Crop Production for Agricultural Improvement

Muhammad Ashraf • Münir Öztürk
Muhammad Sajid Aqeel Ahmad
Ahmet Aksoy
Editors

Crop Production for Agricultural Improvement

Springer

Editors
Muhammad Ashraf
Department of Botany
University of Agriculture
Jail Road
Faisalabad
Pakistan

Münir Öztürk
Center for Environmental Studies
Ege University
Science Faculty Building A-Block
Bornova, Izmir
Turkey

Muhammad Sajid Aqeel Ahmad
Department of Botany
University of Agriculture
Jail Road
Faisalabad
Pakistan

Ahmet Aksoy
Department of Biology
Erciyes University
Rektörlüğü
Melikgazi, Kayseri
Turkey

ISBN 978-94-007-4115-7 ISBN 978-94-007-4116-4 (eBook)
DOI 10.1007/978-94-007-4116-4
Springer Dordrecht Heidelberg New York London

Library of Congress Control Number: 2012937471

Printed on acid-free paper

Springer is part of Springer Science+Business Media (www.springer.com)

Contents

Chapter 1
Crop Improvement Through Different Means: Challenges and Prospects

Muhammad Ashraf, Muhammad Sajid Aqeel Ahmad, Münir Öztürk, and Ahmad Aksoy

Contents

Abstract In the recent years, the looming food scarcity problem has transformed plant sciences as an emerging discipline committed to devise new strategies for enhanced crop productivity. The major factors causing food scarcity are biotic and abiotic stresses such as plant pathogens, salinity, drought, flooding, temperature extremes, nutrient deficiency or excess, etc. which substantially limit crop productivity world-wide. In this scenario, such strategies should be adopted which may be

M. Ashraf (✉) • M.S.A. Ahmad
Department of Botany, University of Agriculture, Faisalabad, Pakistan
e-mail: ashrafbot@yahoo.com; sajidakeel@yahoo.com

M. Öztürk
Centre for Environmental studies, Faculty of Science, Building A Block, Ege University, 35100, Bornova, Izmir, Turkey
e-mail: munirozturk@gmail.com

A. Aksoy
Faculty of Art and Sciences, Department of Biology, Erciyes University, 38039, Kayseri, Turkey
e-mail: aksoy@erciyes.edu.tr

M. Ashraf et al. (eds.), *Crop Production for Agricultural Improvement*,
DOI 10.1007/978-94-007-4116-4_1,

employed to achieve maximum productivity and economic crop returns under such adversaries. Major strategies include pathogen/pest management practices, breeding of new crop varieties, screening and selection of existing crop gene pool, production of genetically modified (GM) crops, exogenous use of osmoprotectants and plant hormones, agronomic and soil reclamation practices, sustainable use of available water supplies, etc. In this book, we have mainly focused on physiological, biochemical, molecular and genetic tools for crop improvement under environmental adversaries. In addition, the adverse effects of different biotic (diseases, pathogens, etc.) and abiotic (salinity, drought, high and low temperatures, metals, etc.) stresses on crop development and the potential strategies to enhance crop productivity under such stressful environments have been critically discussed. Moreover, the role of nutrient, water and soil management in improving crop efficiency is also a part of this book.

Keywords Crop production • Food security • Crop improvement • Stress tolerance • Disease resistance

1 Introduction

The rapidly increasing human population is causing a number of challenges to sustain life on earth. For example, we are losing biodiversity, degrading environment, facing food scarcity, over-exploiting natural resources and performing activities that lead to increased levels of abiotic stresses in our environment. Among these, food scarcity is surely the largest issue that directly or indirectly relates to environmental issues. In this situation, it is imperative to keep updated ourselves with advances in plant production science to meet these scientific challenges and thus overcome the increasing food scarcity and sustain life on earth. For this purpose, we are in need to develop new high yielding and stress tolerant varieties, through modern biotechnological, molecular and genetic tools. We should have enhanced knowledge of stress tolerance mechanisms and should develop methodologies to overcome the stresses. We need to understand our environment and ecosystems in the changing environment and develop methodologies to conserve it. For this purpose, we invited a number of scientists worldwide to review the current scenario of the problems, current development and future prospects of the challenges and their solutions. Their contributions are compiled in this book that is a valuable contribution towards our struggle for improved crop production to meet the demands of the growing human population.

2 Human Population Growth

With the announcement of the United Nations on October 31, 2011 that the World's population has crossed 7 billion, the hard question raised in our minds is "will the current population growth rate be supported by carrying capacity of the earth?" It is a fact that currently, approximately 2.5 babies are being added to the world population per second (US Census Bureau 2010). The current growth rate of world's population is 1.8%.

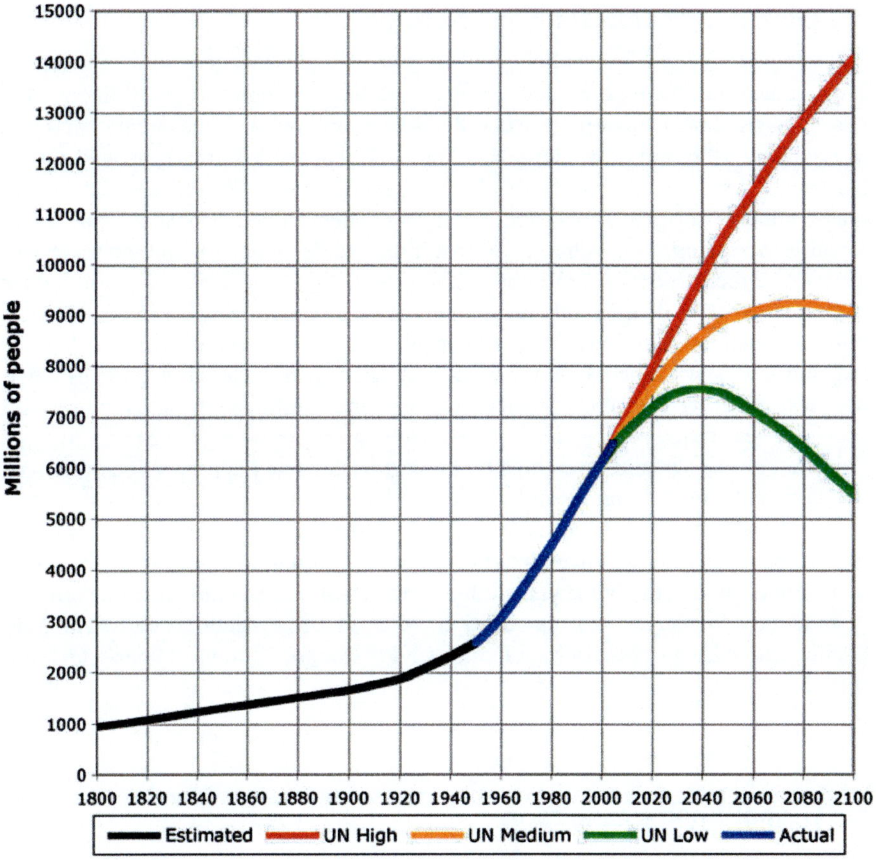

Fig. 1.1 The estimates of world population from years 1800 to 2100. The estimates are based on UN-2004 projections (*red*, *orange*, *green*) and US Census Bureau historical estimates (*black*). If the current growth rate continues, the world population will cross 14 billion at the end of twenty-second century that is almost double the current world population (7 billion) (Sources: UN 2004; U.S. Census Bureau 2010)

It is estimated that more than 120 million people will be added to the planet during the year 2011 while the deaths will be only about 70 million. So, there will be the addition of 50 million people to the planet this year (Population Institute 2011). The UN estimated that at the end of 2025 the human population will cross 8 billion while at the end of 2050 it could be over 10 billion. At the end of twenty-second century, the human population is projected to cross 14 billion (Bongaarts 1997; United Nations 2004). Where the most rapid growth would be? It is estimated to be in Asia and Africa, which will be the most crowded continents on earth. It is estimated that by the year 2025, out of 8 billion world's population, 6 billion will be living only in Asia and Africa. In comparison, the developed countries will be experiencing near zero population growth. Thus, at the end of year 2025, approximately 80% of world population will be living in under-developed countries of Asia and Africa (Fig. 1.1).

3 Global Demands for Food Supply

It is claimed that the food production on the globe is enough to support the world's current population. However, the fact is that a large proportion of the population is still starved. Since the start of twenty-first century, the proportion of malnourished people has been reported to be almost halved in the past 40 years. Nevertheless, recent estimates indicate that the proportion of malnourished/starved people is once again steadily increasing. For example, 843 million people under- or mal-nourished in 1990–1992 increased to 923 million in 2007 (FAO, 2010). In 2009, this figure further increased to 1023, while a little decrease in this figure was reported in 2010 as 925 million under- or mal-nourished people in the world. This shows that the share of malnourished/starved people in the world has steadily increased during the past two decades. If the current trend continues, one can easily estimate the situation of food supply in near future particularly in the developing countries (Figs. 1.2 and 1.3).

Although the statistics presented by FAO indicates that the proportion of hungry people has decreased significantly at the global scale, it is a fact that every day, almost 16,000 children (one child every 5 s) die from nutrition-related causes. According to another estimate, nearly 9 million children died before they reached their fifth birthday only in the year 2008. One third of these deaths were due directly or indirectly to hunger and malnutrition. Most of these deaths occurred in Asia Pacific and African countries including Chad, Congo, Ethopia, Niger and India (Fig. 1.4).

4 Global Food Production

Although, there are 250,000–300,000 known plant species on planet earth, only 150–200 of these are used by humans for dietary purposes. About 75% of the world's food is generated from only 12 plants and 5 animal species (FAO 1999a). Among these, only three crops (rice, maize and wheat) contribute ~60% of calories and proteins obtained by humans from plants, while animals provide about 30% of human requirements for food and agriculture (FAO 1999b). The food production is steadily increasing with the demand. For example, during the year 2011 a record production of cereal grains (2,325 million tonnes) has been estimated by FAO that is 3.7% more than that in the year 2010. Thus, about 507 million tonnes cereal crops have been estimated to be in stock in 2011 by FAO. Overall, there is an increase of 6.0% in wheat, 2.6% growth in the coarse grains and a 3.4% rise for rice production has been estimated by FAO during the year 2011 (FAO 2011a). Nevertheless, a question arises as to whether the people are still hungry worldwide? This is due to the reason that most of the food production is in developed countries, while developing countries experience less increase in food production, resulting in food scarcity and hunger related issues in these regions. Secondly, the World Bank estimates that the increase in global food prices in 2008, accompanied by a global economic

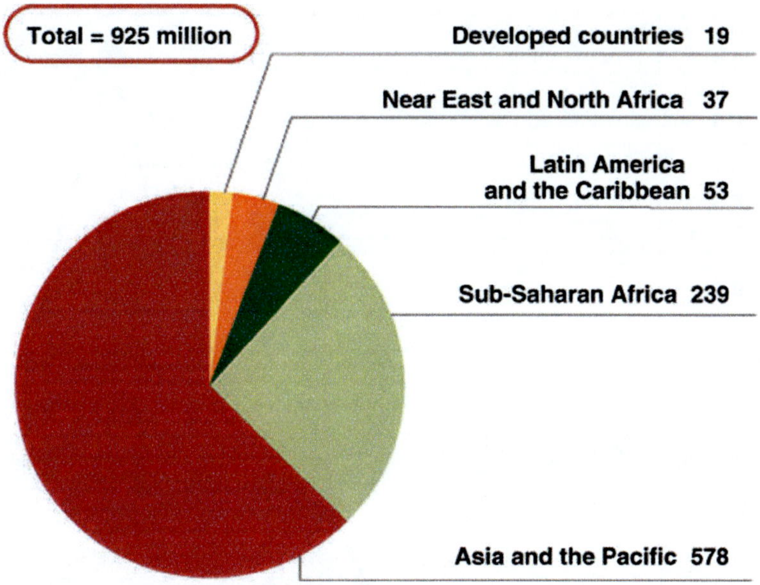

Fig. 1.2 The distribution of hungry people-world wide during 2001. Note that out of 925 million people worldwide, the largest proportion of malnourished people is in the Asia and the Pacific (578 million) followed by Sub-Saharan Africa. Here, the cause of hunger in Asia and the Pacific is population explosion while in the Sub-Saharan Africa is environmental extremes. The least proportion is in developed countries (19 million) that experience almost zero-population growth (Source: World Hunger Education Service 2011; FAO 2010)

depression in 2009 and 2010 has pushed an additional 100–150 million people into poverty worldwide leading to increase in global hunger (Mitchell 2008; Bread for the World 2011) (Fig. 1.5).

5 Is the Population Explosion a Real Problem?

The rapid growth in human population raises a serious question about environmental health and food security issues. The biggest question in our minds is that will the earth be able to support 14 billion people in the year 2100, the population double to the present day (7 billion), with the same limited resources of the present day? Will our requirements of food, health and education and residence be met? Shall we be able to sustain our renewable and non-renewable resources? What would be the situation of biodiversity and croplands? Will our future generations be supplied with clean water and air? Indeed, the policy-makers, economists and ecologists are worried by this situation as these things seem hard to sustain in future, particularly in the developing countries of Asia and Africa.

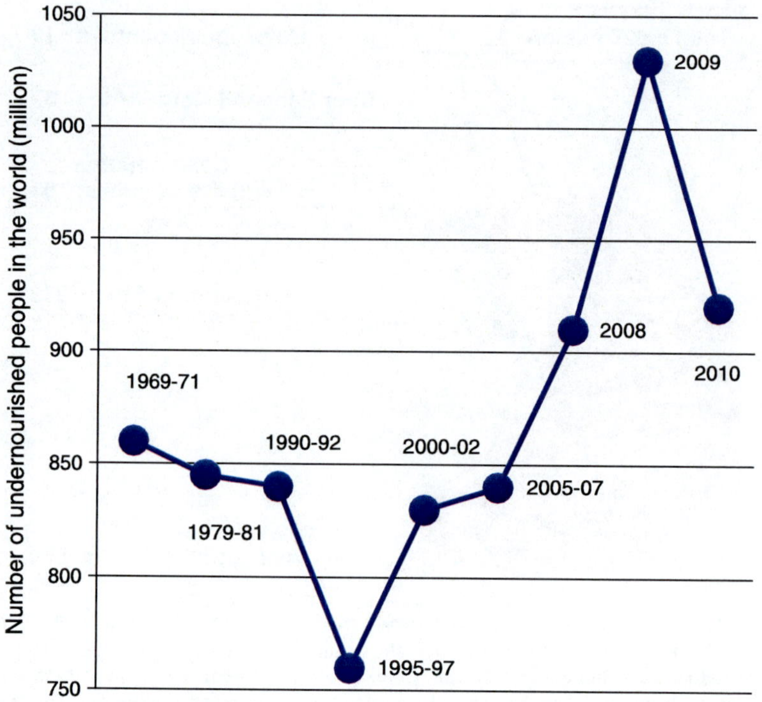

Fig. 1.3 The extent of hungry people from 1969 to 2010. There was a slight decrease in global hunger from 1969 to 1997. However, afterwards, a rapid increase is reported, although there is again a decline in global hunger in 2010 (Sources: World Hunger Education Service 2011; FAO 2010)

The fact is that currently, approximately 434 million people live in areas of either extreme water stress or scarcity. It is estimated that depending on future trends of human population growth, in the year 2025 approximately 2.6–3.1 billion people will be living in areas of water-scarcity (Valerio 2008; US Department of State 2006). Similarly, approximately 600–986 million people will be living in regions where cultivated land will become critically scarce in 2025. Despite the improvement in crop production after Green Revolution aided by technological advances, agricultural experts are worried. The debate is how long crop production will be enough to feed increasing human population. In future, the crops will be produced mostly from today's cropland. Therefore, our current croplands must remain fertile to sustain food production. The minimum amount of land needed to provide the vegetarian diet for one person without the input of any artificial chemical fertilizer and loss of soil nutrients is 0.07 ha. Currently, 415 million people already live in countries having land less than required for a person for this purpose (Population Action International 2011).

Our forested lands are also becoming critically scarce. It is estimated that currently, more than 1.8 billion people live in 36 countries where the forested land is less than

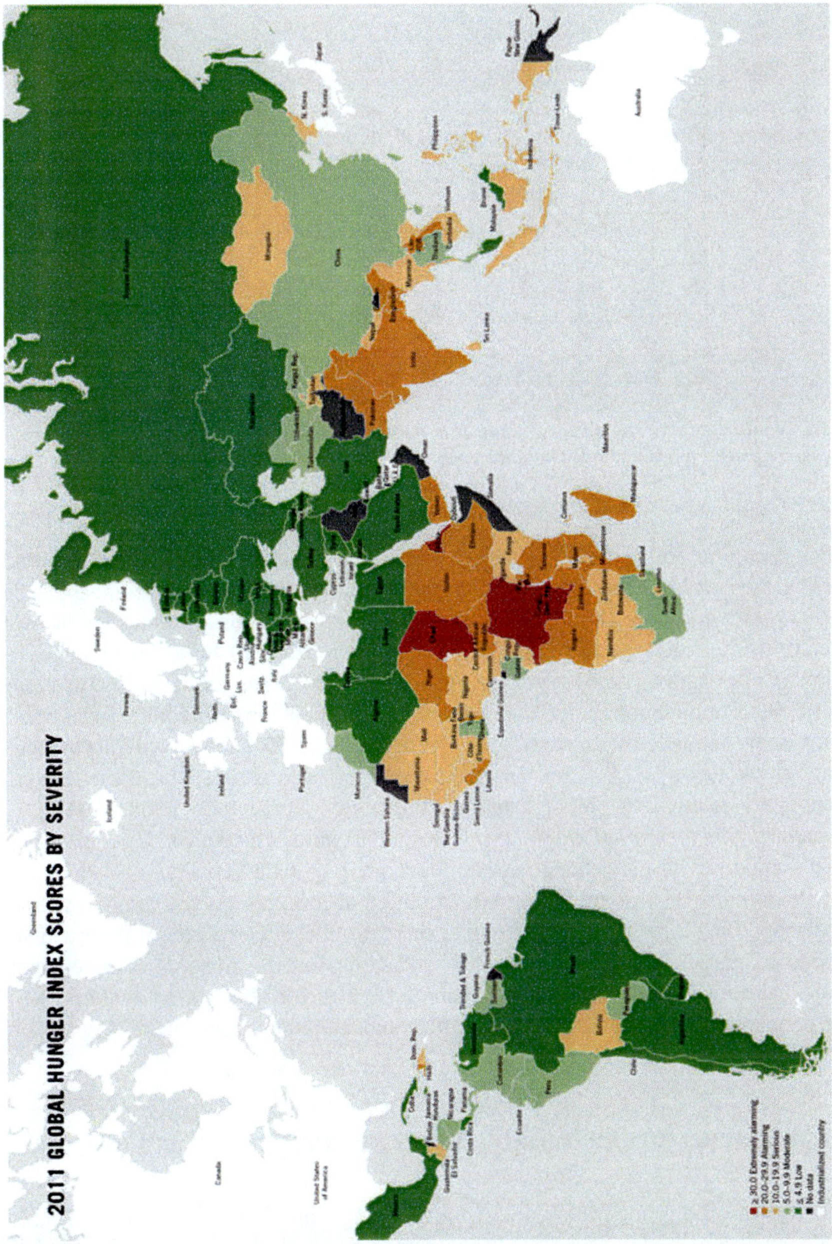

Fig. 1.4 The distribution of global hunger by country (Source: International Food Policy Research Institute 2010; http://www.ifpri.org/ifprimaps/index.php/ghi/v2010)

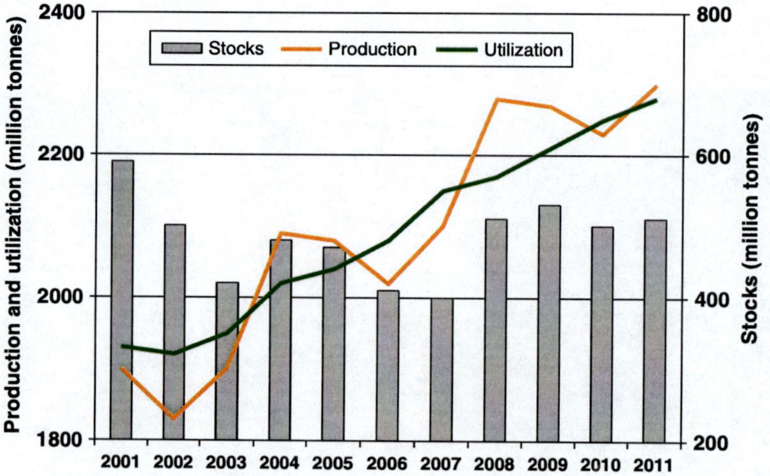

Fig. 1.5 World cereal production and utilization during 2001–2011 (Source: FAO 2011a; http://www.fao.org/worldfoodsituation/wfs-home/csdb/en/)

0.1 ha per capita. This indicates a critically low level of forest cover in these regions. Based on the current deforestation trends and medium population projection, approximately 3 billion people (double to present day) will be living in countries having critically scarce forest land in 2025. At present, more than 1.1 billion people live in biodiversity hot-spots. These regions comprise about 12% of the earth dry-land wherein about 20% of human population is currently living. The alarming fact is that in comparison to the annual growth rate of world's population (1.3%), the growth rate in these biodiversity hotspots is 1.8%, pushing the regions under pressure. Thus, these regions are under severe threat by human activities (Population Action International 2011).

Despite an increase in global crop production is claimed by the FAO, the prices of food commodities have reached to a historical high limit in the year 2011. The average Food Price Index (FPI), a measure in the inflation of food prices, was approximately 100 during 2002–2004. With a consistent increase in the later years, it is now estimated to be more than 200 in the year 2011 indicating that the global food prices have bloomed almost double within only 6 years (FAO 2011b). This has no doubt pushed more people in poverty and made the nutrition related issues more severe (Fig. 1.6).

6 Challenges for Sustainable Crop Production

Currently, the crop production world-wide is facing a number of challenges. These, include, environmental constraints, diseases and pathogens, loss of genetic diversity, and global climate change. Among the abiotic stresses, drought is the most important and most common limitating factor of crop production in arid and semi-arid regions

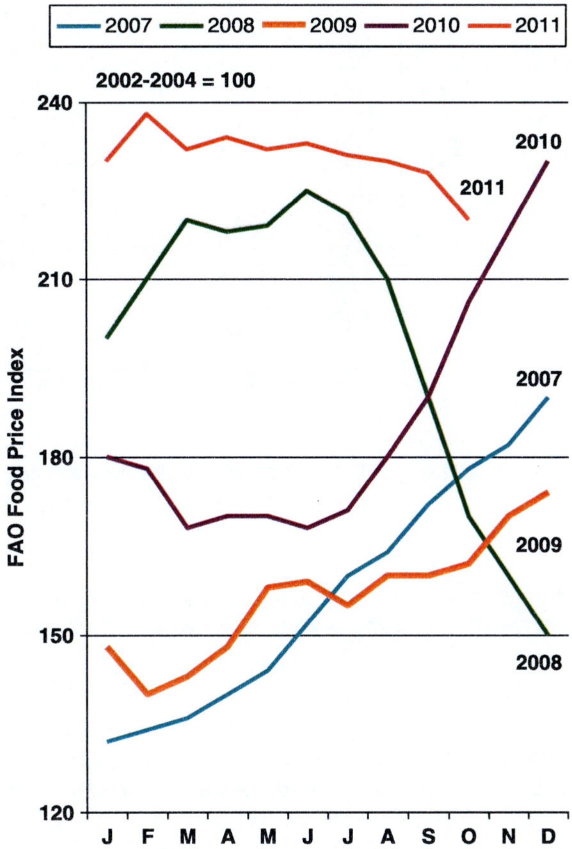

Fig. 1.6 FAO Food Price Index (FPI) from 2002 to 2011. The highest FPI can be seen for the year 2011 (Source: FAO 2011b)

of the world (Saranga et al. 2001). It is estimated that more than 1/4 of total land area is dry and about 1/3 of the world's cultivable land is under water shortage conditions (Kirigwi et al. 2004). The crop quality and production is also seriously influenced by global climatic changes which enhance the frequency and intensity of water shortage thereby making the situation more serious (Hongbo et al. 2005).

Salt stress is the second most prevalent abiotic stress in the world that adversely impacts plant growth (Pessarakli 1991). It is estimated that over 800 million hectares are salt affected in the world either by salinity (397 Mha) or sodicity (434 Mha) which is over 6% of the total land area in the world (FAO 2005). Most of the salinity and all of the sodicity is natural; however, a significant proportion of recently cultivated land has become saline because of land clearing and irrigation. The United Nations Environment Programme (UNEP) and Food and Agriculture Organization (FAO) have estimated that approximately 45 Mha out of 230 Mha of irrigated land in the world are salt affected (FAO 2005). Approximately, 10 Mha of the irrigated land is forced out of cultivation every year due to high salinity (Szabolcs 1989) and one third to half of the irrigated land may be heading towards this fate (Nelson et al. 1998).

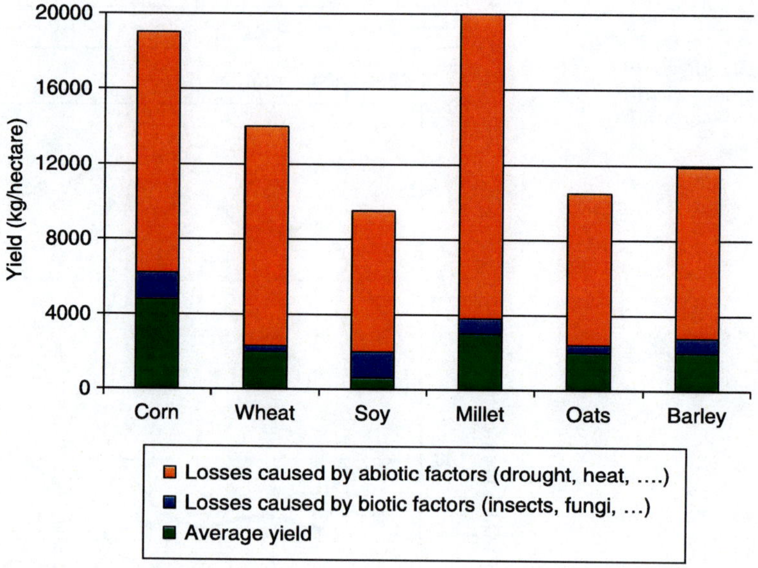

Fig. 1.7 Estimated crop losses due to biotic and abiotic stresses (Bayer Crop Science 2008, http://www.seedquest.com/News/releases/2008/october/23973.htm)

High temperature stress is another major factor that significantly affects plant productivity particularly in arid zones (Bray et al. 2000). Heat stress or heat shock, caused by rise in ambient temperature beyond a threshold level, is a major threat to crop production worldwide (Hall 2001). In general, heat stress is considered when temperature elevates 10–15°C above ambient temperature. However, the probability of its occurrence depends on period of high temperatures occurring during the day and/or the night. Elevated temperatures may lead to alteration in geographical distribution as well as also result in altered growing season of agricultural crops, allowing crop maturity to reach earlier by causing threshold temperature for the start of the season (Porter 2005). Intergovernmental Panel on Climatic Change (IPCC) has estimated that global mean temperature will rise 0.3°C per decade (Jones et al. 1999) and this will reach to 1°C and 3°C by years 2025 and 2100, respectively. The situation becomes worse when heat stress usually combines with drought and salinity stresses, further impeding crop production worldwide.

Other problems of relatively less intensity that hinder crop production include, environmental pollutants such as heavy metals, pesticides, fertilizers, petroleum products, and other organic and inorganic chemicals. Soil mismanagement and loss of soil fertility due to excessive cultivation of crop is also threatening the crop production worldwide. In addition to all these abiotic stresses, biotic stresses such as diseases, pests and pathogens also contribute significantly towards crop losses worldwide, though their contribution is significantly less than that of abiotic stresses (Fig. 1.7).

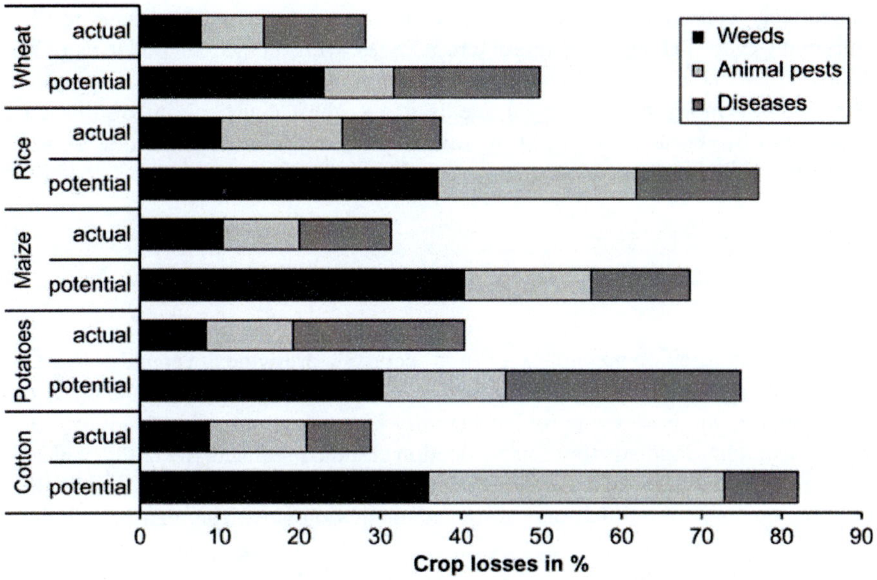

Fig. 1.8 Estimated global crop losses in major crops due to pests and pathogens (Qaim 2011)

7 Crop Losses Due to Biotic and Abiotic Factors

It is estimated that abiotic and biotic stresses collectively contribute more than 50% crop losses worldwide. A survey conducted by Bayer in 2008 indicated that crop losses caused by stresses were significantly greater than the average yield of economically important crops (corn, wheat, soy, millet, oats and barley). They also showed that the abiotic stresses caused significantly higher crop losses than did the biotic ones. For example, the highest crop losses were shown for millet, a crop of the arid regions, where average yield was 2,000 kg/ha and crop losses were 3,800 and 20,000 kg/ha due to biotic and abiotic stresses, respectively. Similarly, the average yield of corn in 2008 was 4,500 kg/ha while the crop losses due to biotic and abiotic stresses were 6,000 and 19,000 kg/ha, respectively. The third highest crop losses were recorded for wheat, another economically crucial crop of third world countries. The average yield of wheat was approximately 1,500 kg/ha, while crop losses were 2,000 kg/ha due to biotic stresses and 14,500 kg/ha due to abiotic stresses. Almost a similar extent of crop losses due to abiotic and biotic factors was reported for barley, oats and soya crops. All these data indicate that crop losses due to abiotic stresses were more severe than those by the biotic ones (Fig. 1.7).

In another report, Qaim (2011) compared the crop losses due to various biotic agents such as disease, weeds and animal pests in five economically important crops, i.e. wheat, rice, maize, potatoes and cotton (Fig. 1.8). He showed that these biotic agents collectively caused approximately 28–40% harvest loss in these economically important crops. Here, the highest crop losses were shown in potatoes (40%) followed by rice (38%) and maize (30%). The harvest losses in wheat and cotton were 28%. Among the biological agents, the highest contribution towards crop losses was by diseases, followed by animal pests and the least was due to weed competition (Fig. 1.8).

8 Strategies for Crop Improvement

In view of the situation prevailing for food security worldwide, it is amply clear that we need to devise concrete methodologies to increase average crop yield. At the first instance, we need to control haphazardly increasing human population so that pressure on our croplands for crop production could be reduced. Secondly, we need to combat environmental adversaries, a major reason of crop losses worldwide, by developing conventional and advanced methodologies. This can be achieved by water management, soil manipulation, nutrient management, screening and selection of the existing gene pool, conventional and molecular breeding, tissue culture, genetic transformations and molecular enhancements. Additionally, we have to manage crop losses arising from biotic agents through disease and pest management.

As discussed earlier, the impact of abiotic stresses on yield losses is more severe than that by the biotic ones. Therefore, we have to combat abiotic stresses in the first instance so as to fulfil our desire to increase crop productivity worldwide. Normally, it is achieved through conventional breeding and selection strategies to select tolerant varieties/lines. Although, such efforts have enduring impact, their development is usually slow and requires a considerable time to succeed (Witcombe et al. 2008). In the recent past, use of various molecular enhancements has shown a promising means to induce short-term resistance to abiotic stresses and have been summarized in various reviews (Ashraf and Foolad 2007; Alcázar et al. 2010; Ashraf 2009; Ashraf et al. 2011). More recently, genetic transformations have also been shown to be another effective and long lasting means to improve crop productivity under stress conditions (Cushman and Bohner 2000; Zhang et al. 2000; Vinocur and Altman 2005; Mittler and Blumwald 2010; Roy et al. 2011). All these reports indicate that there is still a potential to improve crop production under stress conditions in future to overcome the problem of food security of growing human population.

It is a fact that biotic stresses, although have comparatively less damaging impact on harvest losses, most of the genetic modifications to enhance crop productivity have been performed to confer resistance against biotic stresses. For example, Huang et al. (2002) compared the genetic modifications in crop plants against various stresses. They concluded that majority of genetic transformations have been performed for insect resistance (37%), herbicide resistance (29%), stalked traits

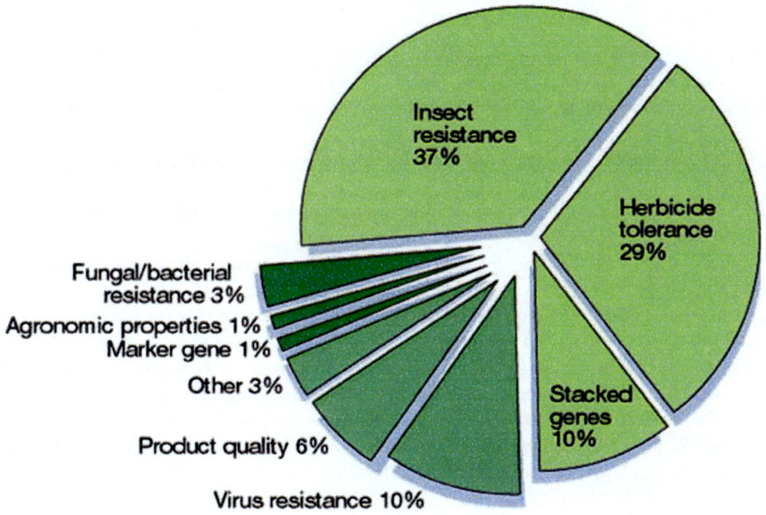

Fig. 1.9 Genetically modified crop traits tested in developed countries from 1987 to 2000 (After Huang et al. 2002)

(10%) and virus resistance (10%). In comparison, a little attention has been paid to agronomic properties (1%), marker genes (1%) and resistance against abiotic stresses that constitutes only 3% of all GM crops tested under field conditions (Fig. 1.9). This shows that we need to focus our efforts to develop GM crops that can perform better under field conditions against abiotic stresses, a major problem for crop production worldwide.

9 Conclusion

It is amply clear from the above discussion that we will be facing food security issues in near future particularly in the developing countries where most of human population will be living. Additionally, the increasing crop losses due to environmental adversaries will amplify food security issues. Majority of crops losses are due to abiotic stresses that cause more than 50% harvest loss. Although, scientists are working hard to increase the average yield of various economically important crop plants, a limited success is achieved due to the increasing extent of abiotic and biotic stresses. Therefore, there is a dire need to devise methodologies to enhance crop production particularly in the stressed-regions of the world.

References

Alcázar R, Altabella T, Marco F, Bortolotti C, Reymond M, Koncz C, Carrasco P, Tiburcio AF (2010) Polyamines: molecules with regulatory functions in plant abiotic stress tolerance. Planta 231:1237–1249

Ashraf M (2009) Biotechnological approach of improving plant salt tolerance using antioxidants as markers. Biotechnol Adv 27:84–93

Ashraf M, Foolad MR (2007) Roles of glycine betaine and proline in improving plant abiotic stress resistance. Environ Exp Bot 59:206–216

Ashraf M, Akram NA, Al-Qurainy F (2011) Drought tolerance: roles of organic osmolytes, growth regulators, and mineral nutrients. Adv Agron 111:249–296

Bayer Crop Science (2008) "We need a second green revolution" – Safeguarding food for a growing world population. http://www.seedquest.com/News/releases/2008/october/23973.htm

Bongaarts J (1997) Future population growth and policy options. A World Bank Report. http://www.info.worldbank.org/etools/docs/library/48424/m1s3bongaarts.pdf

Bray EA, Bailey-Serres J, Weretilnyk E (2000) Responses to abiotic stresses. In: Buchanan BB, Gruissem W, Jones RL (eds) Biochemistry and molecular biology of plants, ASPP, Rockville, pp 1158–1203

Bread for the World (2011) Global Hunger. http://www.bread.org/hunger/global/

Cushmana JC, Bohner HJ (2000) Genomic approaches to plant stress tolerance. Curr Opin Plant Biol 3:117–124

FAO (1999a) Agricultural biodiversity, multifunctional character of agriculture and land conference, Background paper 1. Maastricht, The Netherlands. Sept 1999

FAO (1999b) Women: users, preservers and managers of agrobiodiversity. Food and Agricultural Organization of the United Nations, Rome. http://www.fao.org/sd/nrm/Women%20-%20Users.pdf

FAO (2005) Global network on integrated soil management for sustainable use of salt-affected soils. Rome, Italy. ftp://ftp.fao.org/agl/agll/docs/misc23.pdf

Food and Agriculture Organization (FAO) (2010) The state of Food Insecurity in the World. Food and Agricultural Organization of the United Nations, Rome. http://www.fao.org/docrep/013/i1683e/i1683e.pdf

Food and Agricultural Organization (FAO) (2011a) World food situation – FAO cereal supply and demand brief. Food and Agricultural Organization of the United Nations, Rome. http://www.fao.org/worldfoodsituation/wfs-home/csdb/en/

Food and Agricultural Organization (FAO) (2011b) World food situation – Food Price Index. Food and Agricultural Organization of the United Nations, Rome. http://www.fao.org/worldfoodsituation/wfs-home/foodpricesindex/en/

Hall AE (2001) Crop responses to environment. CRC Press LLC, Boca Raton, Florida.

Hongbo S, Zongsuo L, Mingan S (2005) Changes of anti-oxidative enzymes and MDA content under soil water deficits among 10 wheat (*Triticum aestivum* L.) genotypes at maturation stage. Colloids Surf B Biointerfaces 45:7–13

Huang J, Pray C, Rozelle S (2002) Enhancing the crops to feed the poor. Nature 418:678–684

International Food Policy Research Institute (2010) Global Hunger Index. Consultative Group on International Agricultural Research (CGIAR), USA http://www.ifpri.org/ifprimaps/index.php/ghi/v2010

Jones PD, New M, Parker DE, Martin S, Rigor JG (1999) Surface air temperature and its changes over the past 150 years. Rev Geophys 37:173–199

Kirigwi FM, Van Ginkel M, Trethowan R, Seaes RG, Rajaram S, Paulsen GM (2004) Evaluation of selection strategies for wheat adaptation across water regimes. Euphytica 135:361–371

Mitchell D (2008) A note on rising food crisis. The World Bank Development prospects group. Policy research working paper 4682. The World Bank. http://www-wds.worldbank.org

Mittler R, Blumwald E (2010) Genetic engineering for modern agriculture: challenges and perspectives. Annu Rev Plant Biol 61:443–462

Nelson DE, Shen B, Bohnert HJ (1998) Salinity tolerance--mechanisms, models and the metabolic engineering of complex traits. Genet Eng 20:153–176

Pessarakli M (1991) Dry matter yield nitrogen-15 absorption and water uptake by green bean under sodium chloride stress. Crop Sci 31:1633–1640

Population Action International (2011) Why population matters. Published by Population Action International (PAI), Washington, DC, USA

Population Institute (2011) From 6 billion to 7 billion: how population growth is changing and challenging our world. Population Institute, Washington, DC. www.populationinstitute.org/external/files/reports/from-6b-to-7b.pdf

Porter JR (2005) Rising temperatures are likely to reduce crop yields. Nature 5:435

Qaim M (2011) Genetically modified crops and global food security. In: Beladi H, Choi EK (eds) Frontiers of economics and globalization, vol 10. Emerald Group Publishing Limited, London, ISSN: 1574–8715

Roy B, Noren SK, Mandal AB, Basu AK (2011) Genetic engineering for abiotic stress tolerance in agricultural crops. Biotechnology 10:1–22

Saranga Y, Menz M, Jiang CX, Robert JW, Yakir D, Andrew HP (2001) Genomic dissection of genotype X environment interactions conferring adaptation of cotton to arid conditions. Genome Res 11:1988–1995

Szabolcs I (1989) Salt-affected soils. CRC Press, Florida, Boca Raton

United Nations (2004) World population to 2300. Department of Economic and Social Affairs, Population Division, United Nations, New York

US Census Bureau (2010) International data base, Washington, DC http://www.census.gov/population/international/

US Department of State (2006) Water for the poor act of 2005. Report by Senator Paul Simon (P.L. 109–121) for US Congress, Department of State Publication 11345, Bureau of Oceans and International, Environmental Scientific Affairs, USA

Valerio C (2008) Securing water resources for Water Scarce Ecosystems. United Nations Convention to Combat Desertification (UNCCD), Bonn, Germany. http://www.unccd.int/publicinfo/pagi/docs/UNCCDWaterDocument_30042009PAGI.pdf

Vinocur B, Altman A (2005) Recent advances in engineering plant tolerance to abiotic stress: achievements and limitations. Curr Opin Biotechnol 16:123–132

Witcombe JR, Hollington PA, Howarth CJ, Reader S, Steele KA (2008) Breeding for abiotic stresses for sustainable agriculture. Phil Trans R Soc B 27(363):703–716

World Hunger Education Service (2011) World hunger and poverty facts and statistics. World Hunger Education Service, Washington, DC, USA. http://www.worldhunger.org/articles/Learn/world%20hunger%20facts%202002.htm

Zhang J, Klueva NY, Wang Z, Wu R, David Ho TH, Nguyen HT (2000) Genetic engineering for abiotic stress resistance in crop plants. In Vitro Cell Dev Biol Plant 36:108–114

Part I
Breeding for Crop Improvement

Chapter 2
Bridging Genomic and Classical Breeding Approaches for Improving Crop Productivity

Mehboob-ur-Rahman, Tayyaba Shaheen, Muhammad Ashraf, and Yusuf Zafar

Contents

M. Rahman (✉) • Y. Zafar
National Institute for Biotechnology and Genetic Engineering (NIBGE),
Jhang Road, P.O. Box 577, Faisalabad, Pakistan
e-mail: mehboob_pbd@yahoo.com; mehboob@nibge.org;
y_zafar@yahoo.com

T. Shaheen
Department of Bioinformatics and Biotechnology, GC University, Faisalabad, Pakistan
e-mail: tayaba_pgmb@yahoo.com

M. Ashraf
Department of Botany, University of Agriculture, Faisalabad, Pakistan

Department of Botany and Microbiology, King-Saud University, Riaydh, Saudi Arabia
e-mail: ashrafbot@yahoo.com

M. Ashraf et al. (eds.), *Crop Production for Agricultural Improvement*,
DOI 10.1007/978-94-007-4116-4_2, © Springer Science+Business Media B.V. 2012

Abstract Numerous genomic tools have been used vigorously for studying the inherent genetic polymorphisms which were instrumental in resolving the phylogenies of many crop species, developing genetic maps, initiating marker assisted selection and incorporating genes from distantly related taxa-introduction of Bt genes in cotton, corn etc., these together set a stage for developing crop varieties with improved genetic potential to multiple stresses. Wider adaptation of genomic based breeding in crop improvement programs is impeded due to the narrow genetic base resulting from selection pressures applied during the domestication of many plant taxa, which also can confer genetic vulnerability to crop gene pools. Genomic based breeding may contribute to increasing crop genetic diversity by introgressing novel alleles from feral and or alien species. Association mapping approaches coupled with identifying single nucleotide polymorphisms the most elemental form of polymorphism in the genomes, may facilitate breeding by design. In this article, efforts to advance genomic-based breeding for improving crop species, providing food, feed, fuel and fiber to the world community, will be discussed.

Keywords RFLP • SSRs • SNPs • QTLs • Marker-assisted selection • GM cotton • GM rice • GM wheat • GM soybean • GM maize • GM sorghum

1 Introduction

Molecular markers have been proved vital tools for bridging the genomic tools with the classical breeding procedures for improving the genetic potential of multiple crop species (Rahman et al. 2009). In conventional breeding schemes, various traits of interest are combined in one genotype by hybridizing two genotypes (Beckmann and Soller 1986). Various studies elucidate that pyramiding of complex traits conferred by numerous genes which contribute directly or indirectly to the development of same trait, is really a mammoth task to achieve by deploying classical breeding methods (Beckmann and Soller 1986). Marker-assisted selection (MAS) is an approach which utilizes the tightly linked DNA markers for diagnosing plants having that particular trait of interest (Ribaut and Hoisington 1998). MAS can expedite the process of improved breeding by reducing time for selection of true to type genotype, increasing efficiency in selection procedure and efficient utility of the available resources. In other words MAS is a procedure to merge genomic and conventional resources in a better way (Moose and Mumm 2008). Second strategy for merging the conventional and genomic resources is bringing the deployment of transgenic technology for developing genetically modified (GM) crops. Transgenic crops overcome the limitations of utilizing genetic resources among different species (Qaim and Subramanian 2010). To meet the demands of new era breeding with desired characteristics is unavoidable which is only possible with GM technology (Qaim and Subramanian 2010). In this review economically important crops will be discussed in the context of utilizing the aforementioned technologies to improve their genetic potential of crop plants.

2 Wheat

Among the biotic factors, substantially depressing wheat production, are the rust diseases like leaf rust (Singh et al. 1998), stripe rust (Helguera et al. 2003), and stem rust (Mago et al. 2009). A fungus *Puccinia recondite* causes leaf rust. Two genes *Lr34* and *Lr46* which causes slow rusting have been found effective to combat many disease causing fungi (Singh et al. 1998). All combinations of other Lr genes and *Lr34* genes (Kloppers and Pretorius 1997) have explained the hypersensitive resistance responses. DNA markers have been identified which are linked with the other leaf rust genes (Huang and Gill 2001) and *Lr34* (Suenaga et al. 2003) which have further utility for probing the F_2 wheat plants, and also in succeeding generations, containing the gene(s) which can potentially cause resistance to the disease. In another study resistant genes for leaf rust *Lr47, Lr24, Lr1, Lr9,* were introgressed into bread wheat gentypes (Nocente et al. 2007) using MAS. Similarly, translocation lines 6VS/6AL derived from a cross *Triticum aestivum/Haynaldia villosa which* harbors a gene *Yr26* located on chromosome 1B show resistance to the majority of races of *Puccinia striiformis* f. sp. *tritici* (Pst) causing yellow or stripe rust. DNA markers *Xwe173* and *Xbarc181* were utilized for monitoring the introgression in cultivated wheat varieties (Wang et al. 2008). Another gene *Yr15*, imparts resistance to stripe rust, tagged with two SSR markers *Xgwm413 and Xbarc8*, further these markers served the purpose of diagnosis in all genetic backgrounds except in one (Murphy et al. 2009). Commercialization of the first wheat variety "Patwin" was done by the University of California at Davis (http://www.plantsciences.ucdavis.edu; Helguera et al. 2003) is a master piece example that was developed with the help of diagnostic DNA markers which assisted in introgression of *Yr17* and *Lr37* genes for resistance against stripe rust and leaf rust respectively into one genotype.

T. timopheevii ssp. *Armeniacum,* confers resistance against a recently appeared strain of stem rust (Ug99). The resistant gene *Sr40* was tagged with a closely linked marker *Xwmc344* (0.7 cM), and later two flanking markers *Xgwm374 and Xwmc474* (~2.5 cM) were identified, together can be used in marker-assisted incorporation and pyramiding of *Sr40* to develop superior lines (Wu et al. 2009b). Another gene *Sr39* was introgressed along with *Lr35* gene for resistance against leaf rust into wheat from *Aegilops speltoides.* Mago and Co-workers (2009) induced homoeologous recombinations between the *Ae. Speltoides and* wheat chromosome and developed a set of recombinant lines with reduced *A. speltoides* parts. For the resultant resistant and susceptible genotypes, DNA markers were utilized for conveniently pyramiding of other stem rust resistant genes with enhanced sources of Sr39 which effectively combat the Pgt pathotype TTKSK and its other strains in wheat. Two genes from *Thinopyrum ponticum* (*Sr25* and *Sr26*) were introduced into wheat and proved to be useful against new strains of TTKSK (syn. Ug99) and its types. Co-dominant markers for *Sr25* and *Sr26* were identified which can be potentially used in MAS (Liu et al. 2010).

Powdery mildew is another threat to wheat production. SSR markers linked with genes *Pm4a* and *Pm5e* have been detected (Huang et al. 2003; Ma et al. 2004).

Markers linked with another gene *Pm4b* (STS_241, *Xgwm382,* Me8/Em7_220) were identified which can improve resistance to the powdery mildew disease in wheat (Yi et al. 2008).

A locus Glu-1 has some impact on the wheat's quality of bread making. Coding and promoter regions of this locus were scrutinized for polymorphisms (Radovanovic and Cloutier 2003; Ma et al. 2003). Two specific PCR based markers were confirmed and utilized for alleles identification at *Glu-B1x* locus for its further utility in introgression of cultivated wheat varieties (Xu et al. 2008).

A linkage map of all 14 chromosomes was developed containing 280 SSRs, and also for detection of tan spot resistance associated QTLs. A tetraploid wheat doubled haploid (DH) population was derived by crossing a *T. turgidum* var Lebsock and *T. turgidum* subsp. *carthlicum* (accession PI 94749). A total of five QTLs for tan spot resistance were identified on chromosome arms, 3BL, 7BL, 5AL and 3AS. The out come of this study can facilitate genetic dissection of agronomic traits and marker identification for MAS (Chu et al. 2010).

2.1 GM Wheat

Transgenic studies in wheat have been focused mainly on improvement of grain quality characteristics and effect of expression of endogenous genes on dough quality (Francki 2009). It has been experimentally proven that expression of endogenous gene have positive or negative impact on grain quality and dough characteristics. Impact of genes 1Ax1, 1Dx5, LMW-GS, HMW-GS and pinA have been experimentally determined (Alvarez et al. 2000; Blechl et al. 2007; He et al. 2005; Tosi et al. 2004, 2005; Masci et al. 2002; Martin et al. 2006).

Fusarium graminearum, causes Fusarium head blight which is a challenging disease of wheat globally. Wheat has low resistance against this disease due to narrow genetic diversity in the existing pool. GM wheat containing barley class II chitinase gene was found effective against *F. graminearum* when experimentally tested (Shin et al. 2008).

RNA interference is a sequence specific gene silencing mechanism which can be utilized in determining gene functions. The application of RNAi in wheat has confirmed the function of VRN1, VRN2, SBE11a, SBE11b, EIN2, PDS, GPC and 1Dx5 (Francki 2009).

It has been studied recently that ferulic acid esterase which is derived from *Aspergillus niger* or endo-xylanase (from *Bacillus subtilis*) when expressed under the control of endosperm-specific *1DX5* glutenin promoter have an impact on wheat baking quality (Harholt et al. 2010).

National Institute for Biotechnology and Genetic Engineering has evaluated Arabidopsis AVP1 gene by introducing into tobacco for assessing its role for developing resistance against salinity and drought which are major limiting factors for crop productivity. Arabidopsis AVP1 gene encodes a vacuolar pyrophosphatases that function as proton pump and generates an electrochemical gradient in vacuole activating vacuolar membrane-antiporters including Na^+/H^+ antiporter, which helps in sequestration of Na+ into vacuole as well as overexpression of AVP1 gene promotes vegetative growth by enhancing root development under the influence of auxins. Results of this study elucidate the significance of this gene in salinity and drought tolerance. This gene can further be utilized in economically important crops like wheat (Ibrahim et al. 2009).

3 Rice

All over the world yield of rice is being depressed by a fungal disease called Bacterial Blight (BB). Three genes *xa5*, *xa13* and *Xa21* causing resistance to BB were incorporated in susceptible rice cultivars and were tracked using STS markers flanking these genes (Chunwongse et al. 1993; Huang et al. 1997; Singh et al. 2001). Basmati rice is also highly vulnerable to BB. In another study, pyramiding of two genes *Xa7* and *Xa21* was carried out using MAS for improved resistance for BB in hybrid rice (Zhang et al. 2006). Foreground selection was integrated with background analysis using mapped SSR markers to detect the genes *xa13* and *Xa21* which show resistance against BB and superior quality features while these genes were non-Basmati resource derived. In India an improved Pusa Basmati 1 line, developed through MAS, has been commercialized (Gopalakrishnan et al. 2008). SSR markers were utilized to introgress three major genes for resistance *xa5, xa13 and Xa21* in a superior indica rice variety (Sundaram et al. 2008).

Magnaporthae grisea (fungus) causes a disease blast which is another destructive disease of rice. Three vital genes (*Pi*1, *Piz*-5 and *Pi*ta) control this disease. Utilization of tightly linked RFLP markers has facilitated the pyramiding of these genes and also mapping of these genes on respective chromosomes 11, 6 and 12 (Hittalmani et al. 2000). There are many concerns about a race-specific resistance in many crop plants which can be overcome by non-race-specific resistance that was effectively used in breeding against fungal diseases. Some strains of *Japonica* rice contain a resistant *pi21* allele, is able to improve resistance to the blast disease in rice (Fukuoka et al. 2009).

Tightly linked SSR and RFLP markers with a Waxy gene allele were employed to improve the grain quality of a rice cultivar Zhenshan-97A (Zhou et al. 2003).

Among abiotic stresses, limited water condition is the most detrimental factor for causing substantial reduction in yield. Root traits remained a major focus for tackling this menace. Root length was increased by 12–27% in IR64 by introgressing four QTLs for penetrating roots from Azucena (*japonica* variety) (Shen et al. 2001). Another QTL involved in osmotic adjustment (OA) under drought condition was mapped on chr-8 (Robin et al. 2003) would be helpful in future rice improvement program. Synteny between rice and maize was found for a QTL for OA mapped on chr-3 of rice and chr-1 of maize. This QTL accounts for numerous agronomic and physiological traits contributing tolerance to drought (Zhang et al. 2001). Conservativeness among these regions can pave the way for translating information generated on a well-studied crop species to less-studied crop species. In this case, diagnostic DNA markers have utility in probing individual plants harboring QTLs for useful allele (Nguyen et al. 2004).

Plant type in rice is controlled primarily by a gene *Spk*(t), which was targeted through map-based cloning method (Komori et al. 2009). Both *Spk*(t) and its recessive allele *spk*(t) encode the same 259-aa proteins. However, a SNP was detected in the untranslated part of the gene sequence in 3′-splicing site which is *Spk*(t) allele specific. These findings will be useful in rice breeding program.

At low temperature unavailability of viable pollen causes spikelet sterility in rice. A SNP in alternative oxidase gene (OsAOX1a) and two closely linked QTLs (*Ctb*1 and *Ctb*2) contributing towards tolerance to low temperature in anthers were found tightly linked (Abe et al. 2002). A physical mapping study has revealed that seven putative genes were found for *Ctb*1. the utility of the identified SNP in marker assisted selection for the diagnosis of plants harboring QTLs which contribute for tolerance to cold has been proven (Saito et al. 2004).

MAS was utilized in assembling two major QTLs, one for shoot Na^+ reduction (qSNC-7) located on chr-7 and second for shoot K^+ accumulation (qSKC-1) located on chr-1 were found in three F3 lines developed by crossing Azucena and IR64 which is moderately tolerant (Lin et al. 2004).

3.1 GM Rice

Fifteen years ago first GM rice was developed and since then numerous traits related with yield, nutritional value and quality improvement have been dealt with. These efforts were complemented by sequencing its genome followed by characterization of multiple genes of interest. (Barry 2001; Sasaki and Burr 2000; Goff et al. 2002; Feng et al. 2002; Yu et al. 2002; Sasaki et al. 2002; Delseny 2003; The Rice Chromosome 10 Sequencing Consortium 2003; Sasaki et al. 2005).

Golden rice was developed by introducing daffodil and bacterial genes. The *psy* (phytoene synthase) and *crtl* genes that carry out the four steps required for the production of beta-carotene (Vitamin A in rice endosperm). However this rice variety could not get much attention due to many concerns (Toenniessen 2000).

Biotic factors cause loss to global production of rice 52% annually. Of this insect pests cause 21% of loss. The stable transgenic lines, expressing ASAL (*Allium sativum* leaf agglutinin) gene, were resistant against the sap-sucking pests (Yarasi et al. 2008). Genetic engineering for developing insect resistant rice is a very environment friendly and cost effective technique. *Bt (Cry 1 ac and Cry 2ab)* genes, in *Bacillus thuringiensis* species provide a variety of genes to develop transgenic plants, including rice (de Maagd et al. 2001; Tabashnik et al. 2003).

Three different studies were conducted in Centre of Excellence Molecular Biology regarding genetic engineering of rice. Through particle bombardment GM rice indica varieties Basmati 370 and M 7 were generated which express the novel *cry2A* (Bt) insecticidal gene. This novel *Bt* transgene was analysed for stable integration. Cry 2A protein was found effective against two major rice pests of Indian subcontinent the yellow stem borer and the rice leaf folder.

In a second study through particle bombardment three genes related with insect resistance (the Bt genes *cry1Ac* and *cry2A*, and the snowdrop lectin gene *gna*) were introduced simultaneously into *indica* rice varieties M7 and Basmati 370. The transgene showed stable transmission and expression and significant defense against the most important insect pests yellow stem borer (*Scirpophaga incertulas*), brown plant hopper (*Nilaparvata lugens*) and rice leaf folder (*Cnaphalocrocis medinalis*). This approach led to multi mechanism defense.

In a third study three plasmids containing four genes were used to cotransform Indica rice (*Oryza sativa* L. cvs. Basmati 370 and M7). Two separate vectors were used to transform the Bt genes *cry1Ac* and *cry2A*, while the *hpt* (hygromycin phosphotransferase) and *gna* (snowdrop lectin) genes were transformed through a single, co-integrate vector. It was reported that the introduction of multiple agronomically favourable genes into the rice genome by co-transformation is a feasible approach for engineering elite rice varieties (Maqbool and Christou 1999; Maqbool et al. 1998, 2001).

Transgenic *Bt* hybrid rice has been evaluated in the farmer field in China and was confirmed to be resistant against rice leaf folder (*Cnaphalocrocis medinalis*, RLF) and yellow stem borer (Tu et al. 2000). Numerous transgenic varieties have been produced and field tested for yellow stem borer resistance in Pakistan, India and Mediterranean (Bashir et al. 2005; Ramesh et al. 2004a, b; Breitler et al. 2001).

Transgenic insect resistant plants have been very successful through the years and it has become challenging for opponents of genetic engineering to oppose (Bhattacharya et al. 2006). GM rice against yellow stem borer, leaf folder and sheath blight are really effective (Bhattacharya et al. 2006)

In multiple investigations, genes derived from various plant species have used for introducing in major crops of interest. For example, insecticidal genes such as, protease inhibitors or ribosome inactivating proteins and lectins (Sharma et al. 2004) were introduced in rice. Plant lectins are protective against many organisms. The mannose-specific lectin gene *gna* has been extensively used to develop transgenic rice having resistance against numerous economically important insects (Foissac et al. 2002; Tang et al. 2001; Wu et al. 2002; Nagadhara et al. 2003, 2004) plants. Protease inhibitors which are antimetabolites can protect against a wide range of insect pests. GM rice plants expressing protease inhibitors have been produced already. These are Oryzacystatin, cowpea trypsin inhibitors, bean trypsin inhibitor (Mochizuki et al. 1999), barley trypsin inhibitors (Alfonso-Rubi et al. 2003) and soybean trypsin inhibitors, potato protease inhibitors II (Xu et al. 1996; Sharma et al. 2004). Cowpea trypsin inhibitor (CpTi) transgene in transgenic rice is effective against stem borer (Bentur 2006; Brar and Khush 2007).

Ribosome inactivating proteins have been reported to have antifungal activity in plants. Fungus resistant transgenic maize has been produced using ribosome inactivating proteins (Kim et al. 2003).

4 Cotton

Cotton is one of the most significant natural textile fiber crop worldwide (Rahman et al. 2008). It has been reported that improvement in lint production can be achieved by supplementing the conventional breeding procedures with modern genomic tools for initiating knowledge based breeding program in cotton (Rahman et al. 2009). In this regard, multiple non-conventional research efforts like various cotton databases, sequencing data etc. can help in accelerating the breeding progress (Chen et al. 2007). For initiating MAS in cotton there are many useful DNA markers associated with fiber quality traits (Zhang et al. 2003; Asif 2009). Zhang et al. (2003) identified several QTLs for fiber strength. Two QTLs for fiber strength tagged three SSRs and six RAPDs markers were grouped into one linkage group. One of the RAPDs was converted into sequence characterized amplified regions (SCAR4311920), and was applied for diagnosing the cotton plants containing the main QTL associated with fiber strength (Guo et al. 2003). Recently, this QTL was fine mapped on Chr-24 (D8) (Chen et al. 2009). In another experiment, RFLPs linked with fiber quality traits were found by surveying a population derived from an interspecific cross *G. hirsutum/G. barbadense* (Chee et al. 2005a,b; Draye et al. 2005). Similarly, SSRs originating from *G. barbadense* were utilized to assess the introgression of genomic regions obtained from *G. barbadense* into *G. hirsutum*. It results in increment of 12–20% in fiber length in *G. hirsutum* (Mumtaz 2007). Jixiang and Co-workers (2007) identified AFLP markers linked with fiber and agronomic traits which can be utilized for MAS. Recently, these QTLs linked with agronomic traits and fiber were located on At as well as Dt genomes of tetraploid cotton (Wu et al. 2009a)

Like other crop plants, cotton is also vulnerable to limited water conditions. Globally efforts are underway to identifying QTLs conferring traits help in drought, for using in MAS under water stress condition (Saranga et al. 2001; Paterson et al. 2003). Ullah (2009) detected collectively nine putative QTLs for drought tolerance in cotton. Babar et al. (2009) also detected some QTLs associated with drought conditions.

Fusarium wilt (FW), a disease of fungal origin, causes ultimate death in cotton after yellowing, wilting, defoliation and vascular tissue damage. A number of DNA markers were identified by surveying an F2 population derived from an intraspecific cross (*G. hirsutum*). One of the SSR markers (JESPR304) tightly linked with a FW resistant gene (*FWR*) was mapped on chromosome D3 (c17). Four QTLs were identified of which one QTL was found near marker JESPR304 at proximity of 0.06–0.2 cM. This QTL explained 52.5–60.9% phenotypic variance. This QTL can be further used to develop FW resistant cultivars using MAS (Wang et al. 2009).

Cotton leaf curl a viral disease, has substantially depressed cotton production in Pakistan, and also in neighboring countries like India and China. DNA markers (4 RAPDs and 2 SSRs) were found linked with resistance to the disease (Rahman 2002; Rahman et al. 2006). Two cotton lines NIBGE-2 (Rahman and Zafar 2007b) and NIBGE-115 (Rahman and Zafar 2007a) were developed by utilizing these markers.

4.1 GM Cotton

GM cotton containing Cry 1Ac gene has been commercialized in most of the cotton growing countries, covering ~50% of the total cotton growing area, for imparting resistance to lepidopteron insect pests. The gene was isolated from a soil bacterium species *Bacillus thuringiensis*. Numerous varieties of cotton bio-engineered to produce an insecticidal protein are called Bollgard cotton. It has been commercialized in 1996 by a private seed company in the USA. This product is an effective alternative to chemical insecticides to control the attack of cotton bollworm, *Heliothis virescens*, tobacco budworm and pink bollworm which causes reduction in cotton production costs and insecticide use (Perlak et al. 2001).

To further broaden the spectrum of insects to which the plant is tolerant and to provide an insect resistance management tool to hamper the onset of resistance Bollgard II cotton event 15985 was produced in which the Cry1Ac and Cry2Ab2 proteins has been introduced by genetic modification (Hamilton et al. 2004).

Research efforts are underway for identifying new gene(s) for enhancing the spectrum of insect pests resistance. Recently, a gene was derived from spider venom-toxin that confers resistance to herbivorous insects. The gene has been characterized in tobacco and efforts are on the way to transfer it to cotton (Khan et al. 2006).

The nitrilase gene which confers resistance to herbicide bromoxynil in cotton has been obtained from a bacterial species *klebsiella*. Genes for acetolactate synthase conferring resistance to the imidazolinone and sulfonylurea classes of herbicide were isolated from both the genomes of higher plants and bacteria (Stewart 1994). Herbicide resistant cotton has provided the farmers weed control without damage to cotton which has ultimately increased the income of the farming community (Brookes and Barfoot 2006, 2008).

Virus-resistant and salinity-tolerant GM cotton is at the stage of field evaluation (Zafar 2007). Genetic transformation of cotton with biolistic gene gun was carried out; however, the results were not so successful (Haq et al. 2005). Later, for cotton (*Gossypium hirsutum* L. cv. Coker-312) a silicon carbide whisker-mediated gene transfer system was developed which make recovery of fertile and stable transformants possible (Asad et al. 2008). Antisense RNA technology has also been used to develop transgenic plants resistant to cotton leaf curl disease. The mechanism of resistance was found to be post transcriptional gene silencing (Asad 2004). Transgenic cotton was produced using antisense movement protein gene AV2 in an Indian variety (F864) along with the npt II gene. Transgenic plants were tested for the integration of gene (Sanjaya et al. 2005).

Recently, *Arabidopsis* vacuolar H^+ pyrophosphatase gene (AVP1) was transformed in cotton exhibiting resistance against drought and salt stresses. Also, these transgenic cotton lines have shown improvement in staple length compared to the control. These findings are manifesting the utility of this gene to combat drought and salt stress in cotton (Pasapula et al. 2011)

5 Maize

Maize streak virus (MSV) disease causes 100% yield losses in tropical Africa. QTLs accounting for resistance to MSV in maize have been mapped. Microsatellite markers were utilized to evaluate resistance field trials. Linkage group consisted of 13 microsatellite markers and 3 QTLs were detected associated with resistance to MSV (Lagat et al. 2008).

MAS has been deployed in maize (*Zea mays*) for introgressing genes from wild sources and or novel sources (Ragot et al. 1995), for identifying plants having opaque2 gene (Dreher et al. 2003) and for simple (Ho et al. 2002; Morris et al. 2003) or complex traits enhancement (Bouchez et al. 2002; Willcox et al. 2002). Drought causes delayed silking, causing ~15% yield losses annually, which causes long anthesis-silking interval (ASI). There is a positive correlation between reduced ASI and better yields under drought stress. Genomic regions associated with both yield and ASI (Ribaut et al. 1997) were identified. Out of these, one genomic region showed allelic contribution for short ASI with low grain yield while the others provided alleles for short ASI accounting for high grain yield.

In another study, five genomic regions conferring drought tolerance were introduced in a maize inbred line CML247 from a drought tolerant donor line Ac7643. Performance of some genotypes was 2–4 times better than the control genotype, further development of new cultivars can be based on these genotypes (Ribaut et al. 2004). In another investigation, seedling emergence was increased by monitoring the introduction of QTL, has also shown positive impact of grain yield (Yousef and Juvik 2002). QTLs for grain yield and flowering time were identified in maize (Blanc et al. 2008).

High-quality SNP markers were identified by screening maize inbred lines. It has been demonstrated that during cultivar development genetic diversity has not been utilized extensively. SNPs are potent markers to be utilized in MAS and study of diversity. Two SNPs were identified from a gene putatively linked with diversity within two Chinese heterotic groups. Allelic frequency change at two SNPs and absence of their allele in Brazilian germplasm indicated linkage disequilibrium block of 142 kb (Lu et al. 2009).

5.1 GM Maize

In 1990s, conventional breeding procedures were supplemented by integrating transformation technology for improve the genetic base of corn germplasm. GM-maize was commercialized in mid 1990s containing genes conferring resistant to European corn borer and lepidopteran pests. GM-corn containing insect protection and herbicide resistance was commercialized not only in North America but also in other maize growing areas. In 2008, GM-corn containing at least one biotech trait covered more than 80% of total corn growing area in USA (Mullet 2009).

In the second generation biotechnology era, GM maize having three stacked transgenes or multiple traits for corn rootworm control, herbicide tolerance for more efficient weed control and lepidopteran insect control has been commercialized. It has been cultivated on more than 30 million hectares in 16 different countries by 2007. Transgenic maize hybrids not only increase yield but also contribute to reduction of green house gases, and pesticide use. Although first generation transgenic maize has contributed a lot, moreover next generation under development is even more promising. These traits are designed to enhance maize yield, growth under drought conditions, more efficient nitrogen use, enhancement of protection against insects and pests and improvement of grain quality for food, animal feed and biofuels. Improved farm practices along with maize breeding and biotechnology has remained instrumental in escalating maize productivity worldwide (Mullet 2009).

Transgenic corn containing maize hemoglobin and soybean ferritin were characterized to set a strategy for increasing the iron sink strength in grains. Transgenic kernels were compared to non-transgenic sibling kernels for accumulation of mRNA encoding seed storage proteins and proteins involved in mineral metabolism, as well as mineral content. These experiments help to set stratifies for improving nutritional quality in maize grains (Scott 2010).

In a recent study impact of fall armyworm on GM corn having Cry1Ab and Cry1F insecticidal crystal protein was examined. Field study and bioassays have elucidated Cry1F reduced foliar injury significantly by lowering survivorship of armyworm when compared to that on non-Bt corn tissues; whereas Cry1Ab has limited effects on fall armyworm (Hardke et al. 2011).

6 Sorghum

Sorghum provides at the same time fiber, biofuel, food and feed in the semi arid tropical regions of the world. It is a C_4 grass, and its genome has been sequenced which comprise 730 Mbp. Genetic information generated on sorghum can be utilized in the closely related species like maize and wheat etc. (Paterson et al. 2009).

Shoot fly causes substantial damage at seedling stage of sorghum. A microsatellite marker-based linkage map was constructed using RILs of a cross 296B (susceptible)/IS18551 (resistant) (Satish et al. 2009). Also a total of 29 QTLs conferring multiple traits of interest were identified. These are, seven for oviposition, six for dead hearts, six for abaxial trichome density, four each for leaf glossiness and seedling vigor, and two for adaxial trichome density. Resistance alleles were from IS18551 for most QTLs; however, 296B also contributed alleles to the six QTLs involved in conferring resistance. QTLs identified in this study will offer a firm foundation for initiating marker assisted selection (MAS) for improving resistance to shoot fly in sorghum.

Among diseases, resistance to Anthracnose disease was linked with a RAPD marker OPJ011437 which was converted into SCAR (SCJ01). Correspondence of this region was detected with contig-3966 of the sorghum genome (chr-8) which is useful in resistant plants diagnosis (Singh et al. 2006).

Stay-green mechanism is very important trait for improving drought tolerance in sorghum. Identification of four QTLs (*Stg1*, *Stg2*, *Stg3* and *Stg4*) was carried out utilising different populations (Haussman et al. 2002; Sanchez et al. 2002; Harris et al. 2007). Marker assisted breeding was utilized to develop NILs for *Stg2* (Sanchez et al. 2002). Further, 18 different NILs that were containing introgressed regions of the four important stay-green loci, *Stg1-Stg4* were developed through MAS (Harris et al. 2007).

Several QTLs linked with SSRs for early season cold tolerance were identified. These SSRs were further utilized to initiate MAS for early-season cold tolerance under various environments (Knoll and Ejeta, 2008).

6.1 GM Sorghum

To develop resistance against stalk rot (*Fusarium thapsinum*) the agronomically important gene chi II, which encodes rice chitinase under the constitutive CaMV 35 S promoter, was transferred to sorghum. (Zhu et al. 1998; Krishnaveni et al. 2001).

To improve tolerance to water and NaCl stress, a sorghum cultivar SPV462 was engineered with a *mtlD* gene (derived from *E. coli*) encoding for mannitol-1-phosphate dehydrogenase. The transgenic lines have shown that manipulating mannitol biosynthetic pathway can cause enhanced tolerance to salinity and water deficit in sorghum. The product has not yet commercialized (Maheshwari et al. 2010).

A mutated gene was introduced in sorghum for improving the nutritional quality (Tadesse et al. 2003). This gene encodes a key regulatory enzyme of the lysine pathway (dihydropicolinate synthase). Over expression of this gene produces sorghum lines with elevated lysine content.

7 Barley

Two closely linked QTLs involved in tolerance to low temperature co-occurred with a number of QTLs involved in regulation of cold-regulated (*COR*) genes functions (Vagujfalvi et al. 2003) were detected on barley chr-5 (Francia et al. 2004; Toth et al. 2004).

A gene (*Run8*) contributes in developing resistance against loose smut disease. Marker Assisted Breeding (MAB) was utilized to introgress this gene into a hulless barley cultivar using (double haploid) DH and a line (HB390) was produced and evaluated prior to release commercially in Canada. Another gene *Ruhq* causing resistance to covered smut disease was transferred into hulless barley using the same process (Grewal et al. 2008).

Findings of another study suggested that chr-3, chr-6 and chr-7 are harboring QTLs for malt quality traits and grain yield. However, QTLs located on chr-7 were most useful (Igartua et al. 2000). RFLP markers were used for monitoring the introgression of QTLs associated with yield for developing high yielding NILs. In multi-location trials, one line has shown high yielding property of one parent (Baronesse) while retaining a feature of better malting property of the other parent (Harrington), was recognized (Schmierer et al. 2005).

A new fingerprinting assay "temperature-switch PCR" (TSP) was tested for assessing its efficacy for marker development, its reliability and genotyping accuracy in barley. For assessing gene diversity 87 TSP markers were surveyed. This method has been designed for genotyping codominant SNP on agarose gels. This method has proven to be handy for SNP detection (Hayden et al. 2009).

7.1 GM Barley

Horvath et al. (2001) tested transgenic barley cv. 'Golden Promise', containing β-(1,3–1,4)-glucanase gene, over a period of 3 years in the field trial,. Their study indicated a relatively poor performance of the transgenic barley as compared to the control which might be due to the occurrence of somaclonal variations during regeneration of the transformants. In another study, occurrence of reduced fitness in a field grown transgenic barley has been attributed to somaclonal variations (Manoharan et al. 2006).

Backcrossing transgenic lines having expressing a β-(1,3–1,4)-glucanase, with the original wild type (wt) cv. 'Conlon', to get stable transformants led to plants carrying the transgene and having a wt-like growth habit (Manoharan et al. 2006). In another study, crossing of transgenic lines to elite cultivars resulted in homozygous lines that express the transgene and give a higher 1,000-grain weight and yield than the untransformed cultivar, which elucidated that it is possible to enhance the agronomic features of transgenic barley by cross-breeding with elite cultivars (Horvath et al. 2001; Goedeke et al. 2007).

8 Soybean

Five backcross populations (BC2F4, 468 lines) were derived from a cross *G. max* cv. A2008/*G. soja* acc. 468916 after verification at two different locations for 2 years. One QTL for lodging, four QTLs for yield, four QTLs for maturity, and for plant height five QTLs were identified. Alleles from *G. max* cultivar were showing higher yield potential than alleles from *G. soja* (Wang et al. 2004). In the same study, an SSR marker Sat-107 was found closely linked with the four-seeded pod (4SP) locus which can be effectively utilized for plant selection having this trait (Zhu and Sun 2006).

Marker Assisted Breeding was utilized in six genetic backgrounds to introgress an important QTL conferring high yielding potential (Concibido et al. 2003). The final seed phytoestrogen level is largely controlled by a set of ~6–12 loci (Kassem et al. 2006) making direct selection difficult. Marker assisted selection, enhanced the phytoestrogen content level well above in new soybean genotypes than the level of elite cultivars (Lightfoot 2008).

Charlson and co-workers (2009) identified four QTLs for canopy wilting in three drought environments using RILs population. QTLs for seed yield mechanism (Palomeque et al. 2009a, b), gene *Rps8* conferring resistance to stem and root rot (caused by *Phytophthora sojae*), (Sandhu et al. 2005), *Rsv4* gene conferring soybean mosaic virus resistance (Hwang et al. 2006), *Rag1* gene imparting resistance to soybean aphid (Li et al. 2007) can effectively be utilized in soybean.

8.1 GM Soybean

Transgenic fertile soybean plants were generated capsid polyprotein (pCP) which were resistant to *Bean pod mottle virus* (BPMV). The progeny of the homozygous transgenic line exhibited systemic resistance; these lines could potentially be useful in generating commercial cultivars showing resistance to BPMV (Reddy et al. 2001).

An attempt was made for developing GM soybean that has nutritional properties similar to fish oils (Damude and Kinney 2008). Enriching poultry meat with LC n-3 PUFA is a source of increasing intakes. Studies have shown that the use of oil from transgenic soya in which the fatty acid metabolic pathways have been modified can increase the n-3 VLC-PUFAs of chicken meat. Feeding broilers a soya oil rich in SDA produced meat with approximately 150–200 mg LC n-3 PUFA/100 g meat (Rymer and Givens 2009).

Transgenic soybean lines have been produced which contain the decarboxylase gene (*oxdc*) obtained from a Flammulina spp. using the biolistic process. Further analysis revealed that expression of the *oxdc* gene results in resistance to *S. sclerotiorum* (Cunha et al. 2010).

9 Conclusion

The blend of both classical and genomic breeding approaches has been proven to be very valuable for crop improvement. Still there are many aspects which need attention and have a capacity to progress. Whole genome sequencing and study of syntenic regions have proven the utility of sequence information from one crop to another. In the present world scenario there is a great need of better yield and vigor of vital crops to fulfill the needs of the growing population. Transgenic crops having superb and valuable properties are the most needed entity today. Marker assisted selection and transgenic crops are opening new horizons in the field of genomics. In the coming future further advancement in the field of biotechnology will lead to overcome the present dilemmas.

References

Abe F, Saito K, Miura K, Toriyama K (2002) A single nucleotide polymorphism in the alternative oxidase gene among rice varieties differing in low temperature tolerance. FEBS Lett 527:181–185

Alfonso-Rubi OF, Castanera P, Carbonaro P, Diaz I (2003) Transgenic expression of trypsin inhibitor CMe from barley in indica and japonica rice confers resistance to the rice weevil *Sitophilus oryzae*. Transgenic Res 12:23–31

Alvarez ML, Guelman S, Halford NG, Lustig S, Reggiardo MI, Ryabushkina N, Schewry P, Stein J, Vallejos RH (2000) Silencing of HMW glutenins in transgenic wheat expressing extra HMW subunits. Theor Appl Genet 100:319–327

Asad (2004) Use of antisense RNA technology to suppress cotton leaf curl virus infection in a model plant system. Biochemistry PhD thesis. University of the Punjab, Lahore, p 215

Asad S, Mukhtar Z, Nazir F, Hashmi AJ, Mansoor S, Zafar Y, Arshad M (2008) Silicon carbide whisker-mediated embryogenic callus transformation of cotton (*Gossypium hirsutum* L) and regeneration of salt tolerant plants. Mol Biotechnol 40:161–169

Asif M (2009) Genomic analysis for quality traits in cotton (*Gossypium hirsutum* L) by DNA fingerprinting technology PhD thesis. BZ University, Multan, Pakistan

Babar M, Saranga Y, Iqbal Z, Arif M, Zafar Y, Lubbers E, Chee P (2009) Identification of QTLs and impact of selection from various environments (dry vs irrigated) on the genetic relationships among the selected cotton lines from f6 population using a phylogenetic approach. Afr J Biotechnol 8(19):4802–4810

Barry GF (2001) The use of the Monsanto draft rice genome sequence in research. Plant Physiol 125:1164–1165

Bashir K, Husnain T, Fatima T, Riaz N, Makhdoom R, Riazuddin S (2005) Novel *indica* basmati line (B-370) expressing two unrelated genes of *Bacillus thuringiensis* is highly resistant to two *Lepidopteran* insects in the field. Crop Prot 24:870–879

Beckmann JS, Soller M (1986) Restriction fragment length polymorphisms and genetic improvement of agricultural species. Euphytica 35:111–124

Bentur JS (2006) Host plant resistance to insects as a core of rice IPM. Science, Technology and Trade for Peace and Prosperity (IRRI, ICAR), McMillan India Ltd, pp 419–435

Bhattacharya J, Mukherjee R, Banga A, Dandapat A, Mandal CC, Hossain MA (2006) A transgenic approach for developing insect resistant rice plant types, Science, Technology and Trade for Peace and Prosperity (IRRI, ICAR), Mc Millan India Ltd, pp 245–264

Blanc G, Charcosset A, Veyrieras JB, Gallais A, Moreau L (2008) Marker-assisted selection efficiency in multiple connected populations: a simulation study based on the results of a QTL detection experiment in maize. Euphytica 161:71–84

Blechl A, Lin J, Nguyen S, Chan R, Anderson OD, Dupont FM (2007) Transgenic wheats with elevated levels of Dx5 and/or Dy10 highmolecular- weight glutenin subunits yield doughs with increased mixing strength and tolerance. J Cereal Sci 45:172–183

Bouchez A, Hospital F, Causse M, Gallais A, Charcosset A (2002) Marker-assisted introgression of favorable alleles at quantitative trait loci between maize elite lines. Genetics 162:1945–1959

Brar DS, Khush GS (2007) Breeding rice for resistance to biotic stresses: conventional and molecular approaches. SABRAO J 45:225–234

Breitler JC, Cordero MJ, Royer M, Meynard D, San Segundo B, Guiderdoni E (2001) The –689/+197 region of the maize protease inhibitor gene directs high level, wound inducible expression of the cry1B gene which protects transgenic rice plants from stemborer attack. Mol Breed 7:259–274

Brookes G, Barfoot P (2006) Global impact of biotech crops: socio-economic and environmental effects 1996–2004. AgBioforum 8(2&3):187–196

Brookes G, Barfoot P (2008) Global impact of biotech crops: socio-economic and environmental effects 1996–2006. Ag Bio Forum 11(1):21–38. Available on the World Wide Web: http://www.agbioforum.org

Charlson DV, Bhatnagar S, King CA, Ray JD, Sneller CH, Carter TE Jr, Purcell LC (2009) Polygenic inheritance of canopy wilting in soybean [*Glycine max* (L) Merr]. Theor Appl Genet 119:587–594

Chee PW, Draye X, Jiang C, Decanini L, Delmonte T, Bredhauer B, Smith CW, Paterson AH (2005a) Molecular dissection of interspecific variation between *Gossypium hirsutum* and *G barbadense* (cotton) by a backcross-self approach: I Fiber elongation. Theor Appl Genet 111:757–763

Chee PW, Draye X, Jiang C, Decanini L, Delmonte T, Bredhauer B, Smith CW, Paterson AH (2005b) Molecular dissection of phenotypic variation between *Gossypium hirsutum* and *Gossypium barbadense* (cotton) by a backcross-self approach: III Fiber length. Theor Appl Genet 111:772–781

Chen JZ, Scheffler BE, Dennis E, Triplett B, Zhang T, Guo W, Chen X, Stelly DM, Rabinowicz PD, Town C (2007) Towards sequencing cotton (*Gossypium*) genomes. Plant Physiol 145:1303–1310

Chen H, Qian N, Guo WZ, Song QP, Li BC, Deng FJ, Dong CG, Zhang TZ (2009) Using three overlapped RILs to dissect genetically clustered QTL for fiber strength on ChroD8 in upland cotton. Theor Appl Genet 119(4):605–612

Chu CG, Chao S, Friesen TL, Faris JD, Zhong S, Xu SS (2010) Identification of novel tan spot resistance QTLs using an SSR-based linkage map of tetraploid wheat. Mol Breed 25(2):327–338

Chunwongse J, Martin GB, Tanksley SD (1993) Pre-germination genotypic screening using PCR amplification of half-seeds. Theor Appl Genet 86:694–698

Concibido VC, Vallee BL, Mclaird P, Pineda N, Meyer J, Hummel L, Yang J, Wu K, Delannay X (2003) Introgression of a quantitative trait locus for yield from *Glycine soja* into commercial soybean cultivars. Theor Appl Genet 106:575–582

Cunha WG, Pancotil MLP, Pancoti HL, Ribeiro RE, Aragão FJL (2010) High resistance to *Sclerotinia sclerotiorum* in transgenic soybean plants transformed to express an oxalate decarboxylase gene. Plant Pathol 59(4):654–660. doi:101111/j1365-3059201002279x

Damude HG, Kinney AJ (2008) Engineering oilseed plants for a sustainable, land-based source of long chain polyunsaturated fatty acids. Lipids 42:179–185

De Maagd RA, Bravo A, Crickmore N (2001) How Bacillus thuringiensis has evolved specific toxins to colonize the insect world. Trends Genet 17:193–199

Delseny M (2003) Towards an accurate sequence of the rice genome. Curr Opin Plant Biol 6:101–105

Draye X, Chee P, Jiang C, Decanini L, Delmonte T, Bredhauer B, Smith CW, Paterson AH (2005) Molecular dissection of phenotypic variation between *Gossypium hirsutum* and *G. barbadense* (cotton) by a backcross-self approach. II fiber fineness. Theor Appl Genet 111:764–771

Dreher K, Khairallah M, Ribaut J, Morris M (2003) Money matters (I): costs of field and laboratory procedures associated with conventional and marker-assisted maize breeding at CIMMYT. Mol Breed 11:221–234

Feng Q, Zhang Y, Hao P, Wang S, Fu G, Huang Y, Li Y, Zhu J, Liu Y, Hu X (2002) Sequence and analysis of rice chromosome 4. Nature 420:316–320

Foissac X, Du Edwards MGJP, Gatehouse AMR, Gatehouse JA (2002) Putative protein digestion in a sap-sucking homopteran plant pest (rice brown plant hopper; Nilaparvata lugens, Delphacidae) – identification of trypsin-like and cathepsin – like proteases. Insect Biochem Mol Biol 32:967–978

Francia E, Rizza F, Cattivelli L, Stanca AM, Galiba G, Toth B, Hayes PM, Skinner JS, Pecchioni N (2004) Two loci on chromosome 5H determine low temperature tolerance in a 'Nure' (winter) x 'Tremois' (spring) barley map. Theor Appl Genet 108:670–680

Francki MG (2009) Genomics for wheat improvement: In: Jain SM, Brar DS (eds) Molecular techniques in crop improvement. Springer Science+Business Media B.V., Dordrecht. doi:10.1007/978-90-481-2967-6_12

Fukuoka S, Saka N, Koga H, Ono K, Shimizu T, Ebana K, Hayashi N, Takahashi A, Hirochika H, Okuno K, Yano M (2009) Loss of function of a proline-containing protein confers durable disease resistance in rice. Science 325:998–1001

Goedeke S, Hensel G, Kapusi E, Gahrtz M, Kumlehn J (2007) Transgenic barley in fundamental research and biotechnology. Transgenic Plant J 1:104–117

Goff SA, Ricke D, Lan TH, Presting G, Wang R, Dunn M, Glazebrook J, Sessions A, Oeller P, Varma H (2002) A draft sequence of the rice genome (*Oryza sativa* L ssp *japonica*). Science 296:92–100

Gopalakrishnan S, Sharma RK, Rajkumar KA, Joseph M, Singh VP, Singh AK, Bhat KV, Singh NK, Mohapatra T (2008) Integrating marker- assisted background analysis with foreground selection for identification of superior bacterial blight resistant recombinants in Basmati rice. Plant Breed 127:131–139

Grewal TS, Rossnage BG, Scoles GJ (2008) Validation of molecular markers for covered smut resistance and marker-assisted introgression of loose and covered smut resistance into hulless barley. Mol Breed 21:37–48

Guo W, Zhang T, Shen X, Yu JZ, Kohel RJ (2003) Development of SCAR marker linked to a major QTL for high fiber strength and its usage in molecular-marker assisted selection in Upland cotton. Crop Sci 43:2252–2256

Hamilton KA, Pyla P, Breeze M (2004) Bollgard II cotton: compositional analysis and feeding studies of cottonseed from insect-protected cotton (*Gossypium hirsutum* L) producing the Cry1Ac and Cry2Ab2 proteins. J Agric Food Chem 52:6969–6976

Haq I, Asad S, Zafar Y (2005) Bioloistic-mediated transformation of cotton (*Gossypium hirsutum* L): embryogenic calli as explant. J Plant Biotechnol 7(4):211–218

Hardke JT, Leonard BR, Huang F, Jackson RE (2011) Damage and survivorship of fall armyworm (Lepidoptera: Noctuidae) on transgenic field corn expressing *Bacillus thuringiensis* Cry proteins. Crop Prot 30(2):168–172

Harholt J, Bach IC, Lind-Bouquin S, Nuna K, Brinch-pedersen H, Holm PB, Scheller H (2010) Generation of transgenic wheat (*Triticum aestivum* L) accumulating heterologous endo-xylanase or ferulic esterase in the endosperm. Plant Biotechnol J 8:351–362

Harris K, Subudhi PK, Borrell A, Jordan D, Rosenow D, Nguyen H, Klein P, Klein R, Mullet J (2007) Sorghum stay-green QTL individually reduce post-flowering drought-induced leaf senescence. J Exp Bot 58(2):327–338

Haussmann BIG, Mahalakshmi V, Reddy BVS, Seetharama N, Hash CT, Geiger HH (2002) QTL mapping of stay-green in two sorghum recombinant inbred populations. Theor Appl Genet 106:133–142

Hayden MJ, Tabone T, Mather DE (2009) Development and assessment of simple PCR markers for SNP genotyping in barley. Theor Appl Genet 119(5):939–951

He GY, Jones HD, D'Ovidio R, Masci S, Chen MJ, West J, Butow B, Anderson OD, Lazzeri P, Fido R, Shewry PR (2005) Expression of an extended HMW subunit in transgenic wheat and the effect on dough mixing properties. J Cereal Sci 42:225–231

Helguera M, Khan IA, Kolmer J, Lijavetzky D, Zhong-Qi L, Dubcovsky J (2003) PCR assays for the Lr37-Yr17-Sr38 cluster of rust resistance genes and their use to develop isogenic hard red spring wheat lines. Crop Sci 43:1839–1847

Hittalmani S, Parco A, Mew TV, Zeigler RS, Huang N (2000) Fine mapping and DNA marker-assisted pyramiding of the three major genes for blast resistance in rice. Theor Appl Genet 100:1121–1128

Ho JC, McCouch SR, Smith ME (2002) Improvement of hybrid yield by advanced backcross QTL analysis in elite maize. Theor Appl Genet 105:440–448

Horvath H, Jensen LG, Wong OT, Kohl E, Ullrich SE, Cochran J, Kannangara CG, von Wettstein D (2001) Stability of transgene expression, field performance and recombination breeding of transformed barley lines. Theor Appl Genet 102:1–11

Huang L, Gill BS (2001) An RGA-like marker detects all known Lr21 leaf rust resistance gene family members in Aegilops tauschii and wheat. Theor Appl Genet 103:1007–1013

Huang N, Angeles ER, Domingo J, Magpantay G, Singh S, Zhang G, Kumaravadivel N, Bennett J, Khush GS (1997) Pyramiding of bacterial blight resistance genes in rice: marker-assisted selection using RFLP and PCR. Theor Appl Genet 95:313–320

Huang XQ, Wang LX, Xu MX, Röder MS (2003) Microsatellite mapping of the powdery mildew resistance gene Pm5e in common wheat (*Triticum aestivum* L). Theor Appl Genet 106:858–865

Hwang TY, Moon JK, Yang YK S, Mohankumar S, Yu YH, Lee YH, Kim HS, Kim HM, Maroof MAS, Jeong SC (2006) Application of comparative genomics in developing molecular markers tightly linked to the virus resistance gene Rsv4 in soybean. Genome 49:380–388

Ibrahim M, Khan SA, Zafar Y, Mansoor S, Yusuf A, Mukhtar Z (2009) Expression of a full length Arabidopsis vacuolar H^+-pyrophosphatase (AVP1) gene in tobacco (*Nicotiana tabbacum*) to increase tolerance to drought and salt stresses. J Phytol 1:433–440

Igartua E, Edney M, Rossnagel BG, Spaner D, Legge WG, Scoles GJ, Eckstein PE, Penner GA, Tinker NA, Briggs KG, Falk DE, Mather DE (2000) Marker-based selection of QTL affecting grain and malt quality in two-row barley. Crop Sci 40:1426–1433

Jixiang W, Johnie NJ, McCarty JC, Zhong M, Michael S (2007) AFLP marker associations with agronomic and fiber traits in cotton. Euphytica 153:153–163

Kassem MA, Shultz JL, Meksem K, Cho Y, Wood AJ, Iqbal MJ, Lightfoot DA (2006) An updated 'Essex' by 'Forrest' linkage map and first composite interval map of QTL underlying six soybean traits. Theor Appl Genet 113(6):1015–1026

Khan SA, Zafar Y, Briddon RW, Malik KA, Mukhtar Z (2006) Spider venom toxin protects plants from insect attack. Transgenic Res 15:349–357

Kim JK, Jang IC, Wu R, Zu WN, Boston RS, Lee YH, Alm P, Nahm BH (2003) Co-expression of a modified maizeribosome-inactivating protein and a rice basic chitinase gene in transgenic rice plants confers enhanced resistance to sheath blight. Transgenic Res 12:475–484

Kloppers FJ, Pretorius ZA (1997) Effects of combinations amongst genes Lr13, Lr34 and Lr37 on components of resistance in wheat to leaf rust. Plant Pathol 46:737–750

Knoll J, Ejeta G (2008) Marker-assisted selection for early-season cold tolerance in sorghum: QTL validation across populations and environments. Theor Appl Genet 116:541–553

Komori T, Miyata M, Yamamoto T, Nitta N, Hiei Y, Yano M, Ueki J, Komari T (2009) Isolation and functional analysis of the gene controlling the stubspreading trait in rice (*Oryza sativa* L). Plant Breed 128:568–575

Krishnaveni S, Jeoung JM, Muthukrishnan S, Liang GH (2001) Transgenic sorghum plants constitutively expressing a rice chitinase gene show improved resistance to stalk rot. J Genet Breed 55:151–158

Lagat M, Danson J, Kimani M, Kuria A (2008) Quantitative trait loci for resistance to maize streak virus disease in maize genotypes used in hybrid development. Afr J Biotechnol 7(15):2573–2577

Li Y, Hill CB, Carlson SR, Diers BW, Hartman GL (2007) Soybean aphid resistance genes in the soybean cultivars Dowling and Jackson map to linkage group M. Mol Breed 19:25–34

Lightfoot DA (2008) Soybean genomics: developments through the use of cultivar "Forrest". Int J Plant Genomics 2008:1–22

Lin HX, Zhu MZ, Yano M, Gao JP, Liang ZW, Su WA, Hu XH, Ren ZH, Chao DY (2004) QTLs for Na^+ and K^+ uptake of the shoots and roots controlling rice salt tolerance. Theor Appl Genet 108:253–260

Liu S, Yu L-X, Singh RP, Jin Y, Sorrells ME, Anderson JA (2010) Diagnostic and co-dominant PCR markers for wheat stem rust resistance genes Sr25 and Sr26. Theor Appl Genet 120(4):691–697

Lu Y, Yan J, Guimarães CT, Taba S, Hao Z, Gao S, Chen S, Li J, Zhang S, Vivek BS, Magorokosho C, Mugo S, Makumbi D, Parentoni SN, Shah T, Rong T, Crouch JH, Xu Y (2009) Molecular characterization of global maize breeding germplasm based on genomewide single nucleotide polymorphisms. Theor Appl Genet 120(1):93–115

Ma W, Zhang W, Gale KR (2003) Multiplex-PCR typing of high molecular weight glutenin alleles in wheat. Euphytica 134:51–60

Ma ZQ, Wei JB, Cheng SH (2004) PCR-based markers for the powdery mildew resistance gene Pm4a in wheat. Theor Appl Genet 109:140–145

Mago R, Zhang P, Bariana HS, Verlin DC, Bansal UK, Ellis JG, Dundas IS (2009) Development of wheat lines carrying stem rust resistance gene Sr39 with reduced *Aegilops speltoides* chromatin and simple PCR markers for marker-assisted selection. Theor Appl Genet 119(8):1441–1450

Maheshwari M, Varalaxmi Y, Vijayalakshmi A, Yadav SK, Sharmila P, Venkateswarlu B, Vanaja M, Pardha Saradhi P (2010) Metabolic engineering using *mtlD* gene enhances tolerance to water deficit and salinity in sorghum. Biol Plant 54:647–652

Manoharan M, Dahleen LS, Hohn TM, Neate SM, Yu X-H, Alexander NJ, McCormick SP, Bregitzer P, Schwarz PB, Horsley RD (2006) Expression of 3-OH trichothecene acetyltransferase in barley (*Hordeum vulgare* L.) and effects on deoxynivalenol. Plant Sci 171:699–706

Maqbool SB, Christou P (1999) Multiple traits of agronomic importance in transgenic indica plants, analysis of transgene integration pattern, expression levels and stability. Mol Breed 5:471–480

Maqbool SB, Husnain T, Riazuddin S, Masson L, Christou P (1998) Effective control of yellow stem borer and rice leaf folder in transgenic rice indica varieties Basmati 370 and M7 using the novel *d*-endotoxin cry2A *Bacillus thuringiensis* gene. Mol Breed 4(6):501–507

Maqbool SB, Riazuddin S, Loc NT, Gatehouse AMR, Gatehouse JA, Christou P (2001) Expression of multiple insecticidal genes confers broad resistance against a range of different rice pests. Mol Breed 7(1):85–93

Martin JM, Meyer FD, Smidansky ED, Wanjugi H, Blechl AE, Jiroux MJ (2006) Complementation of the pina (null) allele with the wild type Pina sequence restores a soft phenotype in transgenic wheat. Theor Appl Genet 113:1563–1570

Masci S, Rovelli L, Kasarda DD, Vensel WH (2002) Characterisation and chromosomal localization of C-type low-molecular-weight glutenin subunits in the bread wheat cultivar Chinese Spring. Theor Appl Genet 104:422–428

Mochizuki A, Nishizawa Y, Onodera H, Tabei Y, Toki S, Habu Y, Ugaki M, Ohashi Y (1999) Transgenic rice plants expressing a trypsin inhibitor are resistant against rice stem borers, *Chilo suppressalis*. Entomol Exp Appl 93:173–178

Moose SP, Mumm RH (2008) Molecular plant breeding as the foundation for 21st century crop improvement. Plant Physiol 147:969–977

Morris M, Dreher K, Ribaut JM, Khairallah M (2003) Money matters (II): costs of maize inbred line conversion schemes at CIMMYT using conventional and marker-assisted selection. Mol Breed 11:235–247

Mullet J (2009) Traits and genes for plant drought tolerance. In: Kriz AL, Larkins BA (eds) Molecular genetic approaches to maize improvement biotechnology in agriculture and forestry, vol 63. Springer, Berlin, pp 55–64

Mumtaz II (2007) Identification of structural and functional genomic markers for fiber quality traits in cotton using interspecific population (*G hirsutum* x *G barbadense*). MPhil thesis, QA University, Islamabad, Pakistan

Murphy LR, Santra D, Kidwell K, Yan G, Chen X, Campbell KG (2009) Linkage maps of wheat stripe rust resistance genes Yr5 and Yr15 for use in marker-assisted selection. Crop Sci 49:1786–1790

Nagadhara D, Ramesh S, Pasalu IC, Rao YK, Krishnaiah NV, Sarma NP, Bown DP, Gatehouse JA, Reddy VD, Rao KV (2003) Transgenic indica rice resistant to sap-sucking insects. Plant Biotechnol J 1:231–240

Nagadhara D, Ramesh S, Pasalu IC, Rao YK, Sarma NP, Reddy VD, Rao KV (2004) Transgenic rice the plants expressing snowdrop lectin gene (GNA) exhibit high-level resistance to the whitebacked planthopper (*Sogatella furcifera*). Theor Appl Genet 109:1399–1405

Nguyen TTT, Klueva N, Chamareck V, Aarti A, Magpantay G, Millena ACM, Pathan MS, Nguyen HT (2004) Saturation mapping of QTLs regions and identification of putative candidate genes for drought resistance in rice. Mol Genet Genomics 272:35–46

Nocente F, Gazza L, Pasquini M (2007) Evaluation of leaf rust resistance genes Lr1, Lr9, Lr24, Lr47 and their introgression into common wheat cultivars by marker-assisted selection. Euphytica 155:329–336

Palomeque LC, Jun LL, Li W, Hedges B, Cober ER, Rajcan I (2009a) QTL in mega-environments: I Universal and specific seed yield QTL detected in a population derived from a cross of highyielding adapted 9 high-yielding exotic soybean lines. Theor Appl Genet 119:417–427

Palomeque LC, Jun LL, Li W, Hedges B, Cober ER, Rajcan I (2009b) QTL in mega-environments: II Agronomic trait QTL colocalized with seed yield QTL detected in a population derived from a cross of high-yielding adapted 9 high-yielding exotic soybean lines. Theor Appl Genet 119:429–436

Pasapula V, Shen G, Kuppu S, Paez-Valencia J, Mendoza M, Hou P, Chen J, Qiu X, Zhu L, Zhang X, Auld D, Blumwald E, Zhangl H, Gaxiola R, Payton P (2011) Expression of an Arabidopsis vacuolar H+-pyrophosphatase gene (AVP1) in cotton improves drought- and salt tolerance and increases fibre yield in the field conditions. Plant Biotechnol J 9(1):88–99

Paterson AH, Saranga Y, Menz M, Jiang C, Wright RJ (2003) QTL analysis of genotype x environment interactions affecting cotton fiber quality. Theor Appl Genet 106:384–396

Paterson AH, Bowers JE, Bruggmann R, Dubchak I, Grimwood J, Gundlach H, Haberer G, Hellsten U, Mitros T, Poliakov A, Schmutz J, Spannagl M, Tang H, Wang X, Wicker T, Bharti AK, Chapman J, Feltus FA, Gowik U, Grigoriev IV, Lyons E, Maher CA, Martis M, Narechania A, Otillar RP, Penning BW, Salamov AA, Wang Y, Zhang L, Carpita NC, Freeling M, Gingle AR, Hash CT, Keller B, Klein P, Kresovich S, McCann MC, Ming R, Peterson DG, Rahman M, Ware D, Westhoff P, Mayer KFX, Messing J, Rokhsar DS (2009) The *Sorghum bicolor* genome and the diversification of grasses. Nature 457:551–556

Perlak FJ, Oppenhuizen M, Gustafson K, Voth R, Sivasupramaniam S, Heering D, Carey B, Ihrig RA, Roberts JK (2001) Development and commercial use of *Bollgardâ cotton* in the USA: early promises versus today's reality. Plant J 27(6):489–501

Qaim M, Subramanian A (2010) Benefits of transgenic plants: a socioeconomic perspective. In: Kempken F, Jung C (eds) Genetic modification of plants. Springer, Heidelberg, pp 615–628

Radovanovic N, Cloutier S (2003) Gene-assisted selection for high molecular weight glutenin subunits in wheat doubled haploid breeding programs. Mol Breed 12:51–59

Ragot M, Biasiolli M, Delbut MF, Dell-Orco A, Malgarini L, Thevenin P, Vernoy J, Vivant J, Zimmermann R, Gay G (1995) Marker-assisted backcrossing: a practical example. In: Berville A, Tersac M (eds) Les Colloques, No. 72, techniques et utilizations des marqueurs moleculaires. INRA, Paris, pp 45–56

Rahman M (2002) DNA markers studies in cotton. PhD thesis, Department of Plant Breed Genet. University of Agriculture, Faisalabad, Pakistan

Rahman M, Zafar Y (2007a) Registration of NIBGE-115. J Plant Regist 1:51–52

Rahman M, Zafar Y (2007b) Registration of NIBGE-2. J Plant Regist 2:113–114

Rahman M, Ahmed N, Asif M, Zafar Y (2006) Identification of DNA markers linked with cotton leaf curl disease (CLCD). International Cotton Genome Initiative (ICGI) Workshop, 18–20 Sept Brasilia, Brazil, pp 77–78

Rahman M, Tabassum N, Ullah I, Asif M, Zafar Y (2008) Studying the extent of genetic diversity among *Gossypium arboreum* L genotypes/cultivars using DNA fingerprinting. Genet Res Crop Evol 55:331–339

Rahman M, Zafar Y, Paterson AH (2009) Gossypium DNA markers: types, numbers and uses. In: Paterson AH (ed) Genetics and genomics of cotton. Springer, New York. doi:101007/978-0-387-70810-2_5

Ramesh S, Nagadhara D, Pasalu IC, Kumari AP, Sarma NP, Reddy VD, Rao KV (2004a) Development of stem borer resistant transgenic parental lines involved in the production of hybrid rice. J Biotechnol 111:131–141

Ramesh S, Nagadhara D, Reddy VD, Rao KV (2004b) Production of transgenic indica rice resistant to yellow stem borer and sap-sucking insects, using super-binary vectors of *Agrobacterium tumefaciens*. Plant Sci 166:1077–1085

Reddy MSS, Ghabrial SA, Redmond CT, Dinkins RD, Collins GB (2001) Resistance to bean pod mottle virus in transgenic soybean lines expressing the capsid polyprotein. Phytopathology 91:831–838

Ribaut JM, Hoisington D (1998) Marker-assisted selection: new tools and strategies. Trends Plant Sci 3(6):236–239

Ribaut JM, Jiang C, Gonzales-de Leon D, Edmeades GO, Hoisington DA (1997) Identification of quantitative trait loci under drought conditions in tropical maize. 2: yield components and marker-assisted selection strategies. Theor Appl Genet 94:887–896

Ribaut JM, Bänziger M, Setter T, Edmeades G, Hoisington D (2004) Genetic dissection of drought tolerance in maize: a case study. In: Nguyen H, Blum A (eds) Physiology and biotechnology integration for plant breeding. Marcel Dekker, Inc, New York, pp 571–611

Rice Chromosome 10 Sequencing Consortium (2003) In-depth view of structure, activity, and evolution of rice chromosome 10. Science 300:1566–1569

Robin S, Pathan MS, Courtois B, Lafitte R, Carandang S, Lanceras S, Amante M, Nguyen HT, Li Z (2003) Mapping osmotic adjustment in an advanced back-cross inbred population of rice. Theor Appl Genet 107:1288–1296

Rymer C, Givens DI (2009) The effect of feeding steariodonic acid enriched soya oil to broilers on the fatty acid composition and sensory characteristics of chicken meat. Br Poult Abstr 5:44

Saito K, Hayano-Saito Y, Maruyama-Funatsuki W, Sato Y, Kato A (2004) Physical mapping and putative candidate gene identification of a quantitative trait locus Ctb1 for cold tolerance at the booting stage of rice. Theor Appl Genet 109:515–522

Sanchez AC, Subudhi PK, Rosenow DT, Nguyen HT (2002) Mapping QTLs associated with drought resistance in sorghum (*Sorghum bicolor* L. Moench). Plant Mol Biol 48:713–726

Sandhu D, Schallock KG, Velez NR, Lundeen P, Cianzio S, Bhattacharyya MK (2005) Soybean *Phytophthora* resistance gene Rps8 maps closely to the Rps3 region. J Heredity 96(5):536–541

Sanjaya V, Satyavathi V, Prasad V, Kirthi N, Maiya SP, Savithri HS, Lakshmi Sita G (2005) Development of cotton transgenics with antisense AV2 gene for resistance against cotton leaf curl virus (CLCuD) via *Agrobacterium tumefaciens*. Plant Cell Tissue Organ Cult 81:55–63

Saranga Y, Menz M, Jiang CX, Wright RJ, Yakir D, Paterson AH (2001) Genomic dissection of genotype x environment interactions conferring adaptation of cotton to arid conditions. Genome Res 11:1988–1995

Sasaki T, Burr B (2000) International rice genome sequencing project: the effort to completely sequence the rice genome. Curr Opin Plant Biol 3:138–141

Sasaki T, Matsumoto T, Antonio BA, Nagamura Y (2005) From mapping to sequencing, post-sequencing and beyond. Plant Cell Physiol 46:3–13

Sasaki T, Matsumoto T, Yamamoto K, Sakata K, Baba T, Katayose Y, Wu J, Niimura Y, Cheng Z, Nagamura Y, Antonio BA, Kanamori H, Hosokawa S, Masukawa M, Arikawa K, Chiden Y, Hayashi M, Okamoto M, Ando T, Aoki H, Arita K, Hamada M, Harada C, Hijishita S, Honda M, Ichikawa Y, Idonuma A, Iijima M, Ikeda M, Ikeno M, Ito S, Ito T, Ito Y, Ito Y, Iwabuchi A, Kamiya K, Karasawa W, Katagiri S, Kikuta A, Kobayashi N, Kono I, Machita K, Maehara T, Mizuno H, Mizubayashi T, Mukai Y, Nagasaki H, Nakashima M, Nakama Y, Nakamichi Y, Nakamura M, Namiki N, Negishi M, Ohta I, Ono N, Saji S, Sakai K, Shibata M, Shimokawa T, Shomura A, Song J, Takazaki Y, Terasawa K, Tsuji K, Waki K, Yamagata H, Yamane H, Yoshiki S, Yoshihara R, Yukawa K, Zhong H, Iwama H, Endo T, Ito H, Hahn JH, Kim HI, Eun MY, Yano M, Jiang J, Gojobori T (2002) The genome sequence and structure of rice chromosome 1. Nature 420:312–316

Satish K, Srinivas G, Madhusudhana R, Padmaja PG, Reddy NR, Mohan MS, Seetharama N (2009) Identification of quantitative trait loci for resistance to shoot fly in sorghum [*Sorghum bicolor* (L) Moench]. Theor Appl Genet 119(8):1425–1439

Schmierer DA, Kandemir N, Kudrna DA, Jones BL, Ullrich SE, Kleinhofs A (2005) Molecular marker-assisted selection for enhanced yield in malting barley. Mol Breed 14(4):463–473

Scott MP (2010) Characterization of Transgenic Maize Containing Iron-Rich Proteins in Grain Long Beach Convention Center, Room 308, Seaside Level 296-3

Sharma HC, Sharma KK, Crouch JH (2004) Genetic transformation of crops for insect resistance: potential and limitations. Crit Rev Plant Sci 23:47–72

Shen L, Courtois B, McNally KL, Robin S, Li Z (2001) Evaluation of nearisogenic lines of rice introgressed with QTLs for root depth through marker-aided selection. Theor Appl Genet 103:75–83

Shin S, Mackintosh CA, Lewis J, Heinen SJ, Radmer L, Dill-Macky R, Baldridge GD, Zeyen RJ, Muehlbauer GJ (2008) Transgenic wheat expressing a barley class II chitinase gene has enhanced resistance to Fusarium Head Blight. J Exp Bot 59:2371–2378

Singh RP, Mujeeb-Kazi A, Huerta-Espino J (1998) Lr46: a gene conferring slow-rusting resistance to leaf rust in wheat. Phytopathology 88:890–894

Singh S, Sidhu JS, Huang N, Vikal Y, Li Z, Brar DS, Dhaliwal HS, Khush GS (2001) Pyramiding three bacterial blight resistance genes (xa5, xa13 and Xa21) using marker-assisted selection into indica rice cultivar PR106. Theor Appl Genet 102:1011–1015

Singh M, Chaudhary K, Singal HR, Magill CW, Boora KS (2006) Identification and characterization of RAPD and SCAR markers linked to anthracnose resistance gene in sorghum [*Sorghum bicolor* (L) Moench]. Euphytica 149:179–187

Stewart JM (1994) Potential for crop improvement with exotic germplasm and genetic engineering challenging the future: Proceedings of the World Cotton Research Conference-1, Brisbane Australia, GA Constable and NW Forrester (eds), CSIRO, Melbourne, Australia, pp 313–327

Suenaga K, Singh RP, Huerta-Espino J, William HM (2003) Microsatellite markers for genes Lr34/Yr18 and other quantitative loci for leaf rust and stripe rust resistance in bread wheat. Phytopathology 93:881–890

Sundaram RM, Vishnupriya MR, Biradar SK, Laha GS, Reddy GA, Rani NS, Sarma NP, Sonti RV (2008) Marker-assisted introgression of bacterial blight resistance in Samba Mahsuri, an elite indica rice variety. Euphytica 160:411–422

Tabashnik BE, Carrière Y, Dennehy TJ, Morin S, Sisterson MS, Roush RT, Shelton AM, Zhao J-Z (2003) Insect resistance to transgenic Bt crops: lessons from the laboratory and field. J Econ Entomol 96:1031–1038

Tadesse Y, Sági L, Swennen R, Jacobs M (2003) Optimization of transformation conditions and production of transgenic sorghum (*Sorghum bicolor*) *via* microparticle bombardment. Plant Cell Tissue Organ Cult 75:1–18

Tang K, Zhao E, Sun X, Wan B, Qi H, Lu X (2001) Production of transgenic rice homozygous lines with enhanced resistance to the rice brown planthopper. Acta Biotechnol 21:117–128

Toenniessen GH (2000) Vitamin A deficiency and golden rice: the role of the Rockefeller Foundation. http://www.rockfound.org

Tosi P, D'Ovidio R, Napier J, Bekes F, Shewry PR (2004) Expression of epitope-tagged LMW glutenin subunits in the starchy endosperm of transgenic wheat and their incorporation into glutenin polymers. Theor Appl Genet 108:468–476

Tosi P, Masci S, Giovangrossi A, D'ovidio R, Bekes F, Larroque O, Napier J, Shewry P (2005) Modification of the low molecular weight (LMW) glutenin composition of transgenic durum wheat: effects on lutenin polymer size and glulten functionality. Mol Breed 16:113–126

Toth B, Francia E, Rizza F, Stanca AM, Galiba G, Pecchioni N (2004) Development of PCR-based markers on chromosome 5H for assisted selection of frost-tolerant genotypes in barley. Mol Breed 14(3):265–273

Tu J, Zhang G, Datta K, Xu C, He Y, Zhang Q, Khush GS, Datta SK (2000) Field performance of transgenic elite commercial hybrid rice expressing *Bacillus thuringiensis* –endotoxin. Nat Biotechnol 18:1101–1104

Ullah I (2009) Molecular genetic studies for drought tolerance in cotton. PhD thesis, QA University, Islamabad

Vagujfalvi A, Galiba G, Cattivelli L, Dubcovsky J (2003) The cold-regulated transcriptional activator Cbf3 is linked to the frost-tolerance locus Fr-A2 on wheat chromosome 5A. Mol Genet Genomics 269:60–67

Wang D, Graef GL, Procopiuk AM, Diers BW (2004) Identification of putative QTL that underlie yield in interspecific soybean backcross populations. Theor Appl Genet 108:458–467

Wang C, Zhang Y, Han D, Kang Z, Li G, Cao A, Chen P (2008) SSR and STS markers for wheat stripe rust resistance gene Yr26. Euphytica 159:359–366

Wang P, Su L, Qin L, Hu B, Guo W, Zhang T (2009) Identification and molecular mapping of a Fusarium wilt resistant gene in upland cotton. Theor Appl Genet 119:733–739

Willcox M, Khairallah M, Bergvinson D, Bergvinson D, Crossa J, Deutsch JA, Edmeades GO, Gonza´lez-de-Leo´ ND, Jiang C, Jewell DC, Mihm JA, Williams WP, Hoisington D (2002) Selection for resistance to southwaestern corn borer using marker-assisted and conventional backcrossing. Crop Sci 42:1516–1528

Wu A, Sun X, Pang Y, Tang K (2002) Homozygous transgenic rice lines expressing GNA with enhanced resistance to the rice sap-sucking pest Laodelphax striatellus. Plant Breed 121:93–95

Wu J, Gutierrez OA, Jenkins JN, McCarty JC, Zhu J (2009a) Quantitative analysis and QTL mapping for agronomic and fiber traits in an RI population of upland cotton. Euphytica 165:231–245

Wu S, Pumphrey M, Bai G (2009b) Molecular mapping of stem-rust resistance gene Sr40 in wheat. Crop Sci 49:1681–1686

Xu D, Xue Q, McElroy D, Mawal Y, Hilder VA, Wu R (1996) Constitutive expression of a cowpea trypsin inhibitor gene, CpTi, in transgenic rice plants confers resistance to two major insect pests. Mol Breed 2:167–173

Xu QJ, Liu CL, Chang C, Wang CP, You MS, Lia BY, Liu GT (2008) PCR-based markers for identification of HMW-GS at Glu-B1x loci in common wheat. J Cereal Sci 47:394–398

Yarasi B, Vijaya KS, Pasalu IC, Reddy VD, Rao KV (2008) Transgenic rice expressing *Allium sativum* leaf agglutinin (ASAL) exhibits high-level resistance against major sap-sucking pests. BMC Plant Biol 8:102. doi:10.1186/1471-2229-8-102

Yi YJ, Liu HY, Huang XQ, An LZ, Wang F, Wang XL (2008) Development of molecular markers linked to the wheat powdery mildew resistance gene Pm4b and marker validation for molecular breeding. Plant Breed 127:116–120

Yousef GG, Juvik JA (2002) Enhancement of seedling emergence in sweet corn by marker-assisted backcrossing of beneficial QTL. Crop Sci 42:96–104

Yu J, Hu S, Wang J, Wong GK, Li S, Liu B, Deng Y, Dai L, Zhou Y, Zhang X, Cao M, Liu J, Sun J, Tang J, Chen Y, Huang X, Lin W, Ye C, Tong W, Cong L, Geng J, Han Y, Li L, Li W, Hu G, Huang X, Li W, Li J, Liu Z, Li L, Liu J, Qi Q, Liu J, Li L, Li T, Wang X, Lu H, Wu T, Zhu M, Ni P, Han H, Dong W, Ren X, Feng X, Cui P, Li X, Wang H, Xu X, Zhai W, Xu Z, Zhang J, He S, Zhang J, Xu J, Zhang K, Zheng X, Dong J, Zeng W, Tao L, Ye J, Tan J, Ren X, Chen X, He J, Liu D, Tian W, Tian C, Xia H, Bao Q, Li G, Gao H, Cao T, Wang J, Zhao W, Li P, Chen W, Wang X, Zhang Y, Hu J, Wang J, Liu S, Yang J, Zhang G, Xiong Y, Li Z, Mao L, Zhou C, Zhu Z, Chen R, Hao B, Zheng W, Chen S, Guo W, Li G, Liu S, Tao M, Wang J, Zhu L, Yuan L, Yang H (2002) A draft sequence of the rice genome (*Oryza sativa* L ssp *indica*). Science 296:79–92

Zafar Y (2007) Development of agriculture biotechnology in Pakistan. J AOAC Int 90(5): 1500–1507

Zhang J, Zheng HG, Aarti A, Pantuwan G, Nguyen TT, Tripathy JN, Sarial AK, Robin S, Babu RC, Nguyen BD, Sarkarung S, Blum A, Nguyen HT (2001) Locating genomic regions associated with components of drought resistance in rice: comparative mapping within and across species. Theor Appl Genet 103:19–29

Zhang TZ, Yuan YL, Yu J, Guo WZ, Kohel RJ (2003) Molecular tagging of a major QTL for fiber strength in upland cotton and its marker assisted selection. Theor Appl Genet 106:262–268

Zhang J, Li X, Jiang G, Xu Y, He Y (2006) Pyramiding of Xa7 and Xa21 for the improvement of disease resistance to bacterial blight in hybrid rice. Plant Breed 125:600–605

Zhou PH, Tan YF, He YQ, Xu CG, Zhang Q (2003) Simultaneous improvement for four quality traits of Zhenshan 97, an elite parent of hybrid rice, by molecular marker-assisted selection. Theor Appl Genet 106:326–331

Zhu BG, Sun YR (2006) Inheritance of the four-seeded-pod trait in a soybean mutant and marker-assisted selection for this trait. Plant Breed 125:405–407

Zhu H, Muthukrishnan S, Krishnaveni S, Wilde G, Jeoung JM, Liang GH (1998) Biolistic transformation of sorghum using a rice *chitinase* gene. J Genet Breed 52:243–252

Chapter 3
Breeding for Improved Drought Tolerance

Abazar Rajabi and Eric Schmieder Ober

Contents

Abstract The combination of a changing global climate and an increasing world population requires that crops for food, fibre and fuel need to be more resilient to hotter and drier conditions than they have been in the past. Given the years required to develop and release new varieties, breeders have been working hard to achieve this goal. Various strategies being used to counteract drought stress and some examples of successes will be reviewed. These include improvements in maize yield stability, due in part to improved drought tolerance; new varieties of wheat with increased water use efficiency; upland rice with deeper roots, and pearl millet with better yields in arid areas. While most of the selections are based simply on yield, greater emphasis is being placed on morpho-physiological traits associated with greater stress tolerance. Molecular markers based on QTLs linked to these traits also have

A. Rajabi (✉)
Sugar Beet Seed Institute (SBSI), P.O. Box 31585-4114, Karaj, Iran
e-mail: rajabi@sbsi.ir

E.S. Ober
Rothamsted Research, Harpenden, Hertfordshire, AL5 2JQ, UK
e-mail: eric.ober@rothamsted.ac.uk

M. Ashraf et al. (eds.), *Crop Production for Agricultural Improvement*,
DOI 10.1007/978-94-007-4116-4_3, © Springer Science+Business Media B.V. 2012

been used, although less than the number of papers published in this area would suggest. There is promise in candidate genes that have been highlighted by 'omic studies, and great hope has been placed on transgenic approaches, but few of these have yet borne fruit in the field. The rapidly advancing genomic technologies and ready access to cheap sequence data are accelerating the breeding process. However, the ability to obtain relevant, high quality phenotypic information is the rate-limiting process; new innovations in 'high-throughput' phenotyping may provide solutions. Improved characterisation of test environments in multi-location trials combined with advanced statistical dissection of G x E interactions is helping breeders to improve target varieties. The use of models for predicting gene effects, particularly when combining multiple traits, will find greater application as these tools are developed further. Teams of breeders, physiologists, molecular geneticists, agronomists, pathologists, modellers and statisticians, who can communicate across disciplines, are required to tackle the challenge of producing enough food in a world where production is often constrained by insufficient water resources.

Keywords Drought tolerance • Breeding wheat • Sugar beet • Yield • Stress

1 Introduction

Drought decreases the yield of crops in many production systems throughout the world, especially in arid and semi-arid regions. As water resources for agronomic uses become more limiting, the development of drought tolerant varieties gains greater importance (Barker et al. 2004). For plant breeders and physiologists, drought tolerance is a complex trait. The difficulty of dealing with such complexity has motivated many studies to unravel the genetic and physiological basis of drought tolerance in field crops (Ludlow and Muchow 1990). For any research to have an impact on the genetic improvement of drought tolerance, the effort must be established as a long-term commitment and have clear objectives relevant to field conditions with a strong focus on yield. Historically, most improvements in crop drought tolerance have come from incremental improvements achieved by long-term breeding efforts. It is unlikely that this will change in the foreseeable future. However, recent food security issues, combined with the threats resulting from climate change, impose even greater demands on the breeder to improve yields and yield stability. Increased productivity requires a two-pronged approach: at the top end, strive to increase yield potential with each cycle, and at the other, strive to lift the low yields obtained under poor conditions.

To deal with the latter, this chapter examines certain aspects of how yields can be improved through breeding when water is limiting, using predominantly sugar beet and wheat as examples. This is not a comprehensive review of the literature on the subject, as many good reviews have been written recently (Richards 2006; Reynolds and Tuberosa 2008; Praba et al. 2009; Fleury et al. 2010; Passioura and Angus 2010; Richards et al. 2010; Tardieu and Tuberosa 2010; Ashraf 2010; Blum 2011).

2 Mechanisms of Drought Tolerance

The term 'drought tolerance' is used frequently, but is often ill-defined. It has meaning to molecular biologists, crop physiologists, farmers, and global change modellers, but the context changes with scale from cell to landscape, and therefore, it is interpreted and used differently (Passioura 2007). 'Drought tolerance' in this chapter is the ability to capture the greatest proportion of yield potential (the full expression of genetic potential in the absence of environmental limitations) when water is limiting, and includes any plant process that contributes to this result. Note as well that here 'drought' does not mean Sahara-like conditions, but merely any environment where the supply of moisture is insufficient to support crop growth and development without inhibition. Strictly speaking, drought escape (completion of reproduction before drought), and avoidance (maintenance of high tissue water potentials despite shortage of rainfall) can be differentiated from tolerance, which is the ability of tissues to function despite decreased tissue water potentials induced by water deficit. Mechanisms of drought tolerance include: the accumulation of solutes that help maintain turgor through osmotic adjustment (Boyer et al. 2008); accumulation of compatible solutes such as proline and glycine betaine that also help protect the integrity of cellular machinery (Ashraf and Foolad 2007; Chen and Murata 2011), changes in membrane composition via alteration of constituent lipid and steroid species (Olsson 1995); changes in cell wall properties to allow growth at decreased turgor (Sharp et al. 2004); repair mechanisms that replace damaged macromolecules and detoxify excess reactive species such as superoxide O_2 (Miller et al. 2010). Dehydration tolerance describes the ability of plants to continue metabolism at very low water potentials and to repair and recover after release from desiccation; except for seeds, such extreme conditions are rarely met in agricultural settings.

Plants can produce harvestable yields in water-limited environments through combinations of drought escape, avoidance and tolerance. Thus, many different traits have been proposed as selection targets for indirect selection and genetic analysis (Ludlow and Muchow 1990; Edmeades et al. 2000; Morison et al. 2008; Praba et al. 2009; Reynolds et al. 2009). However, genotype-by-environment interactions commonly confound interpretation of experimental results, and realized genetic gain is often lower than predicted. Thus, there have been few attempts to select for improved drought tolerance via indirect selection based on putative drought tolerance traits, or such efforts often have not contributed to successful commercial improvements of drought tolerance in the varieties grown by farmers. There is a clear urgency for a step-change in the productivity of crops in dry conditions, and improved varieties are a key component for achieving this. In this chapter we highlight where there have been successes, and possible avenues for progress.

In order for the cultivation of a variety to be profitable in water-limited areas, it should possess both improved yield potential and drought tolerance so that it is competitive across a range of environments, in addition to the requisite disease resistance and quality traits expected by the market. Thus, breeding for drought tolerance in elite commercial germplasm focuses on adding novel and improved levels of drought tolerance to genotypes that already demonstrate commercially viable levels of yield in the best environments. To achieve further improvements in drought tolerance, molecular technologies in combination with novel pedigree breeding methods are required.

3 Empirical Breeding

Improving drought tolerance and productivity is one of the most difficult tasks for plant breeders. The difficulty arises from the diverse strategies adopted by plants themselves to combat drought stress depending on the timing, severity of stress and stage of crop growth (Hussain 2006). Breeding for tolerance has been hampered by interactions between genotype and environment resulting from variation and intensity of rainfall from year to year.

There are three main breeding approaches for drought tolerance. The first is to breed for high yield under non-stress conditions (Mitra 2001). In one hand, the maximum genetic potential of yield is expected to be realized under non-stress conditions when expression of genetic variation is maximal (Rosielle and Hamblin 1981). On the other hand, a high positive correlation generally exists between performance in stress and non-stress conditions (Cattivelli et al. 2008; Ober et al. 2010). Therefore, a genotype superior under non-stress conditions will also yield relatively well under drought conditions. However, there is no general agreement on the concept of expression of maximum genetic potential in non-stress conditions (Kumar et al. 2007), as genotype by environment interactions may restrict the high yielding genotype from performing well under drought. The second approach is to breed under actual drought conditions (Hurd 1971; Mitra 2001). The problem of this approach is the high variability in drought intensity from year to year and as a consequence, selection pressure on breeding materials is inconsistent across generations as it is highly dependent on environmental conditions. This problem along with low heritability of yield (Blum 2011) makes the breeding programme complicated and slow.

The third approach is to incorporate morpho-physiological traits related to drought resistance into high-yielding genotypes (Mitra 2001). However, this approach is complicated because the physiological and genetic basis of adaptation to drought conditions has not been well understood. An alternative approach is to improve the yield potential of already resistant material provided that there is genetic variation within such material (Bidinger et al. 2005; Mitra 2001). In order to develop high yielding, drought-tolerant genotypes, simultaneous selection in non-stress environment for yield and in drought condition for stability may be done. As such, the breeding methodology to be applied for drought tolerance is the same as that applied for other purposes.

In general, self-pollinated crops could be improved by pedigree and bulk selection methods and cross-pollinated crops by recurrent selection. Back cross is the appropriate methodology for transferring drought tolerance related traits to a high-yielding genotype. On the other hand, family selection methods such as half-sib and full-sib selections maintain a broad genetic base and make it possible to develop the desired drought tolerant genotype (Yunus and Paroda 1982; Mitra 2001).

There is considerable hope being placed on biotechnology to supply the drought tolerant varieties of the future using transgenic methods. Innovations in this area most likely will come from large multinational seed companies that have the resources to develop, test and licence this material. However, transferring discoveries from the lab bench to farmers' fields is taking longer than expected, and in the meantime, CGIAR centres and the private sector continue to release improved conventional varieties (Gilbert 2010). For the less lucrative or non-hybrid crops, publicly financed breeding will continue to carry the effort.

4 Breeding for Harsh Environments

There is a low probability for simultaneous improvement of yield in both unstressed and severe stress environments (Rosielle and Hamblin 1981). When selection for yield under severe drought stress is performed this produces a genetic shift towards expression of constitutive traits that moderate water use. Such traits include reduced leaf area and leaf area index (Blum and Sullivan 1986; Fukai et al. 1999), constitutively low leaf transpiration rate (Kholov et al. 2010), early flowering (e.g., Blum and Pnuel 1990; Yadav and Bhatnagar 2001) and reduced tillering in cereals.

Low constitutive (potential) total plant dry matter was associated with better water status and yield under drought stress in rice (Fukai et al. 1999). Pearl millet hybrids selected for drought resistance were relatively higher yielding in a series of stress environments (Bidinger et al. 2005; Yadav et al. 2011). However, this gain under stress was achieved at the cost of a lower yield in the non-stressed environments. Stress adaptation was consistent with early flowering, limited tillering and low biomass, traits which are not compatible with a high yield potential phenotype. Secondly, high grain yield cultivars have a large sink, which means a high rate of assimilate demand from the source (Blum 2011). A large sink demand from the source can lead to earlier leaf senescence under stress, compared with a plant with a smaller sink (Khanna-Chopra and Sinha 1988). Genotypes with large yield potential may use water at a faster rate than less productive genotypes, which could be detrimental in situations such as terminal drought in Mediterranean climates where early season soil moisture needs to be conserved for later stages such as grain filling. Thus, genotypes bred for high yield potential may have resource demands that can not be met in severe stress environments.

Where drought stress is severe, a cultivar adapted to the region becomes more specific and less common to other regions. This is well supported by other studies showing that wide adaptation was achieved at the cost of poorer performance in specific stress environments. Thus, a variety with uniform superiority over all environments is a rare occurrence (Blum 2011).

5 Germplasm Resources

Breeders strive to create genetic variation within a pool of breeding germplasm prior to making selections. While there is some concern that in general the genetic base of crop species is narrowing (Tanksley and McCouch 1997), there exists substantial genetic variation for many agronomic, morphological and physiological traits within, for example, sugar beet (Ober et al. 2005) and winter wheat (Ober et al. 2010). Novel sources of pest and disease resistance and tolerance to abiotic stresses are also sought from exotic germplasm such as wild relatives of crop species. One effort to introduce this type of genetic background into elite breeding material is the creation of synthetic wheats (Reynolds et al. 2007). This highlights the importance of maintaining germplasm collections, vital work that is often under-funded in national programmes. There are many examples of how germplasm collections have been used in practice. Venuprasad et al. (2007) found that selection for rice yield under severe drought stress was effective if the population carried high levels of drought tolerance derived from a tolerant donor. This strategy works if the relative drought susceptibility of parental lines is known before making any crosses. In this case, screening founder germplasm for drought tolerance early in the breeding process makes sense. Without this information, evaluating crosses at later stages in the breeding process might be more effective.

In India, pearl millet populations containing various proportions of landrace materials from the dry regions performed better than high yielding cultivars under limited water supply (Yadav and Weltzien 2000). Landrace materials expressed better tillering and stable flowering under drought stress as compared with high yielding cultivars. Landraces were more productive under severe drought than elite materials and expressed less delay in flowering (Yadav 2008). When these drought resistant landraces were top-crossed onto elite materials of large panicles it was possible to achieve greater yield and drought resistance in the hybrids.

A rice breeding program in China (He et al. 2010) used backcross breeding with several donors of drought resistance. A two-round-selection for yield under severe drought together with a two-round-selection for yield under irrigation resulted in 113 BC2F8 introgression lines which included superior yielding lines under both irrigated and drought stress conditions.

6 Yield Stability

For any long-term breeding effort to be successful it is critical to strike the appropriate balance between defining clear breeding objectives and ensuring flexibility within the breeding program. Given the many proposed breeding strategies for improving drought tolerance, the breeder needs a clearly defined set of check hybrids as internal references to decide whether the chosen breeding strategy is achieving adequate progress. This is complicated by the fact that individual check varieties remain static while yields should progress with each breeding cycle. Thus, each introduction of a new check variety should also reflect improvements made in drought tolerance. To improve drought tolerance, the breeder can either target specific combinations of putative drought tolerance traits or simply select for yield in large populations of genotypes in environments that reflect conditions in key market areas. In the former case there is scope to determine which trait combinations have contributed to improved performance and subsequently apply this knowledge to define breeding targets for testing (Podlich et al. 2004; Hammer et al. 2006). Most likely, a combination of both approaches is required. Here we use sugar beet as an example of breeding for drought tolerance, although many of the principles can apply to other hybrid crops and indeed inbreeding crops such as wheat and other cereals.

Improving sugar beet yields in Iran can be used as an example of how breeding for drought tolerance is applied in practice. In Iran, spring sugar beet production is wholly reliant on irrigation as the average annual rainfall in the country is about 250 mm, which mainly occurs during the winter and is not useful for a spring crop such as sugar beet. The average water requirement of the crop to realize the yield potential is approx. 9,000 m^3 ha^{-1}, which is almost entirely provided by irrigation. Due to a high evaporative demand and competition for water, the water table has been greatly lowered and this has increased the pumping charges. Therefore, continued irrigation at this rate does not seem to be a viable long-term solution for supporting future yield increases. One promising method is to develop drought tolerant varieties so that water resources are used more effectively.

The Sugar Beet Seed Institute (SBSI) is one of the leading research institutes working on drought stress in sugar beet, with projects on deficit irrigation, altered sowing pattern, the effects of drought stress on morpho-physiological traits, yield and technological quality, etc. Evaluation of sugar beet hybrids and their parents in stressed and non-stress conditions in Iran indicated that there is heterosis for yield under drought, which could be exploited by development of hybrids (Sadeghian et al. unpublished). This might be the case for other main crops such as maize, sorghum and pearl millet where hybrids are developed towards rainfed and water-limited environments (Blum 2011). There is strong evidence that heterosis in certain tropical maize materials is expressed in effective use of water and improved plant water status, probably as a result of larger root system in the hybrids (Araus et al. 2010). On the other hand, the parental effect is crucial for hybrid performance under drought stress (Blum 2011). This indicates that having one or both drought tolerant parents is the main condition for achieving a drought resistant hybrid. In some cases (e.g. Castiglioni et al. 2008; Araus et al. 2010; Yadav 2010), a recombination of parental drought resistance with high rate of heterosis for yield (namely high yield potential) might be possible.

Family line breeding and progeny selection over 8 years has resulted in improved drought tolerance in sugar beet for Iranian conditions. However, traditional breeding by using yield as a selection index and performing multi-environment yield trials has been costly and slow due to the multigenic nature of the trait and large genotype by environment interactions (Blum 2011). This necessitates using more efficient selection methods such a recurrent selection for further improvement of the level of drought tolerance. In addition, selection for secondary physiological traits plus yield in stress conditions have been shown to be more effective than direct selection for

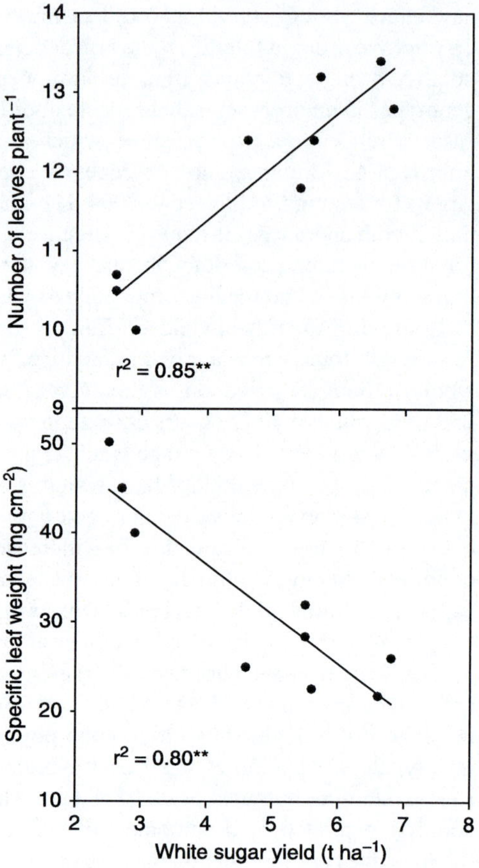

Fig. 3.1 Relationship of number of leaves and specific leaf weight with white sugar yield in three sugar beet populations grown under drought stress condition (Abdollahian Noghabi and Mohammadian 2005)

yield *per se* (Bänziger and Lafitte 1997; Bruce et al. 2002). One of the reasons for using physiological traits in selection is that they may have higher heritability than yield (Cruickshank et al. 2004). However, as each single trait that shows significant genetic correlation with crop performance in water-limited conditions can explain only a fraction of the total variance in yield, therefore one has to consider a combination of traits as selection criteria for improving yield in drought-prone environments.

Abdollahian Noghabi and Mohammadian (2005) studied the effects of varying degrees of drought stress on morpho-physiological traits of three sugar beet populations. They found that among the root and top traits measured, leaf number and specific leaf weight (leaf dry mass per area) had the greatest positive and negative correlations with white sugar yield, respectively (Fig. 3.1). Development and maintenance of the canopy for the capture and efficient use of light is critical for the accumulation of biomass under water-limited conditions as well as irrigated conditions. Rajabi et al. (2009) examined sugar beet genotypes of different genetic backgrounds and established that genetic variation in water use efficiency (dry matter produced per unit water used) of sugar beet could be indirectly exploited by measuring the stable carbon isotope discrimination ratio (Delta) in leaf and root tissue. Rajabi et al. (2008) found

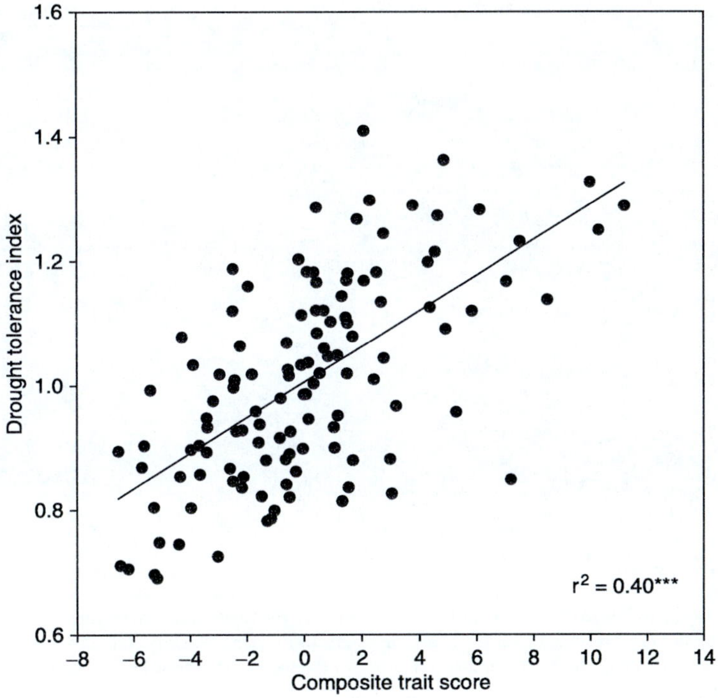

Fig. 3.2 The composite trait score explained 40% of the variation in drought tolerance amongst 120 wheat lines grown under managed drought conditions in the UK in 2008 (Fig. 3.3). Individual traits can explain only a small proportion of the variance in complex traits such as drought tolerance. Therefore, a composite trait score that combines the influence of several key traits was a better descriptor of drought tolerance. This score sums (with equal weighting and according to the sign of the correlation coefficient) standardised values for: green canopy cover, photosynthetic efficiency, flag leaf size, maintenance of ear dry mass at anthesis, leaf porosity, wax score, flag leaf senescence

that Delta, water use efficiency and specific leaf weight were mainly governed by additive gene effects, indicating that a greater genetic gain is expected for these traits whatever type of selection is applied. However, narrow-sense heritability was greater for Delta than specific leaf weight (Rajabi et al. 2008). Despite genetic variability for a range of other traits such as relative water content, osmotic adjustment, cell membrane stability, stomatal density, canopy temperature, ash content, etc. there was poor correlation with yield under drought stress conditions (Ober et al. 2005; Rajabi 2006). There are various ways that multivariate data sets can be handled to produce useful information. Multiple regression techniques and biplot analyses are two approaches that can highlight which combination of morpho-physiological traits best explain the variation in a complex target trait such as yield. The combination of traits describes a phenotypic ideal, or ideotype, that breeders could aim for in selections. Such a set of traits can be described numerically by computing a composite trait score that sums standardised values for each trait. The trait values can be given equal weighting, or weighted according to genetic correlations or heritabilities (Fig. 3.2).

Fig. 3.3 Large polytunnel rainout shelters used to apply managed drought conditions in the field. Shown are plots of 66 wheat genotypes grown at Broom's Barn, Suffolk, UK. The cost-effective facility allows control over the timing and severity of the water deficit with minimal impact on crop microclimate (Ober et al. 2010)

Every breeding strategy for drought-prone environments also has to consider that the timing and intensity of the stress events vary significantly from year to year and plants designed to cope with a specific type of drought may under-perform when the stress conditions are different or absent (Cattivelli et al. 2008). This is particularly true for cereals with sensitive stages of reproduction, but less of a concern for vegetative crops such as sugar beet. Selection under favourable environments also has positive effects when plants are grown in stressed environments (Cattivelli et al. 2008). In a typical Mediterranean environment, selection based on the absolute performance of the genotypes across environments is more successful than selecting for the minimum yield decrease under stress with respect to favourable conditions. An experiment including 46 sugar beet genotypes representing different genetic backgrounds grown in drought and irrigated conditions indicated that sugar beet genotypes with high yielding capacity when irrigated also tended to perform well under drought (correlation coefficient $r=0.64^{***}$), while the genotypes with minimum yield loss under stress did not belong to the group of high yielding genotypes in either irrigated or drought conditions (Ober et al. 2004). Similarly, there was a positive correlation between irrigated and droughted yields across a wide range of wheat genotypes when grown under managed drought conditions in the field (Ober et al. 2010; Fig. 3.3).

7 Conclusion

Increases in yields and yield stability can be achieved with selection and improvement of germplasm on the basis of performance under water-limited conditions that reflect the targeted environment. Selection response may be relatively slow, but should lead to more stable production across environments. Breeding progress depends on the availability of reliable selection tools and facilities where stress can be carefully controlled by managing the level and timing of water deficit. The scope for genetic gains in drought tolerance is remarkable. This can be achieved through judicious choice of selection environments and breeding strategy, rather than through a conventional approach of testing in a random sample of environments, which are often biased towards optimal conditions and less representative of actual farm conditions. Although we have given some specific examples of breeding approaches, each crop species and target environment may have different challenges, and therefore different solutions. These solutions do not have to be complex. While understanding the mechanisms of drought tolerance, the multifarious physiological effects of various allelic combinations and their genetic or epigenetic control can be a complicated phenomenon, this should not daunt the plant breeder. The successful yield gains for stressed environments made so far must continue, and indeed must increase to meet the challenges of a changing climate and growing world population.

Acknowledgements Sugar beet work at Rothamsted was supported by the British Beet Research Organisation, KWS, Strube and Südzucker, and work on wheat by DEFRA through the Sustainable Arable LINK programme (LK0986), with contributions from HGCA, KWS UK, RAGT, and Limagrain. SBSI is supported by the Iranian Ministry of Jihad-e-Agriculture.

References

Abdollahian Noghabi AM, Mohammadian R (2005) Study of relationship of morphological and physiological characteristics of sugar beet with drought tolerance. Research final report (in Persian), Sugar Beet Seed Institute, Iran, 80pp

Araus JL, Sanchez C, Cabrera-Bosquet L (2010) Is heterosis in maize mediated through better water use? New Phytol 187:392–406

Ashraf M (2010) Inducing drought tolerance in plants: recent advances. Biotechnol Adv 28:169–183

Ashraf M, Foolad MR (2007) Roles of glycine betaine and proline in improving plant abiotic stress resistance. Environ Exp Bot 59:206–216

Bänziger M, Laffite HR (1997) Efficiency of secondary traits for improving maize for low-nitrogen target environments. Crop Sci 37:1110–1117

Barker T, Campos H, Cooper M, Dolan D, Edmeades G, Habben J, Schussler J, Wright D, Zinselmeier C (2004) Improving drought tolerance in maize. Plant Breed Rev 25:173–242

Bidinger FR, Serraj R, Rizvi SMH (2005) Field evaluation of drought tolerance QTL effects on phenotype and adaptation in pearl millet [*Pennisetum glaucum* (L.) R. Br.] topcross hybrids. Field Crops Res 94:14–32

Blum A (2011) Plant breeding for water-limited environments. Springer, London

Blum A, Pnuel Y (1990) Physiological attributes associated with drought resistance of wheat cultivars in a Mediterranean environment. Aust J Agric Res 41:799–810

Blum A, Sullivan CY (1986) The comparative drought resistance of landraces of sorghum and millet from dry and humid regions. Ann Bot 57:835–846

Boyer JS, James RA, Munns R, Condon TG, Passioura JB (2008) Osmotic adjustment leads to anomalously low estimates of relative water content in wheat and barley. Funct Plant Biol 35:1172–1182

Bruce WB, Edmeades GO, Barker TC (2002) Molecular and physiological approaches to maize improvement for drought tolerance. J Exp Bot 53:13–25

Castiglioni P, Warner D, Bensen RJ (2008) Bacterial RNA chaperones confer abiotic stress tolerance in plants and improved grain yield in maize under water-limited conditions. Plant Physiol 147:446–455

Cattivelli L, Rizza F, Badeck FW, Mazzucotelli E, Mastrangelo AM, Francia E, Mare C, Tondelli A, Stanca AM (2008) Drought tolerance improvement in crop plants: an integrated view from breeding to genomics. Field Crops Res 105:1–14

Chen THH, Murata N (2011) Glycinebetaine protects plants against abiotic stress: mechanisms and biotechnological applications. Plant Cell Env 34(1):1–20

Cruickshank A, Dowkiw A, Wright G, Rachaputi NC, Nigam S (2004) Heritability of drought-resistance traits in peanut. Proceedings of the 4th international crop science congress. 26 September–1 October, 2004, Brisbane, Australia

Edmeades GO, Baeziger M, Ribaut J (2000) Maize improvement for drought-limited environments. In: Otegui ME, Slafer GA (eds) Physiological basis for maize improvement. Food Products Press, New York

Fleury D, Jefferies S, Kuchel H, Langridge P (2010) Genetic and genomic tools to improve drought tolerance in wheat. J Exp Bot 61:3211–3222

Fukai S, Pantuwan G, Jongdee B (1999) Screening for drought resistance in rainfed lowland rice. Field Crops Res 64:61–74

Gilbert N (2010) Food: inside the hothouses of industry. Nature 466:548–552

Hammer G, Cooper M, Tardieu F, Welch S, Walsh B, van Eeuwijk F, Chapman S, Podlich D (2006) Models for navigating biological complexity in breeding improved crop plants. Trends Plant Sci 11:1360–1385

He YX, Zheng TQ, Hao XB (2010) Yield performances of japonica introgression lines selected for drought tolerance in a BC breeding programme. Plant Breed 129:167–175

Hurd EA (1971) Can we breed for drought resistance? In: Larson KL, Eastin JD (eds) Drought injury and resistance in crops. CSSA Special Publication. No. 2. Crop Science Society of America, Madison, Wisconsin

Hussain SS (2006) Molecular breeding for abiotic stress tolerance: drought perspective. Proc Pakistan Acad Sci 43:189–210

Khanna-Chopra R, Sinha SK (1988) Enhancement of drought-induced senescence by the reproductive sink in fertile lines of wheat and sorghum. Ann Bot 61:649–653

Kholov J, Hash CT, Kakkera A (2010) Constitutive water-conserving mechanisms are correlated with the terminal drought tolerance of pearl millet (*Pennisetum glaucum* L.). J Exp Bot 61:369–377

Kumar R, Venuprasad R, Atlin GN (2007) Genetic analysis of rainfed lowland rice drought tolerance under naturally-occurring stress in eastern India: heritability and QTL effects. Field Crops Res 103:42–52

Ludlow MM, Muchow RC (1990) A critical evaluation of traits for improving crop yields in water-limited environments. Adv Agron 43:107–153

Miller G, Suzuki N, Ciftci-Yilmaz S, Mittler R (2010) Reactive oxygen species homeostasis and signalling during drought and salinity stresses. Plant Cell Env 33:453–467

Mitra J (2001) Genetics and genetic improvement of drought resistance in crop plants. Curr Sci 80:758–763

Morison JIL, Baker NR, Mullineaux PM, Davies WJ (2008) Improving water use in crop production. Philos Trans R Soc Biol Sci 363:639–658
Ober ES, Clark CJA, Le Bloa M, Royal A, Jaggard KW, Pidgeon JD (2004) Assessing the genetic resources to improve drought tolerance in sugar beet. Agronomic traits of diverse genotypes under droughted and irrigated conditions. Field Crops Res 90:213–234
Ober ES, Le Bloa M, Clark CJA, Royal A, Jaggard KW, Pidgeon JD (2005) Evaluation of physiological traits as indirect selection criteria for drought tolerance in sugar beet. Field Crops Res 91:231–249
Ober ES, Clark CJAC, Perry A (2010) Traits related to genotypic differences in effective water use and drought tolerance in UK winter wheat. Asp Appl Biol 105:13–22
Olsson M (1995) Alterations in lipid composition, lipid peroxidation and anti-oxidative protection during senescence in drought stressed plants and non-drought stressed plants of *Pisum sativum*. Plant Physiol Biochem 33:547–553
Passioura J (2007) The drought environment: physical, biological and agricultural perspectives. J Exp Bot 58:113–117
Passioura JB, Angus JF (2010) Improving productivity of crops in water-limited environments. Adv Agron 106:37–75
Podlich DW, Winkler CR, Cooper M (2004) Mapping as you go: an effective approach for marker-assisted selection of complex traits. Crop Sci 44:1560–1571
Praba ML, Cairns JE, Babu RC, Lafitte HR (2009) Identification of physiological traits underlying cultivar differences in drought tolerance in rice and wheat. J Agron Crop Sci 195:30–46
Rajabi A (2006) Carbon isotope discrimination and selective breeding for drought tolerance in sugar beet. PhD dissertation, University of Cambridge, Cambridge, MA
Rajabi A, Ober ES, Griffiths H, Pidgeon JD, Kromdijk W (2008) Genetic characteristics of water-use related traits in sugar beet. Euphytica 160:175–187
Rajabi A, Ober E, Griffiths H (2009) Genotypic variation for water use efficiency, carbon isotope discrimination, and potential surrogate measures in sugar beet. Field Crops Res 112:172–181
Reynolds M, Tuberosa R (2008) Translational research impacting on crop productivity in drought-prone environments. Curr Opin Plant Biol 11:171–179
Reynolds M, Dreccer F, Trethowan R (2007) Drought-adaptive traits derived from wheat wild relatives and landraces. J Exp Bot 58:177–186
Reynolds M, Manes Y, Izanloo A, Langridge P (2009) Phenotyping approaches for physiological breeding and gene discovery in wheat. Ann Appl Biol 155:309–320
Richards RA (2006) Physiological traits used in the breeding of new cultivars for water-scarce environments. Agric Water Manage 80:197–211
Richards RA, Rebetzke GJ, Watt M, Condon AG, Spielmeyer W, Dolferus R (2010) Breeding for improved water productivity in temperate cereals: phenotyping, quantitative trait loci, markers and the selection environment. Funct Plant Biol 37:85–97
Rosielle AA, Hamblin J (1981) Theoretical aspects for selection for yield in stress and non-stress environments. Crop Sci 21:943–946
Sharp RE, Poroyko V, Hejlek LG, Spollen WG, Springer GK, Bohnert HJ, Nguyen HT (2004) Root growth maintenance during water deficits: physiology to functional genomics. J Exp Bot 55:2343–2351
Tanksley SD, McCouch SR (1997) Seed banks and molecular maps: unlocking genetic potential from the wild. Science 277:1063–1066
Tardieu F, Tuberosa R (2010) Dissection and modelling of abiotic stress tolerance in plants. Curr Opin Plant Biol 13:206–212
Venuprasad R, Lafitte HR, Atlin GN (2007) Response to direct selection for grain yield under drought stress in rice. Crop Sci 47:285–293
Yadav OP (2008) Performance of landraces, exotic elite populations and their crosses in pearl millet (*Pennisetum glaucum*) in drought and non-drought conditions. Plant Breed 127:208–210
Yadav OP (2010) Drought response of pearl millet landrace-based populations and their crosses with elite composites. Field Crops Res 118:51–56

Yadav OP, Bhatnagar SK (2001) Evaluation of indices for identification of pearl millet cultivars adapted to stress and non-stress conditions. Field Crops Res 70:201–208

Yadav OP, Weltzien E (2000) Differential response of landrace-based populations and high yielding varieties of pearl millet in contrasting environments. Ann Arid Zone 39:39–45

Yadav RS, Sehgal D, Vadez V (2011) Using genetic mapping and genomics approaches in understanding and improving drought tolerance in pearl millet. J Exp Bot 62:397–408

Yunus M, Paroda RS (1982) Impact of biparental mating on correlation coefficients in bread wheat. Theor Appl Genet 62:337–343

Chapter 4
Breeding for Biotic Stress Resistance/Tolerance in Plants

Carlotta Balconi, Piergiorgio Stevanato, Mario Motto, and Enrico Biancardi

Contents

C. Balconi (✉) • M. Motto
Consiglio per la Ricerca e la Sperimentazione in Agricoltura, CRA – Unità di Ricerca per la Maiscoltura (CRA-MAC), Bergamo, Italy
e-mail: carlotta.balconi@entecra.it;mario.motto@entecra.it

P. Stevanato
Dipartimento di Biotecnologie Agrarie, Università di Padova, Padova, Italy
e-mail: stevanato@unipd.it

E. Biancardi
Consiglio per la Ricerca e la Sperimentazione in Agricoltura, CRA- Centro di Ricerca per le Colture Industriali (CRA-CIN), Rovigo, Italy
e-mail: enrico.biancardi@alice.it

M. Ashraf et al. (eds.), *Crop Production for Agricultural Improvement*,
DOI 10.1007/978-94-007-4116-4_4,

Abstract The long-term goal of crop improvement for biotic stress tolerance in plants is a traditional objective of breeders. Plants must continuously defend themselves against attacks from bacteria, viruses, fungi, invertebrates, and even other plants. This chapter will therefore summarize the benefits and drawbacks of resistance versus chemical protection. Attempts will be made to provide a description on the effective genetic and molecular mechanisms that plants have developed to recognize and respond to infection by a number of pathogens and pests, such as nonhost resistance, constitutive barriers and race-specific resistance, including recent advances in elucidating the structure and molecular mechanisms used by plants to cope with pathogens and pest attacks. This chapter also covers the most relevant problems in breeding for resistance to parasites and will include aspects related to specificity of defense mechanisms, specificity of parasitic ability, inheritance of resistance, gene-for-gene interaction, and durability of resistance. Major considerations in breeding for resistance to parasites, conventional sources of resistance and possible alternatives, namely mutation breeding, genetic manipulations, tissue cultures, and molecular interventions to develop plants resistant to pests and pathogens will also be dealt.

Keywords Defense mechanisms • Genetic basis of resistance • Signal transduction network • Pathogenesis related proteins • Transgenic plants

1 Introduction

Most of the problems facing agriculture in the twenty-first century relate to the growing world population, which is expected to stabilize at around 10–12 billion during the next 70 years (Heszky 2008). The almost doubled population will require a more than proportional increase in food production. During the last decade, world grain yield increased around 0.5% per year, which is three-fold lower than the population growth rate in the same period. The main task for breeders and agronomists will therefore be to increase yields while reducing the use of chemicals. The difficulties of this mission are due to: (i) the limited possibilities of expanding the cropped land area; (ii) the environmental legislation which limits the use of chemicals for disease control; (iii) climate change and the/predicted worsening of biotic and abiotic stresses; (iv) the reduced source of useful traits from crops wild relatives (Cook 2000).

Beacuse more than 42% of the potential world crop yield is lost owing to biotic stresses (15% attributable to insects, 13% to weeds, and 13% to other pathogens), a reduction in this incidence will be one of the more important possibilities for improving plant production (Pimentel 1997). Cook (2000) divided these possibilities as follows: (i) improvement of plant material (breeding for tolerance/resistance); (ii) improvement of root health (e.g. field rotation, soil tillage, soil-borne diseases control); (iii) improvement of irrigation practices (optimal water quality and availability); (iv) protection against airborne hazards (foliar diseases etc.). In this context, the development of tolerant plants to biotic stresses is therefore an important objective of plant breeding strategies with relevant implications for both farmers and the seed and agrochemical industries. In fact genetic resistance has several obvious advantages over the use of chemical pesticides or other methods for parasite control. These include nominal genetic permanency, negligible cost once cultivars are developed, and quite high efficiency. The major downside of genetic resistance to biotic stresses is the fact that selection pressure is placed on parasites populations to develop means of overcoming the resistance, thus practically limiting the time of effectiveness (Table 4.1). In this chapter the genetic, biochemical and molecular mechanisms by which plants defend themselves against attack from pathogens will be examined. In addition breeding approaches towards their improvement will be described.

Table 4.1 Overview of potential and actual losses attributable to fungal and bacterial pathogens, viruses, animals pests and weeds as well as the efficacy of the applied pest control operations in maize, wheat, rice, barley, potatoes, soybean, sugar beet, and cotton

	Pests and pathogens				
	Fungi and bacteria	Viruses	Animal pests	Weeds	Total
Loss potential (%)[a]	14.9	3.1	17.6	31.8	67.4
Actual losses (%)[a]	9.9	2.7	10.1	9.4	32.0
Efficacy (%)[b]	33.8	12.9	42.4	70.6	52.5

Source: Modified from Oerke and Dehne (2004)

[a]As percentage of attainable yields

[b]As percentage of loss potential prevented

2 Fungal and Bacterial Diseases

A plant pathogen is defined as an organism that for a part or all of its life cycle grows inside a plant; this has a detrimental effect on the plant growth and development and ultimately on yield. Several reviews have been published in this field to which the readers are referred for a more in-depth description (e.g. Hammond-Kosack and Jones 2000; Dickinson 2003). The main findings emerging from these studies indicate that pathogens have evolved specialist ways to invade plants: (i) some penetrate the plant surface directly using mechanical pressure or enzymatic attack; (ii) others pass through natural plant openings, for example, stomata or lenticels; (iii) many take advantage of existing wounds. Once inside the plant, three main colonization strategies are deployed by pathogens to use the host plant as substrate for their growth and development: (i) biotrophic organisms ensure the plant cell remains alive; (ii) hemibiotrophic organisms initially keep host plant cells alive, but then kill them at later stages of the infection; (iii) necrotrophic organisms first kill plant cells and then metabolize their contents. In this respect, pathogenesis is the term used to describe the sequence of processes from host and pathogen contact (infection, colonization and plant pathogen reproduction) to the development of the complete syndrome. A pathogen strain that causes disease is termed virulent and its success can be attributed to several factors such as : (i) rapid and high rate of reproduction during the main growing season for plants; (ii) a very efficient dispersal mechanism and long-term survival capacity; (iii) high capacity to generate genetic diversity through haploidy and subsequent sexual reproduction.

2.1 *Plant Defense Mechanisms*

An overview of forms of plant resistance, defined on the basis of innate and acquired resistance, and their mechanisms of response to pathogens is given in Table 4.2. According to Kiraly et al. (2007) innate resistance is exhibited by the plant in two forms: non-specific (nonhost or general) resistance, which is effective against several pathogenic species or several strains (races, biotypes, pathovars) of a single pathogen, and specific resistance. In the latter case, one plant cultivar (variety) can resist infection by one or a few pathogenic strains.

Although plants are in continual contact with potential pathogens, a successful infection is rare. The ability of a particular plant species to prevent successful colonization by a given pathogen species is referred to as nonhost resistance. The molecular basis of nonhost resistance is poorly understood, but presumably relies on both constitutive barriers and inducible responses that involve a large array of proteins and other organic molecules produced, respectively, prior to infection or during pathogen attack (cf. Jones and Dangl 2006; Ferreira et al. 2007). This is in contrast to the vertebrate immune system, in which specialized cells devoted to defense are rapidly mobilized to the infection site, where they either kill the invading organisms or limit their diffusion.

Table 4.2 Overview of forms of plant resistance

Resistance phenomenon	Mechanism
1. ***Innate resistance***	
1.1. Non specific, general resistance	
Non-host resistance	HR; ROS; BAX inhibitor-1; PEN genes
Basal resistance against bacteria	Flagellin/FLS2 interaction; ROS; antimicrobial compounds
Race non-specific mlo resistance and quantitative resistance to fungi	Cell wall thickening; Antimicrobial compounds; ROS
Resistance to necrosis-inducing stresses	High antioxidant capacity
1.2. Specific resistance (cultivar/pathogenic race specificity)	
Extreme resistance-symptomless gene-for-gene resistance	Unknown
Rx-resistance against viruses without HR	
Symptomless reaction to rust pathogens, no visible HR	
Gene-for-gene resistance	ROS; Phytoalexins; Phenol oxidation; Stress proteins
R-gene ↔ Avr-gene interaction associated with the hypertensive response (HR)	
Resistance to pathogen toxins	Enzymatic detoxification; Lack of toxin recept
Gene silencing	Recognition and decomposition of foreign RNAs with ribonucleases
2. ***Acquired resistance***	
After a primary infection and acquired resistance develops against a second infection "Stress memory"	Accumulation of SA; Stimulated antioxidants; Gene silencing; Rhizobacterial induction

Source: Modified from Kiraly et al. (2007)

The typical preformed, constitutive defenses are morphological, structural, and chemical barriers. An example of morphological barrier is the height of lips of stomatal guard cells (Hoch et al. 1987). Certain fungal rust pathogens possess specific detection mechanisms that sense the height of stomatal guard cell lips encountered on susceptible plants. Moreover, waxes, cutin, suberin, lignin, cellulose, callose, and cell wall proteins act as structural barriers that are rapidly reinforced upon the pathogen infection process (Punja 2001). Plants also constitutively produce a variety of secondary metabolites (e.g. phenolics, saponins, terpenoids, steroids and glucosinalates), and antifungal proteins, many of which act as antimicrobial compounds during defense (see Dangl and Jones 2001, for a review). These compounds may be present in their biologically active forms or stored as inactive precursors that are converted to their active forms by host enzymes in response to pathogen attack or tissue damage.

Plants employ two modes of their innate immune system to contrast pathogen infections (see Tsuda and Katagiri 2010, for a recent review). The first mode of immunity is referred to as pattern-triggered immunity (PTI) that is triggered by molecular patterns common to many microbial types. The second mode is triggered by recognition of pathogen effectors and is called as effector-triggered immunity (ETI). At least some cases of PTI and ETI extensively share downstream signaling machinery, that is, PTI and ETI appear to be mediated by an integrated signaling network. However, activated immune responses in ETI are more prolonged and robust than those in PTI. Furthermore, the previous cited authors have reported that network analysis has also revealed that synergistic relationships among the signaling sectors are evident in PTI, which may amplify the signal, whereas compensatory relationships among the sectors dominate in ETI, explaining the robustness of ETI against genetic and pathogenic perturbations. Thus, plants seem to use a common signaling network that differs in PTI and ETI.

There is evidence that induced or acquired resistance includes the hypersensitive response (HR), a form of programmed plant cell death, cell-wall strengthening, and the expression of various defense-related *R* genes (R, resistance; Staskawicz et al. 1995) that mediated recognition of pathogen effectors. The *R* genes activate a series of defense signaling cascades and pathogenesis-related (*PR*) gene expression to generate systemic acquired resistance (SAR); this primes the plant for resistance against a broad spectrum of pathogens (Durrant and Dong 2004; Dangl and Jones 2001). This multicomponent response requires a substantial commitment of cellular resources, including extensive genetic reprogramming and metabolic re-allocation. Thus, defenses are kept under tight genetic control and are activated only if the plant detects a prospective invader.

2.2 *Genetic Basis of Resistance*

Genetic analysis of disease resistance in plants began over 100 years ago when Biffin (1905) reported that resistance in wheat to stripe rust (*Puccinia striiformis*) was inherited as a single recessive Mendelian trait. Since this initial work, many genes conferring resistance to pathogens in crop plants have been characterized, and the genetic basis of pathogenicity (virulence/avirulence) has been studied in many plant pathogens. This knowledge culminated in the development of the gene-for-gene hypothesis by Flor (1971) based on genetic studies of the interaction between flax and the flax rust pathogen, which has provided a framework for much if not all of the work on disease resistance in the years since. In genetic terms, resistance is generally defined by the mode of inheritance, with broad distinctions between oligogenic (controlled by one or few genes of major effect) and polygenic (controlled by many genes of low individual phenotypic effect) resistance.

2.2.1 Qualitative Resistance

Evidence made it clear that many cases of resistance were inherited in a simple way. Most characterized resistance genes are dominant in action; for example the *Hm1* gene of maize conferring resistance to *Cochliobolus carbonum* race 1, a causal agent of northern leaf spot of maize, secretes an HST known as HC-toxin, which interferes with a histone deacetylase (HD) altering host gene expression (Brosch et al. 1995). Indeed, resistant (insensitive to HC-toxin) maize lines contain a dominant resistance gene *Hm1*, which encodes for a NADPH-dependent reductase whose function is to reduce (detoxify) the HC-toxin that the fungus produces to cause disease in susceptible maize (Johal and Briggs 1992). However, some recessive resistance genes have proven important sources of durable resistance – e.g. gene *Sr2* conferring resistance to stem rust in wheat (McIntosh et al. 1995); gene *mlo* for mildew resistance in barley (Jørgensen 1994). The barley *Mlo* gene has been cloned and encodes a transmembrane protein that is a negative regulator of cell death and powered mildew resistance (Büschges et al. 1997). Notably, further research showed that functional *Mlo* genes also exist in *Arabidopsis* (Consonni et al. 2006). Thus, *mlo*-mediated non-specific resistance to powdered mildew might be a more widespread phenomenon among plants than previously hypothesized. There are also many examples of resistance genes that display partial dominance (gene dosage dependence; e.g. the resistance gene *Lr9* in wheat to *Puccinia recondita*; Samborski 1963).

Dominance or recessiveness of resistance genes is, however, not absolute and can even be governed by the attribute used to measure the disease phenotype (Johnson 1992), genetic background, pathogen isolate or environment. Examples of oligogenic resistance are known in which additive and non-additive interaction occurs between genes at separate loci. The genes *Lr13* and *Lr34* in wheat interact in an additive manner to confer resistance to leaf rust, not only with each other, but also with other genes for resistance to leaf rust (Kolmer 1992). Non-additive gene interaction occurs when two genes in the host are only effective when present together. In such cases, the genes in the host are referred to as complementary. Baker (1966) demonstrated complementary action of the genes *Pc3* and *Pc4* conferring resistance to *Puccinia coronata* in the oat cultivar Bond.

Resistance to bacterial infections is not well developed as virus and fungal resistance, partly because bacterial diseases are a main problem only in crop plants like potato, rice, and some fruit trees. Similarly to fungal diseases the most effective type of protection is genetic resistance, which is based on single dominant or semidominant genes. Different classes of *R* genes cloned from various plant species were characterized and tested for their ability in conferring resistance against bacterial pathogens. For example, among these, a map-based cloned *Xa21* gene from rice, gave resistance to bacterial blight , a serious disease in rice caused by *Xantomonas oryzae. Xa21* specify a receptor- like kinase formed bi LRRs in the putative domain and a serine-threonine kinase in the putative intracellular domain pv. *oryzae,* after transferring this gene from a wild rice species to a cultivated indica variety (Wang et al. 1996). Moreover, the last cited authors found on the broad-spectrum resistance of transgenic rice with the *Xa21* gene against 29 diverse isolates, suggesting that a single cloned gene is sufficient to confer multi-isolate resistance. In the same way, the resistance gene *Bs2* from pepper was transferred to tomato, which then had resistance to bacterial disease (Tai et al. 1999). Infiltration of different maize lines with a variety of bacterial pathogens of maize, rice and sorghum has permitted to identify a maize gene, *Rxo1*, which conditions a strong HR to the non-host bacterial pathogen *X. o.* pv. *Oryzicola* (Zhao et al. 2004). The same locus carries a gene (designated *Rba1*) controlling resistance to the maize and sorghum bacterial stripe pathogen *Burkholderia andropogonis*. It was surprising that the same locus controlled resistance to two of only four bacterial pathovars tested. This suggests that this locus may condition defense reactions to other bacterial pathogens.

2.2.2 Quantitative Resistance and QTLs

Quantitative resistance, in contrast to qualitative resistance, is generally considered as partial resistance in a particular cultivar (Young 1996). This type of disease resistance is controlled by multiple loci, referred to as polygenes or quantitative trait loci (QTLs), and does not comply with simple Mendelian inheritance. Examples of such polygenetically inherited resistance are the partial resistance in potato to *Phytopthora infestans*, in maize to *Puccinia sorghi*, and in barley to *Puccinia hordei* (Parlevliet and Zadoks 1977).

Although genetically complex forms of disease resistance are still poorly understood, an effective strategy for studying complex and polygenic forms of disease resistance is known as QTL mapping, which is based on the use of DNA markers (see Young 1996, for a review). With QTL mapping, the roles of specific loci in genetically complex traits can be described; this has also permitted insight to be gained into fundamental questions that have puzzled researchers in the field of plant pathology for decades. Although results of QTL mapping indicate that it is generally not the case, there are examples of several (>10) QTLs involved in quantitative resistance; however, it is much more common to find only three to five loci: frequently, 1 or 2 QTLs predominate.

QTL mapping may also help to determine whether individual QTLs are race-specific or not, and when there is an indication of specificity, the degree to which partial resistance differs between races. For example, quantitative resistance to *P. infestans* in potato was initially described as race-nonspecific (Van der Plank 1982). Dissecting the contributions of individual QTLs, it was clearly demonstrated that loci show distinctly different resistance effects against different pathogen races (Leonards-Schippers et al. 1994). Indeed, only 5 of the 11 statistically significant genomic regions showed no specificity against just two races tested, while the others were significant against just one. Moreover, genetic mapping with DNA markers makes it possible to ask whether homologous resistance genes exist in related plant taxa and may help to test the hypothesis that QTLs are simply variants of qualitative resistance loci that have been (partially) overcome by their respective pathogen. For instance, in rice blast, 3 of the QTLs mapped to the same marker intervals as previously identified qualitative blast resistance genes. It is conceivable that these QTLs represent allelic variants of the known qualitative resistance genes, though only more precise mapping and gene cloning can resolve this definitely. In potato late blight, 1 QTL coincided in location with a dominant, race-specific gene known as *R1*, as well as *Rx2,* a gene for resistance to potato virus (Leonards-Schippers et al. 1994). Moreover, a second late blight QTL mapped to the same region as *Rx1,* a second major resistance gene for potato virus X. Further progress in QTL mapping technology will include the molecular cloning of the underlying genes, including those that confer partial resistance. For example, major genes such as resistance to *Pseudomonas syringae* in tomato and *Arabidopsis* have been isolated and cloned based on map position, while others have been isolated through transposon-tagging (cf. Young 1996). More recently, Hu et al. (2008), to isolate disease resistance QTLs, have established a candidate gene approach that integrates linkage map, expression profile and functional complementation analyses. This strategy has proven applicable in rice for identifying the genes underlying minor resistance QTLs in bacterial blight caused by *Xanthomonas oryzae* pv. *oryzae* and fungal blast caused by *Magnaporthe grisea* systems and it may also help to shed light on disease resistance QTLs of other cereals. The results also suggest that a single minor QTL can be used in rice improvement by modulating the expression of the gene underlying the QTL. Pyramiding 2 or 3 minor QTL genes, whose expression can be managed and that function in different defense signal transduction pathways, may allow the breeding of cultivars that are highly resistant to bacterial blight and blast.

2.3 Plant R Gene-Mediated Disease Resistance

As mentioned above, plants do not have the benefit of a circulating antibody system, so plant cells autonomously maintain constant vigilance against pathogens by expressing vast arrays of *R* genes. These genes have been genetically defined in interactions with all major classes of plant pathogens including fungi, bacteria, and viruses.

In the classic gene-for-gene model – also known as receptor-ligand model – of host pathogen interactions, *R* gene products recognize pathogen elicitor, encoded by avirulence (*Avr*) genes (Fig. 4.1). Resistance gene-mediated resistance is a host-specific defense and can only be activated when both *R* gene and corresponding *Avr* gene are present (Staskawicz et al. 1995); the absence of either component results in disease, which is typically associated with damage and a reduction in yield of the host plant.

Because gene-for-gene are operative in defense response to fungal, bacterial and virus pathogens, and because the host defense responses are similar irrespective of the type of pathogen, common recognition and signal recognition transduction mechanisms are postulated to underlie the gene-for-gene relationship. A single model of pathogen response will therefore be given herein.

Currently, there are two alternative mechanisms to explain this model: direct and indirect interaction (Fig. 4.2). Direct interaction suggests that the pathogen *Avr* effectors interact with plant R proteins directly to trigger *R* gene-mediated resistance signaling. For example, the rice *R* gene *Pi-ta* was initially shown to directly interact with *AVR-Pita* from *Magnaporthe grisea* but no interaction between

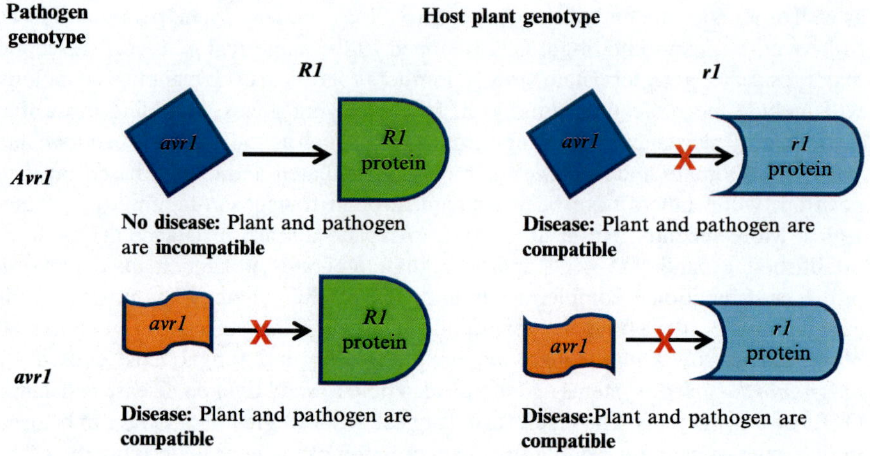

Fig. 4.1 Flor's gene-for-gene model. For resistance (incompatibility) to occur, complementary pairs of dominant genes, one in the host and one in the pathogen, are required. An alteration or loss of the plant resistance gene (*R* changing to *r*) or of the pathogen avirulence gene (*Avr* changing to *avr*) leads to disease (compatibility) (Modified from Hammond-Kosack and Jones 2000)

AVR-Pita and its susceptible allele *Pi-ta* was observed (Jia et al. 2000). In addition, a direct interaction was recently observed between the flax *L* alleles and corresponding flax rust *Avr* genes, which provides evidence for direct, allele-specific interaction between R proteins and diverse Avr proteins (Dodds et al. 2006). Conversely, most studied data prefer the indirect model also called "guard" hypothesis (Jones and Dangl 2006). In this model, R proteins act as "guardees" to monitor the variation/modification of host proteins after coupling with the corresponding Avr effectors. Moreover, evidence suggests that plants possibly employ alternate mechanisms to prevent pathogens from invading in different plant-pathogen interaction systems, which maintain a good balance between different R proteins and Avr effectors to coordinate the conflicts of interaction between different *R* genes and varied Avr effectors in host-pathogen co-evolution (Van der Hoorn et al. 2002). In these two models, R proteins may detect Avr effectors with conserved structure through direct interaction, or could indirectly recognize diverse unrelated pathogen factors after Avr proteins couple with their virulence targets (Chisholm et al. 2006). However, when and how R proteins detect diverse Avr effectors directly or indirectly is unclear and requires in-depth analysis.

More recently, Hann et al. (2010), by reviewing published results on bacterial virulence effectors and their activities, indicated that the major virulence strategy for plant pathogenic bacteria is a deployment of effector molecules within the host cytoplasm. As summarized by these authors (i) each bacterial strain possesses a set of 20–30 effectors which are redundant and interchangeable, and interact promiscuously with host targets, (ii) bacterial effectors have weak, somewhat nonspecific interactions with host molecules, targeting conserved protein domains or molecular

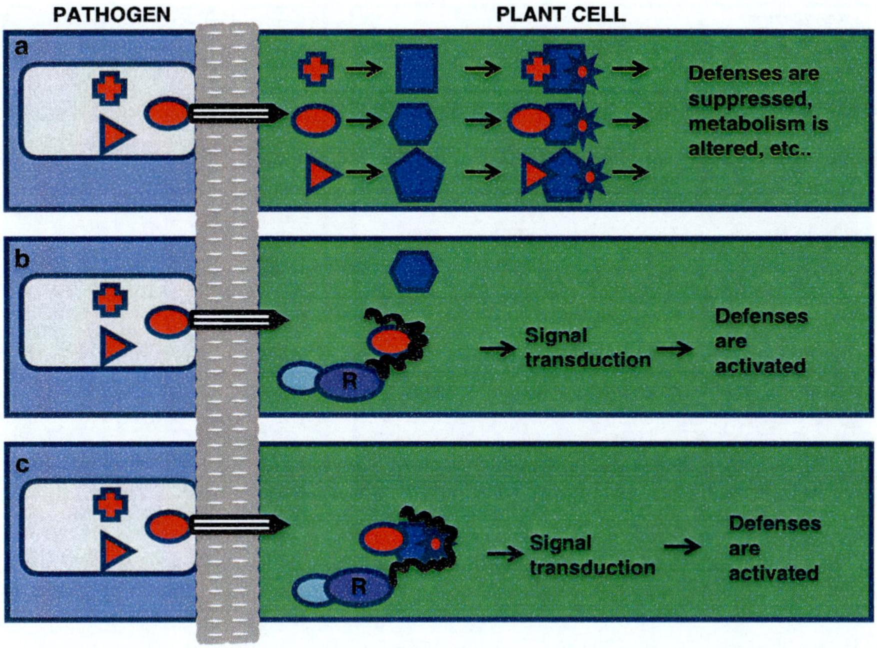

Fig. 4.2 Interactions between pathogen Avr proteins (*red*) and plant R proteins (*blue*). Once inside the plant cell, pathogen Avr proteins target host proteins that control defense responses, metabolism or other plant process that affect pathogen virulence. Panel (**a**): the plant cell does not express an R protein that is capable of recognizing any virulence protein, plant defenses are not activated, disease results from the collective action of the virulence proteins. Panel (**b**): classic receptor-elicitor hypothesis, in which an R protein (*purple*) directly binds a virulence protein, so a complex signal transduction network is activated, which in turn triggers plant defense responses. Panel (**c**): guard hypothesis in which an R protein (*guard*) detects a modified host protein (*guardee, red star*), evenctually as a complex with the "attacking" virulence protein (Modified from McDowell and Woffenden 2003)

structures, (iii) these structures have been coopted by the plant defense machinery as accessory proteins or baits in NB-LRR complexes, and (iv) the link between pathogenicity and immunity apparently lies in the molecular (enzymatic) activities of each effector. Although, the direction of evolution of host immune components with respect to effectors is unclear, some authors have suggested a positive-negative selection model in which positive selection for effectors is balanced with strong negative selection for specific effectors by NB-LRR complexes. The clearest example for such a positive-negative selection scenario is presented by AvrPto and AvrPtoB, which suppress PRR receptor kinases and are recognised by the Prf NB-LRR complex in tomato (Zipfel and Rathjen 2008). Thus, the two levels of pathogen perception may interact to slow pathogen evolution, which is important when recognition specificities are innate and cannot be acquired.

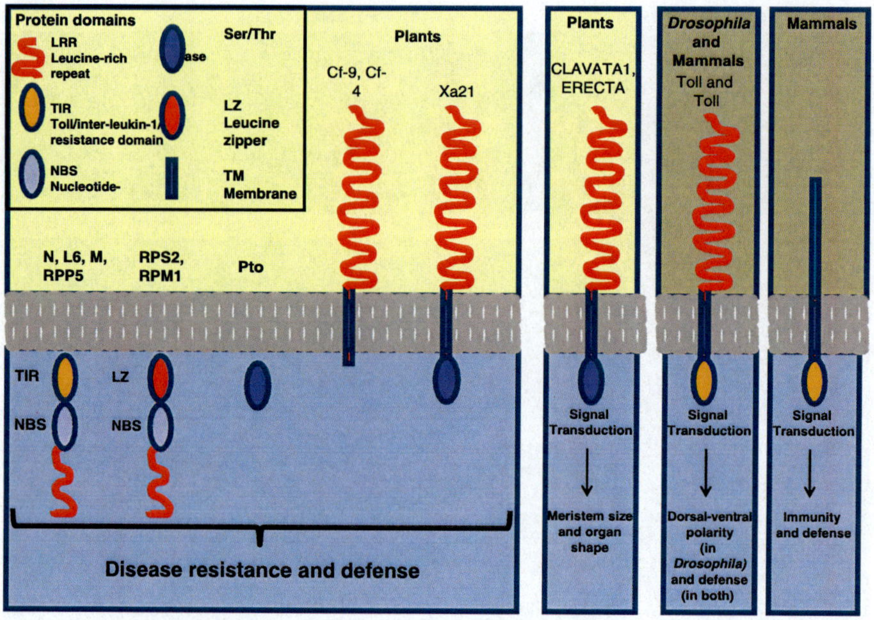

Fig. 4.3 Schematic diagram illustrating several plant resistance proteins and various plant LRR-containing proteins synthesized during the defense response. For comparison are included structurally related proteins that are involved in various aspects of plant development as well as other eukaryotic proteins that coordinate development and the induction of the immune response in animals (Modified from Hammond-Kosack and Jones 2000)

2.3.1 Structure of *R* Genes

The strong phenotypes and natural variability at *R* loci have been exploited by molecular geneticists to clone the *R* genes and investigate their molecular modes of action. To date, over 70 *R* genes have been cloned and some of them have been well characterized (for reviews see Jones and Dangl 2006; Shan et al. 2004; Toyoda et al. 2002). Notably, these studies have not only provided a large body of information on the structure, function and evolution of *R* genes that control resistance to diverse pathogens, but also have generated useful genetic materials to engineer novel resistant cultivars. In addition, some critical defense signaling components, such as NDR1, EDS1, RAR1, and SGT1, have been identified in *R* gene-mediated resistance signaling, which provide important clues to understanding the mechanism of *R* gene-mediated defense signaling (Lin et al. 2007; reviewed by McDowell and Woffenden 2003; Rathjen and Moffett 2003).

Examination of the predicted R protein sequences, based on protein domains that are described in Fig. 4.3, reveals the existence of shared sequence motifs that can be grouped into several superfamilies. The large majority of genes cloned so far belong to the nucleotide-binding site (NBS), leucine-rich repeat (LRR), a motif with homology to the cytoplasmic domains of the *Drosophila* Toll protein and the mammalian interleukin-1 receptor (TIR), a coiled-coil (CC) or leucine zipper (LZ) structure, transmembrane domain (TM), and protein kinase domain (PK). According to these features, at least four principal classes are distinguished among most *R* genes as follows: NBS-LRR, Receptor-like kinase (RLK), LRR-TM and TM-CC (Fig. 4.3). The NBS-LRR genes represent the largest class of *R* genes, and encode proteins with a variable N-terminal domain of approximately 200 amino acids (aa), connected by a predicted NBS domain of approximately 300 aa and a more variable tandem array of approximately 10–40 short LRR motifs. The NBS-LRR genes are further categorized into three subgroups based on the motif within their N-terminus: TIR group, CC or LZ group and non-motif group. Furthermore, studies regarding the NB-LRR signaling pathway have been recently summarized by Eitas and Dangl (2010). The main findings reviewed by these workers indicate that (i) two NB-LRRs can function together to mediate disease resistance against pathogen isolates, (ii) the NB-LRR protein fragments that are sufficient to initiate defense signaling, and (iii) importantly, distinct fragments of different NB-LRRs are sufficient for function. Finally, the cited authors described that accessory proteins (e.g. Pto) and highly related Pto-like kinases have a significant role in regulating the function and down-streaming host genes in NB-LRR signaling.

As more *R* genes have been cloned in recent years, new motifs or structures have been uncovered in R proteins. *RRS1-R* from *Arabidopsis*, conferring resistance to *Ralstonia solanacearum* strain GMI1000 with a type III effector PopP2, which belongs to the YopJ/AvrRxv protein family encodes a typical TIR-NBS-LRR protein, but containing a transcriptional factor WRKY domain in its C-terminus (Deslandes et al. 2002). The WRKY domain is highly conserved and is composed of a region of about 60 aa containing a heptapeptide WRKYGQK in its N-terminus and a zinc-finger motif, which plays a crucial role in regulating plant defense responses (Journot-Catalino et al. 2006).

Recently, the plant resistance gene *Pi-d2*, conferring gene-for-gene resistance to the Chinese rice blast strain, ZB15, encodes a novel type of receptor-like kinase R protein with a predicted extracellular domain of a bulb-type mannose specific binding lectin (B-lectin) and an intracellular serine-threonine kinase domain (Chen et al. 2006). Among all isolated *R* genes, three novel genes do not belong to any of the four types: *Xa5* encoding a TFIIA transcription factor (Jiang et al. 2006), *Xa13* with homologous to nodulin MtN3, and *Xa27* without any hits in the available protein database, *R* genes encode putative receptors that respond to the products of '*Avr* genes' expressed by the pathogen during infection.

2.3.2 Evolution Mechanism of *R* Genes

In plants many *R* genes are located in clusters that comprise several copies of homologous *R* gene sequences arising from a single gene family (simple clusters) or colocalized *R* gene sequences derived from two or more unrelated families (complex clusters), and may also contain unrelated single genes interspersed between the homologs (reviewed in Friedman and Baker 2007). *R* clusters range in size from two tandem paralogs to large complexes spanning several megabases. The largest *R* clusters characterized to date include the maize *Rp1* cluster (~1–52 homologs per haplotype; Smith et al. 2004), the lettuce *Dm3* (aka RGC2) cluster (~12–32 homologs per haplotype), and the potato major late blight resistance (MLB) cluster (~45 homologs per haplotype; Kuang et al. 2005).

Genic and intergenic sequence repeats within *R* clusters, generated by duplications and transposon insertions, provide a structural environment that permits mispairing during recombination, giving rise to unequal crossovers and interlocus gene conversions (McDowell and Simon 2006). Intergenic unequal crossover has the potential to place *R* genes in new structural contexts that may alter expression, whereas intragenic mispairing generates chimeric genes that may encode novel functions. Both types of unequal recombination will also result in altered gene copy number within the cluster (gene duplication on one chromosome and loss on the other) according to the number of genes present in the region between the mispaired recombination sites.

Sequence exchanges (unequal crossovers and/or gene conversions) have been reported in several *R* clusters (Kuang et al. 2005) and are associated with genic diversity, characterized by sequence shuffling and chimeric genes, and haplotypic diversity, characterized by a variable number of *R* homologs within the cluster and a general loss of syntenic/orthologous relationships between haplotypes. Furthermore, unequal recombination, at the *Rp1* cluster and at the *Cf4/9* cluster, has been shown to generate novel *R* haplotypes with resistance specificities that differ from either parent. Interestingly, similar clustering phenomena are seen at (a) virulence loci in multiple, evolutionarily distinct pathogen genomes (Dodds et al. 2006). This accumulated evidence indicates that *R* clusters facilitate rapid evolution via recombinatorial mispairings, generating novel *R* gene sequences that may encode altered specificities or have altered expression patterns.

2.3.3 Signal Transduction Network

The most relevant features of the defense condition indicate that the activation of defense responses in plants is initiated by host recognition of pathogen-encoded molecules called elicitors (e.g., microbial proteins, small peptides and oligosaccharides). A simplified model for signal transduction in plant defense provided by Yang et al. (1997) is given in Fig. 4.4. According to this model the interaction of pathogen elicitors with host receptors (many of which may be encoded by *R* genes) likely activates a signal transduction cascade that involves oxidative burst i.e. reactive oxygen species (ROS), calcium fluxes, ion channel fluxes, G-proteins, nitrogen oxide production, and dephosphorilation of unknown pathogens. Subsequent transcriptional and/or post-translational activation of transcription factors eventually leads to the induction of plant defense genes (Zhu et al. 1996). In addition to eliciting primary defense responses, pathogen signals may be amplified through the generation of secondary plant signal molecules such as salicylic acid (SA) (Durner et al. 1997).

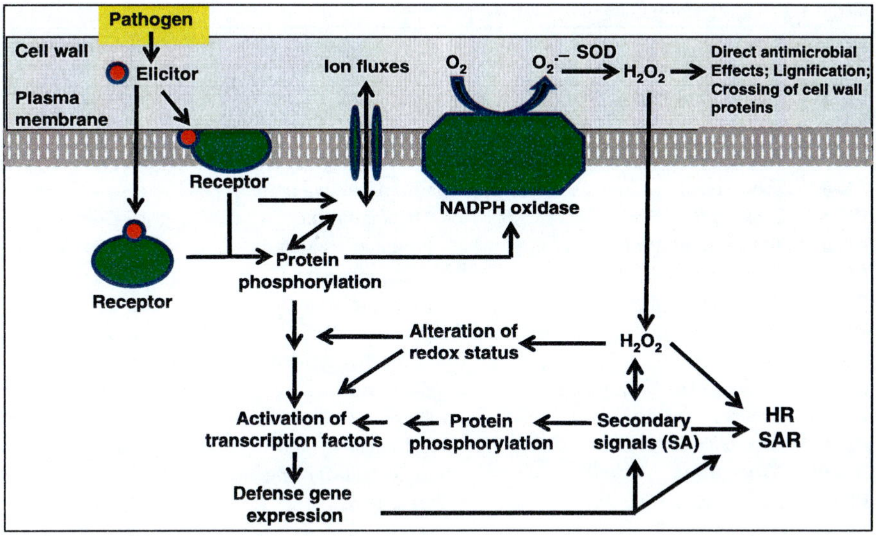

Fig. 4.4 Simplified model for signal transduction in plant defense responses. Host recognition of pathogen elicitors initiates early signaling events such as protein phosphorylation/dephosphorylation, ion fluxes and oxidative burst. Subsequent transcriptional and/or posttranslational activation of transcription factors leads to induction of plant defense genes and biosynthesis of endogenous secondary signals such as SA. Additionally, the activated NADPH oxidase complex generates reactive oxygen species (ROS) such as $O_2^{\cdot}$ – and H_2O_2 that alter the redox status of plant cells and affect defense signaling. SA, ROS, as well as defense genes, all contribute to the development of HR and SAR during plant-pathogen interactions. SOD, superoxide dismutase (Modified from Yang et al. 1997)

Moreover, advances in induced defense signaling research revealed regulators of induced systemic resistance and suggest a model in which jasmonic acid (JA) related transcription factors play a central role in establishing the primed state for enhanced resistance (reviewed in Van der Ent et al. 2009).

Both primary pathogen elicitors and secondary endogenous signals may activate a diverse array of plant protectant and defense genes, whose products include glutathione S-transferases , peroxidases, cell wall proteins, proteinase inhibitors, hydrolytic enzymes (e.g., chitinases and β~-1,3-glucanases), pathogenesis-related (PR) proteins – PR proteins are host-encoded, abundant proteins induced by pathogens and many of them have antimicrobial activity *in vitro* or when overexpressed in transgenic plants, and phytoalexin – phytoalexins are low-molecular-weight, antimicrobial compounds (e.g., phenylpropanoids, terpenoids), whose synthesis is induced following pathogen infection, plus biosynthetic enzymes, such as phenylalanine ammonia lyase and chalcone synthase (reviewed in Yang et al. 1997). Notably, more recently Clay et al. (2009) have identified a metabolic pathway for glucosinates, previously identified as important in avoiding damage by herbivores, as a component of plant defense response against microbial pathogens.

Table 4.3 Families of pathogenesis-related proteins

Family	Type member	Biochemical properties	Molecular mass range (kDa)
PR-1	Tobacco PR-1a	Unknown	15–17
PR-2	Tobacco PR-2	B-1,3-glucanase	30–41
PR-3	Tobacco P,Q	Chitinase class I, II, IV, VI, VII	35–46
PR-4	Tobacco R	Chitin-binding proteins	13–14
PR-5	Tobacco S	Thaumatin-like	16–26
PR-6	Tomato inhibitor I	Proteinase inhibitor	8–22
PR-7	Tomato P69	Endoproteinase	69
PR-8	Cucumber chitinase	Chitinase class III	30–35
PR-9	Tobacco "lignin forming peroxidase"	Peroxidase (POC)	50–70
PR-10	Parsley "PR-1"	"Ribonuclease-like"	18–19
PR-11	Tobacco class V chitinase	Chitinase class V	40
PR-12	Radish Rs-AFP3	Defensins	5
PR-13	Arabidopsis THI-2.1	Thionons	5–7
PR-14	Barley LTP4	Lipid Transfer Proteins	9
PR-15	Barley OxOa (germin)	Oxalate oxidases	22–25
PR-16	Barley OxOLP	Oxalate oxidase-like protein	100 (hexamer)
PR-17	TobaccoPRp27	Unknown	Unknown

Source: Modified from Van Loon et al. (2006)

2.3.4 Pathogenesis Related Proteins (PR)

The concept of PR was introduced in 1980 to designate any protein encoded by the host plant but induced only in pathological or related situations (Antoniw et al. 1980), including viral, fungal or bacterial infections, parasitic attack by nematodes. The main criterion for inclusion among the PR is that the protein concerned is newly expressed upon infection, although not necessarily in all pathological conditions (Van Loon 1999). The term PR-like protein was proposed to accommodate proteins that are present in healthy plants, being induced essentially in a developmentally controlled, tissue-specific manner, and that are not synthesized in response to pathogen infection or related stresses. The distinction between PR proteins and PR-like proteins became blurred by the discovery of specific PR proteins in healthy tissues and the induction of PR-like proteins upon pathogen attack. Recently Van Loon et al. (2006) introduced the general term "inducible defense-related proteins" to include proteins that are mostly non-detectable in healthy tissues and for which induction at the protein level has been demonstrated after pathogen infection. The PR proteins encompass several different groups of structurally and functionally unrelated proteins, which have been grouped into protein families according to coding sequence similarities, serological relationships, and/or enzymatic or biological activities (Tarchevsky 2001; Ferreira et al. 2007). Seventeen classes are now considered, numbered in the order in which they were discovered, from PR-1 to PR-17 (Table 4.3; Van Loon et al. 2006); members of several of these families were demonstrated to have damaging actions on the structures of the parasite in *in vitro* bioassays, thus exhibiting antifungal activity and supporting a possible role for these proteins in plant defense. PR-1 and PR-5 (thaumatin-like proteins and osmotins), are thought to create transmembrane pores and have therefore been termed permatins; PR-2 (β-1,3-glucanases) and PR-3, 4, 8 and 11 (chitinases), which attack β-1,3-glucans and chitin respectively, components of the cell walls in most higher fungi (Honeé 1999). PR-6 proteins (proteinase inhibitors) may target nematodes, whereas the PR-7 protein (an endoproteinase) may be involved in microbial cell wall dissolution (Jordá et al. 2000). The PR-9 family may act in cell wall reinforcement by catalyzing lignifications, leading to enhanced resistance against multiple pathogens (Passardi et al. 2004). Some members of the PR-10 family exhibit a weak ribonuclease activity, suggesting a role in defense against viruses (Park et al. 2004). Members of the PR-12 (defensins), PR-13 (thionins) and PR-14 (lipid transfer proteins) families display antibacterial and antifungal activities (Epple et al. 1997; García-Olmedo et al. 1995; Lay and Anderson 2005). PR-15 (oxalate oxidases) and PR-16 (oxalate oxidase-like proteins) proteins generate hydrogen peroxide that may be toxic to attackers or stimulate plant defense responses (Hu et al. 2003). PR-17 proteins, as yet uncharacterized, have been detected in infected tobacco, wheat and barley (Christensen et al. 2002).

2.4 Breeding Strategies

It is a common notion that if a new character is added to a breeding program, either gains in other characters will suffer (for example yield potential), or the program will have to be expanded by a factor which is dependent on the selection rate. The breeder therefore has to consider whether breeding for disease or pest resistance is economically sustainable. This decision depends mainly on the frequency and extent of disease in the area where the crop is to be grown and on the economic damage caused by the parasite. Moreover, the breeder should identify which type of defense mechanism is most suitable for introduction into the crop. He may choose major-gene resistance with complete expression. Advantages of this type of resistance are: (i) the simple inheritance, which is of course very desirable in a breeding program; (ii) the normally complete protection of the crop from the parasite. A risk in choosing complete major-gene resistance is that this type of resistance may turn out to be transitory. There are, however, cases where major-gene resistance has been durable.

The next step in a breeding program for resistance is the identification of an appropriate source of resistance. The genotypic variation within the genotypes of the crop and often also within related species should be investigated. Source for resistance may be found in taxonomic groups that are more or less distantly related to the crop, such as the cultivar itself, commercial cultivars, landraces, wild progenitors, related species and genera. The potential sources of resistance indicated here are listed in the order in which complications for the breeder increase. The main problems are: (i) failure to secure crosses between the crop and the donor species,

(ii) sterility of the interspecific or intergeneric hybrid, and (iii) poor intrachromosomal recombination (Harlan and De Wet 1971). Many generations of backcrossing are usually needed to remove undesirable traits introduced together with the resistance.

2.4.1 Conventional Methods

In many cases, a single *R* gene can provide complete resistance to one or more strains of a particular pathogen, when transferred to a previously susceptible plant of the same species. For this reason, *R* genes have been used in conventional resistance breeding programs for decades (Austin et al. 2002). *R* gene-mediated resistance has several attractive features for disease control. When induced in a timely manner, the concerted responses can efficiently block pathogen growth with light collateral damage to the plant. No input is required from the farmer and there are no adverse environmental effects. Unfortunately, *R* genes are often quickly defeated by co-evolving pathogens. In this context, it is worth noting that durable resistance is defined as "resistance that remains effective when a cultivar is grown widely in environments favoring disease development" (reviewed by Michelmore 2003). The concept of durable resistance makes no assumptions concerning the mechanisms or genetic control of resistance, and has proved a very useful concept in disease resistance breeding. Although it is now easier to identify and deploy useful *R* genes, the problem of durability remains. Many *R* genes lack durability because they can be defeated by a single loss-of-function mutation in the corresponding *Avr* gene (thereby rendering the pathogen 'invisible'). Because individual *Avr* genes often make only incremental contributions to virulence, pathogens can afford to alter or discard an *Avr* gene with little or no fitness penalty (Leach et al. 2001).

Traditional breeding strategies have used *R* genes 'one at a time' in crop monocultures. Such homogeneous host populations exert strong selection for mutation of the relevant *Avr* gene, and then become extremely vulnerable to the emergent pathogen. As an alternative to single-gene deployment, multiple *R* genes ('pyramids') can be bred into individual plant lines (Pink 2002). In principle, these pyramids require the pathogen to accumulate mutations in multiple *Avr* genes to escape detection. This is unlikely to occur if the mutations have a strong cumulative effect on virulence. Another approach is to grow a mixture of lines, each expressing a different *R* gene(s), in the same plot. A susceptible line can be included in the mixture, to reduce the selection pressure for mutations in *Avr* genes (Mundt 2002). A multiline protocol was tested in a study, with striking success (Zhu et al. 2000). Pyramiding and multiline deployment have not been widely used, owing to the time required for breeding assortments of *R* genes into elite cultivars. However, these strategies will become much more practical as the approaches described earlier are further developed. Furthermore, many *R* genes recognize only a limited number of pathogen strains and therefore do not provide broad-spectrum resistance. Furthermore, introgression of *R* genes into elite cultivars by conventional breeding is a lengthy process. However, recent molecular-level insights into the function of R proteins and downstream signal transduction pathways might provide strategies to remedy these deficiencies.

2.4.2 Applications of Marker-Assisted Breeding

Though polygenically inherited forms of resistance is nearly always durable (Parlevliet 1979) this type of inheritance is more difficult to handle in breeding programs. In particular, backcross programs to introduce polygenes from wild relatives of the crop are heavy or not easily managed. In many agricultural crops a low level of infection is acceptable, and partial resistance may be combined with other control measures, such as the application of pesticides. In the specific case of disease resistance, marker-assisted breeding may have a special role. In this respect, pyramiding several major resistance genes into a valuable genetic background is simplified via the use of marker-based selection (Song et al. 1995). Studies have indicated that pyramiding resistance QTLs can achieve the same level or even a higher level of resistance than that conferred by an *R* gene (Castro et al. 2003a, b; Richardson et al. 2006). This should be especially helpful when screening for one resistance gene interferes with the ability to screen for another, a frequent occurrence in disease resistance breeding. Rather than screen sequentially for the inheritance of single resistance (or simultaneously through progeny screens), individuals that have retained all of the genes of interest can be selected based on DNA marker genotype.

Similarly, gene deployment can be speeded-up via the use of marker assisted breeding. This approach, in which cultivars with complementary sets of resistance genes with differing race-specificities are grown by farmers, aims at achieving durable disease protection. In theory, the capacity to pyramid or deploy genes of interest is not restricted to major, single locus resistance genes. With QTL mapping, partial resistance loci can be treated as Mendelian factors and manipulated just like any major gene. This includes resistance alleles that apparently come from otherwise susceptible parents (Wang et al. 2006), providing the potential for selecting transgressively resistant genotypes. Consequently, QTLs from diverse donors can be quickly introduced into a desirable genetic background or deployed in a set of cultivars. Information about the race-specific (or race-nonspecific) nature of partial resistance loci can obviously play a key role in this process.

2.4.3 Alternative Possibilities for Resistance

It is worth noting that alternative possibilities of obtaining resistance have been exploited. These include mutation breeding and transgenic technologies.

Mutation Breeding

A mutagenic treatment may convert a susceptible genotype into a resistant one. If the mutation is a point mutation, the resistant mutant will be identical to the original cultivar, except for its resistance. Usually, however, there are undesirable side-effects of the mutagenic treatment. Several other genes may also have undergone changes, or the mutation for resistance has undesirable pleiotropic effects. As a consequence, the selection of a resistant mutant should be followed by further breeding efforts (i.e. backcrossing) to produce a commercially acceptable cultivar.

Transgenic Prospects

Currently, there are no fungus-resistant transgenic crops on the market. However, a number of reports have shown promising results in field trials. One example is a potato line that is resistant to late blight (Song et al. 2003). Late blight, caused by the oomycete *Phytophthora infestans*, is infamous as the cause of the Irish potato famine in the nineteenth century and still today causes serious crop losses around the world. The gene that was introduced into the potato line was called *RB* and came from a wild Mexican potato species called *Solanum bulbocastanum*.

There are also prospects for transgenic use of single *R* genes that have previously been proven durable. For example, the pepper gene *Bs2* has provided long-standing resistance against bacterial spot disease, caused by the bacterium *Xanthomonas campestris*. *Bs2* has been cloned from pepper and shown to encode a NB-LRR protein (Tai et al. 1999). *X. campestris* is also a significant pathogen of tomato and a pepper *Bs2* transgene works effectively in tomato against *X. campestris*. Recently cloned *R* genes with potential use against fungal pathogens include the barley *Rpg1* gene and the tomato *Ve1* and *Ve2* genes (Kawchuk et al. 2001). *Rpg1* has provided remarkably durable resistance to stem rust for decades, while *Ve1* and *Ve2* target *Verticillium* species that cause wilt in many different crops. The *Ve* genes can provide resistance to different *Verticillium* species and are functional in potato when expressed as transgenes. The *Rpg1* and *Ve* genes are also interesting from a basic research standpoint because they have novel structural features that distinguish them from previously characterized *R* genes. Additionally, it will be particularly interesting to determine whether these genes can be used as prototypes to identify additional *R* genes by sequence similarity. Additional useful genes might be unearthed through investigations of so-called 'non-host resistance' (Heath 2000). This term refers to interactions in which all varieties of a plant species are resistant to all strains of a particular pathogen species (as opposed to intraspecific variability, which is observed for *R* gene-mediated resistance). Because non-host resistance is not genetically variable, this trait has not been amenable to classical genetic analyses. However, experimental tools now available in model plants (e.g. large-scale forward and reverse genetic screens) have made non-host resistance more accessible to genetic dissection. For example, *Arabidopsis* and tobacco are uniformly resistant to many microbes that plague crops (e.g. *P. infestans*, which caused the Irish potato famine) (Kamoun 2001). Moreover, it is worth noting that certain signal transduction components are used in *R* gene resistance and for some non-host resistances (Peart et al. 2002). Thus, it might be possible to identify effective resistance genes against crop pathogens from model species and transfer them to crops. It will be of great interest to determine whether non-host resistance results from natural pyramiding of *R* genes, and/or from use of *R* genes that recognize virulence factors that are essential for the pathogen to cause disease. Note that non-host resistance might result from several mechanisms (Heath 2000) and it is possible that genetic dissection of non-host resistance will provide unanticipated tools for engineered resistance.

Efforts to transfer *R* genes from model species to crops, or between distantly related crops, could be hampered by a phenomenon termed 'restricted taxonomic functionality' (RTF) (Tai et al. 1999). For example, *Bs2* and several *R* genes from tomato can function as transgenes within related species from the same family (Hulbert et al. 2001) (e.g. tobacco, potato and pepper, which belong to the *Solanaceae*). However, *Bs2* does not function in *Arabidopsis*, nor does the *Arabidopsis RPS2* resistance gene function in tomato (Tai et al. 1999). The molecular basis of RTF is unknown but might reflect an inability of the R protein to interact with signal transduction components that have diverged in the heterologous host (Ellis et al. 2000). It remains to be seen whether RTF is a general attribute of *R* genes. A recent report suggests that it will indeed be possible to transfer certain *R* genes between distantly related species: the *Arabidopsis RPW8* gene provides broad-spectrum resistance to powdery mildew in *Arabidopsis* and in tobacco (Xiao et al. 2003). A solution to the RTF problem might be developed as we gain a deeper understanding of *R* gene signaling.

In bacterial the tomato disease resistance gene *Pto* gives race-specific resistance to *Pseudomonas syringae* pv. Tomato carrying the *avrPto* gene, by overexpressing *Pto* race- non specifical resistance in transgenic tomatoes exhibited a superior tolerance (Tang et al. 1999). Similarly to fungal resistance, overexpression of PR proteins or transfer of PR protein genes, such as the barley lipid transfer protein (LTP2; Molina and García-Olmedo 1997), from other sources has led to increase resistance against bacterial infection. In several plant species, bifunctional enzymes with lysozyme activity have been detected which are hyphotised to be involved in defense bacteria. After transfer of the bacteriophage T4 lysozyme gene, transgenic potatoes had reduced susceptibility toward *Erwinia carotovora atroseptica* infection (Duering et al. 1993). Transfer of the human lysozyme gene resulted to increase resistance against both fungal and bacterial diseases (Nakajima et al. 1997)

In several studies, transgenic plants expressing cereal ribosomal inactivating proteins (RIPs) were used to test defense properties attributed to this group of proteins (reviewed in Balconi et al. 2010). Research in our laboratory showed that transgenic tobacco plants, expressing the maize RIP *b-32* gene driven by the *wun 1* promoter, had increased protection against infection from the soil-borne fungal pathogen *Rhizoctonia solani*. Similarly, other research with wheat transgenic lines, indicated that maize RIP b-32 protein was effective, as anti-fungal protein, in reducing Fusarium head blight (FHB) symptoms. To further explore the antifungal activity of the maize RIP b-32, transgenic maize plants have been developed containing the *b-32* coding sequence driven by a constitutive *35SCaMV* promoter. In this study four homozygous independent maize transgenic lines, with differential ectopic expression of RIP b-32, were evaluated, in comparison with plants expressing RIP *b-32* only in the endosperm, for response to *Fusarium verticillioides* colonization by leaf tissue bioassays. The identification of progenies with a differential RIP b-32 expression in the leaves was useful for setting up pathogenicity experiments. Transgenic progenies expressing RIP b-32 in leaf tissues resulted as less susceptible than the negative control when evaluated for response to *F. verticillioides* attack, showing significantly reduced colony diameter around the inoculated leaves; a good correlation between the RIP b-32 content in the leaves and the level of resistance to Fusarium attack was observed. Collectively, these results confirm that the incorporation of maize RIP *b-32* gene and the ectopical expression of RIP b-32 protein, appears to be an effective and reliable tool in crop disease management programs.

3 Viral Diseases

Viruses are among the most important kinds of plant pathogens causing severe economic losses in many crops. Genetic resistance is one of the different systems to protect crops from virus infection, including also the control of biotic vectors, the use of virus-free plant materials, and practices for avoiding the transmission. If available, the genetic resistances are still considered the most effective mean for avoiding the viral diseases. The study of virus resistance genes implies several questions regarding the molecular mechanisms involved in the plant-virus interaction. Resistance to viral diseases has been divided, similarly for other pathogens, into two principal families: non-host and host-resistance. Host resistance to plant viruses has been more investigated and considers the case where all genotypes within a plant species show resistance or fail to be infected by a particular virus. More than 80% of reported viral resistances is monogenic. The remainder shows quantitative inheritance. About half of monogenic resistances show dominant inheritance. In most but not all cases, dominance has been reported as complete (Fraser 1986). Furthermore, a third important family of host resistance has been identified, initially in studies involving TMV, i.e. SAR. This response can be activated in many plant species by diverse pathogens that cause necrotic cell death (Ross 1961), resulting in diminished susceptibility to later pathogen attack. As SAR has recently been reviewed (Durrant and Dong 2004), this topic is not discussed further here. Virus-induced gene silencing, another induced defense mechanism to virus disease, has also been reviewed recently (Baulcombe 2004).

3.1 Genetic Basis of Virus Resistance

Plants contain many (>200) resistance genes (*R*) that confer resistance to different strains of viruses. The largest class of *R* genes encodes a NB-LRR type of protein. So far all *R* genes that have been isolated conferring resistance to viruses belong to this class. It is tempting to assume that the *R* gene products directly or indirectly interact with other (host or virus-encoded) factors, but this still needs to be demonstrated. Approximately 30% of the *R virus* genes have been tagged with molecular genetic markers that can be exploited for indirect selection via genotype, for locating *R* genes in plant genomes, and for gene isolation. Relatively few QTL for plant viral resistance have been tagged or genetically mapped.

Considerable progresses were also made in the study of *R* gene structure and in the explanation of mechanisms of resistance and viral evolution. The advent of molecular methods has demonstrated that *R* genes may represent different loci with shared or independent evolutionary history, or different alleles at the same locus. There are cases where resistance alleles at two or more loci are required to observe the resistant response. A well-known example is the *bc-u* system in *Phaseolus vulgaris* for resistance to a wide array of BCMV pathotypes. Resistance is observed only when the *bc-u* locus is homozygous recessive and one or more pathotype-specific genes, *bc-1*, *bc-2*, and *bc-3*, are also homozygous at one or more of three additional loci (Drijfhout 1978). In *Capsicum*, for example, full resistance is observed to another potyvirus, *Pepper veinal mottle virus*, only when the resistance alleles *pvr12* (formerly *pvr22*) and *pvr6* are homozygous (Caranta et al. 1996).

One type of *R* gene cluster contains a set of genes, showing similar inheritance and resistance phenotypes that control very closely related viral genotypes. A notable example of this pattern occurs in *Pisum sativum* where recessive resistance has been mapped to two *R* gene clusters on linkage groups II and VI. This type of *R* gene cluster occurs widely in monocots and dicots. For example, the wheat *Bdv1* allele conferring resistance to *Barley yellow dwarf virus* (BYDV) is linked to fungal *R* genes *Lr34* and *Yr18* (Singh 1993). A comprehensive genome-wide analysis of *R* gene clusters and their distribution within a series of crop genomes linked by comparative genetic mapping has been published for the *Solanaceae* (Grube et al. 2000). This study clearly demonstrated that *R* gene clusters often occur at homologous positions in related genomic regions, even in genera that diverged tens of millions of years ago. These clusters may therefore consist of evolutionarily related sequences that diverged to control very different pathogen groups.

The typical R-gene-mediated responses include host-cell death (HR) like that occurring in fungi and bacteria and will not here repeated. The induction of this response is preceded by a specific recognition of the virus, and in many cases this is based on matching (dominant) gene products of the plant (produced from dominant resistance genes, *R* genes) and the virus (avirulence genes). Dominant resistance is frequently associated with the HR response (Fraser 1986), possibly due to the frequent use of HRas a diagnostic indicator for field resistance by plant breeders. HR, induced by specific recognition of the virus, localizes virus spread by rapid programmed cell death surrounding the infection site, which results in visible necrotic local lesions. HR-mediated resistance is a common resistance mechanism for viruses and for other plant pathogens. Because the extent of visible HR may be affected by gene dosage (Collmer et al. 2000), genetic background, environmental conditions such as temperature, and viral genotype, etc., schemes that classify or name virus *R* genes based on presence or absence of HR may obscure genetic relationships. Over the past 10 years significant advances have been made in the understanding of the molecular basis of the HR-mediated resistance. More than 40 plant *R* genes showing monogenic dominant inheritance have been cloned (reviewed in Kuang et al. 2005). Several of these confer resistance to plant viruses (Martin et al. 2003).

Few resistance genes have proved exceptionally durable. Genetic resistance often fails because a resistance-breaking (RB) pathogen genotype increases in frequency.

Based upon data obtained predominantly from plant resistance to fungi, polygenic resistance is often presumed to be more durable than monogenic resistance. The analysis of polygenic resistance traits tends to be much more complex than monogenic or oligogenic traits, so researchers often focus on monogenic resistance because it can be studied and utilized more readily.

3.2 *Resistance Mechanisms*

The natural resistance mechanisms underlying virus resistance in plants have been largely treated in several reviews (Goldbach et al. 2003) and will be here briefly summarized. The main finding emerging from these studies indicates that the genetic material of viruses may be either DNA or RNA, and may be single- or double-stranded. Approximately 77% of characterized plant viruses possess a single plus-(messenger) sense strand of RNA. Infection of plant tissue requires damage to the cell wall and/or plasma membrane which, for insect-borne viruses, is achieved by the penetration by the insect stylet during feeding. Once inside the cell, the virus particle is uncoated to release its nucleic acid, and for at least some plus-stranded RNA viruses, such as tobacco mosaic virus (TMV), uncoating is achieved by cytoplasmic ribosomes which also translate the RNA. Plant virus nucleic acids are not integrated into the host genome. Common translational products amongst most, if not all, viruses, include coat protein, one of more proteins involved in the replication process, and factors involved in the systemic transmission of the virus away from the site of infection. The genomes of cauliflower mosaic virus (CaMV) and tobacco mosaic virus (TMV) contain seven and five open reading frames (ORFs), respectively, which function in the replication and movement of the viral DNA, symptom development, and encapsidation. Genome replication for positive-strand RNA viruses occurs in the cytoplasm, apparently utilizing the translation apparatus of the host. Plant virus movement proteins (MPs), in association with various components of the cytoskeleton of the host cell, facilitate transport of nucleoprotein complexes or virus particles into adjacent cells by way of modified plasmodesmata, channels between plant cells. The processes controlling long-distance transport of virus particles or viral nucleic acids within the phloem are distinct from those controlling movement between mesophyll cells. Once inside the phloem, a rapid movement of virus particles has been documented; for some viruses (e.g. TMV), the coat protein (CP) is necessary for this process; however, for other viruses the CP protein may not be involved.

3.3 *Breeding for Viral Diseases*

3.3.1 Conventional Strategies

One of the most important durable successes of plant breeding for virus resistance was the development of sugar beet with the source of resistance to rhizomania (Fig. 4.5). The disease is caused by the virus BNYVV (Beet Necrotic Yellow Vein Virus)

Fig. 4.5 Section of sugar beet root infected by rhizomania

transmitted by the fungus *Polymyxa betae*. Rhizomania is widespread in many Europeans countries and available data indicate a spread on 60% of total sugar beet cultivated area. Damage to the sucrose production can cause up to a 80–100% yield loss (Biancardi et al. 2002). Three forms of virus have been classified (A, B and P) according to the structure of RNA.

The rhizomania symptoms are evident especially on the roots as: (i) excessive proliferation of the rootlets assuming a beard-like appearance around the tap root; (ii) constrictions of the root tip leading to a wineglass shape; (iii) necrotic rings in the root tip section. Diseased beets, if analyzed, show low sugar content, processing quality, etc. Immunoenzymatic tests (ELISA) performed on the roots can easily quantify the infection.

A and B types there are often associated, while the P is always alone and has been localized only near Phitivier (France). The use of ELISA test for the determination of virus content in the storage root has significantly contributed to the selection of resistant varieties. These genotypes have allowed survival of the sugar beet crop in many cultivated areas. After the discovery of the first resistant materials of

Italian origin, derived from sugar beet progenitor *Beta maritima*, with multigenic resistance, defined "type Alba", more efficient sources of monogenic resistance were introduced, also these derived from *Beta maritima*, allowing optimal productive performance also in infected soils. The sugar beet cultivars with the "Rizor" source of resistance, developed in 1985 by De Biaggi, was the first variety showing an optimum level resistance in heavily infected fields. Later, after some years, the source "Holly" has been isolated on materials of USDA's origin. These two sources of resistance have a good heritability and few cycles of selection are sufficient to improve the resistance trait. Resistance such as "Holly" is classified as monogenic, like that of type "Rizor". The two resistant traits have been mapped very close on several genetic maps. Other sources of resistance have been found recently in these wild beets also belonging to the *Beta vulgaris* L. ssp. *maritima* (L.) Arcang, that is the ancestor of the cultivated beets. Among these, the source named "WB42", developed at the USDA in Salinas (California, USA), is stirring a considerable interest. Studies are still in progress to determine the relationship between the two major sources of resistance (Holly and Rizor).

Genotypes carrying the monogenic sources of resistance frequently exhibit different levels of expression, probably due to the presence of minor genes interacting with the major allele in heterozygous individuals. The resistant varieties used today, when tested in severe disease conditions applied in greenhouses, display no more than 80% resistant plants. Improvement of this percentage should allow better sugar yield even in severely diseased fields. Since the resistance in commercial varieties is usually transmitted by the pollinators, this goal should be possible using varieties in which all plants carry the genes of resistance at least in heterozygous conditions. This result is becoming possible by: (i) using resistant pollinators and seed-bearers; (ii) analyzing with molecular markers for rhizomania resistance genes all pollinating and/or seed-bearing beets employed in seed production; and (iii) discarding the recessive and, when possible, the heterozygous plants. In addition, further sugar yield improvements should be possible combining in the same variety the different sources of resistance. This would be essential where the known sources of resistance appear to be overcome by suspected mutations of BNYVV, or in presence of the more pathogenic strains of the virus (Liu and Lewellen 2007; Panella and Lewellen 2007).

3.3.2 Transgenic Strategies

A knowledge of the molecular biology of aspects of virus function has led to the proposal of three general strategies for plant protection against viruses using genetic engineering techniques: (1) modified cross-protection; (2) the use of satellite nucleic acids; and (3) the use of anti-sense RNA.

Transgenic crop varieties have been successfully deployed to control viral diseases. One of the classical examples is the success with the genetically engineered papaya, which virtually rescued the papaya industry in Hawaii from the threat of the dreaded ring spot disease (Yeh et al. 1998). The transgenic approach would be more appropriate in situations where sufficient levels of resistance to the virus are not available in the related germplasm or the resistance is difficult to transfer by normal crossing techniques because of either reproductive isolation or linkage with other undesirable traits. The production of virus-tolerant transgenic plants has been based on several approaches which follows into two categories: protein-mediated and RNA-mediated protection (reviewed by Prins et al. 2008). In most instances, a gene coding for the complete viral protein or part of a viral protein has been introduced into the crop by transformation.

One strategy used to obtain virus-resistant plants is to transfer genes from the pathogen itself into the plant (pathogen-derived resistance). The most widely used approach is to express the virus coat protein in transgenic plants. In theory, the expression of viral genes disrupts viral infection or symptom development. The first virus-resistant variety to be grown was papaya ringspot virus (PRSV)-resistant papaya (Ferreira et al. 2002; Gonsalves 1998). The GM variety contains a gene that encodes a PRSV coat protein, a strategy that mimics the phenomenon of cross protection. In true cross protection, infection by a mild strain of a virus induces resistance to subsequent infection by a more virulent strain (reviewed in Culver 2002). This approach has been extended to other plants, for example rice (Hayakawa et al. 1992), plum tree (Ravelonandro et al. 1997), tomato (Kaniewski et al. 1999), and peanut (Magbanua et al. 2000). Field trials have also been performed, in the USA, with coat protein-mediated virus-resistant wheat, soybean, sugarcane, sugar beet, cucumber, sweet potato, grapefruit, pineapple, and papaya (USDA 2002).

Another form of pathogen-derived resistance is the use of viral replicase genes (or RNA-dependent RNA polymerase genes), which presumably act by post-transcriptional gene silencing. This technique has been used to confer resistance to potato leafroll virus in potato, to barley yellow dwarf virus in oats, cucumber mosaic virus in tomato, rice tungro, spherical virus in rice, and wheat streak mosaic virus in wheat (Koev et al. 1998; Gal-On et al. 1998; Huet et al. 1999; Sivamani et al. 2000). Because different degrees of virus resistance have been obtained with coat protein-mediated resistance, attempts have been made to ameliorate resistance against cucumber mosaic virus via satellite RNA, especially in tomato (Stommel et al. 1998). This approach has caused controversy, however, because a single-point mutation in the satellite RNA can transform it into a harmful necrogenic form (Tepfer 1993). To protect plants against more than one virus, RIPs, have been expressed in transgenic plants. RIPs are strong inhibitors of protein synthesis and, depending on the plant species from which they originate, they have different levels of toxicity against different hosts. Poke weed antiviral protein (PAP) confers resistance to PVX and PVY in transgenic potatoes and PAPII confers resistance to TMV, PVX, and fungal infections in tobacco (Balconi et al. 2010).

On a more experimental scale are approaches to achieve virus resistance by using antibodies against the virus coat protein. Such antibodies can neutralize virus infection, presumably by interacting with newly synthesized coat protein and disrupting viral particle formation Xiao et al. 2000). Similar to RIPs, broad-spectrum antibodies might be used to protect plants against a wider range of viruses, as has been demonstrated for poty viruses.

Notably, virus-resistant transgenic crops, which offer numerous benefits to growers and consumers, need to be deployed safely after due assessment of safety considerations. However, risk assessment studies need to be realistic to provide valuable assistance to regulatory authorities for the safe and timely release of such crops (Fuchs and Gonsalves 2007).

4 Insect Diseases

Crop losses due to insects and nematodes, estimated at 10–20% for major crops, are a significant factor in limiting crop yields. To overcome this problem modern agriculture uses a wide range of insecticides and nematocides to control pest damage. However, chemical control of pests, in addition to being expensive, frequently results in negative environmental effects. The development of insect- and nematode-resistant plants is therefore an important objective of plant breeding strategies with relevant implications for both farmers and the seed and agrochemical industries. In this section attention will mainly be given to plant response to insects, indicating some specific examples related to nematodes.

4.1 Nature of Plant Resistant Mechanisms

According to Maxwell and Jennings (1980) insect resistance is defined as "those heritable characteristics possessed by the plant which influence the ultimate degree of damage done by insects". Resistance is relative and is measured by using susceptible cultivars of some species as controls. Additionally, host-plant resistance may be the result of a series of interactions between insects and plants which influence the selection of plants as hosts and the effects of plants on insect survival and multiplication. Within this context three mechanisms of plant resistance have been described: (i) non-preference (or antixenosis), (ii) antibiosis and (iii) tolerance (Painter 1958). Tolerance differs from non-preference and antibiosis in its mechanism: non-preference and antibiosis require an active insect response or lack of response. However, tolerance is more subject to variation as a result of environmental conditions than non-preference and antibiosis. The age or size and general vigor of the plant and size of the insect-resistant population also strongly influence the degree of tolerance.

In their long association with pests and pathogens, plants have evolved an impressive arsenal of defensive tools. In this respect, natural pest resistance mechanisms occurring in higher plants have been classified into preformed and inducible resistance mechanisms and throughout the last century agricultural pest control has attempted to harness these mechanisms wherever possible (reviewed in Howe and Jander 2008). Furthermore, plant traits conferring resistance to insect pests may also be classified according to the manner in which they are regulated. Some traits are expressed constitutively under the control of hard-wired developmental programs, irrespective of the insect threat level. For example, reproductive tissues typically accumulate large amounts of defensive proteins and metabolites. In contrast to these preformed barriers, herbivore-challenged plants mount defense responses at the site of tissue damage and, in many cases, systemically in undamaged tissues. Moreover, to induce defensive traits, plants can minimize the fitness consequences of tissue loss by activating physiological processes, such as sequestration of sugars in below-ground tissues, which allow plants to better tolerate insect damage.

4.2 Genetic Bases in Imparting Insect Resistance

At least 30 major or single genes for insect resistance have been tagged or mapped in various crops (e.g. maize, rice, wheat, tomato, mung bean, apple), conferring resistance to species from 5 orders: *Homoptera*, *Hemiptera*, *Diptera*, *Lepidoptera* and *Coleoptera* (reviewed in Yencho et al. 2000). Each gene is known to confer resistance to only one insect species or to closely related species within the same genus. The *Mi* gene from tomato provides an interesting example because it was originally identified as a dominant gene for resistance to a root-knot nematode, *Meloidogyne incognita*. Further studies have shown that it is located at the same locus as that previously known as *Meu-1*, which provides resistance to some isolates of the potato aphid (*Macrosiphum euphorbiae)* and to the silverleaf whitefly *(Bemisia tabaci*) (Nombela et al. 2003). Interestingly, *Mi* is one of the few examples of genes for insect resistance cloned from a plant; it is a member of the nucleotide-binding, leucine-rich (MBS-LRR) repeat family of resistance genes, many members of which have been found to confer isolate-specific resistance to viruses, bacteria, fungi, and nematodes (Hammond-Kosack and Parker 2003). Another NBS-LRR protein, encoded by the melon *Vat* gene, confers increased resistance to both *Aphis gossypii* (cotton aphid) and the transmission of plant viruses by this aphid species (Dogimont et al. 2007). By analogy to plant defense against pathogens, these findings suggest a gene-for-gene interaction between the plant and the insect. However, the presumed avirulence proteins in aphid saliva have not yet been identified. Similarly, research on nematodes has identified at least 15 genes conferring resistance to nematodes in various crop species (reviewed in Jung et al. 1998). For example 2 genes for nematodes resistance have been cloned on the basis of their chromosomal position and identified by genetic complementation. The first was $Hs1^{pro\text{-}1}$ from sugar beet that confers resistance to the beet cyst nematode (*Heterodera schachtii*). The second was *Mi-1* which is responsible for the hypersensitive reaction of tomato roots after infection with *Meloidogyne* spp.

4.3 Gene Mapping and Molecular Markers

Molecular markers have been used to map the above-cited genes for insect resistance in most of the major crop species and to map QTLs for resistance to 11 species of insects from 3 orders (*Homoptera*, *Lepidoptera* and *Coleoptera*) in 6 plant species (reviewed in Yencho et al. 2000). Traits evaluated include direct measures of insect fitness or behavior (e.g. larval weight, population growth, ovipositional preference); plant damage (e.g. scores on scales of 1–9, tunnel length, leaf area defoliated); plant morphology (e.g. trichomes, leaf toughness); and plant chemical content or enzyme activity (e.g. acyl sugars, maysin, polyphenol oxidase). For any single trait scored, the number of QTLs identified varies from 1 to 10, and the percentage of variation explained by any single QTL varies from 1.3% to 58%. More recently, Pfals et al. (2007), by mapping in *Arabidopsis thaliana* QTLs for resistance agents of 2 cruciferan specialist lepidopteron herbivores (*Pieris Brassicae* and *Plutella xilostella*), identified 6 QTLs for resistance against *Pieris* herbivory and found only a weak QTL for *Plutella* resistance. Similarly Omo-Ikerodah et al. (2008), in genetic mapping QTLs affecting resistance to flower bud thrips (*Megalurotrhips sjostedti*) in cowpea, found association between 23 DNA markers and resistance to flower bud thrips. QTLs with effects on resistance were identified in five linkage groups which accounted for 77.5% of the phenotypic variation for resistance. Moreover, molecular markers can greatly speed up the identification of new resistant genes. This aspect is well documented for the Hessian fly (Hf), *Mayetiola destructor* (Say) (*Diptera:Cecidomyiidae*), one of the most destructive pests of wheat worldwide. To date, 31 major *Hf*-resistance genes (named *H1* through *H31*) have been identified from wheat and its relatives (Williams et al. 2003), but distinguishing new genes is difficult by traditional phenotypic differentiation with biotypes. Liu et al. (2005), using molecular markers, have identified a new gene or a new allele of an *H* gene, (tentatively named *Hdic*) on the short arm of wheat chromosome 1A, which confers a high level of resistance to Hf. of a known *H* gene on chromosome 1A. The broad spectrum of resistance conferred by the *Hdic* gene makes it valuable for developing Hf resistant wheat cultivars. Selection for nematode resistance has a long tradition in potato breeding; both polygenic and monogenic types of resistance have been mapped with molecular markers (cf. Jung et al. 1998). These include the *Gro1*, genic resistance to all pathotypes of the root cyst nematode . *Globodera rostochiensis*. Similarly, in soybean cyst nematode different types of resistance to *Heterodera glycine* have been mapped with molecular markers such as the *Rhg1* and the *Rhg4* loci. In barley, the nematode resistance loci *Ha1 and Ha2* have been mapped to chromosome 2, while a new gene, *Ha4* has been mapped to chromosome 5. In *Triticeae*, two loci, *Cre1* and *Cre3*, have been mapped.

In addition to their utility as selectable markers to facilitate breeding efforts, molecular markers can be employed to increase our understanding of the mechanisms of plant resistance to insects. By mapping QTLs encoding for specific plant physical and/or biochemical attributes associated with insect resistance, and comparing the locations of these QTLs with those identified for the phenotypic expression of resistance to a pest species, valuable insights can be obtained into the nature of resistance. Often, these insights have both basic and applied implications that can be used to develop insect-resistant crops more efficiently. The advantages of these techniques is well illustrated by researches carried out by Byrne et al. (1998). These authors have used molecular markers and QTL mapping techniques to unravel the genetic mechanisms of resistance in maize to the corn earworm (CEW), *Helicoverpa zea*, larvae which cause considerable direct yield loss as well as development of kernel-rotting fungi. Moreover, significant negative correlations were reported between maysin concentrations of fresh silks and growth of CEW larvae in dried-silk bioassays (Wiseman et al. 1996). Because C-glycosyl flavones are synthesized via a branch of the well characterized flavonoid biosynthetic pathway, Byrne et al. (1998) hypothesized that loci of that pathway would explain a large portion of the quantitative variation in maysin concentration, and by extension, resistance to CEW. These loci were proposed as "candidate genes" in a series of QTL analyses. (A candidate gene is one that is hypothesized to affect expression of the trait of interest, either *a priori* based on knowledge of trait biology, or *a posteriori*, guided by similar locations of QTLs and genes of known function).

4.4 Direct Defense Responses

Upon attacks by insects, individual plants rely on a matrix-like variety of defense mechanisms, involving physical barriers (leaf toughness and trichomes), toxic or anti-nutritive secondary metabolites, synthesis of defensive proteins, volatile attractants and extrafloral nectars, and/or recruitment of predators and parasitoids, as well as the reallocation of resources upon attack. Additionally, a plant's defense arsenal depends on various genetic, ontogenetic, and environmental factors, which together modulate the complex defensive phenotype and outcome of the interaction.

Although it is known that plants change their primary and secondary metabolism in leaves to resist and tolerate aboveground attack, there is little awareness of the role of roots in these processes (reviewed in Erb et al. 2009). This is surprising given that plant roots are responsible for the synthesis of plant toxins, play an active role in environmental sensing and defense signaling, and serve as dynamic storage organs to allow re-growth. Studying roots is therefore essential for a better understanding of resistance and tolerance to leaf-feeding insects and pathogens. Indeed roots are increasingly recognized to synthesize secondary metabolites implicated in leaf defenses. However, the active role of roots in plant resistance against leaf herbivory implies shoot-root communication. A model of a defensive shoot-root-shoot loop in plant defense reaction has recently been provided by Erb et al. (2009) to which readers are referred for details.

4.4.1 Defensive Metabolites

A remarkably diverse array of over 200,000 low-mass natural products, known as secondary metabolites, are produced by plants. These include alkaloids, furanocoumarins, tannins, saponins, glucosinolates, cyanogenic glycosides, phenolics and benzoaxinoids. This rich diversity results in part from an evolutionary process driven by selection for improving chemical defense against microbial and herbivorous predation. For instance, several terpenoids, the most metabolically diverse class of plant secondary metabolites (>40,000 known structures), play a role in plant defense (Aharoni et al. 2005). The alkaloids, widely distributed secondary metabolites that are best known for their metabolic effects in mammals likely evolved as a defense against insect herbivory.

Benzoxazinoids are also secondary metabolites that are effective in defense and allelopathy. They are abundant in grasses, including the major agricultural crops, i.e. maize and wheat, and other Gramineae, and are synthesized in seedlings and stored as glucosides (glcs) in the vacuole (see Frey et al. 2009, for a recent review). A specific glycosidase, located in the chloroplast, catalyzes the formation of the toxic aglucon when a cell is damaged and disintegrates. DIBOA [2,4-dihydroxy-2H-1,4-benzoxazin-3-one and its C-7-methoxy] derivative DIMBOA are the prevalent representatives of benzoxazinoids in plants. Figure 4.6 gives a schematic representation of the benzoxazinoid biosynthetic pathway in maize as provided by Frey et al. (2009). It has also been shown that DIMBOA is an enzyme inhibitor of α-chymotrypsin, aphid cholinesterase and plasma membrane H^+-ATPase. The correlation between DIMBOA content and protection against insect feeding damage was especially investigated, indicating that DIMBOA can act as a feeding deterrent and reduce the viability of insect larvae, with practical application in developing maize plants with improved insect resistance, as previously suggested by Klun et al. (1970).

Plants also contain significant quantities of various polyphenolic acids, as well as their glycosides and esters. These compounds are implicated in two defense mechanisms: the phenolic fortification of cell walls and the deterrent effect of fiber content (Bergvinson et al. 1995). Free phenols, mainly 4-coumaric and ferulic acid, were implicated as factors contributing to resistance of maize against ECB and the maize weevil (*Sitophilus zeamais*), and recently, to pink stalk borer (*Sesamia nonagrioides*; Santiago et al. 2006). Notably, the transgenic expression of wheat oxalate oxidase in maize significantly increased the phenolic concentrations, mainly ferulic acid. Field testing showed that the transgenic maize exhibited more resistance to ECB than its non-transgenic counterpart. It was suggested that transgenic oxalate oxidase elicits defense responses by generation of H_2O_2 and activating jasmonic acid signaling (Mao et al. 2007).

In addition to possible synergistic effects, metabolic diversity in toxin production by individual plants can also provide defense against multiple herbivores with different feeding styles or resistance mechanisms. Recent work on glucosinolates demonstrates how natural selection for a diverse profile of secondary metabolites can provide defensive specificity. Glucosinolates are found almost exclusively in Brassicales (Hansen et al. 2008); nearly 40 different glucosinolates have been found in *A. thaliana*, and more than 100 breakdown products are likely formed after activation by the enzyme myrosinase. Experiments with four insect herbivores showed

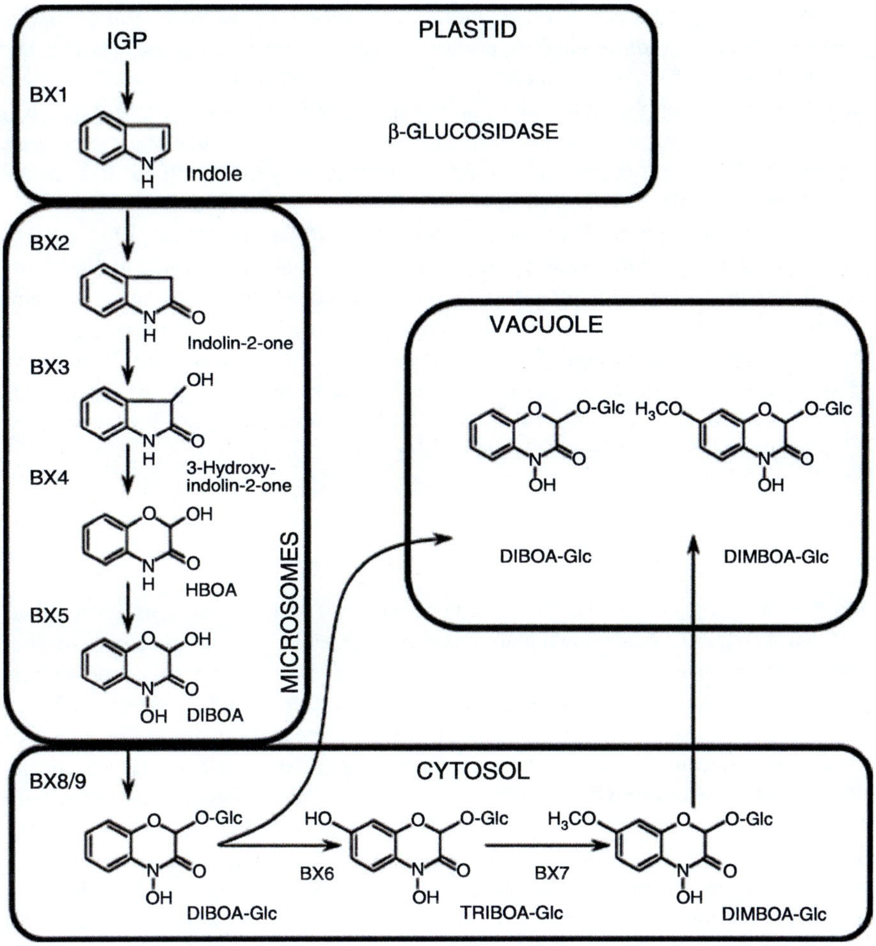

Fig. 4.6 Enzymes and intermediates of benzoxazinoid biosynthesis in maize. In this scheme the BX1 is encoded by the Bx1 gene, a homolog of the Trp synthase a-subunit, catalyses the formation of indole in the first specific pathway. The introduction of four oxygen atoms into the indole moiety that yield DIBOA is catalysed by four cytocrome P450 monooxygenases, termed BX2 to BX5. DIBOA-glc is the substrate of the dioxygenase Benzoxazinless6 (Bx6) and the produced 2,4,7-trihydroxy-2H-1,4-benzoxazin-3-(4H)-one-glc is metabolized by the methyltransferase Bx7 to yield DIMBOA-glc. The enzymatic function of BX1-BX5 is indicated. DIMBOA and DIBOA are accepted as substrates by BX8, while DIMBOA is the preferred substrate of BX9. DIMBOA-glucoside is the predominant benzoxazinoid glucoside in young maize plants. IGP; indole-3-glycerolphosphate, TRIBOA-Glc; TRIBOA-glucoside (Adapted from Frey et al. 2009)

that tryptophan-derived indole and methionine-derived aliphatic glucosinolates have differing effects on Hemiptera and Lepidoptera (Mewis et al. 2005). Indole glucosinolates, which break down in the absence of the activating enzyme myrosinase (Barth and Jander 2006), provide a better defense against *Myzus persicae* than do the more stable aliphatic glucosinolates (Kim and Jander 2007). Almost all genes required for the production of glucosinolates, a diverse class of metabolites found in the model plant *A. thaliana* and other *Cruciferae*, have been identified (Halkier and Gershenzon 2006). As an example of how such knowledge of biochemical pathways can be applied to change plant immunity to herbivory, *A. thaliana* was engineered with three enzymes from grain sorghum to produce the cyanogenic glycoside dhurrin, thereby enhancing resistance to yellow-striped flea beetle (*Phyllotreta nemorum;* Tattersall et al. 2001).

Many defensive compounds are potentially toxic to the plants that produce them. Therefore, the storage of relatively benign precursors that are activated by herbivory is a recurring theme in plant biology. For instance, all three defensive systems mentioned in the previous paragraph include compounds that are sequestered in plants, but not activated until the onset of herbivory. DIBOA is stored as inactive DIBOA-glucoside, glucosinolates are enzymatically activated to produce toxic breakdown products, and the respiratory inhibitor hydrogen cyanide is released from cyanogenic glycosides during herbivory attack.

4.4.2 Defensive Proteins

Insect feeding triggers the expression of plant defensive proteins that exert direct effects on the attackers. The best known plant proteins supposedly involved in defense mechanisms are lectins, ribosome-inactivating proteins (RIPs), inhibitors of proteolytic enzymes, chitinases, and glycohydrolases (reviewed in Carlini and Grossi-de-Sà 2002).

Protease inhibitors (PIs), which impair various mechanistic classes of digestive proteases in the insect midgut, have been thoroughly studied for their role in the active defense response (Ryan 1990). Inhibition of gut proteases by PIs results in amino acid deficiencies that negatively affect the growth and development of the herbivore (Lison et al. 2006; Zavala et al. 2004). The effectiveness of PIs as a defense is often thwarted by the insect's adaptive ability to express digestive proteases that are insensitive to the host plant complement of PIs or that inactivate PIs (e.g. Bayes et al. 2005; Rivard et al. 2004). PIs are synthesized and stored in seeds and tubers of plants and the expression of some *PI* genes is induced in response to mechanical wounding or insect damage. For instance, local and systemic induction of expression of *MPI*, a maize protease inhibitor gene, efficiently inhibits elastase and chymotrypsin-like activities from the larval midgut of *Spodoptera littoralis* (Cordero et al. 1994); this suggests that *MPI* is a factor of maize insect resistance. Similarly, strains of tropical maize germplasm were found to exhibit resistance to Lepidoptera. In these strains, larval feeding led to the induction of a unique cysteine proteinase, *Mir1-CP*; proteinase accumulation was detected at the feeding site, localized predominantly in the phloem of minor and intermediate veins and was correlated with a significant reduction in larval growth (Lopez et al. 2007).

The plant's defensive protein arsenal also includes enzymes that disrupt insect digestive physiology and other aspects of food consumption. Members of the cysteine protease family of enzymes, for example, disrupt the chitin-rich peritrophic membrane that protects the gut epithelium (Konno et al. 2004; Mohan et al. 2006).

Plant lectins and chitinases may also target carbohydrate containing components of the insect gut (Lawrence and Novak 2006; Peumans and Vandamme 1995). Oxidative enzymes such as polyphenol oxidase (PPO) and lipoxygenase (LOX) covalently modify dietary protein through the production of reactive *o*-quinones and lipid peroxides, respectively (Wang and Constabel 2004). Because catalysis by O_2-dependent enzymes is limited by low oxygen levels in the foregut and midgut of some insect species (Thipyapong et al. 1997), an alternative possibility is that PPO and LOX act rapidly (i.e., within seconds) during tissue mastication by insect mouthparts.

The discovery of novel defensive proteins can be facilitated by proteomic analysis of gut content and feces (frass) of insect herbivores. This approach is based on the premise that defensive proteins are relatively resistant to gut proteases and, as a consequence, are highly enriched during passage of the food bolus through the insect. Application of this procedure to the tomato-reared *Manduca sexta* larvae led to the identification of isoforms of arginase and threonine deaminase , which degrade the essential amino acids arginine and threonine, respectively, in the lepidopteran midgut (Chen et al. 2005).

4.4.3 Volatile Defenses

Plants synthesize and emit blends of volatile organic compounds (e.g. terpenoids, green leafy volatiles, and ethylene) in response to damage from herbivorous insects (reviewed in Unsicker et al. 2009). The induced volatiles are proposed to serve a variety of physiological and ecological functions, including the attraction of natural enemies of herbivores, which is termed "indirect defense". Advances in plant biotechnology have allowed investigators to manipulate plant volatile emissions and demonstrate their defensive function in laboratory studies with model plants (Schnee et al. 2006; Kappers et al. 2005). The specificity of this interaction has recently been proved by Degenhart et al. (2009), by restoring the emission of a specific below-ground signal emitted by insect-damaged maize roots. According to these authors, the sesquiterpene (E)-β-caryophyllene is highly attractive to the entomopathogenic nematode *Heterorhabditis megidis*. It was shown that (E)-β-caryophyllene is emitted by ancestral maize and European lines, but most American varieties have lost this ability and do not attract the nematode, which is therefore much less effective as a control agent of the larvae of the western corn rootworm, *Diabrotica virgifera virgifera*, a serious root pest in maize cultivation. To restore nematode attractions, a non-producing maize line was transformed with a *caryophyllene-synthase* gene from oregano, resulting in constitutive emissions of (E)-β-caryophyllene. In root-worm infested field plots, in which they released nematodes, transformed plants received significantly less root damage and had 60% fewer adult beetles emerge than isogenic lines. This demonstration that plant volatile emissions can be manipulated to enhance the effectiveness of biological control agents opens the way for a novel ecologically sound pest control strategy.

4.4.4 Signal Transduction Pathways

There is relatively little information about the signal transduction pathways that connect insect-specific elicitors to the plant defense responses they generate. Evidence indicates that the calcium ion (Ca^{2+}) is involved as a second messenger in many plant signaling pathways, including responses to herbivory (Maffei et al. 2007). Transient increases in cytosolic Ca^{2+} levels activate calmodulin and other calcium-sensing proteins that subsequently promote downstream signaling events, including protein phosphorylation and transcriptional responses. Although no complete mitogen-activated protein kinase (MAPK) signaling cascades (Pitzschke et al. 2009) leading to insect resistance has been identified, there is evidence that such pathways play a role in some plant-insect interactions. In tomato, *Mi-1* mediated resistance was attenuated when expression of certain MAPKs and MAPK kinases was reduced by virus-induced gene silencing (VIGS) (Li et al. 2006). VIGS studies in tomato also showed that at least three MAPKs are required for systemin-mediated defense responses to *Manduca sexta* (tobacco hornworm) (Kandoth et al. 2007).

Many inducible defenses are expressed rapidly (i.e., within hours) in undamaged leaves of herbivore-challenged plants. This systemic response, which has been reported in a wide range of plant species, provides effective resistance to future insect attacks (Karban and Baldwin 1997). Since the discovery of this phenomenon more than 35 years ago (Green and Ryan 1972), research effort has been devoted to the identification of systemic wound signals and the underlying mechanisms by which they are produced, transported, and perceived. In this respect, it was found that systemin, which is a strong peptide elicitor of PI expression in *Solanum lycopersicum*, appears to enhance systemic defenses by amplifying jasmonate synthesis in damaged leaves (Schilmiller and Howe 2005).

The plant hormone jasmonic acid (JA) and related signaling compounds (collectively referred to as jasmonates) appear to be ubiquitous signals for tissue injury and for the subsequent activation of defense responses to many, if not most, insect herbivores (Howe and Jander 2008).

Recent studies with *Nicotiana attenuata* indicate that fatty-acid amino acid conjugates (FACs) in oral secretions of *M. sexta* elicit rapid activation of MAPK activity and defense-related genes in undamaged areas of the attacked leaf (Wu et al. 2007). FAC binding to a hypothetical receptor was proposed to generate a rapidly acting, short-distance mobile signal that triggers MAPK cascades in the damaged leaf. This intraleaf systemic response is followed by the production of a second mobile signal (e.g., jasmonate) that initiates PI expression in distal undamaged leaves. These findings are consistent with the idea that multiple intercellular signals, acting over a range of distances, mediate the complex spatiotemporal responses of plants to herbivory. The fact that both *S. lycopersicum* systemin and FACs positively regulate jasmonate synthesis via a MAPK cascade (Kandoth et al. 2007) suggests that parallel signaling pathways initiated at the plant-insect interface may converge on the jasmonate pathway. In this context, evidence has been provided in the past few years to indicate that the jasmonate family of signaling compounds is involved in endogenous regulation of plant resistance to insects.

4.4.5 Breeding Strategies for Improving Plant Pest Resistance

Although there have been many notable successes in conventional breeding for improved plant resistance to insects, the breeding process is often slow and laborious, and sufficient levels of resistance have not been achieved for some pests. However, recent progress in plant transformation technologies has made it possible to produce new genetically modified cultivars with improved resistance to insect pests by genetic engineering. In addition, with advances in biotechnology, breeding of horizontal resistance, whereby resistance is based on many genes, along with genetically enhanced sustainable pest resistance with fusion genes, offer new strategies in improving plant insect resistance (Wan 2006). Genomic tools are enabling significant progress in the understanding of nematode diseases (reviewed in Bellafiore and Briggs 2010). Genome-wide expression profiling of infected plants has revealed genes that respond to infection and functional tests show they can mediate the interaction with nematodes. Several candidate effectors from nematodes have been identified and functional tests using RNAi have supported their putative roles in pathogenesis. These will increase the possibility to design novel approaches to developing crops resistant to nematode injuries.

Marker Assisted Selection

Once a major gene or QTL has been identified and mapped, marker assisted selection (MAS) and/or mapbased gene cloning can be initiated. Particularly, MAS offers the opportunity of combining different genes for a given pathosystem in a single genotype (gene pyramiding). A prerequisite for gene pyramiding is that loci are not allelic. Moreover, it would be wise to determine if the resistance genes targeted for introgression are indeed potentially durable. In choosing the resistant parent(s) for a mapping population, or choosing among existing mapping populations for a study of insect resistance, knowledge of the mechanisms of resistance involved or prior observation of the durability of a resistant cultivar in the field or in selection experiments can identify cultivars that may be sources of promising major genes or QTLs (Alam and Cohen 1998). The results of a QTL analysis itself will indicate whether the insect resistance in the resistant parent of the mapping population indeed has a polygenic basis. Insight into whether the QTLs influence multiple resistance factors acting on multiple targets within the pest can be gained by analyzing the mapping population for a series of carefully chosen traits.

Selective breeding for QTLs conferring a particular modality of insect resistance (antibiosis, antixenosis), or tolerance (Painter 1958), is another approach to achieving more durable varietal resistance. In this respect, Alam and Cohen (1998), in a QTL analysis of six traits associated with rice resistance to the brown planthopper, found a total of seven QTLs, one of which was predominantly associated with antixenosis and a second with tolerance. Most of the other QTL analyses of insect resistance conducted to date have scored chemical or morphological antibiotic resistance factors, or plant damage ratings under free-choice conditions, the results of which can be influenced by all three resistance modalities (see Yencho et al. 2000, for a review).

Transgenic Plants for Pest Control

Insect-resistant transgenic crops for enhancing insect pest control is currently one of the most important successes of plant biotechnology: more than 30 million hectares are planted worldwide with crops expressing *Bacillus thuringiensis* (Bt) δ-endotoxins (James 2009).

Bt is a soil bacterium that makes crystalline inclusions (Cry proteins) during sporulation (De Maagd et al. 2001; Bravo et al. 2007). These crystals dissolve in the alkaline environment of insect midguts and release protoxin molecules (65–140 kDa) that are processed by midgut proteases to yield active insecticidal proteins (60–70 kDa). These proteins interfere with the ion channel pumps and ultimately lead to the death of insect larva that ingests the crystal. They are quite specific in their host range (determined by ligand-receptor interaction) and this property has been exploited in the development of transgenics tolerant of specific groups of insect pests. The Bt δ-endotoxins are now known to constitute a family of related proteins for which 140 genes have been characterized for Lepidopterans, Coleopterans and Dipterans, and are not toxic to other organisms (Crickmore et al. 1998). Hence, they are safe insecticides and offer an interesting alternative to chemical control agents.

Cry-1 encoding genes have been introduced into several crop species such as maize, rice, cotton, tomato, potato and tobacco, with resistance target insect pests (Hilder and Boulter 1999), the modified varieties are generally referred to as Bt varieties. Transgenic Bt varieties are in several ways better than Bt spray formulation. In Bt transgenic plants, the protein is expressed in all tissues at all times, while the effectiveness of the sprays would be affected by a lack of uniform coverage and instability of the Bt protein, especially on exposure to sunlight.

As an example, several events of transgenic Bt maize have been developed over the past decade and there are currently varieties registered able to control lepidopteran and coleopteran species including the corn borer complex (European Corn Borer- ECB), southwestern corn borer – *Diatraea grandiosella* – and sugarcane borer – *Diatraea saccharalis*-, corn earworm (ECB), fall earmyworm (*Spodoptera frugiperda*), black cutworm (*Agrotis ipsilon*), and Worm Corn Root (WCR) complex (Figs. 4.7 and 4.8; Western, Northern and Mexican rootworms; *Diabrotica virgifera virgifera* LeConte). These insects can cause significant economic damage to maize production and all these transgenic varieties have provided a more effective control than insecticides, with lower cost than traditional insecticide applications and fewer logistical, health, and environmental concerns (Head and Ward 2009). Furthermore, this technology reduces the risk associated with lepidopteran pests like the European corn borer by improving yield stability. The use of multiple Bt proteins in a single product offers the potential for an extended spectrum of pest control and reduced risk of resistance evolving in the target pests.

Fig. 4.7 Damage inflicted by WCR larval feeding on transgenic (MON863) and conventional maize hybrids: the *left* root is a conventional hybrid and has been severely damaged while the root on the *right* side of the frame is protected by event MON863 (Modified from Vaughn et al. 2005)

The benefits of using Bt crops depend on many factors, most obviously the nature or the major insect pests in the area (not all are controlled by Bt) and the insect pressure in a given season (Christou et al. 2006). However, there are concerns regarding the use of Bt transgenic crops, the two major ones being: the effect on non-target organisms, and the possibility of the target insects developing resistance to the Bt protein. In this respect, several studies showed that the effect of maize pollen from Bt crops was negligible on non-target insects, including butterflies, under field conditions (Hodgson 1999). Moreover, though Bt crops have been widely cultivated since 1995, there has been no instance of a pest developing resistance (Ferry et al. 2006). However, given the experience of the diamondback moth having developed resistance to Bt sprays, the development of resistance in the insects cannot be discounted. As a proactive measure, several strategies for insect resistance management have been developed as a package for the cultivation of Bt crops. These strategies include refugia (growing a non-Bt crop on a small proportion of the area along with the Bt transgenic crop), gene pyramiding, and a high dosage of the protein in the plant to prevent any insects escaping from the Bt field (Christou et al. 2006; Ferry et al. 2006).

As alternatives to the Bt *Cry* genes, several candidate genes have been used to develop insect-resistant transgenic plants, such as protease inhibitors (Xu et al. 1996), α-amylase inhibitors (Ishimoto et al. 1996), vegetative insecticidal proteins from Bt (Estruch et al. 1996), cholesterol oxidases (Corbin et al. 1994), and toxins from predators such as mites and scorpions (Barton and Miller 1991). Transgenic tobacco plants expressing chitinase, one of the most important enzymes implicated in insect integument, have shown increased resistance to lepidopteron insects (Ding et al. 1998). Studies on rice (reviewed in Deka and Barthakur 2010) show that some of these candidates appear promising and provide an effective alternative to the Bt

Fig.4.8 Comparison of Bt (**a**) and conventional (**b**) maize hybrids in field trials in Italy (Courtesy CRA-MAC, Bergamo, Italy)

approach. In a transgenic assay in tobacco, enhanced resistance to *Helicoverpa zea* was generated by constitutive expression of the maize ribosome inactivating protein, named RIP- b-32, suggesting that this RIP plays a role in resistance to maize-feeding insects (Dowd et al. 2003). Characteristics of b-32, the developmentally regulated expression and synthesis of a non-toxic precursor, are reminiscent of the concept of phytoanticipine in the chemical defense strategy.

A further interesting solution to the development of insect-resistant plants was provided by Baum et al. (2007) for the control of coleopteran insect pests. Through RNA interference (RNAi) technology, they demonstrated that ingestion of double stranded (ds)RNA supplied in an artificial diet triggers RNAi in several coleopteran species, most notably WCR, which may result in larval stunting and mortality. Interestingly, transgenic maize plants engineered to express WCR dsRNAs show a significant reduction in WCR feeding damage in a growth chamber assay, suggesting that the RNAi pathway can be exploited to control insect pests via *in planta* expression of a dsRNA.

A molecular strategy for establishing nematode resistance in plant species has also been proposed with the development of artificial resistance. This can be achieved by introducing effector genes into the host plant that have a nematocidal impact. Such transgenes can encode enzymatic inhibitors that block physiological processes within the nematodes (e.g. PIs toxins) or degrading enzymes (e.g. collagenases, chitinases), toxic compounds that are ingested (cytotoxins), compounds that bind molecules (e.g. lectins, monoclonal antibodies), enzymes that interact with the nematodes, and substances that cause the breakdown of specific feeding structures (cytotoxins). More recently, RNAi was used to evaluate the role of the 16D10 secretory peptide of *Meloidogyne incognita*, which apparently interacts with the root Scarecrow protein. Expression of dsRNA in *Arabidopsis* to silence the *16D10* gene of infecting nematodes confers resistance to four *Meloidogyne* spp. (*M. incognita, M. javanica, M. arenaria*, and *M. hapla*; Huang et al. 2006). Moreover, RNAi in soybean has been used to target essential genes of *H. glycine*, causing a reduction in the number of females developing on transgenic roots (Klink et al. 2009). Despite this positive finding, results from RNAi experiments should be taken with caution: exposure to dsRNA per se is capable of causing aberrant phenotypes in both cyst and root knot nematodes (Dalzell et al. 2009). To address this problem, a novel design strategy to generate 21 bp siRNAs has been successfully applied to the potato cyst nematode, *G. pallida,* and to the root knot nematode, *M. incognita* (Dalzell et al. 2009).

5 Herbicide Tolerance

Farmers must control weeds that compete with their crops for water, nutrients and sunlight. Depending on the crop and location, weeds can decrease crop yields by 35%–100%. A number of options are available to farmers for minimizing the impact of weeds on crop productivity; one of these is the application of herbicides to the weeds. Indeed, effective weed control is a prerequisite in any crop production system if high yields and good quality are to be achieved, and herbicides have revolutionized weed control in many cropping systems and play an important role in modern agriculture. They provide economical weed control and increase the efficiency of crop production. A number of new herbicides combine high weed killing potency with low- or no-environmental persistence. However, the very effective broad spectrum herbicides available also lack selectivity, thus limiting their use in some cropping operations. On the other hand, the continuous use of the few available selective herbicides is speeding up the development of herbicide resistance in weeds; hence making effective control difficult to achieve in some crops.

5.1 Mode-of-Action and Metabolism

A large amount of knowledge exists on the mechanisms of herbicide mode-of-action and metabolism; these have frequently been described by several authors (e.g. Mazur and Falco 1989; Powles and Shaner 2001) so will not be repeated herein. Briefly, herbicides generally function by disrupting unique and essential processes in plants e.g. photosynthesis, mitosis, pigment biosynthesis or essential amino acid biosynthesis. This in turn has permitted a number of herbicide-tolerant target enzymes naturally existing in different plant species and microorganisms to be identified, as well as a number of herbicide-modifying enzymes leading to herbicide-tolerant organisms. Among the input traits offered to farmers, herbicide resistance has been the most widely adopted.

5.2 Breeding Strategies

Both crops and weeds share essential biochemical processes. Consequently, selectivity is mostly based on differential herbicide uptake between weeds and crops, controlled timing and site of application or rapid detoxification of the herbicide by the crop plants. Reliance on these natural selection processes limits the effective use of potent herbicides; hence mechanisms to impart better herbicide selectivity in crops need to be investigated.

Two approaches can be exploited. The first is the design of specific chemicals with broad selectivity for crops. This approach, however, is expensive and the products thereof may be uneconomical for use by growers, not to mention that it may also increase the already growing chemical load to the environment. Moreover, it has become increasingly difficult to discover new herbicides and even harder to come up with one that has a novel mode of action (Gressel 2002; Tan et al. 2005). The second and more popular approach to crop herbicide selectivity is the development of crop cultivars with tolerance to the already existing effective broad-spectrum herbicides so as to expand the crop options in which they can be used. Two methods can be used to develop crops with resistance to herbicides.

5.3 Conventional Methods

Conventional plant breeding utilizing strains that are known to be tolerant to specific herbicides is one approach that could confer resistance on susceptible crops from closely related species. However, this approach has limitations in that naturally herbicide resistant plants are found more among weed species than in crops. In addition, conventional plant breeding takes a long time to produce a single useful genotype.

5.4 *Biotechnology Techniques*

A faster approach is the use of biotechnology techniques such as *in vitro* cell culture, mutagenesis and selection in physiologically inhibitory concentrations of herbicides (also referred to as brute force selection) or genetic transformation of already existing crop cultivars with genes that confer resistance to herbicides.

5.4.1 Cell Culture and Selection

A number of mutant enzymes have been identified from plant cells in cultures. A trait of agronomic interest that may be expressed by cultured cells is herbicide sensitivity. Herbicides that interfere with basic metabolic activities are expected to inhibit growth of cultured cells as well as of the whole plant. In such instances, herbicide-tolerant mutants can be selected by culturing cells in the presence of a herbicide concentration that is toxic to normal cells, favoring subsequent identification of the herbicide-tolerant target enzyme.

Using cell culture techniques, BASF Inc. produced a maize hybrid that is resistant to the sulfonylurea herbicide, sethoxidim. In their analysis, a mutant cell line (named S2) was identified following continuous culture of maize embryo tissues under high sethoxidim selection pressure. Plants regenerated from this somaclonal mutant line were found to contain a form of the enzyme, acetolactate synthase (ALS, target of sulfonylureas/imidazolinones), which was insensitive to the herbicide. This resistance was subsequently transferred to the commercial hybrid (DK404SR) by backcrossing the S2 line with both of its parental lines. Further investigations showed that the sethoxidim tolerance was inherited as a single partially dominant allele. Similarly, Zambrano et al. (2003) selected a glyphosate-tolerant sugar cane cell line in liquid medium containing 0.8 mM glyphosate and regenerated plants that could tolerate up to five-fold the concentration of glyphosate that killed plants from unselected cells. Cell culture under lethal concentrations of certain herbicides also results in gene amplification in surviving cells that leads to resistance through the overproduction of enzymes targeted by herbicides. For example, a petunia cell line with resistance to glyphosate was selected in this manner and plants regenerated from it survived lethal levels of glyphosate (Steinrucken and Amrhein 1986). This resistance was found to be due to amplification of the gene encoding 5-enolpyruvylshikimate-3-phosphate (EPSP) synthase that caused its overproduction in the cells. Similarly, Caretto et al. (1994) selected carrot cells and subsequently regenerated plants that were resistant to the sulfonylurea herbicide, chlorsulfuron. Resistance in these plants was due to amplification of the *ALS* gene. *In vitro* development of phosphinothricin (PPT) resistant rice has also been reported by inducing plantlet regeneration in explants collected from 7-day old seedlings on medium supplemented with sublethal doses of PPT (Toldi et al. 2000). Other *in vitro* cell selection studies have developed resistance to paraquat in tomato cells (Thomas and Pratt 1982), resistance to glyphosate in carrot and groundnut cells (e.g. Jain et al. 1999) and resistance to a protoporphyrinogen oxidase (PPO) inhibitor in soybean cells (Warabi et al. 2001); however, no viable plant regeneration was reported in these studies.

5.4.2 Mutagenesis

Chemical or physical mutagenesis of seed, microspores or pollen followed by selection under herbicide selection pressure has also been used to develop crop resistance to herbicides. The most common mutagen used is ethyl methanesulfonate (EMS), which is efficient at producing chloroplast mutants (McCabe et al. 1990). In this strategy, seeds or pollen are treated with EMS then grown either *in vitro* or *in vivo* in the presence of a herbicide. Surviving plants are selected and grown to maturity to provide seed that is used for further screening with herbicides. Using this method, Sandhu et al. (2002) developed 21 rice lines that were resistant to glyphosate. Ashfaq-Farooqui et al. (1997) produced atrazine resistant *Solanum melongena* plants by mutagenizing seeds followed by germination and *in vitro* regeneration of plants from the resultant seedling cotyledons. Similarly, Mourad et al. (1993) isolated, by screening seedlings of M2 populations from EMS-treated seeds, a triazolopyrimidine (herbicide) resistant mutant. The resistance was found to be due to a single, dominant, nuclear gene mutation that encodes the ALS enzyme. ALS activity in enzyme extracts from the mutant was about 1,000-fold less sensitive to inhibition by triazolopyrimidine than in extracts from wild-type plants.

Ultra-violet (UV) or EMS treated microspores or pollen can be grown *in vitro* into haploid plantlets whose chromosome number can be doubled to create instant inbred lines bearing a specific herbicide tolerance trait. This method was applied by Ahmad et al. (1991) using microspore UV mutagenesis and haploid culture to develop canola plants that were resistant to chlorsulfuron. Syngenta Seeds Inc. produced the EXP19101T line of imazethapyr-resistant maize using pollen mutagenesis. In that work, EMS mutagenized pollen was used to fertilize the parent line, UE95, progeny plants were screened for tolerance to lethal doses of imazethapyr and resistant ones selected. Tolerance in these plants was found to be the result of a single nucleotide substitution within the ALS encoding gene, which gave a single amino acid change (Ser_{621} to Asn_{621}) in the sequence of the enzyme. This change prevents the binding of the herbicide to the enzyme active site, thus maintaining normal enzyme function. More recently, Venkataiah et al. (2005) reported the production of atrazine-resistant pepper (*Capsicum annuum*) plants regenerated from 3-week-old seedling cotyledons obtained from EMS treated seeds. They also noted maternal inheritance of the atrazine resistance trait. Finally, BASF (Ludwigshafen, Germany) markets non-transgenic CLEARFIELD® imidazolinone-resistant canola, wheat, sunflowers, maize, lentils, and rice, while DuPont (Wilmington, DE) markets non-transgenic STS® soybeans with tolerance to sulfonylurea herbicides. These crops all contain mutagenized versions of the ALS, which are not inhibited by imidazolinone and/or sulfonylurea herbicides (Devine and Preston 2000). Herbicides that inhibit ALS are considered low or very low use-rate herbicides with a good spectrum of weed control and are likely to remain an important part of weed resistance management programs.

5.4.3 Genetic Transformation

Herbicide tolerance is the most common trait in commercial transgenic crops, being part of 82% of all transgenic crops in 2009 (James 2009). Transgenesis for herbicide selectivity involves the identification of a herbicide resistance gene from a plant or microorganism, its isolation and manipulation for efficient plant expression (if it is of microbial origin) and (James 2009) its subsequent delivery, stable integration and expression in the cells of the target crop plant. For the most part, genes encoding for useful herbicide resistance in crops are isolated from herbicide degrading soil microorganisms.

Herbicide tolerance via genetic transformation can be conferred by one or a combination of these four mechanisms:

1. Introduction of a gene(s) encoding for a herbicide detoxifying enzyme(s);
2. Introduction of gene(s) encoding for a herbicide insensitive form of a normal functioning enzyme or over expression of the genes encoding for a herbicide target enzyme such that the normal metabolic functioning is still achieved in the plant even though some of the enzyme is inhibited;
3. Modification of the herbicide target enzyme in such a way that the herbicide molecule does not bind to it and;
4. More recently described engineering of active herbicide efflux from plant cells.

Glyphosate (Monsanto Technology LLC) is one of the most widely used herbicides in the world; it is relatively inexpensive and can be applied after the emergence of resistant crop seedlings. Nearly all broadleaf and grass weeds are eliminated, resulting in reduced competition, higher yields, and cleaner fields at harvest. Adoption of reduced and no-till practices, where dead vegetation is left in the field rather than plowed under, has been a significant unintended feature of herbicide-resistant crops, saving farmers money in fuel costs and reducing soil erosion.

Since 1996, glyphosate-tolerant or Roundup Ready crops have been developed and marketed for soybean and maize. Glyphosate is highly effective against the majority of annual and perennial grasses and broad-leaf weeds and has superior environmental and toxicological characteristics, such as rapid soil binding (resistance to leaching) and biodegradation (which decreases persistence), as well as extremely low toxicity to mammals, birds and fish. Glyphosate resistance is achieved in Roundup Ready® brands by expression of a modified *Agrobacterium* gene encoding for the herbicide insensitive enzyme CP4 enolpyruvyl-shikimate-3-phosphate (Padgette et al. 1996). The GA21 trait for glyphosate-resistant maize relies on a modified maize *epsps* gene, but is largely being replaced by varieties with the NK603 trait which has two copies of CP4 *epsps* with different promoters for better expression in the meristems.

Traits for resistance to three other classes of herbicides have been developed, but have not reached the same level of diffusion as glyphosate resistance. Resistance to oxynil herbicides conferred by the BXN nitrilase from *Klebsiella pneumoniae* (subspecies *ozaenae*) (Stalker et al. 1988) was the first trait engineered in cotton (developed by Calgene, Davis, now Monsanto). Because glyphosate is less expensive and controls more weed species, interest in using the oxynil herbicides has waned and 2004 was the final year of BXN® cotton sales. BXN canola was marketed by Rhone-Poulenc Canada (now Bayer CropScience, Monheim, Germany) and then discontinued.

Phosphinothricin acetyltransferase (PAT or BAR) detoxifies phosphinothricin- or bialaphos-based herbicides (glufosinate) by acetylation of the free NH_2 group of molecules. The *pat* gene is native to *Streptomyces viridichromogenes* and *bar* is from *S. hyhroscopicus* where they act in both the biosynthesis and detoxification of the tripeptide bialaphos (De Block et al. 1987). Like glyphosate, phosphinothricin herbicides control a broad spectrum of weed species and break down rapidly in the soil so that problems with residual activity and environmental impact are greatly reduced. Bayer CropScience markets this trait as Liberty Link® in several species. The *pat* and *bar* genes are also popular plant transformation markers in the research community.

As a technology, herbicide-resistant crops are a valuable tool for efficient weed control. However, doubts remain about the long-term viability of this strategy, particularly the emergence of herbicide-resistant weeds following widespread cultivation of herbicide-resistant crops (Sandermann 2006). Regardless, growers perceive that the benefits of the herbicide resistance characteristic outweigh the risks. It is clear that the widespread adoption of herbicide-resistant cultivars, particularly glyphosate-resistant crops, has dramatically impacted weed communities (Powles and Yu 2010). Weed population shifts to naturally resistant species, species with inherent biological characteristics that make the populations difficult to control, and the evolution of herbicide-resistant biotypes are real, as are the immediate economic issues attributable to the adoption of herbicide-resistant crops and the concomitant use of the herbicide. However, studies have shown the possibility of engineering multiple resistance in plants. In this respect strategies have been suggested to delay the development of herbicide-resistant weeds. These include combined or sequential use of herbicides with different modes of action, crop rotation, integrated weed management.

These studies have opened the avenue for the targeted development of crops that would reduce the environmental chemical load due to the use of different herbicides in crop rotation programs. A greater number of, and more various, modes of resistance have evolved in weeds than in other organisms because herbicides are used far more extensively than other pesticides, and because weed seed output is so prolific. Weeds have evolved unknown mechanisms, even antibiotic, as well as other drug and pesticide resistance. It is also possible that cases of epigenetic resistance may have appeared (Gressel 2009).

6 Conclusions and Future Prospects

Plant pests and diseases have major effects on agricultural production and the food supply. Although application of fungicides and pesticides has helped control plant diseases, chemical control is economically costly as well as environmentally undesirable. The development of new strategies based on a plant's own defense mechanisms for disease control is therefore critical for sustaining agricultural production and improving our environment and health. Basic research on the genetic bases of pest and disease resistance in plants and of host-pathogen interactions has greatly improved the efficiency of manipulating disease resistance genes in practical breeding programs and resulted in the deployment of high-yielding genetically resistant crop cultivars that in some cases have been grown over vast areas, but much remains to be learnt at the interface of the genetics of resistance and crop physiology. The cloning of resistance genes and corresponding avirulence genes has indicated considerable complexity not only in structure but also in the way in which gene products interact and trigger resistance. Hence, our overall understanding of the process is still fragmentary. Furthermore, many gaps remain in our models of the defense signal transduction network and these must be bridged before we can design truly rational strategies to activate the network. Similarly, genetic mapping of plant mutations that alter herbivore resistance, or perhaps responses to purified insect elicitors, will almost certainly lead to the identification of previously unknown defense pathways.

A further point worth noting is that although major genes and QTLs for resistance to numerous pathogens and insect pests have been mapped, the usefulness of this information for MAS breeding programs has not yet been demonstrated. In this respect, the development of new technologies, such as high-throughput DNA sequencing and microarray analysis to facilitate the mapping and cloning of major genes and QTLs for routine use will provide an assemblage of new tools to facilitate the development of crops resistant to pests and pathogens, while analysis of signaling and metabolic pathways will be harnessed to increase the power of MAS and genetic engineering for crop improvement. Furthermore, the complexity of plant-insect interactions makes it difficult to determine which anatomical features, metabolites, and signaling pathways effectively limit pathogen and pest infestations. Genomic information from both host plants and pests and pathogens should accelerate the rate of discovery in this field. The field of genomics will provide powerful tools to investigate these critical factors. Transcript profiling techniques allow the simultaneous examination of thousands of genes, and can be utilized to study changes in gene expression that are transcriptionally regulated. Beyond transcript profiling, genomics also facilitates the functional analysis of genes implicated in resistance and susceptibility. As signaling cascades and metabolic pathways are elucidated in model systems and crop plants, key regulatory genes can be targeted for silencing or overexpression to study the role of these pathways in plant-insect and -pathogen interactions. To achieve a detailed understanding of plant interactions with pathogens and pests, it will ultimately be necessary to combine transcriptomic approaches with proteomic, metabolomic, and mutational analyses. While plant responses have been the focus of most transcriptomic studies, additional levels of complexity can also be analyzed with genomic tools. Investigating changes that occur concurrently within the pathogens and insects is essential to understand the basis of an effective plant defense. Therefore, knowledge accumulated in these studies will help us to establish economical and sustainable strategies to fight insects and diseases of many important crops. Beyond genome sequencing, additional effort should be targeted at identifying pathogens and insect genetic markers, studying natural variation in host plant utilization, and developing methods such as RNA interference for manipulating insect gene expression. The development of such research tools will facilitate studies on both sides of the plant-insect and -pathogen interactions and thereby achieve a more complete understanding of plant response to pathogens and insects.

Since the first transgenic plants appeared almost two decades ago, this technology has contributed to develop new methods of crop protection aiming to increase world food production. It is certain that the methodology developed for creating *Bt* plants will ultimately make the objective of having highly productive, pest- and pathogen-resistant, and environmentally friendly crops, become a reality. The promising alternative of genetic engineering of insect- and pathogen-resistant plants, relying exclusively on the repertoire of plant defense genes, should be thoroughly investigated as it may provide solutions to the problem of increasing plant productivity for future needs. Success in developing transgenic organisms will also benefit from knowledge of the signal transduction pathways that regulate pathogenesis, particularly host range and the availability of a wide range of suitable genes that can be used to increase virulence. Genetic engineering strategies require information on the roles and consequences of these genes, leading to enhanced exploitation of the genetic resources present in plants.

Acknowledgements Part of the research was developed in the projects: "BIORES- Use of bioactive proteins in plant protection against pathogens" funded by CRA- Consiglio per la Ricerca e Sperimentazione in Agricoltura- and in the "IDIAM-Interventions to counterattack the spread and damage from rootworm in maize Italian crop", funded by the MiPAF -Italian Ministry of Agricultural Food and Forestry Policies.

References

Aharoni A, Jongsma MA, Bouwmeester HJ (2005) Volatile science? Metabolic engineering of terpenoids plant. Trends Plant Sci 10:594–602

Ahmad I, Day JP, MacDonald MV, Imgram DS (1991) Haploid culture and UV mutagenesis in rapid-cycling *Brassica napus* for the generation of resistance to chlorsulfuron and *Alternaria brassicola*. Annal Bot 67:521–525

Alam SN, Cohen MB (1998) Detection and analysis of QTLs for resistance to the brown planthopper, *Nilaparvata lugens*, in a doubled-haploid rice population. Theor Appl Genet 97:1370–1379

Antoniw JF, Ritter CE, Pierpoint WS, Van Loon LC (1980) Comparison of three pathogenesis-related proteins from plants of two cultivars of tobacco infected with TMV. J Gen Virol 47:79–87

Ashfaq-Farooqui M, Rao AV, Jayasree T, Sadanandam A (1997) Induction of atrazine resistance and somatic embryogenesis in *Solanum melongena*. Theor Appl Genet 95:702–705

Austin MJ, Muskett P, Kahn K, Jones JDG, Parker JE (2002) Regulatory role of SGT1 in early *R* gene mediated plant defenses. Science 295:2077–2080

Baker EP (1966) Isolation of complementary genes conditioning crown rust resistance in the oat variety Bond. Euphytica 15:313–318

Balconi C, Lanzanova C, Motto M (2010) Ribosome-inactivating proteins in cereals. In: Lord JM, Hartley MR (eds) Toxic plant proteins. Springer, Heidelberg/London/New York pp 149–166

Barth C, Jander G (2006) *Arabidopsis* myrosinases TGG1 and TGG2 have redundant function in glucosinate breakdown and insect defense. Plant J 46:549–562

Barton KA, Miller MJ (1991) Insecticidal toxins in plants. Environ Prot Agency 0431829 A1910612

Baulcombe D (2004) RNA silencing in plants. Nature 431:356–363
Baum JA, Bogaert T, Clinton W, Heck GR, Feldmann P, Ilagano O, Johnson S, Plaetinck G, Munyikwa T, Pleau M, Vaughn T, Roberts J (2007) Control of coleopteran insect pests through RNA interference. Nat Biotechnol 25:1322–1326
Bayes A, Comellas-Bigler M, de la Vega MR, Maskos K, Bode W, Aviles FX, Jongsma MA, Beekwilder J, Vendrell J (2005) Structural basis of the resistance of an insect carboxypeptidase to plant protease inhibitors. Proc Natl Acad Sci U S A 102:16602–16607
Bellafiore S, Briggs SP (2010) Nematode effectors and plant responses to infection. Curr Opin Plant Biol 13:1–7
Bergvinson DJ, Hamilton RI, Arnason JT (1995) Leaf profile of maize resistance factors to European corn borer, *Ostrinia nubilalis*. J Chem Ecol 21:343–354
Biancardi E, Lewellen RT, De Biaggi M, Erichsen AW, Stevanato P (2002) The origin of rhizomania resistance in sugar beet. Euphytica 127:383–397
Biffin RH (1905) Mendel's laws of inheritance and wheat breeding. J Agric Sci 1:4–48
Bravo A, Gill SS, Soberon M (2007) Mode of action of *Bacillus thuringiensis* toxins and their potential for insect control. Toxicon 49:423–435
Brosch G, Ransom R, Lechner T, Walton JD, Loidl P (1995) Inhibition of maize histone deacetylases by HC toxin, the host-selective toxin of *Cochliobolus carbonum*. Plant Cell 7:1941–1950
Büschges R, Hollricher K, Panstruga R, Simons G, Wolter M, Frijters A, Van Daelen R, Van Der Lee T, Diergarde P, Groenenendijk J, Töpsch S, Vos P, Salamini F, Schulze-Lefert P (1997) The barley *Mlo* gene: a novel control element of plant pathogen resistance. Cell 88:695–705
Byrne PF, McMullen MD, Wiseman BR, Snook ME, Musket TA, Theuri JM, Widstrom NW, Coe EH (1998) Maize silk maysin concentration and corn earworm antibiosis: QTLs and genetic mechanisms. Crop Sci 38:461–471
Caranta C, Palloix A, Gebre SK, Lefeb-vre V, Moury B, Daubeze AM (1996) A complementation of two genes originating from susceptible *Capsicum annuum* lines confers a new and complete resistance to Pepper veinal mottle virus. Phytopathology 86:739–743
Caretto S, Giadina MC, Nicolodi C, Mariotti D (1994) Chlorosulfuron resistance in *Daucus carota* cell lines and plants. Involvement of gene amplification. Theor Appl Genet 88:520–524
Carlini C, Grossi-de-Sà MF (2002) Plant toxic proteins with insecticidal properties. A review on their potentialities as bioinsecticides. Toxicon 40:1515–1539
Castro AJ, Capettini F, Corey AE, Filichkina T, Hayes PM, Kleinhofs A, Kudrna D, Richardson K, Sandoval-Islas S, Rossi C, Vivar H (2003a) Mapping and pyramiding of qualitative and quantitative resistance to stripe rust in barley. Theor Appl Genet 107:922–930
Castro AJ, Chen XM, Hayes PM (2003b) Pyramidin quantitative trait locus (QTL) alleles determining resistance to barley stripe rust: effects on resistance at the seedling stage. Crop Sci 43:651–659
Chen H, Wilkerson CG, Kuchar JA, Phinney A, Howe BSGA (2005) Jasmonate-inducible plant enzymes degrade essential amino acids in the herbivore midgut. Proc Natl Acad Sci U S A 102:19237–19242
Chen XW, Shang JJ, Chen DX, Lei CL, Zou Y, Zhai WX, Liu GZ, Xu JC, Ling ZZ, Cao G, Ma BT, Wang YP, Zhao XF, Li SG, Zhu LH (2006) A B-lectin receptor kinase gene conferring rice blast resistance. Plant J 46:794–804
Chisholm ST, Coaker G, Day B, Staskawicz BJ (2006) Host-microbe interactions: shaping the evolution of the plant immune response. Cell 124:803–814
Christensen A, Ho Cho B, Naesby M, Gregersen PL, Brandt J, Madriz-Ordeñana K, Collinge DB, Thordal-Christenen H (2002) The molecular characterization of two barley proteins establishes the novel PR-17 family of pathogenesis-related proteins. Mol Plant Pathol 3:135–144
Christou P, Capell T, Kohli A, Gatehouse J, Gatehouse A (2006) Recent developments and future prospects in insect pest control in transgenic crops. Trends Plant Sci 11:302–308
Clay NK, Adio AM, Denoux C, Jander G, Ausubel FM (2009) Glucosinolate metabolites required for an *Arabidopsis* innate immune response. Science 323:95–101
Collmer CW, Marston MF, Taylor JC, Jahn MM (2000) The *I* gene of bean: a dosage-dependent allele conferring extreme resistance, hypersensitive resistance, or spreading vascular necrosis in response to the potyvirus Bean common mosaic virus. Mol Plant Microbe Interact 13:1266–1270

Consonni C, Humphry ME, Hartmann HA, Livaja M, Durner J, Westphal L, Vogel J, Lipka V, Kemmerling B, Schulze-Lefert P, Somerville SC, Panstruga R (2006) Conserved requirement for a plant host cell protein in powdery mildew pathogenesis. Nat Genet 38:716–720

Cook JR (2000) Advances in plant health management in the twentieth century. Annu Rev Phytopathol 38:95–116

Corbin DR, Greenplate JT, Wong EY, Purcell JP (1994) Cloning of an insecticidal cholesterol oxidase gene and its expression in bacteria and plant protoplasts. Appl Environ Microbiol Biotechnol 44:133–138

Cordero MJ, Raventós D, San Segundo B (1994) Expression of a maize proteinase inhibitor gene is induced in response to wounding and fungal infection: systemic wound-response of a monocot gene. Plant J 6:141–150

Crickmore D, Ziegler R, Fietelson J, Schnepf E, Van Rie J, Lereclus D, Baum J, Dean DH (1998) Revision of the nomenclature for *Bacillus thuringiensis* pesticidal crystal proteins. Microbiol Mol Biol Rev 62:807–813

Culver JN (2002) Tobacco mosaic virus assembly and disassembly: determinants in pathogenicity and resistance. Annu Rev Phytopathol 40:287–308

Dalzell JJ, McMaster S, Johnston MJ, Kerr R, Fleming CC, Maule AG (2009) Non-nematode-derived double-stranded RNAs induce profound phenotypic changes in *Meloidogyne incognita* and *Globodera pallida* infective juveniles. Int J Parasitol 39:1503–1516

Dangl JL, Jones JDG (2001) Plant pathogens and integrated defense responses to infection. Nature 441:826–833

De Block M, Bottermn J, Vandewiele M, Dockx J, Thoen C, Gossele V, Rao Mowa N, Thompson C, Van Montagu M, Leemans J (1987) Engineering herbicide resistance in plants by expression of a detoxifying enzyme. EMBO J 6:2513–2518

De Maagd RA, Bravo A, Crickmore N (2001) How *Bacillus thuringiensis* has evolved specific toxins to colonie the insect world. Trends Genet 17:193–199

Degenhardt J, Hiltpold I, Köllner TG, Frey M, Gierl A, Gershenzon J, Hibbard BE, Ellersieck MR, Turlings TCJ (2009) Restoring a maize root signal that attacts insectkilling nematodes to control a major pest. Proc Natl Acad Sci 106:13213–13218

Deka S, Barthakur S (2010) Overview on current status of biotechnological interventions on yellow stem borer *Scirpophaga incertulas* (*Lepidoptera*: *Crambidae*) resistance in rice. Biotechnol Adv 28:70–81

Deslandes L, Olivier J, Theulieres F, Hirsch J, Feng DX, Bittner-Eddy P, Beynon J, Marco Y (2002) Resistance to *Ralstonia solanacearum* in *Arabidopsis thaliana* is conferred by the recessive *RRS1-R* gene, a member of a novel family of resistance genes. Proc Natl Acad Sci U S A 99:2404–2409

Devine MD, Preston C (2000) The molecular basis of herbicides resistance. In: Cobb AH, Kirkwood RC (eds) Herbicides and their mechanisms of action. Sheffield Academic Press, UK, pp 72–104

Dickinson M (2003) Molecular plant pathology. BIOS Scientific Publishers, London

Ding X, Gopalakrisnan B, Johnson LB, White FF, Wang X, Morgan TD, Kramer KJ, Muthukrishnan S (1998) Insect resistance of transgenic tobacco expressing an insect chitinase gene. Transgenic Res 7:77–84

Dodds PN, Lawrence GJ, Catanzariti A-M, The T, Wang C-IA, Ayliffe MA, Kobe B, Ellis JG (2006) Direct protein interaction underlies gene-for-gene specificity and coevolution of the flax resistance genes and flax rust avirulence genes. Proc Natl Acad Sci U S A 103:8888–8893

Dogimont C, Bendahmane A, Pitrat M, Burget-Bigeard E, Hagen L (2007) Gene resistant to Aphis gossypii. US Patent Application, US 2007/0016977 A1 2007

Dowd PF, Zuo WN, Gillikin JW, Johnson ET, Boston RS (2003) Enhanced resistance to Helicoverpa zea in tobacco expressing an activated form of maize ribosome-inactivating protein. J Agric Food Chem 51:3568–3574

Drijfhout E (1978) Genetic interaction between *Phaseolus vulgaris* and Bean common mosaic virus with implications for strain identification and breeding for resistance. Agric Res Rep 872:1–98

Duering K, Porsch P, Fladung M, Loerz H (1993) Trangenic potato plants resistant to the phytopathogenic bacterium *Erwinia carotovora*. Plant J 3:587–598

Durner J, Shah J, Klessig DF (1997) Salicylic acid and disease rsistance in plants. Trends Plant Sci 2:266–274
Durrant WE, Dong X (2004) Systemic acquired resistence. Annu Rev Phytopathol 42:185–209
Eitas TK, Dangl JL (2010) NB-LRR proteins: pairs, pieces, perception, partners and pathways. Curr Opin Plant Biol 13:1–6
Ellis J, Dodds P, Pryor T (2000) Structure, function and evolution of plant disease resistance genes. Curr Opin Plant Biol 3:278–284
Epple P, Apel K, Bohlamann H (1997) Overexpression of an endogenous thionin enhances resitance of *Arabidopsis* against *Fusarium oxysporum*. Plant Cell 9:509–520
Erb M, Lenk C, Degenhardt J, Turlings CJ (2009) The understimated role of roots in defense against leaf attackers. Trends Plant Sci 12:653–659
Estruch JJ, Warren GW, Mullins MA, Nye GJ, Craig J, Koziel MG (1996) Vip3A, a novel *Bacillus thuringiensis* vegetative insecticidal protein with a wide spectrum of activities against lepidopteran insects. Proc Natl Acad Sci U S A 93:5389–5394
Ferreira SA, Pitz KY, Manshardt R, Zee F, Fitch M, Gonsalves D (2002) Virus coat protein transgenic papaya provides practical control of *Papaya ringspot* virus in Hawaii. Plant Dis 86:101–105
Ferreira RB, Monteiro S, Freitas R, Santos CN, Chen Z, Batista LM, Duarte J, Borges A, Teixeira AR (2007) The role of plant defence proteins in fungal pathogenesis. Mol Plant Pathol 8:677–700
Ferry N, Edwards MG, Gatehouse J, Capell T, Christou P, Gatehouse AM (2006) Transgenic plants for insect pest control: a forward looking scientific perspective. Transgenic Res 15:13–19
Flor HH (1971) Current status of the gene-for-gene concept. Annu Rev Phytopathol 9:275–296
Fraser RSS (1986) Genes for resistance to plant viruses. CRC Crit Rev Plant Sci 3:257–294
Frey M, Schullehner K, Dick R, Fiesselmann A, Gierl A (2009) Benzoxazinoids biosynthesis, a model for evolution of secondary metabolic pathways in plants. Phytochemistry 70:1645–1651
Friedman AR, Baker BJ (2007) The evolution of resistance genes in multi-protein plant resitance systems. Curr Opin Genet Dev 17:493–499
Fuchs M, Gonsalves D (2007) Safety of virus-resistant transgenic plants two decades after their introduction: lesson from realistic field risk assessment studies. Annu Rev Phytopathol 45:173–202
Gal-On A, Wang Y, Faure JE, Pilowsky M, Zelcer A (1998) Trangenic resistance to cucumber mosaic virus in tomato: blocking of long-distance movement of the virus in lines harbouring a defective viral replicase. Phytopathology 88:1101–1107
García-Olmedo F, Molina A, Segura A, Moreno M (1995) The defensive role of nonspecific lipid-transfer proteins in plants. Trends Microbiol 3:72–74
Goldbach R, Bucher E, Prins M (2003) Resistance mechanisms to plant viruses: an overview. Virus Res 92(2):207–212
Gonsalves D (1998) Control of papaya ringspot virus in papaya: a case study. Annu Rev Phytopathol 36:415–437
Green TR, Ryan CA (1972) Wound-induced proteinase inhibitor in plant leaves: a possible defense mechanism against insects. Science 175:776–777
Gressel J (2002) Molecular weed biology. Taylor & Francis, London/New York
Gressel J (2009) Evolving understanding of the evolution of herbicide resistance. Pest Manag Sci 65:1164–1173
Grube RC, Radwanski ER, Jahn MM (2000) Comparative genetics of disease resistance within the *Solanaceae*. Genetics 155:873–887
Halkier BA, Gershenzon J (2006) Biology and biochemistry of glucosinolates. Annu Rev Plant Biol 57:303–333
Hammond-Kosack K, Jones JDG (2000) Response to plant pathogens. In: Buchanan B, Gruissem W, Jones R (eds) Biochemistry & molecular biology of plants. American Society of Plant Physiologists, Rockville
Hammond-Kosack KE, Parker JE (2003) Deciphering plant-pathogen communication: fresh perspectives for molecular resistance breeding. Curr Opin Biotechnol 14:177–193
Hann DR, Selena Gimenez-Ibanez S, Rathjen JP (2010) Bacterial virulence effectors and their activities. Curr Opin Plant Biol 13:388–393

Hansen BG, Kerwin RE, Ober JA, Lambrix VM, Mitchell-Olds T, Gershenzon J, Halkier BA, Kliebenstein DJ (2008) A novel 2-oxiacid-dependent dioxygenase involved in the formation of the goiterogenic 2-hyfroxybut-3-enyl glucosinolate and generalist insect resistance in *Arabidopsis*. Plant Physiol 148:2096–2108

Harlan JR, De Wet JMM (1971) Toward a rational classification of cultivated plants. Taxon 20:509–517

Hayakawa T, Zhu Y, Itoh K, Kimura Y, Izawa T, Shimamoto K, Toriyama S (1992) Genetically engineered rice resistant to rice stripe virus, an insect-transmitted virus. Proc Natl Acad Sci U S A 89:9865–9869

Head G, Ward D (2009) Insect resistance in corn through biotechnology. In: Kriz AL, Larkins BA (eds) Molecular genetic approaches to maize improvement, vol 63. Springer, Berlin/Heidelberg

Heath MC (2000) Nonhost resistance and nonspecific plant defenses. Curr Opin Plant Biol 3:315–319

Heszky L (2008) Challengee of plant breeding early in 21th century. Hungarian Agric Res 4:4–8

Hilder VA, Boulter D (1999) Genetic engineering of crop plants for insect resistance: a critical review. Crop Prot 18:177–191

Hoch H, Staples R, Whitehead B, Comeau J, Wolf ED (1987) Signaling of growth orientation and cell differentiation by surface topography in *Uromyces*. Science 235:1659–1662

Hodgson J (1999) Monarch Bt-corn paper questioned. Nature Biotechnol 17:627

Honeé G (1999) Engineered resistance against fungal plant pathogens. Eur J Plant Pathol 105:319–326

Howe GA, Jander G (2008) Plant immunity to insect herbivores. Annu Rev Plant Biol 59:41–66

Hu X, Bidney DL, Yalpani N, Duvick JP, Crasta O, Folkerts O, Lu G (2003) Overexpression of a gene encoding hydrogen peroxide-generating oxalate oxidase evokes defense responses in sunflower. Plant Physiol 133:170–181

Hu KM, Qiu DY, Shen XL, Li XH, Wang SP (2008) Isolation and manipulation of quantitative trait loci for disease resistance in rice using a candidate gene approach. Mol Plant 1:786–793

Huang G, Allen R, Davis EL, Baum TJ, Hussey RS (2006) Engineering broad root-knot resistance in transgenic plants by RNAi silencing of a conserved and essential root-knot nematode parasitism gene. Proc Natl Acad Sci U S A 103:14302–14306

Huet H, Mahendra S, Wang J, Sivamani E, Ong CA, Chen L, de Kochko A, Beachy RN, Fauquet C (1999) Near immunity to rice tungro spherical virus achieved in rice by replicate-mediated resistance strategy. Phytopathology 89:1022–1027

Hulbert SH, Webb CA, Smith SM, Sun Q (2001) Resistance gene complexes: evolution and utilization. Annu Rev Phytopathol 39:285–312

Ishimoto M, Sato T, Chrispeels MJ, Kitamura K (1996) Bruchid resistance of transgenic azuki bean expressing seed alpha amylase inhibitor of common bean. Entomol Exp Appl 79: 309–315

Jain M, Bhatnagar RK, Sarin NB (1999) Isolation and biochemical diagnosis of cell lines of groundnut (*Arachis hypogaea* L.) selected on glyphosate. Pestic Sci 55:843–849

James C (2009) Global status of commercialized biotech/GM crops: 2009. ISAAA Briefs No 41, Ithaca, NY

Jia YL, McAdams SA, Bryan G, Hershey H, Valent B (2000) Direct interaction of resistance gene and avirulence gene products confers rice blast resistance. EMBO J 19:4004–4014

Jiang GH, Xia ZH, Zhou Y, Wan J, Li DY, Chen RS, Zhai WX, Zhu LH (2006) Testifying the rice bacterial blight resistance gene *xa5* by genetic complementation and further analyzing xa5 (Xa5) in comparison with its homolog TFIIAgamma1. Mol Genet Genomics 275:354–366

Johal GS, Briggs SP (1992) Reductase activity encoded by the HM1 disease resistance gene in maize. Science 258:985–987

Johnson R (1992) Past, present, and future opportunities in breeding for disease resistance, with examples from wheat. In: Robertson R, Jellis GJ (eds) Breeding for disease resistance. Kluwer Academic, Dordrecht

Jones JDG, Dangl JL (2006) The plant immune system. Nature 444:323–329

Jordá L, Conejero V, Vera P (2000) Characterization of *P69E* and *P69F*, two differentially regulated genes encoding new members of the subtilisin-like proteinasi family from tomato plants. Plant Physiol 122:67–74

Jørgensen JH (1994) Genetics of powdery mildew resistance in barley. Crit Rev Plant Sci 13:97–119

Journot-Catalino N, Somssich IE, Roby D, Kroj T (2006) The Transcription factors WRKY11 and WRKY17 act as negative regulators of basal resistance in *Arabidopsis thaliana*. Plant Cell 18:3289–3302

Jung C, Cai D, Kleine M (1998) Engineering nematode resistance in crop species. Trends Plant Sci 3:266–271

Kamoun S (2001) Nonhost resistance to Phytophtora: novel prospects for a classical problem. Curr Opin Plant Biol 4:295–300

Kandoth PK, Stefanie Ranf S, Pancholi SS, Jayanty S, Walla MD, Miller W, Howe GA, Lincoln DE, Stratmann JW (2007) Tomato MAPKs LeMPK1, LeMPK2, and LeMPK3 function in the systemin-mediated defense response against herbivorous insects. Proc Natl Acad Sci U S A 104:12205–12210

Kaniewski W, Ilardi V, Tomassoli L, Mitsky T, Layton J, Barba M et al (1999) Extreme resistance to cucumber mosaic virus (CMV) in transgenic tomato expressing one or two viral coat proteins. Mol Breed 5:111–119

Kappers IF, Aharoni A, van Herpen TW, Luckerhoff LL, Dicke M, Bouwmeester HJ (2005) Genetic engineering of terpenoid metabolism attracts, bodyguards to *Arabidopsis*. Science 309:2070–2072

Karban R, Baldwin IT (1997) Induced responses to herbivory. University of Chicago Press, Chicago

Kawchuk LM, Hachey J, Lynch DR, Kulcsar F, van Rooijen G, Waterer DR, Robertson A, Kokko E, Byers R, Howard RJ, Fischer R, Prufer D (2001) Tomato *Ve* disease resistance genes encode cell surface-like receptors. Proc Natl Acad Sci U S A 98:6511–6515

Kim JH, Jander G (2007) *Myzus persicae* (green peach aphid) feeding on *Arabidopsis* induces the formation of a deterrent indole glucosinolate. Plant J 49:1008–1019

Kiraly L, Barna B, Kiraly Z (2007) Plant resistance to pathogen infection: forms and mechanisms of innate and acquired resistance. J Phytopathol 155:385–396

Klink VP, Kim KM, Martins V (2009) A correction between host-mediated expression of parasite genes as tandem inverted repeats and abrogation of development of female *Heterodera glycines* cyst formation during infection of *Glycine max*. Planta 230:53–71

Klun JA, Guthrie WD, Hallauer AR, Russell WA (1970) Genetic nature of the concentration of 2,4-dihydroxy-7-(2H)-1,4-benzoazin-3(4H)-one and resistance to the European corn borer in a diallel set of eleven maize inbreds. Crop Sci 10.87–90

Koev G, Mohan BR, Dinesh-Kumar SP, Torbert KA, Somers DA, Miller WA (1998) Extreme reduction of disease in oats transformed with the 5′ half of the barley yellow dwarf virus-PAV genome. Phytopathology 88:1013–1019

Kolmer JA (1992) Enhanced leaf rust resistance in wheat conditioned by resistance gene pairs with Lr13. Euphytica 61:123–130

Konno K, Hirayama C, Nakamura M, Tateishi K, Tamura Y, Hattori M, Kohno K (2004) Papain protects papaya trees from herbivorous insects: role of cysteine proteases in latex. Plant J 37:370–378

Kuang H, Wei F, Marano MR, Wirtz U, Wang X, Liu J, Shum WP, Zaborsky J, Tallon L, Rensink W, Lobst SE, Zhang P, Tornqvist CE, Tek A, Bamberg J, Helgeson J, Fry W, You F, Luo MC, Jiang J, Buell CR, Baker BB (2005) The R1 resistance gene cluster contains three groups of independently evolving, type I R1 homologues and shows substantial structural variation among haplotypes of *Solanum demissum*. Plant J 44:37–51

Lawrence SD, Novak NG (2006) Expression of poplar chitinase in tomato leads to inhibition of development in Colorado potato beetle. Biotechnol Lett 28:593–599

Lay FT, Anderson MA (2005) Defensin-components of the innate immune system in plants. Curr Protein Pept Sci 6:85–101

Leach JE, Vera Cruz CM, Bai J, Leung H (2001) Pathogen fitness penalty as a predictor of durability of disease resistance genes. Annu Rev Phytopathol 39:187–224

Leonards-Schippers C, Gieffers W, Schiifer-Pregl R, Ritter E, Knapp SJ, Salamini F, Gebhardt C (1994) Quantitatve resistance to *Phytophtora infestans* in potato: a cause study of QTL mapping in an allogamous plant species. Genetics 137:67–77

Li Q, Xie QG, Smith-Becker J, Navarre DA, Kaloshian I (2006) Mi-1-mediated aphid resistance involves salicylic acid and mitogen-activated protein kinase signaling cascades. Mol Plant Microbe Interact 19:655–664

Lin F, Chen S, Que Z, Wang L, Liu X, Pan Q (2007) The blast resistance hene Pi37 encodes an NBS- LRR protein and is a member of a resistance gene cluster on rice chromosome 1. Genetics 177:1871–1880

Lison P, Rodrigo I, Conejero V (2006) A novel function for the cathepsin D inhibitor in tomato. Plant Physiol 142:1329–1339

Liu HY, Lewellen RT (2007) Distribution and molecular characterization of resistance-breaking isolates of Beet necrotic yellow vein virus in the United States. Plant Dis 91:847–851

Liu XM, Brown-Guedira GL, Hatchett JH, Owuoche JO, Chen MS (2005) Genetic characterization and molecular mapping of a Hessian fly resistance gene transferred from *T. turgidum* ssp. *dicoccum* to common wheat. Theor Appl Genet 111:1308–1315

Lopez L, Camas A, Shivaji R (2007) Mirl-CP, a novel defense cysteine protease accumulates in maize vascular tissues in response to herbivory. Planta 226:517–527

Maffei ME, Mithöfer A, Boland W (2007) Before gene expression: early events in plant- insect interaction. Trends Plant Sci 12:310–316

Magbanua ZV, Wilde HD, Roberts JK, Chowdhury K, Abad J, Moyer JW, WEtzstein HY, Parrott WA (2000) Field resistance to tomato spotted wilt virus in transgenic peanut (Arachis hypogaea L.) expressing an antisense nucleocapsid gene sequence. Mol Breed 6:227–236

Mao J, Burt AJ, Ramputh AI (2007) Diverted secondary metabolism and improved resistance to European corn borer (*Ostrinia nubilalis*) in maize (*Zea mays* L.) transformed with wheat oxalate oxidase. J Agric Food Chem 55:2582–2589

Martin GB, Bogdanove AJ, Sessa G (2003) Understanding the functions of plant disease resistance proteins. Annu Rev Plant Biol 54:2

Maxwell FG, Jennings PR (1980) Breeding plants resistant to insects. Wiley, New York

Mazur BJ, Falco SC (1989) The development of herbicide resistant crops. Annu Rev Plant Mol Biol 40:441–470

McCabe PF, Cslpo A, Timmons AM, Dix PJ (1990) Selection of chloroplast mutants. In: Polland JW, Walker JM (eds) Methods in molecular biology, vol 6, Plant cell and tissue culture. The Humana Press, Totowa

McDowell JM, Simon SA (2006) Recent insights into *R* gene evolution. Mol Plant Pathol 7:437–448

McDowell JM, Woffenden BJ (2003) Plant disease resistance genes: recent insights and potential applications. Trends Biotechnol 21:178–183

McIntosh RA, Wellings CR, Park RF (1995) Wheat rusts: an atlas of resistance genes. CSIRO Publishing, Melbourne

Mewis I, Appel HM, Hom A, Raina R, Schultz JC (2005) Major signaling pathways modulate *Arabidopsis* glucosinolate accumulation and response to both phloem-feeding and chewing insects. Plant Physiol 138:1149–1162

Michelmore RW (2003) The impact zone: genomics and breeding for durable disease resistance. Curr Opin Plant Biol 6:397–404

Mohan S, Ma PWK, Pechan T, Bassford E, Williams WP, Luthe DS (2006) Degradation of the *S. frugiperda* peritrophic matrix by an inducible maize cysteine protease. J Insect Physiol 52:21–28

Molina A, García-Olmedo F (1997) Enhance tolerance to bacterial pathogens caused by the transgenic expression of barley lipid transfer protein LPT2. Plant J 12:669–675

Mourad G, Pandey B, King J (1993) Isolation and genetic analysis of a triazolopyrimidine-resistant mutant of *Arabidopsis*. J Heredity 84:91–96

Mundt CC (2002) Use of multiline cultivars and cultivar mixtures for disease management. Annu Rev Phytopathol 40:381–410

Nakajima H, Murakana T, Ishige F, Akutsu K, Oeda K (1997) Fungal and bacterial disease resistance in transgenic plants expressing human lysozyme. Plant Cell Rep 16:674–679

Nombela G, Williamson VM, Muniz M (2003) The root-knot nematode resistance gene *Mi-1.2* of tomato is responsible for resistance against the whitefly *Bemisia tabaci*. Mol Plant Microbe Interact 16:645–649

Oerke EC, Dehne HW (2004) Safeguarding production-losses in major crops and the role of crop protection. Crop Prot 23:275–285

Omo-Ikerodah EE, Fawole I, Fatokun CA (2008) Genetic mapping of quantitative trait loci (QTLs) with effects on resistance to flower bud thrips (*Megalurothrips sjostedti*) identified in recombinant inbred lines of cowpea (*Vigna inguiculata* L. Walp). African J Biotechnol 7:263–270

Padgette SR, Re DB, Barry GF, Eichholtz DE, Delannay X, Fuchs RL, Kishore GM, Fraley RT (1996) New weed control opportunities: development of soybeans with a Roundup Ready® gene. In: Duke SO (ed) Herbicide-resistant crops: agricultural, economic, environmental, regulatory and technological aspects. Lewis Publishers, Los Angeles, pp 53–84

Painter RH (1958) Insect resistance in crop plants. University Press of Kansas, Lawrence

Panella L, Lewellen RT (2007) Broadening the genetic base of sugar beet: introgression from wild relatives. Euphytica 154:382–400

Park CJ, Kim KJ, Shin R, Park JM, Shin YC, Paek KH (2004) Pathogenesis-related protein 10 isolated from hot pepper functions as a ribonuclease and an antiviral pathway. Plant J 37:186–198

Parlevliet JE (1979) Components of resistance that reduce the rate of epidemic development. Annu Rev Phytopathol 17:203–222

Parlevliet JE, Zadoks JC (1977) The integrated concept of disease resistance. A new view including horizontal and vertical resistance in plants. Euphytica 26:5–21

Passardi F, Penel C, Dunand C (2004) Performing the paradoxical- How plant peroxidases modify the cell wall. Trends Plant Sci 9:534–540

Peart JR, Lu R, Sadanandom A, Malcuit I (2002) Ubiquitin ligase-associated protein SGT1 is required for host and nonhost disease resistance in plants. Proc Natl Acad Sci U S A 99: 10865–10869

Peumans WJ, Vandamme EJM (1995) Lectins as plant defense proteins. Plant Physiol 109:347–352

Pfalz M, Vogel H, Mitchell-Olds T, Kroymann J (2007) Mapping of QTL for resistance against the crucifer specialist herbivore *Pieris Brassicae* in a new *Arabidopsis* inbred line population Da(1)-12xEi-2. PLoS One 2:e578

Pimentel D (1997) Techniques for reducing pesticide use. Wiley, Hoboken

Pink DAC (2002) Strategies using genes for non-durable resistance. Euphytica 1:227–236

Pitzschke A, Schikora A, Hirt H (2009) MAPK cascade signaling networks in plant defense. Curr Opin Plant Biol 12:421–426

Powles SB, Shaner DL (2001) Herbicide resistance and world grain. CRC Press LLC, Boca Raton

Powles SB, Yu Q (2010) Evolution in action: plants resistant to herbicides. Annu Rev Plant Biol 61:317–347

Prins M, Laimer M, Noris E, Schubert J, Wassenegger M, Tepfer M (2008) Strategies for antiviral resistance in transgenic plants. Mol Plant Pathol 9:73–83

Punja ZK (2001) Genetic engineering of plants to enhance resistance to fungal pathogens – a review of progress and future prospects. Can J Plant Pathol 23:216–235

Rathjen JP, Moffett P (2003) Early signal transduction events in specific plant disease resistance. Curr Opin Plant Biol 6:300–306

Ravelonandro M, Scorza R, Bachelier JC, Labonne G, Lery L, Damsteegt V, Callahan AM, Dunez J (1997) Resistance of transgenic *Prunus domestica* to plum pox virus infection. Plant Dis 81:1231–1235

Richardson KL, Vales MI, Kling JG, Mundt CC, Hayes PM (2006) Pyramiding and dissecting disease resistance QTL to barley stripe rust. Theor Appl Genet 113:485–495

Rivard D, Cloutier C, Michaud D (2004) Colorado potato beetles show differential digestive compensatory responses to host plants expressing distinct sets of defense proteins. Arch Insect Biochem Physiol 55:114–123

Ross AF (1961) Systemic acquired resistance induced by localized virus infections in plants. Virology 14:340–358

Ryan CA (1990) Protease inhibitors in plants: genes for improving defenses against insects and pathogens. Annu Rev Phytopathol 28:425–449

Samborski DJ (1963) A mutation in *Puccinia recondita* Rob. Ex Desm. F. sp. Tritici to virulence on transfer, Chinese spring x *Aegilops umbellulata* Zhuk. Can J Bot 41:475–479

Sandermann H (2006) Plant biotechnology: ecological case studies on herbicide resistance. Trends Plant Sci 11:324–328

Sandhu SS, Bastos CR, Azini LE, Neto AT, Colombo C (2002) RAPD analysis of herbicide-resistant Brazilian rice lines produced via mutagenesis. Genet Mol Res 1:359–370

Santiago R, Butrón A, Reid LM, Arnason JT, Sandoya G, Souto XC, Malvar RA (2006) Diferulate content of maize sheaths is associated with resistance to the Mediterranean corn borer *Sesamia nonagrioides* (Lepidoptera: Noctuidae). J Agric Food Chem 54:2274–2279

Schilmiller AL, Howe GA (2005) Systemic signaling in the wound response. Curr Opin Plant Biol 8:369–377

Schnee C, Köllner TG, Held M, Turlings TCJ, Gershenzon J, Degenhardt J (2006) The products of a single maize sesquiterpene synthase form a volatile defense signal that attracts natural enemies of maize herbivores. Proc Natl Acad Sci U S A 103:1129–1134

Shan WX, CaoM DLU, Tyler BM (2004) The Avr1b locus of Phytophora sojae encodes an elicitor and a regulator required for avirulence on soybean plants carrying resistance gene Rps1b. Mol Plant Microbe Interact 17:394–403

Singh RP (1993) Genetic association of gene *Bdv1* for tolerance to barley yellow dwarf virus with genes *Lr34* and *Yr18* for adult plant resistance to rusts in bread wheat. Plant Dis 77:1103–1106

Sivamani E, Brey CW, Dyer WE, Talbert LE, Qu R (2000) Resistance to wheat streak mosaic virus in transgenic wheat expressing the viral replicase (Nib) gene. Mol Breed 6:469–477

Smith SM, Pryor AJ, Hulbert SH (2004) Allelic and haplotypic diversity at the *rp1* rust resistance locus of maize. Genetics 167:1939–1947

Song WY, Wang G-L, Zhu L-H, Fauquet C, Ronald P (1995) A receptor kinase-like protein encoded by the rice disease resistance gene, *Xa21*. Science 270:1804–1806

Song J, Bradeen JM, Naess SK, Raasch JA, Wielgus SM, Haberlach GT, Liu J, Kuang H, Austin-Phillips BCR, Helgeson JP, Jiang J (2003) Gene RB cloned from *Solanum bulbocastanum* confers broad spectrum resistance to potato late blight. Proc Natl Acad Sci U S A 100:9128–9133

Stalker DM, McBride KE, Malyj LD (1988) Herbicide resistance in transgenic plants expressing a bacterial detoxification gene. Science 242:419–423

Staskawicz BJ, Ausubel FM, Baker BJ, Ellis JB, Jones JD (1995) Molecular genetics of plant disease resistance. Science 268:661–667

Steinrucken HC, Amrhein N (1986) The herbicide glyphosate id a potent inhibitor of 5-enolpyruvyl-shikimate-3-phosphate synthase. Biochem Res Commun 94:1207–1212

Stommel JR, Tousignant ME, Wai T, Pasini R, Kaper MJ (1998) Viral satellite RNA expression in transgenic tomato confers field tolerance to cucumber mosaic virus. Plant Dis 82:391–396

Tai TH, Dahlbeck D, Clark ET, Gajiwala P (1999) Expression of the *Bs2* pepper gene confers resistance to bacterial spot disease in tomato. Proc Natl Acad Sci U S A 96:14153–14158

Tan S, Evans RR, Dahmer ML, Singh BK, Shaner DL (2005) Imidazolinone-tolerant crops: history, current status and future. Pest Manag Sci 61:246–257

Tang X, Xie M, Kim YJ, Zhou J, Klessig DF, Martin GB (1999) Overexpression of *Pto* activates response and confer broad resistance. Plant Cell 11:15–29

Tarchevsky IA (2001) Pathogen-induced plant proteins (review). Appl Biochem Microbiol 37:441–455

Tattersall DB, Bak S, Jones PR, Olsen CE, Nielsen JK, Hansen ML, Hoj PB, Moller BL (2001) Resistance to an herbivore through engineered cyanogenic glucoside synthesis. Science 293:1826–1828

Tepfer M (1993) Viral genes and transgenic plants. Biotechnology 11:1125–1132

Thipyapong P, Joel DM, Steffens JC (1997) Differential expression and turnover of the tomato polyphenol oxidase gene family during vegetative and reproductive development. Plant Physiol 113:707–718

Thomas BR, Pratt D (1982) Isolation of paraquat-tolerant mutants from tomato cell cultures. Theor Appl Genet 63:169–176

Toldi O, Toth S, Oreifig AS, Kiss E, Jenes B (2000) Production of phosphinothricin-tolerant rice (*Oryza sativa*) through the application of phosphinothricin as a growth regulator. Plant Cell Rep 19:1226–1231

Toyoda K, Collins NC, Takahashi A, Shirasu K (2002) Resistance and susceptibility of plants to fungal pathogens. Transgenic Res 11:567–582

Tsuda K, Katagiri F (2010) Comparing signaling mechanisms engaged in pattern-triggered and effector-triggered immunity. Curr Opin Plant Biol 13:459–465

Unsicker SB, Kunert G, Gershenzon J (2009) Protective perfumes: the role of vegetative volatiles in plant defense against herbivores. Curr Opin Plant Biol 12:479–485

USDA (United States Department of Agriculture) (2002) Field tests release database for the US. http://www.isb. vt.edu/cfdocs/fieldtests1.cfm

Van der Ent S, Van Wees SCM, Pieterse CMJ (2009) Jasmonate signaling in plant interactions with resistance-inducing beneficial microbes. Phytochemistry 70:1581–1588

Van der Hoorn RAL, De Wit PJGM, Joosten MHAJ (2002) Balancing selection favors guarding resistance proteins. Trends Plant Sci 7:67–71

Van der Plank JE (1982) Disease resistance in plants. Academic, New York

Van Loon LC (1999) Occurrence and properties of plant pathogenesis-related proteins. In: Datta SK, Muthukrishnan K (eds) Pathogenesis-related proteins in plants. CRC Press, New York

Van Loon LC, Rep M, Pieterse CM (2006) Significance of inducible defense-related proteins in infected plants. Annu Rev Phytopathol 44:135–162

Vaughn T, Cavato T, Brar G, Coombe T, DeGooyer T, Ford S, Groth M, Howe A, Johnson S, Kolacz K, Pilcher C, Purcell J, Romano C, English L, Pershing J (2005) A method of controlling corn rootworm feeding using a *Bacillus thuringiensis* protein expressed in transgenic maize. Crop Sci 45:931–938

Venkataiah P, Christopher T, Karampuri S (2005) Selection of atrazine-resistant plants by *in vitro* mutagenesis in pepper (*Capsicum annuum*). Plant Cell Tissue Org Cult 83:7–82

Wan JM (2006) Perspectives of molecular design in breeding in crops. Acta Agron Sin 32:455–462

Wang GL, Song WY, Ruan DL, Sideris S, Ronald PC (1996) The cloned gene *Xa21* confers resistance to multiple *Xantomonas oryzae* pv. oryzae isolates in transgenic plants. Mol Plant Microbe Interact 9:850–855

Wang G, Ding X, Yan M, Qiu D, Li X, Xu C, Wang S (2006) Dual function of rice OsDR8 gene in disease resistance and thiamine accumulation. Plant Mol Biol 60:437–449

Wang JH, Constabel CP (2004) Polyphenol oxidase overexpression in transgenic Populus enhances resistance to herbivory by forest tent caterpillar (*Malacosoma disstria*). Planta 220:87–96

Warabi E, Usui K, Tanaka Y, Matsumoto H (2001) Resistance of a soybean cell line to oxyfluorfen by overproduction of mitochondriak protoporphyrinogen oxidase. Pest Manag Sci 57:743–748

Williams CE, Collier CC, Sardesai N, Ohm HW, Cambron SE (2003) Phenotypic assessment and mapped markers for *H31*, a new wheat gene conferring resistance to Hessian fly (*Diptera*: *Cecidomyiidae*). Theor Appl Genet 107:1516–1523

Wiseman BR, Snook ME, Widstrom NW (1996) Feeding responses of corn earworm larvae (*Lepidoptera*: *Noctuidae*) on corn silks of varying flavone content. J Econ Entomol 89:1040–1044

Wu J, Hettenhausen C, Meldau S, Baldwin IT (2007) Herbivory rapidly activates MAPK signaling in attacked and unattacked leaf regions but not between leaves of *Nicotiana attenuata*. Plant Cell 19:1096–1122

Xiao XW, Chu PWG, Frenkel MJ, Tabe LM, Shukla DD, Hanna PJ, Higgins TJV, Müller WJ, Ward CW (2000) Antibody-mediated improved resistance to ClYVV and PVY infections in transgenic tobacco plants expressing a single-chain variable region antibody. Mol Breed 6:421–431

Xiao S, Charoenwattana P, Holcombe L, Turner JG (2003) The *Arabidopsis* genes RPW8.1 and RPW8.2 confer induced resistance to powdery mildew diseases in tobacco. Mol Plant Microbe Interact 16:289–294

Xu D, Xue Q, McElroy D, Mawal Y, Hilder VA, Wu R (1996) Constitutive expression of a cowpea trypsin inhibitor gene, *CpTi*, in transgenic rice plants confers resistance to two major rice insect pests. Mol Breed 2:167–173

Yang Y, Shah J, Klessig DF (1997) Signal perception and transduction in plant defense responses. Genes Dev 11:1621–1639

Yeh SD, Bau HJ, Cheng YH, Yu TA, Yang JS (1998) Greenhouse and field evaluations of coat-protein transgenic papaya resistant to papaya ringspot virus. Acta Hort 461:321–328

Yencho GC, Cohen MB, Byrne PF (2000) Application of tagging and mapping insect resistance loci in plants. Annu Rev Entomol 45:393–422

Young ND (1996) QTL mapping and quantitative disease resistance in plants. Annu Rev Phytopathol 34:479–501

Zambrano AY, Demey JR, Gonzalez V (2003) *In vitro* selection of a glyphosate-tolerant sugarcane cellular line. Plant Mol Biol Rep 21:365–373

Zavala JA, Patankar AG, Gase K, Hui D, Baldwin IT (2004) Manipulation of endogenous trypsin proteinase inhibitor production in *Nicotiana attenuata* demonstrates their function as antiherbivore defenses. Plant Physiol 134:1181–1190

Zhao BY, Ardales E, Brasset E, Claflin LE, Leach JE, Hulbert SH (2004) The *Rxo1/Rba1* locus of maize controls resistance reactions to pathogenic and non-host bacteria. Theor Appl Genet 109:71–79

Zhu Q, Droge-Laser W, Dixon RA, Lamb C (1996) Transcriptional activation of plant defense genes. Curr Opin Genet Dev 6:624–630

Zhu Y, Chen H, Fan JH, Wang Y, Li Y, Chen J, Fan JX, Yang S, Hu L, Leung H, Mew TW, Teng AS, Wang Z, Mundt CC (2000) Genetic diversity and disease control in rice. Nature 406:718–722

Zipfel C, Rathjen JP (2008) Plant immunity: AvrPto targets the frontline. Curr Biol 18:218–222

Chapter 5
The American Halophyte *Prosopis strombulifera*, a New Potential Source to Confer Salt Tolerance to Crops

Mariana Reginato, Verónica Sgroy, Analía Llanes, Fabricio Cassán, and Virginia Luna

Contents

Abstract Salinity imposes a major environmental threat to agriculture and its adverse impact has become the most serious problem in vast regions of the earth surface. The combination of population growth and land degradation are of such

M. Reginato (✉) • V. Sgroy • A. Llanes • F. Cassán • V. Luna
Laboratorio de Fisiología Vegetal, Departamento de Ciencias Naturales,
Universidad Nacional de Río Cuarto, 5800 Río Cuarto, Argentina
e-mail: mreginato@exa.unrc.edu.ar; vsgroy@exa.unrc.edu.ar; allanes@exa.unrc.edu.ar; acassan@exa.unrc.edu.ar; vluna@exa.unrc.edu.ar

M. Ashraf et al. (eds.), *Crop Production for Agricultural Improvement*,
DOI 10.1007/978-94-007-4116-4_5, © Springer Science+Business Media B.V. 2012

significance that plant salt tolerance improvement has become an urgent need for the future of agriculture. Generation of salt tolerant crops requires a clear understanding of the complex mechanism of abiotic stress resistance in key species. Evolutionary processes under saline conditions gave rise to halophytes which have evolved adaptive traits to cope with the salinity of the environment. They include some specific biochemical, physiological and molecular mechanisms, on one hand, and a capability of natural association with different microorganisms called Plant Stress-Homeo-regulating Rhizobacteria (PSHR), on the other. However, a complete understanding of these processes is still lacking despite the intensive research conducted during the last decade. The genus *Prosopis* includes many important arboreal and shrub-like species that are present in saline zones of the Americas and some of them are considered to be unique terrestrial species due to their combined ability to fix nitrogen and grow under high-salinity conditions. The shrub *Prosopis strombulifera* (Lam) Benth. is distributed from the Arizona desert (U.S.A.) to Patagonia (Argentina) and is especially abundant in the salinized areas of central Argentina. This species showed a halophytic response to NaCl surviving up to 1 M NaCl in *in-vitro* experiments, but in contrast, a strong growth inhibition at lower Na_2SO_4 concentrations. These differential responses to the most abundant salts present in most salinized soils make this species an excellent model to study salt-tolerance mechanisms in halophytic plants. This chapter provides an overview of different salt tolerance mechanisms in the native halophyte *Prosopis strombulifera*, which may be considered a new useful source to improve crop salt tolerance through two biotechnological strategies: considering *P. strombulifera* as a natural gene donor to improve the genetic salt tolerance of low tolerant crops, and considering this species and its rhizosphere as natural sources of PSHR microorganisms capable of physiologically improving salt tolerance in crops.

Keywords Halophyte • NaCl • Na_2SO_4 • Differential tolerance • PSHR microorganisms • Biotechnology

1 Introduction

Nowadays salinity, which has been representing a threat to agriculture in some parts of the world for more than 3,000 years, is affecting more and more lands. Some of the most serious examples of salinity occur in arid and semiarid regions. For example, in Iran, Pakistan, Egypt, and Argentina, out of the total land area of 162.2, 77.1, 99.5, and 237.7 million hectares, about 23.8, 10, 8.7, and 33.1 million hectares are salt-affected, respectively (FAO 2008). As the world population has expanded, so there is a need to grow more food, which requires an increase in the area of land under cultivation as well as in land productivity as yields per hectare (Flowers and Flowers 2005). For this reason, genetic improvement of salt tolerance has become an urgent need for the future of agriculture (Owens 2001; Munns 2007).

The deleterious effects of salinity on plant growth are associated with (1) low water potential of the root medium which causes a water deficit within the plant; (2) toxic effects of ions mainly Na^+ and Cl^-; and (3) nutritional imbalance caused by reduced nutrient uptake and/or transport to the shoot (Munns and Termaat 1986; Marschner 1995; Serrano et al. 1999; Hasegawa et al. 2000; Ashraf and Ahmad 2000; Parida and Das 2005; Munns and Tester 2008; Hariadi et al. 2011). In addition, salt stress can induce oxidative stress, such as generation and/or accumulation of reactive oxygen species, including hydrogen peroxide (H_2O_2), superoxide anion, and hydroxyl radicals (Kovtun et al. 2000; Miller et al. 2010). For that, salt tolerance is a complex trait involving responses to cellular osmotic and ionic stresses, subsequent secondary stresses (*e.g.* oxidative stress), and whole plant coordination of these responses. Various approaches have been advocated to generate salt-tolerant crops, none of which has offered a universal solution (Flowers and Flowers 2005; Munns 2007).

During the last years, it has been proposed that improving our knowledge about the specialized physiology and biochemistry of halophytic plants, which impart their exceptional degree of salt tolerance, would give us new insight into plant salt tolerance, but it still represents a challenge for the scientists. For this purpose, the tolerant relative of *Arabidopsis*, *Thellungiella halophila*, has been considered the alternative halophytic genetic model system (Inan et al. 2004).

Nevertheless, our plant model system *Prosopis strombulifera* exhibits growth promotion up to 500 mM NaCl and its growth is slightly inhibited at higher salt concentrations such as 700 mM NaCl. This exceptional degree of tolerance makes this species a perfect candidate to deepen our knowledge on halophytic salt tolerance mechanisms.

2 *Prosopis strombulifera*, a Halophytic Legume

Within the family Leguminosae, virtually all of the important annual legumes that belong to the subfamily Papilionaceae such as soybeans, beans, peas, and cowpeas are highly salt sensitive and suffer yield reductions from salinities with conductivities as low as 2–3 dS m^{-1} (Richards 1954; Ayers and Westcott 1985). Alfalfa is the most salt tolerant of the commercial legume species showing yield reductions of 50% at salinities of 9.6 dS m^{-1} (Ayers and Westcott 1985). In contrast, the genus *Prosopis*, subfamily Mimosoideae, includes many important arboreal and shrub-like species that are present in saline zones of the America (Burkart 1937, 1952, 1976); these species present a rapid growth at seawater salinity or 45 dS m^{-1} which is nearly 20 times greater than salinities that can be tolerated by annual temperate legumes. The *Prosopis* genus is considered to have unique terrestrial species with the ability to fix nitrogen and grow under high-salinity conditions (Rhodes and Felker 1987).

The spiny shrub *Prosopis strombulifera* (Lam.) Benth. (Burkart 1976) is distributed from the Arizona desert (U.S.A.) to Patagonia (Argentina) and is especially abundant

Fig. 5.1 *Prosopis strombulifera* plants in their natural habit (Saltland in San Luis, Argentina) and plants growing in hydroponic culture

in the salinized areas of central Argentina (Fig. 5.1) (Cantero et al. 1996). The germination of this species is strongly influenced by the nature of the ions in the salt solutions and their interactions (Sosa et al. 2005).

In the salinized soils of southern Cordoba and southwestern San Luis, Argentina where we can find *P. strombulifera* populations, the proportions of NaCl and Na_2SO_4 are similar, although, in some samples, Na_2SO_4 was up to three times more abundant (Sosa et al. 2005). This and other recent studies have suggested an increasing need to compare the effects of Na_2SO_4 and NaCl on plant growth in order to better understand plant responses to the major salts found in the salinized soils of several countries such as Pakistan, China, India and Tunisia (Iqbal 2003; Bie et al. 2004; Shi and Sheng 2005; Naeem and Quereshi 2005; Manivannan et al. 2008; Tarchoune et al. 2010).

Comparative studies of the effects of Cl^- and SO_4^{2-} on *P. strombulifera* germination have demonstrated that SO_4^{2-}-based solutions are considerably more inhibitory than Cl^--based solutions at iso-osmotic concentrations (Sosa et al. 2005; Llanes et al. 2005). Species like *Hordeum vulgare* (Huang and Redmann 1995), *Medicago sativa* (Redmann 1974), *Triticum aestivum* (Hampson and Simpson 1990), *Pinus banksiana* (Croser et al. 2001), and *Ocinum basilicum* (Tarchoune et al. 2010) were found to be more inhibited by Na_2SO_4 than by NaCl. For other species like *Brassica napus* (Huang and Redmann 1995) the reverse was found.

The response of *P. strombulifera* to salinity was different depending on the type of salt used and the osmotic potential in the culture medium. Shoot growth was stimulated up to $\Psi_o = -1.88$ MPa (500 mM) NaCl; this is an interesting halophytic response that differs from other *Prosopis* species (Felker 2007). The genus *Prosopis* occurs worldwide in arid and semiarid regions and includes about 44 species grouped in 5 sections and 8 series (Burkart 1976). Many species of this genus, particularly those belonging to the Algarobia Section, have economic and ecological potential, often being major components of native North and South America ecosystems wherein they offer shade, firewood, food, and forage for wildlife and livestock. Some *Prosopis* species, especially *P. pallida*, *P. juliflora*, *P. tamarugo*, and *P. alba* have individual plants with rapid growth at seawater salinity or 45 dS m^{-1} (Felker 2007). Notwithstanding, according to the literature *P. strombulifera* NaCl-tolerance exceeds the limits described for most *Prosopis* species being closer to tolerance levels observed in some Chenopodiaceae.

On the other hand, this species is much less tolerant to Na_2SO_4, showing strong and sustained shoot growth inhibition from the beginning of the treatment, accompanied by senescence symptoms such as chlorosis, necrosis, and leaf abscission (Reinoso et al. 2005; Reginato 2009). Root growth stimulation was also observed in NaCl-grown *P. strombulifera* plants showing a generalized adaptive mechanism to favor water and nutrient absorption. The SO_4^{2-} anion had a detrimental effect on root growth in comparison with the Cl^- anion, which was also partially alleviated when both anions were present in bisaline solutions.

Similar results were obtained in comparative salinization studies with Na_2SO_4 and NaCl in the halophyte *Chenopodium rubrum* by Warne et al. (1990) who reported full growth inhibition at $\Psi_o = -1.6$ MPa with SO_4^{2-}-based solutions but Cl^--mediated growth inhibition at Ψ_o lower than −2 MPa. These results differ from those obtained by Egan and Ungar (1998), who registered shoot growth stimulation with Na_2SO_4 in *Atriplex prostata* seedlings and growth inhibition in the presence of iso-osmotic NaCl solutions.

Nevertheless, none of these authors analyzed growth responses to iso-osmotic mixtures of both salts. Partial alleviation of SO_4^{2-} toxicity by bisaline solutions is an interesting response in our experiments, confirming previous reports on germination of this species (Sosa et al. 2005; Llanes et al. 2005).

3 Mechanisms of Salt Tolerance in *Prosopis strombulifera*

3.1 Ion Exclusion, Accumulation and Compartmentation

Sodium exclusion (Garcia et al. 1995), K^+/Na^+ discrimination (Asch et al. 2000; Houshmand et al. 2005), and Cl^- exclusion (Rogers and Noble 1992) traits have proven valuable in screening germplasm for salinity tolerance. Some halophytes make an osmotic adjustment by accumulating equivalent amounts of Na^+ and Cl^- in vacuoles (Kefu et al. 2003). *Prosopis strombulifera* does not allow the entrance of great amounts of Na^+ (compared with Cl^- levels) which correlates with the particular characteristics observed in the radical system, such as precocious lignification and suberization of endodermis for salt exclusion (Reinoso et al. 2004, 2005); this observation would imply that *P. strombulifera* falls within the category of salt-excluding plants (Yensen and Biel 2005), which are able to exclude salts from the roots. Similarly, the economically important relatives *Prosopis* species are Na^+ excluders (Villagra et al. 2010).

Nevertheless, Na^+ content in the leaves of *P. strombulifera* was 2–3 fold higher than K^+ content (0.48 and 0.185 mmol g^{-1} DW, respectively) when optimal growth was observed in NaCl treated plants ($\Psi_o = -1.9$ MPa, 500 mM). These results suggest that *P. strombulifera* is a sodium exporter like *Atriplex*. Thus, a good explanation for the unusual salt tolerance of this species may be a combination of exclusion mechanism in the roots and ion accumulation/compartmentation in the leaves. It is remarkable to note that *P. strombulifera* preferentially accumulated Cl^- over Na^+ reaching significant levels of Cl^- anion (2.24 mmol g^{-1} DW) in roots as well as in leaf tissues without toxicity symptoms when optimal growth was observed in NaCl treated plants (Reginato 2009). The total ion content pattern determined at the end of the experiment (48 days) is shown in Fig. 5.2.

In Na_2SO_4-treated *P. strombulifera* plants, there was SO_4^{2-} accumulation in roots and leaves from the beginning of salinization, reaching amounts equivalent to those of Cl^- at the highest salt concentration tested in our experiments (530 mM Na_2SO_4, osmotically equivalent to 700 mM NaCl). Chrispeels et al. (1999) affirmed that SO_4^{2-} causes acidification of the medium, generating membrane depolarization and favoring SO_4^{2-} uptake. The SO_4^{2-} levels in leaves were lower than those of Cl^- in NaCl-treated plants, but they were sufficient to cause metabolic disorders and visible signs of toxicity, as reported above. In bisaline-treated plants a preferential accumulation of Cl^- over SO_4^{2-} was observed, corroborating the fact that SO_4^{2-} is absorbed slowly by the roots of higher plants (Bie et al. 2004). The deleterious effect of SO_4^{2-} on leaf development and root and shoot elongation may be a consequence of several metabolic reactions, such as sulfide formation in the process of sulphate assimilation in the chloroplast, where sulphite reductase catalyzes the reduction of sulphite to sulphide using reduced ferredoxin as an electron donor (De Kok et al. 2005). Free sulphide can be incorporated to L-cysteine by cysteine synthase. If this assimilatory step does not consume all free sulphide, the sulphide could be emitted to the environment or could effectively bind to cytochromes, thus inhibiting mitochondrial respiration (Schmidt 2005).

Some species accumulate NO_3^- anion for charge balance and osmotic adjustment (Martinez-Ballesta et al. 2004). In halophyte *Suaeda salsa* NO_3^- plays an important osmotic role in high salinity (Song et al. 2009). In *P. strombulifera* NO_3^- anion does not seem to have such a role. This species synthesizes compatible solutes like proline and pinitol in high concentrations for charge balance and osmotic adjustment (Llanes 2010).

On the other hand, Na^+ accumulated in *P. strombulifera* leaves at the expense of K^+ in a quantitatively different manner depending on the salinizing agent used. In roots, SO_4^{2-}-based treatment caused greater Na^+ accumulation at the beginning of salinization limiting sodium's transport to the shoot. Kefu et al. (2003) reported that when external Na^+ concentration is very high, it can block K^+ transporters (including high-affinity transporters) causing a decrease in K^+ entrance. Although most plants have an absolute requirement for K^+, and Na^+ is toxic for many biological reactions in the cytoplasm. This does not apply to vacuolar processes, because Na^+ accumulates preferentially in vacuoles being more suitable than K^+ for osmotic adjustment in several species; the replacement of K^+ by Na^+ in the vacuole does not produce toxicity

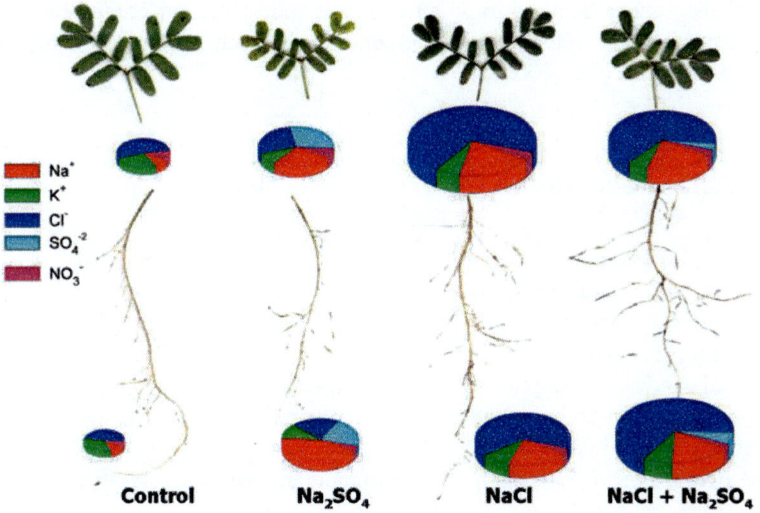

Fig. 5.2 Ion composition in *P. strombulifera* plants. Control plants (48 days); Na_2SO_4, NaCl and NaCl + Na_2SO_4 treated plants ($\Psi_o = -2.6$ MPa, 48 days). The area of the circles designates the total ion content in different organs of the plant. The size of the *circle* in leaves of control plants corresponds to 3.158 mmol/g DW (100%). The sizes of *circles* (and their respective portions) in the other plant organs and treatments are proportionally depicted

(Subbarao et al. 2003). Therefore, the use of inorganic ions is efficient and significantly more economical than the synthesis of compatible organic solutes for plants under salt stress (Song et al. 2009).

The long-distance Na^+ transport and, especially, the mechanism of Na^+ loading into the xylem play a substantial role in the formation of water potential gradient in *Suaeda altissima* (Balnokin et al. 2005); efficiency in long-distance Na^+ transport to stems and leaves is a required mechanism for salt tolerance in halophytes (Patel and Pandey 2007). Then, limitation of sodium's transport to the shoot by SO_4^{2-} anion seems to be one of the reasons for the lack of tolerance of this species to SO_4^{2-} treatment, discarding the possibility that unfavorable plant growth in Na_2SO_4 resulted from Na^+ toxicity due to the higher Na^+ concentration in this solution.

Salt treatments caused a decrease in Ca^{2+} levels at all osmotic potentials tested in our experiments, showing a great difference between controls and treated plants in both leaves and roots (data not shown). The lower Ca^{2+} than Na^+ and K^+ contents in roots may be explained by membrane nonselective cation channels that allow monovalent or divalent cation movement indistinctly. These would constitute the primary way of Ca^{2+} influx to cells (Demidchik et al. 2002).

These results demonstrate that NaCl and Na_2SO_4 differentially affect ion accumulation by plants, generating specific distribution patterns of ions in different organs (Fig. 5.2). This fact suggests an important specific anionic effect on membrane permeability. Increasing concentrations of bisaline solutions allowed higher salt tolerance than increasing concentrations of individual Na_2SO_4 solutions through various phenomena such as ionic antagonism and/or mutual competence, which bring about specific physiological responses, confirming previous results (Sosa et al. 2005; Llanes et al. 2005).

3.2 Water Relations and Water Use Efficiency

Salt stress affects plant water relations by reducing water potential; decreased osmotic potential and increased turgor pressure are common responses in plants exposed to salt. The water use efficiency is addressed by several authors as a relevant mechanism of salt tolerance in plants (Verlues et al. 2006; Grewal 2010). Decreased water consumption per accumulated biomass unit implies a reduced amount of salt absorbed by plants and a reduced energy cost for compartmentation or extrusion of salt itself.

Prosopis strombulifera has the ability to maintain a higher relative water content (RWC) in leaves of seedlings growing in high NaCl concentrations without succulence. Thus, it appears that the extreme tolerance of this species to NaCl would be due to its capability of osmotic adjustment (Fig. 5.3). Also, the high water content in its leaves help ease the impact of Na^+ accumulation in tissues, which is also favored by leaf anatomical adaptations such as the particular distribution of chlorenchyma around the leaf veins (Reinoso et al. 2004) and the modifications on stomatal area and distribution (Reginato 2009) as well as stomatal conductivity that enable *P. strombulifera* seedlings to make a more efficient use of water under these experimental conditions.

Plants treated with high Na_2SO_4 concentrations ($\Psi_o = -2.6$ MPa) failed to make an efficient osmotic adjustment (Fig. 5.3) showing extremely negative osmotic and water potentials in leaves causing water imbalance, which, in comparison with NaCl-treated plants, correlates directly with the marked decrease in individual and total leaf area at $\Psi_o = -1.88$ MPa and lower treatment, reaching final values of 30% and 60% inhibition of these parameters, respectively. An increase in stomatal density and epidermal cell density with smaller stomata, maybe additional efforts of Na_2SO_4-treated plants to survive (Reginato 2009). However, leaf area was not modified in NaCl-treated plants.

3.3 Metabolism of Protection-Compatible Solutes Production

One of the most common stress responses in plants is overproduction of different types of compatible organic solutes (Serraj and Sinclair 2002; Parida and Das 2005; Munns and Tester 2008). Compatible solutes are low molecular weight, highly soluble compounds that are usually nontoxic at high cellular concentrations. Generally, they protect plants from stress through different courses. Multiple functions for these compounds have been suggested. In addition to the conventional role of

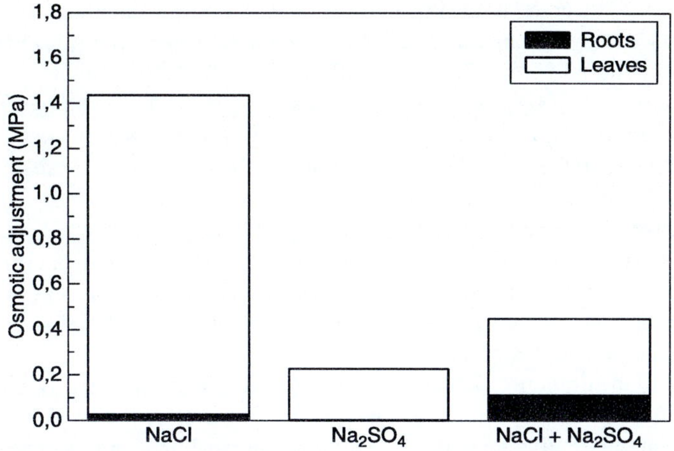

Fig. 5.3 Effects of NaCl, Na_2SO_4 and their iso-osmotic mixture on osmotic adjustment in leaves and roots of *Prosopis strombulifera* plants. Osmotic adjustment was calculated from two linear regressions for all treatments as the difference between osmotic potential (OP) and the estimated OP of tissue ascribed to mere loss of water at each given RWC (OP_0) at RWC of 70%

these compatible solutes in cell osmotic adjustment, they are also reported to act as low-molecular-weight chaperones, stabilizing the photosystem II complex, protecting the structure of enzymes and proteins, maintaining membrane integrity and scavenging ROS (Bray 2002; Manivannan et al. 2008; Türkan and Demiral 2009). Recently, it was also shown that some of these compatible solutes are very efficient in reducing the extent of K^+ loss in response to both salinity and oxidative stress (Cuin and Shabala 2007) in barley and *Arabidopsis* roots.

Compatible solutes fall into three major groups: amino acids (e.g. proline), quaternary amines (e.g. glycine betaine, dimethylsulfoniopropionate) and polyols/sugars (e.g. mannitol, trehalose). Many plant species accumulate significant amounts of glycine betaine, proline, and polyols in response to high salinity (Di Martino et al. 2003; Silveira et al. 2009). Different varieties of a particular plant species exhibit a high degree of variation in salt tolerance (Chen et al. 2007) and a possible causal link between these responses and the differential accumulation of glycine betaine and proline among cereal genotypes has been proposed (Yang et al. 2003). Indeed, the introduction of genes involved in the synthesis of proline, betaines, and polyols in plants contributes to an increased abiotic stress tolerance (Chen and Murata 2002); numerous genetic engineering attempts to manipulate compatible solutes biosynthetic pathways have been made in order to enhance salt tolerance (Wang et al. 2003; Chen et al. 2007; Verbruggen and Hermans 2008).

In *P. strombulifera* seedlings a remarkable proline accumulation occurred under increasing salinity. The proline content increased by 200% since the start of all salt treatments that is in line with the earlier suggestions that proline content in plants increases in a linear relationship with conductivity in the medium (Llanes et al. 2010). In *Medicago truncatula,* proline accumulation in response to NaCl stress was higher in aerial parts (13 fold) than that in roots (8-fold) compared to controls (Armengaud et al. 2004). A similar effect was observed in *P. strombulifera* where proline accumulation in leaves was much more important than in roots. This fact could be related to the high ion accumulation in salt-treated leaves, raising the need for organic molecules production to counteract osmotic balance and/or ion sequestration, scavenging free radicals and buffering cellular redox potential under stress conditions. The results with NaCl and Na_2SO_4 treated plants indicate that proline would be a stress intensity signal but not a salt tolerance indicator, as it increased concomitantly with decrease in osmotic potential, independently of the ion composition of the solution (Llanes et al. 2010).

In view of the literature, there is also a linear relationship between glycine betaine content and salt stress intensity in several species. However, *P. strombulifera* seedlings did not accumulate glycine betaine under any salt treatment in our experiments in line with the statement by Tipirdamaz et al. (2006) that 'species that behaved as proline accumulators contained little betaines and vice versa'.

A number of sugar-alcohols are present in some but not all halophytes from several families, as is the case of pinitol (Ishitani et al. 1996; Murakeozy et al. 2003; Arndt et al. 2004). The phreatophytic but desert-dwelling *Populus euphratica* accumulated the cyclitols chiro-inositol and pinitol to significant concentrations when growing over saline groundwater in the Taklamakan desert in China (Arndt et al. 2004). Sorbitol and mannitol are not widely reported in halophytes, but are present in several families of flowering plants, including the mangrove *Laguncularia racemosa* (Noiraud et al. 2001; Koyro 2006).

Mannitol accumulation in *P. strombulifera* seedlings only occurred in leaves of NaCl-treated plants with moderate to high salinity, while in those treated with sulphate and salt mixture, manitol level was lower than that in controls. By contrast, sorbitol accumulated in greater amounts in sulphate and salt mixture-treated plants, while under NaCl treatment sorbitol level was similar to controls. Thus, it appears that this species uses mannitol + pinitol for osmotic adjustment in the presence of NaCl, while in the presence of the toxic sulfate it synthesizes sorbitol + pinitol. Thus, sorbitol biosynthesis might be a possible manifestation of a carbon metabolism disorder (Llanes et al. 2010).

Mannitol and sorbitol have been targeted for engineering compatible-solute overproduction. Tarczynski et al. (1993) introduced a bacterial gene that encodes mannitol 1-phosphate dehydrogenase into tobacco plants, resulting in mannitol accumulation and enhanced salinity tolerance. In addition, transgenic tobacco plants carrying a cDNA encoding myo-inositol *O*-methyltransferase (*IMT1*) accumulated D-ononitol and, as a result, acquired enhanced photosynthesis protection and increased recovery under drought and salt stress (Shelevela et al. 1997). On the other hand, abnormal phenotypes associated with sorbitol accumulation were also found in transgenic tobacco transformed with *stpd1*, a cDNA encoding sorbitol-6-phosphate dehydrogenase from apple. Plants with low levels of sorbitol (less than 2–3 mmol g^1 fresh weight) developed normally, but necrotic lesions and reduced shoot and root growth, as well as low fertility, were associated with an excessive sorbitol accumulation. The adverse effects observed in these sorbitol overproducers may have resulted from a disturbance in carbohydrate transport and allocation (Shelevela et al. 1998) which confirms our observations in *P. strombulifera.*

More generally, the roles of compatible solutes are still being vigorously debated, especially in relation to crop yield in water-limited environments (Ashraf and Akram 2009; Hamdia and Shaddad 2010). After analyzing a large body of work, many authors suggested that 'the purely osmotic contribution of these metabolites to stress tolerance may not describe their function completely, and that the pathway leading to a particular osmolyte may be more important than accumulation *per se*. Several examples of other significant stress-relieving functions served by either the solute or the associated enzyme pathway have been reported, *e.g.*, free radicals scavenging. Irrespective of the exact function of osmolytes or their biosynthetic pathways, the identification and estimation of their significance to cell turgor maintenance would be an important step forward (Adams et al. 2005; Flower and Colmer 2008).

3.4 Anatomical Modifications

In addition to producing metabolic changes, soil salinity can also affect the plant general growth, particularly its morphology and anatomy (Serrato Valenti et al. 1991; Reinoso et al. 2004, 2005; Tattini et al. 2006; Wang et al. 2008). It is generally known that plants that grow well in highly salinized environments have specific structural adaptations to allow altered physiological and biochemical mechanisms while maintaining their reproductive capacity (Shannon 1994).

Prosopis strombulifera does not possess salt glands in its leaves. When grown under high NaCl salinity some tissues of this plant develop great vacuolization and the radical system acquires particular characteristics such as precocious lignification and (or) suberization of endodermal cells, with casparian strips found much closer to the root tip than that in glycophytes. Consequently, these plants have greater potential to efficiently filter the soil solution to prevent an excessive ion loading onto the xylem (Reinoso et al. 2004). On the other hand, Na_2SO_4 treatment induced structural alterations in the cells and tissues of *P. strombulifera* seedlings, which modified the growth pattern at different levels of organization. These alterations included anatomical and histological differences in roots, stems, and leaflets from Na_2SO_4 -treated plants compared to NaCl -treated plants and those grown without any salt (Reinoso et al. 2005). Although the low water potential and the ion toxic effect had strong incidence on cellular growth in these plants, the photosynthetic apparatus might have not loosened efficiency thus providing the carbon skeletons necessary for neoformed parenchymatic layers in the stem, neo-synthesis of lignin, suberin and tannins, whose concentrations increased proportionally with salinity increase (Reinoso et al. 2005). All these complex compounds of high molecular weight greatly influenced total dry weight of treated seedlings, mainly those treated with Na_2SO_4, because this parameter was not modified in spite of the marked growth inhibition observed in the latter.

An important feature in salt-treated plants is the significant increase in tannin content in all organs, which increased with salt concentration independently of light intensity and temperature conditions (Reinoso et al. 2004, 2005; Reginato 2009). Results of several studies suggest that tannin accumulation occurs in response to different stressing conditions (Fahn and Cutler 1992; Grossoni et al. 1998; Hwang et al. 1995) and may be involved in cell protective roles such as scavenging of reactive oxygen species.

3.5 Antioxidant Defense

Salt stress leads to increased reactive oxygen species (ROS) production in plant cells, which are known to produce a secondary oxidative stress under these conditions (Sreenivasulu et al. 2000). To counteract oxidative stress induced by salinity, plants have developed different strategies among which the stimulation of synthesis of secondary metabolites, particularly polyphenols, may be involved in cell protective roles such as ROS scavenging (Simic and Jovanovich 1994).

As reported earlier, there is an important increase in tannin content in salt-treated *Prosopis strombulifera* plants. Leaf and root content of total phenols, total flavonoids, total flavan-3-ols, condensed tannins (or proanthocyanidins), tartaric acid esters and flavonol analysis showed that salt treatments increased the accumulation of several polyphenols in both leaves and roots (Fig. 5.4).

Sodium sulphate (Na_2SO_4) treatment sharply induced an increase in total polyphenols showing the highest levels of these compounds in our experiments, especially flavonoids and total flavan-3-ols. Bisaline treated plants also accumulated high levels of polyphenols. HPLC analysis showed high levels of rutin, catechin, epicatechin and proanthocyanidine in these plants.

These increases in total flavonoids and flavan-3-ols, when SO_4^{2-} anion is present in the growth solution, may indicate a role for these compounds in counteracting the oxidative damage induced by severe salt stress (Reginato et al. 2010). Agati and Tattini (2010) reported that antioxidant mesophyll flavonoids, at micromolar range, may effectively avoid generation of reactive oxygen forms (e.g. by chelating transition metal ions).

3.6 Changes in Photosynthetic Pigments

Salinity results in both qualitative and quantitative changes in photosynthetic pigment composition depending upon the species tested and quality and quantity of salts used. Chlorophyll concentration in leaves is an important biochemical indicator that represents the potentiality of the photosynthetic apparatus in plants, and an increase in chlorophyll content has been observed in salt tolerant species (Joshi 1976; Reddy et al. 1992).

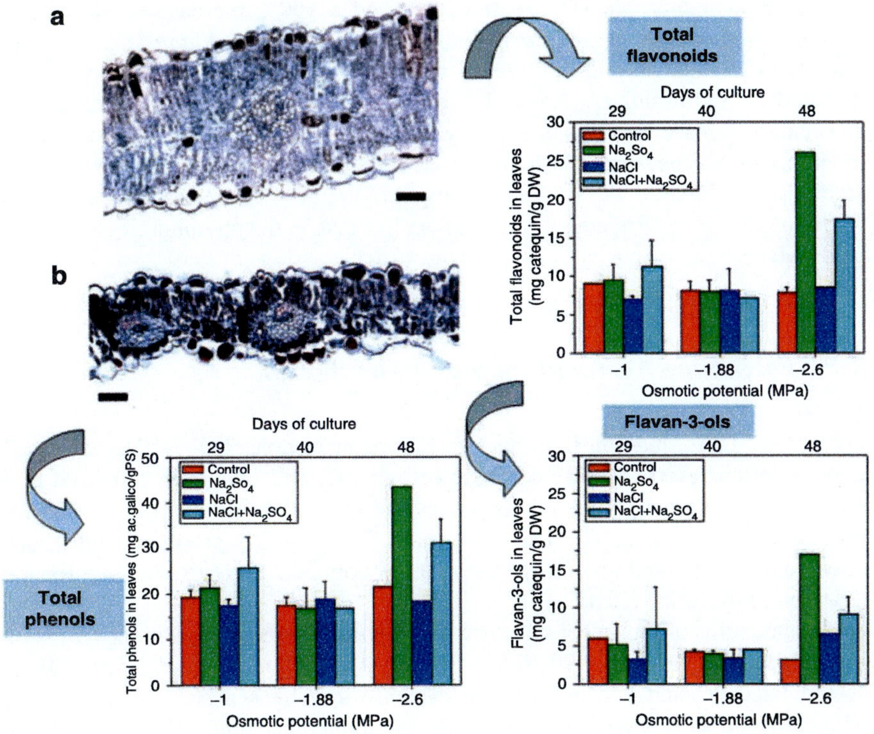

Fig. 5.4 Transverse section of 48 day-old plant leaflets showing the accumulation of condensed tannins under salt stress (Control plants (**a**) and Na_2SO_4 treated plants ($\Psi_o = -2.6$ MPa) (**b**); Scale = 50 μm). The associated graphics show the respective quantification of polyphenols in control and treated plants

The photosynthetic pigment content in *P. strombulifera* was also differentially affected by NaCl, Na_2SO_4, and their mixture. Chlorophyll *a* content remained constant in all treatments. On the contrary, chlorophyll *b* content significantly increased under all treatments of low salinity, with a sharp increase after Na_2SO_4 treatment. Reddy et al. (1997) proposed this increment as a cellular adaptation to satisfy the increasing demand of energy due to salt toxicity. One could speculate that the antennae pigments need to be more effective to deliver sufficient energy to deal with energy-consuming adaptations to salinity. Similar to our results, the halophyte *Aster tripolium* L. showed high tolerance to soil salinity and increased chlorophyll contents to get higher light harvesting capacity and, hence, higher rates of photosynthesis (Ramani et al. 2006).

Carotenoids play three major in the photosystems of thylakoid membranes are light harvest, photoprotection and maintenance of structural stability (Nayak et al. 2002). The effect of salinity on biosynthesis of carotenoids in plants is little known.

In our experiments, Na_2SO_4 treatment caused a 100% increase in carotenoid production from the beginning of salinization. Similarly, *Plantago coronopus* leaves showed salt-induced increase of carotenoid content when plants were cultured in 500 mmol L^{-1} NaCl (Koyro 2006). The proposed function for this carotenoid increase is excess energy dissipation as heat or non-damaging chemical reactions in photosystem I and II (Lu et al. 2003), and chloroplast membranes stabilization (Havaux 1998). Additionally, the increase in the carotenoid pool in *P. strombulifera* is correlated with a marked increase in the phytohormone abscisic acid (ABA).

3.7 Polyamine Accumulation and Metabolism

Polyamines (PAs) play an important role in plant responses. Free PAs are small organic cations essential for eukaryotic cell growth, and have been proposed as a new category of plant growth regulators involved in a wide variety of physiological processes (Liu et al. 2007). The three main PAs in plants are putrescine (Put), spermidine (Spd), and spermine (Spm). Less common PAs are diaminopropane (Dap) and cadaverine (Cad).

Accumulation of PAs under salt stress was reported for mono- and di-cotyledonous species (Bouchereau et al. 1999; Simon-Sarkadi et al. 2002). *Lupinus luteus* seedlings accumulated Put and Spd in leaves in response to increased NaCl (Legocka and Kluk 2005). Salt-tolerant cultivars of rice and tomato accumulated Spd and Spm, whereas salt-sensitive cultivars accumulated Put (Krishnamurthy and Bhagwat 1989; Santa Cruz et al. 1999). PAs are presumed to have protective functions, *e.g.*, scavenging free radicals (Besford et al. 1993), cellular pH maintenance, cellular ionic balance and membrane stabilization (Kakkar and Rai 1997; Aziz et al. 1999; Groppa and Benavides 2008), cationic channels regulation in cells (Liu et al. 2000; Janicka-Russak et al. 2010), prevention of lipid peroxidation (Verma and Mishra 2005; Zhao and Yang 2008) and bioenergetics of photosynthesis (Ioannidis and Kotzabasis 2007). These various processes may occur separately, or be combined as a unified strategy to minimize membrane damage, promote cell growth, or enhance cell survival in response to stress (Liu et al. 2007).

Stressful conditions may alter the content of endogenous PAs, particularly Put, with variation according to type of stress, species, and time of exposure (Kakkar and Rai 1997; Ali 2000). In *P. strombulifera,* leaves of Na_2SO_4-treated plants showed the lowest Put content, which was associated with a strong inhibition of shoot growth and possibility associated with a general metabolic alteration caused by SO_4^{2-} anion (Reginato 2009). The physiological significance of stress-induced Put accumulation has been linked to maintenance of cellular pH, cation/anion balance, and membrane stabilization (Kakkar and Rai 1997). In contrast to leaves, Na_2SO_4-treated roots showed the highest Put contents, together with significant increases in Spd and Spm. PA accumulation has been correlated with adventitious and lateral roots formation

(Altamura et al. 1991) in coincidence with our previous observations in *P. strombulifera* (Reinoso et al. 2005). Perhaps Put is involved in the plasticity of root development, *i.e.*, part of an attempt by Na_2SO_4-treated plants, to optimize ions and water uptake under stressful conditions imposed by this salt.

The low Spd content in *P. strombulifera* leaves, in all treatments, suggest a rapid degradation of this amine to Put, as a part of an interconversion pathway well known in animals and also characterized in higher plants (Duhazé et al. 2002). That part of the high Put content in NaCl-treated plant leaves in our studies could be due to Spd catabolism. On the other hand, plants that accumulated the highest Spd content in leaves were those with higher toxicity and deterioration symptoms caused by Na_2SO_4. Spd accumulation in these Na_2SO_4-treated plants may represent a metabolic disorder, rather than an adaptation mechanism. In addition, Spm content in leaves was not altered by changes in salinity; hence, this PA, at least in its free form, may not play a significant role in salt stress response in this species.

Salinity plant response is often associated with stimulation of PAs oxidation and catabolism (Cona et al. 2006), which play an important role in the regulation of cellular PAs level. Accumulation of Dap and Put could result from the presence of polyamine oxidases (PAOs) acting on Spd, as reported by Duhazé et al. (2002) in various halophytic species of the genus *Limonium*. The specific effect of Dap is poorly understood, but may be involved in biosynthesis of uncommon PAs and/or β-alanine, via oxidative deamination pathway, in some species (Cohen 1998). Biosynthesis of uncommon PAs such as 1.3 Dap, Cad, norspermidine, homospermidine, and norspermine, has been associated with the capacity of some biological systems to grow or function under extreme environmental conditions (Flores 1991). In salinized *P. strombulifera*, several unknown chromatographic peaks were observed, mainly in Na_2SO_4-treated plants at high salinity. The analysis of these peaks, which may correspond to uncommon PAs as above, deserves further research.

The presence of diamine Cad is sporadic and uncommon. Cad has been found in several other genera of Leguminosae, and also in Gramineae and Solanaceae (Flores 1991). Although Cad in roots has been associated with cell elongation and adventitious root formation, similarly to Put (Shevyakova et al. 2001; Carrizo et al. 2001), its accumulation in Na_2SO_4-treated roots of *P. strombulifera* may reflect a stress symptom rather than an adaptative response.

PAs content in leaves and roots of control and salt-treated plants at –2.6 MPa (day 48) is shown in Fig. 5.5 for a more comprehensive understanding of the distribution pattern of these compounds.

In summary, the limited variation in free PAs content in salinized *P. strombulifera* suggests active PA catabolism and conjugation. This could be related to observations that in higher plants subjected to salt stress, PA profile is disturbed in glycophytes but remains stable in halophytes. Internal level of PAs in *P. strombulifera* may remain constant because of equilibrated rates of synthesis and degradation, which are differentially affected by different salts (NaCl, Na_2SO_4, and their iso-osmotic mixture) in roots vs. leaves.

3.8 Hormonal Changes

Hormonal changes in response to environmental abiotic stress have been well-studied in glycophytic plants, but such information is much more limited for halophytic plants. Plants respond to biotic and abiotic stresses by generation of various lipid-based signals. Jasmonates (JAs) are key signaling molecules in plant stress responses and development (for review see Wasternack 2007). This group of compounds includes jasmonic acid (JA), its methyl ester (JAME), its amino acid conjugates, and metabolites such as 12-OH-JA and 11-OH-JA. The octadecanoid cis (+) 12 oxophytodienoic acid (OPDA) is the precursor of JA.

Changes of jasmonate profiles have been reported in response to abiotic stresses such as NaCl, mannitol/sorbitol, and water stress (Kramell et al 2000; Pedranzani et al. 2007). In glycophytes, tomato, barley, and others, involvement of JAs in salt responses has been well established (Pedranzani et al. 2003; Wasternack and Hause 2002).

The same jasmonate family members (OPDA, JA, 12-OH-JA, 11-OH-JA) found in glycophytes are present in the extremely salt tolerant halophyte *Prosopis strombulifera*. In this species, we observed large differences in total amount of JAs between organs, as well as in the ratios of these compounds. Similarly, Miersch et al. (2008) reported differences in the content of JAs among different tissues of various plant species. The major components in *P. strombulifera* roots were OPDA and JA, whereas those in leaves were OPDA and 12-OH-JA, confirming the existence of specific metabolic JAs profiles depending on the organ analyzed, as also observed in glycophytes. Hence, the concept of an "oxylipin signature" (Weber et al. 1997) or a "jasmonate signature" (Miersch et al. 2008), can be applied to the halophyte *P. strombulifera*.

OPDA is synthesized in leaves (in chloroplasts), and there is an evidence of a similar pathway in roots (Hause et al. 2000; Abdala et al. 2003). The high OPDA content in 29-day-old control roots of *P. strombulifera* suggests active biosynthesis in this organ. Contrarily to what it was expected, given the fact that *P. strombulifera* is a halophyte, its roots responded to high salinity by decreasing total content of JAs. OPDA content was higher than that of the other JAs, and was more strongly affected by salt stress. JAs are known to inhibit primary root elongation (Woodward and Bartel 2005; Wasternack 2007). However, a few reports demonstrated that exogenously applied JAs enhance rooting *per se* (Moons et al. 1997; Toro et al. 2003). In *Arabidopsis*, exogenous MeJA positively regulates lateral root formation (Sun et al. 2009). In *P. strombulifera,* NaCl-treated plants showed the lowest OPDA content at −1 MPa, concomitantly with a significant root growth promotion.

The differential response to salt treatments of roots vs. leaves could be attributed to the root's function as a "stress-receptor organ", which responds to stress with more dramatic hormonal changes compared to leaves. Another difference is the higher content of 12-OH-JA in leaves than in roots; hence, the higher content of this compound in leaves may result from active hydroxylative pathway in these organs or transport of root-generated 12-OH-JA. For many years, 12-OH-JA was detected

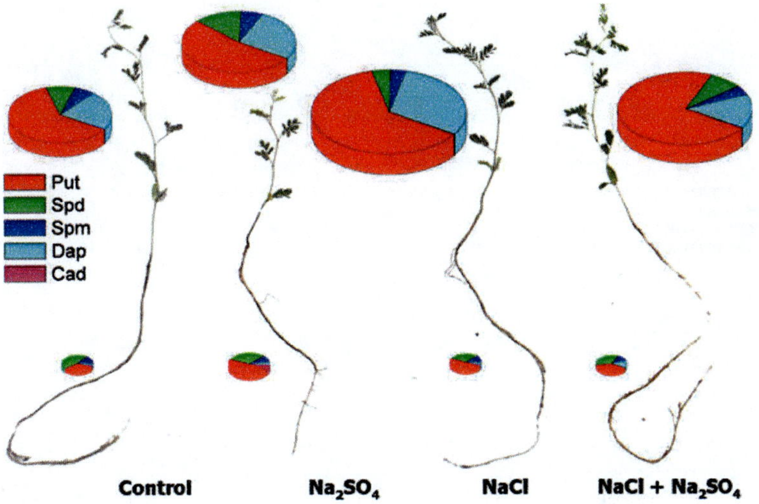

Fig. 5.5 Total PAs levels in *P. strombulifera* plants. Control plants (48 days); Na_2SO_4, NaCl and NaCl+Na_2SO_4 treated plants (Ψ_o=−2.6 MPa, 48 days). The area of the *circles* designates the total PAs levels in different organs of the plant. The size of the circle in leaves of NaCl-treated plants corresponds to 846.66 nanomol/g FW (100%). The sizes of *circles* (and their respective portions) in the other plant organs and treatments are proportionally depicted

only in solanaceous species, although more recently analysis of monocotyledons and dicotyledons revealed the presence of 12-OH-JA in a wide variety of species and organs, including tomato flowers (Hause et al. 2000), *Arabidopsis thaliana* leaves (Gidda et al. 2003), potato tubers and stolons (Cenzano et al. 2005), tomato seeds (Andrade et al. 2005), sunflower seeds (Vigliocco et al. 2007), and soybean seeds (Miersch et al. 2008). Clearly, 12-OH-JA must be involved in physiological processes other than tuberization. On the other hand, 11-OH-JA is usually present in much lower concentration than that of 12-OH-JA, except in the seeds of *Cucurbita pepo* (Miersch et al. 2008). In *P. strombulifera* salt treatments had no effect on both hydroxylates content. As for PAs, Fig. 5.6 shows JAs distribution pattern under salinity.

On the other hand, upon salinity or drought stress, plants accumulate the phytohormone abscisic acid (ABA), which in turn controls many adaptive responses such as salinity-induced stomatal closure and induction of gene expression and tolerance to drought. Results obtained previously in our group (Sosa 2005) demonstrated that optimal growth was observed in *P. strombulifera* seedlings when the salt concentration in the medium was 400–500 mM NaCl (Ψ_o: −1.8 MPa). Thus, higher or lower salt concentrations could be registered as stressful by the plant, which could explain the high ABA levels found in plants growing in the salt-free Hoagland solution (controls). Moreover, a marked decrease in ABA level at Ψ_o: -1.88 MPa in NaCl treated plants compared to controls was found, which could mean that such NaCl

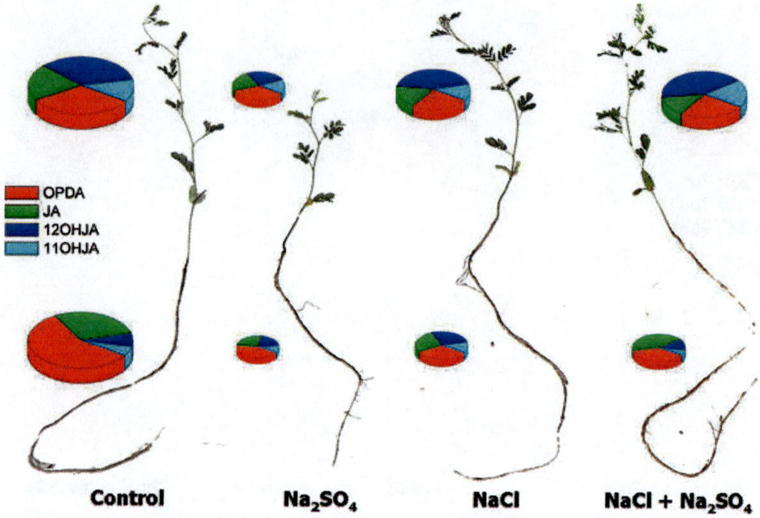

Fig 5.6 Total JAs levels in *P. strombulifera* plants. Control plants (48 days); Na_2SO_4, NaCl and NaCl+Na_2SO_4 treated plants (Ψ_o=−2.6 MPa, 48 days). The area of the *circles* designates the total content of JAs in different organs of the plant. The size of the circle in leaves of control plants corresponds to 12,005 pmol/g DW (100%). The sizes of *circles* (and their respective portions) in the other plant organs and treatments are proportionally depicted

concentration is not a stressful condition for this species, showing a true halophytic response. Higher NaCl concentrations lead to higher ABA accumulation (Sosa 2005; Llanes 2010).

In Na_2SO_4 -treated plant leaves the highest free ABA peaks were quantified, mainly at the highest salt concentration, in coincidence with growth inhibition, toxicity symptoms and significant levels of Spd. As mentioned earlier, a correlation between higher ABA levels and higher PAs accumulation was observed in our experiments. Similarly, Liu et al. (2005) reported ABA-induced PAs synthesis in salt stressed maize plants. Furthermore, Hanzawa et al. (2002) and Urano et al. (2003) informed that exogenous ABA application stimulated transcription and accumulation of genes involved in polyamine biosynthesis.

With respect to ABA metabolism, our results showed that both roots and leaves of *Prosopis strombulifera* seedlings accumulated mainly ABA-GE (glucose conjugated) in relation to phaseic acid (PA) and dihydrophaseic acid (DPA) whose levels were very low. This pattern was especially notable in salinized plants at highest Na_2SO_4 concentration. The greater ABA-GE accumulation recorded in roots under high salt concentrations (−2.6 MPa) could be related to ABA transport to the leaves where highest levels of free ABA were detected. It is noteworthy that in the presence of Na_2SO_4, both roots and leaves showed the highest concentration of ABA-GE as well as free ABA, which correlates with the severe stressful conditions imposed by this salt (Llanes et al. 2008).

4 Biotechnological Approach

4.1 *P. strombulifera as a Natural Gene Donor to Improve Salt Tolerance in Crops*

Development of stress tolerant crop plants requires, among others, knowledge of the physiological mechanisms and genetic control of the contributing traits at different plant developmental stages. In the past two decades, biotechnological research has provided considerable insights into abiotic stress tolerance mechanisms in plants at the molecular level (Munns 2007). Nevertheless, for halophytic species this knowledge is still very incomplete.

The exploration for highly salt-tolerant *Prosopis* species is important for both theoretical and applied considerations. Within the family Fabaceae to which *Prosopis* belongs, virtually all of the important annual legumes of the Papilonoideae subfamily, *i.e.* soybeans, common beans, peas *etc*., are highly salt sensitive to salt and only some alfalfa varieties have a moderate degree of salt tolerance. Thus, identification and full understanding of salt tolerance mechanisms in *Prosopis* species may have relevance to current commercially important legumes (Felker 2007). The search for cDNAs and differentially expressed RNA sequences in this highly salt-tolerant *Prosopis* species, *P. strombulifera,* is being carried and in our lab with the hope that this information would be useful for improving stress tolerance in common annual legumes.

In *P. strombulifera* seedlings four sequences were isolated and identified in our lab, showing homology with proteins encoding genes involved in salinity responses. They were named *PSALl, PSAL3*, *PSAL11* and *PSAL5*, and were selected for further studies of gene expression and functional analysis with the aim of contributing to understand the mechanisms responsible for the extreme salt tolerance of this species. The isolation of *PSAL1* sequence, which has homology with an ABA-dependent atypical LEA protein, is an interesting result related to cellular protection against osmotic stress (Bray 2002; Parida et al. 2007). Preliminary studies of this gene expression showed that *P. strombulifera* roots constitutively accumulate high levels of the corresponding transcript, whereas gene induction was observed only in Na_2SO_4-treated plant roots at $\Psi_o = -1.88$ MPa, which coincides with ABA and ABA-GE accumulation. *PSAL5* sequence encodes a myo-inositol-1-P synthase enzyme, which participates in the formation of the pinitol precursor; its further study will give us valuable information considering that pinitol is the main polyalcohol accumulated in our species under salinity. *PSAL11* sequence shows homology with a potassium and sodium transporter HKT1 from *Mesembryanthemum crystallinum.* Functional analysis of this transporter would provide important information on osmotic potential regulation in *P. strombulifera* seedlings. Similarly, *PSAL3* sequence shows homology with a sodium/proline transporter, which may play a key role in stress tolerance considering that both, sodium and proline, accumulate largely in this species (Llanes 2010).

4.2 *P. strombulifera and Its Rhizosphere as Natural Sources of PSHR Microorganisms Which Are Capable of Improving Salt Tolerance in Crops*

4.2.1 Plant Growth Promoting and Stress Homeo-Regulating Rhizobacteria

Bacteria living in the rhizosphere are commonly named rhizobacteria, and have the ability to associate and colonize a large number of plant species (Garate and Bonilla 2000). Depending on the type of relationship established with plants, they could be classified into *symbiotic* or *free-living* rhizobacteria; whichever, all of them are associated to the endorhizosphere (*i.e.* endophytic inside plant tissues as bacteroids), rhizoplane or ectorhizosphere in a mutualistic relationship (Kloepper et al. 1989). Among the most successful partnerships in nature, there are some members of Order *Rhizobiae* with legumes, and some *free-living* rhizobacteria of the genera *Pseudomonas, Bacillus* and *Azospirillum* with grasses and other non-legumes (Döbereiner and Pedroza 1987). If the rhizobacteria are beneficial for plant growth, they are named plant growth promoting rhizobacteria (PGPR). Cassan et al. (2009a) suggested that microorganisms could be re-classified into three groups, according to their effect on the plant: (a) direct plant growth promotion, (b) indirect plant growth promotion or phytopathogens biocontrol, and (c) regulation of plant growth through homeostasis regulation under abiotic stress conditions.

The first group, PGPR, was proposed by Kloepper and Schroth (1978) and represents rhizobacteria that stimulate plant growth through direct mechanisms such as biological nitrogen fixation, supply or solubilization of essential mineral elements as iron (Fe) or phosphorus (P), and phytohormone production as auxins (Patten and Glick 1996), cytokinins (Tien et al. 1979) and gibberellins (Cassán et al. 2003), as well as certain polyamines particularly cadaverine (Cassán et al. 2009a).

The second group, known as biocontroling-PGPB (Bio-controlling-plant growth promoting bacteria) was defined by Bashan and Holguin (1997) for bacteria (not necessarily rhizospheric ones) that can stimulate plant growth through several biocontrol mechanisms such as biosynthesis of antibiotics or antifungal compounds, which activate under biotic (bacterial or fungal) stressing agents.

The third group, known as PSHR (Plant stress homeo-regulating rhizobacteria), was proposed by Cassan et al. (2009a) for those microorganisms which support or promote plant growth under abiotic stress conditions. The main plant promoting mechanism proposed for the third group is the production of active molecules that regulate abiotic stress tolerance in plants, such as abscisic acid (ABA) (Cohen et al. 2008), jasmonic acid (JA) (Forchetti et al. 2007) certain polyamines as cadaverine (Cassán et al. 2009b), and activities of enzymes such as ACC deaminase (Glick 2005). Another important mechanism, less studied, but of great interest from the ecophysiological point of view, is related to the bacterial ability to immobilize or neutralize toxic compounds, such as heavy metals, salts or ions (rhizoremediation) (Sgroy et al. 2010).

The halophyte *P. strombulifera* has various physiological and biochemical mechanisms that allow optimal growth under high saline conditions as reported earlier. Nevertheless, its adaptive success may depend, at least in part, on its ability to establish and maintain effective associations with plant growth-promoting endophytic or rhizospheric bacteria. Studies carried out in our lab (Sgroy et al. 2010) reported for the first time the isolation and molecular characterization of *P. strombulifera* endophytic bacteria as well as the principal PSHB mechanisms under controlled growth conditions. Twenty-nine endophytic strains were isolated from established *P. strombulifera* plants in their natural habits, with a 69% frequency of positive spore-forming bacilli, including *Bacillus* sp. representing the 80% of gram-positive isolated strains. The genotypic 16 S rDNA sequencing showed a prevalence of 18.75% for *L. fusiformis* (Ps7); 68.75% for *B. subtilis* (Ps8); 6.25% for *B. licheniformis* (Ps14); and 6.25% for *B. pumilus* (Ps19). On the other hand, 31% of isolations were gram-negative bacilli, and genotypic identification showed a prevalence of 55% for *Achromobacter xylosoxidans* (Ps27) and 45% for *Pseudomonas putida* (Ps30).

The physiological evaluation of the potential PHSR mechanisms present in these rhizobacteria showed that all the strains were able to express *ACC deaminase* activity to produce high amounts of several phytohormones functionally associated with the abiotic stress tolerance, such as ABA and JA, and to immobilize the most toxic anion (SO_4^{2-}) for *P. strombulifera* plants grown under controlled conditions (Sgroy et al. 2010).

The current literature has identified a strong relationship between the bacterial ACC-deaminase activity and plant salt-stress tolerance (Ma et al. 2008). Similarly, bacterial production of ABA (Cohen et al. 2008) and JA (Forchetti et al. 2007) was correlated with the microorganism capacity to increase stress tolerance in inoculated plants. Little information is published on the rhizobacterial capacity to promote or regulate plant growth through sulphate immobilization under saline stress conditions. In this regard, our results could be considered the basis for a reconsideration of the rhizoremediation concept under salinity conditions.

The existence of the plant stress homeo regulating mechanisms described earlier in rhizobacteria, could be the result of a co-evolutionary relationship allowing plants to select their most convenient partners for association, so as to improve their natural capacity to face adverse environmental conditions. Our findings suggest the existence of an "endophytic consensus" among all the strains colonizing the same plant, each of them expressing one or more different interacting mechanisms that jointly determine a global answer of plant growth regulation under salt stress.

5 Conclusions and Perspectives

The great opportunity for salinity tolerance research now is the ability to consolidate new molecular techniques with the body of literature on whole plant physiology. This new opportunity in salinity tolerance research provides exciting prospects for ameliorating the impact of salinity stress on plants through genetic engineering, and improving native halophytic species performance for agricultural and environmental sustainability and human health care.

Halophytes have recently been pointed out as a useful approach for generating salt tolerant crops in the future by using them as alternative crops, like *Aster tripolium* and *Salicornia rubra* (Erdei and Bogemans 2009) or as gene donors for genetic manipulation of economically important species, representing a very good model for future research.

The present review is a contribution to a better understanding of the physiological responses of a native halophyte to salinity, taking into account the chemical nature of the major salts present in soils of several countries, not only Argentina. Beside this, a new approach emerging from this work is the consideration of the rhizosphere of halophytes as natural sources of PSHR microorganisms capable to help plants to improve salt tolerance, giving rise to a new biotechnological tool for the inoculants industry.

Finally, we consider that more interdisciplinary work is needed, and the interaction and cooperation between plant physiologists, bacteriologists and molecular biologists should be emphasized.

Acknowledgments This study was supported with funds from CONICET PIP N° 5628, PICTO-ANPCYT-UNRC N° 30093, ICGBE-TWAS Joint Biotechnology Programme, SECYT-Universidad Nacional de Río Cuarto and Ministerio de Ciencia y Tecnología de la Provincia de Córdoba (R N° 1210/2007), Argentina, to V.L.

References

Abdala G, Miersch O, Kramell R, Vigliocco A, Agostini E, Forchetti G, Alemano S (2003) Jasmonate and octadecanoid occurrence in tomato hairy roots. Endogenous level changes in response to NaCl. Plant Growth Regul 40(1):21–27

Adams M, Richter A, Hill A, Colmer T (2005) Salt tolerance in *Eucalyptus* spp.: identity and response of putative osmolytes. Plant Cell Environ 28:772–787

Agatti G, Tattini M (2010) Multiple functional roles of flavonoids in photoprotection. New Phytol 186:786–793

Ali R (2000) Role of putrescine in salt tolerance of *Atropa belladonna* plant. Plant Sci 152:173–179

Altamura M, Torrigiani P, Capitani F, Scaramagli S, Bagni N (1991) *De novo* root formation in tobacco thin layers affected by inhibition of polyamine biosynthesis. J Exp Bot 42:1575–1582

Andrade A, Vigliocco A, Alemano S, Miersch O, Botella M, Abdala G (2005) Endogenous jasmonates and octadecanoids in hypersensitive tomato mutants during germination and seedling development in response to abiotic stress. Seed Sci Res 15:309–318

Armengaud P, Thiery L, Buhot N, Grenier-De March G, Savoure A (2004) Transcriptional *Medicago truncatula* reveals developmental and regulation of proline biosynthesis in environmental specific features. Physiol Plant 120:442–450

Arndt S, Arampatsis C, Foetzki A, Li X, Zeng F, Zhang X (2004) Contrasting patterns of leaf solute accumulation and salt adaptation in four phreatophytic desert plants in a hyperarid desert with saline groundwater. J Arid Environ 59:259–270

Asch F, Dingkuhn M, Dorffling K, Miezan K (2000) Leaf K/Na ratio predicts salinity induced yield loss in irrigated rice. Euphytica 113:109–118

Ashraf M, Ahmad S (2000) Influence of sodium on ion accumulation, yield components and fibre characteristics in salt-tolerant and salt-sensitive lines of cotton (*Gossypium hirsutum* L.). Field Crops Res 66:115–127

Ashraf M, Akram A (2009) Improving salinity tolerance of plants through conventional breeding and genetic engineering: an analytical comparison. Biotechnol Adv 27:744–752

Ayers R, Westcott D (1985) Water quality for irrigation. FAO Irrigation and drainage paper no 29. FAO, Rome

Aziz A, Martin-Tanguy J, Larher F (1999) Salt stress-induced proline accumulation and changes in tyramine and polyamine levels are linked to ionic adjustment in tomato leaf discs. Plant Sci 145:83–91

Balnokin Y, Myasoedov N, Shamsutdinov Z, Shamsutdinov N (2005) Significance of Na^+ and K^+ for sustained hydration of organ tissues in ecologically distinct halophytes of the family chenopodiaceae. Russ J Plant Physiol 52(6):779–787

Bashan Y, Holguin G (1997) *Azospirillum*-plant relationships, environmental and physiological advances. Can J Microbiol 43:103–121

Besford R, Richardson C, Campos J, Tiburcio A (1993) Effect of polyamines on stabilization of molecular complexes in thylakoid membranes of osmotically stressed oat leaves. Planta 189:201–206

Bie Z, Tadashi I, Shinoara Y (2004) Effects of sodium sulfate and sodium bicarbonate on the growth, gas exchange and mineral composition of lettuce. Sci Hort 99:215–224

Bouchereau A, Aziz A, Larher F, Martin-Tanguy J (1999) Polyamines and environmental challenges: recent development. Plant Sci 140:103–125

Bray E (2002) Classification of genes differentially expressed during water-deficit stress in Arabidopsis thaliana: an analysis using microarray and differential expression data. Ann Bot 89:803–811

Burkart A (1937) Estudios morfológicos y etológicos en el género Prosopis. Darwiniana 3:27–47

Burkart A (1952) Las leguminosas argentinas. Acme Agency Press, Buenos Aires

Burkart A (1976) A monograph of the genus *Prosopis* (Leguminosae subfam. Mimosoideae). Catalogue of the recognized species of *Prosopis*. J Arnold Arbo 57:450–525

Cantero J, Cantero A, Cisneros J (1996) La vegetación de los paisajes hidro-halomórficos del centro de Argentina. Ed. Fundación Universidad Nacional de Río Cuarto, Córdoba, Argentina

Carrizo C, Pitta-Alvarez S, Kogan M, Giulietti A, Tomaro M (2001) Occurrence of cadaverine in hairy roots of *Brugmansia candida*. Phytochemistry 57:759–763

Cassán F, Picolli P, Bottini R (2003) Plant growth promoting by *Azospirillum* sp. through gibberelin production. An alternative model to increase crop yield?. In: Albanesi A (ed) Microbiología Agrícola Un aporte de la Investigación Argentina para la sociedad. Universidad Nacional de Santiago del Estero, Argentina. pp 143–158

Cassán F, Maiale S, Masciarelli O, Vidal A, Luna V, Ruiz O (2009a) Cadaverine production by *Azospirillum brasilense* and its possible role in plant growth promotion and osmotic stress mitigation. Eur J Soil Biol 45:12–19

Cassán F, Perrig D, Sgroy V, Masciarelli O, Penna C, Luna V (2009b) *Azospirillum brasilense* Az39 and *Bradyrhizobium japonicum* E 109 promote seed germination and early seedling growth, independently or co-inoculated in maize (*Zea mays* L.) and soybean (*Glycine max* L.). Eur J Soil Biol 45:28–35

Cenzano A, Vigliocco A, Miersch O, Abdala G (2005) Hydroxylated jasmonate levels during stolon to tuber transition in *Solanum tuberosum* L. Potato Res 48:107–115

Chen T, Murata N (2002) Enhancement of tolerance of abiotic stress by metabolic engineering of betaines and other compatible solutes. Curr Opin Plant Biol 5:250–257

Chen Z, Cuin TA, Zhou M, Twomey A, Bodapati PN, Shabala SN (2007) Compatible solute accumulation and stress-mitigating effects in barley genotypes contrasting in their salt tolerance. J Exp Biol 58(15/16):4245–4255

Chrispeels M, Crawford N, Shroeder I (1999) Proteins for transport of water and mineral nutrients across the membranes of plant cells. Plant Cell 11:661–675

Cohen SS (1998) A guide to the polyamines. Oxford University Press, Oxford

Cohen A, Bottini R, Piccoli P (2008) *Azospirillum brasilense* Sp 245 produces ABA in chemically-defined culture medium and increases ABA content in *Arabidopsis* plants. Plant Growth Regul 54:97–103

Cona A, Rea G, Angelini R, Federico R, Tavladoraki P (2006) Functions of amine oxidases in plant development and defence. Trends Plant Sci 11:80–88

Croser C, Renault S, Franklin J, Zwiazek J (2001) The effect of salinity on the emergence and seedling growth of *Picea mariana*, *Picea glauca*, and *Pinus banksiana*. Environ Pollut 115:9–16

Cuin T, Shabala S (2007) Amino acids regulate salinity-induced potassium efflux in barley root epidermis. Planta 225(3):753–761

De Kok L, Castro A, Durenkamp M, Koralewska A, Posthumus S, Stuiver C, Yang L, Stulen I (2005) Pathways of plant sulfur uptake and metabolism-an overview. Landbauforschung Völkenrode, Special Issue 283:5–13

Demidchik V, Davenport R, Tester M (2002) Nonselective cation channels in plants. Annu Rev Plant Biol 53:67–107

Di Martino C, Delfine S, Pizzuto R, Loreto F, Fuggi A (2003) Free amino acids and glycine betaine in leaf osmoregulation of spinach responding to increasing salt stress. New Phytol 158:455–463

Döbereiner J, Pedroza F (1987) Nitrogen-fixing bacteria in non-leguminous crop plants. Science Tech, Madison, pp 1–155

Duhazé C, Gouzerh G, Gagneul D, Larher F, Bouchereau A (2002) The conversion of spermidine to putrescine and 1,3 diaminopropane in the roots of *Limonium tataricum*. Plant Sci 163:639–646

Egan T, Ungar I (1998) Effect of different salts of sodium and potassium on the growth of *Atriplex prostata* (Chenopodiaceae). J Plant Nutr 21(10):2193–2205

Erdei L, Bogemans J (2009) Domestication of halophyte species: from laboratory to production and market. Book of abstracts Primer Congreso de la Red Argentina de Salinidad, pp 20

Fahn A, Cutler D (1992) Xerophytes. Encyclopedia of plant anatomy, vol XIII, 3. Gebruder Borntraeger, Berlin/Stuttgart

FAO (2008) FAO land and plant nutrition management service. http://www.fao.org/ag/agl/agll/spush/

Felker P (2007) Unusual physiological properties of the arid adapted tree legume Prosopis and their applications in developing countries. In: De la Barrera E, Smith W (eds) Perspectives in biophysical plant ecophysiology: a tribute to Park Nobel. Mildred E. Mathias Botanical Garden University of California, Los Angeles, pp 1–41

Flores HE (1991) Changes in polyamine metabolism in response to abiotic stress. In: Slocum R, Flores H (eds) The biochemistry and physiology of polyamines in plants. CRC Press, Boca Raton, pp 214–255

Flower T, Colmer T (2008) Salinity tolerance in halophytes. New Phytol 179:945–963

Flowers T, Flowers S (2005) Why does salinity pose such a difficult problem for plant breeders? Agric Water Manag 78:15–24

Forchetti G, Masciarelli O, Alemano S, Alvarez D, Abdala G (2007) Endophytic bacteria in sunflower (*Helianthus annuus* L.) isolation, characterization, and production of jasmonates and abscisic acid in culture medium. Appl Microbiol Biotechnol 76:1145–1152

Garate A, Bonilla I (2000) Nutrición mineral y producción vegetal. In: Azcón-Bieto Talón E (eds) Fundamentos de fisiología vegetal. McGraw-Hill Interam, Madrid, pp 113–130

Garcia A, Senadhira D, Flowers T, Yeo A (1995) The effects of selection for sodium transport and of selection for agronomic characteristics upon salt resistance in rice (*Oryza sativa* L.). Theor Appl Genet 90:1106–1111

Gidda S, Miersch O, Levitin A, Schmidt J, Wasternack C, Varin L (2003) Biochemical and molecular characterization of a hydroxy-jasmonate sulfotransferase from *Arabidopsis thaliana*. J Biol Chem 278:17895–17900

Glick BR (2005) Modulation of plant ethylene levels by the enzyme ACC deaminase. FEMS Microbiol Lett 251:1–7

Grewal HS (2010) Water uptake, water use efficiency, plant growth and ionic balance of wheat, barley, canola and chickpea plants on a sodic vertosol with variable subsoil NaCl salinity. Agric Water Manag 97:148–156

Groppa MD, Benavides MP (2008) Polyamines and abiotic stress: recent advances. Amino Acids 34:35–45

Grossoni P, Bussotti F, Tani C, Gravano E, Santarelli S, Bottacci A (1998) Morpho-anatomical alterations of leaves of *Fagus sylvatica* L. and *Quercus ilex* L. in different environmental stress conditions. Chemosphere 36:919–924

Hamdia MA, Shaddad MA (2010) Salt tolerance of crop plants. J Stress Physiol Biochem 6:64–90

Hampson C, Simpson G (1990) Effects of temperature, salt and osmotic pressure on early growth of wheat (*Triticum aestivum*). 1. Germination. Can J Bot 68:24–528

Hanzawa Y, Imai A, Michael A, Komeda Y, Takahashi T (2002) Characterization of the spermidine synthase-related gene family in Arabidopsis thaliana. FEBS Lett 527:176–180

Hariadi Y, Marandon K, Tian Y, Jacobsen S, Shabala S (2011) Ionic and osmotic relations in quinoa (*Chenopodium quinoa* Willd.) plants grown at various salinity levels. J Exp Bot 62(1):185–193

Hasegawa P, Bressan R, Zhu J, Bohnert H (2000) Plant cellular and molecular responses to high salinity. Annu Rev Plant Physiol Plant Mol Biol 51:463–499

Hause B, Stenzel I, Miersch O, Maucher H, Kramell R, Ziegler J, Wasternack C (2000) Tissue-specific oxylipin signature of tomato flowers: allene oxide cyclase is highly expressed in distinct flower organs and vascular bundles. Plant J 24:113–126

Havaux M (1998) Carotenoids as membrane stabilizer in chloroplasts. Trends Plant Sci 3:147–151

Houshmand S, Arzani A, Maibody S, Feizi M (2005) Evaluation of salt-tolerant genotypes of durum wheat derived from *in-vitro* and field experiments. Field Crops Res 91:345–354

Huang J, Redmann R (1995) Salt tolerance of Hordeum and Brassica species during germination and early seedling growth. Can J Plant Sci 75:815–819

Hwang S, Lindroth R, Montgomery M, Shields K (1995) Aspen leaf quality affects gypsy moth (*Lepidoptera*: Lymantriidae) susceptibility to *Bacillus thuringiensis*. J Econ Entomol 88:278–282

Inan G, Zhang Q, Li P, Wang Z, Cao Z, Zhang H, Zhang C, Quist T, Goodwin S, Zhu J, Shi H, Damsz B, Charbaji T, Gong Q, Ma S, Fredricksen M, Galbraith D, Jenks M, Rhodes D, Hasegawa P, Bohnert H, Joly R, Bressan R, Zhu JK (2004) Salt Cress. A halophyte and cryophyte Arabidopsis relative model system and its applicability to molecular genetic analyses of growth and development of extremophiles. Plant Physiol 135:1718–1737

Ioannidis N, Kotzabasis K (2007) Effects of polyamines on the functionality of photosynthetic membrane in vivo and *in vitro*. Biochim Biophys Acta 1767:1372–1382

Iqbal RM (2003) Leaf area and ion contents of wheat grown under NaCl and Na_2SO_4 salinity. Pak J Biol Sci 6(17):1512–1514

Ishitani M, Majumder AL, Bornhouser A, Michalowski CB, Jensen RG, Bohnert HJ (1996) Coordinate transcriptional induction of myo-myo-inositol metabolism during environmental stress. Plant J 9:537–548

Janicka-Russak M, Kabaia K, Miodzinska E, Kiobus G (2010) The role of polyamines in the regulation of the plasma membrane and the tonoplast proton pumps under salt stress. J Plant Physiol 167:261–269

Joshi GV (1976) Studies in Photosynthesis under saline conditions. Final report of PL 480 project. Shivaji University Publication, pp 102

Kakkar R, Rai V (1997) Polyamines under salt stress. In: Jaiwal P, Singh R, Gulati A (eds) Strategies for improving salt tolerance in higher plants. Editorial Science Publishers, Enfield, pp 197–203

Kefu Z, Hai F, San S, Jie S (2003) Study on the salt and drought tolerance of Suaeda salsa and Kalanchoe claigremontiana under iso-osmotic salt and water stress. Plant Sci 165:837–844

Kloepper J, Schroth M (1978) Plant growth-promoting rhizobacteria in radish. In: Proceedings of the 4th international conference on plant pathogenic bacteria, vol 2. INRA, Angers, France, pp 879–882

Kloepper J, Lifshitz R, Zablotowicz R (1989) Free-living bacteria inocula for enhancing crop productivity. Trends Biotechnol 7:39–44

Kovtun Y, Chiu W, Tena G, Sheen J (2000) Functional analysis of oxidative stress-activated mitogen-activated protein kinase cascade in plants. Proc Natl Acad Sci U S A 97:2940–2945

Koyro H-W (2006) Effect of salinity on growth, photosynthesis, water relations and solute composition of the potential cash crop halophyte *Plantago coronopus* (L.). Environ Exp Bot 56:136–146

Kramell R, Miersch O, Atzorn R, Parthier B, Wasternack C (2000) Octadecanoid-derived alteration of gene expression and the "oxylipin signature" in stressed barley leaves: implications for different signaling pathways. Plant Physiol 123:177–186

Krishnamurthy R, Bhagwat KA (1989) Polyamines as modulators of salt tolerance in rice cultivars. Plant Physiol 91:500–504

Legocka J, Kluk A (2005) Effect of salt and osmotic stress on changes in polyamine content and arginine decarboxylase activity in *Lupinus luteus* seedlings. J Plant Physiol 162:662–668

Liu J, Bei Q, Luan S (2000) Inward potassium channel in guard cells as a target for polyamine regulation of stomatal movements. Plant Physiol 124:1325–1325

Liu J, Jiang M, Zhou Y, Liu Y (2005) Production of polyamines is enhanced by endogenous absicic acid in maize seedlings subjected to salt stress. J Int Plant Biol 47:1326–1334

Liu J, Kitashiba H, Wang J, Ban Y, Moriguchi T (2007) Polyamines and their ability to provide environmental tolerance to plants. Plant Biotechnol 24:117–126

Llanes A (2010) Indicadores fisiológicos y moleculares de la tolerancia a salinidad en *Prosopis strombulifera*. Su correlación con los niveles endógenos de ABA. PhD thesis, Universidad Nacional de Río Cuarto, Córdoba, Argentina

Llanes A, Reinoso H, Luna V (2005) Germination and early growth of *Prosopis strombulifera* seedlings in different saline solutions. World J Agric Sci 1(2):120–128

Llanes A, Masciarelli O, Luna V (2008) Abscisic acid levels in salinized seedlings of *Prosopis strombulifera*. Biocell 32:173

Llanes A, Reginato M, Palacio G, Luna V (2010) Biochemical indicators of salinity tolerance in the halophyte *Prosopis strombulifera* are differentially affected by NaCl and Na2SO4. In: Öztürk M, Mermut A, Celik A (eds) Urbanisation, land use, land degradation and environment. Daya Publishing House, New Delhi, pp 344–355

Lu C, Jiang G, Wang B, Kuang T (2003) Photosystem II photochemistry and photosynthetic pigment composition in salt-adapted halophyte *Artimisia anethifolia* grown under outdoor conditions. J Plant Physiol 160:403–408

Ma Y, Rajkumar M, Fritas H (2008) Inoculation of plant growth promoting bacterium *Achromobacter xylosoxidans* strain Ax10 for the improvement of copper phytoextraction by *Brassica juncea*. J Environ Manag 90(2):831–837

Manivannan P, Abdul Jaleel C, Sankar B, Kishorekumar A, Murali P, Somasundaram R, Panneerselvam R (2008) Mineral uptake and biochemical changes in *Heliantus annus* under treatment with different sodium salts. Colloids Surf B 62:58–63

Marschner H (1995) Mineral nutrition of higher plants, 2nd edn. Academic, London

Martinez-Ballesta M, Martinez V, Carvajal M (2004) Osmotic adjustment, water relations and gas exchange in pepper plants grown under NaCl or KCl. Environ Exp Bot 52:161–174

Miersch O, Neumerkel J, Dippe M, Stenzel I, Wasternack C (2008) Hydroxylated jasmonates are commonly occurring metabolites of jasmonic acid and contribute to a partial switch-off in jasmonate signaling. New Phytol 177:114–127

Miller G, Susuki N, Yilmaz S, Mittler R (2010) Reactive oxygen species homeostasis and signaling drought and salinity stresses. Plant Cell Environ 33:453–467

Moons A, Prinsen E, Bauw G, Van Montagu M (1997) Antagonistic effects of abscisic acid and jasmonates on salt stress-inducible in rice roots. Plant Cell 9:2243–2259

Munns R (2007) Utilizing genetic resources to enhance productivity of salt-prone land. CAB Rev: Perspect Agric Vet Sci Nutr Nat Res 009:1–11

Munns R, Termaat A (1986) Whole-plant responses to salinity. Aust J Plant Physiol 13:143–160

Munns R, Tester M (2008) Mechanisms of salinity tolerance. Annu Rev Plant Biol 59:651–681

Murakeozy E, Nagy Z, Duhaze C, Bouchereau A, Tuba Z (2003) Seasonal changes in the levels of compatible osmolytes in three halophytic species of inland saline vegetation in Hungary. J Plant Physiol 160:395–401

Naeem MA, Quereshi RH (2005) Rice growth and ionic composition under saline hydroponic conditions: II. Supplemented with Cl^-: SO_4^{2-} ratios. Pak J Agric Sci 42:1–2

Nayak L, Raval M, Biswal B, Biswal U (2002) Topology and photoprotective role of carotenoids in photosystem II of chloroplast: a hypothesis. Photochem Photobiol Sci 1:629–631

Noiraud N, Delrot S, Lemoine R (2001) The sucrose transporter of celery. Identification and expression during salt stress. Plant Physiol 122:1447–1456

Owens S (2001) Salt of the earth. Genetic engineering may help to reclaim agricultural land lost due to salinization. EMBO Rep 2:877–879

Parida AK, Das AB (2005) Salt tolerance and salinity effects on plants: a review. Ecotoxicol Environ Safety 60:324–349

Parida A, George S, Venkataraman G (2007) Identification of stress-induced genes from the drought-tolerant plant *Prosopis juliflora* (Swartz) DC. through analysis of expressed sequence tags. Genome 50:470–478

Patel A, Pandey A (2007) Effect of soil salinity on growth, water status and nutrient accumulation in seedlings of *Cassia montana* (Fabaceae). J Arid Environ 70:174–182

Patten C, Glick BR (1996) Bacterial biosynthesis of indole-3-acetic acid. Can J Microbiol 42:207–220

Pedranzani H, Racagni G, Alemano S, Miersch O, Ramírez I, Peña-Cortés H, Taleisnik E, Machado-Domenech E, Abdala G (2003) Salt tolerant tomato plants show increased levels of jasmonic acid. Plant Growth Regul 41:149–158

Pedranzani H, Sierra de Grado R, Vigliocco A, Miersch O, Abdala G (2007) Pinus pinaster Ait.: abiotic stress and endogenous jasmonates. Plant Growth Regul 52:111–116

Ramani B, Beeck T, Debez A, Stelzer R, Huchzermeyer B, Schmidt A, Papenbrock J (2006) *Aster tripolium* L. and *Sesuvium portulacastrum* L.: two halophytes, two strategies to survive in saline habitats. Plant Physiol Biochem 44:395–408

Reddy RN, Lyle P, Small G (1992) Applications of spectral hole burning spectroscopies to antenna and reaction center complexes. Photosynth Res 31:167–193

Reddy MP, Rao US, Iyengar ER (1997) Carbon metabolism under salt stress. In: Jaiwal PK, Singh RP, Gulati A (eds) Strategies for improving salt tolerance in higher plants. Science Publishers Inc, Enfield, pp 159–181

Redmann RE (1974) Osmotic and specific ion effects on the germination of alfalfa. Can J Bot 52:803–808

Reginato MA (2009) Respuesta de la halófita *Prosopis strombulifera* a diferentes medios salinos. Modificaciones de los parámetros morfofisiológicos y su regulación hormonal. PhD thesis, Universidad Nacional de Río Cuarto, Córdoba, Argentina

Reginato M, Luna V, Becatti E, Castagna A, Ranieri A (2010) Polyphenols Content in Salinized Seedlings of the Halophytic Legume *Prosopis Strombulifera*. Book of abstract First Scientific Meeting COST Action FA0901Putting Halophytes to Work-From Genes to Ecosystems. Naples, Italy, pp 47

Reinoso H, Sosa L, Ramirez L, Luna V (2004) Salt-induced changes in the vegetative anatomy of *Prosopis strombulifera* (Leguminosae). Can J Bot 82:618–628

Reinoso H, Sosa L, Reginato M, Luna V (2005) Histological alterations induced by sodium sulfate in the vegetative anatomy of *Prosopis strombulifera* (Lam.) Benth. World J Agric Sci 1(2):109–119

Rhodes D, Felker P (1987) Mass screening *Prosopis* (mesquite) seedlings for growth at seawater salinity. For Ecol Manag 24:169–176

Richards, LA (Ed.) (1954) Diagnosis and improvement of saline and alkali soils. USDA Agriculture Handbook 60, Washington DC

Rogers ME, Noble CL (1992) Variation in growth and ion accumulation between two selected populations of *Trifolium repens* L. differing in salt tolerance. Plant Soil 146:131–136

Santa Cruz A, Acosta M, Rus A, Bolarin MC (1999) Short-term salt tolerance mechanisms in differentially salt tolerant tomato species. Plant Physiol Biochem 37:65–71

Schmidt A (2005) Metabolic background of H_2S release from plants. Landbauforschung Völkenrode, Special Issue 283:121–129

Serraj R, Sinclair TR (2002) Osmolyte accumulation, can it really help increase crop yield under drought conditions? Plant Cell Environ 25:333–341

Serrano R, Mulet J, Rios G, Marquez J, de Larriona I, Leube P, Mendizabal I, Pascual-Ahuir A, Proft M, Ros R, Montesinos C (1999) A glimpse of the mechanisms of ion homeostasis during salt stress. J Exp Bot 50:1023–1036

Serrato Valenti G, Ferro M, Ferraro D, Rivero F (1991) Anatomical changes in *Prosopis tamarugo* Phill. seedling growing at different levels of NaCl salinity. Ann Bot 68:47–53

Sgroy V, Cassán F, Masciarelli O, Luna V (2010) Endophytic rhizobacteria from the halophyte *Prosopis strombulifera* produce ABA, GA_3, JA and IAA under saline and non-saline culture conditions. In: 20th international conference on plant growth substances (IPGRSA) Tarragona (Spain) June 2010. C0075-Abiotic stress

Shannon MC (1994) The potential for improved salt tolerance of the cultivated soybean. In: Ansari R, Flowers T, Azmi A (eds) Current developments in salinity and drought tolerance of plants. Plant Physiol Div, Atomic Energy Agric Res Centre, Tando Jam, pp 103–113

Shelevela E, Chmara W, Bohnert H, Jensen R (1997) Increased salt and drought tolerance by D-ononitol production in transgenic *Nicotiana tabacum* L. Plant Physiol 115:1211–1219

Shelevela E, Marquez S, Chmara W, Zegeer A, Jensen R, Bohnert H (1998) Sorbitol-6-phosphate dehydrogenase expression in transgenic tobacco: high amounts of sorbitol lead to necrotic lesions. Plant Physiol 117:831–839

Shevyakova N, Rakitin V, Duong D, Sadomov N, Kutnetzov V (2001) Heat shock-induced cadaverine accumulation and translocation throughout the plant. Plant Sci 161:1125–1133

Shi D, Sheng Y (2005) Effect of various salt-alkaline mixed stress conditions on sunflower seedlings and analysis of their stress factors. Environ Exp Bot 54:8–21

Silveira J, Araújo S, Santos Lima J, Almeida Viégas R (2009) Roots and leaves display constrasting osmotic adjustment mechanism in response to NaCl-salinity in *Atriplex nummularia*. Environ Exp Bot 67:1–8

Simic M, Jovanovich S (1994) Inactivation of oxygen radicals by dietary phenolic compounds in anticarcinogenesis. Food phytochemicals for cancer prevention II. American Chemical Society, Washington, DC, pp 20–39

Simon-Sarkadi L, Koesy G, Sebestyen Z (2002) Effect of salt stress on free amino acid and polyamine content in cereals. Act Biol Szeg 46:73–75

Song J, Chen M, Feng G, Jia Y, Wang B, Zhang F (2009) Effect of salinity on growth, ion accumulation and the roles of ions in osmotic adjustment of two populations of Suaeda salsa. Plant Soil 314:133–141

Sosa L (2005) Adaptaciones fisiológicas de *Prosopis strombulifera* a condiciones de salinidad por cloruros y sulfatos. PhD thesis, Universidad Nacional de Río Cuarto, Córdoba, Argentina

Sosa L, Llanes A, Reginato M, Reinoso H, Luna V (2005) Osmotic and specific ion effects on the germination of *Prosopis strombulifera* (Lam.) Benth. Ann Bot 96(2):261–297

Sreenivasulu N, Grimm B, Wobus U, Weschke W (2000) Differential response of antioxidant compounds to salinity stress in salt-tolerant and salt sensitive seedlings of foxtail millet (*Setaria italica*). Physiol Plant 109:435–441

Subbarao G, Ito O, Berry W, Wheeler RM (2003) Sodium: a functional plant nutrient. Crit Rev Plant Sci 22:391–416

Sun J, Xu Y, Ye S, Jiang H, Chen Q, Liu F, Zhou W, Chen R, Li X, Tietz O, Wu X, Cohen JD, Palme K, Li C (2009) *Arabidopsis* ASA1 is important for jasmonate-mediated regulation of auxin biosynthesis and transport during lateral root formation. Plant Cell 21:1495–1511

Tarchoune I, Sgherri C, Izzo R, Lachaal M, Ouerghi Z, Navari-Izzo F (2010) Antioxidative responses of *Ocinum basilicum* to sodium chloride or sodium sulphate salinization. Plant Physiol Biochem 9:772–777

Tarczynsky M, Jensen R, Bohnert H (1993) Stress protection of transgenic tobacco by production of the osmolyte mannitol. Science 259:508–510

Tattini M, Remorini D, Pinelli P, Agati G, Saracini E, Traversi ML, Massai R (2006) Morpho-anatomical, physiological and biochemical adjustments in response to root zone salinity stress and high solar radiation in two Mediterranean evergreen shrubs, *Myrtus communis* and *Pistacia lentiscus*. New Phytol 170:779–794

Tien T, Gaskina M, Hubbell D (1979) Plant growth substances produced by *Azospirillum brasilensis* and their effect on the growth of pearlt millet (*Pennisetum Americanum* L.). Appl Environ Microbiol 37(5):1016–1024

Tipirdamaz R, Gagneul D, Duhazé C, Aïnouche A, Monnier C, Özkum D, Larher F (2006) Clustering of halophytes from an inland salt marsh in Turkey according to their ability to accumulate sodium and nitrogenous osmolytes. Environ Exp Bot 57:139–153

Toro FJ, Martin C, Pelacho AM (2003) Jasmonate promote cabbage (*Brassica oleracea* L. var Capitata L.) root and shoot development. Plant Soil 255:77–83

Türkan L, Demiral T (2009) Recent development in understanding salinity tolerance. Environ Exp Bot 67:2–9

Urano K, Yoshiba Y, Nanjo T, Igarashi Y, Seki M, Sekiguchi F, Yamaguchi-Shinozaki K, Shinozaki K (2003) Characterization of *Arabidopsis* genes involved in biosynthesis of polyamines in abiotic stress responses and developmental stages. Plant Cell Environ 26:1917–1926

Verbruggen N, Hermans C (2008) Proline accumulation in plants: a review. Amino Acids 35:753–759

Verlues PE, Agarwal M, Agarwal SK, Zhu J, Zhu JK (2006) Methods and concepts in quantifying resistance to drought, salt and freezing, abiotic stresses that affect plant water status. Plant J 45:523–539

Verma S, Mishra S (2005) Putrescine alleviation of growth in salt stressed *Brassica juncea* by inducing antioxidative defence system. J Plant Physiol 162:669–677

Vigliocco A, Alemano S, Miersch O, Alvarez D, Abdala G (2007) Endogenous jasmonates during sunflower germination in seeds from plants grown under different soil moisture content. Seed Sci Res 17:91–98

Villagra P, Vilela A, Girodano C, Alvarez J (2010) Ecophysiology of *Prosopis* species from the arid lands of Argentina: what do we know about adaptations to stressful environments? In: Ramawat KG (ed) Desert plants. Springer, Berlin/Heidelberg, pp 321–340

Wang H, Miyazaki S, Kawai K, Deyholos M, Galbraith D, Bohnert H (2003) Temporal progression of gene expression responses to salt shock in maize roots. Plant Mol Biol 52:873–891

Wang Y, Zhang W, Li K, Sun F, Han C, Wang Y, Li X (2008) Salt-induced plasticity of root hair development is caused by ion disequilibrium in *Arabidopsis thaliana*. J Plant Res 121:87–96

Warne P, Guy R, Rollins L, Reid D (1990) The effects of sodium sulphate and sodium chloride on growth, morphology, photosynthesis, and water use efficiency of *Chenopodium rubrum*. Can J Bot 68:999–1006

Wasternack C (2007) Jasmonates: an update on biosynthesis, signal transduction and action in plant stress response, growth and development. Ann Bot 100(4):681–697

Wasternack C, Hause B (2002) Jamonates and octadecanoids: signals in plant stress responses and development. Prog Nucleic Acid Res Mol Biol 72:165–221

Weber H, Vick B, Farmer E (1997) Dinor-oxo-phytodienoic acid: a new hexadecanoid signal in the jasmonate family. Proc Natl Acad Sci U S A 94:10473–10478

Woodward AW, Bartel B (2005) Auxin: regulation, action, and interaction. Ann Bot 95:707–735

Yang W, Rich P, Axtell J, Wood K, Bonham C, Ejeta G, Mickelbart M, Rhodes D (2003) Genotypic variation for glycine betaine in sorghum. Crop Sci 43:162–169

Yensen N, Biel K (2005) Soil remediation via salt-conduction and the hypotheses of halosynthesis and photoprotection. In: Khan M, Weber D (eds) Ecophysiology of high salinity tolerant plants. Springer, Dordrecht, pp 313–344

Zhao H, Yang H (2008) Exogenous polyamines alleviate the lipid peroxidation induced by cadmium chloride stress in Malus hupehensis Rehd. Sci Hort 116:442–447

Chapter 6
Breeding for Biotic Stress Tolerance in Plants

L.F. De Filippis

Contents

L.F. De Filippis (✉)
Centre for Environmental Sustainability (CEnS), School of Environment,
University of Technology, Sydney, P. O. Box 123, Broadway/Sydney, NSW 2007, Australia
e-mail: lou.defilippis@uts.edu.au

M. Ashraf et al. (eds.), *Crop Production for Agricultural Improvement*,
DOI 10.1007/978-94-007-4116-4_6, © Springer Science+Business Media B.V. 2012

Abstract Modern agriculture is concerned with the production of crops used primarily for human and animal food, but in so doing there is often the need (in some cases by law) to protect the environment. In crop production there is also the need to lower production costs, and especially reduce the use of expensive pesticides and fertilizers. It is often an important aim, which is not always fulfilled to apply fertilizers and pesticides only when needed, but in order for this strategy to succeed, a better understanding of biotic stress and associated influences from plant breeding achievements is required. Therefore the impact of biotic stress and injury to plants and plant yield is not only of economic importance to agriculture but is directly related to other biological and environmental questions. For example, biological and economic decision made over the control of biotic stress forms an important part of Integrated Pest Management (IPM). In this chapter, we deal with the latest results and conclusions of yield losses in plant pathology, entomology and weed science, and successful application of breeding approaches to limiting such yield reductions. We intend to cover all biological classes of biotic stressors, and plant breeding methods commonly used for the diversity of organisms involved. It will focus on current knowledge of yield and fitness loss in agricultural ecosystems, and improved approaches in order that crops can better tolerate these biotic factors. Therefore in the first part of the chapter we intend to cover agricultural crops, production and limitations, conventional and molecular breeding, and where DNA-based molecular markers have been used with advantages over traditional phenotype trait selection. Molecular markers can be used to tag biotic resistance genes, and they can serve for improvement of the efficiency of selection in plant breeding, by so called marker assisted selection (MAS). The potential benefits of MAS are discussed, especially with the use of MAS to overcome some of the problems faced by classical phenotypic screening approaches in conventional plant breeding programs. In the second part of the chapter we intend to discuss biotic stress within the context of each biological class of organisms involved in crop losses, and attempt to evaluate the knowledge available in breeding and control of biotic stress damage. Abiotic stress (dealt with elsewhere in the book) will be mentioned from time to time and we certainly make a strong argument for an integrated approach to these two types of stresses in agriculture whenever possible.

Keywords Agricultural crops • Biotic stress • Crop improvement • Crop production • Crop security • Environmental protection • Plant breeding

1 Introduction

Population increase and food security is now an issue of concern. The United Nations have projected that the world population will increase by 25% to 7.5 billion by 2020. On average, an additional 73 million people are added annually to the population of the world (an average of more than two persons every second); of which 95% will live in developing countries. At the moment, nearly 1.2 billion people live in a state of 'absolute poverty' (Oerke and Dehne 2004), of which 800 million people live under uncertain food security, and 160 million pre-school children suffer from malnutrition (Alexandratos 1999). A large number of people also suffer from deficiencies of micronutrients such as iron, zinc and vitamin A. Food insecurity and malnutrition results in serious public health problems, and lost human potential. The amount of land available for crop production is decreasing steadily due to urban growth and land degradation, and the trend is expected to be much more dramatic in developing countries than in developed countries. In 1990, only Egypt, Kenya, Bangladesh, Vietnam, and China had per capita cropland availability below 0.25 ha. However by 2025, countries such as Peru, Tanzania, Pakistan, Indonesia, and the Philippines are likely to join this group (Sharma et al. 2002). The decrease in the amount of land available for crop production and an increase in human population will have major implications for food production and security over the next two to three decades. There had been a remarkable increase in total grain production between 1950 and 1980, but only a marginal increase was realized during 1980–2000 (Myers 1999). Much of the early increase in grain production resulted from an increase in area under cultivation, irrigation, better agronomic practices, and improved cultivars. Yields of several food crops have already reached a plateau in developed countries, and therefore, most of the productivity gains in future will have to be achieved in developing countries.

Rice, maize and wheat provide 60% of the energy consumed as food in the world making them the most important food crops. The global dimension of loss in crop production due to biotic stress is substantial. Oerke et al. (1994) estimated that between 1988 and 1990 from the total attainable production of the eight major crops (wheat, corn, rice, barley, soybean, cotton, potatoes and coffee) worth US$580 billion, about 42% or US$ 240 billion was lost due to insects (total 15%), followed by pathogens (total 13%) and weeds (total 13%). To protect plants from pests and diseases, farmers had spent worldwide in 1998 a total of $34 billion on chemicals, and will spend more in coming years with an annual growth rate in pesticide consumption of 4.4% (Yudelman et al. 1998). However, growing consumer concern about the use of agrochemicals to protect cultivated plants from pests and diseases has led to seriously consider alternatives to keeping plants 'healthy'. Simply prohibiting the use of agro-chemicals would have a devastating effect on agricultural production. Schmidt (2003) estimated that without the use of agricultural chemicals in Germany alone the yield of the major crops would decrease by 35–48%, variation in yield could be high from year to year (32–47%) and farmers' profitability would decline by 32–57%. The most efficient, cost-effective and environment friendly approach to prevent the losses caused by biotic disease epidemics is the development of genetic resistance to these biotic stresses. Various biotic factors inflict a wide range of problems in cultivation. Of these, fortunately only a few are known to cause serious concerns and almost complete loss of agricultural crops, however all have been shown to decrease production. Among them, some are of historical importance, like coffee rust caused by *Hemileia vastatrix*, which caused the catastrophic closure of the coffee plantation industry in Ceylon (Sri Lanka) during the second half of nineteenth century, paving the way for the development of the tea industry. As well, the South American leaf blight (SALB) by *Microcyclus ulei* in rubber, which caused the shifting of rubber cultivation to South and South-East Asia in the late nineteenth century from native

Brazil (Simmonds 1989; Wilson 1999). Biotic stresses therefore can take a heavy toll on crop productivity in agricultural ecosystems.

Most of the biotic stresses in agriculture are seasonal in nature. Field crops, which can also be seasonal may come under the threat of a stress when the periods of the crop and the stress coincide with each other. Whereas, perennial crops are non-seasonal in nature and are subjected to all kinds of biotic stresses irrespective of the seasons. Therefore, perennial crops undergo a sort of natural selection *in situ* against any adverse factors, year after year, and only the fittest must be advanced forward and encouraged to be planted in the years to follow. However, there would be a great deal of geographic variation in the stresses, as well as in the reaction of crops to different kinds of stresses. In practice, many of the biotic factors are controlled in most crops in a seasonal manner, but this method only becomes sustainable in the long term when newly released varieties are tested in different environments and are bred quickly after a 'new' disease arrives. Selection over many environments and for a number of years therefore can be one of the most important breeding strategies utilized in breeding new crop varieties, but this is ever seldom done in a systematic manner.

Crop improvement is important from both a scientific and socio-economic perspective, and the ways in which these factors interact and impact on agricultural development is important; including debates on genetically modified (GM) foods. Horst (2008) described resistance in a host-pathogen relationship as the ability of plants to limit the penetration, development and/or reproduction of invading pathogens. Tolerance of host plants is measured in terms of the ability to maintain growth and yield production, in spite of infection or invasion of pathogens. Although both factors are genetically controlled, the environment of the host and plant can modify expression and production to a certain extent, especially in moderately susceptible or resistant genotypes/varieties. The use of cultivars with single-gene resistance permits the selection of mutations at a single locus to render the resistance effective in a relatively short period of time. However, due to loss of variation and selection pressure, and evolution, new virulent races of a pathogen appear which increases the need to develop durable resistance. Hence, the use of combinations of resistance genes has been suggested as the best method for genetic control of some of the biotic stress organisms (Miklas et al. 2006). This can be achieved by pyramiding effective resistance genes, but expression of individual and group resistance genes is difficult to monitor in the field. Agricultural crops of perennial nature have had few focused attempts in breeding against biotic stress factors. However, the approach is usually to integrate breeding efforts combining all desirable attributes, including yield, quality and resistance (Todorovska et al. 2009). There is a great advantage with these perennial crops, in the way that at any time during the breeding process, emergence of any superior individual can be fixed by vegetative multiplication or cloning in tissue culture.

Crop productivity gains are essential for long-term economic growth, but in the short-term these gains are even more important for maintaining adequate food supplies for a growing world population. It is in this context that biotechnology and plant breeding will play an important role in food production in the near future. In this review, we attempt to take a critical but practical look at the prospects and constraints of various types of plant breeding methods and biotechnologies, and their application in increasing crop production and improving nutritional quality. Within this chapter, we also briefly address the critical issues of biosafety and the impact of genetically engineered crops on the environment. Genetic engineering offers plant breeders access to an infinitely wide array of novel genes and traits, which can be inserted through a short-term single event into high-yielding and locally adapted cultivars. This approach offers rapid introgression of novel genes and traits into elite agronomic backgrounds (Beaver and Osorno 2009). New crop cultivars with resistance to insect pests and diseases, combined with bio-control agents should lead to a reduced reliance on pesticides, and thereby reduce farmers' crop protection costs, while benefiting both the environment and public health. Similarly, genetic modification for herbicide resistance is set to achieve efficient and cost effective weed control to increase farm incomes, while reducing labor demand for weeding and herbicide application. Labor released from agriculture can then be used for other profitable endeavours. In addition, there is an urgent need for less labor-intensive agricultural practices in countries significantly affected by human immune deficiency virus (HIV) and other chronic diseases (e.g. malaria). By increasing crop productivity, agricultural biotechnology can substitute for the need to cultivate new land and thereby conserve biodiversity in areas that are marginal for crop production (Pinstrup-Andersen and Cohen 2000). The potential of these technologies has been extensively tested in the model crop species of temperate and subtropical agriculture. However, there is an urgent need for an increased focus on crops relevant to small farm holders and poorer consumers in the developing countries of the humid and semi-arid tropics. The promise of biotechnology can only be realized by utilising the information and products generated through research on genomics and transgenics to increase the productivity of crops, through enhanced resistance to biotic and abiotic stress factors and improved nutritional quality.

Diseases and pests can cause significant losses to most plant production systems (Beaver and Osorno 2009). Control of these biotic constraints using agrochemicals can increase production costs and create the potential for contamination of both the food and the environment. Resistance to pesticides also represents a valuable disease and pest management tool. Therefore the development of crop cultivars with greater levels of disease and pest resistances is a primary objective of most breeding programs. Examples of plant breeding achievements in individual plants are many and beyond the scope of this chapter, therefore I intend to concentrate on the methods best suited for crop breeding. Finally, conventional breeding methods do have limitations, these will be outlined and this will give way to more discussion and rationale for the use of molecular breeding. Molecular markers can be used in improving the efficiency of selection in plant breeding by marker assisted selection (MAS). The potential benefits of MAS have been widely discussed, especially for solutions to overcome some of the problems faced by classical phenotypic screening approaches in plant breeding programs (Backes and Ostergard 2008).

2 Part A: Crop Breeding Techniques and Production

2.1 Agricultural Crops, Production and Constraints

2.1.1 Introduction

Human population is projected to grow at about 80 million per year to 7.7 billion by 2020, then levelling off at about 10 billion (Tilman 1999; Pinstrup-Andersen and Cohen 2000). This increase population density together with changes in dietary habits and increase use of grains for livestock feed is projected to cause the demand for grain production alone to more than double. However land suitable for agricultural production is limited and most of the soils with high potential production are already under cultivation. Water availability in many regions is restricted and suitable cultivated areas are actually shrinking (Nelson-Smith 1995). Sustainable production for high yield are urgently required, and the availability of fertile soils and the development of high yielding varieties are major challenges to agriculture. Safeguarding crop production from damage from pathogens, pests and weeds are also a major requirement to meet these challenges.

Rice, maize and wheat already use 40% of cropland, and as yields of these three crops, plus some other crops like soybean, cotton and sugar beet respond to high production levels and mechanization more stress will be placed on arable soils. In the last decade world-wide crop production increases have focused on an ever limiting number of plant species and cultivars. Diverse ecosystems have been replaced in many regions by simple agrosystems which are more vulnerable to disease and pests. In order to safeguard productivity of these crops and reach levels of production to meet demand, these crops have to be protected more and more from diseases, pests and weeds. Plant production loss data, including the importance of pathogens and pests, the pest organisms involved and the control and use of pesticides are a prerequisite for economic management and for evaluating efficacy of present practices. Estimates of actual losses in crops worldwide have been difficult to obtain, and since crop production technology and especially crop protection are changing constantly, loss data for all major food and cash crops can be important. However, in the last two decades only revised figures and estimates have been available, and it is with this in mind that most of the loss potential data used below are exactly that (i.e. are only estimates and projections) (Oerke et al. 1994; Oerke and Dehne 2004).

2.1.2 Crop Loss and Trade

(a) **Wheat** – grown in most countries around the world and is the most important cereal crop in the Northern Hemisphere, Australia and New Zealand. The major wheat producing countries are China, India, USA, France and Russia. International wheat trading nations include USA, Canada, France, Argentina and Australia. In wheat, weeds are the most important pests worldwide, and pathogenic rust fungi, smuts and bunts, and some soil-borne pathogens are next. Total losses can vary considerably around regions of the world from 15% to over 35%. Nematodes, rodents, birds, snails and arthropods can cause significant loss in some regions, whereas world-wide losses due to viruses are of minor importance.

(b) **Rice** – production is largely concentrated in Asia, where it is considered a major source of food. Rice is mostly grown for subsistence or local markets, and only about 4% of global production reaches international markets. Bacterial pathogens are very important in rice but there are a diverse set of regional conditions which influences this, and weeds and insects can be more and more important. Weed control, whether by mechanical or chemical means can be very effective in all regions, whereas control of pests and diseases is heavily reliant on synthetic pesticides; and these can have variable effects, from 16% loss to 35% loss.

(c) **Corn** – staple food for human consumption in East Asia, Latin America and parts of Africa, but its production is highest in the Americas and international trade in corn can reach about 12% of production. Weed competition is the most important problem which can depress production the most, followed by animal pests, fungal and bacterial pathogens, with viruses being least significant. In Central Africa and South-East Asia crop protection is largely restricted to weed control, and despite this over 30% of attainable production can be lost.

(d) **Barley** – mostly grown in the Northern Hemisphere, especially in dry areas and areas short of cropping periods. Barley is hardly grown for human consumption anymore but rather for animal feed and malting. The international trade amounts to about 12% of production. Weeds are the cause of most production loss, followed by fungal pathogens, animal pests and viruses. Prevention of barley loss due to pest control measures can be as high as 20% of total production. This however can differ widely amongst regions (from 5% to 15% loss), and under intense cultivation this is mainly due to good weed control.

(e) **Potato** – the production of potatoes has gone through a period of expansion in recent times and is ranked number five in the food crops. Potato can produce more starch than any other crop and is only second in protein production to soybean. Vegetative propagation dominates potato planting therefore all pests and diseases are of importance to control. Major pathogens are bacteria and viruses, and nematodes. Between soil practices, mechanical and chemical control methods, prevention of production losses of near 30% can be achieved, yet despite this, actual losses can approach 40% world-wide.

(f) **Soybean** – an annual member of the legumes and satisfies about half the global needs for vegetable oils and protein. Most important producers are USA, Brazil, China and Argentina. International trade value can be over 25% of world production. Weeds are the most important production dangers followed by fungi, bacteria, viruses and pests. Control measures can protect about 30% of production and attained production can be high, but this is heavily dependent on regions.

Table 6.1 World regional differences in the overall efficacy of actual crop protection practices in eight major crops (wheat, corn, rice, barley, potato, soybean, sugarbeet and cotton)

World area	Percent reduction in total crop loss potential
North Africa	49
Central Africa	32
South Africa	36
North America	59
Central America	46
South America	47
Near East	47
South Asia	41
Southeast Asia	46
East Asia	56
Northwest Europe	70
South Europe	60
Northeast Europe	48
Southeast Europe	46
Oceania	45

Values are estimates for the period 1996–1998 as detailed by Oerke and Dehne (2004)

(g) **Sugar beet** – the most important sugar producing crop after sugar cane, and is mostly a Northern Hemisphere planted crop. It only accounts for about 35% of sugar production world-wide where the most important producers are France, USA and Germany. Seedling production can be slow and the plant can suffer losses from weed competition, but this can vary from 5% to 50% production loss. Pathogens are minor causes of losses in sugar beet crops and actual loss is usually restricted to below 10%.

(h) **Cotton** – not a food crop but one of the most important fibre crops globally. It is grown in almost all tropical and sub-tropical countries. Most important producers are China, USA, India and Pakistan, where in many developing countries it is an important cash crop. Cotton is especially vulnerable to insect attack and then weeds where losses can approach 35–40%. In most countries protection of cotton crops has increased dramatically and can approach on average 50% of production, but this can vary due mainly to the intensity of the growing system and money spent on protection chemicals.

2.1.3 Efficacy of Crop Protection

World-wide, the efficacy of success in crop protection was lowest in wheat, potato and barley (about 45% prevention), moderate in rice, maize and soybean (about 55% prevention) and highest in cotton and sugar beet (about 65% prevention) (Table 6.1). Regional variation in these figures was lowest in sugar beet and cotton,

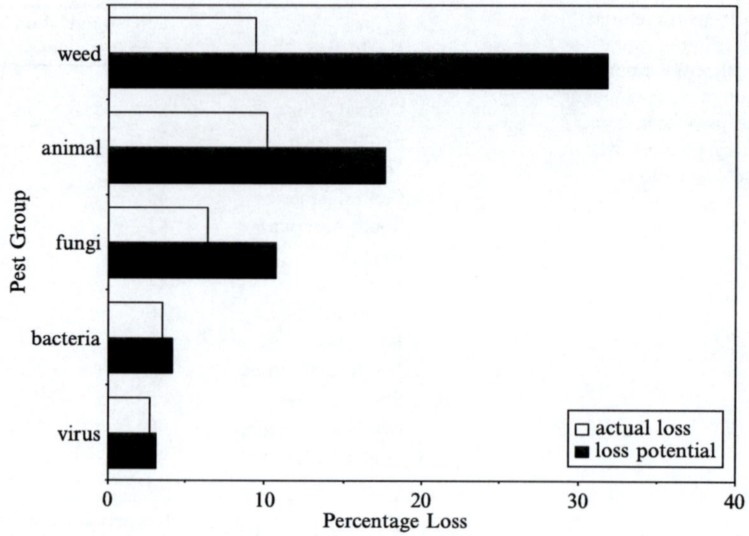

Fig. 6.1 World data on the percentage loss potential estimated, and the actual loss percentage of major agricultural crops according to the five major groups of pests and diseases. Values are estimates for the period 1996–1998 as detailed by Oerke and Dehne (2004)

and highest in potato and wheat. The percentage of loss prevention according to areas around the world ranged from 32% in Central Africa and Eastern Europe to 70% in North West Europe. In East Asia, North America and South Europe it was between 41% and 60% (Table 6.1). In terms of pest control measures by pest group and practice, mechanical or chemical weed control achieved an overall efficacy of almost 70%. Control of animal pests, bacteria and fungi were considerably lower and averaging about 20–45%. Virus control reached an efficacy on average of only about 15% (Fig. 6.1). In order to achieve even these efficacy figures considerable expenditure on herbicides, fungicides and insecticides was necessary, and there was considerable regional differences in such outlays and expenditure on agrochemicals; this is detailed in Fig. 6.2.

2.1.4 Safeguarding Crop Production

Much of the observed yield increase in recent decades can be attributed to better control over biotic stress rather than an increase in yield potential. But progress in improving food security has been uneven, and many developing countries have failed to see such progress. The persistence of food insecurity reflects regional differences rather than to a lack of overall capacity. It is worth noting that at present the world already produces sufficient food, but many people are undernourished because they are poor in terms of agricultural resources, education, technology etc;

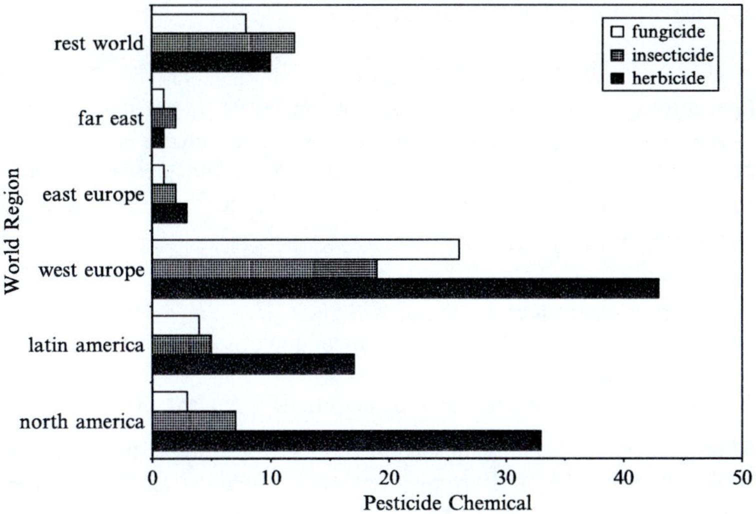

Fig. 6.2 World regional data on the percentage of pesticide used on major agricultural crops according to the three major groups; fungicides, insecticides and herbicides. Values are estimates for the period 1996–1998 as detailed by Oerke and Dehne (2004)

and thus are not able to produce their own food (Alexandratos 1999). The increasing trend of cereal yield increases alone should be sufficient to cope with most of the demographically projected increase in population until about 2025. World trade is expected to increase and expand considerably (Dyson 1999) but after 2025 the growth rate in food crop production of about 1.1% per annum achieved in the past 35 years is below the 2.0% per annum required.

Increase in cultivated land will contribute less than ever before, and in present high crop production systems any rise may be impossible to achieve; e.g. in rice where 70% attainable yield is being achieved in Japan, Korea and parts of China. Further yield increases must now depend on the availability and increase use of genetic high yield potential crop varieties and hybrids (Cassman 1999). However, losses due to weeds, animal pests, pathogens and viruses continue to reduce yields worldwide, and although advances have been made the picture is variable. Development of new compounds that are highly effective against former less controllable pests and the use of GM crops must be considered. Training of farmers by government and non-government organizations have contributed to a better outcome in the past 10 years, especially in Asia and Latin America; the situation is still not favourable in parts of the former Soviet Union and especially in sub-Sahara Africa for example. The problem of food allocation can in part be alleviated to some extent by intensification of crop production where the demand is highest, but this will not always be possible.

2.1.5 Integrated Pest Management (IPM)

Integrated pest management includes various techniques suitable for maintaining pest infestations below economically acceptable levels, rather than attempting to eradicate all pests. Therefore IPM is a management approach that encourages natural control of pest populations by anticipating problems and preventing them from reaching critically high levels. All appropriate techniques can be used in such an approach, such as:

(a) Enhancing numbers of natural enemies
(b) Planting pest resistant crops or cultivars
(c) Adopting appropriate cultural management and practices
(d) Using chemical pesticides judiciously
(e) Using organic 'natural' chemicals as deterrent

IPM programs have been established in various crops around the world and have proven especially suitable in developing countries (Cuyno et al. 2001). Where food supply often suffers from poor crop production technology and crop losses are high due to inadequate pest control, the intensification of food production can only be realized in part by implementation of IPM into current cropping systems. IPM has been used successfully in perennial and annual crops in both temperate and tropical conditions.

2.2 *Conventional Phenotype Breeding*

2.2.1 Introduction

The crop plants we see today are the result of spontaneous mutations followed by millions of years of evolution through natural selection; and more recently human intervention by selection which has artificially accelerated the pace of evolution. Plant breeders have developed desirable traits in plants resulting in the creation of numerous modern crop varieties. Modification of crop plants by man to improve cultivation has persisted for at least 10,000 years. Early farmers produced better crops by simply saving the seeds of their most desirable plants, unknowingly modifying the crops genetic make-up. Since the re-discovery of Mendel's laws in the early 1900s plant breeders have become more rigorous in this approach. Hybrid varieties and selection for increased yield and increased resistance to biotic and abiotic stresses have increased agricultural production consistently (Akhond and Machray 2009).

2.2.2 Backcross Breeding

The method is useful in preserving important cultural and seed traits of plants, yet incorporating simple inherited traits into new lines. However the method is less suitable for quantitatively inherited traits (e.g. tolerance to most abiotic stresses).

Backcrossing is inefficient in removing portion of the genome closely linked to the genes to be targeted in backcrossing. Linkage drag can impede effects to using backcrossing to introgress desirable traits into the recurrent parent. The Inbred Backcross Line (IBL) method developed by Bliss (1993) can produce near heterozygous lines that can be used for replicate testing of traits. One or two backcrosses are made after the initial set of crosses. The backcrosses are followed by a few generations of single seed descent (SSD – see Sect. 2.2.6) to produce an inbred backcross line. This method is well suited to marker assisted selection (MAS) and has been used to identify quantitative trait loci (QTL) conditioning resistance to *Fusarium* root rot (*Fusarium solani*)(Roman-Haviles and Kelly 2005). However, it has many limitations when used in most types of plant breeding, and other methods below appear better.

2.2.3 Gamete Selection

The method has been proposed to simultaneously select plants for multiple traits (Singh 1994). In breeding for multiple traits, gamete selection permits the early generation evaluation of potential resistance in populations. Populations that do not segregate for desired traits in early generation testing can be discarded, this avoids the loss of valuable time and resources. The method can be labour intensive and therefore permits evaluation of few populations, and care is required in selecting parents that possess the degree of resistance required (or desired trait). Gamete selection has been successfully used to develop high-yielding bean lines with resistance to leafhoppers and some diseases (Singh et al. 1998). Asenio-Manzanera et al. (2006) also used gamete selection to select for common bacterial blight and halo-blight resistance. Molecular markers can facilitate gamete selection and the method may be most efficient in pyramiding simple inherited traits or QTL traits to have large effects. The evaluation of large populations can be expensive and time consuming, unless an automated cheap method of screening is available.

2.2.4 Recurrent Selection Breeding

The method permits the accumulation of favourable alleles as a result of recombination in each cycle of selection. Because F_2 plants could be evaluated for plant resistance and seed trait, each cycle could be completed in a shorter period of time than most other methods. Nevertheless at least three cycles at a minimum are required to break up undesirable linkage (Singh 1999). Garcia et al. (2003) used recurrent mass selection to select bean populations with greater resistance to soil-born diseases caused by *Pythium, Rhizoctonia* and *Fusarium*. Breeding lines were developed to produce greater seed yield and higher survival rates than the parents. However, lines developed from inter-gene pool crosses usually have poor performance due to the breakup of some favourable gene complexes within each pool. So, recurrent selection provides additional opportunities for the plant breeder for recombination and formation of new gene complexes that could lead to disease resistance.

2.2.5 Pedigree Breeding

This is a very common method used to develop improved cultivars to diseases. But an important limitation to pedigree selection is the amount of time needed (Fehr 1987). In warm tropical areas, breeders can accelerate cultivar development by planting most of the year round, even in the dry season by using irrigation. Also breeding can be speeded-up by growing additional generations in nurseries or greenhouses. Whilst growing one generation, breeders can screen other breeding lines for disease resistance. Resistance to diseases such as anthracnose and bean common mosaic necrosis virus (BCMNV) have been successful, and in recent years pedigree breeding has also used marker-assisted selection to identify specific genes for disease resistance (Miklas et al. 2001). Exclusion of susceptible lines is necessary and Beaver and Macchiavelli (1998) have noted that screening breeding lines in F_4 or later generations can improve the probability of identifying lines with desirable genotypes; and would reduce the number of lines that need to be evaluated in early generations.

2.2.6 Single Seed Descent (SSD)

The method is recommended when working with crosses between elite lines within a set of varieties, and can provide a way of maintaining genetic variability. SSD can be conducted in the field and target environments, or in nurseries and greenhouses. Macchiavelli and Beaver (2001) noted that breeders often bulk seeds rather than use one single seed. Although bulking reduces genetic variability, on average every third F_6 line would be derived from a different F_2 plant. Concentrating on improvement of a single disease trait favours the use of SSD. SSD is a rapid method to develop recombinant lines for diseases that cannot be phenotyped in early generations. SSD can be costly, therefore it is primarily used in molecular and mapping techniques that have the aim of identifying parents that have increased disease resistance and are useful for later crossing. SSD is quite difficult to manage on a large scale and is most useful at breeding for specific purposes (Fehr 1987).

2.2.7 Bulk Breeding

An old and well established method where multiple generations can be grown each year. It is most appropriate for crosses between elite lines where little segregation for seed type or adaptation would be expected. If the bulked population can be grown in a target environment, some natural selection may occur for traits such as disease resistance, however it is more suited to increasing yields of plants and seeds. Plant breeders would need to advance large samples of bulked populations to avoid genetic drift. Molecular plant breeding techniques might be used to monitor genetic variability and determine an adequate sample size for breeding. Plant breeders must also avoid planting bulked populations in situations where undesirable traits may be favoured, and bulk breeding has been used extensively to obtain disease resistance in lentil to a number of important diseases (Renato-Corte et al. 2002).

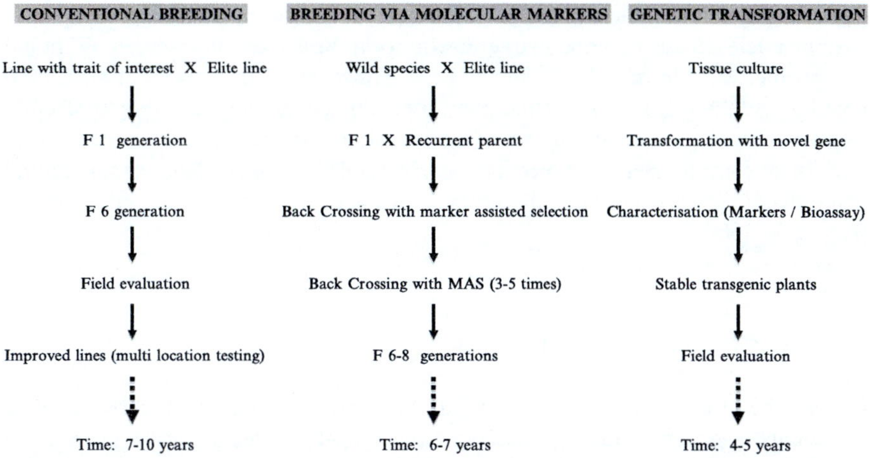

Fig. 6.3 A schematic diagram outlining approaches in crop improvement, comparing the steps required for the most commonly used methods; conventional breeding, breeding via molecular markers and using transgenic technology. Times taken for each method is only an approximation

2.3 Molecular Breeding

2.3.1 Introduction

In a number of situations conventional breeding may be assisted by molecular methodology in order to quicken selections, but also molecular approaches attempt to be more precise and bear lower costs. Figure 6.3 summarises essential differences between conventional breeding and molecular marker breeding with approximate times needed for an improved cultivar to be realised. A number of molecular methods have been employed to improve breeding objectives; these include:

(a) Key enzyme and/or metabolite measurements as a result of expression or even over-expression of genes
(b) Molecular signal changes between the host plant and the pathogen, but these have seldom been used
(c) Quantitative trait loci (QTL) and genomic maps (see Sect. 2.3.4 below)
(d) Molecular markers in MAS (see Sect. 2.3.5)
(e) Association mapping (Muehlbauer et al. 2006)

Using such molecular methods, breeders can screen for genes for disease resistance without running the risk of introducing the disease agent. The risk is that normally disease resistance genes are in clusters, so the breeder must be aware of the risk of introducing susceptibility to a different disease if these genes are linked for example (Michelmore 1995). Other problems include the specificity of some markers to only one plant genome, however as more and more genes are cloned, documented and sequenced it may be possible to design more specific markers to specific alleles. Ideally a good marker useful for routine molecular screening should be specific, reliable, rapid and inexpensive. However as Broughton et al. (2003) noted that at the moment molecular plant breeding techniques (including molecular markers) are only an additional set of tools available for breeders. A suitable combination of conventional and molecular tools is still the best application for agricultural plant breeding, but needs should always be evaluated carefully as should resources (Bernardo 2008).

2.3.2 Key Enzymes and Metabolites

Knowledge of changes in a specific plant metabolic function induced by different treatments or observed during breeding has led to the development of methods to identify genes involved in regulating metabolic pathways and associated physiology and biochemistry. With the availability of tight mutant lines, the use of elegant biochemical screening systems based on this knowledge can be considered a relatively easy approach compared to gene isolation and selection. Many secondary plant metabolites such as flavonols have been implicated in plant pathology, including resistance to biotic stress factors (Sharma et al. 2002). The biochemical pathway to the synthesis of flavonols is well understood, however the use of such biochemical markers has been relatively unsuccessful. Fatty acid analysis and isoenzymes have also had limited success in plant breeding towards biotic stress resistance (Chawla 2009; Davey and Anthony 2010).

2.3.3 Molecular Signals

Responsive Induced Expressed Genes (micro RNA-miRNA and small interfering RNA-siRNA) code for at best small polypeptides that are induced due to biotic stresses. miRNA and siRNA are a class of small size non-gene-coding RNA that can act as regulators of gene expression in both animals and plants. The induction of miRNA may play a role in adaptation and resistance to biotic stress, although abiotic stress related miRNA and their targets are better known. At least a number of miRNA are known to be induced under the action of fungi and bacteria (miRNA 393 and miRNA 159), and more are induced due to viral infection (miRNA 156, miRNA 160, miRNA 164, miRNA 166, miRNA 169 and miRNA 171) (Lin et al. 2009). The mechanism of pest-transcription regulation is not well known or understood. In this regard mi RNA 399 induced during low phosphorus nutrition and miRNA 395 induced during sulphur homeostasis are better understood (Bartel 2004). miRNA and siRNA have been estimated to constitute about 1% of plant genomes, which means that more unknown miRNA and siRNA functions await discovery, and these small RNA structures appear to be another useful tool to enhancing crop resistance or tolerance to pests and diseases.

2.3.4 QTL and Genomic Mapping

When a trait is governed by multiple and/or co-dominant inherited genes, quantitative trait loci (QTL) maps are very informative, which is a more holistic approach to genome interaction. Subsequently molecular markers identified this way are closely linked to the trait under investigation. In this way molecular marker assisted selection (MAS) has possible merit by identifying the important regions of the genome associated with the trait (Torres et al. 2006). Simple inherited traits can be located on genetic maps. It is widely recognized that the development of saturated genetic linkage maps provide a most valuable tool in plant genetics and breeding. Using genetic maps, it is possible to estimate the number of loci controlling genetic variation in a segregating population and to estimate the effects of the loci. Identification of favourable alleles at each locus is then possible for many traits, including disease resistance.

2.3.5 Marker Assisted Selection (MAS)

Some molecular markers available allow the indirect selection of interesting genotypes, and they constitute an essential tool for the development of MAS. Molecular markers fall into a number of categories, each having positive and negative features (also summarized in Table 6.2):

(a) **Restriction Fragment Length Polymorphism (RFLP)** – requires hydrolysis of probe DNA from samples. Can provide high quality data but has severe restrictions on throughput because large amounts of DNA are required and because it is not PCR-based.
(b) **Random Amplified Polymorphic DNA (RAPD)** – is a method based on PCR but uses arbitrary short primers for identity of plant DNA regions. No knowledge of the genome is needed, but by the same token markers can be at many places in the genome. Results can be inconsistent and only dominant genes can be identified.
(c) **Simple Sequence Repeats (SSR)** – are high quality and consistent DNA markers, but they are the most expensive to develop. SSR markers require extensive band sequencing data for each marker developed, and often the markers are species and even cultivar specific.
(d) **Amplified Fragment Length Polymorphism (AFLP)** – requires enzymatic degradation of DNA and careful fragment separation, where only a sub-fraction of the population is sampled by PCR. It can provide too much information at any time. It is more technically demanding and information can be difficult to interpret. It produces very good high quality data which is suitable for high output sources and suitable for automation.
(e) **Single Nucleotide Polymorphism (SNP)** – relies on the fact that the vast majority of differences in eukaryotic organisms are surprising point mutations in their DNA. So there are a vast number of polymorphisms that are SNP.

Table 6.2 The major classes of molecular DNA markers used for indirect selection in marker assisted selection (MAS) in plant breeding

Criteria	RFLP	RAPD	SSR	AFLP	SNP
Principle	Endonuclease restriction and Southern blot	DNA amplification with random primers (in PCR)	PCR of simple sequence DNA repeats	PCR of endonuclease restriction sites	Sequencing part of genome looking for base changes
Polymorphism type	Single base changes (insertion/deletion)	Single base changes (insertion/deletion)	Changes in length of simple repeats	Single base changes (insertion/deletion)	Single base changes (insertion/deletion)
Genomic abundance	High	High/very high	Medium high	High/very high	High
Polymorphism level	Medium	Medium	High	High	Very high
Dominance	Codominant	Dominant	Codominant	Codominant	Need testing
Amount of DNA required	2–10 μg	10–25 ng	50–100 ng	20–50 ng	ng amounts
Sequence information required	No	No	Yes most times	Yes it can help	Yes
Radioactive detection required	Yes/no	No	No	No	No
Development and start-up costs	Medium	Low	High	Medium	Very high
Continual costs per sample	Medium/high	Low	Medium	Medium	High

The list is data compiled from personal experience by the author in areas of criteria, development and costs of the respective methods

The biggest advantage is automation and techniques that do not require electrophoresis to separate fragments.

(f) **Expressed Sequence Tags (EST)** – requires cDNA synthesis to develop and apply, therefore it is the only method listed above which is based on RNA. Preferences for this method may be for plant species where there is already extensive sequencing data.

The use of DNA markers (and indirectly EST markers from RNA) for direct selection offers greater potential gains in breeding for QTL and traits with low heritability, and these can be the most difficult to work with in breeding. However these traits are also amongst the most interesting and the most difficult to develop. The expression of the trait can be greatly affected by 'genotype X environment' interactions and 'epistasis' which can complicate the development of MAS to almost the same extent as compared to traditional field based phenotypic selection (Backes and Ostergard 2008).

2.3.6 Pyramid Multiple Resistance

Pyramiding of multiple resistance genes to foliar fungal pathogens in lentils has been developed (Muehlbauer et al. 2006) and should provide a broader and more durable resistance, as also demonstrated in rice against bacterial blight (Singh 2001). Durable resistance to stem rust in wheat has also been demonstrated and pyramiding resistance genes for several diseases has been reported. Pyramiding multiple resistance in breeding programs now constitutes a major tool for the advancement of MAS breeding (Todorovska et al. 2009).

2.3.7 Association Mapping (AM)

Association or linkage disequilibrium (LD) mapping has been used successfully in lentil, but has been used more successfully for studies of traits in humans for some time now (Flint-Garcia et al. 2003; Muehlbauer et al. 2006). Entire genomes can be examined for markers for quantitative and qualitative traits, and may allow breeders to 'break-out' of the restrictions of F_1-derived mapping populations, and employ other populations available. Gebhardt et al. (2004) summarized the four potential benefits of AM:

(a) Allows assessment of genetic potential of specific genotypes before phenotypic evaluation.
(b) Allows identification of superior trait alleles in germplasm.
(c) Can assist in high resolution QTL mapping.
(d) Can be used to validate candidate genes responsible for individual traits.

However some important issues must be considered before implementation of AM:

(a) Determination of the population structure.
(b) Estimation of nucleotide diversity.

(c) Estimation of haplotype frequencies and LD (ie non-random association of alleles at different loci).
(d) Precise evaluation of phenotypes.

Often nucleotide diversity and LD are not possible on the species investigated, but may be inferred from very closely related species, especially if the two related species have been sequenced. AM also constitutes a major tool for the advancement of plant breeding, but careful attention is required in many crop plants where population structure and precision on phenotypic expression can be poorly documented.

2.4 *Tissue and Cell Culture*

2.4.1 Introduction

The term "tissue culture" is often used to describe all types of aseptic plant culture procedures leading to growth, and falls into various types depending on the growth of plant cells, organs and tissues as end points of the culturing process (Chawla 2009; Neuman and Kumar 2009). Tissue culture can involve any of:

(a) Plant protoplasts (cells without the cell walls).
(b) Callus (loose or compact cell arrangement).
(c) Tissues (such as apical or root meristems, or pith segments).
(d) Organs (such as buds or roots).
(e) Embryos (usually for more specialized use).
(f) Plantlets (usually having both roots and stem/leaves).

Because the growth takes place in a sterile, artificial and often controlled environment we tend to call these techniques *in vitro* techniques. It is fair to point out that the eventual outcome of tissue culture must have in mind always to try to acclimatize these plantlets to greenhouse and eventually open field conditions, otherwise the time, efforts and resources put into tissue culture may not be realized. Details of the various and specific techniques involved in the whole area of tissue culture is beyond the scope of this chapter and will not be dealt with here, but I will refer you to various recent books that address this (Razdan 2003; de Fossard 2007).

Plant tissue culture is not a science in itself. It is merely a range of enabling techniques in which the knowledge of many different scientific disciplines are put to use when a particular type of conventional method cannot be used, or a type of end-user plant product is required; such as:

(a) Stock plant material must be made disease free.
(b) Breeding and/or propagation must be speeded up.
(c) Storage of germplasm and whole genotypes must be preserved.

(d) Plant genotypes must be manipulated, such as it is difficult or impossible *in vivo*.
(e) Increase production of desired plant metabolites.

2.4.2 Use of Tissue Culture

The use of plant tissue culture has had a rapid period of expansion, but recently there are signs of stabilization. In the Netherlands alone in 1988 there were 67 companies producing 62 million plants a year. The number of companies today has decreased from this point due to consolidation and increase in size of operations, but the number of plants produced has in fact increased (de Fossard 2000). Similar consolidation of tissue culture companies is evident in many other countries of the world, and this technology is a major export from developed countries to developing countries in agricultural and horticultural production. It is the emerging developing countries in the area of agriculture and horticulture that are quickly developing such plant tissue culture facilities and technologies. The most important developed countries in tissue culture are the US, Japan, Israel, Italy, Holland, France, Spain, Belgium, UK, India and South Africa. The main developing countries for tissue culture include China, Columbia, Brazil, Kenya and Vietnam (De Filippis 1999). The expansion and use of tissue culture has become very important for some food crops because high quality is very important as one of the most costly aspects of crop production, especially in the planting material used; and especially so if the product is to be exported to international markets. A good example of this is potato where the traditional mode of propagation is by planting tubers or sliced tubers, and this is prone to diseases such as *Erwinia, Rhizoctonia*, leaf roll virus and spindle tuber viroids. The process of accumulation of any of these organisms causes stock degeneration and eventually poor availability of disease-free stocks for further planting. The application of tissue culture for pathogen elimination and superior planting stock and rapid multiplication in potato has lead to its widespread use in both developed and developing countries with excellent results. The use of tissue culture in transformation of plants is now considered essential no matter what transformation system is used, and genetically modified (GM) crops will be dealt with later.

2.4.3 Tissue Culture in Plant Breeding

In vitro techniques can be used very successfully to aid conventional and molecular breeding programs because they have a number of advantages over traditional use of greenhouses and field methods. These include disease elimination and selection of elite breeding lines, and the rapid multiplication of these. Use of 'somaclonal variation' and induced mutation for new genotypes, and screening of breeding stocks is quicker and exact, especially with MAS. There are some disadvantages to consider in that in culture, diseases if present could quickly kill all material, it can be expensive, requires more skills, and material must be conditioned to grow in the field because the plantlets out of tissue culture tend to be metabolically 'weak'; and difficult to grow or establish in unprotected environments (Herman 2009; Davey and Anthony 2010).

2.4.4 Conservation of Germplasm

Conservation of tissue culture material is often easier to protect, but this is very much more of importance for vegetatively propagated plants; ie plants not normally propagated from seeds. The material is easily available in small tubes and can be multiplied quickly. This conservation can be in two forms:

(a) **Cryopreservation** – that is in the frozen state where the techniques developed for some material have been difficult, and water content of the tissue can be a problem, and resolution of this can be costly.
(b) **Lower Temperatures** – this means temperatures above freezing and at a temperature where growth is drastically reduced, and can also be costly on material and space for new culturing conditions.

2.4.5 Embryo Rescue

A method of working and breeding via zygotic embryos in tissue culture where they are usually aborted or become not viable in the plant itself. This is usually due to poor endosperm development or failure, where the media components in tissue culture replace the nutrients normally in the endosperm. Embryo rescue can increase the chance of incorporating new genes into breeding programs and the development of elite lines which otherwise would not be possible. This method has already been used successfully in a number of crops, including potato, adding new and valuable genetic material (Malik 2008).

2.4.6 Microspore, Anther and Ovule Culture

These types of culture material are sometimes called 'haploid cultures' because the number of chromosomes in them has been halved, and in some cases haploid plants can be obtained for breeding purposes. The number of chromosomes can be doubled later by chemical means. This technique is increasing in importance but has limitations in that the method is difficult to master and not successful in many plants, where much depends on the ploidy level of plants; for instance tetraploid plants are very difficult to handle. The advantages of using homozygous diploids after haploid culture is that parents derived this way have lower recombination events and there is a better chance of recovering viable plants afterwards (Vinterhalter et al. 2008b).

2.4.7 Somatic Embryogenesis and Organogenesis

Somatic embryos are independent bipolar structures which are not physically attached to the tissue in culture but are produced from a somatic line of cells. The embryos can grow to complete plantlets. Somatic embryogenesis can even occur spontaneously, and as long as the genetic make-up is stable they can be multiplied very rapidly, and therefore they have great potential. However it is still a method most suitable for breeding rather than rapid multiplication. Organogenesis can be defined as the transformation of a single cell, callus or tissue into an organ-like structure capable of independent growth, but are organ-like and not embryo-like. Plants can eventually be produced from somatic embryos and organs by a sort of 'germination' of them in their own defined culture medium (Murphy 2007; Basha and Sujath 2009).

2.4.8 Somaclonal Variation and Induced Mutations

When plants are regenerated in culture, especially from callus it often happens that new genotypes arise. This phenomenon is called 'somaclonal variation' and there appears to be many mechanisms leading to this, but it occurs in many plants in tissue culture. These genotypic changes can be very useful but may be unstable in culture and must be stabilized otherwise they become useless. In culture it is also easy to induce mutations via conventional means of radiation or mutagenic chemicals. The use of variations in culture and induced mutations have become popular and has produced new stable and useful genotypes for many purposes, including crop breeding (Navratilova et al. 2008; Hoang et al. 2009).

2.4.9 Protoplast and Cell Culture

Protoplasts are true somatic cells which although quite difficult to work with can be grown into a whole plant. Isolation of protoplasts is quite simple now by enzymatic digestion of the cell wall, however culturing protoplasts into a callus, tissue and whole plants is technically demanding, very difficult and perhaps almost impossible for some plants; especially the monocotyledon group. Hybridisation of protoplasts and even cells is relatively simple by fusion, which can be affected by either chemicals (eg PEG) (De Filippis et al. 1996, 2000) or electric fields (electrofusion) (Hampp et al. 1997; De Filippis et al. 2000). Hybridisation of cells and protoplasts has led to new genotypes in which desirable traits have been identified. Hybrids and cybrids (where the cytoplasms only are fused) can be produced and these can be used in many situations, but perhaps the best use of this material is to develop and test genetic markers in MAS.

2.5 *Marker Assisted Selection (MAS)*

2.5.1 Introduction

A molecular genetic marker (sometimes shortened to molecular marker) is a DNA sequence which can be identified by relatively simple biochemical/molecular techniques and localized at a certain position (locus) on the chromosome. In some cases it can be statistically associated with a certain trait through a target gene located at or nearby the position of the marker. So selection based on molecular marker patterns (marker assisted selection – MAS) may complement or replace phenotypic selection for the respective trait (Collard et al. 2005). The main difference between classic phenotypic selection and MAS is the target for selection. In phenotypic selection the direct target is the trait itself and its phenotypic expression in a specific environment, so that alleles of specific genes behind the trait are selected indirectly. In MAS on the other hand the marker allele associated with the favourable trait related gene is the direct target, while the phenotypic traits to be found in a specific environment are the indirect targets.

2.5.2 Development of MAS

Development of markers for MAS includes several steps:

1. The detection of an association between markers used and linked loci affecting the trait. If the trait of interest is qualitative a simple linkage analysis can be used to determine the association between marker and target gene. If on the other hand the trait is quantitative (quantitative trait loci – QTL) a statistical analysis of trait and marker data in segregating populations is needed; ie QTL analysis.
2. The selection of a subsample of markers with expected high usefulness in MAS.
3. The evaluation of the marker in different genetic backgrounds.
4. Eventually a technical optimization of the marker for practical application.

In the case of a qualitative trait with no dominance, phenotypic selection and MAS roughly give the same results, even though incomplete linkage between marker and gene can lead to a certain percentage of errors in prediction of trait expression. The situation is much more complicated in the situation of a quantitative trait where a single trait is influenced by several other genes, and some of these may also influence other traits. Here the combination of MAS and phenotypic selection can have an advantage relative to selection based solely on phenotype; especially traits with modest heritability and high genotype-environment interactions (Moreau et al. 1998).

2.5.3 Application of MAS

In conventional plant breeding, MAS is a well established selection tool (Tuvesson et al. 2007). MAS has its biggest advantage when applied in early stages of the breeding process on single plants. In this case markers can reduce effectively the number of lines to be tested and save planting area and labour in field trials. In breeding of cereals, the success of markers is due to the simultaneous introduction of double haploid technologies (DH). In DH lines the genetic constitution is fixed after haploidisation and selection efficiency is no longer reduced by genetic recombination in the following generations. Thus molecular markers allow for selection that would be difficult, very costly or just not possible when based solely on phenotype; and a good example is disease resistance in the absence of the pathogen (de Oliveira et al. 2005). In perennial plants, MAS can be carried out on seedlings and an earlier selection for traits, and therefore achieves quicker breeding progress (Dilrewanger et al. 2004). Finally, the increased use of automation in marker approaches allows large-scale screening for DNA markers within a short time and has contributed real advantages to this technology (Dayteg et al. 2007).

Up until now the application of MAS in plant breeding has mostly concerned qualitative traits. Well known examples are genes for virus resistance in barley and wheat where molecular markers have been used in pyramiding (stacking) genes against several virus strains (Werner 2005). This effective use when pyramiding several highly effective genes and MAS may be even higher when applied to quantitative traits. Only a few examples of success have been found for MAS in quantitative traits. The reason may be that many issues can complicate or even prevent the use of MAS for QTL (Charcosset and Moreau 2004) such as:

(a) If the target trait is influenced by many QTL with relatively small effects, MAS is inefficient and the sub-population possessing all optimal marker alleles will be small; so a relatively small effect is seen.
(b) Effects of QTL found in analysis can be overestimated due to the influence of undetected QTLs. A possible solution is cross-validation in an independent sample.
(c) Epistatic effects (*i.e.* QTL X QTL interactions) can cause a QTL effect to be highly dependent on genetic background of the respective genotype. This problem can be partially circumvented by the use of several crosses at the same time.
(d) Interactions between the genotypes (*i.e.* QTL) and the environment can be important. This is especially true when the trait is heavily influenced by strong environmental factors. QTL are seldom independent of the environment and success may lie in being able to include genetic environment interaction in analysis (Francia et al. 2005; Christiansen et al. 2006). But a considerable reduction in the number of usable QTL can be expected, especially when the trait is under strong environmental influence.

In vitro selection systems enable an increase rate and a more precise level of MAS. Selection can be affected on numerous individuals and can be done at any stage of plant growth, and this is a real advantage for tissue culture. Therefore MAS in tissue culture has become more important and more efficient, and more exact than any greenhouse or field selection system.

2.6 *Biotechnology and Genetic Manipulated (GM) Crops*

2.6.1 Introduction

Improvements of plants by conventional breeding involves crossing parents holding part of the desired traits while selecting progeny of individuals showing the proper combination of a superior set of traits. Selection however has its limitations and some of these can be removed by molecular biology techniques, but in other cases only the use of transgenic plants can provide success. For a full comparison of steps in conventional breeding, MAS breeding and genetic transformation see Fig. 6.3. Transgenic plants and molecular breeding has opened a number of possibilities:

(a) Isolating and amplifying genes encoding for a particular trait *in vitro.*
(b) Equipping an isolated gene with regulatory signals recognized by the plant.
(c) Introduction and expression of a modified gene into plant cells and tissues.
(d) Regeneration of transgenic plant cells into mature fertile lines of plants.

The main advantage is that it offers the possibility of adding a new trait directly to an existing valuable variety. In such systems potential species barriers to gene incorporation and expression do not exist. Applications using this approach have already been well documented in many dicotyledon plants, and some fewer monocotyledon plants (eg tobacco, potato, tomato, rapeseed, corn, wheat, barley oats, soybean and banana) (Jackson et al. 2006; Akhond and Machray 2009). Genes can be accessed from exotic sources; plants, animals, bacteria, viruses or even humans. The methods used in plants however have some limitations, and there has been limited success because:

(a) Few reliable protocols for introduction and expression of the foreign DNA exist.
(b) Number of documented genes of interest is still limited, although expanding all the time.
(c) Regulation of gene expression is vital for success, but is limited in plants.
(d) Tissue culture methods for plant regeneration are still inadequate.
(e) Long periods of time are required for full technology development and transfer.
(f) Lack of long-term funding for such programs.
(g) Absence of a co-ordinated approach by the scientific community.
(h) Unfavourable public perception towards transgenic plants.

It is interesting to note that GM technology was first developed for plants as early as 1983, but the first GM crops reached the markets only during the mid-1990s. During the next two decades the development of transgenic technology has proceeded rapidly. The global area planted to biotech crops has increased over the last 20 years by an annual rate of 10–12% each year, and the major biotech growing nations are detailed in Table 6.3. However the biotech crops planted in agriculture are still dependent on four GM crops (ie soybean, corn, cotton and canola) in most

Table 6.3 Global area of biotech crops in 2009 by country, and note the top 15 mega-countries that are growing 50,000 ha or more

Rank	Country	Area (million ha)	Biotech crops
1	USA	64.0	Soybean, corn, cotton, canola, squash, papaya, alfalfa, sugarbeet
2	Brazil	21.4	Soybean, corn, cotton
3	Argentina	21.3	Soybean, corn, cotton
4	India	8.4	Cotton
5	Canada	8.2	Canola, corn, soybean, sugarbeet
6	China	3.7	Cotton, tomato, poplar, papaya, sweet pepper
7	Paraguay	2.2	Soybean
8	South Africa	2.1	Corn, soybean, cotton
9	Uruguay	0.8	Soybean, corn
10	Bolivia	0.8	Soybean
11	Philippines	0.5	Corn
12	Australia	0.2	Cotton, canola
13	Burkina Faso	0.1	Cotton
14	Spain	0.1	Corn
15	Mexico	0.1	Cotton, soybean
16	Chile	<0.1	Corn, soybean, canola
17	Colombia	<0.1	Corn
18	Honduras	<0.1	Corn
19	Czech Republic	<0.1	Corn
20	Portugal	<0.1	Corn
21	Romania	<0.1	Corn
22	Poland	<0.1	Corn
23	Costa Rica	<0.1	Cotton, corn
24	Egypt	<0.1	Corn
25	Slovakia	<0.1	Corn

The data was compiled by James (2007)

countries, but the number of countries planting Biotech crops is expanding all the time (Fig. 6.4). The process of genetic transformation of plants involves several distinct stages; namely insertion, integration, expression and stable inheritance of the new DNA, and details of the steps involved are adequately covered in a number of books (Herman 2009; Neuman and Kumar 2009).

2.6.2 *Agrobacterium* System

The plasmid containing soil bacterium *Agrobacterium tumefaciens* can be considered a natural gene transfer system. It is one of the most efficient methods employing the naturally evolved 'crown gall' causing mechanism of DNA transfer. *Agrobacterium* transformation results from the stimulation of plant cell division by gene products encoded by a segment of DNA (T-DNA) which is transferred from the bacterium to the plant.

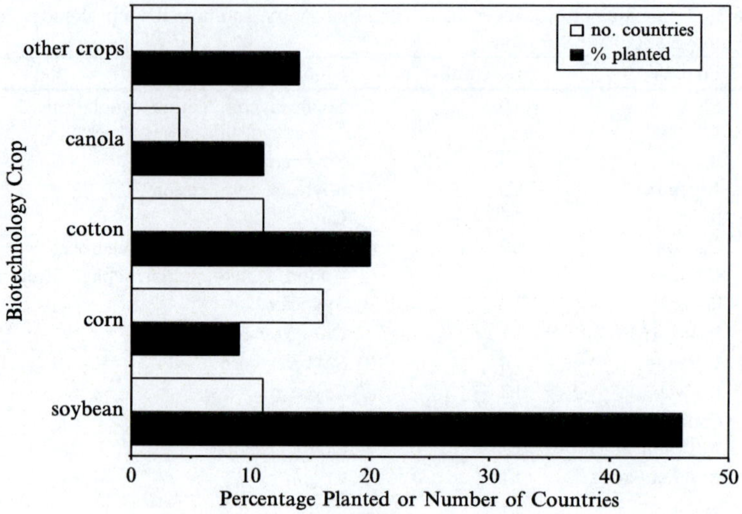

Fig. 6.4 Percentage of the major biotech crops planted, and the number of countries involved for each crop planted in 2009. The data was compiled by James (2007)

The T-DNA and '*vir*' regions required for this transfer are located on the tumor inducing plasmid (Ti). The '*vir*' system will process and transfer any DNA between two flanking regions that delimit the T-DNA, hence its utility. The strain of *Agrobacterium* used as a carrier for DNA transfer into plants has been modified (disarmed) to remove the gall-inducing (tumor producing) ability and engineered to carry the gene of interest, together with a selectable marker. Inside the plant the T-complex is imported into the nucleus of the plant, the T-strand becomes stably inherited into the plant chromosomes, and the genes (including foreign genes) are expressed (Broothaerts et al. 2005).

The *Agrobacterium* system has the ability to integrate into a variety of plants, some monocotyledons have shown some resistance to the method but now has become a routine method even in these crops (Smith and Hood 1995). More important is that the site of integration of the T-DNA is random, it can be rearranged or truncated, but the biggest problem is making sure that the carrier bacterium itself does not persist internally in transgenic plants (Basha and Sujath 2009).

2.6.3 Biolistic or Microprojectile Bombardment

The method was developed in the mid-1980s and has been widely used as a more direct gene transfer system in a variety of plants; including previously hard to transform monocotyledon plants (Southgate et al. 1995). The technique is based on the rapid acceleration of a metal microprojectile to deliver nucleic acids, including specific genes (or DNA) into the intact nucleus or even just into plant cells and tissues. Other biological material and compounds apart from DNA could be delivered, and a commercial particle delivery system for gene transfer has been available on the market (Biolistic PDS-1000) for some time. Due to the genotype-independent physical nature of DNA delivery, the methods employed are simple, efficient and nearly the same regardless of the nature of the foreign DNA or the target cells used. It is therefore a highly valuable and adaptable technique potentially applicable over many different types of cells and tissues. During the last two decades it appears that microprojectile bombardment has become more routine and more reliable a method; so much so that it is replacing the *Agrobacterium* method quickly (Hansen and Wright 1999).

However, there are still problems to be resolved, in particular, more direct and efficient targeting of the nucleus, and tissue culture related totipotency and low regeneration of plants in crops like soybean and others. The results of the biolistic method can be a complex pattern of transgene integration which is undesirable, but can be mitigated by using 'Agrolistic' approaches. Agrolistic transformation allows integration of the gene of interest without the undesired vector sequences. In other words, deliver by biolistic means *Agrobacterium* containing the foreign DNA and only some virulence genes (eg *vir* D1 and *vir* D2) but not *vir* E2 in the T-DNA. It has been demonstrated that only these two '*vir*' genes are required for '*in planta*' stable transformation (Hess and Dressler 1989; Beranova et al. 2008).

2.6.4 Alternative Methods

(a) **Protoplast transformation** – protoplasts can be transformed either by the *Agrobacterium* system or by direct DNA uptake using electrofusion or PEG treatments. The method has been used routinely to study transient expression of transgenes. Transgenic plants have been obtained in very few cases, and only where suitable protocols were already available for culturing of protoplast to whole plants (Shillito 1999; Murphy 2007).

(b) **Chloroplast transformation** – chloroplast transformation has generated much interest as a more 'environmentally friendly' approach to plant genetic engineering. It provides a system where it can avoid out-crossing of transgenes to weeds or other non-target crops; thus reducing the potential toxicity of transgenic pollen to non-targeted species. Biolistic gene delivery is probably the most reliable and reproducible method to use, but the method is only possible in a limited number of species (Daniell et al. 2002; Murphy 2007).

(c) **Floral dip method** – a method that could circumvent the problemsome time consuming steps in previous methods, and is a simple method to master. There are a number of such methods, and they are all placed into the category of 'in plant transformation'. One such method has been well developed in *Arabidopsis* and has achieved high rates (0.5–3%) of transformation, and there are reports of success in a number of other crop plants (eg *Brassica campestris, Medicago trunculata* and radish)(Liu et al. 1998; Trieu et al. 2000; Curtis 2005).

(d) **Pollen mediated transformation** – has attracted significant interest and there have been a number of reports claiming success (Harwood et al. 1996; Heberle-Bors et al. 1996). Basically it involves incubating pollen in DNA solution and electroporation (Hess and Dressler 1989), direct particle bombardment (van der Leede-Plegt et al. 1995) or co-cultivation with *Agrobacterium* (Langridge et al. 1992). Many of the methods were later proven to be not reproducible and only in the report of Smith et al. (1994) have transgenic plants shown to be produced. A report of an enhanced *Agrobacterium* mediated transformation system in flax has been described recently by Beranova et al. (2008).

2.6.5 Future of Plant Transgenics

Introduced DNA in plants usually integrates more or less randomly into the host genome, even when the introduced DNA shows homology with plant endogenous DNA sequences. This is referred to as 'illegitimate recombination' or 'non-homologous end joining'; a very common method for joining DNA ends that are created in cells by double stranded DNA breaks. When an introduced DNA sequence can target an identical sequence in the host genome and integrates itself by homologous recombination (HR) the process is known as 'Gene Targeting' (GT). This event is a prerequisite to application of the most sophisticated tool of 'reverse genetics'; ie gene disruption and allele replacement (Berg 1991b).

In such GT technology virtually any cloned gene, even of unknown function can be specifically mutagenised *in vitro* and re-introduced to its own chromosomal location in order to study its function. But the targeted recombination of incoming DNA offers so many advantages and overcomes many of the problems with gene expression of transgenic plants. In plants only a few reports of GT have been published in *Arabidopsis*, tobacco and maize (Nguyen et al. 2007; Munkacsi et al. 2008). The reported frequency of GT can be low (10^{-3}–10^{-4}) and if it is true that in plants, as in mammals, there appears to be little or no negative consequences to the organism (Drobnik 2008; Vinterhalter et al. 2008a, b) it could be very exciting. More recently large scale GT has been reported in rice using an *Agrobacterium* system (Iida and Terada 2004), and a zinc finger nuclease based technique (Kumar et al. 2006) also shows some promise.

Crucial steps and decisions need to be made in gene targeting (GT) in plants, and these include:

(a) The design of the gene targeting construct.
(b) The choice of tissue to be transformed.
(c) Selection of transformants.
(d) Regeneration of transformants.
(e) Screening for clean, targeted transformants.

During the 1990s and 2000s the genomes of a few different plants have been completely sequenced (Jackson et al. 2006; Amano et al. 2010). Many more plants are well on the way to being sequenced. Nowdays plant molecular research is being directed more and more towards an understanding of gene expression and function. Proteomics with a high through-put is such an approach in this post-genomic period. Reports of proteomic applications in plants and biotic stress are still rather limited compared to other biological and molecular approaches (Xing et al. 2002; Jorrin et al. 2006). But proteomics is becoming important and justifiable because:

(a) Gene function is carried out by proteins and polypeptides.
(b) A large number of genes (an important 80–90%) have no apparent assigned function.
(c) Information obtained by using transcriptone approaches is incomplete.
(d) Correlation between mRNA and protein levels is remarkably and surprisingly low.

2.6.6 RNA Mediated Silencing

Viral coat mediated resistance to some plant viruses have been reported over the last 20 years by conventional breeding, but this technology is fast being replaced by the recently recognized small interfering RNA (siRNA) and micro RNA (miRNA) mediated gene silencing (Ritzenthaler 2005). A recent report for example has demonstrated how artificially constructed miRNA can confer virus resistance in plants (Niu et al. 2006). Transcriptional factors that regulate the expression of several genes related to biotic stress resistance have been described, and belong to the stress signaling pathway. Transcriptional analysis using microarray technology has demonstrated that these are important in abiotic stress tolerance such as drought, cold and salt, but they may also be expressed as a result of biotic stress. For example *Arabidopsis* inoculated with growth promoting *Rhizobacteria* also have enhanced protection against *Erwinia* (Agarwal et al. 2006), and the *'adr*1*'* gene was involved (the so called activated disease resistance 1 gene).

2.6.7 Safety of Bitech (GM) Crops

There has been considerable public debate and concern regarding the safety of biotech (GM) crops. This has lead to an approval protocol in many countries where rigorous testing and safety assessment must be carried out before any such crops are approved for commercial cultivation. The Codex Alimentarius Commission, under FAO and WHO adopted specific guidelines in 2003 for risk assessment of plants derived from biotechnology in the global market (CAC 2003), and these guidelines are quite strict. There appears to be three concerns by the public about genetically engineered crops (Clark 2006):

(a) **Marker genes** – used in biotech crops could introduce toxicity and allergy. Experimental evidence suggests that the genetic approaches and methodology used have little potential for this (Magana-Gomez and de la Barca 2008).

(b) **Perception that there is a safety concern to diet** – gaining access to a wider range of genetic diversity in food crops, crossing species barriers and the introduction of foods with additional proteins are not safe. It is fair to say that most biotech foods are equivalent to existing and traditional foods, and there is no documented proof that any biotech crop has caused toxicity or allergies. Additionally, there is no proof that successive generations of such biotech crops have caused a biologically significant increase in endogenous allerginicity from foods (Goodman et al. 2008).
(c) **Transgenes will spread in the environment** – mathematical models and empirical experimental evidence suggest that genetic approaches used in GM crops have the potential to effectively prevent transgenes from escaping into wild populations, ie those crops that are not reproductively isolated from genetically engineered crops (Lee and Natesan 2006).

3 Part B: Biotic Factors and Crop Production

3.1 Viruses and Viroids

3.1.1 Characteristics and Symptoms

Viruses and viroids are nucleoproteins that have the ability to multiply in living plant cells and cause disease. Viruses can be specific as to infect only one species of plant or infect dozens of different plants, and each plant species may be attacked by many different types of viruses. Viruses always contain nucleic acids (RNA or DNA) plus one or more different kinds of proteins. The composition and structure of the viral proteins is know only for a few plant viruses, and most plant viruses have their nucleic acid in the form of RNA; but about 80–100 plant viruses contain DNA. When a virus infects a plant in the field it moves from one cell to another and multiplies in most of these cells (systemic infection). In all important agricultural and economic infecting viruses they reach the phloem where they are rapidly transported over long distances. Almost all viral diseases appear to cause some degree of dwarfing or stunting of the entire plant; where total crop yields are frequently reduced (Agrios 1997). The most obvious symptom of a viral infected plant usually appears on the leaves, but fruit, stem and roots may also be affected. In the case of latent viruses, infection may proceed without causing visible symptoms. The most characteristic leaf symptoms are grouped into the two categories below:

(a) **Mosaics** – characterised by appearance of light green, yellow or white areas intermingled by normal green tissue on leaves or stems.
(b) **Ringspots** – characterised by appearance of chlorotic or necrotic rings on the leaves, sometimes also on stems and fruit.

A large number of other, less common symptoms have been described, including stunt, dwarf, yellows, streak, tumors, pitting, flattening and distortion, and curling of usually the leaves and fruits of infected plants. The present methods of detecting viruses is transmission of the virus and disease from a diseased plant to a healthy plant by budding, grafting or rubbing sap into a healthy leaf. The most definitive methods however are expensive like electron microscopy, serology and more recently molecular (DNA/RNA) probes.

Viroids are small, low molecular weight ribonucleic acids (RNA) that have the ability to infect plants, replicate themselves in plant cells and cause disease. Viroids are different to viruses in at least two main characteristics:

(a) Size of the RNA in viroids is only between 250 and 370 bases, compared to 4,000–20,000 bases in viruses.
(b) Viruses are always enclosed in a protein coat, while viroids lack this protein coat (only free RNA).

Because of their small size, viroids cannot code for even one small protein due to insufficient number of bases; even for example a replicase-type enzyme essential for viroid replication (Horst 2008).

3.1.2 Important Diseases

Viral and viroid diseases of plants have been loosely categorised into eight classes, primarily based on characteristics of the viruses rather than the plant they infect or the severity of the symptoms, or for that matter affect on production. It is my aim below to just detail the diseases in each of these categories below, and the more important viral diseases of food crop plants are summarised in Table 6.4 and Strange and Scott (2005):

(a) **Rigid Rod-shaped Viruses** – includes tobamoviruses, tobraviruses, furoviruses and hordeiviruses. These viruses are single stranded RNA type and only tobamoviruses and tobraviruses cause serious loss of production in tobacco and some Solonaceae group plants, especially potato, tomato, pepper and pea by damaging leaves, flowers and fruits. Furoviruses at times also reduce production in potato, sugarbeet and peanut however symptoms and losses can be variable. Hordeiviruses are rare and have only been detected in barley, and in general causes only minor losses in production.
(b) **Filamentous Viruses** – includes potexviruses, carlaviruses, capilloviruses, trichoviruses, potyviruses, rymoviruses, bymoviruses, and closteroviruses. These viruses are also RNA viruses but the number of strands is not always certain. Potyviruses and closteroviruses are the main classes that cause serious loss in production and severe diseases in many plants where they appear to be manifested as mosaic, mottling and chlorotic spots on leaves, stems and fruits. Some of the more serious diseases of potyvirus include lettuce mosaic, celery mosaic, bean mosaic, bean yellowing, papaya ringspot, plum pox, potato virus Y, sugarcane mosaic, turnip mosaic, watermelon mosaic and zucchini yellow mosaic. The more serious diseases of closterovirus include beet yellow, citrus

Table 6.4 The world's staple crops and their principal diseases according to the biological class of organism causing the respective diseases

Crop	Biological classes of casual organism
Barley (*Hordeum vulgare*)	**Virus**: Yellow dwarf; Stripe mosaic
	Fungi: Mildews (downy/powdery); Spot blotch and Net blotch; Scald; Scab; Barley stripe
	Nematode: Root-knot; Cyst; Root lesion
Cassava (*Maniot esculenta*)	**Virus**: Cassava mosaic, Geminivirus
	Bacteria: Bacterial blight
	Fungi: Anthracnose
Lentil (*Lens culinaris*)	**Fungi**: Wilt and Vascular wilt; Blight; Rust; Anthracnose
Maize (*Zea mays*)	**Virus**: Chlorotic dwarf; Streak Geminivirus; Yellow dwarf
	Bacteria: Stewart's wilt (*Erwinia*); Corn stunt
	Fungi: Leaf blight; Rust and Smut; Downy mildew; Stalk and ear rots
Common millet (*Panicum miliaceum*)	**Fungi**: Downy mildew
Pearl millet (*Pennisetum glaucum*)	**Fungi**: Ergot
Oats (*Avena sativa*)	**Virus**: Yellow dwarf; Golden stripe; Mosaic potyvirus
	Bacteria: Halo blight
	Fungi: Crown rust and Stem rust; Powdery mildew; Leaf blight/Seedling blight; *Fusarium* root and crown rot; Snow mould; Leaf blotch; *Alternaria*
Potato (*Solanum tuberosum*)	**Virus**: Leafroll; Mosaic virus
	Bacteria: Bacterial wilt; Bacterial soft rot *Erwinia;* Common scab; Bacterial ring rot
	Fungi: Early blight; Black scurf; Late blight; Pink rot
Rice (*Oryza sativa*)	**Virus**: Rice tungro disease; Yellow dwarf
	Bacteria: Bacterial leaf blight
	Fungi: Blast; Brown spot; Sheath blight
Rye (*Secale cereale*)	**Virus**: Yellow dwarf
	Fungi: Snow mold; Brown rust and Stem rust; Ergot; Sharp eyespot; Powdery mildew; Glume blotch/leaf blotch
	Nematode: Eelworm
Sorghum (*Sorgum bicolor*)	**Virus**: Streak disease
	Fungi: Anthracnose; Leaf blight; Zonate leaf spot; Tar spot; Charcoal rot; Rust; Ergot; Downy mildew
Soybean (*Glycine max*)	**Virus**: Mosaic; Yellow mosaic
	Bacteria: Bacterial pustule
	Fungi: Rust; Anthracnose; Purple seed stain; Pod and stem blight; Downy mildew
Sweet potato (*Ipomoea batatas*)	**Virus**: Feathery mottle
	Bacteria: Soil rot; Little leaf
	Fungi: Scab; Fusarium wilt; Black rot; Java black rot; Scurf
	Nematode: Root-knot

(continued)

Table 6.4 (continued)

Crop	Biological classes of casual organism
Wheat (*Triticum aestivum*)	**Virus**: Yellow dwarf **Bacteria**: Bacterial leaf streak **Fungi**: Stem rust and Leaf rust; Stripe or yellow rust; Spot blotch; Head scab; *Fusarium* foot/root rot; Sclerotium foot rot; Tan spot; Powdery mildew; Speckled leaf blotch; Glume blotch; Loose smut; *Rhizoctonia* root rot
Yam (*Dioscorea alata*)	**Virus**: Yam virus complex **Fungi**: Anthracnose; Tuber rots

The list is modified from data compiled from Agrios (1997), Strange and Scott (2005) and Horst (2008)

tristeza and lettuce infectious yellow. The remaining virus diseases in this group can be rare, affect ornamental plants like orchids and carnations, with the exception being the rymovirus and bymovirus which have been shown to have varying severity of effects on grain crops like barley, oats, rice and wheat.

(c) **Isometric Single-stranded RNA Viruses** – includes waikaviruses, luteoviruses, comoviruses, nepoviruses, bromoviruses, cucumoviruses and ilaroviruses. Waikaviruses, luteoviruses, cucumoviruses and nepoviruses are the main classes that cause serious loss in production and severe diseases in many plants where they appear to be manifested as mottling, chlorosis, leaf roll and stunting of plants. Some of the more serious diseases of waikavirus include rice tungro and maize chlorotic dwarf. Some of the more serious diseases of luteovirus include beet western yellow and potato leafroll. The more serious diseases of nepovirus and cucumoviruses include tomato ring spot, cherry leafroll, grapevine fanleaf, raspberry yellow dwarf, ringspot or leafcurl dwarf and cucumber mosaic. The remaining virus diseases in this group are usually rare and do not affect plant production.

(d) **Isometric Double-stranded RNA Viruses** – includes the one group of rheoviruses, which can also infect humans and animals. The more serious diseases cause wound tumor induction except rice dwarf virus. The viruses have a very limited host range amongst grasses, and although the symptoms appear severe the plants are usually young when infected. Therefore overall production loss can be moderate and localised.

(e) **Negative Single-stranded RNA Viruses** – includes diseases caused by rhabdoviruses, which is a large group of 80–100 viruses with a limited host range; but also includes the rare groups of tospoviruses and tenuiviruses. They are mostly viruses in vegetables, weeds and grasses where the disease causes mosaic, vein clearing, yellowing and dwarfing. The most notable are lettuce necrotic yellow, potato yellow dwarf, rice transitory yellowing and wheat striate mosaic. Production losses are quite low.

(f) **Double-stranded DNA Viruses** – includes two groups of caulimoviruses and badnaviruses which are important groups mostly in vegetables and ornamental plants where the diseases cause mosaics and mottles. The most notable are cauliflower mosaic, dahlia mosaic and carnation etch ring where they cause poor growth, poor quality and reduced yields. Badnaviruses affect rice, banana and cacao and in many parts of the world it affects production severely causing large losses.

(g) **Single-stranded DNA Viruses** – includes the large group of geminiviruses which is an important group that has been extensively studied where the disease causes mottles to yellow mosaics with leaves curled and distorted. The most notable diseases are maize streak, beet curly top, African cassava mosaic, bean golden mosaic, squash leaf curl, tomato mottle, tomato yellow leaf curl and banana bunchy top. Geminiviruses affect many plants and can cause devastating losses in many of the crops, particularly in tropical and sub-tropical regions.

(h) **Viroids** – at least 20–30 plant disease have been described with the most important being coconut cadang-cadang, potato spindle tuber, citrus exocortis, avocado sunblotch, apple scar skin, chrysanthemum stunt and chloritic mottle, hop stunt and cucumber pale fruit. How viroids cause disease is largely unknown, but when they do cause a disease, symptoms can be varied and resemble symptoms of true viruses. Viroids are easily spread from plant to plant primarily by mechanical means through sap transfer on hands or tools.

3.1.3 Control Measures

Viral and viroid diseases are difficult to control and the best way is by keeping them out of the area, through quarantine, inspection and certification of crops used. Eradication of diseased plants and control of vectors is good practice, but this can be difficult if not impossible. Periodic virus indexing of crops can help in not introducing viruses but it can be expensive. Incubation of material by heat can be useful and especially hot water dipping of seeds and propagation material. No effective chemical treatments are available for plant viruses. Virus indexing and use of resistant plants are still the best options (Strange and Scott 2005).

3.2 Bacteria

3.2.1 Characteristics and Symptoms

Bacteria and Mycoplasma Like Organisms (MLO) are prokaryotes; single cell microorganisms whose genetic material (DNA) is not bound by a membrane and therefore do not possess a nucleus. In bacteria, the cells are enclosed by both a membrane and a cell wall, however MLO are surrounded by a cell membrane only. Most plant pathogenic bacteria are rod shaped with the exception being *Streptomyces* which are filamentous. Most plant pathogenic bacteria have delicate thread-like flagella, commonly longer than the cells themselves. Based on specific cell wall staining bacteria can be divided into gram-positive or gram-negative bacteria for classification. However if bacteria are allowed to grow (and multiply) on the surface of a solid or semi-solid nutrient medium they will form colonies. Colonies are easier to study and identification of species may be possible based on colony shape, smoothness, elevation from the media, colour and how grainy they appear. Often (but not always) bacteria have single or multiple copies of smaller circular material called plasmids in the cytoplasm. Each plasmid consists of several 'non-essential' genes which can move or be transported between bacteria and even between bacteria and plants; as happens in crown gall disease (see Sect. 2.6.2). Rod-shaped bacteria reproduce by an asexual process known as binary fission or just fission. Bacteria can divide every 20–30 min. Almost all plant pathogenic bacteria develop mostly in the host plant as parasites, on the plant surfaces, in plant debris and in the soil as saprophytes.

Some bacterial pathogens after infection produce most bacteria within their plant host and have reduced numbers present in the soil. These bacteria have evolved very efficient plant-to-plant infection mechanisms often using vectors to sustain high infection rates. Other bacteria are typically soil inhabitants, however they do build-up numbers in plant cells under suitable conditions while still keeping high bacterial numbers in the soil. Dissemination and infection by these bacteria are with the aid of water, insects, animals and humans. Bacterial have flagella but in most cases move only very short distances (Agrios 1997).

The most commonly used method of detecting bacteria is still by serological methods employing metabolites as substrates or fluorescent stains. Another excellent method is by promoting growth and colony identification on selected growth supplemented media, and bacteria can be identified through genetic relatedness or genetic fingerprinting methods analogous to human fingerprinting methods on their DNA. Plant pathogenic bacteria induce as many kinds of symptoms on plants they infect, as do for example pathogenic fungi. They cause leaf spots and blights, soft rots of fruit, roots and storage organs, wilts, excessive growth, scales and cankers (Agrios 1997). What is more important than this is that any one type of symptom can be caused by bacteria belonging to several genera, and each genera may contain pathogens capable of causing quite different symptoms.

3.2.2 Important Diseases

Diseases of plant pathogenic bacteria have been characterised by two different methods below:

(a) **Taxonomic Families** – identification based on correct placement into their bacterial families.
(b) **Disease Symptoms** – families and genera grouped together based on similarity of symptoms.

The more important bacterial diseases of crop plants are summarised in Table 6.4 and Strange and Scott (2005).

(a) **Taxonomy**

1. **Family Enterobacteria** – includes *Erwinia* causing fire blight of pear and apple, Stewart's wilt of citrus and soft rots of fleshy vegetables.
2. **Family Pseudomonads** – includes *Pseudomonas, Rhizobacter, Xanthomonas, Xylophilus*, and manifested as leaf spots, blights, wilts, rots, cankers and galls in many plant species.
3. **Family Rhizobia** – includes *Rhizobium* which causes root nodules in legumes.
4. **Family Firmibacteria** – includes *Bacillus* and *Clostridium*; causing rots of tubers, seeds and stripe in wheat and a number of other cereals.
5. **Family Thallobacteria** – includes *Anthrobacter, Clavibacter, Rhodococcus, Streptomyces*, showing up as blight, wilts, scab and fasciation in a number of plants.

(b) **Disease**

1. **Bacterial Spots and Blights** – symptoms produced mainly by *Pseudomonas* and *Xanthomonas*.
2. **Bacterial Vascular Wilts** – disease produced by *Clavibacter, Erwinia, Pseudomonas* and *Xanthomonas*.
3. **Bacterial Soft Rots** – symptoms produced by Erwinia, Pseudomonas, Bacillus, Clostridium and Streptomyces.
4. **Bacterial Galls** – includes at least 20–30 plant disease primarily due to *Agrobacterium, Pseudomonas, Rhizobacter* and *Rhodococcus*.
5. **Bacterial Cankers** – symptoms produced mainly by *Pseudomonas* and *Xanthomonas*.
6. **Root Nodules of Legumes** – produced by *Rhozobium* and *Frankia*.
7. **Plant Vascular Bacteria** – diseases mainly caused by *Clavibacter* and *Xylella*.
8. **Micoplasma Like Organisms (MLO)** – diseases mostly due from *Mycoplasma*.

3.2.3 Control Measures

Bacteria and MLO diseases of plants are usually difficult to control and a combination of measures are commonly required. Sanitation is essential of equipment and tools. Steam pasteurisation of soil and growing media is one of the best methods to use. Seed sterilisation by hypochlorite soaking is effective and so can hot water treatment. Use of chemicals for control of bacterial diseases is generally less effective than in controlling fungi for example, and foliar spays for bacterial control are the most effective. The use of crop varieties resistant to bacterial diseases is one of the best methods to use (Horst 2008).

3.3 Fungi

3.3.1 Characteristics and Symptoms

Fungi are small, generally microscopic, eukaryotes, usually filamentous, branched, spore-bearing organisms that lack chlorophyll. Fungi have cell walls that contain chitin and glucans embedded in a matrix of polysaccharides and glycoproteins; but a small group (the oomycota) that also cause plant diseases contain cellulose and no chitin. Most of the 100,000 species of fungi are saprophytic, that is they live on dead organic matter which they help to decompose. Very few species cause disease in plants. All plants can be attacked by some kind of fungus, and each of the parasitic fungi can attack one or more kinds of plants. Most plant fungi have a filamentous vegetative body called a mycelium. The mycelium can branch out in all directions. The individual branches are called hyphae. The length of the mycelium may be only a few nanometres in length in some fungi but in others it may reach several metres in length. Some lower fungi however lack true mycelium and produce a system of strands called rhyzomycelium (Agrios 1997). Fungi reproduce chiefly by means of spores; the reproductive bodies. Spores may be formed asexually by budding or as a result of sexual fertilisation. Almost all plant pathogenic fungi spend part of their life cycle on the host plant, in the soil or in plant debris. During the parasitic phase of their life cycle fungi assume various positions and grow in various organs of their host:

(a) Some grow outside their hosts, but send their feeding organs (haustoria) into the plant cells.
(b) Some only grow between the cuticle and epidermal cells.
(c) Some grow only in the intercellular spaces between cells.
(d) Others grow between and within plant cells almost indiscriminantly.

Infection, survival and severity of diseases of most plant pathogenic fungi greatly depend on environmental conditions; as outlined below:

(a) Mycelium survives within a certain range of temperatures and only in contact with moist surfaces.
(b) Spores survive a much broader range of both temperatures and moisture. However spores also require favourable conditions in order to germinate.
(c) Zoospores are the only structures that can move by themselves, but only for very short distances (a few mm).
(d) Most spores depend on their spread from plant to plant, or different parts of the one plant by agents such as wind, water, birds, insects, animals and humans.

Fungi cause local or general symptoms on host plants and such symptoms may occur at any time or may follow a set time pattern. In general, fungi cause local or general necrosis (killing of plant tissue areas) and they often result in reduced growth or stunting of plant organs or the whole plant. Some fungi cause excessive growth in infected plants or plant parts, which can be manifested as growth which is very different and distorted to normal plant growth. Another group of symptoms characteristic of fungal infection include wilting, rusts, smuts and mildew on plant parts (Horst 2008).

3.3.2 Important Diseases

A summary of the most important fungal diseases of crop plants are summarised in Table 6.4 following Strange and Scott (2005).

(a) **Pseudofungi (Fungal-like Organisms)**

1. **Myxomycetes**
   Includes slimemoulds on leaf surfaces in many plants. Clubroot of crucifers, powdery scab of potato, black smut of potato and brown spot of corn. No control is usually necessary or effective, and best to avoid planting such crops in infected areas.
2. **Oomycetes**
   Includes several of the most important plant pathogens known, and these are *Pythium, Phytopthora*, several fungi causing Downy Mildews and white rust (*Albuga*) on crucifers. Disease by Oomycetes are of two types; one which affect plant parts in contact with the soil and another which affect above ground parts, like leaves, stems and fruits.
   Pythium – a major cause of seed rot, seedling damping-off and root rot in all types of plants and seedlings. Occasionally it may cause soft rots of fleshy fruits when in contact with soils.
   Phytopthora – the cause of a late blight of potato, and root rots and blights of many other plants, especially hard to control in forest trees.
   Downy mildews – caused by a number of fungal species in dicotyledon plants which can be destructive in lettuce, tobacco, grape and cucumber. In monocotyledon plants other fungal species are involved in mildews but with similar destructive effects.
   Albugo – can be common but usually exhibit a white rust symptom in crucifers, and is not destructive.

(b) **Ascomycetes (Imperfect Fungi)**
Many Ascomycetes and imperfect fungi cause a variety of diseases in all types of plants. The most important plant pathogens are detailed below.

1. **Sooty Moulds** – appear on leaves and stems of plants as superficial black growths in warm humid weather. Most are not parasitic and live on dew and sugars of leaves. Rarely damage plants and no control is required.
2. **Leaf Curl** – several species of *Taphine* cause curls on leaves, flowers and fruits; even stored fruits. Damage can result to the fruit and even the plant but is easily controlled by a single fungicide spray.
3. **Powdery Mildew** – one of the most common, conspicuous, widespread and easy to recognise plant disease. The fungi causing powdery mildew are

all obligate parasites and they can be so common that they may cause each year losses that surpass any other single type of disease. They seldom kill the plant but by depleting reserve nutrients they may reduce growth the following year of up to 40%. Control is difficult and may involve resistant varieties and systemic fungicide application.

4. **Foliar Diseases** – there are a variety of plant pathogenic diseases of foliage which are economically important due to Ascomycetes. These include *Alternaria*, Rice Blast, Black Rot of Grape, Needle and Blight of Conifers, Banana Leaf Spot (Sigatoka Disease), Cucurbit Stem Blight, *Cernospora* Disease and *Pyrenospora* Disease of Cereals and Grasses.
5. **Stem and Twig Cankers** – poplar canker, Black But of plum and cherry, Chestnut Blight, *Nectria* Canker and *Leucostoma* Canker.
6. **Anthracnose Diseases** – Black Spot of Roses, *Glomerella* disease of annuals and Fruit Rots, Leaf Spot Disease of Gramminacea, *Colletochritum* Disease. Some anthracnose disease can have a serious effect on crops.
7. **Fruit and Seed Diseases** – Ergot of Cereals and Grains, Apple Scab, Brown Rot of Stone Fruit, *Botrytis* are the more important diseases which can cause large production losses.
8. **Vascular Wilts** – *Fusarium* Wilts, *Verticillium* Wilts, Dutch Elm Disease can all be serious.
9. **Root and Stem Rots** – *Gibberella* Stalk, Seedling Blight of Cereals, *Fusarium* Rot, Take-all of Wheat, and some other stem and root rot diseases are examples, and some that can cause production losses.
10. **Postharvest Disease** – *Erwinia* and *Pseudomonas, Rhizobacteria* and *Sclerotium* (Basidiomycetes see below), *Alternaria, Botrytis, Fusarium, Penicillium* also may be involved in serious post-harvest loss of crops.

(c) **Basideomycetes (True Fungi)**

Most Basideomycetes are fleshy fungi, such as the common mushroom. They are mostly saprophytic and/or cause wood decay, which includes root and stem rots of trees. However two very common and very destructive groups of these plant pathogens are present; the rusts and smuts.

1. **Rusts** – cereal rusts, stem rust of wheat and other cereals, Apple Rust, Coffee Rust, Rusts of Fruit Trees, White pine Blister, Fusiform Rust and Rusts of Bean and Roses are of commercial importance.
2. **Smuts** – Corn Smut, Kernel Smut of small grains, Covered Smut, But of Wheat, Sorghum Head Smut, Leaf Smut and Stripe Smut can be considered the most important in crop production loss.
3. **Root And Stem Rots** – *Rhizoctonia* and *Sclerotium* Diseases, *Armillaria* Root Rot can be important.
4. **Wood Rot and Decay** – Can have huge losses in timber and are usually divided into Brown-Rot Fungi and White-Rot Fungi depending on the cell wall components the fungi are preferentially degrading. Some of these fungi are simply wood staining fungi that are just surface moulds, but they can

cause wood discolouration, lines and splits which have some commercial implications in timber sales.

(d) **Mycorrhizal Fungi**
The feeder roots of many flowering plants (Angiosperms) growing in native soils can be infected by these symbiotic fungi. The fungi generally do not cause a root disease but instead can be beneficial to their host plants. There are two types of mycorrhizal fungi described Ectomycorrhizae and Endomycorrhizae. However details of this special plant-fungal relationship is beyond the scope of this chapter.

3.3.3 Control Measures

There are numerous and diverse methods to try to control fungal diseases, and the complexity of many of these methods can appear endless. Few fungal diseases can be controlled just with one method and usually a combination of methods are necessary. Some common methods and practices include the use of resistant plant cultivars, clean tools and machinery, however these are not always possible or effective. But for many plants and their fungal diseases the most effective method of control is the application of chemical sprays and dusts (Strange and Scott 2005). Soil fungi can be controlled by steam pasteurisation. Very often a combination of resistant varieties and good chemical spray programs are essential.

3.4 Nematodes

3.4.1 Characteristics and Symptoms

Nematodes belong to the Animal Kingdom. Nematodes are worm-like in appearance but are quite distinct taxonomically from the true worms. There are several thousand species of nematodes and they live freely in fresh or salt waters or in the soil, where they generally feed on microscopic plants or animals and microorganisms. Only a few hundred species are known to feed on living plants, and only some cause a variety of plant diseases worldwide with considerable economic losses. Plant parasitic nematodes are very small and therefore mostly invisible to the eye. They are easily observed under the microscope as eel-shaped, round in cross section and with longer mostly unsegmented bodies, without legs or appendages. Their bodies are more or less transparent and most penetrate their host plant via a hollow stylet or spear to withdraw nutrients from their host. Their life cycle can be short (2–4 weeks) under optimal conditions but usually will take longer under cool temperatures. When the infection stage is reached they must feed on a suitable host plant otherwise they will die. Lack of suitable hosts may result in complete loss or death of a species within months, but in some species the juvenile stages may dry-up and remain quiescent as the egg can survive for years (Agrios 1997).

Nematodes spread through the soil without aid but generally travel only a few metres at best in one season. They can be faster in more open soil, but they do not survive well if soils are waterlogged. Nematodes are easily spread quickly by anything that moves and can carry soil particles. Farm equipment, irrigation, flooding drainage water, animal feet, birds, dust and humans are primary spreading agents. Plant diseases caused by nematodes can be complex. Root feeding decreases the ability of the plant to take-up water and nutrients from the soil, but also they interact with the plant biochemical mechanism. Although most parasitic nematodes are capable of causing disease on plants by themselves they do operate in the soil where the damaged plant roots are constantly surrounded by fungi and bacteria. Many of these bacteria/fungi can cause plant diseases, and in many cases this association is never made.

Isolation of nematodes from the soil or even from infected plant material is based on a series of filtering and/or sieving methods to allow the nematodes to migrate to an area (usually towards the bottom layer) where they are concentrated and collected for examination under the microscope. Sometimes a sugar flotation and gradient method is used under centrifugation to achieve a larger concentrated layer of nematodes (Horst 2008).

Nematode infection of plants results in the appearance of visible areas and even lesions, or reduce growth:

(a) **Roots** – symptoms may appear as root knots, root galls, lesions, excessive root branching, injured root tips and occasionally root rots; the latter is most likely a secondary effect.
(b) **Above-ground plant** – mostly non-characteristic symptoms but primarily reduced growth, symptoms analogous to nutrient deficiency like yellow foliage, excessive wilting, reduced yield and poor quality.

Direct mechanical injury by nematodes while feeding only slightly damage plants. Most of the damage is caused by a secretion of saliva injected into the plant cells. Some nematodes feed rapidly within seconds and then quickly move on, while others can take several hours. In either case the feeding process results in dead or damaged plant cells, especially at the growing tips of roots where lesions and tissue degradation due to the digestive enzymes and toxins is evident. In some cases this is accompanied by stimulation of cell division uncontrollably and the formation of galls.

3.4.2 Important Diseases

The more important nematode diseases of crop plants are summarised in Table 6.4 and Strange and Scott (2005):

(a) **Root-Knot Nematode (*Meloidogyme*)**
Root-knot nematodes occur throughout the world, but are more prevalent in warm or hot climates with short winters; but can also be widespread in greenhouse plants where growth conditions are ideal. Root-knot nematodes can be easily controlled with steam pasteurisation of soil or growing media. There are also nematicides that have been successful, but are usually expensive to use, and in

the field soil fumigation with chemicals is a more practical and a cheaper method. Damage to plants is usually by lowering crop production, but can also disfigure and reduce market value of crops. This group of nematodes attack over 2,000 species of plants in cultivation and reduces crop production by on average 5%, but can be much higher during heavy infestation.

(b) **Cyst Nematodes (*Heterodera/Globerella*)**
A diverse group of nematodes mostly in temperate parts of the world. Some can have a limited region for infection and some attack only a few crops. Most susceptible plants are tomato, potato and eggplant (for *Globerella*). *Heterodera* on the other hand are most important in cereals, soybean, sugar beet, spinach and crucifers. Soil fumigation is the only practical method of control but may not be economically viable for all crops, and as well nematodes cannot be totally eliminated in most cases. Soybean are the most susceptible crop in North America, Europe, Middle East and Australia where in heavy infested fields it can cause between 25% and 50% loss.

(c) **Citrus Nematode (*Tylenchulus*)**
This nematode is present and common in nearly all parts of the world where citrus is cultivated. In some areas different strains or races are also capable of attacking other crops like grapes and olives in close proximity. Infection results in slow decline in trees, leaves can turn yellow many drop leaves early before some die-back symptoms appear. Citrus nematodes are easily transferred from soil to soil and control of citrus nematode is mostly based on prevention and avoiding re-introduction. Nursery stocks are treated with hot water or a combination of hot water and nematicides before propagation for new plantings.

(d) **Lesion Nematodes (*Pradylenchus*)**
Lesion nematodes occur in all parts of the world where they can attack most field crops, vegetables, fruit trees and many ornamental plants. Plants grow poorly due to root damage and produce lower yields; many plants after some time can die completely. Control is by treatment of the soil with nematicides where control is good, but often not complete. Eventually crop rotation will be necessary.

(e) **Burrowing Nematode (*Rodopholus*)**
Most of these nematodes occur in tropical and sub-tropical regions of the world. Banana plants are particularly susceptible and the disease is called by a few names; banana root rot, blackhead toppling disease or decline of banana. Banana plants are so badly affected where the profitable life of infected banana can be as short as 1 year. The nematodes can also affect severely avocado, tea, black pepper, coconut, coffee, sugar cane, corn, citrus, vegetables, fruit and even ornamental plants. Control can only be achieved by nematode-free plantlets and soil fumigation.

(f) **Stem and Bulb Nematode (*Ditylenchus*)**
This nematode occurs world-wide but is at its most destructive in temperate climates. It can be one of the most destructive of nematodes attacking large numbers of plants including onion, hyacinth, tulip, oat, alfalfa and strawberry.

The nematode is so destructive that it can enter growing young roots after germination at or near the root growing tips (cap), as well as later in growing plantlets via the stomata. Control can be affected through long rotation strategies of crops (at least 2–3 years) and/or fumigation and formaldehyde treatment.

(g) **Seed Gall Nematode (*Aguira*)**
Wheat seed gall nematode is the best known and is prevalent in Europe, Asia, Africa and Australia. The nematode is quite large (3 mm long), overwinters in infected plants or seeds, and galls fall to the soil for re-infection. Fields infested should not be planted with wheat or rye for at least 1 year afterwards.

(h) **Foliar Nematodes (*Aphelenchoides*)**
Nematodes that can feed both ectoparasitically or endoparasitically on above ground parts of plants. Nematodes are most prevalent in ornamental plants like chrysanthemum, aster, dahlia, zinnia and strawberry. Defoliation can occur and if so flowers are affected by reduction in size and quality. Several sanitary practices are required for control, including keeping stems dry and propagation cuttings being treated with hot water.

(i) **Stubby Root Nematodes (*Paratrichedus/Trichedus*)**
Nematodes that occur world-wide and attack cereals, vegetables, shrubs and trees. They can cause stunting, reduced growth, reduced yield and poor quality, but rarely or seldom cause death of plants. Control is mainly through the application of nematicides, but these chemicals are rarely very effective.

(j) **Pine Wilt and Palm Red Ring Disease (*Buraphelunchus*)**
Nematode can cause severe wilt of several tree species, but the nematode has developed a special symbiotic relationship with insects, mainly beetles and weevils which it uses as vectors. Pine wilt can be severe in *Pinus* species and can affect different parts of the tree. Pine nematode vectors reside or burrow into the wood and can overwinter there; new nematodes develop in early spring. Control may involve an insecticide to control the insect vectors, burying any dead wood, however these measures are sometimes very difficult to introduce in pine plantation forests.

3.4.3 Control Measures

Several methods of effective control of nematodes are available, but expense and type of crop may influence the control methods chosen. Control is usually affected through a number of methods, these include cultural practices (ie crop rotation, fallow and cover crops), resistant varieties to nematodes, physical agents like heat treatment and flooding, antagonistic bacteria and fungi and chemicals of various types in the nematocide group. Cultural practices have been successful for decades against nematodes, and as in all cases of pathogen control, exclusion of the nematodes and not introducing them into an area is the best option (Agrios 1997).

3.5 *Protozoa and Insects*

3.5.1 Protozoa

Certain types of flagellate protozoa (*Phytomonas*) have been associated with plant diseases. The taxonomy of the protozoa that have been associated with plant diseases is rather complex except that most, if not all inhabit the phloem sieve tubes of their hosts. It is thought that parasitic protozoa inhibit phloem function, and in this way cause decreased growth, but the mechanism of action is not clear (Agrios 1997). There are four important diseases in this group which do affect some commercially important crops:

(a) **Phloem Necrosis of Coffee** – in Malaya, Guyana, Brazil, San Salvadore and Columbia. The flagellate pathogen is *Phytomonas*, which can be traced from the roots to the trunk where it appears to migrate vertically in the phloem and spreads vertically through the sieve plates.
(b) **Hartrot of Coconut Palm** – in Malaya, Colombia, Ecuador and Trinidad-Tobago. The pathogen is a flagellate *Phytomonas* and is present in the sieve elements of younger leaves, plugging the sieve plates as the elements mature. In this case insects appear to be the vectors for the protozoa.
(c) **Sudden Wilt of Oil Palm** – in South America, especially in Columbia and can spread rapidly in plantations; and is caused by *Phytomonas*. The pathogen is similar to Coconut Hartrot and an insect appears also to be a vector for spread.
(d) **Empty Root of Cassava** – especially prevalent in Brazil where the root system develops poorly, roots remain small and slender. The organism can be transmitted during grafting and spreads rapidly in the field; probably also by an insect vector.

3.5.2 Insects

(a) **Introduction**

Insects are important pests of many plants world-wide because of the damage inflicted by direct feeding, but also because they are vectors or provide infection sites for plant pathogens. The relative importance of insects compared to other biotic stresses (ie pathogens or weeds) depends greatly on the crop and its location. Tropical crops usually can suffer more direct feeding damage from a wider variety of insect pests than the same crops in cold seasonal climates (Berg 1991a). However crops grown in temperate areas or in cold climates are more likely to be attacked by aphids and the viruses they transmit (Heie 1994). The economic importance and damage of insect herbivore will also greatly depend on the life stage and tissues of the plant they attack, and their method of feeding. Insects can feed and damage plants by:

1. Damaging seeds in store or in the field.
2. Damage pods and flowers.
3. Damage leaves and/or may also be sap feeding.

(b) **Defence**

Most plants rely on a suite of defences for protection against insect pests; such as:

1. Accumulation of anti-nutritional compounds during germination (eg lectins).
2. Chemical defence localisation or translocated throughout the plant.
3. Structural defence including simple and glandular hairs (ie trichomes).

All of these defences can act directly on the herbivour pests themselves by:

1. Deterring herbivour feeding (antiserosis).
2. Suppressing herbivour growth and development (antibiosis).
3. Minimising damage symptoms (tolerance).
4. Increase herbivour mortality, directly or indirectly (toxins).

Many of the mechanisms important to insect defence have been bred out of many cultivated crops, mainly because they also detract from the taste and texture of edible plants (Clement 2002). Breeders are only now realising this mistake and are deliberately trying to re-introduce some of these defences in many crops. But achieving pest resistance without reducing agronomic quality has been a huge problem to overcome. An exception to this has been breeding resistance to aphids in alfalfa/lucerne, in part due to the simple and dominant inheritance pattern that many aphid resistance genes have (Edwards and Singh 2006).

(c) **Signalling Pathways**

Similar to plant-pathogen interactions, where defences are induced to improve insect resistance in the plant. However our knowledge of these signalling pathways and downstream changes is not well developed. The best known are chewing insects where potential elicitors have been identified. These include lytic enzymes like β-glucosidase and glycine oxidase, and fatty acid-amino acid conjugates (Kessler and Baldwin 2002). Genomic approaches are being used to study plant responses to chewing insects and primarily have involved clear overlap with pathogen induced and wound induced responses, but some specific effects caused by the insects themselves have been described (Ferry et al. 2004). Voelckel et al. (2004) compared changes brought about by simple, sequential and simultaneous insect attack. They found quite different profiles of metabolites obtained following these different types of attacks, which makes studies of pathways complex.

While concentrating on the plants themselves, additional interactions also open important research areas of what the insects themselves are doing to try to overcome plant resistance (Edwards and Singh 2006). These include insect developed mechanisms like:

1. Contain potent proteinase inhibitors and actively up-regulate proteases.
2. Detoxifying enzymes such as glutathione-S-transferase and cytochrome P450.
3. Minimise and regulate the signals that are part of the plants defence response.

3.6 Higher Plant Taxa (Weeds and Parasitic Plants)

3.6.1 Introduction

Weeds have major economic, environmental and social impacts causing damage to agriculture, natural landscape, waterways and coastal areas. Weeds can impact severely on agriculture by competing with production, contaminating produce and poisoning livestock. Some weeds cause severe allergenic reactions in people. It has been estimated that the cost of weeds to Australian agriculture and horticulture alone is in the vicinity of $4 billion per year. Worldwide weed plants make-up more than one third of the costs to agricultural products (US$350 billion per year) (Pimentel 2002). Once established, weeds pose an ongoing challenge and cost to government, agriculture and the community. Weeds can be imported or enter a country or area by increase of international movements of goods and produce, and this can be legal, illegal or even inadvertently. After establishment in an area the spread or control of this spread can be a major issue to address. Weeds are resilient and very successful in spreading, and establish easily by their very nature. They often respond to changes in environmental conditions very quickly and the causes of these changes in conditions can be varied; including a number of factors such as:

(a) Pathogens and vectors
(b) Pollinators
(c) Storm damage and floods
(d) Fire and frequency of burning
(e) Vegetation clearing
(f) Changes in land-use pattern.

3.6.2 Herbicides

Total reliance on herbicides for weed control is unsustainable with the spread of herbicide resistance and environmental need to reduce the use of pesticides. Herbicide resistance can be a major limitation to crop production around the world, and has spread rapidly to a broad range of weed species (Heap 2004). This rapid development of resistance may have been helped by recent agricultural practices, planting semi-dwarf varieties, poor weed competitive cultivars and a high dependence on herbicides for control. All of these factors are forcing plant breeders to think about different strategies to combat weeds, and even breed for more weed competitiveness in future. Coleman et al. (2001) found that the heritability of important traits for more weed competitiveness was low due to large 'genotype X year' interactions, making genetic gains through phenotypic selection difficult.

However advances are being made in the breeding of strongly competitive crops that have high tolerance to weed pressure and therefore one can maintain high yields in the presence of weeds, and this can be a low cost option. In this regard the newly developed Incremental Crop Tolerance (ICT) measure proposed by Lemerle et al. (2006)

is an interesting tool for breeders. ICT can be applied to existing and new cultivars to reflect increased yield differences between genotypes associated with tolerance to weeds, over and above differences in the underlying yield potential.

3.6.3 Biological Control

In Europe a recent report has concluded that research in biological control of weeds was lagging well behind other countries such as North America, Australia, New Zealand and even South Africa (Sheppard et al. 2006). Classic biological control over the last 200 years or so has in many countries slowly but surely proven itself to be the only low cost, low risk and viable long term control of weeds; but it was not always reliable as a long-term weed control method in the past. Classic biological control uses a co-evolved and specific antagonist of the weed from its native range, which is then screened for effects to any participating non-target native or commercially important species in the new area. The antagonists are then released into the weed population for control. This approach has a long history compared with many other approaches to active pest management (ie agrochemicals).

Classic biological control has had a historically tarnished reputation, largely due to early unscientific and uncontrolled release of mainly invertebrate predators, to the detriment and even decimate native species. In recent cases as well, insect biological control agents have spread beyond their intended targets (Roy et al. 2005). Biological control of weeds now adopt a precautionary approach using the most specific antagonistic invertebrate organisms or even microorganisms against the target. They follow best practice scientific risk analysis and regulator approval before any release is approved (Sheppard et al. 2003). When conditions above are met and practiced, negative effects have proven to be predictable and releases have recently not caused problems.

3.6.4 Parasitic Higher Plants

More than 2,500 species of higher plants are known to live parasitically on other plants. Relatively few of these cause important diseases or loss in production on agricultural crops. However there are a few pests:

(a) **Dodders** – widely distributed in Europe and North America, and crops that suffer losses include alfalfa, corn, sugar beet, several ornamentals and potato. Dodders encircle plant stems and lower leaves and can easily elongate to adjacent plants where they send haustoria within the stem tissue. Infected host plants become weakened, vigour declines, produce poor yields and may kill plants. Prevention is mainly by not introducing dodder seeds and plants into fields, cleaning equipment and tools; however drastic measures like cutting and slashing or burning to kill both the host and dodder may be required. Frequent tilling and herbicides are other methods, especially the use of pre-emergent herbicides to kill dodder seeds.

(b) **Witchweed** – a serious parasite weed in Africa, Asia and Australia, where it can parasatise important economic plants like corn, sugarcane, rice, cowpeas, tobacco and other small grains. Affected plants remain stunted, wilt and turn yellow; death may follow. Witchweed seeds germinate near roots of the host plant and haustoria attack the host roots, and then produces the above ground part of the witchweed plant. Witchweed is difficult to control and best to avoid its introduction at all cost. Plowing material under and use of herbicides are possible, but with variable effects, and the use of resistant cultivars is an additional tool.

(c) **Broomrapes** – occur mainly in warm and dry regions world-wide. They can attack many species of herbaceous dicotyledon crop plants. In some areas up to 70% of crops are lost, but the losses are usually lower. Seeds of broomrapes germinate and grow towards the roots of the host plant and attach to them. Apressoria can grow and extend to the xylem of the roots absorbing nutrients and water, weakening the plant considerably. Control of broomrapes mainly depends on preventing the introduction of seeds into an area.

4 Conclusions/Future Directions

Predicted growth in world population and the likely effects of climate change will pose a serious challenge to crop production and food security, particularly in developing countries. Conventional plant breeding has proven to be effective for the improvement of many traits of economic importance, especially disease resistance. Plant breeders have extended the range of adaptation in crops and improved agronomic traits (eg seed yield potential). Molecular plant breeding techniques have proven to be valuable tools for the improvement of quantitative inherited traits and traits that do not allow phenotypic selection; and some have already proven to be effective for marker assisted selection (MAS) of biotic related stress. Therefore screening of genetic resources coupled with conventional breeding still offers high discovery potential, especially if researchers take full advantage of the landraces and wild species available world-wide. It is therefore clear that an integrated approach involving research involving both the plant and the biotic stress organism, and their biological interactions are still good approaches.

The use of MAS for some complex biotic traits will require a markedly better understanding of the genetic basis of expression of these traits. Genomic mapping of the traits should be a considerable help in the future, as well as a better understanding of the basis for genotype-environment interactions. A reliable, easy and efficient transformation system needs to be further developed. Traits already using genetic manipulation (GM) technology (such as herbicide and insect tolerance) have the potential to lower the cost of crop production and reduce land and soil degradation. At the moment only *Agrobacterium*-mediated and particle bombardment can be considered to be routine. Gene targeting for example is still an elusive goal for the plant biotechnologist. In this regard the ability of a plant to defend against pathogens is important, and contributing to this successful defence are the role of protein kinases and phosphatases (signalling proteins), and the recent development of proteomics which is likely to increase our knowledge of plant-pathogen-signal interactions.

The area of complex metabolic engineering and gene modification is still less successful than conventional and molecular plant breeding. Nevertheless the combination of plant breeding and targeting genes in a biotechnological sense can contribute substantially to human well-being by providing renewable sources of food, feed, chemicals, pharmaceuticals, energy, and for the creation of a sustainable environment for the developed, as well as the developing world. During the last two decades the development and adoption of transgenic technology has progressed rapidly to the point that almost every significant food crop has been successfully transformed. However, the issue of consumer acceptance of such biotech crops should be seriously taken into consideration before development and release of transgenic crops. Finally, a number of important food crops are produced and consumed in developing countries. This means that breeding programs, especially those in developing countries need access to knowledge and methodology that will ensure the use of these enabling technologies of molecular plant breeding in developing countries.

Future impacts of biotechnology in crop production is predicted to be in the areas of:- (i) developing new hybrid crops based on genetic male-sterility, (ii) exploiting transgenic apomixes to fix hybrid vigour in inbred crops, (iii) increase resistance to insect pests, diseases, and abiotic stress, (iv) improve effectiveness of bio-control agents, (v) enhance nutritional value (vitamin A, zinc and iron) of crops and post-harvest quality, (vi) increase efficiency of soil phosphorus uptake and nitrogen fixation, (vii) improve adaptation to soil salinity and metal toxicity, (viii) understanding the nature of gene action and metabolic pathways, (ix) increase photosynthetic activity, sugar and starch production, and (x) production of pharmaceuticals and vaccines.

References

Agarwal PK, Agarwal P, Reddy MK, Sopory SK (2006) Role of DREB transcription factors in abiotic and biotic stress tolerance in plants. Plant Cell Rep 25:1263–1274

Agrios G (1997) Plant pathology, 7th edn. Academic, London

Akhond MAY, Machray GC (2009) Biotech crops: technologies, achievements and prospects. Euphytica 166:47–59

Alexandratos N (1999) World food and agriculture: outlook for the medium and long term. Proc Natl Acad Sci 96:5908–5914

Amano N, Tanaka T, Numa H, Sakai H, Itoh T (2010) Efficient plant gene identification based on interspecies mapping of full-length cDNAs. DNA Res 17:1–9

Asenio-Manzanera MC, Asenio C, Singh SP (2006) Gamete selection for resistance to common and halo-bacterial blights in dry bean intergene pool populations. Crop Sci 46:131–135

Backes G, Ostergard H (2008) Molecular markers to exploit genotype-environment interactions of relevance in organic growing systems. Euphytica 163:523–531

Bartel DP (2004) MicroRNA: genomics, biogenesis, mechanism, and function. Cell 116:281–297
Basha SD, Sujath AM (2009) Genetic analysis of *Jatropha* species and interspecific hybrids of *Jatropha carcas* using nuclear and organelle specific markers. Euphytica 168:197–214
Beaver JS, Macchiavelli R (1998) Breeding strategies for pyramiding genes for resistance. Annu Rep Bean Improv Coop 41:141–142
Beaver JS, Osorno JM (2009) Achievements and limitations of contemporary common bean breeding using conventional and molecular approaches. Euphytica 168:145–175
Beranova M, Rakousk YS, Vaurova Z, Skalicky T (2008) Sonication assisted *Agrobacterium*-mediated transformation enhances the transformation efficiency in flax (*Linum usitatissimum* L.). Plant Cell Tissue Organ Cult 94:253–259
Berg H (1991a) Pesticide use in rice and rice-fish farms in the Mekong Delta, Vietnam. Crop Prot 20:897–905
Berg P (1991b) Reverse genetics: its origins and prospects. Biotechnology 9:342–344
Bernardo R (2008) Molecular markers and selection for complex traits in plants: learning from the last 20 years. Crop Sci 48:1649–1664
Bliss FA (1993) Breeding common bean for improved biological nitrogen fixation. Plant Soil 152:71–79
Broothaerts W, Mitchell HJ, Weir B, Kaines S, Smith LMA, Yang W et al (2005) Gene transfer to plants by diverse species of bacteria. Nature 433:629–633
Broughton WJ, Hernandez G, Blair M, Beebe S, Gepts P, Vandereyen J (2003) Beans (*Phaseolus* spp.) – model food legume. Plant Soil 252:55–128
Cassman KG (1999) Ecological intensification of cereal production systems: yield potential, soil quality, and precision agriculture. Proc Natl Acad Sci 96:5952–5959
Charcosset A, Moreau L (2004) Use of molecular markers for the development of new cultivars and the evaluation of genetic diversity. Euphytica 137:81–94
Chawla HS (2009) Introduction to plant biotechnology (3 rd ed). Science Publishers Incorporated, India
Christiansen MJ, Feenstra B, Skovgaard IM, Andersen SB (2006) Genetic analysis of resistance to yellow rust in hexaploid wheat using a mixture model of multiple crosses. Theor Appl Genet 112:581–591
Clark EF (2006) Environmental risks of genetic engineering. Euphytica 148:47–60
Clement SL (2002) Insect resistance in the wild relatives of food legumes and wheat, In: McComb JA (ed) Plant breeding for the 11th millennium, Proceedings of the 12th Australasia plant breeding conference, Perth, Australia
Codex Alimentarius Commission (CAC) (2003) Joint FAO/WHO food standard programme, 25th session Appendix III Guideline for the conduct of food safety assessment of foods derived from recombinant DNA plants, Codex Alimentarius Commission, Rome
Coleman RK, Gill GS, Rebetze GJ (2001) Identification of quantitative trait loci for traits conferring competitiveness in wheat (*Triticum aestivum* L.). Aust J Agric Res 52:1235–1246
Collard BCY, Jahufer MZZ, Browner JB, Pang ECK (2005) An introduction to markers, quantitative trait loci (QTL) mapping and marker-assisted selection for crop improvement: the basic concepts. Euphytica 142:169–196
Curtis IS (2005) Production of transgenic crops by the floral-dip method. In: Pena L (ed) Transgenic plants: methods and protocols, vol 286, Methods in molecular biology. Humana Press, Totowa
Cuyno LCM, Norton GW, Rola A (2001) Economic analysis of environmental benefits of integrated pest management: a Philippine case study. Agric Econ 25:227–233
Daniell H, Khan MS, Allison L (2002) Milestones in chloroplast genetic engineering: an environmentally friendly era in biotechnology. Trends Plant Sci 7:84–91
Davey MR, Anthony P (2010) Plant cell cultures: essential methods. CABI Publishing, Oxford, UK
Dayteg C, Tuvesson S, Merker A, Jahoor A, Klodinska-Brantestam A (2007) Automation of DNA marker analysis for molecular breeding in crops: practical experience of a plant breeding company. Plant Breed 126:410–415

De Filippis LF (1999) To mark an end and begin a new era in plant tissue culture. In: Johnson KA, McFarlane IJ (eds) Plant tissue culture at the edge of the new millennium. UTS Publication, Sydney

De Filippis LF, Hoffmann H, Hampp R (1996) Identification of somatic hybrids of tobacco generated by electrofusion and culture of protoplasts using RAPD-PCR. Plant Sci 121:39–46

De Filippis LF, Hampp R, Ziegler H (2000) Membrane permeability changes and ultrastructural abnormalities observed during protoplast fusion. J Plant Physiol 156:628–634

de Fossard RA (2000) Commercial micropropagation. CD produced by Magpie Digital Media, Eagle Heights, Queensland

de Fossard RA (2007) Plant tissue culture propagation. CD produced by Magpie Digital Media, Eagle Heights, Queensland

de Oliveira DJ, Alzate-Marin AL, Borem A, Fagundes SD, de Barros EG, Moreira MA (2005) Molecular marker-assisted selection for development of common bean lines resistant to angular leaf spot. Plant Breed 124:572–575

Dilrewanger E, Graziano E, Joobeur T, Garriga-Caldere F, Cosson P, Howad W et al (2004) Comparative mapping and marker-assisted selection in Rosaceae fruit crops. Proc Natl Acad Sci 101:9891–9896

Drobnik J (2008) Time to relax GM regulations in Europe. Plant Cell Tissue Organ Cult 94:235–238

Dyson T (1999) World food trends and prospects to 2025. Proc Natl Acad Sci 96:5929–5936

Edwards O, Singh KB (2006) Resistance to insect pests: what do legumes have to offer? Euphytica 147:273–285

Fehr WR (1987) Principles of cultivar development, vol 1, Theory and techniques. Macmillan Publishing, New York

Ferry N, Edwards MG, Gatehouse JA, Gatehouse AMR (2004) Plant-insect interactions: molecular approaches to insect resistance. Curr Opin Biotechnol 15:155–161

Flint-Garcia SA, Thornsberry JM, Buckler ES (2003) Structure of linkage disequilibrium in plants. Annu Rev Plant Biol 54:357–374

Francia E, Tacconi G, Crosatti C, Barabaschi D, Bulgarelli D, Dall'Aglio E et al (2005) Marker-assisted selection in crop plants. Plant Cell Tissue Organ Cult 82:317–342

Garcia R, Robinson RA, Agfuilar AJ, Sandoval S, Guzman R (2003) Recurrent selection for quantitative resistance to soil-borne diseases in beans in the Mixteca region, Mexico. Euphytica 130:241–247

Gebhardt C, Ballvora A, Walkemeir B, Oberhagemann P, Schuler K (2004) Assessing genetic potential in germplasm collections of crop plants by marker-trait association: a case study for potatoes with quantitative variation of resistance to late blight and maturity type. Mol Breed 13:93–102

Goodman RE, Vieths S, Sampson HA, Hill D, Ebisawa M, Taylor SL et al (2008) Allergenicity assessment of genetically modified crops – what makes sense? Nat Biotechnol 26:73–81

Hampp R, Hoffmann H, Schonherr K, Johann P, De Filippis LF (1997) Fusion and metabolism of plant cells and protoplasts as affected by microgravity. Planta 203:S42–S53

Hansen G, Wright MS (1999) Recent advances in the transformation of plants. Trends Plant Sci 4:226–231

Harwood WA, Chen DF, Creisses GP (1996) Transformation of pollen and microspores. In: Mohan JM, Sopory SK, Veilleux RE (eds) In vitro haploid production in higher plants. Kluwer Academic, Dordrecht

Heap IM (2004) The international survey of herbicide resistant weeds, http://www.weedscience.org

Heberle-Bors E, Stoger E, Touraev A, Zarsky V, Vicente O (1996) *In vitro* Pollen cultures: progress and perspectives. In: Mohapatra SS, Knox RB (eds) Pollen biotechnology, gene expression and allergen characterization. Chapman and Hall, New York

Heie OE (1994) Why are there so few aphid species in the temperate areas of the southern hemisphere? Eur J Entomol 91:127–133

Herman EB (2009) Genetic modification of plants: methods and applications 2005–2009. Agricell Reports Publication, Shrub Oak

Hess D, Dressler K (1989) Tumor transformation of *Petunia hybrida* via pollen co-cultivated with *Agrobacterium tumefaciens*. Bot Acta 102:202–207

Hoang TML, De Filippis LF, Le XT (2009) Salt tolerance and screening for genetic changes in rice mutants after gamma irradiation using RAPD and microsatellite (RAMP) markers. Open J Plant Sci 2:62–69

Horst RK (2008) Westcott's plant disease handbook, 7th edn. Springer, Berlin

Iida S, Terada R (2004) A tale of two integrations, transgene and T-DNA: gene targeting by homologous recombination in rice. Curr Opin Biotechnol 15:132–138

Jackson S, Rounsley S, Prugganan M (2006) Comparative sequencing of plant genomes: choices to make. Plant Cell 18:1100–1104

James C (2007) Global status of commercialised biotech/GM crops: 2007, ISAAA Brief No. 37. ISAAA, Ithaca

Jorrin JV, Rubiales D, Dumas-Gaudot E, Recorbet G, Maldonado A, Castillejo MA, Curto M (2006) Proteomics: a promising approach to study biotic interactions in legumes. A Rev Euphytica 147:37–47

Kessler A, Baldwin IT (2002) Plant responses to insect herbivory. Annu Rev Plant Biol 53:299–328

Kumar S, Allen GC, Thompson WF (2006) Gene targeting in plants: fingers on the move. Trends Plant Sci 11:159–161

Langridge P, Brettschneider R, Lazzeri R, Lorz H (1992) Transformation of cereal via *Agrobacterium* and the pollen pathway: a critical assessment. Plant J 2:631–638

Lee D, Natesan E (2006) Evaluating genetic containment strategies for transgenic plants. Trends Biotechnol 24:109–114

Lemerle D, Smith A, Verbeek B, Koetz E, Lockley P, Martin P (2006) Incremental crop tolerance to weeds: a measure for selecting competitive ability in Australian wheats. Euphytica 149:85–89

Lin HJ, Zhang ZM, Ou SY, Gao SB, Pan GT (2009) Review of plant miRNAs in environmental stressed conditions. Res J Agric Biol Sci 5:803–814

Liu Q, Kasuga M, Sakuma Y, Abe H, Miura S, Yamaguchi-Shinozaki K, Shinozaki K (1998) Two transcriptional factors, DREB1 and DREB, with an EREBP/AP2 DNA binding domain separate two cellular signal transduction pathways in drought- and low-temperature-responsive gene expression, respectively, in *Arabidopsis*. Plant Cell 10:1391–1406

Macchiavelli R, Beaver JS (2001) Effect of number of seed bulked and population size on genetic variability when using the multiple-seed procedure of SSD. Crop Sci 41:1513–1516

Magana-Gomez JA, de la Barca AMC (2008) Risk assessment of genetically modified crops for nutrition and health. Nutr Rev 67:1–16

Malik M (2008) Comparison of different liquid solid culture systems in the production of somatic embryos from *Narcissus* ovary explants. Plant Cell Tissue Organ Cult 94:337–345

Michelmore RW (1995) Molecular approaches to manipulation of disease resistance genes. Annu Rev Phytopathol 33:393–427

Miklas PN, Johnson WC, Delorne R, Gepts P (2001) QTL conditioning, physiological resistance and avoidance to white mould in dry bean. Crop Sci 41:309–315

Miklas PN, Kelly JD, Beebe SE, Blair MW (2006) Common bean breeding for resistance against biotic and abiotic stresses: from classic to MAS breeding. Euphytica 147:105–131

Moreau L, Charcosset A, Hospital F, Gallais A (1998) Marker-assisted selection efficiency in populations of finite size. Genetics 148:1353–1365

Muehlbauer FJ, Cho S, Sarker A, McPhee KE, Coyne CJ, Rajesh PN, Ford R (2006) Application of biotechnology in breeding lentil for resistance to biotic and abiotic stress. Euphytica 147:149–165

Munkacsi AB, Stoxen S, May G (2008) *Ustilago maydis* populations tracked maize through domestication and cultivation in the Americas. Proc R Soc B 275:1037–1046

Murphy DJ (2007) Plant breeding and biotechnology: societal context and the future of agriculture. Cambridge University Press, Cambridge

Myers N (1999) The next green revolution: its environmental underpinnings. Curr Sci 76:507–513

Navratilova B, Luhova L, Petrivalsky M (2008) Effect of UV-C irradiation on mesophyll protoplasts of *Cucumis sativus*. Plant Cell Tissue Organ Cult 94:313–318

Nelson-Smith D (1995) Food or famine politics, economics and science in the world's food supply, Brighton Crop Protection Conference. Weeds 1:3–15

Neuman KH, Kumar AT (2009) Plant cell and tissue culture – a tool in biotechnology. Science Publishers Incorporated, India

Nguyen HT, Wu XL, Vuong TD, Sleper DA, Shannon GJ, Stacey G (2007) Genome maps, genetic diversity and marker-assisted selection for soybean improvement. Mol Plant Breed 5:196–198

Niu QW, Lin SS, Reyes JL, Chen KC, Wu HW, Yeh SD et al (2006) Expression of artificial microRNAs in transgenic *Arabidopsis thaliana* confers virus resistance. Nat Biotechnol 24:1420–1428

Oerke EC, Dehne HW (2004) Safeguarding production – losses in major crops and the role of crop protection. Crop Protect 23:275–285

Oerke EC, Dehne HW, Schonbeck F, Weber A (1994) Crop production and crop protection – estimated losses in major food and cash crops. Elsevier Science, Amsterdam

Pimentel D (ed) (2002) Biological invasions: economic and environmental costs of alien plant, animal and microbe species. CRC Press, Boca Raton

Pinstrup-Andersen P, Cohen M (2000) Modern biotechnology for food and agriculture: risks and opportunities for the poor. In: Persley GJ, Lantin MM (eds) Agricultural biotechnology and the poor. Consultative Group on International Agricultural Research, Washington, DC

Razdan MK (2003) Introduction to plant tissue culture. Science Publishers Incorporated, India

Renato-Corte H, Patto Ramalhol MA, Avelar Goncalves FM, de Barbo CA, Abreu AF (2002) Natural selection for grain yield in dry bean populations bred by the bulk method. Euphytica 123:387–393

Ritzenthaler C (2005) Resistance to plant viruses: old issues, new answers? Curr Opin Biotechnol 16:118–122

Roman-Haviles B, Kelly JD (2005) Identification of quantitative trait loci conditioning resistance to *Fusarium* root rot in common bean. Crop Sci 45:1881–1890

Roy H, Rowland F, Brown P, Ware R, Majerus M (2005) Ecology of the harlequin ladybird, a new invasive species. Br Wildl 16:403–407

Schmidt K (2003) Ergebnisse der Meldungen fur Pflanzenschutzmittel und Wirkstoffe nach des Pflanzenschutzgesetzes fur die Jahre 1999, 2000 und 2001 im Vergleigh zu 1998. Nachr Deut Pflanzenschutzd 55:121–133

Sharma HC, Crouch JH, Sharma KK, Seetharama N, Hash CT (2002) Applications of biotechnology for crop improvement: prospects and constraints. Plant Sci 163:381–395

Sheppard AW, Hill R, DeClerck-Floate RA et al (2003) A global review of risk-benefit-cost analysis for the introduction of classical biological control agents against weeds: a crisis in the making? Biocont News Inform 24:91N–108N

Sheppard AW, Shaw RH, Sforza R (2006) Top 20 environmental weeds for classic biological control in Europe: a review of opportunities, regulations and other barriers to adoption. Weed Res 46:93–117

Shillito R (1999) Methods of genetic transformation: electroporation and polyethylene glycol treatments. In: Vasil I (ed) Molecular improvement of cereal crops. Kluwer, Dordrecht

Simmonds NW (1989) Rubber breeding. In: Webster CC, Baulkill DW (eds) Rubber. Longman Scientific and Technical, Harlow

Singh SP (1994) Gamete selection for simultaneous improvement of multiple traits in common bean. Crop Sci 34:352–355

Singh SP (1999) Production and utilization. In: Singh SP (ed) Common bean improvement in the twenty-first century. Kluwer, Boston

Singh SP (2001) Broadening the genetic base of common bean cultivars: a review. Crop Sci 41:1659–1675

Singh SP, Cardona C, Morales FJ, Pastor-Corrales MA, Voysest O (1998) Gamete selection for upright Carioca bean with resistance to five diseases and a leafhopper. Crop Sci 38:666–672

Smith RH, Hood E (1995) *Agrobacterium tumefaciens* transformation of monocotyledons. Crop Sci 35:301–309

Smith CR, Saunders JA, van Wert S, Cheng JP, Matthews BF (1994) Expression of GUS and CAT activities using electrotransformed pollen. Plant Sci Limerick 104:49–58

Southgate EM, Davey MR, Power JB, Marchant R (1995) Factors affecting the genetic engineering of plants by microprojectile bombardment. Biotechnol Adv 13:631–651

Strange RN, Scott PR (2005) Plant disease: a threat to global food security. Annu Rev Phytopathol 43:83–116

Tilman D (1999) Global environmental impacts of agricultural expansion: the need for sustainable and efficient practices. Proc Natl Acad Sci 96:5995–6000

Todorovska E, Christov N, Slavov S, Christova P, Vassilev D (2009) Biotic stress resistance in wheat – breeding and genomic selection implications. Biotechnol Biotechnol Eq 23:1417–1426

Torres AM, Roman B, Avila CM, Satovic Z, Rubiales D, Sillero JC, Cubero JI, Moreno MT (2006) Fava bean breeding for resistance against biotic stress: towards application of marker technology. Euphytica 147:67–80

Trieu AT, Burleigh SH, Kardailsky IV, Maldonado-Mendoza IE, Versaw WK, Blaylock LA et al (2000) Transformation of *Medicago truncatula* via infiltration of seedlings or flowering plants with *Agrobacterium*. Plant J 22:531–541

Tuvesson S, Dayteg C, Hagberg P, Manninen O, Tanhuanpaa P, Tenhola-Roininen T et al (2007) Molecular markers and double haploids in European plant breeding programmes. Euphytica 158:305–312

van der Leede-Plegt LM, van de Ven BC, Schilder M, Franken J, van Tunen AJ (1995) Development of pollen-mediated transformation for *Nicotiana glutinosa*. Transgenic Res 4:77–86

Vinterhalter B, Jankovic T, Savik K, Nikolic R, Vinterhalter D (2008a) Propagation and xanthone content of *Gentianella austriaca*. Plant Cell Tissue Organ Cult 94:329–335

Vinterhalter B, Savic J, Plastisa J, Raspor M, Ninkovics S, Mitic N, Vinterhalter D (2008b) Nickel tolerance and hyperaccumulation in shoot cultures regenerated from hairy root cultures of *Alyssum murale* Waldst et Kit. Plant Cell Tissue Organ Cult 94:299–303

Voelckel C, Weisser WW, Baldwin IT (2004) An analysis of plant-aphid interactions by different microarray hybridisation strategies. Mol Ecol 13:3187–3195

Werner D (2005) Production and biological nitrogen fixation of tropical legumes. In: Werner D, Newton WE (eds) Nitrogen fixation in agriculture, forestry, ecology, and the environment. Springer, Dordrecht

Wilson KC (1999) Coffee, cocoa and tea. CABI Publishing, Oxford, UK

Xing T, Ouellet T, Miki BL (2002) Towards genomic and proteomic studies of protein phosphorylation in plant-pathogen interactions. Trends Plant Sci 7:224–230

Yudelman M, Ratta A, Nygaard D (1998) Pest management and food production. Looking to the future, food, agriculture and the environment, Discussion paper 25. International Food Policy Research Institute, Washington, DC

Chapter 7
Breeding Wheat for Salt Tolerance and Stem Rust Resistance

Makhdoom Hussain, Aziz ur Rehman, Imran Habib, Mumtaz Hussain, Nadeem Ahmad, Muhammad Arif Khan, Muhammad Hussain, and Faqir Muhammad

Contents

Abstract Fast and effective hydroponics screening technique that could identify physiological variation in salinity tolerance of wheat was applied. A set of 442, previously unexplored wheat varieties/lines representing a wide range of genetic diversity was planted as control in ½ strength Hoagland's nutrient media, whereas two sets of the same material were exposed to salt (NaCl) application under two treatments i.e., 10 dS m^{-1} and 20 dS m^{-1} for the first 2 years (2003–04 and 2004–05).

M. Hussain • A. ur Rehman (✉) • N. Ahmad • M.A. Khan • M. Hussain • F. Muhammad
Wheat Research Institute, AARI, Faisalabad, Pakistan
e-mail: makhdoomhussain@yahoo.com; aziz_kml@yahoo.com; nadeemahmad532@hotmail.com; marif_k@yahoo.com; hussain_pathologist@yahoo.com; faqirmuhammad98@yahoo.com

I. Habib
National Institute for Biotechnology and Genetic Engineering (NIBGE), Faiasalabad, Pakistan
e-mail: imranuaf@hotmail.com

M. Hussain
Plant Pathology Research Institute, Faisalsabad, Pakistan
e-mail: sahu_mumtaz@hotmail.com

M. Ashraf et al. (eds.), *Crop Production for Agricultural Improvement*,
DOI 10.1007/978-94-007-4116-4_7,

For the third year (2005–06), more intensive stress was applied with salinity levels of 12.5 dS m^{-1} and 25 dS m^{-1}. Salinity tolerance was defined as differences in biomass (root-shoot ratio) production in saline versus non-saline conditions over prolonged periods, of 3–4 weeks (seedling to pre booting stage). For this purpose parameters like shoot length, root length, shoot weight and root weight alongwith their relative ratios were measured. As a result of 3 year study 11 common salt tolerant varieties/lines were identified including Pasban-90, LU 26S, V-01078 (Seher 06) and V-01180. Under saline field conditions some of these lines like Gamdow-6, BAV 92//SAP/MON and Lakata-1 produced higher grain yield, whereas, Uqab-2000 was the best yielding in saline field conditions, although it was found non tolerant in hydroponic studies. Line Ning 8319 showed good resistance for stem rust comparable to Parula and Pavon. These selected lines are being used in specific breeding programs for the development of high yielding wheat varieties having high degree of salt tolerance and stem rust resistance.

Keywords Spring wheat • Salinity • Root-shoot ratio and hydroponics • Stem rust

1 Introduction

Vast areas of the world including Pakistan are rendered agriculturally unproductive due to high concentration of salts in the root zone. Out of 20.2 million hectares of cultivated land in Pakistan, 6.8 million hectares are affected with salinity and 2.67 million hectares of the affected area is located in Punjab. The salinity area has been categorized into four major classes (Table 7.1). It has been reported that 1390.03 thousands hectares were strongly saline lands, 804.8 thousands hectares moderately saline, while 472.4 thousands hectares saline areas in the world (Anonymous 2002).

Root zone salinity affects growth of wheat and reduces the number of fertile spikes and spike-bearing tillers (Steppuhn and Wall 1997). The reduced number of tillers, plant height and reduced root growth ultimately affect the yield. A high salt level interferes with the germination of new seeds. Salinity acts like drought on plants, preventing roots from performing their osmotic activity where water and nutrients move from an area of low concentration into an area of high concentration (Greenway and Munns 1980; Munns 1993). Therefore, because of the salt levels in the soil, water and nutrients cannot move into the plant roots.

Salt tolerance is usually defined as the percent biomass production in saline medium relative to plants in non-saline medium, after growth for an extended period of time. As soil salinity levels increase, the stress on germinating seedlings also increases. Perennial plants seem to handle salinity better than annual plants like wheat. In some cases, salinity also has a toxic effect on plants because of the high concentration of certain salts in the soil. Salinity prevents the plants from taking up the proper balance of nutrients they require for healthy growth. Ultimately, salt tolerance of

Table 7.1 Soil salinity rating and electrical conductivity value

Soil depth	Non-saline	Weakly saline	Moderately saline	Strongly saline	Very strongly saline
0–60 cm	<2 dS m^{-1}	2–4 dS m^{-1}	4–8 dS m^{-1}	8–16 dS m^{-1}	>16 dS m^{-1}
60–120 cm	<4 dS m^{-1}	4–8 dS m^{-1}	8–16 dS m^{-1}	16–24 dS m^{-1}	>24 dS m^{-1}

wheat is tested as yield from farmer's fields. However, evaluating field performance under saline conditions is notoriously difficult because of the variability of salinity within fields (Daniells et al. 2001). To address this problem, a rapid and more reliable hydroponics screening program was developed (Iqbal 2003). The germplasm was screened and tolerant lines were evaluated in the field.

A recent outbreak of new wheat stem rust race (Ug99), capable of parasitizing many commercial cultivars has threatened local and world wheat production and potentially 25% of the World's wheat crop is thought to be at risk (Reynolds and Borlauge 2006; Ayliffe et al. 2008). The rust has spread from African nations to Middle East. From where it is expected to reach the wheat growing regions of Asia, raising the possibility of a major epidemic (Stokstad 2007). Due to urgency of stem rust problem, salt tolerant material was got screened from Kenya and a breeding program was initiated to address these problems simultaneously for the development of high yielding wheat varieties.

2 Materials and Methods

2.1 Screening in Hydroponic Culture

Research Site

The research was conducted in the glasshouse of Wheat Research Institute, Faisalabad during the years 2003–04, 2004–05 and 2005–06.

Seed Material

For the year 2003–04, 442 wheat varieties/lines were tested, whereas in the year 2004–05, 496 varieties/lines were included in the salinity screening program. This material includes past and present wheat cultivars of Indo-Pak Subcontinent, promising lines (MTWV and NUWYT), and exotic germplasm (CIMMYT and ICARDA).

Screening Technique

Seed was cultured in sand filled pots. At two leaf stage, four seedlings of each genotype were transferred to thermopore sheets and plugged in holes especially made to hold the seedlings so that they do not drop down in the growth media. Special steel tubs (1 × 1 m) were made having 300-l capacity. Half strength Hoagland's solution (Hoagland and Arnon 1950) was used as nutrient medium. Sodium Chloride (NaCl) was used as a source of salinity.

Electrical Conductivity
Salt tolerance is mostly measured in terms of the stage of plant growth over a range of electrical conductivity (EC) levels. Three treatments with two salinity levels i.e. control, 10 dS m^{-1} and 20 dS m^{-1} (Iqbal et al. 1998) were applied to the wheat seedlings for the first 2 years (2003–04 and 2004–05). For the third year (2005–06), a more intensive selection process was adopted with salinity levels of 12.5 dS m^{-1} and 25 dS m^{-1}. Air pumps were used for proper root aeration and growth. Tap water (EC $\leq$ 2 dS m^{-1}) was used as a main component of the hydroponic medium.

After 3–4 weeks of seedling transplantation (Munns and Richard 2003), plants were plugged out of the thermopore sheets and data for shoot and root lenths and fresh and dry weights were taken for the following parameters:

Statistical Analysis
Data for all the above-mentioned parameters were taken from all four plants/genotypes and mean values calculated. Relative ratios were computed for shoot length, root length, shoot weight, root weight with the following formula (Zulfiqar et al. 2002).

$$\text{Relative salt tolerance} = \frac{\text{Mean performance in stress}}{\text{Mean performance in control}}$$

To calculate relative ratio of shoot length at 10 dS m^{-1} salinity level, shoot length of a genotype at 10 dS m^{-1} of salt was divided by shoot length of the same genotype under control. Similarly, to calculate relative performance of the same genotype at 20 dS m^{-1} salinity level the shoot length of the said genotype was divided by the performance under control. Same calculations were done with all other genotypes. Salt tolerant lines were selected on the basis of their mean relative ratios near or equal to unity (1.0). Genotypes deviating sharply from unity were rejected as they failed to show synchronized performance for the above mentioned parameters.

Frequency distribution of each parameter at two levels of salinity was calculated and population curves of each character were also plotted to study the population dynamics at different levels of salinity.

2.2 *Field Trial*

Sixteen wheat varieties/lines found tolerant in the hydroponic experiment along with four checks, i.e., Inqilab-91, Uqab-2000, AS-2002 and V-02192 were assessed for salt tolerance under field conditions at Biosaline Research Station, Pacca Anna, District T.T. Singh. Randomized Complete Block Design (RCBD) was used having six replications. Plot size was kept at 1.8 m × 3 m (5.4 m^2). There were six rows per plot each having 3 m length and row to row distance of 0.30 m. Soil type was sandy loam. Ground water was applied having EC of 4.8 dS m^{-1}, RSC 21.8 me L^{-1} and

SAR 40 me L^{-1}, whereas soil EC (upper 15 cm) was recorded as 11.0 dS m-1 at sowing and 12 dS m^{-1} at harvest. Sowing was done in dry soil with the help of single row cotton drill and then irrigation applied afterwards. The crop was sown on 15th of November 2006 and harvested on 20th of April 2006 and threshed a week after. Grain yield per plot (g) was recorded. Analysis of variance (ANOVA) was performed using MSTATC (4.1) computer software.

2.2.1 Screening for Stem Rust

The salt tolerant material along with some other lines was sent to KARI, Njoro, Kenya for screening against stem rust race Ug99 through the Wheat Coordinator, NARC, Islamabad during 2006 and 2007 and for its utilization in some breeding programs.

3 Results

The results of Hydroponics experiment and field trial are discussed below:

3.1 Hydroponics Experiment

Data for the characters under study was recorded from all the entries in all the three treatments. The data was compared and relative salt tolerance ratios were calculated for the characters which are discussed character wise as follow:

Shoot Length (cm)
In the first year of screening (2003–04) normal population curve (Fig. 7.1a) was observed at 10 dS m^{-1} salinity level for this character, whereas at 20 dS m^{-1} of salinity the population showed abnormal behavior. Highest frequency of 121 was found at the relative ratio of 0.6–0.9 for 10 dS m^{-1} salinity level. For 20 dS m^{-1} salinity level, highest frequency of 123 was observed in the relative ratio range of 0.3–0.6.

For the second year of screening (2004–05) normal population behavior (Fig. 7.1b) was observed at 10 dS m^{-1} salinity, whereas at 20 dS m^{-1} salinity more abnormal population behavior was recorded. The highest frequency i.e. 185 was observed at the range of 0.6–0.9 at 10 dS m^{-1} salinity, whereas highest frequency of 210 fell in the range of 0.3–0.6 at 20 dS m^{-1}.

For the third year more intense selection criteria were adopted with salinity levels of 12.5 dS m^{-1} and 25 dS m^{-1} to further refine our results. Normal population curve was observed (Fig. 7.1c) for the parameters under study at 12.5 dS m^{-1}, whereas at 25 dS m^{-1} salinity the population curve showed abnormal behavior as it peaked at a value much lower than unity (1.0). Highest frequency of genotypes fell

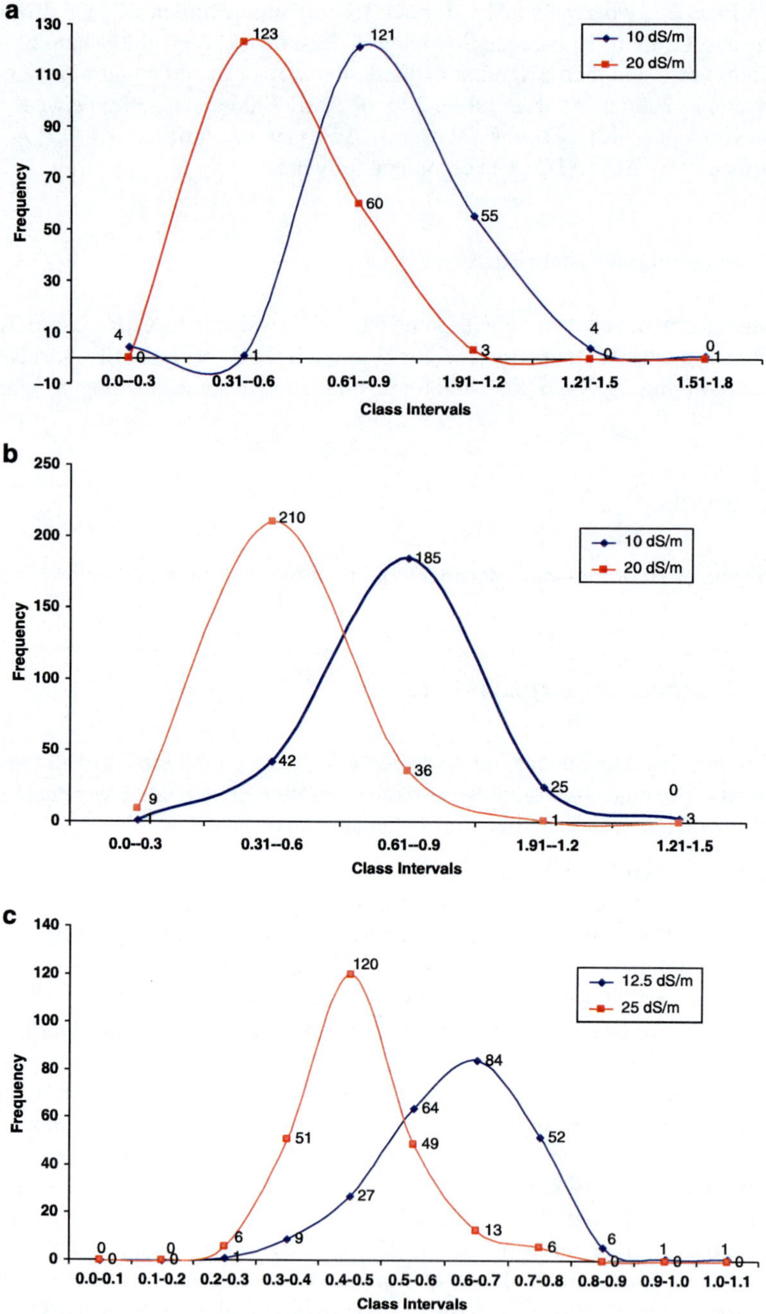

Fig. 7.1 (**a**) Frequency distribution of shoot length at 10 and 20 dS m^{-1} (2003–04). (**b**) Frequency distribution of shoot length at 10 and 20 dS m^{-1} (2004–05). (**c**) Frequency distribution of shoot length at 12.5 and 25 dS m^{-1} (2005–06)

in the range of 0.6–0.7 relative ratio at 12.5 dS m^{-1}. However, at 25 dS m^{-1} salinity, the highest frequency i.e. 120 was observed at much lower range of 0.4–0.5. At the highest salinity level of 25 dS m^{-1}, 7 lines fell in the range of 0.7–1.0 (which is near or equal to unity). These results are strongly in accordance with Cicek and Cacirlar (2002) who also reported that salinity reduced shoot length.

Root Length (cm)
In the first year of screening (2003–04) normal curve for population was observed at 10 dS m^{-1} salinity, whereas at 20 dS m^{-1} abnormal population curve was observed (Fig. 7.2a). Highest frequency of 94 was observed at the range 0.6–0.9 at 10 dS m^{-1} salinity, whereas at 20 dS m^{-1} salinity, the highest frequency of 116 was recorded at the range of 0.3–0.6. For the second year of screening (2004–05), abnormal population curve (Fig. 7.2b) for both salinity levels (10 dS m^{-1} and 20 dS m^{-1}) was observed. Highest frequencies of 152 and 175 were recorded at the range of 0.3–0.6 at salt levels of 10 and 20 dS m^{-1}, respectively.

In the third year of screening (2005–06), genotypes were screened at 12.5 and 25 dS m^{-1} salinity. Wheat genotypes showed abnormal curves at both salinity levels for root length (Fig. 7.2c). Highest frequency (75) was observed for 12.5 dS m^{-1} in the range of 0.8–1.0. In contrast, at 25 dS m^{-1}, highest frequency of 99 was recorded at the range of 0.4–0.6.

Shoot Weight (g)
For the first year (2003–04) of screening, abnormal population behavior at both salinity levels was observed (Fig. 7.3a). 77 and 113 were highest frequencies observed at the range of 0.3–0.6 at 10 and 20 dS m^{-1} respectively. At 10 dS m^{-1} salinity, frequency was distributed between the range of 0.0–0.3. However for 20 dS m^{-1}, the distribution of frequency was in much shorter range of 0.0–2 for this character.

For second year (2004–05) of screening abnormal population curves were observed (Fig. 7.3b) at both salinity levels. Highest frequency of 50 was recorded at the range of 0.4–0.5 at 10 dS m^{-1} salinity. At 20 dS m^{-1} salinity, the highest frequency of 77 was observed at very low range of 0.1–0.2. At 10 dS m^{-1} salinity, the frequency was more evenly distributed within the range of 0.0–2. However at 20 dS m^{-1}, population was unevenly distributed among ranges of 0.0–1.7.

Abnormal population behavior was observed (Fig. 7.3c) at both salt levels of 12.5 and 25 dS m^{-1} for third year (2005–06). At 12.5 dS m^{-1} Salinity, 54 was the highest frequency recorded at the range of 0.2–0.3. Whereas at 25 dS m^{-1}, highest frequency of 99 peaked at much earlier range (0.1–0.2). Frequencies were more evenly distributed among different ranges at 12.5 dS m^{-1}, whereas at 25 dS m^{-1} all the frequencies were unevenly distributed among a lower range of 0.0–1.3. Eleven lines were found in the near to unity range (0.7–1.0) which is considered as selection criteria for salt tolerance.

Root Weight (g)
In the first year of screening (2003–2004) abnormal but almost similar population behavior was observed for both salinity levels (10 and 20 dS m^{-1}) as shown in

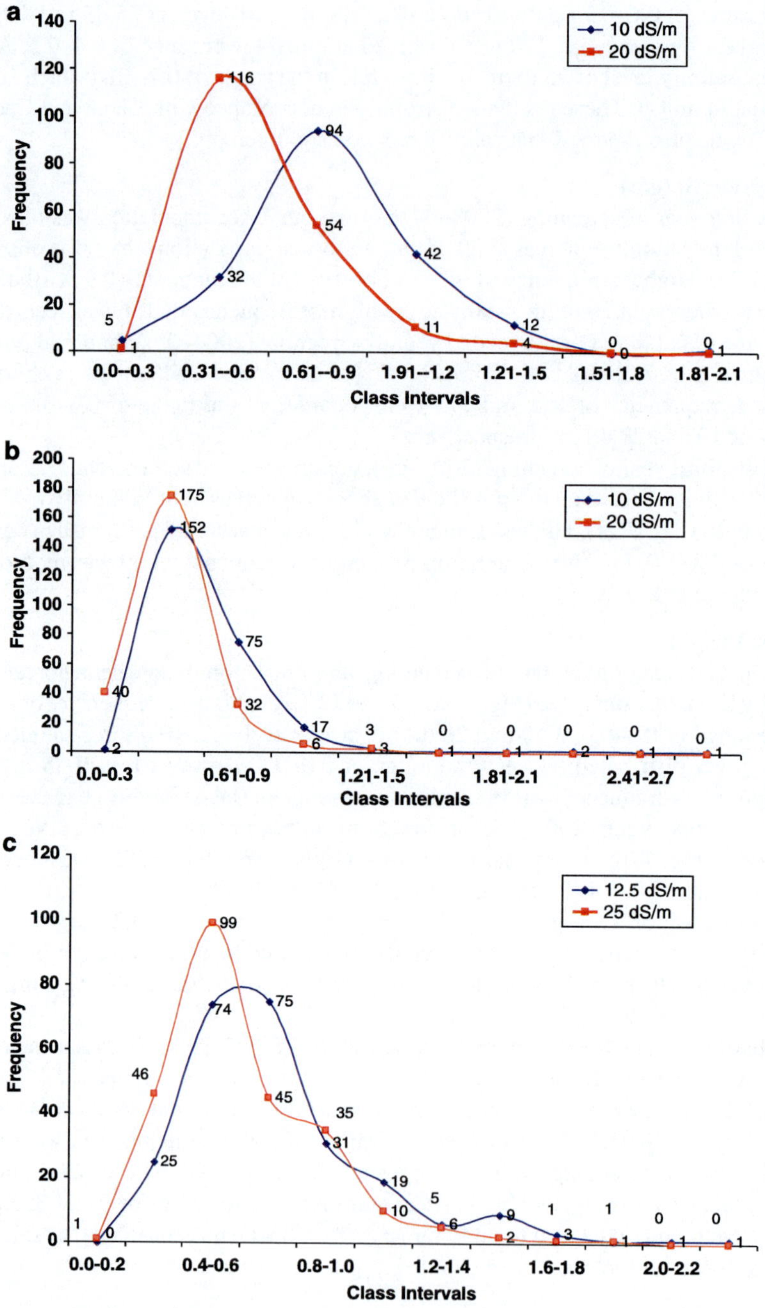

Fig. 7.2 (**a**) Frequency distribution of root length at 10 and 20 dS m^{-1} (2003–04). (**b**) Frequency distribution of root length at 10 and 20 dS m^{-1} (2004–05). (**c**) Frequency distribution of root length at 12.5 and 25 dS m^{-1} (2005–06)

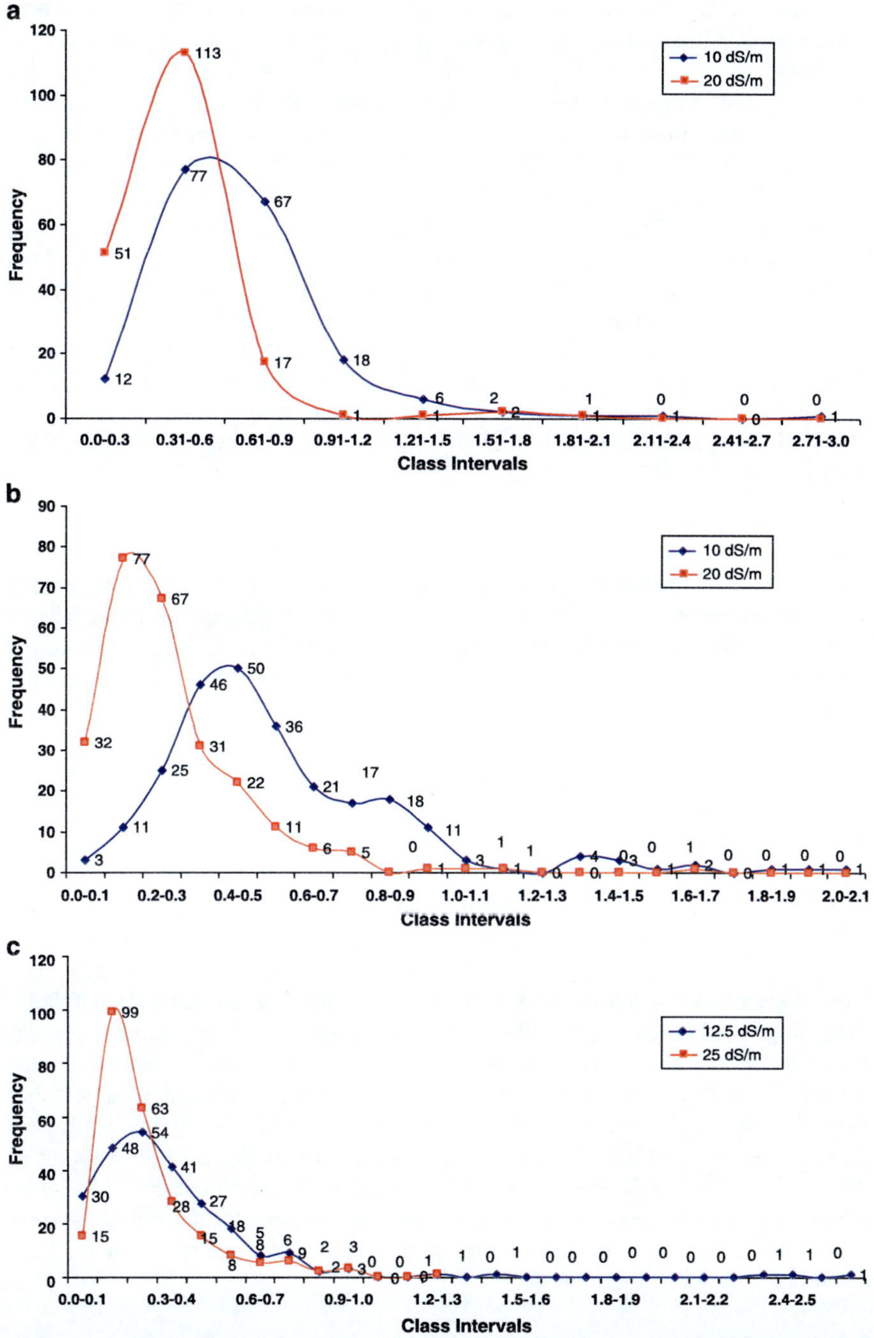

Fig. 7.3 (**a**) Frequency distribution of shoot weight at 10 and 20 dS m^{-1} (2003–04). (**b**) Frequency distribution of shoot weight at 10 and 20 dS m^{-1} (2004–05). (**c**) Frequency distribution of shoot weight at 12.5 and 25 dS m^{-1} (2005–06)

Fig. 7.4a. The highest frequencies of 68 and 72 fell in the range of 0.6–0.9 at both salinity levels (10 and 20 dS m^{-1} respectively).

For the second year (2004–05) of screening again abnormal population curve was noticed at both salt treatments for root weight (Fig. 7.4b). At 10 dS m^{-1}, the frequencies were more evenly distributed with high frequency reaching up to 71 at the range of 0.6–0.9. On the other hand at 20 dS m^{-1}, the frequencies were more unevenly distributed among different class intervals and highest frequency of 98 peaked at relatively lower range of 0.3–0.6.

Like previous 2 years, abnormal population behavior was recorded (Fig. 7.4c) for the third year (2005–06) at both salt levels (12.5 and 25 dS m^{-1}). At the range of 0.4–0.6 the highest frequency (51 genotypes) was recorded at 12.5 dS m^{-1}, whereas at 25 dS m^{-1} frequency of 42 and 44 were closely clustered at the ranges of 0.4–0.6 and 0.8–1.0. At top salinity of 25 dS m^{-1}, 64 lines were found in the 0.7–1.0 range. At 25 dS m^{-1} salinity, relatively less root weight reduction was recorded as compared to lower salinity level of 12.5 dS m^{-1}. Similar findings have also been reported by Grewal et al. (2004).

Salt Tolerant Lines

Eleven common salt tolerant lines were identified from 3 years study. Common salt tolerant lines and their pedigree/parentage are given in Table 7.2. The mean relative ratios of parameters studied were near to unity for these lines as minimum salinity effects were observed for all the above mentioned characters. The relative ratios for root length, shoot length, root weight and shoot weight for these lines were in the range of 0.8–1.

3.2 *Field Trial*

Analysis of variance (Table 7.3) showed that all replications have non significant differences, whereas highly significant differences were found among wheat genotypes for salinity tolerance. Table 7.4 indicates that most of the varieties/lines showed significant differences ($P \geq 0.05$), when subjected to salt affected field. According to Table 7.4, Uqab-2000 out-performed all other genotypes with respect to yield under salt stress. Whereas lines Gamdow-6, V-01171 and LAKATA-1 were ranked after Uqab-2000 as most salt tolerant varieties and these varieties also showed significant higher yield ($P \geq 0.05$) than the other lines including Inqilab-91 under salinity stress. Line FRET-1 was ranked 5th with line VEE/PJN//2*KAUZ at 6th position. Our check Inq-91 was ranked at 7th position with respect to performance in salt stress. Known salt tolerant varieties Pasban-90 and LU26'S were ranked at 17th and 19th positions respectively. These lines are recommended for wheat breeding for salt tolerance.

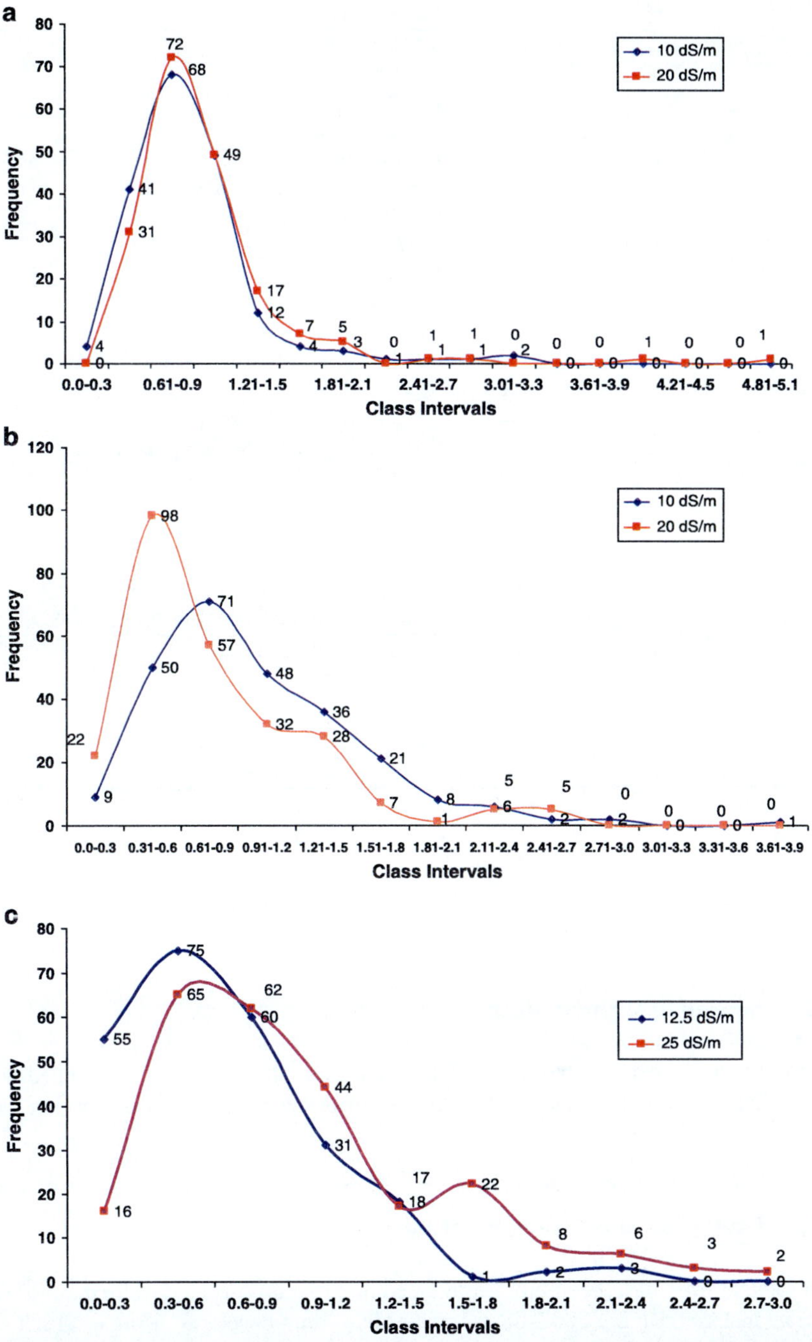

Fig. 7.4 (**a**) Frequency distribution of root weight at 10 and 20 dS m^{-1} (2003–04). (**b**) Frequency distribution of root weight at 10 and 20 dS m^{-1} (2004–05). (**c**) Frequency distribution of root weight at 12.5 and 25 dS m^{-1} (2005–06)

Table 7.2 Salt tolerance lines during 3 years study (2003–06)

*C.B. #	Pedigree/parentage
44	KANCHAN
	OBGD
53	LU 26
	BLS/KHUSHAL 69
72	PASBAN 90
	INIA 66/A. DISTT//INIA 66/3/GEN 81
172	GAMDOW-6
	CM79515-044Y-----
173	LAKTA-1
	ICW91-0233-------
232	PF 70402/ALD'S'//PAT 72/160//ALD'S'/3/PEW'S'
	CM 70474-(1-1)M-1013Y-19M-2Y-1B-0Y
311	NING 8319
414	CHIL/2*STAR=V-98124
	CM 112793-0TOY-22M-20Y-1M-3Y-0M
430	PB-96/87094//MH-97
	V01180
439	V-01171
	BAV92//SAP/MON
450	V-01078
	CHIL/2*STAR/4/BOW/CROW//BUC/PVN/3/...

*CB # according to 05–06 data book

Table 7.3 Analysis of variance for salt tolerance yield trial conducted at Pacca Anna

S.O.V	d.f	Sum of squares	Mean square	F-value	F-tab
Replication	5	1412649.52	282529.90	2.25^{ns}	0.056
Varieties	19	5606269.76	295066.83	2.35**	0.004
Error	95	11949830.41	125787.69		
Total	119	18968749.69			

**Significant at $p \leq 0.01$; ns non significant

3.2.1 Screening for Stem Rust

According to the results obtained (Table 7.5), of the salt tolerant lines screened, line NING 8319 showed desirable level of stem rust resistance like other stem rust resistant lines, Parula and Pavon.

3.2.2 Utilization in Breeding Program

Salt tolerant lines have been utilized in breeding program for the development of salt tolerant material. During 2007–08, 60 crosses have been developed using local salt tolerant lines PASBAN 90, LU 26'S' NING 8319, and some other good local varieties and stem rust resistant lines. During the crop year (2006–07), 40 crosses were developed.

Table 7.4 Field evaluation of the selected lines for salt tolerance at Pacca Anna

Line/variety	Parentage	Ave. yield
Uqab-02	CROW'S'/NAC//BOW'S'	1537.08 a
C.B 172	GAMDOW-6	1325.65 ab
V-01171	BAV 92//SAP/MON	1317.93 ab
C.B 173	LAKATA-1	1313.31 ab
C.B 147	FRET-1	1253.12 abc
C.B 159	VEE/PJN//2*KAUZ	1194.47 abcd
Inq-91		1049.41 bcde
V-98124	CHIL/2* STAR	1040.15 bcdef
V-01180	PB-96/87094//MH-97	1015.46 bcdef
V-01078	CHIL/2*STAR/4/BOW/CROW//BUC/PVN/3	1002.19 bcdef
C.B 226	ESDA 'S'	993.85 bcdef
C.B 330	TRAP # 1/YACO//BAV 92	955.27 bcdef
C.B 232	PF 70402/ALD'S'//PAT 72/160//ALD'S'/3/PEW'S'	939.84 bcdef
V-96015	INQ.91/TANGER	930.58 bcdef
V-02192	SHL88/87094//MH97	896.63 cdef
C.B.44	Kanchan	862.68 cdef
Pasban-90	INIA 66/A. DISTT//INIA 66/3/GEN 81	848.79 cdef
C.B 311	NING 8319	845.70 def
LU26'S	BLS/KHUSHAL 69	787.06 ef
AS-02	WD-97603	638.90 f

LSD value: 406.5 at $P<0.05$

Table 7.5 Stem rust reactions of selected salt tolerant lines along with some other lines at Njoro, Kenya

C.B.#	Variety/line	2006	2007
44	KANCHAN	80S	–
172	GAMDOW	40S	–
173	LAKATA 1	50S	–
232	PF70402/ALD'S'S//PAD72/160//ALDS/3/PEW'S'	40S	–
311	NING 8319	20M	10MR
330	BAV92/SAP//MON	60S	–
430	V-01180	60S	–
414	V 98124	60S	–
–	BB#2/PT//CC/NIA/3/ALD'S'	60S	70S
450	SEHER 06 (V-01078)	60S	–
–	PARULA	30MS	1MSS
–	PAVON	30MSS	–
2	AUQAB	40S	
72	PASBAN90	60S	–
53	LU26'S	70S	70S

C.B. #=Crossing block entry number

4 Discussion

Hydroponic experiment revealed that salt stress reduced the shoot length and the effect becomes more pronounced with the increase in salt concentration. At 10 and 20 dS m^{-1}, normal population behavior was observed. At 20 and 25 dS m^{-1}, the population curve was abnormal which indicated that higher salt concentration reduced the shoot length. Root length also decreased under salt stress. Abnormal population curves indicated that root length was more drastically affected by salt stress as compared to shoot length. This type of behaviors in wheat has previously been reported by Iqbal et al. (1998) and Cicek and Cakirlar (2002).

Salinity stress reduced the shoot weight and shoot weight was severely decreased at higher salinity levels. Only 11 lines were found at the relative value range of 0.7–1.0. This decrease in shoot weight was due to decreased water uptake, reduced photosynthesis and toxicity of sodium and chloride in the shoot cell (Ali et al. 2005), whereas salinity stress increased the root weight of many genotypes as more lines had shown relative ratio near to unity at higher salinity level as compared to lower salinity level. Munns and Termatt (1986); Marcum and Murdoch (1992) also reported increased root weight in plants under salinity stress. The lines that showed relative ratios near to unity for all the characters studied were subjected to field testing in salt affected field along with some other high yielding varieties/lines.

In the field trial, non significant replication mean squares indicated homogenous field conditions. Significant varietal differences for yield performance under saline conditions indicated the presence of genetic variability for grain yield in salt affected fields. Uqab-2000 was the highest yielding line in field conditions, whereas it was not among the best performing lines in aquaculture. Similarly some lines, like KANCHAN, which showed tolerance in lab conditions, were not good as for yield performance in saline field conditions is concerned. Differences in varietal/ lines behavior in the field and aquaculture conditions have been reported in literature (Reynolds and Borlauge 2006). However, lines like GAMDOW-6, V-01171, LAKATA-1, FRET-1 and VEE/PJN//2*KAUZ proved tolerance in laboratory conditions and also produced higher grain yield in saline fields. These lines were sent to Kenya for screening against stem rust. One line (NING 8319) showed good resistance level comparable to known stem rust resistance lines like PARULA and PAVON. These lines were used in the specific breeding programs and the crosses will be further evaluated following shuttle breeding approach. The F2 generation will be sown at Kaghan in summer and F3 in a salt affected field at Faisalabad in normal season. Again F4 at Kaghan and F5 at salt affected field in Faisalabad. The material will be enhanced following modified bulk method. From F5 single heads will be taken and the material will be further evaluated in the normal field in F6 and in the next generation in normal as well as salt affected fields to see their adoptability to normal and saline conditions.

5 Conclusion

Rapid and specific screening methods were adopted for screening wheat germplasm for 3 years and it was concluded that in general terms salinity reduces the biomass of wheat plant but when it reaches a particular level wheat plant increases its root weight to better cope with high salt concentration in the root zone and also to uptake water and nutrients more efficiently. This study helped in selecting 11 salt tolerant varieties/lines on the basis of better and consistent performance for 3 years. Field studies also revealed the superiority of some of these lines for grain yield potential over Inqulab-91, the most popular variety of Pakistan. These genotypes are valuable asset for the wheat breeders and can be used in hybridization program or as commercial cultivars for salt affected areas of Pakistan. Screening for stem rust revealed that NING 8319, PARULA and PAVON can be used as parents for stem rust resistance. The crosses developed during the previous 2 years will be utilized for the development of salt tolerant and stem rust resistant material.

References

Ali Y, Aslam Z, Sarwar G, Hussain F (2005) Genotypic and environmental interaction in advanced lines of wheat under salt-affected soils environment of Punjab. Int J Environ Sci Technol 2:223–228

Anonymous (2002) Agricultural statistics of Pakistan. Government of Pakistan. MINFAL, Economic Wing, Islamabad, pp 83–84

Ayliffe M, Singh RP, Lagudah E (2008) Durable resistance to wheat stem rust needed. Curr Opin Plant Biol 11:187–192

Cicek N, Cakirlar H (2002) The effect of salinity on some physiological parameters in two maize cultivars. Bulg J Plant Physiol 28:66–74

Daniells IG, Holland JF, Young RR, Alston CL, Bernardi AL (2001) Relationship between yield of grain sorghum (*Sorghum bicolor*) and soil salinity under field conditions. Aust J Exp Agric 41:211–217

Greenway H, Munns R (1980) Mechanisms of salt tolerance in non-halophytes. Annu Rev Plant Physiol 31:149–190

Grewal HS, Cornish P, Norrish S (2004) Differential response of wheat cultivars to subsoil salinity/sodicity. In: Proceedings of 4th international crop science congress, Brisbane, Australia, 26 Sept–1 Oct 2004

Hoagland DR, Arnon DI (1950) The water culture method of growing plants without soil. University of California, Berkeley College of Agriculture, Circular no 347

Iqbal RM (2003) Effect of constant and changing salinity environment on different growth parameters in spring wheat. Asian J Plant Sci 2:1112–1117

Iqbal N, Ashraf HY, Javed F, Iqbal Z, Shah GH (1998) Effect of salinity on germination and seedling growth of wheat (*Triticum aestivum* L.). Pak J Biol Sci 1:226–227

Marcum KB, Murdoch CL (1992) Salt tolerance of the coastal salt marsh grass, Sporobolus virginicus L. Kunth. New Phytol 120:281–288

Munns R (1993) Physiological processes limiting plant growth in saline soil: some dogmas and hypotheses. Plant Cell Environ 16:15–24

Munns R, Richard AJ (2003) Screening methods for salinity tolerance: a case study with tetraploid wheat. Plant Soil 253:201–218

Munns R, Termatt I (1986) Whole-plant responses to salinity. Aust J Plant Physiol 13:143–160
Reynolds MP, Borlauge NE (2006) Applying innovations and new technologies from international collaborative wheat improvement. J Agric Sci 144:95–110
Steppuhn H, Wall KG (1997) Grain yields from spring-sown Canadian wheats grown in saline rooting media. Can J Plant Sci 77:63–68
Stokstad E (2007) Deadly wheat fungus threatens worlds's breadbaskets. Science 315:1786–1787
Zulfiqar A, Khan AS, Asad MA (2002) Salt tolerance in bread wheat: genetic variation and heritability for growth and ion relation. Asian J Plant Sci 4:420–422

Chapter 8
The Potential of Breeding Okra (*Abelmoschus esculentus* L.) for Water Stress Tolerance

Abdul Naveed, Asif Ali Khan, and Saeed Rauf

Contents

Abstract Okra plant grows well but produces a few pods under water limited conditions. A linear relationship between okra production and the amount of water supplied is known to exist. Review of previous literature has shown that drought cause damage to okra plant of varying degree during its ontogeny, the reproductive phase being the most prone to the detrimental effects of drought. More specifically, the highest effect on yield was recorded when drought was found to occur during flowering and pod formation. Existence of genetic variability among genotypes is

A. Naveed (✉) • A.A. Khan • S. Rauf
Office of Research, Innovation and Commercialization, Department of Plant Breeding and Genetics, University of Agriculture, Faisalabad, Pakistan
e-mail: ranaveed@gmail.com; asifpbg@hotmail.com; saeedbreeder@hotmail.com

M. Ashraf et al. (eds.), *Crop Production for Agricultural Improvement*,
DOI 10.1007/978-94-007-4116-4_8,

the most important component for the success of a breeding programme for increased drought resistance. Previous reports show the existence of a substantial amount of variability among okra genotypes. However, most of the traits related to drought tolerance are quantitative in nature. Therefore, estimation of type and amount of total genetic variability associated with the target traits is equally important. Determination of the type of genetic variability may help in the formulation of comprehensive breeding programme regarding the further improvement of the target trait. It is also important to find out the target trait that is related to drought tolerance and also shows good relationship with yield. Many traits are measured with complex and time-consuming techniques that are unsuitable for screening large numbers of progeny in breeding programmes. As a result, breeders tend to measure them in small populations. In addition, the stage at which trait allows the evaluation of a plant would also predict its suitability. A trait, which is evaluated earlier in the plant growth and shows freedom of its interaction with growth stages, will speed up the breeding programme. Breeding efforts at molecular level are usually intense for finding drought related molecular markers which helped us in the screening of large population without actually been exposed to drought stress. Furthermore, development in the reverse genetics has led us to isolation of novel drought tolerant genes synthesized by cDNA obtained form mRNA expressed in drought stress and non stress plants. An integrated approach has been presented here for the development of drought tolerance in okra. The increasing strengths in genomics, proteomics and molecular marker systems should be made use of to facilitate breeding of okra genotypes for water stress tolerance.

Keywords Drought tolerance • Genetic variability • Molecular markers • Okra • Susceptibility index

1 Introduction

Okra (*Abelmoschus esculentus* L.) is the only vegetable crop of significance in the Malvaceae family and it is very popular in the Indo-Pak subcontinent. In India, it ranks number one in its consumption but its original home is Ethiopia and Sudan. It is one of the oldest cultivated crops and presently grown in many countries and is widely distributed from Africa to Asia, southern Europe and America (Oyenuga 1969; Hamon 1991; Ariyo 1993; Oyelade et al. 2003). It is a tropical to subtropical crop and is sensitive to frost, low temperature, waterlogging and drought conditions, and the cultivars from different countries have certain adapted distinguishing characteristics specific to the country to which they belong (Siemonsma 1982).

It is an oligo purpose crop, but it is usually consumed for its green tender fruits as a vegetable in a variety of ways. These fruits are rich in vitamins, calcium, potassium and other mineral matters (Camciuc et al. 1981). The mature okra seed is a good source of oil and protein (Karakoltsidis and Constantinidesm 1975; Martin and Ruberte 1979; Oyelade et al. 2003) and has been known to have superior nutritional

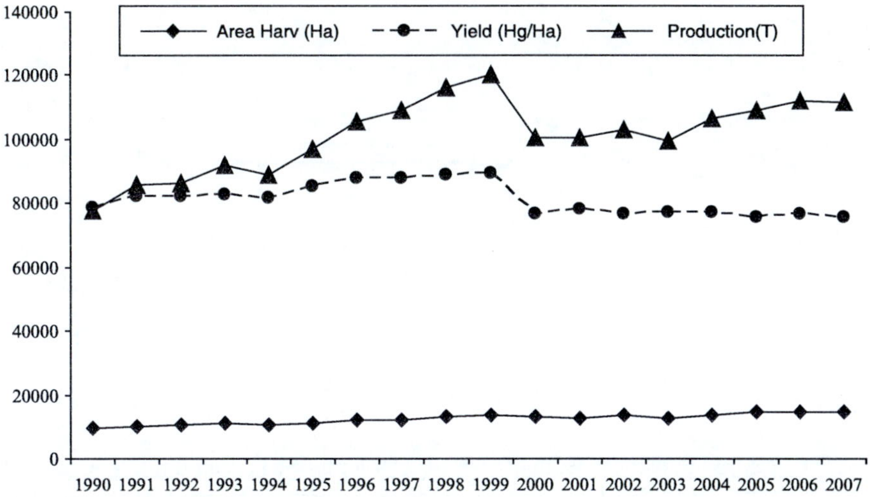

Fig. 8.1 Area (ha), production (Mt) and yield (kg ha^{-1}) of Okra in Pakistan (from 1990 to 2007)

quality. Okra seed oil is rich in unsaturated fatty acids such as linoleic acid (Savello et al. 1980), which is essential for human nutrition. Its mature fruit and stems contain crude fibre, which is used in the paper industry.

In Pakistan, okra is a favourite vegetable cuisine, which resulted in a progressive increase in the area of okra but the last 5-year statistics show that there was a decreasing trend in yield per unit area and production as compared to the last 10 years (Fig. 8.1). A number of factors are considered to cause reduction in yield and production but it is argued that the limiting availability of irrigational water has been a key factor for its reduction.

The rainfall data of 1995–2007 presented in Fig. 8.2 indicates that during the years 1999–2005, Pakistan had been under severe drought. The percentage of rainfall decreased to 66% in Sindh, 52% in Balochistan, 20% in NWFP and 13% in Punjab. In addition, the underground water and water storage reservoirs are rapidly depleting due to a nominal recharging by the rainfalls and continuous silting in the dams. As a result national economy suffered a loss of about $ 2 billion due to water shortage (MinFAL 2006).

In order to face the calamity it is necessary to devise strategies that may improve water use efficiency of crops. There are two strategies to manage drought stressed environment. The first strategy includes the drought stress management by means of agronomic practices i.e. improved soil and water conservation practices associated with the tillage system, weed control, fertility management, optimized plant population, improved forage/ livestock/grains integration and rotation, avoidance of mono-cropping and the diversification of farming (Blum 2005).

The second strategy is to improve the crop drought tolerance. It is the most efficient, cheapest and long-term strategy to manage drought stress (Ludlow 1989) and once it is introduced in a population it will be permanent source to endure stress.

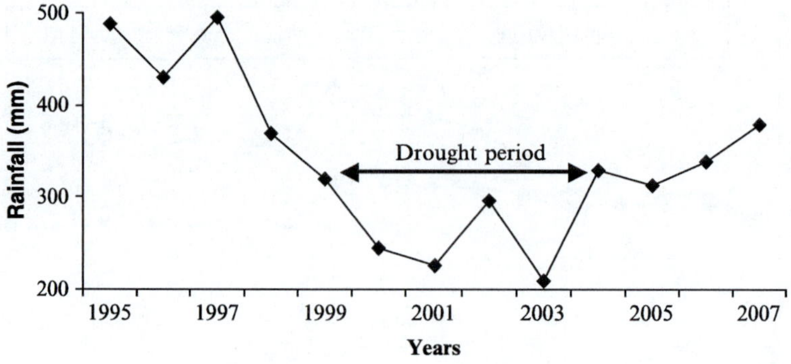

Fig. 8.2 Average annual rainfall (mm) in Pakistan from year (1995–2007)

The progress in plant breeding for drought tolerance has been limited by at least two major problems. First, drought resistance is a complex attribute of a crop community (Blum 1979) and it is not clear how specific mechanisms at the tissue or organ level are integrated in determining yield under water stress (Hsiao and Bradford 1983). The genetics of drought tolerance is complex and not adequately understood (Osmanzai et al. 1987). The complexity arises because of the number of morphological and biochemical systems within a plant, which are related to drought tolerance. Compensation of one system for another and interaction with environment make it even more difficult to correlate physiological and morphological traits with yield under drought conditions.

The importance of okra vegetable in respect of its nutritional value and area under cultivation in Pakistan, and alarming condition of occurrence of drought demands a comprehensive review of the genetic, molecular, physiological and breeding work to facilitate the formulation of a strategy for the development of drought tolerant cultivars in okra.

2 Drought and Its Types

Drought is a condition where soil moisture contents are too low or tightly attracted to soil particles (due to lower osmotic potential) and thus a plant cannot absorb water or quantity of the absorbed water cannot meet its transpiration demands. A plant faces different types of drought stress depending upon its growth stage (Seghatoleslami et al. 2007; Sinaki et al. 2007; Kron et al. 2008). The stress that occurs at seedling stage or during development phase may be called as early drought. Such type of drought usually reduces the crop stand and as a result it damages yield due to lower than optimum plant population (Ashraf et al. 2005). However, farmers usually respond to this drought by replanting their crops (Muasya and Diallo 2001). This type of stress has been proved lethal for early maturing varieties (Muasya and Diallo 2001), which seldom recover from the stress while late maturing varieties have enough time to recover.

Drought occurring during the vegetative growth period is called as vegetative phase drought. This type of drought affects plant assimilatory organs, which usually decrease in number and size, resulting in lower photosynthate production (Kaiser 1987; Chaves 1991; Larcher 1995; Chaves et al. 2002). Consequently the yield decreases due to less amount of assimilates available to the developing pods.

The third type of stress develops during bud formation, flowering and grain filling period. It reduces yield due to abortion of ovule, embryo and sterility of pollen. Review of previous literature shows that this type of drought has the highest detrimental effects on pod yield (Ahmad et al. 2003; Sawadogo et al. 2006; Seghatoleslami et al. 2007). However, the highest effect on yield was recorded when drought was found to occur during flowering and pod formation. Mbagwu and Adesipe (1987) found the greatest reduction (70%) in yield of okra when stress was imposed at flowering and pod filling stage. Ahmad et al. (2003) studied three treatments of drought, (water stress at penta-foliate and bud formation, flowering and pod formation, and seeding and maturity). They recorded maximum reduction in yield when water stress (three consecutive irrigations withheld) was imposed at the flowering and pod formation stages. Sawadogo et al. (2006) observed two types of flowering phases of the okra plant i.e. the period which intervenes during the accelerated growth phase of the main stem (type 1) and the one which starts once the growth of main stem begins to decline (type 2). It was also observed that water deficit during the budding phase causes earlier flowering of type 2 plants and leads to delayed flowering of the type 1 plant.

3 Water Availability and Okra Plant

Although okra is a drought tolerant plant, the availability of water has a significant impact on okra production. It has been found that there was a linear relationship between okra production and the amount of water supplied (Batra et al. 2000; Ahmad et al. 2003). In each season, okra requires supplemental water of 24-acre inch. In okra different studies were carried out on the basis of irrigation regime using ID/CPE ratio where ID is irrigation depth, and CPE is cumulative evaporation from a USWB Class-A pan (Singh and Singh 2005). Batra et al. (2000) used three levels of irrigation based on ID/CPE ratio of 0.6, 0.9 and 1.2 and concluded that irrigation of 1.2 and 0.9 ID/CPE resulted in higher DM production than 0.6 ID/CPE. Verma and Batra (2001) showed that the highest fruit yield could be ensured with moderate intensity of irrigation (ID/CPE 0.9).

4 Plant Adaptation to Drought

Okra is considered to grow well under drought conditions, although plant has shown reduction in yield under drought stress (Ahmad et al. 2003). The drought tolerance of a species may be considered due to its better crop stand rather than yield under drought conditions. However, for sustainable agriculture, better crop stand alone means nothing to farmers until it produces an acceptable yield under drought stress. Therefore drought tolerance may be defined on the basis of yield rather than plant survivability.

Plants have four adaptive mechanisms for drought (Levitt 1972; Mitra 2001) that must be reviewed and subsequently incorporated according to the species intensity and timing of drought. First of them is drought escape. This type of drought tolerance mechanism is achieved by reducing the growth cycle of the crop. Moisture stress occurs at the end of growth cycle and plants usually complete the growth cycle well before the onset of moisture stress (Turner 1986). Early maturing varieties have been equipped with this adaptive mechanism. On the other hand, stress may coincide with the flowering or the grain filling period in case of a late maturing variety. As a result late maturing variety may show reduction in yield. However, this type of mechanism has not been found successful and popular among the breeders due to its negative correlation with yield. The second type of drought resistance mechanism is drought avoidance. In this type a plant does not show the symptoms of drought and usually maintains higher degree of moisture contents through out the crop growth cycle. Several types of traits have been found to be involved in this type of mechanism such as lower transpiration rate, small leaf area, leaf hairs, high root-to-shoot ratio which help the plant to use water slowly and efficiently (Passioura 1976; Turner 1986). Such types of plants are called as water savers. This type mechanism has not been found successful due to its negative correlation with yield. The third mechanism is drought tolerance. This mechanism may be defined as the ability of a genotype to produce higher yield when compared with other genotypes and treated with similar intensity of moisture stress. This type of mechanism is based on the traits such as drought susceptibility index, root length, relative water contents, turgor pressure, and osmotic adjustment that help the plant to absorb more water and from the deeper soil profiles. This type of mechanism has been proved most successful due to its positive correlation with the traits associated with yield. The fourth type is drought recovery. This type of mechanism allows the plant to recover from the deleterious effects of drought when stress occurs during plant development (Kholodova et al. 2007). Different types of traits such as stay green and leaf retention allow the plant to recover after drought (Oosterom et al. 1999; Lenis et al. 2006). Leaf retention and stay green plant traits are indicators of normal functioning of the leaves thus helping the plants to recover from drought by producing more assimilates.

5 Breeding for Drought Tolerance

The genetic improvement of crop plants for drought tolerance, through breeding and selection, involves a favourable combination of plant traits. The success of a drought tolerance breeding programme depends on the following four important features:

1. Presence of sizeable genetic variability for target traits
2. Type of genetic variability associated with the selected traits

3. Relationship of the target traits with yield
4. Cost, stage and speed of measurement
5. Genetic variability for target traits
6. Genetic variability

Success of a drought tolerance breeding programme depends on the existence of genetic variability for the target traits. Genetic variability arises as a result of mutation, inter-variety or inter-specific genetic recombination. These variations tended to accumulate in the germplasm that has been under selection pressure by the target environment. Therefore, local germplasm evolved under target environments would provide the desired variability for selection. However, target traits are to be integrated in superior high yielding cultivars, since land races and wild species have some undesirable features i.e. introgression of target traits may result in deterioration of yield and quality due to linkage drag phenomenon. The basic approach for development of drought tolerant genotypes is to select locally adapted germplasm containing genetic variability for high yield potential and drought adaptive traits (Beck et al. 1990; Vasal et al. 1997). Furthermore, the unpredictable nature of drought dictates that improved genotypes must perform well in both favourable and stressed environments. Thus, combination of stressed and unstressed environments may be used in the selection of genotypes for drought stressed areas.

Existence of genetic variability for drought resistance among the genotype is the most important factor for the success of drought resistance breeding programme. Previous reports showed the existence of a substantial amount of variability among okra genotypes in dry-land for harvest index. This variation in harvest index was due to differences in leaf area, biomass accumulation, osmotic adjustment, photosynthetic efficiency and distribution and in root water extraction (Fereres et al. 1986; Gimenez and Fereres 1986).

In okra variability among the cultivated genotypes for different traits has been reported by Gill et al. (1997). They found considerable variation with respect to vegetative, floral and fruit characters and reaction to diseases and pests. A few distinguishing characters were identified in each variety. Chehda and Fatokun (1991) evaluated the exotic and locally collected germplasm for genetic variability. Significant differences were found between exotic and local material. Three agronomic types (A, B and C) were identified based on flowering date, plant size and fruit yield. Local germplasm belonging to types B (Soudanian or *Abelmoschus esculentus*) and C (Guinean or *A. caillei*) were crossed with a view to transfer drought resistance from type C to B by backcrossing.

Ahmad et al. (2003) screened six okra varieties and showed significant genetic variability in yield and its components and also revealed drought tolerance on the basis of yield in a few varieties. Similarly, Sawadogo et al. (2006) conducted a greenhouse experiment to assess tolerance to water deficit among six okra ecotypes evolved under different conditions. Among these ecotypes, three were chosen for using in okra breeding programmes for tolerance to water deficit.

6 Genotype × Environment Interactions

Progress in developing high-yielding, drought-tolerant cultivars by conventional breeding has been slow, because of difficulties in defining the target environment and complex interactions of drought tolerance with environments (Cooper et al. 1999; Wade et al. 1999; Pantuwan et al. 2002). Breeders appears to have different views that whether selection for drought yield should be carried out under optimum conditions or under targeted drought environment. However, a large number of breeders have shown that drought tolerant genotypes should be selected under a target drought environment (Ceccarelli 1987; Din et al. 1992; Chapman et al. 1997). Ceccarelli (1987) emphasized that selection under stress conditions is expected to be more efficient than selection under favourable conditions when dry areas are the target environment (Ceccarelli 1987). Similarly, Chapman et al. (1997) and Din et al. (1992) concluded that the yield gains under drought would have been unlikely to occur if selection had been done only in well-watered environments.

7 Types of Genetic Variability

Since most of the traits related to drought tolerance are quantitative in nature, therefore, estimation of type and amount of genetic variability associated with the target traits is equally important. Determination of the type of genetic variability may help us in the formulation of comprehensive breeding programme regarding further improvement of the target trait. Additive or non-additive type of genetic variability has been found to be associated with any trait. Additive type of genetic variability arises as a result of cumulative effect of minor alleles while non-additive type of genetic variability arises as a result of dominance and epistasis. Dominance arises as a result of intragenic interaction while epistasis due to intergenic interaction. In addition, these intergenic or intragenic interactions have also been found to be affected by the external stimuli such as drought (Rauf et al. 2007, 2008; Khan et al. 2007; Rauf and Sadaqat 2008a, b).

Due to their quantitative nature, drought related traits couldn't be studied in simpler ways. Specialized biometrical techniques are required to work out the type of genetic variability associated with the traits. These biometrical techniques are dependent on different mating designs such as diallel, line × tester, North Carolina design and generation mean analysis for the estimation of type of genetic variability. Among these mating designs, generation mean analysis has been the most powerful biometrical analysis since it gives additional information about the epistatic interactions. It is based on the mean of five generations i.e. parental, F_1, F_2, BC_1 and BC_2. Information derived from these analyses can be further utilized for the formulation of an effective breeding strategy. A population with preponderance of additive genetic variability can be easily managed and selected through progeny rows or pedigree method. The additive genetic variability has also been used for the estimation of narrow sense heritability (ratio of additive genetic variability to the total genetic variability) and genetic advance. The heritability and genetic advance provide further information on the proportion of genetic variability, which can actually be selected and how much improvement can be brought through selection. Panda and Singh (1997) obtained information on genotypic coefficients of variation, heritability and genetic advance from data of pod yield and seven other traits in 40 F_1s progeny of okra (*Abelmoschus esculentus*) grown at Varanasi under two sowing dates. Number of branches, number of pods and total pod yield per plant had higher genotypic and phenotypic coefficients of variation in all optimum environments. All the characters under study except days to first flower appearance and girth of pod were highly heritable. High heritability coupled with high genetic advance was observed for plant height, number of pods and total pod yield per plant, indicating that these traits are suitable for improvement through selection.

A population high in non-additive genetic variability is difficult to manage and further improvement in trait will be slow. Recurrent selection has been recommended for the population high in non-additive genetic variability in order to break the intergenic or intragenic interaction.

In okra, previous literature has not been able to show a single report on inheritance characteristics of drought related traits. Most of the work was available under optimum conditions. Deo et al. (1997) developed different generations to estimate type of gene action for five quantitative characters, namely number of pods/plant, length of pod, girth of pod, number of seeds/pod and pod yield/plant. They concluded that all type of genetic variability i.e. additive, dominance and epistasis was significant for all traits. However, few reports are available for yield and its related traits, which showed the prevalence of additive type of gene action under drought stress (Ahmad et al. 2004). Ahmad et al. (2004) studied the type of genetic variability in morphological traits through 6×6 diallel technique under normal and drought conditions. They reported additive gene action for fresh fruit yield per plant, days to first flower and seed yield per plant under drought conditions, which suggested selection to be effective in early segregating generations for these traits.

Reports on the inheritance of drought related traits are available in other crop species. Dhanda and Sethi (1998) conducted an experiment on the F1 generation from a half-diallel set of crosses involving two drought tolerant, two moderately tolerant and two sensitive varieties. Additive gene action, in general, played a major role in determining the inheritance of relative water contents. General combining ability (GCA) was the main source of genetic variation among crosses, while specific combining ability (SCA) was negligible. Strong phenotypic correlations existed between *per se* performance and GCA effects in majority of cases, but heterosis was not found important. Genotype-environment interactions and/or differential gene expression appeared to account for different results found between environments and growth stages, respectively. Similarly, Rebetzke et al. (2003) estimated leaf stomatal conductance prior to anthesis on irrigated plants representing different generations of crosses between the low conductance parent and three high conductance varieties. They obtained significant genetic differences between generation means for conductance measured in different crosses and on different days. Gene action was complex with both additive and non-additive (dominance and additive-based epistasis) genetic effects important for expression of leaf conductance. Family-mean and heritability varied, depending on cross and time of sampling. It was suggested that breeders selecting for altered leaf conductance maximise genetic gain by delaying screening of populations until later in the day, and repeat measurements across a minimum of 2 days.

8 Ideal Trait for Evaluation of Drought Tolerance

It is important to find out the target trait that is related to drought tolerance and also shows good relationship with yield. Any traits that improve the drought tolerance by improving its yield may be considered ideal. It may be a physiological, morphological or biochemical trait. Previously, plant breeders have focused on the traits that were related to plant survival, i.e. lower leaf area, lower transpiration, etc. As a result plant breeders continued to evolve varieties that had good crop stand but poor yield in the field (Rauf and Sadaqat 2008a) and these varieties ultimately turned towards failure. Richard (1996) showed that large number of criteria that have been proposed to increase drought resistance of our crops had little impact on improving crop yields in dry environments. They concluded that there were several reasons for this lack of success. Some of these were: (i) criteria proposed have been related more to survival mechanisms under drought than to productivity, (ii) criteria are inappropriate to the target environment, and (iii) criteria are temporal and are therefore likely to have minimal impact on growth and yield over the entire life cycle. Another important reason is that the breeders have never been convinced that the proposed criteria will be successful, as they are too difficult to measure. Therefore it is important to find out the ideal traits for selection. Some of the traits are discussed below:

8.1 Pod Yield

Okra plant is usually grown for its green tender pods that are harvested over multiple times. Therefore, fresh pod yield may be considered as an ideal candidate for selection. However, it has been found in a number of studies that selection through yield and yield components under drought conditions was slow. This may be due to very low heritability of the yield and yield components under drought stress (Quisenberry et al. 1980; Blum 1988; Ntare and Williams 1998; Collaku 1994; Szilagyi 2003). The decrease in the heritability estimates was due to increase in error variance under stress (Hulmel et al. 2005). It was concluded that direct selection for yield and yield components was difficult. Annicchiarico and Pecetti (1998) showed that selection indices based on heading displacement and kernels per spike, was on the average, 20% and 11% more efficient than yield-based selection. Furthermore, genotypes selected under drought conditions on the basis of yield could not take advantage of non-stress year and did not produce any higher yield. Since drought stress is an unpredictable feature, therefore ideal genotypes should produce high pod yield under non-stress conditions while comparable yield under drought stress condition. However, few reports have shown that the most effective selection criterion, among various morphological, physiological, phenological, yield, and yield related traits, for identifying drought resistant genotypes was mean seed yield under drought stress and non stress environments (White et al. 1994; Abebe et al. 1998; Ramırez-Vallejo and Kelly 1998).

8.2 Drought Susceptibility Index

Drought susceptibility index is calculated on the basis of differences between pod yield under non-stress conditions to the stress conditions. Drought susceptibility index can be calculated as under adopting Fischer and Maurer (1978).

$$S = \frac{1 - Y/Y_p}{1 - X_d/X_p}$$

Where Y is the green pod yield per plant of a given genotype under drought, Y_p is the green yield per plant of the same genotype under irrigation, X_d is the mean green yield of all genotypes within group under drought, and Xp is the green yield per plant of all genotypes within group under irrigation. A genotype with higher susceptibility index has been considered as susceptible genotype while genotype with low index has been considered as tolerant (Fischer and Maurer 1978). Drought susceptibility index is calculated on the basis of pod yield thus it reflects yield at whole plant level. Different studies have shown significant relationship of drought susceptibility index with pod yield, thus genotype showing lower susceptibility index may also yield better under drought stress conditions (Fereres et al. 1986; Rauf and Sadaqat 2007a, b). However, in these studies drought susceptibility index was unable to show relationship with yield under non-stress conditions. Thus, selection for lower drought susceptibility index may improve drought tolerance without affecting yield under non-stress conditions. As a result, the genotype may be able to take benefit of non stress years and lower drought susceptibility index may act as a buffer for yield reduction during stress years.

8.3 Root Length

Root length is an important plant trait for the evaluation of genotypes under drought stress, since roots are in direct contact with soil moisture contents and first to be affected by the drought and other abiotic stresses aluminium toxicity (Khan and McNeilly 1998) and soil salinity (Khan et al. 2003). A longer downwards root growth may enable the plant to extract water from deeper profile, while lateral root growth enable it to extract water from its surroundings. However, measuring root length is a difficult and laborious job. Different studies have shown a significant relationship of different traits with root length (Rauf and Sadaqat 2007a, b, 2008a). For example, Rauf and Sadaqat (2008a, b) evaluated sunflower genotypes under drought stress and showed a significant relationship of osmotic adjustment with root length. Therefore, increased osmotic adjustment may be an index of higher root length. Similarly, a positive correlation between root and shoot has been obtained in number of studies. Thus a higher shoot length will be an indication of higher root length. Furthermore root length itself is a morphological trait that depends on the pattern of plant dry matter partitioning. Higher allocation of dry matter to the roots depends upon the level of different of plant growth regulators such as absicisic acid (ABA). It has been found that under drought stress level of ABA sharply increases which triggers the accumulation of dry matter to the roots (Quarrie 1981).

8.4 *Osmotic Adjustment*

Osmotic adjustment (OA) has shown its promise in plant adaptation to drought stress. Many reports have shown a significant relationship of AO with yield. Osmotic adjustment is the ability of a genotype to produce certain osmolytes such as sugars, proline, K^+, Ca^{2+} etc. that are essentially non-toxic and have the ability to reduce the osmotic potential of the plant so that the plant can absorb more water from the soil. Due to higher osmotic adjustment ability, the plants were able to show more root growth, increased translocation of stem reserves to the reproductive parts, thus showing higher harvest index and yield under drought (Ludlow et al. 1990; Moinuddin 2004; Silva et al. 2006). However, a few studies have also shown negative effects of osmotic adjustment on yield and some others have shown no significant effect on yield (Turner et al. 2007). Therefore, beneficial effects of osmotic adjustment have been found to be dependent on the type of crop species or genotype, and specific and environment. Turner et al. (2007) evaluated yield and osmotic adjustment in the F_8 progeny under drought in Australia and India. They found that differences in osmotic adjustment among lines and parents varied from year to year and did not consistently benefit yield when measured in the field under terminal drought. In Australia, differences in osmotic adjustment were not associated with any yield benefit in any year, while in India early flowering resulted in higher yields at three of the four sites, and osmotic adjustment had an inconsistent effect on seed yields. In okra, different studies have shown significant genetic variability among the okra cultivars for osmotic adjustment (Wullschleger and Oosterhuis 1991; Ashraf et al. 2002). Ashraf et al. (2002) studied the osmotic adjustment ability of two cultivars i.e. Sabzpari and Chinese-red, after subjecting them to drought for 30 days. They found significant differences in cultivars for their osmotic adjustment ability. The leaf osmotic adjustment was much lower in Chinese-red than that in Sabzpari. However, overall growth of the two okra cultivars was negatively correlated with the osmotic adjustment.

Different methods have been devised to determine the osmotic adjustment ability of plants. However, simplest method was based on the difference of osmotic potential under controlled conditions to the drought condition at full turgor or after irrigating the field. Babu et al. (1999) compared four different methods for measuring osmotic adjustment in plants. The four methods were: (i) derivation of osmotic adjustment from regressions of leaf relative water content (RWC) on leaf osmotic potential (OP); (ii) estimation of osmotic adjustment from osmotic potential of stressed plants calculated to rehydrated state; (iii) estimation of osmotic adjustment from osmotic potential of stressed plants that have been rehydrated; and (iv) estimation (from data used in Method 1) of osmotic adjustment capacity by the sustained RWC at given osmotic potential of −3.5 MPa. They hypothesized that method 1 was a priori considered as the best estimate. However, they concluded that methods 2 and 3 were less demanding in terms of labour and plant materials than Methods 1 and 4. The results support the use of method 3 (the "rehydration method") as a faster and an economical replacement of method 1.

8.5 *Leaf Relative Water Contents*

Leaf relative water content is a simple technique that can easily determine the water status of the plant. On the basis of relative water contents, plants can easily be discriminated into tolerant or susceptible types. Bhatt and Rao (2005) estimated relative water contents in the irrigated plants of okra, the relative water content (RWC) varied from 72% to 82%, while in the stressed plants, it decreased to 66–75%. The highest RWC was recorded in cv. BO-1 during the stress. Relative water contents have also shown significant relationship with yield. Tahara et al. (1990) studied genetic relationship between grain yield and relative water contents in a random group of plants, selected from an F2 population having the pedigree, TAM W-101/Sturdy. A positive relationship was observed between grain yield and RWC measured during anthesis and mid-grain fill, as the high-yield selections maintained a significantly higher RWC than the low-yield selections. Grain yield and RWC were also positively associated among random selections segregating for both traits. In addition, different studies have suggested that relative water contents can be utilized as a physiological marker of other traits. Sawadogo et al. (2006) also concluded that the best variables for discriminating okra ecotypes during water deficit were the relative water content, and the length and the number of fruits.

8.6 *Stomatal Conductance*

Selection for stomatal conductance has the potential to increase the yield under both irrigated and drought conditions (Rebetzke et al. 2003; Khan et al. 2007). Under irrigated conditions, higher stomatal conductance has been found to be associated with the maintenance of leaf or canopy temperature (Lu et al. 1998) while higher stomatal conductance under water deficit conditions was an indication of some drought tolerance mechanisms that allow the plant to extract more water and transpire (Sashidhar et al. 2000). Stomatal conductance has shown a positive relationship with yield and has been used to evolve high yielding genotypes (Lu et al. 1998). Lu et al. (1998) has shown a significant improvement in the yield of cotton when plants were selected for higher yield. Two types of plant drought tolerance mechanism were observed on the basis of stomatal conductance (Sashidhar et al. 2000). The first mechanism allows lower stomatal conductance with warmer canopies. The lower stomatal conductance inhibits water losses, thus helping the plants to cope with the drought in better way. A term 'water saviours' has been given to these types of plants. However, lower stomatal conductance inhibits the CO_2 fixation, resulting in lower productivity (Jones and Corlett 1992; Cornic 2000; Chaves et al. 2002). The plants showing higher stomatal conductance with cooler bodies form another group (Sashidhar et al. 2000). Higher stomatal conductance under drought stress depends on different mechanisms such as root elongation that allows extraction of more water from the soil. This type of mechanism has shown a significant relationship with yield under stress conditions. Sashidhar et al (2000) evaluated the advantages and disadvantages of both mechanisms. They found that first mechanism resulted in the termination of growth while second mechanism has two disadvantages (a) high demand of energy for root growth (b) danger for the moisture contents to be exhausted quickly. It was concluded that presence of both mechanisms in low intensity i.e. drop in the stomatal conductance by at least 25–30% and uptake of water by lower roots in contact with moist soil help the plants to cope with drought more successfully.

8.7 *Cost, Speed and Stage of Measurement*

Many traits are measured with complex, time-consuming techniques that are unsuitable for screening large numbers of progeny in breeding programmes (Khan et al. 2001). As a result, breeders tend to measure them in small populations. In addition, the stage at which trait allows the evaluation of plants would also predict its suitability. A trait, which is evaluated earlier in the plant growth and shows freedom of its interaction with growth stages, will speed up the breeding programme.

Although morphological traits such as yield contributing traits and reduced plant height are evenly important. However, these traits may be only measured at plant maturity. On the other hand, some of the physiological traits i.e. relative water contents have the potential to evaluate large number of populations during earlier plant growth. While other more complicated and time-consuming techniques such as osmotic adjustment, root length and dry matter partitioning are only useful for screening a small number of genotypes for use as parents.

9 Conclusions and Future Prospects

Stress tolerance in general and drought tolerance in particular is a complex phenomenon, which is affected by many factors and characters and varies in different plant species. Further complexity in breeding crop plants against drought stress arises because of the quantitative nature of the multiple contributing traits. There are some common and specific responses across stresses and plant species (Khan et al. 2000). A common response to stresses is thought to be the relaxation of epigenetic regulation, leading to activation of suppressed sequences and secondary effects as regulatory systems attempt to re-establish genomic order (Madlung and Comai 2004).

Unlike other abiotic stresses, drought is a relatively temporary condition in terms of time, space and plant growth stage. During some years it occurs severely but may be another year there is no drought on account of rainfall in a particular area or in the water catchment areas of a river supporting the irrigating system of the area. Therefore, drought tolerance of a species may be enhanced by selecting traits related to drought tolerance and essentially positively correlated with yield. Independence of correlation with yield under non-stress conditions would help to take advantage of non-stress years. However, the trait(s) under selection should be easily scoreable to handle large segregating population at early growth phase. The molecular markers provide a high throughput system to identify and keep track of the potential trait(s) in large segregating populations. The review has considered the role of various physiological and morphological traits as selection criteria under drought stress. Among the traits leaf relative water contents and osmotic adjustment appeared to have potential to be used as selection criteria for breeding okra genotypes for low moisture conditions. The presence of additive genetic variability and high narrow sense heritability for these traits would increase their responsiveness to the selection. Furthermore, the increasing strengths in genomics, proteomics and molecular marker systems should be integrated in the drought tolerance breeding programmes.

References

Abebe A, Brick MA, Kirkby R (1998) Comparison of selection indices to identify productive dry bean lines under diverse environ mental conditions. Field Crops Res 58:15–23

Ahmad S, Mahmood A, Malik AJ, Karim A, Kumbhar MB (2003) Response of okra to water stress. Sarhad J Agric 19(1):73–79

Ahmad S, Malik AJ, Mahmood A, Kumbhar MB, Karim A (2004) Inheritance studies in okra under drought conditions. Sarhad J Agric 20(1):57–65

Annicchiarico P, Pecetti L (1998) Yield vs. morphophysiological traits based criteria for selection of durum wheat in a semi-arid Mediterranean region (Northern Syria). Field Crops Res 59(3):163–173

Ariyo OJ (1993) Genetic diversity in West African okra (*Abelmoschus caillei*) (*A. Chev.*) Stevels-multivariate analysis of morphological and agronomic characteristics. Genet Resour Crop Evol 40(1):25–32

Ashraf M, Arfan M, Shahbaz M, Ahmad A, Jamil A (2002) Gas exchange characteristics and water relations in some elite okra cultivars under water deficit. Photosynthetica 40:615–620

Ashraf MY, Ashraf M, Sarwar G (2005) Response of okra (*Hibiscus esculentus*) to drought and salinity stresses. In: Dris R (ed) Vegetables: growing environment and mineral nutrition. WFL Publisher, Helsinki, pp 166–177

Babu CR, Pathan MS, Blum A, Nguyen HT (1999) Comparison of measurement methods of osmotic adjustment in rice cultivars. Crop Sci 39:150–158

Batra BR, Mohan I, Arora SK (2000) Effect of irrigation and nitrogen levels on dry matter production by okra (*Abelmoschus esculentus* L. Moench). Har J Hort Sci 29(34):239–241

Beck DL, Vasal SK, Crossa J (1990) Heterosis and combining ability of CIMMYT's tropical early and intermediate maturity maize germplasm. Maydica 35(3):279–285

Bhatt RM, Rao NKS (2005) Influence of pod load on response of okra to water stress. Indian J Plant Physiol 10(1):54–59

Blum A (1979) Genetic improvement of drought resistance in crop plants. A case for sorghum. In: Mussel H, Staples RC (eds) Stress physiology in crop plants. John Wiley and Sons, Inc., New York, pp 430–445

Blum A (1988) Plant breeding for stress environments. CRC Press, Boca Raton

Blum A (2005) Drought resistance, water-use efficiency, and yield potential – are they compatible, dissonant, or mutually exclusive? Aust J Agric Res 56:1159–1168

Camciuc M, Bessifre JM, Vilarem G, Gaset A (1981) Volatile components in okra seed coat. Phytochemistry 48:311–315

Ceccarelli S (1987) Yield potential and drought tolerance of segregating populations of barley in contrasting environments. Euphytica 36:265–273

Chapman S, Crossa J, Basford KE, Karoonenberg PM (1997) Genotype by environment effects and selection for drought tolerance in tropical maize. II. Three mode pattern analysis. Euphytica 95:11–20

Chaves MM (1991) Effects of water deficits on carbon assimilation. J Exp Bot 42:1–16

Chaves MM, Pereira JS, Maroco J, Rodrigues ML, Ricardo CPP, Osório ML, Carvalho I, Faria T, Pinheiro C (2002) How plants cope with stress in the field: photosynthesis and growth. Ann Bot 89:907–916

Chheda HR, Fatokun CA (1991) Studies on okra germplasm in Nigeria. Report of an international workshop on okra genetic resources held at the National Bureau for Plant Genetic Resources, New Delhi, India, 8–12 Oct 1990, pp 21–23

Collaku A (1994) Selection for yield and its components in a winter wheat population under different environmental conditions in Albania. Plant Breed 112:40–46

Cooper M, Rajatasereekul S, Immark S, Fukai S, Basnayake J (1999) Rainfed lowland rice breeding strategies for northeast Thailand. I. Genotypic variation and genotype × environment interactions for grain yield. Field Crops Res 64:131–151

Cornic G (2000) Drought stress inhibits photosynthesis by decreasing stomatal aperture – not by affecting ATP synthesis. Trends Plant Sci 5:183–221

Deo C, Singh KP, Panda PK (1997) Genetic analysis of certain quantitative traits in okra (*Abelmoschue esculentus* (L.) Moench). Adv Plant Sci 10:179–184

Dhanda SS, Sethi GS (1998) Inheritance of excised-leaf water loss and relative water content in bread wheat (*Triticum aestivum* L.). Euphytica 104:39–47

Din NU, Carver BF, Clutter AC (1992) Genetic analysis and selection for wheat yield in drought-stressed and irrigated environments. Euphytica 62(2):89–96

Fereres E, Gimenez C, Fernandez JM (1986) Genetic variability in sunflower cultivar under drought. I. Yield relationships. Aust J Agric Res 37(6):573–582

Fischer RA, Maurer R (1978) Drought resistance in spring wheat cultivars. I. Grain yield response. Aust J Agric Res 29:897–912

Gill SS, Bassi G, Gill BS, Arora SK, Bassi G (1997) Morphological variation and characterisation of okra (*Abelmoschus esculentus* (L.) Moench). Crop Improv 24(1):69–73

Gimenez C, Fereres E (1986) Genetic variability in sunflower cultivars under drought. II. Growth and water relations. Aust J Agric Res 37(6):583–597
Hamon S (1991) Future prospects of the genetic integrity of two species of okra (*Abelmoschus esculentus* and *A. caillei*) cultivated in West Africa. Euphytica 58(2):101–111
Hsiao TC, Bradford KJ (1983) Physiological consequence of cellular water deficits. In: Taylor HM, Jordan WR, Sinclair TR (eds) Limitation to efficient water use in crop production. American Society for Agronomy, Crop Science Society of America, Madison, pp 227–265
Hulmel MB, Heumez E, Pluchard P, Beghin D, Depatureaux C, Giraud A, Le Gouis J (2005) Indirect versus direct selection of winter wheat for low-input or high-input levels. Crop Sci 45:1427–1431
Jones HG, Corlett JE (1992) Current topics in drought physiology. J Agric Sci 119:291–296
Kaiser WM (1987) Effects of water deficit on photosynthetic capacity. Physiol Plant 71:142–149
Karakoltsidis PA, Constantinidesm SM (1975) Okra seeds: a new protein source. J Agric Food Chem 23(6):1204–1207
Khan AA, McNeilly T (1998) Variability in aluminium and manganese tolerance among maize accessions. Genet Resour Crop Evol 45:525–531
Khan AA, McNeilly T, Collins JC (2000) Accumulation of amino acids, proline, and carbohydrates in response to aluminium and manganese stress in maize. J Plant Nutr 23:1303–1314
Khan AA, McNeilly T, Azhar FM (2001) Stress tolerance in crop plants. Int J Agric Biol 3:250–255
Khan AA, Rao SA, McNeilly T (2003) Assessment of salinity tolerance based upon seedling root growth response functions in maize (*Zea mays* L.). Euphytica 131(1):81–89
Khan FA, Azhar FM, Afzal I, Rauf S (2007) Effect of nitrogen regimes on combining ability variation in oil and protein contents in cottonseed (*Gossypium hirsutum* L.). Plant Prod Sci 10(3):367–371
Kholodova VP, Bormotova TS, Semenov OG, Dmitrieva GA, Kuznetsov VV (2007) Physiological mechanisms of adaptation of alloplasmic wheat hybrids to soil drought. Russ J Plant Physiol 54:480–486
Kron AP, Souza GM, Ribeiro RV (2008) Water deficiency at different developmental stages of *Glycine max* can improve drought tolerance. Bragantia Campinas 67(1):43–49
Larcher W (1995) Physiological plant ecology: ecophysiology and stress physiology of functional groups. Springer, New York/Berlin/Heidelberg, p 506
Lenis JJ, Calle F, Jaramillo G, Perez JC, Ceballos H, Cock JH (2006) Leaf retention and cassava productivity. Field Crops Res 95(3):126–134
Levitt J (1972) Responses of plants to environmental stresses. Academic, New York
Lu Z, Percy R, Qualset C, Zeiger E (1998) Stomatal conductance predicts yields in irrigated Pima cotton and bread wheat grown at high temperatures. J Exp Bot 49:453–460
Ludlow MM (1989) Strategies in response to water stress. In: Kreeb IIK, Ritcher H, Hinkley TM (eds) Structural and functional response to environmental stress: water shortage. SPB Academic Press, The Hague, pp 269–281
Ludlow MM, Santamaria JM, Fukai S (1990) Contribution of osmotic adjustment to grain yield in *Sorghum bicolor* (L.) Moench under water-limited conditions. II. Water stress after anthesis. Aust J Agric Res 41(1):67–78
Madlung A, Comai L (2004) The effect of stress on genome regulation and structure. Ann Bot 94:481–495
Martin F, Ruberte R (1979) Milling and use of okra seed meal at the house hold level. J Agric Pureto Rico 63:1–7
Mbagwu JSC, Adesipe FA (1987) Response of three okra (*Abelmoschus esculentus* L. Moench) cultivars to irrigation at specific growth stages. Sci Hort 31:35–43
MinFAL (2006) Ministry of food agriculture and life stock. Government of Pakistan, Islamabad
Mitra J (2001) Genetics and genetic improvement of drought resistance in crop plants. Curr Sci 80:758–763
Moinuddin CRK (2004) Osmotic adjustment in chickpea in relation to seed yield and yield parameters. Crop Sci 44:449–455

Muasya WNP, Diallo AO (2001) Development of early maturing and extra early drought and low nitrogen tolerant varieties using exotic and local germplasm for the dry mid altitude ecology. Proceedings of seventh eastern and southern Africa regional maize conference, pp 253–259
Ntare BR, Williams JH (1998) Heritability of components of a simple physiological model for yield in groundnut under semiarid rainfed conditions. Field Crops Res 58:25–33
Oosterom EJ, Jayachadran R, Bidinger FR (1999) Diallel analysis of the stay green traits and its component in sorghum. Crop Sci 36:549–555
Osmanzai M, Rajaram S, Knapp EB (1987) Breeding for moisture stressed areas. In: Srivastava JP, Porceddu E, Acevedo E, Varma S (eds) Drought tolerance in winter cereals. Wiley, New York, pp 151–162
Oyelade OJ, Ade-Omowaye BIO, Adeomi VF (2003) Influence of variety on protein, fat contents and some physical characteristics of okra seeds. J Food Eng 57:111–114
Oyenuga VA (1969) Nigeria's foods and foodstuffs: their chemistry and nutritive values, 3rd edn. Ibadan University Press, Ibadan
Panda PK, Singh KP (1997) Genetic variability, heritability and genetic advance for pod yield and its contributing traits in okra hybrids. Madras Agric J 84(3):136–138
Pantuwan G, Fukai S, Cooper M, Rajatasereekul S, O'Toole JC (2002) Yield response of rice (*Oryza sativa* L.) genotypes to different types of drought under rainfed lowlands. Part 1. Grain yield and yield components. Field Crops Res 73:153–168
Passioura JB (1976) Physiology of grain yield in wheat growing on stored water. Aust J Plant Physiol 3:559–565
Quarrie SA (1981) Genetic variability and heritability of drought-induced abscisic acid accumulation in spring. Plant Cell Environ 4:147–151
Quisenberry JE, Roark B, Fryrear DW, Kohel RJ (1980) Effectiveness of selection in upland cotton in stress environments. Crop Sci 20:450–453
Ramırez-Vallejo P, Kelly JD (1998) Traits related to drought resistance in common bean. Euphytica 99:127–136
Rauf S, Sadaqat HA (2007a) Sunflower (*Helinathus annuus* L.) germplasm evaluation for drought tolerance. Commun Biomet Crop Sci 2:8–16
Rauf S, Sadaqat HA (2007b) Effects of varied water regimes on root length, dry matter partitioning and endogenous plant growth regulators in sunflower (*Helianthus annuus* L.). J Plant Interact 2:41–51
Rauf S, Sadaqat HA (2008a) Effect of osmotic adjustment on root length and dry matter partitioning in sunflower (*Helianthus annuus* L.) under drought stress. Acta Agric Scand Sect B Soil Plant Sci 58:252–260
Rauf S, Sadaqat HA (2008b) Identification of physiological traits and genotypes combined to high achene yield in sunflower (*Helianthus annuus* L.) under contrasting water regimes. Aust J Crop Sci 1:23–30
Rauf S, Khan TM, Naveed A, Munir H (2007) Modified path to high lint yield in upland cotton (*Gossypium hirsutum* L.) under two temperature regimes. Turk J Biol 31:119–126
Rauf S, Sadaqat HA, Khan IA (2008) Effect of moisture regimes on combining ability variations of seedling traits in sunflower (*Helianthus annuus* L.). Can J Plant Sci 88:323–329
Rebetzke GJ, Condon AG, Richards RA, Farquhar GD (2003) Gene action for leaf conductance in three wheat crosses. Aust J Agric Res 54:381–387
Richards RA (1996) Defining selection criteria to improve yield under drought. Plant Growth Regul 20:157–166
Sashidhar VR, Ankegowda SJ, Mahesh J, Kulkarni M, Srinivas N, Prasad TG, Nagalakshmi U, Devendra R (2000) Should plants keep their (canopy) 'cool' or allow themselves to grow 'warm' under stress: it is a Hobson's choice and plants survive by doing a balancing act. Curr Sci 78:786–792
Savello P, Martin FW, Mill JM (1980) Nutritional composition of okra seed meal. Agric Food Chem 28:1163–1166
Sawadogo M, Zombre G, Balma D (2006) Behaviour of different ecotypes of okra (*Abelmoschus esculentus* L.) under water deficit during budding and flowering phases. Biotechnol Agron Soc Environ 10:43–54

Seghatoleslami MJ, Kafi M, Majidi I, Nour-Mohammadi G, Darvish F (2007) Effect of drought stress at different growth stages on yield and water use efficiency of five proso millet (*Panicum miliaceum*) genotypes. J Sci Technol Agric Nat Res 11:27–231

Siemonsma JS (1982) West African okra – Morphological and cytogenetical indications for the existence of a natural amphidiploids of *Abelmoschus esculentus* (L.) Moench and *A. manihot* (L.) Medikus. Euphytica 31:241–252

Silva H, Copaja SV, Bravo HR, Argandona VH (2006) Relationship between grain yield, osmotic adjustment and benzoxazinone content in *Triticum aestivum* L. cultivars. Zeitschrift Fur Naturforschung 61:704–708

Sinaki JM, Heravan EM, Hossein A, Rad S, Noormohammadi G, Zarei G (2007) The effects of water deficit during growth stages of canola (*Brassica napus* L.). Amer Eur J Agric Environ Sci 2:417–422

Singh BR, Singh DP (2005) Agronomic and physiological responses of sorghum, maize and pearl millet to irrigation. Field Crops Res 42:57–67

Szilagyi L (2003) Influence of drought on seed yield components in common bean. Bulg J Plant Physiol Special Issue:320–330

Tahara M, Carver BF, Johnson RC, Smith EL (1990) Relationship between relative water content during reproductive development and winter wheat grain yield. Euphytica 49:255–262

Turner NC (1986) Adaptation to water deficits: a changing perspective. Aust J Plant Physiol 13:175–190

Turner NC, Abbo S, Berger JD, Chaturvedi SK, French RJ, Ludwig C, Mannur DM, Singh SJ, Yadava HS (2007) Osmotic adjustment in chickpea (*Cicer arietinum* L.) results in no yield benefit under terminal drought. J Exp Bot 58:187–194

Vasal SK, Cordova H, Beck DL, Edmeades GO (1997) Choices among breeding procedures and strategies for developing stress tolerant maize germplasm. In: Edmeades GO, Bänziger M, Mickelson HR, Pena-Valdiva CB (eds) Developing drought and low N-tolerant maize. Proceedings of a smposium, 25–29 Mar 1996, CIMMYT, El Batán, Mexico, pp 336–347

Verma IM, Batra BR (2001) Effect of irrigation and nitrogen on growth and yield in okra. Changing scenario in the production systems of horticultural crops. In: Proceedings of a national seminar, Coimbatore, Tamil Nadu, India, 28–30 Aug 2001. South-Indian-Horticulture, 49:386–388

Wade LJ, McLaren CG, Quintana L, Harnpichitvitaya D, Rajatasereekul S, Sarawgi AK, Kumar A, Ahmed HU, Sarwoto AK, Singh R, Rodriguez J, Sarkarung S (1999) Genotype by environment interactions across diverse rainfed lowland rice environments. Field Crops Res 64:35–50

White JW, Castillo JA, Ehleringer JR, Garcıa JA, Singh SP (1994) Relations of carbon isotope discrimination and other physiological traits to yield in common bean (*Phaseolus vulgaris*) under rainfed conditions. J Agric Sci 122:275–284

Wullschleger D, Oosterhuis D (1991) Osmotic adjustment and the growth response of seven vegetable crops following water-deficit stress. In: Proceedings of the annual meeting on alternative crops research and development programs: strategies to reduce the plight of the farmer, 26(9): 1127–1140

Part II
Biotechnology, Molecular Biology and Genetics

Chapter 9
Biotechnology as an Aid for Crop Improvement to Overcome Food Shortage

Khalid ul Rehman Hakeem, Münir Öztürk, Parvaiz Ahmad, and Abdul Razaque Memon

Contents

K.ul R. Hakeem (✉)
Botany Department, Jamia Hamdard University, New Delhi, India
e-mail: kur.hakeem@gmail.com

M. Öztürk
Science Faculty, Botany Department, Ege University, Bornova-Izmir, Turkey
e-mail: munirozturk@gmail.com

P. Ahmad
Department of Botany, Amar Singh College, University of Kashmir,
Srinagar 190008, Jammu & Kashmir, India
e-mail: parvaizbot@rediffmail.com; parvaizbot@yahoo.com

A.R. Memon
Department of Genetics & Bioengineering, Faculty of Engineering & Information Technologies,
International Burch University, Sarajevo 71000, Bosnia and Herzegovina
e-mail: abdulrezzak.memon@gmail.com; armemon@ibu.edu.ba

M. Ashraf et al. (eds.), *Crop Production for Agricultural Improvement*,
DOI 10.1007/978-94-007-4116-4_9, © Springer Science+Business Media B.V. 2012

Abstract World's population has crossed 6.5 billion with majority of human beings living in developing or under developing countries. Clearly, food security in such countries will be a primary concern over the next few decades. However, options for increased food production to meet this population pressure are limited because most arable land is already under cultivation, and in many areas land use cannot be further intensified without a risk to the long-term productivity. Agricultural land use has been especially intense in recent years because of rapid urbanization and increasing environmental pollution. The ultimate need is to use newer technologies which could help us to curb this food insecurity. Biotechnology is globally recognized as a rapidly emerging, complex and far reaching new technology. It has revolutionized all the fields of life. Recent discoveries and technical innovations in the field of genomics and biotechnology are revealing the full complement of genes in crops, the ability to define genetic variation and use DNA markers to follow chromosome segments with known functions through breeding programmes are leading to new efficiencies in breeding. The ability to isolate and redesign genes and transfer them into different plants also offers the breeder solutions to several key limitations. The convergence of advances in biology-genomics, proteomics, bioinformatics and information technologies is driving the emergence of a new bio-economy. By the usage of this technology we have achieved remarkable success in increasing crop productivity, improving crop quality as well as overcoming food shortage. Additionally the genetically engineered crops have shown a remarkable potential to tackle some of the world's most challenging socioeconomic problems which are more prevalent in the developing world than in the industrialized nations.

Keywords Biotechnology • Food security • Molecular markers • Transgenics • Proteomics • Nanobiotechnology

1 Introduction

The world population has increased by 2.3 billion people in the past 40 years, and by the year 2040, an additional 3.6 billion will be added to it. In fact, in every earth hour about 13,000 new human beings are added up to this globe. Most of this increase is in the developing countries where already one billion people go hungry every day and live in dismal poverty. The African nations of Ethiopia, Nigeria or Egypt each add more people than all of Western Europe combined (World Watch Institute, Washington, DC). India's population is around 1.14 billion with an annual per capita income of less than US $ 1,043 and its population is projected to continue to increase to a total of 1.5 billion by the year 2030 (FAO 2008). The population increase in developing countries constitutes 97% of the global increase (Swaminathan 1995). It is therefore, an intimidating task to feed the ever increasing population in the resource-poor countries where agriculture is already constrained by lack of new arable land, small sized farms, and certain destructive agricultural practices contributing to soil degradation, salinization and ultimately the desertification.

Agriculture is an essential component of societal well-being. It occupies 40% of the land surface, consumes 70% of global water resources and manages biodiversity at genetic, species and ecosystem levels. Intensive use of inorganic fertilizers and pesticides, expansion of irrigation, and capital-intensive farm management has resulted in an unparalleled increase in global agricultural productivity since 1950s. Agriculture, therefore, is and will continue to be central to all strategies for planned socio-economic development of the countries. Despite major advances in agriculture and strong growth in food production in the latter part of the twentieth century, food security for the masses continues to be an area of concern. The application of biotechnological techniques in the agriculture sector can potentially improve food security by raising crop tolerance to adverse weather and soil conditions, by enhancing adaptability of crops to different climates and by improving yields, pest resistance and nutrition, particularly of staple food crops. Biotechnology can, over the next two decades, deliver the next wave of technological change that can be as fundamental and invasive as that brought about by information technology.

Recently the World Food Program (WFP) and the Food and Agriculture Organization (FAO) -reported that 22 countries are experiencing protracted food insecurity, and 17 of these countries are in Africa. There has been a remarkable increase in total grain production between 1950 and 1980, but only a marginal increase has been realized during 1980–1990 (Myers 1999; Ozturk and Uzonur 2004). This increase in grain production has mostly resulted from an increase in area under cultivation, irrigation, better agronomic practices, and improved cultivars. Yields of several crops have already reached a plateau in developed countries, and therefore, most of the productivity gains in the future will have to be achieved in developing countries through better natural resources management and crop improvement. Productivity gains are essential for long-term economic growth, but in the short-term, these are even more important for maintaining adequate food supplies for the growing world population. It is in this context that biotechnology will play an important role in food production during next few decades.

In the present review we attempt to take a practical look at the prospects and constraints of various types of biotechnologies and their application for increasing crop production and improving nutritional quality. Under this context, we also address the important issues of biosafety and impact of the genetically engineered crops on the environment.

2 Biotechnology in Agriculture

The Convention on Biological Diversity (CBD) defines biotechnology as: "any technological application that uses biological systems, living organisms, or derivatives thereof, to make or modify products or processes for specific use". This technology has been incorporated in every part of living system, from disease cure to crop improvement in which new traits are being either introduced or old ones modified to yield better system. Genetic modification of crops is one such method allowing individual characters (gene, factor or trait) to be transferred into crop plants. With the advent of genetic modification through genetic engineering in early 1980s, the natural barrier of only intra-specific exchange of characters was removed and scientists were able to identify and transfer specific genes associated with desirable traits from one organism to the organism of other species that otherwise cannot breed naturally (Ozturk and Uzonur 2004; Shigeto et al. 2006). With these techniques genes from varied class of organisms like bacteria, virus or even animal may be transferred into plants to develop genetically modified plants having exclusively changed characteristics controlled by the specific gene. This gives scientists/ breeders a broader access to desirable traits from any living organism and its possibility of transferring it with much faster rate and greater precision. There are numerous biotechnological approaches through which we can develop effective changes in agricultural fields and yield better crops in terms of their productivity as well as quality. Future impacts of biotechnology in crop production will be in the areas of:

1. Developing new hybrid crops based on genetic male-sterility,
2. Exploit transgenic apomixes to fix hybrid vigour in inbred crops,
3. Increased resistance to insect pests, diseases, and abiotic stress factors,
4. Improved effectiveness of bio-control agents,
5. Enhance nutritional value (vitamin A and iron) of crops and post-harvest quality,
6. Increase efficiency of soil phosphorus and essential micronutrients uptake especially Zn uptake and its translocation in plants,
7. Increase the nitrogen fixation capacity of legumes,
8. Improve adaptation to soil salinity and heavy metal toxicity,
9. Understanding the nature of gene action and metabolic pathways,
10. Increase photosynthetic activity, sugar and starch production, and
11. Production of pharmaceuticals and vaccines in suitable pants (biofarming).

Further understanding the biochemical process(s) at molecular scales and changes in expression levels could be an enormous help with which biotechnology proves beneficial.

3 Fertilizer Usage and Biotechnology

Increase in agricultural productivity during green revolution is largely associated with the augmentation of fertilizers (particularly N-based fertilizers). It is estimated that around 50% of the human population relies on nitrogen (N) fertilizer for food production globally (Pathak et al. 2008). The N fertilizer consumption has grown dramatically in Asia, about 17-fold in the last 40 years (Dobermann and Cassman 2004). However it is surprising to know that only 50% or less of the applied nitrogen is used for producing the aboveground biomass of cereals. The other 50% or more gets dississipated in the wider environment by volatilization, leaching, surface runoff and denitrification (Jeffrey et al. 2002), resulting in the major detrimental impacts on environment, such as eutrophication of fresh water and marine ecosystems (Beman et al. 2005), gaseous emission of oxides reacting with the stratospheric ozone, and the emission of toxic ammonia (Stulen et al. 1998) into the atmosphere. Despite these hazardous impacts on the biosphere the use of N fertilizers has increased 100-folds over the last 100 years, as it is known that the cereal yield and the fertilizer N consumption have increased in a linear fashion during the past 40 years and both are highly correlated (Ladha 2005). At this juncture, the use of nitrogenous fertilizers cannot be reduced because of the pressure of more and more food production to feed the teeming population. Development of crop varieties that can grow and yield well at low nitrogen condition can be a solution to the problem. Tolerance of a crop to low N conditions is a highly desired characteristic for sustainable crop production. Many approaches such as optimal time, rate, and methods of application for matching N supply with crop demand; the use of specially formulated forms of fertilizer, including those with urease and nitrification inhibitors; the integrated use of fertilizers, manures, and/or crop residues; and optimizing irrigation management have been suggested for increasing nitrogen use efficiency (Abrol et al. 1999; Abdin et al. 2005; Raghuram et al. 2006). But their adoption at the farm level has been limited for various reasons in developing countries. Progress has been made in genetic and molecular analysis of low N tolerance and breeding crops for low N conditions. To develop varieties with improved nitrogen use efficiency it is necessary to have high level of genetic diversity for N uptake efficiency. Lian et al. (2005) analyzed the genetic components associated with low N tolerance in rice at the seedling stage. So, through biotechnological approaches we can work out the molecular mechanism of nutrient assimilation in crop plants which will help us to manipulate the important genes (s) which may affect nutrient use efficiency and hence crop yield.

4 Engineering Plants with Other Gene Systems Regulating N Metabolism

Studies with transgenic plants overexpressing genes affecting the N metabolism pathway suggest it is possible to improve or manipulate N metabolism and the growth phenotype of plants, which can improve the nitrogen use efficiency (NEU) of crop plants. In order to identify and understand the regulation of the genes involved in enhancing NUE, proper evaluation of the combined genetic and transgenic approaches to improving NUE are needed as a component of any crop improvement program. The benefits of growing NUE-efficient crops will not be realized until breeders evaluate N metabolism and nitrogen use efficiency in economically important crop plants.

In higher plants, the expression of the NR genes is influenced by several external and endogenous factors and is highly regulated at the transcriptional as well as post-translational levels. The overexpression of either the NR or the NiR gene in plants increases mRNA levels and often affects N uptake. However, the increased uptake of N does not seem to increase the yield or growth of plants, regardless of the N source (Andrews et al. 2004). This is believed to be due, in part, to the complex regulation of both NR and the pathway as a whole. Lea et al. (2006) demonstrated that post-translational regulation of NR strongly affects the levels of free amino acids, ammonium, and nitrate, whereas transcriptional regulation has only minor influence. Plants expressing fully unregulated NR accumulate high concentrations of asparagine and glutamine in leaves; however these transgenic plants grow and develop normally, despite having an NR enzyme that is active during both light and dark periods (Good et al. 2004; Shrawat and Good 2008). Mutants or transgenic plants with altered levels of GS/GOGAT are used to determine the effects of these proteins on plant development and to study the expression of the different members of the GS multigene family. Although several studies demonstrate that an increase in GS activity in transgenic plants has no effect on the phenotype, many researchers show a direct correlation between an enhanced GS activity in transgenic plants and an increase in biomass or yield, upon incorporating a novel *gs1* construct. Transgenic tobacco plants enriched or reduced in plastid glutamine synthetase (GS2, a key enzyme in photorespiration) have been developed (Kozaki and Takeba 1996). Those transgenic plants having twice the normal amount of GS2 had an improved capacity for photorespiration and an increased tolerance to high-intensity light, whereas those with a reduced amount of GS2 had a diminished capacity for photorespiration and were photo-inhibited more severely by high-intensity light compared with the control plants. Ectopic expression of GS1 has been shown to alter plant growth (Oliveira et al. 2002) and the over expression of GS1 in transgenic plants could cause the enhancement of photosynthetic rates, higher rates of photorespiration and enhanced resistance to water stress (El-Khatib et al. 2004). The overexpression of soybean cytosolic GS1 in the shoots of *Lotus corniculatus* was reported to accelerate plant development, leading to early senescence and premature flowering, particularly when plants were grown under conditions of high ammonium (Vincent et al. 1997). Man et al. (2005) provided additional empirical evidence for enhanced nitrogen-assimilation efficiency in GS1 transgenic lines. However, differences in the degree of ectopic GS1 expression have been reported and attributed to positional effects, effectiveness of chimeric constructs, or differences in growth conditions. These differences could account for the lack of correlation between the enhanced expression of GS1 and concomitant growth (Vincent et al. 1997; Ortega et al. 2001). Interestingly, the differences are more striking at a low nitrate concentration. In addition, higher rates of N incorporation into the transgenic plants further demonstrate that the transformed plants have increased NUE (Man et al. 2005).

In comparison to GS, few reports have described the production of transgenic plants overexpressing *GOGAT* genes. Transgenic overexpression and antisense technology have been employed recently to modulate the expression of NADH-GOGAT in alfalfa and rice plants (Schoenbeck et al. 2000; Yamaya et al. 2002). The studies on transgenic rice plants expressing antisense RNA for either GS1 or NADH – GOGAT point towards the possible involvement of GS1 in the export of N via phloem in senescing leaves. On the other hand, in case of developing leaf blades and spikelets, NADH-GOGAT was implicated in the utilization of glutamine transported from senescing organs (Yamaya 2003). While these genes appear to be good candidates for improving NUE in the short run, the degree of improvement may vary with the crop and cropping conditions. Therefore, the utility of transgenic over-expression of N-assimilatory genes for major improvements of NUE remains uncertain, though the possibility that different crops respond differently cannot be ruled out yet.

5 Signalling and Regulation of Nitrogen Metabolism

It is a well known concept in signal transduction that whenever multiple genes are subject to transcriptional regulation by a common signal, it is mediated through a regulatory sequence that exists in all the genes that respond to the signal. These signature sequences, commonly known as response elements, are identified by mutations that abolish their function, and their conserved nature as revealed by homology comparisons. Early experiments in transgenic *Nicotiana* plants using GUS gene fused to NR and NiR promoter sequences clearly demonstrated for the first time that nitrate induction of gene expression requires some sequence(s) associated with the NR and NiR promoters (Rastogi et al. 1993; Quesada et al. 1997). Subsequent studies in transgenic tobacco incorporating the 5′ flanking regions of the two *Arabidopsis thaliana* nitrate reductase genes *NR1* and *NR2* (designated *NP1* and *NP2*) demonstrated that 238 and 330 bp of *NP1*and *NP2* respectively are sufficient for nitrate-dependent transcription (Lin et al. 2005; Lea et al. 2006). These nitrate-responsive elements (NREs) are composed of several copies of a core A[G/C]TCA sequence motif preceded by an ~7-bp AT-rich sequence present in the 5′flanking regions of nitrate reductase (*NR1* and *NR2*) genes. This particular sequence motif was also found to be very well conserved in the 5′ flanking regions of NR and NiR genes from eight other plants (Hwang et al. 1997). Sarkar (2003) compared the flanking sequences of all available plant nitrate responsive genes and found that the NRE core sequence (A[C/G]TCA) was present in multiple copies on both strands in all the known nitrate-responsive genes in many dicots, monocots and cyanobacteria. Though most of the NREs examined contained both the core sequence and a proceeding AT rich sequence, there were some cases which had GC rich regions or did not reveal any AT/GC bias. A more detailed bioinformatic analysis of the entire *Arabidopsis* genome revealed that the proposed NREs are randomly distributed, with no difference between nitrate responsive genes and the presumably nonresponsive genes and intergenic regions in the rest of the genome (Kang et al. 2004; Raghuram et al. 2006). These findings raise doubts on the validity of the proposed NRE as comprising of (A[C/G]TCA) elements preceded by AT-rich sequence. Further work in this area will need a combination of bioinformatic and experimental approaches to redefine the NREs that mediate the expression of all nitrate responsive genes in all plants. The discovery of NREs is important, as it provides an end point for nitrate signal transduction.

For the regulation of nitrate uptake, signals are derived from nitrate, which are involved in triggering widespread changes in gene expression; resulting in reprogramming of N metabolism to facilitate the uptake and assimilation of nitrate and its incorporation into amino acids. The nitrate assimilatory pathway is under tight regulation by the available nitrate and reduced N. In strawberry, increasing external nitrate concentration from 0 to 4 mM markedly increased the cumulative nitrate uptake (Taghavi and Babalar 2007). Several of the LATS- and HATS-related genes, apart from being root specific, are also inducible by nitrate and there is evidence that at least one HATS-related gene, *NpNrt2:1* is also repressible by reduced nitrogen (Quesada et al. 1997). In barley and white spruce, cHATS provides a high affinity, low capacity pathway for nitrate entry in uninduced plants. Nevertheless, cHATS activity is up regulated (approximately three folds) by exposure to nitrate (Trueman et al. 1996). In barley, the fully induced iHATS flux was approximately 30 times higher than that resulting from the cHATS (Quesada et al. 1997). The increase in transcript is accompanied by increased rates of nitrate uptake (Imsande and Touraine 1994). The results on citrus seedlings suggest that LATS is under feedback control by the N status of plant. A decline in uptake rate by the addition of amino acids (Glu, Asp, Asn, Gln) to the external solution has been reported (Cerezo et al. 2000). The use of chemical inhibitors in physiological studies has suggested that protein synthesis is important for nitrate uptake (Aguera et al. 1990) and the transporters may turn over relatively slow. A degradation mechanism for transporter protein in *Arabidopsis* (*AtNrt2:1*) has been suggested (Cerezo et al. 2001). The presence of a number of conserved protein kinase C recognition motifs in the N and C domains of *HvNRT2:1* (Forde 2000) suggests that phosphorylation events are involved in regulating *AtNrt2:1* activity in response to environmental cues. Remans et al. (2006) found that under N-limited conditions, *AtNrt2:1* played a key role as a major NO_3^- uptake system and coordinated lateral root initiation and development with external NO_3^- availability.

The precise mechanism of nitrate sensing and signalling is not yet fully understood. Post-translational regulation of some of the nitrate-responsive enzymes is brought about by 14-3-3 proteins, though they mainly mediate the effect of light and other signals, rather than nitrate. A few elements possibly associated with nitrate signaling are Ca^{2+} and protein kinases/ phosphatases; these have been implicated in mediating the nitrate signal for the expression of NR, NiR and GS2 m RNAs (Sakakibara et al. 1997). In addition to the kinases, Hartwell et al. (1999) described a Ca^{2+} independent PEPCase protein kinase, which is a novel member of the Ca^{2+} calmodulin regulated group of protein kinases. Krapp et al. (2002) described their specific roles in mediating nitrate and other interacting signals. A better understanding of the nitrate-signaling cascade might emerge from the study of mutants related to the signal-transfer cascade from nitrate to the NR gene (Ogawa et al. 2000), revealing more intermediate and potential sites for the manipulation of NUE. Light is an additional signal that regulates the expression of many nitrate-responsive genes, though it has been studied in depth in only a few of them. The role of light in regulation of NR gene expression has often been reviewed (Raghuram and Sopory 1995; Lillo and Appenroth 2001). The effects of light in green plants are signalling and N-use efficiency probably mediated more indirectly, through photosynthesis and sugars (Lillo and Appenroth 2001). At the post-translational level, light acts by modulating the phosphorylation status of the enzyme, in conjunction with 14-3-3 proteins.

Transcriptional regulation of several hundreds of nitrate-responsive genes by nitrate as a signal requires cis-acting regulatory sequences or nitrate responsive elements (Raghuram et al. 2006). ANR1, a putative signalling and N-use efficiency transcription factor homologous to the MADS box family has been reported in *Arabidopsis thaliana* (Zhang and Forde 1998). ANR1 is nitrate inducible and root-specific, and has been shown to be involved in nitrate-dependent stimulation of lateral-root proliferation in transgenic plants (Forde 2002). However, this root-specific transcription factor does not account for the transcription of all the known nitrate responses even in the root, besides being irrelevant for nitrate-responsive gene expression in the shoots. In terms of finding a global target for manipulation of NUE, the successful manipulation of N content by overexpression of the Dof1 transcription factor indicates that unravelling the signalling mechanisms that bring about their coordinated expression of nitrate-responsive genes by N and C metabolites could reveal new targets and approaches for future metabolic-engineering efforts (Lochab et al. 2007). Yanagisawa et al. (2004) generated transgenic *Arabidopsis* lines over expressing Dof1, a maize protein that belongs to Dof family of plant-specific transcription factors known to activate the expression of several C-metabolizing genes associated with organic-acid metabolism. The transformants showed up to 30% higher levels of mRNA and enzyme activities for PEP carboxylase and pyruvate kinase, without any reduction of NR, GS, and GOGAT RNAs. If Dof1 is not nitrate inducible, it means that multiple transcription factors may be involved in the coordinated expression of N and C metabolizing genes.

6 Marker Assisted Selection of High Yielding Varieties

Marker-assisted selection and DNA fingerprinting allow a faster and much more targeted development of improved genotypes for all living species. They also provide new research methods which can assist in the conservation and characterization of biodiversity. The new techniques will enable scientists to recognize and target quantitative trait loci and thus increase the efficiency of breeding for some traditionally intractable agronomic problems such as drought resistance and improved root systems. A genetic marker is a measurable character with Mendelian inheritance. The advancement in DNA marker technology has facilitated in genome analysis and rapid development of many high-density linkage, physical and consensus maps in crops of interest (Gupta and Varshney 2000; Lörz and Wenzel 2005). The most commonly employed markers in crop plants are random amplified polymorphic DNA (RAPD), intersimple sequence repeat (ISSR) markers and amplified fragment length polymorphism (AFLP). These are the markers of choice for crops with inadequate genomic resources, do not require prior sequence information and scan the genome including the repetitive sequences. In fact, the RELP (restriction fragment length polymorphism) approach has been used successfully to identify genetic markers in plants, including rice (Tanksley et al. 1989: Wang et al. 1994). However, the RFLP technique needs specific probes for the target DNA sequences, and use of radioactive elements makes it more costly and tedious. The development of PCR technique has offered a good alternative to the RFLP analysis. The PCR-based RAPD approach using single 10-mer arbitrary primers requires much less DNA, and is technically simple and cheaper compared to the RFLP (Williams et al. 1990). In maize, 42 pairs of proteins showed a 1:2:1 segregation in the F2 population indicative for a monogenic inheritance. Two linkage maps were constructed from RFLP and position-shift loci, which revealed that protein markers were interdispersed between the RFLP markers on all chromosomes (De Vienne et al. 1996). In many cases position-shift variants correspond to the same protein as shown by micro-sequencing (Touzet et al. 1995; Plomion et al. 1997). It is expected that the maps of expressed genes obtained by 2-Dimentional Electrophoresis (2-DE) will be crucial for the candidate gene strategy of quantitative trait loci (QTL) characterization (Thiellement et al. 1999). QTL analysis has been applied to map genes controlling protein quantity for spots on 2-D gels (Touzet et al. 1995) and the loci have been termed PQL for protein-quantity loci (Thiellement et al. 1999). Co-localization of a protein-quantity locus (PQL) and its protein-coding locus would indicate that expression level of the protein is a consequence of allelic differences, whereas co-localization between a PQL and a QTL for a different trait would point to an association of a candidate gene and the variation observed for a trait (Zivy and de Vienne 2000; Thiellement et al. 1999; Consoli et al. 2002; Lian et al. 2005). The level of molecular polymorphism in wheat has been found to be low as compared to many other species (Prasad et al. 2000; Song et al. 2002), thereby limiting studies on variability and diversity using molecular markers. Among different classes of molecular markers, simple sequence repeat (SSR) markers are short (1–6 bp long) tandemly repeated DNA sequences that are highly polymorphic and are useful for a variety of applications in molecular breeding due to its reproducibility, multi-allelic nature, co-dominant inheritance, abundance and high polymorphic information content (PIC), and thus have recently been used to study the genetic variability based on DNA polymorphism in a number of crop species (Morgante and Oliveri 1993; Powell et al. 1996; Gupta et al. 1996; Gupta and Varshney 2000; Prasad et al. 2000; Agarwal et al. 2008). It has also been demonstrated that even limited numbers of SSR markers were adequate to discriminate closely related wheat and barley varieties (Russel et al. 1997; Prasad et al. 2000). The genomes of all eukaryotes contain a class of sequences termed microsatellites (Litt and Luty 1989) or simple sequence repeats (SSRs) (Tautz et al. 1986). Microsattelite are short tandem repeat of 1–6 bp that can repeat up to 100 times (Schloetterer et al. 1991), have emerged as an important source of ubiquitous genetic markers for many eukaryotic genomes (Wang et al. 1994). In plants, it has been demonstrated that microsatellites are highly informative, locus-specific markers in many species (Liu et al. 1996; Moerchen et al. 1996; Smulders et al. 1997), because they are multiallelic, microsatellites have high potential for use in evolutionary studies (Schloetterer et al. 1991; Buchanan et al. 1994) and studies regarding genetic relationships. SSRs show a much higher level of polymorphisms and are informative in hexaploid bread wheat than any other marker system (Ma et al. 1996; Bryan et al. 1997). They can be utilized for several applications in plant genetics mapping, cultivars discrimination and detection of genetic diversity (Gupta and Varshney 2000). SSRs provide an efficient means of detecting genetic diversity as they can detect high number of alleles per assay (Powell et al. 1996).

7 Development of Transgenic Crops

To improve both productivity and sustainability of agriculture, new crop varieties have been introduced by transgenic approach. Over the years, considerable progress has been made in developing transgenic plants in which agriculturally important crops have been improved upon by the incorporation of gene from similar or another species. Much emphasis has been give to develop insect-resistant transgenic crops as the world annual losses from plant pathogens alone are estimated to be 12% (Cook 2006). Transgenic insecticidal crop cultivars are in the process of revolutionizing agriculture and are likely to become a major insect management tactic worldwide. Introducing novel resistant genes into economically important crops can develop insect-resistant crops. This tactic has a potentially key role in integrated pest management of several important pests (Gatehouse and Gatehouse 1998). *Bacillus thuringiensis* (Bt)-, a soil-borne bacterium, based transgenic cultivars have been produced for Cotton (*Gossypium hirsutum*), Potato (*Solanum tuberosum*), maize (*Zea mays*), rice (*Oriza sativa*), tomato (*Lycoperscion esculentum*), alfalfa (*Medicago sativa*) and numerous other crop plants (Hilder and Boulter 1999). The *bar* gene conferring herbicide tolerance was introduced in 1-month-old wheat calli employing both particle bombardment and *Agrobacterium*-mediated transformation strategies (Chugh and Khurana 2003). Herbicide-tolerant soybean was the dominant transgenic crop grown commercially in the USA, Argentina, Canada, South Africa, Romania and Uruguay, occupying 33.3 million hectares in 2001. The current efforts to improve plant stress tolerance by gene transformation have resulted in important achievements. Present engineering strategies rely on the transfer of one or several genes that are either involved in signaling and regulatory pathways, or that encode enzymes present in pathways leading to the synthesis of functional and structural protectants, such as osmolytes and antioxidants, or that encode stress-tolerance-conferring proteins. Transgenic system in indica rice and wheat has been developed. Rice has been transformed with codA/COR47, AtHSP100 and PDC gene while wheat transformed with *hva1* to confer stress resistance. All the transgenics have been characterized at the molecular level for gene integration. Studies have been conducted to transform indica rice plants with *p5cs* gene to make it more tolerant to salinity. The assessment of transformed plants harboring this gene as a single copy showed promising results and further efforts are on to pyramid one more gene for higher tolerance to salinity.

8 Modern Biotechnological Tools and Crop Productivity

During the advent of post genomic era we came to know the newer branches of biotechnology which provides us new insights in understanding the different process(s) and mechanisms that ultimately leads to crop improvement.

8.1 Genomic Approach for Crop Improvement

Genomics research has provided breeders new tools, such as functional molecular markers and informatics, as well as new knowledge about statistics and inheritance phenomena that could increase the efficiency and precision of crop improvement. Currently an impressive number of advances in genetics and genomics have given us enhanced understanding of structural and functional aspects of plant genomes and basic knowledge have been integrated in such a way which can enhance the ability of plant breeders to improve crop plants for our benefit. In the last decade the whole genome sequencing of model plant *Arabidopsis thaliana*, rice, *Sorghum bicolor*, *Medicago truncatula*, *Musa* spp., grape, apple and recently draft sequence of wheat have become available (http://www.ncbi.nlm.nih.gov/genomes/).

The first complete plant genome to be sequenced was that of *Arabidopsis*. The sequenced regions cover 115.4 Mb of the 125-Mb genome and extend into centromeric regions. The genome contains 25,498 genes encoding proteins from 11,000 families (The Arabidopsis Genome Initiative, 2000). *Arabidopsis* contains many families of new proteins but also lacks several common protein families. The complete genome sequence provides the foundation for more comprehensive comparison of conserved processes in all eukaryotes, identifying a wide range of plant-specific gene functions and establishing rapid systematic methods of identifying genes for crop improvement (Tacchini et al. 1995; Varshney et al. 2009; Thakur and Varshney 2010).

Among the most important food crops, rice has the smallest genome (389 Mb) and wheat the largest (15,966 Mb). Arumuganathan and Earle (1991) have grouped other crops into seven classes: *Musa*, cowpea and yam (873 Mb); sorghum, bean, chickpea and pigeonpea (673–818 Mb); soybean (1,115 Mb); potato and sweet potato (1,597–1,862 Mb); maize, pearl millet and groundnut (2,352–2,813 Mb); pea and barley (4,397–5,361 Mb); and oat (11,315 Mb). Genome size is often correlated with plant growth and ecology. The diverse cellular and physiological effects of large genomes may be a function of selection of the major components that contribute to genome size such as transposable elements and gene duplication. The recent advances in genome sequencing, through the development of second generation sequencing technologies and beyond, provide opportunities to develop millions of novel markers, in non-model crop species, as well as identification of genes of agronomic importance. Identification of all genes within a species permits an understanding of how important agronomic traits are controlled, knowledge of which can be directly translated into crop improvement (Chia and Ware 2011). This systematic whole genome sequencing will provide critical information on gene and genome organization and function, which will revolutionize our understanding of crop production and the ability to manipulate those traits contributing to high crop productivity (Pereira 2000).

The use of whole genome information and high-throughout tools has opened up a new field of research called functional genomics. Among its subdisciplines, transcriptomics (the complete set of transcripts produced in a cell) (Zimmerli and Somerville 2004), proteomics (the complete set of proteins produced in a cell) (Roberts 2002) and metabolomics (the complete set of metabolites expressed in a cell) (Stitt and Fernie 2003) have been used by the plant science community.

Recent advances in microarray technology will allow the simultaneous expression and analysis of vast numbers of genes that will elucidate gene function, and the complex multifaceted interactions between genes that result in different phenotypes under varying environmental conditions. These high-throughput or large-scale experimental methodologies combined with statistical and computational analysis (bioinformatics) will give the detailed information about specific gene/genes function linked to specific characters required for crop improvement. These studies will be augmented by more specific investigations based on gene suppression, co-suppression or anti-sensing of a defined sequence (Jain and Barr 2010). Advances in these areas will fuel the mapping of QTL (quantitative trait loci) underlying agronomic traits in less studied crops. The use of QTL markers in crop improvement promises rapid and efficient utilization of novel traits from closely related wild species (Varshney et al. 2011). These new information provided by all the omics disciplines will lead the plant science community to *in silico* simulations of plant growth, development and response to environmental change.

The recent addition of the high quality draft genome of the soybean in the rapidly growing list of crops will not only help the breeders to improve soybean varieties in terms of protein and oil content but will also help to improve nitrogen fixation capacity of many important legumes used for human and animal nutrition. Schmutz et al. 2010 report that the 1.1-gigabase soybean genome-the largest shotgunsequenced plant genome-is predicted to encode 46,000 genes. Two genome duplication events are likely to account for the observation that ~75% of these genes are found in multiple copies. Although the importance of soybean as a source of protein and oil alone testifies to the potential implications of understanding its genetic makeup, this genome will also serve as the reference for ~20,000 leguminous species that play a critical ecological role through their unique ability to fix nitrogen with the help of rhizobial bacteria. Availability of the genome should accelerate the association of quantitative trait loci of nutritional, economic and ecologically important traits with the causal DNA sequences from soybean in the near future. In the longer term, the genome will likely also be leveraged to improve the way in which a range of leguminous subsistence crops are used to both replenish soil nitrogen through crop rotation and meet the expanding needs of developing nations for protein and energy.

Crop improvement can also be carried out by engineering novel RNA interference (RNAi) pathways that create small RNA molecules to alter gene expression in crops and can generate new crop quality traits or provide protection against insects, nematodes and pathogens without introducing new proteins into food and feed products (Auer and Frederick 2009). Although miRNAs are relatively small, they play an important role in gene expression. miRNA is considered as one of the most important post-transcriptional gene regulators (Carrington and Ambros 2003) since it was originally recognized in 2001 (Lagos-Quintana et al. 2001; Lau et al. 2001; Lee and Ambros 2001).

Recent studies have revealed powerful and unexpected roles for small interference RNAs (siRNAs) and microRNAs (miRNAs) in the control of plant growth by the silencing of native genes. Although both of these have pivotal roles in gene silencing, the actions of miRNAs are more extensive and remarkable. Several experiments have demonstrated that many miRNAs regulate almost every aspect of plant growth and development, including leaf morphogenesis and polarity, floral differentiation and development, root initiation and development, vascular development, and the transition from vegetative growth to reproductive growth (Jones-Rhoades et al. 2006; Chuck et al. 2009). Even more intriguingly, it has been discovered that miRNAs play a role in hormone signal transduction (Liu and Chen 2009), the response to environmental stress, and pathogen invasion (Chen et al. 2004; Sunkar et al. 2006). These regulatory miRNAs stimulated the idea of developing artificial miRNAs that can silence specific gene(s). Such targeted gene silencing could permit the direct molecular modulation of plant traits, which could in turn be applied to the breeding of crop species.

8.2 *Proteomics and Crop Improvement*

Proteomics is the fascinating field of research that attracts the thousands of scientific workers over the globe for a wide range of subjects (Tilleman et al. 2005; Bona et al. 2007). The term "Proteome" (PROTEins expressed by genOME) therefore is expected to represent a comprehensive survey of all proteins expressed at a given time, in given conditions. Moreover the expression levels in protein strongly depend on complex regulatory systems; unlike genome, the proteome is highly dynamic (similar to transcriptome). Proteomics is one of the fastest growing areas of biological research. At present, proteomic research aims both identifying new proteins in relation to their function and in unraveling how their expression is controlled within regulatory networks. In the past few years tremendous progress has been made in the field of crop plant proteomics. With the high advancement in technologies and latest bioinformatic tools, studying the plant proteome and its dynamic nature has become comparatively easy and very accurate. Plant physiologists now feel comfortable over this development as many unsolved mysteries about various physiological processes in the plants will now become easy to work out. From seed germination to fruit development/grain filling proteomics research is accelerating by leaps and bounds throughout the globe. Expression of proteins regulating the processes of nutritional as well as hormonal balances has been providing the new insights towards understanding various metabolic processes fully. Plants like other organisms are always under the threat of various stresses. A remarkable achievement has been made in this direction by proteomicists by discovering the functions of various proteins and working out their expression levels.

All living organisms rely on the uptake of nutrients from the environment to sustain energy, metabolism and growth. They have, therefore, evolved numerous alternative programs to adapt to their permanently changing environment. Such programs involve instantaneous responses (changes in intracellular metabolites, activation/inhibition of enzymes by effectors and of proteins through post-translational modifications) as well as slower processes that affect the levels of macromolecules (transcription, translation, mRNA and protein degradation). The availability of complete genome sequences and technologies that allow comprehensive analysis of global mRNA profiles has greatly expanded the ability to monitor the transcriptional reprogramming of cells in response to their environment. However, further studies (often conducted with yeast) indicate that transcripts are imperfect indicators of protein levels and of *in vivo* fluxes (ter Schure et al. 2000; Griffin et al. 2002; Washburn et al. 2003; Daran-Lapujade et al. 2004; Wek et al. 2004; Kolkman et al. 2006), and therefore it brings limited understanding on whole biological systems. The implementation of sensitive and rapid methods for protein identification and the continuous technical improvement of the so far largely descriptive analysis of protein patterns by two-dimensional gel electrophoresis (2-DE) have transformed the combination of both techniques into a powerful tool for functional analysis now also more and more used in plant studies. Proteomics is proving an indispensable tool for examining alterations in the protein profile caused due to gene mutations, introduction or silencing of genes or in response to various stress stimuli in a relatively fast, sensitive and reproducible way. This science is becoming important for generation of information on physiological (e.g. regulatory behaviour and function), biochemical (e.g. metabolic and structural data), genetic (e.g. gene mapping and assigning of the structural gene to the 2D gel map) and architectural (e.g. location of the proteins in the cell) aspects. Proteomics-based approach is proving important for characterization of individuals or lines, estimation of genetic variability within and between different populations, establishment of genetic distances to be used in phylogenetic studies and characterization of mutants with localization of genes encoding revealed proteins (Thiellement et al. 1999). It is becoming a necessity in plant biology for deciphering the function and the role of genes in the on-going plant genome sequencing projects.

Analysis of the *Arabidopsis pasticcino* mutants by 2-DE revealed a considerable percentage of variable spots relative to wild-type controls; evaluation of responses to different hormone treatments indicated that the mutants were affected in cytokinin responses (Faure et al. 1998). A mutant was also used to study cytokinin effects on chloroplast division in the moss *Physcomitrella* at the protein level (Kasten et al. 1997). Comparison of protein patterns in leaves of the late flowering *Arabidopsis* mutant *fy* and wild-type demonstrated qualitative differences. Studies have also demonstrated the capacity of 2-DE to document genetic variability and distinguish between lines and varieties, *e.g.*, when analyzing barley seed and malt (Ostergaard et al. 2002) or wheat grains (Skylas et al. 2005). Positional shifts of proteins were observed in 2-D gel analysis of segregating families of maize, barley, pea, and maritime pine (De Vienne et al. 1996).

Application of proteomics can enormously boost up agricultural production (Dhand 2000; Cánovas et al. 2004; Xu et al. 2006). It is the most promising technique to identify proteins that are induced, repressed, or post-transcriptionally modified during a developmental process as complex as senescence.

8.3 Nanobiotechnology and Crop Improvement

The term 'Nanobiotechnology' was used for the first time by Lynn W. Jelinski (a biophysicist at Cornell University, USA). It is an exciting and rapidly emerging technology allowing us to work, manipulate and create tools and materials at the molecular level that may be of great importance in our day-to-day life. Nature has been performing the "nanotechnological feats" for millions of years. This important technology has the potential to revolutionize the agricultural and food industry with new tools by enhancing the ability of plants to absorb nutrients etc. Smart sensors and smart delivery systems have a great potential to help the agricultural industry by combating viruses and other crop pathogens. Key advances have been made in the ability to make measurements at the sub-cellular level and in understanding the cell as a highly organized, self-repairing, self-replicating, information-rich molecular machine. Single-molecule measurements are shedding light on the dynamics and mechanistic properties of molecular biomachines, allowing the direct investigation of molecular motors, enzyme reactions, protein dynamics, DNA transcription and cell signaling. It has also been possible to measure the chemical composition within a single cell. Nanobiotechnological research and development is likely to facilitate and frame the next stage of development of genetically modified crops, animal production inputs, chemical pesticides and precision farming techniques. While nano-chemical pesticides are already in use, other applications are still in their early stages, and it may be many years before they are commercialized. These applications are largely intended to address some of the limitations and challenges facing large-scale, chemical and capital intensive farming systems. This includes the fine-tuning and more precise micro-management of soils; the more efficient and targeted use of inputs; new toxin formulations for pest control; new crop and animal traits; and the diversification and differentiation of farming practices and products within the context of large-scale and highly uniform systems of production. Nanobiotechnology can have momentous application in plant sciences too for inducting foreign DNA in the cells, an improvement over the transgenics as the gene would not be fully incorporated thereby preventing unintended gene flow into the environment. Nanobiotechnology has made inroads into uncovering fundamental biological processes, including self-assembly, cellular processes, and systems biology. It can also be used to enhance photosynthesis and improve soil management. Removal of heavy metal contamination can be achieved through intense sensing for precision farming. It has also enabled the development of biochips and has a role in green manufacturing. Major applications are in the design of sensors, biofluidics for handling DNA and other molecules, nano-filtration, bioprocessing and traceability of genetically modified food. Nanobiotechnology is evolving as a powerful tool as a result of cross talk between nano scientists and biologists. By operating in the nanoscale realm, at the molecular level, nanotechnology offers a wide range of tools, techniques and applications. Nano biotechnology can stimulate new technologies for studies in cell biology using nano tools, provide opportunities for early detection of diseases through *in-vivo* and *in-vitro* analysis using nano sensing structures with extra ordinary multi-function capabilities and targeted drug delivery. Some of the areas with research priority are: nanoparticles for biosynthesis of nanoparticles, biological templates for nanoparticle assembly, bionano composites, imaging/sensing of nanoparticles/biomolecules, tissue engineering, cell-cell interactions, mammalian/microbial cell development, nano-biosensors with multiple sensing capabilities. The development of genetically encoded molecular sensors, which transduce an interaction of the target molecule with a recognition element into a macroscopic observable, via allosteric regulation of one or more reporter elements, may provide us a chance to understand many physiological processes fully. The recognition element may simply bind the target, bind and enzymatically convert the target, or may serve as a substrate for the target, as in the use of a specific target sequence in the construction of a protease sensor (Nagai and Miyawaki 2004). The most common reporter element is a sterically separated donor-acceptor fluorescence resonance energy transfer (FRET) pair of spectral variants of the green fluorescent protein (GFP; Fehr et al. 2002), although single fluorescent proteins (Doi and Yanagawa 1999) or enzymes (Guntas et al. 2005) are viable as well. Some molecular sensors additionally employ a conformational actuator (most commonly a peptide which binds to one conformational state of the recognition element) to magnify the allosteric effect upon and resulting output of the reporter element.

9 Conclusion

Biotechnology has remarkably affected all the aspects of human life. In agricultural fields too, it has and will surely solve the problems of food insecurity. Understanding the important process(s) of crop production and manipulating the key steps of grain/fruit development have become easier and accurate by this wonderful technology. Developing stress tolerant, high yielding and nutrient efficient crop varieties are some attributes associated with biotechnological tools. One of the tremendous challenges before the breeders/scientists is food insecurity which needs to be addressed timely so as to feed the teeming population. Agricultural lands are vanishing due to the various processes like desertification; salination etc and just overflow of nutrient fertilizers in these fields could not increase crop production as normally crop plants are not nutrient use efficient. Modern biotechnological approaches viz., proteomics, nanobiotechnology etc could also help us to understand the important process(s) of crop plants, thereby improvement in their production.

References

Abdin MA, Dwivedi RK, Abrol YP (2005) Nitrogen in agriculture. In: Singh RP, Shankar N, Jaiwal PK (eds) Nitrogen nutrition and plant productivity. Studium Press LLC, Houston, pp 1–41

Abrol YP, Chatterjee SR, Kumar PA, Jain V (1999) Improvement in nitrogenous fertilizer utilization – physiological and molecular approaches. Curr Sci 76:1357–1364

Agarwal M, Shrivastava N, Padh H (2008) Advances in molecular marker techniques and their applications in plant sciences. Plant Cell Rep 27:617–631

Aguera E, Haba PD, Fontes AG, Maldonando JM (1990) Nitrate and nitrite uptake and reduction by intact sunflower plants. Planta 182:149–154

Andrews M, Lea PJ, Raven JA, Lindsey K (2004) Can genetic manipulation of plant nitrogen assimilation enzymes result in increased crop yield and greater N-use efficiency? An assessment. Anal Appl Biol 45:25–40

Arumuganathan K, Earle ED (1991) Estimation of nuclear DNA amounts of plants by flow cytometry. Plant Mol Biol Rep 9:229–241

Auer C, Frederick R (2009) Crop improvement using small RNAs: applications and predictive ecological risk assessments. Trends Biotechnol 27:644–651

Beman JM, Arrigo K, Matson PM (2005) Agricultural runoff fuels large phytoplankton blooms in vulnerable areas of the ocean. Nature 434:211–219

Bona E, Marsano F, Cavaletto M, Berta G (2007) Proteomic characterization of copper stress response in *Cannabis sativa* roots. Proteomics 7:1121–1130

Bryan GJ, Collins AJ, Stephenson P, Orry A, Smith JB, Gale MD (1997) Isolation and characterization of microsatellites from hexaploid wheat. Theor Appl Genet 94:557–563

Buchanan FC, Adams LJ, Littlejohn RP, Maddox JF, Crawford AM (1994) Determination of evolutionary relationships among sheep breeds using microsatellites. Genomics 22:397–403

Cánovas FM, Eliane DG, Ghislaine R, Jorrin HP, Mock J, Rossignol M (2004) Plant proteome analysis. Proteomics 4:285–298

Carrington JC, Ambros V (2003) Role of microRNAs in plant and animal development. Science 18(336):336–338

Cerezo M, Flors V, Legaz F, Garcia-Agustin P (2000) Characterization of the low affinity transport system for NO_3^- uptake by citrus roots. Plant Sci 160:95–104

Cerezo M, Tillard P, Filleur S, Munos S, Daniel-Vedele F, Gojon A (2001) Major alterations of the regulation of root NO_3^- uptake are associated with the mutation of *NRT2.1* and *NRT2.2* genes in Arabidopsis. Plant Physiol 127:262–271

Chen J, Li WX, Xie DX, Peng JR, Ding SW (2004) Viral virulence protein suppresses RNA silencing-mediated defense but upregulates the role of microRNA in host gene expression. Plant Cell 16:1302–1313

Chia JM, Ware D (2011) Sequencing for the cream of the crop. Nat Biotechnol 29:138–139

Chuck G, Candela H, Hake S (2009) Big impacts by small RNAs in plant development. Curr Opin Plant Biol 12:81–86

Chugh A, Khurana P (2003) Herbicide resistant transgenics of bread wheat (*T. aestivum*) and emmer wheat (*T. dicoccum*) by particle bombardment and *Agrobacterium*-mediated approaches. Curr Sci 84:78–83

Consoli L, Lefevre A, Zivy M, de Vienne D (2002) QTL analysis of proteome and transcriptome variations for dissecting the genetic architecture of complex traits in maize. Plant Mol Biol 48:575–581

Cook J (2006) Towards cropping systems that enhance productivity and sustainability. Proc Natl Acad Sci U S A 103:18389–18394

Daran-Lapujade P, Jansen ML, Daran JM, van Gulik W, de Winde JH, Pronk JT (2004) Role of transcriptional regulation in controlling fluxes in central carbon metabolism of *Saccharomyces cerevisiae*. A chemostat culture study. J Biol Chem 279:9125–9138

De Vienne DBJ, Gerber S, Leonardi A, Le Guilloux M (1996) Two-dimensional electrophoresis of proteins as a source of monogenetic and codominant markers for population genetics and mapping the expressed genome. Heredity 76:166–176

Dhand R (2000) Nature insight: functional genomics. Nature 405:819–865

Dobermann A, Cassman KG (2004) Environmental dimensions of fertilizer N: what can be done to increase nitrogen use efficiency and ensure global food security? In: Mosier AR, Syers JK, Freney JR (eds) Agriculture and the nitrogen cycle: assessing the impacts of fertilizer use on food production and the environment. Island Press, Washington, DC, p 261

Doi N, Yanagawa H (1999) Design of generic biosensors based on green fluorescent proteins with allosteric sites by directed evolution. FEBS Lett 453:305–307

Elkhatib HA, Elkhatib EA, Khalaf Allah AM, El-Sharkawy AM (2004) Yield response of salt-stressed potato to potassium fertilization: a preliminary mathematical model. J Plant Nutr 27:111–122

FAO (2008) FAO yearbook 2008. Food and Agriculture Organization, India

Faure S, Cliqut JB, Thephany G, Boucaud J (1998) Nitrogen assimilation in *Lolium perenne* colonized by the arbuscular mycorrhizal fungus *Glomus fasiculatum*. New Phytol 138:411–417

Fehr M, Frommer WB, Lalonde S (2002) Visualization of maltose uptake in living yeast cells by fluorescent nanosensors. Proc Natl Acad Sci U S A 99:9846–9851

Forde BG (2000) Nitrate transporters in plants: structure, function and regulation. Biochim Biophys Acta 1465:219–235

Forde BG (2002) Local and long-range signalling pathways regulating plant responses to nitrate. Annu Rev Plant Biol 53:203–224

Gatehouse AMR, Gatehouse JA (1998) Identifying proteins with insecticidal activity: use of encoding genes to produce insect-resistant transgenic crops. Pest Sci 52:165–175

Good A, Shrawat A, Muench D (2004) Can less yield more? Is reducing nutrient input into the environment compatible with maintaining crop production? Trends Plant Sci 9:597–605

Griffin TJ, Gygi SP, Ideker T, Rist B, Eng J, Aebersold R (2002) Complementary profiling of gene expression at the transcriptome and proteome levels in *S.* Cerevisiae Mol Cell Proteomics 1:323–333

Guntas G, Mansell TJ, Kim JR, Ostermeier M (2005) Directed evolution of protein switches and their application to the creation of ligand-binding proteins. Proc Natl Acad Sci U S A 102:11224–11229

Gupta PK, Varshney RK (2000) The development and use of microsatellite markers for genetic analysis and plant breeding with emphasis on bread wheat. Euphytica 113:163–185

Gupta PK, Balyan HS, Sharma PC, Ramesh B (1996) Microsatellites in plants: a new class of molecular markers. Curr Sci 70:45–54

Hartwell J, Gill A, Wilkins MB, Jenkins GI, Nimmo HG (1999) Phosphoenol-pyruvate carboxylase kinase is a novel protein kinase regulated at the level of expression. Plant J 20:333–342

Hilder VA, Boulter D (1999) Genetic engineering of crop plants for insect resistance – a critical review. Crop Prot 18:177–191

Hwang CF, Lin Y, D'souza T, Cheng CL (1997) Sequences necessary for nitrate-dependent transcription of Arabidopsis nitrate reductase genes. Plant Physiol 113:853–862

Imsande J, Touraine B (1994) N demand and the regulation of nitrate uptake. Plant Physiol 105:3–7

Jain SM, Barr DS (2010) Molecular techniques in crop improvement, 2nd edn. Springer, London/New York

Jeffrey A, Kaplan I, Zhang D, Shan-Tan ST, Nielsen J (2002) Environmental tracers: identifying the sources of nitrate contamination in groundwater. Soil Sed Waters 6:15–20

Jones-Rhoades MW, Bartel DP, Bartel B (2006) MicroRNAs and their regulatory roles in plants. Annu Rev Plant Biol 57:19–53

Kang JG, Pyo YJ, Cho JW, Cho MH (2004) Comparative analysis of differentially expressed proteins induced by K^+ deficiency in *Arabidopsis thaliana*. Proteomics 4:3549–3559

Kasten B, Buck F, Nuske J, Reski R (1997) Cytokinin affects nuclear and plastome-encoded energy converting plastid enzyme. Planta 201:261–272

Kolkman A, Daran-Lapujade P, Fullaondo A, Olsthoorn MMA, Pronk JT, Monique S, Heck AJR (2006) Proteome analysis of yeast response to various nutrient limitations. Mol Syst Biol 2:1–26

Kozaki A, Takeba G (1996) Photorespiration protects C_3 plants from photooxidation. Nature 384:557–560

Krapp A, Ferrario-Mery S, Touraine B (2002) Nitrogen and signalling. In: Foyer CH, Noctor G (eds) Photosynthetic nitrogen assimilation and associated carbon and respiratory metabolism. Kluwer Academic, Dordrecht, pp 205–225

Ladha JK (2005) Improving the recovery efficiency of fertilizer nitrogen in cereals. J Indian Soc Soil Sci 53:472–483

Lagos-Quintana M, Rauhut R, Lendeckel W, Tuschl T (2001) Identification of novel genes coding for small expressed RNAs. Science 294:853–858

Lau NC, Lim LP, Weinstein EG, Bartel DP (2001) An abundant class of tiny RNAs with probable regulatory roles in *Caenorhabditis elegans*. Science 294:858–862

Lea US, Leydecker MT, Quilleré I, Meyer C, Lillo C (2006) Posttranslational regulation of nitrate reductase strongly affects the levels of free amino acids and nitrate, whereas transcriptional regulation has only minor influence. Plant Physiol 140:1085–1094

Lee R, Ambros V (2001) An extensive class of small RNAs in *Caenorhabditis elegans*. Science 294:862–864

Lian X, Xing Y, Yan H, Xu C, Li X, Zhang Q (2005) QTLs for low nitrogen tolerance at seedling stage identified using a recombinant inbred line population derived from an elite rice hybrid. Theor Appl Genet 112:85–96

Lillo C, Appenroth KJ (2001) Light regulation of nitrate reductase in higher plants: which photoreceptors are involved? Plant Biol 3:455–465

Lin B, White JT, Lu W, Xie T, Utleg AG, Yan X, Yi EC, Shannon P, Khrebtukova I, Lange PH, Goodlett DR, Zhou D, Vasicek TJ, Hood L (2005) Evidence for the presence of disease-perturbed networks in prostate cancer cells by genomic and proteomic analyses: a systems approach to disease. Cancer Res 65:3081–3091

Litt M, Luty JA (1989) A hypervariable microsatellite revealed by in vitro amplification of a dinucleotide repeat within the cardiac muscle actin gene. Amer J Hum Genet 44:397–401

Liu Q, Chen YQ (2009) Insights into the mechanism of plant development: interactions of miRNAs pathway with phytohormone response. Biochem Biophys Res Commun 384:1–5

Liu YG, Biyashev RM, Sagai-Maroof MA (1996) Development of Simple sequence repeats DNA markers and their integration into barley linkage map. Theor Appl Genet 93:869–876

Lochab S, Pathak RR, Raghuram N (2007) Molecular approaches for enhancement of nitrogen use efficiency in plants. In: Abrol YP, Raghuram N, Sachdev MS (eds) Agricultural nitrogen use and its environmental implications. IK International, New Delhi, pp 327–350

Lörz H, Wenzel G (2005) Molecular marker systems in plant breeding and crop improvement. Springer, Berlin/Heidelberg, p 476

Ma Zq, Roder M, Sorrels ME (1996) Frequency and sequence characteristics odd i-, tri-, and tetranucleode microsatellites in wheat. Genome 39:123–130

Man HM, Boriel R, El-Khatib R, Kirby EG (2005) Characterization of transgenic poplar with ectopic expression of pine cytosolic glutamine synthetase under conditions of varying nitrogen availability. New Phytol 167(1):31–35

Moerchen MJ, Cuguen J, Mitchaelis G, Hanni C, Saurnitou-Laprade P (1996) Abundance and length polymorphism of microsatellite repeats in *Beta Vulgaris* L. Theor Appl Genet 92:326–333

Morgante M, Oliveri AM (1993) PCR-amplified microsatellites as markers in plant genetics. Plant J 3:175–182

Myers N (1999) The next green revolution: its environmental underpinnings. Curr Sci 76:507–513

Nagai T, Miyawaki A (2004) A high-throughput method for development of FRET-based indicators for proteolysis. Biochem Biophys Res Commun 319:72–77

Ogawa K, Soutome R, Hiroyama K, Hagio T, Ida S, Nakagawa H, Komamine A (2000) Co-regulation of nitrate reductase and nitrite reductase in cultured spinach cells. J Plant Physiol 157:299–306

Oliveira IC, Brears T, Knight TJ, Clark A, Coruzzi GM (2002) Overexpression of cytosolic glutamine synthetase. Relation to nitrogen, light, and photorespiration. Plant Physiol 129:1170–1180

Ortega JL, Temple SJ, Sengupta-Gopalan C (2001) Constitutive overexpression of cytosolic glutamine synthetase (GS1) gene in transgenic alfalfa demonstrates that GS1 be regulated at the level of RNA stability and protein turnover. Plant Physiol 126:109–121

Ostergaard O, Melchior S, Roepstorff P, Svensson B (2002) Initial proteome analysis of mature barley seeds and malt. Proteomics 2:733–739

Ozturk M, Uzonur E (2004) Genetic engineering and the ecobalance. In: Majali AS, Ergin M, Zoubi M (eds) Biotechnology and genetic engineering for development in the Islamic world. Islamic Academy of Science, National Printing Press, Amman, pp 371–388

Pathak RR, Ahmad A, Lochab S, Raghuram N (2008) Molecular physiology of plant nitrogen use efficiency and biotechnological options for its enhancement. Curr Sci 94:1394–1403

Pereira A (2000) Plant genomics is revolutionising agricultural research. Biotechnol Dev Monit 40:2–7

Plomion C, Costa P, Bahrman N, Frigerio JM (1997) Genetic analysis of needle proteins in maritime pine. 1. Mapping dominant and co-dominant protein markers assayed on diploid tissue, in a haploid-based genetic map. Silvae Genet 46:161–165

Powell W, Morgante M, Andre C, Hanafey M, Vogel J, Tingey S, Rafalski A (1996) The comparison of RFLP, AFLP and SSR (microsatellite) markers for germplasm analysis. Mol Breed 2:225–238

Prasad M, Varshney RK, Roy JK, Balyan HS, Gupta PK (2000) The use of microsatellites for detecting DNA polymorphism, genotype identification and genetic diversity in wheat. Theor Appl Genet 100:584–592

Quesada A, Krapp A, Trueman L, Daneil-Vedele F, Fernendez E, Forde B, Caboche M (1997) PCR-identification of a *Nicotiana plumbaginifolia* cDNA homologous to the high affinity nitrate transporters of the *crnA* family. Plant Mol Biol 34:265–274

Raghuram N, Sopory SK (1995) Light regulation of NR gene expression: mechanism and signal-response coupling. Physiol Mol Biol Plants 1:103–114

Raghuram N, Pathak RR, Sharma P (2006) Signalling and the molecular aspects of N-use efficiency in higher plants. In: Singh RP, Jaiwal PK (eds) Biotechnological approaches to improve nitrogen use efficiency in plants. Studium Press LLC, Houston, pp 19–40

Rastogi R, Back E, Schneiderbauer A, Bowsher C, Moffatt G, Rothstein SJ (1993) A 330 bp region of the spinach nitrite reductase gene prompter directs nitrate-inducible tissue-specific expression in transgenic tobacco. Plant J 4:317–326

Remans T, Nacry P, Pervent M, Filleur S, Diatoff E, Mounier E, Tillard P, Forde BG, Gojon A (2006) The Arabidopsis NRT1.1 transporter participates in the signaling pathway triggering root colonization of nitrate rich patches. Proc Natl Acad Sci U S A 103:19206–19211

Roberts JKM (2002) Proteomics and a future generation of plant molecular biologists. Plant Mol Biol 48:143–154

Russel J, Fuller J, Young G, Thomas B, Taramino G, Macaulay M, Waugh R, Powell W (1997) Discriminating between barley genotypes using microsatellite markers. Genome 40:442–450

Sakakibara H, Kobayashi K, Deji A, Sugiyama T (1997) Partial characterization of signalling pathway of nitrate-dependent expression of the genes for nitrogen assimilatory enzymes using detached maize leaves. Plant Cell Physiol 38:837–843

Sarkar R (2003) Nitrate-induced gene expression and metabolic regulation in rice. M.Sc. thesis. University of Mumbai, Mumbai, India

Schloetterer C, Amos B, Tautz D (1991) Conservation of polymorphic simple sequence loci in cetacean species. Nature 354:313–322

Schmutz J, Cannon SB, Schlueter J, Ma J, Mitros T, Nelson W, Hyten DL, Song Q, Thelen JJ, Cheng J, Xu D, Hellsten U, May GD, Yu Y, Sakurai T, Umezawa T, Bhattacharyya MK, Sandhu D, Valliyodan B, Lindquis E, Peto M, Grant D, Shu S, Goodstein D, Barry K, Futrell-Griggs M, Abernathy B, Du J, Tian Z, Zhu L, Gill N, Joshi T, Libault M, Sethuraman A, Zhang XC, Shinozaki K, Nguyen HT, Wing RA, Cregan P, Specht J, Grimwood J, Rokhsar D, Stacey G, Shoemaker RC, Jackson SA (2010) Genome sequence of the palaeopolyploid soybean. Nature 463:178–183

Schoenbeck MA, Temple SJ, Trepp GB, Blumenthal JM, Samac DA, Gantt SJ, Hernandez G, Vance CP (2000) Decreased NADH- glutamate synthase activity in nodules and flower of alfalfa (*Medicago sativa* L.) transformed with an antisense glutamate synthase transgene. J Exp Bot 51(342):29–39

Shigeto J, Yoshihara S, Adam Suaad EH, Sueyoshi K, Sakamoto A, Morikawa H, Takahashi M (2006) Genetic engineering of nitrite reductase gene improves uptake and assimilation of nitrogen dioxide by *Rhaphiolepis umbellate* (Thunb.) Makino. Plant Biotechnol 23:111–116

Shrawat AK, Good AG (2008) Genetic engineering approaches to improving nitrogen use efficiency. ISB News Report, May 2008

Skylas D, Van Dyk D, Wrigley CW (2005) Proteomics of wheat grain. J Cereal Sci 41:165–179

Smulders MJM, Bredemeijer G, Rus-Kortekaas W, Arens P, Vosman B (1997) Use of short microsatellites from database sequences to generate polymorphism among *Lycopersicon esculentum* and accessions of other *Lycopersicon* species. Theor Appl Genet 97:264–272

Song MT, Lee JH, Cho YS, Jeon YH, Lee SB, Ku JH, Choi SH, Hwang HG (2002) Narrow genetic background of Korean rice germplasm as revealed by DNA fingerprinting with SSR markers and their pedigree information. Korean J Genet 24:397–403

Stitt M, Fernie AR (2003) From measurements of metabolites to metabolomics: an 'on the fly' perspective illustrated by recent studies of carbon-nitrogen interactions. Curr Opin Biotechnol 14:136–144

Stulen I, Pérez-Soba M, De Kok LJ, van der Eerden L (1998) Impact of gaseous nitrogen deposition on plant functioning. New Phytol 139:61–67

Sunkar R, Kapoor A, Zhu JK (2006) Posttranscriptional induction of two Cu/Zn superoxide dismutase genes in Arabidopsis is mediated by downregulation of miR398 and important for oxidative stress tolerance. Plant Cell 18:2051–2065

Swaminathan MS (1995) Population, environment and food security. Issues in Agriculture, No 7. CGIAR, Washington, DC

Tacchini P, Fink A, Thiellement H, Greppin H (1995) Analysis of leaf proteins in late flowering mutants of *Arabidopsis thaliana*. Physiol Plant 93:312–316

Taghavi TS, Babalar M (2007) The effect of nitrate and plant size on nitrate uptake and *in vitro* nitrate reductase activity in strawberry (*Fragaria* X *ananassa* cv. Selva). Sci Hort 112:393–398

Tanksley SD, Young ND, Peterson AH, Bonierbale MW (1989) RFLP mapping in plant breeding-new tools for an old science. Biotechnology 7:258–264

Tautz D, Trick M, Dover GA (1986) Cryptic simplicity in DNA is a major source of genetic variation. Nature 322:652–656

ter Schure EG, van Riel NA, Verrips CT (2000) The role of ammonia metabolism in nitrogen catabolite repression in *Saccharomyces cerevisiae*. FEMS, Microbiol Rev 24:67–83

Thakur V, Varshney R (2010) Challenges and strategies for next generation sequencing (NGS) data analysis. J Comp Sci Syst Biol 3:40–42

The *Arabidopsis* Genome Initiative (2000) Analysis of the genome sequence of the flowering plant *Arabidopsis thaliana*. Nature 408:796–815
Thiellement H, Bharmann N, Damerval C (1999) Proteomics for genetic and physiological studies in plants. Electrophoresis 20:2013–2026
Tilleman K, Deforce D, Elewaut D (2005) Rheumatology: a close encounter with proteomics. Rheumatology 44:1217–1226
Touzet P, Winckler RG, Helentjaris T (1995) Combined genetic and physiological analysis of a locus contributing to quantitative variation. Theor Appl Genet 91:200–209
Trueman LJ, Onyeocha I, Forde B (1996) Recent advances in molecular biology of a family of eukaryotic high affinity nitrate transporters. Plant Physiol Biochem 34:621–627
Varshney RK, Nayak SN, May GD, Jackson SA (2009) Next-generation sequencing technologies and their implications for crop genetics and breeding. Trends Biotechnol 27:522–530
Varshney RK, Bansal KC, Aggarwal PK, Datta SK, Craufurd PQ (2011) Agricultural biotechnology for crop improvement in a variable climate: hope or hype? Trends Plant Sci 16(7):363–371
Vincent R, Fraisier V, Chaillou S, Limami MA, Deleens E, Phillipson B, Douat C, Boutin JP, Hirel B (1997) Over expression of a soyabean gene encoding cytosolic glutamine synthetase in shoots of transgenic *Lotus cornicultus* L. plants triggers changes in ammonium and plant development. Planta 201:424–433
Wang GJ, Costiglione S, Zhang J, Fu RZ, Ma JS, Li WB, Sun YR, Sala F (1994) Hybrid rice (*Oryza sativa* L.): identification and parentage determination by RAPD fingerprinting. Plant Cell Rep 14:112–115
Washburn MP, Koller A, Oshiro G, Ulaszek RR, Plouffe D, Deciu C, Winzeler E, Yates JR III (2003) Protein pathway and complex clustering of correlated mRNA and protein expression analyses in *Saccharomyces cerevisiae*. Proc Natl Acad Sci U S A 100:3107–3112
Wek RC, Staschke KA, Narasimhan J (2004) Regulation of the yeast general amino acid control pathway in response to nutrient stress. In: Winderickx JG (ed) Nutrient-induced responses in eukaryotic cells. Springer, Berlin, pp 171–199
Williams JGK, Kubelik AR, Livak KJ, Rafalski JA, Tingey SV (1990) DNA polymorphisms amplified by arbitrary primers are useful as genetic markers. Nucleic Acids Res 18:6531–6535
Xu K, Xu X, Fukao T, Canlas P, Maghirang-Rodriguez R, Heuer S, Ismail AM, Bailey-Serres J, Ronald PC, Mackill DJ (2006) Sub1A is an ethylene-response-factor-like gene that confers submergence tolerance to rice. Nature 442:705–708
Yamaya T (2003) Biotechnological approaches for modification of nitrogen assimilation in rice. In: Jaiwal PK, Singh RP (eds) Plant gernetic engineering improvement of food crops, vol 2. Science Tech Publishing LLC, Houston, pp 79–88
Yamaya T, Obara M, Nakajima H, Sasaki S, Hayakawa T, Sato T (2002) Genetic manipulation and quantitative-trait loci mapping for nitrogen recycling in rice. J Exp Bot 53:917–925
Yanagisawa S, Akiyama A, Kisaka H, Uchimiya H, Miwa T (2004) Metabolic engineering with Dof1 transcription factor in plants: improved nitrogen assimilation and growth under low nitrogen conditions. Proc Natl Acad Sci U S A 101:7833–7838
Zhang H, Forde BG (1998) An Arabidopsis MADS box gene that controls nutrient-induced changes in root architecture. Science 279:407–409
Zimmerli L, Somerville S (2004) Transcriptomics in plants: from expression to gene function. In: Leister D (ed) Plant functional genomics. Haworth Food Products Press, Binghamton, pp 55–84
Zivy M, de Vienne D (2000) Proteomics: a link between genomics, genetics and physiology. Plant Mol Biol 44:575–580

Chapter 10
Plant Genetic Engineering: Problems and Applications

Bushra Rashid, Tayyab Husnain, and Sheikh Riazuddin

Contents

B. Rashid (✉) • T. Husnain • S. Riazuddin
Centre of Excellence in Molecular Biology, University of The Punjab Lahore,
87-W Canal Bank Road, Thokar Niaz Baig, Lahore 53700, Pakistan
e-mail: bush_rashid@yahoo.com; tayyabhusnain@yahoo.com; riaz@lhr.comsats.net.pk

M. Ashraf et al. (eds.), *Crop Production for Agricultural Improvement*,
DOI 10.1007/978-94-007-4116-4_10,

Abstract Introduction of Biotechnology has lead us to a stage where agricultural progress seems to be much different as before. It has introduced farmers to improve agricultural products while reducing the use of pesticides, and holds increased potential for pharmaceuticals and biomedical requirements. Technology has been improved to use the methods of transformation of agriculturally important crops by using different methods such as *Agrobacterium* and biolistic gun. Genetically engineered/modified crops, by *Agrobacterium*-mediated transformation system are becoming common in many countries. The health concerns on plants, animal and human life due to GM foods is unpredictable. However, plant genetic engineering can be considered either a progress or a threat to mankind depending on the way it is used.

Keywords Transgenic plants • Plant genetic transformation • Tissue culture • Transformation systems • Genetically modified crops • Plant biotechnology

1 Introduction

As the world population increasing, the agriculture sector has the priority to produce maximum foodstuff and other products with good yield in an environmentally sustainable way and cost-effective approach (Ferry et al. 2004). Agricultural production systems are dynamic and vary with time. The increasing awareness of the undesirable side effects of chemical pesticides and development of the insect resistance has forced scientists to look for alternative control measures. Improvements have been made in the crops for introduction of agronomic traits and other agriculturally important characteristics like yield, disease resistance and fibre quality by adopting conventional breeding techniques. This science has continued to offer the farmers with better environmentally friendly varieties. However, it is now improved by the introduction of technology as genetic engineering. Prospectively valuable and genetically superior characteristics known in different plants or yet in non-plants are sometimes unattainable due to the sexual hurdles in the crosses between distinct types. The improvement of technology to introduce and functionally express the foreign genes in plant cells has been adopted in less than two decades and leads to produce the transgenic plants. These transgenic plants show resistance against insects and diseases, better nutritional qualities in seeds and fruits and better adaptation of plants to undesirable environmental/ecological conditions (Danny et al. 1992). There is considerable interest in using transgenic approaches to produce resistance to different pathogens.

The genetically modified (GM) crops have been adapted commercially on an extensive scale since 1996; therefore they have been produced and became a reality in various countries (Dunwell 2000). The major countries initially growing Bt crops are USA, Canada, China, Australia, South Africa, and Mexico. It has been reported that in 2007, 22 countries grew Bt crops on a total area of 42.1 million hectares which accounted for the global area of 37% covered by GM crops (James 2007).

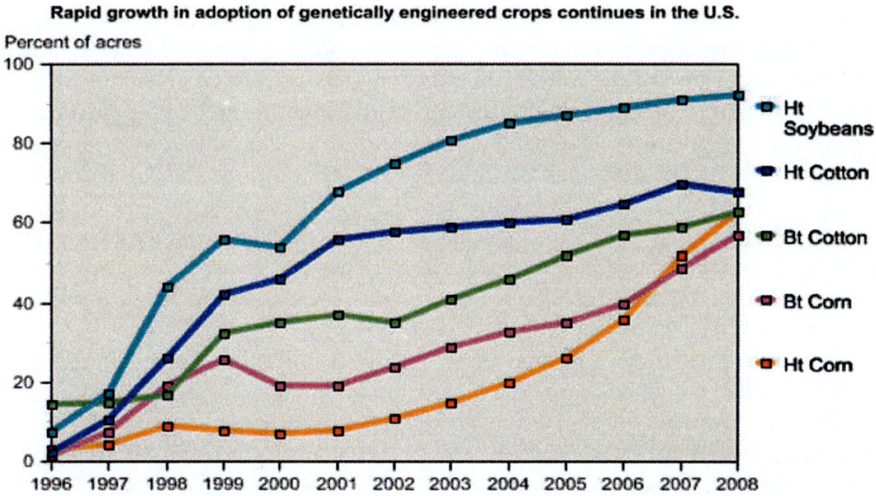

Fig. 10.1 Rapid growth in adoption of genetically engineered crops continues in the U.S. (Source: USDA, Economic Research Service)

Most important crops are; Cotton, Rice, Canola, Soybean, Potato, Corn and vegetables. The main important agronomic traits introduced into the transgenic crops include insect and herbicide resistance (Fig. 10.1). Advances have been made by using different transformation technologies like *Agrobacterium*-mediated transformation (Umbeck et al. 1987; Firoozabady 1989; Cousins et al. 1991; Gould and Magallanes 1998; Satyavathi et al. 2002; Shuangxia et al. 2005; Franklin et al. 2008), Gene Gun or combining both the methods together have also been reported (Firoozabady et al. 1987; Christou et al. 1989; Perlak et al. 1990; McCabe and Martinell 1993; Rajasekaran et al. 1996, 2000; Zapata et al. 1999; Taylor and Fauquet 2002). Sonication assisted transformation (Trick and Finer 1997, 1998; da-Silva and Fukai 2002; Solís et al. 2003; Jiang et al. 2004; Rashid et al. 2008) and Direct DNA uptake into protoplast (Khan and Maliga 1999; Daniell and Dhingra 2002) have also been used to which the researchers are required to get the reported literature/information. In the present review we have identified key issues to develop/adapt the transgenic plant technology and the main problems to be resolved in the practical application of this technology.

2 Confirmation of Transformation

The transformation of introduced genes involves stable integration and expression into the plant nuclear genome. Perlak et al. (1990) presented the idea that the successful integration and expression of a foreign gene in a plant could be predicted to be a result of many different factors. These factors may include the gene product itself, a competent vector, suitable promoter, leader sequences, 3′ non-coding sequences, existence of prospective plant regulators, codon frequency and the secondary structure of the mRNA. Potrykus in 1991 offered an assessment that a combination of phenotypic, genetic and physical records is required for plant transformation technologies which can be based on a strict explanation of evidence of integrative transformation, but this arrangement was not practical to observe the confirmation of transformation in some plants. It is difficult to investigate the population of sexual progeny in some trees that grow and sexually reproduce slowly. The same is also complicated in sexually sterile varieties and some of the crops which are propagated vegetatively like sugarcane which has complex genetics such as polyploidy and aneuploidy.

Sometimes, transformants may have restricted copies of foreign introduced sequences, that would end up in the incorporation of numerous or repeatedly rearranged copies of the relocated DNA. Therefore, an association between substantial and phenotypic facts is not a really crucial attribute of gene transfer technique. Integrative transformation has yet not been clearly demonstrated in such type of cases (Birch 1997). Therefore as competitively commonly more appropriate criteria would be accepted as thorough confirmation of integrative transformation.

Southern blot is the DNA hybridization of various independent genetically modified plants. This method uses a probe for a foreign gene DNA digested with restriction enzymes which finally expect to produce hybridizing fragments of desirable size at required integration sites (Southern 1975). It has been reported that from the integration of T-DNA into plant cells, transforming vector border regions is deleted largely (Tinland 1996; Buck et al. 1999; Chen et al. 2003; Wang and To 2004). For transgenic lines, it is significant to verify that the size of hybridizing fragments including flanking DNA at each integration site should be reproducible within same transgenic line, but differ between independent lines.

This is well-known fact that most of the T-DNA transmited to the plant cell has not been integrated stably. Measurement of the comparative amounts of transient as well as stable expression of reporter gene activity in plant cells infected with *Agrobacterium* has proved this indirectly. Normally, the expression of stably integrated T-DNA is much lesser than the transient expression of T-DNA genes (Castle and Morris 1990; Janssen and Gardener 1990; Nam et al. 1997; Mysore et al. 1998). Single-stranded molecule of T-DNA enters the nucleus (Tinland and Hohn 1995; Yusibov et al. 1994) but it is not apparent if the T-DNA integrates by means of strand invasion of the nearby denatured plant DNA via single-stranded T-strand, followed by second strand synthesis (Tinland and Hohn 1995; Tinland et al. 1995; Rossi et al. 1996; Tinland 1996; Krizkova and Hrouda 1998) or might be the T-strand is changed to an extrachromosomal double-stranded appearance before integration (De-Neve et al. 1997). Switching of the T-strand into a transcriptional competent level involve the production of DNA strand complementary to the T-strand, therefore most of the extrachromosomal T-DNA that primarily enters the nucleus probably becomes double-stranded (Narasimhulu et al. 1996). Thus, for the specific integration on T-strand in the plant chromosome, *vir*D2 has a dynamic role. The energy included in its phosphodiester bond may be provided by the discharge of *vir*D2 protein. This energy may be provided at the Tyr29 residue by means of the first nucleotide of T-strand provided that the 5′-end of the T-strand for ligation to the plant DNA. For nucleophilic 3′-OH from nicked plant DNA, the phosphodiester bond is capable to act as electrophilic substrate (Jayaram 1994). As soon as the mutant *vir*D2 protein attached to the T-strand is transferred, the nucleotides at the 5′-end of the T-strand is lost and the integration process starts to occur (Tinland et al. 1995).

Selection of transformed cells from a number of non-transformed ones is done by the marker genes and for visual selection of transformed tissues a reporter gene like the green fluorescent protein gene is required (Miki and McHugh 2004; Sripriya et al. 2008). Sometimes selectable marker gene affects the plant growth but an optimistic level of antibiotic is essential to reduce the false positives (Zapata et al. 1999; Zhang et al. 2001; Jin et al. 2005; Wu et al. 2006). However, the marker gene is not necessary after the selection of a transgenic plant. Various techniques even published after molecular proof could not be repeated or do not produce same results, therefore it is important that analysis of data at molecular, phenotypic, and genotypic level should be done in a more detailed manner on transformed lines produced for practical or commercial purposes.

The genetic transformation protocols differ from species to species and cultivar to cultivar. It is important to optimize the protocols required for successful transformation of species. There are several factors involved in this technology. First of all, standardization of cells taken from plant regenerable tissues and their interaction with the *Agrobacterium* and secondly, optimization of a suitable method for regeneration of plant organs from transformed cells.

3 Purpose of Plant Transformation

3.1 An Experimental Strategy for Plant Physiology

The advancement of plant genetic engineering technologies has allowed the introduction and functional expression of foreign genes in plant cells. This has further lead to the development of transgenic plants having superior characteristics like insect, disease and herbicide resistance. Plants have been produced for better adaptation to undesirable environmental circumstances and seeds or fruits with improved nutritional traits. Nutrition reduction and physiological injuries have occurred to the plants by sap sucking homopteran insect pests in the course of stylet probing and feeding. They also act as vectors to transmit viral diseases (Rao et al. 1998; Foissac et al. 2000). A homodimeric lectin, ASAL (mannose binding 25 kDa), is identified as a strong control agent against sap sucking insects and has been isolated from leaves of garlic (*Allium sativum*) (Bandyopadhyay et al. 2001; Banerjee et al. 2004; Majumder et al. 2004). It is reported that plants have the ability to adapt their gene expression pattern and activate resistance mechanism in response to control infection and transformation with *Agrobacterium* (Ditt et al. 2001, 2005, 2006). Recently, it has been verified that the expression of *Agrobacterium vir* regulon could be shut down by salicylic acid- a plant signaling compound and it directly influences the infection development (Yuan et al. 2007). These reports present the existence of plant-microbe interaction for physiological improvements.

3.2 A Practical Strategy for Plant Improvement

Potential advantages of this technology include: increased yield, food security, less application of pesticides and practices to reduce soil erosion which ultimately would help out the small farmers for more saving and to improve quality of life and a technology easy to disseminate. Golden rice producers have recently been granted the patent licenses by Monsanto without any cost to meet the requirement of vitamin A in the human body. More crops such as wheat, rice and sugarcane having improved characteristics are now in different phases of inventory as engineering, testing and commercialization of salt-resistance. A large area of farmland presently inappropriate to produce crops will be used for cultivation by this technology. Fruit crops do not have significance as cereal or staple crops such as; wheat, corn, rice and soy in many countries. In Bangladesh papaya fruit as a crop has very essential part in their every day diet after rice. In 1990 Papaya ring-spot virus (PRSV) was menace to demolish the Hawaiian papaya. Therefore, GM papaya (PRSV-resistant cultivar) was produced and since 1998, it has been cultivated commercially very successfully.

4 Requirements for Genetic Transformation

Tissue culture is used in nearly all present practical transformation procedures. It is not a hypothetical requirement but without proper tissue culture procedure it is difficult to achieve an efficient gene transfer, selection and rejuvenation of putative transformants. Therefore improvement in tissue culture systems is desirable to provoke proficient transformation which should not be genotype dependent. Murashige and Skoog in 1962 discussed the comprehensive concerns, alternatives and standardization of tissue culture systems that are practical for plant genetic transformation. They further confer the detailed protocols and addressed the scope of this broad technology. There are advances which have been made by means of biotechnological applications such as ovule culture (McStewart and Hsu 1977; Thengane et al. 1986), protoplast culture (Peeters et al. 1994), somatic embryogenesis and regeneration (Shoemaker et al. 1986; Trolinder and Goodin 1988; Finer 1988; Kumria et al. 2003) and plant gene transfer (Firoozabady et al. 1987; Umbeck et al. 1987; McCabe and Martinell 1993). A number of crops have been transformed through *Agrobacterium* (Perlak et al. 1990; Bayley et al. 1992; Thomas et al. 1995), by gene gun method (Finer and McMullen 1990; Rajasekaran et al. 2000), or by combining both the methods (Majeed et al. 2000). These crops contained different genes known for important agronomic traits. However, the most favorable means for gene delivery is through *Agrobacterium*. This is because it is simple, cost-effective and common for integration of single copy of gene into the host plant genome. It is strongly recommended to apply this system to all crops due to its access to an extensive selection of functional genes in the post-genomic time period. *Agrobacterium* infection diverges to a great extent for plant species, varieties, tissues and cells and amenability to this makes transformation either incompetent or not viable in recalcitrant plant species. Consequently, a plant-pathogen interaction study is needed to investigate the basis of plant recalcitrance towards this bacterium. A highly-efficient transformation procedure is required to develop which should be enormously concise, consistent, genotype independent. Moreover the method needs to be highly embryogenic, reduces the complexity of embryo germination, highly frequent for organogenesis, efficient in rooting of transgenic plantlets, and capable of hardening-off when shifted to soil pots.

5 Recalcitrance and Constraints

Many important plant species (legumes, cereals and woody plants) or best varieties of selected species, still remain recalcitrant to *Agrobacterium* method of transformation and unresponsive to many plant growth regulators (Choffe et al. 2000; Wang and To 2004). Sometimes regeneration is found restricted to only some of the cultivars (Firoozabady and DeBoer 1993; Zhang et al. 2000; Kumria et al. 2005). A number of reports have been published for transformation such as shoot apex using gene gun or *Agrobacterium* (Gould et al. 1991a, b; McCabe and Martinell 1993; Zapata et al. 1999). Particle bombardment of suspension cultures generally gave high-copy-number transformants, but is a time-consuming procedure (Finer and McMullen 1990; Rajasekaran et al. 2000). Therefore, most of the laboratories prefer the *Agrobacterium*-mediated method of genetic transformation followed by the formation of somatic embryos and then organogenesis (Firoozabady et al. 1987; Umbeck et al. 1987; Lyon et al. 1993; Thomas et al. 1995; Jin et al. 2005). But the limitations involved with this method are that: it is laborious, time-intensive procedure, which may take as long as 6–9 months to obtain embryogenic calli from transformed plant cells. It would lead to the development of a few embryos, problems of embryo germination and formation, low percentage and abnormal plantlets which ultimately result in low transformation efficiency and plantlet regeneration. Due to these reasons sometimes it takes 10–12 months to regenerate plantlets for even greatly receptive genotypes. There is evidence that *Agrobacterium* strain, co-cultivation time duration and temperature, bacterium optical density, addition of acetosyringone for the period of co-cultivation and embryogenic calli condition would obviously affect the transformation efficiency (Firoozabady et al. 1987; Fullner and Nester 1996; Li et al. 1996; Jin et al. 2005).

6 Steps to Achieve Transformation

There are many transgenic crops which have been released for commercial production due to considerable progress in plant genetic engineering technology. Progress in the technology has been made due to the improvement in a variety of appropriate tissue culture techniques for regenerating whole plant from single cell or tissues in a large number of plant species along with the development of improved or efficient transformation methods. Hence any type of totipotent plant tissue or the one containing the latent capacity is appropriate for genetic engineering (Sanford et al. 1987; Aragão 2002). Revealing the key events that could effectively be integrated into plant breeding program to develop commercially important transgenic crops, it is significantly important that recovery of fertile transgenic plants may be higher.

6.1 Tissue Culture

Production of virus-free germplasm through tissue culture using shoot meristem or apex cultures is an important discovery. The ornamental nursery industry became very popular after this discovery and fastly growing shoots of many virus infected clones produced the virus free plants (Morel and Martin 1952). Different scientists reported that MS basal medium directly regenerated the plants from inoculated shoots (Murashige and Skoog 1962; Shabde and Murashige 1977). The same medium has been used to regenerate other crops such as *Gossypium* (Gould et al. 1991a), maize (Gould et al. 1991b) and cereals (Hiei et al. 1997). Tissue culture is considered as an essential step for transformation of the majority of plant species, consequently higher priority is needed to establish research procedures that could reduce the somaclonal variation (Birch 1997). Somatic embryogenesis after callus formation would lead the somaclonal variations or genetic mutations but plants produced through shoot apex show very few somaclonal variations or genetic mutations. Because the shoot apex culture lacks the callus formation and somatic embryogenesis step (Gould and Magallanes-Cedeno 1998) which ultimately diminish the activation of retrotransposon activity in cultured plant tissues and avoid creating permanent mutations or somaclonal variations (Hirochika 1993).

Before plant transformation most important thing for the successful tissue culture systems is the accessibility of enough number of cells with latent capacity, because such cells are required to be easily available for genetic transformation treatment. Moreover they are expected to preserve the cell proliferation, regeneration capacity and selection treatment during or after transformation procedures. Somatic embryogenesis and establishment of successive regeneration of plantlets through callus derived lines is complicated and prolonged process. Limitation factor involved with the meristematic cell culture is that the cells are fewer in number and even within those the regenerable cells are very less. As a result the plantlets regenerated or the explants act in response to selection medium are less in quantity

(Firoozabady et al. 1987; Sunilkumar and Rathore 2001). However, once the development of an embryogenic calli from the preliminary material even from a single callus line in any commercial cultivar is established, then it may strengthen the corresponding, several transformation experiments (Rajasekaran et al. 2000; Leelavathi et al. 2004).

6.2 Gene Transfer

Plant biotechnology has enormous potentiality for genetic improvement. Accessibility of appropriate tissue culture system, correct selection of transformants and revival of transformed plants are the pre-requisites for the development of an efficient system for gene transfer technology. Presently the most extensively used techniques for plant genetic engineering or modifications are the *Agrobacterium*-based and particle bombardment through gene. In particle in the bombardment using metal beads as transporter, the DNA is directly incorporated into the host genome. This method has been employed for several years to transform many of the plant species related to cereals, legumes and woody plants (Genga et al. 1991; Aragão et al. 1992; McCabe and Martinell 1993; Zapata et al. 1999; Altpeter et al. 2000).

Dicot crops like cotton, maize have successfully been transformed for many years by applying *Agrobacterium*-mediated method of transformation, but by the passage of time, efficient procedure has been established for monocots (Yu et al. 2000; Tu et al. 2000). Generally, the *Agrobacterium*-based transformation system has benefits over gene gun particle bombardment as it has more chances for stable expression of transgene for next generation and larger insertion frequency for low copy of transgene (Dai et al. 2001; Shou et al. 2004; Travella et al. 2005).

Conventional transport of T-DNA has been involved in the process of *Agrobacterium* infection (Zambryski 1989; Hansen and Chilton 1999). VIP1 and VIP2 are known as the VirE2 interacting proteins (Tzfira and Citovsky 2000). Nuclear import of VirE2 has been promoted by VIP1, a bZIP protein has been expressed in yeast. VIP2 is a protein thought to mediate interaction between chromatin proteins and transcription factor's complex (Frolov et al. 1998; Tzfira et al. 2000) thus providing evidence that intranuclear transport or T-strand integration might be promoted by VIP2. Advanced evidences indicate, VirE3 is also exported into the host cell in addition to bacterial proteins (VirD2 and VirF) and T-DNA. A majority of these proteins are transported discretely from the T-DNA (Vergunst et al. 2000, 2003; Schrammeijer et al. 2003; Cascales and Christie 2004). These *Agrobacterium* proteins participate for key role inside the host cell during genetic transformation. Thus, the function of VirE3 demonstrates that bacterial virulence protein could be replaced by the cellular protein VIP1 inside the host cell. This will interact with and facilitate the nuclear import of VirE2 and this will lead to the successive expression of T-DNA through a-mediated pathway "karyopherin" (Lacroix et al. 2005).

During transformation premature expression of T-DNA corresponds to the stage that instantaneously pursues the nuclear import. Effectiveness of T-DNA expression has been confirmed by inoculating the leaf disks derived from the VIP1 antisense, wild-type, and VIP1 antisense/VirE3 double-transgenic plants transformed with *Agrobacterium* which carries its T-DNA a uidA gene encoding GUS reporter (Tzfira et al. 2001). Total numbers of GUS-expressing areas per transformed leaf disk are determined by the histochemical assay and enzymatic activity in the transformed tissues. The regulatory sequences for expression in bacteria are deficient in the *uid*A gene contained on the T-DNA (Janssen and Gardener 1990), therefore the T-DNA correspond to direct the GUS activity after it is transported to the plant cell other than or except its liable leaky expression in *Agrobacterium* (Tzfira et al. 2001). When comparison was made between the wild-type plants and VIP1 antisense *Agrobacterium* inoculated plants, more or less 25–35% GUS activity was exhibited by the inoculated plants. Encouraging or optimistic outcome has been obtained when the regenerating cells were estimated for transient assay after gene transfer. But based on negative results from transient assay, it is injudicious to anticipate too earlier. It is better to understand the restrictions distressing such type of expression. The level of expression of a transgene is variable and is subjective to various features, for example the site of integration (Matzke and Matzke 1998). Transgene stability also varies among transformants and some plants show a variety of instability even in subsequent generations (Datta et al. 2002).

6.3 Selection

A considerable number of true putative transformants are expected to be regenerated after transformation. For selection of transformants, genes confronting to a selective medium containing an antibiotic (Steinitz et al. 2002; Wang and Tu 2004), genes conferring a phenotype allowing visual or substantial selection (Jefferson 1987; Hadi et al. 1996; Christou 1997; Chen et al. 1998; Sunilkumar and Rathore 2001; Francois et al. 2002) or screening through molecular analyses like PCR to classify plants containing foreign genes (Datta et al. 2002; Jin et al. 2005; Chakraborti et al. 2008) can be used to select putative transformants.

It is reported that production of chimeric primary transformants comprised of transformed germline cells, like the intermediates in the development of homogeneous transformed R_1 progeny plants, the transformation systems usually involve the initial screening to a certain extent but not the toxic selection to get the primary transformants (Gould and Magallanes-Cedeno 1998; Zapata et al. 1999; Datta et al. 2002; Chakraborti et al. 2008). Important factor for *Agrobacterium* transformation in the inoculation of apical meristematic cells and explant is the selection or intensity of an antibiotic adapted to the activity level of promoter This would be helpful to avoid killing the transformed tissues. In R_0 the regenerated plants are ought to be chimeric and presence of transferred genes may be in less than single copy or they may be present as multiple insertion events. Progeny from R_1 or R_2 individuals of self or cross pollinated plants are best to use or to confirm for Southern blot analysis. In this way the germ line revealing the transformation event that was present in the transgenic progeny without chimerism will be selected (Maqbool et al. 2002). In majority of the plant tissues, high level of expression during the selection or all through the life cycle has been produced by the selectable marker genes which are driven by the constitutive promoters. A good example for this is the promoter ASA2 as it drives adequate *hpt* expression which is strong enough for selection on hygromycin containing medium in soybean somatic embryogenic or callus cultures. Moreover, in most of the plant tissues extremely small level of expression is found but in mature maize seeds, there is no expression (Zernova et al. 2008).

A permissive protocol for selection has the key importance to perform the transformation procedures since the plant regeneration system is associated with the viability of the explant. Normally it is expected that, for selection, the promoter along with the *neo* selectable marker resistant gene is required to be active in the meristematic region of the explant. If the meristematic cells are dead or weak, the regeneration is stopped or the cells will not proliferate and the inheritance pattern is modified or come to an end. An optimistic level of kanamycin (10–30 mg/l) is anticipated to be used to select the inoculation of the small shoots for *neo* gene driven by the *nos* promoter in *nptII.* The *CaMV* 35 S promoter does not play dynamic role in the plant shoot apical meristematic cells. Yet if this promoter is used with the selectable marker gene, it will not defend these lively or active cells. Size or dimension and method to prepare the shoot apex explant is very important for the selection of the antibiotic concentration. Small amount of selective agent is suggested for the initial selection process or gradually enhance the selection level if the escapes are extensively produced.

6.4 Transgene Expression

Both the methods of transformation (*Agrobacterium*/gene gun) are known for the stable integration and expression of the foreign genes into the plant cells (Briddon et al. 1998; Finer et al. 1999; Aragão 2002). To observe the transient expression, tissues are bombarded/cocultivated and cultured for a specific duration, generally 24–48 h. Then the tissues are evaluated for the expression of transgene. Qualitative and quantitative measurement of transgene expression is done by reporter gene such as *uidA* (Jefferson 1987), luciferase (Millar et al. 1992) and GFP green fluorescent protein (Davis and Vierstra 1998). It is possible to determine the factors responsible for affecting the gene expression in various plant tissues. This possibility is effective by studying the outcome of mutations, role of introns, codon usage and effectiveness of promoters. Regulatory sequences are modified by deletions or mutations and bombarded into plant cells. Control elements are identified by dissection and promoters are studied for up or down-regulation under that specific physiological condition (Vain et al. 1996; Komari et al. 1998; Finer et al. 1999). Effectiveness of transgenes and promoters are estimated through transient expression assays before commencement of the plant genetic engineering (Jefferson 1987; de-la Riva et al. 1998).

For expression of Vir proteins, plasmid study is also important. Either one of the single plasmid, GFP-VirE3 or GFP-VirE3- mNLS12 from pEGFP-C1-VirE3 or pEGFP-C1-VirE3-mNLS12 were first cloned into the *Nco*I-*BamH*I sites of pRTL2-GUS as a replacement for GUS. After that 35 S promoter-GFP-VirE3-terminator or 35 S promoter-GFP-VirE3-mNLS12-terminator cassettes were moved into the *Sph*I site of pGDR, which already contained a DsRed2 gene driven by the 35 S promoter. This resulted in the construct pGDRGFP- VirE3 and pGDR-GFP-VirE3-mNLS12 (Citovsky et al. 1992; Goodin et al. 2002). VirE3-VirE2 correlation is significantly prone for infection of *Agrobacterium* because the VIP1 antisense plants significantly elevate the susceptibility of VirE3 for transgenic expression by *Agrobacterium* during genetic transformation. Therefore it has been proven that, the function of VIP1 throughout the *Agrobacterium* infection of plant cells is accomplished by VirE3 (Lacroix et al. 2005).

External factors such as temperature and pH also affect the activation of *vir* system. The *vir* genes are not expressed at higher temperatures, since the inactivation of its properties induces a conformational change in the folding of *vir*A (Jin et al. 1993). *Vir*Gc trigger the constitutive expression of *vir* genes. *Vir*Gc is a mutant form of *vir*G which suppress the effect of temperature on *vir*A. However, that temperature does not confer the virulence competence of the mutant for *Agrobacterium*. This may be due to the other proteins that effectively may participate in the transfer of T-DNA process undergoing the folding at high temperature (Fullner and Nester 1996).

Occurrence of gene silencing is a reasonably normal phenomenon for genetic engineering of plants which is wounded up by the presence of multiple copies within the plant genome. Function of cytosine methylation and cosuppression mutually is consequential for down-regulation of transgene expression which is sometimes very incompatible and irregular (Komari et al. 1998; Finer et al. 1999; Iyer et al. 2000; Dong and Li 2006). Methylation of promoter sequences mainly cause the gene silencing at transcriptional level and in this manner chromatin re-modeling proteins are attracted to these sites by interfering with assembly of the transcription factors (Meyer 2000; Wang and Waterhouse 2002). Double stranded RNA is produced by the activation of co-suppression at the RNA level. This would lead to commence the degradation of the target RNA, thus following the gene silencing (Vance and Vaucheret 2001). It is common problem in plant genetic transformation since it avoids sustainable expression of a desired phenotype within transgenic plants. However, while unraveling mechanisms underlying this phenomenon, it is realized that the down-regulation of native genes within the plants is possible with the increasing knowledge.

Recent studies demonstrated that most of the genome is found to be transcribed at individual or multiple phases of cell growth therefore overall patterns of gene expression are vastly active throughout the maturity of epidermal plant cell (Hovav et al. 2008). Advanced skill to collect the particular types of cells in plants and amplification of mRNA for expression profiling has directed for the transcriptome analysis at the cellular level (Galbraith and Birnbaum 2006). Isolating cells via sorting practices or laser microdissection and capture are among the recent technologies known for the determination of changes during differentiation under in vitro or natural systems (Birnbaum et al. 2003; Leonhardt et al. 2004; Casson et al. 2005).

Now microarray is a potent device offered for high throughput analysis of cell differentiation and progress and has made it possible to estimate the participation of thousands of genes in biological processes. Modern applications of this technology consist of wood-forming cambial meristematic tissues (Schrader et al. 2004), pollen transcriptome investigation of *Arabidopsis thaliana* (Pina et al. 2005), the quiescent-center cells of developing roots (Nawy et al. 2005), vascular tissue or epidermis cells of maize (Nakazono et al. 2003), and quickly growing cotton fiber preliminary cells (Wu et al. 2007).

6.5 Integration of Transgene

Genetic transformations of plants take place when tumor-inducing (Ti) plasmid transfer its single stranded copy of T-DNA into the host cell and then integrate that into the genome of the host cell. The current plant genetic engineering is based on this natural gene transfer principle. This phenomenon has been focused by taking out the tumor inducing genes from the bacterium. Then desired genetic sequences are substituted with most of the T-DNA region (Zupan and Zambryski 1997). Recently, using this concept the coat protein genes of several plant viruses have been transported into a number of plant species to develop resistance against viral infection. In this way the organisms engineered with genes derived from a pathogen may exhibit resistance against the pathogen itself (Vasudevan et al. 2008).

After particle bombardment integration of exogenous DNA takes place into plant's genome in the form of simple, single copy or complex multiple copies insertions (Christou 1997; Finer et al. 1999; Smith et al. 2001). This insertion of transgene process is considered to takes place in two-steps. Firstly, the exogenous DNA undergoes homologous recombination and then it is rearranged head to head and/or head to tail after the ligation. Fragmented plasmid and expression cassette's copies can also be interwoven collectively inside these concantamers. This recombination takes place within 30 min of the transformation process. Secondly, after rearrangement the DNA is introduced into the plant's genome. Integration of DNA is supposed to occur by the contribution of native DNA reparation along with the non-homologous recombination at sites of double stranded disruptions in plant's chromosome (Kohli et al. 1998; Puchta 1998).

Stable integration of introduced DNA, decreased somaclonal mutations and unforeseen segregation of transgene in successive generations are the important selection measures for an efficient transformation technology. Therefore, initially a large number of transformed plants should be produced or multiple genes should be introduced for testing or initial screening. Co-transformation technology assists the simultaneous introduction of important multiple genes. This is possible by using single selection marker gene (Francois et al. 2002; Sandhu and Altpeter 2008). Inheritance studies of the rice progeny produced with this technology showed that co-segregation of transgenes occurs altogether in the successive generation. This indicates integration had taken place at the same genetic locus (Taylor and Fauquet 2002). However if the marker and genes of interest are unlinked and they integrate infrequently then this situation relies on the co-transformation methodology (Radhakrishnan and Srivastava 2005). Conversely, removal/modification in the DNA sequence existing in the host may happen by transposition phenomenon (Scholz et al. 2001).

Genes become silent at transcription stage if the same promoter and a polyadenylation signal for different marker genes have been used repetitively (Hohn et al. 2001). Hence it is proposed that the elimination of antibiotic resistant genes from transformed plants or selection for a distinct transgene insertion by segregating by the breeding technology will prove successful. For such situations tight association of the transgenes in the transgenic crops is disadvantageous. So the capacity to control the positions of multiple transgene integrations might be improved by employing the marker gene at restrained sites from the gene(s) of interest for this purpose. Thus, selectable marker gene elimination is considered to be critical for pyramiding of multiple characteristics in transgenic crop. Advances in the technology show that various approaches have effectively been used for exclusion of selectable marker gene from transgenic plants (Vergunst et al. 2000; Luo et al. 2000; Hoa et al. 2002; Zhang et al. 2003; Miki and McHugh 2004; Kopertekh et al. 2004; Kerbach et al. 2005; Radhakrishnan and Srivastava 2005; Sreekala et al. 2005; Cuellar et al. 2006; Darbani et al. 2007). Research in transforming plants need to concentrate on the complications related to the stable integration and consistent expression of the gene after the DNA has been integrated (Aragão 2002).

7 Constraints in Plant Transformation

7.1 *Environment and Public*

Genetically modified inventions in agriculture are not commercial goods marketed as such to the consumer. The research in advancements of genetically modified crops help out the researchers to yield these crops, perhaps along with novel genetic sequences and promoters and/or copyrights on upgraded research methodology. A commercial variety is then released in response to these transitional products (Pray et al. 2005). Since the commercialization of transgenic crops the cultivated land for GM crops has continuously been increasing. None of the negative affect has been observed on biodiversity, therefore many countries consider these crops safe for food and feed purpose. Though, new technologies in this area are in the process to produce the GM crops generating pharmaceutical agents or industrial raw materials. These products will be introduced to the market in near future. For such type of situations, crossing between genetically modified crops and commercial non GM crops is needed to be done under consistent protective processes.

Production of GM crops is one of the possible solutions for food scarcity among poor countries and presently a number of capable improvements are there. US universities and many biotech companies have declared not to ask for patent rights if the GM crop tools would be used to produce crops for developing or poor countries. The genetically modified crops with desirable improved characteristics will be having simply smaller long-lasting impact. Providing farm financial assistance in the developed countries would persistently leave out farmers from the international market (Breithaupt 2003).

It is of enormous concern in crop biotechnology to avoid the flow of transgene from genetically engineered crops to other edible crops and their native species. Male sterility approach is being used for gene imprisonment in addition to decrease the consumers concerns of food purity (Konagaya et al. 2008). Issues relating to Biotech inventions are health (e.g. allergins, toxins, antibiotic resistance), environmental, social, ethical and religious concerns. Public is paying attention to the effects of agricultural applications on environment and human health. In agriculture sector, there is the development of regulatory and motivation programs which may influence the farming systems varying from the selection of agronomic applications to pest management.

The farming division is always being the key resource for improvement in the environment and food security technologies with these characteristics. The agricultural industry may perhaps preserve a small amount of economic profits out of these innovations. One way to do more with less is the theoretical approach to transfer the agricultural system technology from public to the private divisions. The key objectives here are: (i) to communicate the advantages of public research and development to prospective consumers, (ii) to discover the modern ways to execute the organization task in an era of comparatively limited assets, (iii) persuade to direct the technology improvement, (iv) attracting research resources through patent returns. These objectives may be achieved by direct contact between researchers and end users, sharing research findings through publications, utilizing patents and protecting technologies for copy rights and carrying out mutual scientific research (Rubenstein and Heisey 2005).

For recent advances, an online version of "Biosafety regulations of Asia-Pacific countries", published by the FAO, the Asia-Pacific Consortium on Agricultural Biotechnology (APCoAB) and the Asia-Pacific Association of Agricultural Research Institutions (APAARI), is now available. The mechanisms for regulations associated to biosafety of agricultural and food biotechnological products offered in Asian and Pacific countries have been discussed in detail (http://www.apcoab.org/documents/bs_pub.pdf). The prospects of biotech crops seems to be hopeful as the number of countries adopting biotech crops and the area under cultivation, number of crops and the improved traits are estimated to grow twofold during 2006–2015 when large number of farmers will prefer to grow biotech crops and it may possibly be raised up to ten times to 100 million or more than that till 2015. Presently the major limitation to increase the crop production throughout the world is water shortage. By 2011, drought tolerance genes are likely to become accessible principally essential for developing countries suffering comparatively more from drought conditions (James 2007).

7.2 *Intellectual Property Issues*

Transgenic plants have been produced after the efforts of several years but following the overcome to technical limitations, marketable restrictions happen to be more severe obstacle for utilization of genetic engineering. Cultivation and commercialization of transgenic crops in different countries has brought about the considerable economical benefits during the last decade. None of the serious concerns linked to biosafety, environment or pest resistance has been noticed. Modern biotechnology leads to the innovations in products, medicines, agriculture, food, soil rehabilitation, health, environment protection and industry etc. These innovations lead to the advances for inventions in the biotechnological field like: cDNA, mRNA, synthetic DNA, proteins, peptides, fragments, monoclonal antibodies, vaccines, vectors, microorganisms, host cells, cell lines, transgenic and cloned animals, laboratory apparatus, tissue engineering, production methods and gene therapy protocol. For biotechnological inventions patents should relate to the details of the process through which the biological material is prepared, processed or used. Biological material shall mean that the material having genetic sequence competent to reproduce itself or being reproduced while using in biological method.

A monopoly right granted to an inventor is a patent. This right enables the patentee to use his invention as he may use it by making and selling the product or he may authorize others to do so. A Patent remains valid for 20 years subject to renewal. It entails the creative activity and subjected to be used in any kind of industrial as well as agricultural relevance. Patent shall be granted for a new or novel invention and an invention is considered to be novel if it has not been publicized in the prior art. Patents provide benefits like (1) to encourage research and inventions, (2) to disclose new technological discoveries, (3) to disseminate technical information and know-how, (4) to make transfer of technology possible, (5) to create an atmosphere for investments. Policy formulation, administration and enforcement of intellectual property rights are improved by modernising the intellectual property right (IPR) legislation, creating more awareness of the role of IPR for economic growth and trade development and supporting the establishment of a local IPR Organisation. Industrial revolution has modernized the patent structure and modified the system to shelter the mechanical and chemical inventions. There is no management to offer the security to living organisms. In fact, living organisms are supposed to be the breakthrough by nature, but not the scientific innovations (http://www.opendemocracy.net/ecology-foodwithoutfrontiers/article_1817.jsp).

8 Future Needs for Research and Development in Plant Transformation

8.1 Transformation Efficiency

It has been well documented until now that proficient regeneration system is required for the transformed tissues, whatever the method of transformation is used either *Agrobacterium* or biolistic gun, these two process i.e. transformation and regeneration are closely intermingled. Although there are outstanding achievements and novel practical progress has been made in plant biotechnology, but still there are shortcomings for the development of an efficient and genotype independent regeneration protocol to develop genetically engineered plants/crops. A number of factors can provide the boost to the transformation efficiency such as choosing the right strain of *Agrobacterium*, stimulation of *vir* genes with or without acetosyringione, temperature and duration of cocultivation and culture medium etc. (Gawel and Robacker 1990; Kumar et al. 1998; Wilkins et al. 2000; Popelka and Altpeter 2001; Sunilkumar and Rathore 2001; Mishra et al. 2003). Somatic embryogenesis would lead to the regeneration through a probable single-cell origin of the somatic embryo, therefore this method is favorite as compared to organogenesis, consequently reducing the chimerism (Merkle et al. 1995). Conversely the problems associated with regeneration via somatic embryogenesis are: time consuming, rate of recurrence of abnormal embryos is high, problems with embryo germination, problems associated with root and shoot development (Kumria et al. 2003).

The effectiveness of embryogenesis in a cultivar may be assured by adopting a stable and highly competent transformation system that can curtail the time of tissue culture. It was observed that chances of somatic embryogenesis of transformed cells may be increased by using embryogenic calli as explant propagated from a distinct somatic embryogenic callus. Thus it will increase the transformation efficiency and more number of transgenic plants will be produced in less time. Somatic embryogenic calli have effectively been used as explant in plant genetic transformation system (Scorza et al. 1996; Martinelli and Mandolino 2001; Wu et al. 2005). Moreover, stimulating the cell division by phytohormones at the time of cocultivation may enhance the *Agrobacterium* transformation in plant cell cycle at the particular phase (Gelvin 2000). Therefore, adoption of novel methods for cell and tissue culture and genetic transformation acquire considerable assurance to establish innovative genetic variability for progress of agriculturally important crops (Chahal and Gosal 2002; Rashid et al. 2004; Gill et al. 2006).

8.2 Genetic Damages

For the achievement of high number of transformants, an efficient transformation procedure that allows the desired integration of introduced genes and selection of events without severe somaclonal variation and unexpected segregation is needed. Traditional breeding techniques have been used since a long time to introduce the superb varieties of cotton. Nevertheless cotton breeding technology is restricted by being deficient in satisfactory genetic differences in the accessible germplasm collection (Wu et al. 2005) or the absence of various key characteristics like tolerance to insect pests and herbicides as in the genetic pools of many cultivars of sugarcane (Arencibia et al. 1997). Plant genetic engineering assists in the improvement of qualitative and quantitative characteristics and produces a number of agriculturally important crops with improved tolerance/resistance against insects, diseases, microbes, herbicides and abiotic stress. Similar molecular mechanism is employed for the transfer of T-DNA to potential dicot and monocot crops in the presence of an appropriate transformation procedure. This capability is improved by insertion of a selectable marker gene at different position from the desirable gene(s), to control the integration transgene at multiple locations is consequently required to explore the plant biotechnology related research. After laboratory, it is important to investigate the performance of transgenic plants under open field conditions that include evaluating the affect of transgenic plants on other plants in the field and environmental conditions. Furthermore, it is also important to evaluate the stability of foreign genes and issues associated with their interaction with the complicated physiological performance of these transgenic plants.

8.3 *Ideal Transformation Systems*

Pre-requisite for the development of genetically modified plants is the potential for introduction of foreign DNA into the plant genome and after that restoration of fertile plants. It is also important that the transgene(s) may have inherited, integrated and expressed successively into the genome of progeny plants. Particle gun and Agrobacterium- are the most commonly used techniques for this purpose with varying level of accomplishment. Somatic embryogenesis leading the regeneration of transgenic plantlets is laborious and overwhelming which ultimately results for somaclonal variation. Another drawback is that there are insufficient cells for regeneration competence in the meristematic region. This will limit the number of cells to be selected against an antibiotic. Somaclonal variations may be reduced by maintaining the culture condition/time at different stages of the tissue culture procedure. This is possible by taking care of the appropriate phase of cell proliferation and subculturing at proper time. This procedure may be different for each crop or species. Consequently, putative transgenic plants must be selected with substantial efforts and incompetent/ non-integrated copies of the transgene(s) should be removed. Transgenic lines with stable inheritance, integration and expression of the foreign genes are to be screened for future breeding programs and commercialization of transgenic crop. In summary, preparing ourselves and selecting the proper methods would lead us to unravel the troubles involved to achieve the practical application of the aims and objectives.

8.4 *Genetic Transformation vs Plant Breeding and Heritable Assortment*

Plant transformation system independent of genotype is one of the key successes to promote the genetically modified crops technology. For academic purposes or fundamental comprehension related to the gene introduction, integration and expression of transgene in the transgenic plants, model plants of a particular genotype may be used at laboratory level easy. These model plants are easy to study and understand the basic function of gene expression related studies.

As the Agrobacterium-mediated transformation method is species or genotype specific therefore, many issues are required to be considered for optimization of this procedure to establish the efficient transgene integration and expression in different species. These issues may be positive interaction of Agrobacterium and cells to be transformed and a proficient tissue culture procedure to restore the plants after transformation.

In the past, important crop varieties have been genetically modified with improved and desirable traits by conventional breeding programmes. But these conventional breeding techniques are time consuming, laborious and occasionally complicated to generate target tissues for diverse and considerable crops or their species. This is also true with some horticultural or vegetatively propagated plants like potato, cassava and banana to continue with conventional breeding. Single gene or oligogenic plant transformation system direct towards extreme genetic consistency in self as well as cross pollinated plants, but the conventional breeding technology involve the whole genome which is sometimes detrimental. There is need to develop de novo introduction and integration of the transgene for the essential cultivars (Taylor et al. 2002). Therefore it is important to remove undesirable sequences during genetic transformation of divergent genotypes with multiple genes within specie to introduce desirable physical characters. This will help to improve the genetically modified crop plants (Birch 1997). Genetic diversity and its intensity for certain crop specie most likely selected for breeding may be useful to study its evolutionary background (Zhou et al. 2008).

Recently, various agriculturally essential characters affecting the Oat crop were found to be correlated with at AFLP markers studies. This means that the comprehensive studies or investigations would be helpful to improve the genetic diversity. Ultimately marker assisted selections will confirm the selected traits in segregating populations (Achleitner et al. 2008). Further progress in this scenario is that to expand the new molecular markers, mapping of the regions of quantitative loci (QTL) has been proposed. It is documented that the outcome of single or joint QTL alleles is expected to be more perfect in an identical genetic background (Delourme et al. 2008; Holzapfel et al. 2008). Several alleles or quantitative trait alleles at each QTL affect to a particular attribute i.e. for every QTL known, one to three alleles affect that specific trait (Aitken et al. 2008). Moreover molecular marker assisted selection is a consistent technique to validate the additive effect of QTL for the development of the near isogenic lines (NILs) for quantitative trait loci (QTL).

8.5 Marketing of Genetically Modified Plants

Genetically modified plant technology has valuable impact on the recent plant improvement/breeding policies. Basic research involves the laboratory experiments and after that field trials are conducted. These field trials are only conducted for research purpose. Independent homozygous or true to type lines of transgenic plants will be selected. Ultimately this would direct a new variety to public and commercialization. Applied research is must after the basic, to ensure the application of the inventions. Scientific data is mandatory to deal with the issues related to public and ethical matters. Solution of legal or regulatory matters will be helpful to extend the GM products and ultimately this will raise the productivity and improve the quality of foods. In this way people will be benefited all together by having the products which are modified with environmentally friendly traits.

9 Conclusion

The prospective for applications of plant biotechnology to agriculture sector are significant for the assistance to mankind. Presently the adoption of GM crops is recognized at international level. The first commercial transgenic crop was cotton which was engineered for resistance against insects or herbicide. It was first time released in 1996 for commercial production on only 12% of the area for cotton cultivation in United States. By 2007, the countries growing GM crops went up to 23, and among these 12 are developing and 11 developed countries. According to area these countries are in the sequence of USA, Argentina, Brazil, Canada, India, China, Paraguay, South Africa, Uruguay, Philippines, Australia, Spain, Mexico, Colombia, Chile, France, Honduras, Czech Republic, Portugal, Germany, Slovakia, Romania and Poland.

All problems related to food and crop production would not be solved by adoption of agricultural biotechnology. Indeed, these issues may be helped out by improving and adopting these technologies. Judicious classification of food allocation is desirable along with adoption of plant biotechnology which essentially can facilitate to integrated food security approach.

References

Achleitner A, Tinker NA, Zechner E, Buerstmayr H (2008) Genetic diversity among oat varieties of worldwide origin and associations of AFLP markers with quantitative traits. Theor Appl Genet 117:1041–1053

Aitken KS, Hermann S, Karno K, Bonnett GD, McIntyre LC, Jackson PA (2008) Genetic control of yield related stalk traits in sugarcane. Theor Appl Genet 117:1191–1203

Altpeter F, Xu J, Ahmed S (2000) Generation of large numbers of independently transformed fertile perennial ryegrass (*Lolium perenne* L.) plants of forage- and turf-type cultivars. Mol Breed 6:519–528

Aragão FJL (2002) Development of transformation methods toward producing transgenic plants with abiotic stress tolerance. JIRCAS Working Report, pp 35–42

Aragão FJL, Sá MFG, Almeida ER, Gander ES, Rech EL (1992) Particle bombardment-mediated transient expression of a Brazil nut methionine-rich albumin in bean (*Phaseolus vulgaris* L.). Plant Mol Biol 20:357–359

Arencibia A, Vázquez R, Prieto D, Téllez P, Carmona E, Coego A, Hernández L, de la Riva G, Selman-Housein G (1997) Transgenic sugarcane plants resistant to stem-borer attack. Mol Breed 3:247–255

Bandyopadhyay S, Roy A, Das S (2001) Binding of garlic (*Allium sativum*) leaf lectin to the gut receptors of homopteran pests is correlated to its insecticidal activity. Plant Sci 161:1025–1033

Banerjee S, Hess D, Majumder P, Roy D, Das S (2004) The interactions of *Allium sativum* leaf agglutinin with a chaperonin group of unique receptor protein isolated from a bacterial endosymbiont of the mustard aphid. J Biol Chem 279:23782–23789

Bayley C, Trolinder N, Ray C, Morgan M, Quisenberry JE, Ow DW (1992) Engineering 2,4-D resistance into cotton. Theor Appl Genet 83(5):645–649

Birch RG (1997) Plant transformation: problems and strategies for practical application. Annu Rev Plant Physiol Plant Mol Biol 48:297–326

Birnbaum K, Shasha DE, Wang JY, Jung JW, Lambert GM, Galbraith DW, Benfey PN (2003) A gene expression map of the *Arabidopsis* root. Science 302:1956–1960

Breithaupt H (2003) Blame games. EMBO Rep 4(9):819

Briddon RW, Liu S, Pinner MS, Markham PG (1998) Infectivity of African cassava mosaic virus clones to cassava by biolistic inoculation. Arch Virol 143:2487–2492

Buck SD, Jacobs A, Montagu MV, Depicker A (1999) The DNA sequences of T-DNA junctions suggest that complex T-DNA loci are formed by a recombination process resembling T-DNA integration. Plant J 20:295–304

Cascales E, Christie PJ (2004) Definition of a bacterial Type IV secretion pathway for a DNA substrate. Science 304:1170–1172

Casson S, Spencer M, Walker K, Lindsey K (2005) Laser capture microdissection for the analysis of gene expression during embryogenesis of *Arabidopsis*. Plant J 42:111–123

Castle LA, Morris RO (1990) A method for early detection of T-DNA transfer. Plant Mol Biol Rep 8:28–39

Chahal GS, Gosal SS (2002) Principles and procedures of plant breeding. Biotechnological and conventional approaches. Narosa Publishing House, New Delhi

Chakraborti D, Sarkar A, Hossain A, Mondal HA, Schuermann D, Hohn B, Sarmah BK, Das S (2008) Cre/lox system to develop selectable marker-free transgenic tobacco plants conferring resistance against sap sucking homopteran insect. Plant Cell Rep 27:1623–1633

Chen L, Marmey P, Taylor N, Brizard JP, Espinoza SD, Cruz P, Huet H, Zhang S, Dekochko A, Beachy R, Fauquet C (1998) Expression and inheritance of multiple transgenes in rice plants. Nat Biotechnol 16:1060–1064

Chen PY, Wang CK, Soong SC, To KY (2003) Complete sequence of the binary vector pBI121 and its application in cloning T-DNA insertion from transgenic plants. Mol Breed 11:287–293

Choffe KL, Victor JMR, Murch SJ, Saxena PK (2000) In vitro regeneration of *Echinacea purpurea* L.: direct somatic embryogenesis and indirect shoot organogenesis in petiole culture. In Vitro Cell Dev Plant 36:30–36

Christou P (1997) Rice transformation: bombardment. Plant Mol Biol 35:197–203

Christou P, Swain WF, Yang NS, McCabe DE (1989) Inheritance and expression of foreign genes in transgenic soybean plants. Proc Natl Acad Sci 86:7500–7504

Citovsky V, Zupan J, Warnick D, Zambryski P (1992) Nuclear localization of *Agrobacterium* VirE2 protein in plant cells. Science 256:1802–1805

Cousins YL, Lyon BR, Llewellyn DJ (1991) Transformation of an Australian cotton cultivar: prospects for cotton improvement through genetic engineering. Aust J Plant Physiol 18:481–494

Cuellar W, Gaudin A, Solorzano D, Casas A, Nopo L, Chudalayandi P, Medrano G, Kreuze J, Ghislain M (2006) Self-excision of the antibiotic resistance gene *nptII* using a heat inducible Cre-loxP system from transgenic potato. Plant Mol Biol 62:71–82

Dai SH, Zheng P, Marmey P, Zhang SP, Tian WZ, Chen SY, Beachy RN, Fauquet C (2001) Comparative analysis of transgenic rice plants obtained by *Agrobacterium*-mediated transformation and particle bombardment. Mol Breed 7:25–33

Daniell H, Dhingra A (2002) Multigene engineering, dawn of an exciting new era in biotechnology. Curr Opin Biotechnol 13:136–141

Danny J, Llewellyn M, Brown Y, Cousins YL, Hartweck D, Last A, Mathews FM, Thistleton J (1992) The science behind transgenic cotton plants. In: Proceedings of 6th Australian cotton conference, Broadbeach and Queensland, Australia

Darbani B, Eimanifar A, Stewart CN, Camargo WN (2007) Methods to produce marker free transgenic plants. Biotechnol J 2:83–90

da-Silva JAT, Fukai S (2002) Transgene expression following transformation of chrysanthemum by four gene introduction methods. Prop Ornam Plants 2:28–37

Datta K, Baisakh N, Thet KM, Tu J, Datta SK (2002) Pyramiding transgenes for multiple resistance in rice against bacterial blight, yellow stem borer and sheath blight. Theor Appl Genet 106:1–8

Davis SJ, Vierstra RD (1998) Soluble highly fluorescent variants of green fluorescent protein (GFP) for use in higher plants. Plant Mol Biol 36:521–528

De Neve M, De Buck S, Jacobs A, Van Montagu M, Depicker A (1997) T-DNA integration patterns in co-transformed plant cells suggest that T-DNA repeats originate from co-integration of separate T-DNAs. Plant J 11:15–19

De-La Riva GA, González-Cabrera J, Vázquez-Padrón R, Ayra-Pardo C (1998) *Agrobacterium tumefaciens*: a natural tool for plant transformation. Electron J Biotechnol 1(3):118–133

Delourme R, Piel N, Horvais R, Pouilly N, Domin C, Vallée P, Falentin C, Manzanares-Dauleux MJ, Renard M (2008) Molecular and phenotypic characterization of near isogenic lines at QTL for quantitative resistance to *Leptosphaeria maculans* in oilseed rape (*Brassica napus* L.). Theor Appl Genet 117:1055–1067

Ditt RF, Nester EW, Comai L (2001) Plant gene expression response to *Agrobacterium tumefaciens*. Proc Natl Acad Sci 11:10954–10959

Ditt RF, Nester EW, Comai L (2005) The plant cell defense and *Agrobacterium tumefaciens*. FEMS Microbiol Lett 247:207–213

Ditt RF, Kerr KF, Figueiredo PD, Delrow J, Comai L, Nester EW (2006) The *Arabidopsis thaliana* transcriptome in response to *Agrobacterium tumefaciens*. Mol Plant Microbe Interact 19:665–681

Dong HZ, Li WJ (2006) Variability of endotoxin expression in Bt transgenic cotton. J Agron Crop Sci 193(1):21–29

Dunwell J (2000) Transgenic approaches to crop improvement. J Exp Bot 51:487–496

Ferry N, Martin GE, John AG, Angharad MRG (2004) Plant insect interactions: molecular approaches to insect resistance. Curr Opin Biotechnol 15:155–161

Finer JJ (1988) Plant regeneration from somatic embryogenic suspension cultures of cotton (*Gossypium hirsutum* L.). Plant Cell Rep 7:399–402

Finer JJ, McMullen DM (1990) Transformation of cotton (*Gossypium hirsutum* L.) via particle bombardment. Plant Cell Rep 8:586–589

Finer J, Finer K, Ponappa T (1999) Particle bombardment mediated transformation. In: Hammond J, McGarvey PB, Yusibov V (eds) Plant biotechnology. Springer, Berlin, pp 59–80

Firoozabady E (1989) Transformation of cotton. In: Bajaj YPS (ed) Plant protoplasts and genetic engineering, II. Biotechnology in agriculture and forestry 9:140–186

Firoozabady E, DeBoer DL (1993) Plant regeneration via somatic embryogenesis in many cultivars of cotton (*Gossypium hirsutum* L). In Vitro Cell Dev Biol 29:166–173

Firoozabady E, De Boer EL, Merlo DJ, Halk EL, Amerson LN, Rashka KE, Elizabeth EM (1987) Transformation of cotton (*Gossypium hirsutum* L.) by *Agrobacterium tumefaciens* and regeneration of transgenic plants. Plant Mol Biol 10:105–116

Foissac X, Loc NT, Christou P, Gatehouse AMR, Gatehouse JA (2000) Resistance to green leafhopper (*Nephotettix virescens*) and brown planthopper (*Nilaparvata lugens*) in transgenic rice expressing snowdrop lectin (*Galanthus nivalis* agglutinin GNA). J Insect Physiol 46:573–583

Francois I, Broekaert W, Cammue B (2002) Different approaches for multi-transgene-stacking in plants. Plant Sci 163:281–295

Franklin G, Conceição LFR, Kombrink E, Dias ACP (2008) *Hypericum perforatum* plant cells reduce *Agrobacterium* viability during co-cultivation. Planta 227:1401–1408

Frolov MV, Benevolenskaya EV, Birchler JA (1998) Regena (Rga) a *Drosophila* homolog of the global negative transcriptional regulator CDC36 (NOT2) from yeast modifies gene expression and suppresses position effect variegation. Genetics 148(1):317–329

Fullner KJ, Nester EW (1996) Temperature affects the T-DNA transfer machinery of *Agrobacterium tumefaciens*. J Bacteriol 178:1498–1504

Galbraith DW, Birnbaum K (2006) Global studies of cell type specific gene expression in plants. Annu Rev Plant Biol 57:451–475

Gawel N, Robacker CD (1990) Genetic control of somatic embryogenesis in cotton petiole callus cultures. Euphytica 49:249–253

Gelvin SB (2000) *Agrobacterium* and plant genes involved in T-DNA transfer and integration. Annu Rev Plant Physiol Plant Mol Biol 51:223–256

Genga A, Cerjotti A, Bollini R, Bernacchia G, Allavena A (1991) Transient gene expression in bean tissues by high velocity microprojectile bombardment. J Genet Breed 45:129–134

Gill R, Malhotra PK, Gosal SS (2006) Direct plant regeneration from cultured young leaf segments of sugarcane. Plant Cell Tissue Organ Cult 84:227–231

Goodin MM, Dietzgen RG, Schichnes D, Ruzin S, Jackson AO (2002) pGD vectors: versatile tools for the expression of green and red fluorescent protein fusions in agroinfiltrated plant leaves. Plant J 31:375–383

Gould JH, Magallanes-Cedeno M (1998) Adaptation of cotton shoot apex culture to *Agrobacterium*-mediated transformation. Plant Mol Biol Rep 16:1–10

Gould J, Banister S, Hasegawa O, Fahima M, Smith RH (1991a) Regeneration of *Gossypium hirsutum* and *G. barbadense* from shoot apex tissue for transformation. Plant Cell Rep 10:12–16

Gould JMD, Hasegawa O, Ulian EC, Peterson G, Smith RH (1991b) Transformation of *Zea mays* L., using *Agrobacterium tumefaciens* and the shoot apex. Plant Physiol 95:426–434

Hadi M, Mcmullen M, Finer J (1996) Transformation of 12 different plasmids into soybean via particle bombardment. Plant Cell Rep 15:500–505

Hansen G, Chilton MD (1999) Lesson in gene transfer to plants by a gifted microbe. Curr Top Microbiol Immunol 240:21–57

Hiei Y, Komari K, Kubo T (1997) Transformation of rice mediated by *Agrobacterium tumefaciens*. Plant Mol Biol 35:205–218

Hirochika H (1993) Activation of tobacco retrotransposons during tissue culture. EMBO J 12:2521–2528

Hoa TTC, Bong BB, Huq E, Hodge TK (2002) Cre/lox site-specific recombination controls the excision of a transgene from the rice genome. Theor Appl Genet 104:518–525

Hohn B, Levy AA, Puchta H (2001) Elimination of selection markers from transgenic plants. Curr Opin Biotechnol 12:139–143

Holzapfel J, Voss HH, Miedaner T, Korzun V, Häberle J, Schweizer G, Mohler V, Zimmermann G, Hartl L (2008) Inheritance of resistance to Fusarium head blight in three European winter wheat populations. Theor Appl Genet 117:1119–1128

Hovav R, Udall JA, Hovav E, Rapp R, Flagel L, Wendel JF (2008) A majority of cotton genes are expressed in single celled fiber. Planta 227:319–329

Iyer LM, Kumpatla SP, Chandrasekharan MB, Hall TC (2000) Transgene silencing in monocots. Plant Mol Biol 43:323–346

James C (2007) Executive Summary Global Status of Commercialized Biotech/GM Crops. ISAAA Brief 37

Janssen BJ, Gardener RC (1990) Localized transient expression of GUS in leaf discs following cocultivation with *Agrobacterium*. Plant Mol Biol 14:61–72

Jayaram M (1994) Phosphoryl transfer if FLP recombination: a template for strand transfer mechanisms. Trends Biotechnol 19:78–82

Jefferson RA (1987) Assaying chimeric genes in plants: the GUS gene fusion system. Plant Mol Biol Rep 5:387–405

Jiang L, Maoka T, Komori S, Fukamachi H, Kato H, Ogawa K (2004) An efficient method for sonication assisted *Agrobacterium*-mediated transformation of coat protein (CP) coding genes into papaya (*Carica papaya* L.). Acta Biol Exp Sin 37(3):189–198

Jin S, Song Y, Pan S, Nester EW (1993) Characterization of a *virG* mutation that confers constitutive virulence gene expression in *Agrobacterium tumefaciens*. Mol Microbiol 7(4):555–562

Jin S, Zhang X, Liang S, Nie Y, Guo X, Huang C (2005) Factors affecting transformation efficiency of embryogenic callus of Upland cotton (*Gossypium hirsutum*) with *Agrobacterium tumefaciens*. Plant Cell Tissue Organ Cult 81:229–237

Kerbach S, Lorz H, Becker D (2005) Site-specific recombination in *Zea mays*. Theor Appl Genet 111:1608–1616

Khan MS, Maliga P (1999) Fluorescent antibiotic resistance marker for tracking plastid transformation in higher plants. Nat Biotechnol 17:910–915

Kohli A, Leech M, Vain P, Laurie D, Christou P (1998) Transgene organization in rice engineered through direct DNA transfer supports a two-phase integration mechanism mediated by the establishment of integration hot spots. Proc Natl Acad Sci 95:7203–7208

Komari T, Hiei Y, Ishida Y, Kumashiro T, Kubo T (1998) Advances in cereal gene transfer. Curr Opin Plant Biol 1:161–165

Konagaya K, Ando S, Kamachi S, Tsuda M, Tabei Y (2008) Efficient production of genetically engineered, male-sterile *Arabidopsis thaliana* using anther-specific promoters and genes derived from *Brassica oleracea* and B. *rapa*. Plant Cell Rep 27:1741–1754

Kopertekh L, Juttner G, Schiemann J (2004) PVX-Cre-mediated marker gene elimination from transgenic plants. Plant Mol Biol 55:491–500

Krizkova L, Hrouda M (1998) Direct repeats of T-DNA integrated in tobacco chromosome: characterization of junction regions. Plant J 16:673–680

Kumar S, Sharma P, Pental D (1998) A genetic approach to in vitro regeneration of non regenerating cotton (*Gossypium hirsutum* L.) cultivars source. Plant Cell Rep 18:59–63

Kumria R, Sunnichan VG, Das DK, Gupta SK, Reddy VS, Bhatnagar RK, Leelavathi S (2003) High-frequency somatic embryo production and maturation into normal plants in cotton (*Gossypium hirsutum*) through metabolic stress. Plant Cell Rep 21:635–639

Kumria R, Leelavathi S, Bhatnagar RK, Reddy VS (2005) Regeneration and genetic transformation of cotton, present status and future perspectives. www.icgeb.trieste.it/RESEARCH/ND/Reddy.Htm

Lacroix B, Vaidya M, Tzfira T, Citovsky V (2005) The VirE3 protein of *Agrobacterium* mimics a host cell function required for plant genetic transformation. EMBO J 24:428–437

Leelavathi S, Sunnichan VG, Kumria R, Vijaykanth GP, Bhatnagar RK, Reddy VS (2004) A simple and rapid *Agrobacterium*-mediated transformation protocol for cotton (*Gossypium hirsutum* L.), embryogenic calli as a source to generate large numbers of transgenic plants. Plant Cell Rep 22:465–470

Leonhardt N, Kwak JM, Robert N, Waner D, Leonhardt G, Schroeder JI (2004) Microarray expression analyses of *Arabidopsis* guard cells and isolation of a recessive abscisic acid hypersensitive protein phosphatase 2 C mutant. Plant Cell 16:596–615

Li HQ, Sautter C, Potrykus I, Puonti-Kaerlas J (1996) Genetic transformation of cassava (*Manihot esculenta* Crantz). Nat Biotechnol 14:736–740

Luo H, Lyznik LA, Gidoni D, Hodges TK (2000) FLP-mediated recombination for use in hybrid plant production. Plant J 23:423–430

Lyon BR, Cousins YL, Llewell DJ, Dennis ES (1993) Cotton plants transformed with a bacterial degradation gene are protected from accidental spray drift damage by the herbicide 2,4-dichlorophenoxyacetic acid. Transgenic Res 2(3):162–169

Majeed A, Husnain T, Riazuddin S (2000) Transformation of virus resistant genotype of *Gossypium hirsutum* L., with pesticidal gene. Plant Biotechnol 17:105–110

Majumder P, Banerjee S, Das S (2004) Identification of receptors responsible for binding of the mannose specific lectin to the gut epithelial membrane of the target insects. Glycoconj J 20:525–530

Maqbool SB, Zhong H, El-Maghraby Y, Ahmad A, Chai B, Wang W, Sabzikar R, Sticklen MB (2002) Competence of oat (*Avena sativa* L.) shoot apical meristems for integrative transformation, inherited expression, and osmotic tolerance of transgenic lines containing *hva1*. Theor Appl Genet 105:201–208

Martinelli L, Mandolino G (2001) *Genetic transformation in Vitis*. In: Bajaj YPS (ed) Biotechnology in agriculture and forestry. Springer, Heidelberg, 47:325–338

Matzke AJ, Matzke MA (1998) Position effects and epigenetic silencing of plant transgenes. Curr Opin Plant Biol 1:142–148

McCabe DE, Martinell BJ (1993) Transformation of elite cotton cultivars via particle bombardment of meristems. Biotechnology 11:596–598

McStewart JD, Hsu CL (1977) *In ovulo* embryo culture and seedling development of cotton (*Gossypium hirsutum* L.). Planta 137:113–117

Merkle SA, Parrot WA, Flinn BS (1995) Morphogenic aspects of somatic embryogenesis. In: Thorpe TA (ed) In vitro embryogenesis in plants. Kluwer, Dordrecht, pp 155–203

Meyer P (2000) Transcriptional transgene silencing and chromatin components. Plant Mol Biol 43:221–234

Miki B, McHugh S (2004) Selectable marker genes in transgenic plants: applications, alternatives and biosafety. J Biotechnol 107:193–232

Millar AJ, Short SR, Chua NH, Kay SA (1992) A novel circadian phenotype based on firefly luciferase expression in transgenic plants. Plant Cell 4:1075–1087

Mishra R, Wang HY, Yadav NR, Wilkins TA (2003) Development of a highly regenerable elite Acala cotton (*Gossypium hirsutum* cv. Maxxa) a step towards genotype-independent regeneration. Plant Cell Tissue Organ Cult 73:21–35

Morel G, Martin C (1952) Guerison de dahlias atteints dune maladie a virus. C R Séance Acad Sci Paris 233:1324–1325

Murashige T, Skoog F (1962) A revised medium for rapid growth of and bioassays with tobacco tissue cultures. Physiol Plant 15:473–497

Mysore KS, Bassuner B, Deng XB, Darbinian NS, Motchoulski A (1998) Role of the *Agrobacterium tumefaciens* VirD2 protein in T-DNA transfer and integration. Mol Plant Microbe Interact 11:668–683

Nakazono M, Qiu F, Borsuk LA, Schnable PS (2003) Laser-capture microdissection, a tool for the global analysis of gene expression in specific plant cell types: identification of genes expressed differentially in epidermal cells or vascular tissues of maize. Plant Cell 15:583–596

Nam J, Matthysse AG, Gelvin SB (1997) Differences in susceptibility of *Arabidopsis* ecotypes to crown gall disease may result from a deficiency in T-DNA integration. Plant Cell 9:317–333

Narasimhulu SB, Deng XB, Sarria R, Gelvin SB (1996) Early transcription of *Agrobacterium* T-DNA genes in tobacco and maize. Plant Cell 8:873–886

Nawy T, Lee JY, Colinas J, Wang JY, Thongrod SC, Malamy JE, Birnbaum K, Benfey PN (2005) Transcriptional profile of the *Arabidopsis* root quiescent. Plant Cell 17:1908–1925

Peeters M, Willems K, Swennen R (1994) Protoplast to plant regeneration in cotton (*Gossypium hirsutum* L. cv. Coker 312) using feeder layers. Plant Cell Rep 13:208–211

Perlak FJ, Deaton RW, Armstrong TA, Fuchs RL, Sims SR, Greenplate JT, Fischhoff DA (1990) Insect resistant cotton plants. Biotechnology 8:939–943

Pina C, Pinto F, Feijo JA, Becker JD (2005) Gene family analysis of the *Arabidopsis* pollen transcriptome reveals biological implications for cell growth, division control, and gene expression regulation. Plant Physiol 138:744–756

Popelka SC, Altpeter F (2001) Interactions between genotypes and culture media components for improved in vitro response of rye (*Secale cereale* L.) inbred lines. Plant Cell Rep 20:575–582

Pray C, Oehmke FJ, Naseem A (2005) Innovation, dynamic efficiency in plant biotechnology: an introduction to the researchable issues. AgBioForum 8(2 and 3):1

Puchta H (1998) Towards targeted transformation in plants. Trends Plant Sci 3:77–87

Radhakrishnan P, Srivastava V (2005) Utility of the FLP-FRT recombination system for genetic manipulation of rice. Plant Cell Rep 23:721–726

Rajasekaran K, Grula JW, Hudspeth RL, Pofelis S, Anderson DM (1996) Herbicide-resistant *Acala* and *Coker* cottons transformed with a native gene encoding mutant forms of acetohydroxyacid synthase. Mol Breed 2:307–319

Rajasekaran K, Hudspeth RL, Cary JW, Anderson DM, Cleveland TE (2000) High-frequency stable transformation of cotton (*Gossypium hirsutum* L.) by particle bombardment of embryogenic cell suspension cultures. Plant Cell Rep 19:539–345

Rao KV, Rathore KS, Hodges TK, Fu X, Stoger E, Sudhakar S, Williams P, Christou P, Bharathi M, Bown DP, Powell KS, Spence J, Gatehouse A, Gatehouse JA (1998) Expression of snowdrop lectin (GNA) in transgenic plants confers resistance to rice brown plant hopper. Plant J 15:469–477

Rashid B, Husnain T, Riazuddin S (2004) In-vitro shoot tip culture of cotton (*Gossypium Hirsutum*). Pak J Bot 36(4):817–823

Rashid B, Saleem Z, Husnain T, Riazuddin S (2008) Transformation and Inheritance of Bt Genes in *Gossypium hirsutum*. J Plant Biol 51(4):248–254

Rossi L, Hohn B, Tinland B (1996) Integration of complete transferred DNA units is dependent on the activity of virulence E2 protein of *Agrobacterium tumifaciens*. Proc Natl Acad Sci 93:126–130

Rubenstein KD, Heisey PW (2005) Can technology transfer help public-sector researchers do more with less? The case of the USDA's agricultural research service. AgBioForum 8(2 and 3):10

Sandhu S, Altpeter F (2008) Co-integration, co-expression and inheritance of unlinked minimal transgene expression cassettes in an apomictic turf and forage grass (*Paspalum notatum* Flugge). Plant Cell Rep 27:1755–1765

Sanford JC, Klein TM, Wolf ED, Allen N (1987) Delivery of substances into cell tissues using a particle bombardment process. J Particle Sci Technol 5:27–37

Satyavathi V, Prasad V, Gita V, Lakshmi B, Lakshmi SG (2002) High efficiency transformation protocol for three Indian cotton varieties via *Agrobacterium tumefaciens*. Plant Sci 162:215–223

Scholz S, Lorz H, Lutticke S (2001) Transposition of the maize transposable element Ac in barley (*Hordeum vulgare* L.). Mol Gen Genet 264:653–661

Schrader J, Nilsson J, Mellerowicz E, Berglund A, Nilsson P, Hertzberg M, Sandberg G (2004) A high-resolution transcript profile across the wood-forming meristem of poplar identifies potential regulators of cambial stem cell identity. Plant Cell 16:2278–2292

Schrammeijer B, den Dulk-Ras A, Vergunst AC, Jurado JE, Hooykaas PJ (2003) Analysis of Vir protein translocation from *Agrobacterium tumefaciens* using *Saccharomyces cerevisiae* as a model: evidence for transport of a novel effector protein VirE3. Nucleic Acids Res 31:860–868

Scorza R, Cordts JM, Gray DJ, Gonsalves D, Emershad RL, Ramming DW (1996) Producing transgenic Thompson seedless grape (*Vitis vinifera* L.). J Am Soc Hort Sci 12:616–619

Shabde M, Murashige T (1977) Hormonal requirements of excised *Dianthus caryophyllus* L shoot apical meristem *in vitro*. Am J Bot 64:443–448

Shoemaker RC, Couche LJ, Galbraith DW (1986) Characterization of somatic embryogenesis and plant regeneration in cotton. (*Gossypium hirsutum* L). Plant Cell Rep 3:178–181

Shou HX, Frame BR, Whitham SA, Wang K (2004) Assessment of transgenic maize events produced by particle bombardment or *Agrobacterium*-mediated transformation. Mol Breed 13:201–208

Shuangxia J, Xianlong Z, Shaoguang I, Yichun N, Xiaoping G, Chao H (2005) Factors affecting transformation efficiency of embryogenic callus of Upland cotton (*Gossypium hirsutum*) with *Agrobacterium tumefaciens*. Plant Cell Tissue Organ Cult 81:229–237

Smith N, Kilpatrick J, Whitelam G (2001) Superfluous transgene integration in plants. Crit Rev Plant Sci 20:215–249

Solís JIF, Mlejnek P, Studená K, Procházka S (2003) Application of sonication-assisted *Agrobacterium*-mediated transformation in *Chenopodium rubrum* L. Plant Soil Environ 49(6):255–260

Southern EM (1975) Detection of specific sequence among DNA fragments separated by gel electrophoresis. J Mol Biol 98:503–517

Sreekala C, Wu L, Gu K, Wang D, Tian D, Yin Z (2005) Excision of a selectable marker in transgenic rice (*Oryza sativa* L.) using a chemically regulated Cre/loxP system. Plant Cell Rep 24:86–94

Sripriya R, Raghupathy V, Veluthambi K (2008) Generation of selectable marker-free sheath blight resistant transgenic rice plants by efficient co-transformation of a cointegrate vector T-DNA and a binary vector T-DNA in one *Agrobacterium tumefaciens* strain. Plant Cell Rep 27:1635–1644

Steinitz B, Gafni Y, Cohen Y, Perea DJ, Tabib Y, Levski S, Navon A (2002) Insecticidal activity of a cry1A(c) transgene in callus derived from regeneration-recalcitrant cotton (*Gossypium hirsutum* L.). In Vitro Cell Dev Biol Plant 38:247–251

Sunilkumar S, Rathore KS (2001) Transgenic cotton, factors influencing *Agrobacterium*-mediated transformation and regeneration. Mol Breed 8:137–152

Taylor NJ, Fauquet CM (2002) Microparticle bombardment as a tool in plant science and agricultural biotechnology. DNA Cell Biol 21:963–977

Taylor N, Masona M, Schopke C, Fauquet C (2002) Transgenic cassava for food security and economic development. In: Khachatourians G, McHughen A, Scorza R, Nip W-K, Hui Y (eds) Transgenic plants and crops. Mercel Dekker Inc., New York, pp 523–546

Thengane S, Paranjape SV, Khuspe SS, Mascarenhas AF (1986) Hybridization of *Gossypium* species through *in ovulo* embryo culture. Plant Cell Tissue Organ Cult 6:209–219

Thomas JC, Adams DG, Keppenne VD, Wasmann CC, Brown JK, Kanost MR, Bohnert HJ (1995) Protease inhibitors of *Manduca sexta* expressed in transgenic cotton. Plant Cell Rep 14:758–762

Tinland B (1996) The integration of T-DNA into plant genomes. Trends Plant Sci 1:178–184

Tinland B, Hohn B (1995) Recombination between prokaryotic and eukaryotic DNA: integration of *Agrobacterium tumifaciens* T-DNA into the plant genome. In: Setlow JK (ed) Genetic engineering. Plenum, New York

Tinland B, Schoumacher F, Gloeckler V, Bravo-Angel AM, Hohn B (1995) The *Agrobacterium tumifaciens* virulence D_2 protein is responsible for precise integration of T-DNA into the plant genome. EMBO J 14:3585–3595

Travella S, Ross SM, Harden J, Everett C, Snape JW, Harwood WA (2005) A comparison of transgenic barley lines produced by particle bombardment and *Agrobacterium*-mediated techniques. Plant Cell Rep 23:780–789

Trick HN, Finer JJ (1997) SAAT: sonication-assisted *Agrobacterium* mediated transformation. Trans Res 6:329–336

Trick HN, Finer JJ (1998) Sonication-assisted *Agrobacterium*-mediated transformation of soybean [*Glycine max* (L.) Merrill] embryogenic suspension culture tissue. Plant Cell Rep 17:482–488

Trolinder NL, Goodin JR (1988) Somatic embryogenesis in cotton (*Gossypium*). I. Effects of source of explant and hormone regime. Plant Cell Tissue Organ Cult 12:31–42

Tu J, Zhang G, Datta K, Xu C, He Y, Zhang Q, Khush GS, Datta SK (2000) Field performance of transgenic elite commercial hybrid rice expressing *Bacillus thuringiensis* -endotoxin. Nat Biotechnol 18:1101–1104

Tzfira T, Citovsky V (2000) From host recognition to T-DNA integration: the function of bacterial and plant genes in *Agrobacterium* and plant genes in *Agrobacterium*-plant cell interaction. Mol Plant Pathol 1:201–212

Tzfira T, Rhee Y, Chen MH, Kunik T, Citovsky V (2000) Nucleic acid transport in plant microbe interactions: the molecules that walk through the walls. Annu Rev Microbiol 54:187–219

Tzfira T, Vaidya M, Citovsky V (2001) VIP1, an Arabidopsis protein that interacts with Agrobacterium VirE2, is involved in VirE2 nuclear import and Agrobacterium infectivity. EMBO J 20:3596–3607

Umbeck P, Johnson G, Barton K, Swain W (1987) Genetically transformed cotton (*Gossypium hirsutum* L.) plants. Biotechnology 5:263–266

Vain P, Finer KR, Engler DE, Pratt RC, Finer JJ (1996) Intron-mediated enhancement of gene expression in maize (*Zea mays* L.) and bluegrass (*Poa pratensis* L.). Plant Cell Rep 15:489–494

Vance V, Vaucheret H (2001) RNA silencing in plants-defense and counter defense. Science 292:2277–2280

Vasudevan A, Oh TK, Park JS, Lakshmi SV, Choi BK, Kim SH, Lee HJ, Ji J, Kim JH, Ganapathi A, Kim SC, Choi CW (2008) Characterization of resistance mechanism in transgenic Nicotiana benthamiana containing turnip crinkle virus coat protein. Plant Cell Rep 27:1731–1740

Vergunst AC, Schrammeijer B, den Dulk-Ras A, de Vlaam CMT, Regensburg-Tuink TJ, Hooykaas PJJ (2000) VirB/D4-dependent protein translocation from *Agrobacterium* into plant cells. Science 290:979–982

Vergunst AC, van Lier MCM, den Dulk-Ras A, Hooykaas PJ (2003) Recognition of the Agrobacterium VirE2 translocation signal by the VirB/D_4 transport system does not require VirE1. Plant Physiol 133:978–988

Wang H, To K (2004) *Agrobacterium*-mediated transformation in the high-value medicinal plant Echinacea purpurea. Plant Sci 166:1087–1096

Wang M-B, Waterhouse P (2002) Application of gene silencing in plants. Curr Opin Plant Biol 5:146–150

Wilkins T, Rajasekaran K, Anderson DM (2000) Cotton biotechnology. Crit Rev Plant Sci 19:511–550

Wu J, Zhang X, Nie Y, Luo X (2005) High-efficiency transformation of *Gossypium hirsutum* embryogenic calli mediated by *Agrobacterium tumefaciens* and regeneration of insect-resistant plants. Plant Breed 124:142–146

Wu Y, Chen Y, Liang X, Wang X (2006) An experimental assessment of the factors influencing *Agrobacterium*-mediated transformation in tomato. Russ J Plant Phys 53(2):252–256

Wu Y, Liewellyn DJ, White R, Ruggiero K, Al-Ghazi ES, Dennis ES (2007) Laser capture microdissection and cDNA microarrays used to generate gene expression profiles of rapidly expanding fiber initials cells on surface of cotton ovules. Planta 226(6):1475–1490

Yu TT, Skinner DZ, Liang GH, Trick HN, Huang B, Uthukrishnan SM (2000) *Agrobacterium* -mediated transformation of creeping bentgrass using GFP as a reporter gene. Hereditas 133:229–233

Yuan ZC, Edlind MP, Liu P, Saenkham P, Banta LM, Wise AA, Ronzone E, Binns AN, Kerr K, Nester EW (2007) The plant signal salicylic acid shuts down expression of the *vir* regulon and activates quormone-quenching genes in *Agrobacterium.* Proc Natl Acad Sci 104:11790–11795

Yusibov VM, Steck TR, Gupta V, Gelvin SB (1994) Association of single stranded transferred DNA from *Agrobacterium tumefaciens* with tobacco cells. Proc Natl Acad Sci 91:2994–2998

Zambryski P (1989) *Agrobacterium*–plant cell DNA transfer. In: Berg DE, Howe MM (eds) Mobile DNA. American Society for Microbiology, Washington, DC, pp 309–333

Zapata C, Park SH, El-Zik KM, Smith RH (1999) Transformation of a Texas cotton cultivar by using *Agrobacterium* and the shoot apex. Theor Appl Genet 98:252–256

Zernova O, Zhong W, Zhang XH, Widholm J (2008) Tissue culture specificity of the tobacco ASA_2 promoter driving hpt as a selectable marker for soybean transformation selection. Plant Cell Rep 27:1705–1711

Zhang BH, Liu F, Yao CB (2000) Plant regeneration via somatic embryogenesis in cotton. Plant Cell Tissue Organ Cult 60:89–94

Zhang B-H, Liu F, Liu Z-H, Wang H-M, Yao C-B (2001) Effects of kanamycin on tissue culture and somatic embryogenesis in cotton. Plant Growth Regul 33:137–149

Zhang W, Subbarao S, Addae P, Shen A, Armstrong C, Peschke V, Gilbertson L (2003) Cre/lox-mediated marker gene excision in transgenic maize (*Zea mays* L) plants. Theor Appl Genet 107:1157–1168

Zhou HF, Zheng XM, Wei RX, Second G, Vaughan DA, Ge S (2008) Contrasting population genetic structure, gene Xow between *Oryza ruflpogon* and *Oryza nivara.* Theor Appl Genet 117:1181–1189

Zupan J, Zambryski P (1997) The *Agrobacterium* DNA transfer complex. Crit Rev Plant Sci 16:279–295

Web Sites

(http://www.apcoab.org/documents/bs_pub.pdf)

(http://www.opendemocracy.net/ecology-foodwithoutfrontiers/article_1817.jsp)

Chapter 11
Transcriptomics and Proteomics Analysis of Root Nodules of Model Legume Plants

Abdul Razaque Memon

Contents

Abstract Leguminosae is a large and economically important family of plants that, because of their capacity to fix atmospheric nitrogen in symbiosis with rhizobia, are essential components of agricultural ecosystems and an important source in the production of food, feed, forage and other compounds with strong industrial and commercial relevance. During the last decade, the 'omics' technologies of legumes especially the model legumes have provided and still provide unprecedented amount of molecular information that has to be understood in a physiological, developmental and organismal context. New technologies, for example high-throughput sequencing, high multiplexed mapping techniques and rapid development of new bioinformatics tools have provided new information about current model systems and emerging new

A.R. Memon (✉)
Department of Genetics and Bioengineering, International Burch University,
71000 Sarajevo, Bosnia and Herzegovina
e-mail: abdulrezzak.memon@gmail.com;armemon@ibu.edu.ba

M. Ashraf et al. (eds.), *Crop Production for Agricultural Improvement*,
DOI 10.1007/978-94-007-4116-4_11,

models as well. The latest information on the status of development of genomic resources and of related information in model legumes and some other economically important legume crops are summarized in this review. The infection of the plant root by rhizobia triggers several important events in the root cell, resulting in the formation of a nodule – a nitrogen-fixing compartment. Some of the signal molecules involved in the communication between the symbiotic partners have been studied but little is known about the genes and proteins involved in membrane and protein trafficking and targeting towards the symbiosome-the cellular compartment containing the nitrogen-fixing bacteria in the nodule, which is one of the most important processes in nodule formation and development. The central point of this review is to summarize the recent developments in functional genome, transcriptome and proteome of model legume plants specially *Medicago truncatula*, Lotus japonicus and *Glycine max*. A multidisciplinary approach, including plant and bacterial genetics, molecular biology, live cell imaging, biophysics and bioinformatics has been taken to understand the plant-microbe interaction and the biogenesis of root nodules and the genes and proteins involved in nodule formation and efficient nitrogen fixation. Vesicular trafficking plays an important role in rhizobia- root interaction, infection thread formation and the development of root nodules. I have described the role of several small GTP binding proteins in symbiosome formation and root nodule development. Furthermore, there is an increasing need to efficiently convert scientific results into practical applications or products. The economic and environmental costs of the heavy use of chemical N fertilizers in agriculture are a global concern. For this reason legumes could be used as an alternative source for N fertilizers and this would help for cleaning up the environmental pollution caused by chemical fertilizers. Additionally nitrogen-fixing biological systems represent an economically attractive and ecologically sound means of reducing external inputs and improving internal resources. I hope the scientific information given in this review paper could not only be useful to the molecular geneticists and plant breeders but will also assist the agronomists to carry out their research efficiently.

Keywords Model legume plants • Genome • Transcriptome • Proteome • Root nodule • Symbiosis • Small GTP-binding proteins

1 Introduction

The legumes comprise the third largest family of flowering plants with one of the highest economical value in agriculture. Several members of this family, including soybean, alfalfa, peas and beans, have been fundamental to the development of modern agricultural systems. On a worldwide basis, this plant family contributes 33% of humankind's protein intake, while also serving as an important source of fodder and forage for animals, and of edible and industrial oils. Legumes are also distinguished by their unique property of symbiotic nitrogen fixation, providing one of the major sources of available nitrogen in the biosphere (Singh and Jauhar 2009).

It is estimated, that between 40 million and 60 million tons of nitrogen are fixed annually by cultivated legumes (Smil 1999) equivalent to about US$10 billion fertilizer (Graham and Vance 2003). Grain legumes (common bean, pea, lentil, chickpea, and others) are the major, if not the only, source of dietary protein for a large proportion of the population in South Asia, Middle East, North Africa, and other impoverished regions of the world (Singh and Jauhar 2009). Turkey is one of the major producers of grain and fodder legumes. Among the legumes produced in Turkey, chickpeas, lentils and beans hold the most important place and account for 44%, 35% and 18% respectively of the total legume production in Turkey (Anonymus 2000). Legumes also play a positive role in crop rotations with cereals and help replenish soil's nitrogen supply. Turkey is also one of the major gene centers of the legume crops especially for forage legumes (alfalfa and clover). These forage crops are grown as pastures in almost all regions, particularly in eastern and central Anatolia (Anonymus 2000).

Alfalfa has one of the highest nutritional quality as animal feed and contains between 15% and 22% crude protein. It serves as an excellent source of vitamins and minerals. Specifically, alfalfa contains vitamins A, D, E, K, U, C, B1, B2, B6, B12, niacin, panthothanic acid, inositol, biotin, and folic acid. Alfalfa is also high in minerals phosphorus, calcium, potassium, sodium, chlorine, sulfur, magnesium, copper, manganese, iron, cobalt, boron, and molybdenum, as well as trace elements such as nickel, lead, strontium, and palladium (Bickoff et al. 1972). The alfalfa sprouts are directly consumed by humans and its juice is also used in some health food products. Additionally alfalfa and other legumes are reported to be used as a biofuel for the production of ethanol and biodiesel. Alfalfa, clover and broad bean are found to be good candidates for the bioremediation of soils contaminated with high levels of organic compound especially petrol hydrocarbons (Cigdem et al. 2011). Alfalfa is also used for the production of industrial enzymes such as lignin peroxidase, alpha-amylase, cellulase, and phytase (Carroll and Somerville 2009).

2 Legume Genome Sequencing Strategies and Status

Soybean (*Glycine max*), bean (*Phaseolus vulgaris*), pea (*Pisum sativum*) and lentil (*Lens culinaris*) are the most important crop legumes in terms of food nutrition. They are difficult to study because of their large genome and low transformation capacities. The model legume species, *Lotus japonicus* and *Medicago truncatula*, are the accepted model legumes in which root-bacterial interaction, infection process, nodule formation and development and the genes and proteins involved in these processes are widely studied. Both plants have a small diploid genome (470–550 Mp) and provide a rich source of genetic and genomic information which could be utilized in understanding the genomic structure of other important legume crops. These plants have a short life cycle, are self-fertile and can be transformed easily (Cook 1999). *M. truncatula* is closely related to other temperate legumes, such as clover, pea, vetch, chickpea and lentil. *L japonicus* is more distantly related to tropical legumes, such as soybean and bean (*Phaseolid clade*) (Ane et al. 2008). The Phaseoloids are the other major taxonomic group that contains many economically important legume crops including soybean, the legume of the greatest economic importance, for which extensive genomic information is available and is part of a genomic sequencing project (see http://soybase.agron.iastate.edu/). Soybean is proposed as a third model legume in addition to *M. truncatula* and *L. japonicus* because of its economic importance and the phylogenetic proximity to other major crops (Mathesius et al. 2008). In contrast to soybean, the development of the model legume systems in *Lotus* and *Medicago* were given more importance because of their small genome size and relatively easy to transform. *Medicago truncatula* and *Lotus japonicus* are also amenable to efficient molecular, genetic and reverse-genetic analyses, unlike the major crop legumes. This was driven by the discovery of several mutants that are defective in both mycorrhizal and rhizobial symbiosis (Duc et al. 1989; Marsh and Schultze 2001), so that the use of the genetic information from these model species facilitated the genomic sequence project of other economically important grain legumes. In fact this was the objective of the EU framework 6 funded project on ‘Grain Legumes’ (www.eugrainlegumes.org).

The *Medicago truncatula* sequencing project was first initiated by Samuel Roberts Noble Foundation affiliated with University of Oklahoma. The National Science Foundation and the European Union’s Sixth Framework Programme provided funding in year 2003 to complete sequencing of the remaining euchromatic genespace. Six chromosomes are being sequenced by NSF project “Sequencing the Gene Space of the Model Legume, *Medicago truncatula*,” and two are being sequenced by partners in Europe (Ane et al. 2008).

This project used a clone-by clone approach, in which BACs, with average insert size of approximately 120 kb, were sequenced and used to extend BAC-contig tiling paths to produce increasingly large sequence contigs. Contigs are anchor oriented using genetic markers developed from a large proportion of the BAC sequences. These BAC contigs and sequences cover approximately 60% of the major euchromatic regions of the *M. truncatula* genome (Young et al. 2003; Cannon et al. 2005; Young and Udvardi 2009).

A combination of cytogenetic and BAC sequence data show that the *M. truncatula* genome is organized into distinct gene-rich euchromatin separate from repeat-rich pericentromeric regions. This map-anchored, high-quality sequence is extremely valuable as a basis for genomic comparisons with other plant genomes, and laid a foundation for improving many crop and forage legumes. Nevin Young (University of Minnesota), Bruce Roe (ACGT, University of Oklahoma; (chromosomes 1, 4, 6, 8), and Chris Town (JCVI; chromosomes 2, 7) are principal investigators of the U.S. project. In Europe, collaborators include Giles Oldroyd (John Innes Center, UK.) coordinating sequencing of chromosome 3 at the Sanger Center, and Frederic Deballe (INRA-CNRS) coordinating sequencing of chromosome 5 at Genoscope, France. Genetic markers and maps were developed by collaboration between the University of California-Davis, the University of Minnesota, INRA-CNRS Toulouse France, and the Biological Research Center Institute of Genetics in Szeged Hungary. Cytogenetic analysis is underway at Wageningen University. Genome annotation is

Table 11.1 Web based genomic resources of Medicago and related legume species

Medicago genome sequence sites	
Medicago Genome Sequence Resources	http://www.medicago.org
TIGR *Medicago* Database	http://www.tigr.org/tdb/e2k1/mta1/
University of Oklahoma Sequencing Project	http://www.genome.ou.edu/medicago.html
European *Medicago* Genome Database	http://mips.gsf.de/proj/plant/jsf/medi/index.jsp
Medicago functional genomics sites	
Noble Foundation *Medicago* Gene Expression Atlas	http://bioinfo.noble.org/gene-atlas/
Noble Foundation Medicago Mutant Database	http://bioinfo4.noble.org/mutant/
Noble Foundation Medicago Metabolic Pathways	http://mediccyc.noble.org/
Medicago Handbook	http://www.noble.org/MedicagoHandbook/
Comparative legume genomics sites	
Legume Information System (LIS)	http://www.comparative-legumes.org/
INRA/CNRS Integrative Legume Resources	http://www.legoo.org/
Legume Genome Website Links at LIS	http://www.comparative-legumes.org/lis/lis_links.html
Other legume genome sites	
Lotus japonicus Genome Browser	http://www.kazusa.or.jp/lotus/
Lotus japonicus Legume Base	http://www.shigen.nig.ac.jp/bean/lotusjaponicus/top/top.jsp
Phytozome *Glycine max* Genome	http://www.phytozome.net/soybean
	http://www.soybeangenome.org/
	http://rsoy.psc.riken.jp/
	http://compbio.dfci.harvard.edu/tgi/cgi-bin/tgi/gimain.pl?gudb=soybean
Soybase and Soybean Breeders Toolbox	http://soybase.org/
Phaseolus vulgaris (common bean)	http://phaseolus.genomics.purdue.edu/
	http://www.ccg.unam.mx/phaseolusest/
	http://www.phaseolus.net/
	http://beangenes.cws.ndsu.nodak.edu/
Pisum sativum (pea)	http://urgv.evry.inra.fr/UTILLdb
Vigna unguiculata (cowpea)	http://cowpeagenomics.med.virginia.edu/CGKB/
Arachis hypogaea (peanut)	http://peanutgenetics.tamu.edu/cmap
Trifolium pretense (red clover)	http://clovergarden.jp
Tree of life: Fabaceae	http://www.tolweb.org/Fabaceae
International Legume Database	http://www.ildis.org/

being carried out by the International *Medicago* Genome Annotation Group (IMGAG), which involves participants from INRA-CNRS, JCVI/TIGR, NCGR, MIPS, MPIZ, UMN and VIB-Gent. A single uniform set of annotations (with periodic updates) of the gene-rich pseudomolecules, the unanchored BACs and the Illumina assemblies not captured by the BAC-based project are hosted at JCVI and MIPS (www.medicago.org/genome, see also Table 11.1).

Now the international *Medicago truncatuala* (*Mt*) genome sequencing consortium is in its final phase, with ~2,600 unique BACs has already been sequenced. Assembly of these BACs, guided by an optical map of the entire genome, resulted in 8 pseudomolecules spanning ~375 Mb and containing ~250 Mb of sequence, plus ~150 sequenced BACs that do not have chromosomal assignments. Annotation of the unmasked pseudomolecules and orphan BACs resulted in the prediction of around 58,000 gene models of which ~11,500 appear to be transposon-related. Overall, the about 35% of *Medicago euchromatin* appears to be transposon-related. Assembly of Illumina whole genome sequence generated by the Hap-Map project produced an additional ~55 Mb of unique sequence containing ~20,000 gene models (www.medicagohapmap.org/genome). Whole genome comparisons reveal a high degree of synteny with Lotus and soybean, but duplication blocks within the *Medicago* genome are relatively short (~500 kb). Analysis of duplicated genes and expression patterns suggests that a genome duplication early in the 'papilionoid' legume group (containing e.g. soybean, *Medicago*, and peanut) was important for the evolution of the symbiotic nitrogen-fixing nodule in this group of legumes; but also that an even earlier genome duplication may also have played a role in evolution of this capacity (Town 2006; Verdier et al. 2008).

Whole-genome association studies (GWAS) in the model legume *Medicago truncatula* will enable the discovery of genes playing a role in the evolution of symbiosis. To achieve this goal, The International *Medicago truncatula* HAPMAP Consortium, including participants at the University of Minnesota, the National Center for Genome Resources, Boyce Thompson Institute, Hamline University, the University of Southern California, the Noble Foundation, INRA-Montpellier, ENSAT-Toulouse, and INRST-Tunisia, is developing a hapmap GWAS resource for *Medicago*. More than 300 inbred lines spanning the range of *Medicago* diversity are being resequenced using Illumina next generation technology (http://www.jcvi.org/cgi-bin/medicago/overview.cgi).

The *Medicago* genome is highly syntenic with alfalfa and pea and moderately syntenic with soybean. A project funded by the Noble Foundation and the University of Oklahoma is now sequencing 1,000 BAC clones, providing 100 Mbp of sequence and seed points for a minimum tiling path (George et al. 2008; Hougaard et al. 2008; Umezawa et al. 2008). Currently huge efforts have been taken by genetists and plant molecular biologists to develop resources in order to decipher the functions of many *Medicago* genes. These resources include a gene expression atlas/database that can be used to obtain developmental and other types of expression data for most *Medicago* genes, and two large mutant populations, a fast-neutron-bombardment-deletion and a Tnt1-insertion population, which can be used for both forward and high-throughput reverse genetics (Benedito et al. 2008; He et al. 2009). About 15,000 Tnt1 insertion lines have been generated in the R108 genotype and phenotypic screens found 177 of these lines to be defective in SNF. Flanking sequence tag (FST) analysis and a PCR-based approach have identified insertion alleles of the following known symbiotic genes amongst the mutants: DMI1, DMI2, DMI3, NSP1, NSP2, ERN1, NIN, LYK3, SUNN, SST1, HMGR1, SST1, IRE, and EIN2. High-throughput 454 sequencing using a 2D-pooling strategy is being used to accelerate current FST sequencing efforts, which will lead to a comprehensive FST database for all Tnt1 insertion lines. The mutant resources and gene expression atlas have been used to identify novel symbiotic genes, including VAPYRIN, which is required for intracellular accommodation of rhizobia in nodules (Murray et al. 2011).

The *Lotus japonicus* (*Lj*) genome sequencing project is being carried out by the Kazusa DNA Research Institute in Japan (http://www.kazusa.or.jp). This project is also primarily using a clone-by-clone approach, sequencing transformation competent artificial bacterial chromosomes (TACs), with average insert size of approximately 100 kb. The clone-by-clone sequence is also being augmented by a combination of whole genome shotgun (WGS) and low coverage TAC sequencing. The *Lj* sequence coverage has spanned approximately 177 Mbp, or roughly 60% of the euchromatic regions of the *Lj* genome (Young et al. 2005; Cannon et al. 2006).

The *Glycine max* (*Gm*) genome is being sequenced primarily with a Whole Genome Shotgun (WGS) approach, with sequence coming from a combination of random reads, paired fosmid ends, and paired BAC end sequences (http://www.plantgdb.org/GmGDB). This project was initiated by combined effort of the U.S. Department of Energy's Joint Genome Institute (DOE_JGI Community Sequence Program) and the NSF and USDA-ARS (managing mapping and physical mapping components of the project) (http://www.phytozome.net/soybean, Umezawa et al. 2008). The sequence consortium was led by Gary Stacey, Randy Shoemaker, Scott Jackson, Jeremy Schmutz, and Dan Rokhsar. Genome size is around 975 Mb and is captured in 20 chromosomes with a small additional amount of mostly repetitive sequence in unmapped scaffolds. Around 46,430 protein-coding genes have been predicated (Schmutz et al. 2010). The soybean genome is one of the largest high quality draft- whole-genome shotgun sequenced plant genome and will be having a remarkable impact on the genomic research of other economically important legume crops. After more than a decade-long effort, the genome sequences of these three legume species, *Medicago truncatula, Lotus japonicus*, and *Glycine max* (Soybean), seems to be complete. The genome sequences of these three species can be compared with each other and are likely to be predictive of other agriculturally important legume crops (Sato et al. 2008). These three model legumes are all part of the Papilionoideae subfamily of Leguminosae, and their genes are found in large syntenic regions, which should facilitate positional cloning of genes from other related legumes. The impact of these assembled, annotated genomes from all three model plants will be enormous in crop biotechnology. New technologies, for example high-throughput sequencing, high multiplexed mapping techniques and rapid development of new bioinformatics tools will glean new information about current model systems and emerging new models as well (Doyle and Luckow 2003).

The information obtained from genomic data of all three model legumes should be useful for the understanding of traits and the isolation of corresponding genes in target crop species.

3 Genome Resources in Model Legumes

The sequence information of *Medicago truncatula* (as a reference species for legume genomics) will revolutionize legume research, just as the *Arabidopsis* and rice genome sequences have transformed plant research. Groups interested in positional cloning will quickly begin to discover the genes responsible for fundamental biological processes like symbiosis and nitrogen fixation – and this capability will be available to people working both on *Medicago* and related legume species. The researchers working in proteomics will get high quality, full-length sequences as a basis for identifying peptide fragments. Studies in functional genomics will be based on a more solid foundation, where the results of microarray and gene-knockout experiments will be informed by complete sequences for members of relevant gene families. The molecular breeders will be able to discover candidate genes for the QTLs they have mapped in *Medicago* or other legumes and plant pathologists can explore the entire genetic complement of resistance genes and disease response pathways.

This genome sequence of model legume plants opens the door to the crop improvement for other important legumes used for human and animal food production and more recently for biofuel production. As genome sequencing begins in other crop legumes, constructing sequence assemblies will be far more efficient (as rice has aided in the sequencing of maize) (Sato et al. 2008; Young and Udvardi 2009). The latest information on the status of development of genomic resources and of related information in model legumes and some other economically important species are given in Table 11.1.

4 Transcriptomic Analysis

The combination of high-throughput transcript profiling and next-generation sequencing technologies is a prerequisite for genome-wide comprehensive transcriptome analysis. A growing array of genomic tools is being developed for *M. truncatula*. This great interest in transcriptome studies in *Medicago truncatula* is evidenced by the generation and sequencing of more than 70 cDNA libraries, in total yielding more than 250,000 expressed sequence tags (ESTs) deposited in Genbank (Quackenbush et al. 2001). Parallel to the generation of EST data, thousands of oligonucleotide microarrays were hybridized with targets from different biological conditions (Kuster et al. 2007), using layouts such as Mt16kOLI1 (Hohnjec et al. 2005) and Mt16kOLI1Plus (Thompson et al. 2005). The GeneChip® *Medicago* Genome Array was specifically designed to monitor gene expression in *Medicago truncatula, M. sativa*, and the symbiotic organism *Sinorhizobium meliloti*. The *Medicago* Genome Array is particularly useful for molecular biologists and agronomists studying legume genomics and symbiotic relationships between nitrogen fixing bacteria and plants. Affymetrix GeneChip array was designed from the sequence information collected from data sources including the TIGR *M. truncatula* gene index (The Institute for Genomic Research), gene predictions from the International *Medicago* Genome Annotation Group (IMGAG), gene predictions from the *S. meliloti* genome, and *M. sativa* EST information made available by TIGR. This array contains over 61,200 probe sets: 32,167 *M. truncatula* EST/mRNA-based and chloroplast gene-based probe sets; 18,733 *M. truncatula* IMGAG and phase 2/3 BAC prediction-based probe sets; 1,896 *M. sativa* EST/mRNA-based probe sets; and 8,305 *S. meliloti* gene prediction-based probe sets (Frickey et al. 2008) (Affymetrix GeneChip@ *Medicago* Genome array).

This genechip forms the basis for a recently developed gene expression atlas (Benedito et al. 2008; He et al. 2009). For reverse genetics, *M. truncatula* has large and characterized Tnt1 (Tadege et al. 2008; Rogers et al. 2009), fast neutron bombardment (FNB) (http://bioinfo.noble.org/), and targeting-induced local lesions in genomes (TILLING) (Thompson et al. 2005) populations available, as well as RNA interference (RNAi)-based gene silencing (Limpens et al. 2004; Cannon et al. 2009). Recently Affymetrix *Medicago* GeneChips® are focusing on analysis of *Medicago* transcriptomics with precise accuracy. It is because of the availability of wide variety of genomic tools which can be used to do better comparison of gene expression data from a multitude of conditions leading to more accurate results (Benedito et al. 2008; Cannon et al. 2009; Young and Udvardi 2009).

The various sequence and expression datasets are being stored by different organizations which could be downloaded and can be used for further analysis. For example EST libraries around over 100 species have been clustered and assembled at J. Craig Ventor Institute which can be utilized for the identification of species specific gene expression (Quackenbush et al. 2001). *Medicago truncatula* GeneIndex 10.0 are now available at the Dana-Farber Cancer Institute (DFCI) (http://www.dana-faber.org) for further use. These Gene indices allow to relate EST data to the biological conditions which were used to generate cDNA libraries. In addition several statistical methods were developed to assess the differential expression of the gene under a given condition (Stekel et al. 2000; Journet et al. 2002). A range of other different databases such as GEO (Edgar et al. 2002; Barrett et al. 2007), Arrayexpress (Parkinson et al. 2007), PEPR (Chen et al. 2004) the Stanford MicroArray Database (Sherlock et al. 2001), and MedicagoPLEX Experiment Data (PLEXdb) (Wise et al. 2007) store microarray and GeneChip® expression data, offering biologists public access to results from transcriptomics experiments (Henckel and Arnaud 2010). PLEXdb is a MIAME/Plant-compliant database which serves as a public repository for raw and normalized expression data. It also provides annotation for the Affymetrix plant microarrays and for other important microarrays. A popular resource for expression profiles relying on *Medicago* GeneChips® has been developed (Benedito et al. 2008; He et al. 2009). This comprehensive gene expression atlas for *Medicago truncatula* is a rich of information for agronomists and biologists and enables a large scale comparison between the transcriptomes of different plant species. In case of the model legume *Medicago truncatula*, the TRUNCATULIX data warehouse currently integrates five different sequence databases as well as oligonucleotide microarray and GeneChip® expression experiments from different source databases (Henckel et al. 2009). The user can quickly scan the complete database for the expression of genes of interest, but downstream analyses of expression data cannot be performed inside the warehouse. In order to eliminated this problem (lack of an integrated expression analysis), a new analysis tool MediPlEx (*MEDIcago truncatula* multiPLe EXpression analysis) was developed (Henckel and Arnaud 2010). It offers an integrated analysis pipeline for different sets of expression data generated for *Medicago truncatula* (http://www.cebitec.uni-bielefeld.de/mediplex). These resources are complemented by an extensive array of websites and bioinformatic tools (see Table 11.1).

5 Nodule Specific-Transcriptome

It has been shown that, during the nodulation process, there are major changes in plant gene expression (Hogslund et al. 2009). Several hundreds of genes (nodule-specific transcriptome) have been identified that are specifically and strongly up- or down-regulated during the nodulation process. The large majority of these genes are activated at later stages of organogenesis as deduced from the observation that mutants which are affected in the early signaling between the symbiotic partners fail to activate this "nodule-specific transcriptome" (El Yahyaoui et al. 2004). However, the link between the transcriptome activation and the key events of nodule formation – the establishment of a primordium, plant cell infection and cell differentiation of host and endosymbiont – remains unknown. The nodule-specific transcriptome involving several hundred of genes (El Yahyaoui et al. 2004; Benedito et al. 2008) exhibit three major characteristics. First, the repression of plant defense-related genes is probably necessary to avoid repulsion of the infecting rhizobia. Rhizobia possess "microorganism-associated molecular patterns" or MAMPs which are capable of eliciting the innate immune responses of the host plant whereas other rhizobial effectors, notably surface polysaccharides, are thought to attenuate or suppress this defense reaction during infection of root tissues for nodulation (Jones et al. 2007). Second, a transient activation of cell cycle and protein synthesis genes at the incipient stage of nodule development reflects an increase in cell division at this stage of nodule organogenesis. Third, the activation of the secretory pathway and a large number of secretory proteins and/or peptides throughout organogenesis can be understood in the context of a major activity of symbiotic cells in the hosting and maintenance of the symbiosomes. The secretory pathway ensures the correct trafficking of transmembrane or soluble proteins in different compartments of the root cells including in infection threads and symbiosomes of the symbiotic nodule cells. Recent studies with *Medicago truncatula* shows that during infection process secretory proteins are transported to the symbiosome, symbiosome lumen or the bacteroids (Maunoury et al. 2010).

6 Nodule Proteome

Transcriptome approach using microarray and gene expression analysis by real time PCR are useful tools in finding out the abundance of organ or nodule specific transcripts; however it is still a questionable to correlate mRNA abundance with protein levels (Gygi et al. 1999). In contrary, proteomics give relatively a direct assessment of biochemical processes by observing or identifying the actual proteins performing the regulatory and structural functions encoded by the genome and transcriptome. The proteomics studies does not rely on the availability of microarrays, or on extensive data regarding the genomes of interest in that tandem mass spectrometry (MS/MS) allows de novo sequencing and protein identification by homology search. By directly targeting proteins that are key effectors of biotrophic interactions, proteome analysis may thus unravel some actors of the symbiotic molecular dialog, including those of bacterial or fungal origin. Proteomic approaches have been successfully employed to investigate many different legume species including soybean (Komatsu and Ahsan 2009), *M. truncatula* (Bestel-Corre et al. 2002; Watson et al. 2003; Imin et al. 2004; Lei et al. 2005; Colditz and Braun 2010), alfalfa (Watson et al. 2003; Incamps et al. 2005), lupin (*Lupinus albus*) (Brambilla et al. 2009; Tian et al. 2009), *L. japonicus* (Dam et al. 2009), chickpea (*Cicer arietinum*) (Bhushan et al. 2007), pea (*Pisum sativum*) (Wienkoop and Saalbach 2003; Bourgeois et al. 2009), common bean (*Phaseolus vulgaris*) (Lei et al. 2011), and white clover (*Trifolium repens*) (Wilson et al. 2002).

Proteomic studies of the plant symbiosome membrane have shed light on nodule development. A novel 53-kDa late nodulin was isolated from the symbiosome membrane of soybean nodules and found to be regulated by *B. japonicum* inoculation (Winzer et al. 1999; Kistner et al. 2005). Two other known nodulins were identified from the most studied part of symbiosome membrane, the peribacteroid membrane (Panter et al. 2000). In a time-course analysis of protein expression in *M. truncatula* roots upon exposure to *Sinorhizobium meliloti*, two symbiosis related proteins were identified. One is leghemoglobin MtLb2, involved in maintaining oxygen levels, crucial for nitrogenase activity required for nitrogen fixation, and the other is enolase involved in carbon metabolism (Bestel-Corre et al. 2002). The symbiosome membrane protein profile from the model system *Medicago truncatula* and the corresponding bacterium *Sinorhizobium meliloti* was examined using two-dimensional electrophoresis and microcapillary high-performance liquid chromatography (HPLC) tandem mass spectrometry. The identities of 51 proteins were obtained and these proteins were categorized into functional classes to indicate biochemical roles (Catalano et al. 2004). Symbiosome membrane proteins include an H^+-ATPase, ENOD16, ENOD8, nodulin-25, BiP, HSP70, PDI, multifunctional aquaporin, a putative syntaxin, and other proteins of known and unknown identity and function.

Root proteome reference maps have been established for *Medicago truncatula* (Watson et al. 2003; Mathesius et al. 2008) identifying 179 and 40 proteins, respectively; 74 proteins were identified from white lupin; 532 proteins were identified from several organs of soybean (Sakata et al. 2009), 1492 proteins were identified from soybean root hairs (Brechenmacher et al. 2010). In addition, more than 100 proteins excreted from the root tip and root border cells from pea (Wen et al. 2007). A study showed that around 33 proteins were differentially accumulated in meristematic and non-meristematic cells of *M. truncatula* root cells including proteins involved in carbon metabolism, defense and stress response and flavonoid metabolism (Holmes et al. 2008). Most of the previous legume proteomics studies have used the National Center for Biotechnology Information non-redundant (NCBI nr) protein database for protein queries and identifications. The NCBI nr protein database is a very comprehensive and large protein database, consisting of amino acid sequences translated from coding genome sequences from a vast number of different organisms and species including human, animals, plants and microorganisms.

In this data base (NCBI nr protein database), a large number of experimentally determined tentative consensus sequences and expressed sequence tags (i.e., TCs and ESTs) have not been included. Thus, it will be highly beneficial for the legume community to assemble a legume specific protein database that contained a comprehensive compilation of legume sequences. Due to that, a legume specific protein database (LegProt) has been created containing sequences from seven legume species, i.e., *Glycine max, Lotus japonicus, Medicago sativa, Medicago truncatula, Lupinus albus, Phaseolus vulgaris,* and *Pisum sativum* (Lei et al. 2011). The database consists of amino acid sequences translated from predicted gene models and 6-frame translations of tentative consensus (TC) sequences assembled from expressed sequence tags (ESTs) and singleton ESTs. This database (LegProt) significantly increases legume protein identification success rates and the confidence levels compared to the commonly used NCBI nr. These improvements are primarily due to the presence of a large number of legume specific TC sequences in the LegProt database that were not found in NCBI nr and will serve as a valuable resource for legume proteomics (http://bioinfo.noble.org/manuscript-support/legumedb).

6.1 *The Rhizobium – Legume Interaction as a Model for Plant Cell Biology*

Legumes are the key for sustainable agriculture because of their role in elemental nitrogen fixation through a symbiotic association with rhizobia in the roots and make renewable source of nitrogen available to crop plants. The symbiosis between legume plants and soil-living rhizobia provides the largest amount of organic nitrogen in the global nitrogen cycle. The infection of rhizobia to the plant root triggers several important events in the root cell, which finally ends up in a nodule formation – a nitrogen-fixing compartment. Some of the signal molecules like Nod-factor and genes related to the nod factor synthesis, thread and peribacteroid formation have been studied but little is known about the genes and proteins involved in membrane and protein trafficking and targeting in nodule which is one of the most important

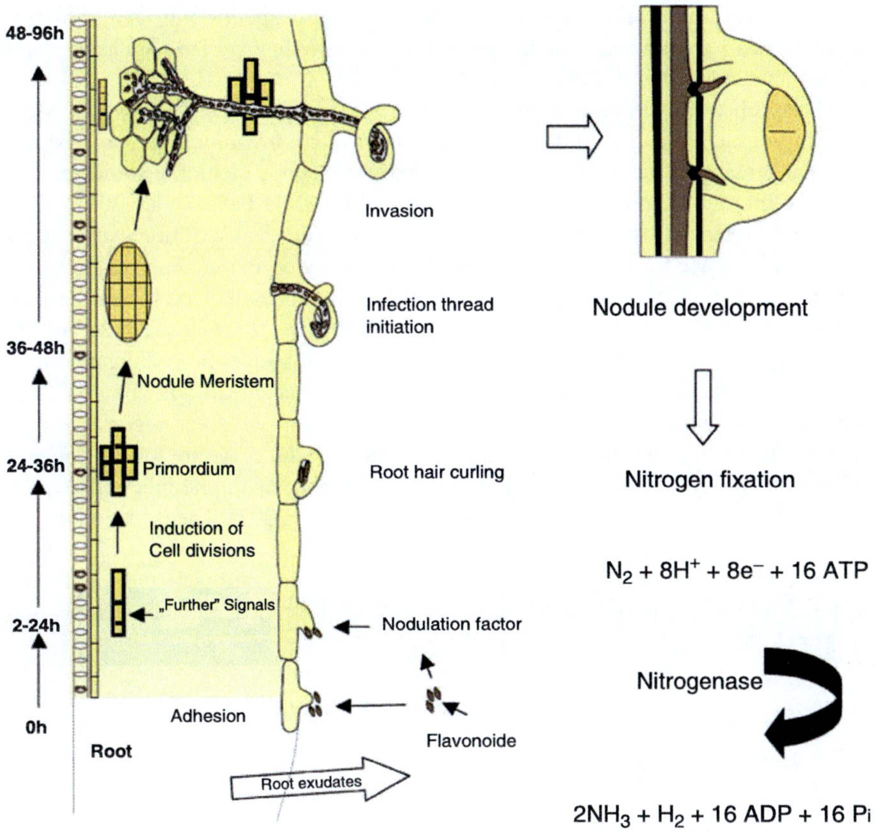

Fig. 11.1 Establishment of the *Rhizobium* – Legume symbiosis. A schematic diagram depict the infection process in early stages of the symbiosis (0–96 h) between *Medicago truncatula* roots and *Sinorhizobium meliloti* (FP7-PEAPOLE-2007-1-ITN; Niehaus, personal communication)

processes in nodule formation and development (Schultze and Kondorosi 1998). Here I will describe recent advances in understanding the molecular mechanisms involved in membrane and protein delivery to the symbiosomes and its effect on nodule development in two legume model systems (*Medicago truncatula* and *Lotus japonicus*).

The symbiosis between soil bacteria of the genus *Rhizobium* and leguminous plants is of outstanding interest to sustainable agriculture and builds one of the best-studied models to understand the molecular biology of plant-microbe interactions (Graham and Vance 2003). During the establishment of this symbiosis plant roots get infected by the microbe leading to cell division and the development of root nodules, specific organs build by the plant and colonized by the invading bacteria (Fig. 11.1). The rhizobia are released into the plant cells, surrounded by a membrane of plant origin, the so-called peribacteroid membrane. This membrane represents an "interface" between the engulfed bacteria and the plant. In a short time these bacteria differentiate into bacteroids and fix atmospheric nitrogen, which is supplied as a nitrogen source to the plant and in return they are fed with assimilates by the plant (Hirsch and Oldroyd 2009).

The symbiotic interaction between legume plant for example *Medicago truncatula* (barrel medic) and their nodulating symbiont, *Shinorhizobium meliloti*, begins with the sensing of flavonoids and some other compounds including some specific proteins in root exudates by bacteria (Fig. 11.1) (Dela-Pena et al. 2008). The flavonoids released by host plants activates *S. meliloti* transcriptional regulators which then induce the expression of about several bacterial *nod* genes required for the biosynthesis of the lipooligosaccharide signaling molecule. Nod factor is essential for nodulation (Kondorosi et al. 1991; Geurts and Bisseling 2002) and its specificity related with each rhizobium strain and plant species is well documented. Rhizobia enters into root cortical cells via an infection thread which starts from infected root hair and then traverses several cell layers to deliver bacteria to root cells in the developing nodule(Gage 2004; Oldroyd and Downie 2004). Bacteria releases from infection thread and enter the cytoplasm of the nodule cells which differentiate into nitrogen-fixing bacteroids (Pessi et al. 2007). The infection thread is a bacteria filled invagination of the root cell wall and its underlying plasma membrane and is formed through fusion of vesicles containing cell wall and membrane material which move in an actin and microtubule dependent manner (Crespi and Frugier 2008).

6.2 Establishment of the Symbiosis

Rhizobia are chemotactically attracted by root exudates and colonize the host plants rhizosphere. Phenolic signal molecules (*nod*-inducers) present in the exudate activate the transcription of the rhizobial *nod* genes. As a result, host-specific lipochitooligosaccharide signal molecules, called Nod-factors, are synthesized by the bacterial symbiont (Schultze and Kondorosi 1998). Rhizobia attach to emerging root hairs – possibly with the help of lectins – forming microcolonies. As a consequence of Nod-factor activity the root tips coil entrapping the rhizobial colony within pockets formed by the walls of the curled root hairs. The hair walls adjacent to the microcolonies are degraded and the rhizobia invade the underlying periplasmic space of the root hair (Fig. 11.1). Plasma membrane derived infection threads, originating from curled root hairs, act as a trigger for rhizobia to enter the root tissues and eventually progress toward the root cortex where a nodule primordium has been initiated (Hirsch and Oldroyd 2009). The release of rhizobia from infection threads into the cytosol of a nodule primordial cells and subsequent cellular localization of bacteria lead to the formation of fully functional nitrogen fixing nodules. Large numbers of symbiosomes are formed and localized in each nodule, which means that enormous amounts of membrane material and proteins must be delivered to the symbiosome membrane for its development and efficient function (Ivanov et al. 2010).

Flavonoides are produced by host roots and Nod factor is synthesized by rhizobia. After nodule development N_2 is fixed in the nodule by the help of nitrogenase enzyme. The establishment of the *Rhizobium* – Legume symbiosis represents an example for a very intimate interaction of two organisms (Fig. 11.1). The cell biology and physiology of both partners is tightly interconnected. The ability of nodule cells to control protein trafficking and targeting to the symbiosome is essential for nodule development and function (Ivanov et al. 2010). The mechanisms by which proteins are targeted to this novel sub-cellular compartment to accommodate its demand for membrane proteins are largely unknown.

Protein targeting to the symbiosome must rely on the plant secretory system, however, very little is known about protein trafficking and targeting in symbiosomes and root nodules. In mammalian and yeast cells, proteins that are destined for the plasma membrane or endosomal organelles travel through the secretory system and are targeted to their final destination in cargo vesicles. Each target organelle in the secretory system must maintain unique biochemical properties for accurate targeting of the cargo vesicles from the *trans* Golgi (Bonifacino and Glick 2004; Gillingham and Munro 2007; Emr et al. 2009). Site-directed vesicle trafficking in plant cells is essential for plant growth and development (Memon 2004; Hanton et al. 2007). Small GTP binding proteins (Rop, Rab, Arf, Sar) and cytoskeletal reorganization are likely to fulfill a unique symbiotic role by contributing to the specialization of the plasma membrane material around the infection thread and specialization of infection droplets and the symbiosome membrane (Memon et al. 1993; Schiene et al. 2004; Yaneva and Niehaus 2005).

6.3 *Recognition of Symbiotic Microbes: Dissecting the Early Signal Transduction Network*

Most of the cultivated legumes are tetraploid and many have large genomes and are recalcitrant to transformation or difficult to regenerate. As a result, two main species, *Medicago truncatula* and *Lotus japonicus*, have been adopted as models for legume research by several research groups. Both legumes are among the best-established systems to study plant-microbe interactions on a molecular level. Mutants of these two model legumes helped to analyze molecular events during the early perception of the Nod-factor. Genes coding for membrane spanning receptors and downstream elements involved in calcium signalling have been identified first in these model plants; however, the chain of events from Nod-factor perception to initiation of plant cell division is not well understood. The model legumes *L. japonicus* and *M. truncatula*, are also able to establish symbiotic interactions with mycorrhizal fungi (Frenzel et al. 2005) with *Mesorhizobium loti* (Stougaard 2000) and *Sinorhizobium meliloti* (Oldroyd et al. 2009) resulting in nitrogen fixing root nodules.

6.4 *Origin and Nature of the Symbiotic Interface*

As shown in Fig. 11.1 rhizobia enters the root epidermis by intracellular infection thread and is first initiated in the root hair cells and concurrently the cells at the root cortex start to differentiate and divide to develop a nodule meristem. This coordinated process between the infection thread formation and elongation and nodule organogenesis is crucial for root nodule formation.

Now question arises that how rhizobial cells are released from the infection thread to the plant cells? And how plant membrane material is delivered to the developing bacteroid? This includes two other principal questions asking: How membrane identity is established? And how the vesicle transport is reprogrammed in infected cells? In this review, a strong focus is given on the role of the small GTP binding proteins in the "endocytotic" uptake of rhizobia by the plant. Bacterial surface polysaccharides are also important "virulence factors" that are needed for a successful infection but how these molecules contribute in the infection process is still not clear. Although isolated bacterial lipopolysaccharides (LPS) and flagellin in disease infected plant cells are taken up by a process that resembles a receptor-mediated endocytosis nothing is known about the uptake of rhizobia by root cells. After the uptake of rhizobia, the bacteria are surrounded by the peribacteroid membranes, which exhibit unique features. It is still not known that how the identity of these membranes is maintained.

Vesicle trafficking plays an important role in root tip growth, infection thread formation and the development of root nodules (Oldroyd and Downie 2008). Several studies have shown that membrane associated small GTPases belonging to the Arf/Sar, Rab, and Rop/Rac families, along with their interacting proteins, playing a vital role in the diverse aspects of root nodule formation and development (Yuksel and Memon 2008, 2009). For example, Rab1, Rab7 and Rab11f are implicated to be essential in the formation of the peribacteroid membrane in *Glycine max*, *Vigna aconitifolia* , *Medicago sativa* and *M. truncatula* (Niehaus et al. 1993; Cheon et al. 1993; Son et al. 2003; Schiene et al. 2004). RAB7 has been reported to play an essential role in vesicular fusion during rhizobial endocytosis and symbiosome formation in soybean (Son et al. 2003). In addition Borg et al. 1997 isolated a total of 33 small GTPbinding genes form a cDNA library of *Lotus japonicus*, the majority of these trascriptes were ubiquitously expressed in all plant tissues with the exception of RAB1, RAB2, RAB5, RAB7, and a RAC which were predominantly expressed in nodules. Similarly, (Schiene et al. 2004) and (Meschini et al. 2008) demonstrated that Rab11 f from *M. sativa* and *Phaseolus vulgaris* is predominantly expressed in the root nodules and possibly involved in targeting vesicles to a symbiotic structure in nodule. Rac-Rop subfamily of the Rho GTPases has been reported to be associated with the plasma membrane at the apex of elongating root hairs and pollen tubes in *Arabidopsis thaliana* and tobacco (Kost 2008). They control tip growth by coordinating actin organization and membrane trafficking.

Although the sorting and targeting of the many secretory proteins from TGN to vacuolar and plasma membrane compartment has been worked out in Arabidopsis, tobacco leaf and BY2 cells but the targeting to the peribacteroid membrane (PBM)

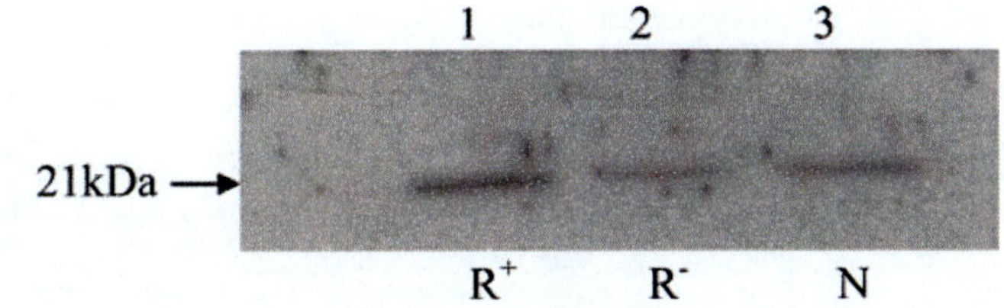

Fig. 11.2 Expression of SAR1 level in the membrane fraction of 3 month old *M. truncatula* infected with *Sinorhizobium meliloti* (R^+) and not infected (R^-) roots, N, nodules. Membrane fractions were isolated at 250,000×g (Yuca and Memon unpublished data)

is still not clearly identified in legumes (Samaj et al. 2004; Richter et al. 2009). One pioneering work of the lab of D.P. Verma showed the involvement of the two GTP-binding proteins, Rab 1 and Rab7 in this process but there were several unresolved questions related to the role of these proteins in peribacteroid formation (Cheon et al. 1993).

The protein and gene expression studies could suggest a nodule-specific expression of small GTP binding proteins especially the expression of several Rabs and Rops (Son et al. 2003; Blanco et al. 2009). The targeting of peribacteroid membrane (PBM) proteins seems to vary and this could be due to the mosaic nature of PBM having properties common to both plasma membrane and vacuoles. Therefore involvement of several GTP binding proteins in PBM biogenesis may be more complex and could be worked out in detail. It is concluded that looking further into the small GTPases of legume plants, more of these genes could be found to be significant in the formation and maintenance of the symbiosome compartment through regulating functions such as shuttling of products between symbiont and host. To realize this we previously made an attempt to retrieve the EST databases of two model legume plants, *Medicago* and *Lotus* and searched the nodule specific expressed transcripts of small GTPases (Yuksel and Memon 2008). As a result of this bioinformatic analyses, two ARL like genes, a group of 10 Rab GTPases and one or two ROP like GTPases were found to be mainly expressed in nodules. The sequences determined as a result of this study, we decided to commence detail expression analysis of these GTP-binding proteins at RNA and protein levels in rhizobium inoculated and non inoculated roots and nodules of *Medicago truncatula*. Our results show three to five times more Sar1 and Arf1 protein content in rhizobium infected roots compared to non inoculated one (Figs 11.2 and 11.3).

Our qReal-time PCR experiments clearly showed about 5–10 times more expression of Sar1 and Arf1, in roots and nodules of inoculated plants compared to the roots of non inoculated ones (see Fig. 11.4a, b). Our results indicate that small GTPases playing a major role in the regulation of membrane and protein transport in root nodule development in legumes.

A schematic diagram (Fig. 11.5) shows the probable role of Small GTPases in various steps of nodule development in legume roots (Yuksel and Memon 2009).

Detailed studies should be carried out to elucidate the role of different GTPases at various steps depicting in the Fig. 11.5. My group and several other groups in

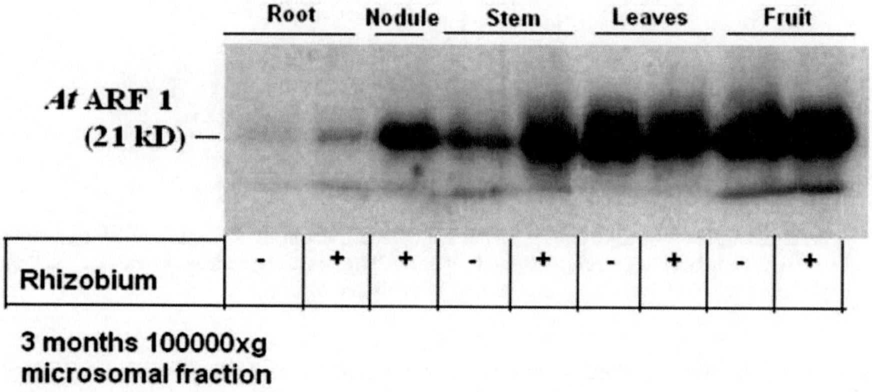

Fig. 11.3 ARF1 expression in the microsomal fraction isolated from different parts of the Medicago *truncatula* grown for 3 months (Yuzbasioglu and Memon unpublished data)

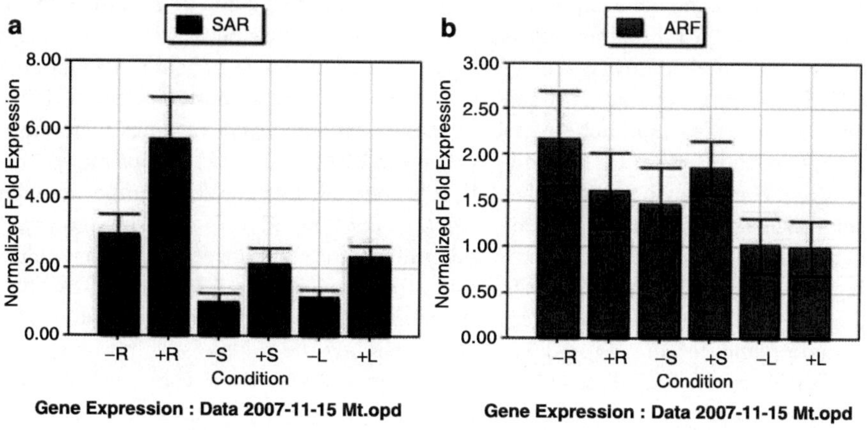

Fig. 11.4 **(a, b)**: The effect of *S. meliloti* infection on SAR and ARF mRNA levels in *M .truncatula* roots after 30 days infection. Ubiqutine was used as an internal control. –R, –S and –L, roots, stems and leaves from uninoculated plants and +R, +S, +L, roots, stems and leaves from inoculated plants (Yuzbasioglu, Keskin, and Memon unpublished data)

Europe (for example Niehaus's group in Bielefeld, Germany, FP7-PEAPOLE-2007-1-ITN) are studying in detail the membrane trafficking from ER to Golgi and from the *trans* Golgi network (TGN) to peribacteroid membranes and/or plasma membranes. To identify the role of small GTP-binding proteins in these processes is one of the important goals of our research, since these proteins are (i) essential and (ii) specific regulators of all vesicle transport events (Samaj et al. 2006). We will take the advantage of the GFP technology and will use GFP/RFP-tagged GTPases as markers to locate the targeting sites and compartments of these proteins and manipulate their specific functions by the use of gain-of-function and loss-of-function mutants of these signal transduction components.

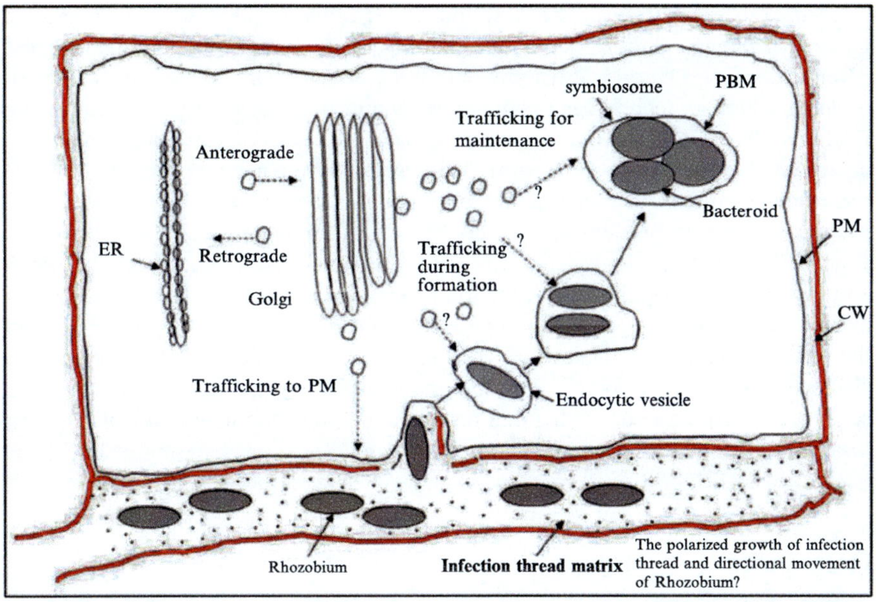

Fig. 11.5 A schematic diagram showing the possible role of various small GTP binding proteins at different steps of membrane trafficking in the formation of symbiosome and nodule development. The *dashed arrows* indicate the possible route of coated –vesicles carrying the cargo to the target compartment. Questions marks describes the involvement of some identified small GTPases in nodule specific trafficking. *PM* plasma membrane, *PBM* peribacteroid membrane, *CW* cell wall, *ER* endoplasmic reticulum

7 Conclusion

Legumes are one of the most economically important crop plants in the world. Because of their unique ability to fix natmospheric nitrogen through a symbiotic relationship with rhizobia, they are a linchpin of sustainable agriculture. This mutallistic interaction between host plant and bacteria provide a plentiful supply of nitrogen to the plants that in turn results in very high protein levels in legumes. They also provide a free and renewable source of usable nitrogen to the subsequent crops grown in rotation. Biological nitrogen fixation (BNF) and the understanding of beneficial plant microbe interactions are essential for a sustainable agriculture. The heavy use of chemical fertilizers especially nitrogen is a global concern for high economic cost and environmental pollution. For sustainable agriculture alternative nitrogen sources should be urgently sorted out. The unique nitrogen fixing features of legumes compare to non nitrogen fixing plants be exploited efficiently to maintain the sustainability in our cropping system. Nitrogen-fixing systems of legumes represent an economically attractive and ecologically sound means of reducing external inputs and improving internal resources. This notion has led to the development of three model legume species, *Lotus japonicus, Medicago truncatula* and *Glycine max* which are providing a rich source of genetic and genomic information.

Medicago truncatula is the first model leguminous plant with a completed genome sequence (http://www.ncbi.nlm.nih.gov/sites/entrez?db=genomeprj&cmd=Retrieve&dopt=Overview&list_uids=9508) available. This allows utilization of molecular biology techniques to study the genomic, transcriptomic, and proteomic relationships of *M. truncatula*. A better knowledge of cellular and molecular processes involved in symbiosome formation may contribute in a long-term to improve the efficiency of BNF by leguminous plants and/or to extend it to other crops. The protein and gene expression studies could suggest a nodule-specific expression of small GTP binding proteins especially the expression of Arf1, Sar1, Rabs and Rops and their possible role in early infection process and nodule development. Recent availability of soybean genome sequence in public domain has not only accelerated the work in finding improved varieties of this economically important crop but will also be used as a key reference for more than 20,000 legume species. This genome sequence opens the door for legume researchers to develop new efficient nitrogen fixing and high quality protein and oil producing crops for a changing world.

References

Ane J, Zhu H, Frugoli J (2008) Recent advances in *Medicago truncatula* genomics. Int J Plant Genomics 256597:1–11

Anonymus (2000) Agricutural strucrureand production statistics Institute Ankara, Turkey. www.die.gov.tr/PROJECTS/GIS/paper1.htm

Barrett T, Troup DB, Wilhite SE, Ledoux P, Rudnev D, Evangelista C, Kim IF, Soboleva A, Tomashevsky M, Edgar R (2007) NCBI GEO: mining tens of millions of expression profiles – database and tools update. Nucleic Acids Res 35:D760–D765

Benedito V, Torres-Jerez I, Murray J, Andriankaja A, Allen S, Kakar K, Wandrey M, Verdier J, Zuber H, Ott T, Moreau S, Niebel A, Frickey T, Weiller G, He J, Dai X, Zhao P, Tang Y, Udvardi M (2008) A gene expression atlas of the model legume *Medicago truncatula*. Plant J 55(3):504–513

Bestel-Corre G, Dumas-Gaudot E, Poinsot V, Dieu M, Dierick J-F, Tuinen Dv, Remacle J, Gianinazzi-Pearson V, Gianinazzi S (2002) Proteome analysis and identification of symbiosis-related proteins from *Medicago truncatula* Gaertn. By two-dimensional electrophoresis and mass spectrometry. Electrophoresis 23(1):122–137

Bhushan S, Pavlov PF, Rudhe C, Glaser E (2007) In vitro and in vivo methods to study protein import into plant mitochondria. Methods Mol Biol 390:131–150

Bickoff E, Kohler G, Smith D (1972) Chemical composition of herbage. In: *Alfalfa* science and technology, vol 15, Agron monog. ASA, CSSA, and SSSA, Madison

Blanco FA, Meschini EP, Zanetti ME, Aguilar OM (2009) A small GTPase of the Rab family is required for root hair formation and preinfection stages of the common bean-rhizobium symbiotic association. Plant Cell 21(9):2797–2810

Bonifacino JS, Glick BS (2004) The mechanisms of vesicle budding and fusion. Cell 116(2):153–166

Borg S, Brandstrup B, Jensen TJ, Poulsen C (1997) Identification of new protein species among 33 different small GTP-binding proteins encoded by cDNAs from Lotus japonicus, and expression of corresponding mRNAs in developing root nodules. Plant J 11(2):237–250

Bourgeois M, Jacquin F, Savois V, Sommerer N, Labas V, Henry C, Burstin J (2009) Dissecting the proteome of pea mature seeds reveals the phenotypic plasticity of seed protein composition. Proteomics 9(2):254–271

Brambilla F, Resta D, Isak I, Zanotti M, Arnoldi A (2009) A label-free internal standard method for the differential analysis of bioactive lupin proteins using nano HPLC-chip coupled with Ion trap mass spectrometry. Proteomics 9(2):272–286

Brechenmacher L, Lei Z, Libault M, Findley S, Sugawara M, Sadowsky MJ, Sumner LW, Stacey G (2010) Soybean metabolites regulated in root hairs in response to the symbiotic bacterium *Bradyrhizobium japonicum*. Plant Physiol 153(4):1808–1822

Cannon S, May G, Jackson S (2009) Three sequenced legume genomes and many crop species: rich opportunities for translational genomics. Plant Physiol 151(3):970–977

Cannon SB, Crow JA, Heuer ML, Wang X, Cannon EK, Dwan C, Lamblin AF, Vasdewani J, Mudge J, Cook A, Gish J, Cheung F, Kenton S, Kunau TM, Brown D, May GD, Kim D, Cook DR, Roe BA, Town CD, Young ND, Retzel EF (2005) Databases and information integration for the *Medicago truncatula* genome and transcriptome. Plant Physiol 138(1):38–46

Cannon SB, Sterck L, Rombauts S, Sato S, Cheung F, Gouzy J, Wang X, Mudge J, Vasdewani J, Schiex T, Spannagl M, Monaghan E, Nicholson C, Humphray SJ, Schoof H, Mayer KF, Rogers J, Quetier F, Oldroyd GE, Debelle F, Cook DR, Retzel EF, Roe BA, Town CD, Tabata S, Van de Peer Y, Young ND (2006) Legume genome evolution viewed through the *Medicago truncatula* and *Lotus japonicus* genomes. Proc Natl Acad Sci U S A 103(40):14959–14964

Carroll A, Somerville C (2009) Cellulosic biofuels. Annu Rev Plant Biol 60(1):165–182

Catalano CM, Lane WS, Sherrier DJ (2004) Biochemical characterization of symbiosome membrane proteins from *Medicago truncatula* root nodules. Electrophoresis 25:519–531

Chen J, Zhao P, Massaro D, Clerch LB, Almon RR, DuBois DC, Jusko WJ, Hoffman EP (2004) The PEPR GeneChip data warehouse, and implementation of a dynamic time series query tool (SGQT) with graphical interface. Nucleic Acids Res 32:D578–D581

Cheon CI, Lee NG, Siddique AB, Bal AK, Verma DP (1993) Roles of plant homologs of Rab1p and Rab7p in the biogenesis of the peribacteroid membrane, a subcellular compartment formed de novo during root nodule symbiosis. EMBO J 12(11):4125–4135

Cigdem E, Ertekin O, Unlu S, Yildizhan Y, Onarici S, Yuksel B, Memon A (2011) Effect of Serbian light crude on alfalfa (*Medicago sativa* L) cultivars. Asian J Chem 23:593–596

Colditz F, Braun H-P (2010) *Medicago truncatula* proteomics. J Proteomics 73:1974–1985

Cook D (1999) *Medicago truncatula* – a model in the making! Curr Opin Plant Biol 2:301–304

Crespi M, Frugier F (2008) De novo organ formation from differentiated cells: root nodule organogenesis. Sci Signal 1(49):re11

Dam S, Laursen BS, Ornfelt JH, Jochimsen B, Staerfeldt HH, Friis C, Nielsen K, Goffard N, Besenbacher S, Krusell L, Sato S, Tabata S, Thøgersen IB, Enghild JJ, Stougaard J (2009) The proteome of seed development in the model legume *Lotus japonicus*. Plant Physiol 149(3):1325–1340

Dela-Pena C, Lei Z, Watson BS, Sumner LW, Vivanco JM (2008) Root microbe communication through protein secretion. J Biol Chem 283(37):25247–25255

Doyle JJ, Luckow MA (2003) The rest of the Iceberg. Legume diversity and evolution in a phylogenetic context. Plant Physiol 131(3):900–910

Duc G, Trouvelot A, Gianinazzi-Pearson V, Gianinazzi S (1989) First report of non-mycorrhizal plant mutants (Myc-) obtained in pea (*Pisum sativum* L.) and fababean (*Vicia faba* L.). Plant Sci 60(2):215–222

Edgar R, Domrachev M, Lash AE (2002) Gene expression omnibus: NCBI gene expression and hybridization array data repository. Nucleic Acids Res 30(1):207–210

El Yahyaoui F, Kuster H, Ben Amor B, Hohnjec N, Puhler A, Becker A, Gouzy J, Vernie T, Gough C, Niebel A, Godiard L, Gamas P (2004) Expression profiling in *Medicago truncatula* identifies more than 750 genes differentially expressed during nodulation, including many potential regulators of the symbiotic program (1 w). Plant Physiol 136(2):3159–3176

Emr S, Glick BS, Linstedt AD, Lippincott-Schwartz J, Luini A, Malhotra V, Marsh BJ, Nakano A, Pfeffer SR, Rabouille C, Rothman JE, Warren G, Wieland FT (2009) Journeys through the Golgi – taking stock in a new era. J Cell Biol 187(4):449–453

Frenzel A, Manthey K, Perlick A, Meyer F, Puhler A, Kuster H, Krajinski F (2005) Combined transcriptome profiling reveals a novel family of arbuscular mycorrhizal-specific *Medicago truncatula* lectin genes. Mol Plant Microbe Interact 18(8):771–782

Frickey T, Benedito VA, Udvardi M, Weiller G (2008) AffyTrees: facilitating comparative analysis of Affymetrix plant microarray chips. Plant Physiol 146(2):377–386

Gage D (2004) Infection and invasion of roots by symbiotic, nitrogen-fixing rhizobia during nodulation of temperate legumes. Microbiol Mol Biol Rev 68(2):280–300

George J, Sawbridge TI, Cogan NO, Gendall AR, Smith KF, Spangenberg GC, Forster JW (2008) Comparison of genome structure between white clover and *Medicago truncatula* supports homoeologous group nomenclature based on conserved synteny. Genome 51(11):905–911

Geurts R, Bisseling T (2002) Rhizobium nod factor perception and signalling. Plant Cell 14(suppl 1):S239–S249

Gillingham AK, Munro S (2007) The small G proteins of the arf family and their regulators. Annu Rev Cell Dev Biol 23:579–611

Graham P, Vance C (2003) Legumes: importance and constraints to greater use. Plant Physiol 131:872–877

Gygi SP, Rist B, Gerber SA, Turecek F, Gelb MH, Aebersold R (1999) Quantitative analysis of complex protein mixtures using isotope-coded affinity tags. Nat Biotechnol 17(10):994–999

Hanton SL, Chatre L, Renna L, Matheson LA, Brandizzi F (2007) De novo formation of plant endoplasmic reticulum export sites is membrane cargo induced and signal mediated. Plant Physiol 143(4):1640–1650

He J, Benedito V, Wang M, Murray J, Zhao P, Tang Y, Udvardi M (2009) The *Medicago truncatula* gene expression atlas web server. BMC Bioinformatics 10(1):441

Henckel A, Arnaud P (2010) Genome-wide identification of new imprinted genes. Brief Funct Genomic 9(4):304–314

Henckel K, Runte KJ, Bekel T et al (2009) BMC Plant Biol 9:19. doi:10.1186/1471-2229-9-19 [Published: 11 FEB 2009 Times Cited: 3 (from Web of Science)]

Hirsch S, Oldroyd GED (2009) Integrated nod factor signaling in plants. In: Baluška F, Mancuso S (eds) Signaling in plants. Springer, Berlin/Heidelberg, pp 71–90

Hogslund N, Radutoiu S, Krusell L, Voroshilova V, Hannah M, Goffard N, Sanchez D, Lippold F, Ott T, Sato S, Tabata S, Liboriussen P, Lohmann G, Schauser L, Weiller G, Udvardi M, Stougaard J (2009) Dissection of symbiosis and organ development by integrated transcriptome analysis of *Lotus japonicus* mutant and wild-type plants. PLoS One 4(8):e6556

Hohnjec N, Vieweg M, Puhler A, Becker A, Kuster H (2005) Overlaps in the transcriptional profiles of *Medicago truncatula* roots inoculated with two different Glomus fungi provide insights into the genetic program activated during arbuscular mycorrhiza. Plant Physiol 137:1283–1301

Holmes P, Goffard N, Weiller G, Rolfe B, Imin N (2008) Transcriptional profiling of *Medicago truncatula* meristematic root cells. BMC Plant Biol 8(1):21

Hougaard BK, Madsen LH, Sandal N, de Carvalho MM, Fredslund J, Schauser L, Nielsen AM, Rohde T, Sato S, Tabata S, Bertioli DJ, Stougaard J (2008) Legume anchor markers link syntenic regions between *Phaseolus vulgaris, Lotus japonicus, Medicago truncatula* and *Arachis*. Genetics 179(4):2299–2312

Imin N, De Jong F, Mathesius U, van Noorden G, Saeed NA, Wang XD, Rose RJ, Rolfe BG (2004) Proteome reference maps of *Medicago truncatula* embryogenic cell cultures generated from single protoplasts. Proteomics 4(7):1883–1896

Incamps A, Hély-Joly F, Chagvardieff P, Rambourg J-C, Dedieu A, Linares E, Quéméneur E (2005) Industrial process proteomics: alfalfa protein patterns during wet fractionation processing. Biotechnol Bioeng 91(4):447–459

Ivanov S, Fedorova E, Bisseling T (2010) Intracellular plant microbe associations: secretory pathways and the formation of perimicrobial compartments. Curr Opin Plant Biol 13(4):372–377

Jones K, Kobayashi H, Davies B, Taga M, Walker G (2007) How rhizobial symbionts invade plants: the *Sinorhizobium-Medicago* model. Nat Rev Microbiol 5:619–633

Journet EP, van Tuinen D, Gouzy J, Crespeau H, Carreau V, Farmer MJ, Niebel A, Schiex T, Jaillon O, Chatagnier O, Godiard L, Micheli F, Kahn D, Gianinazzi-Pearson V, Gamas P (2002)

Exploring root symbiotic programs in the model legume Medicago truncatula using EST analysis. Nucleic Acids Res 30(24):5579–5592
Kistner C, Winzer T, Pitzschke A, Mulder L, Sato S, Kaneko T, Tabata S, Sandal N, Stougaard J, Webb KJ, Szczyglowski K, Parniske M (2005) Seven *Lotus japonicus* genes required for transcriptional reprogramming of the root during fungal and bacterial symbiosis. Plant Cell 17(8):2217–2229
Komatsu S, Ahsan N (2009) Soybean proteomics and its application to functional analysis. J Proteomics 72(3):325–336
Kondorosi E, Pierre M, Cren M, Haumann U, Buiré M, Hoffmann B, Schell J, Kondorosi A (1991) Identification of NolR, a negative transacting factor controlling the nod regulon in *Rhizobium meliloti*. J Mol Biol 222(4):885–896
Kost B (2008) Spatial control of Rho (Rac-Rop) signaling in tip-growing plant cells. Trends Cell Biol 18(3):119–127
Kuster H, Becker A, Firnhaber C, Hohnjec N, Manthey K, Perlick A, Bekel T, Dondrup M, Henckel K, Goesmann A, Meyer F, Wipf D, Requena N, Hildebrandt U, Hampp R, Nehls U, Krajinski F, Franken P, Puhler A (2007) Development of bioinformatic tools to support EST-sequencing, in silico- and microarray-based transcriptome profiling in mycorrhizal symbioses. Phytochemistry 68:19–32
Lei Z, Dai X, Watson BS, Zhao PX, Sumner LW (2011) A legume specific protein database (LegProt) improves the number of identified peptides, confidence scores and overall protein identification success rates for legume proteomics. Phytochemistry In Press, Corrected Proof
Lei Z, Elmer A, Watson B, Dixon R, Mendes P, Sumner L (2005) A two-dimensional electrophoresis proteomic reference map and systematic identification of 1367 proteins from a cell suspension culture of the model legume *Medicago truncatula*. Mol Cell Proteomics 4:1812–1825
Limpens E, Ramos J, Franken C, Raz V, Compaan B, Franssen H, Bisseling T, Geurts R (2004) RNA interference in *Agrobacterium rhizogenes* transformed roots of *Arabidopsis* and *Medicago truncatula*. J Exp Bot 55(399):983–992
Marsh JF, Schultze M (2001) Analysis of arbuscular mycorrhizas using symbiosis-defective plant mutants. New Phytol 150(3):525–532
Mathesius U, Djordjevic MA, Oakes M, Goffard N, Haerizadeh F, Weiller GF, Singh MB, Bhalla PL (2008) Comparative proteomic profiles of the soybean (*Glycine max*) root apex and differentiated root zone. Proteomics 11(9):1707–1719
Maunoury N, Redondo-Nieto M, Bourcy M, Van de Velde W, Alunni B, Laporte P, Durand P, Agier N, Marisa L, Dl V, Delacroix H, Gr D, Ratet P, Aggerbeck L, Kondorosi E, Mergaert P (2010) Differentiation of symbiotic cells and endosymbionts in *Medicago truncatula* nodulation are coupled to two transcriptome-switches. PLoS One 5(3):1–14
Memon AR (2004) The role of ADP-ribosylation factor and SAR1 in vesicular trafficking in plants. Biochim Biophys Acta 1664(1):9–30
Memon AR, Clark GB, Thompson GA Jr (1993) Identification of an ARF type low molecular mass GTP-binding protein in pea (*Pisum sativum*). Biochem Biophys Res Commun 193(3):809–813
Meschini EP, Blanco FA, Zanetti ME, Beker MP, Kuster H, Pueher A, Aguilar OM (2008) Host genes involved in nodulation preference in common bean (*Phaseolus vulgaris*)-Rhizobium etli symbiosis revealed – by suppressive subtractive hybridization. Mol Plant Microbe Interact 21(4):459–468
Murray JD, Muni RRD, Torres-Jerez I, Tang Y, Allen S, Andriankaja M, Li G, Laxmi A, Cheng X, Wen J, Vaughan D, Schultze M, Sun J, Charpentier M, Oldroyd G, Tadege M, Ratet P, Mysore KS, Chen R, Udvardi MK (2011) Vapyrin, a gene essential for intracellular progression of arbuscular mycorrhizal symbiosis, is also essential for infection by rhizobia in the nodule symbiosis of *Medicago truncatula*. Plant J 65(2):244–252
Niehaus K, Kapp D, Puhler A (1993) Plant defence and delayed infection of alfalfa pseudonodules induced by an exopolysaccharide (EPS I)-deficient *Rhizobium meliloti* mutant. Planta 190:415–425

Oldroyd GED, Downie JA (2004) Calcium, kinases and nodulation signalling in legumes. Nat Rev Mol Cell Biol 5(7):566–576

Oldroyd GED, Downie JA (2008) Coordinating nodule morphogenesis with *Rhizobial* infection in legumes. Annu Rev Plant Biol 59(1):519–546

Oldroyd GED, Harrison MJ, Paszkowski U (2009) Reprogramming plant cells for endosymbiosis. Science 324(5928):753–754

Panter S, Thomson R, de Bruxelles G, Laver D, Trevaskis B, Udvardi M (2000) Identification with proteomics of novel proteins associated with the peribacteroid membrane of soybean root nodules. Mol Plant Microbe Interact 13(3):325–333

Parkinson H, Kapushesky M, Shojatalab M, Abeygunawardena N, Coulson R, Farne A, Holloway E, Kolesnykov N, Lilja P, Lukk M, Mani R, Rayner T, Sharma A, William E, Sarkans U, Brazma A (2007) ArrayExpress – a public database of microarray experiments and gene expression profiles. Nucleic Acids Res 35:D747–D750

Pessi G, Ahrens CH, Rehrauer H, Lindemann A, Hauser F, Fischer H-M, Hennecke H (2007) Genome-wide transcript analysis of *Bradyrhizobium japonicum* bacteroids in soybean root nodules. Mol Plant Microbe Interact 20(11):1353–1363

Quackenbush J, Cho J, Lee D, Liang F, Holt I, Karamycheva S, Parvizi B, Pertea G, Sultana R, White J (2001) The TIGR gene indices: analysis of gene transcript sequences in highly sampled eukaryotic species. Nucleic Acids Res 29:159–164

Richter S, Voss U, Jurgens G (2009) Post-Golgi traffic in plants. Traffic 10(7):819–828

Rogers C, Wen J, Chen R, Oldroyd G (2009) Deletion-based reverse genetics in *Medicago truncatula*. Plant Physiol 151(3):1077–1086

Sakata K, Ohyanagi H, Nobori H, Nakamura T, Hashiguchi A, Nanjo Y, Mikami Y, Yunokawa H, Komatsu S (2009) Soybean proteome database: a data resource for plant differential omics. J Proteome Res 8(7):3539–3548

Samaj J, Baluska F, Voigt B, Schlicht M, Volkmann D, Menzel D (2004) Endocytosis, actin cytoskeleton, and signaling. Plant Physiol 135(3):1150–1161

Samaj J, Muller J, Beck M, Bohm N, Menzel D (2006) Vesicular trafficking, cytoskeleton and signalling in root hairs and pollen tubes. Trends Plant Sci 11(12):594–600

Sato S, Nakamura Y, Kaneko T, Asamizu E, Kato T, Nakao M, Sasamoto S, Watanabe A, Ono A, Kawashima K, Fujishiro T, Katoh M, Kohara M, Kishida Y, Minami C, Nakayama S, Nakazaki N, Shimizu Y, Shinpo S, Takahashi C, Wada T, Yamada M, Ohmido N, Hayashi M, Fukui K, Baba T, Nakamichi T, Mori H, Tabata S (2008) Genome structure of the legume, *Lotus japonicus*. DNA Res 15(4):227–239

Schiene K, Donath S, Brecht M, Pühler A, Niehaus K (2004) A Rab-related small GTP binding protein is predominantly expressed in root nodules of *Medicago sativa*. Mol Genet Genomics 272(1):57–66

Schmutz J, Cannon SB, Schlueter J, Ma JX, Mitros T, Nelson W, Hyten DL, Song QJ, Thelen JJ, Cheng JL, Xu D, Hellsten U, May GD, Yu Y, Sakurai T, Umezawa T, Bhattacharyya MK, Sandhu D, Valliyodan B, Lindquist E, Peto M, Grant D, Shu SQ, Goodstein D, Barry K, Futrell-Griggs M, Abernathy B, Du JC, Tian ZX, Zhu LC, Gill N, Joshi T, Libault M, Sethuraman A, Zhang XC, Shinozaki K, Nguyen HT, Wing RA, Cregan P, Specht J, Grimwood J, Rokhsar D, Stacey G, Shoemaker RC, Jackson SA (2010) Genome sequence of the palaeopolyploid soybean. Nature 463(7278):178–183

Schultze M, Kondorosi A (1998) Regulation of symbiotic root nodule development. Annu Rev Genet 32:33–57

Sherlock G, Hernandez-Boussard T, Kasarskis A, Binkley G, Matese JC, Dwight SS, Kaloper M, Weng S, Jin H, Ball CA, Eisen MB, Spellman PT, Brown PO, Botstein D, Cherry JM (2001) The stanford microarray database. Nucleic Acids Res 29(1):152–155

Singh R, Jauhar P (2009) Genetic resources, chromosomes engineering, and crop improvement, vol 5, Grain legum. Taylor and Francis, Boca Raton, London, New York

Smil V (1999) Nitrogen in crop production: an account of global flows. Glob Biogeochem Cy 13(2):647–662

Son O, Yang H-S, Lee H-J, Lee M-Y, Shin K-H, Jeon S-L, Lee M-S, Choi S-Y, Chun J-Y, Kim H, An C-S, Hong S-K, Kim N-S, Koh S-K, Cho MJ, Kim S, Verma DPS, Cheon C-I (2003)

Expression of srab7 and SCaM genes required for endocytosis of *Rhizobium* in root nodules. Plant Sci 165(6):1239–1244

Stekel DJ, Git Y, Falciani F (2000) The comparison of gene expression from multiple cDNA libraries. Genome Res 10(12):2055–2061

Stougaard J (2000) Regulators and regulation of legume root nodule development. Plant Physiol 124(2):531–540

Tadege M, Wen J, He J, Tu H, Kwak Y, Eschstruth A, Cayrel A, Endre G, Zhao PX, Chabaud M, Ratet P, Mysore KS (2008) Large-scale insertional mutagenesis using the Tnt1 retrotransposon in the model legume *Medicago truncatula*. Plant J 54(2):335–347

Thompson R, Ratet P, Kuster H (2005) Identification of gene functions by applying TILLING and insertional mutagenesis strategies on microarray-based expression data. Grain Legumes 41:20–22

Tian L, Peel G, Lei Z, Aziz N, Dai X, He J, Watson B, Zhao P, Sumner L, Dixon R (2009) Transcript and proteomic analysis of developing white lupin (*Lupinus albus* L.) roots. BMC Plant Biol 9(1):1

Town C (2006) Annotating the genome of *Medicago truncatula*. Curr Opin Plant Biol 9(2):122–127

Umezawa T, Sakurai T, Totoki Y, Toyoda A, Seki M, Ishiwata A, Akiyama K, Kurotani A, Yoshida T, Mochida K, Kasuga M, Todaka D, Maruyama K, Nakashima K, Enju A, Mizukado S, Ahmed S, Yoshiwara K, Harada K, Tsubokura Y, Hayashi M, Sato S, Anai T, Ishimoto M, Funatsuki H, Teraishi M, Osaki M, Shinano T, Akashi R, Sakaki Y, Yamaguchi-Shinozaki K, Shinozaki K (2008) Sequencing and analysis of approximately 40 000 soybean cDNA clones from a full-length-enriched cDNA library. DNA Res 15(6):333–346

Verdier J, Kakar K, Gallardo K, Le Signor C, Aubert G, Schlereth A, Town C, Udvardi M, Thompson R (2008) Gene expression profiling of *M. truncatula* transcription factors identifies putative regulators of grain legume seed filling. Plant Mol Biol 67:567–580

Watson B, Asirvatham V, Wang L, Sumner L (2003) Mapping the proteome of barrel medic (*Medicago truncatula*). Plant Physiol 131:1104–1123

Wen J, Parker BJ, Weiller GF (2007) *In Silico* identification and characterization of mRNA-like noncoding transcripts in *Medicago truncatula*. In Silico Biol 7(4–5):485–505

Wienkoop S, Saalbach G (2003) Proteome analysis. Novel proteins identified at the peribacteroid membrane from *Lotus japonicus* root nodules. Plant Physiol 131(3):1080–1090

Wilson KA, McManus MT, Gordon ME, Jordan TW (2002) The proteomics of senescence in leaves of white clover, *Trifolium repens* (L.). Proteomics 2(9):1114–1122

Winzer T, Bairl A, Linder M, Linder D, Werner D, Müller P (1999) A novel 53-kDa nodulin of the symbiosome membrane of soybean nodules, controlled by *Bradyrhizobium japonicum*. Mol Plant Microbe Interact 12(3):218–226

Wise R, Caldo R, Hong L, Shen L, Cannon E, Dickerson J (2007) BarleyBase/PLEXdb. Methods Mol Biol 406:347–363

Yaneva IA, Niehaus K (2005) Molecular cloning and characterisation of a Rab-binding GDP-dissociation inhibitor from *Medicago truncatula*. Plant Physiol Biochem 43(3):203–212

Young N, Udvardi M (2009) Translating *Medicago truncatula* genomics to crop legumes. Curr Opin Plant Biol 12(2):193–201

Young ND, Cannon SB, Sato S, Kim D, Cook DR, Town CD, Roe BA, Tabata S (2005) Sequencing the genespaces of *Medicago truncatula* and *Lotus japonicus*. Plant Physiol 137(4):1174–1181

Young ND, Mudge J, Ellis TH (2003) Legume genomes: more than peas in a pod. Curr Opin Plant Biol 6(2):199–204

Yuksel B, Memon A (2009) Legume small GTPases and their role in the establishment of symbiotic associations with *Rhizobium* spp. Plant Signal Behav 4(4):257–260

Yuksel B, Memon AR (2008) Comparative phylogenetic analysis of small GTP-binding genes of model legume plants and assessment of their roles in root nodules. J Exp Bot 59(14):3831–3844

Chapter 12
Agrobacterium tumefaciens and its Use in Plant Biotechnology

İbrahim İlker Özyiğit

Contents

İ.İ. Özyiğit (✉)
Science & Arts Faculty, Department of Biology, Marmara University, Göztepe-Istanbul, Turkey
e-mail: ilkozyigit@marmara.edu.tr; ilkozyigit@gmail.com

M. Ashraf et al. (eds.), *Crop Production for Agricultural Improvement*,
DOI 10.1007/978-94-007-4116-4_12, © Springer Science+Business Media B.V. 2012

Abstract For years, isolation of foreign genes from one plant and transferring them to another and then observing effects of new genes in transferred plants has only been a dream for a plant biologist. Today, many commercially important species are routinely transformed by different biotechnological methods. Methods available for plant transformation are arranged in three main groups: using biological vectors (virus- or bacteria-mediated transformation), direct DNA transfer techniques (chemical-, electrical-, or laser-induced permeability of protoplasts or cells) and non-biological vector systems (microprojectiles, microinjection or liposome fusion). Today in many countries a number of transgenic important crops such as soybean, maize, cotton, canola, sugarbeet, sugarcane and alfalfa are available and the mostly preferred method is *Agrobacterium*-mediated transformation. In this chapter, some information about this important bacterium and mechanisms of *Agrobacterium*-mediated gene transfer are presented.

Keywords Cotton • Crown-gall disease • Molecular marker techniques • Tissue culture • Transgenic • T-DNA • Vir proteins

1 Characteristics of *Agrobacterium tumefaciens*

The genus *Agrobacterium* is divided into a number of species mostly on the basis of disease symptomology and host range (Gelvin 2003). *Agrobacterium radiobacter* is an "avirulent" species, while *A. tumefaciens* causes crown gall disease; *A. rhizogenes* causes hairy root disease, and *A. rubi* cane gall disease. Recently, a new species *A. vitis* was proposed, which causes the growth of neoplastic tumors on the stem and crown of grapevines and induces necrotic lesions on grape roots and a few other plant species (Otten et al. 1984; Burr et al. 1998; Gelvin 2003; Escobar and Dandekar 2003).

Agrobacterium tumefaciens is a Gram negative, motile, rod shaped soil bacterium, which is non-sporing, and is closely related to the N-fixing rhizobium bacteria, which form root nodules on leguminous plants (Ream 2002). It is also a soil pathogen initiating tumors on plants (DeCleene and DeLay 1976; McCullen and Binns 2006). The bacterium is surrounded by a small number of flagella (Fig. 12.1) (Garrity 2005). The systematic classification of *A. tumefaciens* is given in Table 12.1.

Although today in nomenclature, the name *Agrobacterium tumefaciens* Conn 1942 (Smith and Townsend 1907) is mostly used, there are some synonyms of this bacterium as; *Bacterium tumefaciens* (Smith and Townsend 1907), *Pseudomonas tumefaciens* (Smith and Townsend 1907) Duggar 1909, *Phytomonas tumefaciens*

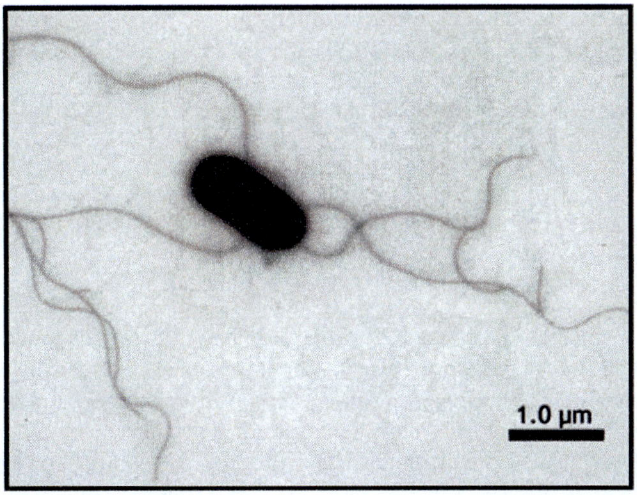

Fig. 12.1 Electron microscopic view of *Agrobacterium tumefaciens* with peritrichous flagella (Modified from Jeon et al. 2008)

Table 12.1 Systematics of *Agrobacterium tumefaciens* (Bergey's Manual of Systematic Bacteriology 2005)

Kingdom	Bacteria
Phylum	Proteobacteria
Class	Alphaproteobacteria
Order	Rhizobiales
Family	Rhizobiaceae
Genus	*Agrobacterium*
Species	*Agrobacterium tumefaciens*

(Smith and Townsend 1907) Bergey et al. 1923, *Polymonas tumefaciens* (Smith and Townsend 1907) Lieske 1928 (LPSN 2011).

Agrobacterium-mediated gene transfer has been of great help in modern plant molecular genetics and genetic engineering (Ream 2002). Nowadays, plant transformation by *A. tumefaciens* has become one of the most used methods for the introduction of foreign genes into plant cells and the subsequent regeneration of transgenic plants (de la Riva et al. 1998; Ziemienowicz 2001; Ream 2002; Özcan et al. 2004; McCullen and Binns 2006). In addition, the genus *Agrobacterium* can transfer a piece of its plasmid DNA into a remarkably broad group of organisms including numerous dicotyledonous and monocotyledonous species (Anderson and Moore 1979; de Frammond et al. 1986; van Wordragen and Dons 1992; Men et al. 2003; Li et al. 2007; Wanichananan et al. 2010) and gymnosperms (Loopstra et al. 1990; Stomp et al. 1990; McAfee et al. 1993; Yibrah et al. 1996; Tang et al. 2001; Grant et al. 2004). Furthermore, *Agrobacterium* can transform fungi (Chen et al. 2000), including yeasts (Bundock et al. 1995; Bundock and Hooykaas 1996; Piers et al. 1996), ascomycetes (de Groot et al. 1998; Abuodeh et al. 2000; Zhiming et al. 2008) and basidiomycetes (Sharma et al. 2006; Okamoto et al. 2010). *Agrobacterium* has been reported to transfer DNA to human cells as well (Kunik et al. 2001; Gelvin 2003).

2 Crown Gall Disease

In nature, *A. tumefaciens* infects the wounded sites in dicotyledonous plants and some monocotyledonous as well, causing the formation of crown gall tumors (de la Riva et al. 1998). These tumors no longer require the continuous presence of the inciting bacterium for proliferation demonstrating that the plant cells have been transformed (White and Braun 1942; McCullen and Binns 2006). Crown gall disease is a common plant disease, affecting more than 600 types of plants. It affects nearly all dicotyledonous plants, woody and herbaceous plants and many commercially important and valuable crops such as brambles, rose, willow, grapes, rice and sugar beet (Burr et al. 1998; Jeon et al. 2008). Although *A. tumefaciens* has a wide host range, many plants are immune to the disease and can be planted in sites with a history of crown gall. Some of these well-known genera, which are resistant or immune to this disease are as follows; *Ailanthus*, *Berberis*, *Betula*, *Catalpa*, *Cedrus*, *Cercis*, *Fagus*, *Ginkgo*, *Larix*, *Magnolia*, *Picea*, *Pyracantha* and *Sambucus* (Lacy and Hansen 2002).

The disease can be identified by the appearance of tumors or galls of varying size and shape on the lower stem where it meets the soil (crown), main roots and sometimes branches of the plant (Fig. 12.2). Infected plant cells contain bacterial genes (plasmid) that replace some of the normal plant cell genes (Khawar and Özcan 2002; Jeon et al. 2008). When they are young, the galls can be white or cream colored and spongy or wart-like; as they age, they become dark and woody (Burr et al. 1998). The tumor usually either appears as a swelling of the plant tissue, or as a separate mass of tissue close to the plant surface, joined only by a narrow neck of tissue (Fig. 12.3). Some of them can reach up to 30 cm in diameter, though 5–10 cm is more common. When plants are infected with the bacterium they become stunted, produce small chlorotic leaves and become more susceptible to extreme environmental conditions such as winter cold and wind. Galls can interfere with the plant's ability to move water and nutrients through the stem, which may result in stunting or decline of the plant (Sule et al. 1995; Jeon et al. 2008).

Plant tumors normally do not kill the plant, but crown gall disease can be fatal if the tumors become too enlarged. Crown gall can also be eradicated using creosote based chemical compounds, copper-based solutions and strong oxidants such as sodium hypochlorite (Utkhede and Smith 1993; Polcaro et al. 2008). However, these are costly to apply in terms of both labor and buying the product. They are also very

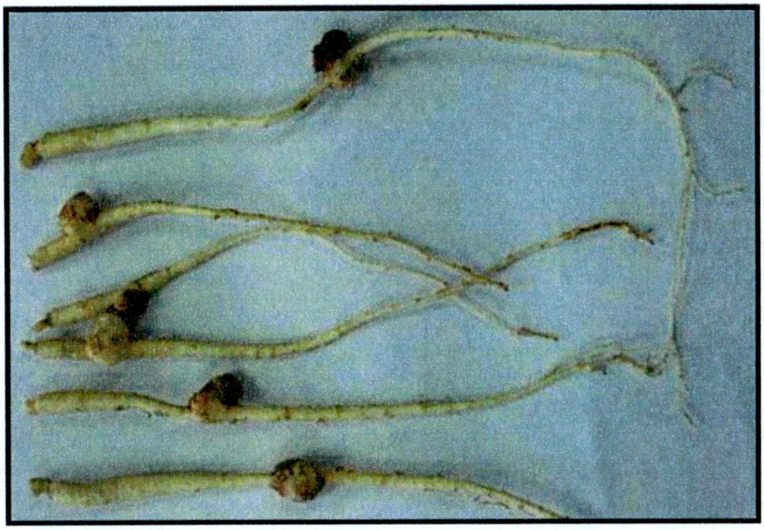

Fig. 12.2 Tumors of Korean ginseng seedling roots collected from a seedbed at the harvest time of seedling roots (Modified from Jeon et al. 2008)

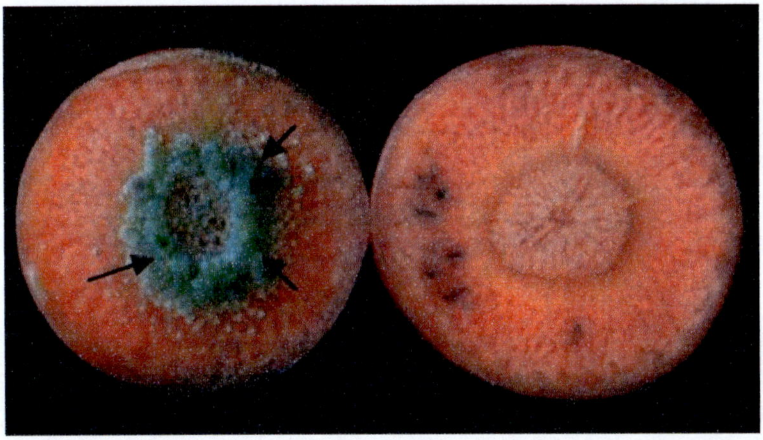

Fig. 12.3 Healthy (*right*) and diseased (*left*) carrot disc showing callus-like swellings (*arrows*) around vascular bundle 2 weeks after *Agrobacterium tumefaciens* inoculation (Modified from Jeon et al. 2008)

harmful to the surrounding environment and accumulation of large quantities of copper in the soil can have a disastrous impact on other plants in the area. Therefore, chemical controls are rarely used against *A. tumefaciens* (Deacon et al. 1988; Zauner et al. 2006).

3 Genome Structure of *Agrobacterium*

The first gene transfer into plants was not realized by scientists, it was realized by *A. tumefaciens*, which has the exceptional ability to transfer a particular DNA segment (T-DNA) of the tumor-inducing (Ti) plasmid into the nucleus of infected plant cells (Fig. 12.4). Therefore, the first genetic engineers are from the genus *Agrobacterium*, not from the genus *Homo*. The molecular basis of genetic transformation of plant cells by *A. tumefaciens* is **(1)** transferring some genes from the bacterium and **(2)** their integration into plant nuclear genome located in a large tumor-inducing Ti plasmid of bacterium (Gelvin 2003; Özcan et al. 2004).

The virulent strain of *A. tumefaciens*, which induces crown gall disease, contains a large mega-plasmid (more than 200 kbp) called Ti plasmid (Fig. 12.5). In genus *Agrobacterium*, Ti plasmids are classified according to opines which are produced and excreted by the tumors they induce (Zhu et al. 2000; Ream 2002; Özcan et al. 2004). During infection, the T-DNA that is a mobile segment of Ti plasmid, is transferred into the plant cell nucleus and integrated into the plant genome (de la Riva et al. 1998; Gelvin 2003). T-DNA region generally represents less than 10% of the Ti plasmid. Some Ti plasmids contain one T-DNA whereas others contain multiple T-DNA regions (Zambryski et al. 1980; Barker et al. 1983; Gelvin 2003). The processing of T-DNA from Ti plasmid and its subsequent export from bacterium into plant cell result in large part from the activity of virulence (*vir*) genes carried by the Ti plasmid (Garfinkel and Nester 1980; Hooykaas et al. 1984; Knauf et al. 1984;

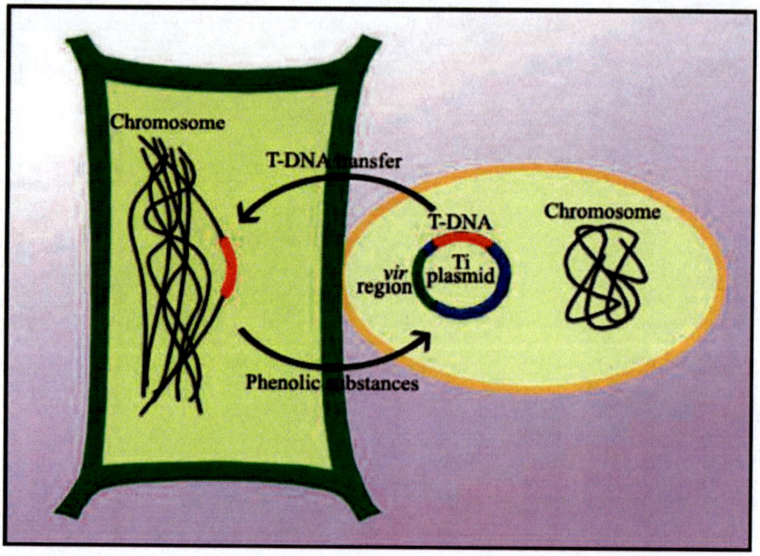

Fig. 12.4 Ti plasmid, T-DNA, *vir* region and T-DNA transfer into the plant cell (Modified from Özcan et al. 2004)

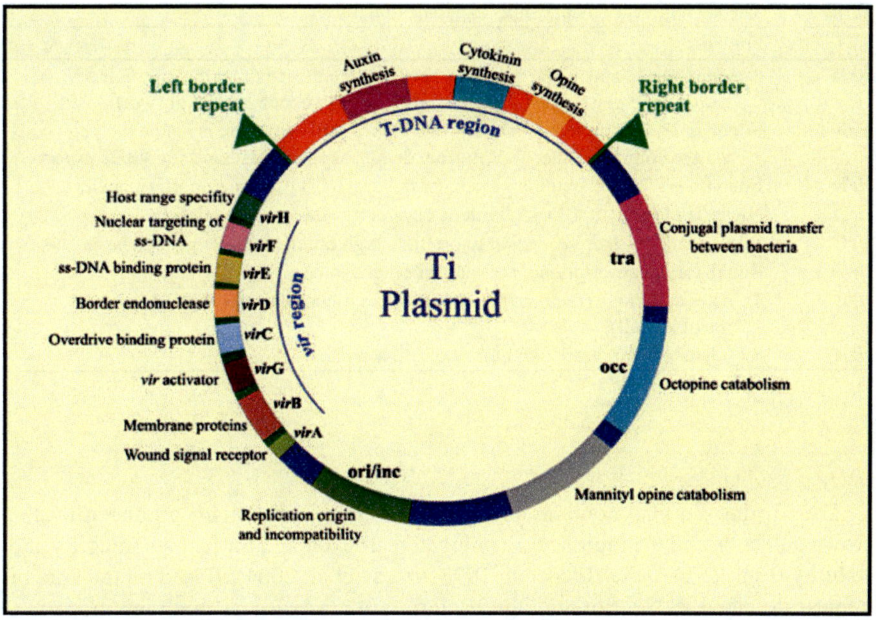

Fig. 12.5 Genetic map of octopine-type Ti plasmid (Modified from Ream 2002 and Özcan et al. 2004)

Lundquist et al. 1984; Horsch et al. 1986; Stachel and Zambryski 1986; Gelvin 2003). In addition, earlier studies have indicated that the Ti plasmid, rather than chromosomal genes, was the major genetic determinant of host range (Loper and Kado 1979; Thomashow et al. 1980).

In *A. tumefaciens*, the T-DNA contains two types of genes; (a) oncogenic genes which are encoding for enzymes involved in the synthesis of auxins and cytokinins that are responsible for tumor formation, and (b) the genes which are encoding synthesis of opines that are responsible for the formation of the novel amino acid-sugar conjugates (de la Riva et al. 1998; McCullen and Binns 2006). The amino acids and sugars are synthesized and excreted by the crown gall cells and consumed by *A. tumefaciens* as carbon and nitrogen sources (de la Riva et al. 1998).

The specific DNA region acting as a *cis* element signal for the transfer apparatus from the Ti plasmid is generated by 25-bp direct repeats. Through the cooperative action of proteins encoded by the genes defined in Ti plasmid virulence region (*vir* genes) and in the bacterial chromosome, the process of T-DNA transfer is carried out (de la Riva et al. 1998; Ream 2002; Gelvin 2000). However, T-DNA transfer and integration into plant genome does not require tumorigenesis or T-DNA encoded proteins (Hoekema et al. 1983; Ream et al. 1983). This situation has allowed the plant scientists to use *A. tumefaciens* to transfer some desired genes into plants in place of T-DNA oncogenes (Ream 2002).

Table 12.2 The *vir* regions of Ti plasmids and their functions

vir Locus	Function
***vir*H**	Encodes VirH1 and VirH2 proteins (they could enhance the transfer efficiency, detoxifying certain plant compounds that can affect bacterial growth)
***vir*F**	Encodes 23 kDa protein (functions once the T-DNA complex is inside the plant cells via the conjugal channel or independently, as it was assumed for VirE2 export)
***vir*E**	Encodes ss-T-DNA binding protein (stabilizes T-DNA during or after transfer)
***vir*D**	Nicks Ti plasmid at T-DNA borders, covalently attaches to T-strand
***vir*C**	Binds to the 'overdrive' region to promote high efficiency T-strand synthesis
***vir*G**	Regulatory (transcriptional activator of other *vir* loci)
***vir*B**	Transfer apparatus (required for export of the T-complex and VirE2 into the plant cell)
***vir*A**	Regulatory (recognizes plant metabolites, activates *vir*G)

The Ti plasmid also contains the genes for opine catabolism produced by the crown gall cells and the regions for conjugative transfer and for its own integrity and stability (Fig. 12.5) (de la Riva et al. 1998; Özcan et al. 2004). Based on the kind of opines produced in the tumors, agrobacteria are classified as octopine, nopaline, succinamopine and leucinopine strains. The biosynthesis of opines is catalyzed by opine synthases, which are encoded by the T-DNA (Ziemienowicz 2001). There are more than 20 opines and each strain induces and catabolizes a specific set of opines. In general, each *A. tumefaciens* strain catabolizes only the opines synthesized by tumors that it will induce (Hooykaas and Shilperoort 1992; Zupan and Zambryski 1997; de la Riva et al. 1998). For example, octopine-type Ti plasmids direct their hosts to synthesize at least eight opines. The *ocs* gene encodes octopine synthase, which reductively condenses pyruvate with either arginine, lysine, histidine or ornithine to produce octopine, lysopine, histopine, or octopinic acid respectively, all of which can be detected in crown gall tumors (Dessaux et al. 1998; Ziemienowicz 2001). Furthermore, some opines induce conjugal transfer of self-transmissible Ti plasmids between strains of *A. tumefaciens* (Petit et al. 1978; Ellis et al. 1982), thereby conferring on other strains the ability to catabolize extant opines. Apparently, *A. tumefaciens* strains create a "niche" (a crown gall tumor-synthesizing particular opines) that offers a favorable environment for growth of the inducing strain (Ream 2002).

There are six operons organized in the *vir* region of Ti Plasmid (30 kb) essential (*vir*A, *vir*B, *vir*D, and *vir*G) or increasing of transfer efficiency (*vir*C and *vir*E) for the T-DNA transfer (Hooykaas and Shilperoort 1992; Zupan and Zambryski 1995; Jeon et al. 1998; de la Riva et al. 1998). In addition, there are also two more operons, which are not necessary for T-DNA transfer called *vir*F and *vir*H (Fig. 12.5) (Özcan et al. 2004). The protein products of these genes, respond to the specific compounds secreted by the wounded plant to generate a copy of the T-DNA and mediate its transfer into the host cell (Table 12.2, Sheng and Citovsky 1996). The *vir*A, *vir*G and *vir*F operons carry one gene, while *vir*E, *vir*C and *vir*H operons carry two genes.

However, *vir*D includes four and *vir*B includes 11 genes (de la Riva et al. 1998; Özcan et al. 2004). *vir* gene expression is induced by chemical signal molecules (phenolic substances and sugar) released from wounded plant cells (Winans et al. 1994). Recognition of these signals by the sensor protein VirA leads to phosphorylation of VirG protein (Stachel and Zambryski 1986; Jin et al. 1990a).

*vir*D gene products are needed for DNA processing which involves recognition and complexing of direct repeat borders of the T-DNA segment (Sheng and Citovsky 1996; Ziemienowicz 2001; Özcan et al. 2004). T-strand production is dependent on the cleavage of the border sequence which is performed by VirD2 in the presence of VirD1 protein. VirD2 cuts the border sequence in a site and strand specific manner and following cleavage it becomes covalently attached to the 5′ end of the nicked DNA (Dürrenberger et al. 1989; Howard et al. 1990; Howard et al. 1992). The nicked DNA is then removed from the plasmid producing single stranded T-DNA (ss-T-DNA). It is believed that by a pilus-like structure containing VirB and VirD4 proteins the T-DNA-VirD2 complex and the VirE2 protein are transferred to the plant (Citovsky et al. 1994; Rossi et al. 1996). In the plant cell, T-DNA becomes associated with the single stranded DNA-binding protein, VirE2 and then the T-DNA-protein complex is transported into the nucleus. Following the transportation, the integration of T-DNA into the nuclear genome occurs (Sheng and Citovsky 1996; Ziemienowicz 2001).

The third component is the cluster of chromosomal virulence (*chv*) genes found on the *Agrobacterium* chromosome. The *chv* genes play important roles in bacterial chemotaxis and attachment to the wounded plant cell (Citovsky et al. 1992; Zambryski 1992; Sheng and Citovsky 1996; de la Riva et al. 1998; Özcan et al. 2004). Different chromosomal-determined genetic element products have exhibited their functional role in the attachment of *A. tumefaciens* to the plant cell and bacterial colonization: the loci *chv*A and *chv*B code for inner membrane proteins essential for the transport of β-1,2 glucan from bacterial cytoplasm into periplasm and in the synthesis of β-1,2 glucan (Cangelosi et al. 1989), the *chv*E is required for the expression of *vir* genes and bacterial chemotaxis (Ankenbauer and Nester 1990; Cangelosi et al. 1990, 1991; de la Riva et al. 1998), the *cel* locus responsible for cellulose fiber synthesis to enable bacterial cells to be firmly adhere to plant cell wall (Matthysse 1983), the *psc*A (*exo*C) locus is involved in the synthesis of both cyclic glucan and acid succinoglycan (Cangelosi et al. 1987, 1991) and the *att* locus acts role in the cell surface proteins (Matthysse 1983; de la Riva et al. 1998).

As an important model system in understanding pathogen-host interactions including recognizing and delivering macromolecules into target cells resulting in disease, *A. tumefaciens* has been established by the intense mechanistic analysis of *Agrobacterium*-mediated transformation (McCullen and Binns 2006). From these studies, a basic model for the cellular transformation of plants by *A. tumefaciens* has been developed (Gelvin 2003; Palmer et al. 2004; Tzfira and Vaidya 2004; Brencic and Winans 2005; Christie et al. 2005; McCullen and Binns 2006). In this chapter, a hypothetical model depicting the most important stages of this process is presented, supported by the most recent experimental data and accepted hypothesis on T-DNA transfer.

4 Molecular Mechanism of Gene Transfer

The process of gene transfer from *A. tumefaciens* into plant cells implies several essential steps: **(1)** bacterial colonization, **(2)** induction of bacterial virulence system, **(3)** generation of T-DNA transfer complex, **(4)** T-DNA transfer and **(5)** integration of T-DNA into plant genome (Fig. 12.6).

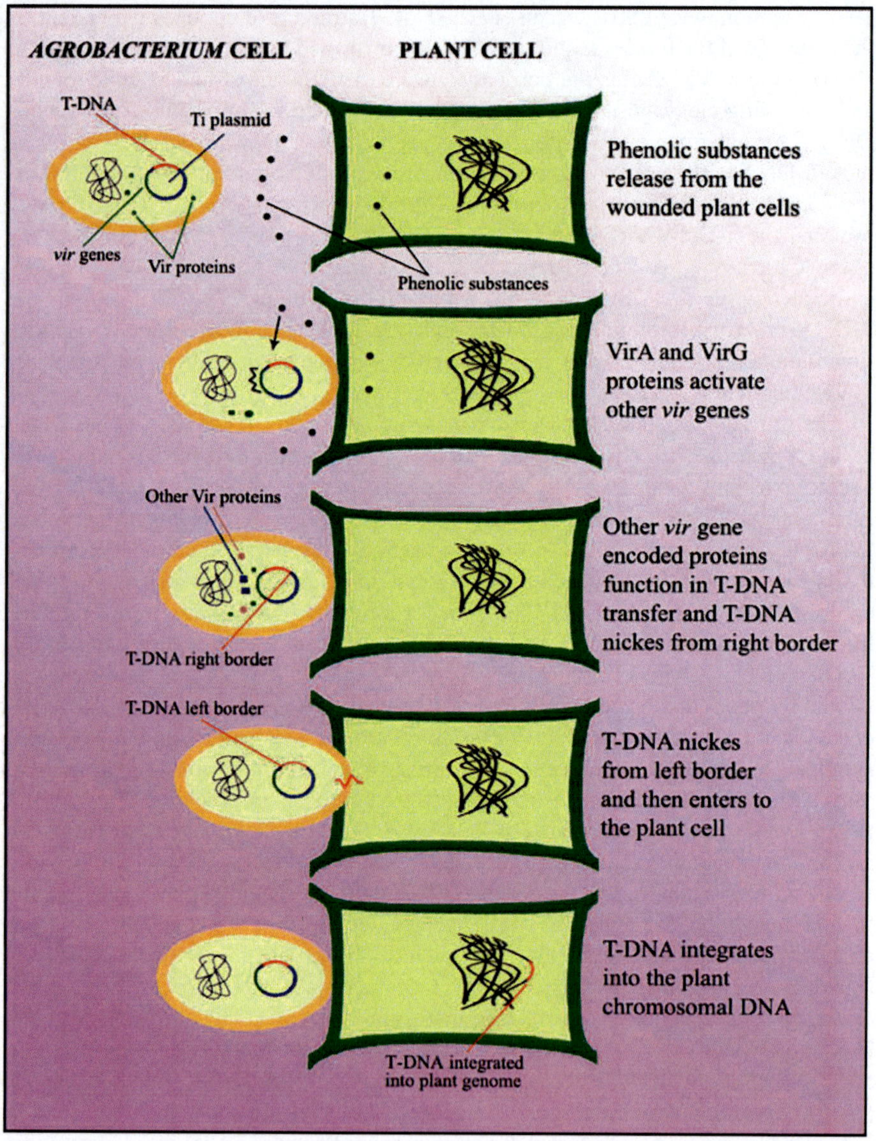

Fig. 12.6 Natural gene transfer from *Agrobacterium tumefaciens* to plant cell (Modified from Watson et al. 1992 and Özcan et al. 2004)

4.1 Bacterial Colonization

Phenolic compounds released to the surface of the host cells during injury act as signals that attract *Agrobacterium* cells to the wounded site in a plant before bacterial colonization (Parke et al. 1987; Ashby et al. 1988; Ziemienowicz 2001). Experiments have shown that some compounds like acetosyringone induce *vir* genes which attract a highly motile strain exhibiting marked pTi-dependent chemotaxis whereas a poorly motile strain does not show any movement. This pTi-dependent chemotaxis requires virulence genes, *vir*A and *vir*G (Shaw et al. 1988). Furthermore, sugars and amino acids also attract *Agrobacterium* (Hawes et al. 1988; Ziemienowicz 2001).

Bacterial colonization is the first step in tumor induction and it occurs when *A. tumefaciens* is attached to the plant cell surface (Özcan et al. 2004). Earlier studies showed that non-attaching *A. tumefaciens* mutants lose their tumor-inducing capacity (Douglas et al. 1982; Cangelosi et al. 1987; Thomashow et al. 1987; Bradley et al. 1997; de la Riva et al. 1998). The cell surface polysaccharides of *A. tumefaciens* are considered to be important in the colonizing process (Bradley et al. 1997). It was reported that the bacterial attachment could be inhibited when preparations of lipopolysaccharides (LPS) of virulent strains are applied to the plant tissue before bacteria (Whatley and Spress 1977; de la Riva et al. 1998).

The LPS are integral components of the outer membrane and contain a lipid moiety, which is embedded in the outer membrane, a conserved core oligosaccharide, and a polysaccharide side chain known as the O-antigen. In addition to LPS, *A. tumefaciens* also produces capsular polysaccharides. Some evidence indicates that capsular polysaccharides have specific functions in the interaction with the host plant (de la Riva et al. 1998). There is evidence that there was a correlation between the production of an acidic polysaccharide and attachment of wild-type bacterium to plant cells (Bradley et al. 1997; de la Riva et al. 1998). Genes involved in the attachment are located in the *Agrobacterium* chromosome, and include *chv*A, *chv*B, *psc*A (or *exo*C) and *att*. Mutations in these loci lead to a loss of virulence towards many plant species (Ziemienowicz 2001).

Experiments related with *chv*A and *chv*B genes showed that mutations in these genes produce non-specific alterations in the bacterial surface and result in absence of β-1,2-glucans in the bacterial membrane, increased production of extracellular polysaccharides, reduced motility and failure to bind to plant cells. The *chv*A gene codes for inner membrane protein essential for the transport of β-1,2-glucans from bacterial cytoplasm into periplasm (Cangelosi et al. 1989; Ziemienowicz 2001) while the *chv*B gene codes for an inner membrane protein likely to be involved in the synthesis of β-1,2-glucan (Zorreguieta et al. 1988). The genes at *att* left side are responsible for molecular signaling events, while the right side genes are involved in the synthesis of fundamental components. The *att* left side includes an operon composed by nine open reading frames (ORF). Four of these ORF exhibit homology to the genes involved in the periplasmic binding protein dependent (or ABC) transport system (Ames et al. 1990; Higgings et al. 1990). A failure observed in a mutant analysis showed that the production and accumulation of specific compounds are necessary for the attachment of bacteria (de la Riva et al. 1998). The ABC transporter genes may be required for the secretion of these substances or some plant-originated activators are introduced into bacteria by the synthesis of compounds specific for the attachment (Matthysse et al. 1996; de la Riva et al. 1998).

After the attachment of bacteria to the plant cell wall, colonization of the plant cell wall begins through cellulose fibers produced by bacteria (Özcan et al. 2004). The *exo*C is required for the synthesis of glucose phosphate isomerase and mutants of the *exo*C are therefore unable to produce extracellular polysaccharides (Uttaro et al. 1990). Furthermore, mutants in the *Agrobacterium* chromosomal *cel* genes (*cel*ABCDE) show deficiency in cellulose synthesis and aggregate formation (Matthysse 1983), thus, these mutants exhibit weak binding capacity to plant wounded sites (Ziemienowicz 2001). In addition to bacterial adhesins and other adhesion factors, several plant factors (plant adhesins) are essential for attachment of *Agrobacterium* to plant cells. These include pectin acceptors, and vitronectin-like and germin-like proteins (Matthysse and Kijne 1998; Ziemienowicz 2001).

4.2 *Induction of Bacterial Virulence System*

The products of the genes encoded by the 30–40 kb *vir* region of the Ti plasmid mediate T-DNA transfer process. The activation of *vir* genes is carried out by two genes, *vir*A and *vir*G, during co-culture with plant cells (Stachel and Nester 1986). Although constitutive expression of *vir*A and *vir*G are basically at low levels but they are also highly stimulated in an autoregulatory fashion (Winans et al. 1988; McCullen and Binns 2006). There is substantial homology between *vir*A and *vir*G and the genes encoding two-component regulatory systems in which a response regulator is activated as a result of controlling the latter's phosphorylation status by a sensor kinase in response to signal input (de la Riva et al. 1998; Wolanin et al. 2002; McCullen and Binns 2006).

VirA is a transmembrane dimeric protein localized on the plasma membrane of the bacterial cell and it acts as a sensor in detection of signal molecules, mainly small phenolic substances, released from wounded plants (Pan et al. 1995; Gelvin 2003). Activation of VirA is achieved by the signals including acidic pH, phenolic substances such as acetosyringone (Winans 1992; Ziemienowicz 2001) and certain group of monosaccharides, which work synergistically with phenolic substances (Ankenbauer and Nester 1990; Cangelosi et al. 1990; Shimoda et al. 1990; Doty et al. 1996; de la Riva et al. 1998). VirA protein consists of three domains: the periplasmic or input domain and two transmembrane domains (TM1 and TM2). The transmembrane domains work as a transmitter (signaling) and receiver (sensor) (Parkinson 1993). The periplasmic domain is required for detection of monosaccharides (Chang and Winans 1992). This structure is a common structure for other transmembrane sensor proteins and the protein is anchored in the inner membrane by a simultaneous alignment with the inner membrane as a result of protein folding

(Seligman and Manoil 1994). The TM2 is the kinase domain and plays a crucial role in the activation of VirA (Huang et al. 1990; de la Riva et al. 1998).

VirA carries out amplification of the transformation system by detection of monosaccharides in the presence of low concentrations of phenolic compounds. This system could be induced only through the periplasmic sugar (glucose/galactose) binding VirE (Ankenbauer and Nester 1990; Cangelosi et al. 1990), which interacts with VirA (Shimoda et al. 1990; Chang and Winans 1992; Shimoda et al. 1993; Turk et al. 1994; de la Riva et al. 1998). In addition, for induction of *vir* gene expression, the phenolic substances are absolutely required whereas the other signals sensitize the bacteria to the phenols (McCullen and Binns 2006).

Activated VirA has the capacity to transfer its phosphate to a conserved aspartate residue of the cytoplasmic DNA binding protein VirG (Jin et al. 1990a, b; Pan et al. 1993). VirG functions as a transcriptional factor regulating the expression of *vir* genes when it is phosphorylated by VirA (Jin et al. 1990a, b; McCullen and Binns 2006). Transcription of *vir*BCDEFGH genes are activated by phosphorylated VirG protein with binding to "*vir* boxes". Except *vir*H gene, all other *vir* genes are involved in T-DNA processing, transport and integration and are essential for the infection (Ziemienowicz 2001).

External factors like temperature and pH also play important roles in activation of *vir* system. The expression of *vir* genes is not possible at temperature higher than 32°C due to a conformational change in the folding of VirA causing the inactivation of its properties (Jin et al. 1990a).

4.3 Generation of T-DNA Transfer Complex

Generation of single-stranded (ss) molecules is produced by activation of *vir* genes representing the copy of the bottom T-DNA strand. The copy number of T-DNA strands is approximately one copy per induced *Agrobacterium* cell and is derived from the coding strand of the T-DNA element (Stachel and Nester 1986; Sheng and Citovsky 1996). Two 25 bp long imperfect direct repeats flank the region known as border sequences. VirD1 and VirD2 proteins from the *vir*D operon recognize and cleave the T-DNA borders (Yanofsky et al. 1986; Stachel and Nester 1986; Filichkin and Gelvin 1993; Ziemienowicz 2001). The nick sites are considered to be the initiation and termination sites for T-strand recovery. After endonucleotidic cleavage by VirD2, the 5′ of the nicked T-DNA gets covalently bound to VirD2. Binding to the 5′ end it prevents 5′ exonuclease activity (Dürrenberger et al. 1989) and distinguishes the 5′-end as the leading end of the T-DNA transfer complex. VirD1 is considered as topoisomerase and it binds to the region where the ss-T-strand is originated. *In vitro* experiments proved that the presence of VirD1 is required for the cleavage of supercoiled stranded DNA by VirD2 (Zupan and Zambryski 1997; Christie 1997). The simultaneous establishment of the excised ss-T-strand is evolutionarily related to other bacterial conjugative DNA transfer processes (Lessl and Lanka 1994; Christie 1997; Zupan and Zambryski 1997; de la Riva et al. 1998; Zupan et al. 2000).

It seems that the right border acts as a starting site for T-strand synthesis, progressing leftwards and terminating at left border even when the left border is mutated or completely absent, although with lower efficiency. The left border can act as an initiation point for ss-T-strand synthesis but the efficiency is much lower (Filichkin and Gelvin 1993).

There is no difference between the right and left border sequences but the right border is more active than the left. The difference between two borders is the presence of an enhancer or "overdrive" sequence close to the right border (Peralta and Ream 1985). It has been found that VirC1 protein specifically recognizes this enhancer (Toro et al. 1989). VirC binds to overdrive region at the right end of the right border of T-DNA (Toro et al. 1989), and helps in unwinding the dsDNA and in activating cleavage of DNA at the right border (Zhu et al. 2000; Ziemienowicz 2001; Gelvin 2003). Deletion of *virC* operon is followed by attenuation of virulence of the *Agrobacterium* strains (Rogowsky et al. 1987; van Haaren et al. 1988; Gelvin 2003).

4.4 The T-DNA Transfer

It is implied that the T-strand from the bacterium passes into the plant cell as a protein-nucleic acid (ss-T-DNA-protein) complex (Sheng and Citovsky 1996) and during translocation of ss-T-DNA-protein complex cross over three membranes, the plant cell wall and cellular spaces (de la Riva et al. 1998). This T-DNA transport is intermediately descripted as the T-complex (Howard and Citovsky 1990) and is a T-strand DNA molecule (carries the genetic information) associated with two proteins, VirD2 and VirE2 (protect the T-strand, help in transporting the DNA across shape and supply specific targeting signals) (Table 12.3) (Sheng and Citovsky 1996). VirE2 is a 69-kDa single stranded DNA binding protein and according to the most commonly accepted model, by binding of VirE2 to the ss-T-DNA-VirD2 complex, a cooperative association arises, preventing from any nuclease digestive activity, stabilizing the DNA into a rigid rod-like structure and making the translocation through membrane channels easier (de la Riva et al. 1998).

VirE2 contains two plant nuclear location signals (NLS) and VirD2 (Tinland et al. 1995). This fact indicates that both proteins play roles in importing proteins to the plant cell nuclei, and a plant import has been shown to mediate nuclear import of VirD2 (Ballas and Citovsky 1997; de la Riva et al. 1998) once the complex enters into the plant cell, the complex is transported to the nucleus (Herrera-Estrella et al. 1988; Shurvinton et al. 1992; Tinland et al. 1995; Rossi et al. 1996; Sheng and Citovsky 1996; Zupan et al. 1996). It is known that VirE1 is required for the export of VirE2 to the plant cell, but other specific functions still await characterization (Binns et al. 1995; Gelvin 2003).

An alternative model has been brought to light for ss-T-DNA complex transfer. This model proposes that covalent attachment of VirD2 to the single strand DNA 5′ end, forms the transfer complex but is uncoated by VirE2. Following the independent transfer of VirE2 into plant cell as the result of natural process, the naked

Table 12.3 The functions of Vir proteins both in *Agrobacterium* and plant cells

Vir protein	Function in *Agrobacterium*	Function in plant
VirA	Sensor for phenolic substances	
VirG	Phenolic response regulator	
VirB1-11	Synthesis of T-pilus, showing ATPase activity and provide the energy required for export of other protein subunits, for T-DNA transport. They thought to form pore for T-DNA export	
VirC1	Possible "overdrive" binding protein; increasing T-DNA transfer	
VirD1	*In vivo* T-DNA process and *in vitro* nicking in T-DNA borders	
VirD2	T-DNA border specific endonuclease activity (ss-T-DNA-VirD2 complex)	Nuclear targeting of the T-strand, protection of the T-strand from 5′exonucleolytic degradation, T-strand integration into the plant genome
VirD4	ATP-dependent linkage of protein complex and necessary for T-DNA translocation	
VirE1	Carrying VirE2 from *Agrobacterium* and its protection	
VirE2	Protecting the T-strand, shape it into a transferable (thin and unfolded) form and supplying specific targeting signals	Formation of a putative "T-complex" in the plant, protection of the T-strand from nucleolytic degradation, unclear targeting of the Ti-strand, passage of the Ti-strand through the nuclear pore complex
VirF	Nuclear targeting of the ss-T-DNA	Host range factor, possible interaction with Skpl proteins to regulate plant cell division cycle
VirH	Possible cytochrome p450 enzyme, plays a role in the host range specificity of bacterial strain	

ss-T-DNA-VirD2 complex is coated by VirE2 inside the plant cell (Lessl et al. 1992; Binns et al. 1995; de la Riva et al. 1998).

It is also possible that the process can be performed by one of the proposed alternatives according to the conditions of infection. Previous studies report that the *vir*B operon products organize suitable cell surface structures for the transfer of ssDNA across the membrane into host cell cytoplasm (Finberg et al. 1995; Stephens et al. 1995; Fernández et al. 1996; Beaupré et al. 1997; Dang and Christie 1997; Rashkova et al. 1997; Zhou and Christie 1997; de la Riva et al. 1998). The VirD4 protein is also required for the ss-T-DNA transport. VirD4 is necessary in the translocation of T-DNA by generating the ATP- dependent linkage of protein complex (Firth et al. 1996). VirB are proteins that present hydropathy characteristics similar to other membrane-associated proteins (Thompson et al. 1988; Ward et al. 1988;

Kuldau et al. 1990; Shirasu et al. 1990; Shirasu et al. 1994; de la Riva et al. 1998). VirD4 is a transmembrane protein but predominantly located at the cytoplasmatic side of the cytoplasmic membrane (Okamoto et al. 1991). Comparative studies show that the genetic organization of transfer regions of broad host range (BHR) plasmids reveals a high degree of homology to the *vir*B operon in genetic organization, nucleotide sequence and protein function (Lessl et al. 1992; Pohlman et al. 1994; de la Riva et al. 1998). Both systems deliver non-self transmissible DNA-protein complex to the recipient host cell. In addition, they are capable for DNA interkingdom delivery (Heinemann and Sprague 1989; Bundock et al. 1995; Piers et al. 1996) suggesting that the T-DNA transfer apparatus and conjugation systems are displaying relatedness and apparently share a common ancestral origin (Christie 1997; Oger et al. 1998). The majority of VirB proteins are organized in a membrane-associated pore-forming structure involving both membranes (Shirasu and Kado 1993a, b; Shirasu et al. 1994; Stephens et al. 1995; Das and Xie 1998). Two VirB proteins, VirB4 and VirB11, peripherally associate with others and located primarily in the cytoplasm, but a small part of VirB4 may span the inner membrane. These proteins serve as ATPases and are thought to provide energy for the physical movement of the DNA into the host cell across the cellular barrier and for export of other protein subunits or both. VirB proteins are thought to form pore for T-DNA export. Recent studies have demonstrated that VirB proteins form pili that show high degree similarity with the conjugative pili (Fullner et al. 1996; Ziemienowicz 2001) and VirB2 constitutes the major subunit of this pili. VirB7, an outer membrane lipoprotein, may facilitate to anchor pilus to the bacterial cell by forming disulfide bonds with the periplasmically localized VirB9 (Ziemienowicz 2001).

*vir*F and *vir*H operons are accessory in the octopine Ti plasmid. One of the *vir*F operon products is a 23 kDa protein that functions once the T-DNA complex is inside the plant cells via the conjugal channel or independently as it was considered for VirE2 export. VirF plays a role identified as assistance in the nuclear targeting of the ss-T-DNA complex (Hooykaas and Shilperoort 1992). The *vir*H operon consists of two genes that code for VirH1 and VirH2 proteins. These Vir proteins are functionally redundant but could enhance the transfer efficiency, detoxifying certain plant compounds that can affect bacterial growth (Kanemoto et al. 1989). If this is the function of VirH proteins, then they play a role in the bacterial attachment to the host plant species (de la Riva et al. 1998).

4.5 Integration of T-DNA into Plant Genome

The *Agrobacterium* T-complex is imported into the host cell nucleus and then the DNA becomes integrated into the host chromosomes (Sheng and Citovsky 1996). The ss-T-DNA complex is directed to the nucleus passing the nuclear membrane inside the plant cell. Both VirD2 and VirE2 proteins of the T-complex have been found to play roles in the integration process and VirF probably plays a minor role in this process (Hooykaas and Shilperoort 1992; Sheng and Citovsky 1996). In addition, the nuclear location signals (NLS) of VirD2 and VirE2 are identified as important players in nuclear targeting of the delivered ss-T-DNA complex, as early described. VirD2 has one functional NLS (de la Riva et al. 1998).

A large nucleoprotein complex called ss-T-DNA complex is created when one molecule of VirD2 is covalently attached to the 5′-end of the T-strand. Nevertheless, the complex is coated with VirE2 molecules and each of them has two NLS. The two NLS of VirE2 are responsible for the continuous nuclear import of ss-T-DNA complex. They probably keep both sides of nuclear pores simultaneously open (de la Riva et al. 1998). The nuclear import process is initiated by certain NLS-binding proteins in the plant cytoplasm. The final step of T-DNA transfer is its integration into the plant genome. The mechanism of the T-DNA integration is largely uncharacterized. It is considered that the integration occurs by illegitimate recombination (Gheysen et al. 1991; Lehman et al. 1994; de la Riva et al. 1998; Puchta 1998).

A sequence at T-DNA 3′-end finds homology with plant DNA resulting in a first contact between the T-strand and plant DNA and generating a gap in 3′–5′ strand of plant DNA. Endonucleases digest the displaced plant DNA at the 3′-end position of the gap and VirD2 attached to the 5′-end accompanies with a nucleotide in the top (5′–3′) plant DNA strand. The 3′ overhanging part of T-DNA and displaced plant DNA are removed away by digesting either with endonucleases or 3′–5′ exonucleases (de la Riva et al. 1998). Then, the 5′ attached to VirD2 end and other 3′-end of T-strand joins the nicks in the bottom plant DNA strand (Tinland et al. 1995). Once T-strand is introduced into the 3′–5′ strand of the plant DNA, a torsional strain would then result in the introduction of a nick into the opposite plant DNA strand. The repair mechanism of the plant cell is activated because of this situation and using the early inserted T-DNA strand as a template, complementary strand is synthesized (Fig. 12.6) (Tinland et al. 1995; de la Riva et al. 1998).

The published data indicates that the function of VirD2 is the precise integration of T-strand into the plant chromosome. VirD2 may provide energy by releasing from its phosphodiester bond, at the Tyr29 residue, with the first nucleotide of T-strand providing the 5′-end of the T-strand for ligation to the plant DNA. This phosphodiester bond is electrophilic center and may act as substrate for nucleophilic 3′-OH from nicked plant DNA (Jayaram 1994). VirD2 is also implicated in other functions such as recruiting plant enzymes involved in DNA repair or recombination to the site of the integration and/or by interaction with some structural chromatin proteins (Ziemienowicz 2001).

5 Use of *Agrobacterium tumefaciens* in Plant Biotechnology

5.1 General View to Plant Biotechnology

Hungarian agricultural engineer Karl Ereky first recorded use of the term "biotechnology" in 1919 (Hopkins 2007). It is defined as "any technological application that uses biological systems, living organisms, or derivatives thereof, to make or modify products or processes for specific use". In fact, the origin of biotechnology could be traced back to prehistoric times. In those times microorganisms had already been used for the processes like fermentation, for production of yoghurt, cheese and vinegar (Chawla 2002). Nevertheless, biotechnology was applied throughout for these products without any real scientific understanding of what was occurring (Lindsay 2002). Biotechnology got boost in the 1970s with the discovery of restriction enzymes, which cut double stranded or single stranded DNA at specific recognition nucleotide sequences (Roberts 1976; Chawla 2002), gene technologies began developing rapidly after that. Today, biotechnology is synonymous with genetic engineering. The term genetic engineering, which was first coined by Jack Williamson, refers to a number of new techniques involving transfer of specific genetic information from one organism to another (Chawla 2002; Stableford 2004).

Plant biotechnology starts with tissue culture techniques (Chawla 2002). Plant cell and tissue cultures are sets of techniques that were designed for growing and cloning any plant parts by using some culture media in aseptic and controlled conditions (Vargas and Flota 2006). With the use of tissue culture techniques, elite plant varieties are clonally propagated, endangered plants are conserved, virus-free plants are produced, germplasm is conserved, and finally secondary metabolites are produced. Additionally, tissue culture serves as an indispensable tool for transgenic plant production (Cardoza 2008). These techniques were developed in the early 1960s and became standard procedures for modern biotechnology. Presently five major areas are recognised in this direction **(1)** large-scale propagation of elite materials, **(2)** generation of genetically modified fertile individuals, **(3)** as a model system for fundamental plant cell physiology aspects, **(4)** preservation of endangered species and **(5)** metabolic engineering of fine chemicals (Vargas and Flota 2006).

Tissue culture techniques have made it possible to transfer foreign genes from one plant to another. Recognition of the biology of *A. tumefaciens* and application of its T-DNA system has been a great step in plant biotechnology. In general, because of these modern gene technological methods, plant biotechnology has grown into a new dimension with putative future possibilities that can hardly be overestimated (Twyman et al. 2002). Today, related with the improved tissue culture and gene transfer methods, plant biotechnology has become one of the most important branches in science. The history of plant biotechnology, especially related to *Agrobacterium* sp. is given below chronologically (Table 12.4).

5.2 *Agrobacterium and Plant Genetic Engineering*

One of the most significant developments in plant biotechnology is stably introducing foreign genes into plant genome. Attempts have been made by using transgenic techniques for manipulating commercially important plants in conjunction with their secondary metabolic pathways for production of a variety of important metabolites (Merkli et al. 1997). The advantages of transformation over conventional cell culture systems are numerous including fast growth and stable high-level production of secondary metabolites making them ideal for biotechnological utilization (Chawla 2002; Twyman et al. 2002).

Table 12.4 The brief history of plant biotechnology related to *Agrobacterium* sp

1853	The first written report of crown gall disease (Fabre and Dunal 1853)
1897	*Agrobacterium vitis* identified as causal agent of crown gall in grape (Cavara 1897)
1902	First attempt of plant tissue culture (Haberlandt 1902)
1907	*A. tumefaciens* identified as causal agent of crown gall in Paris daisy (*Argyranthemum frutescens*) (Smith and Townsend 1907)
1941	*In vitro* culture of crown gall tissues (Braun 1941)
1947	Sterile plant tumor tissue can proliferate indefinitely on hormone-free medium in culture. Tumor cells are proposed to be 'transformed' by an *Agrobacterium*-derived tumor-inducing principle (TIP) (Braun 1947)
1952	First application of micrografting (Morel and Martin 1952)
1954	First plant from single cell (Muir et al. 1954)
1956	Unusual low-molecular weight nitrogenous compounds (opines) are identified exclusively in tumor tissue (Lioret 1956)
1959	Publication of first handbook on plant tissue culture (Gautheret 1959)
1962	Development of Murashige and Skoog nutrient medium (Murashige and Skoog 1962)
1970	Discovery of first restriction endonuclease from *Haemophilus influenzae* Rd. It was later purified and named *HindI* (Smith and Wilcox 1970)
1971	*A. tumefaciens* loses virulence when grown at 37°C. The TIP can be transferred between virulent and avirulent *A. tumefaciens* strains (Hamilton and Fall 1971; Kerr 1971)
	Preparation of first restriction map using *HindI* enzyme to cut circular DNA of SV 40 into 11 specific fragments was prepared (Danna and Nathans 1971)
1974	*A. tumefaciens* virulence depends on the presence of a large 'tumor-inducing' (Ti) plasmid. The TIP is probably a component of the Ti plasmid (Zaenen 1974)
	Biotransformation in plant tissue cultures (Reinhard 1974)
1976	Octopine and nopaline synthesis and breakdown found to be genetically controlled by the Ti plasmid of *A. tumefaciens* (Bomhoff et al. 1976)
1977	The T-DNA region of the Ti plasmid is present in the genome of crown gall tumor cells: the T-DNA is the TIP (Chilton et al. 1977)
1980	The opine concept: the synthesis of opines by transformed cells creates an ecological niche for the infecting strain of *Agrobacterium* (Guyon et al. 1980)
1983	The first plant transformed with a recombinant gene using *A. tumefaciens* as a vector (Zambryski et al. 1983)
1984	T-DNA oncogenes are identified that mediate overproduction of auxin and cytokinin (Klee et al. 1984; Lichtenstein et al. 1984)
	Development of the genetic fingerprinting technique for identifying individuals by analyzing polymorphism at DNA sequence level (Jeffreys et al. 1984)
1985	The *virA*/*virG* two-component regulatory system is identified as a central component of signal perception and transduction in *Agrobacterium* transformation (Stachel and Zambryski 1986)
1987	Isolation of *Bt* gene from bacterium (*Bacillus thuringiensis*) (Barton et al. 1987)
1990	Development of the random amplified polymorphic DNA (RAPD) technique (Williams et al. 1990; Welsh and McClelland 1990)
2001	Publication of the complete genome sequence of two *A. tumefaciens* strains (Goodner et al. 2001; Wood et al. 2001)
2005	220 million acres of biotechnologic crops with herbicide tolerance and/or insect resistance traits were cultivated in 21 countries worldwide (James 2010)
2007	The global area of biotech crops reached 114.3 million hectares (282.4 million acres) worldwide (James 2010)
2010	The global area of biotech crops continued to soar for the fifteenth consecutive year at a sustained growth rate of 10% or 14 million hectares (35 million acres), reaching 148 million hectares or 365 million acres (James 2010)

One of the earliest biotechnology success stories in plant research is the *Agrobacterium* Ti plasmid, which is a classic example of how a useful research tool is created by years of effort combined with chance (Table 12.4) (Mugnier et al. 1986). Elucidation of molecular mechanism of plant genetic transformation via *A. tumefaciens* was made by scientists years ago and Armin Braun proposed the concept of a "tumor-inducing principle" that was stably transferred to and propagated in the plant genome (Braun 1947; Gelvin 2003). As a result of intense research in the 1970s on virulent *Agrobacterium* strains resulted in the identification of large plasmids (Zaenen et al. 1974), although we now know that many strains harboring plasmids are unrelated to virulence. Genetic experiments indicated that the Ti plasmids were involved in tumorigenesis (van Larebeke et al. 1974) and transmission to plant cells and integration into the plant genome is achieved by a specific segment of these plasmids called the T-DNA (Chilton et al. 1977; Zhu et al. 2000; Ream 2002; Özcan et al. 2004). It was thus obvious to suggest that Ti plasmids can be used as a vehicle to transfer and integrate foreign genes into plant cells. This knowledge, "the ability of *A. tumefaciens* to transfer a fragment of its T-DNA" provided the use of *A. tumefaciens* as highly efficient versatile vehicle for introduction of foreign genes into the desired plant genome and provided this bacterium to be used as a powerful tool for plant biotechnology applications (Ziemienowicz 2001; Chawla 2002; Twyman et al. 2002). However, the following properties of Ti plasmids prevent their direct use: **(1)** large size, **(2)** tumor inducing property and **(3)** absence of unique restriction enzyme sites (Chawla 2002).

Some manipulations are required for Ti plasmid before using *Agrobacterium* in genetic engineering. Hormones produced by oncogenes encoded by T-DNA are involved in tumor formation and establishment of a suitable regeneration to the tissue cultured explants can not be accomplished because of these tumors. The hormone imbalance in plant tissues causes disorganized callus growth (Hernalsteens et al. 1980; Twyman et al. 2002). Therefore, an important step in the development of T-DNA vectors is the realization that the transfer of T-DNA to the host plant was dependent on the *vir* genes and the 24-bp direct repeat structures marking the left and right borders of the T-DNA. Within the T-DNA, genes are not essential for transformation, and any sequence could be incorporated therein. This allowed creating disarmed Ti plasmids where all the oncogenes were removed, facilitating T-DNA transfer to plant cells without causing neoplastic growth. Systematically modified a nopaline Ti plasmid, pTiC58, could be used as a vector for the transfer of T-DNA into plant genomes but no longer cause tumors (Zambryski et al. 1983; Wood et al. 2001). This was achieved by deleting all of the T-DNA except for the left and right border sequences, and *nos* coding sequence and replacing this with a region of pBR322. The resulting plasmid pGV3850 was non-oncogenic and disarmed. *Agrobacterium*-carrying pGV3850 were used to infect discs of carrot or potato, or wounded petunia and tobacco plants. Crown gall-like tumors were not produced and callus tissue was only observed when the explants of wounded tissue or discs were cultured in the presence of phytohormones. These results showed that the deletion of internal sequences was required for hormone independent growth (Zambryski et al. 1983). These types of studies offered new promising approaches for the production of transgenic plants where foreign genes could be transmitted through seeds (Barz et al. 2002).

Disarmed Ti plasmids are not preferred for use as primary cloning vectors because of their large size and lack of unique restriction enzyme sites. Therefore, two different types of vector systems called co-integrated and binary vectors have been constructed to be used in transformation carrying the foreign genes inserted into these vectors by plant scientists lately (Özcan et al. 2004).

An incorporation of suitable selectable markers into the T-DNA, Ti plasmids turn into very powerful gene delivery vectors. However, wild-type Ti plasmids are inappropriate for this situation as a result of their large size making them difficult to manipulate *in vitro*. An early approach to cope with this problem required subcloning of the T-DNA into a standard *Escherichia coli* plasmid vector which allowed *in vitro* manipulation by normal procedures, and then integrated into a resident disarmed Ti plasmid in *A. tumefaciens* via homologous recombination (Matzke and Chilton 1981). Although it was an easy system to use, and only required three bacterial strains (triparental matings), the system relied on a complex series of conjugative interactions between *E. coli* and *A. tumefaciens*. However, because the *vir* genes act in *trans* for mobilization and transfer of the T-DNA, it was soon realized that using natural large size of Ti plasmids was unnecessary. Intermediate vectors have been discarded and largely replaced by binary vectors (Bevan 1984), in which the *vir* genes and the T-DNA are cloned on separate plasmids. These can be inserted into *A. tumefaciens* by conjugation with an *E. coli* donor or by freeze-thaw cycles or electroporation. Binary vectors are employed in most contemporary *Agrobacterium*-mediated transformation systems (Barz et al. 2002).

5.2.1 Cointegrate Vectors

Vectors that combine via DNA homology into a resident Ti plasmid are often referred as integrative or co-integrate vectors, or hybrid Ti plasmids (Chawla 2002). For production of an intermediate vector, a conventional *E. coli* plasmid vector was used in subcloning of T-DNA isolated from a parent Ti plasmid for easy manipulation (Matzke and Chilton 1981). These vectors were not capable of replicating themselves in *A. tumefaciens*, and also had no conjugation functions (Fig. 12.7) (Primrose and Twyman 2006).

Transfer capacity for plasmids between different species of Gram-negative bacteria relies on several properties of plasmids. Both *cis*-acting region on the plasmid and *trans*-acting plasmid-encoded molecules are necessary for conjugal transfer. Additionally, many plasmids have a narrow host range for replication and/or they are incompatible with another plasmid in the same cell (Paszkowski et al. 1988; Chawla 2002). Three structures necessary for establishing this system easily are; **(1)** an *E. coli* strain carrying a helper plasmid, which is able to mobilize the intermediate vector in *trans*, **(2)** the *E. coli* strain carrying the recombinant intermediate vector and **(3)** disarmed *Agrobacterium* Ti plasmid (Primrose and Twyman 2006).

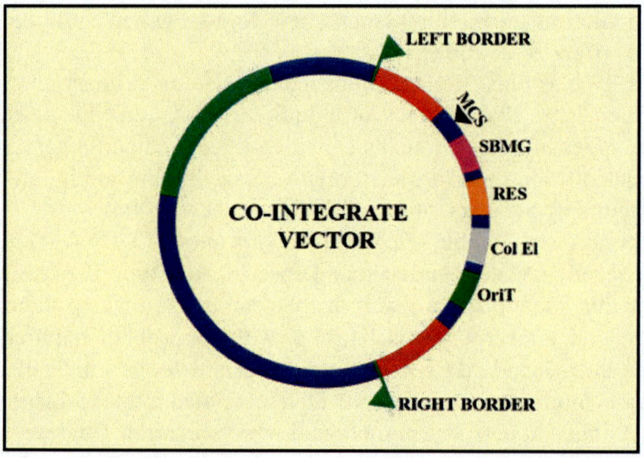

Fig. 12.7 Cointegrate plasmid and its components. *MCS* Multiple cloning site, *SBMG* Selective plant marker gene, *RES* Antibiotic resistance gene, *OriT* Origin of transfer, *Col E1* Replication origin, which belongs to pCol E1 plasmid (Modified from Armitage et al. 1988; Walkerpeach and Velten 1994 and Özcan et al. 2004)

For recombinant T-DNA, maintenance appears to be dependent on recombination. Efficiency of recombination is enhanced if there is a large homologous region shared by the two plasmids. For example, Ti plasmid pGV3850 carries a segment of the pBR322 backbone in its T-DNA (Chawla 2002; Primrose and Twyman 2006). However, a co-integrate plasmid constructed *in vitro* is assembled from parts of different plasmids. Normally these parts are; **(1)** *vir* genes, **(2)** left and right T-DNA borders, **(3)** an exogenous DNA sequence between the two T-DNA borders, and **(4)** plant and bacterial selectable markers (Chawla 2002).

5.2.2 Binary Vectors

The generation of transgenic plants is greatly simplified by T-DNA binary systems. Sophisticated and complex microbial genetic approaches are no longer required to integrate T-DNA regions located on large Ti or Ri-plasmids (Lee and Gelvin 2008). The binary vector systems consist of two autonomously replicating plasmids within *Agrobacterium*; **(1)** a shuttle (binary) vector containing gene of interest between the T-DNA borders and **(2)** a helper Ti plasmid which provides the *vir* genes necessary for facilitating transfer into plant cells (Chawla 2002). A T-DNA is considered disarmed if its tumor-inducing (oncogenic) genes are removed by genetic engineering methods (Chawla 2002; Lee and Gelvin 2008).

A standard binary vector consists of **(1)** multiple cloning site, **(2)** a broad host range origin of replication functional in both *E. coli* and *A. tumefaciens,* **(3)** a selectable marker for both plant and bacteria and **(4)** the left and right T-DNA borders (or at least the right T-border) (Fig. 12.8) (Hoekema et al. 1983; Bevan 1984).

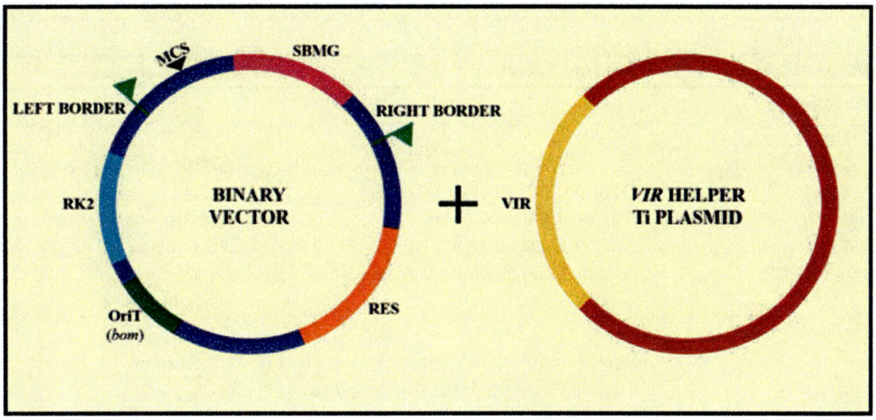

Fig. 12.8 Binary plasmid, its components and *vir* helper Ti plasmid. *MCS* Multiple cloning site, *SBMG* Selective plant marker gene, *RES* Antibiotic resistance gene, *OriT* Origin of transfer and ***bom*** region, *RK2* Replication origin, which belongs to pRK2 plasmid (Modified from Armitage et al. 1988; Walkerpeach and Velten 1994 and Özcan et al. 2004)

In general, the transformation procedure is as follows; **(1)** direct transfer of the recombinant small replicon to *A. tumefaciens* harboring a helper Ti plasmid via bacterial conjugation or direct transfer, **(2)** co-cultivating the plant cells with the *Agrobacterium* for allowing transfer of recombinant T-DNA into the plant genome and **(3)** selecting transformed plant cells under appropriate conditions. The stability of wide host range replicons in *E. coli* and *Agrobacterium* varies considerably, some having possible disadvantages in this type of plasmids. Two different origins of replication known to be active *E. coli* may be unstable depending on the orientation (Lee and Gelvin 2008). There are many advantages of these types of vectors when compared with co-integrated vectors; **(1)** no recombination process takes place between the molecules involved in binary vectors **(2)** high transformation efficiency from *E. coli* to *Agrobacterium* is achieved by using small vectors instead of using large, recombinant, disarmed Ti plasmid (2–3 days versus 4–7 days) **(3)** *Agrobacterium* carrying plant ready genes on a binary vector is easily and efficiently obtained **(4)** the only requirement for binary vectors is an intact plasmid to be inserted into the target bacterium, making the bacterial transformation process more efficient and faster (2–3 days versus 4–7 days) (Chawla 2002; Lee and Gelvin 2008).

5.3 Marker and Reporter Genes

The success rate generally appears to be at a very-low frequency for integration of introduced foreign genes during the genetic transformation of plants (Hopkins 2007). Therefore, the availability of monitoring techniques for analyzing the expression of

Table 12.5 The mostly used marker and reporter genes, their encoded enzymes and conferred resistances

Marker Gene	Enzyme Encoded	Resistance Conferred
Antibiotics		
npt II	Neomycin phosphotransferase	Kanamycin, neomycin, G418, paromycine
hpt or *aph* IV	Hygromycin -phosphotransferase	Hygromycin
dhfr bacterial or mouse	Dihydrofolate reductase	Methotrexate
bla	TEM-1 β-lactamase	Ampicillin
aadA	Association with several transposons (Tn7, Tn21, …)	Streptomycin and spectinomycin
Herbicides		
bar	Phosphinothricin acetyltransferase	Phosphinothricin
aro A	5-enolpyruvylshikimate-3-phosphate synthase	Glyphosate
Modified *als* genes	Acetohydroxyacid synthase (or acetolactate synthase)	Chlorsulfuron, imidazolanones
Reporter Genes		
CAT	Chloramphenicol acetyltransferase	
GUS	β-glucuronidase	
npt II	Neomycin phosphotransferase	
Luc	Luciferase	
bar	Phosphinothricin acetyltransferase	
β-gal	β-galactosidase	

introduced foreign genes in cells and tissues is necessary in gene transfer studies (Galbraith et al. 1995). This can be accomplished by using a marker gene for selection and regenerating plants under the appropriate culture conditions in the genetic transformation (Twyman et al. 2002). For optimization of co-cultivation parameters, culture media, explants and *in vitro* growth conditions for developing efficient transformation protocols in morphogenic tissues and protoplasts, marker genes have been utilized in most of the transformation experiments (Rugini et al. 2000). To achieve the above target, cloning vectors are constructed containing marker genes along with the 'gene of interest'. The marker genes are broadly of two types: **(1)** Selectable markers and **(2)** Reporter genes (Stirn and Lörz 2006; Miki 2008). Selectable marker genes encode easily detectable traits enabling to distinguish marked cells from non-marked cells. The two most widely used selectable marker genes encode proteins that provide herbicide and antibiotic tolerance (Table 12.5) (Galbraith et al. 1995).

A reporter gene is a gene, which encodes a product that can readily be assayed. Reporter genes have been attached to another gene of interest in cell culture, animals or plants (Koo et al. 2007). While the selectable marker genes help the researcher select transgenic tissue, reporter genes usually report which cells are transgenic (Miki 2008). In contrast with selectable marker genes, reporter genes do not confer resistance to selective agents inhibitory to plant development (Stirn and Lörz 2006).

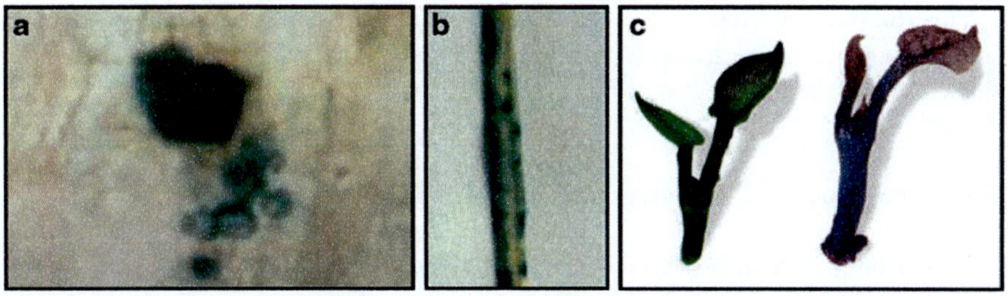

Fig. 12.9 A-*GUS* gene expression in a transversal section of a mature embryo of barley (*Hordeum vulgare* L.) (*Magnification*: 100×) and B-*GUS*-expressing cell clusters in a mature leaf segment from a 3-month-old plantlets of barley (From Gürel and Gözükirmizi 2003) C-Control (*left*) and *GUS* gene expression showing (*right*) young cotton (*Gossypium hirsutum* L.) plantlets (From Özyiğit et al. 2006)

Reporter genes code for products which can be detected directly or catalyze reactions whose products are detectable and as a result, a scientist can detect when and where the gene is active in the genome (Table 12.5) (Schrott 1995; Stirn and Lörz 2006).

The *CAT* (Chloramphenicol acetyl transferase) gene has been employed as a reporter gene in assessing the conditions for polyethylene glycol (PEG)-mediated transfection of kiwifruit protoplasts (Oliveira et al. 1991). A transient expression of *CAT* gene was observed in protoplasts of cv. Harward by using 30% polyethylene glycol 4,000 and by having protoplasts at 45°C for 5 min before transfection. Further research on DNA uptake clarified that transient expression rates of either the *CAT* or the β-glucuronidase (*GUS*) genes were influenced negatively when electroporation was tried without reproducing plasmid uptake (Oliveira et al. 1994; Rugini et al. 2000). The transient *GUS* gene expression was studied in two protoplast populations obtained from *in vitro*-grown leaf material of the same cultivar. One population originating from epidermis and leaf veins was composed of hyaline or light green protoplasts and the other originating from mesophyll was composed of green protoplasts very dense in chloroplasts (Raquel and Oliveira 1996; Rugini et al. 2000). One of the most important systems used to study gene regulatory elements and mechanisms in plants has been the *GUS* gene system (Fig. 12.9) (Miki 2008).

5.4 PCR and Blotting Techniques

Polymerase chain reaction (PCR) and Southern blot analysis can be used in confirmation of the transgenic nature of regenerated plants (Narula et al. 2004). However, screening can be carried out by using Southern blotting or PCR for the confirmation of whether regenerants have integrated the antibiotic resistance gene (and also a gene of interest if this was linked to it within the T-DNA) (Maniatis et al. 1982; van der Meer 2006). Lately, for DNA analysis, northern and western blotting methods have been used (Vargas and Flota 2006). The basic principles of these methods are given here.

5.4.1 Polymerase Chain Reaction (PCR)

The polymerase chain reaction (PCR) is a highly sensitive scientific technique that is widely used to amplify specific DNA sequences requiring appropriate primers (Wu 1995a; Wu et al. 1993, 2004). The PCR technique developed by Kary Mullis in 1983 is now a common and often indispensable technique used in biological research labs for a variety of applications (Bartlett and Stirling 2003). PCR provides the ability to amplify and clone the genomic DNA, cDNA or RNA strands thousands of times, in a matter of minutes, by using the natural ability of an enzyme called polymerase. It is also a relatively rapid and inexpensive procedure (Edwards et al. 1991; Ausubel et al. 1995; Wu 1995a, b; Mohan et al. 1998; Wu et al. 2004). PCR makes studying a certain piece of DNA easier for scientists DNA and RNA expression, genetic diagnosis, detection of mutations and transgene analysis became routine applications by using of this technology (Mohan et al. 1998). Components required to carry out a PCR reaction are; **(1)** the DNA or RNA fragment to be copied, **(2)** two primer fragments, **(3)** the polymerase enzyme and **(4)** a special machine that carefully controls temperature.

With respect to the general principles of PCR procedures, oligonucleotide primers complementary to opposite strands of DNA region are prepared and separated by up to a few hundred base pairs. Denaturing is the word used for separation process (Mohan et al. 1998). *Taq* DNA polymerase purified from thermophile bacterium called *Thermus aquaticus* is a temperature-tolerant enzyme and catalyzes the synthesis of a new DNA strand complementary to a template DNA from 5′ to 3′ direction by using special short sequences called primers during the PCR (Nair 2008). The *Taq* DNA polymerase is stable in high-temperatures (90–96°C) and therefore amplification of target sequences is possible for many cycles using excess primers in a commercial thermocycler. In recent times, high- *Taq* polymerases, such as recombinant polymerase *Tth*, long-span polymerase and high-fidelity PCR polymerase, have been developed (Wu et al. 2004; Nair 2008). The next step involves adding the primers and lowering the temperature to facilitate bonding formation. Since forward and reverse primers are complimentary to the up- and downstream regions of the target DNA, they will bind to those areas. They are employed in the copying process as building blocks to up- and down regions. At this point, the polymerase and deoxyribonucleoside triphosphates are added and temperature is raised to match the optimal working temperature for polymerase. It recognizes the primers and starts to copy. After 20–30 cycles of PCR are performed, hundreds or thousands of copies of the original target sequence are produced (Mohan et al. 1998; Wu et al. 2004; Nair 2008).

The PCR technique has become a valuable tool in the field of molecular genetics because it offers rapid analysis, ease of automation, its relative economy and extraordinary specificity (Mohan et al. 1998). By contrast, the fact that false-negative results may present a problem if positive and negative controls are omitted or do not work in PCR process. Additionally, extra precautions should be taken when assaying putative transformants in PCR screens for eliminating the problem of false positives produced by the plasmid having a source of nonintegrated DNA in the plant cell in the process of *Agrobacterium*-mediated gene transfer. (Cubero and López 2005; Zale 2008). Young plantlets or seedlings regenerated directly after an *Agrobacterium* treatment without an intervening antibiotic treatment should not be used in PCR proceedings, and even then, there might be alive *Agrobacterium* cells existing in plant tissues. An antibiotic such as carbenicillin (Cheng et al. 1997) or cefotaxime (Broothaerts et al. 2005) could be used to control bacterial contamination in regenerated plantlets and whole seedlings (Zale 2008).

5.4.2 Southern, Northern and Western Blotting

There are three different blotting techniques, which are called Southern, northern and western blot. The Southern blot detects DNA (DNA-DNA hybridization), the northern blot detects mRNA (messenger ribonucleic acid, DNA-RNA hybridization) and the western blot detects proteins (protein-protein hybridization such as antigen-antibody binding or protein-ligand binding) (Nair 2008). The blotting techniques are often used to detect genetic abnormalities, detection of mutations and transgene analysis (Mohan et al. 1998). A mixture of DNA, RNA or protein fragments can be separated by gel electrophoresis and the obtained bands can be stained and visualized directly in the gel. However, confirmation and identification of these bands or finding the similarity of one or more of these bands with a known and available molecular probe, it is possible to hybridize these bands with a labeled probe (Wu et al. 2004).

To facilitate this hybridization, the bands are often transferred to a nitrocellulose membrane through a technique described as blotting. When DNA bands are blotted, it is called Southern blotting; when RNA bands are transferred and described as northern blotting and similarly when protein bands are transferred, the technique is described as western blotting.

The Southern Blot

The Southern blot technique was developed by E. M. Southern, begins with the isolation of genomic DNA from cells like peripheral leukocytes or fetal cells (Southern 1975). The technique is used to detect and identify specific DNA fragments on an agarose gel after the separation by electrophoresis (Mohan et al. 1998; Einspanier 2006). This technique consists of six main steps. They are; **(1)** preparation of DNA for analysis **(2)** separation of DNAs by an agarose gel **(3)** denaturation of dsDNAs into ssDNAs **(4)** transfer of DNAs onto a membrane **(5)** hybridization with a non-radioactive or a radioactive probe **(6)** detection of hybridized signals (Wu et al. 2004).

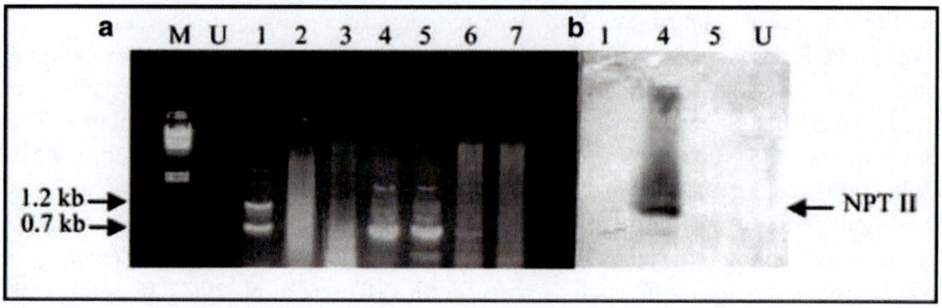

Fig. 12.10 A-PCR and B-Southern blot analysis of barley plants developed from transgenic embryos (From Gürel and Gözükirmizi 2003)

A series of DNA fragments which are reproducible can be obtained by digesting high molecular weight genomic DNA with restriction enzymes. These fragments are separated by electrophoresis in agarose gels based on their molecular weights (Mohan et al. 1998). The resulting dsDNA species are denatured into ssDNAs and then blotted or immobilized onto nylon or a nitrocellulose membrane. The transfer is exact and does not disturb the location and grouping of the DNAs from gel to paper. This process is called "blotting" (Beaudet 1991; Wu et al. 2004). Since there are usually many fragments on a blot, DNA probes are used to localise specific pieces of nucleotide sequences (Fig. 12.10). If the specific labeled probe finds a fit (because of complementarity) on the paper blot, it will bind to the DNA. The membrane is washed to remove unbound probe and then labeled probe is detected by using x-ray film (Beaudet 1991; Mohan et al. 1998).

The Northern Blot

Northern blot hybridization, which was developed by James Alwine, David Kemp and George Stark in 1977, is a procedure in which at different sizes of RNA molecules are separated by agarose gel, fixed onto a membrane following electrophoretic separation, and then incubated with a labeled DNA or RNA probe for hybridization process (Alwine et al. 1977; Wu et al. 2004). The northern blot is similar to the Southern blot except that the northern blot uses mRNA from samples instead of DNA (Wu 1995b; Mohan et al. 1998). One of the most fundamental and powerful tools used in analysis of gene expression is the northern blot technique (Wu 1995b). This technique is a sensitive, reliable and quantitative method for characterization of the steady-state level of RNA transcripts. It consists of five main steps; **(1)** preparation of RNAs for analysis, **(2)** separation of RNAs on an agarose gel, **(3)** transfer of RNAs onto a membrane, **(4)** detection of hybridized signals and **(5)** hybridization with a non-radioactive or a radioactive probe (Wu et al. 2004). RNAs migration through the gel is relative to their size during electrophoresis. Small RNA species move faster than large RNA molecules (Wu 1995b). Northern blot analysis provide information regarding of the species, sizes and expression levels of diversity of RNAs that cannot be revealed by alternative techniques such as the dot/slot blot and RNA protection assays (Wu et al. 2004).

The Western Blot

Proteins translated from mRNA species are regarded as the end products of gene expression. The protein facilitates the specific function of a gene as result of a meaningful gene expression (Wu and Welsh 1996a, b). Therefore, protein expression is one of the most important parts of modern molecular biology. Electrophoresis in which a net charged molecule moves in an electric field is one key approach used for the analysis of proteins. For this aim, the most widely utilized system for separating proteins on the basis of their net charges, sizes and shapes is the SDS-PAGE (sodium dodecyl sulfate poly-acrylamide gel electrophoresis) which was developed in the mid-1960s (Knudsen 1985; Wu 1995b; Wu and Welsh 1996a). Protein antigens can be detected by using another technique called "western blotting" (Burnette 1981). In western blotting, cellular total protein is isolated and separated in a polyacrylamide gel, electroblotted onto a solid PVDF (polyvinylidene difluoride) or nitrocellulose membrane, and then specified with primary and secondary antibodies (Reece 2004; Zale 2008).

The antibodies that are bound to the protein fixed to membrane are detected and analysis of the membrane is carried out following the procedure (Mohan et al. 1998; Wu et al. 2004). Western blot technique is a sensitive, reliable and quantitative method widely employed in the analysis of proteins. The information about the species, sizes and expression levels of diverse proteins can be tracked simultaneously by using this technique. Other alternative techniques cannot provide such information. (Kyhse-Anderson 1984; Wu and Welsh 1996a, b; Wu et al. 2004). This technique consists of five main steps; **(1)** preparation of proteins, **(2)** separation of proteins by SDS-PAGE, **(3)** incubation of the membrane with primary antibodies, **(4)** incubation of the membrane with secondary antibodies, **(5)** detection of specific protein band and **(6)** transfer of proteins onto a membrane (Wu et al. 2004).

The mobilities of proteins on the gel will be influenced by concentration of polyacrylamide gel. Depending on the particular size of the protein of interest, a general chart is displayed below. Normally, lower percentage gels are better for resolving very high molecular weight proteins (Wu et al. 2004). The electrophoretic blot method can be employed for the transfer of proteins from a polyacrylamide gel onto a solid nitrocellulose or nylon membrane. The membrane is incubated with an antibody raised against the transgenic protein, and the antibody binds primarily through hydrogen bonds (Memelink et al. 1994; Sambrook and Russell 2001). During the detection process the protein of interest with an antibody which is linked to a reporter enzyme is detected by using a secondary antibody that drives a colourimetric reaction and produces a colour on film (Memelink et al. 1994). Size approximations are taken by comparing the stained bands to that of the marker loaded during electrophoresis. The amount of target protein can be semiquantified by band intensity (Zale 2008).

5.5 A Model Study, *Agrobacterium*-Mediated Gene Transfer to Meristematic Tissues of Cotton (*Gossypium hirsutum* L.)

Agrobacterium-mediated transformation of plants is now applicable to many dicotyledonous and several monocotyledonous plant species (DeCleene and DeLay 1976; McCullen and Binns 2006). It can be used to transform many different species based on various factors; **(1)** the broad host range of *Agrobacterium*, **(2)** the regeneration responsiveness of many different explant tissues and **(3)** the utility of a wide range of selectable marker genes (van der Meer 2006). In this chapter, *Agrobacterium*-mediated gene transfer to meristematic tissues of cotton is given as a sample gene transfer protocol (Fig. 12.11).

5.5.1 Surface Sterilization

Before sterilization, cottonseeds were kept under flowing tap water for 1 h. They were surface sterilized by immersion in 70% ethanol for 3 min, followed by stirring in 20% commercial bleach (ACE Lever Co.) for 20 min. The surface sterilized seeds were rinsed 3 times with sterile distilled water for 5 min and they were dried onto sterile filter papers. Seed coats were removed with sterile scalpel and pliers prior to germination (Ozyigit et al. 2007; Ozyigit 2008)

5.5.2 Germination

The seeds were germinated on hormone free Murashige Skoog (MS) medium (Murashige and Skoog 1962) containing macro and micro nutrients supplemented with 100 mg L^{-1} myo-inositol, 0.5 mg L^{-1} thiamine-HCl, 0.5 mg L^{-1} nicotinic acid, 0.5 mg L^{-1} pyridoxine-HCl and 3% sucrose. 2.2 g phytagel was added into medium and then pH was adjusted to 5.7 with 1 M NaOH before autoclaving. 20 mL MS medium was poured into Magenta vessels and five seeds were germinated in each Magenta (Özyiğit et al. 2006; Özyiğit and Gözükirmizi, 2008).

5.5.3 Culture Conditions

Seeds were left in growth chamber under a photoperiod of 16 h light (7,500 lux) and 8 h dark, at 25°C and 70% humidity (Ozyigit 2009; Ozyigit and Gozukirmizi 2009).

5.5.4 *Agrobacterium*-Mediated Transformation

A. tumefaciens strain LBA 4404 was used as a transformation vehicle to mediate gene transfer into cotton, cv. Çukurova 1518. Plant expression vectors of pBI 121

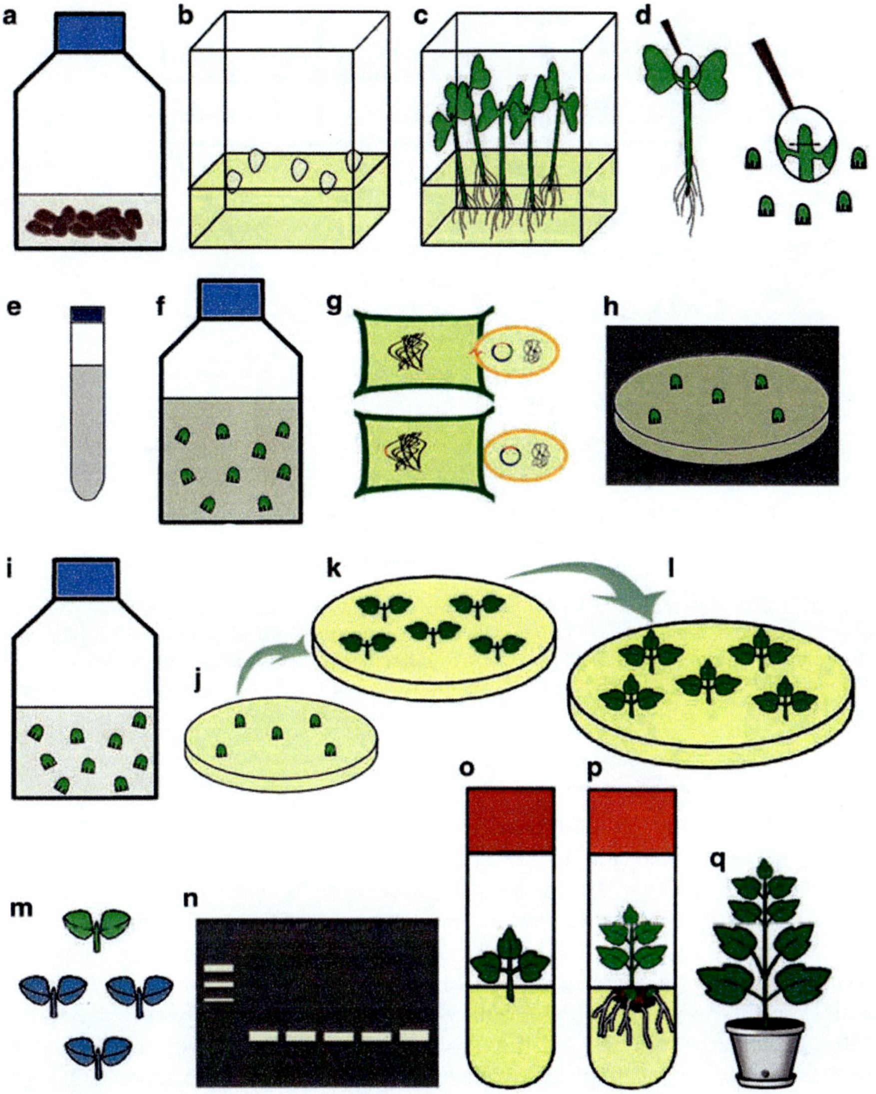

Fig. 12.11 *Agrobacterium*-mediated gene transfer protocol for cotton, (**a**) Surface sterilization, (**b**) Planting seeds, (**c**) 1-week-old young plantlets, (**d**) Isolation of meristematic shoots, (**e**) Breeding of *Agrobacterium*, (**f–g**) Gene transfer to meristematic cells with *Agrobacterium*, (**h**) Culturing explants in darkness, (**i**) Removing *Agrobacterium* by washing with kanamycin, (**j–l**) Culturing explants on selective media, (**m–n**) Histochemical and molecular analysis, (**o–p**) Rooting procedure, (**r**) Adaptation to the soil (Modified from Özyiğit and Gözükirmizi 2008)

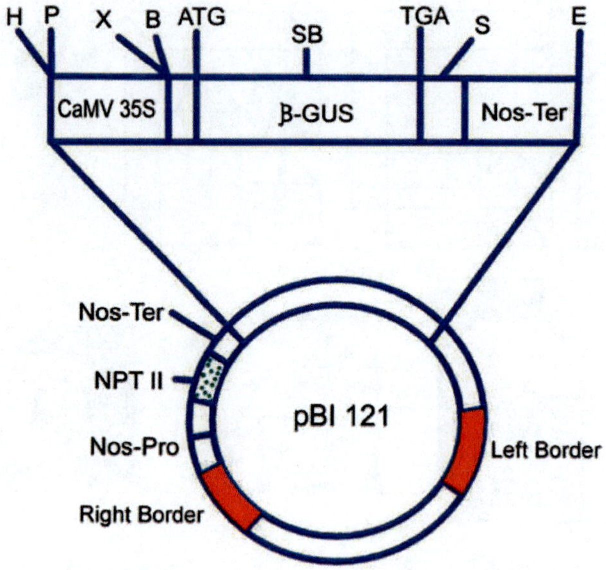

Fig. 12.12 The gene map of pBI-121 plasmid. *Nos-Pro NOS* promoter gene region, *Nos-Ter NOS* gene's terminator region, *CaMV 35S* Cauliflower mosaic virus 35S RNA promoter region, *npt* II Neomycin phosphotransferase II gene, *GUS* β-Glucuronidase gene, Enzyme restriction sites, *H* Hind III, *P* Pst I, *X* Xba I, *B* Bam HI, *SB* Sna BI, *S* Sac I, *E* Eco RI (Modified from Chen et al. 2003)

and pGUS-Int were mobilized into *A. tumefaciens* via triparental mating with pRK 2013 as helper plasmid (Fig. 12.12). The vectors have a selectable marker neomycin phosphotransferase (*npt* II), which confers resistance to kanamycin. The *GUS* and intron-containing *GUS* genes were under the control of CaMV 35S promoter and meristematic tissues carried out the transformation. Meristematic shoot apexes of cotton 1–2 mm long, infected with plasmid-harboring *A. tumefaciens* having final optical density as OD 600 = 0.4. Primary transgenic tissues (TR) were selected on 100 μg/ml kanamycin and 500 μg/ml cefotaxime-containing MS supplemented with 0.1 mg/L KIN + 2 mg/L NAA plates (Uğraş and Gözükirmizi 1999; Özyiğit et al. 2006).

5.5.5 Histochemical GUS Assays

The histochemical GUS assay was also performed with the leaves of 2-weeks-old plantlets. They were washed for 30 min with 50 mM phosphate buffer (pH 7.0) and immersed for 10 min in fixation solution (0.3% formaldehyde, 10 mM MES, 0.3 M mannitol). The samples were put in 1 mM X-gluc (5-bromo-4-choloro-3-indolyl-b-glucuronic acid) solution and incubated at 37°C overnight for blue color development. *GUS* activity was also detected with overnight cultures of *Agrobacterium* harboring pBI 121 and pGUS-Int by histochemical assays (Özyiğit et al. 2006).

5.5.6 PCR Analyses

Genomic DNA was isolated from 2-month-old plant leaves. PCR was run with specific forward and reverse primers ("*GUS* 1" 5′ GGT GGG AAA GCG CGT TAC AAG 3′ and "*GUS* 2" 5′ GTT TAC GCG TTG CTT CCG CCA 3′) of the *GUS* gene and 35 cycles were driven by thermocycler for 60 s at 95°C, 60 s at 36°C and 90 s at 72°C. The PCR was completed for 10 min at 72°C (Uğraş and Gözükirmizi 1999).

5.5.7 Southern Hybridization

It was performed with XbaI digested genomic DNAs (8 μg) to determine the integration of the *GUS* gene into plant genomes and to estimate the number of insertion in the transformants. The overnight digested DNA was fractionated on a 1% agarose gel and transferred to nylon membrane with fixation by UV crosslinking. The non-radioactive digoxigenin (DIG) hybridization system was used for hybridization analysis. The PCR-amplified product of the *GUS* gene was labeled with DIG-dUTP and used as a probe for Southern hybridization. Hybridization was carried at 68°C and immunological detection steps were performed according to manufacturer's instructions (Uğraş and Gözükirmizi 1999).

6 Conclusion

Plant biotechnology has developed into a separate scientific discipline in a very short span of time and *Agrobacterium*-mediated transformation has emerged recently as an efficient method for genetic manipulation of plants, while it was only a dream 25 years ago. Although other direct DNA transfer methods are also being used, *Agrobacterium*-mediated transformation has major advantages over these systems. Many transgenic economically important plant species, or elite varieties have been developed by the *Agrobacterium*-mediated methods. This situation is especially relevant, because plants are the major and most important source of our nutrition. As a result of developments in plant biotechnology and transgene technologies, during the period 1996–2010, biotech crops have been grown on more than 1 billion hectares. As mentioned by James, in 2010 the global area of biotech crops continues to soar for the fifteenth consecutive year at a sustained growth rate of 10% or 14 million hectares, reaching 148 million hectares. United States, Brazil, Argentina, India, Canada, China, Paraguay, Pakistan and South Africa are the principal adopters of biotech crops globally. Although there are many discussions about the transgenic crops and transgene technologies, the amount and variety of transgenic crops will increase in the future parallel to increasing world population and demands.

Acknowledgement I would like to express my special thanks to Professor Sebahattin Özcan (Ankara University, Agronomy Department), Dr. İlhan Doğan (Izmir Institute of Technology, Department of Molecular Biology and Genetics), Dr. Ülkühan Yaşar (Bartin University, Department of Environmental Engineering) and Dr. Göksel Demir (Bahçeşehir University, Department of Environmental Engineering) for a critical reading of this chapter and for their valuable suggestions. My thanks go to İlke Ertem (Biologist) for his assistance in drawing figures, Melike Isci (Marmara University) for rearranging the references. I am also thankful to Dr. Ayse Zeynep Isik for linguistic corrections and Bahçeşehir University, for their technical support.

References

Abuodeh RO, Orbach MJ, Mandel MA, Das A, Galgiani JN (2000) Genetic transformation of *Coccidioides immitis* facilitated by *Agrobacterium tumefaciens*. J Infect Dis 181:2106–2110

Alwine JC, Kemp DJ, Stark GR (1977) Method for detection of specific RNAs in agarose gels by transfer to diazobenzyloxymethyl-paper and hybridization with DNA probes. Proc Natl Acad Sci U S A 74(12):5350–5354

Ames GF, Mimura CS, Shyamala V (1990) Bacterial periplasmic permeases belong to a family of transport proteins operating from *Escherichia coli* to human: traffic ATPases. FEMS Microbiol Rev 75:429–446

Anderson A, Moore L (1979) Host specificity in the genus *Agrobacterium*. Phytopathology 69:320–323

Ankenbauer RG, Nester EW (1990) Sugar-mediated induction of *Agrobacterium tumefaciens* virulence genes: structural specificity and activities of monosaccharides. J Bacteriol 172:6442–6446

Armitage P, Walden R, Draper J (1988) Vectors for the transformation of plant cells using *Agrobacterium*. In: Draper J, Scott R, Armitage P, Walden R (eds) Plant genetic transformation and gene expression: a laboratory manual. Blackwell Scientific Publisher, Oxford, pp 1–67

Ashby AM, Watson MD, Loake GJ, Shaw CH (1988) Ti plasmid-specified chemotaxis of *Agrobacterium tumefaciens* C58 toward *vir*-inducing phenolic compounds and soluble factors from monocotyledonous and dicotyledonous plants. J Bacteriol 170:4181–4187

Ausubel FM, Brent R, Kingston RE, Moore DD, Seidman JG, Smith JA, Struhl K (eds) (1995) Current protocols in molecular biology. Greene Publishing Associates and Wiley-Interscience, John Wiley & Sons, New York

Ballas N, Citovsky V (1997) Nuclear localization signal binding protein from *Arabidopsis* mediates nuclear import of *Agrobacterium* VirD2 protein. Proc Natl Acad Sci U S A 94:10723–10728

Barker RF, Idler KB, Thompson DV, Kemp JD (1983) Nucleotide sequence of the T-DNA region from the *Agrobacterium tumefaciens* octopine Ti plasmid pTi15955. Plant Mol Biol 2:335–350

Bartlett JMS, Stirling D (2003) A short history of the polymerase chain reaction. In: Bartlett JMS, Stirling D (eds) PCR protocols. Humana Press Inc, Totowa, pp 3–6

Barton KA, Whitely HR, Yang NS (1987) *Bacillus thuringiensis* delta-endotoxin expressed in transgenic *Nicotiana tabacum* provides resistance to lepidopteran insects. Plant Physiol 85:1103–1109

Barz WH, Marja K, Caldentey KM (2002) Plant biotechnology – an emerging field. In: Marja K, Caldentey KM, Barz W (eds) Plant biotechnology and transgenic plants. Marcel Dekker Inc, New York, pp 1–20

Beaudet AL (1991) Molecular genetics and medicine. In: Wilson JD, Braunwald E, Isselbacher KJ, Petersdrof RG, Martin JB, Faucci AS, Root RK (eds) Harrison's principles of internal medicine, vol 1. McGraw-Hill Inc, New York, pp 32–46

Beaupré CE, Bohne J, Dale EM, Binns AN (1997) Interactions between VirB9 and VirB10 proteins involved in the movement of DNA from *Agrobacterium tumefaciens* into plant cells. J Bacteriol 179:78–89

Bergey DH, Harrison FC, Breed RS, Hammer BW, Huntoon FM (1923) Bergey's manual of determinative bacteriology, 1st edn. Williams and Wilkins, Baltimore, pp 1–442

Bevan M (1984) Binary *Agrobacterium* vectors for plant transformation. Nucleic Acids Res 12:8711–8721

Binns AN, Beaupre CE, Dale EM (1995) Inhibition of VirB mediated transfer of diverse substrates from *Agrobacterium tumefaciens* by the IncQ plasmid RSF1010. J Bacteriol 177: 4890–4899

Bomhoff G, Klapwijk PM, Kester HCM, Schilperoort RA, Hernalsteens JP, Schell J (1976) Octopine and nopaline synthesis and breakdown genetically controlled by a plasmid of *Agrobacterium tumefaciens*. Mol Gen Genet 145(2):177–181

Bradley LR, Kim JS, Matthysse AG (1997) Attachment of *Agrobacterium tumefaciens* to carrot cells and *Arabidopsis* wound sites is correlated with the presence of a cell-associated, acidic polysaccharide. J Bacteriol 179:5372–5379

Braun AC (1941) Development of secondary tumors and tumor strands in the crown gall of sunflowers. Phytopathology 31:135–149

Braun AC (1947) Thermal studies on the factors responsible for tumor initiation in crown gall. Am J Bot 34:234–240

Brencic A, Winans S (2005) Detection and response to signals involved in host-microbe interactions by plant-associated bacteria. Microbiol Mol Biol Rev 69:155–194

Broothaerts W, Mitchell HJ, Weir B, Kaines S, Smith LM, Yang W, Mayer JE, Roa-Rodriguez C, Jefferson RA (2005) Gene transfer to plants by diverse species of bacteria. Nature 433:629–633

Bundock P, Hooykaas PJJ (1996) Integration of *Agrobacterium tumefaciens* T-DNA in the *Saccharomyces cerevisiae* genome by illegitimate recombination. Proc Natl Acad Sci U S A 93:15272–15275

Bundock P, den Dulk-Ras A, Beijersbergen A, Hooykaas PJJ (1995) Trans-kingdom T-DNA transfer from *Agrobacterium tumefaciens* to *Saccharomyces cerevisiae*. EMBO J 14:3206–3214

Burnette WN (1981) "Western blotting": electrophoretic transfer of proteins from sodium dodecyl sulfate-polyacrylamide gels to unmodified nitrocellulose and radiographic detection with antibody and radioiodinated protein A. Anal Biochem 112:195–203

Burr TJ, Bazzi C, Sule S, Otten L (1998) Crown gall of grape: biology of *Agrobacterium vitis* and the development of disease control strategies. Plant Dis 82:1288–1297

Cangelosi GA, Hung L, Puvanesarajah V, Stacey G, Ozga DA, Leigh JA, Nester EW (1987) Common loci for *Agrobacterium tumefaciens* and *Rhizobium meliloti* exopolysaccharide synthesis and their role in plant interaction. J Bacteriol 169:2086–2091

Cangelosi GA, Martinetti G, Leigh JA, Lee CC, Theines C, Nester EW (1989) Role of *Agrobacterium tumefaciens* chvA protein in export of β-1,2 glucan. J Bacteriol 171:1609–1615

Cangelosi GA, Ankenbauer RG, Nester EW (1990) Sugars induce the *Agrobacterium virulence* genes through a periplasmic binding protein and a transmembrane signal protein. Proc Natl Acad Sci U S A 87:6708–6712

Cangelosi GA, Best EA, Martinetti C, Nester EW (1991) Genetic analysis of *Agrobacterium tumefaciens*. Method Enzymol 145:177–181

Cardoza V (2008) Tissue culture: the manipulation of plant development. In: Stewart CN Jr (ed) Plant biotechnology and genetics. John Wiley & Sons, Inc., Hoboken, New Jersey, pp 113–134

Cavara F (1897) Tubercolosi della vite Intorno alla eziologia de alcune malattie di piante coltivate *Stazoni Sperimentale*. Agrarie Italiane 30:483–487

Chang CH, Winans SC (1992) Functional roles assigned to the periplasmic, linker and receiver domains of the *Agrobacterium tumefaciens* VirA protein. J Bacteriol 174:7033–7039

Chawla HS (2002) Introduction to plant biotechnology, 2nd edn. Science Publishers Inc., Enfield, pp 1–521

Chen X, Stone M, Schlagnhaufer C, Romaine CP (2000) A fruiting body tissue method for efficient *Agrobacterium* – mediated transformation of *Agaricus bisporus*. Appl Environ Microbiol 66:4510–4513

Chen PY, Wang CK, Soong SC, To KY (2003) Complete sequence of the binary vector pBI121 and its application in cloning T-DNA insertion from transgenic plants. Mol Breed 11:287–293

Cheng M, Fry JE, Pang S, Zhou H, Hironaka CM, Duncan DR, Conner TW, Wan Y (1997) Genetic transformation of wheat mediated by *Agrobacterium tumefaciens*. Plant Physiol 115:971–980

Chilton MD, Drummond MH, Merlo DJ, Sciaky D, Montoya AL, Gordon MP, Nester EW (1977) Stable incorporation of plasmid DNA into higher plant cells: the molecular basis of crown gall tumorigenesis. Cell 11:263–271

Christie PJ (1997) *Agrobacterium tumefaciens* T-complex transport apparatus: a paradigm for a new family of multifunctional transporters in eubacteria. J Bacteriol 179:3085–3094

Christie PJ, Atmakuri K, Krishnamoorthy V, Jakubowski S, Cascales E (2005) Biogenesis, architecture, and function of bacterial type IV secretion systems. Annu Rev Microbiol 59: 451–485

Citovsky V, McLean G, Greene E, Howard E, Kuldau G, Thorstenson Y, Zupan J, Zambryski P (1992) *Agrobacterium* plant cell interaction: induction of *vir* genes and T-DNA transfer. In: Verma DPS (ed) Molecular signals in plant-microbe communications. CRC Press, London, pp 169–198

Citovsky V, Warnick D, Zambryski P (1994) Nuclear import of *Agrobacterium* VirD2 and VirE2 proteins in maize and tobacco. Proc Natl Acad Sci U S A 91:3210–3214

Conn HJ (1942) Validity of the genus *Alcaligenes*. J Bacteriol 44:353–360

Cubero J, López M (2005) *Agrobacterium* persistence in plant tissue after transformation. In: Peña L (ed) Methods in molecular biology, transgenic plants: methods and protocols. Humana Press Inc., Totowa, pp 351–364

Dang TAT, Christie PJ (1997) The VirB4 ATPase of *Agrobacterium tumefaciens* is a cytoplasmic membrane protein exposed at the periplasmic surface. J Bacteriol 179:453–462

Danna K, Nathans D (1971) Studies of *sv*40 DNA 1 Specific cleavage of simian virus 40 DNA by restriction endonuclease of *Hemophilus influenzae*. Proc Natl Acad Sci U S A 68(12):2913–2917

Das A, Xie YH (1998) Construction of transposon Tn3phoA: its application in defining the membrane topology of the *Agrobacterium tumefaciens* DNA transfer proteins. Mol Microbiol 27:405–414

de Frammond AJ, Back EW, Chilton WS, Kayes L, Chilton MD (1986) Two unlinked T-DNAs can transform the same tobacco plant cell and segregate in the F1 generation. Mol Gen Genet 202:125–131

de Groot MJA, Bundock P, Hooykaas PJJ, Beijersbergen AGM (1998) *Agrobacterium tumefaciens* mediated transformation of filamentous fungi. Nat Biotechnol 16:839–842

de la Riva GA, González-Cabrera J, Vázquez-Padrón R, Ayra-Pardo C (1998) *Agrobacterium tumefaciens*: a natural tool for plant transformation. Electron J Biotechnol 1(3):118–133

Deacon JW, Macdonald RKW, Fox FM, Lascaris D (1988) Application of alginate gel for protection of wounds against crown gall (*Agrobacterium tumefaciens*). Plant Pathol 37(4):522–528

deCleene M, DeLay J (1976) The host range of crown gall. Bot Rev 42:389–466

Dessaux Y, Petit A, Far SK, Murphy PM (1998) Opines and opine-like molecules involved in plant-*Rhizobiaceae* interactions. In: Spaink HP, Kondorosi A, Hooykaas PJJ (eds) The Rhizobiaceae. Kluwer Academic, Dordrecht, pp 173–197

Doty SL, Yu NC, Lundin JI, Heath JD, Nester EW (1996) Mutational analysis of the input domain of the VirA protein of *Agrobacterium tumefaciens*. J Bacteriol 178:961–970

Douglas CJ, Halperin W, Nester EW (1982) *Agrobacterium tumefaciens* mutants affected in attachment to plant cells. J Bacteriol 152:1265–1275

Duggar BM (1909) Fungous diseases of plants. Ginn and Co, New York, pp 1–508

Dürrenberger F, Crameri A, Hohn B, Koukolíková-Nicola Z (1989) Covalently bound VirD2 protein of *Agrobacterium tumefaciens* protects the T-DNA from exonucleolytic degradation. Proc Natl Acad Sci U S A 86:9154–9158

Edwards JB, Delort J, Mallet J (1991) Oligodeoxyribonucleotide ligation to single-stranded cDNAs: a new tool for cloning 5′ ends of mRNAs and for constructing cDNA libraries by *in vitro* amplification. Nucleic Acids Res 19:5227–5232

Einspanier R (2006) DNA-based methods for detection of genetic modifications. In: Heller KJ (ed) Genetically engineered food methods and detection. 2nd Updated and Enlarged Ed. Wiley-VCH Verlag GmbH & Co., KgaA, Darmstadt, pp 163–185

Ellis JG, Kerr A, Petit A, Tempe J (1982) Conjugal transfer of nopaline and agropine Ti plasmids: the role of agrocinopines. Mol Gen Genet 186:269–274

Escobar MA, Dandekar AM (2003) *Agrobacterium tumefaciens* as an agent of disease. Trends Plant Sci 8(8):380–386

Fabre E, Dunal F (1853) Observations sur les maladies regantes de la vigne. Bull Soc Cent Agric Dep Herault 40:46

Fernández D, Spudich GM, Zhou XR, Berger BR, Christie PJ (1996) *Agrobacterium tumefaciens* VirB7 lipoprotein is required for stabilization of VirB proteins during assembly of the T-complex transport apparatus. J Bacteriol 178:3168–3176

Filichkin SA, Gelvin SB (1993) Formation of a putative relaxation intermediate during T-DNA processing directed by the *Agrobacterium tumefaciens* VirD1, D2 endonuclease. Mol Microbiol 8:915–926

Finberg KE, Muth TR, Young SP, Maken JB, Heitritter SM, Binns AN, Banta LM (1995) Interactions of VirB9, -10 and -11 with the membrane fraction of *Agrobacterium tumefaciens*: solubility studies provide evidence of tight associations. J Bacteriol 177:4881–4889

Firth N, Ippen-Ihler K, Skurray RA (1996) Structure and function of the F factor and mechanism of conjugation. In: Neidhardt FC, Curtis R, Ingraham JL, Lin ECC, Low KB, Magasanik B, Reznikoff WS, Riley M, Schaechter M, Umbarger HC (eds) *Escherichia coli* and *Salmonella*: cellular and molecular biology, 2nd edn. American Society for Microbiology, Washington, DC, pp 2377–2401

Fullner KJ, Lara CJ, Nester EW (1996) Pilus assembly by *Agrobacterium* T-DNA transfer genes. Science 273:1107–1109

Galbraith DW, Bohnert HJ, Bourque DP (eds) (1995) Methods in plant cell biology, vol 50. Academic, New York, pp 1–555

Garfinkel DJ, Nester EW (1980) *Agrobacterium tumefaciens* mutants affected in crown gall tumorigenesis and octopine catabolism. J Bacteriol 144:732–743

Garrity GM (ed) (2005) Bergey's manual of systematic bacteriology, 2nd edn. The Proteobacteria, Originally published by Williams and Wilkins, 1984 2nd edn. Springer, New York

Gautheret RJ (1959) La culture des tissus vegetaux; Techniques et Realisations Masson and Cie Paris, IV, 884pp

Gelvin SB (2000) *Agrobacterium* and plant genes involved in T-DNA transfer and integration. Annu Rev Plant Physiol 51:223–256

Gelvin SB (2003) *Agrobacterium*-mediated plant transformation: the biology behind the "gene-jockeying" tool. Microbiol Mol Biol Rev 67(1):16–37

Gheysen G, Villarroel R, van Montagu M (1991) Illegitimate recombination in plants: a model for T-DNA integration. Gene Dev 5:287–297

Goodner B, Hinkle G, Gattung S, Miller N, Blanchard M, Qurollo B, Goldman BS, Cao Y, Askenazi M, Halling C, Mullin L, Houmiel K, Gordon J, Vaudin M, Iartchouk O, Epp A, Liu F, Wollam C, Allinger M, Doughty D, Scott C, Lappas C, Markelz B, Flanagan C, Crowell C, Gurson J, Lomo C, Sear C, Strub G, Cielo C, Slater S (2001) Genome sequence of the plant pathogen, biotechnology agent *Agrobacterium tumefaciens* C58. Science 294:2323–2328

Grant JE, Cooper PA, Dale TM (2004) Transgenic *Pinus radiata* from *Agrobacterium tumefaciens*-mediated transformation of cotyledons. Plant Cell Rep 22(12):894–902

Gürel F, Gözükirmizi N (2003) Electroporation transformation of barley. In: Jackson JF, Liskens HS (eds) Modern methods of plant analysis, vol 23, Plant transformation. Springer, Berlin, Germany, pp 69–89

Guyon P, Chilton MD, Petit A, Tempé J (1980) Agropine in 'null type' crown gall tumors: evidence for generality of the opine concept. Proc Natl Acad Sci U S A 77:2693–2697

Haberlandt G (1902) Kulturversuche mit isolierten Pflanzenzellen Sitzungsber Akad Wiss Wien Math-Naturwiss Kl. Abt J 111:69–92

Hamilton RH, Fall MZ (1971) The loss of tumor-inducing ability in *Agrobacterium tumefaciens* by incubation at high temperature. Experientia 27:229–230

Hawes MC, Smith LY, Howarth AJ (1988) *Agrobacterium tumefaciens* mutants deficient in chemotaxis to root exudates. Mol Plant Microbe Interact 1:182–186

Heinemann JA, Sprague GF (1989) Bacterial conjugative plasmids mobilize DNA transfer between bacteria and yeast. Nature 340:205–209

Hernalsteens JP, van Vliet F, de Beuckeleer M, Depicker A, Engler G, Lemmes M, Holsters M, van Montagu M, Schell J (1980) The *Agrobacterium tumefaciens* Ti plasmid as a host vector system for introducing foreign DNA in plants. Nature 287:654–656

Herrera-Estrella A, Chen Z, van Montagu M, Wang K (1988) VirD proteins of *Agrobacterium tumefaciens* are required for the formation of a covalent DNA protein complex at the 5′ terminus of T-strand molecules. EMBO J 7:4055–4062

Higgings CF, Hyde SC, Mimmack MM, Gileadi U, Gill DR, Gallagher MP (1990) Binding protein dependent transport systems. J Bioenerg Biomembr 22:571–592

Hoekema A, Hirsch PR, Hooykaas PJJ, Schilperoort RA (1983) A binary plant vector strategy based on separation of *vir* and T-region of the *Agrobacterium tumefaciens* Ti plasmid. Nature 303:179–180

Hooykaas PJJ, Shilperoort RA (1992) *Agrobacterium* and plant genetic engineering. Plant Mol Biol 19:15–38

Hooykaas PJJ, Hofker M, den Dulk-Ras H, Schilperoort RA (1984) A comparison of virulence determinants in an octopine Ti plasmid, a nopaline Ti plasmid and a Ri plasmid by complementation analysis of *Agrobacterium tumefaciens* mutants. Plasmid 11:195–205

Hopkins WG (2007) Plant biotechnology. Infobase Publishing, Philadelphia, pp 1–153

Horsch RB, Klee HJ, Stachel S, Winans SC, Nester EW, Rogers SG, Fraley RT (1986) Analysis of *Agrobacterium tumefaciens* virulence mutants in leaf disks. Proc Natl Acad Sci U S A 83:2571–2575

Howard EA, Citovsky V (1990) The emerging structure of the *Agrobacterium* T-DNA transfer complex. BioEssays 12:103–108

Howard EA, Citovsky V, Zambryski P (1990) The T-complex of *Agrobacterium tumefaciens*. UCLA Symp Mol Cell Biol News Ser 129:1–11

Howard E, Zupan J, Citovsky V, Zambryski P (1992) The VirD2 protein of *A tumefaciens* contains a C-terminal bipartite nuclear localization signal: implications for nuclear uptake of DNA in plant cells. Cell 68:109–118

Huang Y, Morel P, Powell B, Kado CI (1990) VirA, a coregulator of Ti-specified virulence genes, is phosphorylated *in vitro*. J Bacteriol 172:1142–1144

James C (2010) Global status of commercialized biotech/GM crops: 2010 ISAAA Brief No: 42. ISAAA, Ithaca, pp 1–7

Jayaram M (1994) Phosphoryl transfer if FLP recombination: a template for strand transfer mechanisms. Trends Biotechnol 19:78–82

Jeffreys AJ, Wilson V, Thein SW (1984) Hypervariable minisatellite regions in human DNA. Nature 314:67–73

Jeon GA, Eu JS, Sim WS (1998) The role of inverted repeat (IR) sequence of the *vir*E gene expression in *Agrobacterium tumefaciens* pTiA6. Mol Cells 8:49–53

Jeon YH, Park H, Lee BD, Yu YH, Chang SP, Kim SG, Hwang I, Kim YH (2008) First description of crown gall disease on ginseng. Plant Pathol J 24(2):207–210

Jin S, Prusti RK, Roitsch T, Ankenbauer RG, Nester EW (1990a) Phosphorylation of the VirG protein of *Agrobacterium Tumefaciens* by the autophosphorylated VirA protein: essential role in biological activity of VirG. J Bacteriol 172:4945–4950

Jin S, Roitsch T, Ankenbauer RG, Gordon MP, Nester EW (1990b) The VirA protein of *Agrobacterium tumefaciens* is autophosphorylated, is essential for *vir* gene regulation. J Bacteriol 172:525–530

Kanemoto RH, Powell AT, Akiyoshi DE, Regier DA, Kerstetter RA, Nester EW, Hawes MC, Gordon MP (1989) Nucleotide sequence and analysis of the plant-inducible locus *pinF* from *Agrobacterium tumefaciens*. J Bacteriol 171:2506–2512

Kerr A (1971) Acquisition of virulence by non-pathogenic isolates of *Agrobacterium radiobacter*. Physiol Plant Pathol 1:241–246

Khawar KM, Özcan S (2002) *In vitro* induction of crown galls by *Agrobacterium tumefaciens* super virulent strain A281 (pTiBo 542) in lentil (*Lens culinaris* Medik.). Turk J Bot 26:65–170

Klee H, Montoya A, Horodyskit F, Lichtenstein C, Garfinkel D, Fuller S, Flores C, Peschon J, Nester E, Gordon M (1984) Nucleotide sequence of the tms genes of the pTiA6NC octopine Ti plasmid: two gene products involved in plant tumorigenesis. Proc Natl Acad Sci U S A 81:1728–1732

Knauf V, Yanofsky M, Montoya A, Nester E (1984) Physical and functional map of an *Agrobacterium tumefaciens* tumor inducing plasmid that confers a narrow host range. J Bacteriol 160:564–568

Knudsen KA (1985) Proteins transferred to nitrocellulose for use as immunogens. Anal Biochem 147:285–288

Koo J, Kim Y, Kim J, Yeom M, Lee IC, Nam HG (2007) A *GUS*/luciferase fusion reporter for plant gene trapping and for assay of promoter activity with luciferin-dependent control of the reporter protein stability. Plant Cell Physiol 48(8):1121–1131

Kuldau GA, deVos G, Owen J, McGaffrey G, Zambryski P (1990) The *vir*B operon of *Agrobacterium tumefaciens* pTiC58 encodes 11 open reading frames. Mol Gen Genet 221:256–266

Kunik T, Tzfira T, Kapulnik Y, Gafni Y, Dingwall C, Citovsky V (2001) Genetic transformation of HeLa cells by *Agrobacterium*. Proc Natl Acad Sci U S A 98:1871–1876

Kyhse-Anderson J (1984) Electroblotting of multiple gels: a simple apparatus without buffer tank for rapid transfer of proteins from polyacrylamide to nitrocellulose. J Biochem Biophys Methods 10:203–209

Lacy GH, Hansen MA (2002) Crown gall of woody ornamentals *Virginia Cooperative Extension*, Plant Disease, Fact Sheets Publication No: 450–608, pp 1–2. Online at: http: //pubsextvtedu/450/450-608/450-608_pdf.pdf

Lee LY, Gelvin SB (2008) T-DNA binary vectors and systems. Plant Physiol 146:325–332

Lehman CW, Trautman JK, Carroll D (1994) Illegitimate recombination in *Xenopus*: characterization of end-joined junctions. Nucleic Acid Res 22:434–442

Lessl M, Lanka E (1994) Common mechanisms in bacterial conjugation and Ti-mediated T-DNA transfer to plant cells. Cell 77:321–324

Lessl M, Balzer D, Pansegrau W, Lanka E (1992) Sequence similarities between the RP4 Tra2 and the Ti VirB region strongly support the conjugation model for TDNA transfer. J Biol Chem 267:20471–20480

Li HQ, Chen L, Li MR (2007) Establishment of an efficient *Agrobacterium tumefaciens*-mediated leaf disc transformation of *Thellungiella halophila*. Plant Cell Rep 26:1785–1789

Lichtenstein C, Klee H, Montoya A, Garfinkel D, Fuller S, Flores C, Nester E, Gordon M (1984) Nucleotide sequence and transcript mapping of the *tmr* gene of the pTiA6NC octopine Ti plasmid: a bacterial gene involved in plant tumorigenesis. J Mol Appl Genet 2:354–362

Lieske R (1928) Untersuchungen uber die Krebskrankheit bei Pflanzen, Tieren und Menschen" Zentralbl Bakteriol Parasitenk Infektionskr. Hyg Abt I Orig 108:118–146

Lindsay DG (2002) The potential contribution of plant biotechnology to improving food quality. In: Marja K, Caldentey KMO, Barz WH (eds) Plant biotechnology and transgenic plants. Marcel Dekker, New York, pp 195–226

Lioret C (1956) Sur la mise en evidence d'un acide amine non identifie particulier aux tissus de 'crown-gall'. Bull Soc Fr Physiol Veg 2:76–78

Loopstra CA, Stomp AM, Sederoff RR (1990) *Agrobacterium* mediated DNA transfer in sugar pine. Plant Mol Biol 15:1–9

Loper JE, Kado CI (1979) Host range conferred by the virulence specifying plasmid of *Agrobacterium tumefaciens*. J Bacteriol 139:591–596

LPSN (2011) The website of "List of Bacterial names with Standing in Nomenclature" online at: http: //wwwbacteriocictfr/a/agrobacteriumhtml
Lundquist RC, Close TJ, Kado CI (1984) Genetic complementation of *Agrobacterium tumefaciens* Ti plasmid mutants in the virulence region. Mol Gen Genet 193:1–7
Maniatis T, Fritsch FE, Sambrook J (eds) (1982) Molecular cloning: a laboratory manual of Cold Spring Harbor Laboratory. Cold Spring Harbor Laboratory Press, Cold Spring Harbor
Matthysse AG (1983) Role of bacterial cellulose fibrils in *Agrobacterium tumefaciens* infection. J Bacteriol 154:906–915
Matthysse AG, Kijne JW (1998) Attachment of Rhizobiaceae to plant cells. In: Spaink HP, Kondorosi A, Hooykaas PJJ (eds) The Rhizobiaceae. Kluwer Academic, Dordrecht, pp 235–249
Matthysse AG, Yarnall HA, Young N (1996) Requirement for genes with homology to ABC transport system for attachment and virulence of *Agrobacterium tumefaciens*. J Bacteriol 178:5302–5308
Matzke AJM, Chilton MD (1981) Site-specific insertion of genes into T-DNA of the *Agrobacterium* tumour-inducing plasmid: an approach to genetic engineering of higher plant cells. J Mol Appl Genet 1:39–49
McAfee BJ, White EE, Pelcher LE, Lapp MS (1993) Root induction in pine (*Pinus*) and larch (*Larix*) spp using *Agrobacterium rhizogenes*. Plant Cell Tissue Organ Cult 34:53–62
McCullen CA, Binns AN (2006) *Agrobacterium tumefaciens* and plant cell interactions and activities required for interkingdom macromolecular transfer. Annu Rev Cell Dev Bi 22:101–127
Memelink J, Swords KMM, Staehelin LA, Hoge JHC (eds) (1994) Southern, Northern and Western blot analysis. Kluwer Academic, Dordrecht, pp 1–23
Men SZ, Ming XT, Liu RW, Wei CH, Li Y (2003) *Agrobacterium*-mediated genetic transformation of a *Dendrobium* orchid. Plant Cell Tissue Organ Cult 75:63–71
Merkli A, Christen P, Kapetanidis I (1997) Production of diosgenin by hairy root cultures of *Trigonella foenum-graecum* L. Plant Cell Rep 16:632–636
Miki B (2008) Marker genes and promoters. In: Stewart CN Jr (ed) Plant biotechnology and genetics. Wiley, Hoboken, pp 217–244
Mohan V, Mohan A, Hitman GA (1998) Methodology in molecular genetics and its applications to diabetology. Int J Diab Dev Countries 18:46–49
Morel G, Martin C (1952) Guérison de dahlias atteints d'une maladie à virus. C R Acad Sci 235:1324–1325
Mugnier JP, Ready W, Riedel GE (1986) Root culture system useful in the study of biotrophic root pathogens *in vitro*. In: Augustine PC, Danforth HD, Bakst MR (eds) Biotechnology for solving agricultural problems. Nijhof, Dordrecht, pp 147–153
Muir WH, Hildebt AC, Riker AJ (1954) Plant tissue cultures produced from single isolated cells. Science 119:877–878
Murashige T, Skoog F (1962) A revised medium for rapid growth and bioassays with tobacco tissue cultures. Physiol Plant 15(3):473–497
Nair AJ (ed) (2008) Introduction to biotechnology and genetic engineering. Laxmi Publications, Pvt. Ltd, New Delhi, India, pp 1–798
Narula A, Kumar S, Bansal KC, Srivastava PS (2004) Biotechnological approaches towards improvement of medicinal plants. In: Srivastava PS, Narula A, Srivastava S (eds) Plant biotechnology and molecular markers. Anamaya Publishers, New Delhi, pp 78–116
Oger P, Kim KS, Sackett SL, Piper KR, Farrand SK (1998) Octopine-type Ti plasmids code for mannopine-inducible dominant-negative allele of TraR, the quorum-sensing activator that regulates Ti plasmid conjugal transfer. Mol Microbiol 27:277–288
Okamoto S, Toyoda-Yamamoto A, Ito K, Takebe I, Machida Y (1991) Localization and orientation of the VirD4 protein of *Agrobacterium tumefaciens* in cell membrane. Mol Gen Genet 228:24–32
Okamoto T, Yamada M, Sekiya S, Okuhara T, Taguchi G, Inatomi S, Shimosaka M (2010) *Agrobacterium tumefaciens*-mediated transformation of the vegetative dikaryotic mycelium of the cultivated mushroom *Flammulina velutipes*. Biosci Biotechnol Biochem 74(11):2327–2329

Oliveira MM, Barroso J, Pais MS (1991) Direct gene transfer into *Actinidia deliciosa* protoplasts: analysis of transient expression of the *CAT* gene using TLC autoradiography and a GC-MS-based method. Plant Mol Biol 17(2):235–242

Otten L, deGreve H, Leemans J, Hain R, Hooykaas P, Schell J (1984) Restoration of virulence of *vir* region mutants of *Agrobacterium tumefaciens* strain B6S3 by coinfection with normal and mutant *Agrobacterium* strains. Mol Gen Genet 195:159–163

Özcan S, Uranbey S, Sancak C, Parmaksiz İ, Gürel E, Babaoğlu M (2004) *Agrobacterium* araciliğiyla gen aktarimi. In: Özcan S, Gürel E, Babaoğlu M (eds) Bitki Biyoteknolojisi II (Plant biotechnology, II), Genetik Mühendisliği ve Uygulamalari (Genetic engineering and its applications), Cilt II, 2nd edn. SÜ Vakfi Yayinlari, Turkey, pp 112–159

Ozyigit II (2008) Phenolic changes during *in vitro* organogenesis of cotton (*Gossypium hirsutum* L.) shoot tips. Afr J Biotechnol 7(8):1145–1150

Ozyigit II (2009) *In vitro* shoot development from three different nodes of cotton (*Gossypium hirsutum* L.). Not Bot Hort Agrobot 37(1):74–78

Özyiğit İİ, Gözükirmizi N (2008) High efficiency shoot and root formation from cotyledonary nodes of cotton (*Gossypium hirsutum* L.). Pak J Bot 40(4):1665–1672

Ozyigit II, Gozukirmizi N (2009) Efficient shoot and root formation from shoot apices of cotton (*Gossypium hirsutum* L.). Russ J Plant Physiol 56(4):527–531

Özyiğit İİ, Gözükirmizi N, Semiz BD (2006) *Agrobacterium*-mediated transformation of cotton: *Gossypium hirsutum* L "Çukurova 1518" from Turkey. In: Proceedings of the IV international congress of ethnobotany, Istanbul, pp 589–592

Ozyigit II, Kahraman MV, Ercan O (2007) Relation between explant age, total phenols and regeneration response in tissue cultured cotton (*Gossypium hirsutum* L.). Afr J Biotechnol 6(1):3–8

Palmer AG, Gao R, Maresh J, Erbiol WK, Lynn DG (2004) Chemical biology of multihost/pathogen interactions: chemical perception and metabolic complementation. Annu Rev Phytopathol 42:439–464

Pan SQ, Charles T, Jin S, Wu Z, Nester EW (1993) Preformed dimeric state of the sensor protein VirA is involved in plant-*Agrobacterium* signal transduction. Proc Natl Acad Sci U S A 90:39–43

Pan SQ, Jin S, Boulton MI, Hawes M, Gordon MP, Nester EW (1995) An *Agrobacterium* virulence factor encoded by a Ti plasmid gene or a chromosomal gene is required for T-DNA transfer into plants. Mol Microbiol 17:259–269

Parke D, Ornston LN, Nester EW (1987) Chemotaxis to plant phenolic inducers of virulence genes is constitutively expressed in the absence of the pTi plasmid in *Agrobacterium tumefaciens*. J Bacteriol 169:5336–5338

Parkinson JS (1993) Signal transduction schemes of bacteria. Cell 73:857–871

Paszkowski J, Baur M, Bogucki A, Potrykus I (1988) Gene targeting in plants. Embo J 7(13): 4021–4026

Peralta EG, Ream LW (1985) T-DNA border sequences required for crown gall tumorigenesis. Proc Natl Acad Sci U S A 82:5112–5116

Petit A, Tepme J, Kerr A, Holsters M, van Montagu M, Schell J (1978) Substrate induction of conjugative activity of *Agrobacterium tumefaciens* Ti plasmids. Nature 271:570–571

Piers KL, Heath JD, Liang X, Stephens KM, Nester EW (1996) *Agrobacterium tumefaciens* mediated transformation of yeast. Proc Natl Acad Sci U S A 93:1613–1618

Pohlman RF, Genetti HD, Winans SC (1994) Common ancestry between IncN conjugal transfer genes and macromolecular export systems of plant and animal pathogens. Mol Microbiol 14:655–668

Polcaro CM, Brancaleoni E, Donati E, Frattoni M, Galli E, Migliore L, Rapana P (2008) Fungal bioremediation of creosote-treated wood: a laboratory scale study on creosote components degradation by *Pleurotus ostreatus* mycelium. B Environ Contam Toxicol 81(2):180–184

Primrose SB, Twyman RM (eds) (2006) Principles of gene manipulation and genomics, 7th edn. Blackwell Publishing, Malden

Puchta H (1998) Repair of genomic double-strand breaks in somatic cells by one-side invasion of homologous sequences. Plant J 13:331–339

Raquel MH, Oliveira MM (1996) Kiwifruit leaf protoplasts competent for plant regeneration and direct DNA transfer. Plant Sci 121:107–114

Rashkova S, Spudlich GM, Christie PJ (1997) Characterization of membrane and protein interaction determinants of the *Agrobacterium tumefaciens* VirB11 ATPase. J Bacteriol 179:583–591

Ream W (2002) *Agrobacterium* genetics. In: Streips UN, Yasbin RE (eds) Modern microbial genetics, 2nd edn. Wille-Liss Inc., New York, pp 323–348

Ream LW, Gordon MP, Nester EW (1983) Multiple mutations in the T-region of the *Agrobacterium tumefaciens* tumor-inducing plasmid. Proc Natl Acad Sci U S A 80:1660–1664

Reece RJ (ed) (2004) Analysis of genes and genomes. Wiley, Chichester, pp 1–459

Reinhard E (1974) Biotransformations by plant tissue cultures. In: Street HE (ed) Tissue culture and plant science. Academic, London, pp 433–459

Roberts RJ (1976) Restriction endonucleases. Crit Rev Biochem Mol 4(2):123–164

Rogowsky PM, Close TJ, Chimera JA, Shaw JJ, Kado CI (1987) Regulation of the *vir* genes of *Agrobacterium tumefaciens* plasmid pTiC58. J Bacteriol 169:5101–5112

Rossi L, Hohn B, Tinland B (1996) Integration of complete transferred DNA units is dependent on the activity of virulence E2 protein of *Agrobacterium tumefaciens*. Proc Natl Acad Sci U S A 93:126–130

Rugini E, Muganu M, Pesce PG (2000) Transformation of *Actinidia* species (kiwifruit). In: Jain SM, Minocha SC (eds) Molecular biology of woody plants, vol 2. Kluwer Academic, Dordrecht, pp 191–226

Sambrook J, Russell DW (2001) Molecular cloning: a laboratory manual, 3rd edn. Cold Spring Harbor Laboratory Press, Cold Spring Harbor, pp 51–586

Schrott M (1995) Selectable marker and reporter genes. In: Potrykus I, Spangenberg G (eds) Gene transfer to plants. Springer Lab Manual, Berlin, pp 325–336

Seligman L, Manoil C (1994) An amphipathic sequence determinant of membrane protein topology. J Biol Chem 269:19888–19896

Sharma KK, Gupta S, Kuhad RC (2006) *Agrobacterium*-mediated delivery of marker genes to *Phanerochaete chrysosporium* mycelial pellets: a model transformation system for white-rot fungi. Biotechnol Appl Biochem 43(3):181–186

Shaw CH, Ashby AM, Brown A, Royal C, Loake GJ, Shaw C (1988) *vir*A and *vir*G are Ti plasmid functions required for chemotaxis of *Agrobacterium tumefaciens* towards acetosyringone. Mol Microbiol 2:413–417

Sheng J, Citovsky V (1996) *Agrobacterium*-plant cell DNA transport: have virulence proteins, will travel. Plant Cell 8:1699–1710

Shimoda N, Toyoda-Yamamoto A, Nagamine A, Usami S, Katayama M, Sakagami Y, Machida Y (1990) Control of expression of *Agrobacterium tumefaciens* genes by synergistic actions of phenolic signal molecules and monosaccharides. Proc Natl Acad Sci U S A 87:6684–6688

Shimoda N, Toyoda-Yamamoto A, Shinsuke S, Machica Y (1993) Genetic evidence for an interaction between the VirA sensor protein and the ChvE sugar-binding protein of *Agrobacterium*. J Biol Chem 268:26552–26558

Shirasu K, Kado CI (1993a) Membrane location of the Ti plasmid VirB proteins involved in the biosynthesis of a pilin-like conjugative structure on *Agrobacterium tumefaciens*. FEMS Microbiol Lett 111(2–3):287–294

Shirasu K, Kado CI (1993b) The *vir*B operon of the *Agrobacterium tumefaciens* virulence region has sequence similarities to B, C and D open reading frames downstream of the pertussis toxin-operon and to the DNA transfer-operons of broad-host-range conjugative plasmids. Nucleic Acids Res 21(2):353–354

Shirasu K, Morel P, Kado CI (1990) Characterization of the *vir*B operon of an *Agrobacterium tumefaciens* Ti plasmid: nucleotide sequence and protein analysis. Mol Microbiol 4:1153–1163

Shirasu K, Koukolikova-Nicola Z, Hohn B, Kado CI (1994) An inner-membrane-associated virulence protein essential for T-DNA transfer from *Agrobacterium tumefaciens* to plants exhibits ATPase activity and similarities to conjugative transfer genes. Mol Microbiol 11:581–588

Shurvinton CE, Hodges L, Ream W (1992) A nuclear localization signal and the C-terminal omega sequence in the *Agrobacterium tumefaciens* VirD2 endonuclease are important for tumor formation. Proc Natl Acad Sci U S A 89:11837–11841

Smith EF, Townsend CO (1907) A plant-tumor of bacterial origin. Science 24:671–673

Smith HO, Wilcox KW (1970) A restriction enzyme from *Hemophilus influenzae* I. Purification and general properties. J Mol Biol 51:379–391

Southern EM (1975) Detection of specific sequences of DNA fragments separated by gel electrophoresis. J Mol Biol 98(3):503–517

Stableford BM (2004) Historical dictionary of science fiction literature. Scarecrow Press, Lanham, p 133

Stachel SE, Nester EW (1986) The genetic and transcriptional organization of the *vir* region of the A6 Ti plasmid of *Agrobacterium tumefaciens*. EMBO J 5:1445–1454

Stachel SE, Zambryski PC (1986) *vir*A and *vir*G control the plant induced activation of the T-DNA transfer process of *A. tumefaciens*. Cell 46:325–333

Stephens KM, Roush C, Nester EW (1995) *Agrobacterium tumefaciens* VirB11 protein requires a consensus nucleotide-binding site for function in virulence. J Bacteriol 177:27–36

Stirn S, Lörz H (2006) Genetically modified plants. In: Heller KJ (ed) Genetically engineered food methods and detection. 2nd Updated and Enlarged Edition. Wiley- VCH Verlag GmbH & Co KgaA, Weinheim, pp 27–63

Stomp AM, Loopstra C, Chilton WS, Sederoff RR, Moore LW (1990) Extended host range of *Agrobacterium tumefaciens* in the genus *Pinus*. Plant Physiol 92:1226–1232

Sule S, Lehockzy J, Jenser G, Nagy P, Burr TJ (1995) Infection of grapevine roots by *Agrobacterium vitis* and *Meloidogyne hapla*. J Phytopathol 143:169–171

Tang W, Sederoff R, Whetten R (2001) Regeneration of transgenic loblolly pine (*Pinus taeda* L.) from zygotic embryos transformed with *Agrobacterium tumefaciens*. Planta 213(6):981–989

Thomashow MF, Panagopoulos CG, Gordon MP, Nester EW (1980) Host range of *Agrobacterium tumefaciens* is determined by the Ti plasmid. Nature 283:794–796

Thomashow MF, Karlinsey JE, Marks JR, Hurlbert RE (1987) Identification of a new virulence locus in *Agrobacterium tumefaciens* that affects polysaccharide composition and plant cell attachment. J Bacteriol 169:3209–3216

Thompson DV, Melchers LS, Idler KB, Schilperoort RA, Hooykaas PJJ (1988) Analysis of the complete nucleotide sequence of the *Agrobacterium tumefaciens virB* operon. Nucleic Acids Res 16:4621–4636

Tinland E, Schoumacher F, Gloeckler V, Bravo-Angel AM, Hohn B (1995) The *Agrobacterium tumefaciens* virulence D2 protein is responsible for precise integration of T-DNA into the plant genome. EMBO J 14:3585–3595

Toro N, Datta A, Carmi OA, Young C, Prusti RK, Nester EW (1989) The *Agrobacterium tumefaciens vir*C1 gene product binds to overdrive, a T-DNA transfer enhancer. J Bacteriol 171:6845–6849

Turk SCHJ, van Lange RP, Regensburg-Tuink TJG, Hooykaas PJJ (1994) Localization of the VirA domain involved in acetosyringone-mediated vir gene induction in Agrobacterium tumefaciens. Plant Mol Biol 25:899–907

Twyman RM, Christou P, Stoger E (2002) Genetic transformation of plants and their cells. In: Marja K, Caldentey KM, Barz W (eds) Plant biotechnology and transgenic plants. Marcel Dekker, New York, pp 111–141

Tzfira M, Vaidya CV (2004) Involvement of targeted proteolysis in plant genetic transformation by *Agrobacterium*. Nature 431:87–92

Uğraş T, Gözükirmizi N (1999) Expression and inheritance of *GUS* gene in transgenic tobacco plants. Turk J Bot 23:297–302

Utkhede RS, Smith EM (1993) Evaluation of biological and chemical treatments for control of crown gall on young apple-trees in the Kootenay Valley of British-Columbia. J Phytopathol 137(4):265–271

Uttaro AD, Cangelosi GA, Geremia RA, Nester EW, Ugalde RA (1990) Biochemical characterization of avirulent *exo*C mutants of *Agrobacterium tumefaciens*. J Bacteriol 172:1640–1646

van der Meer IM (2006) *Agrobacterium*-mediated transformation of petunia leaf discs. In: Vargas VML, Flota FV (eds) Plant cell culture protocols methods in molecular biology, 2nd edn. Humana Press Inc., Totowa, pp 265–272

van Haaren MJJ, Sedee NJA, de Boer HA, Schilperoort RA, Hooykaas PJJ (1988) Bidirectional transfer from a 24 bp border repeat of *Agrobacterium tumefaciens*. Nucleic Acids Res 16:10225–10236

van Larebeke N, Engler G, Holsters M, van den Elsacker S, Zaenen I, Schilperoort RA, Schell J (1974) Large plasmid in *Agrobacterium tumefaciens* essential for crown gall-inducing ability. Nature 252:169–170

van Wordragen MF, Dons HJM (1992) *Agrobacterium tumefaciens* mediated transformation of recalcitrant crops. Plant Mol Biol Rep 10:12–36

Vargas VML, Flota FV (2006) An introduction to plant cell culture back to the future. In: Vargas VML, Flota FV (eds) Plant cell culture protocols methods in molecular biology, 2nd edn. Humana Press Inc., Totowa, pp 3–8

Walkerpeach CR, Velten J (1994) *Agrobacterium*-mediated gene transfer to plant cells: cointegrate and binary vector system. In: Gelvin SB, Schilperoort RA (eds) Plant molecular biology manual, 2nd edn. Kluwer Academic, Dordrecht, pp 1–19

Wanichananan P, Teerakathiti T, Roytrakul S, Kirdmanee C, Peyachoknagul S (2010) A highly efficient method for *Agrobacterium* mediated transformation in elite rice varieties (*Oryza sativa* L spp *indica*). Afr J Biotechnol 9(34):5488–5495

Ward JE, Akiyoshi DE, Regier D, Datta A, Gordon NP, Nester EW (1988) Characterization of the *vir*B operon from an *Agrobacterium tumefaciens* Ti plasmid. J Biol Chem 263(12):5804–5814

Watson JD, Gilman M, Witkowsky J, Zoller M (1992) Genetic engineering of plants. In: Watson JD, Gilman M, Witkowsky J, Zoller M (eds) Recombinant DNA. WH Freeman and Company, New York, pp 273–276

Welsh J, McClelland M (1990) Fingerprinting genomes using PCR with arbitrary primers. Nucleic Acids Res 18:7213–7218

Whatley MH, Spress LD (1977) Role of bacterial lipopolysaccharide in attachment of *Agrobacterium* to moss. Plant Physiol 60:765–766

White PR, Braun AC (1942) A cancerous neoplasm of plants: autonomous bacteria-free crown-gall tissue. Cancer Res 2:597–617

Williams JGK, Kubelik AR, Livak KJ, Rafalski JA, Tingey SV (1990) DNA polymorphisms amplified by arbitrary primers are useful as genetic markers. Nucleic Acids Res 18:6531–6535

Winans SC (1992) Two-way chemical signaling in *Agrobacterium*-plant interactions. Microbiol Rev 56:12–31

Winans SC, Kerstetter RA, Nester EW (1988) Transcriptional regulation of the *vir*A and *vir*G genes of *Agrobacterium tumefaciens*. J Bacteriol 170:4047–4054

Winans SC, Mantis NJ, Chen CY, Han DC (1994) Host recognition by the VirA, VirG two-component regulatory proteins of *Agrobacterium tumefaciens*. Res Microbiol 145:461–473

Wolanin PM, Thomason PA, Stock JB (2002) Histidine kinases: key signal transducers outside the animal kingdom. Genome Biol 3:1–8

Wood DW, Setubal JC, Kaul R, Monks DE, Kitajima JP, Okura VK, Zhou Y, Chen L, Wood GE, Almeida NF Jr, Woo L, Chen Y, Paulsen IT, Eisen JA, Karp PD, Bovee D Sr, Chapman P, Clendenning J, Deatherage G, Gillet W, Grant C, Kutyavin T, Levy R, Li MJ, McClelland E, Palmieri A, Raymond C, Rouse G, Saenphimmachak C, Wu Z, Romero P, Gordon D, Zhang S, Yoo H, Tao Y, Biddle P, Jung M, Krespan W, Perry M, Gordon-Kamm B, Liao L, Kim S, Hendrick C, Zhao ZY, Dolan M, Chumley F, Tingey SV, Tomb JF, Gordon MP, Olson MV, Nester EW (2001) The genome of the natural genetic engineer *Agrobacterium tumefaciens* C58. Science 294(5550):2317–2323

Wu W (1995a) PCR techniques and applications. In: Kaufman PB, Wu W, Kim D, Cseke L (eds) Handbook of molecular and cellular methods in biology and medicine. CRC Press, Boca Raton, pp 243–262

Wu W (1995b) Electrophoresis, blotting, and hybridization. In: Kaufman PB, Wu W, Kim D, Cseke L (eds) Handbook of molecular and cellular methods in biology and medicine. CRC Press, Boca Raton, pp 87–122

Wu W, Welsh MJ (1996a) Expression of the 25-kDa heat-shock protein (HSP27) correlates with resistance to the toxicity of cadmium chloride, mercuric chloride, *cis*platinum (II)-diammine dichloride, or sodium arsenite in mouse embryonic stem cells transfected with sense or antisense HSP27 cDNA. Toxicol Appl Pharmacol 141:330–339

Wu W, Welsh MJ (1996b) A method for rapid staining and destaining of polyacrylamide gels. Biotechniques 20(3):386–388

Wu LL, Song I, Karuppiah N, Kaufman PB (1993) Kinetic induction of oat shoot pulvinus invertase mRNA by gravistimulation and partial cDNA cloning by the polymerase chain reaction. Plant Mol Biol 21:1175–1179

Wu W, Welsh MJ, Kaufman PB, Zhang HH (eds) (2004) Gene biotechnology, 2nd edn. CRC Press LLC, Boca Raton, pp 1–523

Yanofsky MF, Porter SG, Young C, Albright LM, Gordon MP, Nester EW (1986) The *vir*D operon from *Agrobacterium tumefaciens* encodes a site-specific endonuclease. Cell 7:471–477

Yibrah HS, Gronroos R, Lindroth A, Franzen H, Clapham D, von Arnold S (1996) *Agrobacterium rhizogenes* mediated induction of adventitious rooting from *Pinus contorta* hypocotyls and the effect of 5-azacytidine on transgene activity. Transgenic Res 5:75–85

Zaenen I, van Larebeke N, Teuchy H, van Montagu M, Schell J (1974) Supercoiled circular DNA in crown gall inducing Agrobacterium strains. J Mol Biol 86:109–127

Zale J (2008) Transgenic plant analysis. In: Stewart CN Jr (ed) Plant biotechnology and genetics. Wiley, Hoboken, pp 275–290

Zambryski P (1992) Chronicles from the *Agrobacterium*-plant cell DNA transfer story. Annu Rev Plant Physiol Plant Mol Biol 43:465–490

Zambryski P, Holsters M, Kruger K, Depicker A, Schell J, van Montagu M, Goodman HM (1980) Tumor DNA structure in plant cells transformed by *A tumefaciens*. Science 209:1385–1391

Zambryski P, Joos H, Genetello C, Leemans J, van Montagu M, Schell J (1983) Ti plasmid vector for the introduction of DNA into plant cells without alteration of their normal regeneration capacity. EMBO J 2:2143–2150

Zauner S, Creasap JE, Burr TJ, Ullrich CI (2006) Inhibition of crown gall induction by *Agrobacterium vitis* strain F2/5 in grapevine and *Ricinus*. Vitis 45(3):131–139

Zhiming R, Zheng M, Wei S, Huiying F, Jian Z (2008) Transformation of industrialized strain *Candida glycerinogenes* with resistant gene *zeocin* via *Agrobacterium tumefaciens*. Curr Microbiol 57(1):12–17

Zhou XR, Christie PJ (1997) Suppression of mutant phenotypes of the *Agrobacterium tumefaciens* VcirB11 ATPase by overproduction of VirB proteins. J Bacteriol 179:5835–5842

Zhu J, Oger PM, Schrammeijer B, Hooykaas PJJ, Farrand SK, Winans SC (2000) The bases of crown gall tumorigenesis. J Bacteriol 182:3885–3895

Ziemienowicz A (2001) Odyssey of *Agrobacterium* T-DNA. Acta Biochem Pol 48(3):623–635

Zorreguieta A, Geremia RA, Cavaignac S, Cangelosi GA, Nester EW, Ugalde RA (1988) Identification of the product of an *Agrobacterium tumefaciens* chromosomal virulence gene. Mol Plant Microbe Interact 1:121–127

Zupan JR, Zambryski P (1995) Transfer of T-DNA from *Agrobacterium* to the plant cell. Plant Physiol 107:1041–1047

Zupan JR, Zambryski P (1997) The *Agrobacterium* DNA transfer complex. Crit Rev Plant Sci 16:279–295

Zupan JR, Citovsky V, Zambryski P (1996) *Agrobacterium* VirE2 protein mediates nuclear uptake of ssDNA in plant cells. Proc Natl Acad Sci U S A 93:2392–2397

Zupan JR, Muth TR, Draper O, Zambryski P (2000) The transfer of DNA from *Agrobacterium tumefaciens* into plants: a feast of fundamental insights. Plant J 23:11–28

Chapter 13
Progress and Prospects for Efficient Micropropagation of Woody Plants

Faheem Aftab

Contents

Abstract Despite the progress made so far, tissue culture studies involving woody plants have always remained quite challenging. Amongst several reasons, lack of a suitable explant source still remains a major bottleneck to limit success. The present review describes the progress in micropropagation of difficult to manipulate woody plant species. It discusses the possibility of forcing epicormic or latent buds in a wide range of adult woody plant species of both temperate as well as tropical origin

F. Aftab (✉)
Department of Botany, University of the Punjab, Q. A. Campus, Lahore 54590, Pakistan
e-mail: faheem.botany@pu.edu.pk; drfaheemaftab@yahoo.com

M. Ashraf et al. (eds.), *Crop Production for Agricultural Improvement*,
DOI 10.1007/978-94-007-4116-4_13, © Springer Science+Business Media B.V. 2012

where procurement of clean, juvenile explant material is extremely scanty. Forcing may be accomplished under different environmental conditions such as fog, mist or lab and media including sand, sawdust, perlite, vermiculite and their combinations. The explants derived from such forced shoots have shown good potential for either rooting directly under greenhouse conditions or further manipulation under *in vitro* conditions for axillary bud activation, shoot apex proliferation, multiple shoot formation and somatic embryogenesis. This paper intends to review the progress made so far for an efficient micropropagation of several woody species of economic significance making use of the above technology. Several factors limiting the scope of forced epicormic shoots and possible solutions have also been detailed and discussed herein.

Keywords Adventitious shoots • Bud break • Chilling • Epicormic buds • Micropropagation • Shoot forcing • Somatic embryogenesis

1 Introduction

Over the past few decades, plant tissue culture studies have made considerable progress. Amongst plants belonging to diverse groups and families, reproducible protocols exist for callus induction, its maintenance and regeneration. Owing to the relative ease with which these plants may be maintained under *in vitro* conditions (Aftab et al. 2008a, b), much focus, however, has been on herbaceous and crop plants. It is the possibility of maintaining totipotent plant material *in vitro* that has also paved way for several breakthroughs in the genetic engineering of various crop plants (Aftab et al. 1996; Aftab and Iqbal 1999).

Compared with the above-mentioned plants, woody plants including many tree species have posed special problems and thus considered to be quite recalcitrant under *in vitro* conditions (Aftab and Preece 2007). Lack of suitable explant source and reproducible regeneration protocols, release of polyphenols, slow growth and high rate of contamination have limited success with micropropagation of these plants of economic significance.

Several nursery or greenhouse techniques may now be combined with tissue culture techniques in order to get meaningful results in the propagation of several woody plant species (Preece and Read 2003, 2007). The purpose of this paper is to highlight both the problems as well as the prospects of woody plants' propagation under such an experimental setup. Softwood shoot forcing and forcing epicormic buds has emerged as a great tool resulting in the production of plant material that has shown potential as propagules for further development under greenhouse conditions or as a suitable explant source for the establishment of *in vitro* cultures. Since these techniques have mainly been investigated for the plants of temperate origin, this review also outlines several associated problems and forecasts the potential use of such techniques in the micropropagation of a wide range of woody plants.

2 Forcing Epicormic or Latent Buds

Epicormic or the latent buds may prove to be an ideal source material in plants where clean and juvenile tissue is usually not readily available for either greenhouse propagation or *in vitro* manipulations including micropropagation by tissue culture means. Over the past several years, its use has proven to be of great significance and the resulting forced softwood shoots have either been used directly for rooting purposes or as a source material to obtain nodal or apical explants for *in vitro* culture establishment (Preece and Read 2003). Softwood shoots are usually visible within a week's time on the surface of logs in most species that have earlier been derived or cut from adult trees and further cut to a length of some 40 cm and placed some one-fourth embedded in several possible media (details to follow in the next section) in flats. Several studies have shown the significance of the caliper size on the number of forced shoots (Henry and Preece 1997; Ledbetter and Preece 2003; Fishel et al. 2003), but the log length seems to be relatively less critical a factor influencing softwood shoot forcing (Henry and Preece 1997). Experience with some tropical trees like teak (*Tectona grandis* L.) in this lab have shown that the areas on logs surrounding cut branches have several hidden buds that are usually easier to force and may result in greater number of softwood shoots. This is also true for pecan (*Carya illinoensis* (Wangenh.) C. Koch), a temperate nut-producing tree species of the family juglandaceae (Haroon 2011).

Although the logs have been taken from plants and forced in lab conditions on year-round basis but seasonal influence seems to hold good for the production of larger number of softwood shoots. Epicormic buds on pecan logs thus forced easily and at a greater frequency during the spring season rather than autumn where the logs mostly remained quite dormant and produced either a few shoots or none at all (Aftab and Preece 2007).

Perhaps the most important aspect to consider for softwood shoot forcing is to obtain the branches from within the cone of juvenility (Preece and Read 2003). This allows obtaining softwood shoots with juvenile characteristics that may be rooted easily and thus may be used for large scale propagation. Retaining juvenile characteristics is also important if the aim of the studies is to use softwood shoots as an explant source for *in vitro* culture establishment. Detailed account on juvenility in woody plants can be found in a review article by Preece and Read (2003).

2.1 *Media and Culture Conditions*

Forcing softwood epicormic buds involves placing the logs either in empty flats of suitable size or containing forcing medium. In case of empty flats, it is important that the logs be drenched with water daily in order to retain moisture and keep the buds alive. Drenching is also important in case the medium is dispensed in the flats and logs are placed embedded in the medium in such a manner that some three-fourth of the logs is exposed to the external environment. Some of the media that in particular have been used for a wide variety of plants are perlite and vermiculite (Aftab et al. 2005). Vermiculite adsorbs and retains more water as compared to perlite and hence provides a better opportunity for the logs that have been forced under the lab conditions (Aftab et al. 2005). The choice of medium thus seems to be interacting with the environment. Combinations of perlite and vermiculite have also been used and it is not surprising that epicormic buds have also been forced in empty flats provided the environmental conditions are humid enough (Aftab et al. 2005). From the past experience, it is also known that even sand may prove to be a good forcing medium and may allow forcing of epicormic buds at a reasonable frequency (Aftab and Preece 2007; Akram and Aftab 2007). The use of some other forcing media such as peat moss, sawdust, cocopeat and even cotton have also been evaluated in the author's lab for such forcing experimentation under different environmental conditions, but so far none of these has been found superior to the ones mentioned above.

As pointed out earlier, the media seem to interact with the culture environment. Epicormic buds have been forced in different environmental conditions such as lab, under mist or fog in a greenhouse (Aftab et al. 2005). In all the environmental conditions, temperature and light conditions may be adjusted to suit the plant in question. Greenhouses in the western countries are usually well-equipped with the facilities to regulate temperature, light quality and intensity and produce mist or fog as desired. The intensity of fog or the duration of mist spray and interval in between is also electronically regulated. Both fog as well as mist have shown good results for the epicormic bud forcing in silver maple (*Acer saccharinum* L.) and white ash (*Fraxinus pennsylvanica* Marsh) though the lab conditions, supported relatively lesser number of softwood shoots (Aftab et al. 2005). Use of ambient temperature and light conditions has also been evaluated in teak during the spring season in order to force a large number of epicormic buds (Akram and Aftab 2007). This way of producing softwood shoots may also help in those conditions where a lot of softwood shoot production is desired but greenhouse facilities are non-existent. This may, however, result in lack of reproducibility of results thus limiting its experimental scope.

3 Shoot Forcing

'Shoot forcing' is essentially different from 'forcing epicormic buds' on a medium in flat trays as mentioned above. In the words of Read and Preece (2007), "This technology involves immersing the bases of cut woody stems into solutions containing 2% sucrose and 8-hydroxyquinoline citrate at the rate of 200 mg L^{-1}" for bud break in a number of woody plant species (Yang and Read 1992, 1993; Hamooh and Read 2000). Preece's group at Southern Illinois University at Carbondale, Illinois has conducted research on this aspect for many years. A useful information on several aspects of shoot forcing can be found in the earlier reports by Preece and Read (2003), Read and Preece (2007) and Preece and Read (2007). These review papers and the literature cited therein strongly suggest the possibility of extending this research to woody plant species of tropical regions where work of this nature is almost non-existent till this point in time.

4 Chilling and Bud Break

Ashby et al. (1991) explained that "plants that set buds, become leafless, and do not respond readily to favorable growing conditions are considered dormant". Such plants were said to have a 'chilling requirement' usually of several 100 h before the bud break occurs (Swartz and Powell 1981; Ashby et al. 1991). Vast literature on various aspects of chilling and bud break in woody plant species is available (Weinberger 1950; Kriebel and Wang 1962; Spiers and Draper 1974; Hunter and Lechowicz 1992). The literature hence includes (but not limited to) studies on chilling exposures (Swartz and Powell 1981; Couvillon and Erez 1985), oxidative processes (Shulman et al. 1986), use of bioregulators for bud break (Shulman et al. 1986; Steffens and Stutte 1989) and the associated biochemical changes (Wang and Faust 1988; Wang et al. 1991). This alone necessitates a separate review in order to throw some light on this important physiological aspect of woody plants' propagation. In the present paper, this information only aims at indicating the potential use and importance of such studies to get maximum efficiency in propagation (or micropropagation) through the epicormic buds. The dormant stem sections of woody plants in particular limit the scope of a year-round forcing of epicormic buds and hence the main focus is on plant material that had undergone chilling under natural environment. This is for the same reason that stem sections that are collected during the late spring or early autumn in the temperate region have shown good potential for a greater number of epicormic bud forcing (Aftab et al. 2005). Dormant stem sections collected during other seasons may become an appropriate plant material only if their chilling requirements have been met. This aspect needs further investigation especially in woody plants of tropical regions of the world where a comprehensive information about the chilling requirements of a large number of woody plant species is still lacking.

5 Rooting and Acclimatization

Softwood shoots may be rooted directly under the greenhouse conditions in flats or beds of different sizes containing a suitable medium. This is accomplished usually by cutting softwood shoots of various sizes (usually 4–6 cm but may be longer) from the logs and treating the basal 2 cm cut ends with either indole-3-acetic acid (IAA) or 0.1% indole-3-butyric acid; IBA (Henry and Preece 1997). Softwood shoots after this treatment are transplanted to flat trays of appropriate sizes containing a variety of greenhouse media as just mentioned above. Peat moss is an important medium at this stage. The purpose is to produce roots and acclimatize the plants in the greenhouse environment. Mist or fog conditions also play a vital role during the earlier establishment since the plants cannot afford water loss due to the absence of root system. Even after the establishment of a well developed root system, most plants need humid conditions for quite some time before being transferred to the field conditions. High humidity tents provide an opportunity to maintain the desired humidity level till that time or else may be maintained under the fog or the mist system. The hardening-off or acclimatization of plants derived either from the propagation bench or through micropropagation has been referred to as "the most tricky part of micropropagation" (Preece 2001). This is so because the plants are grown, regenerated and maintained under almost 100% relative humidity. This does not allow a propagator to transfer the plants directly to the field conditions. A stepwise hardening-off or acclimatization process thus needs to be followed and it has been reviewed by Preece (2001) in detail. A successful acclimatization will ensure survival of most plants in the normal greenhouse environment (without mist, fog or frequent watering) but 100% survival is seldom achieved.

6 *In Vitro* Establishment

An important application of softwood shoot forcing is subsequent establishment under *in vitro* conditions. This makes sense in that the plant material is generally cleaner as compared to the one derived from field-grown plants. Though the possibility exists for *in vitro* culture initiation in a large number of plant species, contamination is still a major issue at least for those explants that have been obtained from the shoots grown under mist, fog or high humidity tents. This is because the high humidity condition not only facilitates an optimum growth of plants but also favors the establishment of microbial contamination on the plants' surface. Such contamination is usually so severe that it is difficult to disinfest and culture-contamination during *in vitro* manipulations is usually high. In their experiments, Aftab et al. (2005) have recommended the use of 0.18% H_2O_2 (obtained in the form of a commercial preparation under a trade name of Zerotol; Biosafesystems, USA) on the stems of silver maple and green ash placed under mist or fog in greenhouse. Contamination rate was reduced to 68 and 92.2% in silver maple from the explants derived from stems placed under fog and mist conditions, respectively. The explants derived from shoots forced under the lab condition, however, were readily adapted *in vitro* with only 4% microbial contamination. Consequently, to get the true advantage of softwood shoots in raising *in vitro* cultures, more work is required to find out measures to minimize the microbial contamination of the plant material raised under mist or fog conditions. As a result, if the ultimate objective of forcing softwood shoots is to raise plant material for micropropagation or *in vitro* establishment for somatic embryogenesis, a tissue culturist is usually left with only one option, i.e., forcing softwood shoots under the lab conditions.

6.1 Micropropagation

For micropropagation, the best explant seems to be the nodal segments. Shoot apices may also be used but nodal explants offer potentially more chances of multiple shoot formation that is desired in micropropagation. Additionally, the softwood shoots (usually 2–4 cm) offer a unique opportunity of direct root formation under *in vitro* conditions and may then be hardened-off for further propagation.

6.1.1 Through Nodal Explants

Each nodal explant may provide one, two or more axillary buds that may be developed further under the influence of growth regulators. Multiple shoot formation is of great significance and is obtained using nodal explants subjected to a wide range of hormonal treatments (George and Sherrington 1984; Akram and Aftab 2008). It is important to give attention to details in order to establish nodal explants *in vitro* successfully. Disinfestation of explants must follow trimming at both the cut ends in order to avoid subsequent browning and necrosis of the entire culture. As stated above, the nodal explants, if obtained from the softwood shoots should preferably be forced under the lab conditions. Of the literature available on micropropagation involving a wide variety of plants, nodal explant seems to be the best choice (Quraishi and Mishra 1998; Fraga et al. 2004) amongst others such as hypocotyls (Daysuis and Chagvardieff 1990) and inflorescence (Eizenga and Dahleen 1990), but the choice of an explant may vary from species to species. It is also important that factors governing stock plant growth and physiology be known (Read 1990; Read and Preece 2007).

6.1.2 Through Shoot Apices

Although micropropagation has been reported using shoot apex (Nand et al. 2004; Akram and Aftab 2008), it poses certain disadvantages over the use of nodal explants. One of the most important being a relatively slower growth rate followed by a lesser number of micropropagated plants at the end of an experiment (Akram and Aftab 2008). However, variation in results in different plant species does not allow us to 'label' shoot apices as a second choice and may in fact be of great significance in certain plant groups (Birmeta and Welander 2004; Kadota and Niimi 2004). If enough literature is not available on a certain plant species of interest, its micropropagation should be attempted using all possible explant sources.

6.1.3 Direct Rooting

Direct rooting of softwood shoots under greenhouse conditions is discussed under Sect. 5. For micropropagation, direct rooting of softwood shoots has also been attempted in teak (Akram and Aftab 2009) and Araucaria (Shafiq and Aftab 2008). There are no reports for such an attempt in other plant species. In these studies, softwood shoots (2–4 cm) were surface disinfested and transferred directly to half or full strength MS (Murashige and Skoog 1962), DKW (Driver and Kuniyuki 1984) or WPM (Woody Plant Medium; McCown and Llyod 1981) media containing several doses of IBA. The purpose of such attempts for *in vitro* rooting was simply academic and did not work well at least in teak and Araucaria, but the results with other plant species may be different and the author expects to see some progress in direct rooting of softwood shoots under *in vitro* conditions in the near future.

6.2 Problems Associated with Micropropagated Plants

Phenotypic inconsistency due to somaclonal variation has been attributed to be one of the main problems associated with micropropagated plants (Preece and Trigiano 2001). They have thus suggested the need to avoid adventitious shoots "especially from callus, during micropropagation of woody species". They have further described lack of chlorophyll in "portions of the leaves of some plants" as a result of this variation in some micropropagated plants. "Another fairly common variation in plants *in vitro* is the formation of fasciated stems (flat, wide stems that appear as if two or more stems are fused together lengthwise). Fasciation can have a genetic cause, as the result of a mutation, but it can also be a physiological response of shoots to conditions *in vitro*, such as the cytokinin used" (Preece and Trigiano 2001). Interplay of the growth regulators may be of use to overcome the above-mentioned problems.

It is a well-known fact that the micropropagated plants are usually quite fragile. Since these plants are raised *in vitro* with an ample supply of nutrients and growth regulators and the fact that relative humidity is almost 100%, growth rate is usually optimum. The plants are at the same time quite fragile and cannot cope with the external environment. A successful acclimatization or hardening-off phase is thus a prerequisite to obtain the benefits of micropropagation.

Micropropagation through multiple shoot formation is reported by many workers in a wide variety of plant species (Biroscikova et al. 2004; Ozden-Tokatli et al. 2005; Akram and Aftab 2008). Multiple shoot formation offers production of many shoots within the confines of a limited lab space and thus becomes quite economical. Micropropagation even without a greater number of multiple shoots still is economically favorable in those plants where it is the only means of propagation and conventional methods fail to yield the desirable results. Micropropagation through multiple shoot formation requires standardization of medium for each species and many reports in the available literature may be found focusing on standardization of culture conditions for this purpose.

6.3 Somatic Embryogenesis or Adventitious Shoot Formation – Must Have Option for Certain Recalcitrant Woody Plants

Focus on somatic embryogenesis and adventitious shoot formation in the literature seems to be for two main reasons. Firstly, in those cases where conventional propagation methods fail to yield desirable results and hence need supplemental *in vitro* approaches. Secondly, and perhaps more importantly, development of a reliable plant regeneration system either through somatic embryogenesis or organogenesis including adventitious shoot regeneration is extremely important in those studies where cells, protoplasts or tissues are used as target material to obtain transgenic plants (Aftab and Iqbal 1999). In view of the significance of the later point, a vast literature exists on callus induction, regeneration and somatic embryogenesis in woody plants of significance (Rakesh et al. 1995; Ipekci and Gozukirmizi 2004; Kvaalen et al. 2005; Aftab and Preece 2007). Adventitious shoot formation in particular for micropropagation is an approach that has got somewhat lesser interest of the workers until recently but the continuous advances in the understanding of the role of plant hormones and growth regulators *in vitro* have opened up further possibilities in the development of fast and reliable *in vitro* approaches for the propagation of various recalcitrant woody plant species (Wilhelm 1999; Akram and Aftab 2008).

6.4 Role of Growth Regulators

Generally speaking, exploitation of *in vitro* approaches for the potential development of woody plant species rests mainly on an appropriate use and interplay of various plant hormones and growth regulators; the main focus being on auxins and cytokinins (Vandenbussche and Straeten 2004; Gahan 2007). 'Plant hormone' is a strict term used for only naturally-synthesized signaling molecules that must be synthesized and translocated within the plant body for its physiological action at much lower concentrations of μM or even nM range. Since both naturally-synthesized as well as synthetically-produced signaling molecules are used in tissue culture studies, hence a broader term of 'plant growth regulators' is preferred in this text. The general philosophy for the use of plant growth regulators in woody plant species *in vitro* seems simple. Auxins, e.g., Indole-3-acetic acid (IAA), indole-3-butyric acid (IBA) and 2,4-D (2,4-dichlorophenoxyacetic acid) are generally known to stimulate somatic embryogenesis (Raghavan 2004), rhizogenesis in callus cultures (Wadegaonkar et al. 2006) and root formation of regenerated shoots (Akram and Aftab 2008) and cytokinins promote cell division (Sánchez-Moreiras et al. 2006) or shoot regeneration via organogenesis (Bhagwat and Lane 2004). In most studies, it is the combination of the two or more growth regulators from each class that is reported to influence various growth behaviors *in vitro*. There is an optimum concentration for each growth regulator for a certain growth response or cellular behavior and this optimum may differ slightly or considerably when used in combination with another growth regulator. In general, higher cytokinin-to-auxin ratios have favored shoot initiation in callus cultures. On the other hand, higher auxin-to-cytokinin ratios usually promote rhizogenesis and equal or nearly equal 'moderate levels' of auxins and cytokinins support callus induction. It is essentially the choice of the worker to select a particular concentration or combination of growth regulators on the basis of experimental objectives.

In the language of Huetteman and Preece (1993) "Few innovations in woody plant micropropagation have been as dramatic as the introduction of thidiazuron as a plant growth regulator in culture media". For this reason, thidiazuron (N-phenyl-N′-1, 2, 3-thiadiazol-5-yl urea; TDZ) has been reported as an effective signaling molecule having cytokinin-like activity in most of the literature dealing with many *in vitro* aspects of woody plants during the 1990s and onwards (Huetteman and Preece 1993; Ledbetter and Preece 2004 and Akram and Aftab 2008). In the above-cited review paper of Huetteman and Preece (1993), TDZ's properties have been discussed with reference to chemical activity, axillary shoot formation, callus, adventitious regeneration and rooting besides certain special concerns such as elongation, tolerance, vitrification and shoot fasciation. In fact, in most of the recent literature on temperate woody plants, TDZ's usefulness has been proven in somatic embryogenesis (Bates et al. 1992), axillary shoot formation (Wilhelm 1999; Akram and Aftab 2008), adventitious shoot regeneration (Ledbetter and Preece 2004) and more reports about its 'strong effect' on various growth parameters of woody plants are continuously pouring in. From the personal experience of the author, the use of TDZ has proven to be of great significance in overcoming certain growth and developmental barriers in difficult-to-manipulate woody plants such as *Acer saccharinum* L. and *Fraxinus pennsylvanica* Marsh (Aftab et al. 2005), *Tectona grandis* (Akram and Aftab 2007, 2008), *Carya illinoensis* (Aftab and Preece 2007) and *Araucaria* (Shafiq and Aftab 2008; Batool and Aftab 2008) thus opening doors to the development of essentially rapid and reproducible micropropagation protocols for many related woody plant species. It is, however, surprising that most reports are on plants of temperate origin and its usefulness has not been encashed in the studies pertaining to micropropagation of woody plants of tropical origin (Aftab and Preece 2007; Akram and Aftab 2008). One obvious reason in the developing countries seems its price but effectiveness at very low concentrations may evade its price consideration.

Apart from the use of auxins and cytokinins, gibberellins have also been used in woody plant tissue culture albeit less frequently but its influence in regulating plants' height is well documented (Glocke et al. 2006). A new era on the studies pertaining to brassinosteroids has already started (Brosa 1999; Khripach et al. 2000; Ferrie et al. 2005; Haubrick and Assmann 2006) though its use in woody plants' tissue culture is yet to be determined. Similarly, jasmonic acid and salicylic acid and several other signaling molecules may also be strong candidates for the determination of their morphological effects on various growth and developmental aspects of woody plant species. Further work on these lines will hopefully throw light on these aspects.

7 In Search of Other Novel Candidate Explant Sources

If we see the history of plant tissue culture and the concept of totipotency, every living cell of a plant is considered to have an inherent capability to divide and regenerate into a plant. Based on this, explants have been collected from almost every tissue and organ of a plant. Leaves, nodes, internodes, shoot apices, roots and cambium have been the most common explant sources in wide groups of plant species. However, choice of explants is relatively less in case of recalcitrant woody plant species. The above-mentioned explants in case of adult woody trees are generally hard, lignified and mature enough and pose serious difficulties *in vitro* to revert back or dedifferentiate to meristematic nature. One possible solution is the one mentioned in Sect. 2 and involves the development of epicormic shoots that may be used as an explant source quite effectively. The other approaches have been to use softer tissues such as nucellus obtained from immature fruits in *Mangifera indica* (Rivera-Domínguez et al. 2004), immature seed tissues in *Juglans nigra* (Long et al. 1995) and immature zygotic embryos in *Carya illinoensis* (Mathews and Wetzstein 1993). This approach, however, considerably reduces the period of availability of suitable explants since immature fruits can be collected over a very brief developmental phase of a particular woody plant.

7.1 *Immature Bark Explants and Pith Tissues*

The author in view of the above limitations associated with woody plants suggests trying several uncommon explants as well. In *Carya illinoensis* (Wangenh.) C. Koch (Haroon and Aftab 2008) and *Araucaria heterophylla* (Batool and Aftab 2008), immature bark explants collected directly from adult trees during active growth phase have yielded quite satisfactory results for dedifferentiation leading to an ample callus formation and further morphogenesis. In addition, embryogenic callus from nucellar tissue of *Mangifera indica* spp. has also been obtained on several TDZ levels (Malabadi et al. 2011). Although the availability of epicormic shoots seems to mask the importance of such unusual explants, they may still have a potential role especially in the light of the established usefulness of growth regulators like TDZ in woody plants' micropropagation.

8 Conclusion

Combining relatively newer propagation techniques including shoot forcing and forcing epicormic buds with plant tissue culture holds a promise for the development of rapid and reproducible propagation and micropropagation protocols of many recalcitrant woody plant species. Studies on these lines may potentially ensure propagation of those woody tree species where non-viable seeds or associated problems such as low germination rate and dormancy are frequently encountered or the aim of propagation is to retain genetic fidelity. Since most work cited in this paper is on woody plants of temperate regions of the world, its usefulness in woody plant species including ornamentals and forest trees of tropical origin is yet to be determined and the prospects are wide open. One needs a little caution though since the limitations are many (and many more may be reported in the future) once the work on similar lines is carried out in labs dealing with research on tropical woody plant species.

References

Aftab F, Iqbal J (1999) Plant regeneration from protoplasts derived from cell suspension of adventive somatic embryos in sugarcane (*Saccharum* spp. hybrid cv. CoL-54 and cv. CP-43/33). Plant Cell Tissue Organ Cult 56:155–162

Aftab F, Preece JE (2007) Forcing and *in vitro* establishment of softwood shoots from large stem segments of woody plants. In: Xu Z, Li J, Xue Y, Yang W (eds) Biotechnology and sustainable agriculture 2006 and beyond. Springer, Dordrecht, The Netherlands, pp 437–444

Aftab F, Zafar Y, Malik KA, Iqbal J (1996) Plant regeneration from embryogenic cell suspensions and protoplasts in sugarcane (*Saccharum* spp. hybrid cv. CoL-54). Plant Cell Tissue Organ Cult 44:71–78

Aftab F, Mansouri K, Preece JE (2005) The influence of environment, media and Zerotol on forcing and *in vitro* establishment of softwood shoots from large stem segments of *Acer saccharinum* L. and *Fraxinus pennsylvanica* Marsh. Prop Orn Plants 5:111–116

Aftab F, Akram S, Iqbal J (2008a) Estimation of fixed oils from various explants and *in vitro* callus cultures of jojoba (*Simmondsia chinensis*). Pak J Bot 40:1467–1471

Aftab F, Alam M, Afrasiab H (2008b) *In vitro* shoot multiplication and callus induction in *Gladiolus hybridus* Hort. Pak J Bot 40:517–522

Akram M, Aftab F (2007) Rooting and *in vitro* establishment of softwood shoots of *Tectona grandis* L. In: 5th international conference, 'Propagation of Ornamental Plants', 5–8 Sept 2007, Sofia, Bulgaria, p 36

Akram M, Aftab F (2008) High frequency multiple shoot formation from nodal explants of teak (*Tectona grandis*) induced by thidiazuron. Prop Orn Plants 8:72–75

Akram M, Aftab F (2009) An efficient method for clonal propagation and in vitro establishment of softwood shoots from epicormic buds of teak (Tectona grandis L.). For Stud China 11:105–110

Ashby WC, Bresnan DF, Huetteman CA, Preece JE, Roth PL (1991) Chilling and bud break in silver maple. J Environ Hort 9:1–4

Bates S, Preece JE, Navarrete NE, Van Sambeek JW, Gaffney GR (1992) Thidiazuron stimulates shoot organogenesis and somatic embryogenesis in white ash (*Fraxinus americana L.*). Plant Cell Tissue Organ Cult 31:21–29

Batool N, Aftab F (2008) Thidiazuron induces embryogenesis from mature bark explants in Norfolk Island pine (*Araucaria heterophylla* (Salisb.) Franco). International conference of plant scientists (10th national meeting of plant scientists), 21–24 April 2008. Department of Botany, University of Agriculture, Faisalabad, Pakistan (Abstract, ICPS-193), p 85

Bhagwat B, Lane WD (2004) *In vitro* shoot regeneration from leaves of sweet cherry (*Prunus avium*) 'Lapins' and 'Sweetheart'. Plant Cell Tissue Organ Cult 78:173–181

Birmeta G, Welander M (2004) Efficient micropropagation of *Ensete ventricosum* applying meristem wounding: a three-step protocol. Plant Cell Rep 23:277–283

Biroscikova M, Spisakova K, Liptak S, Pichler V, Urkovic JD (2004) Micropropagation of mature wych elm (*Ulmus glabra* Huds.). Plant Cell Rep 22:640–644

Brosa C (1999) Biological effects of Brassinosteroids. Crit Rev Biochem Mol Biol 34:339–358

Couvillon GA, Erez A (1985) Influence of prolonged exposure to chilling temperatures on bud break and heat requirement for bloom. J Am Soc Hort Sci 110:47–50

Daysuis JM, Chagvardieff P (1990) Plant donor tissue and isolation procedure effect on early formation of embryoids from protoplasts of *Helianthus annuus* L. Plant Cell Tissue Organ Cult 22:183–189

Driver JA, Kuniyuki AH (1984) *In vitro* micropropagation of paradox walnut rootstock. Hort Sci 19:507–509

Eizenga G, Dahleen LS (1990) Callus production, regeneration and evaluation of plants from cultured inflorescences of tall fescue (*Festuca arundinaceae* Schred). Plant Cell Tissue Organ Cult 22:7–15

Ferrie AMR, Dirpaul J, Krochko P, Keller WA (2005) Effects of brassinosteroids on microspore embryogenesis in *Brassica* species. In Vitro Cell Dev Biol Plant 41:742–745

Fishel DW, Zaczek JJ, Preece JE (2003) Positional influence on rooting of shoots forced from the main bole in swamp white oak and northern red oak. Can J For Res 33:705–711

Fraga M, Alonso M, Ellul P, Borja M (2004) Micropropagation of *Dianthus gratianopolitanus*. Hort Sci 39:1083–1087

Gahan PB (2007) Totipotency and the cell cycle. In: Jain SM, Haggman H (eds) Protocols for micropropagation of woody trees and fruits. Springer, Dordrecht, The Netherlands, pp 3–14

George EF, Sherrington PD (1984) Plant propagation by tissue culture. Exegetics Ltd., Eversley, Basingstoke, 478pp

Glocke P, Delaporte K, Collins G, Sedgley M (2006) Micropropagation of juvenile tissue of *Eucalyptus erythronema* × *Eucalyptus stricklandii* cv. 'urrbrae gem'. In Vitro Cell Dev Biol Plant 42:139–143

Hamooh BT, Read PE (2000) Silver thiosulfate alone and in combination with gibberellic acid (GA_3) in the forcing solution influenced bud break and shoot elongation of dormant woody species. Proc Plant Growth Regul Soc Am 26:191–192

Haroon A (2011) Propagation of Pecan (*Carya illinoensis*) using in vitro techniques. PhD thesis, University of the Punjab, Lahore, Pakistan

Haroon A, Aftab F (2008) Immature bark explants: an excellent source for callus induction in Pecan (*Carya illinoensis* (Wangenh.) C. Koch). International conference of plant scientists (10th national meeting of plant scientists), 21–24 April 2008. Department of Botany, University of Agriculture, Faisalabad, Pakistan (Abstract, ICPS-141), pp 63

Haubrick LL, Assmann SM (2006) Brassinosteroids and plant functions: some clues, more puzzles. Plant Cell Environ 29:446–457

Henry PH, Preece JE (1997) Production of shoots from dormant species of *Acer* as influenced by length and caliper of stem sections. J Environ Hort 15:153–156

Huetteman CA, Preece JE (1993) Thidiazuron: a potent cytokinin for woody plant tissue culture. Plant Cell Tissue Organ Cult 33:105–119

Hunter AF, Lechowicz MJ (1992) Predicting the timing of budburst in temperate trees. J Appl Ecol 29:597–604

Ipekci Z, Gozukirmizi N (2004) Indirect somatic embryogenesis and plant regeneration from leaf and internode explants of *Paulownia elongata*. Plant Cell Tissue Organ Cult 79:341–345

Kadota M, Niimi Y (2004) Improvement of micropropagation of Japanese yam using liquid and gelled medium culture. Sci Hort 102:461–466

Khripach V, Zhabinskii V, Groot AD (2000) Twenty years of Brassinosteroids: steroidal plant hormones warrant better crop for the XXI century. Ann Bot 86:441–447

Kriebel HB, Wang CW (1962) The interaction between provenance and degree of chilling in bud break of sugar maple. Silvae Genet 11:125–130

Kvaalen H, Daehlen OG, Rognstad AT, Gronstad B, Egertsdotter U (2005) Somatic embryogenesis for plant production of *Abies lasiocarpa*. Can J For Res 35:1053–1060

Ledbetter DI, Preece JE (2003) Forcing softwood shoots, rooting, and micropropagating *Hydrangea quercifolia* Bartr. Prop Orn Plants 3:3–10

Ledbetter DI, Preece JE (2004) Thidiazuron stimulates adventitious shoot production from *Hydrangea quercifolia* Bartr. leaf explants. Sci Hort 101:121–126

Long LM, Preece JE, Van Sambeek JW (1995) Adventitious regeneration of *Juglans nigra* L (eastern black walnut). Plant Cell Rep 14:799–803

Malabadi RB, Teixeira da Silva JA, Nataraja K, Vijaykumar S, Mulgund GS (2011) Induction of somatic embryogenesis in mango (*Mangifera indica* L.). Int J Biol Technol 2:12–18

Mathews H, Wetzstein HY (1993) A revised protocol for efficient regeneration of somatic embryos and acclimatization of plantlets in pecan, *Carya illinoensis*. Plant Sci 91:103–108

McCown BH, Llyod G (1981) Woody plant medium (WPM). A mineral nutrient medium formulation for microculture of wood plant species. Hort Sci 16:453

Murashige T, Skoog F (1962) A revised medium for rapid growth and bioassays with tobacco tissue cultures. Physiol Plant 15:373–397

Nand N, Drew RA, Ashmore S (2004) Micropropagation of two Australian native fruit species, *Davidsonia pruriens* and *Davidsonia jerseyana* G. Harden & J.B. Williams. Plant Cell Tissue Organ Cult 77:193–201

Ozden-Tokatli Y, Ozudogru EA, Akcin A (2005) *In vitro* response of pistachio nodal explants to silver nitrate. Sci Hort 106:415–426

Preece JE (2001) The most tricky part of micropropagation: establishing plants in greenhouse and fields. Comb Proc Int Plant Prop Soc 51:300–303

Preece JE, Read PE (2003) Novel methods in micropropagation. Acta Hort 616:71–76

Preece JE, Read PE (2007) Forcing leafy explants and cuttings from woody species. Prop Orn Plants 7:138–144

Preece JE, Trigiano RN (2001) Tissue culture of woody plants. In: Jones RK, Benson DM (eds) Diseases of woody ornamentals and trees in nurseries. The American Phytopathological Society, APS Press, St. Paul, pp 451–456

Quraishi A, Mishra SK (1998) Micropropagation of nodal explants from adult trees of Cleistanthus collinus. Plant Cell Rep 17:430–433

Raghavan V (2004) Role of 2,4-dichlorophenoxyacetic acid (2,4-D) in somatic embryogenesis on cultured zygotic embryos of *Arabidopsis*: cell expansion, cell cycling, and morphogenesis during continuous exposure of embryos to 2,4-D. Am J Bot 91:1743–1756

Rakesh M, Minocha SC, Simola LK (1995) Somatic embryogenesis and polyamines in woody plants. In: Jain S, Gupta P, Newton R (eds) Somatic embryogenesis in woody plants, vol I. Kluwer Academic, Dordrecht, pp 337–359

Read PE (1990) Environmental effects in micropropagation. In: Ammirato P, Evans DA, Sharp WR, Bajaj YPS (eds) Handbook of plant cell culture, vol 5. McGraw Hill, New York, pp 95–125

Read PE, Preece JE (2007) Micropropagation of ornamentals: the wave of the future? Prop Orn Plants 7:150–159

Rivera-Domínguez M, Manzanilla-Ramírez MA, Robles-González M, Gómez-Lim MA (2004) Induction of somatic embryogenesis and plant regeneration of 'Ataulfo' mango (*Mangifera indica*). Plant Cell Tissue Organ Cult 79:101–104

Sánchez-Moreiras AM, de la Peña TC, Roger MJR (2006) Cell cycle analyses for understanding growth inhibition. In: Reigosa MJ, Pedrol N, González L (eds) Allelopathy, a physiological process with ecological implications. Springer, Dordrecht, pp 141–156

Shafiq M, Aftab F (2008) The influence of environment and log diameter on forcing and establishment of softwood shoots from large stem segments of *Araucaria heterophylla.* International conference of plant scientists (10th national meeting of plant scientists), 21–24 April 2008. Department of Botany, University of Agriculture, Faisalabad, Pakistan (Abstract, ICPS-066), 30pp

Shulman Y, Nir G, Lavee S (1986) Oxidative processes in bud dormancy and the use of hydrogen cyanamide in breaking dormancy. Acta Hort 179:141–148

Spiers JM, Draper AD (1974) Effect of chilling on bud break in rabbiteye blueberry. J Am Soc Hort Sci 99:398–399

Steffens GL, Stutte GW (1989) Thidiazuron substitution for chilling requirement in three apple cultivars. J Plant Growth Regul 8:301–307

Swartz HJ, Powell LE (1981) The effect of long chilling requirement on time of bud break in apple. Acta Hort 120:173–178

Vandenbussche F, Straeten DVD (2004) Shaping the shoot: a circuitry that integrates multiple signals. Trends Plant Sci 09:499–506

Wadegaonkar PA, Bhagawat KA, Rai MK (2006) Direct rhizogenesis and establishment of fast growing normal root organ culture of Withania somnifera Dunal. Plant Cell Tissue Organ Cult 84:223–225

Wang SY, Faust M (1988) Changes of fatty acids and sterols in apple buds during bud break induced by a plant bioregulator, thidiazuron. Physiol Plant 72:115–120

Wang SY, Jiao HJ, Faust M (1991) Changes in the activities of catalase, peroxidase, and polyphenol oxidase in apple buds during bud break induced by thidiazuron. J Plant Growth Regul 10:33–39

Weinberger JH (1950) Chilling requirements of Peach varieties. Proc Am Soc Hort Sci 56:122–128

Wilhelm E (1999) Micropropagation of juvenile sycamore maple via adventitious shoot formation by use of thidiazuron. Plant Cell Tissue Organ Cult 57:57–60

Yang QG, Read PE (1992) Pre-forcing treatments influence bud break and shoot elongation in forced woody species. J Environ Hort 10:101–103

Yang QG, Read PE (1993) *In vitro* culture of 'Van-houtte's spirea explants from "secondary cultures" and dormant stems forced in solutions containing plant growth regulators. Plant Cell Tissue Organ Cult 33:25–30

Chapter 14
Novel Methods in Micropropagation of Pistachio

Engin Tilkat, Yelda Ozden Çiftçi, Hulya Akdemir, Ahmet Onay, and Emine Ayaz Tilkat

Contents

Abstract The focus of this paper is to describe the novel methods developed for the different stages of micropropagation: installation of mature apical shoot tips and elimination of browning exudates; forcing hardwood shoots from the lignified stem sections; forcing axenic leaves; initiation of embryogenic masses (EMS); encapsulation of somatic embryos and cryopreservation of axillary buds for storage, and the facilitation of rooting. These developed methods are not being used by commercial micropropagation laboratories yet. Exudation of phenolics is inhibited

E. Tilkat (✉) • A. Onay • E.A. Tilkat
Faculty of Science and Literature, Department of Biology, University of Batman, 72080 Batman, Turkey
e-mail: etilkat@gmail.com; ahmeto@dicle.edu.tr; emine.ayaz@batman.edu.tr

Y.O. Çiftçi • H. Akdemir
Faculty of Science, Department of Molecular Biology and Genetics, Gebze Institute of Technology, 41400 Kocaeli, Turkey
e-mail: yelda75@yahoo.com; pinarakdemir@gmail.com

M. Ashraf et al. (eds.), *Crop Production for Agricultural Improvement*,
DOI 10.1007/978-94-007-4116-4_14, © Springer Science+Business Media B.V. 2012

when the disinfested and rinsed explants are cut at the base of the shoot tips and washed twice for 1 h by shaking them in sterile distilled water on a shaker at 100 rpm. For shoot initiation, 15–20 cm long terminal stem sections are cut into 3–4 cm sections and placed in pots filled with peat, perlite and soil. Two or three weeks later, the developed softwood shoots are excised; or freshly harvested three to nine apical tips of 1–2 cm are disinfested and used as explants. Leaves excised from axenic shoot cultures were also used to induce organogenesis on a Murashige and Skoog (MS) medium with Gamborg vitamins supplemented with combinations of different concentrations of 6-benzylaminopurine (BA) and indole-3-asetic acid (IAA). Calcium alginate gel was used to encapsulate somatic embryos to produce synthetic seeds. The encapsulated somatic embryos can be stored under refrigerated conditions for short-term conservation. For long-term conservation, some axillary buds can also be stored by using vitrification and one-step freezing technique; however, other cryogenic techniques should also be tested. With these improved stages, application of *Pistachio vera* L. micropropagation in commercial clonal orchards will be feasible as an alternative to traditional propagation in the near future.

Keywords *Pistacia vera* L. • Leaf forcing • Shoot forcing • Synthetic seeds • Somatic embryo • *In vitro* germplasm conservation

1 Introduction

The pistachio is usually propagated outside of Turkey by budding selected scions onto seedling rootstock of the same species or other pistachio species such as *P. atlantica, P. terebinthus*, *P. mutica*, *P. khinjuk*, *P. integerrima* or hybrids such as *P. atlantica*, *P. integerrima* (PGI), *P. atlantica* x *P. integerrima* (PGII and UCB-1) (Ferguson et al. 2002). Many of the institutions that currently produce plants for production are successful and have developed their protocols based on past research. Usage of pistachio as a rootstock results in a great variability in the performance of the grafted cultivars affecting diverse traits such as fungal resistance, shell splitting, blank nut production or yields (Hormaza and Wünsch 2007). Consequently, a greater uniformity is desirable and could be obtained by using vegetative propagated rootstock (Crane 1984). There are some techniques that are currently being developed that have the potential to impact commercial micropropagation in a positive manner. Breakthroughs are largely due to organogenesis and embryogenesis of the commercial cultivars, but may also be utilized in the micropropagation of stock cultivars.

The focus of this paper is concerned with the newer techniques of *in vitro* micropropagation of pistachio, such as the installation of mature apical shoot tips and the elimination of browning exudates; the forcing of the lignified stem sections and axenic leaves; the initiation of embryogenic masses; the encapsulation of somatic embryos; cryopreservation of axillary buds for storage; and the facilitation of rooting. None of these techniques is in widespread use in commercial micropropagation in an effective manner, but all appear to have a great potential and may become adopted by labs in the near future.

2 Revitalization of Pistachio

Clonal propagation is not always successful with older material due to poor rooting. Plants in the juvenile phase of growth generally are easier to propagate than plants that have reached the adult phase. The change from the juvenile to the adult phase is considered to be the most serious constraint to root cuttings (Howard 1990). As far as the rooting of adult *P. vera* material is concerned, rooting was attempted by several researchers (Joley 1960; Joley and Opitz 1971; Sakoury 1976; Bustamante-Garcia 1984) *in vitro* using a mist system. Unfortunately, no more than 5% rooting was achieved. Perhaps the major dilemma with rooting of cuttings of pistachios is the rapid loss in rooting capacity with the increasing age of the parent tree. Seedling explants are usually most readily established *in vitro* and proliferate more rapidly with tree species where micropropagation of adult material is often difficult. The majority of juvenile growth occur when the plant is young and still exhibiting juvenile characteristics. As plant ages, the older and lower parts retain their juvenile traits and an adult phenotype appears on the newest growth (Preece and Read 2003). In the case of pistachio *in vitro* culture, the origins of explants, in particular their positions on the mother plants, have been reported to influence their establishment and growth (Barghchi and Alderson 1985; Abousalim 1990; Onay 1996). We have been conducting research for the revitalisation of adult Turkish pistachio cultivars for years. At first, studies were carried out to investigate whether the degree of juvenility or maturity of an apical meristem depends on its distance (along the trunk or along the branches) from the root-shoot junction (Onay 1996). A diagram of the sampled gradients of the apical meristems is shown in Fig. 14.1, and the data reported are given in Table 14.1.

Depending on the time the explants were harvested, the position of the explants did not significantly influence the survival rate obtained by the explant positions A, B and C with 90%, 90%, and 100%, respectively (Table 14.1). However, the greatest length of shoots was obtained using the explants harvested from position C. There were significant differences between the positions tested. Explants from position C (terminal tips) produced most buds, and the mean was also significantly greater than the explants harvested from the other positions. This older, but more juvenile portion of the plant, is called "the cone of juvenility" (Preece and Read 2003). The challenge is to collect freshly sprouted shoot tips from the different positions of a mature tree, or if such juvenile shoot tips are not available, one may even go to the extreme measure of cutting down a tree to encourage juvenile shoots to grow from the stump, and then forcing them, especially during the dormant season, in a forcing solution indoors.

3 Shoot Forcing

When outdoor plants were used, it is difficult to remove microflora growing on them with using surface disinfestion techniques (Preece and Read 2003). The microbes are often in tiny cracks and crevasses and the disinfestant may not penetrate these areas. Many woody species are either too large or growing in the ground and cannot

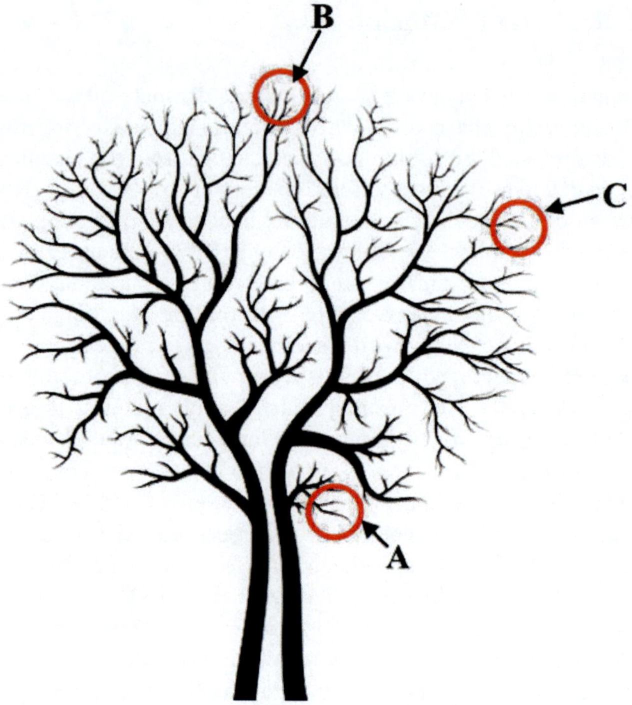

Fig. 14.1 Diagram of the sampled shoot tip gradients: (**a**) fresh sprouts from the trunk, (**b**) apical tips from the orthotropic far end of the root-shoot junctions, and (**c**) plagiotropic apical tips from far part of the root-shoot junction

Table 14.1 The effects of explant position harvested from different parts of a mature tree on viable culture and shoot growth

Position of the explant	Viable culture rate (%)	Mean shoot length (mm)	Mean shoot number
A	90	6.5 ± 0.61 a	4.2 ± 0.36 b
B	90	4.3 ± 0.50 b	3.8 ± 0.35 b
C	100	6.8 ± 0.82 a	5.3 ± 0.33 a
χ^2 (2 df)	P>0.05		

be removed easily to greenhouse or growth room. Therefore, it is often necessary to remove parts of the plant such as branch tips or segments, usually during the dormant season, and force new shoot growth indoors or in a greenhouse environment. This new growth can then be excised, successfully surface disinfected and used for explant material. We have been conducting research on forcing shoots for years.

Much of the focus has been done on shoot tips harvested from mature trees during the dormant season. Terminal lignified stem sections, 3–4 cm long were collected from 30-year-old trees and then the cut ends were immersed in a 10 mg L^{-1} BA solution for 24 h before placing them in a greenhouse medium. Subsequently, the cuttings were potted in a sterile 1:1 mixture of sand and soil and placed in a greenhouse (65±5% R.H.) at 25°C/24 h. The photoperiod was supplied by mercury fluorescent lamps (25 mmol photons m^{-2} s^{-1}) in order to force the axillary buds to sprout. After 3 weeks, the newly formed leafy shoots from the forced buds were excised and surface sterilized by a 2-min dip in absolute ethanol followed by a 10-min soak in a 20% (v/v) NaOCl solution. Similarly, 20–25 cm long stems (Read and Yang 1991) were harvested from deciduous trees and then soaked for 15 min in a 0.78% NaOCl solution with Tween 20 (Yang and Read 1993). Following the bleach treatment, a fresh cut was made at the base. The lower cut ends of the stems were placed in containers with water, 200 mg L^{-1} 1,2,8-hydroxy-quinoline citrate (8-HOC), sucrose and, sometimes plant growth regulators. Read and Yang (1991) reported that faster bud break and greater bud and shoot elongation were promoted if the stems received the 15 min soak in the bleach solution prior to the forcing when compared to those that did not. Because these shoots were excised from the tips of branches of the most adult portion of a plant, they can be prone to be somewhat difficult to propagate. The effects of plant growth regulators in the forcing solution are, at least somewhat, species and plant growth regulator specific.

4 *In Vitro* Installation of Mature Pistachio Shoot Tips and Exudation of Phenolics

Juvenile explants are usually more readily established *in vitro*, grow and proliferate at a more rapid rate, and are more rootable than adult material (Preece and Read 2003). This is especially true with pistachio species where micropropagation of adult material is often difficult. Eradication of fungal and bacterial contaminants and browning were encountered as major problems in the establishment of pistachio tissue culture reported by many pistachio researchers (Bustamante-Garcia 1984; Barghchi and Alderson 1985; Abousalim 1990; Onay 1996; Mederos-Molina and Trujillo 1999; Ozden Tokatli et al. 2005). Data on the effectiveness of a sterilization method has not been published by these researches, although Abousalim (1990) and Onay (1996) gave optimum conditions for certain materials. Especially in the case of explants from mature pistachio trees, disinfestation was best achieved when the explants were obtained from actively growing meristem tip cultures of 50-year-old pistachio. (*P. vera* L.) (Onay 1996). However, when nodal bud segments and meristem tips were used for the initiation of a culture (Tilkat et al. 2008), a very high level of media browning was frequently observed. Browning, especially during explant establishment, appears to be the main limiting factor in the establishment of tissue cultures of pistachio species. There are a few interesting recent reports on the use of anti-oxidants, glutation, ascorbic acid, citric acid, PVP, L-cysHCL, Na-DEDC, and Rasmonal, but they did not make a visible difference on the absorption of phenolic compounds. To obtain satisfactory results, we have developed a new method that allows a faster and safer initiation and establishment of adult material. In our protocol, apical shoot tips were excised from 25-year-old mature trees. The branches were cut into 10–15 cm shoot tips. The softwood shoot tips were excised into 1–2-cm long tips. The explants were surface disinfested with 10% sodium hypochlorite (NaOCl) for 30 min and rinsed three times with sterile distilled water, then the base of the shoot tips was cut. After this, these explants were washed twice for 1 h by shaking them in sterile distilled water on a shaker at 150 rpm. The exudation of phenolics was not observed when the explants were kept in twice-sterile distilled water for 1 h immediately after the excision of the basal ends, which were then blotted with sterile filter paper. During preparation, polyphenolic compounds are oxidized by polyphenol oxidates (PPO), inducing a strong browning on the cut surface and media and leading to the subsequent death of the explant (Tilkat 2006). Applying the new method described above, actively growing shoot tips of pistachio were cleaned and disinfested, dissected to 4–6 mm and placed on a modified MS medium of 1 mg L^{-1} BA. A yield of 100% sterile and non-browned explants was obtained in comparison to the formerly nearly 100% losses of non-washed shoot tips. The reason for obtaining a high survival rate from the treatment described above is that the cultured shoot tips could accumulate Na^+ ion from NaOCl which was used for surface sterilization. Therefore, phenolics were released from the basal-cut-ends of the apical shoot tips; they could react with Na^+ ion and form salt. This salt possibly dissolved in the water and the phenolic compounds lost their specific properties (Onay et al. 2007a; Tilkat et al. 2008).

5 Forcing Axenic Leaves

This technique is very different from forcing stem sections onto shoot tips, because already regenerated shoots can be used in a BA and IAA forcing solution. The details of the method are as follows: Healthy mature tips of 30-year-old male *P. vera* L. "Siirt" were collected from the pistachio research institute in Gaziantep province, a province of southeastern Turkey, and were used as the source of material (Tilkat and Onay 2009). To establish aseptic shoot cultures, apical shoot tips were disinfested by the protocol described above, and cultured on a modified Murashige and Skoog (MS) (Murashige and Skoog 1962) medium (initiation medium) with Gamborg vitamins (Gamborg et al. 1968) supplemented with 1.0 mg L^{-1} BA, 30 gL^{-1} sucrose and 5.5 gL^{-1} agar. Adventitious shoot buds from the initiation medium were subcultured on the fresh initiation medium every 3–4 weeks. The regenerated adventitious shoot buds from the *in vitro* cultures were maintained, and proliferated on the initiation medium for more than 1 year before the forcing studies. Semi-mature leaves along with a portion of the petiole were excised and cultured on an agarified MS medium with their petioles and abaxial sides in contact with the medium. The MS basal medium was supplemented with 3% (w/v) sucrose, 6-benzylaminopurine (BA) thidiazuron (TDZ) and kinetin (Kin) (each at 1 mg L^{-1}) and different concentrations of BA or TDZ (0,0, 1,0, 2,0, 3,0 and 4,0 mg L^{-1}) and IAA (0,5, 1,0 and 2,0 mg L^{-1}) were tested on supplements on the MS basal medium for shoot regeneration experiments. The formation of direct shoot buds from different woody species requires more exacting culture conditions of light and temperature, and complex medium composition, especially with regard to plant growth regulators (Tomar and Gupta 1988; Gharyal and Maheshwari 1990). Gharyal and Maheshwari (1990) reported 36% direct shoot regeneration from petioles on a B5 medium supplemented with 0.5 mg L^{-1} IAA and 1 mg L^{-1} BA. However, 31.9% of direct shoot regeneration was obtained in female cultivar ("Siirt") from culturing leaflet explants on a MS medium supplemented with 1 mg L^{-1} IAA and 2 mg L^{-1} BA (Tilkat and Onay 2009). Shoots subcultured on the 1:2 IAA:BA media also proliferated rapidly, but the medium regenerated hyperhydric shoots. Once shoot buds were obtained from axenic leaves, the use of only BA on the proliferation medium proved to be more beneficial than the combination of IAA and BA. The results reported by Tilkat and Onay (2009) and Tilkat and co-workers (2009b) showed consistency with other studies where BA and IAA promoted the proliferation and elongation of shoots in buds from *Ladebouria hyacinthiana* (Turakhia and Kulkarni 1988) leaves. The protocol for forcing the regenerated leaves for organogenesis summarized above on the female cultivar "Siirt" showed consistency with the pistachio male cultivar "Atli" (Tilkat and Onay 2009; Tilkat et al. 2009b). These results show that a combination of BA and IAA was crucial for direct shoot regeneration.

6 Initiation of Embryogenic Mass

Somatic embryogenesis in *P. vera* L. has been reported in a number of studies (Onay et al. 1995, 1996, 1997; Onay and Jeffree 2000; Onay 2000a, b; Onay et al. 2005). In these reports initiation of embryogenic cultures from immature kernels ranged from 0.0% to 10%. More recently, Onay et al. (2007b) obtained somatic embryogenic lines with five open pollinated cultivars at frequencies of between 7.0% and 19%. Previous studies have suggested that somatic embryo formation was also influenced by the type of media and explant and by the type of cytokinin used in the culture medium (Onay et al. 1995; Onay and Jeffree 2000). In pistachio we have standardized the efficient protocols for direct somatic embryogenesis from immature fruits (Onay et al. 1995, 1996; Onay and Jeffree 2000), mature zygotic embryos (Onay 2000a; Onay et al. 2007b), and juvenile leaf explants (Onay and Namli 1998). The protocol involves multiple stages, including embryogenic mass formation, maturation, germination, and plantlet development.

Another factor pertinent to somatic embryogenesis (SE) in pistachio is the composition of the initiation medium. Various media have been used to culture immature fruits of pistachio, the most common of which is MS (Murashige and Skoog 1962). The only plant growth regulator for inducing direct embryogenic tissue from mother tissue has been BA. We have also demonstrated that TDZ was also another effective agent for the induction of callus from leaf explants of pistachio, but BA was again essential for the expression of somatic embryogenesis (Onay and Namli 1998).

The color of EMS varied depending on the genotype from yellowish green to deep green with some red pigmentation. Proliferative ability was influenced by all genotypes. Once an EMS was initiated in a liquid medium containing 3% sucrose, with or without (after a few subcultures) 0.5 mg L^{-1} BA, it could be maintained successfully in liquid or the agarified medium (if the agarified medium was used for maintenance with or without BA, the proliferation of EMS takes a longer time). Therefore, subculture on the initiation medium with or without BA is a prerequisite for the maintenance of embryogenic tissues.

7 Cryopreservation and Encapsulation of *In Vitro* Grown Pistachio Regenerants

7.1 *In Vitro Conservation and Cryopreservation of Pistachio Germplasm*

As the diversity of wild pistachio types is destroyed by severe anthropogenic pressures, such as land clearance, charcoal burning, and over-grazing (Barghchi and Alderson 1989; Padulosi and Hadj-Hassan 2001), both wild and cultivated pistachio germplasm are today under threat of genetic erosion and a comprehensive gene bank for the pistachio has not been established yet. Thus, it is important to develop suitable conservation strategies for the medium and long-term preservation of pistachio germplasm in which slow-growth storage strategies can be carried out at low temperatures (above-freezing) in the former, whereas the material can be cryopreserved in liquid nitrogen (LN) at ultra-low temperatures (−196°C) in the latter. However, interestingly, up to now only a few reports can be found in the literature about both slow-growth storage and the cryopreservation of pistachio germplasm (Ozden Tokatli et al. 2008, 2010; Akdemir et al. 2012).

7.2 *Slow-Growth Storage*

Slow-growth storage of plant materials aims to reduce growth with increasing intervals between subcultures and, therefore, it is used as a strategy for germplasm conservation and exchange. In slow growing species standard culture conditions can also be used for medium-term conservation; however, in most cases, modified environmental conditions and/or a culture medium are preferred to be used in association with a decrease in light intensity or even its complete suppression for the induction of growth reduction (Engelmann 1997).

The potential of slow-growth storage of *P. vera* microshoots has also been investigated. An attempt to reduce growth by incorporating different concentrations of growth reducing substance like abscisic acid (ABA, 0.25–4 mg L^{-1}) or an osmotic inhibitor of growth like mannitol (2.5–40 mg L^{-1}) to the media in combination with a decreasing of the incubation temperature from 26°C to 4°C and/or by using different photoperiods and light intensities was tried (Barghchi 1986). Using this approach some cultures have been stored for nearly 18 months, and shoots taken from these stored cultures were able to resume growth after their transfer to normal proliferation conditions.

Apart from the usage of microshoots for the conservation of germplasm, propagules can also be preserved in synthetic seeds (synseeds) which can mimic the formation and germination of natural seeds. The concept of encapsulation was first applied to somatic embryos (Murashige 1977), and then, extended to other vegetative propagation units (Bapat et al. 1987) by using buds and nodal segments as encapsulated explants. In time, several methods of encapsulation have been tested for the production of synseeds, including gelation, complex coacervation, and interfacial polymerization (Redenbaugh et al. 1987). Since the gelation technique proved to be the most efficient, different gels were compared to form sufficiently hard capsules and to guarantee the maintenance of somatic embryo viability (Redenbaugh et al. 1991). Among them, Na-alginate is today by far the most commonly used compound, due to its ability to form hydrogels in the presence of divalent cations (Ca^{+2}), its nontoxicity and its low cost. These features have led to the technology being applied to a wider range of species and explants (Piccioni and Standardi 1995; Lambardi et al. 2006).

When both axillary/apical buds or somatic embryos are used for encapsulation, important advantages of this technology rely on its potential to improve not only the mass propagation of plants, but that they are also suitable for short-term conservation studies at cold (above-freezing) conditions with the ability of germplasm storage, and reducing the need for transferring and subculturing during off-season periods (West and Preece 2006), a condition which diminishes the labor costs of commercial micropropagation laboratories. Last but not least, the synseeds can also be used for the long-term preservation of germplasm with recently developed "encapsulation-dehydration" and "encapsulation-vitrification" techniques (Panis and Lambardi 2006).

In the case of pistachio, synseeds were prepared by using both axillary buds (Ozden Tokatli et al. 2008) and somatic embryos (Onay et al. 1996) with using Na alginate as the gelling and $CaCl_2.2H_2O$ as the complexing agent. In the case of encapsulated axillary buds, more than 75% of the synseeds were able to convert to shoots after 4 weeks of culture in 2 mg L^{-1} BA containing MS media, whereas 14% of the somatic embryos showed germination in 4% (w/v) sucrose and 1 mg L^{-1} BA enriched MS media. Moreover, with encapsulation of somatic embryos, it was possible to conserve the propagules up to 2 months at 4°C with a 14% recovery rate which may confirm that encapsulation is a practical procedure for short-term storage of embryogenic pistachio tissue and may be applicable to the preservation of desirable elite genotypes (Onay et al. 1996).

7.3 Cryopreservation

Cryopreservation refers to the storage of biological specimens at ultra-low temperatures (196°C) in liquid nitrogen (LN) as a cryogenic medium (Withers and Engelmann 1997).

As at this ultra-low temperature, all cellular divisions and metabolic processes are stopped, it is the only sound technique that allows the conservation of specimens for a theoretically unlimited period of time (Engelmann 2004). For pistachio cryopreservation, protocols were developed initially for seeds by using dehydration and one-step freezing techniques (Ozden Tokatli et al. 2007). Using this technique, a maximum of 90% germination was obtained following 8 h of drying in silica gel (corresponding to 11.7% moisture content on a FW basis) and direct immersion in LN. Then, more recently, a different cryogenic approach, vitrification and one-step freezing, was applied for the cryopreservation of pistachio axillary buds (Ozden Tokatli et al. 2008) in which excision of buds from *in vitro* grown and cold-acclimated shoots were followed by their pre-treatment with a sucrose enriched medium for 2 days at 4°C. This freezing procedure, referred to as vitrification, comprises a loading treatment with a cryoprotectant solution (2 M glycerol and 0.4 M sucrose, Matsumoto et al. 1994) at room temperature (RT), followed by the exposure of explants to a plant vitrification solution (PVS2; 30% glycerol (w/v) + 15% ethylene glycol (w/v) + 15% DMSO (w/v) in a culture medium with 0.4 M sucrose, Sakai et al. 1990) at 25°C or 0°C for various incubation periods (15, 30, 60, 90 min), rapid cooling and warming, and final removal of the vitrification solution by washing samples with an unloading solution consisting of a liquid culture medium supplemented with 1.2 M sucrose. After 4 weeks of culture in 2 mg L^{-1} BA containing MS medium, both vitrified (control) and vitrified/ cryopreserved (LN+) axillary buds were analyzed. The results showed that prolonged exposure of buds to the PVS2 solution up to 90 min resulted in reduction in the survival and the proliferation percentages of the control samples in both applied temperatures. The initial survival percentage of axillary buds (100%) were decreased to 75.8% with exposure in 0°C and to 56.2% in 25°C; whereas, the proliferation rates were also decreased from 90% (initial) to 72.7% and 66.7%, respectively. These results clearly indicate that prolonged exposure of samples to a vitrification solution leads to detrimental effects due to its toxicity. However, it was also possible to recover some axillary buds after cryopreservation when PVS2 was applied for 60 min in 0°C and 90 min both in 0°C and 25°C prior to their direct immersion in LN. It should be noted that other cryogenic procedures, including encapsulation-dehydration, encapsulation-vitrification and droplet-vitrification techniques, should be carried out on pistachio in order to develop more efficient long-term conservation techniques.

7.4 *Facilitation of Rooting*

The *in vitro* initiation of roots for the regenerated pistachio shoots was found to be especially dependent upon the auxin type and the concentration (Barghchi and Alderson 1989; Parfitt and Almehdi 1994; Onay 1996; Onay 2000b; Ozden Tokatli et al. 2005; Tilkat 2006). Successful reports on the tissue culture of adult *P. vera* L. produced somatic plantlets, but obstacles pertaining to the low incidence of the rooting of plantlets reproducibility during the rooting regime and the poor establishment of *in vitro* propagated plants after acclimatization as a result of fungal infections have been reported (Onay 2000b; Tilkat et al. 2008). Abnormality *in vitro* morphologies related to the callus production of cultured plant tissues is also a hindrance to commercial *in vitro* plant propagation as losses of up to 50% can be observed (Tilkat 2006). Although a high rate of rooting (84%) of *P. vera* cv. Atli was obtained on a medium supplemented with 2.0 mg L^{-1} IBA (Tilkat et al. 2008), the plantlets were at times prone to severe callusing and hyperhydricity.

Very recently, a new protocol was described by Tilkat and Onay (2009a). The washing of the cut ends of the microshoots before the rooting treatment had an effect on the rooting. There was evidence of a significant difference in the frequencies of rooted shoots between the results obtained from shoots excised from the washed and unwashed, which is not a fact that has been reported by other researchers concerned with the pistachio. This method provides a clean environment for microcutting by solving the released phenolics and other metabolites which hinder rhizogenesis. Improvements obtained in rooting percentages (from 67.5% to 87.5%) were produced when cloned shoots were used (Tilkat and Onay 2009a); this finding has been previously reported for seedling material of the pistachio (Onay 1996). This indicates that not only growth substances, but also some other factors may limit root induction, and this merits attention before any practicable application. The rooting responses are significantly influenced by the applied dipping IBA concentration and its duration, a maximum of 90.2% shoot-forming roots was obtained with 2 mg L^{-1} IBA after a 5-mi dipping in 10 mg L^{-1} IBA (Tilkat and Onay 2009a). Even though high levels of rooting were obtained, a dipping protocol was not suitable for future routine use as roots associated with callus are often vulnerable to pathogen attack once plantlets are transferred to soil. A quick dip approach for the rooting of pistachio under *in vitro* conditions has also been reported by another pistachio researcher (Abousalim 1990). The rooting of the regenerated 4-year-old *P. vera* L. material was tested using the quick auxin-dip approach. Up to 50% of the plantlets were rooted with a mean of three roots. A quick dip approach for in vivo rooting was also tested. After 5 weeks of culture, an 85% rooting was obtained on mature softwood cuttings of *P. vera* under mist, but only after treatment with an extremely high level of IBA (35,000 mg L^{-1}) (Al Barazi and Schwabe 1982).

The use of longer shoots (≥4 cm long) for rooting also showed a strong effect on root regeneration. The enhancement of root formation in mature male pistachio on 2 mg L^{-1} IBA support is new and of interest. Similar to these results, 80% of the shoots forming roots have also been reported by Tilkat and Onay (2009) using the regenerated shoots from mature female trees. Although a high percentage (90%) of shoots forming roots has been reported by Parfitt and Almehdi (1994) using juvenile material (*P. vera* "Kerman" seedlings), our percentage of shoots forming roots was the highest (92%) of these obtained by others working with *P. vera* (Barghchi and Alderson 1989; Parfitt and Almehdi 1994; Onay 2000b; Ozden Tokatli et al. 2005; Tilkat 2006). Once healthy and strongly rooted plantlets were obtained, acclimatization of the pistachio was easy (Onay et al. 2007a; Tilkat and Onay 2009).

8 Conclusions

In this chapter, the novel methods developed for the different stages of micropropagation of *Pistacia vera* L. have been reviewed, with a special attention paid to recent reports. Pistachio is one of the most important nut crops and its world production has increased significantly over the past 30 years mainly due to the growing demand in the world markets and the future appears to be more optimistic for the pistachio nut. However, traditional methods used for pistachio propagation and improvement are apparently not adequate to meet these demands. Application of biotechnological methods, such as *in vitro* micropropagation and conservation, to pistachio can create new opportunities in breeding, cloning, disease control and germplasm conservation. As regards micropropagation, mature pistachio cultivars have been extremely difficult to establish due to the high incidence of contamination and the oxidation of phenolic compounds. However, the successful protocols concerned with the different stages of micropropagation were described: installation of mature apical shoot tips and elimination of browning exudates; forcing hardwood shoots from the lignified stem sections; forcing axenic leaves; initiation of embryogenic masses (EMS) and the facilitation of rooting. The studies carried on both mature pistachio female "Siirt" and male "Atli" cultivars showed that a combination of BA and IAA (2 mg L^{-1} and 1 mg L^{-1}, respectively) should be incorporated to MS medium in order to force direct organogenesis from leaves, whereas BA and/or TDZ should be added to the media for induction of somatic embryogenesis from immature fruits and juvenile leaf explants. At present, somatic embryogenesis is restricted only to juvenile tissues as explants from mature trees, when cultured, have not demonstrated any embryogenic potential. To extend the expression of embryogenic potential to the explants from mature trees, it is necessary to identify cell or tissue types which are physiologically identical to these juvenile tissues. In the case of conservation, both slow-growth storage and cryopreservation methods were established for short and long-term preservation, respectively. It was shown that pistachio microshoots and encapsulated somatic embryos could be stored for a period in above freezing temperatures in former, while it was possible to cryopreserve both pistachio seeds and axillary buds with using appropriate cryogenic methods in ultra low temperatures, in latter. Cryobanks for seeds and synthetic seeds for somatic embryos of pistachio are a reality and investigations based upon cryopreservation of plumules will have a great impact on storage and management of genetic resources. Continued researches into the use of biotechnological methods will contribute to improve the vegetative propagation of the pistachio and ultimately should play an important role in future breeding programs for pistachio and other important woody species.

Acknowledgements The authors thank Mrs. Linda Thain-Ali for valuable advice and critical reading of the manuscript. We would like also to thank the anonymous reviewers for their helpful comments and suggestions

References

Abousalim A (1990) Micropropagation and micrografting of Pistachio (*P. vera* L. and *Pistacia atlantica* Desf.). PhD thesis, Department of Horticulture, Wye College, University of London, UK

Al Barazi Z, Schwabe WW (1982) Rooting softwood cutting of adult *Pistacia vera*. J Hort Sci 57:247–252

Akdemir H, Süzerer V, Tilkat E, Yildirim H, Onay A, Çiftçi YO (2012) In vitro conservation and cryopreservation of mature pistachio (*Pistacia vera* L.) germplasm. J Plant Biochem Biotechnol (in press)

Bapat VA, Mhatre M, Rao PS (1987) Propagation of *Morus indica* L. (mulberry) by encapsulated shoot buds. Plant Cell Rep 46:393–395

Barghchi M (1986) In vitro storage of plant genetic resources. In: Plant Physiology Division Biennial Report. Palmerston North, Department of Scientific and Industrial research, p 52

Barghchi M, Alderson PG (1985) In vitro propagation of *P. vera* L. and commercial varieties of Ohadi and Kallegochi. J Hort Sci 60:423–440

Barghchi M, Alderson PG (1989) Pistachio (*Pistacia vera* L.). In: Bajaj YPS (ed) Biotechnology in agriculture and forestry, vol 5, Trees II. Springer, Berlin/Heidelberg, pp 68–98

Bustamante-Garcia MA (1984) Micropropagation and Rejuvenation of *Pistacia* species and the mechanism by which light influences root initiation. PhD thesis, University of California, Davis, USA

Crane JC (1984) Pistachio production problems. Fruit Var J 38:74–85

Engelmann F (1997) In vitro conservation methods. In: Ford-Lloyd BV, Newburry HJ, Callow JA (eds) Biotechnology and plant genetic resources conservation and use. CABI, Wallingford, pp 119–162

Engelmann F (2004) Plant cryopreservation: progress and prospects. In Vitro Cell Dev Biol Plant 40:427–433

Ferguson L, Beede RH, Reyes H, Metheney P (2002) California pistachio rootstocks evaluations. Acta Hort 591:63–66

Gamborg OL, Miller RA, Ojima K (1968) Nutrient requirements of suspension culture of soybean root cells. Exp Cell Res 50:151–158

Gharyal PK, Maheshwari SC (1990) Differentiation in explants from mature leguminous trees. Plant Cell Rep 8:550–553

Hormaza JI, Wünsch A (2007) Pistachio Genome mapping and molecular breeding in plants. In: Kole C (ed) Fruits and nuts, vol 4, Chap. 12. Springer, Berlin/Heidelberg, pp 243–251

Howard B (1990) Rooting potential in cuttings. Garden 115:608–611

Joley LE (1960) Experiment with propagation of the genus *Pistacia*. Proc Inter Prop Soc 10:287–292

Joley LE, Opitz KW (1971) Further experiments with propagation of *Pistacia*. Proc Inter Plant Prop Soc 21:67–76

Lambardi M, Benelli C, Ozudogru EA, Ozden-Tokatli Y (2006) Synthetic seed technology in ornamental plants. In: Teixeira da Silva J (ed) Floriculture, ornamental and plant biotechnology, vol II. Global Science Books, London, pp 347–354

Matsumoto T, Sakai A, Yamada K (1994) Cryopreservation of *in vitro*-grown apical meristems of wasabi (*Wasabi japonica*) by the vitrification and subsequent high plant regeneration. Plant Cell Rep 13:442–446

Mederos-Molina S, Trujillo MI (1999) Elimination of the browning and *in vitro* development of shoots in *Pistacia vera* L. cv. 'Mateur' and *P. atlantica* Desf. culture. Acta Societatis Botanicorum Poloniae 68:21–24

Murashige T (1977) Plant cell and organ cultures as horticultural practices. Acta Hort 78:17–30

Murashige T, Skoog F (1962) A revised medium for rapid growth and bioassay with tobacco cultures. Physiol Plant 15:473–497

Onay A (1996) In vitro organogenesis and embryogenesis of Pistachio, *Pistacia vera* L. PhD thesis, University of Edinburgh, UK

Onay A (2000a) Somatic embryogenesis in cultured kernels of pistachio, *Pistacia vera* L. cv Siirt. In: Proceedings of the 2nd Balkan botanical congress, İstanbul, 14–16 May. Plants of the Balkan Peninsula: into the next Millennium, vol 11, no:109–115

Onay A (2000b) Micropropagation of pistachio from mature trees. Plant Cell Tissue Organ Cult 60:159–162

Onay A, Jeffree CE (2000) Somatic embryogenesis in pistachio. In: Jain SM, Gupta PK, Newton RJ (eds) Somatic embryogenesis in woody plants. Chapter 10, Section B. Kluwer Academic, Dordrecht, Netherlands

Onay A, Namli S (1998). Antepfıstığinin (*Pistacia vera* L.) yapraklarında somatik embriyogenesis. XIV. Ulusal Biyoloji Kongresi, Samsun, 7–10 Eylül 1998, Cilt II, Bitki Fizyolojisi-Bitki Anatomisi ve Hidrobiyoloji Seksiyonları, Samsun, pp 44–53

Onay A, Jeffree CE, Yeoman MM (1995) Somatic embryogenesis in cultured immature kernels of pistachio, P. vera L. Plant Cell Rep 15:192–195

Onay A, Jeffree CE, Yeoman MM (1996) Plant regeneration from encapsulated embryoids and an embryogenic mass of pistachio, *Pistacia vera* L. Plant Cell Rep 15:723–726

Onay A, Firat MZ, Namli O (1997) An improved method for embling production in Pistachio, Pistacia vera L. using liquid medium matured somatic embryos. Turk J Biol 21:159–174

Onay A, Işikalan Ç, Adiyaman F (2005) *Pistacia vera* pistachio. In: Litz RE (ed) Biotecnology of fruit and nut crops, vol 29, Biotechnology in agriculture series. CABI, Wallingford, pp 62–72

Onay A, Tilkat E, Akturk Z, Yildirim H (2007a) Micropropagation of male pistachio (*P. vera* L. cv. 'Atli') and cherry (*Prunus avium* L.) trees. Agriculture, Forestry and Food Technologies Research Grant Committee (TOVAG), Project Final Report, #TOVAG-3355. The Scientific and Technical Research Council of Turkey (TUBITAK)

Onay A, Tilkat E, Yildirim H, Suzerer V (2007b) Indirect somatic embryogenesis from mature embryo cultures of Pistachio, *Pistacia vera* L. Prop Ornamen Plants 7:68–74

Ozden Tokatli Y, Ozudogru EA, Akcin A (2004) In vitro regeneration of pistachio (*Pistacia vera* L.) through organogenesis: effect of silver nitrate, polyvinylpyrrolidone and citric acid. In: Proceedings of "2004 World Congress on In Vitro Biology", San Francisco, California, USA, 22–26 May 2004, vol 40, p 46-A

Ozden Tokatli Y, Ozudogru EA, Akcin A (2005) In vitro response of pistachio nodal explants to silver nitrate. Sci Hort 106:415–426

Ozden Tokatli Y, Ozudogru EA, Gumusel F, Lambardi M (2007) Cryopreservation of *Pistacia* spp. seeds by dehydration and one-step freezing. CryoLetters 28:83–94

Ozden Tokatli Y, Ozudogru EA, Akdemir H, Gumusel F (2008) Application of cryopreservation technology for pistachio germplasm conservation. Acta Hort (ISHS) 839:245–252

Ozden Tokatli Y, Akdemir H, Tilkat E, Onay A (2010) Current status and conservation of *Pistacia* germplasm. Biotechnol Adv 28:130–141

Padulosi S, Hadj-Hassan A (2001) Towards a comprehensive documentation and use of *Pistacia* genetic diversity in central and W. Asia, N. Africa, Europe, The International Plant Genetic Resources Institute (IPGRI), Rome

Panis B, Lambardi M (2006) Status of cryopreservation technologies in plants (crops and forest trees). In: Ruane J, Sonnino A (eds) The role of biotechnology in exploring and protecting agricultural genetic resources. FAO, Rome, pp 61–78

Parfitt DE, Almehdi AA (1994) Use of a high CO_2 atmosphere and medium modifications for the successful micropropagation of pistachio. Sci Hort 56:321–329

Piccioni E, Standardi A (1995) Encapsulation of micropropagated buds of six woody species. Plant Cell Tissue Organ Cult 42:221–226

Preece JE, Read PE (2003) Novel methods in micropropagation. Acta Hort 616:71–76

Read PE, Yang G (1991) Plant growth regulator effects on rooting of forced sotwood cuttings. Acta Hort 300:197–200

Redenbaugh K, Slade D, Viss P, Fujii JA (1987) Encapsulation of somatic embryos in synthetic seed coats. Hort Sci 22:803–809

Redenbaugh K, Fujii J, Slade D, Viss P, Kossler M (1991) Artificial seeds-encapsulated somatic embryos. In: Bajaj YPS (ed) Biotechnology in agriculture and forestry, High-Tec micropropagation. Springer, Heidelberg, pp 395–416

Sakai A, Kobayashi S, Oiyama I (1990) Cryopreservation of nucellar cells of navel orange (*Citrus sinensis* Osb. Var. Brasiliensis Tanaka) by vitrification. Plant Cell Rep 9:30–33
Sakoury K (1976) Propagation of Pistachio by cuttings. M.Sc. thesis, California State University Presno, USA
Tilkat E (2006) Micropropagation of male *P. vera* L. via apical shoot tip culture. PhD thesis, Institute of Science, University of Dicle, Turkey, 142pp
Tilkat E, Onay A (2009) Direct shoot organogenesis from in vitro derived mature leaf explants of pistachio. In Vitro Cell Dev Biol Plant 45(1):982–998
Tilkat E, Onay A, Yildirim H, Ozen HC (2008) Micropropagation of mature male pistachio *Pistacia vera* L. J Hort Sci Biotechnol 83:328–333
Tilkat E, Onay A, Ozden Tokatli Y (2009a) In Vitro rooting improvement of adult pistachio, Pistacia vera L. 'Atlı'. Acta Hort (ISHS) 839:215–221
Tilkat E, Onay A, Yildirim H, Ayaz E (2009b) Direct plant regeneration from mature leaf explants of pistachio *Pistacia vera* L. Scientia Hort 121(3):361–365
Tomar UK, Gupta SC (1988) Somatic embryogenesis and organogenesis in callus cultures of a tree legume *Albizzia richardiana* King. Plant Cell Rep 7:70–73
Turakhia DV, Kulkarni AR (1988) In vitro regeneration from leaf explant of *Ladebouria hyacinthiana* (*Scilla indica*). Curr Sci 57:214–216
West TP, Preece JE (2006) Use of acephate, benomyl and alginate encapsulation for eliminating culture mites and fungal contamination from in vitro cultures of hardy hibiscus (*Hibiscus moscheutos* L.). In Vitro Cell Dev Biol Plant 42:301–304
Withers L, Engelmann F (1997) In vitro conservation of plant genetic resources. In: Altman A (ed) Agricultural biotechnology. Marcel Dekker, Inc., New York, pp 57–88
Yang G, Read PE (1993) In vitro culture of 'Vanhoutte's' spirea explants from "secondary cultures" and dormant stems forced in solutions containing plant growth regulators. Plant Cell Tissue Organ Cult 33:25–30

Chapter 15
Crop Productivity and Water Use Efficiency: The Role of Carbon Isotope Discrimination Technique

Javed Akhter and Philippe Monneveux

Contents

J. Akhter (✉)
Nuclear Institute for Agriculture and Biology, P.O. Box 128, Jhang Road, Faisalabad 38000, Pakistan
e-mail: javedakhterniab@yahoo.com; akhjniab@yahoo.com

P. Monneveux
Generation Challenge Programme, c/o CIMMYT, P. 6-641, 06600 Mexico D.F., Mexico
e-mail: pmommeveux@yahoo.fr

M. Ashraf et al. (eds.), *Crop Production for Agricultural Improvement*,
DOI 10.1007/978-94-007-4116-4_15, © Springer Science+Business Media B.V. 2012

Abstract At a time when global population continues to increase, global agriculture now accounts for 70% of the amount of water used on earth and crop losses due to water loss are much higher than crop losses due to any other causes. Reliable selection of crop varieties or genotypes with increased grain yield is essential both under irrigated and rain-fed conditions, due to water shortage commonly experienced in many environments. This can be obtained by increasing crop water use efficiency (WUE) either by breeding, or by soil, water and crop management. WUE can be enhanced by reducing water losses (runoff, drainage, evaporation), thus increasing the proportion of the water used by the crop. There is, indeed, a clear relationship between the amount of water transpired and yield across a diverse range of crop species. The carbon isotope discrimination technique has been proposed by several authors as a mean of selecting cultivars with improved WUE. Carbon isotope discrimination (Δ) of plant tissues (leaf or grain) has been shown to correlate negatively with transpiration efficiency and positively with grain yield in a wide range of crops. In the present chapter, recent progress in improving crops for yield and water-use efficiency using the carbon isotope discrimination technique are reviewed and some possible avenues for making further advances are discussed. The review describes the basis of carbon discrimination in C_3 and C_4 crops with a special emphasis on cereals, dominant food crops grown in irrigated and rain-fed areas.

Keywords Carbon isotope discrimination • Isotopic fractionation • Water scarcity • Water-use efficiency • Transpiration efficiency • Drought tolerance • Crop productivity

1 Introduction

Humanity is dependent on freshwater for life, and water is increasingly becoming a strategic commodity. In several countries, including Pakistan, water resources current availability and future water supplies are at risk as a result of population increase and urbanization. Increasing water scarcity and high population growth have already increased the food imports in arid and semi-arid countries (UK meteorological Report 2006), the local agricultural sector becoming less and less able to produce staple crops in sufficient quantity. As the world population continues to increase, more people will require more water for the cultivation of food, fibers and industrial crops. The Comprehensive Assessment of Water Management in Agriculture (2007) reveals that around 1.2 billion people, or almost one-fifth of the world's population, live in areas of physical scarcity, while 500 million people are approaching this situation.

In many regions, rainfall, a major source of freshwater, is expected to be also severely affected by changes in global climate, caused by higher concentration of greenhouse gases. The International Panel for Climate Change (IPCC 2007), in its third Assessment Report on climate change, emphasized that global warming is happening at a much faster rate than expected. Global temperatures are rising nearly twice as fast as previously thought. The same report predicts that temperatures could rise by as much as 5.8°C by the end of the century and that rainfall will increase in regions that already have sufficient rains, while dry regions will become even drier. Climate change is expected to account for about 20% of the global increase in water scarcity. Countries that already suffer from water shortages will be hit hardest. Significantly, there will be major increases in water scarcity even if the water impacts of climate change prove to be neutral or enhance of the world's hydrological budget. With neither being reasonably expected to happen, the impact of a changing climate will affect not only bulk water availability but will also worsen the extremes of drought and floods. A 2006 study by the UK Meteorological Office concluded that, with no mitigation of climate change, the severe droughts that now occur only once every 50 years will occur every other year by 2100. Globally, it is estimated that food and feed crop demand will nearly double in the coming 50 years (UN Water 2007).

2 Agriculture and Water Scarcity

Today, agriculture accounts for 70% of all water use globally, up to 95% in several developing countries. Increasing the efficiency of water use and enhancing agricultural water productivity at all levels of the production chains are becoming priorities in a growing number of countries. Water use efficiency in this sector is very poor, not exceeding 45%. Thereby, enormous quantities of water could be saved in the agricultural sector, and increasing water-use efficiency of both irrigated and rain-fed crops is an urgent imperative (Hamdy et al. 2003). Several strategies are required to improve water use in irrigated and rain-fed agriculture (Wang et al. 2002). Breeding crop varieties that are more efficient in water use is one strategy (Condon et al. 2004; Condon and Hall 1997; Rebetzke et al. 2006). Others include better management of the water resource and changes in crop management (Gregory 2004). None of these strategies should be seen as operating in isolation. Rather, it is likely that the greatest gains will be obtained through complementary approaches involving each of them. A systematic approach to agricultural water productivity requires actions at all levels, from crops to irrigation schemes, in order to ameliorate inter-sectoral competition for water resources and optimize social and economic outcomes (Bacon 2004). To keep pace with the growing demand for food and taking into consideration an increase in water productivity, it is estimated that 14% more freshwater will need to be withdrawn for agricultural purposes by 2030 so as to obtain the 55% increase in food production needed.

3 Water Use Efficiency in Agriculture

Water use efficiency is a measure of the amount of water used by plants per unit of plant material produced. The term can be applied at the leaf, whole-plant, and ecosystem levels (Jones 2004). At the leaf level, it is more precisely referred to as "instantaneous water use efficiency" or "transpiration efficiency", TE (Farquhar et al. 1982) and represents the moles of CO_2 taken up by photosynthesis (A) divided by the moles of water lost through transpiration in a unit of time and per unit leaf area (T). At the whole-plant level, water use efficiency referred to as the units of dry matter synthesized divided by the units of water lost (Gregory et al. 1997). At the ecosystem level, WUE (often referred to as "agronomic WUE") can be calculated as the grams of dry weight gained by plants during the growing season per unit land area divided by the millimeters of water lost (including evaporation directly from the soil). Agronomic water use efficiency, or yield of harvested product achieved from the water used by the crop through all means (i.e., precipitation and/or irrigation) may be important for farmers and agronomists. The prospect of improving agronomic water-use efficiency by manipulating leaf-level water-use efficiency (i.e., TE) has long been an attractive one (Condon et al. 2004).

According to Gregory et al. (1997), the ratio of total biomass produced against total available water may be expressed as:

$$\mathrm{WUEB} = \mathrm{M}/(\mathrm{Es}+\mathrm{T}+\mathrm{R}+\mathrm{D}) \tag{15.1}$$

where M, Es, T, R and D are the biomass produced, the evaporation from the soil surface, the transpiration during the growing season, the runoff, and the drainage below the root zone, respectively.

Equation 15.1 can be rearranged to give:

$$\mathrm{WUEB} = (\mathrm{M}/\mathrm{T})/\left[1+\left[(\mathrm{Es}+\mathrm{R}+\mathrm{D})/\mathrm{T}\right]\right] \tag{15.2}$$

Agronomic water use efficiency on a grain basis, the ratio of production of total biomass against total available water can be expressed as:

$$\mathrm{WUEG} = \mathrm{HI} * (\mathrm{M}/\mathrm{T})/\left[1+\left[(\mathrm{Es}+\mathrm{R}+\mathrm{D})/\mathrm{T})\right]\right] \tag{15.3}$$

where HI is the harvest index, i.e., the ratio of grain yield to above-ground biomass.

The potential role of carbon isotope discrimination (Δ) for improving water use efficiency.

Equation 15.3 makes clear that improved WUEG can be obtained by either an increase of M/T, a reduction of Es/T, or an increase of HI. It has been postulated by several authors (Jones 1997; Impa et al. 2005) that Δ, the carbon isotope discrimination by the crop, can be used to estimate or predict WUEG. This assumption is mainly based on the significant association repeatedly observed between transpiration efficiency and Δ (Farquhar et al. 1982; Condon and Richards 1992) and on the fact that Δ was found largely determined by stomata opening and transpiration (Morgan et al. 1993).

3.1 Biochemical Basis and Variability of Δ in Plants

Isotopes are atoms of the same element that differ in mass. Carbon is made up of isotopes with masses 12, 13 and 14 (written ^{12}C, ^{13}C, ^{14}C, respectively). ^{14}C is radioactive and decay to produce isotopes of a different element over time. ^{12}C and ^{13}C are stable isotopes and do not experience radioactive decay.

During physical, chemical and biological processes, the heavy and light isotopes partition differently. This phenomenon is called isotope fractionation. In almost all reactions the light isotopes react faster than the heavy ones. Heavier isotopes have lower mobility, lower diffusion velocity and collision frequency with other molecules, and generally higher binding energies, compared to lighter isotopes. The differences in binding energy and reaction rates are proportional to the mass difference between isotopes. Thus, light elements are more likely to exhibit isotopic fractionation than heavy ones. For example, the isotopes ^{12}C and ^{13}C of the relatively light C atom have an 8% mass difference and undergo stable isotope fractionation. In contrast, the isotopes ^{87}Sr and ^{86}Sr of the heavy Sr atom have a 1.1% mass difference and do not exhibit detectable mass fractionation. Almost 98.9% of atmospheric CO_2 contains the ^{12}C and only a small part (1.1%) contains ^{13}C. Carbon, the major building block of carbohydrate and proteins in plant tissues contains both light (C^{12}) and heavy (C^{13}) stable isotopes.

Organic matter in plants is depleted in ^{13}C and consequently have lower $^{13}C/^{12}C$ ratio compared to atmospheric CO_2 (Craig 1954; Bender 1968), because plants fractionate carbon isotopes during photosynthesis. The magnitude of fractionation depends upon photosynthesis type, environment, genotypes and many other factors. The biochemical and physical phenomena underlying isotope fractionation are well recognized now and have been used to address a large range of problems in plant physiology and ecology (Rundel et al. 1979; Troughton 1979; Vogel 1980; O'Leary 1988; Farquhar et al. 1989).

3.2 Measurement of Carbon Isotope Discrimination

Measurements of ^{13}C contents of CO_2 are carried out with an isotope ratio mass spectrometer specially designed for high precision measurement of ratio R defined as:

$$R = {}^{13}CO_2 / {}^{12}CO_2 \tag{15.4}$$

R values are converted to carbon isotope composition (δ) as:

$$\delta^{13}C\ (‰) = \left[(\text{Rsample} / \text{Rstandard}\ -1) \times 1000\right] \tag{15.5}$$

The standard is the CO_2 obtained from a limestone from Pee Dee Belmenite "PDB" formation in South Carolina, USA with absolute $^{13}C/^{12}C$ (Rstandard) of 0.0112372. The units are 'per mil' or ‰ and $\delta^{13}C$ has negative values. In the

Table 15.1 Carbon isotope fractionations associated with photosynthesis

Process	Fractionation (‰)[a]	Reference
Solubility of CO_2 in water	1.1	O'Leary (1984)
Hydration of CO_2	–9.0	Mook et al. (1974)
CO_2 diffusion in air	4.4	Hersterberg and Siegenthaler (1991)
CO_2 diffusion in aqueous solution	0.7	O'Leary (1984)
Spontaneous hydration of CO_2	6.9	Marlier and O'Leary (1984)
Carbonic anhydrase – catalyzed hydration of CO_2	1.1	Paneth and O'Leary (1985)
Carboxylation PEP	2.0	O'Leary (1984)
Carboxylation of RuBP	29.0	Roeske and O'Leary (1984)

[a]Positive values in this table indicate that the product is depleted in ^{13}C compared to the starting state; negative values indicate enrichment

absence of industrial activity the CO_2 in air has $\delta^{13}C$ value of –8‰ and become more negative after each year as result of fossil fuel combustion and deforestation (Keeling et al. 1979). The values are converted to carbon isotope discrimination (Δ) as the isotopic difference between source and product using formula (Farquhar et al. 1989):

$$\Delta(\text{‰}) = \left[(\delta_a - \delta_p)/(1 + \delta p)\right] \times 1000 \tag{15.6}$$

where δp is $\delta^{13}C$ of samples and δa, –8.00‰ the $\delta^{13}C$ of atmospheric CO_2 (Keeling et al. 1979).

3.3 Components of Isotope Fractionation

Isotope fractionations values of different physical and chemical processes involved in photosynthesis are listed in Table 15.1.

3.3.1 Thermodynamic and Kinetic Fractionations

Two main types of isotope fractionations have been recognized, thermodynamic and kinetic. Thermodynamic fractionation reflects differences in equilibrium constants and will cause one isotope to concentrate more in one component of a reversible system that is in equilibrium. If the heavier isotope concentrates in a component, then that component is commonly referred to as enriched or heavy. If it is the light isotope that concentrates, then the component is referred to as depleted or light. In most circumstances the heavy isotope concentrates in the component in which the element is bound more strongly and thus equilibrium isotope effects usually reflect relative differences in the bond strengths of the isotopes in the various components of the system.

Kinetic fractionation reflects differences in rate constants for isotopic species and occurs when one isotope reacts more rapidly than the other in an irreversible system or a system in which the products are swept away from the reactants before they have an opportunity to come to equilibrium. Normally, the lighter isotope will react more rapidly than the heavy isotope and thus the product will be lighter than the reactant. The isotope fractionation will only occur in systems in which there is both an isotope effect and a reaction that does not proceed to completion. Thus, even in the presence of an isotope effect, there will be no isotope fractionation if all the reactant goes to a single product because all the atoms have reacted. The ratio of the heavy to the light isotope must be the same in the product as it was in the reactant. The magnitude of an isotope effect is expressed as a fractionation factor, represented as f and defined as the ratio of the heavy to light isotope in the product divided by the ratio of the heavy to light isotope in the reactant (f=[(heavy/light) product]/[(heavy/light) reactant]). When f is greater than 1, the product is heavy or enriched. When it is less than 1, the product is light or depleted. Most fractionation factors lie between 0.9 and 1.1. A fractionation factor of 1.050 is often referred to as a 5% isotope effect (for further basic information on isotope fractionation, see Galimov 1985 and Hayes et al. 1982).

3.3.2 Isotopic Fractionation in Physical Processes

During photosynthesis diffusion processes are involved at different stages and components of isotopic fractionations. The heavier isotopes diffuse slowly than lighter ones. The diffusion of $^{12}CO_2$ and $^{13}CO_2$ in air shows a fractionation of 4.4‰ (O'Leary 1981; Hersterberg and Siegenthaler 1991) while diffusion of dissolved CO_2 in water is slow and shows fractionation of 0.7‰.

3.3.3 Isotopic Fractionation in Enzymatic Processes

Isotopic fractionation associated with enzyme-catalyzed reactions has been measured for a wide variety of enzymatic systems. The enzyme Ribulose 1,5-bisphosphate carboxylase/oxygenase (RuBisCO) which catalyzes the first step in CO_2 fixation in C_3 plants is a major source of isotopic fractionation in C_3 plants. A consistent value of 29‰, compared with dissolved CO_2 at 25°C, was found by several investigators using a variety of methods (Cleland 1987; O'Leary 1988; Cook 1991). The initial carboxylation in C_4 plants is carried out by PEP carboxylase, and the measured isotopic fractionation is 2‰ (Whelan et al. 1973; O'Leary 1981). However, this is the fractionation starting from HCO_3^-, whereas in plants the reaction begins with CO_2. When referred to gaseous CO_2 the isotope fractionation becomes 5.7‰.

3.3.4 Environmental Effects on Fractionation

The values of fractionation can slightly change with species, CO_2 concentration, or other natural variables. Isotopic fractionation in physical and diffusive processes should not be affected by the range of environmental conditions or CO_2 concentration encountered during normal plant growth. The isotope fractionation due to RuBisCO is independent of CO_2 concentration and almost independent of pH (Roeske and O'Leary 1984). The fractionation associated with the CO_2 hydration equilibrium has a small temperature effect (Mook et al. 1974). The species effects on the fractionation due to RuBisCO is still controversial. The enzymes from *Rhodospirillum rubrum* (Roeske and O'Leary 1985) and *Anacystis nidulans* (Guy et al. 1987) show fractionation significantly different from spinach enzyme.

4 Carbon Isotope Discrimination and Photosynthetic Types in Plants

Following the first observation of slight depletion of heavy ^{13}C in plants with respect to inorganic carbonaceous material (Nier and Gulbransen 1939), Murphey and Nier (1941) reported a wide variation in ^{13}C among different plant species. From early surveys of $^{13}C/^{12}C$ in plants, Wickman (1952) suggested that ^{13}C depletion was a cyclic process and that the degree of ^{13}C depletion in plants depended on the amount of recycling of soil derived CO_2. Craig (1954) disagreed and suggested that fractionation processes within leaves accounted for the variation in ^{13}C composition and that the magnitude of fractionation was environment-dependent. Meanwhile, Baertschi (1953) showed that no fractionation occurred during the respiration of bean seedlings and found 26‰ ^{13}C depletion during CO_2 assimilation. The similarity of the fractionation factor found by Urey (1948) (29‰) and by Weigl and Calvin (1948) (27‰) suggested that fractionation process was almost the same for different plants and growth conditions. The values reported by these authors incidentally corresponded closely to the maximum fractionation factor for photosynthesis in nature. Park and Epstein (1961) proposed a three step model to explain the observed differences in ^{13}C composition in leaves relative to atmospheric $CO_{2,}$ suggesting some additional photosynthetic and diffusion fractionations.

Wickman (1952) found an aquatic grass with high ^{13}C content compared to known values. The connection between ^{13}C contents in terrestrial plants and C_4 syndrome became clear when Kortshak et al. (1965) and Hatch and Slack (1966) discovered the C_4 pathways in tropical grasses. Based upon isotopic studies, Bender (1968, 1971) and Smith and Epstein (1971) postulated a parallel evolution for C_3 and C_4 plants rather than one common origin. At that time some succulents with anomalous ^{13}C values (Bender 1971) having the ability to fix CO_2 at night were linked to crassulacean acid metabolism (CAM), a separate photosynthetic type group (Bender et al. 1973).

C_3 and C_4 processes involve different isotopic fractionation. C_3 plants convert atmospheric CO_2 to a phosphoglycerate compound with three C atoms. Carbon isotopes are strongly fractionated by photosynthesis. C_3 plants show $\delta^{13}C$ values ranging from −32‰ to −20‰ with an average value of −27‰. C_4 plants convert CO_2 to dicarboxylic acid, a four-C compound and have higher $\delta^{13}C$ values than C_3

plants, ranging from −17‰ to −9‰ with a mean of −13‰ relative to PDB. The difference in $^{13}C/^{12}C$ ratio between C_3 and C_4 is correlated with isotopic fractionation present between the ribulose biphosphate carboxylase (RuBP) activity in C_3 plants and phosphoenolpyruvate (PEP) carboxylase activity in C_4 plants. RuBP discriminates more against ^{13}C than PEP (Christeller et al. 1976). Plants exhibiting C_3 pathways are consequently the most depleted in 13C, and plants endowed with C_4 pathways are the least depleted. Plants with crassulacean acid metabolism show intermediate value (Bender et al. 1973). The ^{13}C value is a standard method for distinguishing the C_3 and C_4 plant groups.

4.1 *Discrimination in C_3 Plants*

Over 80% of crop plants are C_3 and most terrestrial plants are C_3. All forest communities and most temperate zone plant communities are dominated by C_3 plants. The native crops of North America and Europe are almost exclusively C_3.

There are several processes that contribute to isotope fractionation measured in plant dry matter of C_3 species (Farquhar et al. 1989; Brugnoli and Farquhar 2000) and carbon isotope discrimination can be calculated as:

$$\Delta^{13}C = [a + (b - a)C_i / C_a - d] \quad (15.7)$$

where a is the discrimination due to diffusion (4.4‰), b is the discrimination due to carboxylation (30‰ when corrected for the equilibrium effect on CO_2 dissolution), C_i is the internal gas-phase pressure of CO_2, C_a is the external CO_2 pressure and the term d involves contributions from respiration, liquid-phase diffusion, isotopic changes due to carbon export, CO_2 fixation in C_3 plants by PEP carboxylase and a variety of other factors (Farquhar et al. 1989). The value of d is usually small and variations in d can be generally ignored. The respired carbon is isotopically different from leaf carbon but most available evidence suggests that the IF associated with respiration is small (O'Leary 1981; Farquhar et al. 1989). The IF associated with photorespiration in soybean (*Glycine max*) 7‰ (Rooney 1988).

The following approximate expression has consequently been developed by Farquhar and Richards (1984) to rely Δ and C_i/C_a:

$$\Delta^{13}C = a + (b - a)C_i / C_a \quad (15.8)$$

It means that the isotope fractionation in a C_3 plant mainly occurs in two steps. In the first step, CO_2 diffuses through the stomata to the site of carboxylation; in the second step, this CO_2 is taken up irreversibly by the appropriate carboxylase. The fractionation associated with diffusion is 4.4‰ and with carboxylation is 29‰ in C_3 plants (30‰ when CO_2 dissolution is included). Fractionation may be lower than 4.4‰ when internal CO_2 concentration is low (e.g., when stomata are nearly closed) and $\delta^{13}C$ can approach −2‰. If stomata are open, the internal CO_2 concentration approaches the external CO_2 concentration, and the observed fractionation approaches the carboxylation fractionation, and leaf $\delta^{13}C$ would then approach −38‰.where *a* is the fractionations associated with diffusion of CO_2 into the intercellular air spaces ($a=4.4$) and *b* the fractionation associated with carboxylation of CO_2 into the first products of photosynthesis by Rubisco ($b=28$). Substituting the numerical values indicated into Eq. 15.8 generates the following:

$$\Delta^{13}C \approx 4.4 + 23.6 C_i / C_a \tag{15.9}$$

This simple equation shows that $\Delta^{13}C$ is positively related to the ratio C_i/C_a.

4.2 *Relationship Between Carbon Isotope Discrimination and TE in C_3 Plants*

TE, the ratio of net photosynthesis to transpiration (at the leaf level) can be mathematically expressed as below (Condon et al. 2004):

$$A / T \approx 0.6\, Ca\left(1 - Ci\right) / Ca) / \left(Wi - Wa\right) \tag{15.10}$$

Equation 15.10 indicates that *TE* depends on two factors. One is C_i/C_a, which decrease enhance the value of $(1 - C_i/C_a)$ and consequently *TE*. The other is $(W_i - W_a)$. Higher the gradient driving transpirational water loss, higher is *TE* (Condon et al. 2004).

It is now accepted that relative differences in C_i/C_a, at least within C_3 species, may be evaluated indirectly by measuring carbon isotope composition of plant dry matter (Farquhar et al. 1989).

However, earlier it was noted that *A*/*T* was negatively related to C_i/C_a (Eq. 15.7). Therefore, $\Delta^{13}C$ and *TE* are negatively associated in C_3 plants (Hall et al. 1994; Hubick and Farquhar 1989).

4.3 *Discrimination in C_4 Plants*

C_4 plants are characteristically found in hot, arid environments. A selective advantage of C_4 photosynthesis is a more efficient use of water. Some crops of big economical importance are C_4 plants: maize, sorghum, millet, and sugar cane.

Carbon isotope discrimination in C_4 plants can be calculated as:

$$\Delta = a + (b4 + b\, 3\phi - a)\, C_i / C_a \tag{15.11}$$

where a is the fractionation due to diffusion (4.4‰), b4 is the fractionation from gaseous CO_2 through PEP carboxylase (−6.7‰), b3 is the fractionation due to

RuBisCo (30‰), and ϕ is the fraction of CO_2 released in the bundle sheath that leaks to the mesophyll, where it may be either taken up by PEP carboxylase or released to the atmosphere. Stomatal opening causes opposite effect in C_3 and C_4 plants (compare above two equations). Increasing relative humidity under controlled environments causes an increase in $\delta^{13}C$ in C_4 plants but a decrease in C_3 plants (Madhavan et al. 1991).

The bases of carbon isotope discrimination in C_4 plants are consequently more complex than in C_3 plants. The variation in carbon isotopic discrimination in C_4 plants depends not only on C_i/C_a (as in C_3 plants), but also on leakiness, ϕ (Farquhar 1983), the proportion of CO_2 released into the bundle sheath which is not fixed by Rubisco and leaks back to the mesophyll. ϕ is a measure of the extent to which PEP carboxylation exceed Rubisco carboxylation (Henderson et al. 1998) and has been estimated using a variety of techniques (Hatch et al. 1995). Some authors suggested that for a variety of C_4 species Φ could range from 0.2 to 0.6 (Sandquist and Ehleringer 1995; Saliendra et al. 1996; Meinzer and Zhu 1998). Henderson et al. (1992) and Williams et al. (2001) however found Φ to be constant and equal to 0.2 in *Sorghum bicolor* (L.) Moench, cultivated under a wide range of irradiances, temperatures, and partial pressures of CO_2. Krall and Edwards (1990) also found ϕ to be 0.2 in a number of C_4 species. Since there is less variation in ϕ in a given species, Δ is linearly related to C_i/C_a as in C_3 species, and consequently positively related to TE (Henderson et al. 1992, 1998). Since the slope of the relationship between Δ and C_i/C_a is lower than in C_3 plants (Henderson et al. 1998) genetic variation of C_i/C_a lead to much smaller variation in Δ than in C_3 plants. On the other hand, temporal and environmental variations in C_i/C_a give smaller variations in Δ than in C_3 species.

4.4 Discrimination in CAM Plants

The succulents that have the ability to utilize the crassulacean acid metabolism (CAM) have the ability to fix CO_2 at night as malate along the C_4 pathway. The assimilated carbon is then further processed by ribulose bisphosphate (RuBP) carboxylase on the following day, when the stomata can remain closed and restrict transpiration. Those succulents that have a strong tendency to operate in the CAM mode and assimilate only small amounts of CO_2 during the day via the C_3 pathway have isotope rations similar to those of C_4 plants. On the other extreme, such plants that have only a weak affinity for the CAM syndrome tend to have isotope ratios in the C_3 range. The latter group can usually be induced to fix carbon at night by subjecting the plants to water stress (Osmond et al. 1976) and the isotope ratios then become intermediate. The genus Aloe with 63 species has C_3-like and C_4-like species with some intermediate ones (Vogel 1980). CAM plants are distinguished from C_4 plants by anatomy (succulence) or physiology (diurnal malic acid cycling).

5 Processes Affecting Carbon Isotope Composition in Plants

5.1 Carbon Isotope Composition of the Atmosphere

Measurements at various stations show that the $\delta^{13}C$ of atmospheric CO_2 is currently close to −8‰ (Goodman and Francey 1988). This value is slowly decreasing (Keeling et al. 1979) at an average rate of 0.028‰ per year as the atmosphere becomes depleted in $^{13}CO_2$ relative to $^{12}CO_2$. This depletion occurs as a consequence of the anthropogenic burning of fossil fuels and deforestation and isotopic composition typical of C_3 plant matter (−26‰). The ice core record shows that δa was about −6.4‰ prior to the industrial revolution (Friedly et al. 1986). A latitudinal gradient in δa with 0.2‰ more negative at 60°N than at 60°S has been reported (Keeling et al. 1979). Seasonal variation in δa are also observed, being the greatest at high northern latitudes and these changes in δa correlate well with seasonal changes in the atmospheric partial pressure of CO_2.

5.2 Isotope Composition of Air Within Plant Canopies

Variations in δ_a may occur within or just above, plant canopies. The canopy layer does not need to be precisely defined in each case, but it must be close enough to vegetation for diurnal cycles in nearby plants and soil to be measured.

5.3 Flow of CO_2 Across the Laminar Boundary Layer and Within the Leaf

The theoretical value for boundary layer discrimination is 2.9‰ (Kays 1966). Diffusion across the stomata and the discrimination against $^{13}CO_2$ in free air has a theoretical value of 4.4‰.

5.4 Transport of CO_2 Within the Leaf

Discrimination against $^{13}CO_2$ in free air has a theoretical value of 4.4‰. Diffusion across the stomata varies and at low stomatal apertures (less than about 0.5 μm) the nature of the diffusion of gases is markedly different than that at greater apertures of in free air. For stomatal pores whose diameter is much greater than this value, the rate of diffusion is dependent on the concentration gradient and on the frequency and type of molecular collision and conventional isotope discrimination. The discrimination associated with this diffusion is 11.3‰, which is markedly different from the 4.4‰

value for continuous flow. Diffusion through intercellular air spaces and isotope fractionations associated with diffusion from sub-stomatal cavity to the cell wall is significant. Vitousek et al. (1990) found increases in diffusion resistance as a consequence of increases in leaf thickness.

The pathway of flow from the cell wall surface to the sites of carboxylation, sometimes referred to as the liquid phase, involves several resistances in series and total fractionation associated with this pathway is normally taken as the sum of the equilibrium fractionation as CO_2 enters solution (1.1‰ at 25°C; Mook et al. 1974; Evans et al. 1986) and the discrimination occurring during diffusion of dissolved CO_2 in water (0.7‰) (O'Leary 1984). The carbonic anhydrase interconverts CO_2 and bicarbonate and isotope fractionation for the equilibrium reaction is temperature dependent, being –9‰ at 25°C (Mook et al. 1974).

5.5 *Fractionation Associated with Carboxylation*

A major kinetic fractionation (30‰) occurs during the fixation of CO_2 by Ribulose –1,5 Bisphosphate Carboxylase Oxygenase (RuBisCO) (Roeske and O'Leary 1984). Phosphoenolpyruvate (PEP) discriminates against ^{13}C and at 25°C, fractionation is usually considered to be ~5.7‰ (O'Leary 1981). Fractionation associated with PEP must be taken into account for C_3 as well as for C_4 plants (Farquhar and Richards 1984), as a proportion of the carbon fixed is associated with anaplerotic reactions. Plants contain many other carboxylases, most of which function in biosynthetic pathways. However, these carboxylases probably contribute less than 0.05% of the total plant C (Raven and Farquhar 1990).

6 Carbon Isotope Discrimination and Its Relationship with Grain Yield

6.1 *C_3 Crops*

Carbon isotope discrimination (Δ) has been proposed by several authors as a predictive selection criterion for grain yield of wheat under drought stress (Monneveux et al. 2005). Wheat grain yield was found to be positively correlated to grain Δ under the conditions of South Australia (Condon et al. 1987), California (Ehdaie et al. 1991), Italy (Morgan et al. 1993), Morocco (Bazza 1996), Spain (Araus et al. 1998), South of France (Merah et al. 2001b), Greece (Tsialtas et al. 2001), Pakistan (Akhter et al. 2008) and Algeria (Hafsi et al. 2001).

It has been observed that the association between grain Δ and yield was quite stable over growing seasons under those Mediterranean-type climatic conditions (Merah et al. 2001b; Monneveux et al. 2005), characterized by post-anthesis water deficit and referred by Rajaram et al. (1995) as wheat mega-environment ME4A. Few exceptions were reported by Hafsi et al. (2003) and Araus et al. (2003), under very severe drought conditions, in Algeria and southern Spain, respectively. In some cases, grain yield was also positively correlated to leaf Δ, assessed at anthesis (Merah et al. 1999, 2001c). Under semi arid climate Akhter et al. (2008) reported significant positive correlations of grain yield and WUE with both grain Δ and early leaf Δ in bread wheat, grown under different water regimes.

Different hypotheses can explain the association between grain Δ and grain yield and its stability under terminal drought conditions. Firstly, high grain Δ reflects an ability of the plant to maintain open stomata after anthesis (Morgan et al. 1993; Sayre et al. 1995; Merah et al. 1999, 2001a). The association often reported between grain yield and leaf Δ (Merah et al. 1999, 2001c) agrees with this hypothesis. Secondly, high grain Δ characterizes genotypes more dependent on pre-anthesis reserves for grain filling. Under severe post-anthesis water stress, photosynthesis is more reduced than translocation. For translocation, plants mainly use assimilates from preanthesis reserves accumulated during periods of reduced stress (Loss and Siddique 1994) and which Δ values are higher. The significant correlation generally reported between leaf and grain Δ supports this hypothesis. Finally, high grain Δ under postanthesis water stress can be a consequence of earliness (Condon et al. 2002). Strong negative associations have been found between Δ and days to heading or anthesis in studies conducted with cereals in Mediterranean, terminal-drought environments, with low Δ genotypes flowering later than high Δ genotypes (Craufurd et al. 1991; Sayre et al. 1995).

The association between grain yield and Δ was found to be weaker and less stable under pre-anthesis water stress by Xu et al. (2007) and under residual soil moisture conditions by Misra et al. (2006). In these two environments, referred to as mega-environments ME4B and ME4C, respectively (Rajaram et al. 1995), the association between grain yield and Δ highly depended on the quantity of water stored in the soil before sowing and on the evaporative demand during the growth cycle (Monneveux et al. 2005).

Little is known concerning the association between grain yield and Δ under pre-anthesis water stress. In contrast to ME4A, the association is generally weaker and highly depends on environment. Condon et al. (2002) postulated that the dependence on stored assimilates for grain filling increases as the degree of post-anthesis water stress increases. This suggests that in ME4B conditions, a small proportion of starch originates from reserves accumulated in vegetative organs and that most of the C products are synthesized after anthesis, under more optimal water conditions. This hypothesis is supported by the weak or not significant correlations between leaf and grain Δ reported under his drought scenario (Monneveux et al. 2005; Xu et al. 2007). Under this drought scenario, yield is likely to be related to pre-anthesis growth, which may influence the source and sink capacity. Maximizing crop transpiration, in such conditions, can be achieved by a higher stomatal conductance when plants are suffering drought (i.e. before anthesis) and/or a rapid growth after water recovering (i.e., after anthesis). Consistent genotypic variation was noted in Δ during the vegetative period in this environment (Monneveux et al. 2005), reflecting differences among genotypes in their ability to maintain open stomata (Morgan et al. 1993). The lack of correlation between grain yield and leaf Δ at anthesis in this mega-environment (Monneveux et al. 2005; Xu et al. 2007) suggests that differences in transpiration rate among genotypes become less marked with time, and poorly explain final yield differences. The correlation between grain Δ and grain yield becomes stronger as the pre-anthesis water stress is weaker, suggesting that the strength of the correlation between Δ and yield is highly dependent on the quantity of water stored in the soil at sowing.

Under residual moisture stress (ME4C), the relationship between Δ and yield was found to be weaker and less stable than under ME4A (Misra et al. 2006). According to Condon and Richards (1993), high-Δ genotypes tend to grow faster than low-Δ genotypes. By covering the ground more quickly, they are more efficient in reducing soil evaporation. Their higher biomass at anthesis and greater reserves enable them to translocate a large amount of stored assimilates to the grain. High Δ may also reflect higher stomatal conductance, particularly after anthesis, when soil moisture decreases and water stress becomes stronger. The weak association between grain Δ and grain yield suggests that differences in transpiration rate had little influence on yield differences among genotypes. At anthesis soil water depletion may have been greater for high − Δ genotypes than for low − Δ genotypes, which were more conservative in their water use. In some cases, excessive soil water depletion by high − Δ genotypes can offset the advantages described above. This could explain the negative relationships between Δ and grain yield reported by Condon et al. (2002) in the northern Australian wheat belt. In this environment, transpiration represents a high proportion of total crop water use and high TE (low Δ) can lead to better soil water conservation thus sustaining better growth until the end of the growth cycle.

Under irrigated conditions, no association is generally found between Δ and grain yield. This suggests that the potential decrease in Ci associated with increased photosynthetic capacity was already largely offset by the increase in stomatal aperture thereby restricting the variation in Δ and reducing the correlation between Δ and grain yield. Few crops are however grown in a total absence of water stress.

6.2 C_4 Crops

In C_4 crops, information concerning Δ variation in different organs of the plant and its possible association with biomass or grain yield is scarce. Substantial variation in Δ has been reported in several species, as maize (Samejima 1984), *Panicum coloratum* (Ohsugi et al. 1988), sorghum (Hubick et al. 1990) and sugarcane (Meinzer et al. 1994). The variation for Δ with abiotic stresses and the relationship between Δ and yield under drought have been poorly investigated in C_4 crops. Bowman et al. (1989) and Meinzer et al. (1994) reported an increase of Δ in leaves with salinity in *Andropogon glomeratus* and sugarcane, respectively. A comparison of two sugarcane cultivars highly differing for salinity tolerance revealed that Δ was significantly lower in the tolerant than in the susceptible cultivar (Meinzer et al. 1994).

In grasses Akhter et al. In *Atriplex confertifolia*, water stress did not affect Δ. In pearl millet, Bruck et al. (2000) found an increase of Δ with water stress but did not find differences between two cultivars highly contrasting for drought tolerance. Δ analysis conducted by Monneveux et al. (2007) at flowering, under drought and well-watered conditions, in different organs of maize (*Zea mays* L.) inbred lines differing for drought tolerance showed that leaf Δ was higher than ear and silk Δ. Drought stress significantly increased Δ in all organs. Under drought, tolerant inbred lines showed significantly higher Δ than susceptible ones. There was a significant positive correlation between leaf, ear and silk Δ, and ear dry weight at flowering, a trait closely associated to grain yield. Δ was then analyzed at flowering in leaves of a set of drought tolerant maize hybrids and checks. Drought tolerant hybrids had significantly higher grain yield and Δ than the used checks. No correlation was found, however, between Δ and grain yield within tolerant hybrids. The use of Δ appears consequently accurate for a first screening of lines or hybrids highly contrasting for drought tolerance, but not for a more advanced selection among tolerant hybrids. More studies are needed in C_4 crops to establish the causes of Δ variation and precise which organ sampled and time of sampling may allow the better precision in the screening process.

7 Carbon Isotope Discrimination and Water Use Efficiency (WUE)

As mentioned above, and as experimentally confirmed by Farquhar et al. (1982), Farquhar and Richards (1984), Heitholt (1989), Johnson et al. (1990) and Ritchie et al. (1990), Δ and TE are negatively related in C_3 crops. Agronomic water-use efficiency could be improved by exploiting variation in leaf-level water-use efficiency (Morgan and LeCain 1991). Negative correlation between Δ and WUE has been reported in several species. Çagirgan et al. (2005) reported higher WUE in barley cultivars with lower Δ values. A negative correlation was reported between WUE and Δ in rice (Impa et al. 2005) under drought conditions. Negative correlation between Δ and WUE has also been reported in other crops as peanut (Hubick et al. 1986), tomato (Martin and Thorstenson 1988), common bean (*Phaseolus vulgaris* L.), for a range grass (*Agropyrum desertorum*) (Ehleringer et al. 1990), and cowpea (Ismail and Hall 1992). A significant negative correlation has been obtained in wheat between Δ and M/T (Condon et al. 1990; Ehdaie et al. 1991). In some grasses (*Leptochloa fusca* (L.) Kunth and *Sporobolus arabicus* Boiss), leaf Δ was found to be significantly and negatively correlated with Water-use efficiency under semi-arid conditions (Akhter et al. 2003). In rice, Impa et al. (2005) also found a negative association between leaf Δ and WUE. However, due to the difficulty of directly assessing WUE components, these studies were carried out in pot experiments. Runoff and drainage were negligible and soil surface was generally covered by a plastic film to minimize soil evaporation. WUE was consequently close to M/T (see Eq. 15.2). Kirda et al. (1992), by using a water balance model (Hatfield 1990) to evaluate evapo-transpiration and by measuring soil water with a neutron probe, reported a highly significant negative association between Δ and WUE in durum wheat brown in Austria. Condon et al. (1993), however, did not found in bread wheat cultivated in Australia a significant correlation between Δ and WUE, estimated by using a soil water balance and neglecting run off and drainage.

In conclusion, carbon isotope discrimination of plant tissues reflects the variation of photosynthesis to stomatal conductance ratio over a considerable length of time and under variable environmental conditions, and consequently represents a long-term integrative estimate of transpiration efficiency. As emphasized by Chaves et al. (2003) this may be quite different from long-term WUE defined, according to Gregory et al. (1997), as the units of dry matter synthesized divided by the units of water lost (M/T).

It is even more difficult to use *Δ* for predicting variation for agronomic WUE. WUE variation is manly driven by factors that poorly rely to plant characteristics (e.g., runoff, drainage, soil evaporation). Transpiration, which depends on the crop, generally represents a small part of the water available. Allen (1990), for example, found that transpiration represented less than 35% of total water use for barley in Syria. Similarly, in Niger, Wallace et al. (1993) reported that transpiration was less than evaporation from the soil surface. Such finding led to the conclusion that Δ poorly informs about agronomical WUE that can be improved mainly by crop management techniques. Rather, Δ can be used for improving yield of C_3 crops under drought prone conditions, with the limitations that have been previously mentioned. The systematic use of easy-to-use water balance models (e.g., the *budget* model developed by Raes et al. 2006a), would be very useful to simultaneously (i) describe the drought scenario faced by the crop and (ii) estimate the different components of agronomical WUE. This would permit, respectively, (i) to better anticipate the conditions and mega-environments in which Δ can be efficiently used as a selection trait to improve yield and to test the effect of different crop management techniques on agronomical WUE.

8 Conclusion

Many studies conducted across the globe suggest that carbon isotope discrimination has great potential to select high water-use efficient cultivars in different mega (water) environments. Correlation between Δ of plant tissues (leaf or grain) is negative with transpiration efficiency and positive with grain yield in a wide range of crops. Increasing water scarcity under erratically changing climatic conditions demands a reliable and fast screening tool for suitable plants with high WUE. The present review highlights the use of Δ in C_3 and C_4 plants. It also provides the basis of carbon discrimination in common food crops grown in irrigated and rain-fed areas. Although this technique has provided the better selection criteria to the physiologist, agronomist and breeders, still studies are required to establish a relationship of Δ (grain/leaf or specific tissue) with yield and/or WUE in the target environment for different crop plants.

References

Akhter J, Mahmood K, Tasneem MA, Naqvi MH, Malik KA (2003) Comparative water use-efficiency of *Sporobolus arabicus* and *Leptochloa fusca* and its relation with carbon-isotope discrimination under semi-arid conditions. Plant Soil 249:263–269

Akhter J, Sabir SA, Lateef Z, Ashraf MY, Haq MA (2008) Relationships between carbon isotope discrimination and grain yield, water-use efficiency and growth parameters in wheat (*Triticum aestivum* L.) under different water regimes. Pak J Bot 40:1441–1454

Allen SJ (1990) Measurement of evaporation from soil under sparse barley crops in Northern Syria. Agric For Meteorol 49:291–309

Araus JL, Amaro T, Casadesus J, Asbati A, Nachit MM (1998) Relationship between ash content, carbon isotope discrimination and yield in durum wheat. Aust J Plant Physiol 25:835–842

Araus JL, Villegas D, Aparicio N, Garcia del Moral LF, El Hani S, Rharrabti Y, Ferrio JP, Royo C (2003) Environmental factors determining carbon isotope discrimination and yield in durum wheat under Mediterranean conditions. Crop Sci 43:170–180

Bacon MA (2004) Water use efficiency in plant biology. Blackwell Publishing-CRC Press, Boca Raton, p 327

Baertschi P (1953) Die Fractionierung der natürlichen Kohlenstoffisotopen im Kohlendioxydstoffwechsel grüner. Pflanzen Helv Chim Acta 36:773–781

Bazza M (1996) Carbon-13 discrimination as a criterion for identifying water use efficiency cultivars under water deficit conditions. In: Isotope studies on plant productivity. IAEA-TECDOC-889, Vienna, pp 123–130

Bender MM (1968) Mass spectrometric studies of carbon 13 variations in corn and other grasses. Radiocarbon 10:468–472

Bender MM (1971) Variations in the $^{13}C/^{12}C$ ratios of plants in relation to the pathway of photosynthetic carbon dioxide fixation. Phytochemistry 10:1239–1244

Bender MM, Rouhani I, Vines HM, Black CC (1973) $^{13}C/^{12}C$ ratio changes in crassulacean acid metabolism. Plant Physiol 52:427–430

Bowman WD, Hubick KT, von Caemmerer S, Farquhar GD (1989) Short term changes in leaf carbon isotope discrimination in salt and water stressed C_4 grasses. Plant Physiol 90: 162–166

Brück H, Payne WA, Sattelmacher B (2000) Effects of phosphorus and water supply on yield, transpirational water-use efficiency, and carbon isotope discrimination of Pearl millet. Crop Sci 40:120–125

Brugnoli E, Farquhar GD (2000) Photosynthetic fractionation of carbon isotopes. In: Leegood RC, Sharkey TD, von Caemmerer S (eds) Photosynthesis: physiology and metabolism, vol 9, Advances in photosynthesis. Kluwer Academic, Dordrecht, pp 399–434

Çagirgan MI, Özbas MO, Heng L, Afza R (2005) Genotypic variability for carbon isotope discrimination in the mutant and improved lines of barley. Isotopes Environ Health Stud 41(3):229–235

Chaves MM, Maroco JP, Pereira JS (2003) Understanding plant responses to drought: from genes to the whole plant. Funct Plant Biol 30:239–264

Christeller JT, Laing WA, Troughton JH (1976) Isotope discrimination of ribulose-1,5 diphosphate carboxylase. No effect of temperature or HCO concentration. Plant Physiol 57:580–582

Cleland WW (1987) The use of isotope effects in the detailed analysis of catalytic mechanisms of enzymes. Bioorg Chem 15:283–302

Condon AG, Hall AE (1997) Adaptation to diverse environments: genotypic variation in water-use efficiency within crop species. In: Jackson LE (ed) Agricultural ecology. Academic, San Diego, pp 79–116

Condon AG, Richards RA (1992) Broad sense heritability and genotype x environment interaction for carbon isotope discrimination in field-grown wheat. Aust J Agric Res 43:921–934

Condon AG, Richards RA (1993) Exploiting genetic variation in transpiration efficiency in wheat: an agronomic view. In: Ehleringer JR, Hall AE, Farquhar GD (eds) Stable isotopes and plant carbon-water relations. Academic, New York, pp 435–695

Condon AG, Richards RA, Farquhar GD (1987) Carbon isotope discrimination is positively correlated with grain yield and dry matter production in field grown wheat. Crop Sci 27:996–1001
Condon AG, Farquhar GD, Richards RA (1990) Genotypic variation in carbon isotope discrimination and transpiration efficiency in wheat. Leaf gas exchange and whole plant studies. Aust J Plant Physiol 17:9–22
Condon AG, Richards RA, Farquhar GD (1993) Relationship between carbon isotope discrimination, water use efficiency and transpiration efficiency for dryland wheat. Aust J Agric Res 44:1693–1711
Condon AG, Richards RA, Rebetzke GJ, Farquhar GD (2002) Improving water-use efficiency and crop yield. Crop Sci 42:122–132
Condon AG, Richards RA, Rebetzke GJ, Farquhar GD (2004) Breeding for high water use efficiency. J Exp Bot 55:2447–2460
Cook PF (1991) Enzyme mechanism from isotope effects. CRC Press, Inc, Boca Raton. ISBN 10: 0849353122
Craig H (1954) Carbon-13 in plants and the relationship between carbon-13 and carbon-14 variation in nature. J Geol 62:115–149
Craufurd PQ, Austin RB, Acevedo E, Hall MA (1991) Carbon isotope discrimination and grain yield in barley. Field Crops Res 27:301–313
David M (2007) Water for food, water for life: a comprehensive assessment of water management in agriculture: International Water Management Institute Earthscan, cop. London
Ehdaie B, Hall AE, Farquhar GD, Nguyen HT, Waines JG (1991) Water-use efficiency and carbon isotope discrimination in wheat. Crop Sci 31:1282–1288
Ehleringer JR, White JW, Johnson DA, Brick M (1990) Carbon isotope discrimination, photosynthetic gas exchange, and transpiration efficiency in beans and range grasses. Acta Oecol 11:611–625
Evans JR, Sharkey TD, Berry JA, Farquhar GD (1986) Carbon isotope discrimination measured concurrently with gas exchange to investigate CO_2 diffusion in leaves of higher plants. Aust J Plant Physiol 13:281–292
Farquhar GD (1983) On the nature of carbon isotope discrimination in C_4 species. Aust J Plant Physiol 10:205–226
Farquhar GD, Richards RA (1984) Isotopic composition of plant carbon correlates with water-use efficiency of wheat genotypes. Aust J Plant Physiol 11:539–552
Farquhar GD, O'Leary MH, Berry JA (1982) On the relationship between carbon isotope discrimination and inter cellular carbon dioxide concentration in leaves. Aust J Plant Physiol 9:121–137
Farquhar GD, Ehleringer JR, Hubick KT (1989) Carbon isotope discrimination and photosynthesis. Annu Rev Plant Physiol Plant Mol Biol 40:503–537
Friedly H, Lotscher H, Oeschger H, Siegenthaler U, Stauffer B (1986) Ice record of the C^{13}/C^{12} ratio of atmospheric CO_2 in the past 2 centuries. Nature 324:237–238
Galimov EM (1985) The biological fractionation of isotopes. Academic, New York
Goodman HS, Francey RF (1988) $^{13}C/^{12}C$ and $^{18}O/^{16}O$ in baseline CO_2. In: Forgan BW, Fraser PJ (eds) Baseline Atmospheric Program CSIRO, Canberra, Australia, pp 54–58
Gregory PJ (2004) Agronomic approaches for increasing water use efficiency. In: Bacon MA (ed) Water use efficiency in plant biology. Blackwell Publishing-CRC Press, Boca Raton, pp 142–171
Gregory PJ, Simmonds LP, Warren GP (1997) Interactions between plant nutrients, water and carbon dioxide as factors limiting crop yields. Philos Trans R Soc Lond 352:987–996
Guy RD, Fogel ML, Berry JA, Hoering TC (1987) Isotope fractionation during oxygen production and consumption by plants. Prog Photosy Res 9:597–600
Hafsi M, Monneveux P, Merah O, Djekoun A (2001) Discrimination isotopique du carbone et rendement du blé dur dans les Hautes-Plaines Sétifiennes. Algérie Sécheresse 12:1–7
Hafsi M, Pfeiffer WH, Monneveux P (2003) Flag leaf senescence, carbon content and carbon isotope discrimination in durum wheat grown under semi-arid conditions. Cereal Res Commun 31:161–168
Hall AE, Richards RA, Condon AG, Wright GC, Farquhar GD (1994) Carbon isotope discrimination and plant breeding. In: Janick J (ed) Plant breeding reviews, vol 12. Wiley, New York, pp 81–113
Hamdy A, Ragab R, Scarascia-Mugnozza E (2003) Coping with water scarcity: water saving and increasing water productivity. Irrig Drain 52:3–20

Hatch MD, Slack CR (1966) Photosynthesis in sugarcane leaves: a new carboxylation reaction and the pathway of sugar formation. Biochem J 101:103–111

Hatch MD, Agostino A, Jenkins CLD (1995) Measurements of the leakage of CO_2 from bundle sheath cells of leaves during C_4 photosynthesis. Plant Physiol 108:173–181

Hatfield JL (1990) Measuring plant water stress with an infrared thermometer. HortScience 25:1535–1538

Hayes JM, Klein PD, Klein ER (1982) Fractionation: an introduction to isotopic measurements and terminology. Spectra 8(4):3–8

Heitholt JJ (1989) Water use efficiency and dry matter distribution in nitrogen- and water-stressed winter wheat. Agron J 81:464–469

Henderson SA, von Caemmerer S, Farquhar GD (1992) Short term measurements of carbon isotope discrimination in several C_4 species. Aust J Plant Physiol 19:263–285

Henderson SA, von Caemmerer S, Farquhar GD, Wade L, Hammer G (1998) Correlation between carbon isotope discrimination and transpiration efficiency in lines of the C_4 species *Sorghum bicolor* in the glasshouse and the field. Aust J Plant Physiol 25:111–123

Hersterberg R, Siegenthaler U (1991) Production and stable isotopic composition of CO_2 in a soil near Bern, Switzerland. Tellus 43B:197–205

Hubick K, Farquhar GD (1989) Carbon isotope discrimination and the ratio of carbon gained to water lost in barley cultivars. Plant Cell Environ 12:795–804

Hubick KT, Farquhar GD, Shorter R (1986) Correlation between water-use efficiency and carbon isotope discrimination in diverse peanut (*Arachis*) germplasm. Aust J Plant Physiol 13:803–816

Hubick KT, Hammer GL, Farquhar GD, Wade LJ, von Caemmerer S, Henderson SA (1990) Carbon isotope discrimination varies genetically in C_4 species. Plant Physiol 91:534–537

Impa SM, Nadaradjan S, Boominathan P, Shashidhar G, Bindumadhava H, Sheshshayee MS (2005) Carbon isotope discrimination accurately reflects variability in WUE measured at a whole plant level in rice. Crop Sci 45:2517–2522

International Panel for Climate Change (2007) Fourth assessment report on climate change working group II report impacts, adaptation and vulnerability. www.ipcc.ch/publications/publications_and_data_reports.shtml

Ismail AM, Hall AE (1992) Inheritance of carbon isotope discrimination and water use efficiency in cowpea. Crop Sci 33:498–503

Johnson DA, Asay KH, Tieszen LL, Ehleringer JR, Jefferson PG (1990) Carbon isotope discrimination: potential in screening cool-season grasses for water-limited environments. Crop Sci 30:338–343

Jones HG (1997) New concepts in plant water relations: relevance to horticultural production. In: Proceedings of 2nd international symposium on irrigation of horticultural crops, Chartzoulakis KS (ed) Acta Hort 449:371–378

Jones HG (2004) What is water use efficiency? In: Bacon MA (ed) Water use efficiency in plant biology. Blackwell Publishing-CRC Press, Boca Raton, pp 27–41

Kays WM (1966) Convective heat and mass transfer. McGraw-Hill, New York

Keeling CD, Mook WG, Tans PP (1979) Recent trends in the $^{13}C/^{12}C$ ratio of atmospheric carbon dioxide. Nature 277:121–123

Kirda C, Mohamed ARAG, Kumarasinghe KS, Montenegro A, Zapata F (1992) Carbon isotope discrimination at vegetative stage as an indicator of yield and water use efficiency of spring wheat (*Triticum turgidum* L. var durum). Plant Soil 147:217–223

Kortschak HP, Hartt CE, Burr GO (1965) Carbon dioxide fixation in sugarcane leaves. Plant Physiol 40:209–213

Krall JP, Edwards GE (1990) Quantum yields of photosystem II electron transport and carbon dioxide fixation in C_4 plants. Aust J Plant Physiol 17:579–588

Loss SP, Siddique KHM (1994) Morphological and physiological traits associated with wheat yield increases in Mediterranean environments. Adv Agron 52:229–276

Madhavan S, Treichel I, O'Leary MH (1991) Effects of relative humidity on carbon isotope fractionation in plants. Botan Acta 104:292–294

Marlier J, O'Leary MH (1984) Carbon kinetic isotope effects on the hydration of carbon dioxide and dehydration of bicarbonate ion. J Am Chem Soc 106:5054–5057

Martin B, Thorstenson YR (1988) Stable carbon isotope composition ($\delta^{13}C$), water use efficiency, and biomass productivity of *Lycopersicon esculentum*, *Lycopersicon pennellii*, and the F_1 hybrid. Plant Physiol 88:213–217

Meinzer FC, Zhu J (1998) Nitrogen stress reduces the efficiency of the C_4 CO_2 concentrating system, and therefore quantum yield in *Saccharum* (sugarcane) species. J Exp Biol 49:1227–1234

Meinzer FC, Plaut Z, Saliendra NZ (1994) Carbon isotope discrimination, gas exchange, and growth of sugarcane cultivars under salinity. Plant Physiol 104:521–526

Merah O, Deléens E, Monneveux P (1999) Grain yield, carbon isotope discrimination, mineral and silicon content in durum wheat under different precipitation regimes. Physiol Plant 107:387–394

Merah O, Deléens E, Souyris I, Monneveux P (2001a) Ash content might predict carbon isotope discrimination and grain yield in durum wheat. New Phytol 149:275–282

Merah O, Deléens E, Souyris I, Nachit M, Monneveux P (2001b) Stability of carbon isotope discrimination and grain yield in durum wheat. Crop Sci 41:677–681

Merah O, Deléens E, Teulat B, Monneveux P (2001c) Productivity and carbon isotope discrimination in durum wheat organs under a Mediterranean climate. Crit Rev Acad Sci Paris 324:51–57

Misra SC, Randive R, Rao VS, Sheshshayee MS, Serraj R, Monneveux P (2006) Relationship between carbon isotope discrimination, ash content and grain yield in wheat in the Peninsular Zone of India. J Agron Crop Sci 192:352–362

Monneveux P, Reynolds MP, Trethowan R, González-Santoyo H, Peña RJ, Zapata F (2005) Relationship between grain yield and carbon isotope discrimination in bread wheat under four water regimes. Eur J Agron 22:231–242

Monneveux P, Sheshshayee MS, Akhter J, Ribaut JM (2007) Using carbon isotope discrimination to select maize (*Zea mays* L) inbred lines and hybrids for drought tolerance. Plant Sci 173:390–396

Mook WG, Bommerson JC, Staverman WH (1974) Carbon isotope fractionation between dissolved bicarbonate and gaseous carbon dioxide. Earth Planet Sci Lett 22:169–176

Morgan JA, LeCain DR (1991) Leaf gas exchange and related leaf traits among 15 winter wheat genotypes. Crop Sci 31:443–448

Morgan JA, Zerbi G, Martin M, Mujahid MY, Quick JS (1993) Carbon isotope discrimination and productivity in winter wheat. J Agron Crop Sci 171:289–297

Murphey BF, Nier AO (1941) Variations in the relative abundance of the carbon isotopes. Phys Rev 59:771–772

Nier AO, Gulbransen EA (1939) Variations in the relative abundance of the carbon isotopes. J Am Chem Soc 61:697–698

O'Leary MH (1981) Carbon fractionation in plants. Phytochemistry 20:553 567

O'Leary MH (1984) Measurement of the isotope fractionations associated with diffusion of carbon dioxide in aqueous solutions. J Phys Chem 88:823–825

O'Leary MH (1988) Carbon isotopes in photosynthesis. Bioscience 38:328–335

Ohsugi R, Samejima M, Chonan N, Murata T (1988) $\delta^{13}C$ Values and the occurrence of suberized lamellae in some *Panicum* species. Ann Bot 62:53–59

Osmond CB, Bender MM, Burris RH (1976) Pathways of CO_2 fixation in the CAM plant *Kalanchoe daigremontiana*: III Correlation with $\delta^{13}C$ value during growth and water stress. Aust J Plant Physiol 3:787–799

Paneth P, O'Leary MH (1985) Carbon isotope effect on dehydration of bicarbonate ion catalyzed by carbonic anhydrase. Biochemistry 24:5143–5147

Park R, Epstein S (1961) Metabolic fractionation of C_{13} and C_{12} in plants. Plant Physiol 36:133–138

Raes D, Geerts S, Kipkorir E, Wellens J, Sahli A (2006a) Simulation of yield decline as a result of water stress with a robust soil water balance model. Agric Water Manage 81:335–357

Rajaram S, van Ginkel M, Fischer RA (1995) CIMMYT's wheat breeding mega-environments In: Li ZS, Wix ZY (eds) Proceedings of the 8th international wheat genetics symposium, Beijing, China, pp 1101–1106

Raven JA, Farquhar GD (1990) The influence of N metabolism and organic acid synthesis on the natural abundance of C isotopes in plants. New Phytol 116:505–529

Rebetzke GJ, Richards RA, Condon AG, Farquhar GD (2006) Inheritance of carbon isotope discrimination in bread wheat (*Triticum aestivum* L.). Euphytica 150:97–106

Ritchie SW, Nguyen HT, Holaday AS (1990) Leaf water content and gas exchange parameters of two wheat genotypes differing in drought resistance. Crop Sci 30:105–111

Roeske CA, O'Leary MH (1984) Carbon isotope effects on the enzyme-catalyzed carboxylation of ribulose bisphosphate. Biochemistry 23:6275–6284

Roeske CA, O'Leary MH (1985) Carbon isotope effect on carboxylation of ribulose bisphosphate catalyzed by ribulose bisphosphate carboxylase from *Rhodospirillum rubrum*. Biochemistry 24(7):1603–1607

Rooney MA (1988) Short-term carbon isotope fractionation by plants. Ph.D. thesis, University of Wisconsin

Rundel PW, Rundel JA, Ziegler H, Stichler W (1979) Carbon isotope ratios of central Mexican Crassulaceae in natural and greenhouse environments. Oecologia 38:45–50

Saliendra NZ, Meinzer FC, Perry M, Thom M (1996) Associations between partitioning of carboxylase activity and bundle sheath leakiness to CO_2, carbon isotope discrimination, photosynthesis, and growth in sugarcane. J Exp Bot 47:907–914

Samejima M (1984) Intraspecific variation of ^{13}C discrimination in C_3 and C_4 species. In: Genetical approach in photosynthetic research of higher plants Bulletin of Green Energy Program Group II No 5, Ministry of Agriculture, Forestry and Fisheries, Tokyo, pp 70–80 [In Japanese]

Sandquist DR, Ehleringer JR (1995) Carbon isotope discrimination in the C_4 shrub *Atriplex confertifolia* along a salinity gradient. Great Basin Nat 55:135–141

Sayre KD, Acevedo E, Austin RB (1995) Carbon isotope discrimination and grain yield for three bread wheat germplasm groups grown at different levels of water stress. Field Crops Res 41:45–54

Smith BN, Epstein S (1971) Two categories of $^{13}C/^{12}C$ ratios for higher plants. Plant Physiol 47:380–384

Troughton JH (1979) $\delta^{13}C$ as an indicator of carboxylation reactions. In: Gibbs M, Latzko E (eds) Photosynthesis II Photosynthetic carbon metabolism and related processes. Springer, Berlin/Heidelberg/New York

Tsialtas JT, Tokatlidis IS, Tamoutsidis E, Xynias IN (2001) Grain carbon isotope discrimination and ash content of cv Nestos bread wheat selected for high and low yield in absence of competition. Cereal Res Commun 29:391–396

UK Meteorological Office (2006) Climate change. 2006 Report. McCarthy MP, Willett K (eds) http://enquire.hertscc.gov.uk/qol/2006/climatechange06.cfm

UN Water (2007) Coping with water scarcity: a strategic issue and priority for system-wide action UN Water first annual report 07 http://www.unwater.org/flashindex.html

Urey HC (1948) Oxygen isotopes in nature and in the laboratory. Science 108(2813):602–603

Vitousek PM, Field CB, Matson PA (1990) Variation in foliar $\delta^{13}C$ values in Hawaiian *Metrosideros polymorpha*: a case of internal resistance? Oecologia 84:362–370

Vogel JC (1980) Fractionation of the carbon isotopes during photosynthesis. Springer, Heidelberg/New York, pp 1–29

Wallace JS, Allen SJ, Gas JHC, Holwill J, Lloy CR (1993) Components of the energy and water balance at the HAPEX-Sahel southern super-site. IAHS 212:365–371

Wang HX, Liu CM, Zhang L (2002) Water-saving agriculture in China: an overview. Adv Agron 75:135–171

Weigl J, Calvin M (1948) An isotope effect in photosynthesis. J Chem Phys 17:210

Whelan T, Sackett WM, Benedict CR (1973) Enzymatic fractionation of carbon isotopes by phosphoenolpyruvate carboxylase from C_4 plants. Plant Physiol 51:1051–1054

Wickman FE (1952) Variations in the relative abundance of the carbon isotopes in plants. Geochim Cosmochim Acta 2:243–254

Williams DG, Gempko V, Fravolini A, Leavitt SW, Wall GW, Kimball BA, Pinter PJ, LaMorte R, Ottman M (2001) Carbon isotope discrimination by *Sorghum bicolor* under CO_2 enrichment and drought. New Phytol 150:285–293

Xu X, Yuan H, Li S, Trethowan R, Monneveux P (2007) Relationship between carbon isotope discrimination and grain yield in spring wheat cultivated under different water regimes. J Integ Plant Biol 49:1497–1507

Chapter 16
Behaviour of Plant Pathogens for Crops Under Stress During the Determination of Physiological, Biochemical, and Molecular Approaches for Salt Stress Tolerance

Murat Dikilitas and Sema Karakas

Contents

Abstract Agricultural technology is the only way to meet the nutritional needs of the world's population. In the last century, crop production has significantly increased and has reached its plateau, coincident with the increase in world population. However, a large percentage of the human population still does not have enough food, therefore many are underfed and face malnourishment. The world population will continue to grow along with poverty, environmental health concerns, issues surrounding the availability of clean water sources, etc. By the middle of this century it may reach 10 billion. Therefore, scientists need to try to keep the relationship between crop production and population on an upward slope. To fully understand trends and to determine increases in crop production, many marginal areas need to be included into agricultural lands. If environmental pollution, drought, salinity, disease, and insect problems, as well as the use of irrigation water of marginal

M. Dikilitas (✉)
Department of Plant Protection, Faculty of Agriculture, Harran University, S. Urfa, Turkey
e-mail: m.dikilitas@gmail.com

S. Karakas
Department of Soil Science, Faculty of Agriculture, Harran University, S. Urfa, Turkey

M. Ashraf et al. (eds.), *Crop Production for Agricultural Improvement*,
DOI 10.1007/978-94-007-4116-4_16, © Springer Science+Business Media B.V. 2012

quality, are taken into account along with newly added low fertile agricultural lands, many crop plants, for a major part of their growth periods, are grown under adverse conditions. Therefore, now and in the future, scientists have to find physiological, biochemical, and molecular approaches to overcome negative production issues. However, the behaviour and characteristics of plant pathogens have also changed. In this chapter, recent developments in agriculture regarding the production of crops under stress conditions and the behaviour of plant pathogens are evaluated and discussed.

Keywords Plant-salt interactions • Pathogen-salt interactions • Abiotic stress tolerance • Microorganisms • Crop production • Crop improvement

Abbreviations

CMC	Carboxymethylcellulase
FOV	*Fusarium oxysporum* f.sp. *vasinfectum*
GA	Gibberellic Acid
GM	Genetically Modified
IAA	Indole Acetic Acid
NAR	Net Assimilation Rate
PAL	Phenylalanine Lyase
PG	Polygalacturanase
PL	Pectate Lyase
QTL	Quantitative Trait Locus
RAPD	Random Amplification of Polymorphic DNA.
RFLP	Restriction Fragment Length Polymorphism
RGR	Relative Growth Rate
VW	Verticillium Wilt

1 Introduction

The production of safe and sufficient amount of food of high quality in sustainable agricultural systems is a new trend in agriculture. For achieving high quality food goals, new biochemical, molecular, and physiological treatments, combined under the name of biotechnology, are the most promising approaches. In many countries around the world, organizations operating under various names such as Agriculture 2020 or 2020-European Vision for Plant Science, etc., have been working on the current major challenges in biology and plant sciences by investigating how plants of the future can provide solutions to numerous problems such as increasing demand for food, competition for land use, climate change, etc. (Cavalieri 2005; Dikilitas and Karakas 2011; www.arabidopsis.org). Ameliorating these types of issues will partly be dependent on increasing our knowledge of plant biology and environmental sciences.

Over the last century, advances in plant sciences have been spectacular. However, in the next 20 years, in order to obtain more tolerant plant species, information regarding biochemical and molecular processes should be converted into predictive modelling. To achieve this goal, many new technologies and analytical techniques should be developed. Since abiotic factors such as cold, drought, and salt interact with biotic stress factors such as pathogens, insects, and beneficial microorganisms, their complex metabolisms need to be fully understood. The salt, heat, drought, and cold tolerant plants could play important roles in areas where crops are lost, and could be important for increasing crop yields (Ashraf and Foolad 2007). In the past, agricultural improvements, through the use of traditional breeding methods, managed to keep pace with increasing food demands. Using traditional methods, plant breeders developed new strains of crops that significantly increased yields by 3–6% a year, with method that required many years to bring together desirable and to eliminate undesirable traits (Abumhadi and Atanassov 2010). Over the next decade, scientists will use genomics and other biochemical approaches (Atanassov 2009).

Plant biotechnology offers significant improvements in every area of crop production. Recent advances in plant molecular biology have allowed us to discover and to isolate important genes that regulate yields and that tolerate environmental stress (Abumhadi and Atanassov 2010). Intensive studies in plant biotechnology began in the 1980s with the use of *Agrobacterium* (Bevan et al. 1983) and these studies are still carried out using the same organism (Harfouche et al. 2011). Recent biotechnological advances in direct gene transfer have helped engineers to develop more resistant plants with new traits from unrelated organisms such as fungi, viruses, or bacteria that are otherwise very difficult to transfer using classical breeding programs (Abumhadi and Atanassov 2010; Jauhar, 2006). Plant biotechnology using molecular markers, genomics, and proteomics are new and promising tools for improving crop plants. For example, tree genetic engineering has advanced to the point at which genes for desirable traits can now be introduced and expressed efficiently; examples include biotic and abiotic stress tolerance, improved wood properties, root formation, and phytoremediation (Harfouche et al. 2011). Also, through genetic engineering, the biological components of plants can be maintained and crop production can be increased when chemicals or irrigation water or nutrients are minimized (Sridhara 2006; Fageria et al. 2008). However, stable gene expressions are key issues that need to be resolved before transgenic plants can be used commercially.

Diseases may have an impact on crop loss, and result in high costs associated with their control through protective chemicals and drug resistance. Various high-tech approaches have been proposed to protect plants from such type of stresses. Virus resistant transgenic plants, especially, have gained attention, and fungi-, bacteria-, and nematode-resistant plants are currently in the development stage. Very few reports exist regarding the loss of tolerance or pathogenicity after a period of time in living organisms (Krokene and Solheim 2001). Engineering plants to contain increased disease resistances or stress tolerances that are durable can be a challenge for biotechnologists. Despite much effort, most attempts using genetic engineering to develop crops that are resistant to fungal and bacterial diseases or salt tolerances have been unsuccessful in the long term. The most common reason for this is that the success of resistance (*R*) genes relies upon recognition by the encoded *R* protein of a specific avirulence (Avr) protein in the pathogen. Therefore, given the gene-for-gene relationship underlying this type of resistance, *R* genes are extremely vulnerable to a single loss-of-function mutation in the corresponding pathogen *Avr* gene, which leads to a loss of resistance (Rougon-Cardoso and Zipfel 2010).

In this chapter we attempt to underline important issues regarding the stress mediated aspects using new approaches and methods. We will emphasize the pathological impact of microorganisms such as fungi or bacteria under stress conditions, whether they evolve during the stress period and adapt themselves to conditions of crop plants, are suppressed by stress and cannot tolerate stress as crop plants, or can surpass defence barriers without exerting remarkable force. Attempt is made here to understand the possible behaviours of plant pathogens, as well as recent advances in crop production for crops that grow under stress conditions.

2 Improvements in Salinity and Disease Tolerance in Crop Plants

Salinity affects over 6% of the world's land and 20% of the world's irrigated land, posing a major threat to our future (http://www.fao.org). Increased incidences of salinity and disease stress affecting the productivity in principal crops has been common in arid and semi-arid areas. Therefore, diseases and other abiotic stress factors are interconnected in saline soils, and their combined impacts, are very complex. Strategies for obtaining maximum crop yields under stressed conditions include the following: (1) the breeding of new crop varieties, (2) the screening and the selection of existing germplasm in potential crops, (3) the production of GM crops, and (4) the exogenous use of osmoprotectants (Fageria et al. 2008; Athar and Ashraf 2009). Since conventional breeding programs are limited, molecular biology techniques have been introduced for inducing stress tolerance in economically important crops. As for other stress tolerance mechanisms, molecular biology techniques, aimed at crop improvements in salt tolerant species, have followed common biochemical and physiological pathways. However, due to toxic and osmotic effects, such as selective ion accumulation or exclusion, the control of sodium uptake and its distribution within plants, and the compartmentation of ions at the cellular or the entire plant level are important issues that should be considered (Munns and Tester 2008; Athar and Ashraf 2009).

In the first studies on molecular approaches salt tolerant genes controlling salt uptake and transport, or regulating osmotic functions in plants were considered first. However, disease resistance mechanisms and crop improvements for disease resistance were considered separately. Understanding stress metabolism in detail is crucially important since many stresses are involved and such a situation could possibly become more complex due to the involvement of biotic stress factors in physiologically and biochemically improved salt tolerant plants (Dikilitas 2003). Therefore, it is important to identify key genes and their functions at the cellular as well as at the entire plant level using advancements in functional genomics (Athar and Ashraf 2009). However, when developing transgenic plants for salt stress tolerance, disease incidence should be considered. At present, in order to increase plant defence mechanisms and salt tolerances under adverse conditions, it is logical and appropriate to apply exogenous chemicals, plant growth regulators, or compatible solutes or microorganisms. As a result of its complexity, salinity tolerance is difficult to measure. Not only are there a number of genes controlling salinity tolerance whose impacts interact with environmental conditions, but there are also two major and distinct components of salinity tolerance that are often difficult to distinguish (Munns 2002). Screening based on growth needs to allow for two distinct mechanisms for salinity tolerance are as follows: (1) tolerance to the osmotic effect of the saline soil solution, and (2) tolerance to the salt-specific nature of the saline solution.

Significant technological advances in the field of molecular biology have been made during the past decade. For achieving salinity tolerance, molecular techniques designed to selectively introduce desired genes may provide alternative means to classical plant breeding. These techniques, based on specific traits that are controlled by one gene (e.g. a transcription factor or an important ion channel), will benefit the development of salinity-tolerant cultivars. Blumwald et al. (2000) reported the development of a salinity tolerant transgenic tomato plant for which the over-expression of the vacuolar Na^+/H^+ antiporter showed dramatic improvement in vegetative growth and fruit yield. The antiporter is the only known transporter that compartmentalizes Na^+ within the vacuole, and the extrusion of Na^+ from the cell, whereas Na^+ has little chance of toxic effects on metabolism, or transportation to younger leaves and fruits. Such studies indicate a great potential for transgenic approaches. The use of molecular markers in breeding programs is increasing rapidly. Marker-assisted selection is non-destructive and can provide information on the genotype of a single plant without exposing the plant to stress. The technology is capable of handling a large number of samples. In order to use marker-assisted selection in breeding programs, markers must be closely linked to the trait and must operate across different genetic backgrounds. In recent years the efficiency of genetic mapping has also greatly improved with the advent of high-density maps that incorporate microsatellite markers, which are overtaking tedious RFLP markers and unreliable RAPD markers. Molecular approaches for transferring genes from one plant to another have also gained popularity. Munns et al. (2003) reported that two major genes control Na^+ uptake in durum wheat. One gene retrieves Na^+ from the xylem in roots, one enhances K^+ loading in the xylem, while another retrieves Na^+ from the xylem in both the root and the shoot (James et al. 2006). Details and comprehensive information regarding crop productivity under saline conditions can be obtained from the work of Munns (2009), and Athar and Ashraf (2009). For plant disease resistance, detailed information can be obtained from the work of Knox and Clarke (2007). However, major improvements for crop production generally involve either improvements for disease resistance or abiotic stress tolerance and these two stress factors should be handled together while crop production improvements are considered.

3 Impact of Plant Pathogens Under Salt Stress Conditions

Environmental factors such as drought, excess irrigation, and high temperatures interact with microorganisms in the air or in the soil fauna. Therefore, these factors may increase or decrease the virulence of pathogens. In a similar manner, salinity may also interact with fungi in soil fauna and may reduce or increase the effectiveness of fungi. Since many semi-arid and arid areas of the world are characterized by salinity, such a case cannot be inevitable. In the past, researchers have directed their attention either to the relationship between plants and abiotic stresses, or between plants and pathogens. Relationships also exist between abiotic stresses such as drought, heat, salinity, etc., and microorganisms. Therefore, a newer concept is the interaction between plants, pathogens, and abiotic stresses. The exposure of plants to salt stress generally begins with the exposure of roots to stress. Salt stress leads to changes in the growth, morphology, and physiology of roots that, in turn, change water and ion uptake, and the production of signals (hormones) that communicate information to the shoot. The entire plant is then affected when roots are grown in a salt medium (Grover et al. 2011). Since any plant pathogen living under these conditions may be a potential danger to crops, both stress factors should be considered together in relation to crop improvement.

Improving salt tolerance in plants becomes more complex as plants are exposed to additional environmental and biotic stress factors. Interactions of heat and drought, water logging, heavy metal toxicity, and biotic stresses have been studied (Singh et al. 1997; Obuekwe et al. 2005; Jin et al. 2010). The stress responses of many plants have been identified as multigenic traits. The largest threat to molecular and biochemical techniques that aim to obtain resistant or tolerant crops under stress conditions is adaptation of their biotic enemies, that eventually reduce crop production. Additionally, more virulent races of pathogens can possibly emerge and additional destructive consequences can be inevitable.

In general, due to the presence of other environmental factors, plant responses, in addition to salt stress, could interact with biotic stress factors (Dikilitas 2003). In nature, such environmental complexities are a usual occurrence. Both abiotic stress factors such as drought and salinity, or heat and salinity, or heavy metal toxicity and salinity lead to osmotic stress in plants and cause many similar metabolic responses, although salinity may induce additional responses (Singh et al. 1997; Dikilitas 2003; Dimartino et al. 2011). Further progressive symptoms of these stresses can make plants more prone to additional abiotic and biotic stress factors. Salt stresses have not only been studied in semi-arid and arid areas, but also in heavy metal contaminated soils. Plants near industrial vicinities, especially in developing countries, suffer from both stressors. For example, an additive effect of NaCl (6 EC) and lead acetate (1.0 mM) on biomass accumulation has been observed in *Vigna radiata* (L.) Wilczek cv. Pusa baisakhi (Singh 1995; Xu et al. 2010).

Plants growing in the saline environments of arid zones may suffer from high salt stress and high temperature during summer and low temperatures during winter. Under these conditions, plants show an accumulation of low molecular organic osmolytes, changes in polyamine levels, the synthesis and release of ethylene, and other types of metabolite adaptations to high salt and high temperatures. In many studies, heat, high temperature, drought, and salt have resulted in the increased accumulation of carbohydrates such as sucrose, sorbitol, mannitol, glycerol, and nitrogenous compounds such as proline, putrescine, betaine, glycine, choline, and organic acids (Ramagopal 1993; Gamalero et al. 2009); and have indicated the existence of a common defense system that may operate in two or multiple stress systems. Although combined stress factors are responsible for a high loss of agricultural productivity, the development of crops resistant to combined stress factors have, so far, resulted in limited success due to the difficulties of multigene transfer. Most studies have concentrated on the cloning of genes that are responsible for the synthesis of enzymes, leading to the accumulation of osmolytes. However, the regulation of genes and gene products responsible for the responses have not been extensively studied (Ramagopal 1993). Interactions between two kinds of stresses may cause significant alterations in normal physiological processes in all plant organisms, including economically important crops, and may alter the regulation of common transport systems and metabolic events in the cell. Eventually the induction of stress adaptation mechanisms for the survival of stressed plants occurs. However, a wide range of adaptations are required to cope with stress factors.

The adaptations of microorganisms, as well as crop plants, to stress are a complex multilevel regulatory process in which many genes may be involved (Srivastava et al. 2008). In certain species such as halophilic or thermophilic organisms living in adverse conditions, enzymatic activities are normally produced without any stress symptoms (Madigen 1999; Palmero et al. 2010). However, many microorganisms develop different adaptation mechanisms in order to combat the stress (Grover et al. 2011). The susceptibility of plants to insects and diseases may be affected by the presence of nutrients in the soil. For example, salinity induces several metabolic changes in diseased plants, such as accumulation of proline and glycine betaine, and antioxidant enzymes (Bray et al. 1991; Jbir et al. 2001; Dikilitas 2003). Under the combined stress of salinity, and with the wilt pathogen *Verticillium albo-atrum*, metabolic changes in lucerne plants, such as the accumulation of proline and antioxidant enzymes, were significantly effected and their accumulation drastically and unexpectedly decreased (Dikilitas 2003). These findings clearly indicate that the pathogen managed to survive and to infect the host plant after a period of adaptation. Therefore, genetic manipulation of improved crop cultivars for salt tolerance should be coupled with improved disease resistance if we really want to improve crop production under such conditions. The possible interactions between plants and microbes and their possible behaviours under stress conditions are presented in Fig. 16.1.

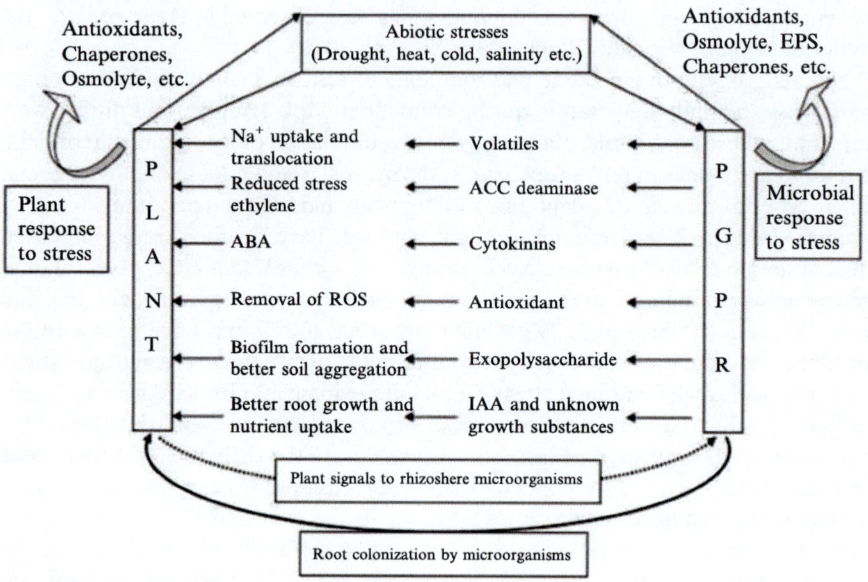

Fig. 16.1 A conceptual diagram of plant-microbe interactions under abiotic stress (Grover et al. 2011)

3.1 *Interactive Effects of Salinity and Plant Diseases on Crop Productivity*

The irrigation waters rich in salts in nature effect the distribution and severity of diseases. One of the earliest studies regarding interactions between salt and pathogens was made by Holmes (1976) on maple trees; where young trees were badly affected by high NaCl stress while low salt concentration alone only induced a foliar scorch. However, with the combined effect of NaCl and *V. albo-atrum*, the limitation of crop production was more pronounced. Similar reports were published by Nachmias et al. (1993, 1994) on potato plants irrigated with a salt solution ($NaCl:CaCl_2$; 4:1) equivalent to 5 mmhos cm^{-1}. Drought stress was also induced by reducing irrigation to 25%. The treatments increased symptom severity and eventually reduced crop production, while increasing pathogen growth in the vascular system. Again, Orenstein et al. (1995) simulated field conditions in potato crop plants by irrigating with a salt solution equivalent to 10.2 mmhos cm^{-1} under a drought stress induced by 10% polyethylene glycol (PEG). Potato growth was significantly reduced while the conidi forming unit (c.f.u) count of *V. dahliae* increased when a comparison was made with that of the control. Similarly, irrigation with saline water (EC=3.2) resulted in a build up of a *Fusarium oxysporum* f.sp. *radicis-lycopersici* population at the root zone of tomatoes in soil (Triky-Dotan et al. 2005). Howell et al. (1994) reported that *V. albo-atrum* inoculation alone; on disease resistant lucerne plants (cv NK-89786); had a minimal impact on foliage yield. However, when the resistant cultivar was treated with salt at 3, 5, and 7.6 dS m^{-1}, the yield was reduced when compared with non-inoculated controls. *In vitro* studies on *V. dahliae*, cultured on potato-based solid or liquid media containing 0.05–0.35 M NaCl, showed that the growth of the fungus was depressed with an increase in the NaCl concentration, but that the production and behaviour of the conidia were not impaired, and conidial differentiation and germination ability gradually increased as the concentration of salt increased (Danti and Broggio 1997; Goudarzi et al. 2008). High soil salinity or saline irrigation has also been reported to increase the severity of stem rot caused by *Phytophthora citrophthora* in citrus roots (Sulistyowati and Keane 1992; Grattan and Grieve 1999; Syvertsen and Levy 2005). Although it did not stimulate growth of the pathogen *in vitro*, increases in disease under saline conditions were attributed to a direct effect in reducing resistance in the host. Besri (1990) and Triky-Dotan et al. (2005) also reported that high salt stress in irrigation water resulted in the breakdown of resistance in tomato cultivars. Similar results on the effect of salinity stress on development of plant diseases have been reported for phytophthora root rot in chrysanthemum and citrus (Blaker and MacDonald 1986; MacDonald, 1984); for pythium blight in penncross creeping bentgrass (*Agrostis palustris* Huds. 'Penncross') (Rasmussen and Stanghellini 1988); for olive trees infected with *Verticillium dahliae* (Levin et al. 2003); and for pistachio trees infected with *V. dahliae* (Mohammadi et al. 2007). Increases in disease severity under saline conditions have also sometimes been attributed to an increase in the virulence of a pathogen (Ragazzi et al. 1994, 1996). For example, salinity increased the development of the motile zoosporangia *Perkinsus* sp., and developmental rates increased with increasing salinity (Ahn and Kim 2001). Similar findings were made by Ragazzi et al. (1994, 1996), who reported that *Fusarium oxysporum* f.sp. *vasinfectum* chlamydospores and their germ tubes grew better in a saline medium and that the pathogen was more virulent when it had been cultured on a saline-enriched medium.

The results of Dikilitas (2003) showed that the growth and sporulation of *V. albo-atrum* isolates, from lucerne (V1-strong pathogen) and tomato (V2-weak pathogen), decreased after the threshold level of 150 mM NaCl. Under non-saline conditions, the fastest growing isolate V2 was the most affected isolate under saline conditions, especially when accounting for sporulation. Above 150 mM NaCl fungal isolates still continued to grow and to produce an ample amount of spores that were capable of infecting host plants. It has been suggested that the highly virulent isolate (V1), showing a high salt tolerance, also has a high adaptation capacity in relation to salinity, since the appearance of the colonies of both isolates, especially the V1 isolate, became more feathery with increasing NaCl concentrations (Dikilitas 2003). At low osmotic potentials, the morphological switch to feathery patterns in some mycelial fungi has been attributed to their ability to adapt to drought conditions (Griffith and Boddy 1991; Llamas et al. 2008). For example, Hasan (2002) stated that the mycelial growth of *F. oxysporum* significantly increased with 1–10% (0.17 – 1.70 M) concentrations of NaCl, although the production of GA (Gibberellic acid) and IAA (Indole acetic acid) by the fungus decreased above 1% NaCl. On the other hand, Amir et al. (1996) reported that concentrations of NaCl over 1% (170 mM) negatively influenced the radial growth of *F. oxysporum* f. sp. *lini*. In general, a narrower range of environmental conditions are required for sporulation than for mycelial growth. Gao and Shain (1995) reported that the conidial germination of *Cryphonectria parasitica* (Murr.), chestnut blight fungus, was more sensitive than mycelial growth to NaCl at a –2.0 MPa osmotic potential.

A study containing the salt-adapted wilt fungus *V. albo-atrum* showed that the growth of isolates of the non-salt-adapted (V1 and V2) and salt-adapted strains (V1-, V2-150; V1-, V2-200), continuously grown by sub-culturing (in their respective conditions) under NaCl stress for at least 8 months, showed similar growth patterns in Petri plates although the salt-adapted strains of V1-200 and V2-200 produced less spores (Dikilitas 2003). When the non-salt-adapted lines (V1 or V2) were cultured on 150 and 200 mM NaCl containing media, their growth was lower than those of the corresponding salt-adapted lines. Although salt-adaptation resulted in the ability to maintain better growth rates under elevated concentrations of NaCl than strains that had not been adapted, adaptation did coincide with a decrease in the ability to form spores, in all of the salt-adapted isolates the effect however, was more marked in V1- and V2-200 than in the corresponding -150 strains, which may impact the ability of *V. albo-atrum* to infect plants under saline conditions. The lower rate of sporulation under saline conditions may still be efficient enough to cause pathogenicity, and pathogen may eventually develop an adaptation mechanism. For example, findings of Llamas et al. (2008) have indicated that the marine strains of *F. solani* have a physiological mechanism that permits survival in environments with low water potential. Many fungal species, such as *Aspergillus*, *Penicillium*, *Rhizopus*, and *Fusarium*, produce increased amount of hormones, such as GA or IAA, or both, under salinity (0.5% and 1%, NaCl) stress, suggesting that an increase in the level of hormones may act as an adaptive response that maintains the stability of the fungus. However, at higher concentrations of NaCl (above 1%), the content of GA and IAA decreased over time.

Salinity not only exerts stress on fungal metabolism but also reduces the resistance of cultivars. Under such circumstances a lower germination rate for conidia may still result in pathogenicity. MacDonald (1984) reported that the root systems of hydroponically grown chrysanthemums exposed to salinity following inoculation with *Phytophthora cryptogea* developed more severe symptoms than plants that were not exposed to salinity. Also, the hyphae of *P. cryptogea* rapidly colonized salinity-stressed roots causing extensive necrosis after 12 h of inoculation, indicating a salt-induced change from resistant to a highly susceptible condition. Similarly, Turco et al. (2002) reported that a nutrient solution with an EC of 20 mS cm^{-1} increased the severity of *Fusarium* wilt by approximately 34% in the cotton cultivars 'Coker 304' and 'Acala SJ2', as compared to the non-saline test group. Recently, Al-Sadi et al. (2010) reported that increasing irrigation-water salinity from 0.01 to 5.0 dS m^{-1} significantly increased mortality in cucumber seedlings inoculated with *Pythium. aphanidermatum. In vitro* tests in culture media amended with various concentrations of NaCl also showed that the growth of *P. aphanidermatum*, *P. spinosum*, and *P. splendens* isolates were either stimulated or unaffected at salinity levels that were stressful for cucumber (electrical conductivity = 5 dS m^{-1}). The isolates of

P. aphanidermatum from greenhouses with no salinity problems were as tolerant to salinity as the isolates obtained from salinity-affected greenhouses, suggesting that increased mortality in cucumber seedlings at higher salinity levels may imply a synergistic interaction between salinity stress and salinity tolerant *Pythium* species in cucumber seedlings, resulting in greater seedling losses.

A reduction in the growth of plants under saline conditions can largely be attributed to reduced water absorption caused by reduced water potential in the root environment. Under saline conditions, Na^+ and Cl^- ions are taken up at high rates, and may lead to excessive accumulation in tissues, which may inhibit the uptake of other ions into the root and their transportation to the shoot (Mer et al. 2000). As reported by Rasmussen and Stanghellini (1988) (who stated that increased salinity levels predisposed Penncross creeping bentgrass to cotton blight caused by *Pythium aphanidermatum* and accelerate the onset and development of disease) may predispose the condition of crop plants. The combined effect of both salt (even low concentrations) and the pathogen may cause more serious problems than those of the salt or the pathogen alone. For such conditions, additional energy may be required in order to cope with disease stress (Dikilitas et al. 2009). The resistance of plants to a pathogen may depend on the speed and the extent of synthesis of the enzymes induced in the host by the pathogen. PAL (Phenylalanine ammonia lyase) is a key enzyme in the production of the basic molecule used for the biosynthesis of most phenolics, including phytoalexins and lignin (Tang 2001; Neves et al. 2010). The accumulation of PAL in response to *V. albo-atrum*, isolate V2, and/or NaCl was studied in cell suspension cultures of lucerne cv. Kabul. Isolate V2, or NaCl, or both, were effective in inducing PAL accumulation in cell cultures *in vitro* (22 °C). The combined effect of NaCl (50 mM) and of the elicitor, derived from the isolate V2 of *V. albo-atrum* (0.05 or 0.1 mg ml^{-1}) resulted in a further increase in PAL activity. However, PAL activity was inversely correlated with a further increase in the concentration of NaCl (200 mM) and the elicitor (0.1 mg ml^{-1}). Similarly, Jbir et al. (2001) reported that salt stress increased PAL activity in wheat plants. On the other hand, Dunn et al. (1998) reported that after 30 days of high salinity (0.1 M NaCl), citrus plants grew more slowly and produced a lower PAL activity and, as a result, became more susceptible to nematode (*Tylenchulus semipenetrans*) attacks. The recovery of tomato plants from the effects of *V. albo-atrum* or salt was quicker and earlier than the combined effect of salt and fungus. Therefore, the results suggest that *V. albo-atrum* was still pathogenic under saline conditions (50 mM) and delayed the recovery of plants, regardless of their age. This is in good agreement with the results of Swiecki and MacDonald (1991) who reported that the exposure of tomato plants (*L. esculentum* Mill.) to salinity stress, either before or after inoculation with *Phytophthora parasitica*, increased root and crown rot severity relative to non-stressed controls. The synergy between salinity and *P. parasitica* was most pronounced on young (pre-bloom) plants and less pronounced on older (post-bloom) plants. For example, the interaction between salinity and *P. parasitica* significantly increased root necrosis, reduced top weight, and caused a higher incidence of crown infection than for the corresponding non-stressed plants. A well known outcome is that plants exposed to salinity show marginal Ca^{2+} deficiency (Kostandi and Soliman 1998;

El-Iklil et al. 2002). Under such conditions, the membrane becomes fragile, permitting the continuous out flow of assimilates (sugars and amino acids) that facilitate fungal growth (Hancock and Huisman 1981).

Although pathogenic fungi living under adverse conditions act as pathogens, they may still be pathogens when adverse conditions turn into optimal conditions for pathogen growth. Salt-adapted strains of V2 (V2-150 or V2-200) were still pathogenic to tomatoes even under non-saline conditions, suggesting that the fungus did not lose its pathogenicity over a period of time, even when sub-cultured under saline conditions (Dikilitas 2003). A tomato isolate of *Phytophthora. parasitica* has been recovered from saline soil and was found to be more tolerant to salinity than one recovered from non-saline soils (Bouchibi et al. 1990). Similarly, Ragazzi et al. (1994, 1996) reported that mycelia of *F. oxysporum* f. sp. *vasinfectum* (FOV) from non-saline medium and from saline-enriched medium both produced wilt symptoms. However, symptoms for a medium-resistant cotton, cv. GSC 20, appeared earlier and advanced rapidly with mycelium from the saline-enriched culture. Furthermore, Ragazzi and Vecchio (1992) reported that chlamydspore viability and the pathogen virulence of FOV was enhanced when sub-cultured on NaCl-enriched media. Turco et al. (1999) also reported that FOV had a greater enzymatic activity in saline environments, especially with regard to pectate lyase (PL) and polygalacturonase (PG) enzymes. However, high salt concentrations (85–250 mM) may strongly inhibit the enzymatic activities (cellulase, PL, and PG) of soil-borne fungi, as well as growth-promoting fungi, as is the case with *Cladosporium cladosporioides*, *F. moniliforme*, *F. oxysporum*, *F. solani*, *Rhizoctonia solani*, and *Macrophominia phaseolina* on the nodules of *Vicia faba* L (Omar and Abd-Alla 2000). The crop plants in areas where soil salinity is a problem could undergo an increase in susceptibility to pathogens and suffer a greater yield loss, as a result of combined stresses or increased pathogenicity due to the adaptation of the fungus.

Although plants may not show stress symptoms and may metabolize normally under low to moderate salinity levels, additional energy may be required to cope with disease, due to the stress caused by microorganisms such as fungi. For example, *Verticillium* wilt (VW) disease generally depends on the conditions in which plants are grown, and it has been reported that the irrigation regime may increase the incidence of wilt (Jefferson and Gossen 2002). If irrigation water is associated with salinity, detrimental effects on plants would be inevitable. In fact, *Verticillium* wilt has now become more prevalent in many areas where an increase in the salinity of irrigation water poses its greatest threat (Nachmias et al. 1993; Saadatmand et al. 2008). It is known that the reaction to infection by disease can be modified by environmental factors, the potential for interactions between salinity and disease is a realistic possibility that must be considered. It has been reported that abiotic stresses may increase symptoms by directly impacting the pathogen or its host (Dimartino et al. 2011). Similar results have been reported by Snapp and Shennan (1994) who found that a tomato cv. UC82B became vulnerable to infection by *Phytophthora. parasitica* when subjected to salt stress, and produced thinner roots and higher root-senescence rates when compared to the non-stress control group. Various researchers also report that soil salinity markedly increased the incidence of *Phytophthora* root rot in tomato and chili peppers under field and greenhouse conditions (Snapp et al. 1991; Sanogo 2004). The combination of salinity and enhanced disease severity may also lead to significant reductions in fresh fruit yields, fruit size, and total above-ground biomass. On the other hand, salinity decreases the resistance of crop plants against soil-borne saprophytes such as *Pythium ultimum.*

The effect of salinity on microorganisms is quite similar to that of crop plants. High salinity concentrations lower the production of mycelia and the sporulation of fungi, like reduced germination capacity and rooting ability of crop plants. A reduction in germination under saline conditions may be attributed both to the toxic and osmotic effects of NaCl. Chandler et al. (1994) reported that reduced osmotic potentials decrease germination and radial extension in *Verticillium* species. Similar findings have been recorded by McQuilken et al. (1992) and Goudarzi et al. (2008), who found that a decrease in osmotic potential caused a reduction in mycelial growth and the germination of oospores of *Pythium oligandrum* and *Macrophomina phaseolina* respectively. A decrease in the growth of *Phytophthora* sp. caused by salinity was also reported by Wilkens and Field (1993) and DiLeo et al. (2010). The lower sporulation rate resulting from higher concentrations of NaCl may affect the pathogenicity of the fungus, and another factor may also influence infectivity of the germination rate of those spores that are produced, since germination rate either decreases or increases depending on the concentration of NaCl.

It is not clear how salinity stress has such a pronounced influence on disease severity in plants, while having relatively less effect on non-inoculated plants. The mechanism is not fully understood, but may result from an impairment of normal host defence mechanisms. The plants respond to pathogenic invasion in numerous ways that function to block, slow, or prevent the pathogen from successful establishment or spread in the host's tissue. Many defence responses involve the biosynthesis of compounds that are toxic to invading pathogens. However, under salinity stress many defense responses and symptoms, including changes in membrane permeability, necrotic leaf margins, leaf drop, wilt, the ultra structure of organelles, and the synthesis of DNA, RNA, proteins, carbohydrates, and ultra low molecular weight antimicrobial substances, such as the accumulation of phytoalaexins, are severely affected (Jbir et al. 2001; Sanogo 2004). The host is predisposed to disease by the decreased availability of water and the accumulation of toxic ions in tissues is also a possibility. As a result, the plant may have lower photosynthetic activity and a lower daily growth (Brugnoli and Bjorkman 1992; Saadatmand et al. 2008). Inhibitions of these processes in dehydrated tissues have been reported to reduce plant resistance, through the inhibition of protein synthesis that may contribute to increased host susceptibility by preventing the synthesis of important enzymes and the low molecular weight of antimicrobial substances for disease resistance in the host (Turco et al. 2007). However, little work has been done on the enzyme synthesis for disease resistance under salinity or dehydrating conditions (Boyer 1995; Turco et al. 2007).

Under saline conditions fungi may become more aggressive. For example, El-Abyad et al. (1992) reported that, when compared to non-saline conditions, the activities of the cell wall enzymes (xylanase and galactanase) of *Sclerotium rolfsii* increased with increased salt concentrations. Dikilitas (2003) suggested that low levels of NaCl (50 mM) have a positive influence on fungal growth. Turco et al. (1999, 2002) determined that the virulence of the pathogen, surmised by the increased production of cell wall enzymes and conidia, increased under 50 mM NaCl. Under non-saline conditions, the impact of the fungus was more destructive than pathogen infection. Similarly, Regragui et al. (2003) stated that tomato plants of var Marmande became highly susceptible to infection by the virulent Moroccan isolate P80 of *Verticillium albo-atrum*, and was slightly susceptible to the avirulent isolate P(3) A of the same fungus. Repeated watering of infected plants with saline solution enriched by NaCl (80 mM) induces an aggravation of *Verticillium* symptoms caused by the aggressive isolate, and a gain of new pathogenic aptitude for the non-pathogenic isolate. The impact of salinity on the fungus developed on CMC (carboxymethylcellulase) medium produced a slight decrease in mycelial growth and a significant increase in CMC activity, *in vitro*. Dikilitas (2003) has reported that even non-pathogens may act as pathogens under saline conditions. For example, weak pathogenic isolates (VL, VS, and VF) of *V. albo-atrum* cause significant reductions in the relative growth rate (RGR), the net assimilation rate (NAR), and the height of lucerne plants under 50 mM NaCl (Dikilitas 2003).

3.2 The Response of Salt Tolerant Plants to the Effect of Plant Pathogens Under Normal or Salinity Stress

Determination of the degree of resistance of crop plants under saline conditions, as well as testing salt tolerant crop plants for resistance to diseases, is very important. Using this approach, we can know the threshold level of salt tolerance or disease resistance under combined stress. The breaking point of disease resistance or salt tolerance under combined stress is also possible. A study carried out on the pathogenicity of *V. albo-atrum*, isolate V1, on various disease-resistant and salt-tolerant plants (13R Supreme, Vertus, and Bilensoy-80); and disease susceptible and salt susceptible (Rambler) or disease susceptible and salt tolerant (R-350-N) cultivars of lucerne has revealed that cvs. 13R Supreme, Vertus, and Bilensoy-80 showed resistance to V1 under non-saline and saline conditions (50 mM NaCl), whereas Rambler and R-350-N both showed susceptibility under saline or non-saline conditions, as indicated by height, RGR, and NAR measurements, compared to corresponding controls. This result indicates that a low level of salinity is not enough to alter the resistance of disease-resistant and salt tolerant cvs. 13R Supreme, Vertus, and Bilensoy-80 to the pathogenicity of *V. albo-atrum*. However, when cultivars are tested under high salinity stress (100 mM NaCl) against *V. albo-atrum*, the resistance of cultivars is surpassed (Dikilitas 2003). Similarly, Besri (1990) reported that high salt levels in irrigation water in Morocco caused a total breakdown in the resistance of tomato cultivars that were normally resistant to race 1 of *V. dahliae*. He also reported that cultivars that were resistant to race 2 became susceptible with increasing soil salinity.

Regenerating salt tolerant plants through biochemical or biotechnological means is remarkable. However, when salt tolerant plants are regenerated, one should keep in mind that these plants can face other stress factors, in nature, other than salinity. If the stress factor is biotic, complex biochemical mechanisms may emerge. For example, biochemically and physiologically salt-tolerant improved lucerne *M. media* cv. Rambler displayed the highest degree of susceptibility to *V. albo-atrum*, under both normal and saline conditions, and the susceptibility of salt tolerant plants increased with an increasing level of tolerance to salinity. The salt tolerant-generated lucerne cultivar (*M. media* strain R-350-N); which thrives under 350 mM NaCl; showed a great susceptibility to the pathogen *V. albo-atrum* under normal and even low levels of salinity (Dikilitas 2003). The salt tolerant or disease resistant, or both salt tolerant and disease resistant, plants may lose their potential tolerance or resistance mechanisms to stress agents is important. However, it is critical to point out that under adverse conditions plant pathogens increase their virulence, while alone, they become more aggressive and produce more virulent strains and races in non-polluted environments. Also, disease-susceptible cultivars, whether they are salt tolerant or not, under low saline conditions, show a susceptibility to *V. albo-atrum*. Therefore, if we need to improve crop production under stress conditions, we definitely cannot ignore the weakest side of crop plants with regard to the potential danger of such stress. Although much work has been done in the past regarding breeding crop plants with a tolerance to salinity, we suggest that salt tolerant and disease resistant cultivars need to be included in breeding programs in order to minimize further yield declines in areas where the two factors occur together.

Salinity may have a positive effect on fungal growth, and may increase the development of motile zoosporangia, spores, conidia, and the germ tubes of fungi. In such a case, fungi may grow better and become more virulent. After an adaptation period, halophytic races of fungi may appear and seriously threaten crop production in non-saline soils. Palmero et al. (2010) has reported that the increase of mycelium growth of *Fusarium oxysporum* strains under moderate saline conditions was simulated with KCl or NaCl. These findings indicate that *F. oxysporum* was well adapted to exist in moderate saline conditions. Although this feature is not directly related to the pathogenicity of the isolate, the experimental results indicate that the pathogen has the capacity of growing at low osmotic pressures, which could explain the prevalence of the *Fusarium* genus in soils in semiarid areas.

On the other hand, the negative effect of salinity could prevent pathogen propagules from reaching root infection sites. In this manner, the effectiveness of microorganisms in saline soils may also be reduced. Amir et al. (1996) have reported that salinity induced soil suppressiveness to vascular fusariois. Salinity reduced the sporulation and germination of *Fusarium oxysporum*. Engel and Grey (1991) stated that chloride fertilizers increased the yield of winter wheat and reduced the severity of root diseases caused by *Fusarium culmorum*. Similar findings were made by Kostandi and Soliman (1998), who found that the effect of saline irrigation water containing NaCl or Na2SO4 reduced susceptibility to smut by 22.7% and 10.8% respectively. The wheat and barley seedlings irrigated with saline water in the Arnon and Hoagland (1940) solution, showed an accumulation of Na^+ and Cl^- ions, as well as metabolites such as glycine betaine, in leaves, decreasing the population and the growth rate of the aphids *Schizaphis graminum* and *Rhopalosiphum padi* feeding on the plants (Araya et al. 1991). Araya et al. (1991) concluded that the accumulation of NaCl in the plants was the real cause for the reduction in the survival and reproduction of aphids. However, *in vitro* studies indicate that the accumulation of higher glycine-betaine was not deleterious to the insects. Therefore, salt tolerant barley accumulating higher amounts of glycine-betaine under salt stress made the barley prone to insect attacks. However, it should be noted that crop plants displaying a resistance to pathogen or insect attack should be very tolerant to high salt stress. The low or high concentrations of salt tolerated by attacking microorganisms or insects will have drastic negative consequences on crop plants.

The negative effect of salt on microorganisms has inspired many workers and led them to use salt as a fungicide. For example, the studies of Mecteau et al. (2002; 2008) and Mill et al. (2004) showed that several salts (0.2 M) completely inhibited mycelial growth and the spore germination of *Fusarium sambucinum*, a casual agent of potato dry rot under *in vitro* conditions. The authors suggested that controlling dry rot may be possible with the application of various salts such as sodium benzoate, sodium metabisulfite, potassium sorbate, and ammonium salts since the post-harvest application of thiabendazole, an effective fungicide, has become less effective to many strains of *F. sambucinum*, which has become resistant to this fungicide. Among the salts tested, aluminium chloride and sodium metabisulfite were able to significantly reduce the development of potato dry rot. Recently, similar reports have been made by El-Mougy and Abdel-Kader (2009) for potato early blight disease caused by *Alternaria solani*. However, for the effective application of salts, there is a need to understand the chemical and biochemical basis of pathogen-plant-salt interactions. However, the fungi may also develop a resistance to salts, which have curative or preventive impacts, as they have developed a resistance to fungicides.

4 Reducing the Impacts of Salinity and Disease Incidence via Bioamelioration for Crop Improvement

Many studies related to understanding the stress metabolism of plants have been designed to keep one stress in mind and to concentrate on its biochemical and molecular metabolism. In nature, however, many abiotic and biotic stress factors are interconnected. Under saline conditions the impacts of pathogens have more detrimental effects than that of each stress factor separately. In order to reduce the impact of both salt stress and pathogens, crop plants resistant to both or more stress factors obtained via molecular or biochemical techniques are thought to be the only solution. However, the reduction of at least salt stress via bio-amelioration, using either halophytes or microorganisms, could partly reduce the impact of salt stress. Additionally, microorganisms that have high bio-control potentials under saline stress could well be the best and the least expensive solution for the combined effects of stress. In most cases, excessive salt concentrations in soils cannot be reduced with time by routine irrigation and crop management practices. Such soils should be ameliorated, for agricultural purposes, in order to meet increasing food demands. The movement of excess soluble salts via leaching, the surface flushing of salts from soils, and the biological reduction of salts from the soil by halophytes or microorganisms are some commonly used methods (Qadir et al. 2000; Dikilitas and Karakas 2010). In recent years, there has been increased interest in the bio-control of pathogens and the phytoremediation of saline soils, especially in arid and semi-arid regions of the world as a result of increase in pollution and soil salinity via excessive irrigation. To date, the control of plant pathogens or the remediation of saline soils have been considered to be separate issues. However, these two stress factors may be present in one field or even on one single crop plant. Considering these two stress issues together through physiological, biochemical, or molecular approaches is important for crop improvement. But every approach related to crop improvement should be evaluated based on cost and efficacy, as well as practical application.

Various plant and microorganism species exist for bioremediation. In fact, for the bio-control of various important plant pathogens, competitive microorganisms also exist. The important issue here is to test and to increase the efficacy of completive microorganisms under adverse conditions. The potential bio-control agents or fungi isolated from saline soils are better adapted to such conditions and, therefore, are more efficient in plant growth promotion, and may better control plant pathogens under salt stress (Gamalero et al. 2009; Urja and Meenu 2010). Many microorganisms; especially nitrogen fixers; have been shown to change the properties of salt-affected soils, resulting in the bioremediation of saline soils (Zahran 1991; Al-Abed et al. 2004; Egamberdieva 2011). The major groups of salt tolerant microorganisms reported from many salt-affected soils include free nitrogen fixing bacteria (e.g. *Azotobacter, Alcaligens, Azospirillum*); cyanobacteria (e.g. *Anabaena, Nodularia, Nostoc*, etc.); and symbiotic *Rhizobia* species (Zahran 1991). Recently Urja and Meenu (2010) reported that inoculation with fungal isolates of *Aspergillus sp.* (S-11) and *Penicillium sp.* (S12B) on chickpea plants led to an increase in plant growth, total chlorophyll content, and the fresh and dry weight of plants with 2% NaCl salinity as compared to non-inoculated controls. Obuekwe et al. (2005) reported hydrocarbon degradation by desert fungi in crude oil-affected areas, and found that fungi such as *Fusarium lateritium* and *Drechslera sp.* grew in the presence of 10% NaCl, while *Papulaspora sp.* grew in 5% NaCl. However, growth was dependent on the nutritional status of the growth medium. Similarly, Al-Abed et al. (2004) found that the EC level of soil was markedly reduced with aerobic bacteria. The result could be a good sign for the use of bio-control agents which cannot tolerate high salt levels. The bio-control agents such as *Trichoderma harzianum*; obtained at low or moderate salt concentrations; have been applied for the biological control of plant pathogens on saline soils (Regraui and Lahlou 2005). In areas where soil salinity aggravates plant pathogens, research on new strains of *Trichoderma* will help to select more salt tolerant strains. Determining the adaptive conditions of these strains to saline soils and their interactions with other abiotic and biotic factors will be useful. Recently, a new approach for developing improved bio-control strains of fungi by increasing salinity tolerance phenotypes using mutation techniques has been determined (Mohammad and Haggag 2006). Such an approach could be useful for enhancing salt tolerance, metabolic production and bio-control ability against plant pathogens under non-saline and saline conditions, and may efficiently protect crop plants, as well as increase crop production. Mohamed and Haggag (2006) stated that exposing a wild-type culture of *Trichoderma harzianum* to gamma irradiation induced two stable salt-tolerant mutants (Th50M6 and Th50M11). Under saline conditions, both mutants greatly surpassed their wild type strain in growth rate, sporulation, and biological proficiency against *Fusarium oxysporum*, the causal agent of tomato wilt disease. *Trichoderma* mutants that produce a higher content of chitinases, cellulases, as well as antibiotics, significantly reduce wilt disease incidence and improve the yield and mineral content of tomato plants under both saline and non-saline soil conditions. In some crops such as onions and bell peppers, vesicular-arbuscular mycorizhal (VAM) fungi also seem to increase salt tolerance (Borde et al. 2011). However, to date, the use of mycorrhizae is still controversial.

5 Future Prospects

The development of crop varieties with an increased salt tolerance is crucial for providing a long-term solution to the problem of salinization. However, the problem of improving salt tolerance in plants becomes more complex when plants are exposed to other environmental and biotic factors. The disease-resistant plants should be generated from suspension cells that have been adapted to higher salt concentrations. It is possible to regenerate 50 mM NaCl-tolerant plants from a disease resistant crop plant. However, it is not known whether or not an increase in the level of salt tolerance in the resistant cultivar will decrease the resistance of the plants towards fungi under non-saline and saline conditions. Commercially available drought or salt tolerant plants should be considered in order to maximize crop production and marketable fruit yields. Changes in gene expression have been reported in various crops and microorganisms when they were subjected to various environmental stresses (Katsuhara and Kawasaki, 1996; personal communication with Dr. Tinley Basset, Swansea University, Wales-UK), as such, salt-adapted strains of the fungus should be examined on a molecular level to determine if any changes resulting from the accumulation of NaCl have been made in the gene-sequence. If changes have been made, the genes should be characterized.

A further objective for future research should be to assess the long-term effects of salt accumulation within the soil profile and ground water. In order to achieve the proposed future objectives for long-term improvements in crop production in saline environments, research must be carried out in a manner that promotes close associations between scientists, commercial growers, retailers, and consumers. Plant scientists, including plant physiologists, molecular biologists, geneticists, and plant pathologists should work together to improve agricultural productivity.

Potential bio-control agents for plant diseases should be improved, for controlling diseases under normal and saline conditions, through biochemical or molecular means because efficient resource management and crop improvements for evolving better breeds can help us overcome diseases to some extent. However, such strategies take time and are cost intensive. Therefore, in the short-term, there is a need to develop simple and low cost biological methods for the management of abiotic and biotic stresses (Grover et al. 2011). Using plant growth promoting bacteria or beneficial fungi under salt stress conditions instead of GM plants is advantageous when considering a large number of plants and their cultivars, as well as the multiplicity of genes that would need to be engineered into plants in order to ensure salt tolerance (Gamalero et al. 2009). The technology is easier to access for farmers. In fact, generating more salt tolerant halophytes or microorganisms with high bio-control ability would be much easier and logical than generating genetically improved salt tolerant and disease resistant plants. The increased levels of salinity are frequently associated with the irrigation practices of agricultural land which pose a threat to crop production, especially where water quality is marginal. Under such conditions of poor water quality, plants are not only stressed but may also be more susceptible to various pathogens. One strategy for maintaining yields in such situations or increasing yields in marginal areas is to develop salt tolerant crop strains. In this environment, such strains should also be resistant to pathogens, including halophytic races of pathogens that may have adapted to living in higher salt concentrations. Therefore, understanding the interactions between crop plants and potential pathogens under salinity is an important part of any project aimed at developing strategies for disease control for crops grown in saline soils, which may have a minimal impact in stressed environments.

Now-a-days, in many countries, regenerated or modified plants with resistances to stress are available for increasing crop production even on marginal land. Although these developments are promising, only one factor is considered to be important when plants are developed against stress. However, by acting synergistically, many stress factors can deteriorate plant development. In near future, due to a number of reasons, irrigated lands may be exposed to salinity. Therefore, salt-tolerant plants that tolerate high levels of salt may be economically important (Dikilitas 1997, 2003; Al-Rawahy 2000). However, stress factors other than salinity may still have the potential to damage plants that tolerate the effects of salt. Dikilitas and Smith (1999) have shown that salt tolerant lucerne strains generated for saline conditions, when attacked by microorganisms, become more susceptible than those of non-salt tolerant parental cultivars. Since a living organism is involved, adaptation, or in the long term, the evolution of the pathogen may create larger problems for agricultural lands; or, in the short term, the tolerance level of salt by regenerated plants may be less than that of the plant pathogen. So, the problem is not yet resolved. Since the use of arable land is decreasing, it is important to understand the physiological responses and the defense mechanisms of plants required to be grown under stress conditions. The economically important crop plants should be bred for their resistance to disease and non-pathogenic stress agents, and should be made commercially available.

6 Conclusion

The conclusion that can be drawn from this overview is that when saline water is used to irrigate crop plants in areas where infecting soil microorganisms may be present, soil salinity should be maintained and controlled in order to prevent infecting microorganisms from reaching a degree of virulence that would put even medium resistant cultivars at risk. For crop plants, different stressors may interact, both in their occurrence and impact. So, real progress in crop improvement is likely to depend on a better understanding of these interactions and their consequences for plants and ecosystems. We must keep in mind that there are two contributing factors, resistance of the plant and virulence of the pathogen. Furthermore, studies at the cellular and molecular level are needed in order to understand the overall response of plants. The plant pathogenic fungi develop an enormous array of strategies under salt stress conditions in order to infect their host as discussed above, such as production of extracellular enzymes, conidia, and mycelium growth.

Acknowledgements We dedicate this chapter to the memory of Prof John Gallon (University of Wales, Swansea-UK).

References

Abumhadi NM, Atanassov AI (2010) Future challenges of plant biotechnology and genomics. Rom Biotechnol Lett 15(2):127–142

Ahn KJ, Kim KH (2001) Effect of temperature and salinity on *in vitro* zoosporulation of *Perkinsus sp*. in Manila clams *Ruditapes philippinarum*. Dis Aquat Organ 48(1):43–46

Al-Abed N, Amayreh J, Al-Afifi A, Al-Hiyari G (2004) Bioremediation of a Jordanian saline soil: a laboratory study. Commun Soil Sci Plant Anal 35(9–10):1457–1467

Al-Rawahy SH (2000) Study of the adaptive mechanisms evolved through selecting NaCl tolerant cells and plants of alfalfa (*M. media* cv. Rambler). PhD thesis, University of Wales, Swansea

Al-Sadia AM, Al-Masoudia RS, Al-Habsib N, Al-Saida FA, Al-Rawahy SA, Ahmed M, Deadman MD (2010) Effect of salinity on Pythium damping-off of cucumber and on the tolerance of *Pythium aphanidermatum*. Plant Pathol 59:112–120

Amir H, Mair A, Riba A (1996) Role de la microflore dans la resistance a la Fusariose vascularie induite par la salinite dans un sol de palmeraie. Soil Biol Biochem 28(1):113–122

Araya F, Abarca O, Zunica GE, Corcuera LJ (1991) Effects of NaCl on glycine-betaine and on aphids in cereal seedlings. Phytochemistry 30(6):1793–1795

Arnon DI, Hoagland DR (1940) Crop production in artificial culture solutions and in soils with special reference to factors influencing yield absorption of inorganic nutrients. Soil Sci 50:463–483

Ashraf M, Foolad MR (2007) Roles of glycine betaine and proline in improving plant abiotic stress resistance. Environ Exp Bot 59:206–216

Atanassov A (2009) Genomics and the global challenges, Biotechnol & Biotechnol Eq. 23/2009/ SE, pp 448–451, Special edition: 120 years of Academic Education in Biology, XI Anniversary Scientific Conference

Athar HR, Ashraf M (2009) Strategies for crop improvement against salinity and drought stress: an overview. In: Ashraf M, Ozturk M, Athar HR (eds) Salinity and water stress improving crop efficiency, vol 44, Tasks for vegetation sciences. Springer, Germany, pp 1–16

Besri M (1990) Effect of salinity on the development of tomato Verticillium wilt in Morocco. In: Proceedings of the fifth international Verticillium symposium, Leningrad, p 61

Bevan M, Flavell R, Chilton M (1983) A chimaeric antibiotic resistance gene as a selectable marker for plant cell transformation. Nature 304:184–187

Blaker NS, MacDonald JD (1986) The role of salinity in the development of Phytophthora root rot of citrus. Phytopathology 76:970–975

Blumwald E, Aharon GS, Apse MP (2000) Sodium transport in plant cells. Biochim Biophys Acta 1465:140–151

Borde M, Dudhane M, Jite PJ (2011) Growth photosynthetic activity and antioxidant responses of mycorrhizal and non-mycorrhizal bajra (*Pennisetum glaucum*) crop under salinity stress condition. Crop Prot 30(3):265–271

Bouchibi N, van Bruggen AHC, MacDonald JD (1990) Effect of ion concentration and sodium: calcium ratio of a nutrient solution on Phytophtora root rot of tomato and zoospore motility and viability of *Phytophtora parasitica*. Phytopathology 80:1323–1329

Boyer JS (1995) Biochemical and biophysical aspects of water deficits and the predisposition to disease. Annu Rev Phytopathol 33:251–274

Bray L, Chriqui D, Gloux K, Lerudulier D, Meyer M, Peduzzi J (1991) Betaines and free amino acids in salt stressed vitro plants and winter resting buds of *Populus trichocarpa* x *deltoides*. Physiol Plant 83(1):136–143

Brugnoli E, Bjorkman O (1992) Growth of cotton under continuous salinity stress: influence of allocation pattern, stomatal and non-stomatal components of photosynthesis and dissipation of excess light energy. Planta 187:335–347

Cavalieri AJ (2005) Biotechnology and agriculture in 2020: a report of the project on technology futures and global power, wealth and conflict. II Centre for Strategic and International Studies (Washington, DC) III

Chandler D, Heale JB, Gillespie AT (1994) Effect of osmotic potential on the germination of conidia and colony growth of *Verticillium lecanii*. Mycol Res 98(4):384–388

Danti S, Broggio M (1997) Impact of salinity stress on the biometric responses of an isolate of *Verticillium dahliae* Kleb. From potato. In: Proceedings of the seventh international Verticillium symposium, Cape Sounion, Athens, Greece, p 53

Dikilitas M (1997) A study of interactions between the phytopathogenic fungus *Verticillium albo-atrum* (Reinke & Berth) and tomato (*Lycopersicon esculentum*) and various cultivars of lucerne (*Medicago sativa* L. & *M. media*). M.Phil. thesis, University of Wales, Swansea, UK

Dikilitas M (2003) Effect of salinity & its interactions With *Verticillium Albo-Atrum* on the disease development In Tomato (*Lycopersicon Esculentum* Mill.) and Lucerne (*Medicago Sativa* & *M. Media*) Plants. Ph.D. thesis, University of Wales, Swansea

Dikilitas M, Karakas S (2010) Salts as potential environmental pollutants, their types, effects on plants and approaches for their phytoremediation. In: Ashraf M, Ozturk M, Ahmad MSA (eds) Plant adaptation and phytoremediation. Springer, London, pp 357–381

Dikilitas M, Karakas S (2011) Agriculture in 2020: what is the target? Harran Univ J Fac Agric 14(3):1–2

Dikilitas M, Kocyigit A, Yigit F (2009) A molecular-based fast method to determine the extent of DNA damages in higher plants and fungi. Afr J Biotech 8(14):3118–3127

Dikilitas M, Smith CJ (1999) Pathogenicity of *Verticillium albo-atrum* to salt tolerant lucerne strains. Biological Control Agents in Crop & Animal Protection, 24–28 Aug 1999, University of Wales, Swansea

DiLeo MV, Pye MF, Roubtsova TV, Duniway JM, MacDonald JD, Rizzo DM, Bostock RM (2010) Abscisic acid in salt stress predisposition to Phytophthora root and crown rot in tomato and chrysanthemum. Phytopathology 100:871–879

Dimartino M, Panebianco S, Vitale A, Castello I, Leonardi C, Cirvilleri G, Polizzi G (2011) Occurrence and pathogenicity of *Pseudomonas fluorescens* and *P. putida* on tomato plants in Italy. J Plant Pathol 93(1):79–87

Dunn DC, Duncan LW, Romeo JT (1998) Changes in arginine, Pal activity, and nematode behaviour in salinity-stressed citrus. Phytochemistry 49(2):413–417

Egamberdieva D (2011) Survival of *Pseudomonas extremorientalis* TSAU20 and *P. chlororaphis* TSAU13 in the rhizosphere of common bean (*Phaseolus vulgaris*) under saline conditions. Plant Soil Environ 57(3):122–127

El-Abyad MS, Abu-Taleb AM, Khalil MS (1992) Impact of salinity stress on soil-borne fungi of sugar beet. Plant Soil 143:75–83

El-Iklil Y, Karrou M, Mrabet R, Benichou M (2002) Salt stress effect on metabolic concentrations of *Lycopersicum esculentum* and *Lycopersicum cheesmanii*. Can J Plant Sci 82(1):177–183

El-Mougy NS, Abdel-Kader MM (2009) Salts application for suppressing potato early blight disease. J Plant Prot Res 49(4):353–361

Engel RE, Grey WE (1991) Chloride fertilizer effects on winter wheat inoculated with *Fusarium culmorum*. Agron J 83(1):204–208

Fageria NK, Baligar VC, Li YC (2008) The role of nutrient efficient plants in improving crop yields in the twenty first century. J Plant Nutr 31:1121–1157

Gamalero E, Graziella B, Bernard RG (2009) The use of microorganisms to facilitate the growth of plants in saline soils. In: Khan MS, Musarrat AZJ (eds) Microbial strategies for crop improvement. Springer, Berlin/Heidelberg, pp 1–22

Gao S, Shain L (1995) Effects of osmotic potential on virulent and hypovirulent strains of the chestnut blight fungus. Can J For Res 25(6):1024–1029

Goudarzi A, Banihashemi Z, Manouchehr M (2008) Effect of water potential on sclerotial germination and mycelial growth of *Macrophomina phaseolina*. Phytopathol Mediterr 47:107–114

Grattan SR, Grieve CM (1999) Salinity-mineral nutrient relations in horticultural crops. Sci Hortic 78:127–157

Griffith GS, Boddy L (1991) Fungal decomposition of attached angiosperm twigs. III. Effect of water potential and temperature on fungal growth, survival and decay of wood. New Phytol 117:259–269

Grover M, Ali Sk Z, Sandhya V, Rasul A, Venkateswarlu B (2011) Role of microorganisms in adaptation of agriculture crops to abiotic stresses. World J Microbiol Biotechnol 27:1231–1240

Hancock JG, Huisman OC (1981) Nutrient movement in host-pathogen systems. Annu Rev Phytopathol 19:309–331

Harfouche A, Meilan R, Altman A (2011) Tree genetic engineering and applications to sustainable forestry and biomass production. Trends Biotechnol 29(1):9–17

Hasan HAH (2002) Gibberellin and auxin production by plant root-fungi and their biosynthesis under salinity-calcium interaction. Rostlinna Vyroba 48(3):101–106

Holmes EW (1976) Verticillium wilt of salt-injured sugar maple-preliminary study. Proc Am Phytopathol Soc 3:305–306

Howell AB, Francois L, Erwin DC (1994) Interactive effect of salinity and *Verticillium albo-atrum* on Verticillium wilt disease severity and yield of two alfalfa cultivars. Field Crops Res 37:247–251

James RA, Davenport R, Munns R (2006) Physiological characterization of two genes for Na^+ exclusion in wheat: *Nax1* and *Nax2*. Plant Physiol 142:1537–1547

Jauhar P (2006) Cytogenetic architecture of cereal crops and their manipulation to fit human needs: opportunities and challenges. In: Singh RJ, Jauhar PP (eds) Genetic resources, chromosome engineering, and crop improvement, vol 2, Cereals. CRC Taylor & Francis Press, Boca Raton, pp 1–25

Jbir N, Ammar S, Chaibi W, Ayadi A (2001) PAL activity and ionic contents of two wheat species differing in their sensitivity to NaCl, in response to salt stress. J Trace Microprobe Tech 19(3):447–450

Jefferson PG, Gossen BD (2002) Irrigation increases Verticillium wilt incidence in a susceptible alfalfa cultivar. Plant Dis 86(6):588–592

Jin X, Hengxia Y, Xiaojing L, Xia L (2010) Salt affects plant Cd-stress responses by modulating growth and Cd accumulation. Planta 231:449–459

Katsuhara M, Kawasaki T (1996) Salt stress induced nuclear and DNA degradation in meristatic cells of barley roots. Plant Cell Physiol 37:169–173

Knox RE, Clarke FR (2007) Molecular breeding approaches for enhanced resistance against fungal pathogens. In: Punja ZK, De Boer SH, Sanfacon H (eds) Biotechnology and plant disease management. CAB International, Oxfordshire, pp 321–357

Kostandi SF, Soliman MF (1998) The role calcium in mediating smut disease severity and salt tolerance in corn under chloride and sulphate salinity. J Phytopathol 146:191–195

Krokene P, Solheim H (2001) Loss of pathogenicity in the blue-stain fungus Ceractocystis polonica. Plant Pathol 50(4):497–502

Levin AG, Lavee S, Tsror (Lahkim) L (2003) Epidemiology of *Verticillium dahliae* on olive (cv. Picual) and its effect on yield under saline conditions. Plant Pathol 52:212–218

MacDonald JD (1984) Salinity effects on susceptibility of chrysanthemum root rot to *Phytophthora cryptogea*. Phytopathology 74:621–624

Madigen O (1999) Thermophilic and halophilic extremophiles. Curr Opin Microbiol 2:265–269

McQuilken MP, Whipps JM, Cooke RC (1992) Effects of osmotic and matric potential on growth and oospore germination of the biocontrol agent *Phythium oligandrum*. Mycol Res 96(7):588–591

Mecteau MR, Arul J, Russell JT (2002) Effect of organic and inorganic salts on the growth and development of *Fusarium sambucinum*, a causal agent of potato dry rot. Mycol Res 106(6):688–696

Mecteau MR, Arul J, Tweddell RJ (2008) Effect of different salts on the development of *Fusarium solani* var. *coeruleum*, a causal agent of potato dry rot. Phytoprotection 89(1):1–6

Mer RK, Prajith PK, Pandya DH, Pandey AN (2000) Effects of salts on germination of seeds and growth of young plants of *Hordeum vulgare*, *Triticum aestivum*, *Cicer arietinum* and *Brassica juncea*. J Agron Crop Sci 185:209–217

Mills AAS, Platt HW, Hurta RAR (2004) Effect of salt compounds on mycelial growth, sporulation and spore germination of various potato pathogens. Postharvest Biol Technol 34:341–350

Mohammad H, Haggag WF (2006) Biocontrol potential of salinity tolerant mutants of *Trichoderma harzianum* against *Fusarium oxysporum*. Braz J Microbiol 37:181–191

Mohammadi AH, Banihashemi Z, Maftoun M (2007) Interaction between salinity stress and Verticillium Wilt disease in three pistachio rootstocks in a calcareous soil. J Plant Nutr 30:241–252

Munns R (2002) Comparative physiology of salt and water stress. Plant Cell Environ 25:239–250

Munns R (2009) Strategies for crop improvement in saline soils. In: Ashraf M, Ozturk M, Athar HR (eds) Salinity and water stress improving crop efficiency, vol 44, Tasks for vegetation sciences. Springer, Berlin, pp 99–110

Munns R, Tester M (2008) Mechanisms of salinity tolerance. Annu Rev Plant Biol 59:651–681

Munns R, Rebetzke GJ, Husain S, James RA, Hare RA (2003) Genetic control of sodium exclusion in durum wheat. Aust J Agric Res 54:627–635

Nachmias A, Kaufman Z, Livescu L, Tsror L, Meiri A, Caligari PDS (1993) Effects of salinity and its interactions with disease incidence on potatoes grown in hot climates. Phytoparasitica 21(3):245–255

Nachmias A, Livescu L, Tsror L, Ben Hador G, Orenstein J, Sabbah S, Orion D (1994) Enhancement of *Verticillium* infection in potato under stress. In: 6th international Verticillium symposium, Dead Sea, Israel, Phytoparasitica 23:51

Neves GYS, Marchiosi R, Ferrarese MLL, Siqueira-Soares RC, Ferrarese-Filho O (2010) Root growth inhibition and lignification induced by salt stress in soybean. J Agron Crop Sci 196:467–473

Obuekwe CO, Adekunle MB, Esmaeil A, Mulder JL (2005) Growth and hydrocarbon degradation by three desert fungi under conditions of simultaneous temperature and salt stress. Int Biodeter Biodegr 56:197–205

Omar SA, Abd-Alla MH (2000) Physiological aspects of fungi isolated from root nodules of faba bean (*Vicia faba* L.). Microbiol Res 154(4):339–347

Orenstein J, Sabbah S, Nachmias A (1995) Enhancement of *Verticillium* dahliae infection of potato plants under stress conditions. In: 6th international Verticillium symposium, Dead Sea, Israel, Phytoparasitica 23:51

Palmero Llamas D, De Cara GM, Iglesias Gonzalez C, Ruiz Lopez G, Tello Marquina JC (2008) Effects of water potential on spore germination and viability of *Fusarium* species. J Ind Microbiol Biotechnol 35:1411–1418
Palmero D, De Cara M, Moreno MM, Iglesias C, Tello JC (2010) Stimulation of mycelial growth of pathogenic and seabed isolates of Fusarium oxysporum in presence of salts. Afr J Microbiol Res 4(17):1859–1861
Qadir M, Ghafoor A, Murtaza G (2000) Amelioration strategies for saline soils: a review. Land Degrad Dev 11(6):501–521
Ragazzi A, Vecchio V (1992) Behaviour of chlamydospores of *Fusarium oxysporum* f. sp. *vasinfectum* in substrates containing sodium chloride. Phytopathol Mediterr 31:85–87
Ragazzi A, Vechio V, Dellavalle I, Cucchi A, Mancini F (1994) Variations in the pathogenicity of *Fusarium oxysporum* f.sp. *vasinfectum* in relation to the salinity of the nutrient medium. J Plant Dis Prot 101(3):263–266
Ragazzi A, Broggio M, Soldi R, Moricca S, Dellavalle I, Sabatini S (1996) Biometry and pathogenicity of *Fusarium oxysporum* f. sp. *vasinfectum* on a Na2SO4 enriched medium. J Plant Dis Protect 103(3):272–278
Ramagopal S (1993) Advances in understanding the molecular biology of drought and salinity tolerance in plants-the first decade. In: Lodha ML, Mehta SL, Ramagopal S, Srivastava GP (eds) Adv Plant Biotech Biochem. Indian Soc Agril Biochemists, Kanpur, pp 38–39
Rasmussen SL, Stanghellini ME (1988) Effect of salinity stress on development of *Phythium* blight in *Agrostis palustris*. Phytopathology 78:1495–1497
Regragui A, Rahouti M, Et Lahlou H (2003) Effet Du Stress salin sur *Verticillium albo-atrum*: pathogénécité et production d'enzymes cellulolytiques *in vitro* Cryptogamie. Mycology 24(2):167–174
Regraui A, Lahlou H (2005) Effect of salinity on *in vitro Trichoderma harzianum* antagonism against *Verticillium dahliae*. Pak J Biol Sci 8(6):872–876
Rougon-Cardoso A, Zipfel C (2010) A new approach to confer broad-spectrum disease resistance in plants. Information Systems for Biotechnology News Report, 5–8 June
Saadatmand AR, Banihashemi Z, Sepaskhah AR, Maftoun M (2008) Soil salinity and water stress and their effect on susceptibility to Verticillium Wilt disease, ion composition and growth of pistachio. J Phytopathol 156:287–292
Sanogo S (2004) Response of chile pepper to *Phytophthora capsici* in relation to soil salinity. Plant Dis 88:205–209
Singh RP (1995) Differential response of growth and nitrate assimilation in sesame and mungbean seedlings to heavy metal stress. Proc Acad Environ Biol 4:215–220
Singh RP, Choudhary A, Gulati A, Dahiya HC, Jaival PK, Sengar RS (1997) Response of plants to salinity in interaction with other abiotic and biotic factors. In: Jaiwal PK, Singh RP, Gulati A (eds) Strategies for improving salt tolerance in higher plants. Science Publishers Inc, Enfield, pp 25–39
Snapp SS, Shennan C (1994) Salinity effects on root growth and senescence in tomato and the consequences for severity of Phytophthora root rot infection. J Am Soc Hort Sci 119(3):458–463
Snapp SS, Shennan C, van Bruggen AHC (1991) Effects of salinity on severity of infection by *Phytophthora parasitica* Dast., ion concentrations and growth of tomato, *Lycopersicon esculentum* Mill. New Phytol 119:275–284
Sridhara S (2006) Vertebrate pests in agriculture-the Indian scenario. Scientific, Jodhpur
Srivastava S, Yadav A, Seem K, Mishra S, Chaudhary V, Srivastava CS (2008) Effect of high temperature on *Pseudomonas putida* NBRI0987 biofilm formation and expression of stress sigma factor RpoS. Curr Microbiol 56(4):453–457
Sulistyowati L, Keane PJ (1992) Effect of soil salinity and water content on stem rot caused by *Phytophthora citrophthora* and accumulation of phytoalexin in citrus rootstocks. Phytopathology 82(7):771–777
Swiecki TJ, MacDonald JD (1991) Soil salinity enhances Phytophthora root rot of tomato but hinders asexual reproduction by *Phytophthora parasitica*. J Am Soc Hortic Sci 116(3):471–477

Syvertsen J, Levy Y (2005) Salinity interactions with other abiotic and biotic stresses in citrus. HortTechnology 15(1):100–103

Tang M (2001) Elicitor-induced defence responses and mechanisms of signal transduction in *Medicago sativa* L. Ph.D. thesis, University of Wales, Swansea

Triky-Dotan S, Yermiyahu U, Katan J, Gamliel A (2005) Development of crown and root rot disease of tomato under irrigation with saline water. Phytopathology 95:1438–1444

Turco E, Broggio M, Ragazzi A (1999) Enzymes produced by *Fusarium oxysporum* f. sp. *vasinfectum* compared on a saline and a non-saline substrate. J Plant Dis Prot 106:333–341

Turco E, Naldini D, Ragazzi A (2002) Disease incidence and vessel anatomy in cotton plants infected with *Fusarium oxysporum* f.sp. *vasinfectum* under salinity stress. J Plant Dis Prot 109(1):15–24

Turco E, Vizzuso C, Franceschini S, Ragazzi A, Stefanini FM (2007) The *in vitro* effect of gossypol and its interaction with salts on conidial germination and viability of *Fusarium oxysporum* sp. *vasinfectum* isolates. J Appl Microbiol 103(2007):2370–2381

Urja P, Meenu S (2010) Role of single fungal isolates and consortia as plant growth promoters under saline conditions. Res J Biotechnol 5(3):5–9

Wilkens S, Field CD (1993) Effect of varying sea-water salinity on growth kinetics of *Phytophthora polymorphica*. Mycol Res 97(9):1135–1139

Xu Y, Sun X, Jin J, Zhou (2010) Protective effect of nitric oxide on light induced oxidative damage in leaves of tall fescue. J Plant Physiol 167:512–518

Zahran HH (1991) Conditions for successful *Rhizobium* legume symbiosis in saline environments. Biol Fertil Soils 12(1):73–80

Chapter 17
Biochemical and Molecular Aspects of Drought Tolerance in Wheat *Triticum* L. Genotypes

Irada Mammad Huseynova, Samira Mahammadrahim Rustamova, Saftar Yusif Suleymanov, and Jalal Alirza Aliyev

Contents

Abstract Wheat (*Triticum aestivum* L.) cultivars contrasting in genetic makeup and differing in drought-resistance were grown in field conditions in a wide area under normal water supply and severe water deficit. One of the genotypes (Azamatli-95) was short-stemmed, with vertically oriented small leaves and drought-tolerant while

I.M. Huseynova (✉) • S.M. Rustamova • S.Y. Suleymanov • J.A. Aliyev
Institute of Botany, Azerbaijan National Academy of Sciences,
40, Patamdar Shosse, Baku AZ 1073, Azerbaijan
e-mail: i_guseinova@mail.ru; babaliyevasamira@mail.ru; suleymanov_s@mail.ru; aliyev-j@botany-az.org

M. Ashraf et al. (eds.), *Crop Production for Agricultural Improvement*,
DOI 10.1007/978-94-007-4116-4_17,

the other genotype (Giymatli-2/17) was short-stemmed, with broad and drooping leaves and drought-sensitive. It was found out that the content of CP I (M_r 115 kD) and apoprotein of P700 with M_r 63 kD, also LHC II polypeptides insignificantly increased in the drought-resistant cv. Azamatli-95 under extreme water supply condition while their content decreased in drought-sensitive cv. Giymatli-2/17. The intensity of synthesis of α- and β-subunits of CF_1 (55 and 53.5 kD) and 33–30.5 kD proteins also decreased in the sensitive genotype. The intensity of short wavelength peaks at 687 and 695 nm sharply increased in the fluorescence spectra (77 K) of chloroplasts from Giymatli-2/17 under water deficiency and there was a stimulation of the ratio of fluorescence band intensity F687/F740. After exposure to drought, cv Giymatli-2/17 showed a larger reduction in the actual PS II photochemical efficiency of chloroplasts than cv. Azamatli-95. Activities of catalase, ascorbate peroxidase, superoxide dismutase and glutathione reductase, as well as photochemical activity of photosystem I and photosystem II were studied in leaves of durum and bread wheat genotypes in ontogenesis. It was found out that dynamics of catalase and ascorbate peroxidase functioning in well-watered plants through ontogenesis practically did not change both among durum and among bread wheat cultivars. Functioning of these enzymes during ontogenesis under water deficit differed. Catalase activity increased in all stressed genotypes: in durum wheat cultivars maximal activity was observed in the milk ripeness and in bread wheat cultivars at the end of flowering. Ascorbate peroxidase activity also increased under water deficit: in tolerant wheat genotypes maximal activity occurred at the end of flowering, and in the sensitive ones at the end of ear formation. The maximum activity of glutathione reductase both as in the control, as well as in drought-subjected plants was observed in the anthesis stage. Superoxide dismutase activity was lower than the control during ontogenesis, except in the last stages. It should be noted that PS I and PS II photochemical activities were also high in genotypes subjected to drought both at the end of ear formation and flowering stages. Drought resistance was checked by RAPD-PCR as a quick and easy method for durum (*Triticum durum* L.) and bread (*Triticum aestivum* L.) wheat genotypes contrasting in tolerance. P6 primer (5′ TCGGCGGTTC 3′) produced 920 bp band mainly in drought tolerant genotypes. It was found that P7 (5′ TCGGCGGTTC 3′) primer produced 750 bp band was not absolutely universal for *Triticum* L. genotypes. In order to identify *DREB1* genes in these genotypes PCR-analysis was carried out using functional markers specific for A, B, and D genomes. It was found that *DREB 1* gene was located on chromosome 3A in all genotypes, excepting one semi-tolerant genotype Tale-38. In comparison with other genotypes, a 717 bp PCR product of *DREB -B1* gene was located on B genome in drought-tolerant Barakatli-95. The results reported here provide an entry point and a reference to future analysis of gene expression during drought. In addition, these results can suggest possible targets for the enhancement of stress tolerance in crops by genetic engineering. The data presented here might be used for monitoring environmental stresses in field-grown plants and selecting stress-resistant varieties for growth under unfavorable conditions.

Keywords Wheat genotypes • Antioxidant enzymes • Thylakoid membrane polypeptides • *DREB* genes • Functional markers • PCR analysis • Chlorophyll • Fluorescence • Photosystems

Abbreviations

CAT	Catalase
Chl	Chlorophyll
DCIP	2,6-dichlorophenolindophenol
DREB	Drought resistance element binding
F	Fluorescence
LHC	Light-harvesting chlorophyll a/b-protein complex
MV	Methyl viologen
PCR	Polymerase Chain Reaction
POD	Peroxidase
PS I	Photosystem I
PS II	Photosystem II
RAPD	Random Amplified Polymorphic DNA
ROS	Reactive oxygen species
RWC	Relative water content
SDS	Sodium dodesyl sulphat
SOD	Superoxiddismutase

1 Introduction

Plants are subjected to a range of abiotic and biotic stresses that affect their growth and development. In particular, it is predicted that water deficit will continue to be a major abiotic factor affecting global crop yields (Sharma and Lavanya 2002). One third of the world's population resides in water-stressed regions, and with elevated CO_2 levels in the atmosphere and climatic changes predicted in the future, drought could become more frequent and severe.

Wheat is a staple food crop for more than 35% of the world population and also one of the widely cultivated crops in Azerbaijan, where drought is the main abiotic stress limiting its grain yield. Wheat is counted the 'big three' cereal crops, with over 600 million tons being harvested annually. Wheat is unrivalled in its range of cultivation, from 67° in Scandinavia and Russia to 45° in Argentina, including elevated regions in the tropics and sub-tropics (Shewry 2009). In view of a projection by Rajaram (2001) more than 50% of the 237 million ha area in the world under wheat cultivation is affected by periodic drought. So study of the wheat anti-drought mechanism is of great importance wheat production and biological breeding for the sake of coping with abiotic and biotic conditions. Much research is involved in this hot topic, but the pace of progress is not so large because of drought resistance being a multiple-gene-control quantitative character and wheat genome being larger (16,000 Mb) than those of most other crops (Shao et al. 2005). However, despite all the recent technological breakthroughs, the overall contribution of genomics-assisted breeding to the release of drought-resilient wheat cultivars has so far been marginal. The elucidation of genomic regions associated with the expression of traits involved in drought adaptation, the novel genes discovery or the determination of their expression patterns in response to drought stress will provide the basis of effective engineering strategies leading to enhanced wheat germplasm for specific agro-ecological niches. For any molecular assessment to be performed, it is paramount to firstly establish the plant adaptation strategy to overcome drought (Zhao et al. 2008).

To cope with highly variable environmental stresses plants have to set a series of adaptation mechanisms ranging from cellular metabolism to physiological and developmental responses (Zaidi et al. 2010). These may be classified into three groups: drought escape, drought avoidance and drought tolerance (Turne et al. 2001).

In response to stress, plants activate a number of defense mechanisms that function to increase tolerance to adverse conditions. The response to drought stress, which involves a number of biochemical-molecular mechanisms, is complex (Seki et al. 2003, 2007; Zheng et al. 2010). The application of this emerging understanding to the genetic engineering of food crops has already led to examples of improved drought tolerance and increased yield under drought (Hu et al. 2006, 2008; Nakashima et al. 2007).

Extreme circumstances can limit CO_2 fixation and enhance the generation of reactive oxygen species (ROS), such as superoxide radical (O_2^-), hydrogen peroxide (H_2O_2) and hydroxyl radical ($OH^\bullet$). Increased levels of ROS cause damage to various cellular components, such as enzyme inhibition, protein degradation, DNA and RNA damage, and membrane lipid peroxidation, which ultimately culminate in cell death (Ishikawa et al. 2010). The plants have developed different scavenging mechanisms to control the level of ROS. Antioxidant resistance mechanisms may provide a strategy to enhance tolerance, and processes underlying antioxidant responses to stress must be clearly understood (Ashraf 2009).

Production of ROS and other radicals increases dramatically during water deficiency and in plants enhanced levels of reactive oxygen species are generated in various intracellular compartments that may cause oxidative damage or act as signals (Apel and Hirt 2004; Gechev et al. 2006; Laloi et al. 2004). Although ROS in plants are produced under normal growth conditions, their concentration remains low (Polle 2001). Water stress causes stomatal closure, which reduces the CO_2/O_2 ratio in leaves and inhibits photosynthesis. These conditions increase the rate of ROS, particularly in chloroplast and mitochondria, via enhanced leakage of electrons to oxygen (Moussa and Abdel-Aziz 2008). The superoxide radicals and their dismutation product, hydrogen peroxide, can directly attack membrane lipid and inactivate SH containing enzymes (Sairam et al. 2001; Mishra and Singhal 1992). The enhanced production of ROS in chloroplasts and peroxisomes has been correlated with drastic changes in nuclear gene expression (Gadjev et al. 2006), that reveals the transfer of $^1O_2$-derived signals from the plastids to the nucleus. Many of the $^1O_2$-responsive genes are different from those activated by superoxide (O_2^-) or H_2O_2, suggesting that O_2^-/H_2O_2- and $^1O_2$-dependent signaling occurs via distinct pathways. These pathways could act independently or may interact with each other (Baruah et al. 2009). Plants protect cellular and sub-cellular systems from the cytotoxic effects of active oxygen radicals with anti-oxidative enzymes such as SOD, POD and CAT as well as metabolites like glutathione, ascorbic acid, tocopherols and carotenoids (Moussa and Abdel-Aziz 2008).

The most important phenomenon in drought stress is the deleterious effect that can be seen in the photosynthetic machinery of the plant cells. Photosynthetic electron transfer reactions that take place in thylakoid membranes have been shown to be remarkably resistant to dehydration (Cornic et al. 1992; Kaiser 1987), suggesting that higher plants develop adaptive mechanisms conferring dehydration tolerance to the photosynthetic apparatus. In response to drought, the adaptation shown by many plants could partly be due to changes in membrane composition and phase behavior, which optimizes the fluidity (Navari-Izzo et al. 1993). Indeed, models for thylakoid membrane function require mobility of protein components and redox carriers. Membrane proteins are particularly important for the functionality of the photosynthetic apparatus (Duxbury et al. 1991; Friso et al. 2004). The function of membrane proteins are influenced by the lipids matrix in which they are embedded (Horvath et al. 1989). Within the thylakoids the memrane lipids have an important role to play in stabilizing the structural arrangement and, via the lipid-protein complexes and possibly in maintaining their spatial distribution (Li et al. 1989). Alterations in bulk membrane lipids perturb all functions by inducing changes in the structure and function of thylakoid membrane protein complexes (Caldwell and Whitman 1987). Until now, very few drought-induced chloroplastic proteins have been identified in detail and the molecular basis of drought-stress tolerance is still unknown.

To develop crop plants with enhanced tolerance of drought stress, a basic understanding of gene regulatory network is essential (Shinozaki and Yamaguchi-Shinozaki 2002; Valliyodan and Nguyen 2006; Zhao et al. 2008). Responses to drought stress are extremely different according to the complexity of the genetic background of most crops (Rampino et al. 2006). Hundreds of genes and their products respond to these stresses at transcriptional and translational level. Understanding the functions of these stress-inducible genes helps unravel the possible mechanisms of stress tolerance (Sreenivasulu et al. 2007). Transcriptome analysis using microarray technology (Bohnert et al. 2001; Seki et al. 2001, 2002; Zhu et al. 2001) has revealed that genes induced by stress could be categorized into two groups according to the functions of their products. The first group consists of functional proteins such as membrane proteins that maintain water movement through membranes (water channel proteins and membrane transporters); key enzymes for osmolyte biosynthesis (proline, betaine and sugars, etc.); the detoxification enzymes enabling cellular, physiological or biochemical metabolism to maintain a normal level (glutathione *S*-transferase, hydrolase, catalase, superoxide dismutase and ascorbate peroxidase, etc.); and other proteins for the protection of macromolecules (LEA protein, osmotin, antifreeze proteins, chaperons and mRNA binding protein, etc.). Tolerance to drought or high salinity can be improved by introduction of genes encoding LEA proteins, proline synthetase or betaine synthetase, etc. The second group comprises regulatory proteins, i.e. transcription factors (bZIP, MYC, MYB and *DREB*, etc.), protein kinases (MAP kinase, CDP kinase, receptor protein kinase, ribosomal-protein kinase and transcription-regulation protein kinase, etc.) and proteinases (phosphoesterases and phospholipase C, etc.) involved in the regulation of signal transduction and gene expression (Agarwal et al. 2006).

Many genes whose expressions are induced by drought have been isolated and characterized at the cellular and molecular level (Wei et al. 2009; Zhu 2002; Shinozaki et al. 2002). Genetic engineering of plants for tolerance to extreme abiotic stresses could be achieved by the regulated expression of stress-induced transcription factors, which in turn would regulate the expression of a large number of relevant downstream genes. Thus, transcription factors are powerful tools for genetic engineering as their overexpression can lead to the up-regulation of a whole array of genes under their control (Agarwal et al. 2006). Dehydration-responsive element binding (*DREB*) proteins constitute a large family of transcription factors that are induced by abiotic stresses. They regulate a large number of functional genes related to drought, high-salinity and low temperature (Wei et al. 2009). The *DREB* transcription factors could be dichotomized as *DREB1* and *DREB2*, which are involved in two separate signal transduction pathways. To date, full-length sequences of *DREB* genes have been cloned from wheat (Shen et al. 2003a), rice (Chen et al. 2003), corn (Qin et al. 2007), Arabidopsis (Liu et al. 1998), and the halophyte *Atriplex hortensis* (Shen et al. 2003b).

Marker-assisted selection (MAS) provides a strategy for accelerating the process of wheat breeding (Ashraf 2010). Through marker-assisted breeding (MAB) it is now possible to examine the usefulness of thousands of genomic regions of a crop germplasm under water limited regimes, which was, in fact, previously not possible. Marker-assisted breeding approach is a prospective alternative to traditional breeding, because of being less time-consuming and labor- and cost-effective.

The Random Amplified Polymorphic DNA-PCR (RAPD-PCR) technology is a powerful tool in quickly identifying markers related to drought tolerance and RAPD technique was found to be quite effective in determining the genetic variation among wheat genotypes and could be utilized as DNA fingerprinting for variety identification and for the establishment of plant breeder rights (Iqbal et al. 2007).

However, conventional markers, such as restriction fragment length polymorphisms (RFLPs), random amplified polymorphic DNAs (RAPDs), amplified fragment length polymorphisms (AFLPs) and simple sequence repeats (SSRs), used in common wheat, are usually not developed from the genes themselves, because the cloning of genes in wheat is complicated by its allohexaploid ($2n=6x=42$) nature and large genome size. In contrast, functional markers (FMs) are usually designed from polymorphisms within transcribed regions of functional genes. Such markers are completely correlated with gene function (Andersen and Lübberstedt 2003). Therefore, FMs can dramatically facilitate accurate selection of target genes (Wei et al. 2009).

The aim of this study was to investigate structural and functional characteristics of thylakoid membrane and antioxidant enzymes that ensure resistance of plant organisms to water deficit. For this purpose wheat (*Triticum* L.) genotypes with contrasting genetic architecture and different genetically stipulated sensitivities to drought were used. Such approach allows to identify not only precise limits of variation of plant reaction to stress, but also to reveal specific features typical for high-resistant genotypes that may be taken into consideration in crop breeding practice for development of drought tolerant varieties. At the same time, RAPD and functional markers were used to confirm drought tolerance in 12 wheat genotypes.

2 Materials and Methods

2.1 Plant Material and Dehydration Conditions

A total of 12 wheat genotypes, three tetraploid (*Triticum durum* L., AABB, $2n=4x=28$) and nine hexaploid cultivars (*Triticum aestivum* L., AABBDD, $2n=6x=42$) including tolerant, moderately-tolerant and non-tolerant to drought were used (Table 17.1). Different sensitivities of these cultivars to drought had been determined during a few years in different regions of Azerbaijan based on grain yield (Aliev 1998, 2001). The plants were provided by the Experimental Station of the Research Institute of Agriculture (Baku, Azerbaijan). All genotypes were grown in field conditions in wide area under normal water supply and dryland conditions from November to May. A group of plants was cultivated under optimum irrigation condition (control), and another set of plants was subjected to water deficit. Dehydration was imposed by withholding water supply. Samples were collected from control and stressed plants at grain filling period up 9:30 a.m. to 10:30 a.m. Roots and shoots were separated, their fresh weights recorded and then these samples were dried in an oven for dry weight measurements. Three different samples for each treatment were taken and analyzed twice.

2.2 Relative Water Content

Leaf relative water content (RWC) was estimated gravimetrically following Tambussi et al. (2005).

2.3 Isolation of Thylakoid Membranes

Leaves were homogenized with a Waring blender at full speed four times for 20 s each in an ice-cold grinding chloroplast isolation medium (1:6 w/v) containing 0.4 M sucrose, 20 mM Tris, 10 mM NaCl, 1 mM EDTA (sodium salt), 5 mM sodium ascorbate, and 0.1% polyethylene glycol, at pH 7.8 following the procedure of Aliyev et al. (1992). The homogenate was filtered through four layers of cheesecloth twice. The filtrate was centrifuged at 200 x*g* for 5 min and then the supernatant

Table 17.1 Wheat genotypes and their drought tolerance status

Genotype name	Reaction to drought
Triticum durum L. (Tetraploid AABB)	
Barakatli-95	Tolerant
Garagylchyg-2	Sensitive
Gyrmyzy bugda	Tolerant
Triticum aestivum L. (Hexaploid AABBDD)	
Azamatli-95	Tolerant
Giymatly-2/17	Semi-tolerant
Gobustan	Tolerant
Gyrmyzy gul	Semi-tolerant
Tale 38	Semi-tolerant
Ruzi 84	Tolerant
12 nd FAWWON No 97 (130/21)	Sensitive
4 tn FEFWSN No 50 (130/32)	Semi-tolerant
Saratovskaya	Tolerant

centrifuged at 1,000 x*g* for 10 min. The chloroplast pellet was suspended for 30 min in a hypotonic buffer consisting of 5 mM Tris-HCl (pH 8.0) and 1 mM $MgCl_2$, and centrifuged at 5,000 x*g* for 20 min. The pelleted thylakoid membranes were resuspended in 5 mM Tris-HCl (pH 8.0). All steps were executed at 4°C.

2.4 *Chlorophyll and Protein Determinations*

The chlorophyll concentration was determined in 80% acetone extract (Mc-Kinney 1941). Protein determination was performed according to Sedmak and Grossberg (1977). All samples were frozen in liquid nitrogen and stored at –80°C until required.

2.5 *Electrophoresis Analysis*

For polypeptide analysis, the samples of thylakoid membranes were separated under denaturing conditions at 2–3°C in the presence of 0.1% (w/v) SDS using a 10–25% (w/v) linear gradient polyacrylamide gel (acrylamide: methylenebisacrylamide ratio = 30:0.8) in combination with the Laemmli buffer system (Laemmli 1970) as described previously (Guseynova et al. 2001). To each slot, 20–45 μl of samples (an equal Chl content) were applied. The gels were stained for 30 min with 0.04% (w/v) Coomassie brilliant blue G-250 (France) prepared in 3.5% perchloric acid ($HClO_4$). Immediately after electrophoresis the gels were scanned using an Ultroscan 2202 densitometer (LKB, Sweden) with a 633 nm laser as the light source. A set of standard proteins (kD) consisting of bovine serum albumin (66), glyceraldehyde-3-phosphate dehydrogenase (36), carbonic anhydrase (29), trypsinogen (24), trypsin inhibitor (20.1), and lactalbumin (14.2) (Sigma, USA) were used for the determination of molecular masses of the polypeptides.

2.6 *Fluorescence of Chloroplasts*

The measurements of fluorescence (F) at 77 K were performed using a Hitachi-850 (Japan) fluorescence spectrophotometer as reported previously (Asadov et al. 1986). Fluorescence emission spectra were corrected for the spectral sensitivity of the spectrophotometer using rhodamine B. Chlorophyll fluorescence was excited by dark blue light with wavelength of 440 nm. The samples on quartz glass fiber were quickly frozen at 77 K by dipping the glass fiber into liquid nitrogen.

2.7 *Assay of Electron Transport Activities*

Electron transport activities of chloroplasts isolated from control and drought-stressed plants were determined polarographically as O_2 evolution or uptake at 20°C using a water-jacketed Clark type oxygen electrode chamber under illumination with saturating white actinic light (850 μE m^{-2} s^{-1}), according to Aliyev et al. (1992). Chloroplast concentrations equivalent to 100 μg Chl were used for all measurements. Artificial donors and acceptors were added immediately before or during illumination. The following electron transport activities were assayed in μmol $O_2 \cdot mg^{-1}$ $Chl \cdot h^{-1}$. PS II activity ($H_2O \rightarrow K_3Fe(CN)_6$) was measured in a medium containing 330 mM sorbitol, 40 mM Hepes-NaOH, pH 7.6, 10 mM NaCl, and 5 mM $MgCl_2$ using 0.5 mM $K_3Fe(CN)_6$ as terminal electron acceptor. PS I activity was assayed in the reaction mixture contained in 2 ml, 80 mM sucrose, 30 mM Tris-HCl, pH 8.0, 10 mM NaCl, 10 mM $MgCl_2$, 1 mM sodium ascorbate, and 2 μM 3-(3-4-dichlorophenyl)-1,1-dimethylurea (in order to block electron transport from PS II), using 0.3 mM 2,6-dichlorophenolindophenol as electron donor and 50 μM methylviolgen as electron acceptor.

Measurements of photoinduced changes of fluorescence yield from F_0 level to F_{max} were carried out at room temperature using laboratory-built set-up as described earlier (Klimov et al. 1982). Potential to the formula:

$$\Phi_p = F_v/F_m = (F_m - F_o)/F_m$$

2.8 *Enzyme Extraction and Activity Determination*

Enzyme extract was prepared by homogenizing leaf material (1 gf. wt.) with a pestle in an ice-cold mortar with 0.05 M Na_2HPO_4/NaH_2PO_4 (pH 7.0) buffer. The homogenates were filtered through four layers of cheesecloth and then centrifuged at 4°C. The supernatant was collected and used for the assays of enzymatic activities.

2.9 Catalase (CAT)

The activity of catalase was determined as a decrease in absorbance at 240 nm for 1 min following the decomposition of H_2O_2 as described by Kumar and Knowles (Kumar and Knowles 1993). The reaction mixture contained 50 mM phosphate buffer (pH 7.0) and 15 mM H_2O_2 and the reaction was initiated by adding enzyme extract.

2.10 Ascorbat Peroxidase (APO)

The activity of ascorbat peroxidase was assayed according to Nakano and Asada (Nakano and Asada 1981). The assay mixture consisted of 0.05 mM ASA, 0.1 mM H_2O_2, 0.1 mM EDTA, 50 mM sodium phosphate buffer (pH 7.6), and 0.3 mL enzyme extract. The activity was measured as a decrease in absorbance at 290 nm for 30 s.

2.11 Glutathione Reductase (GR)

Glutathione reductase activity was determined at 340 nm for 10 min in the reaction mixture containing 100 mM potassium phosphate buffer (pH 7,8), 1 mM EDTA, 0,2 mM NADPH and 0.5 mM GSSG (Yannarelli and Fernandez-Alvarez 2007).

2.12 Superoxide Dismutase (SOD)

Superoxide dismutase activity was estimated by using SOD Assay Kit-WST (Sigma-Aldrich, USA). The absorbance was recorded at 450 nm and one enzyme unit of SOD activity was defined as the amount of enzyme required to cause 50% inhibition of the rate of NBT reduction. Protein content was determined according to Sedmak (Sedmak and Grossberg 1977) using bovine serum albumin as a standard.

2.13 DNA Extraction

DNA was extracted from leaves using CTAB method (Murray and Thompson 1980) with some modifications. In the growth room, 5–7 cm long pieces of fresh leaf material was cut from the plants and the leaf tissues were ground in a preheated 2×CTAB extraction buffer (100 mM Tris, pH 8, 1.4 M sodium chloride, and 20 mM EDTA, pH 8.0). Liquid nitrogen ground samples were also processed with CTAB buffer. The samples were incubated for 60 min in 60°C water bath with occasional vigorous shaking. The samples were mixed gently after adding 400 μl of chloroform and placed on an orbital shaker for 20 min at room temperature. After centrifugation at 5,000 rpm, an equal volume of cold absolute isopropanol was added to the supernatant. The solution was well mixed and incubated for 60 min at 20°C. The sample was centrifuged for 5 min at 5,000 rpm to pellet the DNA, which was followed by washing with 70% alcohol and then dried at 56°C for 5 min. DNA was resuspended by adding 300 μl TE buffer (10 mM Tris, 1 mM EDTA at pH 8.0).

2.14 DNA Quantification

After diluting, the DNA was quantified by taking the optical density (OD) at $\lambda = 260$ nm with a spectrophotometer ULTROSPEC 3300 PRO ("AMERSHAM", USA). The purity of genomic DNA was determined by the A260/A280 absorbance ratio. The quality was also examined by running the extracted DNA samples on 0.8% agarose gel stained with 10 mg/ml ethidium bromide in 1×TBE (Tris base, Boric acid, EDTA) buffer. The gel was visualized and photographed under UV light.

2.15 Polymerase Chain Reaction Conditions for RAPD

PCR was carried out essentially as described by Williams (Williams et al. 1990). Two 10-mer oligonucleotide primers (Eurogentec S.A., Belgique) were used for DNA amplification (Table 17.2). Amplifications were performed in "Applied Biosystems 2720 Thermal Cycler" as follows: first 4 min at 94°C followed by 10 cycles of: 1 min at 94°C, 1 min at 36°C and 1 min at 72°C. After that, for next 35 cycles, 0.2°C was added to annealing temperature. After the final cycle, samples were incubated at 72°C for 15 min and then held at 4°C prior to analysis.

2.16 Polymerase Chain Reaction Conditions for Functional Markers

PCR was carried out essentially, as described by Wei et al. (2009). Five pairs of genome-specific primers designed for the wheat *DREB 1* genes using the Primer Premier 5.0 software (http://www.premierbiosoft.com) were used for DNA amplification (Table 17.3). Genome-specific PCR was performed in a total volume of 20 μl containing 80 ng of genomic DNA, 1 × PCR reaction buffer, 0.25 μM of each primer, 0.45 mM of each deoxyribonucleotide, 4.0 mM $MgCl_2$ and 1.6 U of Taq DNA polymerase (Sigma, USA).

The PCR was carried out using "Applied Biosystems 2720 Thermal Cycler" as follows: initial denaturation at 94°C for 3 min; 34 cycles of 94°C for 1 min, an annealing step at variable annealing temperatures depending on the primer pairs for

Table 17.2 Primer nucleotide sequence used to amplify DNA

Primer designation	Sequence 5′→3′
P1	TCGGCGGTTC
P2	CTGCATCGTG

Table 17.3 Genome-specific primers of the wheat *DREB 1* genes used for PCR reactions

Primers	Sequences (5′→3′)	Chromosome location	Expected size (bp)	Ann. temp. (°C)
P18F	CCCAACCCAAGTGATAATAATCT	3B	717	50
P18R	TTGTGCTCCTCATGGGTACTT			
P20F	TCGTCCCTCTTCTCGCTCCAT	3D	1,193	63
P20R	GCGGTTGCCCCATTAGACATAG			
P21F	CGGAACCACTCCCTCCATCTC	3A	1,113	63
P21R	CGGTTGCCCCATTAGACGTAA			
P22F	CTGGCACCTCCATTGCCGCT	3D	596	63
P25F	CTGGCACCTCCATTGCTGCC	3A	596	57
PRa[a]	AGTACATGAACTCAACGCACAGGACAAC			

[a]PR is a public primer matched with P22F and P25F, respectively

1 min, 72°C for 1.5 min; and a final extension at 72°C for 10 min and then held at 4°C prior to analysis. The PCR products were electrophoresed on 2.5% agarose gels, stained with ethidium bromide and visualized under UV light by"Gel Documentation System UVITEK'.

3 Results and Discussion

The investigated genotypes responded to water deficit through various changes in physiological and biochemical processes. Significant differences in relative water content (RWC) were observed between normally irrigated plants and those subjected to water stress (Fig. 17.1). Genotype Giymatli-2/17 grown in normal water supply condition showed the higher RWC in the leaves. Drought-stress conditions induced a slightly larger decrease in RWC in the sensitive cv. Giymatli-2/17 than in the tolerant cv. Azamatli-95; dehydration decreased the RWC by 14% in comparison with fully irrigated plants. The rate of water loss during drought was low in Azamatli-95. The RWC lowered from 83.9% to 72.1% following stress. Exposure to drought caused a reduction in dry weight accumulation in Giymatli-2/17 plants, whereas it had smaller, insignificant effects in cv. Azamatli-95, even though both cultivars showed a certain drop in RWC.

A reduction in the total chlorophyll content and Chl a/b ratio occurred during drought stress (Fig. 17.2). This pattern of change was not evident in tolerant genotype Azamatli-95, in which these parameters did not change statistically, whereas

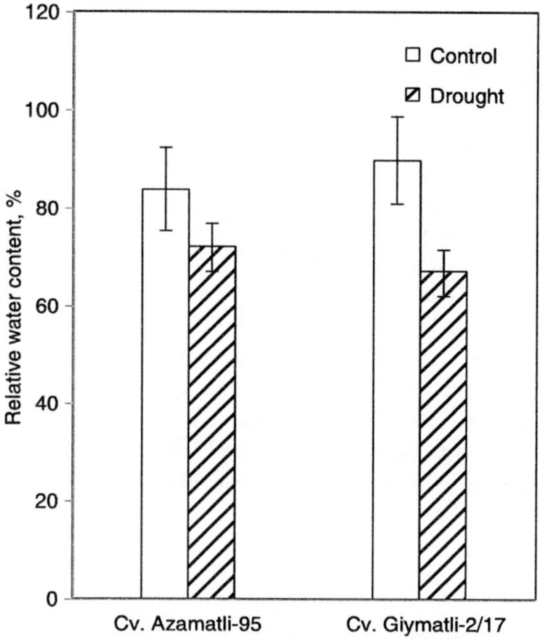

Fig. 17.1 Effect of water stress on relative water content of leaves in normally irrigated and drought stressed plants of *Triticum aestivum* wheat. Vertical bars represent SE of mean, $n=6$

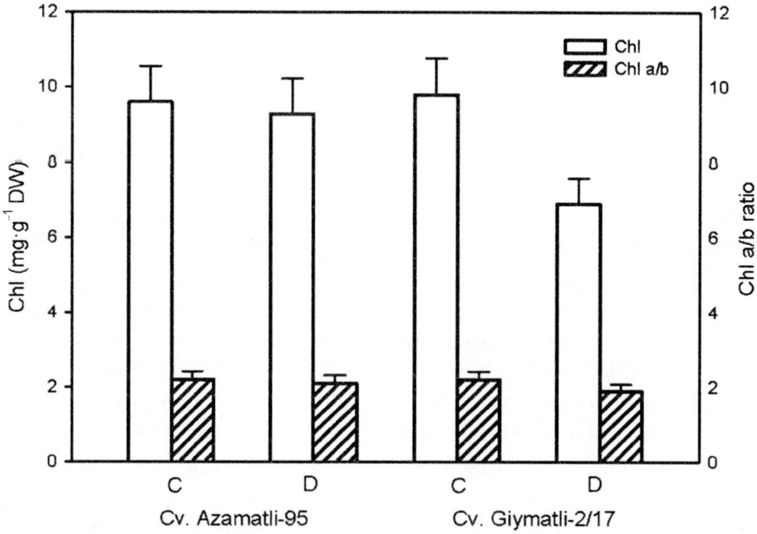

Fig. 17.2 Chlorophyll content of leaves of the wheat genotypes Azamatli-95 and Giymatli-2/17 during water deficit; C-control; D drought stress. Results are the means of three repetitions of two independent experiments SE $\pm$ $(n=6)$

the difference was significant in sensitive cv. Giymatli-2/17. A drought-induced decrease in pigment contents was previously reported in several plant species, including pea (Moran et al. 1994), durum wheat (Loggini et al. 1999) and *Boea hydroscopica* (Navari-Izzo et al. 2000). The drought-sensitive cv. Giymatli-2/17 showed a slight increase in the pool size of xantophyll-cycle components, but such effect was not shown in the tolerant cv. Azamatli-95, which may be explained by its higher rate of electron transport compared with cv. Giymatli-2/17 (Guseynova et al. 2006).

Total protein synthesis was slightly reduced by water deficit in these experiments. The decrease in thylakoid proteins observed during dehydration may be associated with degradation of lipoprotein thylakoid membrane structure. In addition, the photosynthetic apparatus may show acclimation responses such as changes in the relative proportion of stacked and unstacked membrane domains (Anderson and Aro 1994). At the ultrastructural level the thylakoid system of hydrated chloroplasts was organized in several well-defined and regularly distributed grana connected by parallel stroma lamellae. The increased thylakoid stacking in dried chloroplasts could be a consequence of membrane and/or environmental changes leading to a weakening of the repulsive force between the membrane surfaces. Another influential factor might be the rise due to water loss in the stroma ionic charge screening the repulsive force between thylakoids (Barber 1982).

The protein profiles of thylakoid membranes in nonstressed and water-stressed plants were analysed. Figure 17.3 shows density patterns from Coomassie blue staining SDS-PAGE analysis of membrane proteins of two wheat genotypes with differential tolerance to drought. As shown in Fig. 17.3, thylakoid membranes isolated from the wheat genotypes grown under normal water supply appeared to have about 26 polypeptides with M_r from 115 to 11 kD. It was found that Giymatli-2/17 genotype with broad and lodging leaves and drought-sensitive is characterized by low content of chlorophyll a-protein of photosystem (PS) I core (CP I) and β-subunit of CF_1 ATP-synthase complex, the high content of proteins in the 33–30.5 kD region and the relative high amount of polypeptides of light-harvesting complex (LHC) under normal irrigation in comparison with drought-tolerant genotype Azamatli-2/17, having vertically oriented small leaves. Drought stress caused significant changes in the content and composition of thylakoid membrane proteins. The content of CP I (115 kD) and apoprotein of P700 (63 kD) were maintained at relatively high levels in tolerant genotype Azamatli-95, but were slightly little affected by drought in more sensitive cv. Giymatli-2/17. It is interesting to note that the intensity of 60 kD polypeptide strongly increased (about 2-fold higher) in the drought-resistant genotype Azamatli-95. However, a detection of this polypeptide was not available in the experiments with seedlings of wheat grown in a growth chamber under controlled environment conditions (Guseynova et al. 2006). On the basis of the results of the present study and literature data, it is possible to suggest that this protein is related to dehydrins (PCA 60). Seasonal expression of dehydrins has been noted in several species (Wisnewski et al. 1999). The dehydrin family of proteins is induced by environmental stresses that result in cellular dehydration (Close 1997). All these protein groups are characterized with high hygrophilous

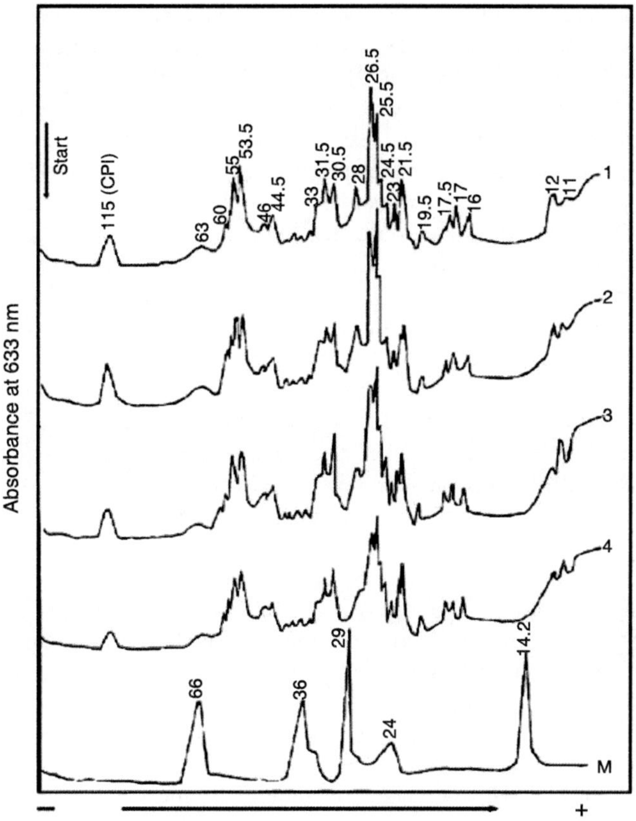

Fig. 17.3 Density patterns from Coomassie blue staining SDS-PAGE (10–25% gel) analysis of thylakoid membrane proteins from wheat plants grown in field conditions under normal water supply (Azamatli-95) (*1*) and Giymatli-2/17 (*3*) and drought stress (Azamatli-95) (*2*) and Giymatli-2/17 (*4*). M, standard proteins (kD): bovine serum albumin (66), glyceraldehydes-3-phosthate dehydrogenase (36), carbonic anhydrase (29), trypsinogen (24), and α-lactalbumin (14.2). Samples on an equal Chl content (25 μg) were placed on the gel. The samples were applied to each slot corresponding to 50 μg of protein

protein molecules. During dehydration of cells they prevent water loss on account of high hydrophilic capacity and stabilize cell proteins. PCA 60 was freely distributed in the cytosol, plastid, and nucleus. Although the functional role of dehydrins remains speculative, the data support the hypothesis that it plays a role in preventing denaturation of proteins exposed to dehydrative stresses in a manner similar to chaperones.

The synthesis of α- and β-subunits of CF_1 ATP synthase complex (55 and 53.5 kD, respectively) tended to increase slightly in stressed plants of Azamatli-95 and decrease in cv. Giymatli-2/17. The low content of β-subunits of CF_1 ATP synthase complex has been also shown in pea plants subjected to water deficit at high light exposure (Giardi et al. 1995; Guseynova et al. 2006). Steady-state levels of the core antenna of PS II (CP 47 and CP 43) serving as the connecting antenna between the main light harvesting complex LHC II and reaction center of PS II remained more or less unchanged in both genotypes. These results agree with the data reported earlier (Masojidek et al. 1991; Giardi et al. 1995; Guseynova et al. 2006).

The most striking change was the appearance of protein with molecular mass of 40.5 kD in tolerant genotype Azamatli-95. It is absent in leaves from nonstressed plants, but at a lower level it was detected in only tolerant genotypes, subjected to water deficit. According to the current literature, C 40.4 protein shares high sequence homology with CDSP (termed CDSP for chloroplastic drought-induced stress protein), the previously described accumulation of a 34 kD thylakoid protein in tomato in response to drought (Pruvot et al. 1996). Substantial increases in CDSP 34 transcript and protein abundance were also observed in potato plants, subjected to high illumination (Gillet et al. 1998). The accumulation of two chloroplastic nuclear-encoded proteins in water-stressed *Solanum tuberosum* plants were reported (Quartacci et al. 1995). A stromal protein of 32 kD related to thioredoxins, was suggested to maintain the redox state of chloroplastic proteins upon drought stress (Rey et al. 1998). Another protein of 34 kD, named CDSP 34 protein is proposed to participate in structural stabilization of thylakoids upon environmental constraints and prevent damage resulting from osmotic or oxidative stress (Pruvot et al. 1996). It is supposed that C40.4 protein is closely bound to LHC II and has functional role by modeling photosynthetic effectiveness and light dissipation of excess absorbed light energy inside antenna complex (Monte et al. 1999).

At the same time, in the sensitive genotype Giymatli-2/17 there was a considerable decrease in the amount of proteins in 33–30.5 kD region. The decrease in the content of 31.5 kD protein in thylakoid membrane from water-stressed plants (especially in cv. Giymatli-2/17) seems to be due, in part to its enhanced degradation rate (Guseynova et al. 2006; Sippola et al. 1998). High rate of D_1-protein turnover provides stability of thylakoid membranes and their electron transport chain to damaging action of free radical forms under stress conditions. On the other hand, thylakoid membranes from stressed plants showed an increased level of LHC polypeptides (28–24.5 kD) in tolerant cv. Azamatli-95 compared to Giymatli-2/17, at which the levels of these units decreased.

A slight increase in 21.5 kD polypeptide (according to literature it is related to WSCP-water-soluble chlorophyll proteins) was also observed in both genotypes under drought. Such effect was found in our previous studies with durum wheat seedlings under water stress (Guseynova et al. 2006). It is supposed that this protein might involve in the decrease of protease activity in leaf senescence.

Drought also caused a decrease in the synthesis of low molecular weight polypeptides of 17.5–12 kD in both genotypes under extreme conditions of water supply. The intensity of 11 kD polypeptide slightly increased in cv. Azamatli, but significantly decreased in cv. Giymatli-2/17. Correlation between tolerance and overexpression of some proteins including 60, 40.5, and 28–24.5 kD assumes that changes in expression of these polypeptide genes can be functionally involved in the ability of plants to survive and grow under water deficiency.

Table 17.4 Photosynthetic membrane proteins from wheat chloroplasts subjected to changes under drought stress

Samples	Molecular mass of proteins, kD[a]					
Azamatli-95 (control)	CP I, 115 (PS I core)	60	55 and 53.5 α and β sub. CF_1	33–30.5	28–24.5 (proteins of LHC)	21.5
Azamatli-95 (drought)	0	+	+	0	+	+
Giymatli-2/17 (control)	CP I, 115 (PS I core)	60	55 and 53.5 α and β- sub. CF_1	33–30.5	28–24.5 (proteins of LHC)	21.5
Giymatli-2/17 (drought)	–	0	–	–	–	+

[a](+) protein content is increasing, (–) content is decreasing, 0 not changes

According to the current literature, there is a cycle of PS II reparation during which the most damaged 32 kD protein (D_1) of reaction center of PS II is replaced (Melis 1998). Selective proteolysis is involved, inactive form of D_1-protein is removed, and newly synthesized D_1-polypeptide is integrated into the PS II holocomplex (Sippola et al. 1998; Chaloub et al. 2003). High rate of D_1-protein turnover provides stability of thylakoid membranes and their electron-transport chain to damaging action of free radicals formed under stress conditions (Guseynova et al. 2006). Thus, the literature and our results suggest that the biochemical response at the level of D_1-turnover and intensive synthesis of polypeptides 60, 40.5 and 28–24.5 kD could act as a general adaptation signal for the plant in response to water stress. Table 17.4 presents the data of membrane proteins in which quantitative changes are significant under drought stress.

In parallel the fluorescence emission spectra (77 K) of chloroplasts from normally irrigated and drought-stressed plants were also measured. As shown in Fig. 17.4, chloroplasts from drought-sensitive genotype Giymatli-2/17 has more intensive fluorescence at 740 nm from PS I under normal water supply. The F687/F740 ratio of control (non-drought stressed) chloroplasts of genotype Azamatli-95 was close to 0.38 and for genotype Giymatli-2/17-0.35. The shift of the main peak from 742 to 740 (in Azamatli-95) and from 740 to 738 nm (in Giymatli-2/17) was observed in both genotypes grown under water deficit. According to the data on the contents of pigments in leaves with normal irrigation and in the plants subjected to water deficit, a short wavelength shift of the main maximum in the fluorescence spectra was coupled with a decrease in the amount of chlorophyll in PS I antenna (see Fig. 17.2). The fluorescence intensity at 740 also slightly increased. The short wavelength peaks at 687 and 695 nm (fluorescence from the PS II core complex CP 47 and CP 43) remained and their fluorescence intensities started to increase sharply under water deficit. It is especially observed in drought-sensitive genotype Giymatli-2/17. At the same time, in chloroplasts from stressed plants, the F687/F740 ratio increased compared with the control plants; the lowest value was that of the Azamatli-95 (F687/F740=0.45) and the highest of Giymatli-2/17 (F687/F740=0.77), suggesting again that the most detrimental influence of drought stress occurred in Giymatli-2/17. The results suggest that antenna system of the photosynthetic

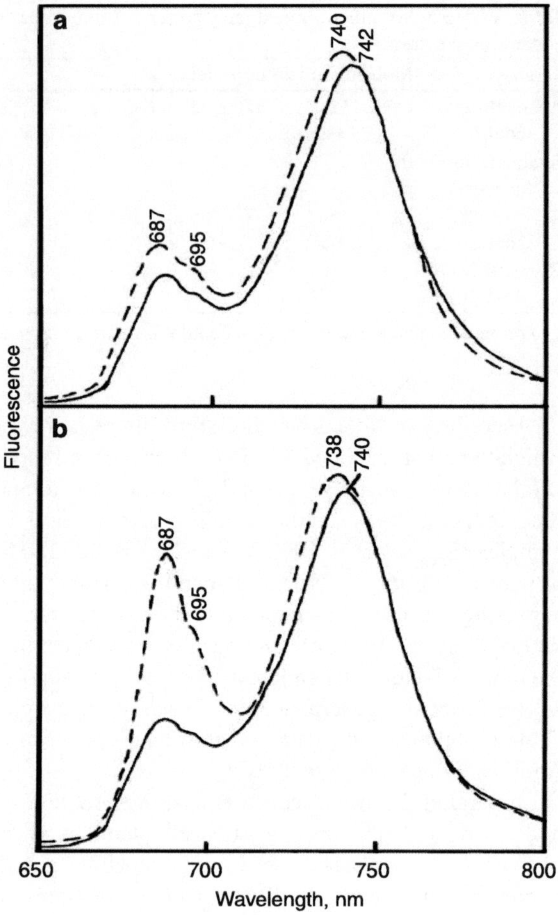

Fig. 17.4 Fluorescence emission spectra at 77 K of chloroplasts from drought-tolerant Azamatli-95 (**a**) and drought-sensitive Giymatli-2/17 (**b**) genotypes grown under normal water supply (*solid curves*) or drought conditions (*dashed curves*)

apparatus in the drought-tolerant genotype Azamatli-95 was rapidly reorganized, and plants began to adapt to environmental stress. More frequently changes in F687/740 ratio may be explained by redistribution of excitation light energy between PS II and PS I. This rise in short wave fluorescence intensity may have been due to the lower content of RC′s of PS II synthesized under water deficit.

Significant differences were found in functional activity of photosynthetic apparatus at the level of photochemical reactions of chloroplasts in comparative studies of genotypes distinguishing by the genetic architecture and drought resistance. In our experiments, the highest PS II activity (oxygen evolution rate) of irrigated plants was found in drought-sensitive genotype Giymatli-2/17 with broad and drooping leaves (Table 17.5). Drought stress caused a significant change in the photochemical activity of chloroplasts in both genotypes. The electron transport activities of all stressed plants were lower than those in the control plants. However, the activity of PS II was significantly affected by dehydration in cv. Giymatli-2/17 only 41% of

Table 17.5 The Photosystem II and Photosystem I activity in chloroplasts from wheat genotypes subjected to drought stress (μmol O_2·mg^{-1} chlorophyll·h^{-1})

Genotypes	Photosystem II $H_2O \rightarrow K_3Fe(CN)_6$	In %	Photosystem I DCIP·H→MV	In %
Azamatli-95 (control)	45±4	100	250±12	100
Azamatli-95 (drought)	35±3	78	225±9	90
Giymatli-2/17 (control)	85±7	100	190±8	100
Giymatli-2/17 (drought)	35±4	41	150±4	79

control values. In drought-stressed cv. Azamatli-95 leaves the photochemical activity of PS II was about 78% of the control value. The case of PS II inactivation in both genotypes may be due to the suppression of synthesis of 32 kD protein (D_1-protein of RC of PS II), which is carrier of photochemical active form of Chl a P680, or breach of electron transfer from pheophytin intermediate electron carrier on quinone acceptor (Q_A) in non-cyclic transport of electrons. It is likely that, desiccation inhibited the energy transfer from the Chl molecules anchor to PS II core complexes. PS I activity (O_2 uptake rate), however, was affected much less under drought stress. It can be caused by a higher ability of PS I to adapt to dehydration.

Concerning the other drought stress actions, several authors reported that PS II photochemistry is predisposed by drought stress to photoinhibitory damage (Björkman and Powles 1984; Peltier et al. 1995). In contradiction, Genty et al. (1987) concluded that PS I-mediated electron transport was inhibited by drought, whereas PS II electron transport remained the unaffected. During rehydration, PS II activities recovered slowly, but PS I complexes recovered their functional forms very quickly (within 1 min) (Hirai et al. 2004).

It is known that fluorescence yield is minimal (F_o), when primary electron acceptor of PSII, i.e. plastoquinone (Q_A), is oxidized. Reduction of Q_A results in rise of chlorophyll fluorescence, approximately 3–5 times, up to F_m level. Rise in chlorophyll fluorescence yield from initial (F_o) to maximal (F_m) level, i.e. appearance of variable fluorescence (F_v, where $F_v = F_m - F_o$), reflects process of accumulation of reaction centers (RC) of PSII in "closed" state with reduced primary quinone acceptor (Q_A). Values of fluorescent parameters that characterize the functional state of photosynthetic apparatus of winter wheat plants grown under different conditions of water regime are shown in Table 17.6. Potential quantum yield of photochemical reactions of PSII (F_v/F_m ratio) in chloroplasts from control (non-drought stressed) plants was 0.74 for Azamatli-95 and 0.81 for Giymatli-2/17, that is typical for normally grown plants. As it is evident from Table 17.6, the state of PSII in dehydration process was significantly changed. Potential yield of photochemical reactions of PSII underwent appreciable changes in comparison with control plants; the highest value of F_v/F_m was in Azamatli-95 ($F_v/F_m = 0.71$) and the lowest in Giymatli-2/17 ($F_v/F_m = 0.69$). It is interesting to note, that chloroplasts from non-drought-tolerant genotype Giymatli-2/17 had higher value of photochemical efficiency of PSII under regular irrigation conditions of growth. However, low ratio of F_v/F_m again confirms that strong effect of drought appeared in genotype Giymatli-2/17 (genotype

Table 17.6 Change of parameters of chlorophyll fluorescence in chloroplasts isolated from wheat leaves after drought. Fluorescence components: F_o – constant fluorescence; F_v – variable fluorescence; F_m – maximal fluorescence[a]

Variant	Control	Drought	% from control
Azamatli-95			
F_o	29.0±1.2	30.0±2.8	103
F_v	85.0±6.1	74.5±5.4	87
F_v/F_m	0.74	0.71	96
Giymatli-2/17			
F_o	27.0±1.1	28.5±2.9	106
F_v	118.0±6.5	63.0±4.3	54
F_v/F_m	0.81	0.69	85

[a]Average arithmetic and standard mistakes from three independent experiments, each of which was carried out in double biological frequency are shown in the table

Giymatli-2/17 is strongly affected by drought). Decreasing of a photochemical efficiency (F_v/F_m) under severe drought can be considered as the main reason of damage of photosynthetic reaction centers.

Both Q_B-reducing and Q_B-non-reducing complexes of PSII make a contribution in variable fluorescence (F_v). Charge separation is realized in Q_B-non-reducing complexes of PSII, but electrons are not transported to plastoquinone pool. Q_B-reducing complexes of PSII in active state are able to realize electron transport between Q_A and Q_B. They lose this ability when D_1-protein is damaged and turn to Q_B-non-reducing complexes (Pshibytko et al. 2003). In optimal conditions due to reactions of reparation cycle the constant ratio between these types of complexes of PSII is supported. Probably, dehydration induces disruption of reactions at the acceptor side of PSII, expressed in increasing a number of Q_B-non-reducing centers.

Under water deficit, linear electron transport is suppressed by the accumulation of plastoquinones caused from difficulty of lateral diffusion of plastoquinones because of increased viscosity of lipid bilayer (Hirai et al. 2004), that could be caused by increasing of reduced level of plastoquinone pool, damage of Q_B-binding site with D_1-protein and deterioration of conditions of damaged D_1-protein reparation (Melis 1998).

The results presented here show that functioning dynamics of CAT and APX in well-watered plants through ontogenesis practically did not change both among durum and among bread wheat cultivars. In Barakatli-95 and Garagylchyg-2 both enzymes exhibited a maximal activity in the end of flowering, in Azamatli and Giymatli-2/17 – in the end of ear formation (Fig. 17.5). Functioning of these enzymes during ontogenesis under water deficit differed: CAT activity increased in all stressed genotypes as compared with control: in durum wheat cultivars maximal activity was observed in the milk ripeness and in bread wheat cultivars at the end of flowering. In drought-tolerant genotypes Barakatli-95 and Azamatli-95 CAT activity increased more substantially as compared to that in the sensitive ones. The increase in CAT activity in plants has been reported in other studies under water stress (Mukherjee and Choudhuri 1983; Quartacci and Navari-Izzo 1992). It is

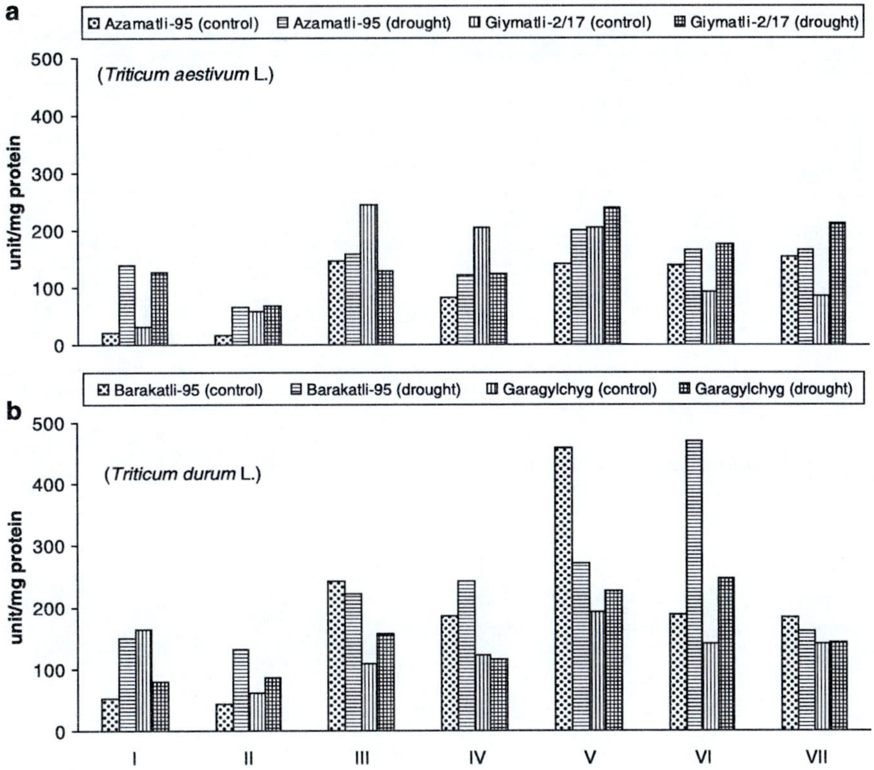

Fig. 17.5 Effect of water stress at different ontogenesis stages on CAT activity (unit/mg protein). (*I* – stalk emergence, *II* – beginning of earing, *III* – end of earing, *IV* – flowering, *V* – end of flowering, *VI* – milky ripeness, *VII* – wax ripeness)

known that catalase reacts with H_2O_2 directly to form water and oxygen (Smirnoff 1993). The decrease in CAT activity in the end of ontogenesis could indicate its inactivation by the accumulated hydrogen peroxide by water shortage and could be explained partly by photoinactivation of the enzyme. When plants are not exposed to water stress, resynthesis of CAT compensates for the loss of total activity caused by irradiance. Inhibition of protein synthesis induced by water stress (Badiani et al. 1990) conceivably could impair resynthesis and partly account for the marked decrease in CAT activity in plants subjected to water stress in the light.

APX activity also increased under water deficit: in tolerant wheat genotypes maximal activity occurred at the end of flowering, and in the sensitive ones in the end of ear formation (Fig. 17.6). An increase in POD activity was also observed by different authors during drought and salt stress (Badiani et al. 1990; Siegel 1993). It indicates the formation of large amounts of H_2O_2 during water stress. Increased activity can explain with some assumptions: elevated H_2O_2 concentrations could release POD from membrane structures, with which it is normally associated. POD

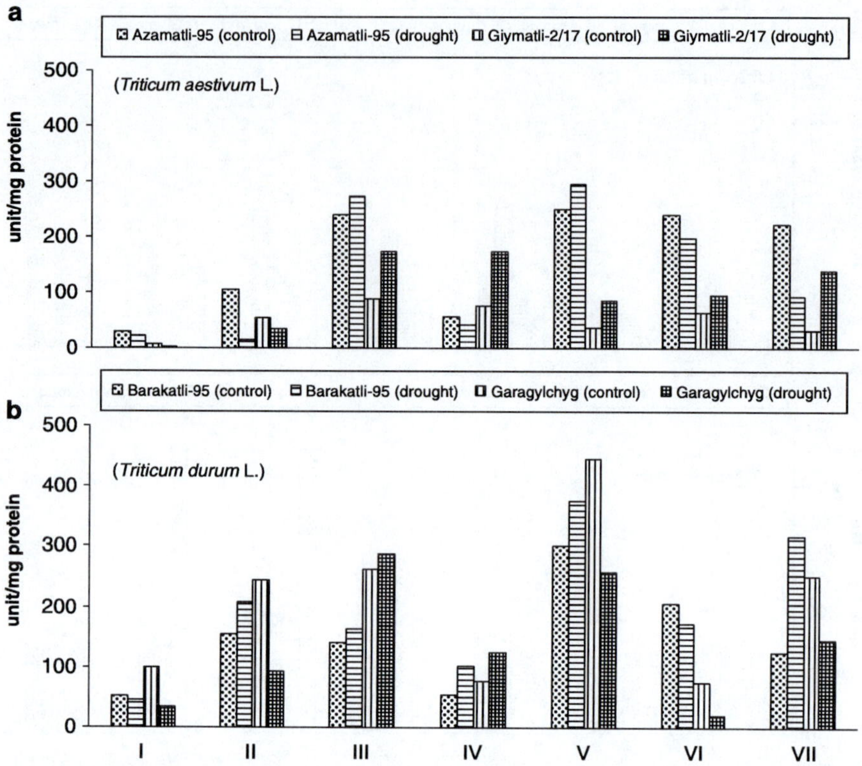

Fig. 17.6 Effect of water stress at different ontogenesis stages on APX activity (unit/mg protein). (*I* – stalk emergence, *II* – beginning of earing, *III* – end of earing, *IV* – flowering, *V* – end of flowering, *VI* – milky ripeness, *VII* – wax ripeness)

could be synthesized *de novo* at least in some cases. Water stress could increase the accumulation of POD substrates, such as glutathione, ascorbate, and phenolic compounds, which, in turn, are scavengers of activated oxygen species (Winston 1990). Also it is known, that H_2O_2 participates in signal transduction at development of oxidizing stress, inducing genes of cytosolic POD. The maximum activity of GR both in the control, as well as in drought-subjected plants was observed at the anthesis stage. GR activity in drought-tolerant durum wheat Barakatli-95 and resistant bread wheat Azamatli-95 was higher than the control in all stages of ontogenesis (Fig. 17.7).

SOD functioning dynamics through ontogenesis differed from CAT and APX. Interestingly, SOD activity was lower than the control during ontogenesis, except at the last stages. In drought-tolerant Barakatli-95 and Azamatli-95 it increased against the control only at wax ripeness stage, when the effect of drought was the highest (Fig. 17.8). Many authors specify the key role SOD in antioxidative protection (Raychaudhuri 2000; Alscher et al. 2002). However, the results reported here on

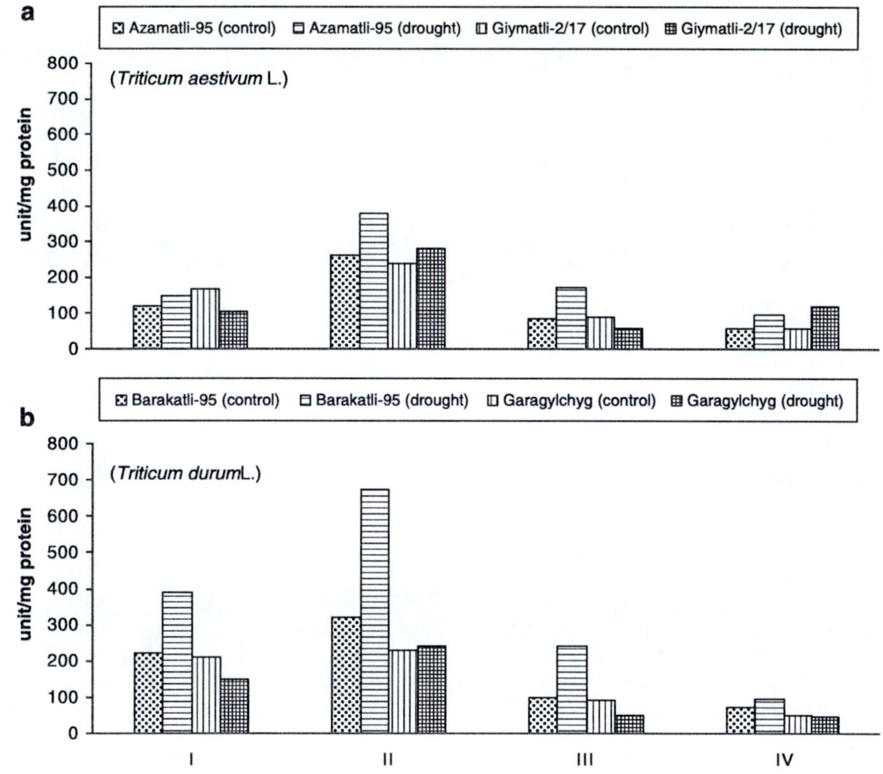

Fig. 17.7 Effect of water stress at different ontogenesis stages on GR activity (unit/mg protein). (*I* – earing, *II* – flowering, *III* – milky ripeness, *IV* – wax ripeness)

SOD activity differed partly from the literature. Nevertheless, it is known that plant cells contain a little isoform of SOD which probably unequally reacts to water deficiency. The study of these SOD isoforms during water stress induction revealed differential regulation of their activities: MnSOD and FeSOD activities increased rapidly while Cu/ZnSOD activities decreased in cowpea plants (Brou et al. 2007). Also direct correlation between APX activity and carotenoid content was observed through ontogenesis. It should be noted that in plants subjected to drought PS II and PS I activities were also high both at the end of ear formation and flowering stages (Tables 17.7 and 17.8).

RAPD-PCR analysis was carried out using seedlings of 12 wheat genotypes *Triticum* L. with different levels of drought tolerance (Table 17.1). For this purpose, P6 and P7 RAPD markers associated with drought tolerance were used (Pakniyat and Tavakol 2007). According to the literature data in tolerant genotypes these primers should produce appropriate fragments.

Figure 17.9 shows the electrophoretic patterns of PCR products, obtained by applying P6 primer (5′ TCGGCGGTTC 3′). This primer produces 920 bp band. As

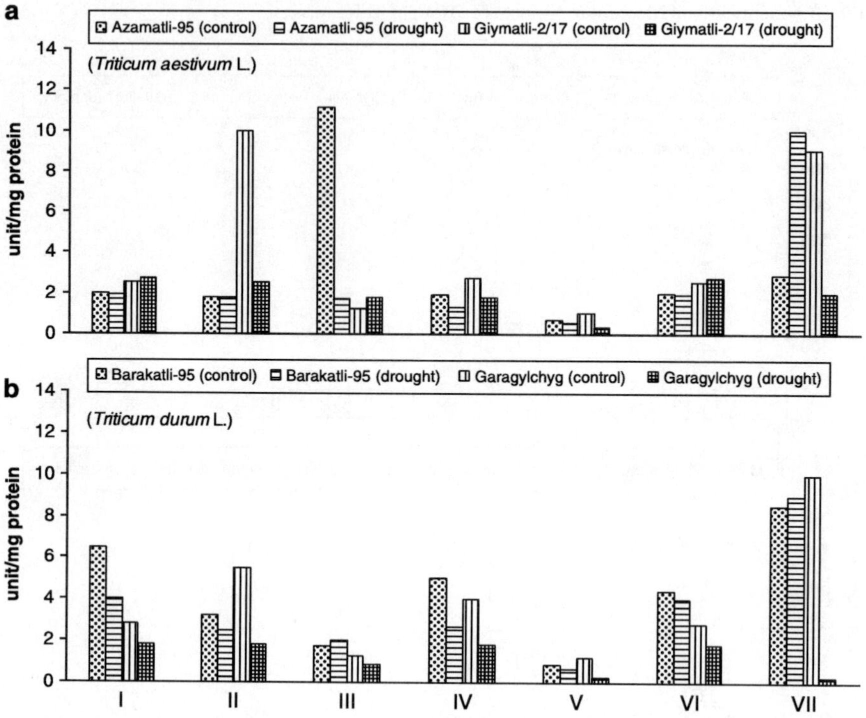

Fig. 17.8 Effect of water stress at different ontogenesis stages on SOD activity (unit/mg protein). (*I* – stalk emergence, *II* – beginning of earing, *III* – end of earing, *IV* – flowering, *V* – end of flowering, *VI* – milky ripeness, *VII* – wax ripeness)

Table 17.7 Photosystem II activity in chloroplasts from wheat genotypes at different stages of ontogenesis (μmol $O_2 \cdot mg^{-1}$ chlorophyll$\cdot h^{-1}$)

Genotypes	Stalk emergence	Beginning of earing	End of earing	Flowering	End of Flowering	Milky ripeness	Wax ripeness
Barakatli-95 (control)	20	24	120	100	84	40	12
Barakatli-95 (stress)	60	60	160	92	100	45	26
Garagylchyg (control)	80	108	84	120	100	24	20
Garagylchyg (stress)	72	72	120	112	108	27	24
Azamatli-95 (control)	88	100	120	88	100	28	40
Azamatli-95 (stress)	104	72	136	120	92	16	8
Giymatli-2/17 (control)	140	72	140	112	120	20	12
Giymatli-2/17 (stress)	120	96	160	104	120	22	8

Table 17.8 Photosystem I activity in chloroplasts from wheat genotypes at different stages of ontogenesis (μmol $O_2 \cdot mg^{-1}$ chlorophyll·h^{-1})

Genotypes	Stalk emergence	Beginning of earing	End of earing	Flowering	End of flowering	Milky ripeness	Wax ripeness
Barakatli-95 (control)	200	600	820	600	740	680	560
Barakatli-95 (stress)	540	400	840	500	520	500	280
Garagylchyg (control)	420	300	540	500	460	360	280
Garagylchyg (stress)	320	100	560	440	440	360	280
Azamatli-95 (control)	400	180	640	400	500	400	240
Azamatli-95 (stress)	400	180	640	440	520	440	160
Giymatli-2/17 (control)	400	240	740	420	440	440	120
Giymatli-2/17 (stress)	360	180	760	420	440	440	160

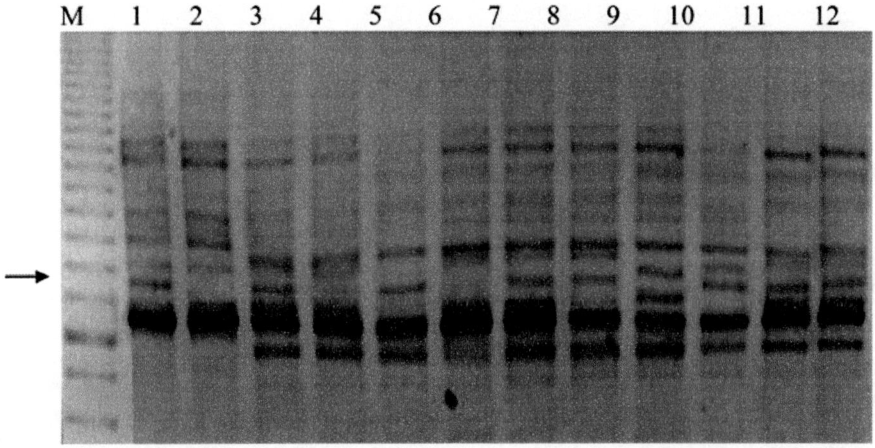

Fig. 17.9 PCR amplification profiles of *Triticum* L. wheat genotypes using P6 primer (5′ TCGGCGGTTC 3′). The arrow shows 920 bp DNA fragment, present in drought tolerant varieties and absent in non-tolerant. *M* – DNA ladder 100. *1* – Barakatli-95, *2* – Garagylchyg-2, *3* – Azamatli-95, *4* – Giymatli-2/17, *5* – Gyrmyzy bugda, *6* – Gyrmyzy gul, *7* – Tale 38, *8* – Ruzi 84, *9* – 12 nd FAWWON No 97 (130/21), *10* – 4 tn FEFWSN No 50 (130/32), *11* – Nurlu-99, *12* – Gobustan

it follows from figure, in drought-sensitive durum wheat genotype Garagylchyg-2 and moderately-tolerant bread wheat genotypes Giymatli-2/17 and Qyrmyzy gul 920 bp bands were not revealed. This locus is well seen on PCR amplification profiles from all other genotypes supporting their drought tolerance. It is noticeable

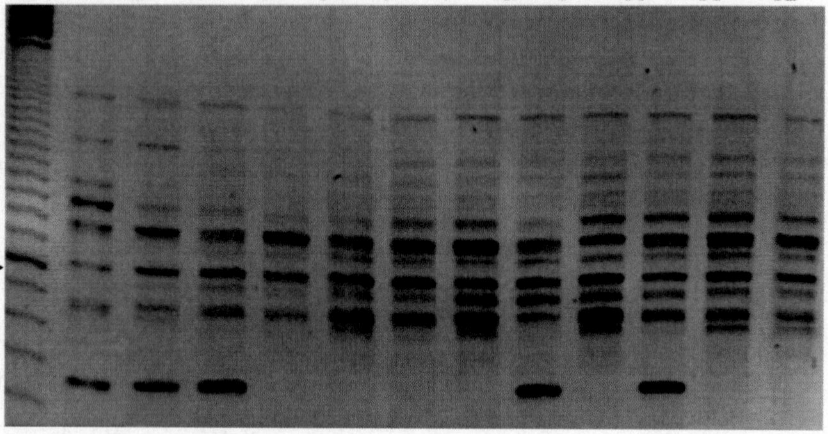

Fig. 17.10 PCR amplification profiles of wheat genotypes *Triticum* L. using a primer P7 (5′ TCGGCGGTTC 3′). The arrow shows 750 bp DNA fragments. *M* – DNA ladder 100. *1* – Barakatli-95, *2* – Garagylchyg-2, *3* – Azamatli-95, *4* – Giymatli-2/17, *5* – Gyrmyzy bugda, *6* – Gyrmyzy gul, *7* – Tale 38, *8* – Ruzi 84, *9* – 12 nd FAWWON No 97 (130/21), *10* – 4 tn FEFWSN No 50 (130/32), *11* – Nurlu-99, *12* – Gobustan

that 12nd FAWWON No 97 (130/21) is estimated as non-tolerant to drought. However, as the electrophoretic patterns indicate, 920 bp locus is also present in 12nd FAWWON No 97 (130/21). In terms of genetics it must have a potential for tolerance, but for some reasons expression of these genes not come true. The second RAPD primer in our experiments was P7 marker (5′ TCGGCGGTTC 3′). This primer produces 750 bp band. Figure 17.10 demonstrates that this band has occurred neither in tolerant variety Barakatli-95, nor in non-tolerant varieties Garagylchyq-2 and Giymatli-2/17. Meanwhile, in drought sensitive variety Gyrmyzy gul this locus has occurred. Therefore from the PCR results carried out with this primer it can be conclude that P7 RAPD marker is not absolutely universal for drought tolerance.

At the same time PCR analysis was carried out to identify *DREB 1* genes, which was responsive for drought tolerance in 12 wheat *Triticum* L. genotypes with different levels of resistance to drought. For this purpose functional markers for *DREB 1* genes, especially synthesized for A, B and D wheat genomes, were used. P25F/PR was designed to amplify a 596-bp DNA fragment downstream of *DREB-A1* in the A genome. P21F/P21R was selected to amplify an upstream region (1113-bp DNA fragment) of the same gene. Similarly, P22F/PR and P20F/P20R were designed to amplify sequences from the D genome, with the amplifications resulting in 596 and 1193-bp DNA fragments, respectively. The P18F/P18R primers, which amplify a 717-bp DNA fragment, were designed as a B genome-specific primer pair (Table 17.3).

Figure 17.11 demonstrates gel electrophoresis of PCR profiles of amplified DNA from *Triticum* L. using the primer pair P21F/P21R. As shown in Fig. 17.11, fragment amplified with this marker in 1113-bp region is present in the high drought

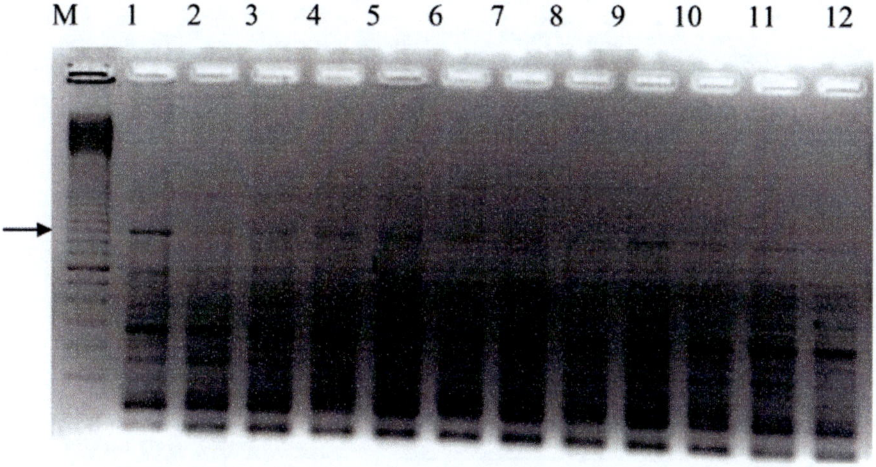

Fig. 17.11 PCR amplification profiles of wheat genotypes *Triticum* L. using an A genome-specific primer pair P21F/P21R. The *arrow* shows 1113 bp DNA fragments. *M* DNA ladder 100. *1* – Barakatli-95, *2* – Garagylchyg-2, *3* – Gyrmyzy bugda, *4* – Azamatli-95, *5* – Giymatli-2/17, *6* – Gobustan, *7* – Gyrmyzy gul, *8* – Tale-38, *9* – Ruzi-84, *10* – 12nd FAWWON No 97 (130/21), *11* – 4tn FEFWSN No 50 (130/32), *12* – Saratovskaya

tolerant durum wheat genotype Barakatli-95. Except in moderately-tolerant bread wheat genotype, *DREB 1*-responsive fragment was obtained in both drought tolerant and sensitive genotypes. This indicates that in these genotypes *DREB 1* gene responsive for tolerance to drought as well as other abiotic stresses is in the third chromosome of A genome (Wei et al. 2009). The second primer for *DREB 1* gene in A genome is a P25F/PR that amplifies fragments of 596-bp.

As it seemed from Fig. 17.12, 596-bp fragments were not synthesized in selected genotypes. Absence of these fragments can be explained by some mutations that, probably, took place in *DREB 1* gene region, complementary to this primer. Thus the presence of *DREB 1* gene in A genome is confirmed by PCR results using the primer pair P21F/P21R.

The results obtained using primer pair P18F/P18R, specific for *DREB 1* gene in B genome, are shown in Fig. 17.13. As it seemed from Fig. 17.3, fragment pair 717–789 bp revealed only in Barakatli-95. This shows that *DREB 1* gene also occurred in Barakatli-95 B genome. It should be noted that in our experiments this genotype also shows its high drought tolerance in other parameters (Aliev 1998, 2001). As to the absence of this pair of fragments in other wheat genotypes it can be explained from the literature that the *DREB1* proteins showed the most specific variations in the B genome, including three single amino acid mutations (amino acids 46, 140 and 200) and a deletion of 24 amino acids in a region rich in Ser and Thr in the orthologous A and D genomes.

PCR analysis with P22F/PR and P20F/P20R primers specific for D genome was also carried out. As it is known D genome occurs in hexaploid genotypes *Triticum*

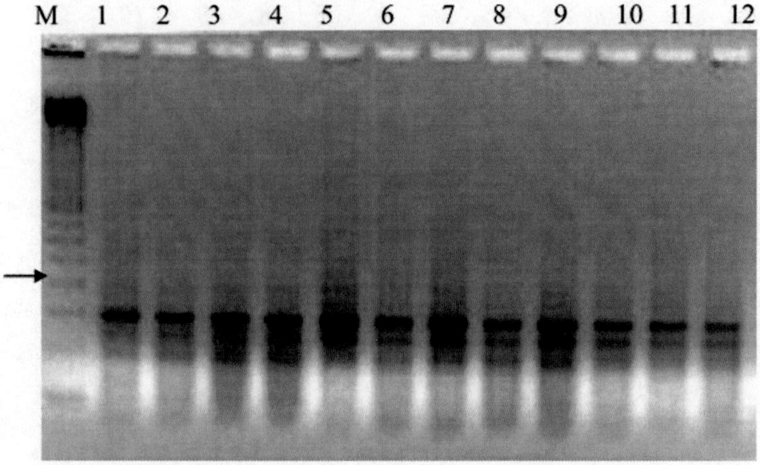

Fig. 17.12 PCR amplification profiles of wheat genotypes *Triticum* L. using an A genom-specific primer pair P25F/PR. The *arrow* shows 596-bp DNA fragment. *M* DNA ladder 100. *1* – Barakatli-95, *2* – Garagylchyg-2, *3* – Gyrmyzy bugda, *4* – Azamatli-95, 5 – Giymatli-2/17, *6* – Gobustan, *7* – Gyrmyzy gul, *8* – Tale-38, *9* – Ruzi-84, *10* – 12nd FAWWON No 97 (130/21), *11* – 4tn FEFWSN No 50 (130/32), *12* – Saratovskaya

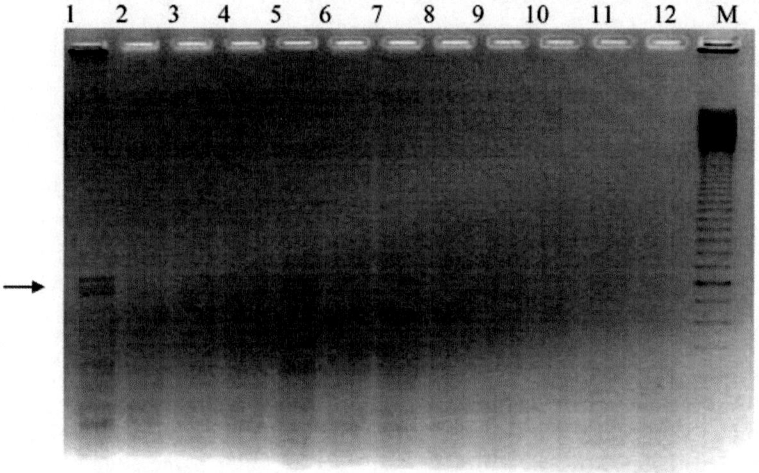

Fig. 17.13 PCR amplification profiles of wheat genotypes *Triticum* L. using a B genome-specific primer pair P18F/P18R. The *arrow* shows 717–789 bp DNA fragment. *M* DNA ladder 100. *1* – Barakatli-95, *2* – Garagylchyg-2, *3* – Gyrmyzy bugda, 4 – Azamatli-95, *5* – Giymatli-2/17, *6* – Gobustan, *7* – Gyrmyzy gul, *8* – Tale-38, *9* – Ruzi-84, *10* – 12nd FAWWON No 97 (130/21), *11* – 4tn FEFWSN No 50 (130/32), *12* – Saratovskaya

aestivum L. In our experiments there were no appropriate fragments responsive to these markers. This allows us to suppose that the hexaploid wheat varieties used here were possibly nulliosomics, i.e. chromosome pair possibly a *DREB 1* gene is absent from the D genomes.

4 Conclusion

So, dehydration involves many changes resulting in disrupted membrane integrity, suppression of many photosynthetic genes, decreased activities of photosystems, etc. The higher water content and its better distribution in the stressed cv. Azamatli-95 permitted the plants to retain a higher turgor in comparison with cv. Giymatli-2/17, which resulted in the maintenance of growth. Drought induced photoinhibition and photodestruction of pigments and pigment-protein complexes and destabilization of photosynthetic membrane. In cv. Azamatli-95 the lack of changes in pigment content and composition following drought indicated the capacity to preserve the photosynthetic apparatus. At the same time, the decline in PS II activity induced by water deficit was more marked in sensitive cv. Giymatli-2/17 than in tolerant cv. Azamatli-95. The drought-sensitive genotype Giymatli-2/17 responded to a period of stress by reducing photosynthetic efficiency and biomass accumulation. In this genotype, the defense mechanisms prevent plants from suffering irreversible damages during drought. Therefore, in cv. Azamatli-95 the photosynthetic electron transport was probably sufficient to preclude the buildup of excess energy in PS II. On the other hand, drought tolerant genotype Azamatli-95 seems able to avoid drought stress by maintaining a high photosynthetic activity, and does not suffer an oxidative stress high enough to trigger the defense mechanisms active in the genotype Giymatli-2/17. Dynamics of changes of anti-oxidative enzymes activity in wheat genotypes under normal water supply and deficiency of water in all phases of ontogenesis reveal that drought differently changes a balance between production of free radicals and enzymes involved in antioxidative defence mechanism.

In the drought-affected tolerant wheat the lower degree of changes in membrane in comparison with those in the sensitive genotype, is probably due to the presence of a more fluid bilayer, since neither non-bilayer-forming lipids of thylakoid membrane nor free fatty acids (FFA) accumulated following drought. Changes in fatty acid saturation are required to preserve an appropriative balance of bilayer and non-bilayer forming lipids in the membrane. In drought adaptation, it is probably the occurrence of bilayer/nonbilayer transformations and their influence on the packaging of proteins that are of primary importance. The larger capacity of the charged lipids to swell with water may also have increased the ability to bind water in the tolerant cultivar.

Thus, results obtained from RAPD-PCR analysis are promising beginning for further research of plant tolerance to water stress on molecular-genetic basis. Marker-assisted selection based on genotype will greatly increase breeding efficiency (Manavalan et al. 2009). Recent advances in wheat research, ranging from breeding programs to genome sequencing and genomics technologies, provide unprecedented opportunities to understand global patterns of gene expression and their association with the development of specific phenotypes, as well as promising tools for the genetic improvement of plants cultivated in adverse environments by molecular breeding or transgenic approaches. The presence of DREB 1 gene in A genome of 3 tetraploid and 8 hexaploid wheat genotypes was identified using functional markers. Unlike other genotypes, this gene was also found to be present in B genome of tetraploid genotype Barakatli-95. Understanding of the functions of these stress-inducible genes helps unravel the possible mechanisms of stress tolerance. Marker-assisted selection was also employed to improve the staygreen trait involved in drought tolerance of wheat. The results presented here open an excellent opportunity to develop stress tolerant crops in future.

Acknowledgement This work was financially supported by grant from the Science Development Foundation under the President of the Republic of Azerbaijan (EIF-2010-1(1)-40/24-3).

References

Agarwal PK, Agarwal P, Reddy MK, Sopory SK (2006) Role of *DREB* in abiotic and biotic stress tolerance in plants. Plant Cell Rep 25:1263–1274

Aliev JA (1998) Importance of photosynthesis of various organs in protein synthesis in grain of wheat genotypes under water stress. In: Garab G (ed) Photosynthesis: mechanisms and effects. Proceedings of the XIth international congress on photosynthesis, Budapest (Hungary). Kluwer Academic, Dordrecht/Boston/London. v.5, pp 3829–3832

Aliev JA (2001) Physiological bases of wheat breeding tolerant to water stress. In: Bedo Z, Lang L (eds) Wheat in a global environment. Proceedings of the 6th international wheat conference, Budapest, Hungary. Kluwer Academic, Dordrecht/Boston/London, v.9, pp 693–698

Aliyev JA, Suleymanov SY, Guseinova IM, Asadov AA, Ismailov MA (1992) Effect of specific translation inhibitors on polypeptide composition and spectral characteristics of wheat thylakoid membrane. Biochemistry 57:679–686

Alscher RG, Erturk N, Heath LS (2002) Role of superoxide dismutases (SODs) in controlling oxidative stress in plants. Exp Bot 53:1331–1341

Andersen JR, Lübberstedt T (2003) Functional markers in plants. Trends Plant Sci 8(554):560

Anderson JM, Aro EM (1994) Grana stacking and protection of photosystem II in thylakoid membranes of higher plant leaves under high irradiance: a hypothesis. Photosynth Res 41:315–326

Apel K, Hirt H (2004) Reactive oxygen species: metabolism, oxidative stress, and signal transduction. Annu Rev Plant Biol 55(373):399

Asadov AA, Zulfugarov IS, Suleymanov SY, Aliyev JA (1986) Pigment organization in light-harvesting complex and pigment-protein complex of photosystem I of spinach chloroplasts. DAN SSSR 287:444–447

Ashraf M (2009) Biotechnological approach of improving plant salt tolerance using antioxidants as markers. Biotechnol Adv 27:84–93

Ashraf M (2010) Inducing drought tolerance in plants: recent advances. Biotechnol Adv 28:169

Badiani M, De Biasi MG, Colognola M, Artemi F (1990) Catalase, peroxidase and superoxide dismutase activities in seedlings submitted to increasing water deficit. Agrochimica 34:90–102

Barber J (1982) Influence of surface charges on thylakoid structure and function. Annu Rev Plant Physiol 33:261–295

Baruah A, Simkova K, Apel K, Laloi C (2009) Arabidopsis mutants reveal multiple singlet oxygen signaling pathways involved in stress response and development. Plant Mol Biol 70:547–563

Björkman O, Powles SB (1984) Inhibition of photosynthetic reactions under water stress: interaction with light level. Planta 161:490–504

Bohnert HJ, Ayoubi P, Borchert C, Bressan RA, Burnap RL, Cushman JC, Cushman MA, Deyholos M, Fisher R, Galbraith DW, Hasegawa PM, Jenks M, Kawasaki S, Koiwa H, Kore-eda S, Lee B-H, Michalowski CB, Misawa E, Nomura M, Ozturk N, Postier B, Prade R, Song C-P, Tanaka Y, Wang H, Zhu JK (2001) A genomic approach towards salt stress tolerance. Plant Physiol Biochem 39(295):311

Brou YC, Zeze A, Diouf O, Eyletters M (2007) Water stress induces over-expression of superoxide dismutases that contribute to the protection of cowpea plants against oxidative stress. Afr J Biotechnol 6(17):1982–1986

Caldwell CR, Whitman CE (1987) Temperature-induced protein conformational changes in barley root plasma membrane-enriched microsomes. I. Effect of temperature on membrane protein and lipid mobility. Plant Physiol 84:918–923

Chaloub RM, Silva LM, Roodriges MA, Santos CPD (2003) Phase transition of thylakoid membranes photoinhibition in the cyanobacterium *Anabaena siamensis*. Photosynth Res 78:143–152

Chen JQ, Dong Y, Wang YJ, Liu Q, Zhang JS, Chen SY (2003) An AP2/EREBP-type transcription-factor gene from rice is cold-inducible and encodes a nuclear-localized protein. Theor Appl Genet 107:972–979

Close TJ (1997) Dehydrins: a commonality in the response of plants to dehydration and low temperature. Physiol Plant 100:291–296

Cornic G, Ghashghaic J, Genty B, Briantais JM (1992) Leaf photosynthesis is resistant to a mild drought stress. Photosynthethica 27:295–309

Duxbury CL, Legge RL, Paliyath G, Barber RF, Thompson JE (1991) Alterations in membrane protein conformation in response to senescence-related changes. Phytochemistry 30:63–68

Friso G, Giacomeli L, Ytterberg AJ, Peltier J-B, Rudella A, Sun Q (2004) In-depth analysis of the thylakoid membrane proteome of *Arabidopsis thaliana* chloroplasts: new proteins, new functions, and a plastid proteome database. Plant Cell 16:478–499

Gadjev I, Vanderauwera S, Gechev TS, Laloi C, Minkov IN, Shulaev V, Apel K, Inze D, Mittler R, Van Breusegem F (2006) Transcriptomic footprints disclose specificity of reactive oxygen species signaling in *Arabidopsis*. Plant Physiol 141(2):436–445

Gechev TS, Breusegem FV, Stone JM, Denev I, Laloi C (2006) Reactive oxygen species as signals that modulate plant stress responses and programmed cell death. BioEssays 28(1091):1101

Genty B, Briantais J-M, Viera Da Silva JB (1987) Effects of drought on primary photosynthetic processes of cotton leaves. Plant Physiol 83:360–364

Giardi MT, Cona A, Kucera T, Masojidek J, Mattoo AK (1995) D_1 turnover as an adaptation signal for plants to recover from drought. In: Proceedings of the international congress on integrated studies on drought tolerance of higher plants "Inter Drought 95", pp 1–5

Gillet B, Beyly A, Peltier G, Rey P (1998) Molecular characterization of CDSP34, a chloroplastic protein induced by water deficit in *Solanum tuberosum* L. plants, and regulation of CDSP 34 expression by ABA and light illumination. Plant J 16:257–262

Guseynova IM, Suleymanov SY, Aliyev JA (2001) Regulation of chlorophyll-protein complex formation and assembly in wheat thylakoid membrane. J Biochem Mol Biol 34:496–501

Guseynova IM, Suleymanov SY, Aliyev JA (2006) Protein composition and native state of pigments of thylakoid membrane of wheat genotypes differently tolerant to water stress. Biochemistry 71:173–177

Hirai M, Yamakawa R, Nishio J, Yamaji T, Kashino Y, Koike H, Satoh K (2004) Deactivation of photosynthetic activities as triggered by loss of a small amount of water in a desiccation-tolerant cyanobacterium, Nostoc commune. Plant Cell Physiol 45:872–878

Horvath G, Droppa M, Hideg E, Rozsa Z (1989) The role of phospholipids in regulating photosynthetic electron transport activities: treatment of chloroplasts with phospholipase A_2. J Photochem Photobiol Biol 3:515–527

Hu H, Dai M, Yao J, Xiao B, Li X, Zhang Q, Xiong L (2006) Overexpressing a NAM, ATAF, and CUC (NAC) transcription factor enhances drought resistance and salt tolerance in rice. Proc Natl Acad Sci U S A 103:12987–12992

Hu H, You J, Fang Y, Zhu X, Qi Z, Xiong L (2008) Characterization of transcription factor gene SNAC2 conferring cold and salt tolerance in rice. Plant Mol Biol 67:169–181

Iqbal A, Khan AS, Khan IA, Awan FS, Ahmad A, Khan AA (2007) Study of genetic divergence among wheat genotypes through random amplified polymorphic DNA. Genet Mol Res 6:476–481

Ishikawa T, Takahara K, Hirabayashi T, Matsumura H, Fujisawa S, Terauchi R, Uchimiya H, Kawai-Yamada M (2010) Metabolome analysis of response to oxidative stress in rice suspension cells overexpressing cell death suppressor Bax inhibitor-1. Plant Cell Physiol 51(1):9–20

Kaiser WM (1987) Effects of water deficit on photosynthetic capacity. Physiol Plant 116:173–181

Klimov VV, Allakhverdiev SI, Shuvalov VA, Krasnovsky AA (1982) Effect of extraction and re-addition of manganese on light reactions of Photosystem II preparations. FEBS Lett 148:307–312

Kumar CN, Knowles N (1993) Changes in lipid peroxidation and lipolytic and free-radical scavenging enzyme during aging and sprouting of potato (*Solanum tuberosum* L.) seed-tubers. Plant Physiol 102:115–124

Laemmli UK (1970) Cleavage of structural proteins during the assembly of the head of Bacteriophage. Nature 227:680–685

Laloi C, Apel K, Danon A (2004) Reactive oxygen signalling: the latest news. Curr Opin Plant Biol 7(323):328

Li G, Knowles PF, Murphy DJ, Nishida I, Marsh D (1989) Spin-label EPR studies of lipid-protein interactions in thylakoid membranes. Biochemistry 28:7446–7452

Liu Q, Kasuga M, Sakuma Y, Abe H, Miura S, Yamaguchi-Shinozaki K (1998) Two transcription factors, DREB1 and DREB2, with an EREBP/AP2 DNA binding domain separate two cellular signal transduction pathways in drought- and low-temperature-responsive gene expression, respectively in Arabidopsis. Plant Cell 10:1391–1406

Loggini B, Scartazza A, Brugnoli E, Navari-Izzo F (1999) Antioxidative defense system, pigment composition, and photosynthetic efficiency in two wheat cultivars subjected to drought. Plant Physiol 119:1091–1098

Manavalan LP, Gutticonda SK, Tran LP, Nguyen TH (2009) Phisiological and molecular approaches to improve drought resistance in soybean. Plant Cell Physiol 50:1260–1276

Masojidek J, Trivedi S, Halshaw L, Alexiou A, Hall DO (1991) The synergistic effect of drought and light stresses in sorghum and pearl millet. Plant Physiol 96:198–207

Mc-Kinney G (1941) Absorption of light by chlorophyll solutions. J Biol Chem 140:315–322

Melis A (1998) Dynamics of photosynthetic membrane composition and function. Biochim Biophys Acta 1052:87–106

Mishra RK, Singhal GS (1992) Function of photosynthetic apparatus of intact wheat leaves under high light and heat stress and its relationship with pereoxidation of thylakoid lipids. Plant Physiol 98:1–6

Monte E, Ludevid D, Prat S (1999) Leaf C40.4: a carotenoid associated protein involved in the modulation of photosynthetic efficiency. Plant J 19:399–410

Moran JF, Becana M, Iturbe-Ormaetyl I, Frechilla S, Klucas RV, Aparicio-Tejo P (1994) Drought induced oxidative stress in pea plants. Planta 194:346–352

Moussa HR, Abdel-Aziz SM (2008) Comparative response of drought tolerant and drought sensitive maize genotypes to water stress. Aust J Crop Sci 1:31–36

Mukherjee SP, Choudhuri MA (1983) Implications of water stress-induced changes in the leaves of endogenous ascorbic acid and hydrogen peroxide in *Vigna* seedlings. Physiol Plant 58:166–170

Murray MG, Thompson WF (1980) Rapid isolation of high molecular weight plant DNA. Nucleic Acids Res 8:4321–4325

Nakano Y, Asada K (1981) Hydrogen peroxide is scavenged by ascorbate-specific peroxidase in spinach chloroplasts. Plant Cell Physiol 22:867–880

Nakashima K, Tran LSP, Nguyen VD, Fujita M, Maruyama K, Todaka D, Ito Y, Hayashi N, Shinozaki K, Yamaguchi-Shinozaki K (2007) Functional analysis of a NAC-type transcription factor OsNAC6 involved in abiotic and biotic stress-responsive gene expression in rice. Plant J 51:617–630

Navari-Izzo F, Quartacci MF, Melfi D, Izzo R (1993) Lipid composition of plasma membranes isolated from sunflower seedlings grown under water stress. Physiol Plant 87:508–514

Navari-Izzo F, Quartacci MF, Pinzino C, Rascio N, Vazzana C, Sgherri CLM (2000) Protein dynamics in thylakoids of the desiccation-tolerant plant *Boea hydrosconica* during dehydration and rehydration. Plant Physiol 124:1427–1436

Pakniyat H, Tavakol E (2007) RAPD markers associated with drought tolerance in bread wheat (*Triticum aestivum* L.). Pak J Biol Sci 10:3237–3239

Peltier G, Tourneux C, Cournac L, Dimon B, Rumeau D (1995) Effects of water deficit on the photosynthetic activity of C_3 plants measured using ^{18}O and mass spectrometry. In: Proceedings of international congress of integrated studies on drought tolerance of higher plants "Inter Drought 95". Aug 28 – Sept 2, Montpellier, France, VIII B

Polle A (2001) Dissecting the superoxide dismutase ascorbate glutathione pathway in chloroplasts by metabolic modeling. Computer simulations as a step towards flux analysis. Plant Physiol 126:445–462

Pruvot G, Cuine S, Peltier G, Rey P (1996) Characterization of a novel drought-induced 34 kDa protein located in the thylakoids of *Solanum tuberosum* L. plants. Planta 198:471–479

Pshibytko NL, Kalitukho LN, Kabashnikova LF (2003) Vliyaniye vodnogo deficita i visokoy temperaturi na sostoyaniye fotosistemi II v listyakh *Hordeum vulgare* raznogo vozrasta. Russ J Plant Physiol 50:51–58

Qin F, Kakimoto M, Sakuma Y, Maruyama K, Osakabe Y, Tran LSP (2007) Regulation and functional analysis of ZmDREB2A in response to drought and heat stress in *Zea mays* L. Plant J 50:54–69

Quartacci MF, Navari-Izzo F (1992) Water stress and free radical mediated changes in sunflower seedlings. J Plant Physiol 139:621–625

Quartacci MF, Pinzino C, Sgherri CLM, Navari-Izzo F (1995) Lipid composition and protein dynamics in thylakoids of two wheat cultivars differently sensitive to drought. Plant Physiol 108:191–197

Rajaram S (2001) Prospects and promise of wheat breeding in the 21st century. Euphytica 119:3–11

Rampino P, Pataleo S, Gerardi C, Perotta C (2006) Drought stress responses in wheat: physiological and molecular analysis of resistant and sensitive genotypes. Plant Cell Environ 29(2143):2152

Raychaudhuri SS (2000) The role of SOD in combation oxidative stress in higher plants. Bot Rev 66:89–98

Rey P, Pruvot G, Becuwe N, Eymery F, Rumeau D, Peltier G (1998) A novel thioredoxin-like protein located in the chloroplast is induced by water deficit in *Solanum tuberosum* L. plants. Plant J 13:97–107

Sairam RK, Chandrasekhar V, Srivastava GC (2001) Comparison of hexaploid and tetraploid wheat cultivars in their responses to water stress. Biol Plant 44(1):89–94

Sedmak JJ, Grossberg SE (1977) A rapid, sensitive, and versatile assay for protein using Coomassie brilliant blue G 250. Anal Biochem 79:544–552

Seki M, Narusaka M, Abe H, Kasuga M, Yamaguchi-Shinozaki K, Carninci P, Hayashizaki Y, Shinozaki K (2001) Monitoring the expression pattern of 1300 *Arabidopsis* genes under drought and cold stresses by using a full-length cDNA microarray. Plant Cell 13:61–72

Seki M, Narusaka M, Ishida J, Nanjo T, Fujita M, Oono Y, Kamiya A, Nakajima M, Enju A, Sakurai T, Satou M, Akiyama K, Taji T, Yamaguchi-Shinozaki K, Carninci P, Kawai J, Hayashizaki Y, Shinozaki K (2002) Monitoring expression profile of 7000 *Arabidopsis* genes under drought, cold-and high-salinity stresses using a full-length cDNA microarray. Plant J 31(279):292

Seki M, Kamei A, Yamaguchi-Shinozaki K, Shinozaki K (2003) Molecular responses to drought, salinity and frost: common and different paths for plant protection. Curr Opin Biotechnol 14:194–199

Seki M, Umezawa T, Urano K, Shinozaki K (2007) Regulatory metabolic networks in drought stress responses. Curr Opin Plant Biol 10:296–302

Shao HB, Liang ZS, Shao MA, Sun Q (2005) Dynamic changes of anti-oxidative enzymes of 10 wheat genotypes at soil water deficits. Colloids Surf B Biointerfaces 42(187):195

Sharma KK, Lavanya M (2002) Recent developments in transgenic for abiotic stress in legumes of the semi-arid tropics. In: Ivanaga M (ed) Genetic engineering of crop plants for abiotic stress. JIRCAS Working Report, JIRCAS, Tsukuba, Japan 23:61–73

Shen YG, Zhang WK, He SJ, Zhang JS, Liu Q, Chen SY (2003a) An EREBP/AP2-type protein in *Triticum aestivum* was a DRE-binding transcription factor induced by cold, dehydration and ABA stress. Theor Appl Genet 106:923–930

Shen YG, Zhang WK, Yan DQ, Du BX, Zhang JS, Liu Q, Chen SY (2003b) Characterization of a DRE-binding transcription factor from a halophyte *Atriplex hortensis*. Theor Appl Genet 107:155–161

Shewry PR (2009) Wheat. J Exp Bot 60(6):1537–1553

Shinozaki K, Yamaguchi-Shinozaki K (2002) Molecular responses to dehydration and low temperature: differences and cross-talk between two stress signaling pathways. Curr Opin Plant Biol 3:217–223

Shinozaki K, Yamaguchi-Shinozaki K, Seki M (2002) Regulatory network of gene expression in the drought and cold stress responses. Curr Opin Plant Biol 6:410–417

Siegel BZ (1993) Plant peroxidases: an organism perspective. Plant Growth Regul 12:303–312

Sippola K, Kanervo E, Murata N, Aro E-M (1998) A genetically engineered increase in fatty acid unsaturation in *Synechoccoccus* sp. PCC 7942 allows exchange of D_1 proteins forms and sustenance of photosystem II activity at low temperature. Eur J Biochem 251:641–648

Smirnoff N (1993) The role of active oxygen in the response of plants to water deficit and desiccation. New Phytol 125:27–58

Sreenivasulu N, Sopory SK, Kavi Kishor PB (2007) Deciphering the regulatory mechanisms of abiotic stress tolerance in plants by genomic approaches. Gene 388:1–13

Tambussi EA, Nogues S, Araus JL (2005) Ear of durum wheat under water stress: water relations and photosynthetic metabolism. Planta 221:446–458

Turne NC, Wrigh GC, Siddique KHM (2001) Adaptation of grain legumes (pulses) to water limited environments. Adv Agron 71:193–223

Valliyodan B, Nguyen HT (2006) Understanding regulatory networks and engineering for enhanced drought tolerance in plants. Curr Opin Plant Biol 9:189–195

Wei B, Jing R, Wang CH, Chen J, Mao X, Chang X, Jia J (2009) *DREB*1 Genes in wheat (*Triticum aestivum* L.): development of functional markers and gene mapping based on SNPs. Mol Breed 23:13–22

Williams JG, Kubelik KJ, Livak JA, Tingey SV (1990) DNA polymorphisms amplified by arbitrary primers are useful genetic markers. Nucleic Acids Res 18:6531–6535

Winston GW (1990) In: Alscher RG, Cumming JR (eds) Stress responses in plants: adaptation and acclimation mechanisms. Wiley, New York, p 57

Wisnewski M, Webb R, Balsamo R, Close TJ, Yu XM, Griffith M (1999) Purification, immunolocalization, cryoprotective, and antifreeze activity of PCA60: a dehydrin from peach (*Prunus persica*). Physiol Plant 105:600–608

Yannarelli GG, Fernandez-Alvarez AJ (2007) Glutatione reductase activity and isoforms in leaves and roots of wheat plants subjected to cadmium stress. Phytochemistry 68:505–512

Zaidi I, Ebel C, Touzri M, Masmoudi K, Hanin M (2010) TMKP1 is a novel wheat stress responsive MAP Kinase phosphatase localized in the nucleus. Plant Mol Biol 73:325–338

Zhao CX, Guo LY, Cheruth AJ, Shao HB, Yang HB (2008) Prospectives for applying molecular and genetic methodology to improve wheat cultivars in drought environments. Curr Rev Biol 331(579):586

Zheng J, Fu J, Gou M, Huai J, Liu Y, Jian M, Huang Q, Guo X, Dong Z, Wang H, Wang G (2010) Genome-wide transcriptome analysis of two maize inbred lines under drought stress. Plant Mol Biol 72:407–423

Zhu JK (2002) Salt and drought stress signal transduction in plants. Annu Rev Plant Biol 53:247–273

Zhu T, Budworth P, Han B, Brown D, Chang HS, Zou G, Wang X (2001) Towards elucidating the global expression patterns of developing *Arabidopsis*: parallel analysis of 8300 genes by a high-density oligonucleotide probe array. Plant Physiol Biochem 39(221):242

Chapter 18
Molecular Basis of Disease Resistance in Cereal Crops: An Overview

Hadi Bux, Muhammad Ashraf, Awais Rasheed, Dipak Sharma Poudyal, Alvina Gul Kazi, and Muhammad Afzaal

Contents

H. Bux (✉)
Institute of Plant Sciences, University of Sindh, Jamshoro 76080, Pakistan
e-mail: hadiqau@gmail.com

M. Ashraf • A.G. Kazi
Atta-ur-Rahman School of Applied Biosciences (ASAB), National University of Science and Technology (NUST), Islamabad 44000, Pakistan
e-mail: ashrafjahanian@yahoo.com;alvina_gul@yahoo.com

A. Rasheed • M. Afzaal
Department of Plant Sciences, Quaid-i-Azam University, Islamabad 44000, Pakistan
e-mail: awaispbg@gmail.com; afzaalshahidhussain@hotmail.com

D.S. Poudyal
Department of Plant Pathology, Washington State University, Pullman, WA 99164-6430, USA
e-mail: ds.poudyal@gmail.com

M. Ashraf et al. (eds.), *Crop Production for Agricultural Improvement*,
DOI 10.1007/978-94-007-4116-4_18,

Abstract Deep insight into the molecular mechanism of disease resistance in plants is essential for devising sophisticated breeding strategies leading towards crop protection. Most of the disease resistance (*R*) genes show genetic and structural similarity even though their target pathogens are extremely diverged in strain types, mode of action, virulence and target plant part. The search for a unified model operating for disease resistance in cereals may provide improved control of devastating pathogens which is critical for global food security. A super family of '*R*' genes encoding NBS-LRR (Nucleotide binding site- Leucine rich repeats) with its several sub-classes, recently identified ABC (ATP binding cassette) transporter and kinase START genes confer resistance to numerous pathogens. These molecular motifs activate a wide array of metabolic responses under tight genetic control as the plant detects a prospective invader. Some of the '*R*' genes require additional genes for the expression of resistance. Investigations into the structure of '*R*' genes, protein products, their location and mechanism of interaction with pathogen elicitor molecules and future prospects are discussed in this review.

Keywords Cereals • '*R*' genes • Defense mechanism • Guard hypothesis

1 Introduction

Cereal crops are an important source of food for population. Currently, protecting crops from different diseases and environmental stresses is a pressing issue. Different approaches are applied to reduce the effects of pathogens and environmental stresses. Chemical control measure is common in this regard. Having serious environmental and health hazards, this approach is discouraged. However, use of chemicals in agriculture is common in developed countries, but it is not affordable in developing countries. Unlike chemical method, use of host genetic resistance is convenient, affordable and environment-friendly. In recent times, much effort has been directed towards understanding defense mechanism of diseases in cereal crops. Findings of Flor (1956) leading to gene-for-gene hypothesis provided a basis for predicting the molecular basis of disease resistance. In view of molecular interpretation of the Flor's findings, avirulent (*Avr*) genes encode signal transduction that is perceived by the products of plant *R* genes. These are regarded as foundation concept in disease resistance. The *R*-gene/avirulence factor complex is thought to instigate a series of signaling cascades leading to disease resistance. Rapid oxidative bursts, cell wall strengthening, induction of defense gene expression and rapid cell death at the site of infection are the downstream cellular events that confer resistance state (Morel and Dangl 1997). In a more elaborated form, the direct or indirect recognition of pathogens by host '*R*' genes leads to a resistance response known as effector-triggered immunity (ETI), which includes localized programmed cell death (PCD), known as the hypersensitive response (HR) which ultimately restrict pathogen growth (Dangl et al. 1996). R proteins serve to recognize pathogen effectors either through direct interaction or as guards for target molecules, and are known to confer resistance to bacteria, viruses, nematodes, oomycetes, insects, and biotrophic fungi (Martin et al. 2003). Genes for resistance, their protein products, and underlying mechanisms are being investigated (Hammond-Kosack and Jones 1996). Sufficient progress has been made in these aspects that will facilitate developing effective control strategies.

2 *R* Gene Classes

Several classes of *R* genes investigated in cereals are depicted in Table 18.1. Brief description about their mechanics and structural features is given below.

2.1 NBS-LRR Genes

The majority of *R* genes which had been investigated so far by gene cloning encode nucleotide binding site (NBS) and a leucine-rich-repeat (LRR) region. These NBS-LRR types of genes are abundant in plant species (Meyers et al. 1998). These plant *R* genes encode proteins that have a putative amino-terminal signaling domain, a nucleotide binding site and a series of carboxy-terminal leucine repeats (Meyers et al. 2005). There are two classes of NBS-LRR proteins. One major class had an amino-terminal TIR (Toll/interleukin receptor) domain also called TIR-NBS-LRR or TNL proteins. Other class includes the genes which encode an amino-terminal coiled-coiled motif (CC-NBS-LRR or CNL proteins). The mechanism of resistance induced by *Pm3a* and its other allelic forms in wheat (Feng et al. 2010) and by *Pb1* against rice blast clearly exemplifies that the single amino acid residue at the final position of the kinase-2 motif is the characteristic of coiled-coil (CC) motif while the tryptophan (W) and aspartic acid (D) are the characteristics of TIR-type proteins. The details of the molecular functions of these protein domains and their interacting partners are still being established. However, the consistent identification of this class of proteins across diverse plant species demonstrates that the NBS-LRR genes are a pillar of plant defense against pathogens. The majority of the *R* genes in *Arabidopsis* are TNL genes; however, they have not yet been reported in cereals. Approximately, 1,500 NBS coding sequences encoding NBS gene analyzed in rice have no TIR binding domain (Zhou et al. 2004). *Lr1* from wheat encoding resistance against *Puccinia triticina* has a CNL type domain. Same is the case of *Lr10, Lr21,* and *Pmb3b*, from wheat *R*-genes cloned to date (Feuillet et al. 2003; Huang et al. 2003; Yahiaoui et al. 2004).

2.2 Protein Kinase

The other smaller, class of *R* proteins are those composed of serine/threonine protein kinase (S/TPK) domains such as the barley Rpg1 stem rust *R* gene

Table 18.1 Comprehensive information of cereal disease resistance genes cloned so far

Crop	Disease	Pathogen	Gene	Proteins encoded	Reference
Wheat	Leaf rust	*Puccinia triticina*	*Lr10*	NBS-LRR	Feuillet et al. (2003)
			Lr1	NBS-LRR	Cloutier et al. (2007)
			Lr21	NBS-LRR	Huang et al. (2003)
			Lr34	ABC transporter	Karttinger et al. (2009)
	Powdery mildew	*Blumeria graminis*	*Pm3*	NBS-LRR	Yahiaoui et al. (2004)
			Pm3a,b,d,f	NBS-LRR	Feng et al. (2010)
	Stripe rust	*Puccinia striiformis*	*Yr36*	Kinase START	Fu et al. (2009)
			Yr18	ABC transporter	Karttinger et al. (2009)
	Cereal cyst nematodes	*Heterodera avenae*	*Cre3*	NBS-LRR	Lagudah et al. (1997)
Rice	Bacteria blight	*Xanthomonas oryzae*	*Xa1*	NBS-LRR	Yushimura et al. (1998)
			Xa21	Receptor like protein kinase	Song et al. (1995)
			Xa26	Receptor like protein kinase	Sun et al. (2004)
	Rice blast	*Magnaporthe grisea*	*Pi-b*	NBS-LRR	Wang et al. (1999)
			Pi-9	NBS-LRR	Qu et al. (2006)
			Pi-37	NBS-LRR	Lin et al. (2007)
			Pi-36	NBS-LRR	Liu et al. (2007)
			Pb-1	NBS-LRR	Hirabayashi et al. (2010)
			OsWAKI1	Wall-associated receptor like protein kinase	Li et al. (2009)
			Pi-ta	NBS-LRR	Bryan et al. (2000)
Barley	Powdery mildew	*Blumeria graminis*	*mlo*	Mutant seven transmembrane	Freialdenhoven et al. (1996)
			Mla1	NBS-LRR	Zhou et al. (2001)
			Mla6	NBS-LRR	Halterman et al. (2001)
	Stem rust	*Puccinia graminis*	*Rpg1*	Protein kinase	Brueggeman et al. (2002)
			Rpg4/Rpg5	NBS-LRR and protein kinase	Brueggeman et al. (2008)
Maize	Leaf rust	*Puccinia sorghi*	*Rp1-D*	NBS-LRR	Collins et al. (1999)
			Rp3	NBS-LRR	Webb et al. (2002)
	Southern corn leaf blight	*Cochliobolus carbonum*	*Hm1*	HC toxin reductase	Johal and Briggs (1992)

(Brueggeman et al. 2002). Resistance to bacterial pathogen *Pseudomonas syringae* conferred by tomato *pto* gene also encodes the same protein kinase but it additionally requires an NBS-LRR gene, *prf*, for defense activation (Salmeron et al. 1996). The barley Rpg5 stem rust *R* gene characterizes a unique structure among *R* genes in that it contains an N-terminal NBS-LRR and a C-terminal S/TPK (Brueggeman et al. 2008). This receptor-like protein kinase (RLK) super-family consists of trans-membrane proteins that recognize stimuli by their extracellular domains and transmit the signals through their cytoplasmic kinase domains (Morris and Walker 2003). These distinctive structural features give RLKs the mandate to play significant roles in plant defense activities. In fact, a few plant RLK members have been found involved in defense activation. In rice, two genes *Xa21* and *Xa26* act as resistance (*R*) genes and play a key role in plant cultivar-specific *R* protein-dependent resistance mediated by a receptor like protein kinase (Song et al. 1995; Sun et al. 2004). Recent advancements have suggested that RLKs are ubiquitous throughout the plant kingdom (Shiu et al. 2004), but the precise function for most of the RLKs has not been established. Cell wall is the first line of plant defense in response to external stimuli, but little is known about what communication might occur between the cell wall and plasma membrane. Additionally, when a pathogen attacks it is not clearly known which types of proteins are integrated in mediating contact between the cell wall and plasma membrane and the subsequent signal transduction. Recently, an identified member of cell wall-associated receptor kinases (WAKs) gene in rice designated as *OsWAK1* has been reported to play an important role in rice blast disease resistance (Li et al. 2009). WAKs differ from other kinds of plant RLKs in the epidermal growth factor (EGF) like repeats in the extracellular domain (Shiu and Bleecker 2001). WAKs physically link the plasma membrane to the cell wall matrix through the extracellular domain and have the potential to directly signal cellular events through their cytoplasmic kinase domain.

2.3 *Kinase START Gene*

START (steroidogenic acute regulatory protein-related lipid transfer) domain proteins in humans have been recognized to participate in lipid trafficking, metabolism, and sensing; and their binding with sterols and ceramides and they cause protein conformational changes (Alpy and Tomasetto 2005). The recent development by the positional cloning of the high-temperature stripe rust resistance gene *Yr36* revealed that two pairs of duplicate genes are responsible to encode a START kinase (Fu et al. 2009) This domain had the ability to bind lipids from PST at high temperature and change its conformation resulting in initiation of signaling cascade leading to programmed cell death. The first pair of duplicated genes had two short putative genes (*1BR1 and 1BR2*) with an in-between ring finger domain. The other pair of duplicated genes designated as WKS1 and WKS2 encode 86% identical proteins that have a predicted kinase domain followed by a predicted START domain. This combination of the kinase and START domains in *WKS1* apparently is the result of a novel domain shuffling, because these two domains are not found together in other organisms (Fu et al. 2009). The *WKS1* kinase has high resemblance to several *Arabidopsis* WAK-like kinases but *WKS1* lacks the additional domains characteristic of WAK-like kinases (Verica and He 2002). The *WKS1* kinase belongs to the non-RD kinases, which are frequently involved in the early steps of the innate immune response (Dardick and Ronald 2006).

2.4 ATP Binding Cassette Transporter

Resistance to multiple fungal pathogens in cereals through durable/non-race specific genes, have saved the cereal production since decades. Sequence analysis of the most important durable rust resistance gene *Lr34* revealed the presence of a gene rich island containing eight open reading frames predicted to encode proteins with homologies to a hexose carrier, an ATP-binding cassette (ABC) transporter, two cytochromes P450, two lectin receptor kinases, a cysteine proteinase, and a glycosyl transferase (Krattinger et al. 2009). The nucleotide sequence of *Lr34* spanned about 11,805 bp. *Lr34* consisted of 24 exons as revealed by sequencing of full length cDNA. The predicted 1401-amino acid protein belongs to the pleiotropic drug resistance subfamily of ABC transporters. Pleiotropic drug resistance transporters share a common basic structure containing two cytosolic nucleotide binding domains and two hydrophobic trans-membrane domains. In the genome of *Arabidopsis*, 15 pleiotropic drug resistance like genes have been diagnosed and additionally 23 members have been described in rice. The closest *Lr34* homolog in rice is *OsPDR23*, which is closest homolog of *Lr34* in rice showed 86% amino acid similarity and in *Arabidopsis*; the closest homologs are 56% identical to *Lr34* at the amino acid level. The knowledge about substrate specificity of these pleiotropic drug resistance transporters is very limited but these are known to confer resistance to various drugs (Rogers et al. 2001). In *Arabidopsis*, it has previously been reported that PEN3/PDR8 contributes to resistance toward non-host pathogens (Stein et al. 2006). The described model suggested that PEN3 may be involved in translocating toxic compounds derived from glucosinolates into the apoplast (Lipka et al. 2008).

2.5 HC Toxin Reductase

The maize *Hm1* gene which confers resistance against southern corn leaf blight caused by the fungal pathogen *Cochliobolus carbonum* represent a very different class of cereal resistance gene (Johal and Briggs 1992). Unlike the other *R* genes, *Hm1* encodes the enzyme HC toxin reductase which detoxifies a specific cyclic tetrapeptide toxin produced by the fungus (HC toxin) which is essential for pathogenicity. The cyclic tetrapeptide HC toxin produced by the pathogen had structure cyclo (D-Pro-L-Ala-D-Ala-L-Aeo), where *Aeo* stands for 2-amino-9,10-epoxy-8-oxodecanoic acid. The maize *Hm* gene encodes a carbonyl reductase. This reductase product uses NADPH as a cofactor to reduce the 8-carbonyl moiety of the side chain of *Aeo* to produce the biologically inactive 8-hydroxy derivative (Meeley and Walton 1991). HC toxin reductase is found in extracts only of maize of genotype with heterozygous or homozygous dominant condition and is not detectable in maize of genotype hm/hm (Meeley et al. 1992). Therefore, detoxification of HC toxin is the basis of the specific reaction of maize to *corn* leaf blight. Therefore, cereal *R* genes are seen to encode a range of different proteins that in some cases obviously have very different functions.

3 Location of the Proteins

The site of these proteins in dicotyledonous species has been recognized as the cell cytoplasm. Few proteins products have also been identified residing on the cytoplasmic face of the cell membrane (Boyes et al. 1998). In barley resistance to stem rust operated by Rpg5 encode NBS-LRR-S/TPK domain. The location of trans-membrane domains, if functional, suggests that LRR domain may reside outside the cell and act as the pathogen receptor, whereas the NBS and PK domains are intracellular and transmit the disease resistance signaling (Brueggeman et al. 2008).

4 Mechanism of Resistance

Different mechanisms have been identified that carry out the defense process in cereals. Most important models are described as below.

4.1 Receptor- Ligand Interaction

This model proposes the mechanism based on the direct interaction between *R* proteins and *Avr* proteins. Investigations into the fungal pathogen of rice blast (*Magnaparthe grisea*), have revealed that the rice *Pi-ta* gene encodes an NBS-LRR protein that recognizes *Magnaporthe* pathotypes in a race-specific manner (Bryan et al. 2000). The pathogen produces a metalloprotease protein as a corresponding pathogen avirulence gene product (AVR-Pita) (Orbach et al. 2000). This recognition process activates the subsequent defense process. Such direct interaction between the plant resistance protein and the pathogen avirulence protein has been reported *in vivo* and *in vitro* (Jia et al. 2000). However, such evidence of receptor-ligand interaction is not common in cereals.

4.2 The Guard Hypothesis

Lack of direct interaction between the pathogen avirulence product and *R* proteins have led to the recognition of other processes involved in defense mechanism.

The model known as 'guard hypothesis' proposed by Van-der-Biezen, and Jones (1998) suggests that the avirulence products of the pathogen come into contact with the non-R factors and modify them. This interference when sensed by the *R* gene products triggers the defense mechanism (Bogdanove 2002). Furthermore, this model envisages that in cereals interaction between R proteins produced by host plant and another protein known as 'guardee' occurs. Upon invasion of the 'guardee' protein by the pathogen, the host plant activates its defense mechanism. Subsequently, the cascade signal transduction initiates to protect the host. Although direct interaction between R and Avr proteins, may not essentially happen. Evidence of this model has recently been reported in *Arabidopsis* (Mackey et al. 2002). The process has not been thoroughly understood and is under investigation (Schneider 2002; Van der Hoorn et al. 2002).

5 Durability of Disease Resistance

Durable resistance against pathogen encoded by the minor genes had attracted the researchers due to their effectiveness to protect crops from pathogens. One of the examples of durable resistance is of *Lr34/Yr18* which offer resistance against leaf and yellow rust and has been exploited by the breeders since the last 50 years. The mechanism of durable, non-race specific resistance is primarily elicited by highly conserved pathogen-associated molecular patterns (PAMPs). Plants have acquired this PAMP-triggered immunity (PTI) during co-evolution with pathogens (Zellerhoff et al. 2010). Quantitative or non-host resistance is assumed to be durable but, in the case of quantitative host resistance, it is attenuated by pathogen effector molecules that reprogram host cells for the pathogen's demand leading to susceptibility triggered by effectors known as effector-triggered susceptibility (Jones and Dangl 2006). In many cultivars the quantitative host resistance is not very effective due to weaker combinations of desirable alleles encoding PTI components that determine the accessibility of the plant for pathogen effectors or the strength of host responses. Molecular basis of PAMP-triggered immunity (PIT) in the model plant *Arabidopsis thaliana* is very well understood (Boller and Felix 2009). However, this knowledge is meager for *Triticeae,* and cereals need to be investigated for PTI. Bischof et al. (2011) reviewed the global transcriptional patterns that modify in *Triticeae* during interactions (compatible or non-compatible) with biotrophic and cell death-inducing fungal pathogens. The key factors to be identified include factors responsible for resistance and susceptibility and transcriptome-based identification of developing biological function during disease development in host plant. Among transcriptionally stress activated genes are members of the wheat induced resistance 1 (*WIR1*) family which includes the members of transcriptionally stress activated genes. On the evidence based on genomic, genetic and transcriptome, the role of *WIR1* genes in modulating quantitative resistance to powdery mildew on barley is presented by Douchkov et al. (2011). They also proposed a way beyond the description of gene-expression patterns towards gene-function analysis in *Triticeae*. Along these lines, transgenic expression or knockdown of PTI or ETS related genes is a powerful way to demonstrate gene function in durable/quantitative resistance and might support future breeding for durable resistance in *Triticeae*. The current technical possibilities of spatial, temporal and quantitative control of transgene expression facilitated by promoters, other un-translated regulatory elements and protein signal peptides and motifs can be used for studying subcellular localization of transgene products (Hensel et al. 2011). Aghnoum and Niks (2011) developed two experimental barley lines; one had extremely low resistance to powdery mildew and the other had extremely high quantitative resistance. These barley lines which are likely to have stacked favorable alleles for quantitative resistance, for high and low efficiency of PTI might be a valuable tool for future investigations into the mechanisms of quantitative resistance and susceptibility.

6 Additional Genes Required for the Expression '*R*' Genes

In addition to *R* genes, additional genes have been reported to complement the defense mechanism operated by *R* genes. Investigations into susceptible mutants revealed that a number of additional genes are required for '*R*' gene activity. The additional genes suggest that multiple genes are actively involved in operations of multiple resistance signaling pathways. In barley, a family of *Mla* genes encodes a large number of NBS-LRR proteins. Each protein has a defense capability against a specific powdery mildew pathotype. Although, both *Mla1* and *Mla6* genes encode 91% identical proteins. The latter requires two additional genes *Rar1* and *Sgt1* to activate the defense response. The former encodes a zinc-binding protein containing a highly conserved cysteine- and histidine-rich domain (CHORD) (Shirasu et al. 1999); and *Mla6* (Azevedo et al. 2002) encodes a subunit of the SCF ubiquitin ligase complex. On the other hand, *Mla1* gene operates independently (Halterman et al. 2001; Azevedo et al. 2002). Homologues of *Rar1* and *Sgt1* gene have been observed in dicotyledonous species and known to show additional interaction with HSP90 proteins (Takahashi et al. 2003).

7 Conclusion and Future Prospects

The understanding of the molecular basis of resistance in cereals is dire need of the time to design potential measures to save the crops from insects and pathogens. Co-evolving strains of the pathogens have made cereals vulnerable threatening global food security. Although many novel insights into the defense mechanism triggered in plants to cope with the dangerous pathogens have emerged in present time, thorough understanding of the phenomena is still to be done. Facts about the structure of the '*R*' genes, protein complexes and mechanism of defense signaling are to be discovered to project the picture clearly. Discovery and candidate positional cloning of more '*R*' genes, understanding of their protein products will have effective biotechnological applications in the field of plant protection. The application of functional genomics to disease resistance will significantly accelerate the pace of discovery and provide new insights into *Avr-R* interactions. Moreover, a detailed understanding of the structural basis of recognition will assist researchers to approach the holy grail of designing *R* proteins that recognize essential virulence factors. Considering a variety of strategies that pathogens utilize and their ability to rapidly adapt it, would be rash to predict the development of a magic bullet for durable, broad-spectrum resistance. However, it is reasonable to expect a novel array of sophisticated omics technologies that will provide efficient protection in certain situation, when they are judiciously integrated with other control measures.

References

Aghnoum R, Niks NN (2011) Transgressive segregation for extremely low and high level of quantitative resistance to powdery mildew in barley. J Plant Physiol 168:45–50

Alpy F, Tomasetto C (2005) Give lipids a START: the StAR-related lipid transfer (START) domain in mammals. J Cell Sci 118:2791–2801

Azevedo C, Sadanandom A, Kitagawa K, Freialdenhoven A, Shirasu K, Schulze-Lefert P (2002) The RAR1 interactor SGT1, an essential component of *R* gene-triggered disease resistance. Science 295:2073–2076

Bischof M, Eichmann R, Hückelhoven R (2011) Pathogenesis-associated transcriptional patterns in Triticeae. J Plant Physiol 168:9–19

Bogdanove AJ (2002) Protein-protein interactions in pathogen recognition by plants. Plant Mol Biol 50:981–989

Boller T, Felix G (2009) A renaissance of elicitors: perception of microbe-associated molecular patterns and danger signals by pattern-recognition receptors. Annu Rev Plant Biol 60:379–406

Boyes DC, Nam J, Dangl JL (1998) The *Arabidopsis thaliana RPM1* disease resistance gene product is a peripheral plasma membrane protein that is degraded coincident with the hypersensitive response. Proc Natl Acad Sci 95:15849–15854

Brueggeman R, Kudrna RN, Kudma N, Kilian H, Han F, Chen J, Druka A, Steffenson B, Kleinhofs A (2002) The barley stem rust resistance gene *Rpg1* is a novel disease resistance gene with homology to receptor kinases. Proc Natl Acad Sci 99:9328–9333

Brueggeman R, Druka A, Nirmala J, Cavileer T, Drader T, Rostoks N, Mirlohi A, Bennypaul H, Gill U, Kudrna D (2008) The stem rust resistance gene *Rpg5* encodes a protein with nucleotide-binding site, leucine-rich, and protein kinase domains. Proc Natl Acad Sci 105:149–170

Bryan GT, Wu KS, Farrall L, Jia Y, Hershey HP, McAdams SA, Faulk KN, Donaldson GK, Tarchini R, Valent B (2000) A single amino acid difference distinguishes resistant and susceptible alleles of the rice blast gene *Pi-ta*. Plant Cell 12:2033–2046

Cloutier S, McCallum BD, Loutre C, Banks TW, Wicker T, Feuillet C, Keller B, Jordan MC (2007) Leaf rust resistance gene *Lr1*, isolated from bread wheat (*Triticum aestivum* L.) is a member of the large *psr567* gene family. Plant Mol Biol 65:93–106

Collins N, Drake J, Ayliffe M, Sun Q, Ellis J, Hulbert S, Pryor T (1999) Molecular characterization of the maize *Rp1-D* rust resistance haplotype and its mutants. Plant Cell 11:1365–1376

Dangl JL, Dietrich RA, Richberg MH (1996) Death don't have no mercy: cell death programs in plant-microbe interactions. Plant Cell 8(10):1793–1807

Dardick CD, Ronald PC (2006) Plant and animal pathogen recognition receptors signal through non-RD kinases. PLoS Pathog 2:1–15

Douchkov D, Johrde A, Nowara D, Himmelbach A, Lueck S, Niks R (2011) Convergent evidence for a role of WIR1 proteins during the interaction of barley with the powdery mildew fungus *Blumeria graminis*. J Plant Physiol 168:20–29

Feng DS, Ma X, Lin AL, Wang HG, Tian JC (2010) Isolation of resistance gene analogues to powdery mildew resistance sequences in hexaploid wheat. Biol Plant 54(3):551–555

Feuillet C, Travella S, Stein N, Albar L, Nublat A, Keller B (2003) Map-based isolation of the leaf rust disease resistance gene *Lr10* from the hexaploid wheat (*Triticum aestivum* L.) genome. Proc Natl Acad Sci 100:15253–15258

Flor H (1956) The complementary genic systems in flax and flax rust. Adv Genet 8:29–54

Freialdenhoven A, Peterhansel C, Kurth J, Kreuzaler F, Schulze-Lefert P (1996) Identification of genes required for the function of non-race-specific mlo resistance to powdery mildew in barley. Plant Cell 8:5–14

Fu D, Uauy C, Distelfeld A, Blechl A, Epstein L, Chen X, Sela H, Fahima T, Dubcovsky J (2009) A Kinase-*START* gene confers temperature-dependent resistance to wheat stripe rust. Science 323:1357–1360

Halterman D, Zhou F, Wei F, Wise RP, Schulze-Lefert P (2001) The *MLA6* coiled-coil, NBS-LRR protein confers *AvrMla6*-dependant resistance specificity to *Blumeria graminis f. sp. hordei* in barley and wheat. Plant J 25:335–348

Hammond-Kosack KE, Jones JDG (1996) Inducible plant defense mechanisms and resistance gene function. Plant Cell 8:1773–1791

Hensel G, Himmelbach A, Chen D, Douchkov K, Kumlehn J (2011) Transgene expression systems in the Triticeae cereals. J Plant Physiol 168:30–44

Hirabayashi H, Sato H, Nonoue Y, Kuno-Takemoto Y, Takeuchi Y, Kato H, Nemoto H, Ogawa T, Yano M, Imbe T, Ando I (2010) Development of introgression lines derived from *Oryza rufipogon* and *O. glumaepatula* in the genetic background of japonica cultivated rice (*O. sativa* L.) and evaluation of resistance to rice blast. Breed Sci 60:604–612

Huang L, Brooks SA, Li WL, Fellers JP, Trick HN, Gill BS (2003) Map-based cloning of leaf rust resistance gene *Lr21* from the large and polyploidy genome of bread wheat. Genetics 164:655–664

Jia Y, McAdams SA, Bryan GT, Hershey HP, Valent B (2000) Direct interaction of resistance gene and avirulence gene products confers rice blast resistance. EMBO J 19:4004–4014

Johal GS, Briggs SP (1992) Reductase activity encoded by the *HM1* disease resistance gene in maize. Science 258:985–987

Jones JDG, Dangl JL (2006) The plant immune system. Nature 444:323–329

Krattinger SG, Lagudah ES, Spielmeyer W, Singh RP, Huerta-Espino J, McFadden H, Bossolini E, Selter LL, Keller B (2009) A putative ABC transporter confers durable resistance to multiple fungal pathogens in wheat. Science 323:1360–1363

Lagudah ES, Moullet O, Appels R (1997) Map-based cloning of a gene sequence encoding a nucleotide binding domain and a leucine-rich region at the *Cre3* nematode resistance locus of wheat. Genome 40:659–665

Li H, Zhou SY, Zhao WS, Su SC, Peng YL (2009) A novel wall-associated receptor-like protein kinase gene, *OsWAK1*, plays important roles in rice Blast disease resistance. Plant Mol Biol 69:337–346

Lin F, Chen S, Que Z, Wang L, Liu X, Pan Q (2007) The blast resistance gene *Pi37* encodes a nucleotide binding site leucine-rich repeat protein and is a member of a resistance gene cluster on rice chromosome 1. Genetics 177:1871–1880

Lipka U, Fuchs R, Lipka V (2008) Arabidopsis non-host resistance to powdery mildews. Curr Opin Plant Biol 11:404–411

Liu X, Lin F, Wang L, Pan Q (2007) The in silico map-based cloning of *Pi36*, a rice coiled-coil nucleotide-binding site leucine-rich repeat gene that confers race-specific resistance to the blast fungus. Genetics 176:2541–2549

Mackey D, Holt BF, Wiiq A, Danql JL (2002) *RIN4* interacts with *Pseudomonas syringae* type III effector molecules and is required for *RPM1*-mediated resistance in *Arabidopsis*. Cell 108:743–754

Martin GB, Bogdanove AJ, Sessa G (2003) Understanding the functions of plant disease resistance proteins. Annu Rev Plant Biol 54:23–61
Meeley RB, Walton JD (1991) Enzymatic detoxification of HC-toxin, the host selective cyclic peptide from *Chochliobolus carbonum*. Plant Physiol 97:1080–1086
Meeley RB, Johal GS, Briggs SP, Walton JD (1992) A biochemical phenotype for a disease resistance gene of maize. Plant Cell 4:71–77
Meyers BC, Chin DB, Shen KA, Sivaramakrishnan S, Lavelle DO, Zhang Z, Mitchelmore RW (1998) The major resistance gene cluster in lettuce is highly duplicated and spans several megabases. Plant Cell 10:1817–1832
Meyers BC, Kaushik S, Nandety RS (2005) Evolving disease resistance genes. Curr Opin Plant Biol 8:129–134
Morel J, Dangl JL (1997) The hypersensitive response and the induction of cell death in plants. Cell Death Different 4:671–683
Morris ER, Walker JC (2003) Receptor-like protein kinases: the keys to response. Curr Opin Plant Biol 6:339–342
Orbach MJ, Farrall L, Sweigard JA, Chumley FG, Valent B (2000) A telomeric avirulence gene determines efficacy for the rice blast resistance gene *Pi-ta*. Plant Cell 12:2019–2032
Qu S, Liu G, Zhou B, Bellizzi M, Zeng L, Dai L, Han B, Wang GL (2006) The broad-spectrum blast resistance gene *Pi9* encodes a nucleotide-binding site-leucine-rich repeat protein and is a member of a multi-gene family in rice. Genetics 172:1901–1914
Rogers B, Decottignies A, Kolaczkowski M, Carvajal E, Balzi E, Goffeau A (2001) The pleiotropic drug ABC transporters from *Saccharomyces cerevisiae*. J Mol Microbiol Biotechnol 3:207–214
Salmeron JM, Oldroyd GE, Rommens CM, Scofield SR, Kim HS, Lavelle DT, Dahlbeck D, Staskawicz BJ (1996) Tomato *Prf* is a member of the leucine-rich repeat class of plant disease resistance genes and lies embedded within the *Pto* kinase gene cluster. Cell 86:123–133
Schneider DS (2002) Plant immunity and film Noir: what gumshoe detectives can teach us about plant-pathogen interactions. Cell 109:537–540
Shirasu K, Lahaye T, Man-Wah T, Zhou F, Azevedo C, Schulze-Lefert P (1999) A novel class of eukaryotic zinc-binding proteins is required for disease resistance signaling in barley and development in *C. elegans*. Cell 99:355–366
Shiu SH, Bleecker AB (2001) Receptor-like kinases from *Arabidopsis* form a monophyletic gene family related to animal receptor kinases. Proc Natl Acad Sci U S A 98:10763–10768
Shiu SH, Karlowski WM, Pan R, Tzeng YH, Mayer KF, Li WH (2004) Comparative analysis of the receptor-like kinase family in *Arabidopsis* and rice. Plant Cell 16:1220–1234
Song W-Y, Wang G-L, Chen L-L, Kim H-S, Pi L-Y, Holsten T, Gardner J, Wang B, Zhai WX, Zhu LH (1995) A receptor kinase-like protein encoded by the rice disease resistance gene, *Xa21*. Science 270:1804–1806
Stein M, Dittgen J, Sanchez-Rodriguez C, Hou BH, Molina A (2006) *Arabidopsis PEN3/PDR8*, an ATP binding cassette transporter, contributes to non-host resistance to inappropriate pathogens that enter by direct penetration. Plant Cell 18:731–746
Sun X, Cao Y, Yang Z, Xu C, Li X, Wang S, Zhang Q (2004) *Xa26*, a gene conferring resistance to *Xanthomonas oryzae pv. oryzae* in rice, encodes an LRR receptor kinase-like protein. Plant J 37:517–527
Takahashi AA, Casais C, Ichimura K, Shirasu K (2003) HSP90 interacts with *RAR1* and *SGT1* and is essential for RPS2-mediated disease resistance in *Arabidopsis*. Proc Natl Acad Sci U S A 100:11777–11782
Van der Biezen EA, Jones JDG (1998) Plant disease-resistance proteins and the gene-for-gene concept. Trends Biochem Sci 12:454–456
Van der Hoorn RA, De Wit PJ, Joosten MH (2002) Balancing selection favors guarding resistance proteins. Trends Plant Sci 7:67–71
Verica JA, He ZH (2002) The cell wall-associated kinase (WAK) and WAK-like kinase gene family. Plant Physiol 129:455–459

Wang ZX, Yano M, Yamanouchi U, Iwamoto M, Monna L, Hayasaka H, Katayos Y, Sasaki T (1999) The *Pib* gene for rice blast resistance belongs to the nucleotide binding and leucine-rich repeat class of plant disease resistance genes. Plant J 19:55–64

Webb CA, Richter TE, Collins NC, Nicolas M, Trick HN (2002) Genetic and molecular characterization of the maize *rp3* rust resistance gene. Genetics 162:381–394

Yahiaoui N, Srichumpa P, Dudler R, Keller B (2004) Genome analysis at different ploidy levels allows cloning of the powdery mildew resistance gene *Pm3b* from hexaploid wheat. Plant J 37:528–538

Yoshimura S, Yamanouchi U, Katayose Y, Toki S, Wang ZX, Kono I, Kurata N, Yano M, Iwata N, Sasaki T (1998) Expression of *Xa1*, a bacterial blight-resistance gene in rice, is induced by bacterial inoculation. Proc Natl Acad Sci U S A 95:1663–1668

Zellerhoff N, Axel H, Wubei D, Stephane B, Ulrich S, Schweizer P (2010) Non-host resistance of barley to different fungal pathogens is associated with largely distinct, quantitative transcriptional responses. Plant Physiol 152:2053–2066

Zhou F, Kurth J, Wei F, Elliot C, Vale G, Yahiaoui N, Keller B, Somerville S, Wise R, Schulze-Lefert P (2001) Cell-autonomous expression of barley *Mla1* confers race-specific resistance to the powdery mildew fungus via a *Rar1*-independent signaling pathway. Plant Cell 13:337–350

Zhou T, Wang Y, Chen J, Araki H, Jing Z, Jiang K, Shen J, Tian D (2004) Genome-wide identification of NBS genes in japonica rice reveals significant expansion of divergent non-*TIR NBS-LRR* genes. Mol Genet Genomics 271:402–415

Chapter 19
Polyamines: Role in Plants Under Abiotic Stress

Parvaiz Ahmad, Ashwani Kumar, Aditi Gupta, Xiangyang Hu, Khalid ul Rehman Hakeem, Mohamed Mahgoub Azooz, and Satyawati Sharma

Contents

P. Ahmad (✉)
Department of Botany, Amar Singh College, Srinagar 190008, Jammu & Kashmir, India
e-mail: parvaizbot@rediffmail.com; parvaizbot@yahoo.com

A. Kumar • A. Gupta • S. Sharma
Biochemistry Laboratory, CRDT, Indian Institute of Technology, Hauz Khas, New Delhi 110016, India
e-mail: ashwaniiitd@hotmail.com; 29.aditi@gmail.com; satyawatis@hotmail.com

X. Hu
Key Laboratory of Biodiversity and Biogeography, Kunming Institute of Botany, Institute of Tibet Plateau Research at Kunming, Chinese Academy of Sciences, Kunming 650204, China
e-mail: Huxiangyang@mail.kib.ac.cn

K. ul Rehman Hakeem
Molecular Ecology Laboratory, Department of Botany, Faculty of Science, Jamia Hamdard, New Delhi 110062, India
e-mail: Kur.hakeem@gmail.com

M.M. Azooz
Department of Biological Sciences, Faculty of Science, King Faisal University, Saudi Arabia

Department of Botany, Faculty of Science, South Valley University, 83523 Qena, Egypt

M. Ashraf et al. (eds.), *Crop Production for Agricultural Improvement*,
DOI 10.1007/978-94-007-4116-4_19,

Abstract Environmental changes, irrespective of source, cause a variety of stresses in plants. These stresses affect the growth and development and trigger a series of morphological, physiological, biochemical and molecular changes in plants. Abiotic stress is the primary cause of crop loss worldwide. The most challenging job before the plant biologists is the development of stress tolerant plants and maintenance of sufficient yield of crops in this changing environment. Polyamines can be of great use to enhance stress tolerance in such crop plants. Polyamines are small organic polycations present in all organisms and have a leading role in cell cycle, expression of genes, signaling, plant growth and development and tolerance to a variety of abiotic stresses. High accumulation of polyamines (putrescine, spermidine and spermine) in plants during abiotic stress has been well documented and is correlated with increased tolerance to abiotic stress. Genetic engineering of PA biosynthetic genes in crop plants is the way to create tolerance against different stresses. The present review throws light on the role of polyamines in plants.

Keywords Abiotic stress tolerance • Polyamines • ADC • ODC • SAMDC

1 Introduction

Plants are exposed continuously to a variety of adversely changing environmental factors such as heat, cold, light, drought, acidity, alkalinity, oxidative damage and metal damage, which affect plant distribution, growth, development and productivity (Ahmad et al. 2008, 2010a, b). These stressful conditions are associated with the losses in the productivity of many of the agriculturally important crops and therefore, affect the economic returns of the country. Thus, concerted efforts are underway worldwide to understand the mechanism of plant resistance against these stresses. There are several natural ways of self-defense in the plants to cope with these stressful conditions: they can induce several functional or regulatory genes (Bartels and Sunkar 2005) or can undergo different physiological or biochemical changes. The accumulation of some functional substances, such as compatible solutes and protective proteins, is an important element of the physiological and biochemical response of plants to the stressful conditions (Liu et al. 2007; Ahmad and Sharma 2008; Ahmad et al. 2010a, b, 2011). In addition to these responses by the plants, molecules known as 'polyamines', have also been known to be an integral part of plant stress response (Bouchereau et al. 1999; Walters 2003a, b; Alcázar et al. 2006b).

Polyamines (putrescine, spermine, spermidine and cadaverine), are the widely distributed of N containing organic molecules, which were discovered more than 100 years ago and hold their significance from the minutest bacteria to multicellular pants, animals and mammals. In addition to their stabilizing effects, which they confer by binding to the intracellular anions (DNA, RNA, chromatin and proteins), they are also known to possess several regulatory functions (Igarasahi and Kashiwagi 2000; Alcázar et al. 2006b, 2010; Kusano et al. 2008). In plants, they have been associated with regulating many physiological processes, such as organogenesis, embryogenesis, floral initiation and development, leaf senescence, fruit development and ripening, and abiotic and biotic plant stress responses (Galston and Kaur-Sawhney 1990; Kumar et al. 1997; Walden et al. 1997; Malmberg et al. 1998; Bouchereau et al. 1999; Bagni and Tassoni 2001; Alcázar et al. 2006b, 2010; Kusano et al. 2008).

Several changes in concentrations of polyamines in plant cells take place while responding to the stressful conditions (Bouchereau et al. 1999; Alcázar et al. 2006b, 2010; Groppa and Benavides 2008). The importance of this process can be exemplified by the fact that the levels of Put may account for 1.2% of the dry matter, representing at least 20% of the nitrogen (Galston 1991) under stressful conditions. Though the exact mechanism of involvement of polyamines during stressful conditions is not fully understood, studies are ongoing to study the molecular mechanisms (Liu et al. 2007; Alcazar et al. 2010). Evaluating the complete genome sequence of Arabidopsis has facilitated the use of global 'omic' approaches in the identification of target genes in polyamine biosynthesis and signaling pathways (Alcazar et al. 2010). The advantages of the progress made in these directions have made possible the generation of Arabidopsis transgenic plants, which are resistant to various stresses (Alcazar et al. 2010). Efforts can be made towards the development of such varieties for the agriculturally important crops as well. Such studies add to the economic potential from the agricultural sector touched by the biotechnological advances and hence, further research in these directions is noteworthy.

2 Polyamine Biosynthesis in Different Organisms

The biosynthetic pathways of polyamines have been established for many organisms ranging from bacteria to plants to mammals (Kusano et al. 2007). The synthesis essentially starts from the two amino acid precursor molecules, L-arginine and L-methionine. An overview of the general pathway is given in Fig. 19.1.

In mammals and fungi, putrescine (Put, 1,4-diaminobutane) is produced by a single pathway catalyzed by ornithine decarboxylase (ODC, EC 4.1.1.17) whereas, in plants two alternative pathways operate, namely the ODC-catalyzed reaction, as in mammals and the second is from arginine (Arg), as a result of the action of Arg decarboxylase (ADC, EC 4.1.1.19), via agmatine. A few plant species, including *Arabidopsis thaliana*, lack the ODC pathway (Hanfrey et al. 2001; Kusano et al. 2007). The polyamines in

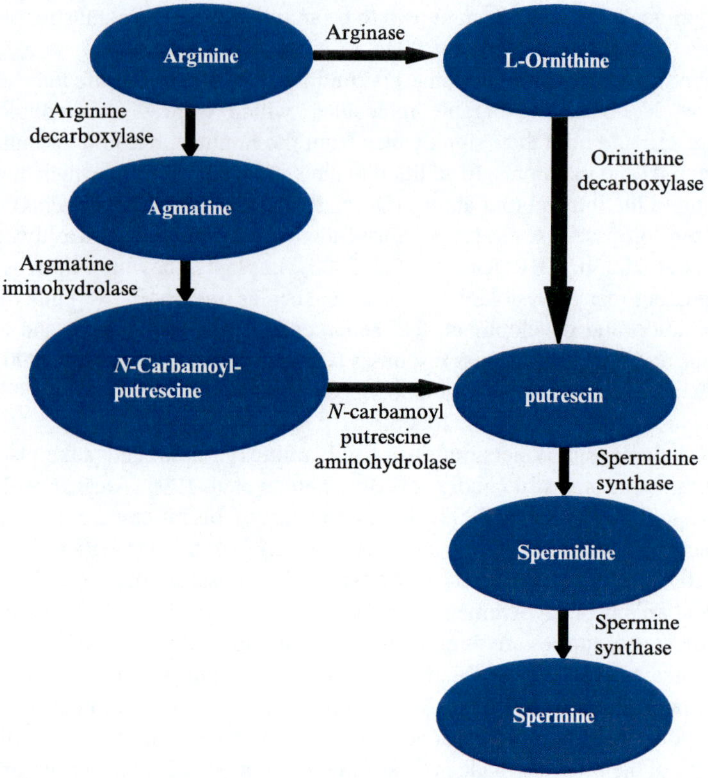

Fig. 19.1 Biosynthetic pathway of polyamines in plants

plants are not only found in the cytoplasm, but also in certain organelles like mitochondria, chloroplasts and vacuoles (Kumar et al. 1997; Kusano et al. 2008). The genes encoding enzymes for the polyamine biosynthesis pathway have been cloned and characterized from various plant species (Bell and Malmberg 1990; Michael et al. 1996; Bagni and Tassoni 2001; Liu et al. 2007; Kusano et al. 2008).

Briefly, starting from arginine, the diamine putrescine is synthesized via ornithine by arginase (EC 3.5.3.1) and ornithine decarboxylase (ODC, EC 4.1.1.17). Putrescine can also be synthesized via agmatine by three sequential reactions catalyzed by arginine decarboxylase (ADC, EC 4.1.1.19), agmatine iminohydrolase (AIH, EC 3.5.3.12), and *N*-carbamoylputrescine amidohydrolase (CPA, EC 3.5.1.53), respectively (Kusano et al. 2008). Putrescine is further transformed to Spd and Spm by successive transfers of aminopropyl groups from decarboxylated *S*-adenosylmethionine (dSAM) catalysed by specific Spd and Spm synthases. The aminopropyl groups are derived from methionine, which is first converted to *S*-adenosylmethionine (SAM) by methionine adenosyltransferase (EC 2.5.1.6), and then decarboxylated in a reaction catalyzed by *S*-adenosylmethionine decarboxylase (SAMDC, EC 4.1.1.50). The resulting decarboxylated SAM is utilized as an aminopropyl donor (Fig. 19.1).

The use of the polyamine inhibitors have helped in the evaluation of their respective roles. Four commonly used inhibitors of PA synthesis are: (1) difluoromethylornithine (DFMO), an irreversible inhibitor of ODC; (2) difluoromethylarginine (DFMA), an irreversible inhibitor of ADC; (3) methylglyoxyl- bis-guanylhydrazone (MGBG), a competitive inhibitor of *S*-adenosyl-methionine decarboxylase (SAMDC); and (4) cyclohexylamine (CHA), a competitive inhibitor of spermidine synthase. Common oxidases are diamine oxidase and polyamine oxidase (PAO), Each PA has been found to be catabolized by a specific oxidase (Kaur-Sawhney et al. 2003).

3 Polyamine Catabolism

The concentrations of the polyamines in the cells is also maintained by the catabolic pathways (Bagni and Tassoni 2001; Cona et al. 2006). Copper containing diamine oxidases (CuAO, EC 1.4.3.6) and flavine-containing polyamine oxidases (PAO, EC 1.5.3.11) catalyse the oxidative de-amination of PAs. CuAO, which prefers diamine substrates, oxidizes Put and cadaverine (1,5-diaminopentane) with concomitant production of pyrroline, NH_3 and H_2O_2, and the resulting aldehyde is further metabolized to γ-aminobutyric acid via Δ^1-1-pyrroline (Bagni and Tassoni 2001; Kusano et al. 2008). On the other hand, PAO oxidizes Spd and Spm, producing 4-aminobutanal and *N*-(3-aminopropyl)-4-aminobutanal, respectively, in addition to 1,3-diaminopropane and H_2O_2 (Kusano et al. 2007). This means that plant PAOs are involved in the terminal catabolism of polyamines (Kusano et al. 2008). These enzymes are associated with the cell walls of tissues, where lignification, suberization and wall stiffening occur (Slocum 1991). Spermine oxidase (SMO), a FAD-dependent amine oxidase, which directs the back-conversion of spermine to spermidine with concomitant production of 3-aminopropanal and H_2O_2, was initially identified in mammalian cells (Wang et al. 2001; Vujcic et al. 2002; Cervelli et al. 2003; Kusano et al. 2008). Diaminopropane can be converted into β-alanine, whereas pyrroline can be further catabolized to γ-aminobutyric acid (GABA) in a reaction catalysed by pyrroline dehydrogenase (PDH). The γ-aminobutyric acid is subsequently transaminated and oxidised to succinic acid, which is incorporated into the Krebs cycle. Thus, this pathway ensures the recycling of carbon and nitrogen from Put. Far from being only a means of eliminating cellular PAs, the enzymes involved in PA catabolism and the products deriving from their action, have been demonstrated to be involved in important physiological processes (Bouchereau et al. 1999). A simple illustration of the catabolic pathways has been shown in Fig. 19.2.

4 Role of Polyamines in Plants

The first reference of the polyamines in plants can probably be dated back to 1911 when Ciamician and Ravenna demonstrated the presence of putrescine in *Datura stramonium* (Bagni and Tassoni 2001). In plant cells, the diamine putrescine (Put),

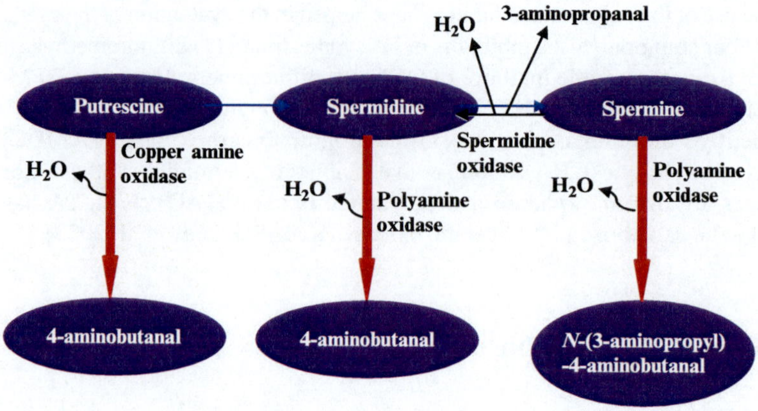

Fig. 19.2 Polyamine degradation in plants

triamine spermidine (Spd) and tetramine spermine (Spm) constitute the major PAs. Cadaverine is also present in legumes. These occur either in the free form or as conjugates bound to phenolic acids and other low molecular weight compounds or to macromolecules such as proteins and nucleic acids owing to their positive charge (Kaur-Sawhney et al. 2003). Besides stimulating DNA replication, transcription and translation, they have contributed to various biological processes in plant morphogenesis, growth, embryogenesis, organ development, leaf senescence, abiotic and biotic stress response and infection by fungi and viruses (Kumar et al. 1997; Walden et al. 1997; Malmberg et al. 1998; Bouchereau et al. 1999; Liu et al. 2000, 2010; Alcázar et al. 2006a, b; Groppa et al. 2008; Kusano et al. 2007, 2008). Their biological activity has been attributed to their cationic nature. Plant polyamines also contribute towards several characteristics of agro-economical importance, such as phytonutrient content, fruiting and fruit quality, vine life, flowering and carnation plants (Kakkar and Rai 1993; Mehta et al. 2002; Piqueras et al. 2002; Matto et al. 2006; Paschalidis and Roubelakis-Angelakis 2005).

Some of the observations suggest that PAs can act by stabilizing membranes, scavenging free radicals, affecting nucleic acids and protein synthesis, RNAse, protease and other enzyme activities, and interacting with hormones, phytochromes, and ethylene biosynthesis (Slocum et al. 1984; Galston and Tiburcio 1991). Because of these numerous biological interactions of PAs in plant systems, it has been difficult to determine their precise role in plant growth and development (Kaur-Sawhney et al. 2003). However, recent use of genomic and proteomic approaches will lead to a better understanding of the functioning in the plants (Kaur-Sawhney et al. 2003; Franceschetti et al. 2004).

The mechanisms involved in the polyamine interactions have been unveiled, at least to some extent with the help of specific polyamine inhibitors, thus explaining their physiological roles in plant growth and development. Clearly, PAs are involved in many plant developmental processes, including cell division, embryogenesis, reproductive organ development, root growth, tuberization, floral initiation and development, fruit development and ripening as well as leaf senescence and abiotic stresses as mentioned above (Evans and Malmberg 1989; Galston et al. 1997; Bais and Ravishankar 2002; Kaur-Sawhney et al. 2003; Cona et al. 2006; Rhee et al. 2007; Groppa et al. 2008; Alcazar et al. 2010). During these developmental processes, changes in the concentrations of the free and conjugated PA's and the enzymes associated with their biosynthesis (ADC, ODC, SAMDC, etc.) take place. Many authors have reported that the increase in the PA levels and the associated biosynthetic enzymes are associated with the rapid cell division in many plant systems e.g., carrot embryogenesis (Montague et al. 1978; Feirer et al. 1984), tomato ovaries (Heimer and Mizrahi 1982; Neily et al. 2011), tobacco ovaries (Slocum and Galston 1985; Franceschetti et al. 2004), and fruit development (Kakkar and Rai 1993; Paschalidis and Roubelakis-Angelakis 2005; Falasca et al. 2010). It has been observed that cells undergoing division (apical shoots, meristems, flowers, etc) contain higher levels of PAs whereas cells undergoing expansion and elongation contain lower levels of PA synthesized via ADC (Kaur-Sawhney et al. 2003). This has been further exemplied as follows: higher levels of endogenous PAs were found in flowers and siliques as compared to their levels in leaves and bolts of certain strains of Arabidopsis; addition of the PA biosynthetic inhibitors, DFMA and CHA to the culture medium, at time of seed germination, inhibited bolting and flower formation and this was partially reversed by addition of exogenous Spd (Applewhite et al. 2000; Kaur-Sawhney et al. 2003). These results clearly show that Spd is involved in flower initiation and development. Similar results have been reported in other plants also (reviewed by Galston et al. 1997; Bais and Ravishankar 2002). The regulation of many important plant hormones such as auxins, gibberellins, ethylene, etc., which play a vital role in plant growth and developmental processes has been correlated with the changes in the PA metabolism. Of the important plant hormones, ethylene is of particular interest as PAs and ethylene are said to play antagonist roles (Kaur-Sawhney et al. 2003). While PAs inhibit senescence of leaves (Kaur-Sawhney et al. 1982), cell cultures of many monocot and dicot species (Muhitch et al. 1983) and fruit ripening (Kakkar and Rai 1993), ethylene promotes all these processes. The most commonly held view is that ethylene is an effective inhibitor of ADC and SAMDC, the key enzymes in PA biosynthetic pathway, and on the other hand, PAs tend to inhibit ethylene synthesis from SAM (Kaur-Sawhney et al. 2003). Plants are exposed to continuous and rapid changing environmental factors (biotic and abiotic) such as light, temperature, water, nutrient availability, and water. These have a major impact on plant growth and productivity and PAs play an important role in these stresses as briefly discussed below.

5 Role of Polyamines in Plant Tolerance to Abiotic Stress

Richards and Coleman (1952) observed the presence of a predominant unknown ninhydrin positive spot that accumulated in barley plants when exposed to potassium starvation. This compound was identified as putrescine. Later on, it was shown that K-deficient shoots fed with L-^{14}C-arginine produced labeled Put in a more rapid way compared to feeding with labeled ornithine (Alcazar et al. 2010). These results suggested that decarboxylation of arginine was the main way of accumulation of Put under K deficiency (Smith and Richards 1962). The relevance of the ADC pathway in plant responses to abiotic stress was later on established by Galston et al. at Yale University (Flores and Galston 1982). It has been observed that polyamines accumulate in plants during various stressful conditions (see Bouchereau et al. 1999; Alcázar et al. 2006b, 2010; Groppa and Benavides 2008). These all reports support the fact that polyamines do play a protective role during the stressful conditions. Several examples have been quoted by Alcazar et al. (2010) in which genetic modification of the genes involved in PA biosynthetic pathway have proven useful in developing plant tolerance against abiotic stresses. The different stress factors have been briefly discussed below.

5.1 *Mineral Deficiency*

This is one of the most common stress related factors affecting plants almost everywhere. However, studies related to this type of stress are often performed on leaves and/or seedlings, as the external symptoms of deficiency become acute. The accumulation of Put in leaves of K-deficient barley plants was first reported by Richards and Coleman (1952) and subsequent studies by others have established that specific role of Put in maintaining a cation- anion balance in plant tissues. As a result of K starvation, this diamine accumulation (via ADC activation), is widespread among mono- and di-cotyledonous species and may well be a universal response (Bouchereau et al. 1999). The exact reason behind the increase in Put is unclear. The induced high levels of Put might be the cause of the stress injury. Put might also be beneficial for plants. Alternatively, high levels of Put could be one of the many physiological changes induced by mineral nutrient deficiency without any specific significance (Bouchereau et al. 1999). There are several other examples listing the changes in the polyamine content while responding to the mineral deficiencies (Geny et al. 1997). However, the changes differed according to the tissue and the stage of development.

5.2 *Cold Stress*

The injury due to cold causes alteration in the membrane structure, and the chilling injury involves phase transition in the molecular ordering of membrane lipids (Raison and Lyons 1970). This can cause several deleterious effects like increased membrane permeability and alteration of the activity of membrane proteins. Cold treatment has been reported to increase the levels of Put, and this correlates with the increase in the induction of arginine decarboxylase (ADC) genes (ADC1, ADC2 and SAMDC2) (Urano et al. 2003; Cuevas et al. 2008, 2009). On the other hand, levels of free Spd and Spm remain constant or even decrease in response to cold treatment (Alcazar et al. 2010). The absence of correlation between enhanced *SAMDC2* expression and the decrease Spm levels may be a result of increased Spm catabolism (Cuevas et al. 2008; Alcazar et al. 2010). Boucereau et al. (1999) reported that in the chilling-tolerant-cultivar, chilling induced an increase in free abscisic acid (ABA) levels first, then ADC activity and finally free Put levels. Fluridone, an inhibitor of ABA synthesis, inhibited the increase of free ABA levels, ADC activity and free Put levels in chilled seedlings of a chilling-tolerant cultivar. These effects resulted in a reduced tolerance to chilling and could be reversed by the pre-chilling treatment with ABA. All these results suggest that Put and ABA are integrated in a positive feedback loop, in which ABA and Put reciprocally promote each other's biosynthesis in response to abiotic stress (Fig. 19.1). This highlights a novel mode of action of polyamines as regulators of ABA biosynthesis (Alcazar et al. 2010).

5.3 *Thermal Stress*

When exposed to heat stress, plants have the ability to synthesize uncommon long chain PAs (caldine, thermine). The levels of free and bound PAs, as well as ADC and polyamine oxidases (PAO) activities, were higher in tolerant than in sensitive cultures of different crop. (Kuehn et al. 1990; Philipps and Kuehn 1991; Roy and Ghosh 1996; Bouchereau et al. 1999). The increased activities of the transglutaminases indicated the high content of the polyamines. This indicates a correlation between heat-stress tolerance, ADC, PAO and transglutaminase activities (Bouchereau et al. 1999).

5.4 *Drought Stress*

Certain plants during water scarcity tend to accumulate putrescine (Put) which is supported by the fact that transcript profiling under these conditions induces the expression of certain genes involved in the biosynthetic pathway. The expression of some of these genes is also induced by ABA treatment (Perez-Amador et al. 2002; Urano et al. 2003; Alcazar et al. 2010). This throws light upon the fact that up- regulation of PA-biosynthetic genes and accumulation of Put under water stress are mainly ABA-dependent responses (Alcazar et al. 2010).

5.5 *Salt Stress*

Differences in PA (Put, Spd, Spm) response under salt-stress have been reported among and within species. For example, according to Prakash and Prathapsenan (1988), endogenous levels of PAs (Put, Spd and Spm) decreased in rice seedlings under NaCl stress, whereas Basu et al. (1988) reported that salinity resulted in accumulation of these compounds in the same material (Bouchereau et al. 1999). Santa-Cruz et al. (1997) reported that the (Spd + Spm):Put ratios increased with salinity in the salt-tolerant tomato species (*Lycopersicon pennellii*, Carrel D'Arcy) but not in the salt-sensitive tomato species (*L. esculentum*). In both species, stress treatments decreased the levels of Put and Spd. The Spm levels did not decrease with salinity in *L. pennellii* over the salinization period, whereas they greatly decreased in *L. esculentum*. The effects of different NaCl concentrations on maize embryogenic cells derived from immature embryo cultures of a salt-sensitive inbred line (cv. w64) and a resistant hybrid (cv, Arizona) have also been reported where increased salt concentration remarkably decreased the growth of the calluses and showed a significant increase in the total PA (Put, Spd) content, especially caused by a rise in Put. It has been reported by Bouchereau et al. (1999) that using the inhibitors of Put synthesis, the ADC pathway in tomato plants operates in both stress and control conditions, whereas the ODC pathway is stimulated only under the stress conditions. These findings are further supported by the studies of Urano et al. (2003) who concluded that the expressions of the arginine decarboxylase 2(ADC 2) and spermine synthases (SPMS) during the 24 h stress treatment maintained and hence, increased the levels of Put and Spm. Yamaguchi et al. (2006) also suggested the protective role of Spm when its addition suppressed the salt sensitivity in Spm deficient mutants. Bouchereau et al. (1999) suggested that polyamine responses to salt stress are also ABA-dependent, since both *ADC2* and *SPMS* are induced by ABA. In fact, Alcazar et al. (2006a) reported that stress-responsive, drought responsive (DRE), low temperature-responsive (LTR) and ABA-responsive elements (ABRE and/or ABRE-related motifs) are present in the promoters of the polyamine biosynthetic genes. This also reinforces the view that in response to drought and salt treatments, the expression of some of the genes involved in polyamine biosynthesis are regulated by ABA (Alcazar et al. 2010). The study of the *Arabidopsis thaliana* flowers by Tassoni et al. (2010) has also supported the hypothesis that polyamine levels (mainly Spm) increase with the increase in the salt concentration and therefore, contribute to plant tolerance during the stressful conditions.

5.6 Osmotic Stress

Osmotic treatments using sorbitol induced high levels of Put and ADC in detached oat leaves (Flores and Galston 1984), whereas, Spd and Spm show a dramatic decrease. Bouchereau et al. (1999) reported that osmotica with widely different assimilation routes, such as sorbitol, mannitol, proline, betaine and sucrose, all induce a rise in Put. These changes are coincident with measurable signs of a stress, such as wilting and protein loss. Tiburcio et al. (1995) reported that when peeled oat leaves are incubated with sorbitol in the dark, they lose chlorophyll and senescence rapidly. Senescence could be delayed by including Spm in the incubation medium.

The senescence-retarding effect of Spm was correlated with increase in the incorporation of labeled precursors into proteins, RNA and DNA. They also concluded that osmotic shock in the dark induces an activation of the pathway catalyzed by ADC. Borrell et al. (1996) have reported the regulation of ADC synthesis by Spm in osmotically-stressed oat leaves using a polyclonal antibody to oat ADC and a cDNA clone encoding oat ADC. Treatment with Spm in combination with osmotic-stress resulted in increased steady-state levels of ADC mRNA, yet the levels of ADC activity decreased. This absence of correlation has been explained by the fact that Spm inhibits processing of the ADC proenzyme, which results in increased levels of this inactive ADC form and a subsequent decrease in the ADC-processed form (Bouchereau et al. 1999). They also showed that in osmotically-stressed oat leaves, degradation of cytochrome thylakoid proteins and the enzyme rubisco can be avoided by addition of Spm to the incubation medium. Thus post-translational regulation of ADC synthesis by Spm may be important in explaining its anti-senescence properties. Interestingly, Masgrau et al. (1997) concluded that the over-expression of oat ADC in tobacco resulted in similar detrimental effects to those observed by ADC activation induced by osmotic-stress in the homologous oat leaf and stem (chlorosis and necrosis). Therefore, optimum levels of polyamines are necessary for the proper growth and development of plants (Bouchereau et al. 1999). Recently, Liu et al. (2010) have investigated the changes in the content and the form of polyamines (PAs) in the leaves of two wheat (*Triticum aestivum* L.) cultivars seedlings, differing in drought tolerance, under the osmotic stress by polyethylene glycol (PEG) treatment. The results suggested that free-Spd, -Spm and PIS-bound Put (perchloric acid insoluble bound putrescine) facilitated the osmotic stress tolerance of wheat seedlings. The important roles of reactive oxygen species in the relationship between ethylene and polyamines (PAs) have also been investigated in the leaves of spring wheat seedlings under root osmotic stress (Li et al. 2010).

5.7 Hypoxia

There has been a lot of work done by Reggiani's group on the role of polyamines under the hypoxic stress conditions. Reggiani et al. (1990) reported that there are many examples available where plant shoots and seedlings of different Gramineae species, when subjected to lack of oxygen, provide evidence of an association between tolerance and the capacity to accumulate Put. Species such as rice and barnyard grass which are adapted to germinate in an oxygen deprived environment, showed a greater capacity of Put accumulation than the anoxia-intolerant species (Reggiani and Bertani 1989). This consideration supports the hypothesis for a role of Put as a protective compound against hypoxia (Reggiani and Bertani 1990; Bouchereau et al. 1999). Reggiani et al. (1989) have reported that Put is required for the anaerobic elongation of rice coleoptiles, but it has no effect on aerobic elongation of rice coleoptiles where auxin is active. This group has also concluded that with a decrease in oxygen concentration, the conjugated Put became predominant in comparison with the free forms (80% at 0.3% oxygen) and there is a negative correlation between Put accumulation (specially under conjugated forms) and shoot elongation (Reggiani and Bertani 1989; Bouchereau et al. 1999). On the other hand, the results of Lee et al. (1996) have indicated that increase in the activities of ADC and ODC, and Put levels are essential for the elongation of *Scirpus* shoots grown under submergence.

5.8 *Ozone Stress*

Ozone, the protective gas in the upper atmosphere, is known to protect us from the harmful UV rays of the sun. But it is known to have serious effects on the vegetation. Experiments are ongoing throughout the world in this respect. According to Heagle (1989) O_3-stress can lead to a significant decline in net photosynthesis, cause leaf injury and accelerate senescence, even when applied at low levels. Reaction to this stress triggers many biochemical changes in plants such as increase in ABA, peroxidases, phenolic compounds, ethylene and polyamines, which form a part of the plant self-defense mechanism. Rowland-Bamford et al. (1989) observed that the ADC activity in the ozone treated barley leaves increased before the damage became apparent. Many more examples have been quoted by Bouchereau et al. (1999) supporting the protective role of the polyamines during the ozone damage. Though the exact mechanism is not clear, there can be a possibility of PAs being involved in the free radical scavenging (Bors et al. 1989). This is also supported by the fact that the levels of superoxide radical formed enzymatically with xanthine oxidase or chemically from riboflavin or pyrogallol were inhibited *in vitro* by Put, Spd or Spm at 10–50 mM (Drolet et al. 1986). Also, superoxide radical protection was inhibited by PAs when added to microsomal membrane preparations. These findings have been also supported by the fact that PAs tend to inhibit lipid peroxidation (Tang and Newton 2005; Zhao and Yang 2008). These conclusions were, however, disputed by the findings of Langebartels et al. (1991) as mentioned by Bouchereau et al. (1999). Leaf injury, caused by O_3 in the tobacco cultivar Bel W3 could be prevented by feeding Put, Spd or Spm through the root. These exogenous treatments were correlated with a two to three-fold increase in soluble conjugated Put and Spd (monocaffeoyl forms). Conjugated Put and Spd associated with cell wall and membrane fractions were increased four to six-fold. When free PAs were assayed *in vitro* for their radical-scavenging properties, very low rate constants were found. On the other hand, PA conjugates had relatively high rate constants. It was thus concluded that free PAs could not account for the protection against O_3 damage. But assuming their role in the ozone damage, it was suggested that the protective effect of exogenous free PAs was mediated by their prior conversion to conjugated forms. Consistent with this hypothesis, it was found that monocaffeoyl Put, an effective scavenger of oxyradicals, was present in the apoplastic fluid of tobacco leaves exposed to O_3 (Dat et al. 2003). The results of Navakoudis et al. (2003) also support these findings showing that the enhanced atmospheric ozone is the accumulation of polyamines, generally observed as an increase in putrescine level, and in particular its bound form to thylakoid membranes. A study by Schraudner et al. (1990) also discovered a relationship between ethylene emission and PA biosynthesis was found in O_3-treated potato and tobacco plants, the leaves of which show early senescence in response to the pollutant. In the presence of O_3, all compounds of ethylene biosynthetic pathway in tobacco leaves were up-regulated. Put and Spd levels also increased, as did ornithine decarboxylase (ODC) activity (Bouchereau et al. 1999).

6 Polyamine Biosynthetic Genes and Stress Tolerance

The expression of genes responsible for the PA synthesis has benefited the plants to withstand environmental stresses. The over-expression or the down-regulation of the genes for PA metabolism in transgenic plants have been reported by many workers during environmental stress (Kumar et al. 1997; Walden et al. 1997; Malmberg et al. 1998; Capell et al. 1998; Rajam et al. 1998; Roy and Wu 2001; Bhatnagar et al. 2002). The genes which have been reported to be involved in the PA metabolism are *ODC*, *ADC* or *SAMDC*. Bhatnagar et al. (2002) have demonstrated that the cellular levels of Put increases by overexpressing *ODC* or *ADC* cDNA. Panicot et al. (2002a) have also reported that overexpression of *ODC* or *ADC* cDNA increases the Put levels in plants. Cheng et al. (2009) reported that transformation of yeast *SAMDC* in tomato increased Spm and Spd under high temperature stress. Overexpression of *SPDS* in Arabidopsis (Kasukabe et al. 2004); tobacco (Franceschetti et al. 2004) and sweet potato (Kasukabe et al. 2006) plants have conferred tolerance to multiple stresses. Polyamines have been proved to act as antioxidants and protect the plants from oxidative damage and maintain homeostasis in plant cells (Rodrıguez-Kessler et al. 2006). Accumulation of polyamines during environmental stresses in plants has been associated with increase in the levels of antioxidant enzyme activities like SOD, CAT, etc. Increase in MDA content has been observed during temperature stress in tomato, which leads to lipid peroxidation (Cheng et al. 2009). Overexpression of *ySAMDC* in transgenic tomato increases the Spm and Spd levels, which in turn decreases MDA content (Cheng et al. 2009). The overexpression of *SAMDC* gene in transgenic rice and tobacco showed increased levels of PA and confers tolerance to drought and salinity (Roy and Wu 2002; Waie and Rajam 2003). Table 19.1 provides further information about the PA transgenics.

7 Integration of Polyamines with Other Molecules During Stress Conditions

Polyamines affect several physiological processes in plants by activating the biosynthesis of signaling molecules like NO, H_2O_2; they affect ABA synthesis and signaling and are involved in Ca^{2+} homeostasis and ion channel signaling during the abiotic stress conditions. Figure 19.1 summarizes this information.

Table 19.1 Polyamine genes that can be expressed in plants for abiotic stress tolerance

Gene overexpressed	Plant	Response	Reference
ADC	*Oryza sativa*	Salt tolerance	Roy and Wu (2001)
ADC	*Brassica juncea*	Chilling and salt	Mo and Pua (2002)
ADC1, ADC2	*Oryza sativa*	Drought tolerance	Capell et al. (2004)
ADC1, ADC2	*Arabidopsis thaliana*	Freezing	Cuevas et al. (2008)
ADC	*Malus domestica*	Chilling, Salt and Dehydration	Hao et al. (2005)
ADC	*Oryza sativa*	Chilling	Akiyama and Jin (2007)
At ADC2	*Arabidopsis thaliana*	Salt tolerance	Urano et al. (2004)
ADC	*Solanum melongena*	Chilling, Salt and Dehydration	Prabhavathi and Rajam (2007)
ADC	*Zea maize*	Salt	Jimenez-Bremont et al. (2007)
MdADC	*Malus sylvestris* (L.) Mill. var. domestica	Salt tolerance	Liu et al. (2006)
PaADC2	*Pringlea antiscorbutica*	Chilling and salt	Hummel et al. (2004)
Mouse *ODC* cDNA	*Populous nigra* X maximowiczii cells	Stress tolerance	Bhatnagar et al. (2001)
Mouse *ODC* cDNA	tobacco	Salt stress	Kumria and Rajam (2002)
MdSAMDC2	*Malus sylvestris*	Cold and salt	Hao et al. (2005)
MdSAMDC2	*Pyrus communis*	salt	He et al. (2008)
SPDS cDNA from Cucurbita ficifolia	Arabidopsis	Chilling, salinity, drought	Kasukabe et al. (2004)
SPDS cDNA from Cucurbita ficifolia	Sweet potato	Increase in Spd	Kasukabe et al. (2006)
MdSPDS1	*Pyrus communis*	Salt, Heavy metal and osmotic stress	Wen et al. (2008)

Abscisic acid (ABA) is an anti-transpirant that reduces water loss through stomatal pores on the leaf surface in response to water deficit, resulting in the redistribution and accumulation of ABA in guard cells and finally closure of the stomata (Bray 1997). Many authors (Liu et al. 2000; An et al. 2008; Alcazar et al. 2010) have reported that Put, Spd and Spm also regulate stomatal responses by reducing their aperture and inducing closure, and Put modulates ABA biosynthesis in response to abiotic stress. Thus, polyamines are involved in the ABA mediated stress responses which affect the stomatal closure. Polyamines are also linked with reactive oxygen species (ROS) and NO signaling as amino oxidases during the catabolic process generate H_2O_2 which is a ROS (associated with plant defense and abiotic stress) and also there is evidence in which polyamines are reported to enhance the production of NO (Tun et al. 2006). NO is also known to enhance the salt stress tolerance in plants by regulating the content and proportions of the different types of free polyamines. According to Neill et al. (2008), both H_2O_2 and NO are involved in the regulation of stomatal movements in response to ABA, in such a way that NO generation depends on H_2O_2 production. Thus, altogether, the available data indicate that polyamines, ROS (H_2O_2) and NO act synergistically in promoting ABA responses in guard cells (Alcazar et al. 2010).

Polyamines are positively charged compounds, which can interact electrostatically with negatively charged proteins, including ion channels. Indeed, polyamines at their physiological concentration block the fast-activating vacuolar (FV) cation channel in a charge-dependent manner ($Spm^{4+} > Spd^{3+} >> Put^{2+}$), at both whole-cell and single-channel levels, thus indicating a direct blockage of the channel by polyamines (Bruggemann et al. 1998). According to Alacazar et al. (2010), in response to different abiotic stresses, such as potassium deficiency, Put levels are increased drastically (reaching millimolar concentrations), whereas the levels of Spd and Spm are not significantly affected, and this increase of Put may significantly reduce FV channel activity. Bruggemann et al. (1998) have also reported that all PA levels increase in amount, and the enhanced Spm concentration probably blocks FV channel activity under salinity stress. These observations can be explained by the fact that polyamines in plants may thus modulate ion channel activities through direct binding to the channel proteins and/or their associated membrane components (Delavega and Delcour 1995; Johnson 1996; Alcazar et al. 2010). Phosphorylation and dephosphorylation of ion channel proteins are closely related to their activities. Thus, polyamines could also affect protein kinase and/or phosphatase activities to regulate ion channel functions (Bethke and Jones 1997; Michard et al. 2005; Alcazar et al. 2010). However, Zhao et al. (2007) points out that for elucidating the molecular mechanisms underlying polyamine action, identification of ion channel structural elements and/or receptor molecules regulated by polyamines would be of great importance.

Polyamines also tend to maintain Ca^{2+} homeostasis. Several examples have been reported by Alcazar et al. (2010). Yamaguchi et al. (2006, 2007) proposed that the protective role of Spm against high salt and drought stress is a consequence of altered control of Ca^{2+} allocation through regulating Ca^{2+}permeable channels. The increase in cytoplasmic Ca^{2+}results in prevention of Na^+/K^+ entry into the cytoplasm, enhancement of Na^+/K^+ influx to the vacuole or suppression of Na^+/K^+ release from the vacuole, which in turn increases salt tolerance (Yamaguchi et al. 2006; Kusano et al. 2007; Alcazar et al. 2010). Thus, polyamines have a definite role in calcium homeostasis during stress conditions.

8 Conclusions and Future Prospects

Considerable evidence shows that polyamines (PAs) are involved in a myriad of plant processes including DNA regulation, gene transcription, organ development, fruit ripening, leaf senescence and various environmental stresses. The use of the genetic approaches, proteomic approaches and various analytical techniques have made it possible to further understand their mechanisms of action, binding, interaction, transport, signaling, homeostatic control of their metabolic pathways and their defensive role in biotic and abiotic stress conditions, although the exact reasoning is still difficult to interpret. Nevertheless, even this lack of information does not hamper further research into polyamines as they now constitute one of the widely distributed groups of organic molecules in nature with an important contribution towards maintaining plant growth and development, increasing crop production, defensive actions during stress conditions, combating various diseases and more recently acting as biomarkers for cancer detection. Thus, a spectral range of their applications in plants, animals and mammals offer a wide scope into their further research.

Polyamines have now been considered as secondary messengers in addition to being known as vital plant regulators (Liu et al. 2007). Although the exact mechanism of action of polyamines during the stressful conditions is not known, genetic tools have been found useful; traditional quantitative trait loci (QTL) mapping (Alonso-Blanco et al. 2009) and genome-wide association mapping (Nordborg and Weigel 2008) can be used for the identification of the genes underlying the mode of action and regulation of polyamines (Alcazar et al. 2010). Cloning of these genes would be another added advantage as these could be used in the same way as from chemicals to alleviate or mitigate stress derived injury for crop protection. Transfer of such technology to the other crops will help create germplasm which would be better adapted to the harsh stressful conditions and thus contributing to enhanced agricultural productivity.

References

Ahmad P, Sharma S (2008) Salt stress and phyto-biochemical responses of plants. Plant Soil Environ 54:89–99

Ahmad P, Sarwat M, Sharma S (2008) Reactive oxygen species, antioxidants and signaling in plants. J Plant Biol 51:167–173

Ahmad P, Jaleel CA, Salem MA, Nabi GH, Sharma S (2010a) Roles of enzymatic and non-enzymatic antioxidants in plants during abiotic stress. Crit Rev Biotechnol 30:161–175

Ahmad P, Umar S, Sharma S (2010b) Mechanism of free radical scavenging and role of phytohormones during abiotic stress in plants. In: Ashraf M, Ozturk M, Ahmad MSA (eds) Plant adaptation and phytoremediation. Springer, Dordrecht/Heidelberg/London/New York, pp 99–108

Ahmad P, Ashraf M, Younis M, Hu X, Kumar A, Akram NA, Al-Qurainy F (2011) Role of transgenic plants in agriculture and biopharming. Biotechnol Adv. doi:10.1016/j.biotechadv.2011.09.006

Akiyama T, Jin S (2007) Molecular cloning and characterization of an arginine decarboxylase gene up-regulated by chilling stress in rice seedlings. J Plant Physiol 164:645–654

Alcázar R, Cuevas JC, Patrón M, Altabella T, Tiburcio AF (2006a) Abscisic acid modulates polyamine metabolism under water stress in *Arabidopsis thaliana*. Plant Physiol 128:448–455

Alcázar R, Marco F, Cuevas JC, Patrón M, Ferrando A, Carrasco P, Tiburcio AF, Altabella T (2006b) Involvement of polyamines in plant response to abiotic stress. Biotechnol Lett 28:1867–1876

Alcazar R, Altabella T, Marco F, Bortolotti C, Reymond M, Koncz C, Carrasco P, Tiburcio AF (2010) Polyamines: molecules with regulatory functions in plant abiotic stress tolerance. Planta 231:1237–1249

Alonso-Blanco C, Aarts MGM, Bentsink L, Keurentjes JJB, Reymond M, Vreugdenhil D, Koornneef M (2009) What has natural variation taught us about plant development, physiology, and adaptation? Plant Cell 21:1877–1896

An ZF, Jing W, Liu YL, Zhang WH (2008) Hydrogen peroxide generated by copper amine oxidase is involved in abscisic acid-induced stomatal closure in *Vicia faba*. J Exp Bot 59:815–825

Applewhite PB, Kaur-Sawhney R, Galston AW (2000) A role for spermidine in the bolting and flowering of *Arabidopsis*. Physiol Plant 108:314–320

Bagni N, Tassoni A (2001) Biosynthesis, oxidation and conjugation of aliphatic polyamines in higher plants. Amino Acids 20:301–317

Bais HP, Ravishankar GA (2002) Role of polyamines in the ontogeny of plants and their biotechnological applications. Plant Cell Tissue Organ Cult 69:1–34

Bartels D, Sunkar R (2005) Drought and salt tolerance in plants. Crit Rev Plant Sci 24:23–58

Basu R, Maitra N, Ghosh B (1988) Salinity results in polyamine accumulation in early rice (*Oryza sativa* L.) seedlings. Aust J Plant Physiol 15:777–786

Bell E, Malmberg RL (1990) Analysis of a cDNA encoding arginine decarboxylase from oat reveals similarity to the *Escherichia coli* arginine decarboxylase and evidence of protein processing. Mol Gen Genet 224:431–436

Bethke PC, Jones RL (1997) Reversible protein phosphorylation regulates the activity of the slow-vacuolar ion channel. Plant J 11:1227–1235

Bhatnagar P, Glasheen BM, Bains SK, Long SL, Minocha R, Walter C, Minocha SC (2001) Transgenic manipulation of polyamine metabolism in poplar (*Populus nigra* X maximoviczii) cells. Plant Physiol 125:2139–2153

Bhatnagar P, Minocha R, Minocha SC (2002) Genetic manipulation of polyamines in poplar cells. The regulation of putrescine catabolism. Plant Physiol 128:1455–1469

Borrell A, Besford RT, Altabella T, Masgrau C, Tiburcio AF (1996) Regulation of arginine decarboxylase by spermine in osmotically-stressed oat leaves. Plant Physiol 98:105–110

Bors W, Langebartels C, Michel C, Sandermann H Jr (1989) Polyamines as radical scavengers and protectants against Ozone damage. Phytochemistry 28:1589–1595

Bouchereau A, Aziz A, Larher F, Martin-Tanguy J (1999) Polyamines and environmental challenges: recent development. Plant Sci 140:103–125

Bray EA (1997) Plant responses to water deficit. Trends Plant Sci 2:48–54

Bruggemann LI, Pottosin II, Schonknecht G (1998) Cytoplasmic polyamines block the fast-activating vacuolar cation channel. Plant J 16:101–105

Capell T, Escobar C, Liu H, Burtin D, Lepri O, Christou P (1998) Over-expression of oat arginine decarboxylase cDNA in transgenic rice (*Oryza sativa* L.) affects normal development pattern *in vitro* and results in putrescine accumulation in transgenic plants. Theor Appl Genet 97:246–254

Capell T, Bassie L, Christou P (2004) Modulation of the polyamine biosynthetic pathway in transgenic rice confers tolerance to drought stress. Proc Natl Acad Sci U S A 101:9909–9914

Cervelli M, Polticelli F, Federico R, Mariottini P (2003) Heterologous expression and characterization of mouse spermine oxidase. J Biol Chem 278:5271–5276

Cheng L, Zou Y, Ding S, Zhang J, Yu X, Cao J, Lu G (2009) Polyamine accumulation in transgenic tomato enhances the tolerance to high temperature stress. J Integ Plant Biol 51:489–499

Cona A, Rea G, Angelini R, Federico R, Tavladoraki P (2006) Functions of amine oxidases in plant development and defence. Trends Plant Sci 11:80–88

Cuevas JC, López-Cobollo R, Alcázar R, Zarza X, Koncz C, Altabella T, Salinas J, Tiburcio AT, Ferrando A (2008) Putrescine is involved in Arabidopsis freezing tolerance and cold acclimation by regulating abscisic acid levels in response to low temperature. Plant Physiol 148:1094–1105

Cuevas JC, López-Cobollo R, Alcázar R, Zarza X, Koncz C, Altabella T, Salinas J, Tiburcio AT, Ferrando A (2009) Putrescine as a signal to modulate the indispensable ABA increase under cold stress. Plant Signal Behav 4:219–220

Dat JF, Pellinen R, Van De Cotte B, Langebartels C, Kangasjarvi J, Inze D, Van Breusegem F (2003) Changes in hydrogen peroxide homeostasis trigger an active cell death process in tobacco. Plant J 33:621–632

Delavega AL, Delcour AH (1995) Cadaverine induces closing of *Escherichia coli* porins. EMBO J 14:6058–6065
Drolet G, Dumbroff EB, Legge RL, Thompson JE (1986) Radical scavenging properties of polyamines. Phytochemistry 25:367–371
Evans PT, Malmberg RL (1989) Do polyamines have a role in plant development? Annu Rev Plant Physiol Plant Mol Biol 40:235–269
Falasca G, Franceschetti M, Bagni N, Altamura MM, Biasi R (2010) Polyamine biosynthesis and control of the development of functional pollen in kiwifruit. Plant Physiol Biochem 48:565–573
Feirer RP, Mignon G, Litvay JD (1984) Arginine decarboxylase and polyamines required for embryogenesis in the wild carrot. Science 223L:1433–1435
Flores HE, Galston AW (1982) Analysis of polyamines in higher plants by high performance liquid chromatography. Plant Physiol 69:701–706
Flores HE, Galston AW (1984) Osmotic stress-induced polyamine accumulation in cereal leaves. I. Physiological parameters of the response. Plant Physiol 75:102–109
Franceschetti M, Fornale S, Tassoni A, Zuccherelli K, Mayer MJ, Bagni N (2004) Effects of spermidine synthase over-expression on polyamine biosynthetic pathway in tobacco plants. J Plant Physiol 161:989–1001
Galston AW (1991) On the trail of a new regulatory system in plants. New Biol 3:450–453
Galston AW, Kaur-Sawhney R (1990) Polyamines in plant physiology. Plant Physiol 94:406–410
Galston AW, Tiburcio AF (eds) (1991) Lecture course on polyamines as modulators of plant development 257. Fundacion Jaun, Madrid
Galston AW, Kaur-Sawhney R, Altabella T, Tiburcio AF (1997) Plant polyamines in reproductive activity and response to abiotic stress. Bot Acta 110:197–207
Geny L, Broquedis M, Martin-Tanguy J, Soyer JP, Bouard J (1997) Effects of potassium nutrition on polyamine content of various organs of fruiting cuttings of *Vitis vinifera* L. cv. Cabernet Sauvignon. Am J Enol Vitic 48:85–91
Groppa MD, Benavides MP (2008) Polyamines and abiotic stress: recent advances. Amino Acids 34:35–45
Groppa MD, Rosales EP, Iannone MF, Benavides MP (2008) Nitric oxide, polyamines and Cd-induced phytotoxicity in wheat roots. Phytochemistry 69:2609–2615
Hanfrey C, Sommer S, Mayer MJ, Burtin D, Michael AJ (2001) Arabidopsis polyamine biosynthesis: absence of ornithine decarboxylase and the mechanism of arginine decarboxylase activity. Plant J 27:551–560
Hao YJ, Zhang Z, Kitashiba H, Honda C, Ubi B, Kita M, Moriguchi T (2005) Molecular cloning and functional characterization of two apple S-adenosylmethionine decarboxylase genes and their different expression in fruit development, cell growth and stress responses. Gene 350:41–50
He L, Ban Y, ınoue H, Matsuda N, Liu J, Moriguchi T (2008) Enhancement of spermidine content and antioxidant capacity in transgenic pear shoots overexpressing apple spermidine synthase in response to salinity and hyperosmosis. Phytochemistry 69:2133–2141
Heagle AS (1989) Ozone and crop yield. Ann Rev Phytopathol 27:397–412
Heimer YM, Mizrahi Y (1982) Characterization of ornithine decarboxylase of tobacco cells and tomato ovaries. Biochem J 201:373–376
Hummel I, Gouesbet G, El Amrani A, Aïnouche A, Couée I (2004) Characterization of the two arginine decarboxylase (polyamine biosynthesis) paralogues of the endemic subantarctic cruciferous species Pringlea antiscorbutica and analysis of their differential expression during development and response to environmental stress. Gene 342:199–209
Igarasahi K, Kashiwagi K (2000) Polyamines: mysterious modulators of cellular functions. Biochem Biophys Res Commun 271:559–564
Jiménez-Bremont JF, Ruiz OA, Rodríguez-Kessler M (2007) Modulation of spermidine and spermine levels in maize seedlings subjected to long-term salt stress. Plant Physiol Biochem 45:812–821
Johnson TD (1996) Modulation of channel function by polyamines. Trends Pharmacol Sci 17:22–27

Kakkar RK, Rai VK (1993) Plant polyamines in flowering and fruit ripening. Phytochemistry 33:1281–1288

Kasukabe Y, He L, Nada K, Misawa S, Ihara I, Tachibana S (2004) Overexpression of spermidine synthase enhances tolerance to multiple environmental stresses and up-regulates the expression of various stress-regulated genes in transgenic *Arabidopsis thaliana*. Plant Cell Physiol 45:712–722

Kasukabe Y, He L, Watakabe Y, Otani M, Shimada T, Tachibana S (2006) Improvement of environmental stress tolerance of sweet potato by introduction of genes for spermidine synthase. Plant Biotechnol 23:75–83

Kaur-Sawhney R, Shih Flores HE, Galston AW (1982) Relation of polyamine synthesis and titer to aging and senescence in oat leaves. Plant Physiol 69:405–410

Kaur-Sawhney R, Tiburcio AF, Altabella T, Galston AW (2003) Polyamines in plants: an overview. J Cell Mol Biol 2:1–12

Kuehn GD, Rodriguez-Garay B, Bagga S, Phillips GC (1990) Novel occurrence of uncommon polyamines in higher plants. Plant Physiol 94:855–857

Kumar A, Altabella T, Taylor M, Tiburcio AF (1997) Recent advances in polyamine research. Trends Plant Sci 2:124–130

Kumria R, Rajam MV (2002) Ornithine decarboxylase transgene in tobacco affects polyamine metabolism, *in vitro* morphogenesis and response to salt stress. J Plant Physiol 159:983–990

Kusano T, Yamaguchi K, Berberich T, Takahashi Y (2007) The polyamine spermine rescues Arabidopsis from salinity and drought stresses. Plant Signal Behav 2:251–252

Kusano T, Berberich T, Tateda C, Takahashi Y (2008) Polyamines: essential factors for growth and survival. Planta 228:367–381

Langebartels C, Kerner KJ, Leonardi S, Schraudner M, Trost M, Heller W, Sanderman H Jr (1991) Biochemical plant response to ozone. I. Differential induction of polyamine and ethylene biosynthesis in tobacco. Plant Physiol 91:882–887

Lee TM, Shieh YJ, Chou CH (1996) Role of putrescine in enhancing shoot elongation in *Scirpus mucronatus* under submergence. Plant Physiol 96:419–424

Li CZ, Jiao J, Wang GX (2010) The important roles of reactive oxygen species in the relationship between ethylene and polyamines in leaves of spring wheat seedlings under root osmotic stress. Plant Sci 166:303–315

Liu K, Fu H, Bei Q, Luan S (2000) Inward potassium channel in guard cells as a target for polyamine regulation of stomatal movements. Plant Physiol 124:1315–1326

Liu JH, Nada K, Honda C, Kitashiba H, Wen XP, Pang XM, Moriguchi T (2006) Polyamine biosynthesis of apple callus under salt stress: importance of the arginine decarboxylase pathway in stress response. J Exp Bot 57:2589–2599

Liu JH, Kitashiba H, Wang J, Ban Y, Moriguchi T (2007) Polyamines and their ability to provide environmental stress tolerance to plants. Plant Biotechnol 24:117–126

Liu HP, Dong BH, Zhang YY, Liu ZP, Liu YL (2010) Relationship between osmotic stress and the levels of free, conjugated and bound polyamines in leaves of wheat seedlings. Plant Sci 166:1261–1267

Malmberg RL, Watson MB, Galloway GL, Yu W (1998) Molecular genetic analyses of plant polyamines. Crit Rev Plant Sci 17:199–224

Masgrau C, Altabella T, Farras R, Flores P, Thompson AJ, Besford RT, Tiburcio AF (1997) Inducible overexpression of oat arginine decarboxylase in transgenic tobacco plants. Plant J 11:465–473

Matto AK, Sobolev AP, Neelam A, Goyal RK, Handa AK, Segre AL (2006) Nuclear magnetic resonance spectroscopy-based metabolite profiling of transgenic tomato fruit engineered to accumulate spermidine and spermine reveals enhanced anabolic and nitrogen-carbon interactions. Plant Physiol 142:1759–1770

Mehta RA, Cassol T, Li N, Ali N, Handa AK, Matto AK (2002) Engineered polyamine accumulation in tomato enhances phytonutrient content, juice quality, and vine life. Nat Biotechnol 20:613–618

Michael AJ, Furze JM, Rhodes MJ, Burtin D (1996) Molecular cloning and functional identification of a plant ornithine decarboxylase cDNA. Biochem J 314:241–248

Michard E, Dreyer I, Lacombe B, Sentenac H, Thibaud JB (2005) Inward rectification of the AKT2 channel abolished by voltage dependent phosphorylation. Plant J 44:783–797

Mo H, Pua EC (2002) Up-regulation of arginine decarboxylase gene expression and accumulation of polyamines in mustard (*Brassica juncea*) in response to stress. Physiol Plant 114:439–449

Montague M, Koppenbrink J, Jaworski E (1978) Polyamine metabolism in embryogenic cells of *Daucuscarota*. I. Changes in intracellular content and rates of synthesis. Plant Physiol 62:430–433

Muhitch MJ, Edwards LA, Fletcher JS (1983) Influence of diamines and polyamines on the senescence of plant suspension cultures. Plant Cell Rep 2:82–84

Navakoudis E, Lutz C, Langebartels C, Lutz-Meindl U, Kotzabasis K (2003) Ozone impact on the photosynthetic apparatus and the protective role of polyamines. Biochim Biophys Acta 1621:160–169

Neill S, Barros R, Bright J, Desikan R, Hancock J, Harrison J, Morris P, Ribeiro D, Wilson I (2008) Nitric oxide, stomatal closure, and abiotic stress. J Exp Bot 59:165–176

Neily MH, Matsukura C, Maucourt M, Bernillon S, Deborde C, Moing A, Yin YG, Saito T, Mori K, Asamizu E, Rolin D, Moriguchi T, Ezura H (2011) Enhanced polyamine accumulation alters carotenoid metabolism at the transcriptional level in tomato fruit over-expressing spermidine synthase. J Plant Physiol 168:242–252

Nordborg M, Weigel D (2008) Next-generation genetics in plants. Nature 456:720–723

Panicot M, Masgrau C, Borrell A, Corderio A, Tiburcio AF, Altabella T (2002) Effects of putrescine accumulation in tobacco transgenic plants with different expression levels of oat arginine decarboxylase. Physiol Plant 114:281–287

Paschalidis KA, Roubelakis-Angelakis KA (2005) Spatial and temporal distribution of polyamine levels and polyamine anabolism in different organs/tissues of the tobacco plant. correlations with age, cell division/expansion, and differentiation. Plant Physiol 138:142–152

Perez-Amador MA, Leon J, Green PJ, Carbonell J (2002) Induction of the arginine decarboxylase *ADC2* gene provides evidence for the involvement of polyamines in the wound response in Arabidopsis. Plant Physiol 130:1454–1463

Philipps GC, Kuehn GD (1991) Uncommon polyamines in plants and other mechanisms. In: Slocum RD, Flores HE (eds) Biochemistry and physiology of polyamines in plants. CRC Press, Boca Raton, pp 121–133

Piqueras A, Cortina M, Serna MD, Casas JL (2002) Polyamines and hyperhydricity in micropropagated carnation plants. Plant Sci 162:671–678

Prabhavathi VR, Rajam MV (2007) Polyamine accumulation in transgenic eggplant enhances tolerance to multiple abiotic stresses and fungal resistance. Plant Biotechnol 24:273–282

Prakash L, Prathapsenan G (1988) Effect of NaCl salinity and putrescine on shoot growth, tissue ion concentration and yield of rice (*Oryza sativa*). J Agron Crop Sci 160:325–334

Raison JK, Lyons JM (1970) Oxidative activity of mitochondria isolated from plant tissues sensitive and resistant to chilling injury. Plant Physiol 45:386–389

Rajam MV, Dagar S, Waie B, Yadav JS, Kumar PA, Shoeb F, Kumria R (1998) Genetic engineering of polyamine and carbohydrate metabolism for osmotic stress tolerance in higher plants. J Biosci 23:473–482

Reggiani R, Bertani A (1989) Effect of decreasing oxygen concentration on polyamine metabolism in rice and wheat shoots. J Plant Physiol 135:375–377

Reggiani R, Bertani A (1990) Polyamines and oxygen availability. In: Flores HE, Arteca RN, Shannon JC (eds) Polyamines and ethylene: biochemistry, physiology and interactions. Rockville, MA: American Society of Plant Physiologists, pp 394–396

Reggiani R, Hochkoeppler A, Bertani A (1989) Polyamines and anaerobic elongation of rice coleoptiles. Plant Cell Physiol 30:893–898

Reggiani R, Giussani P, Bertani A (1990) Relationship between the accumulation of putrescine and the tolerance to oxygen deficit stress in Graminae seedlings. Plant Cell Physiol 31:489–494

Rhee HJ, Kim E-J, Lee JK (2007) Physiological polyamines: simple primordial stress molecules. J Cell Mol Med 11:685–703

Richards FJ, Coleman EG (1952) Occurrence of putrescine in potassium deficient barley. Nature 170:460–461

Rodrıguez-Kessler M, Alpuche-Solís AG, Ruiz OA, Jiménez-Bremont JF (2006) Effect of salt stress on the regulation of maize (*Zea mays* L.) genes involved in polyamine biosynthesis. Plant Growth Regul 48:175–185

Rowland-Bamford AJ, Barland AM, Lea PJ, Mansfield TA (1989) The role of arginine decarboxylase in modulating the sensitivity of barley to ozone. Environ Pollut 61:93–99

Roy M, Ghosh B (1996) Polyamines, both common and uncommon, under heat stress in rice (*Oryza sativa*) callus. Plant Physiol 98:196–200

Roy M, Wu R (2001) Arginine decarboxylase transgene expression and analysis of environmental stress tolerance in transgenic rice. Plant Sci 160:869–875

Roy M, Wu R (2002) Overexpression of S-adenosylmethionine decarboxylase gene in rice increases polyamine level and enhances sodium chloride-stress tolerance. Plant Sci 163: 987–992

Santa-Gruz A, Pérez-Alfocea MA, Bolarin C (1997) Changes in free polyamine levels induced by salt stress in leaves of cultivated and wild tomato species. Plant Physiol 101:341–346

Schraudner M, Trost M, Kerner K, Heller W, Leonardi S, Langebartels C, Sanderman H Jr (1990) Ozone induction and function of polyamines in ozone-tolerant and ozone-sensitive tobacco cultivars. In: Flores HE, Arteca RN, Shannon JC (eds) Polyamines and ethylene: biochemistry, physiology and interactions. Rockville, MA: American Society of Plant Physiologists, pp 394–396

Slocum RD (1991) Tissue and subcellular localisation of polyamines and enzymes of polyamine metabolism. In: Slocum RD, Flores HE (eds) Biochemistry and physiology of polyamines in plants. CRC Press, Boca Raton, pp 22–40

Slocum RD, Galston AW (1985) Changes in polyamine biosynthesis associated with post-fertilization growth and development in tobacco ovary tissues. Plant Physiol 79:336–343

Slocum RD, Kaur-Sawhney R, Galston AW (1984) The physiology and biochemistry of polyamines in plants. Arch Biochem Biophys 325:283–303

Smith TA, Richards FJ (1962) The biosynthesis of putrescine in higher plants and its relation to potassium nutrition. Biochem J 84:292–294

Tang W, Newton RJ (2005) Polyamines reduce salt-induced oxidative damage by increasing the activities of antioxidant enzymes and decreasing lipid peroxidation in Virginia pine. Plant Growth Regul 46:31–43

Tassoni A, Franceschetti M, Bagni N (2010) Polyamines and salt stress response and tolerance in *Arabidopsis thaliana* flowers. Plant Physiol Biochem 46:607–613

Tiburcio AF, Besford RT, Borrell A, Mace M (1995) Metabolism and function of polyamines during osmotically induced senescence in oat leaves and protoplasts. In: Wallsgrove RM (ed) Amino acids and their derivatives in higher plants. Cambridge University Press, Cambridge, UK, pp 205–225

Tun NN, Santa-Catarina C, Begum T, Silveira V, Handro W, Floh EIS, Scherer GFE (2006) Polyamines induce rapid biosynthesis of nitric oxide (NO) in *Arabidopsis thaliana* seedlings. Plant Cell Physiol 47:346–354

Urano K, Yoshiba Y, Nanjo T, Igarashi Y, Seki M, Sekiguchi F, Yamaguchi-Shinozaki K, Shinozaki K (2003) Characterization of Arabidopsis genes involved in biosynthesis of polyamines in abiotic stress responses and developmental stages. Plant Cell Environ 26:1917–1926

Urano K, Yoshiba Y, Nanjo T, Ito T, Yamaguchi-Shinozaki K, Shinozaki K (2004) Arabidopsis stress-inducible gene for arginine decarboxylase AtADC2 is required for accumulation of putrescine in salt tolerance. Biochem Biophys Res Commun 313:369–375

Vujcic S, Diegelmann P, Bacchi CJ, Kramer DL, Porter CW (2002) Identification and characterization of a novel flavin-containing spermine oxidase of mammalian cell origin. Biochem J 367:665–675

Waie B, Rajam MV (2003) Effect of increased polyamine biosynthesis on stress response in transgenic tobacco by introduction of human S-adenosylmethionine gene. Plant Sci 164:727–734

Walden R, Cordeiro A, Tiburcio AF (1997) Polyamines: small molecules triggering pathways in growth and development. Plant Physiol 113:1009–1013

Walters DR (2003a) Resistance to plant pathogens: possible roles for free polyamines and polyamine catabolism. New Phytol 159:109–115

Walters DR (2003b) Polyamines and plant disease. Phytochemistry 64:97–107

Wang Y, Devereux W, Woster PM, Stewart TM, Hacker A, Casero RA Jr (2001) Cloning and characterization of a human polyamine oxidase that is inducible by polyamine analogue exposure. Cancer Res 61:5370–5373

Wen XP, Pang XM, Matsuda N, Kita M, Inoue H, Hao YJ et al (2008) Overexpression of the apple spermidine synthase gene in pear confers multiple abiotic stress tolerance by altering polyamine titers. Transgenic Res 17:251–263

Yamaguchi K, Takahashi Y, Berberich T, Imai A, Miyazaki A, Takahashi T, Michael A, Kusano T (2006) The polyamine spermine protects against high salt stress in *Arabidopsis thaliana*. FEBS Lett 580:6783–6788

Yamaguchi K, Takahashi Y, Berberich T, Imai A, Takahashi T, Michael AJ, Kusano T (2007) A protective role for the polyamine spermine against drought stress in Arabidopsis. Biochem Biophys Res Commun 352:486–490

Zhao H, Yang H (2008) Exogenous polyamines alleviate the lipid peroxidation induced by cadmium chloride stress in *Malus hupehensis* Rehd. Sci Hort 116:442–447

Zhao FG, Song CP, He JQ, Zhu H (2007) Polyamines improve K^+/Na^+ homeostasis in barley seedlings by regulating root ion channel activities. Plant Physiol 145:1061–1072

Part III
Crop Management

Chapter 20
Crop Diversification Practices in Saskatchewan, Canada

Ahmet Ruhi Mermut

Contents

Abstract Agriculture in the province of Saskatchewan is currently undergoing a significant structural change in response to changing economic, environmental, and political conditions. Diversification is becoming an agricultural strategy to reduce economic risk on the farm. Deciding on the type of diversification is not easy. The strategy is to integrate environmental concerns into the development process without fettering development itself. It has to improve environmental conditions and to increase productivity, especially of degraded lands. The Canadian Prairies pause to celebrate the centennial year of agriculture. The farmers and scientists look to a bright future as they continue to bring innovative ideas and technologies to crop producers. The farmers are increasingly adopting extended and diversified crop rotations together with conservation tillage practices. Many of these newer cropping systems are recognized as being more environmentally sustainable; however, often there is a conflict between achieving the long-term goal of resource sustainability

A.R. Mermut (✉)
Department of Soil Science, University of Saskatchewan,
Saskatoon, S7N 5A8SK, Canada
e-mail: mermut@harran.edu.tr; mermut@pedology.net; ar.mermut@gmail.com

M. Ashraf et al. (eds.), *Crop Production for Agricultural Improvement*,
DOI 10.1007/978-94-007-4116-4_20, © Springer Science+Business Media B.V. 2012

and the short-term goal of economic viability. Adoption of technologies to improve nitrogen use efficiency will have the potential to significantly reduce energy use and Green House Gases (GHG). Application of all nitrogen at seeding or split within crop application will reduce the risk of loss of income due to under- or over-fertilization. Including oilseed and pulse crops in rotations that have traditionally been monoculture and cereal based, and reducing the frequency of summer fallow, contributes to higher net farm incomes in most regions, despite the higher production costs. In general, the profitability of cereal–oilseed–cereal pulse systems is > cereal–oilseed > monoculture cereal rotations in the more humid regions.

Keywords Crop diversification • Canadian agriculture and resources • Water management • Conservation tillage

1 Introduction

Centennial celebration of agriculture is now taking place in the Canadian Prairies. Farmers look to a bright future as they continue to bring innovative ideas and technologies to help the agricultural industry in the region to grow and prosper. Monocultures or cropping systems with a small number of crop components were the cropping systems and highly simplified in early stage of agriculture in western Canada.

Climatic conditions in North America are similar to those in Eurasian Steppes. Cold semi-arid regions of the Northern Great Plains (Prairies) receive about 350–400 mm rain per year. These numbers show that the Canadian Prairies are analogous to those in the Eurasian Steppes. Rainfall in Saskatchewan varies between 250 and 450 mm per year with a growing season of 100–120 days year^{-1} (Cameron and Oram 1994). Great similarities in soils, landscapes, geology, hydrology, and ecology exist between Canadian Prairies and northern Kazak Plateau, including the cultivation history. Large areas were opened for agriculture after the World War II in both regions.

One of the fundamental arguments is that, about 50% of soil organic matter (SOM) is lost in the top soil, due to intensive agricultural practices in North America (Lal et al. 1999). Uncultivated soils were in equilibrium with the native vegetation and accumulated large SOC reserves and cultivation has disrupted the steady state equilibrium. The lost carbon primarily can be returned to the soil. With good management practices it may be possible to exceed the original native SOM content of many soils. Reduced grain price in the last 15 years has forced Canadian farmers to diversify agriculture and test new crops in the prairies. These include canola, chick peas, lentil, pigeon pea etc.

Research around the world has proven that soil fertility and organic matter is declining due to long-term use of conventional tillage techniques (Putterbaugh 1993; Acton and Gregorich 1995). The loss of soil fertility and organic matter are the two major factors that cause the reduction of soil quality. Commercial fertilisers can be applied to correct the soil fertility, but this will not directly influence the soil organic matter. Loss of organic matter will adversely affect the soil erosion and increased erosion will cause a further decline of soil fertility and overall soil quality. Low fertility will produce less biomass, contributing faster decomposition of soil organic matter.

2 Land Resources

Farmland is concentrated in the south, reflecting severe limitations of climate and soils further north. Most productive land is located within 200 km of the U.S.A. border, in the agricultural heart lands of Ontario and Quebec, extending considerably farther north in the wheat land of Saskatchewan and adjacent parts of Alberta and Manitoba. About 7% of the land mass is used for agriculture, which is about 67.8 million ha. Of this land area:

1. Only 33 million ha (49%) were actually cultivated cropland.
2. The total number of farms is about 293 000 with an average size of 231 ha.
3. Today population employed directly in agriculture is a little over 3%.

About 47.6 million ha of land alone occur in the Canadian Prairie Provinces. The amount and distribution of land masses in different soil zones are shown in Table 20.1 and Figs. 20.1 and 20.2. Despite the long distance from markets and poor market structures, Canadian agriculture is so highly integrated within international trade through the *greater liberalization of international trade*. The results of this policy are yet to be seen.

3 Water Resources Management

1. Basic ideas of the water resource managements include:
2. Surface irrigation where it is possible
3. Groundwater management
4. Watershed development and management-participatory approach
5. Every drops counts: Ground water recharge/ rainwater harvesting a movement in the farmer's field- peoples' participation
6. Convergence approach

4 Crop Diversification Strategy

According to Agriculture Canada (1989), crop diversification strategy has four main pillars as shown in Fig. 20.3. These include: sustainability, diversification, market responsiveness, and self-reliance. This strategy includes the following principles:

1. More crops per drops
2. Productivity and yield gap syndrome- a human development setting

Table 20.1 Percent of land mass distribution in different soil zones of Prairie Provinces of Canada (Agriculture Canada 1989)

Soil zone	Manitoba	Saskat	Alberta	Total
Brown	–	25	17	19
Dark brown	–	35	19	25
Black	77	30	19	25
Dark grey	14	8	16	11
Grey luvisol	9	2	9	6
Total	11.1	45.1	24.2	80.4

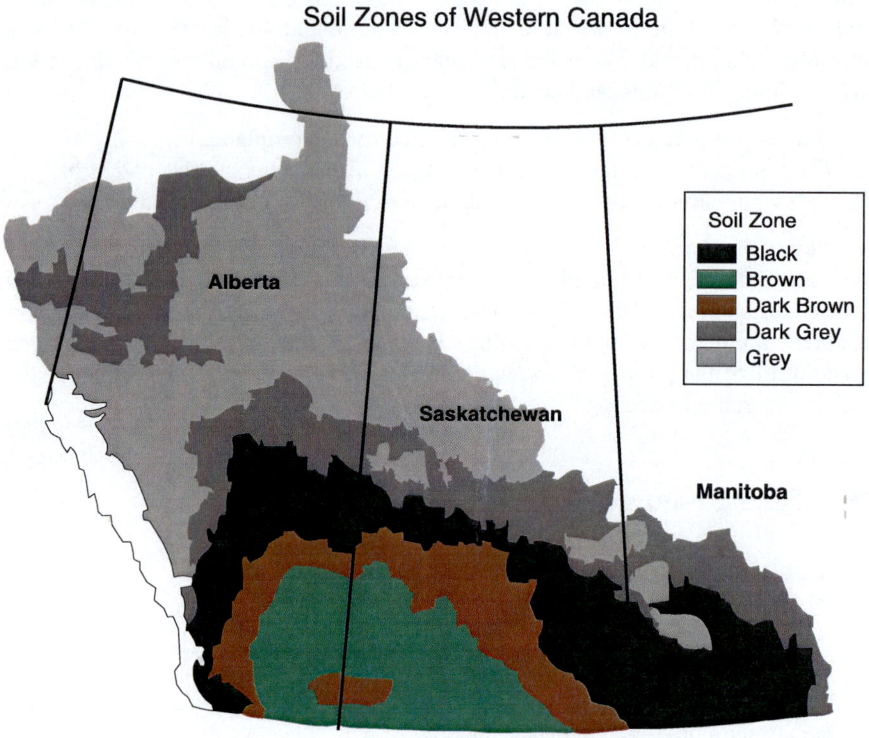

Fig. 20.1 Distribution of soil zones in three western provinces of Canada

3. Grey into green (Wasteland development)
4. Contract Farming
5. Reach the unreached in real time (Revamping of agricultural extension system)
6. Precision farming
7. Agricultural marketing – Post harvest management- in a global setting

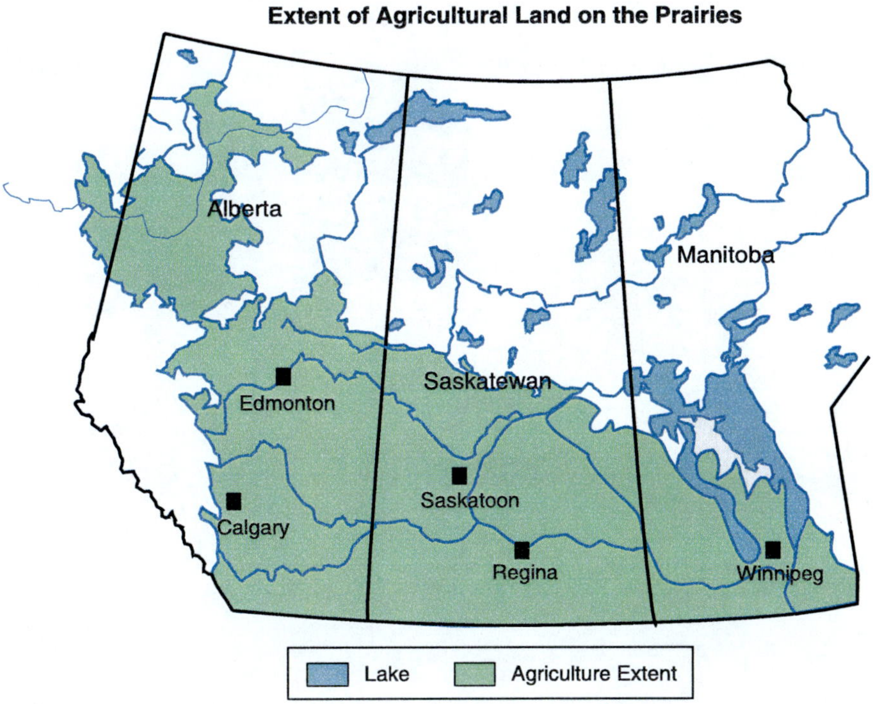

Fig. 20.2 Agricultural land area, in Manitoba, Saskatchewan, and Alberta (Agriculture Canada 1989)

5 Non-renewable Energy

This includes both as direct energy such as diesel, gasoline and indirect energy (fertilizer). Direct on-farm energy expenditure for mechanical power accounts for over $1 billion. This makes about 10% of total farm operating cost in the three Prairie Provinces. Gasoline, diesel and other fuels met over 70% of this requirement; electricity and natural gas constituting the remainder. Primary agricultural production also requires indirect energy embodied in machinery, fertilizer and pesticides. Nitrogen fertilizer can account for up to 70% of the energy used in crop production.

The types of crops produced and the amount of crop inputs will change in response to changes in climate and changes in the relative cost of inputs as they are affected by Green house gas (GHG) reduction policies and climate adaptation strategies in other sectors.

6 Project on Crop Diversification Scenarios

High fertilizer and energy efficiency of dry beans and chickpeas makes them significant crops in Western Canada. After a century of wheat farming Canadian farmers turn to tropical chickpeas, lentils etc., the increase in pulse and oilseed

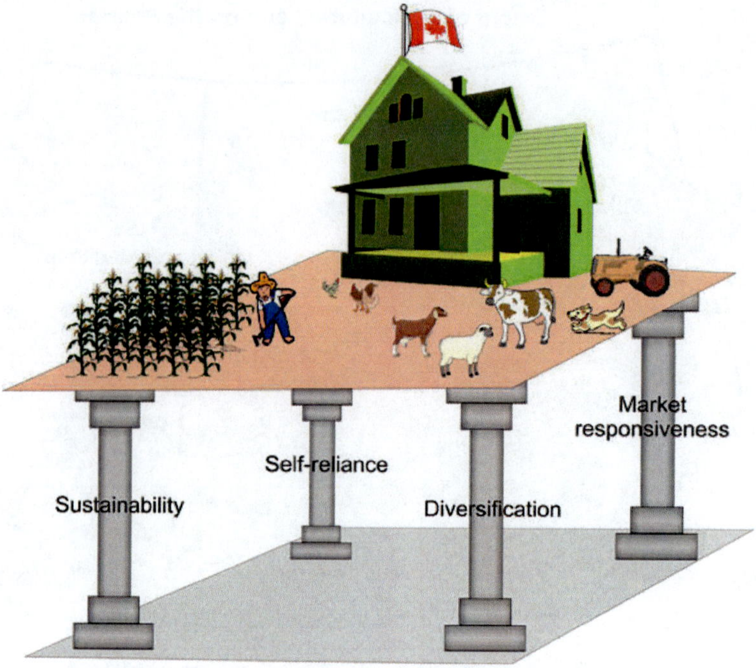

Fig. 20.3 A vision of agriculture in Canada, conceptualizing crop diversification as one of the four essential supporting pillars

Table 20.2 A comparison of pulse crops seeded in 2000 and 2001 (000 ha) (Statistic Canada 2002)

Crop	Year 2000	Year 2001	2000–2001 Change %
Dry peas	1,240	1,461	+18
Lentils	689	732	+5
Chickpeas	295	502	+70
Dry beans	169	163	–4

crops area for each province predicted to be in the year 2005 were 41%, 56%, and 51% for Alberta, Saskatchewan and Manitoba, respectively (Table 20.2).

7 Summer Fallow

In the past two decades, the area under summer fallow decreased in all soil zones, but the decline was small in the drier Brown soil zone (Statistic Canada 1988). The largest decline in fallow frequency occurred in the Black, Dark Gray, and Gray soil zones (subhumid region) where fallow area dropped 63% from 3.6 Mha in 1976 to 1.3 Mha in 1998. Over the same period, fallow area has decreased approximately

50% in the Dark Brown soil zone from 3.4 to 1.7 Mha. In the Brown soil zone, fallow remained constant at about 2.9 Mha until the period between 1996 and 1998 when fallow area dropped 16% to 2.4 Mha.

8 No Tillage or Conservation Tillage

No-till systems, do not use tillage for establishing a seedbed. Crops are simply planted into the previous year's crop residue. No-till planters are equipped with coulters that slice the soil, allowing a double disc opener to place the seed at a proper depth. The slot is closed with a spring press wheel. Herbicides are typically used as the sole means for weed control in no-till systems.

The term *conservation tillage* refers to a number of strategies and techniques for establishing crops in a previous crop's residues, which are purposely left on the soil surface. The principal benefits of conservation tillage are improved water conservation and the reduction of soil erosion. Additional benefits include: reduced fuel consumption, reduced compaction, planting and harvesting flexibility, reduced labor requirements, and improved soil tilth.

9 Conclusions

1. Most grain growers are well aware of the relationship between no-till and increased use of herbicides and the consequent exposure to herbicide resistance risk. This problem is causing a substantial number of growers to reduce their use of no-till. However, with the soil conservation benefits a key driver, no-till adoption is expected to increase over the next 5 years.
2. The adoption of minimum tillage, stubble retention, direct drill, and no-till sowing systems have allowed greater cropping intensity and been associated with reduced risk of soil degradation.
3. Including oilseed and pulse crops in rotations that have traditionally been monoculture and cereal based, and reducing the frequency of summer fallow, contributes to higher net farm incomes in most regions, despite the higher production costs.
4. In general, the profitability of cereal–oilseed–cereal pulse systems > cereal–oilseed > monoculture cereal rotations in the more humid regions.

References

Acton DF, Gregorich LJ (1995) The health of our soils: towards sustainable agriculture in Canada. Centre for Land and Biological Resources Research, Research Branch Agriculture and Agri-Food Canada, Central Experimental Farm, Ottawa, ON. Publication 1906/E

Agriculture Canada (1989) Agriculture Canada Bi-weekly Bulletin Agriculture and Agrifood Canada, Ottawa

Cameron D, Oram P (1994) Minimum and reduced tillage: its use in North America and Western Europe and its potential application in Eastern Europe, Russia, and Central Asia. International Food Policy Research Institute, Washington DC, p 121

Lal R, Follet RF, Kimble JM, Cole VR (1999) Managing U.S. cropland to sequester carbon in soil. J Soil Water Cons 54:374–381

Putterbaugh TL (1993) The geography, climate and soils of the Former Soviet Union. In: Science, agriculture and environment in the Former Soviet Union. American Association for the Advancement of Science, Washington, DC, pp 1–30

Statistic Canada (1988) infostats@statcan.gc.ca www.statcan.gc.ca, 150 Tunney's Pasture Driveway Ottawa, ON. K1A 0T6

Statistic Canada (2002) infostats@statcan.gc.ca www.statcan.gc.ca, 150 Tunney's Pasture Driveway Ottawa, ON. K1A 0T6

Chapter 21
Invasive Weed Species – A Threat to Sustainable Agriculture

Ghazala Nasim and Asad Shabbir

Contents

Abstract There are two main threats to biodiversity in the world. One of them is the direct inadvertent destruction of habitats by people which is done basically through inappropriate resource use or pollution the other highly serious but under

G. Nasim (✉)
Institute of Agricultural Sciences, University of the Punjab,
Quaid-e-Azam Campus, Lahore 54590, Pakistan
e-mail: ghazalanasim@hotmail.com

A. Shabbir
Institute of Agricultural Sciences, University of the Punjab,
Quaid-e-Azam Campus, Lahore 54590, Pakistan

School of Land Crop and Food Sciences, The University of Queensland,
Brisbane, QLD 4072, Australia
e-mail: asad@uq.edu.au

M. Ashraf et al. (eds.), *Crop Production for Agricultural Improvement*,
DOI 10.1007/978-94-007-4116-4_21, © Springer Science+Business Media B.V. 2012

estimated problem is the threat to natural and semi-natural habitats by invasion of alien organisms it is the lasting threat as compared to the first one because when exploitation stops the existing aliens do not disappear but continue to spread and consolidate. Our planet is quickly being taken over by hundreds of hardy aggressive species spread either knowingly or accidentally by human hands these are considered the second most significant threat to biodiversity after habitat destruction. These weeds are causing damages worth billions of dollars world wide and Pakistan being no exception. The countries rich repository of 6,000 of vesicular plants is highly threatened by a number of alien weeds among those some are worth mentioning such as Parthenium weed (*Parthenium hysterophorus*), Lantana (*Lantana camara*), Paper mulberry (*Brossonetia papyrifera*), Mesquet (*Prosopis juliflora*), Wild tamarind (*Leucaena leucocephala*), Khakiweed (*Alternanthera pungens*) etc. These weeds especially *P. hysterophorus* are highly noxious plant species and threatening the natural and agriculture ecosystem of the country. *Parthenium* is allergenic to both humans and animals and after invading Punjab is quickly moving towards NWFP. This is primarily because of unawareness about this plant and secondly that it is being constantly used in floral bouquets. Similarly there are many more in the list developed by IUCN in the year 2000 for Pakistan. Though it is difficult to manage these aggressive alien organisms but success stories have been documented regarding biological/chemical control and other integrated approaches. More over a careful revision of legislation regarding quarantine and enforcement of the same may prove useful to mitigate the threats of these weeds.

Keywords Invasive weeds • Biological pollution • Mycotrophy • Mycorrhiza • Phytosanitation • Management • Biological control

1 Introduction

A plant considered undesirable, unattractive, or troublesome, especially one growing where it is not wanted, as in a garden or any plant that is a hazard, nuisance, or causes injury to man, his animals, or his desired crops is called weed. Etymologically, "weed" derives from the Old English word for "grass" or "herb," but during the Middle Ages the meaning has changed to indicate an undesirable plant that grows where it is not wanted, especially among agricultural plots. This has historically been the primary meaning of the word, although in the nineteenth century, American writers grew increasingly aware that calling a plant a "weed" was an arbitrary human judgment, as there is no natural category of weeds. In the words of Ralph Waldo Emerson, a weed "is a plant whose virtues have not yet been discovered." Today, biologists tend to share that opinion, since many of the plants that are designated as weeds are, in fact, closely related to popular crops. Indeed, "weed" has fallen out of usage among biologists, although those who study agriculture still find the term useful in discussions of weed control and management (Radosevish et al. 2007).

Weeds are native or non-native plants that are unwanted in a particular area at a particular time. Change the area and the time and the plant might be desirable or even cultivated. When weeds become so wide-spread that they threaten crops, livestock, or native species, they may become more than just a "weed". They may then find themselves on a state list of plants to be attacked in a methodical manner with state support. They might then be termed "noxious weed", "invasive species", "exotic species", "alien species", or some similar term as set forth in law by each governing body. An invasive species id defined as a foreign species whose introduction does, or is likely to, cause economic or environmental harm or harm to human health. These noxious weeds find their way into new areas via boats, planes, soles of shoes, imported animals and agricultural products, etc. With increasing frequency, these most dangerous of weeds are causing the extermination of native flora and fauna from larger and larger areas. In the USA alone the invasive plants, animals and aquatic organisms have significantly reduced the economic productivity and ecological balance of the U.S. agriculture and natural resources. On land under cultivation, weeds compete with crops for water, light, and nutrients. On rangelands and in pastures, weeds are those plants that grazing animals dislike or that are poisonous. Many weeds are hosts of plant disease organisms or of insect pests. Some originally unwanted plants later were found to have virtues and came under cultivation, while some cultivated plants, when transplanted to new climates, escaped cultivation and became weeds in the new habitat. Density of single or many weed species can be changed depending on some factors during a long period. Because of globalization of economies and subsequent movement of people and materials, invasion by alien plant species is going to be the second important factor causing loss of biodiversity next to land use pattern. Changes in atmospheric composition and climate are regarded as long term factors influencing weed invasion, increasing in relative importance over time (Adkins 1995).

An invasive species is an introduced species that spreads out and often causes ecological effects on other species of ecosystems. These invasive exotic plants can cause significant displacement of native vegetation. But exactly what makes these exotic plants invasive? To begin with let's start with some basic definition: The vegetation historically found in a local area is termed native vegetation. These plants have traditionally been found in the area and are well-suited to maintain themselves in their environments. Exotic plants are those plants found in a particular area, but which originate from another continent or country. These plants can also be referred to as non-native. However non-native plants are not always exotic. Non-native plants may also be native elsewhere in the same country, but not found in the local area. For example, redwoods are native to California and would be non-native in Tennessee, but not exotic. Invasive plants are plants, native, non-native, exotic, that can cause significant ecological or economic damage. Invasive plants can displace more ecologically and economically valuable plants. These plants are typically characterized by a rampant rate of spread. This rate of dispersal is due to both human activity and vegetative characteristics. Humans assist in the dispersal of plants by moving plants materials to and from locations. Humans also cause significant disturbance to native

Table 21.1 A list of characters which enable weedy plants to become so competitive, persistent, and pernicious

1. **Number of seeds per plant**: Weeds are producers of large number of seeds. Example: Barnyard grass, 7,000 seeds; Common Purslane, 52,000 seeds; Common Lambs quarters, 72,000 seeds; Redroot Pigweed, 117,000 seeds; Russian Thistle, 200,000 seeds.
2. **Dormancy and longevity**: Dormancy is the ability of seeds to remain viable in the soil for extended periods of time. For Example: Johnson grass, 20 years; Field Bindweed, 20+ years; Common Lambs quarter, 40 years; Redroot Pigweed, 40 years.
3. **Special Adaptations or Appendages**: Plants have developed means to assist in their spread and distribution, these include: Hooks and spines – sandbur, puncture vine; Pappus (parachutes) – musk thistle, milkweeds; Good Looks – spotted knapweed, white top, Dalmatian toadflax.
4. **Vegetative Reproductive Capabilities**: Vegetative reproductive structures are those asexual portions of the plant which allow for new plants to arise without the fertilization of the flower. Examples include the following: Roots with adventitious buds Ex.: Leafy Spurge and Canada Thistle; Rhizome Ex. Johnson grass and Bermuda grass; Tubers Ex.: Yellow and Purple Nutsedge; Crowns Ex.: Dandelions and Plantains; Stolons Ex.: Bermuda grass; Special Characteristics: "Weeds have the greater will to live".
5. **Weapon of Mycotrophy**: Weeds owe their success towards the ease with which they establish symbiosis with the native soil inhabiting arbuscular mycorrhizal fungi however they do not entirely rely on the association for their survival.

plant communities that provide opportunities for the establishment of exotic and/or non-native invasives. The seeding and sprouting characteristics of the plants also significantly contribute to the rate of dispersal. Though a negative term, "invasive" plants can also be native plants that we usually do not consider to be a problem (Anderson 1996).

The introduction of many non-native or exotic plants has occurred both intentionally, as with ornamental plantings, and accidentally, as a by-product of commerce. For example, the movement of hey, carrying seeds of invasive plants, can introduce exotic and/or non-native invasive plants to new areas. Intentional plantings occur in both ornamentals, whore foliage, form and/or flowers make them attractive additions to landscaping, and other plants introduced to be used as erosion control or wildlife forage. Often, these plants exhibit reproduction traits that enable them to spread rapidly (Radosevish et al. 2007).

One reproductive tool of many invasive plants is the production of thousands of light, wind-blown seeds per plant that can rapidly disseminate to other areas (Table 21.1). Other types of invasive plants have seeds with very hard seed coatings. This seed coat protects the seed and allows it to remain viable for several years after being dropped by the parent plant. Invasive species with hard seed coats can lay dormant in the soil for years, waiting for the seed coat to wear away and for appropriate weather and site conditions to occur for germination. Conditions that facilitate seed sprouting can include several consecutive wet growing seasons, site disturbance that increase light availability and/or clears the soil surface of other vegetation, or droughts (Barrett and Richardson 1986; Barrett 1988).

Scarification of the seed coat, by animals or other means of abrasion, can also stimulate sprouting. Scarification occurs when the seed passes though an animal's digestive track. The fruits and seeds produced by the invasive plants are usually very attractive to wildlife as a food source, which aids in subsequent seed dispersal. Birds, in particular, play a large role in disseminating seeds over broad areas. Another reproductive characteristic of invasive plants is the ability to sprout from the roots and root nodded along with the stems. If the above ground vegetation of an invasive plant is killed, the roots can rapidly send up a replacement sprout. Some plants reproduce by roots nodes located along the above aboveground stem. When one of the root nodded comes into contact with the soil surface, a new root system begins to develop and eventually forms a new plant (Baskin and Baskin 1988).

Invasive plants quickly colonize areas that are frequently disturbed, such as road embankments and construction areas. They can also be found along field edges, riparian areas and other managed places. With a few exceptions these types of plants are not found in shaded areas, such as the interior of forests. Most invasive plants are better suited for sunny sites and are considered intolerant of shade. However, species such as privet and Japanese grass can present major problems in forested and shaded areas of the United States. Forest disturbances, such as ice storms, fires, insect and disease infestations, harvests or canopy-removing event, can open forested areas to exotic plant invasions (Radosevish et al. 2007).

2 Weeds: Types and Classification

Weeds are really just one type of plant that we have decided shouldn't be growing in one particular place. It's just your point of view as to what makes a weed a weed. Some weed-type plants are invasive and fast growing. Their growth habit overtakes our cultivated turf plants, depriving them of food and water. Other weeds are extremely noxious and cause problems for humans if they get close to them. Weeds may be classified in various ways. The classification may be Based on economic importance, Lifestyle, leaf morphology, their likeness with grasses, herbs or shrubs, based on taxonomic position, based on place of occurrence (habitat), relationship with the host, nature of stem, and other such parameters (Radosevish et al. 2007).

3 Biological Pollution

Weeds are pollutants to our environment. Their characteristics are: they appear attractive, desirable, or harmless, weaken and may kill native vegetation, can be toxic, painful, or otherwise injurious to humans. can increase (or decrease) soil erosion and associated water quality problems, multiplies itself exponentially and can generate tons of itself in a matter of months, spreads naturally in water, wind, or soil, is also spread by wildlife, livestock, and recreationists. Effects are not usually apparent until the spread is already out of control because it can lay dormant and undetected underground for decades, then reappear and spread (Marwat et al. 2010).

Biological invasion by alien invasive species is now recognized as one of the major threats to native species and ecosystems. The unknown invader will cower in crates, mope in machines, and snuggle in ships. It can hide inside fruits, vegetables and meat. Sometimes it can even convince people to carry it across the border in their coat pockets. Parthenium weed (*Parthenium hysterophorus* L.), an alien invasive weed species, has spread in most parts of the world through one or a combination of above mentioned routes. It is now spreading rapidly throughout Pakistan. Worldwide, it has been designated as one of the most troublesome weed species. The adverse effects of this weed on human beings, livestock, crop production, and biodiversity are well-documented. As a result of a lack of information on its spread in Pakistan since its invasion, a an internal linkage project was designed in which a comprehensive phytosociological survey, with special reference to *Parthenium* weed management, was carried out in Punjab and Khyber Pkhtoonkhowa, Pakistan in later years of the past decade (Marwat et al. 2010).

4 What Makes Weeds Invasive

Biological invasion by alien species is now recognized as one of the major threats to native species and ecosystems, yet awareness of the problem is alarmingly low. The effects on biodiversity are immense and often irreversible. Invasive species occur outside their natural range. They are non-native plants and animals that harm or endanger native plants and animals or other aspects of biodiversity. Alien invasive species occur in all groups of plants and animals. They include competitors, predators, pathogens and parasites (McNeely et al. 2001). They have invaded almost every type of native ecosystem and caused hundreds of extinctions (Joshi 2001).

Through increased volume of trade and international transport over the past few centuries, natural barriers such as oceans and mountains that once prevented the movement of species have now become ineffective, ending millions of years of biological isolations. Introductions of alien species can happen deliberately or unintentionally, for example, by organism's transport in containers, ships, cars or soil. Tourists and homeowners often unwillingly introduce alien plants into wilderness areas, for example, by planting imported ornamental species in their gardens which then flourish and out-compete native species (McNeely et al. 2001).

Human exploration, colonization, and commercial trade have dramatically increased the diversity, scale, and impact of the invasion. Introduced species often find no natural enemies in their new habitat and therefore spread quickly and easily. Invasive species are a real threat to our environment and economy. Invasive species are a real threat to our environment and economy. Invasive species threaten biodiversity, habitat quality, and ecosystem function. They are the second greatest threat to native species, behind habitat destruction. Introduced species are also presenting an ever increasing threat to foot and fiber production. Most troublesome weeds in Pakistan are exotic in origin and they have arrived here sometimes in the past mainly through human activities (Tables 21.2, 21.3, 21.4, and 21.5). Once established in

Table 21.2 Common weed species of Pakistan

Scientific names	Family	Common name	Local name
Achyranthes asper	Amaranthaceae	Prickly chaff flower	Puth kanda, chirchita
Alhagi maurorum	Fabaceae	Prickly clover	Juvansa
Amaranthus spinosus	Amaranthaceae	Spiny pigweed	Cholai, Kandiali cholai
Amaranthus viridis	Amaranthaceae	Pig weed	Jangli cholai, tandulia
Anagalis arvensis	Primulaceae	Poison weed	Billi booti, chadder
Argemone maxicana	Papaveraceae	Prickly poppy	Sialkanta
Arundo donax	Poaceae	Thatch grass	Narra
Asphodelus tenuifolius	Lialiaceae	Jungle onion	Piazi, bhokat
Astragalus sp	Fabaceae	Milk vetch	Rotphullai
Avena fatua	Poaceae	Wild oats	Jangli jai
Atriplex crassifolia	Chenopodiaceae	Salt bush	Lani
Bidens biternata	Asteraceae	Black jack	Dipmal, Phutium
Boerhavia procumbens	Nyctaginaceae	Spreading hogweed	Biskhapra
Bracharia ramose	Poaceae	Signal grass	Bajra grass, bandri
Bracharia reptens	Poaceae	Bracharia	Chhota madhana
Brassica nigra	Brassicaceae	Black mustard	Jangli sarsoon
Bromus japonicus	Poaceae	Brome grass	Slai ghass
Calotropis procera	Asteraceae	Caltrope	Aak
Cannabis sativa	Cannabinaceae	Neck weed	Bhang
Carthamus oxycantha	Asteraceae	Wild safflower	Kandiari
Cenchrus ciliaris	Poaceae	Buffel grass	Dhaman
Cenchrus pennesetiformis	Poaceae	Burgrass	Kutta ghass
Centauria iberica	Asteraceae	Corn flower	Pohla otthkanda
Centaurium pulchellum	Asteraceae	Showy century	Rattan jot
Cerastium vulgatum	Caryophyllaceae	Field chickweed	Phullan booti
Chenopodium album	Chenopodiaceae	Common goose foot	Bathu
Chenopodium ambrosiodes	Chenopodiaceae	Alkali weed	Lani bathu
Chenopodium murale	Chenopodiaceae	Fathen	Talla bathu, krund
Cichorium intybus	Asteraceae	Blue daisy	Kasni
Cirsium arvense	Asteraceae	Creeping thistle	Leh, bhur bhur
Cleome viscose	Capparidaceae	Spider flower	Bogra, hul hul
Commmelina benghalensis	Commelinaceae	Asiatic dayflower	Kamlina
Convolvulus arvensis	Convolvulaceae	Common bindweed	Lehli
Convolvulus pluricaulis	Convolvulaceae	Summer bindweed	Hirankhuri
Corchorus antichorus	Tiliaceae	Jew's mallow	Baphali
Corchorus tridens	Tiliaceae	Wild jute	Jangli patsan
Conyza ambigua	Asteraceae	Tall fleabane	Loosen booti
Coronopus didymus	Brassicaceae	Swine cress	Naskari, gandi booti
Cuscuta campestris	Convolvulaceae	dodder	Amar bale, akash bale
Cynodon dactylon	Poaceae	Barmuda grass	Talla, khabbal
Cyprus difformis	Cypraceae	Umbrella plant	Ghoin
Cyprus iria	Cypraceae	Umbrella sedge	Buro-choocha
Cyprus rotundus	Cypraceae	Nut sedge	Deela, mork, motha

(continued)

Table 21.2 (continued)

Scientific names	Family	Common name	Local name
Dactyloctaenium aegypticum	Poaceae	Coast buttongrass	Madhana ghass
Datura fastuosa	Solanaceae	Jimson weed	Sufaid dhatura
Desmostachya bipinnat	Poaceae	Deep root grass	Dhabb
Dicanthium annulatum	Poaceae	Dicanthium	Bra jerga, phelwan
Digera muricata	Amaranthaceae	Digera	Tandla tendulia
Digitaria adscendens	Poaceae	Finger grass	Moti khabbal
Diplachne fusca	Poaceae	Beetle grass	Lumb ghass
Echinocloa colona	Poaceae	Millet rice	Kala swank
Echonocloa crusgali	Poaceae	Barnyard grass	Dhiddan
Eclipta prostrate	Asteraceae	False daisy	Daryai booti
Eichhnornia crassipis	Pontenderiaceae	Water orchid	Kalali
Eleusine indica	poaceae	Finger millet	Madhani
Eleusine flagellifera	Poaceae	Crowfoot grass	Chhimber ghaas
Eragrostis pilosa	poaceae	Soft love grass	Chiri ghaas
Euphorbia dracunculoides	Euphorbiaceae	Green spurge	Kanghi dodhak
Euphorbia helioscopia	Euphorbiaceae	Sun spurge	Chatter booti
Euphorbia granulata	Euphorbiaceae	Snake weed	Laal dodhak
Euphorbia hirta	Euphorbiaceae	Garden spurge	Hazardani
Festuca cristata	Poaceae	Winter grass	Domb ghaas
Fimbristylis dichotoma	Cypraceae	Hoora grass	Kalooro
Fumaria indica	Papaveraceae	Fumitory	Shahtra
Gallium aperine	Rubiaceae	Catch weed	Warri booti
Gnaphalium indicum	Asteraceae	Cud weed	Bairaksha
Heliotropium indicum	Boraginaceae	Indian heliotrope	Oont chara
Hydrilla verticillata	Hydrocharitaceae	Hydrilla	Jala
Ipomoea camea	Convolvulaceae	Railway creeper	Besharmi booti
Lactuca serriola	Asteraceae	Chinese lettuce	Jangli salad
Lantana camara	Verbenaceae	Lantana	Panj phulli
Lathyrus aphaca	Fabaceae	Crow pea	Jangli matri
Lathyrus sativus	Fabaceae	Grass pea	Kaseri
Launaea nudicaulis	Asteraceae	Yellow purge	Bathal
Lepedium sativum	Brassicaceae	Common cress	Halon
Laptochloa panicea	Poaceae	Henbit	Lamb ghaas
Lippia nudiflora	Verbinaceae	Prostrate vervane	Bukken booti
Lolium temulentum	Poaceae	Poison rye grass	Dhanak
Malva neglecta	Malvaceae	Dwarf mallo	Sonchal
Marsilia minuta	Marsiliaceae	Paper wort	Chopatti
Medicago polymorpha	Fabaceae	Bur clover	Maina
Melilotus alba	Fabaceae	White sweet clover	Sufaid senji
Melilotus	Fabaceae	Yellow sweet clover	Zard senji
Mentha longifolia	Lamiaceae	Wild mint	Jangli podina
Mukia maderaspatana	cucurbitceae	Wild cucurbit	Chibbher
Nelumbo nucifera	Nympheaceae	Water lily	Kanwal
Nicotiana plumbagifilia	Solanaceae	Wild tobacco	Jangli tambacco
Nonnea pulla	Boraginaceae		Lwien booti

(continued)

Table 21.2 (continued)

Scientific names	Family	Common name	Local name
Nymphaea nouchali	Nympheaceae	White water lily	Kutta kammi
Opuntia nigricans	Cactaceae	Cactus	Thohar
Orobanche aegyptica	orobanchaceae	Broom rape	Haddah
Oxalis corniculata	Oxalidaceae	Wood sorrel	Khatti booti
Oxalis corymbosa	Oxalidaceae	Wood sorrel	Khatkal
Oxalis pescarpae	Oxallidaceae	Bermuda	Khatti booti
Panicum antidotale	poaceae	Blue panic	Bansi ghaas
Panicum italicum	Poaceae	Foxtail millet	Kangni
Panicum ramosum	Poaceae	Brown top millet	Bandari
Panicum glaucum	Poaceae	Yellow foxtail	Lommar ghaas
Parthenium hysterophorus	Asteraceae	Parthenium weed	Gajar booti
Paspalidium flavidum	Poaceae		
Paspalum paspaloides	Poaceae	Knot grass	Naroo ghaas
Peganum hermala	Zygophyllaceae	Wild rue	harmal
Pennisetum purpureum	Poaceae	Elephant grass	Hathi ghaas
Phalris minor	Poaceae	Bird's seed grass	Dumbi sittee
Physalis minima	Solanaceae	Cape goose berry	Bhambool
Pistia stratiotes	Arucaceae	Duck salad	Aabi slad
Poa annua	Poaceae	Annual blue grass	Poa ghaas
Polygonum plebejum	polygonaceae	Smart weed	Dranak
Polypogon monspeliensis	Poaceae	Winter grass	Dumb ghaas
Portulaca oleracea	Portulacaceae	Parsely	Qulfa
Ranunculus muricatus	Ranunculaceae	Rough seeded	Chambal
Rhynchosia minima	Fabaceae	Rhyncosia	Maini
Rumex dentatus	Polygonaceae	Broad leaf dock	Jangli palak
Saccharum benghalensis	Poaceae	Tiger grass	Sarkanda
Saccharum spontaneum	Poaceae	Giant reed	Kahi
Sagittaria guayanensis	Alismataceae	Arrowhead	Chiri napay
Salsola baryosma	Chenopodiaceae	Soapwort	Kala takla
Saponaria vaccaria	Caryophyllaceae	Cow herb	
Schoenoplectus juncoides	Juncaceae	Softrush	Bhookal booti
Scirpus maritimus	Cypraceae	bulrush	Deela
Sesbania bispinosa	Fabaceae	Dunder fiber	Dhancha
Sesbania sesban	Fabaceae	Common giant	Jantar
Setaria pumila	Poaceae	Yellow foxtail	Zard loomer ghaas
Setaria italic	Poaceae	Ittalian millet	Kangni
Setaria viridis	Poaceae	Green foxtail	Bandri
Setaria verticillata	Poaceae	Bristly foxtail	Lahdra
Silene conoidea	Caryophyllaceae	Catchfly	Bhoora takla
Sylbium marianum	Asteraceae	Milk thistle	Kandiali
Sisymbrium irio	Brassicaceae	Hedge mustard	Jangli sarson
Solanum nigrum	Solanaceae	Black nightshade	Makko
Solanum xanthocarpum	Solanaceae	Spinay weed	Kandiari
Sonchus arvensis	Asteraceae	Perennial sowthistle	Peeli dodhak
Sonchus asper	Asteraceae	Spiny sowthistle	Kandiali dodhak
Sonchus oleraceous	Asteraceae	Annual sowthistle	Dodhak

(continued)

Table 21.2 (continued)

Scientific names	Family	Common name	Local name
Sorghum halepense	Poaceae	Arabian millet	Baru ghaas
Spergula arvensis	Caryophyllaceae	Corn spurry	Kalri booti
Sphenoclea zeylanica	Sphenocleaceae	Goose weed	Mirch booti
Sporobolus pallidus	Poaceae	Sprorobolous	Lamb ghaas
Stellaria media	Caryophyllaceae	Chick weed	Phullan booti
Striga lutea	Scrophulariaceae	Witch weed	Angari booti
Trianthema portulacastrum	Aizoaceae	Desert horse purslane	Itsit
Trianthema monogyna	Aizoaceae	Horse purslane	Itsit
Tribulus terrestris	Zygophyllaceae	Puncture vine	Bhakhra
Trigonella foenum-graecum	Fabaceae	Fenugreek	Methi
Trigonella monantha	Fabaceae	Trefoil	Maini
Typha domingensis	Typhaceae	Cattail	Dib
Veronica agrostis	Scrophulariaceae	Field speedwell	Veronica
Verbascum Thapsus	Scrophulariaceae	Common mullein	Gidar tambakoo
Vicia hirsuta	Fabaceae	Two-seeded vetch	Revari khurd
Vicia sativa	Fabaceae	Common vetch	Chhtri matri
Vicia tetrasperma	Fabaceae	Four-seeded vetch	Revari
Withania somnifera	Solanaceae	Winter cherry	Aksan
Xanthium strumarium	Asteraceae	Common cocklebur	Mohabbat booti

favourable habitats, these aliens spread like forest fire and threatened the environment and economy of the country. Invasive species threaten biodiversity, habitat quality, and ecosystem function. They are the second greatest threat to native species, behind habitat destruction. Introduced species also present an ever-increasing threat to food and fiber production (Marwat et al. 2010, 2011).

Most troublesome weeds in Pakistan are exotic in origin and they have arrived here sometimes in past mainly through human activities (Tables 21.2, 21.3, 21.4, and 21.5). Once established in favorable habitats, these aliens spread like forest fire and threatens the environment and economy. Pakistan has a rich repository of vascular plants and about 6,000 species have been reported. According to an estimate 8% of these species are endemic to Pakistan. According to Ali and Qaiser (1986) number of species per genus is much lower as compare to global average, reflecting a high diversity at generic level. Some of the world's worst alien invasive species have already been invaded country deteriorating the beautiful landscapes, agricultural lands plant biodiversity. Among those following are worth mentioning; Parthenium weed (*Parthenium hysterophorus*), Lantana (*Lantana camara*), Paper mulberry (*Brossunetia paperifera*), Mesquite (*Prosopis juliflora*) Wild tamarind (*Leucaena leucocephala*), Khakiweed (*Alternanthera pungens*) etc.

P. hysterophorus a noxious alien invasive weed worldwide is threatening the natural and agricultural ecosystems of the country. This weed is becoming a challenge for scientists as with in a short span of time it has invaded a significant portion of Punjab and now switching to NWFP. This noxious weed is allergenic to both humans and livestock but unfortunately florists are using this weed in bouquets showing lack of

Table 21.3 Grass land weeds at Quaid-e-Azam Campus, University of the Punjab, Lahore

Sr. #	Weed species	Sr. #	Weed species
1.	*Achyranthes asper*	27.	*Imperata cylindrical*
2.	*Ageratum conizoides*	28.	*Inula vestita*
3.	*Anagalis arvensis*	29.	*Lippia nodiflora*
4.	*Atriplex crassifolia*	30.	*Lamium sp.*
5.	*Abutilon indicum*	31.	*Launia nudicaulis*
6.	*Boerhavia diffusa*	32.	*Oligomeris glaucoscens*
7.	*Cynodon dactylon*	33.	*Mazus rugosus*
8.	*Calotropis procera*	34.	*Ncotiana plumbegiflora*
9.	*Convolvulus arvensis*	35.	*Oligomeris glaucoscens*
10.	*C. pleuricaulis*	36.	*Panicum antidotale*
11.	*Cenchrus pennisetiformis*	37.	*Polypogon monplensis*
12.	*C. setigerous*	38.	*Polygonum plebejum*
13.	*Cassis sp.*	39.	*Pullicaria crispa*
14.	*Croton sparsiflora*	40.	*Sporobolous pallidus*
15.	*Cyprus rotundus*	41.	*Rumax dentatus*
16.	*Dicanthium annulatum*	42.	*Solanum nigrum*
17.	*Dalbergia sisso*	43.	*Sonchus asper*
18.	*Dactyloctaenium aegypticum*	44.	*Suaeda fruticosa*
19.	*Desmostachya bipinnata*	45.	*Saccharum spontaneum*
20.	*D. lupinvola*	46.	*Setaria italica*
21.	*Eclipta alba*	47.	*Tribulus terrestris*
22.	*Euphorbia pilolifera*	48.	*Trigonella foenum-graecum*
23.	*E. prostrata*	49.	*Stellaria media*
24.	*Eleusine flagellifera*	50.	*Withania somnifera*
25.	*Eleusine flagellifera*	51.	*Xanthium strumarium*
26.	*Kochia indica*	52.	*Zizyphus numularia*

awareness at general public and municipal government level (Shabbir 2000). Its presence in metropolitan areas is threat to private and public property. Huge infestation in suburbs is also a potential threat to urban agriculture (Plates 21.1, 21.2, and 21.3a, b; Fig. 21.1).

Paper mulberry (*Broussonetia paperifera* Vent.) is an invasive alien weed tree became a menace in twin cities of Rawalpinia-Islamabad where thousands of residents are susceptible to its pollen allergy. Unfortunately, this weed is spreading to new localities like Lahore and Gujranwala; a huge population can be seen along the main drainage courses like of *Zafar Ali Road* in Lahore. In capital Islamabad where situation is worst, most of the citizens left the city and moved to pollen free areas. *Lantana cammra* a serious environmental weed and well known for its invasiveness globally, present in most parts of the country. Unfortunately awareness about this weed is little and people still using it as a hedge plant there by assisting its spread to newer localities. Mesquite (*Prosopis juliflora*) is another serious tree weed threatening the biodiversity of arid and semiarid environments of southern Punjab and Sind where it has reduced the primary productivity of the region. Mesquite was first introduced by British people in nineteenth century to green the Thar Desert but

Table 21.4 List of invasive weeds of Pakistan

Scientific name	Family	English name	Local name	Origin	Worst affected areas
Broussonecia papyrifera	Moraceae	Paper mulberry	Gul toot	South East Asia	From Lahore to Peshawar especially Islamabad, Rawalpindi and Northern Pakistan
Prosopis juiliflora	Mimosaceae	Mesquite	Kiker	West Indies & Mexico	Sindh
Eichhornia crassipes	Pontederiaceae	Water hyacinth	Gul-e-bakauli	Amazon Basin, South America	Water bodies of Sindh & Punjab
Salvinia molesta	Salvinaceae	Water fern	Not available	South America	Wetlands and irrigation channels of Thatta
Lantana camara	Verbinaceae	Lantana	Punch phuli	America	Islamabad
Parthenium hysterophorus	Compositae	White top	Booti	México, Central America	Islamabad & upper Punjab
Cannabis sativa	Cannabinaceae	Hemp	Bhang	Central and Western Asia	Waste areas of northern Punjab and NWFP
Pistia stratiotes	Araceae	Water cabbage	Jai kumbi	Old and New World Tropics	On the edges of old lakes
Ipomoea carnea	Convolvulaceae	Railway creeper	Railway creeper	Tropical America	In southern Sindh and Indus delta
Emex spinosa	polygonaceae	Prickly dock	Kafir kanda	Mediterranean region	In cooler parts of the country
Galium aprine	Rubiaceae	Catchweed bedstraw	Galium	Europe	Distributed in Pakistan from plains to 12,000 ft
Xanthuium stromarium	Compositae	cocklerbur	Puth kando	A New World Specie	Most of the rangelands of Pakistan
Leucanea leucocephala	Mimosaceae	Ipil ipil	As adopted local	Central America	Parts of Pakistan-in Punjab
Lolium temulentim	Graminae	Rye grass	Dhanak	Mediterranean region	Throughout Pakistan from plains to 2,000 ft

Modified from PARC (1999)

Table 21.5 Weeds of family Asteraceae from Pakistan (Khalid 1995)

Sr. #	Weed species	Sr. #	Weed species
1.	*Achillea millefolium*	28.	*Gnaphalium affine*
2.	*Achillea wilhelmsii*	29.	*Ifloga spicata*
3.	*Adenostemma lavenia*	30.	*Jurinea modesta*
4.	*Ageratum houstonianum*	31.	*Koelpinia linearis*
5.	*Artemesia scoparia*	32.	*Lactuca dissacta*
6.	*Bidens pilosa*	33.	*Lactuca serriola*
7.	*Caesulia axillaris*	34.	*Laggera aurita*
8.	*Calendula arvensis*	35.	*Launaea nudicaulis*
9.	*Carduus edelbergii*	36.	*Launea resedifolia*
10.	*Carthamus oxyacantha*	37.	*Matricaria aurea*
11.	*Carthamus lanatus*	38.	*Pentanema vestitum*
12.	*Centaurea iberica*	39.	*Picris hieraciodes*
13.	*Cichorium intybus*	40.	*Reichardia tingitana*
14.	*Cirsium arvense*	41.	*Saussurea heteromalla*
15.	*Cirsium wallichii*	42.	*Sigesbackia orientalis*
16.	*Conyza aegyptiaca*	43.	*Silybum marianum*
17.	*Conyza bonariensis*	44.	*Sonchus asper*
18.	*Conyza canadensis*	45.	*Sonchus lachnocephalus*
19.	*Conyza Sp.*	46.	*Sonchus oleraceus*
20.	*Conyzanthus squamatus*	47.	*Sphaeranthus indicus*
21.	*Cousinia minuta*	48.	*Taraxacum wallichii*
22.	*Echinops echinatus*	49.	*Tragopogon gracillis*
23.	*Eclipta alba*	50.	*Tridax procumbens*
24.	*Epilasia bungei*	51.	*Veronia anthelmintica*
25.	*Filago hurdwarica*	52.	*Xanthium strumarium*
26.	*Francoeuria undulata*	53.	*Youngia japonica*
27.	*Galinsoga parviflora*		

unfortunately this plant has become a weed and threat to unique desert flora of the country (Shabbir 2002).

A list of alien invasive species of Pakistan was developed by IUCN in 2000 but it is suspected that there would be many more to add as most of them are undocumented. Invasive species are very difficult to mange as these are equipped with weedy charters that native species are not. In developed countries like Australia millions of dollars are spent annually to eradicate invasive weeds. An exotic species is a species introduces into an area from somewhere else, often a different continent. Because the species is not native to the new area, it is frequently successful in establishing a viable population and quietly disappears. This is the fate of many pet birds, reptiles, and fish that escape or are deliberately released from their native habitats. Occasionally, however, an introduced species finds the new environment very much to its liking and can become an invasive species, thriving, spreading out, and perhaps eliminating native species native species by predation or competition for space or food. Exotic species are major agents in driving exotic species to extinction and are responsible for an estimated 39% of all animal extinctions since 1600 (Shabbir 2002).

Plate 21.1 Some grass and non grass weeds of Pakistan; (a) Germinating weed seeds to evaluate soil seed bank of seeds; (b) *Dactyloctaenium aegypticum*; (c) *poa annua*; (d) *Sporobolus pallidus*; (e) *Euphorbia prostrate*; (f) *Cynodon dactylon*; (g) *Pullicaria crispa*; (h) *Saccharum spontaenium*; (i) *Imperata cylindrical*; (j) *Parthenium hysterophorus*; (k) *Heliotropium* sp; (l) *Verbena* sp

Other than this, global warming directly reflects on rising sea levels due to melting of ice caps and natural expansion of sea water as it becomes warmer. Consequently, areas adjoining the coast and wetlands could be frequently flooded and the distribution pattern of monsoon rains gets altered with more intense downpours, storms and hurricanes. Phytosociological survey of floristic composition of weeds reveals the recent invasion of rice fields by alien invasive weeds *Leptochloa chinensis* and

Plate 21.2 (a–d) Some invasive weeds of Pakistan; Invasion of *L. cammara* in Islamabad; (**A**) Mixed communities of invasive weeds in Islamabad; (**B**) Brousunetia paperifera; (**C**) Parthenium weed with Z. *bicolrata*

Marsilea quadrifolia. These two weed species dominated over the native weeds such as *Echinochloa* sp. and others by virtue of their amphibious adaptation to alternating flooded and residual soil moisture conditions prevalent during recent years in this region (Yaduraju and Kathiresan 2003; Kathiresan 2005).

5 Invasive Weeds of Pakistan

Beautiful flowers, interesting foliage, tough constitution, and fast growth are plant qualities that appeal to gardeners and landscapers. When these desirable horticultural characteristics occur in plants that are not native to a particular place, these exotics could escape from maintained landscapes, invade natural areas, and damage native plant communities of the area. Non-native plants that readily spread in natural areas, either vegetative or via seeds, pose a significant threat to the health and welfare of rich biodiversity of the area. These plants may also be considered as exotic invasive pests (Radosevish et al. 2007).

While in contrast to this category of exotic plants, Endemic plants are those which occur within a specified locality; not introduced and have a natural distribution confined to a particular geographic area. Endemic, in a broad sense, can mean "belonging" or

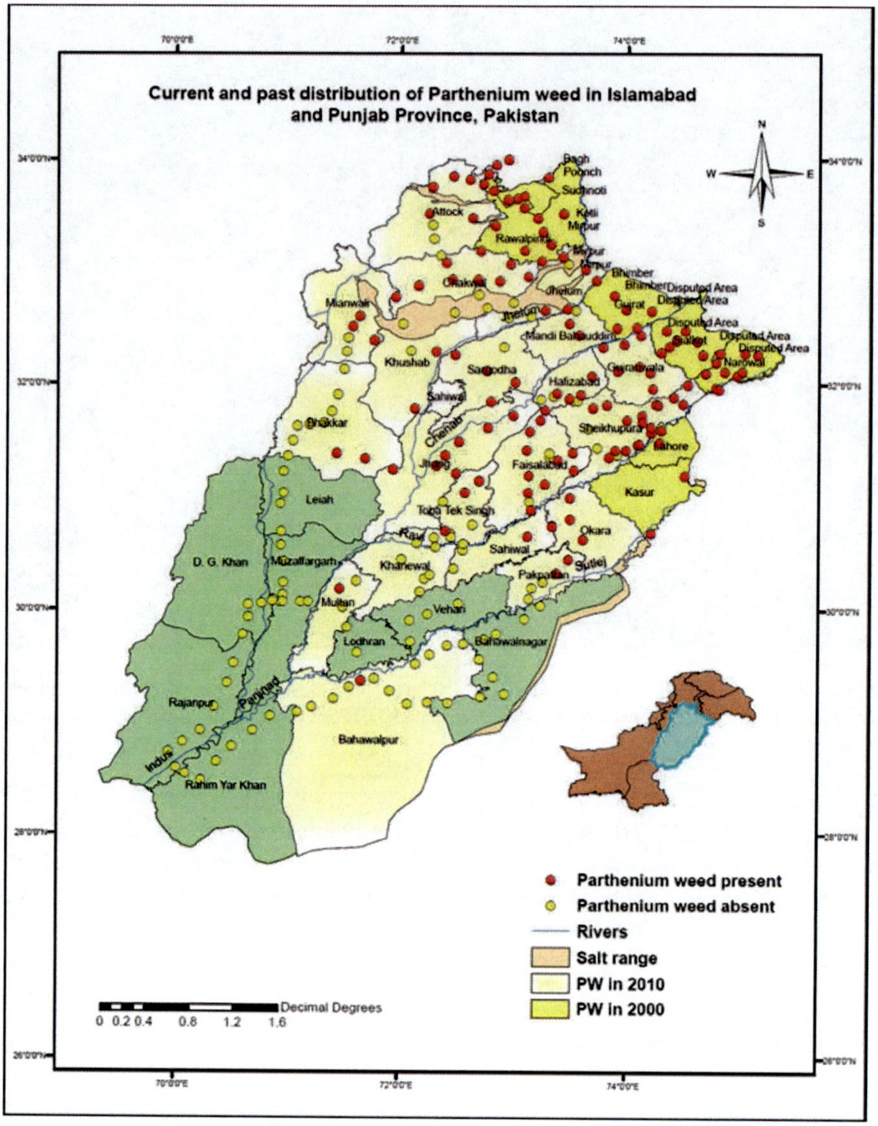

Plate 21.3 (**a**) Distribution of Parthenium weed in Punjab. (**b**) Distribution of Parthenium weed in Khyber Pukhtunkhwa

"native to", "characteristic of", or "prevalent in" a particular geography, race, field, area, or environment; native to an area or scope. The native and/or endemic plants have evolved through geological time in its native geographic location, developing a strong connection to the land, each other and the wildlife (Radosevish et al. 2007).

In contrast to it, some weeds find their way into new areas via boats, planes, soles of shows, imported animals and agricultural products, etc are called noxious or alien weeds.

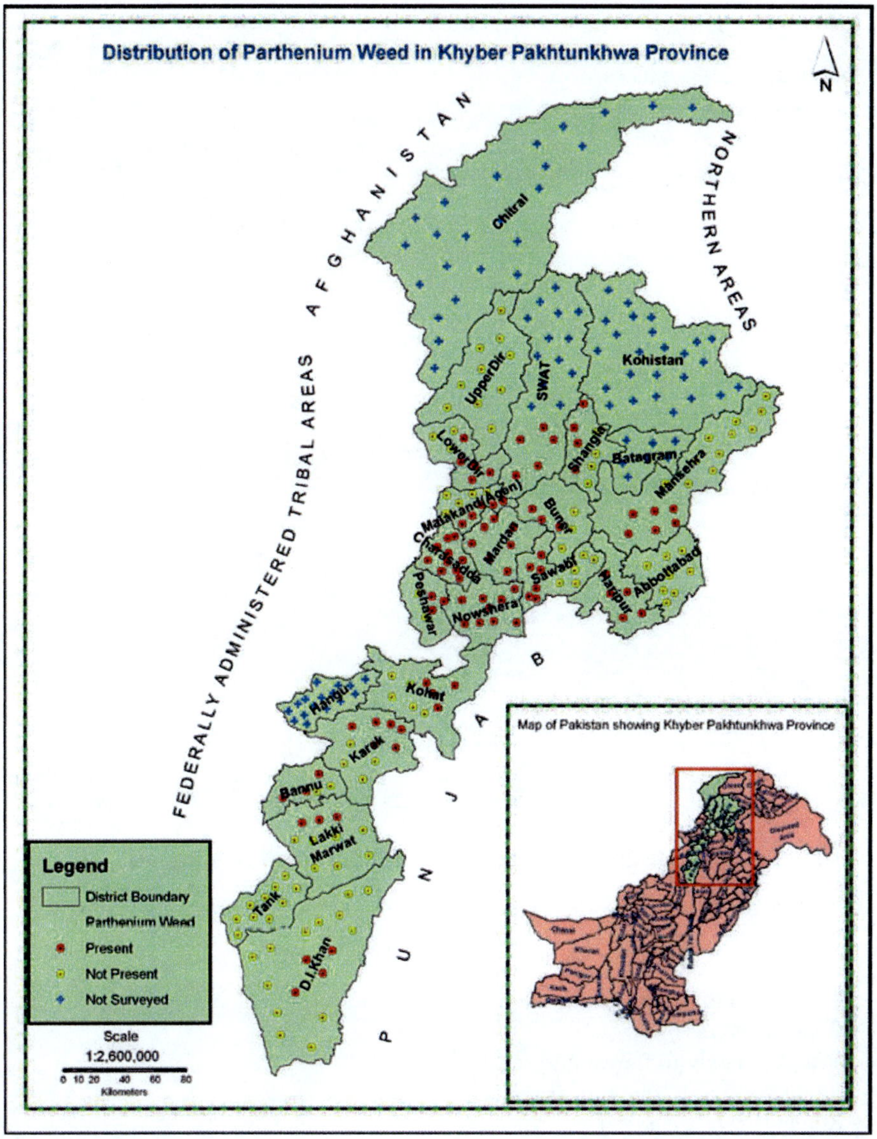

Plate 21.3 (continued)

Some of the plants introduced in Pakistan from other parts of the world are an important part of gardening and landscaping. Most of these plants are well behaved and rarely stray beyond the garden wall e.g. *Lantana camara*. Only about one percent of these non-natives readily escape into wild and become invasive in natural areas. Invasive plants exhibit certain traits (Marwat et al. 2011).

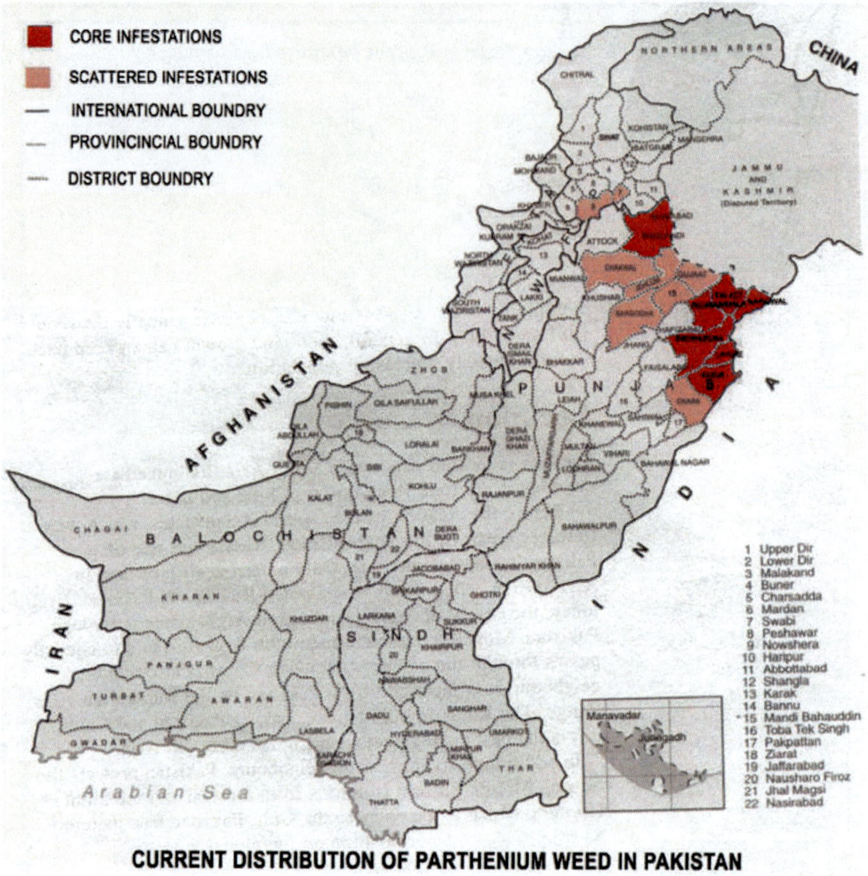

Fig. 21.1 Map of Pakistan showing Parthenium distribution

- Adaptation to local climate
- Rapid growth
- Mature quickly to flower and set seed
- Produce copious amounts of seeds
- Effective seed dispersal
- Rampant vegetative spread
- No major pests or disease problems

These traits can give exotic invasive plants undue advantage in wild habitats like forests, wetlands, marshes and grasslands. Exotic species can overwhelm native plants depriving them of nutrients, water, light and space and many totally displace native species, replacing a diverse ecosystem with a near sterile monoculture and resulting introduction of biodiversity, loss of endangered species and their habitats, loss of habitat and food sources for wildlife, and disruption of native plant-animal associations. Exotic invasive plants threaten the health and stability of natural heritage of an area and needs to establish a handsome amount of funds for research to find proper control options (Pimental et al. 2004).

Some exotic invasives are agricultural pests-forage grasses and noxious weeds. Some were introduced to modify climate like paper mulberry while some hitched rides on boats or got dumped out of homes from floral bouquets or aquaria. Many are horticultural. Unfortunately some of the invasive traits listed above can increase a plant's horticultural desirability (Marwat et al. 2010).

The worldwide search for "new" and exciting landscape plants has played a role in the dramatic shift of plant species' distribution around the globe, and the introduction of potentially invasive species is a real danger. A research conducted by Reichard (1999) in the United States has shown that 85% of the non-native invasive woody plants in North America were originally brought in as ornamental or landscape plants. Gardner's, nursery owners, landscapers, and design professionals should be aware of this potential and understand the consequences.

6 Parthenium Weed: Spread Status and Management– A Case Study from Pakistan

Parthenium hysterophorus L. (Asteraceae), commonly known as parthenium weed, is an annual herb native to the Gulf of Mexico and central South America, and has become widespread in North America, South America, the Caribbean and many parts of Africa, Asia and Australia (Navie et al. 1996; Tamado 2002; Mahadevappa 1997). Parthenium weed is not native to Pakistan and it probably arrived in late 1990s and mainly occurs in central and upper Punjab, some parts of NWFP and Azad Jammu Kashmir (Shabbir 2002, 2006; Shabbir and Bajwa 2006; Adkins and Navie 2006) affecting the natural and agro ecosystems (Fig. 21.1; Plate 21.3a, b). Parthenium weed is potential threat to cultivated and grazing parts of country, In Ethiopia and India this weed has become a problem in some of major crops (Khosla and Sobti 1979; Nath 1981; Tomado and Milberg 2000) and range lands in Australia (Adkins et al. 1996, 2001; Navie et al. 1996). Where ever this weed has invaded, it resulted in severe economic, health and environmental problems (Chippendale and Panetta 1994; McFadyen 1995; Tamado 2002; Nadeem et al. 2005). Parthenium weed can be seen every where especially during summer in non-cropped areas of central and upper districts of Punjab. This weed is emerging as a major weed in wastelands, plotted housing colonies, degraded areas, rocky crevices, along water channels, roadsides and railway tracks (Shabbir and Bajwa 2006). It has recently also been reported in cultivated lands (Shabbir 2002, 2006; Shabbir and Bajwa 2006).

In addition to losses incurred on crop production, this noxious weed can severely affect animal and human health and biodiversity. Human health hazards of this weed are well documented by McFadyen (1995) from Australia. In Pakistan, this weed is growing concern in municipal areas where, unfortunately, florists are using excessively in bouquets in Pakistan. A multiphase collaboration study of Institute of Agricultural Sciences, Punjab University, Lahore and Department of Dermatology King Edward Medical University Lahore revealed that an appreciable number of allergy patients in Lahore are sensitive to parthenium weed. It has caused a major outbreak of airborne contact dermatitis. The impact of *P. hysterophorus* on livestock production is also significant, both directly and indirectly, and it affects grazing lands, animal health, milk and meat quality (Tudor et al. 1982; Chippendale and Panetta 1994), and the marketing of pasture seed and grain. Very little, or sometimes no other, vegetation can be seen in *P. hysterophorus* dominated areas. Wherever it invades, it forms a territory of its own by replacing the indigenous natural flora including medicinal herbs utilised by man as a source of medicine (Shabbir 2002).

Since times immemorial a variety of natural herbs has become the basis of the traditional systems of medicine of treatment of many diseases in many parts of the globe. The traditional systems have been existed since remote past, in countries such as China, India and Pakistan. The traditional medicinal system continues to be of vital importance. It has been estimated by world health organization (WHO) that nearly 80% of human population depends upon traditional systems for primary health care. Amongst the world's ancient civilizations, Pakistan has been recognized as rich repository of natural herbs, which are being used in different indigenous systems of medicine i.e., Hikmat, Tib and Homeopathy. A rich heritage preserved the diversity of natural herbs widely distributed in different phyto-ecological zones of the country. The medicinal herbs viz, *Piccrorhiza* sp., *Aconitum* sp., *Artemisia* sp. etc., occur abundantly in western Himalayas, under arid and sub arid conditions. *Calotropis* sp. *Aloe barbidensis, Tribulus terristris and Withania sominefera* etc are very common. The medicinal herbs are adapted to wide range of agro-climatic regions and soil textures. Some of them also prefer to grow under degraded soils and wastelands. However, *P. hysterophorus* is quickly capturing these areas. They therefore, undergo continuous struggle/interference with *P. hysterophorus* in which this weed definitely becomes the winner, thus existence of economic herbs in these very areas in under threat (Shabbir 2002).

Parthenium hysterophorus grows well in almost all soil types although growth is more luxuriant in black as compared to latrite soil. It prevails all the years around due to its day neutral habit, temperature insensitivity, drought tolerance and absence of seed dormancy. Under arid conditions the weed has a very high survivability due to low photorespiration and emerges throughout the year, where other plants/herbs hardly withstand. Its seeds germinate in a wide range of soil pH that is 2.5–10. High regenerative and reproductive potential of the weed are the other favorable reasons for its wide spread. Root, stumps, petiole and even detached midribs have the ability to regenerate when in contact with soil. However, most of the natural herbs are not equipped with these versatile characters and as a result they cannot withstand in close competitions with it and their population is under gradual decline in Pakistan. Moreover, national parks and forest reserves are also a rich repository of many economic herbs, however, these areas were observed to be heavily infested with *P. hysterophorus* posing a serious threat to this useful flora (Shabbir 2002).

7 Management of Invasive Weeds

7.1 Prevention of Spread

Globalisation through increased trade, transport, travel and tourism will inevitably increase the intentional or accidental introduction of organisms to new environments, and it is widely predicted that climate change will further increase the threat posed by invasive species (Table 21.6). Enforcement of strict quarantine in developing countries is imperative to tackle the problem of invasive species. The establishment of wash-down sites for vehicles and implements, for thorough cleaning at international and national interstate borders will prevent much spread to newer localities (Radosevish et al. 2007).

Manual and mechanical methods: The control of invasive weed through manual and mechanical means is sometimes neither effective nor cost effective due to reasons that some weeds are allergic nature. Some of alien invasive weed species like parthenium weed is highly allergenic and prolonged exposures may result in serious contact dermatitis (McFadyen 1995) In Australia parthenium weed is a serious problem in perennial grasslands and some summer crops, where it has reduced beef production by an estimated Aus $16.5 million annually. Manual and mechanical control to such large property is economically not possible additionally such program also risks human health. If effective measures are not taken to manage the spread of this weed, a similar situation could emerge in Pakistan. In some cases mechanical control methods have shown success in controlling alien invasive weed. *Lantana cammra* an invasive weed in many parts of the world can be managed by hand pulling and bulldozing. Water hyacinth can also be managed mechanically with Hyacinth destruction boat in Florida. Manual and mechanical methods for control of invasive weed species are not effective in all situations. Some of other disadvantages of mechanical control of weeds are labour intensiveness, high cost and regrowth issues

Chemical control: Chemical control is not only partially effective and is expensive but also require repeated applications and likely to be environmentally toxic. Parthenium weed infestations along roadsides are the major sources of weed spread. One of the primary means of weed management for small, local populations is the use of Atrazine, or an Atrazine plus 2, 4-D mix, or Dicamba, or Hexazinone, or Metsulfuron or a Picloram plus 2, 4-D mix. Any one of these treatments could give good results depending upon the situation in which the weed was present (Adkins and Navie 2006). Glyphosate and Isoproturon were checked for their efficacy in controlling parthenium weed in wild situations. Glyphosate was found to be more effective in control of this noxious weed (Shabbir 2002). Chemical control of some of woody alien invasive species like paper mulberry is not possible due to its hardy nature of the plant to tolerate extreme condition

Competitive displacement ability: Several beneficial plants are known to behave highly competitive with many weeds. Some native plants can regain competitive

Table 21.6 Principles of weed management

Prevention of spread

Globalisation through increased trade, transport, travel and tourism will inevitably increase the intentional or accidental introduction of organisms to new environments, and it is widely predicted that climate change will further increase the threat posed by invasive species. Enforcement of strict quarantine in developing countries is imperative to tackle the problem of invasive species. The establishment of wash-down sites for vehicles and implements, for thorough cleaning at international and national interstate borders will prevent much spread to newer localities.

Field scouting

Field scouting is a key component of an Integrated Weed Management (IWM) system. It involves the systematic collection of weed and crop data from the field (weed distribution, growth stage, population, crop stage, etc.). The information is used, in the short term, to make immediate weed management decisions to reduce or avoid economic crop loss. In the long term, field scouting is important in evaluating the success or failure of weed management programs and for making sound decisions in the future.

Manual and Mechanical Control

The control of weeds through manual and mechanical means is sometimes neither effective nor cost effective due to reasons that some weeds are allergic nature. Manual and mechanical methods for control of invasive weed species are not effective in all situations. Some of other disadvantages of mechanical control of weeds are labour intensiveness, high cost and regrowth issues. The mechanical control of weeds involves:

- **Burial**: This method is most effective on annual weeds in which all the growing points can be buried. Burial is usually less effective on perennial weeds which have underground stems and roots and are capable of re-growth from these underground storage organs.
- **Cultivation**: The main objective in cultivation is to cut the root system of weeds and deep cultivation should usually be avoided due to damage to the crop roots. Deep cultivation may also bring more weed seed to the surface where they will germinate.
- **Mowing**: is another method of mechanical control. Mowing is usually the most effective on tall growing annuals, and not as effective on short growing plants or perennials.

Nitrogen fertility

Nitrogen fertilizer can affect the competition between crops and weeds, and in the subsequent crops. For example, nitrate promotes seed germination and seed production in some weed species. Nitrogen fertilization may increase weed growth instead of increasing crop yield. Selective placing nitrogen in a band can favour the crop over the weed. Using legume residues as opposed to chemical nitrogen fertilizer to supplement nitrogen needs of the crop can enhance weed suppression. Legume residues release nitrogen slowly with less stimulation of unwanted weed growth.

Mulches

Mulches may be of the following three types:

- Black plastic
- Clear plastic
- Infrared-transmitting plastic

Crop Competition

Crop competition is usually one of the cheapest and best methods of weed control; however, it is often one of the most overlooked methods. Weeds compete with crops for space, light, moisture, nutrients, and carbon dioxide. Usually the plant which starts first and is growing under ideal conditions will have the competitive advantage. Factors such as planting date, row spacing, seeding rate, planting depth, soil moisture, soil fertility, and soil pH have an influence on the competitive advantage of the crop or weed.

(continued)

Table 21.6 (continued)

Competitive Displacement Ability
Several beneficial plants are known to behave highly competitive with many weeds. Some native plants can regain competitive edge if grazing pressure is managed.

Crop Rotation
If the same crop is planted in the same field year after year, there usually will be some weed or weeds which are tolerant and favored by the cultural practices and herbicides used on that crop. By rotating to other crops, many of the cultural practices and herbicide programs are changed.

Biological Control
Biological control by means of insects and pathogens has been used in several parts of the world to control the exotic weeds. This exploits coevolved natural enemies from the centre of the origin of the target pest species offering practical and sustainable solutions for the management of weeds.

Chemical Control
Chemical control is not only partially effective and is expensive but also require repeated applications and likely to be environmentally toxic. Chemical control can be applied at three stages i.e., preplant, preemergence, and postemergence.

Prevention
If effective weed control has been achieved using other methods, one further step should be considered. This is preventing weeds from re-infesting the area. Knowledge of how weeds enter the field is important. Weed seed may be distributed in crop seed, hay, straw, by wind, water, animals, machinery, and other ways. The use of certified, registered, and foundation seed, or clean planting material cannot be over emphasized in preventing weeds from infesting fields. It is also important to clean equipment before entering fields or when moving from one field to another.

Tillage
Tillage and cultivation are the most traditional means of weed management in agriculture. Both expose bare ground, which is an invitation for weeds to grow. Bare ground also encourages soil erosion, speeds organic matter decomposition, disturbs soil biology, increases water runoff, decreases water infiltration, damages soil structure, and costs money to maintain (for fuel and machinery or for hand labor).

Watch the topsoil
If establishing a landscape that requires the addition of topsoil, consider where that topsoil is coming from. It may be coming from an area that is heavily infested with perennial weeds, which means those vegetative structures may be getting a free ride into your yard.

edge if grazing pressure is managed. Parthenium weed is allelopathic (Adkins and Sowerby 1996) with root and shoot leachates capable of reducing growth and germination of numerous crops. The successful spread of the weed, in part, may be attributed to these allelopathic properties (Mersie and Singh 1987). Shabbir and Bajwa (2005) noticed in the field that *Senna occidentalis* is replacing *P. hysterophorus* gradually in patches in the capital city of Islamabad. Aqueous extract of *S. occidentalis* in different ratios were effective to check the germination and early growth of *P. hysterophorus*. *S. occidentalis* at different concentrations showed least germination of *P. hysterophorus* and a significant gradual depression in fresh biomass accumulation. *Senna occidentalis* and *P. hysterophorus* both are competitive weeds of wastelands. In view of health hazards and likely threats to biodiversity due to *P. hysterophorus*, it is probably advisable to promote *S. occidentalis* growth, which is harmless medicinal plant in Parthenium infested areas in Pakistan. In a similar kind of study, Shafique et al. (2005) conducted aqueous extracts bioassays to evaluate the allelopathic potential of five tree species iz. *Azadirachata indica* L. *Mangifera indica* L., *Syzygium cumini* (L) Skeels., *Ficus benghalensis* L. and *Melia azdarach* L. for their use in *Parthenium hysterophorus* control. Aqueous extracts of 8 and 10% concentrations of all the test species invariably and significantly suppressed the germination and early growth of Parthenium weed. In few separate studies conducted by Javaid et al. (2005) revealed that some indigenous grasses like *Desmostachya bipinnata* and *Imprata cylinderica* had suppressive effect on distribution of parthenium weed, however, Shabbir (2002) observed an opposite trend when studied the distribution pattern of *D. bipinnata*. These results, obtained using a wide range of potentially useful pasture plants, create a foundation on which future field-based studies can now be undertaken to identify a useful plant or plants that can competitively displace parthenium weed (O'Donnell and Adkins 2005).

Fire and weed management: Fire has historically been an important part of the ecology of many of ecosystems. According to Keely (2001) burning may be effective at controlling noxious weeds but unless accompanied by revegetation with native species, is unlikely to diminish the alien dominance. A study carried out by Shabbir (2006) in Lahore revealed that when crop residues were burnt, the subsequent emergence of parthenium seedlings was significantly reduced in next coming crop. Wheat and rice fields, when watered after harvest showed that fire has significantly reduced the parthenium weed with no emergence in bare areas. On the other hand residues that escaped fire were stuffed with parthenium weed. This study suggests that parthenium weed may be a potential threat in those zero tillage rice – wheat cropping systems. In some other studies in Australia, smoke and heat did not significantly stimulate parthenium weed germination. Fire did, however, result in one-off increases in parthenium weed densities, which after subsequent fires rapidly declined (Vogler et al. 2002). In other studies smoke has been shown to stimulate parthenium weed seed germination and in some cases seedling emergence (Adkins et al. 2000, 2001).

Biological control: Biological control by means of insects and pathogens has been used in several parts of the world to control the exotic weeds. This exploits coevolved natural enemies from the centre of the origin of the target pest species offering practical and sustainable solutions for the management of alien invasive species (Plate 21.4). Biological control has been shown to be more cost effective than chemical means for parthenium weed. Encouraging results have been reported using biotrophic fungi (rusts) and several insect groups as agents for biological control of Parthenium weed (Dhileepan 2001, 2003a, b). *Zygogramma bicolorata* Pallister a beetle released as biocontrol agent against this weed in India and Australia has been reported by the author in Pakistan from Changa manga forest reserves (Shabbir

Plate 21.4 Two important weeds, *Euphorbia helioscopia* and *Euphorbia prostrate* and their biocontrol options: (**a** and **b**) *E. helioscopia* and its leaf rust; (**c** and **d**) *Euphorbia prostrate* and its leaf rust

2003; unpublished data) There is need to introduce some fungal pathogens along with some other insects as biocontrol agents in Pakistan to control its rapid spread. Unfortunately there exist no regulations regarding the release of fungal pathogens however; CABI-Pakistan has the prosvision to release the insect as biocontrol agents only. Moreover, a mealy bug has been reported on parthenium which seems to be very effective biocontrol agent but unfortunately not host specific and found attacking many wild and ornamental plants.

7.2 Mycotrophy in Relation to Ecological Success

Mycotrophy is the phenomenon of developing a mutualistic symbiotic relationship between plant roots/underground portions and fungi belonging to a specific group, Glomeromycota. The relationship facilitates a bilateral beneficent coordination. The fungus provides nutrients to plants ans in turn gets prepared food materials from plants. Two major types include ectomycorrhizae and arbuscular mycorrhizae. Of these two, later is the most ubiquitous and more than 50% of land plants are obligatory involves in this relationship. The relationship has existed millions of years ago when plants came to land. It is well established that plants which are ecologically predominant are able to establish strong mycotrophic relationship (Nasim 2010).

The invasion of alien plants can cause a serious threat to native ecosystems and economics (Pimentel 2002; Pimentel et al. 2004), and is the second cause of biodiversity losses after habitat destruction (Vitousek et al. 1997). Species invasion are favoured by the vulnerability of the invaded ecosystems (Sakai et al. 2001), the multiplication of introductions by human exchanges (Williamson 1996), or due to their own genetic, biological, physiological, and ecological attributed (Roy 1990; Prinzing et al. 2002). The investigation focusing on ecological attribute base invasion success of the alien plants on an enemy release hypothesis, where introduced species escape their natural enemies, thus giving them an advantage (Maron and Vila 2001; Kean and Crawley 2002). Other studies (Richardson et al. 2000; Klironomos 2002; Rudgers et al. 2005) have dealt with the potential role of the arbuscular mycorrhizal (AM) symbiosis to facilitate plant invasion in new areas. Most invasive plants are mycotrophic and thus preadapted to successfully establish in habitats containing AM fungi. Keeping in view the wide spread distribution of AM fungi and low host specificity, these mutualistic associations could favour invasion of many ecosystems by alien plants (Richardson et al. 2000; Fumanal et al. 2006). Mycorrhizal interactions could be a key point in invasive plant process, not only by facilitating local adaptation and/or reducing environmental stress, but also through their direct or indirect effects on interplant competition. Moreover, AM are now recognized as a major factor in structuring plant communities (Richardson et al. 2000).

Weeds also develop strong Mycorrhizal relationship to facilitate them for accelerated acquisition of the nutrients and water. *Parthenium hysterophorus* is an example of a notorious weed which is newly introduced in Pakistan. Besides exhibiting other "weedy" characteristics it is able to develop strong arbuscular mycorrhizal association. This relationship enables plants to out compete in ecologically diverse habitats. The year round active phase of the plant (except 1 ½ month in winter) enables AM fungi to actively absorb nutrients and save its energy from being concentrated to form spores and other perennating structures. Plant to plant colonization is also reported facilitating an easy spread of the weed (Nasim 2010).

7.3 *Management Through Awareness Partnership and Participation*

To address the issue of invasive weed species, public awareness and community participation and involvement are of the prime importance. Partnership between public and private sectors in activities to address IAS has shown great success in many countries. Promotion of awareness of invasive weeds issues by convening workshops and seminars, as well as conducting publicity events and media campaigns resulted in effective management program globally. Department of Mycology and Plant Pathology University of Punjab organized a national symposium on the awareness of Parthenium weed from 6 to 7th August 2004. Scientists, teachers and students from various parts of the country participated in this symposium and recommendations were formulated to create awareness about this dreaded weed in scientific and non-scientific community. PARTHENIUM NEWS: a biannual news letter is publishing form department of Mycology and Plant Pathology to create awareness about this weed (Plate 21.5). Ensure the sustainability of IAS prevention and management activities in the region by developing long-term programs of action.

Florists are excessively using parthenium weed, one of the most dangerous weeds that causes different diseases among humans, in the preparation of bouquets (Shabbir 2002). This misuse is not only threat to health of general public but also a source of spread to new localities. The rapid spread is probably due to its use in bouquets. An action group named Pakistan Parthenium Action Group (PPAG) was formed in 2002. This group ensures that the weed receives adequate publicity to create awareness that helps early identification when Parthenium weed germinate in newer locality. Another objective of this Action group is to establish protocols to reduce the movement and spread of parthenium weed particularly in municipal areas of metropolitan cities of Punjab and raise community's ability to recognize parthenium weed, know of its impacts and skills required for its management.

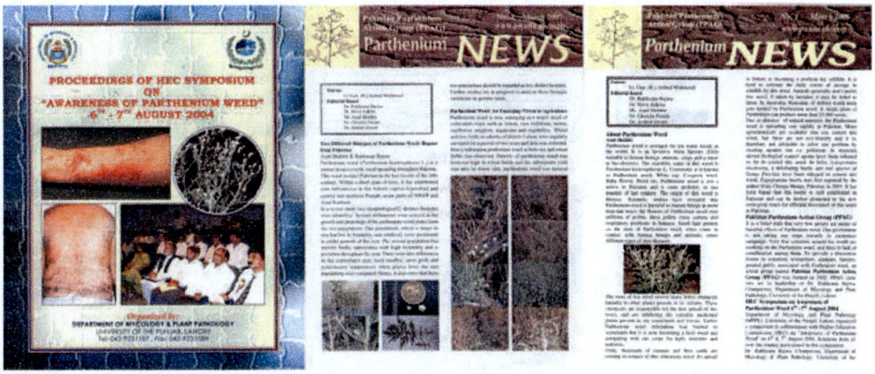

Plate 21.5 Face copies of publications of IMPP on awareness program of Invasive weeds of Pakistan

8 Recent Trends in Weed Management and Modeling

Weeds still represent an important constraint to crop production in the world. Agricultural practice has demonstrated that the same philosophy of integrated management used for insect control also needs to be adopted for weed control. Because of the vast number of serious weed problems in all regions of the world, just one method alone can no longer be relied on. The developments in weed management have taken place mainly through the application of new approaches and methods, rather than the introduction of new herbicides. Moreover, agriculture has promoted new methods, which also directly or indirectly affect weed control. These methods are original *per se* and demand a new approach for weed control. For example, organic agriculture does not permit the use of chemical herbicides, therefore cultural and biological control is the only possible means left to cope with weeds (Marwat et al. 2010).

Presently, the major problems in wheat fields are associated with herbicide-resistant weeds. In this context, one of the most serious incidences is the one associated with isoproturon-resistant grass *Phalaris* minor in North India. The entire rice-wheat area of 10 million ha is affected by the resistant biotype of *Phalaris* in the Indo-Gangetic plains. The problem is more serious in a 5 million ha area (Malik 2002, personal communication). To overcome this problem, Indian specialists have adopted an integrated approach for cropping wheat in rotation with direct-seeded rice. There are already more than 10,000 ha of wheat planted with the no zero till system in India, and it has brought about several benefits, such as improving water-use efficiency, preventing soil erosion and compaction, reducing the need for applying herbicides, and increasing wheat yields. The studies on long-term trials on zero till have shown that *Phalaris minor* stand decreased over the 3-year period because of the combined effect of the herbicide clodinafop and zero tillage (Rice-Wheat Consortium and CIMMYT 2003).

Herbicide-resistant crops offer the potential for simpler weed control, more effective management of problematic and resistant weeds, more timely weed control with potential to employ critical period, increased usage of minimum or zero tillage and avoidance of yield loss caused by current 'selective' herbicides (FAO 1998). Although the benefits are certain, herbicide-resistant crops (HRCs) may also pose direct risks to human health, and may cause indirect, ecological, and evolutionary risks, which have not received enough attention from regulatory agencies (Denison 2001).

9 Instruments of Quarantine, Environmental Law and Legislation

Some exotic plant pests leave immediate evidence of their presence. Signs of disease, crop damage or weed growth appear almost instantly. Other foreign pests however can go undetected for months or even years in the absence of proper surveillance.

Without early detection, these insidious organisms can become established in a new geographical range and permanently damage agriculture and natural resources of the area.

9.1 An Endangered Ecosystems Act

The endangered species act formally recognizes the importance of saving species that are in serious trouble. The act has enjoyed some success but is considered seriously flawed by many environmentalists because it does not directly address the major reasons species become endangered: the loss of crucial ecosystems that are their habitats, alien invasive species, pollution, climate change, disease etc. Thus, ecosystem diversity, a level of diversity higher than biodiversity, is also important to maintain (Wright 2005).

Ecosystem diversity is based on the idea that there should ne a variety of ecosystems in a given area, be it local national or global. By protecting many different types of ecosystems, many endangered species will be protected, and a bulk of other species will be prevented from becoming endangered. Thus protecting ecosystem diversity requires both the qualitative and quantitative preservation of ecosystem. In the United States only a number of ecosystems have been degraded from 85% to 98%. What has happened to these ecosystems? In large measures, they have been altered considerably by urban development, agriculture, exploitation of their resources, and pollution including the introduction of alien weeds. In response to such real and continuous losses, many scientists and environmentalists have called for a national Endangered Ecosystem Act (Wright 2005).

Implementation of endangered Ecosystem Act would involve steps similar to those taken to protect endangered species – that is. Taking an inventory of the status of ecosystem to identify those in trouble, protecting endangered ecosystems from further activities that would damage them, establishing recovery plans for many of the most critical ecosystems, and promoting research and monitoring to maintain surveillance and to gain knowledge for better management and protection (Wright 2005).

9.2 Plant Protection and Quarantine

The best and most cost-effective solution for mitigating agricultural damage is preventing the entry of exotic plant pests or invasive weeds into the country. Plant protection and quarantine (PPQ) is a regulatory program of USA, it is responsible for establishing effective regulations and policies to protect the country's agriculture through the exclusion of exotic species. PPQ determines what plants and plant products can be imported into this country and what products pose a high risk and, therefore, should be kept out. Based on varying levels of risk, PPQ's regulations provide multiple layered of protection ranging from product bans to commodity treatments or other actions that lessen pest risk (e.g., restricting imports to certain geographical regions or certain seasons) (Shabbir 2002).

Plants and seeds requiring permits for propagation must enter the United States through specific ports-of-entry, where such items can be inspected and certified free of invasive pests and disease. At these plant inspection stations, PPQ also enforces the rules and regulations that apply to the import of plant species protected by the endangered species of Wild Fauna and Flora (CITES). PPQ currently operates 17 plant inspection stations nationwide and has been instrumental in surveying for invasive organisms, that do not exist in the United States but have been intercepted at U.S. port of entry. These pest potentially affect urban and natural areas and the Nations lumber and agricultural industry. The proactive work helps ensure that if these pests are found, they are found quickly before that can become established here. Early detection saves million of dollars in control ans eradication costs, protects natural resources, and presents the economic disruption of agricultural industry (Shabbir 2002).

9.3 Pakistan Environmental Protection Act (PEPA)

Pakistan Environmental Protection Act (PEPA), enacted in 1997, is an elaborate, comprehensive act to provide for the protection, conservation, rehabilitation and improvement of the environment, for the prevention and control of pollution and promotion of sustainable development. It rectifies CITES (Convention of International trade in endangered species of wild fauna and flora 1973); Ramsar (Conventional on wetlands of international importance 1971; 1975; Vienna convention for the protection of ozone layer, Montreal Protocol on Substance that deplete ozone, 1985; Stockholm Declaration on Human Environment, 1972; World's conservation strategy, 1980; World Charter for Nature, 1982; Man and Biosphere Program, 1970; Brundtland report, 1987). In all these conventions and declarations it indirectly addresses the issue of biodiversity conservation but no direct reference is made to the invasive species and not a single sub-section is focusing to tackle the problem. The new biosafety and quarantine rules have been laid down recently and are waiting for their enforcement and implementation. PEPA has badly failed due to illiteracy and unawareness prevailing in the masses, besides paucity of funds and corruption. An intensive revision of PEPA is therefore recommended keeping in view the recent development taking place in the developed world with special focus on the role of invasive species on economy and biodiversity of the region.

10 Conclusion

Weeds are a menace for agriculture in the world. These unwanted plants parasitize on the nutrients and other resources available for the crop plants. Billion of dollars are spent on weedicides to tackle with the problem. In developing countries like

Pakistan the damages caused by weeds are a nasty blow to the struggling economy of the country. In view of use of these weedicides and anthropogenic activities the problem become multifold with the invasion of alien weeds like *Parthenium hyterophorus*. Investigations are in progress to manage these weeds using biological methods or with competitor indigenous plants. It is also needed to strengthen our national biosafety regulations and bring them at par with the international criteria particularly for transportation of grains and other biological materials and soil. To avoid the problem, quarantine and phytosanitary measures should also strictly be reinforced.

Acknowledgements The authors are grateful to Professor Dr. Steve Adkins, School of Land Crop and Food Sciences, University of Queensland, Brisbane, Australia for his valuable suggestions. The grant from Higher Education Commission (HEC) of Pakistan under International Linkage Program is also acknowledged for some segments of the chapter.

References

Adkins S (1995) Weed science lecture guide. University of Queensland, Brisbane

Adkins SW, Navie SC (2006) Parthenium weed: a potential major weed for agroecosystems in Pakistan. Pak J Weed Sci Res 12(1–2):19–36

Adkins SW, Sowerby MS (1996) Allelopathic potential of *parthenium hysterophorus* L. in Australia. Plant Prot 1:20–23

Adkins SW, Navie SC, McFadyen RE (1996) Control of parthenium weed: a centre for tropical pest management team effort. In: Proceedings of the 11th Australia weeds conference, Melbourne, Australia

Adkins SW, Davidson PJ, Matthew L, Navie S, Wills DA, Taylor IN, Bellairs SM (2000) Smoke and germination of arable and rangeland weeds. In: Black M, Bradford K, Vazquez-Ramor J (eds) Seed biology: advances and applications. CAB International, Cambridge, pp 347–359

Adkins SW, Navie SC, McFadyen RE, Tomley A, Dhileepan K, Chamberlain J, Graham GC, Adamson D, Goldsworthy D, Dearden S (2001) Parthenium weed: the recent research effort. Proceedings of the 6th Queensland weeds symposium, Caloundra, Australia, pp 130–135

Ali SI, Qaiser M (1986) A phytogeographical analysis of the phanerogams of Pakistan and Kashmir. Proc R Soc Edinb 89B:89–101

Anderson WP (1996) Weed science: principles and applications, 3rd edn. West, St. Paul

Barrett SCH (1988) Genetics and evolution of agriculture weeds. In: Altieri MA, Liebman M (eds) Weed management in agroecosystems: ecological approaches. CRC Press, Boca Raton, pp 57–75

Barrett SCH, Richardson BJ (1986) Genetic attributes of invading species. In: Groves R, Burdon JJ (eds) Ecology of biological invasions - an Australian perspective. Australian Academy of Sciences, Canberra, pp 21–33

Baskin CC, Baskin JM (1988) Seeds: ecology, biogeography and evolution of dormancy and germination. Academic, New York

Chippendale JF, Panetta FD (1994) The cost of *Parthenium* weed to the Queensland cattle industry. Plant Prot 9:73–76

Denison RF (2001) Meeting report: ecologists and molecular biologists find common ground in the rhizosphere. Trends Ecol Evol 16:535–536

Dhileepan K (2001) Effectiveness of introduced biocontrol insects on the weed *Parthenium hysterophorus* (Asteraceae) in Australia. Bull Entomol Res 91:167–176

Dhileepan K (2003a) Current status of the stem-boring weevil *listronotus setosipennis* (coleoptera: curculionidae) introduced against the weed *parthenium hysterophorus* (asteraceae) in Australia. Biocont Sci Technol 13:3–12

Dhileepan K (2003b) Seasonal variation in the effectiveness of the leaf-feeding beetle *zygogramma bicolorata* (coleoptera: chrysomelidae) and stem-galling moth *epiblema strenuana* (Lepidoptera: tortricidae) as biocontrol agents on the weed *parthenium hysterophorus* (asteraceae). Bull Entomol Res 93:393–401

FAO (1998) Technical meeting on benefits and risks of transgenic herbicide-resistant crops. Rome, Italy, 16–18 Nov 1998, 38pp

Fumanal C, Plenchette B, Chauvel B, Bretagnolle F (2006) Which role can arbuscular mycorrhizal fungi play in the facilitation of *Ambrosia artemisiifolia* L. invasion in France. Mycorrhiza 17:25–35

Javaid A, Anjum T, Bajwa R (2005) Biological control of *parthenium* II: allelopathic effect of *desmostachya bipinnata* on distribution and early seedling growth of *parthenium hysterophorus* L. Int J Biol Biotechnol 2(2):459–463

Joshi D (2001) Alien invasive species: a global threat to biodiversity. The Katmandu Post 23 Dec

Kathiresan RM (2005) Case study on Habitat management and rehabilitation for the control of alien invasive weed (*Prosopis juliflora*), Report submitted to Water Resource Organization, Public Works Department, Tamilnadu, India

Kean RM, Crawley MJ (2002) Exotic plant invasion and the enemy release hypothesis. Trends Ecol Evol 17:164–170

Keely JE (2001) Fire and invasive species in Mediterranean-climate ecosystems of California. In: Galley KEM, Wilson TP (eds) Proceedings of the invasive species workshop: the role of fire in the control and spread of invasive species. Fire conference 2000: The first national congress on fire ecology, prevention, and management. Miscellaneous publication no. 11. Tall Timbers Research Station, Tallahassee, FL, pp 81–94

Khalid S (1995) Weeds of Pakistan, Asteraceae, Published by National herbarium, National Agricultural Research Centre (NARC), Park Road, Islamabad

Khosla SN, Sobti SN (1979) Parthenium – a national health hazard, its control and utility – a review. Pesticides 13:121–127

Klironomos JN (2002) Feedback with soil biota contributes to plant rarity and invasiveness in communities. Nature 417:67–70

Mahadevappa M (1997) Ecology, distribution, menace and management of parthenium. In: Mahadevappa M, Patil VC (eds) Proceedings of the 1st international conference on parthenium management, 6–8 October 1997. University of Agricultural Sciences, Dharwad, India, pp 1–12

Maron JL, Vila M (2001) When do herbivores affect plant invasion? Evidence for the natural enemies and biota resistance hypothesis. Oikos 95:361–373

Marwat KB, Hashim S, Ali H (2010) Weed management: a case study from north-west Pakistan. Pak J Bot 42:341–353

Marwat KB, Hashim S, Nawab K, Khattak AM, Khan MA (2011) Integrated weed management in wheat. Pak J Bot 43(1):625–633

McFadyen RE (1995) *Parthenium* weed and human health in Queensland. Aust Fam Physician 24:1455–1459

McNeely JA, Mooney HA, Neville LE, Schei PJ, Waage JK (eds) (2001) A global strategy on invasive alien species. IUCN, Gland/Cambridge

Mersie W, Singh M (1987) Allelopathic effect of parthenium (*parthenium hysterophorus* L.) extract and residue on some agronomic crops and weeds. J Chem Ecol 13:1739–1747

Nadeem M, Rani Z, Aman S, Kazmi AM, Shabbir A (2005) Parthenium weed: a growing concern in Pakistan. J Pak Assoc Dermatol 15(1):4–8

Nasim G (2010) Glomalean Spore Flora of Pakistan, Monograph published by HEC, Islamabad, 333pp

Nath R (1981) Note on the effect of Parthenium extract on seed germination and seedling growth in crops. Ind J Agric Sci 51(8):601–603

Navie SC, McFadyen PE, Panetta FD, Adkins SW (1996) The biology of Australian weeds 27, *parthenium hysterophorous* L. Plant Prot 11:76–88

O'Donnell C, Adkins SW (2005) Management of parthenium weed through competitive displacement with beneficial plants. Weed Biol Manag 5:77–79

Pakistan Agriculture Research Council (1999) Invasive species of Pakistan. IUCN/CABI-Bioscience, University of Karachi, Karachi

Pimental D, Zuniga R, Morrison D (2004) Update on the environmental and economic costs associated with alien-invasive species in the united states. Ecol Econ 52:273–288

Pimentel D (ed) (2002) Biological invasions: economic and environmental cost of alien plant, animal, and microbe species. CRC Press, New York

Pimentel D, Berger B, Filiberto D, Newton M, Wolfe B, Karabinakis B (2004) Water resources: agricultural and environmental issues. BioScience 54(10):909–918

Prinzing A, Durka W, Klotz S, Brandl R (2002) Which species become aliens. Evol Ecol Res 4:385–405

Radosevish SR, Holt JS, Ghersa CM (2007) Ecology of weeds and invasive plants, 3rd edn. Wiley, Hoboken

Reichard SH (1999) A method for evaluating plant invasiveness. Public Garden 14(2):18–21

Rice-Wheat Consortium for the Indo-Gangetic Plains and CIMMYT (2003) Research report, New Delhi, India

Richardson DM, Allsopp N, D'Antonio CM, Milton SJ, Rejmanek M (2000) Plant invasions – the role of mutualism. Biol Rev 75:65–93

Roy J (1990) In search of characters of plant invaders. In: DiCastri AJ, Debussche M (eds) Biological invasions in Europe and the Mediterranean basin. Kluwer, Dordrecht, pp 335–352

Rudgers J, Mattingly W, Koslow J (2005) Mutualistic fungus promotes plant invasion into diverse communities. Oecologia 144:463–471

Sakai AK, Allendorf FW, Holt JS, Lodge DM, Molofsky J, With KA, Baughman S (2001) The population biology of invasive species. Annu Rev Ecol Syst 32:305–332

Shabbir A (2000) Allelopathic effects of *Parthenium hysterophorus* on germination and early growth of different weeds. M.Sc. thesis, University of Arid Agriculture, Rawalpindi

Shabbir A (2002) *Parthenium hysterophorus* L. An exotic weed threatening the biodiversity and agricultural lands of Islamabad and adjoining districts. M.Phil thesis, Department of Botany, University of the Punjab, Lahore, Pakistan

Shabbir A (2006) Parthenium weed (*Parthenium hysterophorus* L.) a new emerging threat to agricultural lands of Pakistan. Paper presented in 2nd international weed science conference organized by Weed Science Society of Pakistan held from 20 to 22 Mar 2006 at University of Arid Agriculture, Rawalpindi, Pakistan

Shabbir A, Bajwa R (2005) *Senna occidentalis*: a native plant to restore the natural vegetation of Islamabad. World conference on ecological restoration – A global challenge held from 12 to 18 Sept, Zaragoza, Spain, Europe

Shabbir A, Bajwa R (2006) Distribution of Parthenium weed (*Parthenium hysterophorus*) I: an alien invasive weed species threatening the biodiversity of Islamabad. Weed Biol Manag 6(2):89–95

Shafique S, Bajwa R, Javaid A, Shafique S (2005) Biological control of *Parthenium* IV: suppressive ability of aqueous leaf extracts of some allelopathic trees against germination and early seedling growth of *Parthenium hysterophorus* L. Pak J Weed Sci Res 11:75–79

Tamado T (2002) Biology and management of Parthenium weed (*Parthenium hysterophorus* L.) in Eastern Ethiopia. Ph.D. dissertation. Swedish University of Agricultural Sciences, Acta Universitatis Agriculturae Sueciae, Agraria 311, Uppsala, Sweden, 106pp

Tomado T, Milberg P (2000) Weed flora in arable fields of eastern Ethopia with emphasis on occurrence of *Partheniurn hysterophorus*. Weed Res 40:507–521

Tudor GD, Ford AL, Armstrong TR, Bromage EK (1982) Taints in meat from sheep grazing *Parthenium hysterophorus*. Aust J Exp Agric Animal Husb 22:43–46

Vitousek PM, D'Antonio CM, Loope LL, Rejmanek M, Westbrooks R (1997) Introduced species: a significant component of human caused global change. New Zealand J Ecol 21:1–16

Vogler W, Navie S, Adkins S, Setter C (2002) Use of fire to control parthenium weed. A report for the Rural Industries Research and Development Corporation. University of Queensland, Brisbane, p 41

Williamson MH (ed) (1996) Biological invasion. Chapman and Hall, London

Wright RT (2005) Environmental science, 9th edn. Pearson Prentice Hall, Pearson Education Inc., New Jersey

Yaduraju NT, Kathiresan RM (2003) Invasive Weeds in the Tropics. In: Proceedings of 19th Asian Pacific Weed Science Society Conference, Manila, Philippines, vol I, pp 59–68

Chapter 22
An Overview of Plant Growth Promoting Rhizobacteria (PGPR) for Sustainable Agriculture

Rifat Hayat, Iftikhar Ahmed, and Rizwan Ali Sheirdil

Contents

Abstract Soil bacteria beneficial to plant growth usually referred to as plant growth promoting rhizobacteria (PGPR), are capable of promoting plant growth by colonizing the plant root. The mechanisms of PGPR-mediated enhancement of crop growth includes (i) a symbiotic and associative nitrogen fixation; (ii) solubilization and mineralization of other nutrients; (iii) production of hormones e.g. auxin i.e. indole acetic acid (IAA), abscisic acid (ABA), gibberellic acid and cytokinins; (iv) production of ACC-deaminase to reduce the level of ethylene in crop roots thus enhancing root length and density; (v) ability to produce antagonistic siderophores, β-1-3-glucanase, chitinases, antibiotics, fluorescent pigment and cyanide

R. Hayat (✉) • R.A. Sheirdil
Department of Soil Science and SWC, PMAS Arid Agriculture University, Rawalpindi, Pakistan
e-mail: hayat@uaar.edu.pk; rizwan_sheirdil52@yahoo.com

I. Ahmed
Plant Biotechnology Program, National Agricultural Research Centre, Park Road, Islamabad, Pakistan
e-mail: iftikharnarc@hotmail.com

M. Ashraf et al. (eds.), *Crop Production for Agricultural Improvement*,
DOI 10.1007/978-94-007-4116-4_22, © Springer Science+Business Media B.V. 2012

against pathogens and (vi) enhanced resistance to drought and oxidative stresses by producing water soluble vitamins niacin, thiamine, riboflavin, biotin and pantothenic acid. Increased crop production through biocontrol is an indirect mechanism of PGPR that results in suppression of soil born deleterious microorganisms. Biocontrol mechanisms involved in pathogen suppression by PGPR include substrate competition, antibiotic production, and induced systemic resistance in the host. PGPR can play an essential role in helping plants to establish and grow in nutrient deficient conditions. Their use in agriculture can favour a reduction in agro-chemical use and support ecofriendly crop production. Trials with rhizosphere-associated plant growth-promoting P-solubilizing and N_2-fixing microorganisms indicated yield increase in rice, wheat, sugar cane, maize, sugar beet, legumes, canola, vegetables and conifer species. A range of beneficial bacteria including strains of *Herbaspirillum, Azospirillum* and *Burkholderia* are closely associated with rhizosphere of rice crops. Common bacteria found in the maize rhizosphere are *Azospirillum sp., Klebsiella sp., Enterobacter sp., Rahnella aquatilis, Herbaspirillum seropedicae, Paenibacillus azotofixans,* and *Bacillus circulans*. Similarly, strains of *Azotobacter, Azorhizobium, Azospirillum, Herbaspirillum, Bacillus* and *Klebsiella* can supplement the use of urea-N in wheat production either by BNF or growth promotion. The commonly present PGPR in sugarcane plants are *Azospirillum brasilense, Azospirillum lipoferum, Azospirillum amazonense, Acetobacter diazotrophicus, Bacillus tropicalis, Bacillus borstelensis, Herbaspirillum rubrisubalbicans* and *Herbaspirillum seropedicae*. Symbiotic N_2-fixing bacteria collectively known as Rhizobia are currently classified into six genera; Rhizobium, Allorhizobium, Azorhizobium, Bradyrhizobium, Mesorhizobium and Sinorhizobium and 91 species. Their inoculation may increase nodulation and N_2-fixation in legumes. All these Rhizobiumn spp. can minimize chemical N fertilizers by BNF, but only if conditions for expression of N_2-fixing activity and subsequent transfer of N to plants are favourable. In this Chapter, PGPR role has been discussed in the process of crop growth promotion, their mechanisms of action and their importance in crop production on sustainable basis.

Keywords PGPR • BNF • P-solubilization • Phytohormones • Biocontrol • Cereals • Sugarcane and legumes crops

1 Introduction

Plant growth promoting rhizobacteria (PGPR) are capable of promoting plant growth by colonizing their roots and can play an essential role in helping plants to establish and grow in nutrient deficient conditions. Their use in crop production can reduce the agro-chemical use and support ecofriendly sustainable food production. Plant growth promotion by PGPR is due to root hair proliferation, root hair deformation and branching, increase in seedling emergence, early nodulation, nodule functioning, enhanced leaf surface area, vigor, biomass, increasing indigenous plant hormones levels, mineral and water uptake, promoted accumulation of carbohydrates and yield in various plant species (Podile and Kishore 2006). The demand of PGPR biofertilizers has been increasing day by day with increase in the importance of organic agriculture with minimum inputs of chemicals. The population of PGPR in rhizospheric soil varies and depends largely on crop species and soil health (Tilak et al. 2005). Numerous studies clearly indicated the positive effect of PGPR on growth of different crops at different agroecological boundaries. Trials with soil rhizospheric N_2-fixing and P-solubilizing *Bacillus* species showed yield increase in wheat (de Freitas 2000), rice (Sudha et al. 1999), maize (Pal 1998), sugar beet (Çakmakçi et al. 2006), canola (de Freitas et al. 1997), and conifer species (Bent et al. 2002). One of the most important and oftenly reported PGPR is *Bacillus polymyxa,* also known as *Paenibacillus polymyxa* (Timmusk et al. 1999) with a range of properties, including P-solubilization (de Freitas et al. 1997), nitrogen fixation (Coelho et al. 2003); production of cytokinin (Timmusk et al. 1999), antibiotic (Rosado and Seldin 1993), hydrolytic enzymes (Nielson and Sorenson 1997), colonization hair and cortical cells (Shishido et al. 1999), and increased root and shoot growth of crops (Sudha et al. 1999). Some strains of *Rhodobacter* are known to fix N_2 (Drepper et al. 2002), but extensive studies are needed to validate its ability to fix N_2. *Pseudomonas* sp. effectively adapt to new environments and colonize winter wheat roots (Misko and Germida 2002) thus significantly increasing root dry weight in spring wheat (Walley and Germida 1997), yield in sugar beet (Çakmakçi et al. 2001), and promote the growth of spinach (Urashima and Hori 2003).

Sustainable agriculture production can be achieved by emphasizing the use of PGPR as biofertilizer inoculants (Schippers et al. 1995). Generally, bacteria promote plant growth in three different ways (Glick 1995, 2001): synthesizing growth promoting harmones for the plants (Dobbelaere et al. 2003), facilitating the uptake of nutrients from the soil (Çakmakçi et al. 2006), and lessening or preventing the plants from diseases (Saravanakumar et al. 2008). The indepth mechanisms involved in plant growth by PGPR are yet to be investigated (Dey et al. 2004). However, the possible explanations include (i) solubilization of mineral phosphates and mineralization of other nutrients (Richardson 2001; Banerjee and Yasmin 2002); (ii) a biological nitrogen fixation (Kennedy et al. 2004a, b); (iii) ability to produce hormones like auxin i.e. indole acetic acid (IAA) (Patten and Glick 2002), abscisic acid (ABA) (Dobbelaere et al. 2003), gibberellic acid and cytokinins (Dey et al. 2004); (iv) ability to produce ACC-deaminase to reduce the level of ethylene in root of developing plants thereby increasing the root length and growth (Li et al. 2000; Penrose and Glick 2001); (v) antagonism against phytopathogenic bacteria by producing siderophores, β-1,3-glucanase, antibiotics, chitinases, fluorescent pigment and cyanide (Glick and Pasternak 2003); and (vi) mediated resistance to drought (Alvarez et al. 1996) and oxidative stresses (Stajner et al. 1995, 1997) and production of water soluble vitamins thiamine, niacin, pantothenic acid, biotin and riboflavin (Revillas et al. 2000).

2 Biological Nitrogen Fixation (BNF)

Nitrogen is the most important nutrient and its deficiency severely affects crop yields. Most of the soils around globe are deficient in nitrogen and applications of nitrogenous fertilizer are essential for achieving maximum crop yield. Nitrogen is required for all living organisms. Although 78% of the atmosphere consists of dinitrogen, but it cannot be used by most organisms and consequently the availability of nitrogen in a form suitable for assimilation is often a major limiting factor for crop growth. The production of chemical nitrogen fertilizers not only depletes non-renewable energy resources but also poses human and environmental hazards, besides being very expensive. Urea is the cheapest and readily available N source, but unfortunately less than 50% of the applied urea is used by plants. This low efficiency of use is mainly caused by NH_3 volatilisation and denitrification, and losses from leaching. Leaching of NO3-N causes ground water toxicity. Volatilisation and denitrification pollute the atmosphere through the evolution of greenhouse gases like N2O, NO and NH3. BNF not only complements and substitutes the mineral fertilizers but can be an economically beneficial and ecologically sound alternative (Glick et al. 1999).

BNF is a major source of nitrogen for plants as a part of environment friendly agricultural practices. It contributes for 65% of the total nitrogen currently utilized in crop production and will be important contributor for future agriculture (Matiru and Dakora 2004). This fixation is an important biochemical reaction next to plant photosynthesis for life on earth and occurs through symbiotic N_2-fixing bacteria possessing the nitrogenase enzyme and in association with legumes (and some woody species) that converts atmospheric elemental dinitrogen (N_2) into ammonia (Shiferaw et al. 2004). The potential of nitrogen fixation for most legumes species is in the range of 200–300 kg N ha^{-1} $crop^{-1}$ (Peoples et al. 1995).

In the recent years, use of biological inoculants for sustainable crop production is attaining popularity in various parts of the world and biological nitrogen fixation represents the major source of N input in agricultural soils including those in arid regions. Symbiotic nitrogen-fixing bacteria include the cyanobacteria, the genera *Rhizobium*, *Azorhizobium*, *Bradyrhizobium*, *Sinorhizobium*, *Allorhizobium*, *Mesorhizobium* and *Frankia* (Paul and Clark 1996; Brock et al. 2000). The mechanisms of symbiotic N_2-fixation between *Rhizobium* and legumes has been studied intensively. The symbiosis between *Frankia* and non-leguminous actinorhizal plants, is also been investigated now a days. Symbiotic system is the major N_2-fixation system, playing a significant role in improving the fertility and maximizing productivity of low-N soils. Biological N_2-fixed through rhizobium-legume symbiosis can also be beneficial to cereals growing in intercrops or to subsequent crops rotated with legumes. In many natural grassland systems, the grasses use N_2 fixed by their legume counterparts for their nitrogen requirement and the protein available through this association enhances the forage quality for livestock production (Paynel et al. 2001). In addition to symbiotic N_2-fixation in legumes, rhizobia as PGPR are also capable of contributing to growth promotion in non-legume species (Hoflich 2000). To act as PGPR, rhizobia naturally produce molecules (auxins, abscisic acids, cytokinins, riboflavin, lumichrome, lipo-chitooligosaccharides and vitamins) that promote crop growth, and their colonization and infection of cereal root would be expected to increase vigor and grain yield (Matiru and Dakora 2004). Other PGPR role of *Rhizobium* includes their ability to produce phytohormones (Arshad and Frankenberger 1998), solubilization of inorganic phosphorus (Chabot et al. 1996), siderophore release (Plessner et al. 1993; Jadhav et al. 1994) and antagonism against plant pathogenic microorganisms (Ehteshamul-Haque and Ghaffar 1993). Application of rhizobium inoculants under rainfed conditions on legumes like guar (*Cyamopsis tetragonoloba* L. Taub), moth (*vigna acontifolia*), mung (*vigna radiate*), and mash (*vigna mungo*) give up to 10–25% yield benefits (Rao 2004; Hayat et al. 2008a, b).

Free living bacteria as well as rhizobial strains can promote the growth of cereals by contributing to N-economy through their ability to fix N_2 (Zahir et al. 2004). It is reported that biological N_2-fixed by the diazotrophic PGPR may be a contributing factor of rice growth promotion in addition to other mechanisms (Biswas et al. 2000). Kennedy and Islam (2001) also explained the possible role and mechanisms of non-symbiotic bacteria to crop growth from BNF view point. Providing the plant with essential nutrients, e.g. NH_4-N through atmospheric nitrogen fixation or aiding the plant in nutrient uptake is also considered as direct plant growth promotion. The important free living and associative nitrogen fixing genera are; *Azospirillum, Azotobacter, Acetobacter* (also known as *Gluconacetobacter*) *Azoarcus, Achromobacter, Bacillus, Burkholderia, Clostridia, Citrobacter, Enterobacter, Herbaspirillum, Kelbsiella, Mycobacterium, Pseudomonas, Rhodobacter* and *Serratia*. This list is increasing day by day.

3 Mineral-Solubilization and Uptake

PGPR directly contribute to the plant growth by facilitating plant nutritions through solubilization of inorganic phosphates and production of iron chelating siderophores thus increasing phosphorous and iron uptake.

3.1 Phosphorus Solubilization

Phosphate solubilizing bacteria (PSB) could play an important role in supplying phosphate to plants in a more environment friendly and sustainable manner. The naturally abundant PSB solubilize Calcium-bound phosphatic compounds in an alkaline soil environment and convert the insoluble phosphatic compounds into soluble forms and make them available to crop plants. PSB are widely applied in agronomic practices in order to increase the productivity of crops while maintaining the health of soils. The beneficial effects of PSB on plant growth vary significantly depending on environmental conditions, bacterial strains, plant and soil conditions (Şahin et al. 2004; Çakmakçi

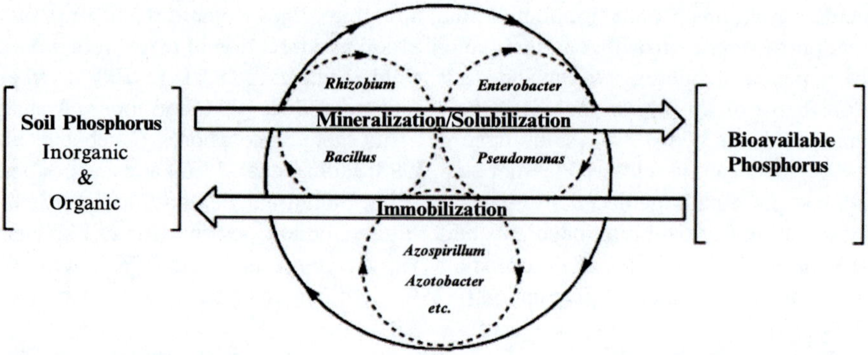

Fig. 22.1 Schematic diagram of soil phosphorus mobilization and immobilization by bacteria (Khan et al. 2009)

et al. 2006). Various bacterial species in the genera *Bacillus, Rhizobium* and *Pseudomonas,* have proven to be the most powerful phosphate solubilizing bacteria (Banerjee et al. 2006). There are also reports of phosphate solubilization by non-symbiotic nitrogen fixer, *Azotobacter* (Kumar et al. 2001). The phosphate-solubilizing activity of *Rhizobium* (e.g., *Rhizobium/Bradyrhizobium*) is associated with the production of 2-ketogluconic acid, indicating that phosphate-solubilizing activity of the organism is entirely due to its ability to reduce pH of the medium (Halder and Chakrabarty 1993). The phosphate-solubilizing ability also depends on nature of nitrogen source used in the media, with greater solubilization in presence of ammonium salts than when nitrate is used as nitrogen source. This has been attributed to extrusion of protons to compensate for ammonium uptake, leading to a decreased extra-cellular pH (Roos 1984). In some cases, however, ammonium can lead to decrease in phosphorus solubilization (Reyes et al. 1999). Several strains of P-solubilizers have been identified in vitro as *Bacillus brevis, B. megaterium, B. polymyxa, B. sphaericus, B. thuringiensis* and *Xanthomonas maltophilia* (de Freitas et al. 1997). P-solubilizing bacteria of *Bacillus* and *Paenibacillus* (formerly *Bacillus*) sp. have been applied to soils to successfully enhance the phosphorus status of plants (van Veen et al. 1997).

Since 1950s it is reported that P-solubilizing bacteria release phosphorus from organic and inorganic soil phosphorus pools through mineralization and solubilization (Fig. 22.1; Khan et al. 2009). Lowering of soil pH by microbial production of organic acids such as acid phosphatases, lactate, citrate, and succinate gluconic and keto gluconic acids etc. (Goldstein 1995; Deubel et al. 2000) and proton extrusion is the main principal mechanism of mineralization of organic form of phosphorus. Release of phosphorus from insoluble and adsorbed forms is also an important aspect of phosphorus solubilizing bacteria regarding phosphorus availability in soils. Phosphorus solubilizing bacteria transform soil phosphorus to forms that can easily be taken up by crops. Bacteria assimilate soluble phosphorus, and make it available by preventing it from adsorption (Khan and Joergensen 2009). Bacteria also enhance phosphorus availability to crops by solubilizing precipitated forms of phosphorus (Chen et al. 2006).

Fe - and Al - P and Ca - P are the main precipitated forms in acidic and alkaline soils respectively and can be solubilized involving organic anions and associated protons. Complex form of adsorbed phosphorus my be released by chelating metal ions through ligand exchange reactions. Soil bacteria use carbon and serve as a sink for phosphorus by immobilizing it from soil (Bünemann et al. 2004). Phosphorus solubilizing bacteria mainly depend on soil phosphorus and organic matter contents and release phosphorus from their cells for crops (Kim et al. 1998a). Agricultural and range land soils are main habitat of phosphorus solubilizing bacteria (Yahya and Azawi 1998) and their use in crop production can reduce chemical P by 50% without any significant reduction of crop yield (Yazdani et al. 2009). PSB biofertilizers have great potential for sustaining crop yield along with optimized phosphorus fertilization. Examples of P-solubilizing PGPR include *Azotobacter chroococcum* (Kumar and Narula 1999) and *Bacillus circulans* in wheat (Singh and Kapoor 1998), *Enterobacter agglomerans* in tomato (Kim et al. 1998b), *Pseudomonas chlororaphis* or *Pseudomonas putida* in soybean (Catellan et al. 1999) and other prominent P-solubilizing genera are *Klebsiella, Balkhurdaria* and *Azospirillum* etc. The availability of nutrients i.e. C, N and metals ions are the determining factors of P-solubilizing ability of PGPR. The activity of P-solubilizing rhizobacteria varies with nitrogen source and increases in the presence of low levels of Ca^{2+} and EDTA (Nautiyal et al. 2000). Soil bacteria could increase the P nutrition of plants through increased solubility of Ca-phosphates. Phosphate solubilization is the result of combined effect of pH decrease and production of organic acids (Fankem et al. 2006). Bacteria through secretion of different types of organic acids e.g. carboxylic acid and rhizospheric pH lowering mechanisms (He and Zhu 1988) dissociate the bound forms of phosphate like $Ca_3(PO_4)_2$. Nevertheless, buffering capacity of the medium reduce the effectiveness of PSB in releasing P from tricalcium phosphates (Stephen and Jisha 2009). Acidification due to proton substitution/excretion of H^+ (accompanying greater absorption of cations than anions) or release of Ca^{2+} by microbial population releases P from apatite (Goldstein 1994; Villegas and Fortin 2002). The reverse phenomenon become operative when uptake of anions increases than that of cations, with excretion of OH^-/HCO_3^- (Tang and Rengel 2003). PSB produce carboxylic anions, which have high affinity to calcium. This phenomenon results in more solubilization of phosphorus than acidification alone (Staunton and Leprince 1996). Complexation favoured by organic acids results in cations complexing structures in P solubilization, which is an important mechanism. Decrease in pH as well as synthesis of carboxylic acids results in the release of calcium phosphate (Ca-P); however, release of protons cannot be the single mechanism (Deubel et al. 2000). Crop yields improve with increased solubilization of the fixed soil P as well as the applied phosphates by PSB (Gull et al. 2004).

Most annual crops often do not get benefitted by direct application of phosphate rock in a short time. Phosphatic rocks can be solubilized by acid producing microorganisms (Gyaneshwar et al. 2002) to release more P for plant uptake. The inorganic P-solubilizing activities by PSB ranges between 25 and 42 μg P mL^{-1}, whereas the organic P mineralization may occur between 8 and 18 μg P mL^{-1} (Tao et al. 2008). The application of P fertilizers can be reduced by 25% and 50%, by using PSB inocula along with single super phosphate and rock phosphate, respectively

(Sundara et al. 2002). It is reported that 29–62% P can be released by *Pseudomonas putida*, *P. fluorescens* Chao and *P. fluorescens* Tabriz; along with highest value of 0.74 mg P/50 mL from Fe2O3 (Ghaderi et al. 2008). *Pseudomonas fluorescens* is very effective and can solubilize 100 mg P L^{-1} containing $Ca_3(PO_4)_2$ or 92 and 51 mg P L^{-1} containing $AlPO_4$ and $FePO_4$, respectively (Henri et al. 2008). Rock phosphatic minerals are often insoluble to provide sufficient P for plant uptake; however, using PSBs can release P from the fixed insoluble minerals and thus help to increase crop yields (Zaidi 1999). PGPR not only improves BNF but also contribute in increasing the availability of soluble P and thus, enhance plant growth (Ponmurugan and Gopi 2006). It is reported that PSB enhanced seedling length of *Cicer arietinum* (Sharma et al. 2007) and increase sugarcane yield (Sundara et al. 2002). Co-inoculation of PSB and PGPR can reduce application of P fertilizers by 50% without affecting corn yield (Yazdani et al. 2009). Grain yield of wheat increased 20–40% by the application of inoculation along with P fertilizers against sole P fertilization (Afzal and Bano 2008). The indigenous PSB populations often are not effective in releasing P from bound phosphates mainly due to high buffering capacity of soil; however, inoculants of PSB biofertilizers may contribute considerably in more plant P uptake. Plant available P increased by the activity of PSB especially belonging to the genera: *Bacillus, Pseudomonas* and *Enterobacter*. So, PSB has enormous potential in biofertilizer formulations to be exploited in increasing crop yields by increasing fixed P in the soil, as well as by making good use of natural reserves of phosphate rocks.

3.2 Iron Uptake

Iron being essential micronutrient of plants plays a key role in several enzymes with redox activity. Its role is important in photosynthesis, NO_2 and SO_4 reduction and N_2 assimilation, and is therefore indispensable for chlorophyll biosynthesis (Rashid 1996). Iron is essential component of different steps involved in biosynthetic pathways and formation of chlorophyll required physiologically available iron to the plant (Lopez-Millan et al. 2001). Catalase and peroxidase are considered as important protoheme enzymes which can be used as biological markers in iron acquisition studies. Around 30% of the agronomic crops growing around the globe are facing the problem of chlorosis due to iron deficiency (Imsande 1998). Total soil Fe is always in excess of crop requirements, which is present in highly insoluble form of ferric hydroxide. In nature, various forms of iron are present in abundance but remain unavailable to plants due to their different solubility and bioavailability behaviour. The amount of soluble Fe in soil is much less than the total Fe contents thus iron remains unavailable for crops even in iron rich soils and contributes little in crop production (Podile and Kishore 2006). When iron availability to crops is inadequate for growth, leaves become pale green, yellow or white and eventually brown (Brittenham 1994). The availability of iron in soil solution is 10–18 M, a concentration even not sufficient for sustaining microbial growth. A large number of soil bacteria produce siderophores that bind Fe^{3+} and help in iron uptake (Podile and Kishore 2006). These siderophores can be used by rhizosphere bacteria and crops can absorb bacterial Fe^{3+} siderophore complexes. This mechanism is involved in iron absorption by crops especially in calcareous soils. Different kind of PGPR produce different kinds of siderophores. Pseudomonads are leading siderophore producers among PGPR and siderophore producing strains of *fluorescent psuedomonas* may be inoculated in calcareous soils for solubilization of non-available forms of iron (Sharma and Johri 2003).

Siderophores are iron chelating compounds having low molecular weight which are released under iron limited rhizospheric soils. These siderophores possess high binding affinity and specificity for iron (III) and help to facilitate its bioavailability into the biological cell (Schalk et al. 2001). The transportation of ferric siderophore complexes is a heterologous uptake and mediated by specific receptor proteins (Meyer et al. 2000). PGPR with the ability to produce siderophore play a vital role for iron acquisition in rhizo spheric soil. It is reported that under non-sterile soil system, crops show no iron-deficiency symptoms and have more iron uptake in roots as compared to crops grown in sterile soil (Masalha et al. 2000). This supports the possible role of PGPR in iron acquisition. Iron is present in different complexes each having different solubilities in natural system; therefore, the bioavailability of iron depends upon the potential of siderophores to chelate the iron from its complexes (Hersman et al. 2001). A cold-tolerant mutant of *Pseudomonas fluorescens* with 17-fold increase in siderophore production enhances colonization and PGP effect on mungbean (Katiyar and Goel 2004). Similarly, seed inoculation of maize with siderophore-producing *Pseudomonas Pseudomonas chlororaphis* enhances seed germination and root shoot biomass (Sharma and Johri 2003). Siderophores producing PGPR suppressed root pathogens by consuming available soil iron (Kloepper et al. 1980) and with the addition/availability of iron even in acidic soils ($pH<6.0$), these siderophores becomes less effective (Neilands and Nakamura 1991). Different bacterial proteins are involved in iron uptake and transport. This uptake by different bacterial species depends on the available concentration of soil iron.

4 Plant Growth Regulators

PGPR synthesize and export phytohormones also called as plant growth regulators (PGRs), that may be synthesized in defined organs of plant and can be translocated to other sites, where these trigger specific biochemical, physiological, and morphological role in plant growth and development (Hayat et al. 2011). PGRs are organic in nature that promote physiological processes of plants even at low concentrations and also take part in the development of tissues where they are produced. Among PGRs, gibberellins, cytokinins, auxins, abscisic acid and ethylene are well documented. The most common and active auxin in plants is indole-3-acetic acid (IAA), that regulates many aspects of plant growth and development throughout the plant cell cycle, from cell division, cell elongation and differentiation to root initiation, apical dominance, tropistic responses, flowering, fruit ripening, senescence and stimulation of plant growth. Regulation of these processes by auxin is believed to involve auxin-induced changes in gene expression (Guilfoyle et al. 1998). In addition to IAA, bacteria such as *Paenibacillus polymyxa* and *Azospirillum* also release other compounds in rhizosphere that could indirectly contribute to plant growth promotion like indole-3-butyric acid (IBA), Trp and tryptophol or indole-3-ethanol (TOL).

Cytokinins too are important phytohormones usually present in small amounts in biological samples (Vessey 2003). These enhance cell division, root development and root hair formation (Frankenberger and Arshad 1995), and are also involved in the processes such as photosynthesis or chloroplast differentiation. They are also known to induce opening of stomata, suppress auxin-induced apical dominance, and inhibit senescence of plant organs, especially in leaves (Crozier et al. 2001). More than 30 growth promoting compounds of cytokinin group are reported that are produced by plant associated PGPR. Cytokinin producing bacteria *Azotobacter chroococcum* and cytokinin precursor's isopentyl alcohol (IA) and adenine (ADE) have been tested on maize crop under controlled and field conditions by Nieto and Frankenberger (1991). They observed improvement in crop growth. PGPR also produce widely recognized gibberellic acid (GA) and gibberellins (GAs).Over 89 GAs are known to date, most of which are primarily responsible for stem elongation (Dobbelaere et al. 2003). GAs also affect reproductive processes in a wide range of plants (Crozier et al. 2001). PGPR like *Azospirillum* and *Pseudomonas* sp. produce cytokinins and gibberellins (gibberellic acid). In addition to IAA, abscisic acid (ABA) has also been detected by radio-immunoassay or TLC in supernatants of *Azospirillium* sp. and *Rhizobium* sp. cultures (Dobbelaere et al. 2003).

Ethylene is a plant growth regulator synthesized by almost all species of bacteria (Primrose 1979). It acts as a ripening hormone, promotes adventitious root and root hair formation, stimulates germination, breaks dormancy of seeds; promote plant growth, development, and senescence. However, if ethylene concentration remains high after germination, root elongation, as well as symbiotic N2 fixation in leguminous plants is inhibited. PGPR produces enzyme like 1-aminocyclopropane-1-carboxylate (ACC) deaminase, which hydrolyzes ACC and lowers the level of ethylene in crop rhizosphere. The product of this hydrolysis, ammonia and α-ketobutyrate, can be used by the bacterium as a source of nitrogen and carbon for growth. In this way the bacterium acts as a sink for ACC and thus lowers the ethylene level in plants, preventing some of the potentially deleterious consequences of high ethylene concentrations (Glick et al. 1998). PGPR with ACC-deaminase characteristics improve crop growth and yield and may be included bio-fertilizer biotechnology (Shaharoona et al. 2006). Role of PGPR in the production of phosphatase, β-glucanase, dehydrogenase and antibiotics has also been recognized. Another recently identified mechanism for plant growth promotion is due to production of volatiles by PGPR. Ryu et al. (2004) discussed in detail the role of bacterial volatiles in plant growth promotion *in vitro*. PGPR release different volatile blends and the differences in these volatile blends stimulate the plant growth. Volatile compounds like 3-hydroxy-2-butanone (acetoin) and 2,3-butanediol, produced by *Bacillus subtilis* and *B. amyloliquefaciens* stimulated the growth of *Arabidopsis thaliana* in *in-vitro* experiments. The volatile-mediated growth promotion by PGPR is by activation of cytokinin-signaling pathways.

5 Role of PGPR in Biocontrol Aspects

PGPR indirectly help in plant growth by suppression of deleterious microorganisms that inhibit plant growth or root pathogens through antibiosis, parasitism, competition for nutrients and space within the vicinity of plant roots, and/or activation of host defense responses (Podile and Kishore 2006). The strains of *Bacillus subtilis* are the most widely used PGPR due to their disease reducing and antibiotic producing capabilities (Kokalis-Burelle et al. 2006). Fluorescent pseudomonads are also known to suppress soil born fungal pathogens by producing antifungal metabolites and by sequestering iron in rhizosphere through the release of iron-chelating siderophores, rendering it unavailable to other organisms (Dwivedi and Johri 2003).

Suppression of deleterious microorganisms by PGPR is mainly by parasitism, by competing for available nutrients, production of enzymes or toxins and inducing resistance by activating plant defense response against pathogens (Podile and Kishore 2006). Fluorescent pseudomonads establish themselves on plant roots and sink the available nutrients, thus limiting the available nutrients required for the growth of pathogen (Walsh et al. 2001). PGPR also compete for nutrients with native rhizosphere microbes for elimination of pathogens. Siderophore production by PGPR, sequester most of available Fe^{3+} in the rhizosphere and force the pathogens for iron starvation, thus is a major contributor for pathogen suppression (O'Sullivan and O'Gara 1992). Suppression of *Fusarium wilt* of radish by *Pseudomonas* strain WCS358 through siderophore-mediated competition for iron (Costa and Loper 1994) is another example. Some PGPR's degrade the cell wall of pathogenic fungi through production of hydrolytic enzymes (chitinase, β-1,3-glucanase). Purified chitinases of *Bacillus subtilis* AF 1 (Manjula et al. 2004), *Serratia marcescens* (Kishore et al. 2005b; Ordentlich et al. 1988) and *S. plymuthica* (Frankowski et al. 2001) are highly antifungal. β-1,3-glucanase producing strain of *Pseudomonas cepacia* inhibits the rhizosphere proliferation of various phytopathogenic fungi including *Rhizoctonia solani*, *Sclerotium rolfsii* and *Pythium ultimum* (Fridlender et al. 1993) and synergistic action of the two hydrolytic enzymes chitinases and β-1,3-glucanases was more effective in the inhibition of fungal pathogens than either enzyme alone (Tanaka and Watanabe 1995). Hydrolytic enzyme produced by pathogenic fungi includes pectolytic enzymes (polygalacturonases, pectin, lysis), cellulases and cutinase. PGPR's inhibit (Podile and Kishore 2006) these pathogenic fungi and thus activity of these enzymes can be reduced with the reduction in virulence (Beraha et al. 1983). *Bacillus megaterium* B 153-2-2 inhibits the activities of extracellular enzymes, like cellulase, pectin lyase and pectinase produced by *Rhizoctonia solani,* by producing an extracellular endoproteinase (Bertagnolli et al. 1996). A groundnut seed endophytic bacterium *Pseudomonas aeruginosa* GSE 18, effective in control of groundnut stem rot disease, when applied as seed treatment or soil amendment, inhibits the production of cell wall-degrading enzymes (CWDE) polygalacturonase and cellulase by the pathogen *Sclerotium rolfsii* tested in vitro (Kishore et al. 2005c).A few strains of rhizobacteria activate plant defense responses against a broad spectrum of plant pathogens, termed as induced systemic resistance (ISR). Rhizobacteria-mediated ISR has been demonstrated in many plant-pathogen systems wherein the bacterium and the challenging pathogen remained spatially separated, and these observations indicate that ISR is genetically determined (Pieterse et al. 2001).

Inoculation of different strains of *Pseudomonas fluorescens* strains reduce the seedling mortality caused by *Aspergillus niger* (Dey et al. 2004) and show strong inhibition to *Sclerotium rolfsii* by reducing the incidence of stem rot severity. They also produce siderophores and antifungal metabolites. Production of antifungal metabolites by fluorescent *Pseudomonads* has been found to suppress soil-borne fungal pathogens on many occasions (Dey et al. 2004; Catellan et al. 1999). Similarly, *Pseudomonas putida* produces siderophores which convert a fusarium-conducive soil into a fusarium-suppressive soil for growth of different crops. Improvement in plant growth and disease resistance to a broad array of plant pests can be accomplished using PGPR (Kloepper et al. 2004). It is well recognized that PGPR can influence plant growth through creating resistance against pathogens. *Bacillus subtilis* is one such commercialized PGPR and it acts against a wide variety of pathogenic fungi (Banerjee et al. 2006).

6 Role of PGPR in Crop Production

Rice, wheat and maize are the three major staple food crops for world's population. A variety of PGPR's participate in interaction with C3 and C4 plants and can significantly increase their yield (Kennedy et al. 2004a, b). Rice crop removes around 16–17 kg N to produce 1 t dry weight of rice including straw. Wheat crop requires about 26–28 kg N to produce 1 t of grain including straw (Angus 2001). Maize plants require 9–11 kg N to produce 1 t biomass (Kennedy et al. 2004a, b). The N requirement of cereals is normally met by fertilization at a rate depending on soil fertility with chemical urea (Scharf 2001). PGPR inoculant biofertilizers can, in principle, be used to supplement or reduce the use of urea-N (Kennedy et al. 2004a, b). Those closely associated with rice rhizosphere are *Azospirillum*, *Burkholderia* and *Herbaspirillum*. A free living heterotrophic diazotroph like *Azotobacter (A. vinelandii* and *A. chroococcum)* uses C from sugar as energy source (Kennedy and Tchan 1992). There are obligatory anaerobic heterotrophs like *Clostridia* which are only capable of fixing N_2 in the complete absence of oxygen (Kennedy and Tchan 1992; Kennedy et al. 2004a, b) and are usually isolated from rice fields (Elbadry et al. 1999). Their activity in rice may be enhanced with the addition of organic source like straw (Kanungo et al. 1997), presumably as a result of microbial breakdown of cellulose into cellobiose and glucose. Yield of rice can be increased with the application of *Azotobacter, Azospirillum lipoferum* and *Azospirillum brasilense,*

*Azospirillum (*Reis et al. 2000). Similarly, *Burkholderia vietnamiensis* increases rice grain yields significantly up to 8 t ha^{-1} (Tran Vân et al. 2000) by supplementing 25–30 kg N ha^{-1} as synthetic fertilizer. Family *Enterobacteriaceae* has several diazotrophs like *Klebsiella* (*K. pneumonia*), *Enterobacter* (*E. cloacae*), *Citrobacter* (*C. freundii*), and *Pseudomonas* (*Ps. putida* or *Ps. fluorescens*) isolated from rice rhizosphere with plant growth promoting traits (Kennedy et al. 2004a, b). Another rice endophyte *Herbaspirillum seropedicae* (James et al. 2000) can fix upto 45% of total plant N in rice from the atmosphere (Baldani et al. 2000). The N_2- fixation range by *Herbaspirillum* was assessed upto 58 mg $tube^{-1}$ under aseptic conditions (Reis et al. 2000). *Azoarcus* sp. is also endophytic Gram-negative N_2-fixing bacterium firstly isolated from Kallar grass *(Leptochloa fusa* Kunth*)* growing in saline-sodic soils of Pakistan (Reinhold-Hurek et al. 1993). *Azoarcus* colonise rice under both laboratory and field condition (Hurek et al. 1994; Reinhold-Hurek et al. 2002). Multi-strain biofertilizer (BioGro) containing three different PGPR like *Pseudomonas* (*Ps. fluorescens*, *Ps. putida*), *Klebsiella* (*K. pneumonia*) and *Citrobacter* (*C. freundii*) isolated from rice rhizospheric soil of Hanoi, Vietnam (Nguyen et al. 2003) significantly increased rice grain yield in Vietnam and Australia (Williams and Kennedy 2002). National Institute for Biotechnology and Genetic Engineering (NIBGE) of Pakistan also prepared a multi-strain rice biofertilizer rice with the commercial name "BioPower" (Malik et al. 2002) and similar product is also available in Egypt (Hegazi et al. 1998). These inoculants are claimed to give similar yield increases on rice farms upto 20% (Kennedy et al. 2004a, b).

Similarly, strains of *Azospirillum, Azotobacter, Azorhizobium, Bacillus, Herbaspirillum* and *Klebsiella* can supplement the use of urea-N in wheat production either by BNF or growth promotion (Kennedy and Islam 2001). The N requirement of wheat is higher than that for rice, because of its higher grain protein content. Wheat yields vary widely from 1 to 7 t ha^{-1} dependent on inherent soil fertility, amount of applied fertilizer, wheat variety, diseases, other management practices and environmental conditions (Angus 2001).Thus, estimated amount of N removed by wheat crop varies between 26 and 200 kg N ha^{-1}, depending on the yield (Reeves et al. 2002). To maximize wheat yields in soils that are not capable of supplying enough N, chemical fertilizers such as urea are used to enhance N supply. Biofertilizers are also being used to supplement the use of urea worldwide. The estimated amount of BNF by such wheat-bacterial associations was between 10 and 30 kg N ha^{-1} (Kennedy and Islam 2001) or about 10% of their total-N requirement. Wheat transformed about 30% of carbon assimilates into the process of rhizodeposition and part of this below ground translocated C is incorporated by rhizosphere micro-organisms (Kuzyakov and Domanski 2000). Studies indicate that PGPR may act as natural elicitor for improving the growth and yield of wheat. Important PGPR responsible for increased wheat yield in different parts of world are *Pseudomonas* (*ps. cepacia, ps. aeruginosa, ps. fluorescens* and *ps. putida, ps. chlororaphis*), *Bacillus (B. cereus), Azospirillum (A. brasilense, A. lipoferum) and Herbaspirillum (H. seropedicae).* Common PGPR species found in rhizosphere of maize are *Enterobacter sp., Rahnella aquatilis, Paenibacillus azotofixans, Herbaspirillum seropedicae, Bacillus circulans, Klebsiella sp.* and *Azospirillum sp.* (Chelius and Triplett 2000). The positive effects of *Azospirillum* on maize growth are mainly derived from physiological changes of the inoculated plant roots, which enhance water and mineral nutrient uptake (Okan and Kapulnik 1986). Both *Azospirillum brasilense* and *Azospirillum irakense* are used as inoculant biofertilizer for maize. Other species of *Azospirillum* capable of increasing the yield of maize are *A. lipoferum,* and *A. indigens. Azorhizobium caulinodans* is also capable of giving such beneficial effects (Riggs et al. 2001). *Herbaspirillium seropedicae* can improve the ability of maize plant to use fertilizer N more efficiently and the yield increase due to *H. seropedicae* was up to 19.5% (Riggs et al. 2001). PGPR strains *Burkholderia (B. cepacia), Pseudomonas (P. fluorescens), Serratia (S. proteamaculans,* and *S. liquefaciens), Rhizobium (R. etli* bv. *Phaseoli, R. leguminosarum* bv. *Trifolii) and Sinorhizobium* sp. increase corn growth, plant height and grain yield of maize in different agro-ecological zones. The diazotrophs commonly present in sugarcane plants are *Acetobactor* also known as *Gluconacetobacter (A. diazotrophicus), Azospirillum (A. brasilense, A. lipoferum, A. amazonense), Burkholderia (B. brasilensis, B. tropicalis),* and *Herbaspirillum (H. seropedicae, H. rubrisubalbicans)* (Kennedy and Islam 2001). Sugarcane requires approximately 1.45 kg N ha^{-1} to produce 1 t moist biomass (Bhuiyan 1995) or about 7 kg N ha^{-1} for 1 t of dry cane. Generally 150–250 kg urea-N ha^{-1} is applied for sugarcane cultivation depending on soil fertility, genotype and the targeted yield. Evidence from Brazil indicates fertilizer-N of sugarcane can be reduced to half by exploiting BNF systems, claimed to be based on diazotrophic PGPR such as A*cetobacter* (*Gluconacetobacter*) and *Herbaspirillum* (Döbereiner 1997; Döbereiner and Baldani 1998). More than 70% of sugarcane N (200 kg N ha^{-1} y^{-1}) was derived from biological fixed N2 by *Azospirillum* (*A. diazotrophicus*) (Boddey et al. 1991). Similarly, *Acetobacter* (with *nif*H^{+}) -sugarcane system has also been well established (Lee et al. 2002). *Azotobacter sp.* and *Klebsiella mobilis* are reported for improving potato yield. Similarly *Pseudomonas fluorescent* and *Achromobacter piechaudii* increase tomato yield. Positive effect of PGPR (*Bacillus cereus; Pseudomonas putida; Pseudomonas fluorescens; Serratia liquefaciens; Arthrobacteer cetreus; Escherichia coli, Mesorhizobium loti* and *Delftia acidovorans*) inoculation on the growth and yield of rapeseed have been reported by many researchers. *Burkholderia cepacia* alone or in combination with *Enterobacter* spp. and *Pseudomonas fluoroscens* have also been tested for its ability to promote growth of sorghum (*Sorghum bicolar*) (Chiarini et al. 1998). Inoculation with effective bacterial strains (*Pseudomonas alcaligenes, Pseudomonas denitrificans, Bacillus polymyxa, Azospirillum brasilense* and *Mycobacterium phlei)* increases the root and shoot growth of cotton (Anjum et al. 2007). Çakmakçi et al. (2007) also conducted a study on barley under greenhouse conditions in order to investigate seed inoculation with five different N_2-fixing (*Bacillus licheniformis*, *Rhodobacter capsulatus*, *Paenibacillus polymyxa*, *Pseudomonas putida*, and *Bacillus* spp. OSU-142) and two different phosphate solubilising (*Bacillus megaterium* and Bacillus spp. M-13) bacteria in comparison to control and mineral fertiliser (N and P) application. Co-inoculation of legumes with PGPR and *Rhizobium* has received increasing attention in recent years. Co-inoculation with symbiotic bacteria and rhizosphere bacteria may increase nodulation through a variety of mechanisms. PGPR strains, from a range of genera, enhance legume growth, nodulation and nitrogen fixation when co-inoculated with their effective rhizobia. Examples of these are *Azospirillum* (*A. lipoferum, A. brasilense*), *Azotobacter* (*A. chroococcum*), *Bacillus* (*B. cereus, B. endophyticus, B. pumilus, B. subtilis, B. firmis, B. megaterium*), *Paenibacillus* (*P. lautus, P. macerans, P. polymyxa,*), *Pseudomonos* (*Ps. fluorescens, Ps. putida, Ps. aeruginosa*) Serratia (*S. lequefacians, S. proteamaculans*), *Aeromonas hydrophia* and, *Streptomyces* (Pal et al. 2004; Kishore et al. 2005a; Figueiredo et al. 2007) . All these bacteria including cyanobacteria can supplement urea-N by BNF, but only if conditions for expression of N_2-fixing activity and subsequent transfer of N to plants are favourable. In general, it is believed that PGPR are more effective in promoting plant growth under limited supply of nutrients; however, present scenario does not allow to compromise on actual potential of crop productivity by reducing use of chemical fertilizers. Hence, it is of prime importance to isolate such PGPR strains that could be effective even under optimum nutrition. The biosynthesis of ethylene in plant roots is significantly affected by concentration of NO_3-N present around the roots (Ligero et al. 1999). Higher levels of NO_3^- in rooting medium stimulate ACC-oxidase activity leading to increased ethylene production, which is generally believed to be inhibitory to root growth (Glick et al. 1998). The nitrogenous fertilizer applied in ammonical form is readily oxidized to NO_3-N under aerobic conditions. It is very likely that NO_3-N present in the vicinity of roots reduces the efficiency of PGPR by inducing ethylene synthesis. However, PGPR containing ACC-deaminase minimizes the NO_3^- induced ethylene synthesis.

7 Conclusion

PGPR are able to enhance significantly the yield of cereals, legumes and sugarcane crops etc. Soil and plant health can be enhanced through PGPR-crops rhizospheric interactions. From the agricultural viewpoint, the aim will be to increase the yield on sustainable basis, while maintaining soil health. To achieve this, knowledge about indigenous PGPR for different crop rhizospheric environment must be improved. It is also necessary to adopt and maintain minimal standards for production of PGPR inoculants/biofertilizers. PGPR biofertilizers not only promote crop growth but also enhance the resistance against pathogens. Integrated use of PGPR along with inorganic and organic nutrient sources provides a sustainable agricultural ecosystem.

References

Afzal A, Bano A (2008) *Rhizobium* and phosphate solubilizing bacteria improve the yield and phosphorus uptake in wheat (*Triticum aestivum* L.). Int J Agric Biol 10:85–88

Alvarez MI, Sueldo RJ, Barassi CA (1996) Effect of *Azospirillum* on coleoptile growth in wheat seedlings under water stress. Cereal Res Commun 24:101–107

Angus JF (2001) Nitrogen supply and demand in Australian agriculture. Aust J Exp Agric 41:277–288

Anjum MA, Sajjad MR, Akhtar N, Qureshi MA, Iqbal A, Jami AR, Hassan M (2007) Response of cotton to plant growth promoting rhizobacteria (PGPR) inoculation under different levels of nitrogen. J Agric Res 45(2):135–143

Arshad M, Frankenberger WT Jr (1998) Plant growth regulating substances in the rhizosphere. Microbial production and function. Adv Agron 62:46–51

Baldani VLD, Baldani JI, Dobereiner J (2000) Inoculation of rice plants with the endophytic diazotrophs *Herbaspirillums seropidicae*. Biol Fertil Soils 30:485–491

Banerjee MR, Yasmin L (2002) Sulfur oxidizing rhizobacteria: an innovative environment friendly soil biotechnological tool for better canola production. Proceeding of Agroenviron, 26–29 Oct 2002, Cairo, Egypt, pp 1–7

Banerjee MR, Yesmin L, Vessey JK (2006) Plant growth promoting rhizobacteria as biofertilizers and biopesticides. In: Rai MK (ed) Handbook of microbial biofertilizers. Haworth Press, Inc., New York

Bent E, Breuil C, Enebak S, Chanway CP (2002) Surface colonization of lodgepole pine (*Pinus contorta* var lati folia [Dougl. Engelm.]) roots by *Pseudomonas fluorescens* and *Paenibacillus polymyxa* under gnotobiotic conditions. Plant Soil 241:187–196

Beraha L, Wisniewski V, Garber ED (1983) Avirulence and reduced extracellular enzyme activity in *Geotrichum candidum*. Bot Gazette 144:461–465

Bertagnolli BL, Soglio FKD, Sinclair JB (1996) Extracellular enzyme profiles of the fungal pathogen *Rhizoctonia solani* isolate 2B-12 and of two antagonists, *Bacillus megaterium* strain B153-2-2 and *Trichoderma harzianum* isolate Th008. I. Possible correlations with inhibition of growth and biocontrol. Physiol Mol Plant Pathol 48:145–160

Bhuiyan NI (1995) Intensive cropping and soil nutrient balance in Bangladesh. In: Hussain MS, Huq SMI, Iqbal M, Khan TH (eds) Improving soil management for intensive cropping in the tropics and sub-tropics. Bangladesh Agricultural Research Council, Dhaka, pp 61–71

Biswas JC, Ladha JK, Dazzo FB (2000) *Rhizobia* inoculation improves nutrient uptake and growth of lowland rice. Soil Sci Soc Am J 64:1644–1650

Boddey RM, Urquiaga S, Reis V, Döbereiner J (1991) Biological nitrogen fixation associated with sugar cane. Plant Soil 137:111–117

Brittenham GM (1994) New advances in iron metabolism, iron deficiency and iron overload. Curr Opin Hematol 1:549–556

Brock JL, Albrecht KA, Tilbrook JC, Hay MJM (2000) Morphology of white clover during development from seed to clonal populations in grazed pastures. J Agric Sci 135:103–111

Bünemann EK, Bossio DA, Smithson PC, Frossard E, Oberson A (2004) Microbial community composition and substrate use in a highly weathered soil as affected by crop rotation and P fertilization. Soil Biol Biochem 36:889–901

Çakmakçi R, Kantar F, Şahin F (2001) Effect of N_2-fixing bacterial inoculations on yield of sugar beet and barley. J Plant Nutr Soil Sci 164:527–531

Çakmakçi R, Dönmez F, Aydın A, Şahin F (2006) Growth promotion of plants by plant growth-promoting rhizobacteria under greenhouse and two different field soil conditions. Soil Biol Biochem 38:1482–1487

Çakmakçi R, Erat M, Erdoğan ÜG, Dönmez MF (2007) The influence of PGPR on growth parameters, antioxidant and pentose phosphate oxidative cycle enzymes in wheat and spinach plants. J Plant Nutr Soil Sci 170:288–295

Catellan AJ, Hartel PG, Fuhrmann JJ (1999) Screening for plant growth rhizobacteria to promote early soybean growth. Soil Sci Soc Am J 63:1670–1680

Chabot R, Antoun H, Cescas MP (1996) Growth promotion of maize and lettuce by phosphate-solubilizing *Rhizobium leguminosarum* biovar phaseoli. Plant Soil 184:311–321

Chelius MK, Triplett EW (2000) Diazotrophic endophytes associated with maize. In: Triplett EW (ed) Prokaryotic nitrogen fixation: a model system for the analysis of a biological process. Horizon Scientific Press, Wymondham, pp 779–791

Chen YP, Rekha PD, Arun AB, Shen FT, Lai WA, Young CC (2006) Phosphate solubilizing bacteria from subtropical soil and their tricalcium phosphate solubilizing abilities. Appl Soil Ecol 34:33–41

Chiarini L, Bevivino A, Tabacchioni S, Dalmastri C (1998) Inoculation of *Burkholderia cepacia*, *Pseudomonas fluorescens* and *Enterobacter* sp. on *Sorghum bicolor*: root colonization and plant growth promotion of dual strain inocula. Soil Biol Biochem 30:81–87

Coelho MRR, Weid I, von der Zahner V, Seldin L (2003) Characterization of nitrogen-fixing *Paenibacillus* species by polymerase chain reaction-restriction fragment length polymorphism analysis of part of genes encoding 16S rRNA and 23S rRNA and by multilocus enzyme electrophoresis. FEMS Microbiol Lett 222:243–250

Costa JM, Loper JE (1994) Characterization of siderophore production by the biological control agent *Enterobacter cloacae*. Mol Plant-Microb Interact 7:440–448

Crozier A, Kamiya Y, Bishop G, Yokota T (2001) Biosynthesis of hormones and elicitors molecules. In: Buchanan BB, Grussem W, Jones RL (eds) Biochemistry and molecular biology of plants. American Society of Plant Biologists, Rockville, pp 850–900

de Freitas JR (2000) Yield and N assimilation of winter wheat (*Triticum aestivum.*, var Norstar) inoculated with rhizobacteria. Pedobiologia 44:97–104

de Freitas JR, Banerjee MR, Germida JJ (1997) Phosphate-solubilizing rhizobacteria enhance the growth and yield but not phosphorus uptake of canola (*Brassica napus* L.). Biol Fertil Soils 24:358–364

Deubel A, Gransee A, Merbach W (2000) Transformation of organic rhizodeposits by rhizoplane bacteria and its influence on the availability of tertiary calcium phosphate. J Plant Nutr Soil Sci 163:387–392

Dey R, Pal KK, Bhatt DM, Chauhan SM (2004) Growth promotion and yield enhancement of peanut *(Arachis hypogaea* L) by application of plant growth promoting rhizobacteria. Microbiol Res 159:371–394

Dobbelaere S, Vanderleyden J, Okon Y (2003) Plant growth-promoting effects of diazotrophs in the rhizosphere. Crit Rev Plant Sci 22:107–149

Döbereiner J (1997) Biological nitrogen fixation in the tropics: social and economic contributions. Soil Biol Biochem 29:771–774

Döbereiner J, Baldani VLD (1998) Biological nitrogen fixation by endophytic diazotrophs in non-leguminous crops in the tropics. In: Malik KA, Mirza MS, Ladha JK (eds) Nitrogen fixation with non-legumes. Kluwer Academic Publishers, Dordrecht, pp 3–7

Drepper T, Raabe K, Giaourakis D, Gendrullis M, Masepohl B, Klipp W (2002) The *Hfq*-like protein Nr*f*a of the phototrophic purple bacterium *Rhodobacter capsulatus* controls nitrogen fixation via regulation of ni*f*A and an*f*A expression. FEMS Microbiol Lett 215:221–227

Dwivedi D, Johri BN (2003) Antifungals from fluorescent pseudomonads: biosynthesis and regulation. Curr Sci 12:1693–1703

Ehteshamul-Haque S, Ghaffar A (1993) Use of *Rhizobia* in the control of root diseases of sunflower, okra, soybean and mungbean. J Phytopathol 138:157–163

Elbadry M, El-Bassel A, Elbanna K (1999) Occurrence and dynamics of phototrophic purple non-sulphur bacteria compared with other asymbiotic nitrogen fixers in rice fields of Egypt. World J Microbiol Biotechnol 15:359–362

Fankem H, Nwaga D, Deubel A, Dieng L, Merbach W, Etoa FX (2006) Occurrence and functioning of phosphate solubilizing microorganisms from oil palm tree (*Elaeis guineensis*) rhizosphere in Cameroon. Afr J Biotechnol 5:2450–2460

Figueiredo MVB, Martinez CR, Burity HA, Chanway CP (2007) Plant growth-promoting rhizobacteria for improving nodulation and nitrogen fixation in the common bean (*Phaseolus vulgaris* L.). World J Microbiol Biotechnol 24(7):1187–1193

Frankenberger WTJ, Arshad M (1995) Photohormones in soil: microbial production and function. Deker, New York, p 503

Frankowski J, Lorito M, Scala F, Schmid R, Berg G, Bahl H (2001) Purification and properties of two chitinolytic enzymes of *Serratia plymuthica* HRO-C48. Arch Microbiol 176: 421–426

Fridlender M, Inbar J, Chet I (1993) Biological control of soilborne plant pathogens by a β-1,3-glucanase-producing Pseudomonas cepacia. Soil Biol Biochem 25:1211–1221

Ghaderi A, Aliasgharzad N, Oustan S, Olsson PA (2008) Efficiency of three *Pseudomonas* isolates in releasing phosphate from an artificial variable-charge mineral (iron III hydroxide). Soil Environ 27:71–76

Glick BR (1995) The enhancement of plant growth by free-living bacteria. Can J Microbiol 41:109–117

Glick BR (2001) Phytoremediation: synergistic use of plants and bacteria to cleanup the environment. Biotechnol Adv 21(3):83–393

Glick BR, Pasternak JJ (2003) Plant growth promoting bacteria. In: Glick BR, Pasternak JJ (eds) Molecular biotechnology – principles and applications of recombinant DNA, 3rd edn. ASM Press, Washington, DC, pp 436–454

Glick BR, Penrose DM, Li J (1998) A model for lowering plant ethylene concentration by plant growth promoting rhizobacteria. J Theor Biol 190:63–68

Glick BR, Patten CL, Holguin G, Penrose DM (1999) Biochemical and genetic mechanisms used by plant growth promoting bacteria. Imperial College Press, London, 267pp

Goldstein AH (1994) Involvement of the quinoprotein glucose dehydrogenises in the solubilization of exogenous phosphates by gram-negative bacteria. In: Torriani A, Yagil GE, Silver S (eds) Phosphate in microorganisms: cellular and molecular biology. ASM Press, Washington, DC, pp 197–203

Goldstein AH (1995) Recent progress in understanding the molecular genetics and biochemistry of calcium phosphate solubilization by Gram-negative bacteria. Biol Agri Hortic 12:185–193

Guilfoyle TJ, Ulmasov T, Hagen G (1998) The ARF family of transcription factors and their role in plant hormone responsive transcription. Cell Mol Life Sci 54:619–627

Gull M, Hafeez FY, Saleem M, Malik KA (2004) Phosphorus uptake and growth promotion of chickpea by co-inoculation of mineral phosphate solubilizing bacteria and a mixed rhizobial culture. Aust J Exp Agric 44:623–628

Gyaneshwar P, Kumar GN, Parekh LJ, Poole PS (2002) Role of soil microorganisms in improving P nutrition of plants. Plant Soil 245:83–93

Halder AK, Chakrabarty PK (1993) Solubilization of inorganic phosphate by *Rhizobium.* Folia Microbiol 38:325–330

Hayat R, Ali S, Amara U, Khalid R, Ahmed I (2011) Soil beneficial bacteria and their role in plant growth promotion. A Review. Ann Microbiol 60:579–598

Hayat R, Ali S, Siddique MT, Chatha TH (2008a) Biological nitrogen fixation of summer legumes and their residual effects on subsequent rainfed wheat yield. Pak J Bot 40(2):711–722

Hayat R, Ali S, Ijaz SS, Chatha TH, Siddique MT (2008b) Estimation of N_2-fixation of mung bean and mash bean through xylem uriede technique under rainfed conditions. Pak J Bot 40(2):723–734

He ZL, Zhu J (1988) Microbial utilization and transformation of phosphate adsorbed by variable charged minerals. Soil Biol Biochem 30:917–923

Hegazi NA, Faye M, Amin G, Hamza MA, Abbas M, Youssef H, Monib M (1998) Diazotrophs associated with non-legumes grown in sandy soil. In: Malik KA, Mirza MS, Ladha JK (eds) Nitrogen fixation with non-legumes. Kluwer Academic Publishers, Dordrecht, pp 209–222

Henri F, Laurette NN, Annette D, John Q, Wolfgang M, François-Xavier E, Dieudonné N (2008) Solubilization of inorganic phosphates and plant growth promotion by strains of *Pseudomonas fluorescens* isolated from acidic soils of Cameroon. Afr J Microbiol Res 2:171–178

Hersman LE, Folsythe JH, Ticknor LO, Maurice PA (2001) Growth of *Pseudomonas mendocina* on Fe (III) (hydr) oxides. Appl Environ Microbiol 67:4448–4453

Hoflich G (2000) Colonization and growth promotion of non-legumes by *Rhizobium* bacteria. Micobial biosystems: new prontiers. In: Bell CR, Brylinsky M, Johnson-Green P (eds) Proceedings of the 8th international symposium on microbial ecology, Atlantic Canada Society for Microbial Ecology, Halifax, Canada, pp 827–830

Hurek T, Reinhold-Hurek B, Van Montagu M, Kellenberger E (1994) Root colonization and systemic spreading of *Azoarcus* sp. strain BH72 in grasses. J Bacteriol 176:1913–1923

Imsande J (1998) Nitrogen deficit during soybean pod fill and increased plant biomass by vigorous N_2 fixation. Eur J Agron 8(1–2):1–11

Jadhav RS, Thaker NV, Desai A (1994) Involvement of the siderophore of cowpea *Rhizobium* in the iron nutrition of the peanut. World J Microbiol Biotechnol 10:360–361

James EK, Gyaneshwar P, Barraquio WL, Mathan N, Ladha JK (2000) Endophytic diazotrophs associated with rice. In: Ladha JK, Reddy PM (eds) The quest for nitrogen fixation in rice. International Rice Research Institute, Los Ban˜os, pp 119–140

Kanungo PK, Panda D, Adhya TK, Ramakrishnan B, Rao VR (1997) Nitrogenase activity and nitrogen fixing bacteria associated with rhizosphere of rice cultivars. J Sci Food Agric 73:485–488

Katiyar V, Goel R (2004) Siderophore-mediated plant growth promotion at low temperature by mutant of fluorescent pseudomonad. Plant Growth Regul 42:239–244

Kennedy IR, Islam N (2001) The current and potential contribution of asymbiotic nitrogen requirements on farms: a review. Aust J Exp Agric 41:447–457

Kennedy IR, Tchan Y (1992) Biological nitrogen fixation in no leguminous field crops: recent advances. Plant Soil 141:93–118

Kennedy IR, Choudhury AIMA, KecSkes ML (2004a) Non-symbiotic bacterial diazotrophs in crop-farming systems: can their potential for plant growth promotion be better exploited? Soil Boil Biochem 36(8):1229–1244

Kennedy N, Brodie E, Conolly J, Clipson N (2004b) Impact of lime, nitrogen and plant species on bacterial community structure in grassland microcosms. Environ Microbiol 6:1070–1080

Khan KS, Joergensen RG (2009) Changes in microbial biomass and P fractions in biogenic household waste compost amended with inorganic P fertilizers. Bioresour Technol 100:303–309

Khan AA, Jillani G, Akhtar MS, Naqvi SMS, Rasheed M (2009) Phosphorus solubilizing bacteria: occurrence, mechanisms and their role in crop production. J Agric Biol Sci 1(1):48–58

Kim KY, Jordan D, McDonald GA (1998a) Effect of phosphate-solubilizing bacteria and vesicular-arbuscular mycorrhizae on tomato growth and soil microbial activity. Biol Fert Soils 26: 79–87

Kim KY, Jordan D, McDonald GA (1998b) *Enterobacter agglomerans,* phosphate solubilizing bacteria, and microbial activity in soil: effect of carbon sources. Soil Biol Biochem 30:995–1003

Kishore GK, Pande S, Podile AR (2005a) Phylloplane bacteria increase seedling emergence, growth and yield of field-grown groundnut (*Arachis hypogaea* L.). Lett Appl Microbiol 40:260–268

Kishore GK, Pande S, Podile AR (2005b) Biological control of late leaf spot of peanut (*Arachis hypogaea* L.) with chitinolytic bacteria. Phytopathology 10:1157–1165

Kishore GK, Pande S, Rao JN, Podile AR (2005c) *Pseudomonas aeruginosa* inhibits the plant cell wall degrading enzymes of *Sclerotium rolfsii* and reduces the severity of groundnut stem rot. Eur J Plant Pathol 113(3):315–320

Kloepper JW, Leong J, Teintze M, Schroth MN (1980) Enhanced plant growth by siderophores produced by plant growth-promoting rhizobacteria. Nature 286:885–886

Kloepper JW, Ryu CM, Zhang S (2004) Induced systemic resistance and promotion of plant growth by *Bacillus* spp. Phytopathology 94(11):1259–1266

Kokalis-Burelle N, Kloepper JW, Reddy MS (2006) Plant growth-promoting rhizobacteria as transplant amendments and their effects on indigenous rhizosphere microorganisms. Appl Soil Ecol 31(1–2):91–100

Kumar V, Narula N (1999) Solubilization of inorganic phosphates and growth emergence of wheat as affected by *Azotobacter chroococcum* mutants. Biol Fert Soils 28:301–305

Kumar V, Behl RK, Narula N (2001) Establishment of phosphate solubilizing strains of *Azotobacter chroococcum* in the rhizosphere and their effect on wheat cultivars under greenhouse conditions. Microbiol Res 156:87–93

Kuzyakov Y, Domanski G (2000) Carbon input by plants into the soil. J Plant Nutr Soil Sci 163:421–431

Lee S, Pierson B, Kennedy C (2002) Genetics and biochemistry of nitrogen fixation and other factors beneficial to host plant growth in diazotrophic endophytes. In: Vanderleyden J (ed) Proceedings of the ninth international symposium on nitrogen fixation with nonlegumes, Katholique Universiteit, Leuven, Belgium, pp 41–42

Li J, Ovakin DH, Charles TC, Glick BR (2000) An ACC deaminase minus mutant of *Entreobacter cloacae* UW4 no longer promotes root elongation. Curr Microbiol 41:101–105

Ligero F, Poreda JL, Gresshoff PM, Caba JM (1999) Nitrate inoculation is in enhanced ethylene biosynthesis in soybean roots as a possible mediator of nodulation control. J Plant Physiol 154:482–488

Lopez-Millan AF, Morales F, Abadia A, Abadia J (2001) Iron deficiency-associated changes in the composition of the leaf apoplastic fluid from field-grown pear (*Pyrus communis*) trees. J Exp Bot 52:1489–1498

Malik KA, Mirza MS, Hassan U, Mehnaz S, Rasul G, Haurat J, Bauy R, Normanel P (2002) The role of plant associated beneficial bacteria in rice-wheat cropping system. In: Kennedy IR, Chaudhry ATMA (eds) Biofertilisers in action. Rural Industries Research and Development Corporation, Canberra, pp 73–83

Manjula K, Kishore GK, Podile AR (2004) Whole cells of *Bacillus subtilis* AF 1 proved effective than cell free and chitinase-based formulations in biological control of citrus fruit rot and groundnut rust. Can J Microbiol 50:737–744

Masalha J, Kosegarten H, Elmaci O, Mengel K (2000) The central role of microbial activity for iron acquisition in maize and sunflower. Biol Fert Soils 30:433–439

Matiru VN, Dakora FD (2004) Potential use of rhizobial bacteria as promoters of plant growth for increased yield in landraces of African cereal crops. Afr J Biotechnol 3(1):1–7

Meyer SLF, Massoud SI, Chitwood DJ, Roberts DP (2000) Evaluation of *Trichoderma virens* and *Burkholderia cepacia* for antagonistic activity against root-knot nematode, *Meloidogyne incognita*. Nematology 2:871–879

Misko AL, Germida JJ (2002) Taxonomic and functional diversity of pseudomonads isolated from the roots of field-grown canola. FEMS Microbiol Ecol 42:399–407

Nautiyal CS, Bhadauria S, Kumar P, Lal H, Mondal R, Verma D (2000) Stress induced phosphate solubilization in bacteria isolated from alkaline soils. FEMS Microbiol Lett 182:291–296

Neilands JB, Nakamura K (1991) Detection, determination, isolation, characterization and regulation of microbial iron chelates. In: Winkelmann G (ed) CRC handbook of microbial iron chelates. CRC Press, London

Nguyen TH, Deaker R, Kennedy IR, Roughly RJ (2003) The positive yield response of field-grown rice to introduction with a multistrain biofertiliser in the Hanoi area, Vietnam. Symbiosis 35:231–245

Nielson P, Sorenson J (1997) Multi-target and medium independent fungal antagonism by hydrolytic enzymes in *Paenibacillus polymyxa* and *Bacillus pumilus* strains from barley rhizosphere. FEMS Microbiol Ecol 22:183–192

Nieto KF, Frankenberger WT Jr (1991) Influence of adenine, isopentyl alcohol and *Azotobacter chroococcum* on the vegetative growth of *Zea mays*. Plant Soil 135:213–221

O'Sullivan DJ, O'Gara F (1992) Traits of fluorescent *Pseudomonas* spp. involved in suppression of plant root pathogens. Microbiol Rev 56:662–676

Okan Y, Kapulnik Y (1986) Development and function of *Azospirillum* inoculated roots. Plant Soil 90:3–16

Ordentlich A, Elad Y, Chet I (1988) The role of chitinase of *Serratia marcescens* in biocontrol of *Sclerotium rolfsii*. Phytopathology 78:84–88

Pal SS (1998) Interactions of an acid tolerant strain of phosphate solubilizing bacteria with a few acid tolerant crops. Plant Soil 198:169–177

Pal M, Karthikeyapandian V, Jain V, Srivastava AC, Raj A, Sengupta UK (2004) Biomass production and nutritional levels of berseem (*Trifolium alexandrium*) grown under elevated CO_2. Agric Ecosyst Environ 101:31–38

Patten CL, Glick BR (2002) Role of Pseudomonas putida indole-acetic acid in development of the host plant root system. Appl Environ Microbiol 68:3795–3801

Paul EA, Clark FE (1996) Soil microbiology and biochemistry, 2nd edn. Academic, London, 340pp

Paynel F, Murray PJ, Cliquet B (2001) Root exudates: a pathway for short-term N transfer from clover and ryegrass. Plant Soil 229:235–243

Penrose DM, Glick BR (2001) Levels of 1-aminocyclopropane-1-carboxylic acid (ACC) in exudates and extracts of canola seeds treated with plant growth promoting bacteria. Can J Microbiol 47:368–372

Peoples MB, Ladha JK, Herridge DF (1995) Enhancing legume N_2 fixation through plant and soil management. Plant Soil 174:83–101

Pieterse CMJ, Van Pelt JA, Van Wees SCM, Ton J, Leon-Kloosterziel KM, Keurentjes JJB, Verhagen BWM, Knoester M, Sluis IV, Bakker PAHM, Van Loon LC (2001) Rhizobacteria-mediated induced systemic resistance: triggering, signaling and expression. Eur J Plant Pathol 107:51–61

Plessner O, Klapach T, Guerinot ML (1993) Siderophore utilization by *Bradyrhizobium japonicum*. Appl Environ Microbiol 59:1688–1690

Podile AR, Kishore GK (2006) Plant growth promoting rhizobacteria. In: Gnanamanickam SS (ed) Plant associated bacteria. Springer, Amsterdam, pp 195–230

Ponmurugan P, Gopi C (2006) Distribution pattern and screening of phosphate solubilizing bacteria isolated from different food and forage crops. J Agron 5:600–604

Primrose SB (1979) Ethylene and agriculture: the role of the microbe. J Appl Bacteriol 46:1–25

Rao AV (2004) Microbial biotechnology for sustainable agricultural production in arid soils. In: Ray RC (ed) Soil microbial biotechnology for sustainable agricultural production. Oxford and IBH Publishing Co., New Delhi

Rashid A (1996) Secondary and micronutrients. In: Rashid A, Memon KS (eds) Soil science. National Book Foundation, Islamabad, pp 341–385

Reeves TG, Waddington SR, Ortiz-Monasterio I, Banziger M, Cassadey K (2002) Removing nutritional limits to maize and wheat production: a developing country perspective. In: Kennedy IR, Choudhury ATMA (eds) Biofertilizers in action. Rural Industries Research and Development Corporation, Canberra, pp 11–36

Reinhold-Hurek B, Hurek T, Gillis M, Hoste B, Vancanneyt M, Kersters K, DeLey J (1993) *Azoarcus* gen. Nov., nitrogen fixing proteobacteria associated with roots of Kallar grass *(Leptochloa fusca (L.)* Kunth) and description of two species, *Azoarcus indigens* sp. Nov. and *Azoarcus communis* sp. Nov. Int J Syst Bacteriol 43:574–584

Reinhold-Hurek B, Egener T, Hurek T, Martin D, Sarkar A, Zhang L, Miche L (2002) Regulation of nitrogen fixation and assimilation of *Azoarcus sp.* BH72 new approaches to study biodiversity of grass endophytes. In: Vanderleyden J (ed) Proceedings of the ninth international symposium on nitrogen fixation with non-legumes. Katholique Universideit Leuven, Leuven, p 48

Reis VM, Baldani JI, Baldani VLD, Döberener J (2000) Biological dinitrogen fixation in the graminae and palm trees. Crit Rev Plant Sci 19:227–247

Revillas JJ, Rodelas B, Pozo C, Martinez-Toledo MV, Gonzalez LJ (2000) Production of B-group vitamins by two *Azotobacter* strains with phenolic compounds as sole carbon source under diazotrophic and adiazotrophic conditions. J Appl Microbiol 89:486–493

Reyes E, Garcia-Castro I, Esquivelm F, Hornedo J, Cortes-Funes H, Solovera J, Alvarez-Mon M (1999) Granulocyte colony-stimulating factor (G-CSF) transiently suppresses mitogen-stimulated T-cell proliferative response. Br J Cancer 80(1/2):229–235

Richardson AE (2001) Prospects for using soil microorganisms to improve the acquisition of phosphorus by plants. Aust J Plant Physiol 28:897–906

Riggs PG, Chelius MK, Iniguez AL, Kaeppler SM, Triplet EW (2001) Enhanced maize productivity by inoculation with diazotrophic bacteria. Aust J Plant Physiol 28:829–836

Roos W (1984) Relationship between proton extrusion and fluxes of ammonium ions and organic acids in *Penicillium cyrlopium*. J Gen Microbiol 130:1007–1014

Rosado AS, Seldin L (1993) Production of potentially novel anti-microbial substance by *Bacillus polymyxa*. World J Microbiol Biotechnol 9:521–528

Ryu CM, Farag MA, Hu CH, Reddy MS, Kloepper JW, Paré PW (2004) Bacterial volatiles induce systemic resistance in *Arabidopsis*. Plant Physiol 134:1017–1026

Şahin F, Çakmakçi R, Kantar F (2004) Sugar beet and barley yields in relation to inoculation with N_2-fixing and phosphate solubilizing bacteria. Plant Soil 265:123–129

Saravanakumar D, Lavanya N, Muthumeena B, Raguchander T, Suresh S, Samiyappan R (2008) *Pseudomonas fluorescens* enhances resistance and natural enemy population in rice plants against leaf folder pest. J Appl Entomol 132:469–479

Schalk IJ, Hennard C, Dugave C, Poole K, Abdallah MA, Pattus F (2001) Iron-free pyoverdin binds to its outer membrane receptor FpvA in *Pseudomonas aeruginosa*: a new mechanism for membrane iron transport. Mol Microbiol 39:351–360

Scharf PC (2001) Soil and plant tests to predict optimum nitrogen rates for corn. J Plant Nutr 24:805–826

Schippers B, Scheffer RJ, Lugtenberg JJ, Weisbek PJ (1995) Biocoating of seed with plant growth promoting rhizobacteria to improve plant establishment. Outlook Agric 24:179–185

Shaharoona B, Arshad M, Zahir ZA, Khalid A (2006) Performance of *Pseudomonas* spp. containing ACC-deaminase for improving growth and yield of maize (*Zea mays* L.) in the presence of nitrogenous fertilizer. Soil Biol Biochem 38:2971–2975

Sharma A, Johri BN (2003) Growth promoting influence of siderophore-producing *Pseudomonas* strains GRP3A and PRS9 in maize (*Zea mays* L.) under iron limiting conditions. Microbiol Res 158:243–248

Sharma K, Dak G, Agrawal A, Bhatnagar M, Sharma R (2007) Effect of phosphate solubilizing bacteria on the germination of *Cicer arietinum* seeds and seedling growth. J Herb Med Toxicol 1:61–63

Shiferaw B, Bantilan MCS, Serraj R (2004) Harnessing the potential of BNF for poor farmers: technological policy and institutional constraints and research need. In: Serraj R (ed) Symbiotic nitrogen fixation: prospects for enhanced application in tropical agriculture. Oxford and IBH publishing Co. Pvt. Ltd., New Delhi, p 3

Shishido M, Breuil C, Christopher PC (1999) Endophytic colonization of spruce by plant growth promoting rhizobacteria. FEMS Microbiol Ecol 29:191–196

Singh S, Kapoor KK (1998) Inoculation with phosphate-solubilizing microorganisms and a vesicular-arbuscular mycorrhizal fungus improves dry matter yield and nutrient uptake by wheat grown in a sandy soil. Biol Fert Soils 28:139–144

Stajner D, Gasaić O, Matković B, Varga SZI (1995) Metolachlor effect on antioxidants enzyme activities and pigments content in seeds and young leaves of wheat (*Triticum aestivum* L.). Agr Med 125:267–273

Stajner D, Kevreaan S, Gasaić O, Mimica-Dudić N, Zongli H (1997) Nitrogen and Azotobacter chroococcum enhance oxidative stress tolerance in sugar beet. Biol Plant 39:441–445

Staunton S, Leprince F (1996) Effect of pH and some organic anions on the solubility of soil phosphate: implications for P bioavailability. Eur J Soil Sci 47:231–239

Stephen J, Jisha MS (2009) Buffering reduces phosphate solubilizing ability of selected strains of bacteria. World J Agric Sci 5:135–137

Sudha SN, Jayakumar R, Sekar V (1999) Introduction and expression of the cry1Ac gene of *Bacillus thuringiensis* in a cereal-associated bacterium, *Bacillus polymyxa*. Curr Microbiol 38:163–167

Sundara B, Natarajan V, Hari K (2002) Influence of phosphorus solubilizing bacteria on the changes in soil available phosphorus and sugarcane yields. Field Crops Res 77:43–49

Tanaka H, Watanabe T (1995) Glucanases and chitinases of *Bacillus circulans* WL-12. J Ind Microbiol 114:478–483

Tang C, Rengel Z (2003) Role of plant cation/anion uptake ratio in soil acidification. In: Rengel Z (ed) Handbook of soil acidity. Marcel and Dekker, New York, pp 57–81

Tao G, Tian S, Cai M, Xie G (2008) Phosphate solubilizing and mineralizing abilities of bacteria isolated from soils. Pedosphere 18:515–523

Tilak KVBR, Ranganayaki N, Pal KK, De R, Saxena AK, Nautiyal CS, Mittal S, Tripathi AK, Johri BN (2005) Diversity of plant growth and soil health supporting bacteria. Curr Sci 89:136–150

Timmusk S, Nicander B, Granhall U, Tillberg E (1999) Cytokinin production by *Paenibacillus polymyxa*. Soil Biol Biochem 31:1847–1852

Tran Vân V, Berge O, Ke SN, Balandreau J, Heulin T (2000) Repeated beneficial effects of rice inoculation with a strain of *Burkholderia vietnamiensis* on early and late yield components in low fertility sulphate acid soils of Vietnam. Plant Soil 218:273–284

Urashima Y, Hori K (2003) Selection of PGPR which promotes the growth of spinach. Jpn J Soil Sci Plant Nutr 74:157–162

Van Veen JA, Van Overbeek LS, Van Elsas JD (1997) Fate and activity of microorganisms introduced into soil. Microbiol Mol Biol Rev 16(2):121–135

Vessey JK (2003) Plant growth promoting rhizobacteria as biofertilizers. Plant Soil 255:571–586

Villegas J, Fortin JA (2002) Phosphorus solubilization and pH changes as a result of the interactions between soil bacteria and arbuscular mycorrhizal fungi on a medium containing NO_3 as nitrogen source. Can J Bot 80:571–576

Walley FL, Germida JJ (1997) Response of spring wheat (*Triticum aestivum*) to interactions between *Pseudomonas* species and *Glomus clarum* NT4. Biol Fertil Soils 24:365–371

Walsh UF, Morrissey JP, O'Gara F (2001) Pseudomonas for biocontrol of phytopathogens: from functional genomics to commercial exploitation. Curr Opin Biotechnol 12:289–295

Williams RL, Kennedy IR (2002) A model for testing the effectiveness of biofertiliser for Australian rice production. In: Choudhury ATMA, Kennedy IR (eds) Biofertilisers in action. Rural Industries Research and Development Corporation, Canberra, pp 112–114

Yahya A, Azawi SKA (1998) Occurrence of phosphate solubilizing bacteria in some Iranian soils. Plant Soil 117:135–141

Yazdani M, Bahmanyar MA, Pirdashti H, Esmaili MA (2009) Effect of Phosphate solubilization microorganisms (PSM) and plant growth promoting rhizobacteria (PGPR) on yield and yield components of Corn (*Zea mays L.*). Proc World Acad Sci Eng Technol 37:90–92

Zahir A, Muhammad A, Frankenberger WT Jr (2004) Plant growth promoting rhizobacteria: applications and perspectives in agriculture. Advan Agron 81:97–168

Zaidi A (1999) Synergistic interactions of nitrogen fixing microorganisms with phosphate mobilizing microorganisms. Ph.D. thesis, Aligarh Muslim University, Aligarh, India

Chapter 23
Arbuscular Mycorrhizae for Sustainable Agriculture

Ghazala Nasim

Contents

Abstract Arbuscular mycorrhizal (AM) associations have now been recognized having a cosmopolitan and ubiquitous occurrence forming with the roots and underground portions of plants. AM associations have been reported to enhance plant vigor by enabling them to absorb more nutrients. The thread like hyphae of the fungi work as conduits for pumping essential nutrients into the plant body, thereby imparting plant ample amount of resistance to combat with the soil born and other diseases. Due to these promising features of AM, they are being widely used as

G. Nasim (✉)
Institute of Agricultural Sciences, University of the Punjab,
Lahore 54590, Pakistan
e-mail: ghazalanasim@hotmail.com

M. Ashraf et al. (eds.), *Crop Production for Agricultural Improvement*,
DOI 10.1007/978-94-007-4116-4_23,

biological fertilizer at commercial level in many countries of the world and it has considerable practical significance in sustainable agricultural practices.

Keywords Arbuscular Mycorrhiza • Plant nutrition • Biocontrol agent • Metal toxicity • Micropropagated plantlets • Transgenic plants • Responsiveness • Sustainable agriculture

1 Introduction

The term mycorrhiza, which literally means fungal root, was first coined by AB Frank in 1985 during his study of endophytic fungi (Frank 1985). Since then, plethora of literature has mounted on various aspects of mycorrhizae. This is a unique example of symbiosis between two eukaryotes, soil fungi and plants. This association induces important morphological and/or physiological changes in each partner that lead to reciprocal benefits, mainly in terms of nutrient supply (Balestrini and Lanfranco 2006). Mycorrhizae comprise several distinctive groups that are recognized based on the type of host, fungus and the structures formed by the fungus. Seven such groups of mycorrhizae have been recognized out of which arbuscular mycorrhizae and ecto-mycorrhizae are the most common and widely distributed ones (Allen et al. 2003). Arbuscular mycorhizal (AM) fungi are the obligate mycotrophs and are the most ubiquitous and cosmopolitan in nature. More than 80% of land plants form arbuscular mycorrhizae (Brundrett 2002). These not only include plants belonging to angiosperms but algae, gymnosperms, pteridophytes, bryophytes and mosses are also beneficiaries of this relationship (Fonseca et al. 2006; Zhang et al. 2004; Nasim 2010). Recent studies have shown that AM fungi are much more widely distributed than had been considered before (Nasim and Iqbal 1991a, b; Iqbal and Nasim 1991). There are now reports on the occurrence of AM fungi in portions other than roots (Nasim 1990a, b). The submerged leaves of vascular plants and algae (Bagyaraj et al. 1979), senescent leaves of mosses, rhizoids, thalli of bryophytes and pteridophytes (Nasim et al. 1987; Nasim 1990a, b; Nasim 2010), decaying peanut leaves (Taber and Trappe 1982), pegs and nodules of the peanut plant (Graw and Rehm 1977; Nasim and Shaheen 1992), decaying leaf litter and leaves of *Dalbergia sisso* and *Termanalia arjuna* (Nasim and Iqbal 1989; Nasim et al. 1996a, b.; Nasim 2010) and scale leaves and rhizomes of a number of plants have been reported to harbor AM structures through several investigations.

Ectomycorrhizae are distinguished from arbuscular mycorrhizae on the basis of the presence of a Hartig net (Brundrett 2004). Hartig net consists of labyrinthine hyphae between root cells and is meant for the exchange of nutrients between the two partners. The thickness of the Hartig net may be variable depending upon the host-fungus combination (Burgess et al. 1994). The presence of an outer hyphal sheath called mantle confers protection from the pathogens by creating a physical barrier. While the hyphae protruding from the sheath enhance host nutrient supply by tremendously increasing the surface area. The fungi forming ectomycorrhizae belong to Ascomycota and Basidiomycota. On the other hand, AM fungi which belong to Glomeromycota (Schüßler et al. 2001), form branched tree-like structures as the primary sites of nutrient exchange. Arbuscules are ephemeral structures which are often absent or hard to observe due to poor staining or host age and pigmentation. Vesicles are the storage structures formed at later stages of the host growth. The vesicles may serve as propagules for vegetative reproduction besides azygospores which are formed extramtrically by the fungus (Brundrett 2004).

Mycorrhiza is mutually beneficial relationship and the organisms involved in linking plant and soil, transporting mineral nutrients to the plants and C compounds to the soil and its biota (Bago et al. 2003). Mycorrhizal associations involve three-way interactions between host plants, mutualistic fungi and soil factors. They are therefore both agents of plant nutrition and soil nutrition (Hock 2001; van der Heijden and Sanders 2002; Allen et al. 2003; Querejeta et al. 2003a, b.). The hyphae of the arbuscular mycorrhizal fungi transport resources such as HPO_4, NH_4, Ca, S, K, Zn, Cu, H_2O to the host (Treseder and Allen 2002; Chen et al. 2004; Goicoechean et al. 2004). Reports on the effect of the arbuscular mycorrhizae on uptake of cesium and some other cations are also available (Berreck and Haselwandter 2001; George et al. 1994). Arbuscular mycorrhizal (AM) fungi contribute to the formation of relatively stable aggregate structures. Fibrous roots and AM fungal hyphae can be viewed as a "sticky-string bag" that contributes to the entanglement and enmeshment of soil particles to form macro-aggregates (Sylvia et al. 1997), a basic building block of soil structure. Furthermore, glomalin a glycoprotein, produced by AM fungi is deposited on their outer hyphal walls and on adjacent soil particles. Glomalin appears to be a rather stable hydrophobic glue that might reduce macro-aggregate disruption during wetting and drying events by retarding water movement into the pores within the aggregate structure (Miller and Jastrow 2000). In a nutrient poor environment such as sand dunes or rugged mountain slopes, AMF contributes not only to plant nutrition, but also to the process of dune stabilization by binding sand grains into wind-resistant aggregates (Forster and Nicolson 1981; Nasim and Bajwa 2001). They do so by binding the soil particles together (Miller and Jastrow 1992) and may serve as a substrate for other polysaccharide-producing microorganisms. The bacterial polysaccharides in fact cement the soil particles together (Tisdall and Oades 1979).

Land degradation due to salinity, waterlogging, erosion, etc., are serious and growing problems all over the world (Bell 1988; Scott 1992). These fungi also play a vital role in alleviating the effects of salinity (Al-Karaki et al. 2001). By improved nutrient acquisition, AM fungi compensate for the nutritional imbalances imposed by salinization (Sylvia and Williams 1992). Some other environmental stresses such as drought (Neumann and George 2004; Auge et al. 2003, 2004a, b.), micronutrient imbalances and industrial effluents (Oliveira et al. 2001), heavy metal toxicity (Chaudhry et al. 1999; Andrade et al. 2004), biocide treatment (Heggo et al. 1990), slurry application (Chistie and Kilpatric 1992), sulfur dioxide fumigation (Clappert and Reid 1990), and wild fire recovery (Puppi and Tartnlini 1991), involve the use of AM fungi (Barea et al. 1993). Some AM fungi are adapted to adverse conditions so they can benefit plants under a variety of environmental stresses (Mosse et al.

1981). AM can also reduce the toxicity of certain metals for plants, while at non-toxic or such optimal level, their acquisition is enhanced by symbioses (Bethlenfalvay 1992; Khan et al. 2000; Kochian et al. 2004). AM also plays a positive role in protecting plants from pH extremes (Sylvia and Williams 1992).

2 AM in Relation to Plant Nutrition

Arbuscular mycorrhizal associations are mutually beneficial relationships in which both the partners achieve advantage. The fungal partners which are the obligate symbionts rely on carbon provided by their plant hosts to complete their life cycle. In return, the fungus provides essential nutrients like phosphorus and nitrogen to the plant host. This nutrient exchange takes place in between root cortical cells and arbuscules. Arbuscules never penetrate the host plasmalemma and cytoplasm rather are enveloped by the newly synthesized host membrane. They remain in this interfacial space where the exchange process occurs (Balestrini and Bonfante 2005).

Mycorrhizal development initiates with the germination of AM fungal spores present in the soil. Hyphae that enter the vicinity of the suitable host start to ramify more vigorously (Akiyama et al. 2005). When it comes in contact with the host root surface, a hyphopodium is formed and the host epidermis is penetrated by the fungus (Genre et al. 2005). After getting into the host root cortex, the hyphae start forming highly branched tree-like arbuscules (Smith and Read 1997). Arbuscules are short-lived strictures and begin to degenerate after 4–10 days of activity (Strack et al. 2003). After the complete senescenece of arbuscules, the host cell may be ready for recolonization. AM colonization is a continuous process and a mature plant root may have cells at various stages of colonization. The extent to which the plant root is colonized depends upon various host, fungal and environmental factors. Even the genetic makeup of the host plays an important role in its responsiveness towards mycotrophy (Maeda et al. 2006; Javot et al. 2007a).

The whole process of nutrient exchange takes place in the interfacial apoplast (Balestrini and Bonfante 2005). During arbuscule development the peri-arbuscular membrane continues to produce cell wall materials, including pectins, xyloglucans, polygalactrunans and hydrxyproline rich glycoproteins. The components are secreted into the interfacial apoplast without being assembled into a wall (Balestrini and Bonfante 2005). Van Buurnan et al. (1999) and Balestrini et al. (1997) identified transcripts encoding proline rich proteins in cells of certain plants colonized with arbuscules. The permeability and nutrient transfer properties of the arbscular membranes are further enhanced by the presence of aquaporin proteins (Uehlein et al. 2007). These proteins are also known to regulate water relations in plants (Javot and Maurel 2002). The expression of aquaporin genes has been reported to be variable in drought and salt stressed conditions (Porcel et al. 2006; Ouzaid et al. 2005).

Phosphate is an essential nutrient and is limiting for plant growth in many environments (Bucher 2007). Phosphate is present in the soil in the form of inorganic orthophosphate and is readily sequestered by cations like iron, aluminum and calcium resulting into reduced mobility of this phosphate form. This ultimately results into the establishment of a localized depletion zone (Bucher 2007). To combat deficiency in the soil tons of phsopahte fertilizers are added globally as soil supplement. This results into to an increasing cost of extraction of this component and eutrophication of surface and subsurface water bodies.

The mycorrhizal associated pathway of P-uptake is operated through specific phosphate transporter genes encoding members of the phosphate transporter 1 (Pht 1) family (Bucher 2007; Javot et al. 2007b). Pht 1 proteins are proton:phosphate symporters belonging to the major facilitator superfamily and play a key role in the transport of phosphate across the peri-arbuscular membrane (Karandashove and Bucher 2005; Javot et al. 2007b)

Arbuscular mycorrhizal fungi are heterotrophic organisms and obligate symbionts totally relying on host for the supply of carbon rich compounds. Studies have shown that the form of carbon transferred from host plant to fungus is hexose sugar (Ravnskov et al. 2003.). However, it still needs to be verified that whether arbuscules are the point of hexose delivery or it takes place from elsewhere. The most prevalent form of sugar translocated in the plants is sucrose. It needs to be broken down into hexoses under the action of invertase and sucrose synthase. AM fungi have not been reported to release such enzymes rather mycorrhiza-stimulated production of synthase and invertase genes have been identified in various plant hosts including maize (Ravnskov et al. 2003) and tomato (Schaarschmidt et al. 2006). Once in the intraradical hyphae, much of the carbon is converted into storage lipids predominantly trigylycerides. Lipids not only act to store carbon but are also the main form of carbon translocated from intra- to extra-radical hyphae, where they provide the major respiratory substrate (Pfeffer et al. 1999).

AM fungi extend a number of benefits to their host plants. In addition to enhanced mineral nutrition, they can increase tolerance to water and salt stress, induce greater resistance to pathogens, and reduce sensitivity to toxic substances or pollutants present in the soil (Smith and Read 1997). The host plant, however, has to pay a fairly high price for all these services, with up to 20% of the host fixed carbon being delivered to the microbial symbionts.

3 AM as a Biocontrol Agent

Arbuscular Mycorrhizal associations between roots of higher plants and fungi have been reported to exist for long ago. Since most of the land plants are able to form mycorrhizae, it is logical to attempt to relate the mycorrhizae to the incidence and severity of plant diseases. Such efforts were also needed to find alternative means of controlling diseases without the use of expensive and potentially harmful chemicals in order to achieve the goal of sustainable agriculture (Bethlenfalvay and Linderman 1992). Rhizosphere microbial changes occur when mycorrhizae are formed. When these associations are formed, the fungi live both within root tissue and external to those tissues. They could have direct interactions with other soil organisms or they could influence those organisms indirectly by changing host plant physiology, especially root physiology and, in turn, pattern of exudation into the '*mycorrhizosphere*'. Similarly, the behavior of AM fungi is also indirectly affected by other soil microbes capable of influencing the host physiology.

A number of investigations have been carried out to draw conclusions regarding the role of mycorrhizae in plant-pathogen interactions (Azcón-Aguilar and Barea 1996; Bagyaraj 1984; Caron 1989; Dehne 1982; Hooker et al. 1994; Ingham 1988; Jalali and Jalali 1991; Linderman 1988; Linderman and Paulitz 1990; Schenck 1983, 1989; Schoenbeck 1979; Zak 1964; Akhtar and Siddiqui 2008). Various authors seem to suggest different mechanisms regarding the role of mycorrhizae in plants disease control. The common agreement is on the enhanced nutrition, competition for host photosynthates and infection sites, morphological changes in roots and root tissues, changes in chemical constituents of plant tissues, reduction of abiotic stresses and microbial changes in the mycorhizosphere (Linderman 2000).

It has been reported that where arbuscular mycorrhizae reduce disease, the same effects could be reproduced by correcting the nutrient deficiency particularly for phosphorus. However, the enhanced plant vigor provides better substrate for obligate pathogens and pathogens causing foliar diseases to infect and multiply (Meyer and Dehne 1986). While some other investigations report that P-induced changes in root exudation could affect spore germination by the pathogens (Graham and Menge 1982). On the other hand, in some of the reports it has been shown that P-nutrition is not involved in disease reduction (Caron et al. 1985, 1986a, b, c.).

It is now well established that AM formation enhances the rate of photosynthesis in the host plant, thus, compensating the C withdrawn by the fungus for its use (Dauds et al. 2000). Competition may exist between the AM fungus and the pathogen for the same nutrients or the infection site. It has been indicated by Dehne (1982) that the pathogen infects plant root tissue adjacent to those colonized by the AM fungus. This further testifies that there is no competition for nutrients or space. For nematode pathogens it has been documented that host nutrients are required for reproduction and development and a direct competition between nematodes and AM fungi may result in reduced reproduction of nematodes (Hussey and Roncadori 1982).

There are very few reports available in the literature suggesting that localized morphological changes occur in mycorrhizal roots (Becker 1976; Dehne and Schoenbeck 1979; Nasim et al. 2012). These authors have suggested that the changes may include increased lignifications of the root cells of the endodermis or may cause hypertrophy and swelling in plant root tissue. Other workers (Wick and Moore 1984) showed an increased wound-barrier formation that inhibited the black root rot pathogen of holly. Such studies indicate that AM colonization in some plants may create physical barrier for the entry of the pathogen (Linderman 2000).

Changes in specific chemical components in plants that could deter pathogens have been reported, such as increased levels of arginine in tissues suppressing sporulation of *Thielaviopsis* (Baltruschat and Schoenbeck 1975), increase in chitinase in mycorrhizal plants that could be detrimental to pathogens with chitinous walls (Dumas-Gaudot et al. 1992), or accumulation of phenolics in mycorrhizal plants

(Grandmaison et al. 1993). There is also considerable research concerning the mycorrhiza-induced defense mechanisms to explain reduced responses of AM plants to infection by specific pathogens (Cordier et al. 1999; Vierheilig et al. 1994; Dumas-Gaudot et al. 2000). These studies have examined AM and non-AM plants for changes in defense reactions and have shown that arbuscular mycorrhizae may induce slight, often transient, but potentially significant activation of metabolic and morphological changes related to disease resistance mechanisms. These' changes are thought to be elicited, and are often reversed unless challenged by a pathogen. However, the magnitude of the plant reaction is generally low compared to pathogen-induced reactions, and there is still no proof or demonstration that the mycorrhiza-induced reactions are directly involved in reducing plant disease (Dassi et al. 1998; Smith and Read 1997). Even the reported increase in phytoalexin-like isoflavanoid compounds, such as reported by Harrison and Dixon (1994), and Sundaresan et al. (1993) have not been implicated directly in the suppression of a specific disease. Benhamou et al. .(1994) showed the activation of plant defense mechanisms, but their study was conducted on transformed carrot roots in an *in vitro* root organ culture system, far removed from a real agrosystem or natural ecosystem where plants are growing in soil and mycorrhizae are associated with a myriad of other soil microbes.

There are plant diseases that definitely are influenced by environmental stresses such as drought or nutrient deficiency or excess. The literature is extensive regarding the improved nutrition and tolerance to soil drought by arbuscular mycorrhizae (George 2000; Auge 2000). These effects could be very real in natural and agro-ecosystems, and could play a major part in preventing, suppressing, or reducing the severity of diseases. If nutrient stress weakens the plant, making it more susceptible to pathogen ingress or spread, then indirectly arbuscular mycorrhizae have influenced the disease by compensating for the nutrient deficiency. Form of nitrogen has been shown to influence root disease incidence by providing a kind that plants can use but pathogens cannot. Different forms of nitrogen have been shown to alter the bacterial populations that can suppress the pathogen. However, mycorrhizae were not evaluated in such studies, so their role in inducing the microbial shift cannot be assessed.

There is now substantial indication in the literature that mycorrhizae induce changes in the microbial populations in the mycorrhizosphere, some of which could directly or indirectly influence the incidence and/or severity of plant diseases. Recent research, however, has focused on other mechanisms of interaction that largely relate to the host response to AM fungus colonization that could limit development of the pathogen and disease. Many researchers believe that host defense responses are not of sufficient magnitude or timely enough to have a significant impact on disease (Filion et al. 1999). This underscores the need to search for more universal and acceptable explanations for why plants with mycorrhizae usually tolerate or resist diseases better than plants without mycorrhizae. Furthermore, since plants in undisturbed ecosystems seem to tolerate diseases, or seem fully resistant or immune to them, one might surmise that they benefit from the full complement of mechanisms that function in tandem to suppress pathogens, even when they are introduced. An underlying feature of these mycorrhizal plants that could contribute to their success in resisting disease, as well as to many other mycorrhiza-induced *traits,* is that they have a balanced and fully-stocked mycorrhizosphere microbial composition.

The literature suggests that arbuscular mycorrhizae alter the composition of functional groups of microbes in the mycorrhizosphere, including the numbers and/ or activity of pathogen antagonists (Secilia and Bagyaraj 1987). Yet a number of other papers provide results, showing of potential increased antagonistic activity against the pathogen, even though that potential was not acknowledged. Results of Caron et al. (1986a), and St-Arnaud et al. (1994) suggest that microbes in the mycorrhizosphere might have influenced the disease or pathogen level in studies on *Phytophthora* root rot, *Fusarium* tomato root rot, and marigold *Pythium* root rot, respectively. The natural tendency is to consider other mechanisms to explain the results. Usually, this is because one expects control treatments consisting of a microbial filtrate from the mycorrhizal fungus inoculum (excluding AM fungal propagules), which would compensate for any effects associated with microbes contributing to the measured response. While those microbes might have been added to all the experimental units equally, the further population enhancement of certain components only occurs when mycorrhizae are present due to changes in root exudation and by hyphal exudation. Thus, the level of antagonism might not be the same for the treatments with or without mycorrhizae.

A number of studies have been carried out in Pakistan to monitor the effect of pre-inoculation of arbuscular mycorrhizae on the incidence of plant diseases (Table 23.1). In these investigations it has been observed that presence of AM in plant roots confers resistance against root born plant pathogens. The host and AM-pathogen combinations had been variable. Some of the studies were survey works that established that the mycorrhizal plants were more resilient to the incidence of diseases.

4 AM Ameliorating the Adverse Effects of Nutritional Stresses and Metal Toxicity

The modern agriculture demands a reduction of chemical biocides to control plant diseases and insect pests. It actually lies within the premise of sustainable agriculture that the production inputs should be reduced without significant losses to productivity. Although the mainstay of Pakistan's economy is agriculture, the crop production in the country is confronted with various constraints, among these, low soil fertility status being the most common. On an average, the agricultural soils of province Punjab contain less than 1% organic matter (Farooq-e-Azam 1988). It necessitates the application of chemical fertilizers. Although chemical fertilizers are effective in increasing the yield, these may cause some problems such as soil structure deterioration, ground water pollution and high nitrates in vegetables.

Table 23.1 An overview of studies conducted in Pakistan

Host/AM fungi	Pathogenic fungus	Effect	References
Brassica napus	*Rhizoctonia solani*	VA mycorrhiza as a deterrent to pathogenic infection	Iqbal et al. (1977)
Brassica napus	*Rhizoctonia solani*	VA mycorrhiza as a deterrent to pathogenic infection	Iqbal and Mehmood (1982)
Brassica napus	*Rhizoctonia solani*	VA mycorrhiza as a deterrent to pathogenic infection	Iqbal and Mehmood (1982)
Brassica napus	*Rhizoctonia solani*	VA mycorrhiza as a deterrent to pathogenic infection	Iqbal and Mehmood (1986)
Brassica oleraceae/Glomus mosseae	*Rhizoctonia solani* + *Fusarium moliniforme*	VA mycorrhiza as a deterrent to mixed pathogenic infection	Iqbal et al. (1987)
Brassica oleraceae/Glomus mosseae	*Fusarium moliniforme*	VA mycorrhiza as a deterrent to pathogenic infection	Iqbal et al. (1987-I)
Brassica oleraceae/Glomus mosseae	*Rhizoctonia solani*	VA mycorrhiza as a deterrent to pathogenic infection	Iqbal et al. (1987-II)
Brassica oleraceae/Glomus mosseae	*Rhizoctonia solani*	VA mycorrhiza as a deterrent to pathogenic infection	Iqbal and Nasim (1988-IV)
Chlorophytum commosum	*Thermomyces stellatus*	Behaviour of vesicular arbuscular mycorrhizal fungi against a pathogen in the scales of *Chlorophytum commosum* Baker	Nasim (1994)
Solanum tuberosum/glomus mossease + *Glomus fasciculatum*	*Rhizoctonia solani*	VA mycorrhiza as a deterrent to mixed pathogenic infection	Nasim et al. (1996a, b)
Lycopersicon esculentum, Brassica oleracea/Glomus mossease + *Glomus fasciculatum*	*Rhizoctonia solani* + *Fusarium moniliforme* + *F. oxysporum* + *Drechselera sorokiniana*	VA mycorrhiza as a deterrent to mixed pathogenic infection	Nasim (2005)
Capsicum annum/Mixed AM inoculum	*Macrophomina phaseolina*	Occurrence of vesicular arbuscular mycorrhizal association in chillies infected with root borne pathogens.	Iqbal et al. (1990)
Cicer areitinum/mixed inoculum	*Ascochyta rabiei*	Effect of foliar application of Ascochyta rabiei on growth and vesicular arbuscular mycorrhizal status of eight chick pae varieties	Nasim et al. (2003)
Lycopersicon esculentum/ mixed inoculum	*Fusarium* spp	Effect of variable soil texture and VAM preinoculation on the occurrence of damping off caused by *Fusarium* spp. on tomato.	Sheikh (1991)
Lycopersicon esculentum/ mixed inoculum	*Fusarium* spp	Effect of preinoculation of tomato seedlings with VAM, *Trichoderma viride* and *Aspergillus flavus* on the pathogenicity of *Fusarium* spp.	Naz (1991)

Similarly, severe insect/pest attack has led to the over-use of pesticides. The residues from these pesticides persist both in food and the environment, posing menace to human and animal health on one side, while they can interrupt natural processes and ecosystems on the other. There is a strong conviction among researchers to probe into alternatives of chemical based agriculture.

It is now generally recognized that AM enhances the uptake of nitrogen and of relatively less mobile nutrients such as phosphorus, sulfur, copper, zinc and boron (Giasson et al. 2008). AM fungi increase the plant contact area with the soil for up to 47-folds (Smith and Read 1997). Nutrients are taken up via the fungal hyphae by specific uptake systems and can be mobilized and transported to the plant via extra-metrical and intramatrical hyphae and arbuscules.

Heavy metals occur naturally on the soil. This term is often used for metals and metalloids that have been associated with contamination and potential toxicity (Duffus 2002). The term "heavy metal" or "trace metals" refers to the metal ions like arsenic, cadmium, chromium, copper, manganese, mercury, molybdenum, nickel, lead, selenium and zinc. In the agroecosystems, the main sources of metal contamination are increased use of chemical fertilizers and biocides, from the application of metal containing wastes such as sewage sludge, animal manure, coal and wood ashes to soil, and from atmospheric deposition (Mhatre and Pankhurst 1997; Kabata-Pendias 2001; Kabata-Pendias and Mukherjee 2007). Although some of these metals are essential nutrients for plants and are needed for growth and development of plants, but some are toxic pollutants like Cd, Pb and As. The presence of high concentration of these heavy metals in the agricultural soils is a threat to the quality of soils and surface waters (Kabata-Pendias and Mukherjee 2007). There are certain factors which greatly influence the uptake of metal ions by plants or microbes (Kloke et al. 1984). These include: (1) nature of soil types; (2) nature of the metal ion and its interaction with soil colloids and other soil components; (3) concentration and chemical form of the metal entering the soil; (4) minerological composition; (5) sorptive properties of soils and their binding; (6) physical, chemical and biological soil properties; (7) biological activity of the rhizosphere; (8) duration of contact with the surface binding of these metals; (9) chemical composition of the soil solution; (10) plant type and plant exudates.

Metal tolerance of arbusculer mycorrhizal fungi has been determined using several methods including AM spore number, root colonization and extent of root colonization (Weissenhorn et al. 1993). Unfortunately, such methods did not provide sufficient information on conditions, limitations, and threshold values ensuring the survival and growth of AM fungi, or about the genetic basis for multi-metal resistance and tolerance. However, AM fungi co-exist with other microbial communities and plant roots that can tolerate and accumulate metals, and this could confound the real interactions between AM fungi and metals in the medium (Giasson et al. 2008). Several reports have shown that AM fungi from metal contaminated soils have developed tolerance against metal toxicity and are well adapted (Weissenhorn et al. 1993, 1994; Toler et al. 2005; Sudova et al. 2007). At present, potential interaction mechanism between AM fungi and metals, and the cellular and molecular mechanism of heavy metal tolerance in AM fungi is poorly understood (Leyval and Joner 2001;

Martin et al. 2007). Metal transporters and plant encoded transporters are involved in the tolerance and uptake of heavy metals (Gohre and Paszkowski 2006; Hildebrandt et al. 2007) from the extracellular media, or in their mobilization from intracellular stores (Gaither and Eide 2001). In general, mobilization of metals in the soil by soil microbes can be achieved by protonation, chelation and chemical transformation (Gadd 2005). The exudates such as citric acid and other organic compounds, released from both plants and microbes are very effective in solubilizing and releasing metals from soil components. AM fungi have been reported to sequester and accumulate metals in their biomass as well as in the roots of host plants (Burke et al. 2000; Gadd 2005). Arbuscular mycorrhizal fungi have great potential in the remediation of disturbed land and low fertility soils through heavy metal mobilization, uptake, metal hyperaccumulation, phytoextraction, metal stabilization and phytovolatilization.

5 Soil Aggregation and Role of AM

Soil is the natural covering on most of the earth's land surface. It is a natural medium for the growth of plants. Soils support plants by holding down their root masses, although strong winds or saturated soils cause withdrawal of some trees. Soils hold enough water to supply plant requirements for several days, sometimes even weeks between rains or irrigation (Kapulnik and Douds 2000).

Soils on the basis of their particle size and composition may be classified as sand, gravel, loam, silt and clay loam (Miller and Donahue 1995). According to this classification, loams are soils with both fine (silt and clay) and coarse size classes well represented. The effects of soil texture on its moisture relations are rather complicated. However, fine textured soils have a great advantage in holding much water. The excess of moisture is held in the upper layers, which are highly vulnerable to drying. The water is not admitted in readily and is mostly lost as run off. Root penetration is also discouraged in such soils so that seedlings may not be able to approach the deeper moisture before surface soil dries. Moreover, these soils tend to be poorly aerated especially in deeper layers. Such conditions of anoxia lead to shallow rooting which results in drought susceptible plants (Doubenmire 1979).

In moderately coarse textured or in heavy well aggregated soil, there exists large interstitial spaces which facilitate the diffusion of gases. As a result, CO_2 produced by respiring organisms and roots, is able to escape rather more easily and oxygen used up in the process also moves into the soil with corresponding ease. In heavy soils, especially those that are not well aggregated or poorly aggregated, the deficiency of O_2 and toxicity of excess of CO_2 become limiting factors for many plants. Moreover, in these coarse soil water readily percolates down into deeper layers and becomes unavailable in upper soils layers. Nelson (1987) in a careful review concluded that most effects of arbuscular mycorrhizae on water relations of their host are probably due to improved phosphorus nutrition.

Higher moisture levels and congested plant growth are the conducive conditions for the development of damping off (Strange 2003). The moisture level of a soil is very much dependent upon its water holding capacity. This water holding capacity depends strongly upon a variety of soil characteristics such as its texture, size of constituent mineral particles, the manner in which these particles are arranged and the amount of organic matter (Linderman 2001).

Soil fertility largely depends upon soil humus, which is the end product of degradative and metabolic activities in soil. Soil humus, the complex array of substances left after extensive chemical and biological breakdown of fresh plant and animal residues make up 60–70% of the total organic carbon in soils and are quite resistant to further microbial decomposition. Because of its complexity, humus is often divided by solubility separations into *humic acid* (HA), *fulvic acid* (FA) and *humin.* The three compounds resemble structurally but differ in molecular weight and functional group content (Schnitzer and Khan 1972). Both fulvic and humic acids are soluble in dilute sodium hydroxide solution, but humic acid is larger and will precipitate out (be insoluble) when the solution is made acidic. Humin is the portion of humus that is insoluble in dilute sodium hydroxide (Miller and Donahue 1995). These compounds have beneficial effects both on soil structure and plant productivity. They enhance the uptake of ions by chelate formation and hence an increase in growth is caused by establishing positive effect on root enzymes and soil O'Donell (1973) regarded soil humus as a plant food. It enhances the water holding capacity and nutrient level of the soil. The high-molecular-weight humic substances, humic acids and humates, alter the physical characteristics of the soil, while the low-molecular-weight humic substances, fulvic acids and fulvates, are involved in chemical reactions in the soil that influence plant's metabolic processes. Fulvic acids can form through the enzymatic and/or chemical oxidation of humic acids. Both fulvic acids and humic acids found in soil result from the chemical and biological degradation of dead organisms. The formation of these substances may come about by the oxidative changes of organic fragments, microbial synthesis, or chemical condensation after biological breakdown or self-digestion of humic biomass (Miller and Donahue 1995).

Soil aggregation is of prime importance for the growth of plants and determines the quality and fertility of field soils. This soil characteristic is based on the spatial and temporal actions of various organomineral binding agents which may be classified into three classes (Miller and Jastrow 2000). The first category of these binding agents composed of microbial and plant-derived polysaccharides and mucigels undergo decomposition quite rapidly and are referred to as transient binding agents. The second category referred to as temporary agents is composed mainly of living or dead fibrous roots and fungal hyphae and can normally persist through a growing season or even longer in perennial systems. The third category considered to be persistent binding agents is composed of decayed or more rendered materials having humic acid moieties in association with clays, amorphous mineral complexes, or both. Thus, among these categories, AM plays an important role as temporary stabilizing agents of macroaggregates. It can be conceptualized by viewing three dimensional network of fibrous roots and AM fungal hyphae as a "sticky string bag" that physically entangles or enmeshes smaller aggregates and particles creating rather stable macroaggregates (Miller and Jastrow 2000). In fect, recent studies have shown that a hydrophobic glycoprotein occurs on the surface of AM fungi. This glycoprotein is named as glomalin and is actually exuded from both intra- and extra-radical hyphae and deposited on soil surface. It has also been established that it is produced by majority of AM strains, though they may differ considerably in the amount they produce (Miller and Jastrow 2000). Hyphae of AM fungi also help create the conditions that are conducive to the formation of microaggregates. This is achieved through the exudation of organic substances from the hyphae which favour the establishment of microbial communities. In the presence of clays and silts, these substrates along with hyphal fragments can act as nucleating points for the formation of microaggregates.

6 AM and Micropropagated and Transgenic Plants

The group of fungi forming arbuscular mycorrhizae is now being placed in Glomeromycota. This is a small group comprising seven genera and 160 species. The genus *Glomus* comprising 110 species is the most important genus of order Glomerales (Schüßler et al. 2001.). However, the fungi are ubiquitous in nature and are world-wide in distribution forming the mutualistic symbiotic association with over 95% of the plant species (Krishna 2005; Johnson et al. 2006). During the last 20 years, these fungi are reported to be of much wider occurrence than were thought 3–4 decades ago. The literature is ample confirming their presence in portions other than roots (Nasim and Iqbal 1991a, b; Iqbal and Nasim 1991). Colonization of roots by mycorrhizal fungi has been shown to improve growth and productivity of several field crops (Javaid et al. 1993; Cavagnaro et al. 2006) by increasing nutrient element uptake (Al-Karaki 2002). Mycorrhizal symbiosis is frequently associated with increased photosynthetic rates of mycorrhizal plants (Martins et al. 1997, 1999; Wu and Xia 2006). Mycorrhizae also impart other benefits to plants including production of certain secondary metabolites (Schliemann et al. 2008), enhancement of nitrogen fixation by symbiotic or associative N_2-fixing bacteria (Javaid et al. 1994), osmotic adjustment under drought stress (Augé et al. 2008; Ruiz-Lozano 2003), increased resistance to pests (Khaosaad et al. 2007), tolerance to various abiotic stress factors (Takeda et al. 2007), and improving soil aggregation (Wu et al. 2008). Thus they improve soil physical properties and stability. Studies have shown that AM fungi alleviate allelopathic stress and improve crop growth and yield (Bajwa et al. 1999, 2003; Javaid 2007).

6.1 *AM and Micropropagated Plants*

The significance of AM fungi for a tissue culture raised plantlets is of special concern. The plantlet mass multiplied through the aseptic technique has to face vicissitudes of the environment when are transplanted in the unsterilized field soil. Root born plant diseases are one of them. Over 50% plantlet mortality has been reported at this stage. This hampers the utility of this technology for human well-being. The plant tissue culture experts have been using chemicals for acclimatization of these fragile plantlets (Short 1990, 1991). In view of the escalating cost of chemicals like antitranspirents and synthetic waxes and their unfriendliness to the environment, AM fungal preinoculation may serve as a vital tool to reduce the mortality percentage. However, to fulfill the objective, selection of the right organism or consortia of microbes is needed to be done carefully as it has been evidenced by Bougher et al. (1990) that the response of plants to mycorrhization does not only depend upon nutrient availability, but also on the fungus species or even strains of the same species.

The abilities of mycorrhizal species and strains to promote plant growth has opened new perspectives for the use of these fungal inoculums in nurseries and forestries. Inoculations of forestry species were performed with different fungal-host combinitions and inoculums types under different conditions. The results of these studies have shown that the plants grow better, have more extended root systems and both roots and shoots have increased dry weights, although the ratios between the dry weight of roots and shoots were lower for mycorrhizal plants (Smith and Read 1997). The total number of short roots of mycorrhizal plants has been higher than for non-mycorrhizal plants. A study conducted by Nasim et al. (1992) and Nasim and Bajwa (2008) the wheat plants responded differently to various types of fungal inocula. The inoculum types included spore concentrates of AM fungi, AM root fragments, scale-like leaves of underground plant portions, earthworm fecal pellets, and cycas rhizosphere soil. Root colonization by mycorrhizal plants can result in lower plant growth rates if fungal compatibility, nutrient availability, light intensity or temperature is not suitable for plant development (Table 23.2; Plate 23.1; and Fig. 23.1) (Conjeaud et al. 1996).

In another study, the fungi forming AM with four tissue culture raised potato varieties have been listed and their dynamics has been studied in the field/tunnel system (Nasim 2010). The spore number in the rhizosphere of four potato varieties cultivated in Pakistan has been recorded so that the suitable associative fungi may be selected carefully. The major objective of the present study was to indicate and characterize AM fungus or a group of fungi potentially beneficial for the crop of tissue culture raised plantlets in potato growing area (Nasim 2010). Micropropagated plants are adversely affected by water stress, either due to low absorption capacity of roots or due to stomata, which are deficient in regulation of water loss (Bonga 1977; Flick et al. 1983). Acclimatization of micropropagated plants corresponds to transition period when roots become adapted to a substrate with less available nutrients, and to an autotrophic condition. At this stage, the presence of mycorrhizae could increase the availability of limiting nutrients such as phosphorus and nitrogen facilitating the absorption. Water stress can be responsible for the high mortality of many micropropagated plants (Martins 2008).

Tissue culture raised plants develop under high moisture and low light intensity, often with low lignifications levels and decreased functionality of the root system have low survival rates during hardening. Mycorrhizal inoculation of such tissue

Table 23.2 List of Glomaromycota spores isolated from soils of potato growing fields, Sahiwal district, Pakistan

Species name	Percentage of spores (100 g^{-1} soil)
Acaulospora bireticulata Rothwell & Trappe	+++
Acaulospora sp. (Unidentified)	++
Gigaspora sp. (Unidentified)	++
Glomus albidum Walker & Rhodes	+
G. claroideum Schenck & Smith	+
G. dimorphicum Boyetchko & Tewari	+++
G. deserticola Trappe, Bloss & Menge	+
G. etunicatum Becker & Gerd.	++
G. fasciculatum (Thaxter Sensu Gerd.) Gerd. & Trappe	++++[a]
G. halonatum Rose & Trappe	++
G. heterosporum Smith, Schenck & Tewari	++
G. macrocarpum Tul & Tul	+++
G. microaggregatum Koske, Gemma & Olexia	++++
G. microcarpum Tul & Tul	+++
G. monosporum Gerd & Trappe	++
G. mosseae (Nicol. & Gerd.) Gerd. & Trappe	++++[a]
G. multicaule Gerd & Bakshi	+
G. pallidum Hall	++
Glomus sp. I (Unidentified)	+
Glomus sp. II (Unidentified)	+
Glomus sp. III (Unidentified)	++
Glomus sp. IV (Unidentified)	+
Sclerocystis sp. (Unidentified)	+++
Scutellospora sp. I (Unidentified)	++
Scutellospora sp. II (Unidentified)	+
Spore Type I (Unidentified)	+
Spore Type II (Unidentified)	++
Spore Type III (Unidentified)	++
Spore Type IV (Unidentified)	+

Key: +=0–25%, ++=50–75% and +++=75–100%

[a]Highly abundant

culture raised plantlets before acclimatization increases survival thereby enhancing the functionality of the root system and the plant mineral nutrition (Martins 2004, 2008; Martins et al. 1996).

6.2 AM and Transgenic Plants

Most plant species except those belonging to certain plant families, are involved in a symbiotic association with mycorrhizal fungi (Trappe 1987). Most crop species form arbuscular type of mycorrhizae which are important in uptake of diffusion

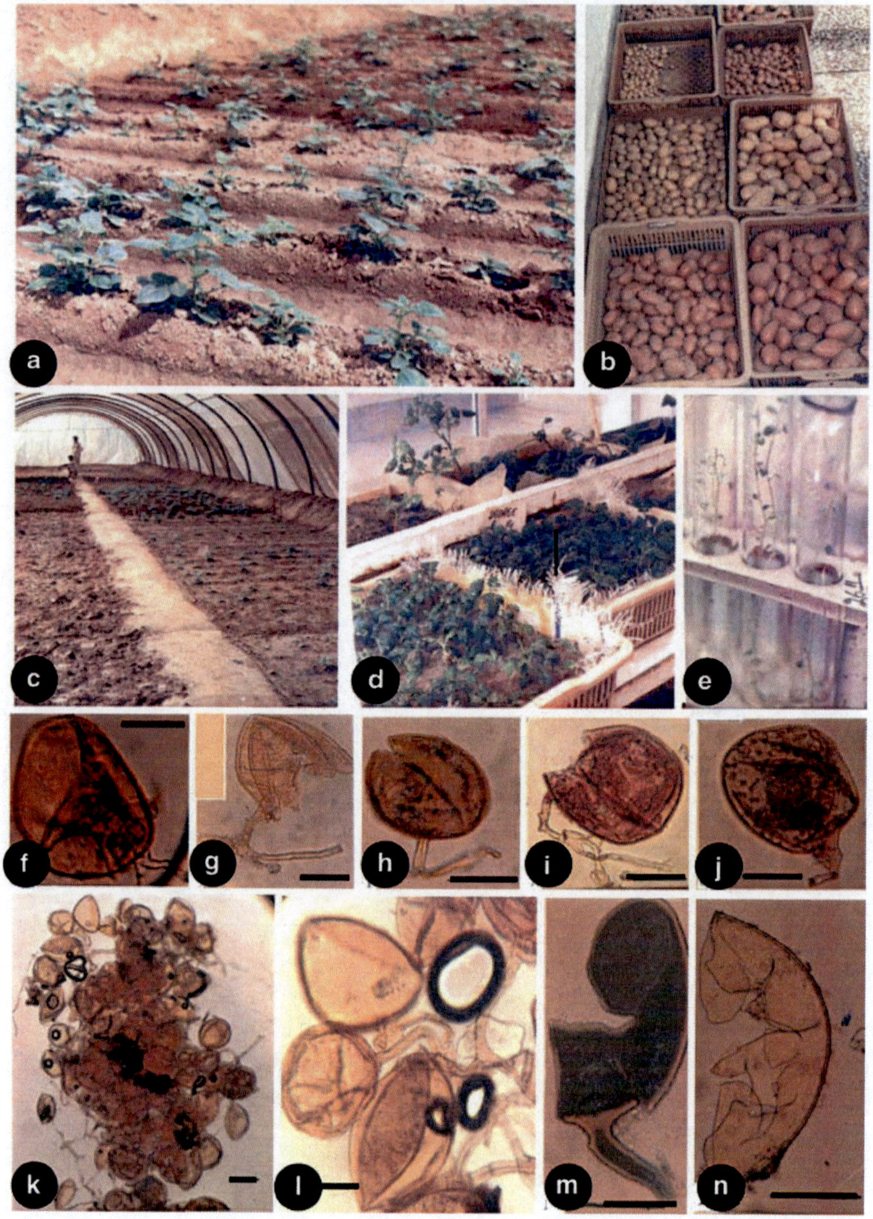

Plate 23.1 (**a–e**) Tissue culture raised potato plantlets in growth tunnels, test trays and tubes in Sahiwal area of Pakistan. (**f**) *Glomus albidum*. (**g** and **h**) *Glomus claroideum*; (**i**) *Glomus microcarpum*; (**j**) *Glomus multicaule*; (**k** and **l**) *Glomus mosseae*; (**m**) *Glomus monosporum*; (**n**) *Glomus* sp. (bar = 30 μm)

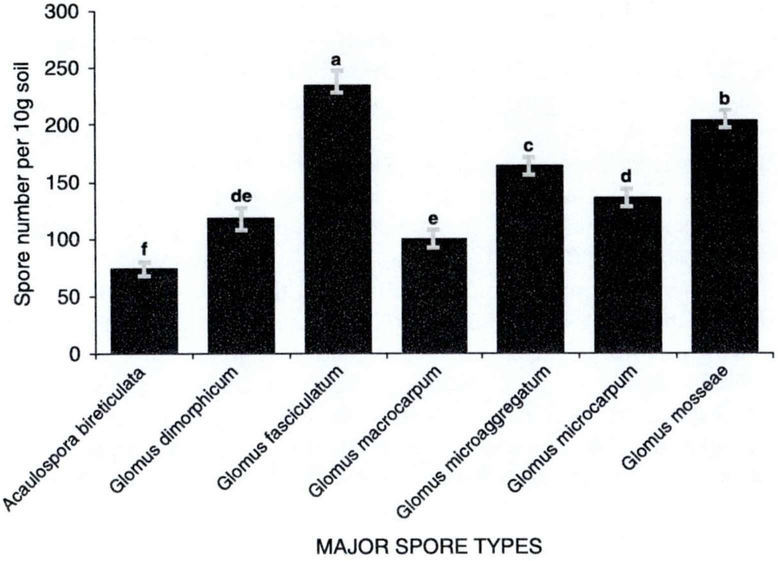

Fig. 23.1 Spore counts for seven major AM species in the rhizosphere of tissue cultured potato plants. Means sharing same lowercase letters are non-significant at 5% level

limited soil nutrients (Smith and Read 1997). They increase the host resistance to pathogens (Linderman 2000), contribute to soil aggregate stability (Miller and Jastrow 2000), and influence plant water relations (Auge 2000). Arbuscular Mycorrhizae (AM) are host non-specific that means one fungal species can form association with more than one plant species and one plant species may be host to many AM fungus species. Never-the-less variation exists in the extent of colonization and host growth benefits among plant species and even among cultivars within a species, and preferential associations between mycorrhizal fungi and particular hosts are frequently observed. Colonization of roots by AM fungi and the formation of structures that characterize the AM symbiotic association, i.e., appresoria, extra-radical and intraradical hyphae, arbuscules and vesicles, involve a complex series of events. Scientists have raised plant mutants that show a block at some stage of the colonization process and are useful in determining the factors involved in the interaction between hyphae and root cells at each step in that process.

6.2.1 Differences Among Plant Species and Mycorrhizal Species

Preferential association of mycorrhizal fungus species and plant species are indicated by differences in spore densities in different cropping systems. Johnson et al. (1992), found that spore density of *Glomus aggregatum* and *Glomus occultum* was higher in fields with maize history and *Glomus microcarpum* higher in field with soybean history. These differences were found in fields which had been planted to the same crop for 5 consecutive years. Dodd et al. (1990) also found that spore densities are dependent upon the dry or wet weather, and also upon the cropping rotation of sorghum, cassava, cowpea and kudzu. *Glomus occultum* spore density increased marledly with the sorghum crop. *Acaulospora mellea, A. marrowae* and *Entrophospora colombiana* spore density was greatest with the cow pea crop. The fact that *G. occultum* spore density increased under the maize (Johnson et al. 1992) and sorghum crops (Dodd et al. 1990) indicates that mycorrhizal species preferences for a specific host may extend across related plant species. Bethlenfalvay et al. (1982) found differential interaction of mycorrhizal species when used to infect soybean and sorghum and further showed that soil type was important in determining the extent of the response. The results of these studies indicate that plant and mycorrhizal fungus species interactions are important in the ecology of agricultural system. A study conducted by Miller et al. (1985) on assessing the comparative response of seven *Glomus* species on apple showed a twofold difference. Interestingly, the AM fungus treatment resulted in the greatest response and the least response was with two different isolates of *Glomus occultum*. Hetrick et al. (1992), found substantial differences among AM fungus species for their ability to colonize and promote the growth of wheat and wheat relatives. Different isolates of *Glomus etunicatum* provided different growth benefits. These studies further support the observation that variation in plant response to an isolate or a mycorrhizal fungal species can be as great as variation in response to various species of the fungi. Little is understood about the genetic differences between effective vs. ineffective isolates, because it is not possible to manipulate them genetically (Park and Kaeppler 2000).

It is likely that isolates and species differ with regard to optimal temperature (Baon et al. 1993) and soil conditions for mycorrhiza formation. Although AM fungi are remarkable for their ability to colonize a wide host range, the combined results of both retrospective spore density analysis and controlled experimental studies clearly show that some level of host preference or functional capability (Gianinazzi-Pearson 1984; Giovannetti and Mosse 1980) may occur between mycorrhizal genotypes and certain plant species.

6.2.2 Differences Among Cultivars and Isolates

There are numerous examples of cultivars within plant species that differ in the extent of colonization by mycorrhizal fungi and their responsiveness to them. These include *Arachis hypogaea* (Kesava-Rao et al. 1990.), *Hordeum vulgare* (Baon et al. 1993), *Medicago sativa* (Lackie et al. 1988), *Oryza sativa* (Dhillion 1992), *Pennisetum americanum* (Krishna et al. 1985), *Triticum aestivum* (Hetrick et al. 1996), *Vigna unguiculata* (Mercy et al. 1990), and *Zea mays* (Toth et al. 1990).

Mycrorrhizal fungi are especially important in absorption and P uptake of plants under P-limiting conditions, but colonization is generally suppressed by P sufficiency of the host (Smith and Read 1997). Baon et al. (1993) compared eight cultivars of barley that differed in P efficiency for their responsiveness to inoculation with

Glomus etunicatum. The results showed that the cultivars least able to grow at low P were most responsive to added P or mycorrhizae and the cultivars which attained the greatest growth at low P were the least responsive to P amendment or mycorrhizae. Their results provided evidence that cultivars best able to grow at low P are not likely to be the most efficient under high P conditions. Mycorrhizal responsiveness of ten wheat cultivars was compared at three P regimes (Hetrick et al. 1996). At the lowest P level, six of the cultivars responded positively to colonization with AM fungi, and four cultivars had no response or responded negatively to inoculation. Mycorrhizal responsiveness declined with increasing P for the six responsive cultivars, but was unaffected for the four non-responsive cultivars. Previous investigations on wheat have shown that landraces and cultivars released before 1950s, before the era of chemical fertilization, were more responsive to mycorrhizal fungi than more recently released cultivars (Hetrick et al. 1992, 1993, 1995). From this observation it was possible to conclude that selection under high fertilization may result into the loss of alleles which allow colonization and control responsiveness to mycorrhizal fungi.

Toth et al. (1984) have shown that in the case of maize breeding and selection for efficiency of P uptake may have resulted in a change in the genetic potential for colonization. High P inbreds were colonized to a greater extent than low P inbreds ones. In this case, selection of maize plants for P efficiency may have resulted in the inadvertent selection of maize susceptible to mycorrhizal fungi, but other selection criteria may have resulted in an opposite trend, namely of selecting plants "resistance" to infection by mycorrhizal fungi (Toth et al. 1990).

In another study with 28 inbred lines of maize in low P soil, the mycorrhizal responsiveness varied from 66% to 653% in the greenhouse (Kaeppler et al. 1998). There was also a twofold difference in root colonization in root colonization among these lines which was not related to mycorrhizal responsiveness. These workers have also found a significant difference among mycorrhizal fungal species in enhancing maize growth; *G. etunicatum* providing a consistent positive response across three inbred lines; *G. intraradices* showing differential response across lines and several species; and *Acaulospora morrowae* which caused no significant response in any of the lines studies.

The above cited studies on barley, wheat and maize exemplify the range of variation present among genotypes within a plant species and within species of mycorrhzal fungi for the ability to form a mutually beneficial interaction. These genetic interactions, coupled with environmental factors such as soil fertility, light intensity, variation in the density of mycorrhizal populations in soil and microbial competition within the rhizosphere have made it challenging to harness the mycorrhizal interaction in agronomic applications.

6.2.3 Genetic Basis in Host Determination of Colonization and Responsiveness

The genetic basis of the plant-micorrhizal fungus interaction has not yet been fully characterized. Genetic control of the interactions between AM fungi and host-plant exists both within host as well as symbiont species. Genetic analysis of variations among AM fungal isolates has been nearly impossible since these fungi propagate sexually and have multinucleate spores and cells (Park and Kaeppler 2000). Information on host genes controlling the interactions is now becoming available both through the analysis of mutants (Duc et al. 1989; Peterson and Guinel 2000) and through the studies of quantitative variation in populations. The genetic relationship between AM symbiosis phenotypes (e.g. percent of root colonized) and agronomic traits such as yield, biomass accumulation, and disease resistance is of particular interest from a plant improvement perspective.

Toth et al. (1990) have observed a positive correlation between disease susceptibility and colonization in 13 inbred lines of maize. Their results have suggested that selection for resistance to fungal pathogens may also result in decreased AM fungus colonization. Hetrick et al. (1995) provided the first data on the chromosomal location of genes controlling mycorrhizal responsiveness in wheat. Kaeppler et al. (1998) mapped quantitative trait loci (QTL) controlling growth at low P in the absence of mycorrhiza and mycorrhizal responsiveness in an inbred population of maize.

These studies futher define genetic factors affecting the host symbiont interaction and indicate that the inbred lines developed in the era of chemical fertilization contain alleles that allow a positive interaction with AM fungi.

The debate that current cultivars contain alleles for mycorrhizal responsiveness has important implications for future agricultural production systems. Cultivars for sustainable systems will likely need to maximize microbial interactions to replace chemical inputs. If current elite cultivars retain alleles for this function, then it is reasonable to expect high yielding genotypes for reduced input systems to be bred in a limited amount of time. However if breeders reach back to less developed germplasm to develop these cultivars, years of selection for disease resistance, quality, and other agronomic charecteristics present in current hybrids will be sacrificed.

7 Culturing of AM Fungi for Commercial Purposes

AM fungi have not been grown *in vitro* and therefore it is difficult to produce large quantities of fungal biomass for biochemical and genetical studies. Numerous attempts have been made to culture these fungi, but these have met with little or no success (Hepper 1984; Burggraff and Beringer 1989) because of the obligate nature of these fungi. The most common method employed for culturing AM fungi is trap culture, where a bait plant is grown in diluted soil from the field. This usually results into a mixer of species. Mosse (1962) was the first to establish that surface sterilized spores could infect plant roots under axenic conditions. Subsequently, various methods were developed to successfully infect a host plant (Menge and Timmer 1982) for starting a pure line of an AM fungus under sterilized conditions (Hepper 1981). AM fungi can be maintained pure and viable for long-term periods on a suitable synthetic growth medium or substrate in association with excised roots.

The current practice of inoculating plants involves the use of soil/turface-based pot-culture inoculum, produced on a suitable host plant grown under sterilized growth conditions for a period of about 2–3 months (Augé et al. 2008.). To break the spore dormancy, the inoculum is incubated in refrigerator or at room temperature for another month prior to application (Mehrotra 2005). To produce standard AM inoculum, the plant species used for raising AM inoculum are *Zea mays, Pennisetum typhoides, Allium porrum, Sorghum bicolor, Glycine max, Paspalum notatum, Cenchrus ciliaris, Chloris guyana* and *Panicum maximum*. The commercial inoculum is available in USA, India, Canada, Japan, and UK and contains several kinds of propagules, including spores, mycorrhizal root fragments and living mycelium (Mehrotra 2005). Time constraints are, however, there and it takes almost 1 year for the inoculum to get ready.

Although AM fungi have been reported to exhibit host non-specificity, however, a number of studies have shown that plant genetic makeup and AM responsiveness are closely related (Krishna 2005). The variations exist between species and genera of the same family, cultivars of the same crop species and different transgenic lines of the same plant.

Genetic control of the interactions between AM fungi and host plant exists both within host as well as symbiont species. However, the genetic basis of the plant-micorrhizal fungus interaction has not yet been fully characterized. The reason that genetic analysis of variations among AM fungal isolates has been nearly impossible as these fungi propagate asexually and have multinucleate spores and cells (Park and Kaeppler 2000).

In a study conducted by Nasim et al. (2010) it has been indicated that some of the transgenic lines of tobacco responded remarkably well to the inoculation by *Glomus intraradices*. It addresses the hypothesis that the transgenic events in tobacco has enhanced the responsiveness of *G. intraradices*. Moreover it also attempts to compare the rate of multiplication of fungual inoculum in transgenic tobacco roots with that of wild type Xanthi.

The data on fresh and dry biomass of tobacco lines at three different harvests have shown a significant difference of some of the genetic lines as compared to wild types (Nasim, G. Unpublished data). However, the determination of mycorrhizal parameters has indicated a multifold increase in colonization and sporulation by *G. intraradices*. The first harvest of experiment I was conducted after three and a half months considering for the provision of enough time to allow the fungal hyphae to invade the plant root system and initiate sporulation. The monstrous sporocarps were first observed at that stage. The second experiment was then planned keeping in view the summary of the results of the first experiment. Eight genetic lines were then short-listed including seven synthetics and one wild type. The modified seed lines were picked on the basis of their interaction with the test fungus. In this experiment, first harvest was taken after a period of 6 weeks allowing the minimum time for colonization by the AM fungus. The results were very exciting and even at this stage of host growth, the colonization and sporulation by *G. intraradices* was remarkably high (Plate 23.2). The spore aggregates/sporocarps were much bigger (Plate 23.3). The maximum size reported was 2.1 × 2.8 cm. These were not only

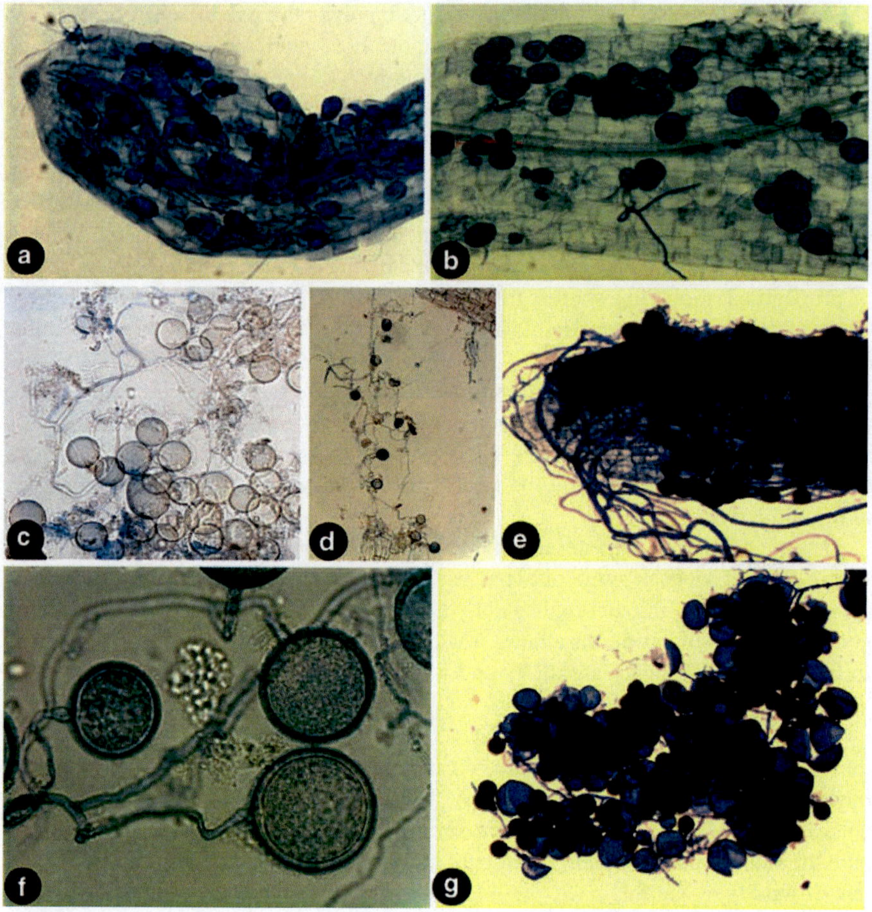

Plate 23.2 (**a**) Vesicles in tobacco root inoculated with *Glomus intraradices*; (**b**) Vesicles in tobacco root inoculated with *G. diaphanum*; (**c–d**) Extrammatrical azygospores formed by *G. intraradices* with in association with tobacco roots; (**e**) Vesicles ans extraradical hyphae; (**f**) Spores enlarged; (**g**) Extramatrical vesicles

formed along with the plant root system but also developed on the mesh which is generally used to block the wholes present at the bottom of the pot (Plate 23.4). The results when viewed in terms of number of such monstrous sporocarps and the mesh squares colonized have indicated that *m-orange* and *gbr-15* are the two constructs with a potential to produce maximum number of such structures (Fig. 23.2). Therefore, these findings may be of interest for mass multiplication of AM inoculum under controlled conditions. Practical steps may be initiated to manipulate genes specifically targeting the responsiveness of host plants towards a particular AM fungus or a consortium of these fungi. Furthermore, the results of this study are in line with those of Wenke and Lianfeng (2008) who have stated that when *Bt* transgenic crops are planted continuously, it may lead to a surprisingly weird results as regards

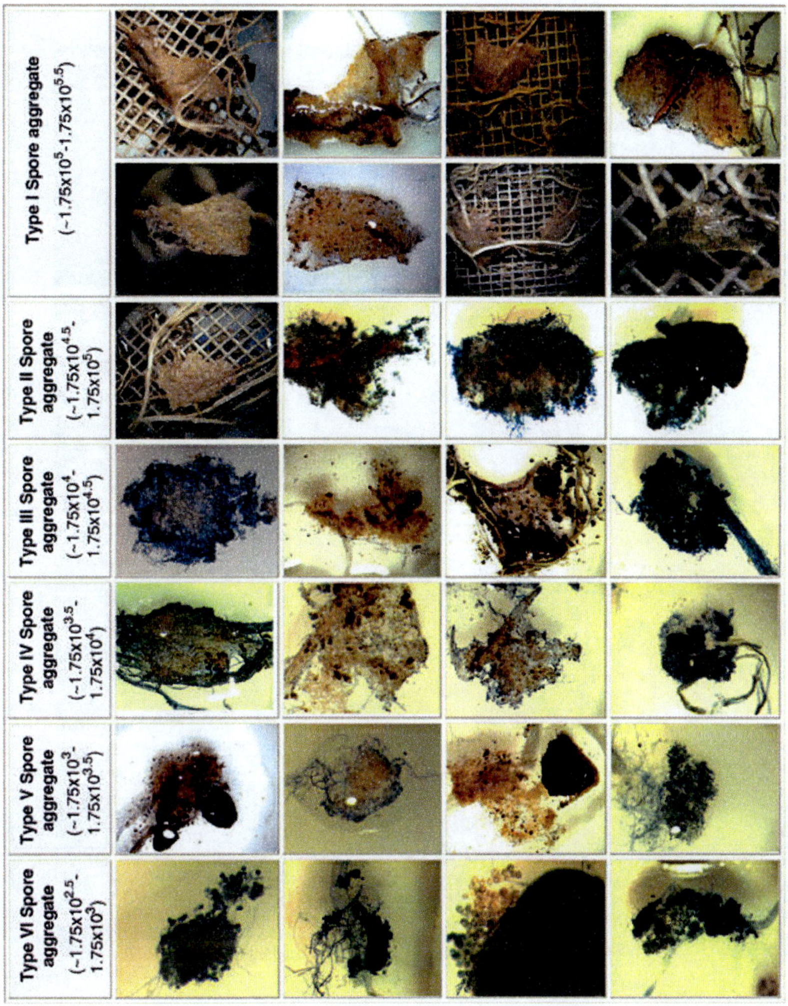

Plate 23.3 Different categories of spore aggregates depending upon the number of spores per aggregate. These categories were used to determine differences within the transgenic lines

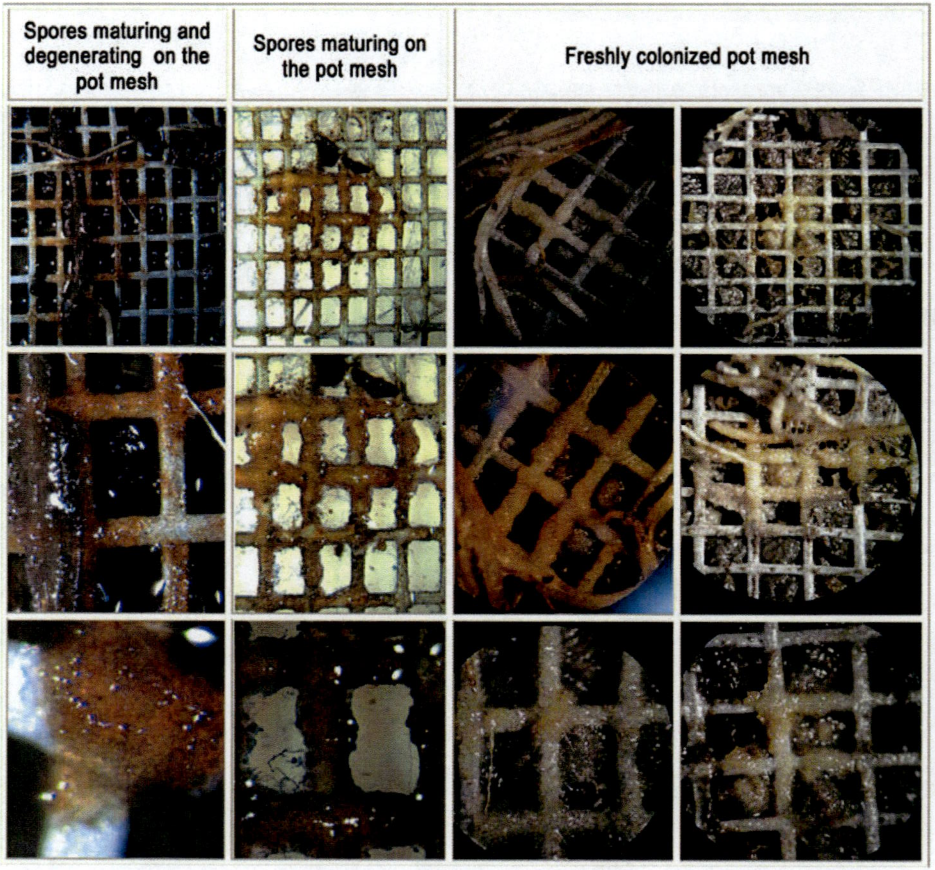

Plate 23.4 Different stages of spore colonization and maturity on pot mesh in transgenic tobacco lines. The number of squares (2×3 mm) colonized was adopted as a parameter to differentiate extent of colonization in various genetic lines

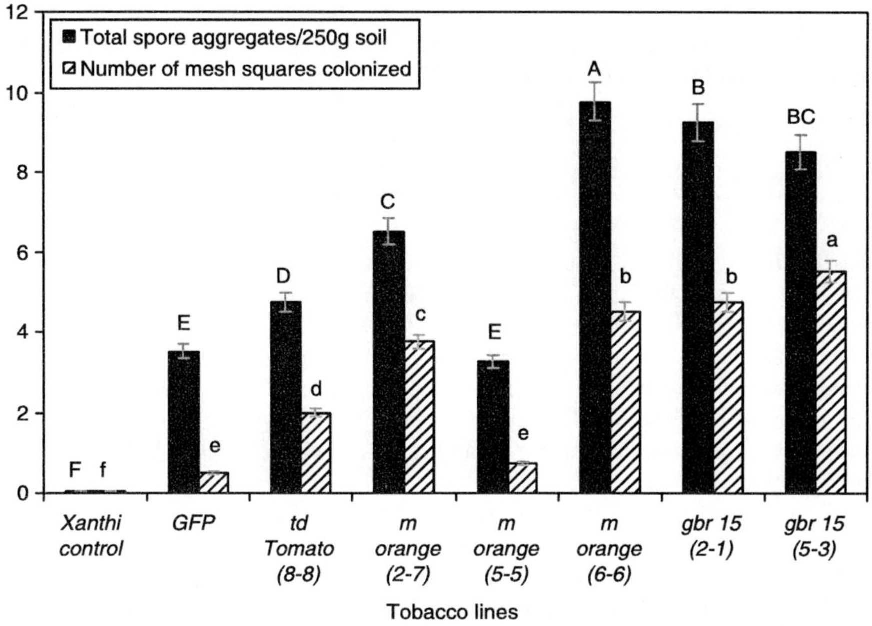

Fig. 23.2 Arbuscular mycorrhiza (AM) data of eight selected genetic lines of tobacco at first harvest after 6 weeks of inoculation. Means sharing same uppercase or lowercase letters are non-significant at 5% level

specific abundance and diversity of AM fungi. They concluded that interaction between AM fungi and *Bt* transgenic crops at individual and community level are an urgent soil ecological issue. Similarly, the results of the present study show that growing transgenic crop plants without proper risk assessment may have interesting ecological and agricultural implications.

The study presented in this session strongly suggests that the response of a species of AM fungi, *Glomus intraradices* is highly variable when inoculated to some of the fluorescent transgenic lines of tobacco, *Nicotiana tobaccum*, cv. Xanthi. The AM fungus formed monstrous sporocarp-like spore aggregates up to 2.8 cm in dimension. In view of these findings and accelerating costs of chemical fertilizers along with the environmental impact of excessive use of these agrochemicals, it is been suggested to develop transgenic strains which can promote the growth and sporulation of these beneficial fungi in pot culture for further utilization as biofertilizer for economically important plants.

These host-AM combinations may be exploited as biofertilizers for field applications. Biofertilizers are living fertilizer compounds consisting of microbial inoculants which are able to fix atmospheric nitrogen, solubilize phosphorus, decompose organic material or oxidize sulphur in the soil. On application, it enhances the growth and yield of plant yield and also improves soil fertility and reduces pollution. Biofertilizers are an alternative to mineral fertilizers for increasing soil productivity and plant growth in sustainable agriculture. Bio-inoculants like arbuscular mycorrhizae (AM), free living and symbiotic N-fixers, phosphorus solubilizing bacteria (PSB), and plant growth promoting rhizobacteria (PGPR), could be regarded as wide spectrum biofertilizers (Gianinazzi et al. 2002).

8 Role of AM in Sustainable Agriculture

Sustainable agricultural systems employ natural processes to achieve acceptable levels of productivity and food quality while minimizing adverse environmental impacts (Harrier and Watson 2004). The concept of sustainability in agriculture emphasizes that it must be ecologically sound or environment-friendly, economically viable and socially just or simply 'sustainable agriculture' is the practice of farming using principles of ecology. It relies on the principle that we must meet the needs of the present without compromising the ability of future generations to meet their own needs. Sustainable agriculture relies on long-term solutions using proactive measures rather than reactive measures at system level (Siddiqui and Pichtel 2008). Several factors contribute towards the development of agriculture on sustainable basis. Soil fertility is very important in this regard. Soil fertility can be regulated through the control of soil-borne diseases, increasing microbial activity leading to increased competition and parasitism within the rhizosphere (Jawson et al. 1993; Knudsen et al. 1995).

The fundamental objective of agriculture is human food security. However, in the past few decades the intensification of agricultural activity has resulted into tremendous environmental degradation. Moreover, the use of agricultural chemicals has produced significant effects on the natural ecosystem. Such constraints of modern-day agriculture have given birth to new concepts in farming, such as organic farming, natural farming, biodynamic agriculture, do-nothing agriculture, eco-farming, etc., (Panwar et al. 2008). This back to natural farming or sustainable agriculture is the only option which encompasses soil and crop productivity by integration of agricultural management technology at the same time maintaining and enhancing the farm profitability and environmental quality (Panwar et al. 2008). In the concept of sustainable agriculture, plant microbe interactions particularly with the beneficial soil fungi like AM have played a significant role. AM fungal associations occur widely throughout the plant kingdom including most of the agricultural and horticultural crops (Gerdemann 1968). AM fungi are ubiquitous soil microbes, but little is known of natural ecology of these fungi. The conventional agronomic practices may adversely affect the efficiency of these microbes. The impact of soil factors such as soil pH, soil moisture contents (Redhead 1977) and light intensity (Hayman 1974) on AM fungi is also well documented. The soil conditions like salinity, water-logging, erosion, soil types, water holding capacity, soil porosity, fertility status, and vegetation greatly influence AM fungal associations (Panwar et al. 2008).

Mycorrhizal fungi are clearly instrumental in augmenting plant nutrient availability particularly in nutrient stressed soils. Research has shown that the growth increment due to AM fungi is basically through an increased surface area for the absorption of nutrients. The increased uptake of phosphate, nitrogen, sulphur, potassium, calcium, zinc, iron, copper, and water transfer, are most commonly reported (Allen 1991). Moreover, the inter-plant movement of nutrients has also been reported in between plants connected with AM fungi (Read et al. 1989). Mycorrhizal inoculation enhances nodulation in leguminous plants resulting into manifold increase in nitrogen fixation (Carling et al. 1978). Arbuscluar mycorrhizal fungi play an important role also in plant-water relations. The associations improve the unsaturated hydraulic conductivity of the roots either by modifying root morphology and root anatomy or indirectly by hormonal or structural changes in the host plant (Auge 2000; Augé et al. 2008.).

The role of AM fungi in improving soil quality and aggregation is also well established (Miller and Jastrow 2000). They have suggested that AM hyphae form and stabilize aggregates of soil through three distinct processes: (1) The AM hyphae physically entangle primary particles of soil; (2) roots and AM hyphae create conditions that enable microorganisms to proliferate in the soil; and (3) roots and AM hyphae enmesh and bind microaggregates and smaller macroaggregates into larger macroaggregates. This improvement in soil structure has a direct impact on the indigenous microbial community through aeration and moisture infiltration and indirectly through stimulation of plant root growth. This may make mycorrhizal plants particularly useful for reclamation of soils with problems of surface crusting and sand dunes. Glomalin, a glycoprotein produced by AM fungal hyphae plays a major role in aggregate stabilization (Wright and Upadhyaya 1998). Glomalin has been considered to enhance ecosystem productivity and carbon sequestration in soil ecosystems (Wright et al. 2000; Rilling 2004). AM fungi also contribute towards microbial biomass resulting into improvement of soil quality.

In problematic soils, the AM fungi have been reported to regulate the uptake of heavy metals and salts (Diaz et al. 1996; Augé et al. 2008.). This may be due to high metal sorption capacity of these fungi, which could 'filter' metal ions during uptake (Joner et al. 2000).

For large scale application in agricultural systems on sustainable basis, it is necessary to evaluate the efficacy of consortia of organisms ensuring better chances of survival of plants and which are able to withstand competition with the indigenous fungi. It is further needed to conserve the diversity of AM fungal flora to ensure the restoration of plant community in abandoned agricultural lands. For commercialization, it is necessary to identify genetic basis for effective nutrient transport for increasing efficacy and adaptation to crop vigor and yield. The research is needed to be done on biotechnological application of AM fungi, technical and commercial aspects of their application for micropropagated or genetically modified plants in sustainable agricultural systems.

9 Conclusion

Arbuscular mycorrhiza plays a vital role in the life of a plant. There are innumerable advantages for the plant harbouring this relationship. AM technoly may be merged with modern biotechnologies to tap greater advantage in terms of sustainable agriculture and food security. Cultivars or genetic lines with greater responsiveness towards mycotrophy may be evolved through integrated research efforts.

References

Akhtar MS, Siddiqui ZA (2008) Arbuscular mycorrhizal fungi as potential bioprotectants against plant pathogens. In: Siddiqui ZA, Akhtar MS, Futai K (eds) Mycorrhizae: sustainable agriculture and forestry. Springer, Dordrecht, pp 321–336

Akiyama K, Matsuzaki K, Hayashi H (2005) Plant sesquiterpenes induce hyphal branching in arbuscular mycorrhizal fungi. Nature 435:824–827

Al-Karaki GN (2002) Benefit, cost and phosphorus use efficiency of mycorrhizal field grown garlic at different soil phosphorus levels. J Plant Nutr 25:1175–1184

Al-Karaki GN, Hammad R, Rusan M (2001) Response of two tomato cultivars differing in salt tolerance to inoculation with mycorrhizal fungi under salt stress. Mycorrhiza 11:43–47

Allen MF (1991) The ecology of mycorrhiza. Cambridge University Press, Cambridge, p 184

Allen MF, Swenson W, Querejeta JI, Egerton-Warburton LM, Treseder KK (2003) Ecology of mycorrhiza: a conceptual framework for complex interactions among plants and fungi. Annu Rev Phytopathol 41:271–303

Andrade SAL, Abreu CA, de Abreu MF, Silveira APD (2004) Influence of lead additions on arbuscular mycorrhiza and *Rhizobium* symbioses under soybean plants. Appl Soil Ecol 26(2):123–131

Auge RM (2000) Stomatal behavior of arbuscular mycorhizal plants. In: Kapulnik Y, Dauds DD Jr (eds) Arbuscular mycorrhizas: physiology and function. Kluwer, Dordrecht, pp 201–238

Auge RM, Moor JL, Cho K, Stutz JC, Sylvia DM, Al-Agely AK, Sexton AM (2003) Relating foliar dehydration tolerance of mycorrhizal *Phaseolus vulgaris* to soil and root colonization by hyphae. J Plant Physiol 160:1147–1156

Auge RM, Moor JL, Sylvia DM, Cho KH (2004a) Mycorrhizal promostion of host stomatal conductance in relation to irradiance and temperature. Mycorrhiza 14(2):85–92

Auge RM, Sylvia DM, Park S, Buttery BR, Sexton AM, Moor JL, Cho KH (2004b) Partitioning mycorrhizal influence on water relations of *Phaseolus vulgaris* into soil and plant components. Can J Bot 82(4):503–514

Augé RM, Toler HD, Sams CE, Nasim G (2008) Hydraulic conductance and water potential gradients in squash leaves showing mycorrhiza induced increases in stomatal conductance. Mycorrhiza 18:115–121

Azcón-Aguilar C, Barea JM (1996) Arbuscular mycorrhizas and biological control of soil-borne plant pathogens – an overview of the mechanisms involved. Mycorrhiza 6(6):457–464

Bago B, Pfeffer PE, Abubaker J, Jun J, Allen JW, Brouillette J, Douds DD, Lammers PJ, Shachar Y (2003) Carbon export from arbuscular mycorrhizal roots involves the translocation of carbohydrates as well as lipid. Plant Physiol 131(3):1496–1507

Bagyaraj DJ (1984) Biological interactions with VA mycorrhizal fungi. In: Conway Powell LL, Bagyaraj DJ (eds) VA mycorrhiza. CRC Press, Boca Raton, pp 131–153

Bagyaraj DJ, Manjunath AK, Patil RB (1979) Interaction between vesicular arbuscular mycorrhiza and *Rhizobium*: their effect on soybean in the field. New Phytol 82:141–145

Bajwa R, Akhtar J, Javaid A (2003) Role of VAM in alleviating allelopathic stress of *Parthenium hysterophorus* on maize. Mycopathologia 1:15–30

Bajwa R, Javaid A, Haneef B (1999) EM and VAM technology in Pakistan V: response of chickpea (*Cicer arietinum* L.) to co-inoculation with effective microorganisms (EM) and VA mycorrhiza under allelopathic stress. Pak J Bot 31:387–396

Balestrini R, Bonfante P (2005) The interface compartment on arbuscular Mycorrhizae: a special type pf plant cell wall? Plant Biosyst 139:8–15

Balestrini R, Jose-Estanyol M, Puigdomenech P, Bonfante P (1997) Hydroxyproline rich glycoprotein mRNA accumulation in maize root cells colonized by the arbuscular mycorrhizal fungus as revealed by in situ hybridization. Protoplasma 198:36–42

Balestrini R, Lanfranco L (2006) Fungal and plant gene expression in arbuscular mycorrhizal symbiosis. Mycorrhiza 16(8):509–524

Baltruschat H, Schoenbeck F (1975) Studies on the influence of endotrophic mycorrhiza on the infection of tobacco by *Thielaviopsis basicola*. Phytopathol Z 84:172–188

Baon JB, Smith SE, Alston AM (1993) Mycorrhizal responses of barley cultivars differeing in P efficiency. Plant and Soil 157:97–105

Barea JM, Azcon R, Azcon-Aquilar C (1993) Mycorrhiza and crops. Adv Plant Pathol 9:167–189

Becker WN (1976) Quantification of onion, VAM and their resistance to *Pyrenochacta terristeris*. Ph.D. thesis, University of Illionois, Urbana

Bell A (1988) Trees, water balance, and salt – a fine balance. Ecosystem 58:2–8

Benhamou N, Fortin JA, Hamel C, St Arnaud M, Shatilla A (1994) Resistance responses of mycorrhizal Ri T-DNA-transformed carrot roots to infection by *Fusarium oxysporum* f. sp. chrysanthemi. Phytopathology 84:958–968

Berreck M, Haselwandter K (2001) Effect of the arbuscular mycorrhizal symbiosis upon uptake of cesium and other cations by plants. Mycorrhiza 10(6):275–280

Bethlenfalvay GJ (1992) Vesicular arbuscular mycorrhizal fungi in nitrogen fixing legumes. Problem and prospects. In: Norris JR, Read DJ, Varma AK (eds) Techniques for the study of Mycorrhizae. Academic, New York

Bethlenfalvay GJ, Linderman RG (1992) Mycorrhizae in sustainable agriculture. ASA special publication no. 54. Agronomy Society of America, Madison. 124pp

Bethlenfalvay GJ, Pacovsky RS, Brown MS (1982) Parasitic and mutualistic relation between a mycorrhizal fungus and soybean: development of the endophyte. Phytopathology 72:894–897

Bonga JM (1977) Application of tissue culture in forestry. In: Reiner J, Bajaj YPS (eds) Applied and fundamental aspects of plant cell, tissue, and organ culture. Springer, New York, pp 93–107

Bougher NL, Grove TS, Malajczuk N (1990) Growth and phosphate acquisition of Karri (*Eucalyptus diversicolor* F. Muell.) seedlings inoculated with ectomycorrhizal fungi in relation to phosphate supply. New Phytol 114:77–85

Brundrett MC (2002) Coevolution of roots and mycorrhizas of land plants. New Phytol 154:275–304

Brundrett MC (2004) Diversity and classification of mycorrhizal associations. Biol Res 79:473–495

Bucher M (2007) Functional biology of plant phosphate uptake at root and mycorrhiza interfaces. New Phytol 173:11–26

Burgess T, Dell B, Malajczuk N (1994) Variations in mycorrhizal development and growth stimulation by 20 *Pisolithus* isolates inoculated on to *Eycalyptus grandis* W. Hill ex Maiden. New Phytol 127:731–739

Burggraff AJP, Beringer JE (1989) Absence of nuclear DNA synthesis in vesicular-arbuscular mycorrhizal fungi during invitro development. New Phytol 111:25–33

Burke SC, Angle JS, Chaney RL, Cunningham SD (2000) Arbuscular mycorrhizae effects on heavy metal uptake by corn. Int J Phytoremid 2:23–29

Carling DE, Riehle WG, Brown MF, Tinker PB (1978) Effects of vesicular arbuscular mycorrhizal fungus on nitrate reductase and nitroginase activities in nodulating and non-nodulating soybeans. Phytopathology 68:1590–1596

Caron M (1989) Potential use of mycorrhizae in control of soil-borne diseases. Can J Plant Pathol 11:177–179

Caron M, Fortin JA, Richard C (1985) Influence of substrate on the interaction of *Glomus intraradices* and *Fusarium oxysporum* f sp. *radicis-lycopersici* on tomatoes. Plant and Soil 87:233–239

Caron M, Fortin JA, Richard C (1986a) Effect of *Glomus intraradices* on infection by *Fusarium oxysporum* f sr. *radicis-lycopersici* on tomatoes over a 12-week period. Can J Bot 64:552–556

Caron M, Fortin JA, Richard C (1986b) Effect of inoculation sequence on the interaction between *Glomus intraradices* and *Fusarium oxysporom* f. sp. *radicis-lycopersici* in tomatoes. Can J Plant Pathol 8:12–16

Caron M, Richard C, Fortin JA (1986c) Effect of phosphorus concentration and *Glomus intraradices* on *Fusarium* crown and root rot of tomatoes. Phytopathology 76:942–946

Cavagnaro TR, Jackson LE, Six J, Ferris H, Goyal S, Asami D, Scow KM (2006) Arbuscular mycorrhizas, microbial communities, nutrient availability, and soil aggregates in organic tomato production. Plant and Soil 282:209–225

Chaudhry TM, Hill L, Khan AG, Kuek C (1999) Colonization of iron and zinc contaminated dumped filter cake waste by microbes, plants and associated mycorrhizae, Chap. 27. In: Wong MH, Wong JWC, Baker AJM (eds) Remediation and management of degraded land. CRC Press, Boca Raton, pp 275–283

Chen HYH, Kayahara GJ, Klinka K (2004) Effects of light and substrate on planted conifer seedlings in coastal British Columbia. Northwest Sci 78(2):101–110

Chistie P, Kilpatric DJ (1992) Vesciular arbuscular infection in cutgrass land following long-term slurry application. Soil Biol Biochem 24:325–330

Clappert MJ, Reid DM (1990) Effects of sulfur dioxide fumigation on *Phleum pratense* and vesicular arbuscular mycorrhizal fungi. New Phytol 115:465–469

Conjeaud C, Scheromm P, Mousain D (1996) Effects of phosphorus and ectomycorrhiza on maritime pine seedlings (*Pinus pinaster*). New Phytol 133:345–351

Cordier C, Pozo MJ, Barea JM, Gianinazzi S, Gianinazzi-Pearson V (1999) Cell defense responses associated with localized and systemic resistance to *Phytophthora parasitica* induced in tomato by an arbuscular mycorrhizal fungus. Mol Plant Microbe Interact 11:1017–1028

Dassi B, Dumas-Guadot E, Gianinazzi S (1998) Do pathogenesis-related (PR) proteins play a role in bioprotection of mycorrhizal tomato roots towards *Phytophthora parasitica?* Physiol Mol Plant Pathol 52:167–183

Dauds DD Jr, Pfeffer PE, Shachar-Hill Y (2000) Carbon partitioning, cost and metabolism of arbuscular mycorrhizas. In: Kapulnick Y, Douds DD Jr (eds) *Arbuscular mycorrhizas*: physiology and function. Kluwer, Dordrecht, pp 107–130

Dehne HW (1982) Interaction between vesicular-arbuscular mycorrhizal fungi and plant pathogens. Phytopathology 72:1115–1119

Dehne HW, Schoenbeck F (1979) Untersuchungen zurn einfluss der endotrophen mycorrhiza auf pflanzenkrankheiten ll. Phenolstotlwechsel und ligniftzierung. Phytopathol Z 95:210–216

Dhillion S (1992) Host-endophyte specificity os vesicular-arbuscular mycorrhizal colonization of the *Oryza sativa* L. at the pretransplant stage in low or high phosphorus soil. Soil Biol Biochem 24:405–411

Diaz G, Azcon-Aguilar C, Honrubia M (1996) Influence of arbuscular Mycorrhizae on heavy meta (Zn and Pb) upatake and growth of *Lygium spartum* and *Anthyllis cytisoides*. Plant and Soil 180:241–249

Dodd JC, Arias I, Koomen I, Hayman DS (1990) The management of populations of vesicular arbuscular mycorrhizal fungi in acid-infertile soils of a savanna ecosystem. I. The effect of precropping and inoculation with VAM-fungi on plant growth and nutrition in the field. Plant and Soil 122:229–240

Doubenmire C (1979) Plant environment. McGraw Hill, New York, p 240

Duc G, Trouvelot A, Gianinazzi-Pearson V, Gianinazzi S (1989) First report of non-mycorrhizal plant mutants (*myc-*) obtained in pea (*Pisum stivum* L.) and faba bean (*Vicia faba* L.). Plant Sci 6:215–222

Duffus JH (2002) "Heavy metals" a meaningless term. Pure Appl Chem 74:793–807

Dumas-Gaudot E, Furlan V, Grenier J, Asselin A (1992) New acidic chitinase isoforrns induced in tobacco roots by vesicular-arbuscular mycorrhizal fungi. Mycorrhiza 1:133–136

Dumas-Gaudot E, Gollotte A, Cordier C, Gianinazzi S, Gianinazzi-Pearson V (2000) Modulation of host defence systems. In: Kapulnik Y, Dauds DD Jr (eds) Arbuscular Mycorrhizas: physiology and function. Kluwer, Dordrecht, pp 173–200

Farooq-e-Azam (1988) Studies on organic matter fractions of some agricultural soils of Pakistan. Sarhad J Agric Res 4(3):364–369

Filion M, St-Arnaud M, Fortin JA (1999) Direct interactions between the arbuscular mycorrhizal fungus *Glomus intraradices* and different rhizosphere microorganisms. New Phytol 141:525–533

Flick CE, Evans DA, Sharp WR (1983) Organogenesis. In: Evans DA, Sharp WR, Ammirato PV, Yamada Y (eds) Handbook of plant cell culture, vol I, Techniques for propagation and breeding. McMillan, New York, pp 13–81

Fonseca HMAC, Berbara RLL, Pereira ML (2006) *Lunaria cruciata*, a potential in vitro host for *Glomus proliferum* and *G. intraradices*. Mycorrhiza 16:503–508

Forster SM, Nicolson TH (1981) Microbial aggregation of sand in maritime dune succession. Soil Biol Biochem 13:205–208

Frank AB (1985) Über die auf Wurzelymbiose beruhende Ernährung gewisser Bäume durch unterirdische Pilze. Berichte der Deutschen Botanischen Gessellschaft 3:128–145

Gadd GM (2005) Microorganisms in toxic metal-polluted soils. In: Buscot F, Varma A (eds) Microorganisms in soils: roles in genesis and function. Part V. Book series in soil biology. Springer, Berlin/Heidelberg, pp 325–356

Gaither LA, Eide DJ (2001) Eukaryotic zinc transporters and their regulation. Biometals 14:251–270

Genre A, Chabaud M, Timmers T, Bonfante P, Barker DG (2005) Arbuscular mycorrhizal fungi elicit a novel intracellular apparatus in *Medicago truncatula* root epidermal cells before infection. Plant Cell 17:3489–3499

George E (2000) Nutrient uptake. In: Kapulnick Y, Douds DD Jr (eds) Arbuscular Mycorrhizas: Physiology and function. Kluwer, Dordrecht, pp 307–344

George E, Romhneld V, Marchner H (1994) Contribution of mycorrhizal fungi in micronutrient uptake by plants. In: Manthey JA, Crowley DE, Luster DG (eds) Biochemistry of metal nutrients in the Rhizosphere. CRC Press, Florida, Boca Raton, pp 93–109

Gerdemann JW (1968) VA mycorrhizal and plant growth. Annu Rev Phytopathol 6:397–418

Gianinazzi S, Schüepp H, Barea JM, Haselwandter K (2002) Mycorrhiza technology in agriculture: from genes to bioproducts. Birkhäuser, Basel

Gianinazzi-Pearson V (1984) The host fungus specificity, recognition and compatibility in mycorrhizae. In: Verma DPS, Hohn T (eds) Genes involved in microbe-plant interactions. Springer Verlag, Vienna, pp 223–253

Giasson E, Figueiredo SR, Tornquist CG, Clarke RT (2008) Digital soil mapping using logistic regression on terrain parameters for several ecological regions in Southern Brazil. In: Hartemink AE, Mcbratney A, Mendonca-Santos ML (eds) Digital soil mapping with limited data. Springer, Newyork, pp 225–231

Giovannetti M, Mosse B (1980) An evaluation of techniques for measuring vesicular arbuscular mycorrhizal infection in roots. New Phytol 84:489–500

Gohre V, Paszkowski U (2006) Contribution of arbuscular symbiosis to heavy metal phytoremidiation. Planta 223:1115–1122

Goicoechean N, Sanchez-Diaz M, Saez R, Iraneta J (2004) The association of barley with AM fungi can result in similar yield and grain quality as a long term application of P or P-K fertilizers by enhancing root phosphate activity and sugars in leaves at tilllering. Biol Agric Hortic 2(1):69–80

Graham JH, Menge JA (1982) Influence ofvesicular-arbuscu1ar mycorrhizae and soil phosphorus on take-all disease of wheat. Phytopathology 72:95–98

Grandmaison I, Olah GM, Van Calsteren MR, Furlan V (1993) Characterization and localization of plant phenolics likely involved in the pathogen resistance expressed by endomycorrhizal roots. Mycorrhiza 3:155–164

Graw D, Rehm S (1977) Vesicular arbuscualr mycorrhiza in den fruchattaragern von *Arachis hypogeal* L. Zeitschrift fun Aker-und Plfenzenban 145:75–78

Harrier LA, Watson CA (2004) The potential role of arbuscular mycorrhizal (AM) fungi in the bioprotection of plants against soil-born pathogen in organic and/or other sustainable farming systems. Pest Manag Sci 60:149–157

Harrison MJ, Dixon RA (1994) Spatial patterns of expression of flavonoid/isoflavonoid pathway genes during interactions between roots of *Medicago truncatula* and the mycorrhizal fungus *Glomus vesiforme*. Plant J 6:9–20

Hayman DS (1974) Plant growth reponse to vesicular arbuscular mycorrhiza. VI: effect of light and temperature. New Phytol 74:71–80

Heggo A, Angle JS, Chny RL (1990) Effects of vesicular arbuscular mycorrhizal fungi on heavy metal uptake by soyabeans. Soil Biol Biochem 22:265–269

Hepper CM (1981) Techniques for studying the infection of plants by vesicular arbuscular mycorrhizal fungi under axenic conditions. New Phytol 88:641–647

Hepper CM (1984) Isolation and culture os VA mycorrhizal fungi. In: Powell CL, Bagayaraj DJ (eds) VA mycorrhiza. CRC Press, Boca Raton, pp 95–112

Hetrick BAD, Wilson GWT, Cox TS (1992) Mycorrhizal dependence of modern wheat varieties, land races, and ancestors. Can J Bot 70:2032–2040

Hetrick BAD, Wilson GWT, Cox TS (1993) Mycorrhizal dependence of modern wheat varieties and ancestors: a synthesis. Can J Bot 71:512–518

Hetrick BAD, Wilson GWT, Gill BS, Cox TS (1995) Chromosomal location of mycorrhizal responsive genes in wheat. Can J Bot 73:891–897

Hetrick BAD, Wilson GWT, Todd TC (1996) Mycorrhizal response in wheat cultivars: relationship to phosphorus. Can J Bot 74:19–25

Hildebrandt U, Regvar M, Bothe H (2007) Arbuscular mycorrhiza and heavy metal tolerance. Phytochemistry 68:139–146

Hock B (2001) Fungal associations. Springer, Berlin

Hooker JE, Jaizme-Vega M, Atkinson D (1994) Biocontrol of plant pathogens using arbuscular mycorrhizal fungi. In: Schuepp H, Gininazzi S (eds) Impact of arbuscular mycorrhizas on sustainable agriculture and natural ecosystem. Birkhduser Verlag, Basel, pp 190–200

Hussey RS, Roncadori RW (1982) Vesicular-arbuscularmycorrhizaemay limit nematode activity and improve plant growth. Plant Dis 66:9–14

Ingham RE (1988) Interactions between nematodes and VA mycorrhizae. Agric Ecosyst Environ 24:169–182

Iqbal SH, Mehmood T (1982) Occurrence of mycorrhizae and damping off caused by *Rhizoctonia solani* in relation to age of *Brassica napus* seedlings. Biologia 2:41–58

Iqbal SH, Mehmood T (1986) Vesicular arbuscular mycorrhiza as a deterrent to damping off caused by *Rhizoctonia solani* in *Brassica napus*. Biologia 33:122–127

Iqbal SH, Nasim G (1991) Are under ground non-root portions of tropical plants vesicular arbuscular mycorrhizal? Trans Mycol Soc Jpn 32:467–476

Iqbal SH, Nasim G, Niaz M (1987) VA mycorrhiza as a deterrent to mixed pathogenic infection caused by *Rhizoctonia solani* and *Fusarium moliniforme* in *Brassica oleraceae*. Biologia 33(2):201–207

Iqbal SH, Nasim G (1988-IV) VA mycorrhiza as a deterrent to damping off caused by *Rhizoctonia solani* in *Brassica oleraceae*. Biologia 34(1):215–222

Iqbal SH, Qureshi KS, Ahmed JS (1977) Influence of mycorrhiza on damping off caused by *Rhizoctonia solani* in *Brassica napus*. Biologia 23:197–208

Iqbal SH, Zahoor R, Nasim G, Khalid AN (1990) Occurrence of vesicular arbuscular mycorrhizal association in chillies infected with root borne pathogens. Sci Int 2(3):233–238

Jalali BL, Jalali I (1991) Mycorrhiza in plant disease control. In: Arora DK, Rai B, Mukelji KG, Knudsen GR (eds) Handbook of applied mycology, soil and plants, vol 1. Marcel Dekker, New York, pp 131–154

Javaid A (2007) Allelopathic interactions in mycorrhizal associations. Allelopathy J 20(1):29–42

Javaid A, Hafeez FY, Iqbal SH (1993) Interaction between vesicular arbuscular (VA) mycorrhiza and *Rhizobium* and their effect of biomass, nodulation and nitrogen fixation in *Vigna radiata* (L.) Wilczek. Sci Int 5:395–396

Javaid A, Iqbal SH, Hafeez FY (1994) Effect of different strains of *Bradyrhizobium* and two types of vesicular arbuscular mycorrhizae (VAM) on biomass and nitrogen fixation in *Vigna radiata* (L.) Wilczek var. NM 20–21. Sci Int 6:265–267

Javot H, Maurel C (2002) The role of aquaporinsin root water uptake. Ann Bot 90:301–313
Javot H, Penmetsa RV, Terzaghi N, Cook DR, Harrison MJ (2007a) A *Medicago truncatula* phosphate transporter indespensible for the arbuscular mycorrhizal symbiosis. Proc Natl Acad Sci U S A 104:1720–1725
Javot H, Pumplin N, Harrison MJ (2007b) Phosphate in the arbuscular mycorrhizal symbiosis: transport properties and regulatory roles. Plant Cell Environ 30:310–322
Jawson MD, Franzlubbers AJ, Galusha DK, Aiken RM (1993) Soil fumigation within monoculture ans rotations-response of corn ans Mycorrhizae. Agron J 85:1174–1180
Johnson D, Leake J, Read DJ (2006) Role of arbuscular mycorrhizal fungi in carbon and nutrient cycling in grassland. In: Gadd GM (ed) Fungi in biogeochemical cycles. Cambridge University Press/British Mycological Society, Cambridge
Johnson NC, Copeland PJ, Crookston RK, Pfleger FL (1992) Mycorrhizae: possible explaination for yield decline with continuous corn and soybean. Agron J 84:387–390
Joner EJ, Briones R, Leyval C (2000) Metal binding capacity of arbusculal mycorrhizal fungi. Plant and Soil 226:227–234
Kabata-Pendias A (2001) Trace element in soils and plants, 3rd edn. CRC Press, Boca Ratoon
Kabata-Pendias A, Mukherjee AB (2007) Trace elements: from soil to human. Springer, Berlin/Heidelberg/New York
Kaeppler SM, Phillips RL, Olhoft P (1998) Molecular basis of heritable tissue culture-induced variations in plants. In: Jain AM, Brar DS, Ahloowalia BS (eds) Somaclonal variations ans induced mutations in crop improvement. Kluwer, Dordrecht, pp 467–486
Kapulnik Y, Douds DD Jr (2000) Arbuscular Mycorrhizas: physiology and function. Kluwer, Dortrecht, p 384
Karandashove V, Bucher M (2005) Symbiotic phosphate transport in arbuscular mycorrhizas. Trends Plant Sci 10:22–29
Kesava-Rao PS, Tilak KVBR, Arunachalam V (1990) Genetic variation for VA mycorrhiza-dependent phosphate mobilization in groundnut (*Arachis hypogaea* L.). Plant and Soil 122:137–143
Khan AG, Kuek C, Chaudhry TM, Khoo CS, Hayes WJ (2000) Role of plants, mycorrhizae and phytochelators in heavy metal contaminated land remediation. Chemosphere 41:197–207
Khaosaad T, Garcia-Garrido JM, Steinkellner S, Vierheilig H (2007) Take-all disease is systemically reduced in roots of mycorrhizal barley plants. Soil Biol Biochem 39:727–734
Kloke A, Saurbeck D, Vetter H (1984) The contamination of plants and soils with heavy metals and the transport of metals in terrestrial food chains. In: Nriagu JO (ed) Changing metal cycles and human health. Springer, Berlin/New York
Knudsen IMB, Debosz K, Hockenhull J, Jensen DF, Elmholt S (1995) Suppressiveness of organically and conventionally managed soils towards brown foot rot of barley. Appl Soil Sci Ecol 12:61–72
Kochian LV, Hoekenga OA, Pineros MA (2004) How do crop plants tolerate acid soils? Mechanisms of alluminum tolerance ans phosphorus efficiency. Annu Rev Plant Biol 55:459–493
Krishna KR (2005) Mycorrhizas: a molecular analysis. Science Publishers, Inc, Enfield
Krishna KR, Shetty KG, Dart PJ, Andrews DJ (1985) Genotype dependent variation in mycorrhizal colonization and response to inoculation of pear millet. Plant and Soil 86:113–125
Lackie SM, Bowley SR, Peterson RL (1988) Comparison of colonization among half-sib families of *Medicago sativa* L. by *Glomus versiforme* (Daniels and Trappe) Berch. New Phytol 108:477–482
Leyval C, Joner EJ (2001) Bioavailability of heavy metals in the mycorrhizosphere. In: Gobran GR, Wenzel WW, Lombi E (eds) Trace elements in the rhizosphere. CRC Press, Boca Raton, pp 165–185
Linderman RG (1988) Mycorrhizal interactions with the rhizospheric mucroflora: the mycorrhizosphere effect. Phytopathology 78:366–371
Linderman RG (2000) Effect of mycorrhizas on plant tolerance to diseases. In: Kapulnik Y, Dauds DD Jr (eds) Arbuscular mycorrhizas: physiology and function. Kluwer, Dordrecht, Boston, pp 201–238
Linderman RG (2001) Vesicular-arbuscular mycorrhiza and plant growth response to soil amendment with composted grape pomace or its water extract. Hortic Technol 11(3):446–450

Linderman RG, Paulitz TC (1990) Mycorrhizal-rhizobacterial interactions. In: Hornby D (ed) Biological control of soil-born plant pathogens. CAB International, Wallingford
Maeda D, Ashida K, Lguchi K, Chechetka SA, Hijikata A, Okusako Y, Deguchi Y, Izui K, Hata S (2006) Knockdown of an arbuscular mycorrhiza-inducable phosphate transporter gene of *Lotus japonicas* suppresses mutualistic symbiosis. Plant Cell Physiol 47:807–817
Martin F, Perotto S, Bonfante P (2007) Mycorrhizal fungi: a fungal community at the interface between soil ans roots. In: Pinton R, Varanini Z, Nannipieri P (eds) The rhizosphere biochemistry and organic substances at the soil plant interface. Marcel Dekker, New York, pp 201–236
Martins A (2004) Micorrizacao controlada de *Castanea sativa* Mill. Aspectos fisiologicos da micorrizacao in vitro e ex vitro. Dissertacao de doutoramento em biologia/Biotechnologia Vegetal. Faculdade de Ciencias de Lisboa, Universidade Classica de Lisboa, p 506
Martins A (2008) In vitro mycorrhization of micropropagated plants: studies on *Castanea sativa* Mill. In: Siddiqui ZA, Akhtar MS, Futai K (eds) Mycorrhizae: sustainable agriculture and forestry. Springer, Dordrecht, pp 321–336
Martins A, Barroso J, Pias MS (1996) Effect of ectomycorrhizal fungi on survival and growth of micropropagated *Castanea sativa* Mill. Mycorrhiza 6:265–270
Martins A, Casimiro A, Pias MS (1997) Influence of mycorrhization on physiological parameters of *Castanea sativa* Mill micropropagated plants. Mycorrhiza 7:161–165
Martins A, Santos MM, Santos H, Pias MS (1999) A P31 NMR study of phosphate levels in roots of ectomycorrhizal and non-mycorrhizal *Castanea sativa* Mill. Trees 13:168–172
Mehrotra VS (2005) Mycorrhiza: a premier biological tool for managing soil fertility. In: Mehrotra VS (ed) Mycorrhiza: role and applications. Allied publishers Pvt. Limited, New Delhi, pp 1–65
Menge JA, Timmer LW (1982) Procedures for inoculation of plants with vesicular arbuscular mycorrhizae in the laboratory, greenhouse ans field. In: Schenck NC (ed) Methods and principles of mycorrhizal research. American Phytopathological Society, St. Paul, pp 59–68
Mercy MA, Shivashankar G, Bagyaraj DJ (1990) Mycorrhizal colonization in cowpea is host dependent and heritable. Plant and Soil 121:292–294
Meyer J, Dehne HW (1986) The influence of VA mycorrhizae on biotrophic leaf pathogens. In: Physiological and genetical aspects of mycorrhizae. Proceedings of the first European symposium on Mycorrhizae, edited by National Institute of Agronomic Research, Dijon, 1–5 July 1985, pp 781–786
Mhatre GN, Pankhurst CE (1997) Bioindicators to detect contaimination of soils with special reference to heavy heavy metals. In: Pankhurst CE, Doube BM, Gupta VVSR (eds) Biological indicators of soil health. CAB International, New York, pp 349–369
Miller DD, Domoto DA, Walker C (1985) Colonization and efficacy of different endomycorrhizal fungi with apple seedlings at two phosphorus levels. New Phytol 100:393–402
Miller RM, Jastrow JD (1992) The role of mycorrhizal fungi in soil conservation. In: Bethlenfalvey GJ, Lindermanv (eds) Mycorrhizae in sustainable agriculture. ASA special publication no. 54. Agron Soc Am, Crop Sci Soc Am, and Soil Sci Soc Am, Madison, pp 29–44
Miller RM, Jastrow JD (2000) Mycorrhizal fungi influence soil structure. In: Kapulnik Y Jr, Dauds DD (eds) Arbuscular Mycorrhizas: physiology and function. Kluwer, Dordrecht, Boston, pp 3–18
Miller RW, Donahue RL (1995) Soils in our environment, 7th edn. Prentice Hall Inc, Upper Saddle River
Mosse B (1962) The establishment of vesicular arbuscular mycorrhizae under aseptic conditions. J Gen Microbiol 27:509–520
Mosse B, Stribley DP, Le Tacon F (1981) The ecology of mycorrhizae and mycorrhizal fungi. In: Alexender M (ed) Advances in microbial biology. Plenum Press, New York, pp 137–210
Nasim G (1990a) VA mycorrhiza in Pteridophytes. Vesicular arbuscular mycorrhiza in *Marselia quardifolia* (short communication). Sci Int 2(1):47–50
Nasim G (1990b) Vesicular arbuscular mycorrhizal endophyte in roots, scale leaves and epidermal cells of the rhizomatous tissues of *Colocasia antiqourum* (L.). Schott Sci Khyber 3(2):183–194
Nasim G (1994) Behaviour of vesicular arbuscular mycorrhizal fungi against a pathogen in the scales of *Chlorophytum commosum* Baker. Pak J Phytopathol 6(1):1–6

Nasim G (2005) Role of vesicular arbuscular mycorrhiza ans other soil fungi as biocontrol and bioindicator agents for some economically important crops of Pakistan. Ph.D. Thesis, University of the Punjab, Pakistan

Nasim G (2010) Glomalean spore flora of Pakistan. Published by Higher Education Commission of Pakistan (HEC), pp 365

Nasim G, Ali M, Mehmood N (2010) Molecular analysis of *Ganoderma lucidum* isolates from Lahore. Pak J Bot 42(5):3307–3315

Nasim G, Auge RM, Stewart NC, Toler HD (2012) Gene manipulation of tobacco affects colonization by *Glomus intraradices* in green house conditions. Afr J Biotechnol (in press)

Nasim G, Bajwa R (2001) Role of symbiotic soil fungi in controlling road-side erosion and in the establishment of plant communities. Poster presentation in global change open science conference "Challenges of a changing earth", Amsterdam, The Netherlands, 10–13 July 2001

Nasim G, Bajwa R (2008) Evaluation of inoculum potential of fifteen different inocula for inducing arbuscular mycorrhiza in wheat (*Tritium aestivum* L. cv. Pak 81). Paper presented in international conference of plant scientists, University of Agriculture, Faisalabad, 21–24 April 2008

Nasim G, Din M, Ali S, Shabbir A, Ilyas MB (2003) Effect of foliar application of *Ascochyta rabiei* on growth and vesicular arbuscular mycorrhizal status of eight chickpea varieties. Mycopathologia 1(1):85–94

Nasim G, Fatima A, Shaheen M, Bajwa R (1992) Effect of six different inocula on growth and VAM status of wheat (*Triticum aestivum* L. Pak 81). J Syst Exp Biol 2(1):47–68

Nasim G, Iqbal SH (1989) Vesicular arbuscular mycorrhizal fungi in angiospermic leaf litter. Sci Int 1(6):388–391

Nasim G, Iqbal SH (1991a) Fate of endogonaceous spores in soil. Trans Mycol Soc Jpn 32:517–522

Nasim G, Iqbal SH (1991b) Species of *Glomus* associated with non-root portions of some rhizomatous plants and characteristics of their mycorrhizae. Trans Mycol Soc Jpn 32:541–545

Nasim G, Iqbal SH, Bhutta AA (1987) *Equisetum:* an excellent host for vesicular arbuscular mycorrhiza. Biologia 33(1):97–103

Nasim G, Shah T, Zahoor R (1996a) Vesicular arbuscular mycorrhiza as a deterrent to pathogenic infections caused by *Rhizoctonia solani* in potato. Sci Khyber 9(2):63–70

Nasim G, Shaheen M (1992) Seasonal dynamics and interaction of seven endogonaceous spore types associated with peanut (*Arachis hypogaea* L.). J Syst Exp Biol 2(1):37–46

Nasim G, Zahoor R, Shaheen M, Saeed S (1996b) Colonisation patterns of *Glomus mosseae* on decayed leaves of *Dalbergia sisso* Roxb. and *Terminalia arjuna* Whight and Arn. Sci Khyber 9(10):47–60

Naz N (1991) Effect of preinoculation of tomato seedlings with VAM, *Trichoderma viride* and *Aspergillus flavus* on the pathogenicity of *Fusarium* spp. M.Sc. Thesis, University of the Punjab

Nelson CE (1987) The water relations of vesicular arbuscular systems. In: Safir GR (ed) Ecophysiology of VA mycorrhizal plants. CRC Press, Boca Ratoon, Florida, pp 71–91

Neumann E, George E (2004) Colonization with the arbuscular mycorrhizal fungus *Glomus mosseae* (Nicol. and Gerd.) enhanced phosphorus uptake from dry soil in *Sorghum bicolor* (L.). Plant and Soil 261(1–2):245–255

O'Donell RW (1973) The auxin like effects of humic preparations from leonardite. Soil Sci 116:100–112

Oliveira RS, Dodd JC, Castro PML (2001) The mycorrhizal status of *Pharagmites australis* in several polluted soils and sediments of an industrialized regions of northern Portugal. Mycorrhiza 10(5):241–247

Ouzaid F, Hildebrandt U, Schmelzer E, Bothe H (2005) Differential gene expression on arbuscular mycorrhizal-colonized tomato grown under heavy metal stress. J Plant Physiol 162:634–649

Panwar J, Yadav RS, Yadav BK, Tarafdar JC (2008) Arbuscular mycorrhizae: a dynamic microsymbiont for sustainable agriculture. In: Siddiqui ZA, Akhtar MS, Futai K (eds) Mycorrhizae: sustainable agriculture and forestry. Springer, Dordrecht, pp 159–176

Park JL, Kaeppler SW (2000) Effects of genetic differences among crop species and cultivars upon the arbuscular mycorrhizal symbiosis. In: Kapulnik Y, Dauds DD Jr (eds) Arbuscular mycorrhizas: physiology and function. Kluwer, Dordrecht, pp 131–146

Peterson RL, Guinel FC (2000) The use of plant mutants to study regulation of colonization by AM fungi. In: Kapulnik Y, Dauds DD Jr (eds) Arbuscular mycorrhizas: physiology and function. Kluwer, Dordrecht, pp 201–238

Pfeffer P, Douds D, Becard G, Shachar-Hill Y (1999) Carbon uptake ans the metabolism and transport of lipids in an arbuscular mycorrhiza. Plant Physiol 120:587–598

Porcel R, Aroca R, Azcon R, Ruiz-Lozano JM (2006) PIP aquiporins gene expression in arbuscular mycorrhizal *Glycine max* and *Lactuca sativa* plants in relation to drought stress tolerance. Plant Mol Biol 60:389–404

Puppi G, Tartnlini N (1991) Mycorrhizal types in three mediterranean communities affected by fire to different extent. Acta Oecol 12:295–304

Querejeta JI, Barea JM, Allen MF, Caravaca F, Roldan A (2003a) Differential response of delta C-13 and water use efficiency to arbuscular mycorrhizal infection in two aridland woody plant species. Oecologia 135(4):510–515

Querejeta JI, Egerton-Warburton LM, Allen MF (2003b) Direct nocturnal water transfer from oaks to their mycorrhizal symbionts during severe soil drying. Oecologia 134(1):55–64

Ravnskov S, Wu Y, Graham J (2003) Arbuscular mycorrhizal fungi differentially affect expression of genes coding for sucrose synthases in maize roots. New Phytol 157:539–545

Read DJ, Leak JR, Langdale AR (1989) The nitrogen nutrition of arbuscular mycorrhizal fungi and their host plants. In: Body I, Marchant R, Read DJ (eds) Notrogen, phosphorus and sulphur utilization by fungi. Cambridge University Press, Cambridge, pp 181–204

Redhead JF (1977) Endotrophic mycorrhizas in Nigeria. Species of the endogonaceae and their distribution. Trans Br Mycol Soc 69:275–280

Rilling MC (2004) Arbuscular mucorrhizae, glomalin, and soil aggregation. Can J Soil Sci 84:355–363

Ruiz-Lozano JM (2003) Arbuscular mycorrhizal symbiosis and alleviation of osmotic stress. New perspectives for molecular studies. Mycorrhiza 13:309–317

Schaarschmidt S, Roitsch T, Hause B (2006) Arbuscular mycorrhiza induces gene expression of the apoplastic invertase *LIN6* in tomato (*Lycopersicon esculentum*) roots. J Exp Bot 57:4015–4023

Schenck NC (1983) Can mycorrhizae control root diseases? Plant Dis 65:230–234

Schenck NC (1989) Vesicular arbuscular mycorrhizal fungi and the control of fungal root diseases. In: Chet I (ed) Innovative approaches to plant disease control. Wiley, New ork, pp 179–191

Schliemann W, Ammer C, Strack D (2008) Metabolite profiling of mycorrhizal roots of *Medicago truncatula*. Phytochemistry 69(1):112–146

Schnitzer M, Khan SU (1972) Humic substances in the environment. Marcel Dekker, Inc., New York, p 327

Schoenbeck F (1979) Endomycorrhiza in relation to plant diseases. In: Schippers B, Gams W (eds) Soil borne plant pathogens. Academic Press, London, New York, San Francisco, pp 271–280

Schüβler A, Schwarzott D, Walker C (2001) A new fungal phylum, the Glomeromycota: phylogeny and evolution. Mycol Res 105:1413–1421

Scott PR (1992) Why we need trees in low rainfall agricultural areas. In: Bicknell D (ed) Low rainfall agroforestry. Department of Agriculture, Perth, pp 23–33

Secilia J, Bagyaraj DJ (1987) Bacteria and actinomycetes associated with pot cultures of vesicu1ar-arbuscu1ar mycorrhizas. Can J Mcrobiol 33:1069–1073

Sheikh MA (1991) Effect of variable soil texture and VAM preinoculation on the occurrence of damping off caused by *Fusarium* spp. on tomato. M.Sc. Thesis, University of the Punjab, Lahore, Pakistan

Short KC (1990) Application of in vitro techniques for the production and the improvement of horticultural plants. In: Sangwan RS, Sangwan BS (eds) Impact of biotechnology in agriculture. Kluwer, Dordrecht, pp 15–27

Short KC (1991) The physiology of cultured plantlets and methods for facilitating their transfer to field conditions. In: Conservation of plant genetic resources through in vitro methods. American Society of Agronomy, Madison, pp 65–73

Siddiqui ZA, Pichtel J (2008) Mycorrhizae: an overview. In: Siddiqui ZA, Akhtar MS, Futai K (eds) Mycorrhizae: sustainable agriculture and forestry. Springer, Dordrecht, pp 1–36

Smith SE, Read DJ (1997) Mycorrhizal symbiosis, 2nd edn. Academic, San Diego

St-Arnaud M, Hamel C, Caron M, Fortin JA (1994) Inhibition of *Pythium ultimum* in roots and growth substrate of mycorrhizal *Tagetes patula* colonized with *Glomus intramdices*. Can J Plant Pathol 16:187–194

Strack D, Fester T, Hause B, Schliemann W, Walter MH (2003) Arbuscular sycorrhiza: biological, chemical, and molecular aspects. J Chem Ecol 29:1955–1979

Strange RN (2003) Introduction to plant pathology. Wiley, Chichester

Sudova R, Jurkiewicz A, Turnau K, VosatKa M (2007) Persistance of heavy metal tolerance of the arbuscular mycorrhizal fungus *Glomus intraradices* under different cultivation regimes. Symbiosis 43:71–81

Sundaresan P, Raja NU, Gunasekaran P (1993) Isolation and accumulation of phytoalexins in cowpea roots infected with a mycorrhizal fungus (*Glomus fasciculatum*) and their resistance to *Fusarium* wilt disease. J Biosci 18:291–301

Sylvia DM, Fuhrmann JJ, Hartel PG, Zuberer DA (1997) Principles and applications of soil microbiology. Prentice-Hall, Berlin

Sylvia DM, Williams SE (1992) Vesicualr arbuscular mycorrhizae and environmental stress. In: Bethlenfalvay GT, Linderman RD (eds) Mycorrhiza in sustainable agriculture. American Society of Agronomy, Madison, pp 101–124

Taber RA, Trappe JM (1982) Vesicular arbuscular mycorrhiza in rhizomes, scalelike leaves, roots and xylem of ginger. Mycologia 74:156–161

Takeda N, Kistner C, Kosuta S, Winzer T, Pitzschke A, Groth M, Sato M, Kaneko T, Tabata S, Parniske M (2007) Proteases in plant root symbiosis. Phytochemistry 68:111–121

Tisdall JM, Oades JM (1979) Stabilization of soil aggregates by the root segments of ryegrass. Aust J Soil Res 17:429–441

Toler HD, Morton JB, Cumming JR (2005) Growth and metal accumulationof mycorrhizal sorghum exposed to elevated copper and zinc. Water Air Soil Pollut 164:155–172

Toth R, Page T, Castleberry R (1984) Differences in mycorrhizal colonization of maize selections for high and low ear leaf phosphorus. Crop Sci 24:994–997

Toth R, Toth D, Stark D (1990) Vesicualr arbuscular colonization in zea mays affected by breeding for resistance to fuangl pathogens. Can J Bot 68:1039–1044

Trappc JM (1987) Phylogenic and ecological aspects of mycotrophy in angiosperms from an evolutionary standpoint. In: Safir GR (ed) Ecophysiology of VA mycorrhizal plants. CRC Press, Boca Raton

Treseder KK, Allen MF (2002) Evidence for direct N and P limitation of arbuscular mycorrhizal fungi. New Phytol 155:507–515

Uehlein N, Fileschi K, Eckert M, Bienert GP, Bertl A, Kaldenhoff R (2007) Arbuscular mycorrhizal mycorrhizal symbiosis and plant aquaporin expression. Phytochemistry 68:122–129

Van Buurnan ML, Maldonado-Mendoza IE, Trieu AT, Blaylock LA, Harrison MJ (1999) Novel genes induced during an arbuscular mycorrhizal (AM) symbiosis formed between *Medicago truncatula* ans *Glomus versiforme*. Mol Plant Microbe Interact 12:171–181

van der Heijden MGA, Sanders IR (2002) Mycorrhizal ecology: ecological studies analysis and synthesis, vol 157. Springer, Berlin

Vierheilig H, Alt M, Mohr U, Boller T, Wiernken A (1994) Ethylene biosynthesis and activities of chitinase and beta-1,3-glucanase in the roots of host and non-host plants ofvesicular-arbuscular mycorrhizal fungi after inoculation with *Glomus mosseae*. J Plant Physiol 143:337–343

Weissenhorn I, Glashoff A, Leyval C, Berthelin J (1994) Differential tolerance to Cd and Zn of arbuscular mycorrhizal (AM) fungal spores isolated from heavy metal polluted and unpolluted soils. Plant and Soil 167:189–196

Weissenhorn I, Leyval C, Berthelin J (1993) Cd tolerant arbuscular mycorrhizal (AM) fungi from heavy metal-polluted soils. Plant and Soil 157:247–256

Wenke L, Lianfeng D (2008) Interaction between Bt. Transgenic crops ans arbuscular mycorrhizal fungi: a new urgent issue of soil ecology in agroecosystems. Acta Agric Socand Sect B 58(2):187–192

Wick RL, Moore LD (1984) Histology of mycorrhizal and non-mycorrhizal *Ilex crenata* 'Helleri' challenged by *Thielaviosis basicola.* Can J Plant Pathol 6:146–150

Wright SF, Rilling MC, Nicholas KA (2000) A soil protein important in carbon sequestration. In: Proceedings of American Chemical Society annual meeting symposium, pp 721–725

Wright SF, Upadhyaya A (1998) A survey of soil for aggregate stability and glomalin. A glycoprotein produced by hyphae of arbuscular mycorrhizal fungi. Plant and Soil 198:97–107

Wu OS, Xia RX (2006) Arbuscular mycorrhiza influence growth, osmotic adjustment and photosynthesis of citrus under well-watered and water stress conditions. J Plant Physiol 163:417–425

Wu OS, Xia RX, Zou YN (2008) Improved soil structure and citrus growth after inoculation with three arbuscular mycorrhizal fungi under drought stress. Eur J Soil Biol 44:122–128

Zak B (1964) Role of mycorrhizae in root disease. Annu Rev Phytopathol 2:377–392

Zhang Y, Guo LD, Liu RJ (2004) Arbuscular mycorrhizal fungi associated with common pteridophytes in Dujiangyan, southwest China. Mycorrhiza 14:25–30

Chapter 24
A Site-Specific Potassium Fertilization Approach to Overcome Sporadic Response of Crops

Abdul Hannan, Muhammad Arif, Muhammad Arif, and Muhammad Waqas

Contents

A. Hannan (✉)
Directorate of Land Reclamation (Irrigation Department), Canal Bank Mughal Pura, Lahore, Punjab, Pakistan
e-mail: ahannan_alro@yahoo.com; ahannan70@gmail.com

M. Arif
Institute of Soil and Environmental Sciences, Department of Agriculture, Faisalabad, Pakistan
e-mail: marifjnw@gmail.com

M. Arif
Department of Mathematics and Statistics, University of Agriculture, Faisalabad, Pakistan
e-mail: arif_uaf@hotmail.com

M. Waqas
Directorate of General of Extension, Department of Agriculture, Davis Road, Lahore, Punjab, Pakistan
e-mail: muhwaqas_786@yahoo.com

M. Ashraf et al. (eds.), *Crop Production for Agricultural Improvement*,
DOI 10.1007/978-94-007-4116-4_24, © Springer Science+Business Media B.V. 2012

Abstract As soils vary in adsorption characteristics depending upon their physical, chemical and mineralogical composition, it is difficult to identify a single critical value of potassium (K) concentration for a variety of crops on different textured soils. In general long experimental trials need to be conducted to develop fertilizer recommendations for a particular region in a country. Even then these proposals are often insufficient to get desired yield targets. However, fertilizer recommendations based on soil solution concentration level using sorption isotherm technique provide a good indicator of soil fertility. To this end Freundlich and Langmuir adsorption models are widely used to monitor nutrient behavior in the soil and for predicting site specific and crop specific fertilizer requirements. This approach requires short experimentation; say for example, only one trial at selected sites. Before starting the experiment, target soil solution levels are set in the laboratory based on soil characteristics using above said adsorption models. After completing the trial and to arrive at precise yield target, yield data are further evaluated using different yield response models (linear plus plateau, quadratic, exponential/Mitscherlich, Extended logistic, and Boltzman sigmoid models). In this study maize fodder crop was taken using above said approach. Total green fodder yield of 61.91 Mg ha^{-1} was obtained at a target soil solution level of 14.50 mg L^{-1} equivalent to 144 kg K ha^{-1}. However, to get maximum output with respect to fodder quality traits, a higher soil solution level need to be adjusted. This would definitely lead to higher fertilizer rate.

Keywords Freundlich • Langmuir • Fertilizer recommendation • Soil solution • Yield target

1 Introduction

Potassium is one of the essential major nutrients required both for plants and animals together with nitrogen as well as phosphorus (Oborn et al. 2005). Total K contents in soils range between 3 and 100 Mg ha^{-1} in the upper 200 cm of the soil profile. More than 95% K is bound in the mineral form, whereas less than 3% is in soil extractable phases (Sparks 1987). The essentiality of K to plant growth was first established by Von Liebig in 1840 (Sparks 2000) and since then this subject has been studied extensively (Du et al. 2004; Askegaard et al. 2005). Regardless of much study, the fundamental fate, movement, and availability to plants that govern its chemical and physical phenomena have not been investigated completely. Earlier studies have revealed that K behaviour has a good relationship to various soil and environmental components (Munson 1985), such as the types of soil minerals present, moisture regime (Malla 2002; Surapaneni et al. 2002), cropping and fertilizer history (Askegaard et al. 2005), temperature variations (Barber 1995), and soil weathering (Cakmak 2002; Askegaard et al. 2005). No coordinated explanation has been arrived at that could link all the said variables in a logical way. That is why soil tests based K fertilization mostly remained ineffective on various types of soils resulting in unpredictable yield for various crops (Munson 1985).

For predicting fertilizer requirements, Freundlich and Langmuir models have been widely used to monitor nutrient adsorption in soil which involves development of sorption isotherm as it quantitatively predicts the exact amount of fertilizer required for maximum crop yield. It predicts in a better way how the nutrient status of a soil will change upon cropping. By knowing the critical solution level for a particular crop, the exact amount of fertilizer can be computed using nutrient-sorption isotherm technique equivalent to soil solution level needed for maximum crop yield on various types of soils. Different soil solution K levels serve as a guideline of K availability which depends upon the rate of removal by the plants as well as rapidity at which K can be desorbed from the adsorbed phase. Thus K concentration in soil solution provides a good indicator of soil fertility. K fertilization therefore, is important factor to meet the K requirements of a particular cropping system. Very few studies have been reported with respect to potassium adsorption characteristics of alkaline and calcareous soils. Keeping this in view, present overview has been prepared for: (a) to evaluate the adsorption isotherm technique for optimum yield and quality of crops, (b) prepare site-specific and crop-specific K fertilizer recommendations for a particular region. In this regard, maize fodder crop in potato-maize cropping sequence was evaluated as a test case.

2 K Categorization in Soil

Potassium in soil is found in solution, exchangeable, non-exchangeable (fixed) and inert (structural) K forms. These four forms can be easily identified with the help of appropriate laboratory methods, however, no clear-cut margins exist in the field among these forms (Sharpley 1990). The kinetics and equilibrium reactions among these four forms ultimately affect its availability to plants.

Extractable (Exchangeable + non-exchangeable) K level constitutes only a minor portion of total K, whereas the major portion is present in mineral fraction (Sparks and Huang 1985). The different forms of K availability to plants and microbes are in the order; solution > exchangeable > fixed (non-exchangeable) > mineral (Sparks and Huang 1985; Sparks 1987, 2000).

The relative amount of these four forms depends upon soil factors like types and amount of clay, K input and output from the field, leaching losses as for instance in sandy soils and K fixation in particularly in the illitic dominated soils. A brief description of these four K forms is given below:

2.1 Solution K

The solution K is the form which is not only taken up by the plants and microbes directly but also subjected to leaching losses in soil. The replenishment as well as depletion from exchangeable and non-exchangeable K sites at any given time determines the K content of this fraction. The processes that cause the reduction in the solution K concentration are also a cause for the release of K from non-exchangeable sites. In normal agricultural soils the level of solution K may vary from 3 to 30 mg KL^{-1} (During 1984; Haby et al. 1990). Soil moisture content, bivalent cations in both solution and exchangeable sites and reactions occurring between different forms of K leading to equilibrium and kinetic reactions affect the level of K in the soil solution (Sparks and Huang 1985; Sparks 2000). Similarly, clay minerals and soil buffering capacity cause variations in the amount of K level in the solution. Weakly buffered allophanic clays replenish solution K slowly while well-buffered clay minerals like micas and vermiculites maintain the level of solution K rapidly and fairly unchanged (Parfitt 1992).

2.2 Exchangeable K

It is the portion of extractable K which occupies exchange sites of soil colloids (Malavolta 1985). Exchangeable K is held with different bonding forces at particular adsorption sites (planer, edge, wedge, cracks etc.) of the clay minerals. Although fairly a constant number of exchange sites are present on clay particles through isomorphic substitution yet, the exchange sites on humus colloids and amorphous clay minerals increase as pH increases mainly due to the dissociation of H^+ from weak acid groups. The kinetic and thermodynamic factors are responsible for holding a particular amount of K^+ at exchange sites of clay minerals (Parfitt 1992). The displacement of exchangeable K^+ by Ca^{2+} is particularly important in soils where Ca^{2+} is added in large quantities through phosphatic fertilizers and gypsum. These additions of Ca^{2+} may be responsible for leaching of K^+ due to the selective adsorption of Ca^{2+} at exchange sites (Edmeades 1982). Exchangeable K is generally less than 2% of total soil K and ranges between 10 and 400 mg kg^{-1}.

2.3 Non-exchangeable or Fixed K

Non-exchangeable or interlayer fixed K differs from structural K as in the former K is not a part of the crystal structure of minerals. It is held between adjacent tetrahedral layers of dioctahedral and trioctahedral micas, vermiculites, and intergrade clay minerals such as chloritized vermiculite (Sparks and Huang 1985; Sparks 1987). Fixation of potassium on clay colloids depends upon the relative differences in forces between K^+ ions with clay surfaces and hydration forces among K^+ ions and dominancy of the former causes its fixation. This results in a partial collapse of the crystal structures and the K^+ ions are physically trapped to varying degrees, making K release a slow, diffusion controlled process (Sparks 1987). The weathered wedge zones of micas and vermiculites are also source of fixed (non-exchangeable) $NH_4^{\ +}$ and H_3O^+ ions, due to their size similar to K^+, can exchange K from wedge zones only, while hydrated Ca^{2+} and Mg^2 cations whose sizes are larger than K^+ are unable to exchange K from wedge zones. This fixed (non-exchangeable) K may also get released to exchangeable form when a reduction in exchangeable/soil solution K occurs as a result of crop removal, leaching or increased microbial activity (Sparks et al. 1980; Sparks 2000). Non-exchangeable K is fairly to moderately available to plants (Mengel 1985; Sparks and Huang 1985; Sparks 1987).

2.4 Structural K

Structural K is known by a number of alternative names like mineral K, un-weathered K, native K, matrix K, or inert K. It generally makes up the bulk of the total K in most soils, however soil parent material as well as formation stage determine the exact amount of total K in soils (Sparks and Huang 1985). In K-bearing minerals such as biotite/muscovite micas, orthoclase/microcline feldspar and volcanic glasses it is mostly bonded covalently within their crystal structures (Metson 1968a). These minerals are generally found in the coarser fractions because of their partial weathering/alteration during soil formation stages, as the particle size decreases the degree of alteration increases. The replacement of K^+ with Mg^{2+}, Ca^{2+} or Al^3 from the interlayer spaces of micas results in the commencement of weathering of micas which in turn lead to the formation of illite, vermiculite, smectite and other interstratified minerals. For example, smectites may form with the complete removal of K^+ from interlayer spaces of mica in an environment of abundant Ca^{2+} and Mg^{2+}, whereas vermiculite may form if environment is changed to more acidic condition.

3 General Fertilizer Recommendations

Soil testing is an important tool for preparing fertilizer recommendations, but it is little used by farmers due to the lack of supportive research. Though standard soil testing methods are good tools to monitor soil fertility status but it often fails to predict the precise requirement of fertilizer for a particular crop. In order to predict the response of a particular crop to any fertilizer based on soil tests require thorough understanding of the mechanism of bioavailability of nutrient present in soil as well as target crop yield. The significant variations due to soil conditions at different sites do not allow to develop a strong relationship between available nutrient pool and crop yield. For this purpose people generally use the critical value i.e., that amount of available nutrient above which the probability of a crop response to applied fertilizer is poor and below which the reverse occurs. Recommendation of nutrient fertilizers merely on the basis of soil test demands strongly to find the critical amount of nutrient present in the soil because it varies with crop type. Fertilizer recommendations in general are obtained from simple experiments and extrapolated on generalized soil properties but when made by using the adsorption technique with Freundlich or Langmuir models, it results in higher yields, greater economic return, and balanced fertilization in alkaline calcareous soils (Hannan 2008). Thus, there is an urgent need for more site specific nutrient recommendations that can be readily transferred to and which meet farmer's production goals and resources. The response of crop to applied K is erratic due to lack of refined critical level of soil K and this unpredictability is further aggravated due to different adsorption characteristics of various soils for K. The fate of added K as fertilizer depends upon the initial level of soil K and concentration of K in soil solution depends upon the rate of removal by the plants and the rapidity at which K can be desorbed from the adsorbed phase, whereas adsorption equilibrium solution K levels serve as an index of K availability. This depicts that equilibrium K concentration provide a better index of soil fertility using sorption isotherm technique.

As adsorption characteristics vary from soil to soil so knowledge about the factors which affect the K adsorption characteristics of the soils is necessary to predict the fate of added K fertilizers and to arrive at precise K fertilizer recommendations (Sparks and Huang 1985). As stated by Grewal et al. (1992), the yield of potato can be increased up to 50% through an improvement in nutrient management. Thus, big chances exist to enhance crop yield and quality by improving nutrient management.

4 Types of Exchange Sites on the Clay Minerals

A number of researchers have explained the exchange behaviour of K with respect to Mg and Ca on illite by assuming the presence of three types of exchange sites for K, each with its own exchange coefficient (Sumner and Bolt 1962; Bolt et al. 1963). These are (a) planner sites that occur on the outer surface of the crystal structures, (b) edges sites, and (c) interlattice sites situated between layers of the minerals. Bolt et al. (1963) showed that interlattice sites showed the highest Gapon selectivity coefficient (KG), where as planar sites exhibit the lowest kG. Goulding (1987) postulated various types of adsorption for K on clay minerals such as planar, edge, interlayer, wedge, crack, and step potassium exchange sites in 2:1 layer silicate minerals.

5 Potassium Sorption Isotherms Versus Nutrient Bioavailability

Sorption isotherm describes the relation between the amount of substance that is adsorbed by the solid phase of the soil and its equilibrium amount in soil solution. Among numerous chemical processes occurring in soil, sorption is one of most important processes which describe the fate and mobility of nutrients in the soil as well. A number of essential elements like P, K, S, Cu, Zn, can be studied practically through sorption isotherm (Hunter 1980). Sorption isotherm further explores whether applied plant nutrients react, fix or make complex with the soil. Without having a comprehensive knowledge regarding the characteristics of nutrient sorption isotherm of a particular experimental site it is inevitable to establish fertilizer optima for enhancing crop productivity, therefore understanding the nutrient sorption characteristics of the soil of the experimental site is inevitable. This approach helps to envisage a safe level of any one of the nutrients in the soil because it can be adjusted in the available pool of the soil at a level in which it is neither deficient nor toxic to the crops (Hunter 1980). This makes the potassium sorption isotherms a part of quantity-intensity approach (Neiderbudde 1986), that is used to evaluate soil K supply to plants and explain the exchange of K by other ions mostly Ca (Bedrossian and Singh 2002) and to determine interchangeable potassium in order to understand its dynamics in soil (Yunda et al. 1997).

Of all the equations, Freundlich and Langmuir adsorption models give a closer description of the real adsorption phenomenon in the soil system.

5.1 Amount of Nutrient Adsorbed

$$\frac{x}{m} = [Ci - Cf]V \times 10 / W \tag{24.1}$$

Where:

$\frac{x}{m}$ is the amount of the nutrient adsorbed per unit weight of soil (mg kg^{-1})

Ci is the initial concentration of the nutrient added (mg L^{-1}),
Cf is the concentration of that nutrient in the equilibrium condition (mg L^{-1}),
V is the volume of solution taken (ml),
W is the weight of soil sample used (g)

The theoretical modifications made in the Eqs. (24.1, 24.2, 24.3, and 24.4) for the fertilizer application are given as under: Freundlich equation:

$$\frac{x}{m} = aC^{b} \tag{24.2}$$

Where x/m and C are same as discussed above.

By rearranging:

$$\frac{x}{m} = \log a + b \log C \tag{24.3}$$

$$\frac{x}{m} = \log a + b \log C + \log Tc \quad (24.4)$$

Where Tc is the target nutrient concentration which will be adjusted by adding the amount of fertilizer corresponding to its specific soil solution level.

$$\text{Fertilizer dose}\left(\text{mg kg}^{-1}\right) = \text{antilog}\left(\frac{x}{m} = \log a + b \log C + \log Tc\right) \quad (24.5)$$

$$\text{Fertilizer dose}\left(\text{kg ha}^{-1}\right) = \text{Fertilizer dose}\left(\text{mg kg}^{-1}\right) \times 2 \quad (24.6)$$

Finally the Eq. 24.6 is used to compute fertilizer rates against specific soil solution K levels. The 'a' and 'b' values are taken from the intercept and slope of the Freundlich model, with b values of $0<n<1$. These parameters have no physical meaning, although Sposito (1980) presented 'b' as measure of heterogeneity of adsorption sites on the adsorbent surface. Equation 24.3 estimates the total amount of nutrient needed to be added just to satisfy adsorptive surfaces of the soil. The amount added in addition to that (Eq. 24.4) will contribute to that which will contribute to its concentration that will come into soil solution. By developing adsorption isotherm in the laboratory, different levels of solution (say for example for potassium 0, 3, 6, 9, 12, 15, 18, 21, 24, and 27 mg L^{-1}) can be worked out theoretically using Freundlich model and then equivalent to these specific solution levels K fertilizer doses can be calculated as given above. For micronutrients or over narrow range of nutrient concentration Langmuir equation can be used in the same way.

Langmuir equation:

$$\frac{x}{m} = \frac{\kappa cb}{1+\kappa c} + Tc \quad (24.7)$$

$$\text{Fertilizer dose}\left(\text{mg kg}^{-1}\right) = \text{antilog}\left(\frac{x}{m} = \frac{\kappa cb}{1+\kappa c} + Tc\right) \quad (24.8)$$

$$\text{Fertilizer dose}\left(\text{kg ha}^{-1}\right) = \text{Fertilizer dose}\left(\text{mg kg}^{-1}\right) \times 2 \quad (24.9)$$

These equations help in resolving sporadic nutrient response in the agricultural soils of Asia and to find out specific soil solution level required for maximum yield for a given crop over a given set of environmental conditions. In spite of certain limitations; as these provide rough estimate of the buffering capacity of the soil which might not reflect the true situation in the field due to mineralization of organic pool, rhizophere and mycorrhizal effects; but on overall basis these effects contribute to a very limited extent because in adsorption isotherm the net change in the charge of the soil either due to fluctuations in pH or the presence of different organic acids has also been counted for (Xu et al. 2005). Several examples prove the capacity of this model to predict adsorption phenomena of various elements by soils, such as Pal et al. (1999) for K, Erdem et al. (2004) for Cu, Co, Zn, and Mn, Imtiaz et al. (2006) for Zn; and Fontes and Coelho (2005) for Mo.

6 Yield Response Models

High K application rates are unwanted because of economic reasons and environmental pollution problems, for instance, higher K rate does not improve yield in potato (Hannan 2008; Lal and Singh 1983; Rhue et al. 1986, and Chapman et al. 1992). Negative affects are assigned to salt damage (Bilski et al. 1988; Maier et al. 1994), and physiological changes (Westermann et al. 1994), while both reduction in potato yield and small-sized tubers are attributed to insufficient K supply (McDole 1978; Satyanarayana and Arora 1985). Fertilizer recommendations thus must optimize crop yield and quality for maximizing profit, and would cut down the risk of environmental pollution; therefore different yield response models can be evaluated to identify the optimum K rate for maximum output. A brief summary of commonly used yield response models is given below:

6.1 Linear Plus Plateau Model

$$Y = a + bx \quad \text{if } x < c$$

$$Y = P \quad \text{If } x \geq c$$

Where 'Y' is the yield (Mg ha^{-1}) and x is K rate (kg ha^{-1}); 'a' represents intercept, 'b' denotes linear coefficient, 'c' is the optimum K rate at the juncture of the linear response and the plateau lines, while 'P' is the plateau yield.

6.2 Quadratic Model

$$Y = a + b\,x + cx^2$$

Where 'Y' is the yield (Mg ha^{-1}) and 'x' is K rate (kg ha^{-1}); 'a' is intercept, 'b' is linear coefficient and 'c' is quadratic coefficient.

6.3 The Exponential/Mitscherlich Model

$$Y = M\left(1 - \exp^{-c(x+b)}\right)$$

Where 'Y' is the yield in Mg ha^{-1}, 'x' is K rate in kg ha^{-1}, 'M' is the highest yield. The 'c' stands for Mitscherlich constant and the 'b' constant calculates the quantity of extractable K.

6.4 Extended Logistic Model

$$Y = Y_{max} / (1 + \exp(b - c * x))$$

Where $'Y'_{max}$ is the highest yield in Mg ha^{-1} and 'x' is the K rate in kg ha^{-1}, 'a' is intercept, 'b' is the linear coefficient and 'c' stands for logistic coefficient. R^2 values for models were calculated through regression analysis.

6.5 Boltzman Sigmoid Model

$$Y = \frac{Y\,\mathrm{min} + (Y\,\mathrm{max} - Y\,\mathrm{min})}{(1 + \exp(Y50 - X) / b)}$$

Where $'Y'_{max}$ is the maximum and $'Y'_{min}$ is the minimum yield (Mg ha^{-1}), 'X' is K rate in kg ha^{-1}, 'b' is linear coefficient, $'Y'_{50}$ is the yield halfway between $'Y'_{max}$ and $'Y'_{min.}$ Coefficients of determination (R^2 values) were calculated through regression analysis.

7 Example of Site-Specific K Fertilization – A Case Study

7.1 Potassium Adsorption

Three soils were selected before sowing at 0.3 m depths and their physical and chemical attributes were determined (Table 24.1). The potassium adsorption data of the selected soils best fitted to Freundlich ($R2 \geq 0.95^{**}$) as given in Table 24.2. The reason might be that Freundlich model assumes unlimited sorption sites of heterogeneous medium which give better correlation due to mixed mineralogy of the soil with illite as the dominant clay mineral. These findings are supported by the results obtained by Pal et al. (1999). The data shows that though all soils had altogether different fertilizer requirements to maintain the same soil solution level, in this study Freundlich adsorption isotherm technique was thus used to predict the quantity of added K required for the adjustment of soil solution to achieve prescribed isotherm based soil solution K levels because K concentration in the soil solution has a more universal applicability. Such studies can provide a suitable basis for determining K fertilizer requirements, especially on soils with varying K adsorption characteristics. In terms of fertilization (Table 24.3), Typic Ustochrepts soil possessed greater K adsorption capacity and required large K application

Table 24.1 Physical and chemical characteristics of selected soils

Parameters	Typic Halorthids	Typic Camborthids	Typic Ustochrepts
EC (dS m^{-1})	1.02	0.94	1.33
pH	8.16	8.29	8.22
O.M (%)	0.49	0.81	0.43
Extractable P (mg kg^{-1})	6.45	7.89	7.26
Extractable K (mg kg^{-1})	242	81	65
CEC ($cmol_c$ kg^{-1})	6.26	6.78	8.78
$CaCO_3$ (%)	7.93	9.15	9.78
Clay (%)	16	14	31
Textural class	Sandy clay loam	Sandy loam	Silty clay loam
FAO world map	Orthic solonchaks	Cacaric gleysols	Cacaric cambisols

Table 24.2 Parameters of Freundlich equations for the adsorption data of three soils

Site	a	b	R^2
Typic Ustochrepts	0.4253	1.1681	0.96*
Typic Halorthids	0.2732	0.9731	0.98*
Typic Camborthids	0.345	1.024	0.95*

*Significant at $P=0.01$

Table 24.3 Model based K rates[a] for the four soils against different soil solution levels

Target soil solution K level	Typic Ustochrepts	Typic Halorthids	Typic Camborthids
(mg L^{-1})	K rate kg ha^{-1}		
0	Control	Control	Control
3	19	11	14
6	43	21	28
9	69	32	42
12	97	42	56
15	126	52	71
18	156	62	85
21	187	73	100
24	218	83	115
27	250	93	129

[a]Rates calculated with Freundlich model

i.e. 250 kg ha^{-1} . The other two soils had almost similar clay contents and other properties but differed only in available K. This factor again leads to varied fertilizer requirement to maintain the same soil solution level.

7.2 *Maize Fodder Yield and K Concentration*

The model based K rates applied to first and second maize fodder crops (Fig. 24.1), revealed that interaction was non-significant with respect to yield, dry matter and crude protein for first and second crop, therefore the data of two crops was pooled.

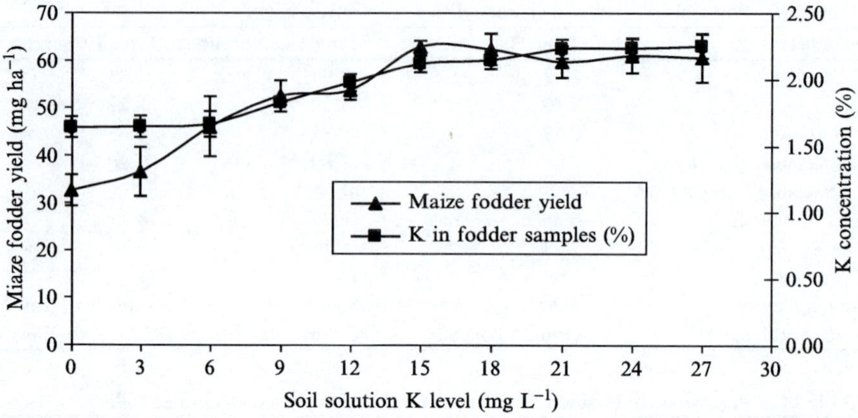

Fig. 24.1 Maize fodder yield and K concentration in plant samples as affected by various soil solution K levels

Green fodder yield increased significantly ($P<0.05$) in response to K application attaining maximum yield of 62.76 Mg ha^{-1} with T_6 where K was applied at 144 (average of 145–152) kg ha^{-1}. Adjusted soil solution K level for maximum yield was 15 mg L^{-1}. This treatment resulted in 47.99% higher green fodder yield as compared to control. Further increase of K rate could not improve maize green fodder yield and treatments remained statistically at par. A closer look on the data indicated that green fodder yield was over 60 Mg ha^{-1} with T_6 to T_{10} where corresponding K fertilizer rates were 145–192 kg ha^{-1}. This increase in fodder yield with K is a big achievement in a country where average green fodder yield is very low i.e. 22 Mg ha^{-1} (Economic Survey of Pakistan 2007–2008), and the existing livestock is facing 54% fodder shortage at national level.

Response of maize fodder to K application was found linear (Fig. 24.1), in which K concentration increased from 1.64% (please check) to the level best of 2.20% on highest K rate of 206 kg ha^{-1}. K concentration was 2.04% in T_6 when K was used at 152 kg ha^{-1}, coinciding with the maximum fodder yield (62.76 Mg ha^{-1}). From T_7 to T_{10} fodder yield remained statistically at par but dry matter content decreased from 26.87% (T_7) to 26.11% (T_{10}) when K concentration increased from 2.07% to 2.20%. Potassium is known to support plant turgidity in the guard cells surrounding stomata followed by an influx of K that reduces the aperture of the stomata, thus contributing to the regulation of water uptake and rate of transpiration; this factor might explain the fact of stagnant green fodder yield (T_7 to T_{10}) and decreasing trend of dry matter content with increasing level of K concentration in plant samples. The results are also comparable to the findings of Vicente-Chandler et al. (1962) who described that maximum dry matter yield of maize fodder was observed at 450 kg K ha^{-1} year^{-1} but K uptake continued to increase up to 900 kg ha^{-1} year^{-1} indicating luxury consumption which is not desirable as livestock particularly cattle need K in low concentration (NRC 1984). Luxury consumption of K may be avoided by split application yearly. McLeod (1965) however reported that K uptake increased from 0.8% to 3.4%

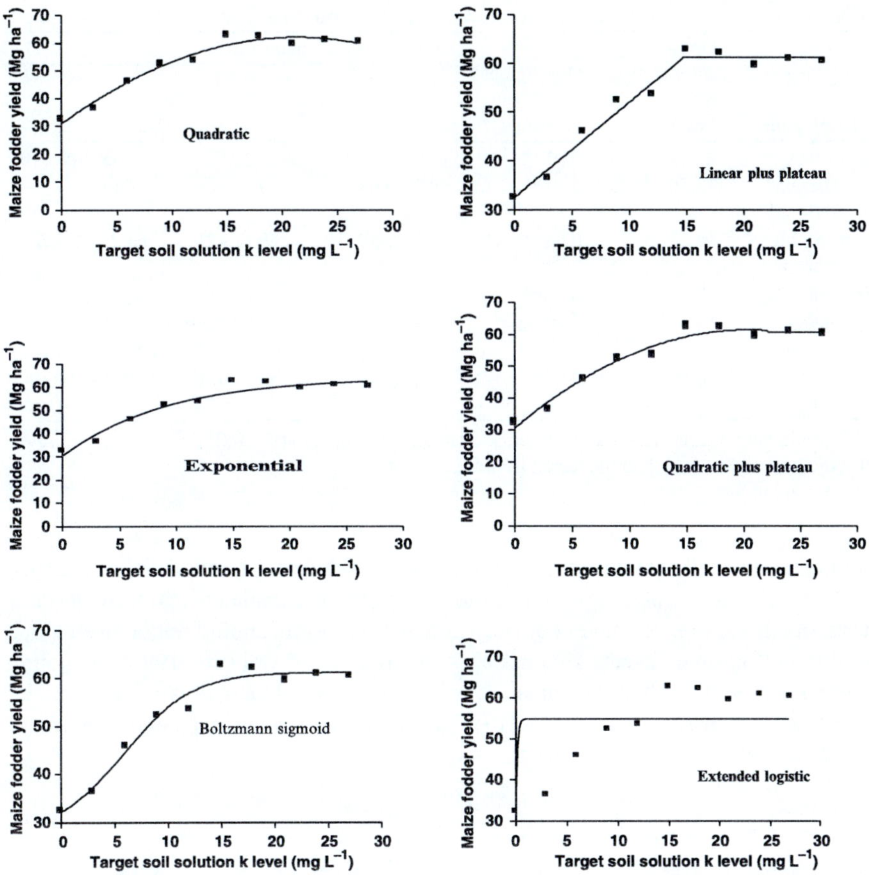

Fig. 24.2 Evaluation of maize fodder yield by different yield response models

with increased K rate but it could not affect forage yield. In addition to K supply the relatively high forage yields of the fertilized maize fodder crops were due to the factors like adequate supply of good quality canal water throughout the growing season, optimum supply of other plant nutrients, favorable temperature etc. As soil was deficient in available K status, the response of maize to K fertilization was anticipated.

7.3 Evaluation of Yield Response Models to Predict Optimum K Rate

In order to find out optimum K rate, maize fodder yield data was evaluated by various yield response models (Fig. 24.2 and Table 24.4). The optimum K rates varied greatly among the tested models. Slight difference among all models' R^2 was

Table 24.4 Optimum soil solution K levels predicted by various models

Model name	Optimum soil solution K level (mg L^{-1})	Coefficient of determination (R^2)	Coefficients of equations a	b	c
Quadratic	21.33	0.97**	30.80 (1.77)	2.92 (0.30)	−0.068 (0.010)
Boltzmann sigmoid	18.50	0.97**	27.58(6.68)	61.27 (1.28)	505 (201)
Linear plus plateau	14.50	0.98**	1.94 (0.20)	32.49 (1.43)	61.19 (0.83)
Exponential	20.50	0.95**	–	5.74 (1.76)	0.11 (0.030)
Extended logistic	[a]	0.40 ns	−0.65 (0.77)	7.30 (5.16×10^8)	–
Quadratic plus plateau	18.50	0.97**	30.51 (2.01)	3.03 (0.045)	−0.07 (0.02)

*Values in parenthesis are standard error of estimate (significant at $P = 0.05$)
[a]Predicted rate was less than the tested minimum K rate

observed ($R^2 \geq 0.92$**) with the exception of extended logistic model (Table 24.4). With similar R^2 values, however, large variations in calculated optimum K rates were obtained. The coefficient of determination therefore should not be used alone to find out optimal K rate. This matches the findings of Colwell (1994). Although models differed in their estimated standard error, yet linear plus plateau model expressed less biasness with respect to R^2 and S.E., therefore it was used for optimum K rate calculation in all other parameters.

7.4 Effect of K Rate on Dry Matter (DM) and Crude Protein

The DM increased to the plateau level of 26.87% with T_7 (160 kg K ha^{-1}) (Fig. 24.3) but it then decreased thereafter possibly due to dilution effect in both the crops. The DM content (%) was slightly higher in second than that of first crop. This might be the result of elevated residual K coupled with higher K fertilizer rate, which goes against the findings of Lioveras et al. (2001) who observed no significant effect of different K fertilizer rate (0, 42, 83, 166 and 332 kg ha^{-1}) on alfalfa DM yield under Mediterranean climate. Pravinchandra and Kotecha (2006), however, reported an enhancement in DM to the tune of 13.3% at 300 kg K ha^{-1}.

Crude protein (CP) is a key nutrition that must be considered in adequate amounts in various animal diets. Crude protein varied from 9.37% (T_1) to statistically significant level of 12.06% with T_9 while it declined to 12.03% in T_{10}. Increase upto 22.28% higher than control was observed with T_9 where 169 kg K ha^{-1} was added equivalent to solution K level of 24 mg L^{-1} (Fig. 24.3). CP concentration with both T_9 and T_{10} approached 12%, which is acceptable limit for dry cow rations (Thomas et al. 1998). For maize fodder, the plateau yield of CP to K rate was agreed with

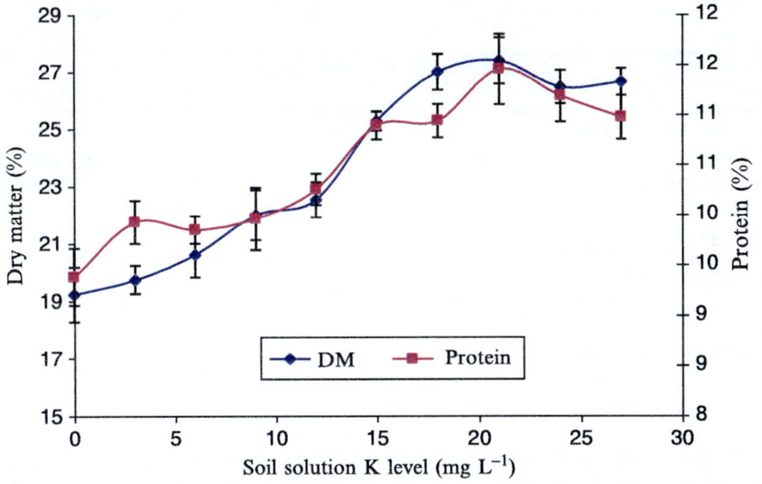

Fig. 24.3 Dry matter and crude protein content as affected by various soil solution K levels

work of Pravinchandra and Kotecha (2006) who reported 17.6% higher CP in Lucerne when 300 kg K ha^{-1} was used.

7.5 Optimum Soil K Solution Level for Maximum Output

For maximum green fodder yield a target soil solution K level required was found to be 14.50 mg L^{-1} (Table 24.5) which could be adjusted by adding 144 kg K ha^{-1} and predicted yield corresponding to 14.50 mg L^{-1} level was 61.19 Mg ha^{-1}. As no significant variation was observed regarding dry matter and crude protein contents both for soil solution and K rate, these were pooled together for the sake of convenience. To get maximum output with respect to dry matter and crude protein, a soil solution level of 20 mg L^{-1} would be sufficient and its equivalent K rate was 165 kg ha^{-1}. A different K rate is required for yield and quality parameters (dry matter and crude protein contents).

8 Conclusions

1. K fertilizer application based on soil solution level worked out through adsorption model provided the basis in resolving sporadic K response of crops on soils of mixed mineralogy as adsorption based K rates were different for the four selected soils
2. Linear plus plateau model predicted economic K rate with greater accuracy
3. Owing to the high K fertilizer prices, integrated approach of using adsorption and yield response models proved very useful in optimizing K fertilizer application

Table 24.5 Optimum Soil solution K levels for maximizing various maize fodder quality parameters

Variables	Soil solution K level (mg L^{-1})	Optimum K rate (kg ha^{-1})[a]	R^2	Predicted value*
Maize fodder yield (Mg ha^{-1})	14.50[a]	144[b]	0.98*	61.19[c]
Dry matter (%)	19.25[a]	164[b]	0.98*	26.83[c]
Crude protein content (%)	20.75[a]	167[b]	0.95*	11.19[c]
Cumulative average of dry matter and protein content	20	165	–	–

*Significant at $P=0.01$

[a]Average soil solution level for fodder yield, dry matter and crude protein content using adsorption model

[b]Values determined by linear plus plateau yield response model

[c]Predicted yielded parameters

in alkaline calcareous soils, for instance it required only 144 kg ha^{-1} (soil solution level of 14.50 mg KL^{-1}) to maximize fodder yield, however, considerably higher rate would be needed to harvest crop with maize fodder of best quality in terms of dry matter and crude protein, (soil solution level vary from 19.25 and 20.75 mg L^{-1})

4. The growers would have to decide one option out of quality or quantity depending upon their marketing system because one cannot have both maximum quantity and supreme quality at the same time.

References

Askegaard M, Hansen HCB, Schjoerring JK (2005) A cation exchange resin method for measuring long-term potassium release rates from soil. Plant Soil 271:63–74

Barber BA (1995) Barber soil nutrient bioavailability, 2nd edn. Wiley, New York

Bedrossian ST, Singh B (2002) Premature senescence in cotton in relation to potassium availability in soil. In: Proceedings of the 11th cotton conference, Brisbane, Convention and Exhibition Centre. 13–15 Aug 2002

Bilski JJ, Nelson DC, Conlon RL (1988) The response of four potato cultivars to chloride salinity, sulfate salinity and calcium in pot experiments. Am Potato J 65:85–90

Bolt GH, Sumner ME, Kamphorst A (1963) A study between three categories of potassium in an illitic soil. Soil Sci Soc Am Proc 27:294–299

Cakmak I (2002) Plant nutrition research: priorities to meet human needs for food in sustainable ways. Plant Soil 247(1):3–24

Chapman KSR, Sparrow LA, Hardman PR, Wright DN, Thorp JRA (1992) Potassium nutrition of Kennebec and Russet Burbank potatoes in Tasmania: effect of soil and fertilizer potassium on yield, petiole and tuber potassium concentrations, and tuber quality. Aust J Exp Agric 32(4):521–527

Colwell JD (1994) Estimating fertilizer requirements. A quantitative approach. CAB International, Wallingford

Du Y, Hayashi JS, Xu YF (2004) Some factors controlling the adsorption of potassium ions on clayey soils. Appl Clay Sci 27:209–213

During C (1984) Fertilizers and soils in New Zealand, 3rd rev edn. Government Printer, Wellington, New Zealand, 355pp

Economic Survey of Pakistan (2007–2008) http://www.finance.gov.pk/survey/home.htm
Edmeades DC (1982) Effects of lime on effective cation exchange capacity and exchangeable cations on a range of New Zealand soils. NZ J Agric Res 25:7–33
Erdem E, Karapinar N, Donat R (2004) The removal of heavy metal cations by natural zeolites. J Coll Interf Sci 280:309–314
Fontes RLF, Coelho HA (2005) Molybdenum determination in Mehlich-1 and Mehlich-3 soil test extracts and molybdenum adsorption in Brazilian soils'. Commun Soil Sci Plant Anal 36(17):2367–2381
Goulding KWT (1987) Potassium fixation and release. Proceedings of the Colloquium of the International Potash Institute 20:137–154
Grewal JS, Sharma RC, Saini SS (1992) Agrotechniques for intensive potato cultivation in India. Indian Council of Agricultural Research, New Delhi
Haby VA, Russelle MP, Skogley EO (1990) Testing soils for potassium, calcium, and magnesium. In: Westerman RL (ed) Soil testing and plant analysis. Soil Science Society of America, Madison, pp 181–228
Hannan A (2008) Evaluation of sorption isotherm based soil solution potassium concentration levels for maximizing crop yields. Ph.D. thesis, Institute of Soil and Environmental Sciences, University of Agriculture, Faisalabad
Hunter AH (1980) Laboratory and green house techniques for nutrient survey studies to determine the soil amendments required for optimum plant growth. Agro Services International, Orange City
Imtiaz M, Alloway B, John A, Memon M, Yousuf M, Parvez K, Saleem-ul-Haq S, Hussain SK (2006) Zinc sorption in selected soils. Commun Soil Sci Plant Anal 37(11):1675–1688
Lal B, Singh KB (1983) Effect of potassium on the quality of potato (*Solanum tuberosum* L.). Agric Sci Digest 3:30–33
Lioveras J, Ferran J, Boixadera J, Bonet J (2001) Potassium fertilization effects on alfalfa in a Mediterranean climate. Agron J 93:139–143
Maier NA, Dahlenburg AP, Williams CMJ (1994) Effects of nitrogen, phosphorus, and potassium on yield, specific gravity, crisp colour, and tuber chemical composition of potato (*Solanum tuberosum* L.) cv. Kennebec. Aust J Exp Agric 34:813–824
Malavolta E (1985) Potassium status of tropical and subtropical region soils. In: Munson RD (ed) Potassium in agriculture. American Society of Agronomy, Madison, pp 163–200
Malla PB (2002) Vermiculites. In: Dixon JB, Schulze DG (eds) Soil mineralogy with environmental applications. Soil Science Society of America, Madison, pp 501–529
McDole RE (1978) Potassium fertilizer tails with potatoes in coarse-textured soils in southern-eastern Idaho. Am Potato J 55:161–170
McLeod LB (1965) Effect of nitrogen and potassium on the yield and chemical composition of alfalfa, bromegrass, orchardgrass, and timothy grown as pure species. Agron J 57:261–266
Mengel K (1985) Dynamics and availability of major nutrients in soils. Adv Soil Sci 2:65–131
Metson AJ (1968a) Potassium. In: Soils of New Zealand. Part 2. New Zealand Soil Bureau bulletin 26. Wellington, New Zealand, Government Printer, pp 82–95
Munson RD (1985) Potassium in agriculture (ed) ASA, CSSA, Madison
Neiderbudde EA (1986) Factors affecting potassium release and fixation in soil. In: 13th international congress on soil science, Hamburg, FRG VI:1155–1167
NRC (1984) National Research Council. Nutrient requirements of beef cattle, 6th rev edn. National Research Council, National Academy Press, Washington, DC
Oborn I, Andrist-Rangel Y, Askekaard M, Grant CA, Watson CA, Edwards AC (2005) Critical aspects of potassium management in agricultural systems. Soil Use Manag 21:102–112
Pal Y, Wong MTF, Gilkes RI (1999) The forms of potassium and potassium adsorption in some virgin soils from South-Western Australia. Aust J Soil Res 37:695–709
Parfitt RL (1992) Potassium-calcium exchange in some New Zealand soils. Aust J Soil Res 30:145–158
Pravinchandra CP, Kotecha AV (2006) Effect of phosphorus and potassium application on forage yield, quality, nutrient uptake and persistence of lucerne (*Medicago sativa* L.). In: 18th world congress of soil science, 9–15 July 2006. Philadelphia, Pennsylvania, USA

Rhue RD, Hensel DR, Kidder G (1986) Effect of potassium fertilization on yield and leaf nutrient concentrations of potatoes grown on a sandy soil. Am Potato J 63:665–681

Satyanarayana V, Arora PN (1985) Effect of nitrogen and potassium on yield and yield attributes of potato (var. Kufri Bahar). Indian J Agron 30:292–295

Sharpley AN (1990) Reaction of fertilizer potassium in soils differing mineralogy. Soil Sci 149:44–51

Sparks DL (1987) Potassium dynamics in soil. Adv Soil Sci 6:1–63

Sparks DL (2000) Bioavailability of soil potassium D-38-D-52. In: Sumner ME (ed) Handbook of soil science. CRC Press, Boca Raton

Sparks DL, Huang PM (1985) Physical chemistry of soil potassium. In: Munson RD (ed) Potassium in agriculture. Soil Science Society of America, Madison, pp 201–276

Sparks DL, Zelazny LW, Martens DC (1980) Kinetics of potassium exchange in a Paleudult from the coastal plain of Virginia. Soil Sci Soc Am J 44:37–40

Sposito G (1980) Derivation of the Freundlich equation for ion-exchange reactions in soils. Soil Sci Soc Am J 44:652–654

Sumner ME, Bolt GH (1962) Isotopic exchange of potassium in an illite under equilibrium conditions. Soil Sci Soc Am Proc 26:541–544

Surapaneni A, Palmer AS, Tillman RW, Kirkman JH, Gregg PEH (2002) The mineralogy and potassium supplying power of some loessial and related soils of New Zealand. Geoderma 110(3–4):191–204

Thomas ED, Sniffen CJ, Allshouse RD, Ballard CS, Majewski CJ (1998) Fertilization of forage grasses for dry dairy cows. Proceeding V 7. American Forage and Grassland Council, Indianapolis, pp 134–138

Vicente-Chandler J, Pearson RW, Abruna F, Silva S (1962) Potassium fertilization of intensively managed tropical grasses under humid tropical conditions. Agron J 54:450–453

Westermann DT, Tindall TA, James DW, Hurst RL (1994) Nitrogen and potassium fertilization of potatoes: yield and specific gravity. Am Potato J 71:417–431

Xu, Ren-Kou, An-Zhen Z, Guo-Liang J (2005) Effect of low molecular weight organic anions on adsorption of potassium by variable charge soils. Comm Soil Sci Plant Anal 36:1029–1039

Yunda AL, Guzman G, Ferrucho A, Shelter LR (1997) Isotherms of potassium adsorption in a level oxisol of the eastern zones of Colombia. Colomb Mag Chem 26:1–7

Chapter 25
Optimal Supply of Micronutrients Improves Drought Tolerance in Legumes

Muhammad Yasin Ashraf, Khalid Mahmood, Muhammad Ashraf, Javed Akhter, and Faqir Hussain

Contents

Abstract Legumes are of considerable importance for providing food and feed world over. In comparison to cereal grains, legume seeds are rich in protein and thus provide highly nutritive food. Legumes are grown on a wide range of soils varying in texture and fertility. Most of the soils of arid and semi-arid regions, being low in soil moisture content, are also low in fertility. So to maximize plant productivity, proper supply of macro- and micro-nutrients to crops is essential. As a general practice, optimal supply of macronutrients to crops is usually ensured but that of

M.Y. Ashraf (✉) • K. Mahmood • J. Akhter • F. Hussain
Nuclear Institute for Agriculture and Biology (NIAB), Faisalabad, Pakistan
e-mail: niabmyashraf@gmail.com; kmahmoodniab@yahoo.com; javedakhterniab@yahoo.com; fhussainfsd@yahoo.com

M. Ashraf
Department of Botany, University of Agriculture, Faisalabad, Pakistan

Department of Botany and Microbiology, King-Saud University, Riaydh, Saudi Arabia
e-mail: ashrafbot@yahoo.com

M. Ashraf et al. (eds.), *Crop Production for Agricultural Improvement*,
DOI 10.1007/978-94-007-4116-4_25,

micronutrients is ignored. In view of a plethora of literature, it is now well established that application of micronutrients is effective in alleviating the adverse effects of abiotic stresses such as salinity and drought. The involvement of micronutrients in different physiological and biochemical activities of the legume plants is well documented because correlations between micronutrient supply and crop growth and productivity have been often observed. Use of micronutrients like zinc (Zn), iron (Fe), boron (B), manganese (Mn), molybdenum (Mo), copper (Cu), cobalt (Co) and nickel (Ni) has now become a common practice to increase crop yield especially under adverse environmental conditions. Plants deficient in micronutrients may become susceptible to diseases and abiotic stresses. Rapid leaching of acids in sandy soils tends to produce a deficiency of tightly held nutrients such as Zn, Fe, Cu or B. Therefore, problem soils such as acid, alkaline or sandy soils are often deficient in one or more micronutrient elements. Micronutrient application not only improves the stress tolerance potential indirectly (because micronutrients deficient plants exhibit an impaired defense response) but also results in improving a number of metabolic phenomena. Thus, application of micronutrients as foliar or soil amendment is recommended to achieve optimum crop productivity from the soils having inherent micronutrient deficiency and low moisture contents. In this chapter, the role of micronutrient management in improving the drought tolerance potential and productivity of legumes is reviewed and critically discussed.

Keywords Micronutrients • Legumes • Drought tolerance • Metabolic changes and plant productivity

1 Introduction

Economically effective and sustainable modern agriculture requires high yielding crops necessary to get reasonable profit so as to meet the high demand for food of the growing population of the world. Although scientists have developed high yielding crop varieties, deficiency of food is still there, which is attributed to unavailability of sufficient irrigation water for crops. Water shortage is a worldwide problem, which affects the growth and productivity of crops through reducing the availability of macro- and micro-nutrients necessary for photosynthesis and other metabolic phenomena (Mahajan and Tuteja 2005; Miletic et al. 2010). Mineral nutrients play an active role in minimizing the photo-oxidative damages caused by generation of reactive oxygen species (ROS) in plants growing under drought conditions (Cakmak 2005; Wang et al. 2005). Therefore, limited supply of water adversely affects plant growth and metabolic processes of the plants (Ashraf et al. 1998) and proper application of macro- and micro-nutrients leads to higher plant productivity under drought conditions (Hussain et al. 2004). Many reports are available which prove strong relationships between micronutrient availability, plant growth and productivity, and recommend proper supply of micronutrients such as zinc (Zn), manganese (Mn), copper (Cu), molybdenum (Mo), iron (Fe), and boron (B) (Mann et al. 2002; Bhadoria et al. 2003; Rashid and Ryan 2004; Welch and Graham 2004; Gupta 2005; He et al. 2005; Liang et al. 2007; Waraich et al. 2011). Several studies have shown that proper supply of micro-nutrients to crops resulted in more vigorous seedlings, lower vulnerability to plant diseases, and improved crop drought resistance/ tolerance (Frossard et al. 2000; Bouis 2003; Welch and Graham 2004). Introduction of high yielding crop varieties and liberal applications of irrigation water and fertilizers containing major nutrients (NPK), have increased the prevalence of micro-nutrient deficiencies in soils, and consequently, in human nutrition (Imtiaz et al. 2010).

In Pakistan, 22 Mha of cultivated area comprises predominantly alluvial soils which are alkaline, calcareous and deficient in micronutrients and organic matter (Imtiaz et al. 2010). Due to calcareous nature and high pH of soils, nutrients are fixed and become less available to plants. The most widespread deficiency is of Zn which is observed in 70% of the soils of Pakistan. Boron deficiency is another major nutritional disorder which severely affects growth and crop productivity in Pakistan (Rashid 2005). The third commonly prevalent disorder is Fe chlorosis which has been recorded in peanut, chickpea, mungbean and other legumes (Rourke et al. 2007). Copper and Mn deficiencies are also observed in some areas of Pakistan but the problem is of localized nature (Rashid 2005; Imtiaz et al. 2010).

Leguminous crops are grown all over the world for food (Franca et al. 2000). They contain considerable amount of protein, starch, oils and vitamins necessary for human growth and health. Legumes are mostly grown in the arid and semi-arid regions, where availability of water for crops is limited and yield losses due to water stress are very high (Farias-Rodriguez et al. 1998). Many reports (e.g. El-Hamdaoui et al. 2003; Ashraf and Iram 2005; Arrese-Igor et al. 2011) have shown that nodulation and nitrogen fixation of the legumes were severely affected by the water stress. Photosynthesis and stomatal conductance were also reduced due to the drought stress resulting in decreased biomass and productivity of legumes (Flexas et al. 2002; Taiz and Zeiger 2002; Purcell 2009). This chapter discusses the influence of micronutrients, particularly Zn, Fe, B, Mn, Cu, Mo, Co and Ni, on improvement in drought tolerance and productivity of plants for sustainable agriculture in drought-prone areas of the world.

2 Zinc

Zinc (Zn) is the most deficient micronutrient in the soils of many countries including Bangladesh, India, Nepal and Pakistan (Nayyar et al. 2001; Imtiaz et al. 2010), so its application either as foliar or soil amendment is recommended. Foliar application of Zn could be of great importance to plants grown under sandy soil conditions as most sandy soils suffer from micronutrients deficiency particularly that of Zn. Sandy soils in many semi-arid regions are known to limit mobility and availability of soil-Zn to plant roots (Marschner 1993; Imtiaz et al. 2006). Foliar application with Zn showed a significant effect on growth, yield, and yield components and seed quality of groundnut under sandy soil conditions (Gobarah et al. 2006).

Earlier, Karaman et al. (1999) reported that biomass production increased with increasing Zn concentrations applied to bean (*Phaseolus vulgaris* L.) plants. Foliar application of Zn or urea and their combination on mungbean improved the plant height by producing the tallest plants and increased the number of branches per plant as a result of which yield per plant improved (Abd-El-Lateef et al. 1998; Malakouti 2007; Rajaiea et al. 2009). Krishna (1995) also found that foliar application of Zn significantly enhanced the dry biomass, seed and straw yield, and seed protein contents in mungbean.

Foliar application of Zn showed a positive effect on growth, yield and yield components in mungbean grown under stress environments (Thalooth et al. 2006). Kassab (2005) indicated that foliar application of Zn, Mg, Mn and Fe significantly increased growth, yield and yield components of mungbean plants grown in soils with low moisture contents. Under arid condition, foliar application of Zn was effective in enhancing the growth and yield of mungbean (Gupta et al. 2003; Kassab 2005). Foliar application of Zn and K was found effective in improving the leaf area, number of leaves, biomass, number of pods and seed weight in mungbean grown in soils having limited moisture contents (Thalooth et al. 2006). Earlier, Basole et al. (2003) conducted an experiment on hormones and nutrient management of soybean and recorded a significant improvement in growth and yield of soybean when treated with Zn or Zn + hormones. Ali and Mowafy (2003) reported that application of Zn (2%) as foliar spray improved yield and seed quality of peanut when grown in sandy soils. Similarly, there are many reports which indicated that foliar application of Zn had a significant effect on growth, yield and seed quality of groundnut under sandy soil conditions (Gobarah et al. 2006; El-Tohamy and El-Greadly 2007). The above discussed findings of different workers clearly indicate that proper application of Zn to either soil or foliage is effective in enhancing the stress tolerance potential in different legume crops as a result of which enhanced crop yield can be achieved. The recommendations of various workers regarding Zn application in legumes to obtain optimal crop productivity are summarized in Table 25.1.

3 Iron

Iron (Fe) is essential micronutrient for a number of key enzymes of the nitrogenase complex in legume plants, and also for the electron carrier, i.e. ferredoxin (Rashid 2005; Imtiaz et al. 2010). Legumes also have a relatively higher iron requirement for the synthesis of heme, the component of haemoglobin, and nodule formation. Under iron deficiency, nodule mass, leghaemoglobin content, number of bacteroids and nitrogenase activity are drastically reduced (Tang et al. 1992; Scherer et al. 2008). Under water deficit conditions, iron deficiency caused chlorosis in peanut and nodule formation also failed until adequate iron supply through foliar applications. Parkpian and Boonkerd (1989) reported that nodule development in groundnut was decreased under iron deficiency in black calcareous soils. However, population of bacteria (*Bradyrhizobium*) in the soil and rhizosphere was not

Table 25.1 Recommendations and impact of Zn application on improving growth, yield and other parameters of legumes under drought conditions

Legume species	Recommended rate of Zn	Effect of Zn fertilizers	Reference
Chickpea (*Cicer arietinum*)	2.5 µg Zn g^{-1} soil	Improved biomass production, water relations and osmotic adjustment	Khan et al. (2004)
Snap bean (*Phaseolus vulgaris*)	Zn (chelated form 13%), 0.3–0.5 g L^{-1} foliar application	Increased chlorophyll and carbohydrate contents in pods and reduced fibers	El-Tohamy and El-Greadly (2007)
Lentil (*Lens culinaris*) and chickpea (*Cicer arietinum*)	12–15 kg $ZnSO_4$ ha^{-1} through soil	Improved growth and yield	Ramakrishna et al. (2000)
Lentil (*Lens culinaris*) and chickpea (*Cicer arietinum*)	25 kg $ZnSO_4$ ha^{-1} through soil	Improved nodulation, root growth, yield and uptake of Zn, B, Fe and P	
Mungbean (*Vigna radiata*)	0.5% Fe, Zn and Mn applied as foliar spray	Improved chlorosis caused by abiotic stress and increased number of pods and seed yield	Unkasem and Tawonsok (1988)
Mungbean (*Vigna radiata*)	300 mg kg^{-1} Zn-EDTA as foliar spray	Improved yield and yield components	Thalooth et al. (2006)
Groundnut (*Arachis hypogaea*)	Foliar spraying with Zn, tap water (control), 0.50, 0.75 and 1.00 g Zn L^{-1}	Increased oil percentage, protein and seed yield	Gobarah et al. (2006)
Groundnut (*Arachis hypogaea*)	1,000 mg L^{-1} $ZnSO_4$ as foliar spray	Increased oil and protein contents in seeds and improved overall yield	Darwish et al. (2002)

influenced by Fe deficiency and root infection by *Bradyrhizobium* was also not affected. Foliar application of Fe resulted in greater number of nodules and thus increased the nodule mass in plants. As a result of Fe application, nodules in sprayed plants contained 200 times higher bacteroids per unit weight, and 14 times higher leghaemoglobin content. Deficiency of Fe delayed the nitrogenase activity in all the tested groundnut varieties (Marschner 2011).

Iron deficiency in crop plants can be overcome commonly by foliar application of inorganic salts or chelated compounds of Fe. Five foliar applications of 0.5% $FeSO_4$ solution at 10-day intervals starting at 10 days after emergence was found a very effective measure in alleviating iron chlorosis in peanut plants with low moisture contents, however, yield was also substantially improved (Ratanarat et al. 1990). Chlorotic symptoms are sometimes commonly observed in mungbean cultivars grown on calcareous soils. Mungbean plants readily recovered from chlorosis after a foliar application of nutrient solution containing 0.5% Fe, Zn and Mn and also produced greater number of pods (Unkasem and Tawonsook 1988). The method of foliar applications has the drawback that translocation of applied Fe within the plant is poor. The translocation rate is species-dependent and hardly exceeds 50% of the Fe applied to a given leaf or leaflet. Under field conditions, spraying is often required at frequent intervals to provide adequate Fe to the developing canopy, because the Fe translocation from previously treated areas is insufficient (Chen and Barak 1982). Mahmoudi et al. (2005) conducted studies with lentil and chickpea in solution culture experiment to examine the influence of iron on growth and chlorosis caused due to Fe deficiency. They reported a significant improvement in growth and reduction in chlorosis by the application of 30 μM Fe in both legumes. However, the effect was more pronounced in chickpea than in lentil. Production of legumes especially groundnut was reduced due to the Fe deficiency when grown on black calcareous soils of Thailand. Although these are considered fertile soils, due to $CaCO_3$ and alkaline pH, deficiency of Fe may result in legume crops (Osotsapar 2000). Therefore, all the earlier mentioned studies prove that the Fe-deficient or calcareous soils with high pH reduce the uptake of Fe which results in reduced crop photosynthesis, growth and yield. So, a proper supply of iron is necessary to have optimal legume crop yield from soils with low moisture contents (Table 25.2).

4 Boron

Boron (B) is an essential micronutrient for plants because it is involved in the activation of enzymes necessary for starch synthesis and hence for cellulose production (Blevins and Lukaszewski 1998; El-Hamdaoui et al. 2003; Hänsch and Mendel 2009). It plays an active role in the transport of sugars from the site of synthesis to meristem regions of stem and roots (Nable et al. 1997; Bolaños et al. 2006). So, the transport of sugars is retarded in boron-deficient plants as a result of which plant growth is reduced (O'Neill et al. 2004). Boron is also involved in cell multiplication and nitrogen metabolism (Sinha and Chatterjee 2003). Nutrient

Table 25.2 Recommendations and impact of Fe application on improving growth, yield and other parameters of legumes under drought conditions

Legume species	Recommended rate of Fe	Effect of Fe fertilizers	Reference
Lentil (*Lens culinaris*)	0.5 kg $FeSO_4$ ha^{-1} through soil	Improved yield and nutrient uptake	Ramakrishna et al. (2000)
Lentil (*Lens culinaris*) Chickpea (*Cicer arietinum*)	Application in growth medium, 30 μM Fe	Decreased chlorosis and enhanced growth of both the legumes	Mahmoudi et al. (2005)
Mungbean (*Vigna radiata*)	0.5% Fe, Zn and Mn applied as foliar spray	Reduced chlorosis caused by water stress and increased number of pods and seed yield	Unkasem and Tawonsok (1988)
Groundnut (*Arachis hypogaea*)	Foliar spray of 0.1% aqueous solution of Fe, 2 kg Fe ha^{-1} as iron citrate or Iron sulfate 400 g ha^{-1}	Improved yield and yield components	Singh and Joshi (1997)
Groundnut (*Arachis hypogaea*)	Foliar spray of 0.5% $FeSO_4$	Alleviated chlorosis caused by water stress in plants and improved yields	Ratanarat et al. (1990)
Peanut (*Arachis hypogaea*)	Foliar application of 0.5% $FeSO_4$	Reduced chlorosis and increased growth	Parkpean et al. (1986)

uptake is generally reduced in soils having low moisture contents. Application of B activates the absorption of other nutrients, photosynthetic efficiency, phosphorus uptake, hormone synthesis and fat metabolism (Dell and Huang 1997; Tewari et al. 2009). All these metabolic processes are helpful in sugar transport throughout the plant body (Brown and Shelp 1997; Boaretto et al. 2008; Hänsch and Mendel 2009). Boron availability in soils varies and is related with the type and degree of weathering of the parent material. Sandy soils are often deficient in B while clay soils with high organic matter contain enough amount of B (Barber 1995). Crops suffer boron deficiency commonly during the periods of drought probably because the supply of B to the root is reduced as a result of reduction in water flow to the roots. Decrease in soil moisture may cause a proportionate reduction in the rate of B diffusion to the root (Barber 1995; Boaretto et al. 2008; Tewari et al. 2009). Boron deficiency in legumes is common under drought conditions. Rerkasem et al. (1993) reported that deficiency of B decreased the seed yield up to 45% in peanut, 60% in soybean and 93% in chickpea. They also noted the depression resembling to hollow heart on the inner surface of cotyledons of soybean and peanut seeds, which was 50% in peanut and 17% in soybean seeds while black gram seeds did not show these types of symptoms. Application of B significantly reduced or even eliminated these symptoms in some varieties of the above-mentioned legume species (Rerkasem et al. 1993; Marsalis et al. 2009).

Extensive studies were conducted on uncovering the effect of boron (B) on peanut and soybean on a Typic Tropaqualf in Northern Thailand. For a comparison, sunflower, green gram, black gram, wheat, and rice were also included in B rate trial. Boron omission induced the hollow heart symptom in 10% of peanut kernels; the incidence of symptoms was correlated with B contents of the affected kernels. On the other hand, omission of B did not affect the appearance of soybean seed or the grain yield of both crops (soybean and peanut). In the experiment on B application rate, omission of B reduced grain yield by 50% in sunflower and by 40–80% in black gram. B deficiency symptoms in green gram and the hollow heart symptom in peanut kernels were observed in the B omitted treatment, but the grain yield of soybean, peanuts, rice, or wheat was not affected. Reduction in grain yield in black gram and green gram under B deficiency was apparently due to delay or inhibition of reproductive development and thus reduced pod set (Rerkasem et al. 1993; Osotsapar 2000).

Keerati-Kasikorn et al. (1987) reported boron deficiency in 28% of peanut crops studies in dry season in Khon Kaen province of Thailand. Hollow heart symptom was recorded in about 50% seeds of the B deficient crops during the dry season and 30% of seeds in wet season. Ninety-five percent of normal seeds had higher B content (>12 μg B g^{-1}), whereas 70% of the seeds with hollow heart had lower B content. The average B content of normal peanuts was 16.4 μg B g^{-1} compared with the seeds with hollow heart having 10.6 μg B g^{-1}. A comparative study on the susceptibility to B deficiency of various soybean, peanut and black gram cultivars indicated a wide range of variation related to the genotypic response. In B-deficient soils, depression in seed yield was 30–60% in soybean, 45% in peanut and 93% in black gram. Boron deficiency also induced a localized depression on the internal surface of one or both cotyledons of some soybean seeds, resembling the symptom of hollow heart in peanut seeds. It induced 50% hollow heart in peanut, 17% in soybean, while black gram seeds had no symptoms. The addition of boron reduced or eliminated the symptoms (Rerkasem et al. 1993; Marsalis et al. 2009).

El-Hamdaoui et al. (2003) investigated the effect of different levels of B (from 9.3 to 93 μM B) and Ca (from 0.68 to 5.44 mM Ca) on growth, nitrogen fixation, and mineral composition of pea (*Pisum sativum* L., cv. Argona) grown in saline conditions. Addition of B and Ca in the growth medium alleviated the adverse effects of salinity (75 mM NaCl) on plant growth, nodulation and nitrogen fixation. The increase in salt tolerance of symbiotic plants mediated by B and Ca can be correlated with the availability of nutrients because both B and Ca contents reduced in shoots and in nodulated roots due to salinity. Increase in K and Fe was also recorded in shoot and root of pea plants due to addition of B and Ca. So, application of B and Ca is beneficial in improving the growth and nutrient uptake in pea under saline conditions. Recommendations of different workers regarding B for enhancing crop yield and stress tolerance potential in different legumes are summarized in Table 25.3.

5 Manganese

Manganese (Mn) is a very important micronutrient because it catalyses various enzymatic activities and plays an active role in electron transport. Supplemental Mn stimulates the N_2-fixation in soybean under water deficit conditions (Purcell et al. 2000; Vadez et al. 2000; Sinclair et al. 2003; Khoshgoftarmanesh et al. 2010). It is also involved in antioxidative metabolism (Polle et al. 1992; Khoshgoftarmanesh et al. 2007). Manganese deficiency is very common in arid or semi-arid soils, and the cultivation of leguminous crops is severely affected on these soils. However, Mn supplementation is effective in improving crop yield (Purcell et al. 2000) through regulating nitrogen fixation because Mn is also important in the ureide degradation in leaves of soybean. Increases in the N_2-fixation by Mn application in soybean under water deficit was also reported by Vadez et al. (2000) and Sinclair et al. (2003). Drought sensitive cultivars required higher amount of Mn than that did the tolerant ones (Todd et al. 2006; Khoshgoftarmanesh et al. 2007). In a drought sensitive soybean cultivar, xylem flow reduced due to low water potential resulting in decrease in Mn delivery in leaves necessary for degradation of ureide enzymes as a result of which ureides build-up takes place. Use of Mn for effectively enhancing growth and yield of leguminous crops suggested by various researchers is summarized in Table 25.4.

Table 25.3 Recommendations and impact of B application on improving growth, yield and other parameters of legumes under drought conditions

Legume species	Recommended rate of B	Effect of B fertilizers	Reference
Peanut (*Arachis hypogaea*)	Soil application 0.25–0.50 kg B ha^{-1}	Increased pod and seed number as well as seed yield per plant	Keerati-Kasikorn et al. (1987)
Lentil (*Lens culinaris*) and chickpea (*Cicer arietinum*)	0.5–1.0 kg B ha^{-1} as soil application	Improved yield and nutrient uptake	Ramakrishna et al. (2000)
Black gram (*Vigna mungo*)	Foliar spray of 50 g ha^{-1} borax at the time of flowering and at pod setting	Improved seed size and yield	Rerkasem (1989)
Green gram (*Vigna radiata*) and black gram (*Vigna mungo*)	Soil application of 4–32 kg ha^{-1} borax in case of green gram and 4–16 kg ha^{-1} borax in black gram	Increased seed yield	Rerkasem et al. (1990)
Pea (*Pisum sativum* cv. Argona)	Foliar spray of 55.8 μM B	Increased number of nodules and nitrogen fixation in pea under water deficit conditions induced by salinity	El-Hamdaoui et al. (2003)
Pea (*Pisum sativum*)	9.3–93 μM B	Improved germination and seedling development	Bonilla et al. (2004)
Pea (*Pisum sativum*)	Foliar spray of 55.8 μM B	Improved nitrogen fixation and Fe uptake	Bolaños et al. (2006)
Soybean (*Glycine max*)	Foliar application of 30 μM B	Improved growth, biomass, pod and seed yield, concentration of boron and reduced ribonuclease peroxidase, acid phosphatase and starch phosphorylase activities	Sinha and Chatterjee (2003)

Table 25.4 Recommendations and impact of Mn and Mo application on improving growth, yield and other parameters of legumes under drought conditions

Legume species	Recommended rates of Mn or Mo	Effect of micronutrient fertilizers	Reference
Soybean (*Glycine max*)	Foliar application of 52.8 mM Mn	Improved growth and nitrogen fixation under water stress conditions	Vadez et al. (2000)
Soybean (*Glycine max*)	Addition of 90 μM Mn in nutrient solution	Decreased ureides accumulation and improved growth and N_2-fixation	Izaguirre-Mayoral and Sinclair (2005)
Faba bean (*Vicia faba*)	Foliar application of 0.1% Mn	Improved growth and productivity under adverse conditions created by NaCl salinity	El-Fouly et al. (2001)
Soybean (*Glycine max*)	Soil application of 390–780 g Mo ha^{-1}	Increased nodulation and number of effective nodules per plant	Tenywa (1997)
Cowpea (*Vigna unguiculata*)	Soil application of 0.5 kg Mo ha^{-1}	Increased nitrogen fixation	Yakubu et al. (2010)
Bambara groundnut (*Vigna subterranean*)			
Groundnut (*Arachis hypogaea*)			
Common bean (*Phaseolus vulgaris*)	Inoculation with 12 g Mo kg^{-1} of seeds	Improved number of pods per plant, number of seeds per plant, 100-seed weight and seed yield	Bambara and Ndakidemi (2010)
Groundnut (*Arachis hypogaea*)	Foliar application of 100 g Mo ha^{-1} as MoO_3	Increased number of nodules, nodule weight and nitrogenase activity	Hafner et al. (1992)

6 Molybdenum

Molybdenum (Mo) regulates the enzymatic action in oxidation reduction reactions in plants (Mendel and Hansch 2002; Tan et al. 2010). It is also involved in N_2-fixation in leguminous plants (Williams and Frausto da Silva 2002). Molybdenum deficiency resulted in poor growth and decrease in chlorophyll content in plants (Marschner 1995; Kaiser et al. 2005). Application of Mo in deficient soils improved the nodulation and N_2-fixation by enhancing the nitrogenase activities (Adams 1997; Bambara and Ndakidemi 2010). So, deficiency of Mo resulted in deficiency of N which can be observed from the N deficiency symptoms and application of NH_4 or Mo is beneficial to regulate all the above mentioned metabolic activities. Similarly, supplemental application of Mo is necessary for active nodule system and optimal plant growth and productivity. Chongpraditnun et al. (1997) reported that the foliar application of Mn increased Mo uptake and translocation by reducing the adsorption of Mo to the surface of cell wall or cell membrane of leaves which may increase the movement of Mo into the phloem. Foliar application of ammonium molybdate significantly increased biomass and N_2-fixation in clover (El-Bably 2002) and in cowpea (Ndakidemi 2005). Molybdenum is an essential component of two major enzymes, nitrogenase and nitrate reductase in plants (Beevers and Hagenman 1983; Khan et al. 1990; Ashraf et al. 1995; Westermann 2005). Nitrogen fixation is regulated by nitrogenase which also helps the bacteria in fixing N_2 (Mengel and Kirkby 1987; Thibaund 2005; Westermann 2005) and for these activities Mo is necessary. Hafner et al. (1992) reported that Mo application increased the growth and symbiotic N_2-fixation in groundnut (*Arachis hypogaea* L.). Application of fertilizer containing 0.1% Mo was effective in enhancing the pod yield from 37% to 86% and total N from 53 to 108 kg N ha^{-1}. Foliar spray of Mo (100 g Mo ha^{-1} as MoO_3) increased nitrogenase activity by 2–4 times as compared to non-sprayed control plants. The increase in nitrogenase activity was mainly due to increase in nodule dry weight and to a lesser extent to increase in specific nitrogenase activity per unit nodule dry weight. The recommendations of various workers regarding Mo application through soil or as foliar spray are given in Table 25.4.

7 Copper

Calcareous or sandy soils are usually deficient in copper (Cu) because they have low moisture contents (Aref 2011). As most of the plants require very low amount of Cu, soil scientists do not recommend its application which may be the reason of its deficiency. The effect of Cu deficiency on crops is a significant reduction in yield and sometimes complete failure of a crop (Nascimento et al. 2003). However, copper application resulted in yield increase under stress environments (Khoshgoftar and Hajimozaffari 2006; Khoshgoftarmanesh et al. 2007). It was recorded that 53–62% of Cu uptake in some legumes is due to uptake by mycorrhizal hyphae (Barber 1995).

Peanut plants have been reported to depend on vesicular abuscular mycorrhizae for Cu supply in degraded soils (Bell et al. 1990; Quilambo 2003). Its concentration, 1–1.5 mg kg^{-1} (dry weight basis), in the shoot tips promotes vegetative growth and yield of peanut (Bell et al. 1990). Higher rates of phosphate fertilizer application may enhance the Cu deficiency (Tiaranan et al. 1985). Marked reduction in nodule development and N_2-fixation was recorded in the Cu-deficient legumes (Snowball and Robson 1980). It was also observed that foliar application of Cu increased dry matter yield and water use efficiency, respectively (El-Bably 2002). The level of Cu in soil solution decreases with increasing pH due to stronger adsorption (Reddy et al. 1995; Georgaka and Spanos 2010). If the deficiency is due to pH imbalance, addition of micronutrients can adversely affect the plant and metabolic activities because the level of individual micronutrients may affect the level of other micronutrients in the plant through antagonism. To get benefits from Cu application, recommendations are given in Table 25.5. On the other hand, toxicity can occur when micronutrients such as Cu are applied in excess.

8 Cobalt

Cobalt (Co) is considered to be very essential element for legumes because its proper supply is necessary for nodule formation and N_2-fixation processes (Bonilla and Bolaños 2010). Bolacer et al. (2003) reported its active role in the symbiotic N_2-fixation mechanism. The presence of cobamide coenzymes in the nodules of many legumes and nitrogen fixing non-legumes also confirmed its involvement in N_2-fixation (Gad 2006). Earlier, Castro et al. (1997) reported that *Phaseolus* seeds treated with Co showed a significant effect on nodulation, dry weight, physiological quality, vigour, protein and nutrient content of seeds. Due et al. (1999) and Tenywa (2003) observed that Co application enhanced nodulation and number of effective nodules in soybean plants and N contents in shoots. Sharma and Bhandari (2002) showed that seed inoculation with *Bradyrhizobium*, farmyard manure (FYM) and minerals significantly increased the number of nodules, their fresh weight, leghaemoglobin content of nodules and chlorophyll content in cowpea plants. The highest performance of all these parameters was recorded in Co alone or Co+FYM treatments. Abd El-Moez and Gad (2002) reported that the addition of Co along with inorganic and organic fertilizers increased the uptake of macro- and micro-nutrients in cowpea plants. Howell and Skoog (1955) observed that Co promoted stem coleoptile elongation, opening of hypocotyle hooks, and leaf expansion and bud development in peas. Gad (2006) conducted pot and field experiments to examine the effect of Co on growth and nutrient uptake in pea plants and noted that application of Co (8 mg L^{-1}) significantly enhanced the plant height, number of branches and leaves, leaf area, stem diameter, and root length. Cobalt application improved nodule formation process and increased the efficiency of organic and inorganic nitrogen fertilizers (urea and ammonium nitrate) and peanut compost in pea plants (Tenywa 2003). Addition of Co in the growth media significantly improved nutrients

Table 25.5 Recommendations and effects of Cu, Co and Ni application on improving growth, yield and other parameters of legumes under drought conditions

Legumes species	Recommended rate of micronutrients	Effect of micronutrients fertilizers	Reference
Peanut (*Arachis hypogaea*)	20 kg $CuSO_4$ ha^{-1}	Total dry matter and seed yields were greatly improved	Chew et al. (1979)
Peanut (*Arachis hypogaea*)	40 kg $CuSO_4$ ha^{-1}	Increased pod yield , haulm yield and oil yield	Desai (1994)
Soybean (*Glycine max*)	Copper ore tailings 600 g ha^{-1}	Increased seed and oil yield and uptake of all nutrients	Virupaksh (1995)
Chickpea (*Cicer arietinum*)	Copper ore tailings 600 g ha^{-1}	Improved chickpea grain yield and nutrient contents/uptake	Jadi (1997)
Soybean (*Glycine max*)	454 g Co ha^{-1}	Increased nodulation and number of effective nodules per plant	Tenywa (1997)
Pea (*Pisum sativum*)	8 mg Co L^{-1} through fertigation	Increased N fertilizer use efficiency, growth, biomass and pod size and seed yield	Gad (2006)
Peanut (*Arachis hypogaea*)	50 mg Co kg^{-1} of soil	Altered the activities of antioxidant enzymes i.e. catalase, peroxidase and polyphenol oxidase	Jaleel et al. (2008)
Groundnut (*Arachis hypogaea*)	Soil application, 0.21 kg Co ha^{-1}	Increased 10% kernel yield	Basu and Bhodaria (2008)
Soybean (*Glycine max*)	Fertigation with 5–8 mM Ni	Increased growth, urease and hydrogenase activities under Ni deficient environments	Klucas et al. (1983)
Soybean (*Glycine max.*)	Ni 1 μg L^{-1} in nutrient solution	Improved growth, urease and hydrogenase activities under adverse environmental conditions	Eskew et al. (1983)

uptake in plants even under low moisture availability conditions. Its application can reduce the requirement of organic and inorganic fertilizers up to 67% and 25%, respectively, of the recommended doses. It also enhanced the fresh and dry weight of shoots and roots, nodule number and weight, uptake of N, P, K, Fe, Mn and Zn as well as yield, number of pods and seeds quality (Tenywa 2003). Recommendations regarding foliar or soil application of Co are summarized in Table 25.5.

9 Nickel

Nickel (Ni) is an essential element for several biological processes, like H_2 oxidation and urea hydrolysis in plants (Bromilow et al. 1993; Zahran 1999; Coutinho and Mazo 2005; Zobiole et al. 2010). Klucas et al. (1983) reported that the application of Ni @ 5 to 8 mM is effective in increasing the 7- to 10-fold urease activity in soybean grown with nitrate or symbiotically; it significantly increased the hydrogenase activity in leaves and isolated nodule bacteroids. Zobiole et al. (2010) also reported that free-living *Rhizobium japonicum*, cultured under chemolithotrophic conditions, needed Ni application for growth and activation of hydrogenase activity. Ni is essential micronutrient element for the proper growth of soybean plant and *R. japonicum* (Ureta et al. 2005). Eskew et al. (1983), Taiz and Zeiger (1998) and Ureta et al. (2005) reported that soybean plants deficient in Ni showed higher accumulation of urea and as a result necrotic lesions appeared on their leaflet tips. Nickel deficiency resulted in delayed nodulation and reduction in early growth. Application of Ni @ 1 $\mu g\ L^{-1}$ to the nutrient media prevented urea accumulation, necrosis, and reduction in growth. This proved that Ni is essential for soybeans and possibly for all legumes. The available recommendations regarding Ni application for crop improvement are given in Table 25.5.

10 Conclusion

Micronutrients are essential for legumes grown in soils with low moisture contents. Their application is recommended for most legume species particularly to counteract deleterious effects of drought stress on growth, yield and metabolic activities. Application of micronutrients increases plant growth, nitrogen fixation, dry matter and seed yield of legumes. Moreover, they are essential component of enzymes, photoelectron chain and activate various metabolic activities in legumes in response to drought stress. Overall, application of micronutrients through foliage or soil is necessary to obtain optimal productivity of crops grown in soils with low moisture contents. Proper application of micronutrients to crops may thus help meet the food demands of ever-growing population of the world.

References

Abd-El-Lateef EM, Ashour NI, Farrag AA (1998) Effect of foliar spray with urea and some micronutrients on mungbean (*Vigna radiata* (L.) Wilczek) growth, yield and seed chemical composition. Bull Nat Res Center Cairo 23(2):219–232

Abd El-Moez MR, Gad N (2002) Effect of organic cotton compost and cobalt application on cowpea plants growth and mineral composition. Egypt J Appl Sci 17:426–440

Adams JF (1997) Yield response to molybdenum by field and horticultural crops. In: Gupta UC (ed) Molybdenum in agriculture. Cambridge University Press, Cambridge

Ali AAG, Mowafy SAE (2003) Effect of different levels of potassium and phosphorus fertilizer with the foliar application of zinc and boron on peanut in sandy soil. Zigzag J Agric Res 30(2):335–358

Aref F (2011) Iron, copper and manganese concentration in maize leaf as influenced by soil and foliar application of zinc sulfate and boric acid. Int J Acad Res 3:1080–1087

Arrese-Igor C, Gonzalez EM, Marino E, Gil-Quintana E (2011) Physiological responses of legumes nodules to drought. Plant Stress 5:24–31

Ashraf M, Iram A (2005) Drought stress induced changes in some organic substances in nodules and other plant parts of two potential legumes differing in salt tolerance. Flora 200:535–546

Ashraf MY, Azmi AR, Khan AH, Naqvi SSM, Ala SA (1995) Effect of water stress on different enzymatic activities in wheat. Acta Physiol Plant 17(4):315–320

Ashraf MY, Ala SA, Bhatti AS (1998) Nutritional imbalance in wheat (*Triticum aestivum* L.) genotypes grown under soil water deficit. Acta Physiol Plant 20:307–311

Bambara S, Ndakidemi PA (2010) *Phaseolus vulgaris* response to *Rhizobium* inoculation, lime and molybdenum in selected low pH soils in Western Cape, South Africa. Afr J Agric Res 5(14):1804–1811

Barber SA (1995) Soil nutrient bioavailability: a mechanistic approach. Wiley, New York

Basole VD, Deotale RD, Ilmulwar SR, Raut SS, Kadwe SB (2003) Effect of hormone and nutrient on morpho-physiological characters and yield of soybean. J Soils Crops 13:135–139

Basu M, Bhadoria PBS (2008) Performance of groundnut (*Arachis hypogaea* L) under nitrogen fixing and phosphorus solubilizing microbial inoculants with different levels of cobalt in alluvial soils of eastern India. Agron Res 6(1):15–25

Beevers L, Hagenman RH (1983) Uptake and reduction of nitrate: bacteria and higher plant. In: Läuchli A, Bieleski RL (eds) Inorganic plant nutrition. Encyclopedia of plant physiology. New series, vol. 15A. Springer, Berlin, pp 351–375

Bell PW, Rerkasem B, Keerati-Kasikorn P, Phetchawee S, Hiunburana N, Ratanarat S, Pongsakul P, Loneragan JF (1990) Mineral nutrition of food legumes in Thailand, with particular reference to micronutrients. Australian Centre for International Agricultural Research (ACIAR). Technical report no. 16, 52pp

Bhadoria PBS, Prakash YS, Rakshit A (2003) Relative efficacy of organic manures on rice production in lateritic soil. Soil Use Manag 19:80–82

Blevins DG, Lukaszewski KM (1998) Functions of boron in plant nutrition. Ann Rev Plant Physiol Plant Mol Biol 49:481–500

Boaretto RM, Quaggio JA, Assis F, Filho AM, Giné MF, Boaretto AE (2008) Absorption and mobility of boron in young citrus plants. Commun Soil Sci Plant Anal 39:2501–2514

Bolacer D, Nagarajan P, Gunasekaran S (2003) Effect of organic amendments on nodulation and yield of blackgram in acid soil. Legume Res 26:192–195

Bolaños L, Martín M, El-Hamdaoui A, Rivilla R, Bonilla I (2006) Nitrogenase inhibition in nodules from pea plants grown under salt stress occurs at the physiological level and can be alleviated by B and Ca. Plant Soil 280:135–142

Bonilla I, Bolaños L (2010) Mineral nutrition for legume-rhizobia symbiosis: B, Ca, N, P, S, K, Fe, Mo, Co, and Ni: a review. Sustain Agric Rev 1:253–274

Bonilla I, El-Hamdaoui A, Bolaños L (2004) Boron and calcium increase *Pisum sativum* seed germination and seedling development under salt stress. Plant Soil 267:97–107

Bouis HE (2003) Micronutrient fortification of plants through plant breeding: can it improve nutrition in man at low cost? Proc Nutr Soc 62:403–411

Bromilow RH, Chamberlain K, Tench AJ, Williams RH (1993) Phloem translocation of strong acids: glyphosate, substituted phosphonic, and sulfonic acids in *Ricinus communis* L. Pestic Sci 37:39–47

Brown PH, Shelp J (1997) Boron mobility in plants. Plant Soil 193:85–101

Cakmak I (2005) Role of potassium in alleviating the detrimental effects of abiotic stress plants. J Plant Nutr Soil Sci 168:521–530

Castero AMC, Bouretto AE, Nakagawa J (1997) Treatments of seeds of *Phasolous Vulgaris* L. with cobalt, molybdenum and vitamine B_{12}. Revista-Brasileire de Sementes 16:26–30

Chen Y, Barak P (1982) Iron nutrition of plants in calcareous soils. Adv Agron 35:217–240

Chew WY, Joseph KT, Ramlt K (1979) Influence of applied copper and other micronutrients on groundnuts and sorghum on Malaysian digotropic peat. Trop Agric 56(1):25–32

Coutinho CFB, Mazo LH (2005) Complexos meta' licos com o herbicida glyphosate: Revisao. Quımica Nova 28:1038–1045

Darwish DS, El-Gharreib EG, El-Hawary MA, Rafft OA (2002) Effect of some macro and micronutrients application on peanut production in a saline soil in El-Faiyum Governorate. Egypt J Appl Sci 17:17–32

Dell B, Huang L (1997) Physiological response of plants to boron. J Plant Soil 193:103–120

Desai RM (1994) Effect of copper sulphate, sulphur and copper ore tailings on groundnut and their residual effect on bengalgram. M.Sc. (Agri.) thesis, University of Agricultural Sciences, Dharwad

Due IS, Sharma RL, Rao AV (1999) Effect of foliar application of molybdenum and cobalt on soyabean *Rhizobium* symbiosis. Legume Res 1:97–100

El-Bably AZ (2002) Effect of irrigation and nutrition of copper and molybdenum on Egyptian clover (*Trifolium alexandrnium* L.). Agron J 94:1066–1070

El-Fouly MM, Zeinab MM, Zeinab SA (2001) Miconutrient spray is a tool to increase tolerance of faba bean and wheat to salinity. In: Horst WJ (ed) Plant nutrition, food security and sustainability of agro-ecosystem. Kluwer Academic, Dordrecht, pp 222–223

El-Hamdaoui A, Redondo-Nieto M, Torralba B, Rivilla R, Bonilla I, Bolaños L (2003) Influence of boron and calcium on the tolerance to salinity of nitrogen-fixing pea plants. Plant Soil 251:93–103

El-Tohamy WA, El-Greadly NHM (2007) Physiological responses, growth, yield and quality of snap beans in response to foliar application of yeast, vitamin E and zinc under sandy soil conditions. Aust J Basic Appl Sci 1(3):294–299

Eskew DL, Welch RM, Cary EE (1983) Nickel: an essential micronutrient for legumes and possibly all higher plants. Science, 11 Nov, pp 621–623

Farias Rodriguez R, Robert BM, Carlos A, Pena-Cabriales JJ (1998) The accumulation of trehalose in nodules of several cultivars of common bean (*Phaseolus vulgaris*) and its correlation with resistance to drought stress. Physiol Plant 102:353–359

Flexas J, Bota J, Escalona JM, Sampol B, Medrano H (2002) Effects of drought on photosynthesis in grapevines under field conditions: an evaluation of stomatal and mesophyll limitations. Func Plant Biol 29:461–471

Franca MGC, Pham-Thi CAT, Pimentel ROP, Rossiello Y, Fodil Z, Laffray D (2000) Differences in growth and water relations among *Phaseolus vulgaris* cultivars in response to induced drought stress. Environ Exp Bot 43:227–237

Frossard E, Bucher M, Mächler F, Mozafar A, Hurrell R (2000) Potential for increasing the content and bioavailability of Fe, Zn and Ca in plants for human nutrition. J Sci Food Agric 80:861–879

Gad N (2006) Increasing the efficiency of nitrogen fertilization through cobalt application to pea plant. Res J Agric Biol Sci 2(6):433–442

Georgaka A, Spanos N (2010) Study of the Cu(II) removal from aqueous solutions by adsorption on titania. Global NEST J 12:239–247

Gobarah ME, Mohamed MH, Tawfik MM (2006) Effect of phosphorus fertilizer and foliar spraying with zinc on growth, yield and quality of groundnut under reclaimed sandy soils. J Appl Sci Res 2(8):491–496

Gupta AP (2005) Micronutrient status and fertilizer use scenario in India. J Trace Elemen Med Biol 18:325–331

Gupta PK, Sharma NN, Acharya HK, Gupta SK, Mali GC, Henry A, Kumar D, Singh NB (2003) Response of mungbean to zinc and iron on vertisols in south-eastern plain of Rajasthan. Adv Arid Legume Res, pp 259–262

Hafner H, Ndunguru BJ, Bationo A, Marschner H (1992) Effect of nitrogen, phosphorus and molybdenum application on growth and symbiotic N_2-fixation of groundnut in an acid sandy soil in Niger. Nutr Cycl Agroecosyst 31:69–77

Hänsch R, Mendel RR (2009) Physiological functions of mineral micronutrients (Cu, Zn, Mn, Fe, Ni, Mo, B, Cl). Curr Opin Plant Biol 12:259–266

He ZL, Yang XE, Stoffella PJ (2005) Trace elements in agroecosystems and impacts on the environment. J Trace Elemen Med Biol 19:125–140

Howell RW, Skoog F (1955) Effect of adenine and other substances on growth of excised *Pisum epicotyls* cultured *in vitro*. Am J Bot 42:356–360

Hussain A, Chaudhry MR, Wajad A, Ahmed A, Rafiq M, Ibrahim M, Goheer AR (2004) Influence of water stress on growth, yield and radiation use efficiency of various wheat cultivars. Int J Agric Biol 6:1074–1079

Imtiaz M, Alloway BJ, Aslam M, Memon MY, Khan P, Siddiqui SH, Shah KH (2006) Zinc absorption in selected soils. Commun Soil Sci Plant Anal 23:1675–1688

Imtiaz M, Rashid A, Khan P, Memon MY, Aslam M (2010) The role of micronutrients in crop production and human health. Pak J Bot 42(4):2565–2578

Izaguirre-Mayoral ML, Sinclair TR (2005) Soybean genotypic difference in growth, nutrient accumulation and ultrastructure in response to manganese and iron supply in solution culture. Ann Bot 96:149–158

Jadi MB (1997) Residual effect of copper, sulphate, sulphur and copper ore tailings on growth and yield of Bengal gram. M.Sc. (Agri.) thesis, University of Agricultural Sciences, Dharwad

Jaleel CA, Jayakumar K, Chang-Xing Z, Azooz MM (2008) Effect of soil applied cobalt on activities of antioxidant enzymes in *Arachis hypogaea*. Global J Mol Sci 3(2):42–45

Kaiser BN, Gridley KL, Brady JN, Phillips T, Tyerman SD (2005) The role of molybdenum in agricultural plant production. Ann Bot 96:745–754

Karaman MR, Brohi AR, Inal A, Taban S (1999) Effect of iron and zinc applications on growth and mineral nutrient concentrations of beans (*Phaseolus vulgaris* L.) grown in artificial siltation soils. Turk J Agric Forest 23(2):341–348

Kassab OM (2005) Soil moisture stress and micronutrients foliar application effects on the growth and yield of mungbean plants. J Agric Sci Mansoura Univ 30:247–256

Keerati-Kasikorn P, Panya P, Plasket D, Bell RW, Loneragan JF (1987) A field survey of boron deficiency in peanut grown in Khon Kaen Province. In: Proceedings of the sixth Thailand national groundnut meeting. Prince of Songkla University, Songkla, Thailand (In Thai with English abstract)

Khan AH, Ashraf MY, Azmi AR (1990) Effect of NaCl on growth and nitrogen metabolism of sorghum. Acta Physiol Plant 12(3):233–238

Khan HR, McDonald GK, Rengel Z (2004) Zinc fertilization and water stress affects plant water relations, stomatal conductance and osmotic adjustment in chickpea (*Cicer arietinum* L.). Plant Soil 267:271–284

Khoshgoftar AH, HajiMozaffari E (2006) Approaches to enhance iron concentration in wheat grain produced in Qom Province. In: Iran 2nd Central Asian cereals conference, 13–16 June 2006, Cholpon-Ata, Issyk Kul Lake, Kyrgyz Republic

Khoshgoftarmanesh AH, Sharifi HR, Mirzapour MH, Schulin R (2007) Plant genotype and Zn fertilization effects on nutritional quality of wheat grain produced in saline soils. In: 9th International Conference of the Biochemistry of Trace Elements (ICOBTE), July 2007, Beijing, China

Khoshgoftarmanesh AH, Schulin R, Chaney RL, Daneshbakhsh B, Afyuni M (2010) Micronutrient-efficient genotypes for crop yield and nutritional quality in sustainable agriculture. A review. Agron Sustain Dev 30:83–107

Klucas RV, Hans FJ, Russell SA, Evans HJ (1983) Nickel: a micronutrient element for hydrogen-dependent growth of *Rhizobium japonicum* and for expression of urease activity in soybean leaves. Proc Natl Acad Sci 80:2253–2257

Krishna S (1995) Effect of sulphur and zinc application on yield, S and Zn uptake and protein content of mung (green gram). Legume Res 18:89–92

Liang YC, Sun WC, Zhu YG, Christie P (2007) Mechanisms of silicon-mediated alleviation of abiotic stresses in higher plants, a review. Environ Pollut 147:422–428

Mahajan S, Tuteja N (2005) Cold, salinity and drought stresses. Arch Biochem Biophys 444:139–158

Mahmoudi H, Ksouri R, Gharsalli M, Lachaâl M (2005) Differences in responses to iron deficiency between two legumes: lentil (*Lens culinaris*) and chickpea (*Cicer arietinum*). J Plant Physiol 162:1237–1245

Malakouti MJ (2007) Zinc is a neglected element in the life cycle of plants: a review. Middle East Russ J Plant Sci Biotechnol 1:1–12

Mann EN, de Resende PM, Mann RS, De Carvalho JG, Von Pinho EVD (2002) Effect of manganese application on yield and seed quality of soybean. Agropecuaria Brasileira 37:1757–1764

Marsalis MA, Puppala N, Goldberg NP, Ashigh J, Sanogo S, Trostle C (2009) New Mexico peanut production. Circular-645, p 16, Cooperative Extension Services, College of Agriculture, Consumer and Environmental Science, NM State University, USA

Marschner H (1993) Zinc uptake from soils. In: Robson AD (ed) Zinc in soils and plants. Kluwer Academic, Dordrecht, pp 59–77

Marschner H (1995) Mineral nutrition of higher plants, 2nd edn. Academic, London

Marschner P (2011) Marschner's mineral nutrition of higher plants. Academic/Elsevier, San Diego/London

Mendel RR, Hansch R (2002) Molybdoenzymes and molybdenum cofactor in plants. J Exp Bot 53:1689–1698

Mengel K, Kirkby EA (1987) Principles of plant nutrition, 4th edn. International Potash Institute, Basel

Miletic RPR, Dodig D, Milutinovic S, Mihajlovic I, Nikodijevic SM (2010) Strategies for solving the problem of drought in Eastern Serbia. http://www.wgcrop.icidonline.org/40doc.pdf

Nable RO, Bafiuelos GS, Paull JG (1997) Boron toxicity. Plant Soil 193:181–198

Nascimento CWA, Fontes RLF, Cesar A (2003) Copper availability as related to soil copper fractions in oxisols under liming. Sci Agric 60:167–173

Nayyar VK, Arora CL, Kataki PK (2001) Management of soil micronutrients deficiencies in the rice-wheat cropping system. J Crop Prod 4:87–131

Ndakidemi PA (2005) Nutritional characterization of the rhizosphere of symbiotic cowpea and maize plants in different cropping systems. Doctoral degree thesis. Cape Peninsula University of Technology, Cape Town, South Africa, 150pp

O'Neill MA, Ishii T, Albersheim P, Darvill AG (2004) Rhamnogalacturonan II: structure and function of a borate cross-linked cell wall pectic polysaccharide. Ann Rev Plant Biol 55:109–139

Osotsapar Y (2000) Micronutrients in crop production in Thailand. Extension Bulletins, Food Fertilizer Technology Centre, 5 F.14 Wenchow St., Taipei 10616, Taiwan R.O.C

Parkpean P, Ratanarat S, Phetchawee S (1986) Diagnosis of iron chlorosis in soybean and peanut using orthophenanthroline method. In: Annual report, Soil Science Division, Department of Agriculture, Thailand (In Thai with English abstract)

Parkpian P, Boonkerd N (1989) Involvement of iron in nodulation and symbiolic nitrogen fixation in groundnut. In: The proceedings of national groundnut conference, Roi-et., Thailand, 3–5 May 1989

Polle A, Chakrabarti K, Chakrabarti S, Seifert F, Schramel P, Rennenberg H (1992) Antioxidants and manganese deficiency in needles of Norway spruce (*Picea abies* L.) trees. Plant Physiol 99:1084–1089

Purcell LB (2009) Physiological responses of N_2 fixation to drought and selecting genotypes for improved N_2 fixation. In: Emerich DW, Krishnan HB (eds) Nitrogen fixation in crop production. ASA, CSSA and SSSA, Madison, pp 211–238

Purcell LB, King CA, Ball RA (2000) Soybean cultivar differences in ureides and the relationship to drought-tolerant nitrogen fixation and manganese nutrition. Crop Sci 40:1062–1070

Quilambo ON (2003) The vesicular-arbuscular mycorrhizal symbiosis. Afr J Biotechnol 2:539–546

Rajaiea M, Ejraieb AK, Owliaiec HR, Tavakolid AR (2009) Effect of zinc and boron interaction on growth and mineral composition of lemon seedlings in a calcareous soil. Int J Plant Prod 3:39–50

Ramakrishna A, Gowda CLL, Johansen C (2000) Management factors affecting legumes production in the Indo-Gangetic plains. In: Johansen C, Duxbury JM, Virmani SM, Gowda CLL, Pande S, Joshi PK (eds) Legumes in rice and wheat cropping systems of the Indo-Gangetic plain: constraints and opportunities. ICRISAT/Cornell University, Patancheru/New York, pp 156–165

Rashid A (2005) Establishment and management of micronutrient deficiencies in soils of Pakistan. A review. Soil Environ 24:1–22

Rashid A, Ryan J (2004) Micronutrient constraints to crop production in soils with Mediterranean type characteristics: a review. J Plant Nutr 27:959–975

Ratanarat S, Masangsan W, Vadeesirisak P, Ditsantia C, Phanuvas V (1990) Effects of ion foliar spray and specific *Rhizobium* strains on yield components of three peanut cultivars. In: Proceedings, 8th national groundnut research meeting, 3–5 May 1989, Roi-Et, Thailand

Reddy KJ, Wang L, Gloss SP (1995) Solubility and mobility of copper, zinc and lead in acidic environments. Plant Soil 171:53–58

Rerkasem B (1989) Boron deficiency in food legumes in Northern Thailand. Thai J Soil Fertil 16:130–152

Rerkasem B, Bell RW, Loneragan JP (1990) Effects of seed and soil boron on early seedling growth of black and green gram. In: Beusichem ML (ed) Plant nutrition – physiology and applications. Kluwer Academic, Dordrecht, pp 281–285

Rerkasem B, Bell RW, Lodkaew S, Loneragan JF (1993) Boron deficiency in soybean [*Glycine max* (L.) Merr.], peanut (*Arachis hypogaea* L.) and black gram [*Vigna mungo* (L.) Hepper]: symptoms in seeds and differences among soybean cultivars in susceptibility to boron deficiency. Plant Soil 150:289–294

Rourke JAO', Charlson DV, Gonzalez DO, Vodkin LO, Graham MA, Cianzio SR, Grusak MA, Shoemake RC (2007) Microarray analysis of iron deficiency chlorosis in near-isogenic near-isogenic soybean lines. BMC Genomics 8:476–488

Scherer HW, Pacyna S, Spoth KR, Schulz M (2008) Low levels of ferredoxin, ATP and leghemoglobin contribute to limited N_2 fixation of peas (*Pisum sativum* L.) and alfalfa (Medicago sativa L.) under S deficiency conditions. Biol Fertil Soils 44:909–916

Sharma SR, Bhandari SC (2002) Effect of organic manure and mineral nutrients on the symbiotic efficiency in cowpea. Indian J Pulses Res 15:156–160

Sinclair TR, Vadez V, Chenu K (2003) Ureide accumulation in response to Mn nutrition by eight soybean genotypes with N_2 fixation tolerance to soil drying. Crop Sci 43:592–597

Singh AL, Joshi YC (1997) Prevention and correction of iron-deficiency chlorosis of groundnut in India. In: Ando T et al (eds) Plant nutrition for sustainable food production and environment. Kluwer Academic, Tokyo, pp 271–272

Sinha P, Chatterjee C (2003) Influence of low boron on yield and seed quality of soybean. Indian J Agric Res 37(3):193–197

Snowball K, Robson AD (1980) The effect of copper on nitrogen fixation in subterranean clover *(Trifolium subterraneum)*. New Phytol 85:63–72

Taiz L, Zeiger E (1998) Plant physiology. Sinauer Associates, Sunderland, 719 pp

Taiz L, Zeiger E (2002) Plant physiology, 3rd edn. Sinauer Associates, Sunderland

Tan YF, O'Toole N, Taylor NL, Millar AH (2010) Divalent metal ions in plant mitochondria and their role in interactions with proteins and oxidative stress-induced damage to respiratory function. Plant Physiol 152:747–761

Tang C, Robson AD, Dilworth MJ, Kuo J (1992) Microscopic evidence on how iron deficiency limits nodule initiation in *Lupinus angustifolius* L. New Phytol 121:457–467

Tenywa JS (1997) Influence of molybdenum and cobalt fertilization on symbiotic nitrogen fixation indicators in an oxisol in Uganda. Afr Crop Sci J 5:87–92

Tenywa JS (2003) Influence of cobalt fertilization on symbiotic nitrogen fixation indicators in an oxisol in Uganda. Afr Crop Sci J 5:87–92

Tewari RK, Kumar P, Sharma PN (2009) Morphological and oxidative physiology of boron-deficient mulberry plants. Tree Physiol 30:68–77

Thalooth AT, Tawfik MM, Mohamed HM (2006) A comparative study on the effect of foliar application of zinc, potassium and magnesium on growth, yield and some chemical constituents of mungbean plants grown under water stress conditions. World J Agric Sci 2(1):37–46

Thibaund GR (2005) Molybdenum relationships in soils and plants. KwaZulu-Natal Department of Agriculture and Environmental Affairs, Cedara College, Private Bag X9059, Pietermaritzburg, 3200, South Africa

Tiaranan N, Pimsarn S, Claimon S, Punpruk P (1985) Correction of nutrient deficiencies of legumes in Thailand. In: Persley GL (ed) Tropical legume improvement, ACIAR proceedings no 8, pp 54–55

Todd CD, Tipton PA, Blevins DG, Piedras P, Pineda M, Polacco JC (2006) Update on ureide degradation in legumes. J Expt Bot 57:5–12

Unkasem B, Tawonsook J (1988) Micronutrient deficiency of mungbean in Takhli soil. In: Proceedings of the 3rd Mungbean conference, Kanchanaburi, 21–23 Nov 1988, Thailand

Ureta AC, Imperial J, Ruiz-Argueso T, Palacios JM (2005) *Rhizobium leguminosarum* biovar viciae symbiotic hydrogenase activity and processing are limited by the level of nickel in agricultural soils. Appl Environ Microbiol 71:7603–7609

Vadez V, Sinclair TR, Serraj R (2000) Asparagine and ureide accumulation in nodules and shoots as feedback inhibitors of N_2 fixation in soybean. Physiol Plant 110:215–223

Virupaksh DM (1995) Effect of copper sulphate, sulphur and copper ore tailings on growth and yield of soybean in vertisols. M.Sc. (Agri.) thesis, University of Agricultural Sciences, Dharwad

Wang ZY, Ge YX, Main R, Baker J (2005) Development of highly tissue culture responsive lines of *Lolium temulentum* by anther culture. Plant Sci 168:203–211

Waraich EA, Amad R, Ashraf MY, Saifullah AM (2011) Improving agricultural water use efficiency by nutrient management. Acta Agric Scandi – Soil Plant Sci 61(4):291–304

Welch RM, Graham RD (2004) Breeding for micronutrients in staple food crops from a human nutrition perspective. J Exp Bot 55:353–364

Westermann DT (2005) Nutritional requirements of potatoes. Am J Potato Res 82:301–307

Williams RJP, Frausto da Silva JJR (2002) The involvement of molybdenum in life. Biochem Biophys Res Commun 292:293–299

Yakubu H, Kwari JD, Sandabe MK (2010) Effect of molybdenum on N_2-fixation by some grain legumes varieties in Sudano- Sahelian zone of north eastern Nigeria. Am Eur J Agric Environ Sci 8(1):109–115

Zahran HH (1999) *Rhizobium*-legume symbiosis and nitrogen fixation under severe conditions and in an arid climate. Microbiol Mol Biol Rev 63(4):968–989

Zobiole LHS, Oliveira RS Jr, Kremer RJ, Constantin J, Yamada T, Castro C, Oliveira FA, Oliveira A Jr (2010) Effect of glyphosate on symbiotic N_2 fixation and nickel concentration in glyphosate-resistant soybeans. Appl Soil Ecol 44:176–180

Chapter 26
Potential of Rhizobia for Sustainable Production of Non-legumes

Ijaz Mehboob, Muhammad Naveed, Zahir Ahmad Zahir, and Muhammad Ashraf

Contents

I. Mehboob (✉) • Z.A. Zahir
Institute of Soil & Environmental Sciences, University of Agriculture,
Faisalabad 38040, Pakistan
e-mail: zaeem_methabacha@yahoo.com; zazahir@yahoo.com

M. Naveed
Institute of Soil & Environmental Sciences, University of Agriculture,
Faisalabad 38040, Pakistan

AIT – Austrian Institute of Technology GmbH, Bioresources Unit,
Konrad-Lorenz-Strasse 24, 3430 Tulln, Austria
e-mail: mnaveeduaf@yahoo.com

M. Ashraf
Department of Botany, University of Agriculture, Faisalabad 38040, Pakistan

Department of Botany and Microbiology, King Saud University,
Riyadh, Saudi Arabia
e-mail: ashrafbot@yahoo.com

M. Ashraf et al. (eds.), *Crop Production for Agricultural Improvement*,
DOI 10.1007/978-94-007-4116-4_26,

Abstract Rhizobia are familiar as the symbiotic associates of legumes, forming N_2-fixing nodules. However, it has been proven that these bacteria also have the ability to survive and colonize the roots of non-legumes as efficiently as they colonize the roots of their legume host. Although example of N_2-fixation in non-legume parasponia has been observed, the researchers are yet unable to extend this nitrogen-fixation symbiosis to major cereals of the planet. Only nodule-like structures or hypertrophies or outgrowths have been recorded with rhizobia on the roots of non-legumes yet without significant N_2-fixation. Generally, three types of rhizobial interactions with non-legumes have been demonstrated i.e. interactions which result in growth and yield promotion of the interacting non-legumes, those which result in poor or detrimental effect on the growth and yield of inoculated non-leguminous plants and those which cannot result in any increase or decrease (missing effect/no effect/neutral effect) in the plant growth. Like other PGPR, rhizobia can affect the non-legumes beneficially by solubilizing sparingly soluble organic and inorganic phosphates, by releasing phytohormnes, enzymes, siderophores, lumichromes, lipo-chito-oligosaccharides, exo-polysaccharides and riboflavins. They can also promote the growth of non-legumes by inhibiting the growth of pathogens by sequestering the iron in the rhizosphere with siderophore production, by releasing the antibiotics and/or by the production of cell wall degrading enzymes. They can also play a significant role in alleviating the deleterious effects of various environmental stresses. Even more, they can improve the growth of non-legumes by changing the host-plant susceptibility by releasing different bio-stimulatory agents. However, plant variety, cultural conditions, native micro-flora, soil and other ecological factors have been reported affecting the degree to which rhizobial association benefit the non-legumes. But it has also been recognized that the potential of PGPR strains of rhizobia (regarding colonization to a variety of plants, adoption under variable soil and environmental conditions and against the pathogens that can attack the host plants) can be improved further through dual/mix inoculation with other beneficial microorganisms. Finally, it is recognized that competent rhizobial strains could be used as biofertilizers, stress regulators, and as biocontrol agents to non-legumes to increase their production. However, much more research efforts are needed to develop rhizobial strains which can effectively improve crop productivity under a variety of environments.

Keywords Rhizobia • Biofertilization • Phytostimulation • Bioprotection • Phytoremediation • N_2-fixation • Non-legumes

1 Introduction

Worldwide, the population depends on agricultural plants as they are food for humans and feed for animals. Demand for food and feed is gradually increasing day-by-day. To meet the demand, farmers apply additional inputs in soil in the form of chemical fertilizers, pesticides, herbicides, etc. But these chemicals disturb the environment, subvert ecology, degrade soil and mismanage water resources. Other anthropogenic activities (i.e. manufacturing industries, mining of natural resources and energy production) further accelerate these biospheric pollutions. Hence, soil fertility maintenance and remediation of contaminated soils was ever in need of an effective and affordable technological solution. Current trends in agriculture are focused on reduction in the use of chemical pesticides and inorganic fertilizers, compelling the search for alternatives that enhance environmental quality (Haggag 2002). Improving the soil fertility and crop production without the use of chemical fertilizers is a difficult task (Ray et al. 2000; Bera et al. 2006). Plant growth promoting rhizobacteria (PGPR) - mediated remediation of contaminated soils (Jing et al. 2007; Khan et al. 2009), improvement in soil fertility, enhancement in plant growth, control of plant pests and diseases is an alternative eco-friendly approach (Kloepper 1993).

Originally, the "rhizobacteria" that exert a beneficial effect on plant growth (referred to as PGPR) when reintroduced as plant inoculant in a soil containing competitive micro-flora were restricted to only free-living bacteria. But this restriction was uplifted during the last couple of decades when the work of a lot of researchers clearly indicated that rhizobia can associate with the roots of non-legumes also, without forming true nodules, and can promote their growth by using various direct and indirect mechanisms of actions. The original definition of beneficial rhizobacteria restricted to only free-living bacteria is now changed as "any root-colonizing bacteria". Generally, growth promotion as a result of N_2-fixation in legume nodules by rhizobia is not considered as a PGPR mechanism of action, but when these associative bacteria do not exhibit morphological modification of the host plant are considered as PGPR.

Recently, PGPR have been classified as phytostimulators, biofertilizers, biopesticides and rhizoremediators depending on their activities (Somers et al. 2004). Rhizobia, like other PGPR, also exert their beneficial effects on the growth of non-leguminous plants (Hoflich et al. 1995; Noel et al. 1996; Yanni et al. 1997) by producing phytohormones (i.e. auxins, abscisic acid, gibberellic acid and cytokinins, etc.), increasing solubility and availability of nutrients (i.e. mineral P solubilization through organic acid production and mineralization of organic P through release of acid phosphatases). They are also helpful in inhibiting the growth of plant pathogens via competition (i.e. out-competing the pathogens by sequestering the iron in the rhizosphere with siderophore production), antibiosis (i.e. production of an antibiotic namely peptide trifolitoxin whose spectrum of activity includes plant pathogen bacteria) and parasitism (i.e. production of enzymes like chitinase which usually degrade the cell wall of the plant pathogenic fungi). Rhizobia play a significant role in changing the host-plant susceptibility by producing various biostimulatory agents (i.e. lipopolysaccharide production) or in interaction with other beneficial microorganisms. Also, rhizobia are able to protect plants from the deleterious effects of various environmental stresses by producing different metabolites (i.e. rhizobia can mitigate the effects of water stress by decreasing leaf stomatal conductance via abscisic acid and lumichrome production). In addition, rhizobia can produce other beneficial chemicals which can promote the growth of inoculated plants directly or indirectly e.g. lipochito-oligosaccharides (LCOs), lumichrome, rhiboflavin, enzymes, siderophores or exopolysaccharides in the rhizosphere (Mehboob et al. 2009). In addition, certain rhizobial strains containing ACC deaminase activities could be used to ameliorate the stresses of phytoremediator plants (Glick and Pasternak 2003; Glick 2010) which they experience during remediation of contaminated soils. Conclusively, it could be inferred that rhizobial species are very helpful in stimulating plant growth, protecting them from different pathogenic attacks, environmental stresses, in sustaining fertility level and reviving the health of contaminated soil through remediation without disturbing/degrading the environment. However, proper exploitation of these organisms is a pre-requisite.

2 Rhizobia-non-legume Association: An Overview

Rhizobia belong to *Rhizobiaceae* family. *Allorhizobium, Azorhizobium, Bradyrhizobium, Mesorhizobium, Rhizobium* and *Sinorhizobium* are their common genera (Tan et al. 2001a; Sessitsch et al. 2002; West et al. 2002; Mia and Shamsuddin 2010) which are collectively referred to as rhizobia (Chaintreuil et al. 2000; Shaukat and Siddqui 2003; Dakora 2003; Okazaki et al. 2004; Matiru and Dakora 2005a, b). By nature they are gram-negative, chemo-organotrophic or chemo-lithotrophic organisms (Werner 1992). Generally, they are famous for their unique ability to establish specific associative interaction with leguminous plants and form nodules in which they fix atmospheric nitrogen (Schloter et al. 1997). However, nonspecific associative interactions of rhizobia with roots of non-legumes have also been investigated (Reyes and Schmidt 1979; Chabot et al. 1996).

Generally, plant-associated bacteria are of two types: those that are closely associated and those that are loosely associated with plants. They may be endophytic (which reside in the internal tissue of the plant), phyllospheric (which reside on the above-ground plant parts i.e. leaves, stems, blossoms and fruits) or rhizospheric (which reside within the rhizosphere) (Weyens et al. 2009). Of these interactions, there are associative interactions between plant roots and bacteria, which exert effects on growth and yield of many plants (Lemanceau 1992; Yanni et al. 1997). These interactions may be stimulating or inhibiting depending upon the rhizosphere survival and colonizing ability of the bacteria and their phyto-effective metabolic characteristics (Hoflich et al. 1994; Antoun et al. 1998).

The ability of rhizobia to survive and colonize the roots of non-legumes as efficiently as they colonize the roots of their legume host has been proven (Gaur et al. 1980; Ladha et al. 1989; Al-Mallah et al. 1990; Yanni et al. 1995; Chabot et al. 1996; Antoun and Prevost 2000; Bhattacharjee et al. 2008). Moreover, rhizobia are now considered endophytes or rhizobacteria of non-leguminous plants (Sessitsch et al. 2002). There is plethora of literature available in which rhizobial inoculants have been recognized as endophytes [those microorganisms that live within host plants for at least part of their life and do not cause apparent symptoms of diseases (Mei and Flinn 2010)] in the roots of non-legumes (McInroy and Kloepper 1995;

Prayitno et al. 1999; Sturz et al. 1999; Biswas et al. 2000a; Matiru et al. 2000; Hilali et al. 2001; Yanni et al. 2001; Reiter et al. 2002; Peng et al. 2002; Lupwayi et al. 2004; Zahir et al. 2004; Singh et al. 2006) such as rice (Balley et al. 1983; Roger and Watanabe 1986; Ladha et al. 1989; Khush and Bennett 1992; Natalia et al. 1994; Ueda et al. 1995; Yanni et al. 1997; Chaintreuil et al. 2000; Engelhard et al. 2000; James et al. 2000; Perrine et al. 2001; Tan et al. 2001b; Peng et al. 2002; Anyia et al. 2004; Matiru and Dakora 2004; Naidu et al. 2004; Singh et al. 2005; Mano and Morisaki 2008; Bhattacharjee et al. 2008; Senthilkumar et al. 2008, 2009), wheat (Sabry et al. 1997; Biederbeck et al. 2000; Anyia et al. 2004; Kaci et al. 2005; Afzal and Bano 2008), maize (Schloter et al. 1997; Gutierrez-Zamora and Martinez-Romero 2001; Reiter et al. 2002; Rosenblueth and Martinez-Romero 2004; Hossain 2007; Mehboob et al. 2008; Cassan et al. 2009), lettuce (Chabot et al. 1993, 1996; Noel et al. 1996), cotton (McInroy and Kloepper 1995; Martinez-Romero et al. 2000; Egamberdiyeva et al. 2004; Hafeez et al. 2004), barley and canola (Lupwayi et al. 2000; Peix et al. 2001; Humphry et al. 2007), mustard (Chandra et al. 2007), oilseed rape (Lupwayi et al. 2000), sorghum, millet, Sudan grass (Matiru et al. 2005), sunflower and okra (Ehteshamul-Haque and Ghaffar 1993; Alami et al. 2000; Sheikh et al. 2006), potato (Sturz et al. 1999), and tobacco (Ji et al. 2010).

A strong evidence exists which indicates that in the presence of germinating seed and root developing systems, rhizobia could flourish in a similar way with legume and non-legumes (Pena-Cabriales and Alexander 1983). Under particular conditions, nodule-like structures or hypertrophies formed by rhizobia on the roots of non-legumes without significant N_2 fixation has also been reported by a number of researchers (Al-Mallah et al. 1989; Bender et al. 1990; De Bruijn et al. 1995; Jing et al. 1992; Li et al. 1991; Rolfe and Bender 1990; Ridge et al. 1992; Trinick and Habdodas 1995; Naidu et al. 2004). Furthermore, for the analysis of rhizobial attachment to the roots of rice and wheat seedlings in terms of dynamic equilibrium model such as Langmuir adsorption isotherm was reported by Shimshick and Hebert (1979) in which the maximum number of binding sites was $8 \times 10^9/g$ of fresh root (gfr) at 22°C. As an evidence, Wiehe and Hoflich (1995) reported a range of more than log 5 to less than log 3 CFU/gfr regarding maize root colonization by rhizobia. Whereas, Chabot et al. (1996) during investigation on the potential of maize and lettuce root colonization by rhizobia reported that root population of rhizobia was averaged log 4.1 CFU/gfr on maize roots 4 weeks after seeding and log 3.7 CFU/gfr on lettuce roots 5 weeks after seeding. They verified the competence of rhizobia to colonize and survive on maize and lettuce roots. Wiehe et al. (1994) reported that rhizobia can colonize not only the rhizoplane but also the cortex and root cap intercellular spaces on the non-leguminous plants. To characterize the associative colonization of the non-leguminous plants by rhizobia, a time-course study was also performed by Schloter et al. (1997) in which they indicated the presence of rhizobia in lysed cells of the root cortex as well as in intracellular spaces of central root cylinder cells. Endophytic invasion of rhizobia in rice roots through cracks in the epidermis and fissures created during emergence of lateral roots was also indicated by Reddy et al. (1997). Whereas, Sabry et al. (1997) reported invasion of rhizobia between cells of the cortex, within the xylem and the root meristem of wheat.

Surface colonization of the rhizoplane followed by endophytic colonization within roots and then ascending endophytic migration into the stem base, leaf sheath and leaves of rice have been reported by Chi et al. (2005). An evidence also exists that for the crack entry invasion of emerging lateral roots of wheat, rice, maize and oil-seed rape by rhizobia (Cocking et al. 1994; Gough et al. 1996; Webster et al. 1997; O'Callaghan et al. 2000; Hoflich 2000; Anyia et al. 2004). Rhizobial attachment to asparagus, oat, rice and wheat has been described by Shimshick and Hebert (1979), Ladha et al. (1989) and Terouchi and Syono (1990). Whereas, Wiehe and Hoflich (1995) demonstrated the multiplication and survival of rhizobia in the rhizosphere of wheat, corn, rape and *Brassica napus*. Rhizobial presence at the root epidermal surfaces as well as inside the tissues of inoculated sorghum and millet plants has been shown by Matiru et al. (2005). Pena and Reyes (2007) detected *Rhizobium* as rhizosphere colonizer of lettuce. Prayitno et al. (1999) studied the root colonization pattern of *Rhizobium* sp. and reported that the rhizobia preferentially colonize rice root surface, or at the emerging lateral root zones and at the root tips. Also, Singh et al. (2006) demonstrated that rhizobial cell can go inside and colonize the rice root interiors. Even so, Perrine-Walker et al. (2007b) showed that *Rhizobium* spp. can infect rice roots and colonize the intercellular spaces of the rice roots.

As far as phyto-effective metabolic characteristics are concerned, rhizobia have been recognized to release many powerful phyto-effective metabolites through which they affect fundamental processes in non-leguminous plant development. There are many metabolites of rhizobia recognized so far such as cytokinins (Phillips and Torrey 1972; Puppo and Riguard 1978; Upadhyaya et al. 1991; Hoflich et al. 1995; Noel et al. 1996), abscisic acid (Phillips and Torrey 1970; Hirsch et al. 1997; Law and Strijdom 1989; Atzorn et al. 1988; Minamisawa et al. 1996), indole-3-acetic acid (Law and Strijdom 1989; Hunter and Kuykendall 1990; Minamisawa et al. 1996; Antoun et al. 1998; Vessey 2003; Naidu et al. 2004; Dazzo et al. 2005; Matiru et al. 2005; Boiero et al. 2007; Chandra et al. 2007; Pandey and Maheshwari 2007; Pena and Reyes 2007; Venieraki et al. 2011), gibberellic acid (Atzorn et al. 1988; Yanni et al. 2001; Humphry et al. 2007; Chi et al. 2005; Afzal and Bano 2008), ethylene (Boiero et al. 2007), lipo-chito-oligosaccharides (Lopez-Lara et al. 1995; Dyachok et al. 2000; Prithiviraj et al. 2000; Smith et al. 2002), lumichrome (Phillips et al. 1999; Yang et al. 2002; Beveridge et al. 2003; Dakora 2003; Matiru and Dakora 2005a), siderophore (Schwyn and Neilands 1987; Chabot et al. 1993; Derylo et al. 1994; Dudeja et al. 1997; Antoun et al. 1998; Carson et al. 2000; Arora et al. 2001; Deshwal et al. 2003; Hossain and Martensson 2008), rhiboflavin (West and Wilson 1938; Phillips et al. 1999; Dakora et al. 2002; Dakora 2003), exopoly-saccharide (Alami et al. 2000; Reitz et al. 2000; Kaci et al. 2005; Santaella et al. 2008; Berge et al. 2009), phenolic acids (Mishra et al. 2006), nod factors (Spaink 1992; Spaink and Lugtenberg 1994; Xie et al. 1995; Zhang and Smith 2001; Smith et al. 2002), enzyme ACC-deaminase (Belimov et al. 2001; Trott et al. 2001; Yasutu et al. 2001; Kaneko et al. 2002; Sullivan et al. 2002; Ma et al. 2003a, b; Okazaki et al. 2003; Belimov et al. 2005; Hafeez et al. 2008; Duan et al. 2009), antibiotics (Chakraborty and Purkayastha 1984; Ehteshamul-Haque and Ghaffar 1993; Breil et al. 1993), and lipopolysaccharides (Reitz et al. 2000) have been known for a long time. In general, the metabolites produced as a result of beneficial interaction between rhizobia and non-leguminous plants have been shown to cause increase in seed germination, rate of radical elongation, seedling vigor, root architecture (length, branching, biovolume, surface), shoot growth, photosynthetic activity, leaf area, chlorophyll contents, stomatal conductance, harvest index, grain yield, nutrient uptake, protein content, tolerance to abiotic stresses (Hilali et al. 2001; Hafeez et al. 2004; Siddiqui 2007; Reimann et al. 2008; Hossain and Martensson 2008).

Rhizobial interaction with non-legumes does not always lead to the desired results. Their interaction, at certain time or site, event may result in poor, missing or detrimental effect on the growth and yield of inoculated plants (Mehboob et al. 2009). As evidence, Antoun et al. (1998) have reported deleterious effects of some of the rhizobial species due to the overproduction of IAA and related compounds. Whereas, O'Sullivan and O'Gara (1992) and Alstrom and Burns (1989) showed harmful effects due to the production of HCN by rhizobial inoculants. Similarly, Perrine-Walker et al. (2007a) reported inhibitory effects of rhizobial inoculants on growth and development of non-leguminous plants due to the production of high concentration of auxin and nitrate. Also, suppression of non-legumes with rhizobial inoculation due to the production of growth inhibitor by the strain used has been described by El-Tarabily et al. (2006), whereas Perrine-Walker et al. (2009) reported rice growth inhibition by a rhizobial strain (i.e. *Sinorhizobium meliloti* 1021) which produces bacteriocin-like substance.

It is now widely accepted that the degree to which these associations between rhizobia and non-leguminous plants benefit their growth varies with plant variety, inoculated bacteria, cultural conditions, native microflora, soil and other ecological factors (Lynch 1990; O'Sullivan and O'Gara 1992; Antoun et al. 1998; Biswas et al. 2000b; Hilali et al. 2001; Dobbelaere et al. 2003; Depret et al. 2004; Mehboob et al. 2008; Hussain et al. 2009). Overall, rhizobial strains have been classified into three general groups; strains that promote growth, those that inhibit growth, and strains having no influence in their association with non-leguminous plants (Prayitno et al. 1999; Perrine et al. 2001, 2005).

Finally, the above-cited research evidences have not only provided the better understanding about rhizobial interaction with non-legumes, but also recognized that competent rhizobial strains could be applied to non-legumes to increase their production. However, much more research efforts are needed to develop rhizobial strains specific to crop, variety, soil and environment.

3 Ecophysiology of Rhizobia-Plant Interaction

3.1 Plant Growth Enhancement – Mode of Action

Microbial population in the soil around the roots [Hiltner (1904) for the first time called this soil "rhizosphere"] is 19–32 times more than the root free soil (Bodelier et al. 1997) because of the release of organic compounds such as amino acids, sugars, vitamins, organic acids, auxins and flavonoides by the plant roots, which can be rapidly utilized by the microorganisms (Somers et al. 2004; Dardanelli et al. 2008, 2009; Raaijmakers et al. 2009). Microorganisms are attracted to this nutritious environment and use these plant root exudates for growth and multiplication on the root surface and in the adjacent rhizosphere soil (Lynch and Whipps 1991; Dakora and Phillips 2002). Actually, the complex exchanges that take place around the plant roots lead to the establishment of rhizosphere interaction. In other words, this complex molecular signaling regulates all the beneficial, deleterious and neutral relationships between microorganisms and the plant roots. The study of these systems has led to the discovery that plants and bacteria communicate by using chemical signals, which are involved in a successful interaction (Bolton et al. 1986; Dardanelli et al. 2008, 2009). Generally, the most abundant microorganisms in the rhizosphere are bacteria. The competent bacteria that aggressively colonize the plant roots are termed as "rhizobacteria" (Antoun and Kloepper 2001) which can have a neutral, deleterious or beneficial effect on plant growth. Deleterious microorganisms are injurious to the plant, but beneficial microorganisms can save the plant from biotic and abiotic stresses, whereas the neutral microorganisms neither harm the plant nor benefit it. Of these types of rhizobacteria, the beneficial could be only 2–5%, which can have beneficial effect on the growth of plants and have been termed as plant growth promoting rhizobacteria (PGPR) (Kloepper and Schroth 1978). Plant growth promoting rhizobacteria can affect plant growth either directly by providing plants with a compound synthesized by the bacterium or by facilitating the uptake of certain nutrients from the environments or indirectly by decreasing or preventing the deleterious effects of one or more phytopathogenic organisms (Glick et al. 1995). Originally, only free-living bacteria were included in the PGPR definition and not nitrogen fixing symbiotic rhizobia (Kloepper 1993). But now rhizobia are also considered as PGPR (Chandra et al. 2007) when they do not exhibit morphological modification of the host plant. It is widely reported that rhizobia, like other PGPR, can also behave like PGPR with non-legume plants as well (Hoflich et al. 1995; Noel et al. 1996; Yanni et al. 1997; Antoun et al. 1998; Rodriguez and Fraga 1999; Sessitsch et al. 2002) as they exert their positive effects on plant growth through various mechanisms different from nitrogen fixation (Dobbelaere et al. 2003). The well known PGPR strains of rhizobia include members of the genera *Allorhizobium, Azorhizobium, Bradyrhizobium, Mesorhizobium, Rhizobium and Sinorhizobium*. PGPR strains of rhizobia modulate the growth of non-legumes by direct or indirect modes of action described as under:

3.1.1 Direct Growth Promoting Activities

Generally, rhizobacteria cause direct growth promotion of plants in the absence of pathogens by producing phyto-effective metabolites (Lugtenberg and Kamilova 2009). Rhizobia can promote the growth of non-leguminous plants directly by producing plant hormones like auxins, gibberellins, abscisic acid and cytokinins, lowering of plant ethylene levels by producing the enzyme 1-aminocyclopropane-1-carboxylic acid (ACC) deaminase, providing bio-available phosphorus for plant uptake, sequestering iron for plants by siderophores and other useful compounds like lipo-chito-oligosaccharides (LCOs), lumichrome and rhiboflavin, etc. (Mehboob et al. 2009). Direct growth promotion of non-legumes by rhizobia has also been earlier reported (Hoflich et al. 1995; Chabot et al. 1996; Noel et al. 1996; Yanni et al. 1997).

Phytohormones are the substances that stimulate plant growth at micromolar concentration (or even lower). Also, they regulate essentially all physiological and developmental processes during the life cycle of a plant (Chiwocha et al. 2003). Phytohormones that are produced by the rhizobia in response to seed or root inoculation with non-legumes, such as indole-3-acetic acid, cytokinins, gibberellins and abscisic acid can stimulate the growth and yield of non-leguminous plants. The most studied phytohormone released by rhizobia during interaction with non-legumes is indole-3-acetic acid (IAA). This adds to plant growth and development by improving root/shoot growth and seedling vigor. Rhizobial strains producing IAA include *Azorhizobium caulinodans, Bradyrhizobium japonicum, Bradyrhizobium elkanii, Mesorhizobium loti, Rhizobium japonicum, Rhizobium leguminosarum, Rhizobium lupine, Rhizobium meliloti, Rhizobium phaseoli, Rhizobium trifolii* and *Sinorhizobium* spp. (Dullaart 1970; Badenoch-Jones et al. 1982; Wang et al. 1982; Kaneshiro et al. 1983; Garcia-Rodriguez et al. 1984; Wheeler et al. 1984; Ernesten et al. 1987; Kittell et al. 1989; Hoflich et al. 1995; Minamisawa et al. 1996; Noel et al. 1996; Antoun et al. 1998; Biswas et al. 2000a; Yanni et al. 2001; Naidu et al. 2004; Dazzo et al. 2005; Boiero et al. 2007; Chandra et al. 2007; Afzal and Bano 2008; Weyens et al. 2009; Senthilkumar et al. 2009; Vargas et al. 2009; Chi et al. 2010). Although rhizobia produce other plant hormones as well such as cytokinins, gibberellins and abscisic acid, much less information is available in this regard in the literature. Cytokinins usually stimulate cell division and cell enlargement. So far, strains of only *Rhizobium* spp. have been recognized for cytokinin production (Phillips and Torrey 1970, 1972; Newcomb et al. 1977; Puppo and Riguard 1978; Wang et al. 1982; Upadhyaya et al. 1991; Hoflich et al. 1994; Caba et al. 2000; Senthilkumar et al. 2009). Also, gibberellin production by rhizobia has been less widespread yet the strains of *Rhizobium phaseoli, Sinorhizobium meliloti and Rhizobium leguminosarum* have been isolated and demonstrated that these can promote the growth and yield of non-leguminous plants via gibberellin production (Atzorn et al. 1988; Yanni et al. 2001; Chi et al. 2005; Boiero et al. 2007; Humphry et al. 2007; Afzal and Bano 2008). Very little information regarding enhancement in growth of non-legumes through the production of abscisic acid is available. However, Dangar and Basu (1987) and Boiero et al. (2007) have reported *Bradyrhizobium japonicum* as growth promoter of non-legumes via abscisic acid production.

Lowering of plant ethylene levels by producing 1-aminocyclopropane-1-carboxylic acid (ACC) deaminase enzyme is also considered a growth promoting mechanism in non-leguminous plants by rhizobial inoculation. Actually, these rhizobia take up the immediate precursor of the ethylene i.e. ACC and convert it into α-ketobutyrate and NH_3 in order to use it as a source of nitrogen and carbon. Generally, inoculation with ACC-deaminase producing rhizobial strains may result in longer roots and may relieve several forms of stresses such as heavy metals, pathogens, drought, and salinity, etc. Many rhizobial strains have been found to produce ACC-deaminase enzyme such as *Rhizobium leguminosarum* bv. *viciae, Rhizobium hedysari, Rhizobium japonicum, Bradyrhizobium japonicum, Bradyrhizobium elkani, Rhizobium gallicum, Mesorhizobium loti* and *Sinorhizobium meliloti* (Kaneko et al. 2000; Sullivan et al. 2002; Ma et al. 2003a, b, 2004; Okazaki et al. 2004; Uchiumi et al. 2004; Madhaiyan et al. 2006; Hafeez et al. 2008; Duan et al. 2009).

Increase in growth and yield development by providing bio-available phosphorus uptake via rhizobial inoculation is also a well studied mechanism. In fact, a large portion of soil phosphorus as well as of applied phosphorus immobilized either in organic or inorganic form and becomes unavailable to plants depending upon the type and pH of soil. Some rhizobial strains have the ability to solubilize such bound phosphorus from either organic form through releasing phosphatases (Abd-Alla 1994) or inorganic form by releasing organic acids such as 2-ketogluconic acid, glutamic acid, sulphuric acid, nitric acid and carbonic acids (Halder et al. 1990; Halder and Chakraborty 1993; Chabot et al. 1996; Antoun et al. 1998; Dazzo et al. 2000; Arora et al. 2001; Mikanova and Novakova 2002; Alikhani et al. 2006; Jayasinghearachchi and Seneviratne 2006; Afzal and Bano 2008) thereby increasing its uptake and facilitating plant growth. *Rhizobium* genera have been recognized to have the most powerful phosphate solubilizing strains (Abd-Alla 1994; Rodriguez and Fraga 1999). However, the rhizobial strains that have been investigated to solubilize or mineralize phosphate in the rhizosphere of non-legumes include *Rhizobium leguminosarum, Rhizobium meliloti, Mesorhizobium mediterraneum, Bradyrhizobium japonicum* and *Bradyrhizobium* sp. (Wood and Cooper 1984; Halder et al. 1990; Halder and Chakraborty 1993; Chabot et al. 1996; Rodriguez and Fraga 1999; Antoun et al. 1998; Peix et al. 2001; Vessey 2003; Egamberdiyeva et al. 2004; Hara and de Oliveira 2004; Puente et al. 2004; Rodriguez et al. 2006; Fernandez et al. 2007; Afzal and Bano 2008)

Rhizobial sequesterations of iron by producing siderophores are assumed to serve as a source of plant available iron under iron deficient environment. Actually, siderophores are low-molecular-weight compounds produced to sequester Fe^{3+} as they have very high Fe^{3+} affinity constants (10^{25}–10^{52}) (Matzanke et al. 1989) and are used to mobilize iron (Plessner et al. 1993). Iron which is needed by plants for chlorophyll synthesis is often present in the highly insoluble form of ferric hydroxide. Many rhizobia such as *Rhizobium meliloti, Rhizobium tropici, Rhizobium leguminosarum* bv. *viciae, Rhizobium leguminosarum* bv. *trifolii, Rhizobium leguminosarum* bv. *phaseoli, Sinorhizobium meliloti and Bradyrhizobium* sp. produce siderophores that can bind insoluble Fe^{3+}, and reduce it to soluble Fe^{2+}, which is preferred by plants (Schwyn and Neilands 1987; Carson et al. 1992; Guerinot 1991, 1994; Chabot et al. 1993; Plessner et al. 1993; Derylo et al. 1994; Jadhav et al. 1994; Dudeja et al. 1997; Antoun et al. 1998; Carson et al. 2000; Arora et al. 2001).

Lipo-chito-oligosaccharides (LCOs) are bacteria-to-plant signal molecules produced by rhizobia during rhizobia-legume symbiosis (Spaink 1992). In the absence of auxins and cytokinins, LCOs have the ability to restore or resume cell division and embryogenesis in non-legumes (De Jong et al. 1993; Dyachok et al. 2000). Also, rhizosphere application of LCOs at low concentration (10^{-7} to 10^{-9} M) can promote seed germination, early seedling, root mass and root length, whereas leaf spray with micromolar concentration of LCOs (10^{-6}, 10^{-8} or 10^{-10}) can increase photosynthate production and grain yield of non-legumes (Smith et al. 2001, 2002; Miransari and Smith 2009). Rhizobia of the genera *Azorhizobium, Bradyrhizobium, Rhizobium, Mesorhizobium and Sinorhizobium* have been observed to improve seed germination and seedling development in non-legumes. These are also known for increase in foliar photosynthetic rates, induction in expression of flavonoid genes, cell division and embryogenesis via releasing LCOs in the rhizosphere (Spaink 1992; Spaink and Lugtenberg 1994; Xie et al. 1995; Zhang and Smith 2001; Smith et al. 2002).

Likewise, lumichrome is also a signaling compound that can encourage growth of non-legumes and has been identified from the culture filtrate of *Sinorhizobium meliloti* cell (Phillips et al. 1999; Yang et al. 2002; Dakora 2003). It can stimulate seedling development and also help in alleviation of water stress (Phillips et al. 1999; Matiru and Dakora 2005a).

Riboflavin is a vitamin which enzymatically or photochemically may be changed into lumichrome (Yagi 1962; Yanagita and Foster 1956; Phillips et al. 1999) that in turn stimulates plant growth (Dakora et al. 2002) via motivating root respiration (Phillips et al. 1999). Only, few researchers i.e. West and Wilson (1938), Rodelas et al. (1993), Sierra et al. (1999), Phillips et al. (1999), Dakora et al. (2002), Yang et al. (2002) have recognized the riboflavin as growth promoter of non-legumes produced by *Rhizobium, Sinorhizobium meliloti* and *Rhizobium leguminosarum* bv. *viciae*.

3.1.2 Indirect Growth Promoting Activities

Rhizobia promote the growth of non-leguminous plants indirectly by lessening or preventing the injurious effects of one or more phytopathogenic organisms through a variety of mechanisms. These include the ability to produce anti-pathogen metabolites such as antibiotics, fungal cell wall-degrading enzymes or production of volatiles such as hydrogen cyanide, which suppress the growth of pathogens. Rhizobia can produce siderophores that chelate iron, making it unavailable to pathogens. They may also have ability to successfully compete with pathogens for nutrients, the ability to stimulate host plant defense mechanism through induced systemic resistance (ISR), and the ability to increase root adhering soil (RAS).

To prevent the proliferation of phytopathogens via synthesizing antibiotics is one of the most effective mechanisms of PGPR. Rhizobia are capable of producing antimicrobial compounds known as antibiotics, as well as fungal cell-wall-degrading enzymes that are inhibitory to the growth and/or activities of non-legume pathogens. Many rhizobia such as *Rhizobium leguminosarum* bv. *trifolii, Rhizobium leguminosarum* bv. *viciae, Rhizobium meliloti, Rhizobium trifolii, Sinorhizobium meliloti and Bradyrhizobium japonicum* have been reported to secrete antibiotics and cell-wall-degrading enzymes that can inhibit/kill the pathogens of non-leguminous plants (Schwinghamer and Belkengren 1968; Antoun et al. 1978; Tu 1978, 1979;

Parveen and Ghaffar 1991; Breil et al. 1993; Parveen et al. 1993; Ehteshamul-Haque and Ghaffar 1992, 1993; Ehteshamul-Haque et al. 1996; Siddiqui et al. 1998, 2000; Siddiqui and Mahmood 2001; Ozkoc and Deliveli 2001; Shaukat and Siddqui 2003; Bardin et al. 2004; Chandra et al. 2007).

Rhizobial strains have also been found capable of displacing pathogens through better competition for nutrients. For example, rhizobia can inhibit a widely occurring plant pathogen *Macrophomina phaseolina* by producing high-affinity iron binding siderophores (Arora et al. 2001) under iron limiting conditions which sequester iron in the rhizosphere and make it less available to deleterious microorganisms. Rhizobia thus out-compete the pathogens for available iron, thus causing death of pathogens or limit their growth.

Furthermore, rhizobia can inhibit pathogens or deleterious microorganisms by producing HCN (Wei et al. 1996; Antoun et al. 1998; Vidhyasekaran and Muthamilan 1999; Chandra et al. 2007). Some reports show that the rhizobial strains such as *Rhizobium leguminosarum* bv. *phaseoli, Rhizobium leguminosarum* bv. *trifolii, Rhizobium leguminosarum* bv. *viciae* and *Mesorhizobium loti* are able to produce HCN (Antoun et al. 1998; Chandra et al. 2007).

Defensive capacity of plants can be enhanced when they are stimulated appropriately. Use of selected PGPR strains could trigger resistance in plant parts upon inoculation. Non-leguminous plants such as rice, sunflower, okra and potato can be protected against viral, bacterial or fungal attack through physiological immunity which can be induced with the colonization of their roots by certain rhizobial species (Ehteshamul-Haque and Ghaffar 1993; Abdelaziz et al. 1996; Nautiyal 1997, 2000). This process has been termed as induced systemic resistance (ISR). In ISR process, it is considered that the inducing rhizobacteria in the plant roots generate signal, which spreads systemically inside the plant and boosts the defensive capacity of the distant tissues from the succeeding infection by the pathogens. Various rhizobial species such as *Rhizobium etli, Rhizobium leguminosarum* bv. *phaseoli* and *Rhizobium leguminosarum* bv. *trifolii* can induce systemic resistance in different non-legumes by producing various biostimulatory agents (Yanni et al. 2001; Peng et al. 2002; Mishra et al. 2006; Singh et al. 2006).

The importance of root adhering soil (RAS) is well established because of the uptake of water and nutrients required by the plants for their growth uptaken from the soil. Therefore, around the root system, soil structure and aggregate stability is even more important. *Rhizobium alamii* sp. nov., *Rhizobium phaseoli, Rhizobium trifolii, Rhizobium meliloti, Rhizobium* and *Rhizobium leguminosarum* are able to produce compounds such as exopolysaccharides (EPS) (Martens 1982; Alami et al. 1998, 2000; Kaci et al. 2005; Berge et al. 2009). These rhizobial EPS can improve plant growth by increasing RAS through better control of water and nutrient uptake.

Conclusively, growth and yield of non-leguminous plants could be triggered by rhizobial inoculation directly via producing phyto-effective metabolites such as plant hormones, enzymes, lipo-chito-oligosaccharides, siderophores, lumichrome, riboflavin, etc. by increasing the uptake of nutrients or by increasing resistance in them against biotic and abiotic stresses. Whereas, rhizobia can indirectly promote the growth of non-leguminous plants by inhibiting the phytopathogens via producing antibiotics, fungal cell wall-degrading enzymes and hydrogen cyanide or through starvation via siderophore production. Also, rhizobia can increase the growth and yield of non-legumes indirectly through improving their defense mechanism or by increasing root adhering soil.

3.2 *Single and Mixed Population of Microbes*

No doubt, the exploitation of single microbe is easy and economical. Also, the commercial formulations of single inoculants are more stable than consortium inoculants. But it is also a proven fact that a single PGPR agent may sometimes account for the inconsistent performance because of its inadequate colonization, inadaptability under variable soil environments or inability against all pathogens that attack the host plant, as reported by Raupach and Kloepper (1998) that a single PGPR agent may not perform well at all times in all kinds of soil environment. In contrast, mixture of PGPR agents that need diverse pH, moisture and temperature conditions, may adapt better to the environmental changes that occur throughout the growing season, may colonize roots more aggressively, may protect the plant against a broader range of pathogens and utilize a wider array of plant growth mechanisms. Earlier, Hoflich et al. (1994) reported that a combination of microorganisms with different metabolic capacities (N_2-fixation, P-mobilization; production of phytohormones and antibiotics) can partly surpass the effect of single inoculations, or can produce a positive effect where single inoculations are ineffective. In other words, it could be said that plant benefits derived from the co-inoculation/mixed inoculation will be superior to that of plant inoculated with single PGPR agent (Xavier and Germida 2002; Adesemoye et al. 2008). Conclusively, despite exploiting their individual plant growth promoting capacity, the potential of selected rhizobia can be improved further through dual/mix inoculation with other microorganisms for additive and/or synergistic effects (Dobbelaere et al. 2003). In general, the advantages of mixture of PGPR agents include broad spectrum of actions, improved efficacy, reliability and allowance of combination of various traits without genetic engineering (Janisiewicz 1996). But sometimes the incompatibility of mixture inoculants may arise because of which inhibition of each other as well as target pathogens may occur. Hence, for the development of mixture inoculants, compatibility of microorganisms is a prerequisite.

Rhizobial potential to promote the growth of non-legumes can be improved further through dual inoculation with other PGPR. For example, Berg (2009) has reported that on the basis of beneficial effects of PGPR and rhizobia, studies using inoculant mixtures were very promising. Also, for providing a more balanced nutrition to plants and to improve the performance of plants in nutrient-deficient and degraded habitats, a compatible combined inoculation of rhizobia, arbuscular mycorrhizal (AM) fungi and PGPR strains is more effective than a single microorganism (Algawadi and Gaur 1988; Belimov et al. 1995; Requena et al. 1997; Galleguillos et al. 2000; Prasad and Chandra 2003; Gunasekaran et al. 2004). Similarly,

Valdes et al. (1993) revealed that rhizobia and VAM fungi when used together synergistically stimulate plant growth. A number of researchers have reported improved yields of many non-leguminous crops such as sorghum, barley, rice, maize, wheat, sugar beet, okra and lettuce by using rhizobia in combination with other PGPR (Algawadi and Gaur 1992; Chabot et al. 1993; Hoflich et al. 1994; Belimov et al. 1995; Yanni et al. 1997; Galal 2003; Sahin et al. 2004; Han and Lee 2005; Sheikh et al. 2006; Afzal and Bano 2008).

Different combinations of rhizobia with other PGPR which have been studied include *Rhizobium*, and *Azotobacter* (Barea et al. 1975), *Rhizobium* with *Azospirillum* or with *Azotobacter* (Burns et al. 1981; Iruthayathas et al. 1983; Plazinski and Rolfe 1985a, b; Sarig et al. 1986; Rodelas et al. 1996; Burdman et al. 1998, 2000). *Bradyrhizobium japonicum* and *Aeromonas* and *Pseudomonas* spp. (Polonenko et al. 1987), *Rhizobium leguminosarum* and *Pseudomonas putida, Pseudomonas fluorescens, Pseudomonas fonticola* (Chanway et al. 1989), *Rhizobium leguminosarum* and *Azospirillum* (Vidhyasekaran and Muthamilan 1995), *Bradyrhizobium japonicum, Rhizobium* sp. and mycorrhizal fungus *Glomus mosseae* (Xie et al. 1995), *Rhizobium etli, Rhizobium tropici* and *Azospirillum brasilense* (Burdman et al. 1996), *Bradyrhizobium* and *Pseudomonas proteamaculans, Pseudomonas liquefaciens* (Zhang et al. 1997), *Enterobacter* and *Bradyrhizobium* sp. (Gupta et al. 1998), *Bradyrhizobium japonicum* and *Serratia proteamaculans* (Dashti et al. 1998), *Sinorhizobium meliloti* and *arbuscular mycorrhizal fungi* i.e. *Glomus intraradices* (Galleguillos et al. 2000), *Mesorhizobium* and *Pseudomonas* (Goel et al. 2002; Sindhu et al. 2002), *Fluorescent Pseudomonas* and *Rhizobium* (Kumar et al. 2001; Chebotar et al. 2001), *Bradyrhizobium japonicum* and *Bacillus thuringiensis* (Bai et al. 2003), *Rhizobium leguminosarum* (*Rhizobium* sp. strain Thal-8/SK8) and phosphorus solubilizing bacteria (*Pseudomonas* sp. strain 54RB) (Afzal and Bano 2008).

Finally, the applicability of rhizobia in combination with other beneficial bacteria to improve growth of non-legumes has been proven more effective, however, the research is very rare in this regard. Hence, there is a need to explore a wide range of combinations of rhizobia with other PGPR that interact synergistically in order to result promising yields of various important non-legumes.

3.3 Nitrogen Fixation in Non-legumes

Fixation of atmospheric nitrogen in leguminous plants by symbiotic bacteria is a popular phenomenon. But the extension of nitrogen-fixing symbiosis to cereals has been a long-standing goal in the field of biological nitrogen fixation. So, among the efforts, there are some investigations which have shown that the N_2-fixing microorganisms are fixing N_2 through exogenous or endogenous symbiosis with some of non-leguminous plants. The most recognized example is of Parasponia (a non-leguminous plant) in which *Rhizobium* forms nodules and fixes N_2 in a similar fashion as in leguminous plants (Trinick 1979; Werner 1992) and of oilseed rape nodulation stimulation by *Rhizobhium parasponium* RP 501 and *Bradyrhizobium* CP 283 (Cocking et al. 1992). Moreover, nodule-like structures or hypertrophies or outgrowth on roots formed by rhizobia have also been observed in oilseed rape, rice, *Arabidopsis thaliana*, and other non-legumes (Al-Mallah et al. 1989, 1990; Bender et al. 1990; Rolfe and Bender 1990; Jing et al. 1990, 1992; Li et al. 1991; Ridge et al. 1992; Spencer et al. 1994; De Bruijn et al. 1995; Trinick and Habdodas 1995). Furthermore, Velazquez et al. (2005) reported that coexistence of symbiosis and pathogenicity-determining genes has been shown to occur in strains of *Rhizobium rhizogenes* which enabled them to induce nodules or tumors depending on plant species. Also, Sabry et al. (1997) observed increased dry weight and nitrogen contents as a result of nitrogenase activity of *Azorhizobium caulinodans* in inoculated wheat grown in pot under controlled conditions and demonstrated the involvement of symbiotic nitrogen fixation. Naidu et al. (2004) also observed nitrogenase activity in rice plants inoculated with *Azorhizobium caulinodans* in a hydroponic study. Low level of nitrogen fixation was also observed by Chen et al. (1992), Yu and Kennedy (1995) and Cocking et al. (1995) in wheat inoculated with rhizobia. These discoveries encouraged to make attempts to extend the host range of *Rhizobium* from legumes to non-legumes. But a number of inoculation experiments failed to show a substantial contribution of biological nitrogen fixation to plant growth in most cases. For example, Spencer et al. (1994) did not detect nitrogenase activity in studies of the interaction between rhizobia and cultured potato tissues. Riggs et al. (2001) have reported that inoculation with different strains of diazotrophs did not relieve the N-deficiency symptoms of unfertilized maize in either field or greenhouse assays. However, Quispel (1991) has suggested that only in endophytic systems are the prerequisites for effective nitrogen fixation likely to be fulfilled in interactions between non-legumes and diazotrophic bacteria.

Presently, the endophytic diazotrophs for which evidence exists that their plant stimulation effect is related to their ability to fix N_2 include the *Azoarcus* sp., *Burkholderia* sp., *Gluconacetobacter diazotrophicus* and *Herbaspirillum* sp. (Vessey, 2003). Several other studies have also indicated the contribution of biological nitrogen fixation (BNF) in the growth promotion of non-leguminous plants through associative interaction with diazotrophs. For example, in a study conducted by Ladha et al. (1987) it was reported that out of total needed nitrogen by the rice plants, 20–25% was supplied by associative nitrogen fixation. Higher contribution in required nitrogen level to rice plants via biological nitrogen fixation (BNF) was reported by Baldani et al. (2000) when rice seedlings were inoculated with strains of *Burkholderia* spp. Whereas, Sevilla and Kennedy (2000) have revealed the involvement of the biologically fixed nitrogen in the observed growth promotion of rice upon inoculation with *Acetobacter diazotrophicus*. Christiansen-Weniger (1997) also reported the colonization of 2,4-D-induced para-nodules of rice by *Azospirillum*. Some other diazotrophs such as *Pantoea* sp., *Ochrobactum* sp., *Herbaspirillum* sp., *Pantoea agglomerans* YS19 and *Azoarcus* sp. have been proposed to be responsible for the supply of biologically fixed nitrogen to their host rice plant (Ladha et al. 1997; Verma et al. 2004; You et al. 2005; Feng et al. 2006; Reinhold-Hurek et al. 2006). Furthermore, Gantar and Elhai (1999) elaborated the possibility of colonization of wheat para-nodules by Cyanobacterium *Nostoc* spp. strain 2S9B and of N_2 fixation. Also, increase in fixed nitrogen in 2,4-D induced para-nodulating maize plants by *Azospirillum* have been stated by Christiansen-Weniger and Vanderleyden (1994). Moreover, enhanced nitrogenase activity in wheat with *Azospillum brasilense* and *Azospirillum lipoferum* was reported by Kennedy and Tchan (1992), whereas the same activity was noticed by Saikia et al. (2004) in maize plants treated with *Azospirillum*. Regarding various attempts which have been made so far, to extend the host range of *Rhizbium* from legumes to non-legumes through plant genetic manipulation have seen little or no success in achieving the induction of symbiosis between cereals and diazotrophs (Saikia and Jain 2007), however, various N_2-fixing diazotrophs found in close association with non-leguminous plant roots such as *Azospirillum lipoferum*, *Azospirillum brasilense* and *Azospirillum amazonense* have been isolated from the roots of rice (Ladha et al. 1982; Pereira et al. 1988).

No doubt, the evidence regarding biological nitrogen fixation in non-legumes has been reviewed in the above discussion, yet the work is in its infancy and requires much more advances.

4 Practical Applications

4.1 Biofertilization

The use of plant growth promoting rhizobacterial inoculants, as biofertilizers, is now common in many regions of the world with the aim of reducing the application of potentially polluting industrial fertilizers and pesticides (Burdman et al. 2000). Whereas, "Biofertilizer" designates the biological product or a substance which contains living microorganisms which, when applied to seed, plant surfaces, or soil, the microorganisms colonize the rhizosphere or the interior of the plant and promote growth by increasing the supply or availability of primary nutrients to the host plant (Vessey 2003). So, a biofertilizer increases crop growth through combinations of biological nitrogen fixation, growth promoting or hormonal substances, and increased availability of soil nutrients (Cocking 2003). Hence, the importance of biofertilizers lies in their ability to supplement/mobilize soil nutrients with minimal use of non-renewable resources. The utility of rhizobial strains as biofertilizers has been demonstrated by Bardin et al. (2004) and Chi et al. (2005) so, as a biofertilizer, rhizobia should promote the growth of plants by supplying symbiotically fixed nitrogen, phytohormones and by increasing the availability/solubility of insoluble/fixed nutrients. Furthermore, Canbolat et al. (2006) recognize the biofertilizers as an alternative to mineral fertilizers, whereas Galal et al. (2001) demonstrated that biofertilization technology can minimize the costs of production and at the same time, avoid the environmental hazards. There are many investigations which are confirming the ability of rhizobia to colonize the non-legumes (Gaur et al.

1980; Ladha et al. 1989; Al-Mallah et al. 1990; Yanni et al. 1995; Hoflich et al. 1995; Chabot et al. 1996; Schloter et al. 1997; Webster et al. 1997; Prayitno et al. 1999; Antoun and Prevost 2000; Gopalaswamy et al. 2000; Bhattacharjee et al. 2008; Ji et al. 2010; Chi et al. 2010), produce phyto-effective hormones (Phillips and Torrey 1970, 1972; Puppo and Riguard 1978; Atzorn et al. 1988; Law and Strijdom 1989; Hunter and Kuykendall 1990; Upadhyaya et al. 1991; Hoflich et al. 1995; Noel et al. 1996; Minamisawa et al. 1996; Hirsch et al. 1997; Antoun et al. 1998; Yanni et al. 2001; Vessey 2003; Naidu et al. 2004; Chi et al. 2005; Dazzo et al. 2005; Matiru et al. 2005; Boiero et al. 2007; Chandra et al. 2007; Pandey and Maheshwari 2007; Pena and Reyes 2007; Humphry et al. 2007; Afzal and Bano 2008; Naher et al. 2009; Senthilkumar et al. 2009), and increase uptake of nutrients (Amara and Dahdoh 1997; Khokhar and Qureshi 1998; Alami et al. 1999; Biswas et al. 2000a; Yanni et al. 2001; Hafeez et al. 2004; Etesami et al. 2009a, b). Rodriguez and Fraga (1999) showed the commercial production of biofertilizers containing the strains of *Rhizobium* and their usage with different crops. Improvements in the growth and yield parameters of different non-leguminous crops (Table 1) as a result of rhizobial biofertilization through production of phyto-effective metabolites and by mobilization or increased uptake of nutrients, which are discussed below:

The efficacy of various rhizobial strains to enhance the growth and yield of wheat through hormone production or by increasing uptake of nutrients has been assessed in a number of studies (Hoflich et al. 1994; Amara and Dahdoh 1997; Galal et al. 2001; Anyia et al. 2004; Egamberdiyeva et al. 2004; Lupwayi et al. 2004; Afzal and Bano 2008; Etesami et al. 2009a, b). These authors have reported that rhizobial strains used as a biofertilizer on wheat promoted the growth via increased uptake of nutrients (such as N, P, K, Na, Zn, Fe and Cu), whereas, in some other studies it has been reported that the increased growth and yield when wheat was biofertilized with rhizobia was via production of phyto-effective metabolites (Kavimandan 1985; Law and Strijdom 1989; Hoflich 2000; Monteleone et al. 2003).

The growth promoting ability of rhizobial inoculants in maize has been examined in a number of studies via increased uptake of nutrients (such as N, P, Mg, Fe, Cu) (Gaur et al. 1980; Fyson and Oaks 1990; Chabot et al. 1993, 1996, 1998; Pineda et al. 1994; Hoflich et al. 1994; Schloter et al. 1997; Hoflich 2000; Prevost et al. 2000; Peng et al. 2002; Egamberdiyeva et al. 2004; Matiru and Dakora 2005b). In some other studies, increased maize growth and yield has been reported via hormone (i.e. auxin, cytokinine) production by using rhizobial biofertilizers (Hoflich et al. 1994; Gutierrez-Zamora and Martinez-Romero 2001; Souleimanov et al. 2002).

Likewise, a number of researchers have reported an increased growth and yield in rice upon rhizobial inoculation via hormone production (Reddy et al. 1997; Yanni et al. 1997, 2000; Biswas et al. 2000a, b; Perrine et al. 2001; Chi et al. 2005) such as auxin, and gibberellic acid. Whereas, in some other studies increased growth and yield of rice plants has been reported by rhizobial inoculation via improved uptake of nutrients such as N, P, K, Ca, Mg, Zn, Na, Mo and Fe (Bashan et al. 1990; Yanni et al. 2001; Biswas 1998; Biswas et al. 2000a; Peng et al. 2002; Singh et al. 2006).

Improvement in growth of canola and lettuce with rhizobial inoculation has been observed by Chabot et al. (1993) possibly with the production of indole-3-acetic

Table 26.1 Yield improvement by *Rhizobium* inoculation and proposed mechanism(s) of action in non-legumes

Host plant	Rhizobium	Growing conditions	Proposed mechanism(s) of action	% increase over control	Reference
Rice	*Rhizobium leguminosarum* bv. trifolii strain E11 and E12	Field trial	Production of indole-3-acetic acid and gibberellin (GA7)	Grain yield by 46 and 42%	Yanni et al. (1997)
	Azorhizobium caulinodans strains (TAL-1926 and IBRG-42)	Pot trial	Not described	Shoot dry weight by 120 and 58%, %N by 28 and 19 and total N by 140 and 68%	Khokhar and Qureshi (1998)
	Rhizobium sp.	Laboratory and greenhouse	Root colonization, N_2-fixation	Bacterization promoted rice plant growth	Prayitno et al. (1999)
	R. leguminosarum bv. trifolii E11, *Rhizobium* sp. IRBG 74 and *Bradyrhizobium* sp. IRBG271	Pot trial	Indole-3-acetic acid	Grain yield up to 22%, straw yield up to 19% and NPK uptake up to 28%	Biswas et al. (2000a)
	R. leguminosarum bv. trifolii E11, *Rhizobium* sp. IRBG74	Laboratory and greenhouse	Production of indole-3-acetic acid	Grain yield by 17 and 12% while straw yield by 20 and 15%, respectively.	Biswas et al. (2000b)
	Bradyrhizobium spp.	Greenhouse	Root colonization, N_2-fixation, auxin production	Grain yield and shoot growth by 20%	Chaintreuil et al. (2000)
	Rhizobium leguminosarum bv. trifolii E11	Axenic conditions	Production of indole-3-acetic acid	Significant increase in plant biomass	Dazzo et al. (2000)
	Azorhizobium caulinodans	Gnotobiotic and field	Colonization and N_2-fixation	Plant height (13%), shoot dry weight by 8%, fresh weight by 7%.	Nieuwenhove et al. (2000)
	Rhizobium leguminosarum bv. trifolii, *Rhizobium leguminosarum* bv. viciae	Laboratory and greenhouse	Production of auxin and nitrate, and root colonization	Inoculation could either promote, inhibit or have no influence on plant growth	Perrine et al. (2001)

Rhizobium leguminosarum bv. trifolii strain E12, IRBG74, *Bradyrhizobium* sp. strain IRBG271	Pot trial	Not described	Grain yield up to 16%, photosynthetic rate up to 12%	Peng et al. (2002)
Rhizobium sp.	Pot trial	Production of indole-3-acetic acid, gibberellin and root colonization	Grain yield by 20-69%, shoot dry mass 11-43%	Chi et al. (2005)
Rhizobium leguminosarum bv. phaseoli (RRE6), *Rhizobium leguminosarum* bv. trifolii (ANU 843)	Greenhouse	Biocontrol/production of phenolics	Root/shoot dry weight up to 26%, grain yield up to 15%	Mishra et al. (2006)
Rhizobium sp.	Greenhouse	N_2-fixation and root colonization	Strains RRE3, RRE5, and RRE6 increased biomass up to 21%, grain yield up to 13%	Singh et al. (2006)
Rhizobium leguminosarum bv. trifolii strain R4	Magenta jars	N_2-fixation	Strain R4 improved plant shoot and root dry mass compared to other strains	Perrine-Walker et al. (2007a)
Rhizobium phaseoli, Rhizobium leguminosarum, Mesorhizobium ciceri	Pot trial	Not described	*R. phaseoli* strain LSI 29 increased the paddy yield and biomass by 43 and 18%, respectively	Hussain et al. (2009)
Bradyrhizobium strains UPMR48, UPMR29	Laboratory assay	Not described	Dry biomass by 6 and 11%, vigor index by 21 and 20%, respectively	Mia and Shamsuddin (2009)
Rhizobium sp.	Axenic trial	Rhizosphere and root colonization, and N_2-fixation	Plant biomass by 36%, tissue N by 4.47%	Naher et al. (2009)

(continued)

Table 26.1 (continued)

Host plant	Rhizobium	Growing conditions	Proposed mechanism(s) of action	% increase over control	Reference
	Azorhizobium caulinodans strain ORS 571	Gnotobiotic and hydroponics	Production of indole-3-acetic acid, cytokinins and nitrogenase activity	Biomass and total N increased up to 70 and 50%, respectively	Senthilkumar et al. (2009)
	Rhizobium sp.	Axenic condition	Production of IAA, siderophores and P-solubilization	Dry seedling weight and germination vigor increased up to 58 and 43%, respectively	Vargas et al. (2009)
	Sinorhizobium meliloti 1021	Axenic and greenhouse	Nutrient uptake and production of indole-3-acetic acid	Shoot growth was increased up to 44%	Chi et al. (2010)
Wheat	*R. leguminosarum* bv. Trifolii strain R39	Greenhouse and field trials	Production of auxins and cytokinins	Strain R39 resulted in significantly higher wheat yield by 8%	Hoflich et al. (1994)
	Rhizobium sp.	Greenhouse	Nutrient solubilization/ uptake	Grain yield by 106%	Amara and Dahdoh (1997)
	Rhizobium leguminosarum bv. trifolii	Greenhouse	Not described	Strain IAT168 increased the shoot dry matter by 24%	Hilali et al. (2001)
	Rhizobium strain (KYGT207)	Greenhouse	Production of exopolysaccharides	Strain KYGT207 increased shoot and root dry mass by 85 and 56%, respectively	Kaci et al. (2005)
	Rhizobium leguminosarum strain (Thal 8)	Pot trial	P-solubilization	Grain yield up to 30%	Afzal and Bano (2008)
	Rhizobium leguminosarum var. phaseoli, Rhizobium leguminosarum var. viciae	Greenhouse trial	Production of indole-3-acetic acid and nutrient solubilization	Root dry weight was increased up to 24%	Etesami et al. (2009a)
	Rhizobium sp.	Pot trial	Production of indole-3-acetic acid	Shoot and root weight was increased up to 37%	Etesami et al. (2009b)

Maize/Corn	*Rhizobium* sp.	Laboratory and Field trials	P-solubilization, Siderophores and auxins production	*Rhizobium* sp. increased lettuce yield only by 11%	Chabot et al. (1993)
	R. leguminosarum bv. trifolii strain R39	Greenhouse and field trials	Production of auxins and cytokinins	Inoculation increased maize yield up to 10%	Hoflich et al. (1994)
	Rhizobium leguminosarum bv. phaseoli	Laboratory and pot trials	P-solubilization	Strain P31 increased the maize shoot dry matter by 8% with full dose of P fertilizer, strain P31 increased lettuce dry matter by 15% with half P dose	Chabot et al. (1996)
	R. leguminosarum bv. Phaseoli	Pot trial	P-solubilization and root colonization	Inoculation improved maize yield up to 30% in high P soil	Chabot et al. (1998)
	Bradyrhizobium japonicum	Greenhouse	Not described	Strain 532C increased the shoot dry matter yield by 8.66%	Prevost et al. (2000)
	Rhizobium strain NCCB 100053, 100054	Greenhouse	Low nutrient solubilization	Plant growth not affected or was inhibited	El-Tarabily et al. (2006)
	Mesorhizobium ciceri, Rhizobium leguminosarum, Rhizobium phaseoli	Axenic conditions	Not described	*Rhizobium* sp. increased the root/shoot length and seedling biomass up to 48, 21 and 61%, respectively	Mehboob et al. (2008)
Barley	*Bradyrhizobium japonicum*	Hydroponic conditions	Not described	Barley root length and fresh weight by 12.9 and 6.3%, respectively	Carletti et al. (1994)

(continued)

Table 26.1 (continued)

Host plant	Rhizobium	Growing conditions	Proposed mechanism(s) of action	% increase over control	Reference
	R. leguminosarum bv. Trifolii strain R39	Greenhouse and field trials	Production of auxins and cytokinins	Strain R39 resulted in higher yield i.e. up to 16%	Hoflich et al. (1994)
	Mesorhizobium mediterraneum	Growth chamber	P-solubilization	Plant biomass by 56% and P uptake by 100%	Peix et al. (2001)
	Rhizobium radiobacter strains 204	Laboratory and pot trials	Production of indole-3-acetic acid and gibberellic acid	Shoot and root length by 63 and 37%, respectively	Humphry et al. (2007)
	Bradyrhizobium japonicum strain 532C	Laboratory	Lipo-chitooligosaccharides and Gibberellins	Barley germination was increased up to 44%	Miransari and Smith (2009)
Brassica campes-tris/napus	*Rhizobium leguminosarum*	Gnotobiotic conditions	Production of indole-3-acetic acid and cytokinin	Inoculation promoted seedling growth of *Brassica*	Noel et al. (1996)
	Mesorhizobium loti strain MP6	Laboratory and field trials	Biocontrol, production of Hydrocyanic acid, indole-3-acetic acid, and P-solubilization	Grain yield by 52%, *Sclerotinia sclerotiorum* was reduced by 75%	Chandra et al. (2007)
Lettuce	*Rhizobium* sp.	Laboratory and field trials	P-solubilization, Siderophores and auxins production	*Rhizobium* sp. increased lettuce yield by 11%	Chabot et al. (1993)
	Rhizobium sp. strains 33 and 45	Laboratory and greenhouse	Indole-3-acetic acid and P-solubilization	Inoculation showed higher weights of lettuce plantlets	Pena and Reyes (2007)

Sorghum	*Bradyrhizobium japonicum* strain (Tal 110), *Sinorhizobium meliloti* strain 1, R.I. bv. viciae Cn6 and R.I. bv. viciae strain 30.	Leonard jars	Production of indole-3-acetic and nutrient solubilization	Inoculation promoted sorghum shoot fresh growth by 11-51% and shoot dry weight by 8-55%. Shoot P and K by 17-250%	Matiru et al. (2005)
Cotton	*Rhizobium leguminosarum* bv. trifolii (E11)	Growth room conditions	Production of indole-3-acetic acid	Strain E11 increased root dry weight, shoot dry weight, biomass and N uptake by 248, 48, 75 and 57%, respectively	Hafeez et al. (2004)
Sunflower	*Rhizobium* sp. strain (YAS34)	Pot trial	Production of exopolysaccharides	Strain YAS34 showed up to 50 and 70% increase in shoot and root diameter in normal and water stressed- conditions, respectively	Alami et al. (2000)
Radish	*Rhizobium, Bradyrhizobium*	Growth cabinets/ Greenhouse	Production of indole-3-acetic acid and siderophores, P-solubilization	On an average *Rhizobium* and *Bradyrhizobium* improved 25% radish dry matter	Antoun et al. (1998)
Potato	*Rhizobium etli* strain G12	Pot trial	Biocontrol	Growth of potato improved, Nematode infection reduced by 43%	Reitz et al. (2000)
Tomato	*Bradyrhizobium japonicum*	Hydroponic conditions	Not described	Total root length of tomato increased by 87%	Carletti et al. (1994)

acid and cytokinin (Noel et al. 1996; Chabot et al. 1996; Pena and Reyes 2007) and with the increased mineral uptake (Han and Lee 2005). Hoflich et al. (1994) assessed the growth improvement of barley plants by rhizobial inoculation via mobilization of phosphorous by the inoculants. Similarly, Peix et al. (2001) reported a significant improvement in barley yield and uptake of P, N, K, Ca and Mg upon inoculation with *Mesorhizobium mediterraneum* strain. Growth improvements in barley have also been observed by Lynch and Clark (1984) and Humphry et al. (2007) via growth promoting molecules (such as gibberellic acid) released by rhizobia upon inoculation. Inoculation of sunflower with *Rhizobium* sp. promoted its growth via enhanced uptake of nitrogen (Alami et al. 2000). Increase in radish dry matter yield with *Bradyrhizobium japonicum* was recorded by Antoun et al. (1998). Increased growth of sugar beet via indol-3-acetic acid has been reported by Loper and Schroth (1986) when they used rhizobia as inoculant to these plants. Induced growth and yield of *Brassica campestris* by inoculation with *Mesorhizobium loti* MP6 strains has been demonstrated by Chandra et al. (2007). They recorded the production of indole acetic acid and phosphate solubilization ability of the strains. *Bradyrhizobium japonicum* and *Rhizobium leguminosarum* promoted cotton growth (Hafeez et al. 2004) through efficient uptake of nutrients (Egamberdiyeva et al. 2004). Matiru et al. (2005) investigated growth promotion of sorghum and finger millet seedlings as cultured aseptically in Leonard jars as a result of inoculation with *Rhizobium* spp. They also reported the production of indole acetic acid by the strains examined.

Finally, it could be inferred that rhizobia possess the characteristics on the basis of which it could be used safely and economically as biofertilizer and hence can improve growth and yield of non-legumes through biological nitrogen fixation, hormonal substances and increased availability of plant nutrients.

4.2 Bioprotection

Any condition under which a practice whereby survival or activity of a pathogen is reduced through the agency of another living organisms (except by man himself) resulting in the reduction of incidence of disease caused by pathogens, is termed as biocontrol (Garrette 1965). Whereas, Handelsman and Stabb (1996) defined the biocontrol as a process which involves harnessing of disease-suppressive microorganisms to improve plant health.

More simply, it could be said that biocontrol is a process by which one living organism (which is beneficial/PGPR) is used to limit the growth and propagation of another (which is undesirable one/pathogen) or a pathogenic organism is maintained at low inoculum density or is controlled or eradicated by beneficial organisms.

Several PGPR strains of rhizobia serve as potential biocontrol agents and exhibit numerous mechanisms of biocontrol. These mechanisms include competition (Arora et al. 2001), production of metabolites, which affect the pathogen directly such as antibiotics (Schwinghamer and Belkengren 1968; Antoun et al. 1978; Chakraborty and Purkayastha 1984; Deshwal et al. 2003; Bardin et al. 2004;

Chandra et al. 2007), cell wall degrading enzymes (Ehteshamul-Haque and Ghaffar 1993; Ozkoc and Deliveli 2001), siderophores (Schwyn and Neilands 1987; Derylo et al. 1994; Dudeja et al. 1997; Carson et al. 2000; Deshwal et al. 2003), and HCN (Wei et al. 1996; Antoun et al. 1998; Vidhyasekaran and Muthamilan 1999; Chandra et al. 2007). Some of the PGPR strains of rhizobia do not produce metabolites against the pathogens rather they induce systemic resistance in the host plant (Yanni et al. 2001; Peng et al. 2002; Mishra et al. 2006; Singh et al. 2006) which account for the disease protection.

A variety of rhizobial strains have been used as biocontrol agents for disease/pathogen suppression of various non-leguminous plants. *Bradyrhizobium japonicum, Rhizobium meliloti* and *Rhizobium leguminosarum* were used to control okra and sunflower pathogen such as *Macrophomina phaseolina, Rhizoctonia solani, Fusarium solanik* and *Fusarium solani* (Ehteshamul-Haque and Ghaffar 1993). Whereas, Siddiqui et al. (1998, 2000) used *Rhizobium trifolii* and *Bradyrhizobium* sp. to control the root rot of the same crops i.e. okra and sunflower. Reitz et al. (2000) used *Rhizobium etli* strain G12 as biocontrol agent against the Cyst nematode (*Globoderma pallida*) of potato. Bardin et al. (2004) used *Rhizobium leguminosarum* bv. *viciae* as biocontrol agent against sugar beet pathogen *Pythium* causing damping-off disease. Likewise, Chandra et al. (2007) demonstrated that *Mezorhizobium loti* can be used to reduce the incidence of white rot disease of *Brassica campestris* when applied as a biocontrol agent. Also, the ability of *Sinorhizobium meliloti* and *Rhizobium trifolii* to be used as a biocontrol agent against root rot/knot diseases and a pathogen *Fusarium oxysporum* of sunflower and tomato have been demonstrated (Antoun et al. 1978; Siddiqui et al. 1998; Shaukat and Siddqui 2003). Whereas, *Bradyrhizobium japonicum, Rhizobium leguminosarum* and *Sinorhizobium meliloti* have also been reported as successful biocontrol agents against fungal pathogens belonging to the genera *Fusarium, Macrophomina* and *Rhizoctonia* (Ehteshamul-Haque and Ghaffar 1993; Ozkoc and Deliveli 2001). Arora et al. (2001) reported inhibition of *Macrophomina phaseolina,* a widely occurring plant pathogen, by the siderophores produced by rhizobia. Whereas Tu (1978), Ehteshamul-Haque and Ghaffar (1992) and Siddiqui et al. (2000) revealed that *Bradyrhizobium japonicum* can control a disease of mustard and sunflower known as root rot when used as biocontrol agents, whereas Parveen and Ghaffar (1991), Parveen et al. (1993) and Ehteshamul-Haque et al. (1996) elaborated that *Rhizobium meliloti* can control root knot nematodes in okra. Likewise, Mishra et al. (2006) used *Rhizobium leguminosarum* bv. *phaseoli* strain RRE6 and *Rhizobium leguminosarum* bv. *trifolii* strain ANU843 for the control of rice plant pathogen known as *Rhizoctonia solani.*

The ability of rhizobia to protect the non-leguminous plants through induction of systemic resistance against pathogens has also been recognized by various researchers. Mishra et al. (2006) have reported the *Rhizobium leguminosarum* bv. *phaseoli* and *Rhizobium leguminosarum* bv. *trifolii* who protected the rice plants during pathogenic stresses. A large number of researchers have used rhizobial inoculants to control fungal diseases via induction of systemic resistance in various non-leguminous plants such as okra, sunflower and potato (Ehteshamul-Haque and Ghaffar 1993; Nautiyal 1997, 2000; Reitz et al. 2000).

Conclusively, rhizobia could be used as biocontrol agent against some pathogens of non-legumes. However, there is a need to test this potential of rhizobia against a wider range of phytopathogens.

4.3 Phytoremediation

Land fill, thermal treatment, excavation, acid leaching and electroreclamation are the methods generally used to remediate the contaminants. The great problems in their practical application are large destruction of soil structure and fertility, their high dependence on the contaminants of concern, soil properties, site conditions, their high cost, low efficiency, and so on. Hence, for the heavy metal contaminated soils, the development of phytoremediation strategies is necessary (Chaney et al. 2000; Jing et al. 2007; Khan et al. 2009; Shao et al. 2010), because these are *in situ*, cost-effective, non-intrusive, environment caring, aesthetically pleasing and socially usual (Alkorta and Garbisu 2001; Garbisu et al. 2002).

Actually, phytoremediation includes processes (physical, chemical and biological) by which the plants sequester, extract, and/or detoxify the contaminants (Cunningham et al. 1996; Wenzel et al. 1999). Phytoremediation technologies can be applied to clean up and/or stabilize both organic and inorganic contaminants but several barriers (such as evapotranspiration of volatile contaminants, phytotoxicity of the contaminants and/or degradation intermediates of the contaminants) are still there in field-scale use of these technologies (Gerhardt et al. 2009; Weyens et al. 2009). To overcome these barriers plant growth promoting rhizopheric and endophytic bacteria having appropriate characteristics can be used (Zhuang et al. 2007). Moreover, plant growth promoting rhizobacteria are important component of phytoremediational technology (Khan et al. 2009; Glick 2010) and provide high efficiency for phytoremediation when used in combination with plants, because they increases the availability and mobility of the contaminants via redox changes, acidification, phosphate solubilization and through releasing chelating agents (Abou-Shanab et al. 2003) on one hand, while on the other hand, they exert beneficial effects on the growth and development of plants (Hoflich et al. 1995; Noel et al. 1996; Yanni et al. 1997) by producing phytohormones, increasing solubility and availability of nutrients, inhibiting the growth of plant pathogens via competition, antibiosis, parasitism and by changing the host plant susceptibility by producing various biostimulatory agents (Breil et al. 1993; Dazzo et al. 2000; Arora et al. 2001; Ozkoc and Deliveli 2001; Dakora 2003; Matiru et al. 2005; Van loon 2007). Furthermore, literature suggests that certain rhizobacterial strains containing ACC deaminase activities could be used successfully to ameliorate the stresses of phytoremediator plants which they experience during remediation of contaminated soils and promote their growth (Glick and Pasternak 2003; Glick 2010). Also, Dell'Amico et al. (2008) demonstrated that bacterial strains can alleviate the inhibitory effects of heavy metals through the production of IAA, siderophores and ACC deaminase activity. In another study, Burd et al. (1998, 2000) stated that ACC deaminase and siderophore producing plant growth promoting bacteria can help plants to overcome many of the effects of high levels of metal. So, as rhizobial strains have the ability to improve the biomass through the production of IAA, siderophores, increasing nutrient uptake and via ACC deaminase activity of the inoculated plants explained as above in section-II hence could be tried in association with phytoremediator plants to improve their phytoremediation efficiency.

4.4 Stress Regulation

Environmental stresses such as salinity, drought, and heat affect significantly the crop yields (Mendelsohn and Rosenberg 1994; Kibblewhite et al. 2008). Likewise, heat stress is a severe threat to world agriculture (Mendelsohn and Rosenberg 1994). An accelerated production of ethylene in soil and plants have been reported under various types of stresses such as heavy metals (Reed and Glick 2005), waterlogging (Bradford and Yang 1980), drought (Apelbaum and Yang 1981), and salt (Nadeem et al. 2009, 2010). Plant growth promoting rhizobacteria having ACC deaminase activity are effective in promoting plant growth under these stress conditions by lowering the ethylene whose higher levels have inhibitory effect on root and shoot growth (Wang et al. 2000; Grichko and Glick 2001; Nie et al. 2002; Mayak et al. 2004a, b; Nadeem et al. 2007, 2009; Arshad et al. 2008; Zahir et al. 2009). There are several reports of ACC-deaminase activity in *Rhizobium leguminosarum* and *Mesorhizobium loti* (Belimov et al. 2001, 2005; Ma et al. 2003a, b; Sullivan et al. 2002). Also, it has been reported that rhizobial inoculation can alleviate the effect of water stress by altering leaf stomatal conductance, transpiration, photosynthetic capacity and root morphology of the inoculated non-legume plants (Phillips and Torrey 1970; 1972; Phillips et al. 1999; Figueiredo et al. 1999; Alami et al. 2000; Anyia et al. 2004; Matiru and Dakora 2005a; Chi et al. 2005).

No doubt, the above few demonstrations revealed that rhizobia having ACC-deaminase can protect non-legumes against various stresses, yet the extensive research is required to explore the details of the mechanisms involved.

5 Conclusion and Future Prospects

The rhizobial secretions are an important way for non-leguminous plants to respond and alter their environment. Over the last several years, research and technical advances have provided a better understanding of rhizobial strains association and promotion in non-legumes growth. These advances could be applied to agricultural systems to enhance production by increasing crop growth through combinations of biological nitrogen fixation, releasing phyto-effective substances, increasing defense responses against soil-borne pathogens, improving resistance in plants against biotic and abiotic stresses and/or for further favoring in adequate colonization, adaptability under variable soil environments or against all pathogens that attack the host plant through association with other beneficial soil microbes. Also, remediation efforts have further contributed in improving understanding of rhizobial role under contaminated environments. Hence, switching from growing non-leguminous plants to the cultivation of plant-rhizobial communities in order to reach a high productivity under minimal energy and chemical investments and with minimal pressures on the environment is an applicable technology. Accordingly, more research is needed on genetics, molecular biology and ecology of rhizobia and non-leguminous plants to improve the performance and use of rhizobia as bacterial inoculants of non-legumes as well as for attaining sustainable site-specific rhizobial-based agro technologies. So, future research should be focused on:

1. The study of signal-exchange between rhizobia and non-leguminous plants for the understanding of mechanisms affecting colonization.
2. As such, attempts on mechanisms clearing up the synergistic interaction are required so as to discover the biochemical basis of interactions between rhizobia and non-leguminous plants.
3. Additionally, the approach to improve plant growth promoting characters of rhizobial PGPR strains by genetic manipulation through recombinant DNA technology offer charm for investigations.
4. The use of potent and crop-specific promoters that could be activated under the specific soil and environmental conditions is another important approach to successful gene expression in engineered strain that invites the researchers towards its study.
5. Also, the approach of selection of mutants with increased production of phytoeffective molecules by classical genetic methods cannot be underestimated.
6. Attention should also be given to investigations and application of new combinations of rhizobia and other PGPR for improved results.
7. The conditions are yet to be determined which can help in extending the ability of rhizobia for endophytic nitrogen fixation in non-legumes.
8. Rigorous research is also required on the assortment of native or genetically manipulated rhizobial strains showing both ACC-deaminase and other contaminant-specific genes for evolving more valuable phytoremediation schemes.
9. Even more, work on the option of strains useful equally to non-legumes and legumes, especially in the crop rotation systems is also needed.

References

Abd-Alla MH (1994) Use of organic phosphorous by *Rhizobium leguminosarum* bv. *viciae* phosphatases. Biol Fertil Soils 8:216–218

Abdelaziz RA, Radwansamir MA, Abdel-Kader M, Barakat MA (1996) Biocontrol of faba bean root-rot using VA mycorrhizae and its effect on biological nitrogen fixation. Egypt J Microbiol 31:273–286

Abou-Shanab RA, Angle JS, Delorme TA, Chaney RL, van Berkum P, Moawad H, Ghanem K, Ghozlan HA (2003) Rhizobacterial effects on nickel extraction from soil and uptake by *Alyssum murale*. New Phytol 158:219–224

Adesemoye AO, Torbert HA, Kloepper JW (2008) Enhanced plant nutrient use efficiency with PGPR and AMF in an integrated nutrient management system. Can J Microbiol 54:876–886

Afzal A, Bano A (2008) *Rhizobium* and phosphate solubilizing bacteria improve the yield and phosphorus uptake in wheat (*Triticum aestivum*). Int J Agric Biol 10:85–88

Alami Y, Heulin T, Milas M, de Baynast R, Heyraud A, Villain A (1998) Polysaccharide microorganism and method for obtaining same composition containing it and application. European Patent No 97–1624970212

Alami Y, Champolivier L, Merrien A, Heulin T (1999) Role de *rhizobium* sp., rhizobactérie productrice d'exopolysaccharide dans l'agrégation du sol rhizosphérique du tournesol: conséquences sur la croissance de la plante et la résistance de la plante à la contrainte hydrique. OCL – Oleagineux Corps gras Lipides 6:524–528

Alami Y, Achouak WA, Marol C, Heulin T (2000) Rhizosphere soil aggregation and plant growth promotion of sunflower by an exopolysaccharide producing *Rhizobium* sp. strain isolated from sunflower roots. Appl Environ Microbiol 66:3393–3398

Algawadi AR, Gaur AC (1988) Associative effect of *Rhizobium* and phosphate-solubilizing bacteria on the yield and nutrient uptake of chickpea. Plant Soil 105:241–246

Algawadi AR, Gaur AC (1992) Inoculation of *Azospirillum brasilense* and phosphate-solubilizing bacteria on yield of sorghum (*Sorghum bicolor* L. Moench) in dry land. Trop Agric 69:347–350

Alikhani HA, Saleh-Rastin N, Antoun H (2006) Phosphate solubilization activity of rhizobia native to Iranian soils. Plant Soil 287:35–41

Alkorta I, Garbisu C (2001) Phytoremediation of organic contaminants in soils. Bioresource Technol 79:273–276

Al-Mallah MK, Davey MR, Cocking EC (1989) Formation of nodular structures on rice seedlings by rhizobia. J Exp Bot 40:473–478

Al-Mallah MK, Davey MR, Cocking EC (1990) Enzyme treatment, PEG, Bioting and nodule stimulation in white clover by *Rhizobium trifolii*. J Plant Physiol 15:225–258

Alstrom S, Burns RG (1989) Cyanide production by rhizobacteria as a possible mechanism of plant growth inhibition. Biol Fertil Soils 7:232–238

Amara MAT, Dahdoh MSA (1997) Effect of inoculation with plant-growth promoting rhizobacteria (PGPR) on yield and uptake of nutrients by wheat grown on sandy soil. Egypt J Soil Sci 37:467–484

Antoun H, Kloepper JW (2001) Plant growth-promoting rhizobacteria (PGPR). In: Brenner S, Miller JH (eds) Encyclopedia of genetics. Academic Press, New York, pp 1477–1480

Antoun H, Prevost D (2000) PGPR activity of *Rhizobium* with nonleguminous plants. http://www.ag.auburn.ed/~mlowens/argentina/pdf%20manuscripts/prevost.pdf. Accessed on 23 Dec 2010

Antoun H, Bordeleau LM, Gagnon C (1978) Antagonisme entre *Rhizobium melilotiet Fusarium oxysporum* en relation avec lefficacite symbiotique. Can J Plant Sci 58:75–78

Antoun H, Beauchamp CJ, Goussard N, Chabot R, Lalande R (1998) Potential of *Rhizobium* and *Bradyrhizobium* species as plant growth promoting rhizobacteria on non- legumes: effect on radishes (*Raphanus sativus* L.). Plant Soil 4:57–68

Anyia AO, Archambault DJ, Slaski JJ (2004) Growth promoting effects of the diazotroph *Azorhizobium caulinodans* on Canadian wheat cultivars. Environmental Technologies, Alberta Research Council, Vegreville, Alberta, T9C 1T4, Canada. Available on: http://www.arc.ab.ca/. Cited 25 Dec 2010

Apelbaum A, Yang SF (1981) Biosynthesis of stress ethylene induced by water deficit. Plant Physiol 68:594–596

Arora NK, Kang SC, Maheshwari DK (2001) Isolation of siderophore-producing strains of *Rhizobium meliloti* and their biocontrol potential against *Macrophomina phaseolina* that causes charcoal rot of groundnut. Curr Sci 8:673–677

Arshad M, Shaharoona B, Mahmood T (2008) Inoculation with *Pseudomonas* spp. containing ACC deaminase partially eliminates the effects of drought stress on growth, yield, and ripening of pea (*Pisum sativum* L.). Pedosphere 18:611–620

Atzorn R, Crozier A, Wheeler CT, Sandberg G (1988) Production of gibberellins and indol-3-acetic acid by *Rhizobium phaseoli* in relation to nodulation of *Phaseolus vulgaris* roots. Planta 175:532–538

Badenoch-Jones J, Summons RE, Djordjervic MA, Shine J, Letham DS, Rolfe BG (1982) Mass spectrometric quantification of indole-3-acetic acid in *Rhizobium* culture supernatants: relation to root curling and nodule initiation. Appl Environ Microbiol 44:275–280

Bai YM, Zhou X, Smith DL (2003) Enhanced soybean plant growth resulting from coinoculation of *Bacillus* strains with *Bradyrhizobium japnicum*. Crop Sci 43:1774–1781

Baldani VLD, Baldani JI, Dobereiner J (2000) Inoculation of rice plants with the endophytic diazotrophs *Herbaspirillum seropedicae* and *Burkholderia* spp. Biol Fertil Soils 30:485–491

Balley R, Thomas-Bauzon D, Heulin T, Balandreau J, Richard C, De Ley U (1983) Determination of the most frequent N_2-fixing bacteria in the rice rhizosphere. Can J Microbiol 36:881–887

Bardin SD, Huang HC, Pinto J, Amundsen EJ, Erickson RS (2004) Biological control of *Pythium* damping-off of pea and sugar beet by *Rhizobium leguminosarum* bv. Viceae. Can J Bot 82:291–296

Barea JM, Azcon R, Hayman DS (1975) Possible synergistic interactions between endogyme and phosphate-solubilizing bacteria in low phosphate soils. In: Sanders FE, Mose B, Tinker PB (eds) Endomycorrhizas. Academic Press, London, pp 409–417

Bashan Y, Harrison SK, Whitmoyer RE (1990) Enhanced growth of wheat and soybean plants inoculated with *Azospirillum brasilense* is not necessarily due to general enhancement of mineral uptake. Appl Environ Microbiol 56:769–775

Belimov AA, Kojemiakov AP, Chuvarliyeva CV (1995) Interaction between barley and mixed cultures of nitrogen fixing and phosphate-solubilizing bacteria. Plant Soil 17:29–37

Belimov AA, Safronova VI, Sergeyeva TA, Engorova TV, Metveyeva AA, Tsyganov VE, Borisov AY, Tikhonovich IA, Kluge C, Preisfeld A, Dietz K, Stepanok VV (2001) Characterization of plant growth promoting rhizobacteria isolated from polluted soils and containing 1-aminocyclopropane-1-carboxylate deaminase on nutrient status of the plant. Can J Microbiol 47:642–652

Belimov AA, Hontzeas N, Safronova VI, Demchinskaya SV, Piluzza G, Bullitta S, Glick BR (2005) Cadmium-tolerant plant growth-promoting bacteria associated with the roots of Indian mustard (*Brassica juncea* L. Czem.). Soil Biol Biochem 37:241–250

Bender GL, Preston L, Barnard D, Rolfe BG (1990) Formation of nodule-like structures on the roots of the non-legumes rice and wheat. In: Gresshoff PM, Roth LE, Stacey G, Newton W (eds) Nitrogen fixation: achievements and objectives. Chapman and Hall, London, p 825

Bera R, Seal A, Bhattacharyya P, Das TH, Sarkar D, Kangjoo K (2006) Targeted yield concept and a framework of fertilizer recommendation in irrigated rice domains of subtropical India. J Z Univ Sci B 7:963–968

Berg G (2009) Plan-microbe interactions promoting plant growth and health: perspectives for controlled use of microorganisms in agriculture. Appl Microbiol Biotechnol 84:11–18

Berge O, Lodhi A, Brandelet G, Santaella C, Roncato MA, Christen R, Heulin T, Achouak W (2009) *Rhizobium alamii* sp. Nov., an exopolysaccharide-producing species isolated from legume and non-legume rhizosphere. Int J Syst Evol Microbiol 59:367–372

Beveridge CA, Gresshoff PM, Rameau C, Turnbull CGN (2003) Additional signaling compounds are required to orchestrate plant development. J Plant Growth Regul 22:15–24

Bhattacharjee RB, Singh A, Mukhopadyay SN (2008) Use of nitrogen-fixing bacteria as biofertilizer for non-legumes: prospects and challenges. Appl Microbiol Biotechnol 80:199–209

Biederbeck VO, Lupwayi NZ, Hanson KG, Rice WA, Zentner RP (2000) Effect of long-term rotation with letils on rhizosphere ecology and on endophytic rhizobia in wheat. In: Proceedings of the 17th American conference on symbiotic nitrogen fixation. Universite de Laval, Quebec, p 80

Biswas JC (1998) Effect of nitrogen fixing bacteria on growth promotion of lowland rice (*Oryza sativa* L.). PhD thesis, Department of Soil Science, University of Philippines, Los Banos

Biswas JC, Ladha JK, Dazzo FB (2000a) Rhizobial inoculation improves nutrient uptake and growth of lowland rice. Soil Sci Soc Am J 64:1644–1650

Biswas JC, Ladha JK, Dazzo FB, Yanni YG, Rolfe BG (2000b) Rhizobial inoculation influences seedling vigor and yield of rice. Agron J 92:880–886

Bodelier PLE, Wijhuizen AG, Blom CWPM, Laanbroek HJ (1997) Effects of photoperiod on growth of and denitrification by *Pseudomonas chlororaphis* in the root zone of *Glyceria maxima*, studied in a gnotobiotic microcosm. Plant Soil 190:91–103

Boiero L, Perrig D, Masciarelli O, Penna C, Cassan F, Luna V (2007) Phytohormone production by three strains of *Bradyrhizobium japonicum* and possible physiological and technological implications. Appl Microbiol Biotechnol 74:874–880

Bolton GW, Nester EW, Gordon MP (1986) Plant phenolic compounds induce expression of the *Agrobacterium tumefaciens* loci needed for virulence. Science 232:983–985

Bradford KJ, Yang SF (1980) Xylem transport of 1-aminocyclopropane-1-carboxylic acid, an ethylene precursor, in waterlogged tomato plants. Plant Physiol 65:322–326

Breil BT, Ludden PW, Triplett EW (1993) DNA sequence and mutational analysis of genes involved in the production and resistance of the antibiotic peptide trifolitoxin. J Bacteriol 175:3693–3702

Burd GI, Dixon DG, Glick BR (1998) A plant growth promoting bacteria that decreases nickel toxicity in seedlings. Appl Environ Microbiol 64:3663–3668

Burd GI, Dixon DG, Glick BR (2000) Plant growth-promoting bacteria that decrease heavy metal toxicity in plants. Can J Microbiol 46:237–245

Burdman S, Volpin H, Kigel J, Kapulnik Y, Okon Y (1996) Promotion of nod gene inducers and nodulation in common bean (*Phaseolus vulgaris*) roots inoculated with *Azospirillum brasilense*. Appl Environ Microbiol 62:3030–3033

Burdman S, Jurkevitch E, Schwartsburd B, Hampel M, Okon Y (1998) Aggregation in *Azospirillum brasilense*: effects of chemical and physical factors and involvement of extracellular components. Microbiology 144:1989–1999

Burdman S, Jurkevitch E, Okon Y (2000) Recent advances in the use of Plant Growth Promoting Rhizobacteria (PGPR) in agriculture. In: Subba Rao NS, Dommergues YR (eds) Microbial interactions in agriculture and forestry, vol II. Science Publishers Inc., Plymouth, pp 229–250

Burns TA, Bishop PE Jr, Israel DW (1981) Enhanced nodulation of leguminous plants roots by mixed culture of *Azotobacter vinelandii* and *Rhizobium*. Plant Soil 62:399–412

Caba JM, Centeno ML, Fernandez B, Gresshoff PM, Ligero F (2000) Inoculation and nitrate alter phytohormone levels in soybean roots: differences between a supernodulating mutant and the wild type. Planta 211:98–104

Canbolat MY, Barik KK, Cakmarci R, Sabin F (2006) Effects of mineral and biofertilizers on barley growth on compacted soil. Acta Agric Scand 56:324–332

Carletti S, Caceres ER, Llorent B (1994) Growth promotion by PGPR on different plant species growing in hydroponic conditions. In: Improving plant productivity with rhizosphere bacteria. Proceedings of the 3rd International Workshop on Plant growth-promoting rhizobacteria, March 7–11, 1994, Adelaide, Australia

Carson KC, Holliday S, Glenn AR, Dilworth MJ (1992) Siderophore and organic acid production in root nodule bacteria. Arch Microbiol 157:264–271

Carson KC, Meyer JM, Dillworth MJ (2000) Hydroxamate siderophores of root nodule bacteria. Soil Biol Biochem 32:11–21

Cassan F, Perrig D, Sgroy V, Mascoarelli O, Penna C, Luna V (2009) *Azospirillum brasilense* Az39 and *Bradyrhizobium japonicum* E109, inoculated singly or in combination, promote seed germination and early seedling growth in corn (*Zea mays* L.) and soybean (*Glycine max* L.). Eur J Soil Biol 45:28–35

Chabot R, Antoun H, Cescas MP (1993) Stimulation de la croissance du maïs et de la laitue romaine par des microorganismes dissolvant le phosphore inorganique. Can J Microbiol 39:941–947

Chabot R, Antoun H, Cescas MP (1996) Growth promotion of maize and lettuce by phosphate-solubilizing *Rhizobium leguminosarum* biovar *phaseoli*. Plant Soil 184:311–321

Chabot R, Beauchamp CE, Kloepper JW, Antoun H (1998) Effect of phosphorus on root colonization and growth promotion of maize by bioluminescent mutants of phosphate solubilizing *Rhizobium leguminosarum* biovar *phaseoli*. Soil Biol Biochem 30:1615–1618

Chaintreuil C, Giraud E, Prin Y, Lorguin J, Ba A, Gillis M, De laiudie P, Dreyfus B (2000) Photosynthetic *bradyrhizobia* are natural endophytes of the African wild rice *Oryza breviligulata*. Appl Environ Microbiol 66:5437–5447

Chakraborty U, Purkayastha RP (1984) Role of Rhizobitoxine in protecting soybean roots from *Macrophomina phaseolina* infection. Can J Microbiol 30:285–289

Chandra S, Choure K, Dubey RC, Maheshwari DK (2007) Rhizosphere competent *Mesorhizobium loti* MP6 induces root hair curling, inhibits *Sclerotinia sclerotiorum* and enhances growth of Indian mustard (*Brassica csmpestris*). Braz J Microbiol 38:128–130

Chaney RL, Brown SL, Stuczynski RI, Daniels WL, Henry CL, Li YM, Siebielec G, Malik M, Angle JS, Ryan JA, Compton H (2000) Risk assessment and remediation of soils contaminated by mining and smelting of lead, zinc and cadmium. Rev Int Contam Ambient 16:75–192

Chanway CP, Hynes RK, Nelson LM (1989) Plant growth-promoting rhizobacteria: effects on growth and nitrogen fixation of lentil (*Lens esculenta Moench*) and pea (*Pisum sativum L.*). Soil Biol Biochem 21:511–517

Chebotar VKK, Asis CA, Akao S (2001) Production of growth-promoting substances and high colonization ability of rhizobacteria enhance the nitrogen fixation of soybean when coinoculated with *Bradyrhizobium japonicum*. Biol Fertil Soils 34:427–432

Chen CT, Maundu J, Cavaness J, Dandurand LM, Orser CS (1992) Characterization of salt-tolerant and salt-sensitive mutants of *Rhizobium leguminosarum* biovar, *viciae* strain C12046. FEMS Microbiol Lett 90:135–140

Chi F, Shen SH, Cheng HP, Jing YX, Yanni YG, Dazzo FB (2005) Ascending migration of endophytic rhizobia, from roots to leaves, inside rice plants and assessment of benefits to rice growth physiology. Appl Environ Microbiol 71:7271–7278

Chi F, Yang P, Han F, Jing Y, Shen S (2010) Proteomic analysis of rice seedlings infected by *Sinorhizobium meliloti* 1021. Proteomics 10:1861–1874

Chiwocha SD, Abrams SR, Ambrose SJ, Cutler AJ, Loewen M, Ross AR, Kermode AR (2003) A method for profiling classes of plant hormones and their metabolites using liquid chromatography-electrospray ionization tandem mass spectrometry: an analysis of hormone regulation of thermodormancy of lettuce (*Lactuca sativa* L.) seeds. Plant J 35:405–417

Christiansen-Weniger C (1997) Ammonium excreting *Azospirillum brasilense* C3: GUS A inhabiting induced root tumors along stem and roots of rice. Soil Biol Biochem 29:943–950

Christiansen-Weniger C, Vanderleyden J (1994) Ammonia excreting *Azospirillum* sp. become intracellularly established in maize (*Zea mays*) para-nodules. Biol Fertil Soils 17:1–8

Cocking EC (2003) Endophytic colonization of plant roots by nitrogen-fixing bacteria. Plant Soil 252:169–175

Cocking EC, Srivastava JS, Kothari SL, Davey HR (1992) Invasion of non-legume plants by diazotrophic bacteria. In: Khush GS, Bennett J (eds) Nodulation and nitrogen fixation in rice: potential and prospects. International Rice Research Institute, Philippines, pp 119–121

Cocking EC, Webster G, Batchelor CA, Davey MR (1994) Nodulation of non-legume crops: a new look. Agro-Food-Industry. Hi-Tech. Jan 21–24

Cocking EC, Kothjari SL, Batchelor CA, Jain S, Webster G, Jones J, Jotham J, Davey MR (1995) Interaction of *Rhizobium* with non-legume crops for symbiotic nitrogen fixation. In: Fendrik I, Del Gallo M, Vanderleyden J, de Zamaroczy M (eds) *Azospirillum* VI and related microorganisms, NATO ASI Series, Fl G37. Springer, Berlin, pp 197–205

Cunningham SD, Anderson TA, Schwab AP, Hsu FC (1996) Phytoremediation of soil contaminated with organic pollutant. Adv Agron 56:55–114

Dakora FD (2003) Defining new roles for plant and rhizobial molecules in sole and mixed plant cultures involving symbiotic legumes. New Phytol 58:39–49

Dakora FD, Phillips DA (2002) Root exudates as mediators of mineral acquisition in low-nutrient environments. Plant Soil 245:35–47

Dakora FD, Matiru V, King M, Phillips DA (2002) Plant growth promotion in legumes and cereals by lumichrome, a rhizobial signal metabolite. In: Finan TM, O'Brian MR, Layzell DB, Vessey JK, Newton W (eds) Nitrogen fixation: global perspectives. CABI Publishing, Wallingford, pp 321–322

Dangar TK, Basu PS (1987) Studies on plant growth substances, IAA metabolism and nitrogenase activity in root nodules of Phaseolus aureus Roxb. var. mungo. Biol Plantarum 29:350–354

Dardanelli MS, Fernandez FJ, Espuny MR, Rodriguez MA, Soria ME, Gil Serrano AM, Okon Y, Megias M (2008) Effect of *Azospirillum brasilense* coinoculated with *Rhizobium* on *Phaseolus vulgaris* flavonoids and Nod factor production under salt stress. Soil Biol Biochem 40:2713–2721

Dardanelli MS, Manyani H, Gonzalez-Barroso S, Rodriguez-Carvajal MA, Gil-Serrano AM, Espuny MR, Lopez-Baena FJ, Bellogin RA, Cegias M, Ollero FJ (2009) Effect of the presence of the plant growth promoting rhizobacterium (PGPR) *Chryseobacterium balustinum* Aur9 and salt stress in the pattern of flavonoids exuded by soybean roots. Plant Soil 328:483–493

Dashti N, Zhang F, Hynes R, Smith DL (1998) Plant growth promoting rhizobacteria accelerate nodulation and increase nitrogen fixation activity by field grown soybean [*Glycine max* (L.) Merr.] under short growing seasons. Plant Soil 200:205–213

Dazzo FB, Yanni YG, Rizk R, de Bruijn FJ, Rademaker J, Squartini A, Corich V, Mateos P, Martinez-Molina E, Velazquez E, Biswas JC, Hernandez RJ, Ladha JK, Hill J, Weinman J, Rolfe BG, Vega-Hernandez M, Bradford JJ, Hollingsworth RI, Ostrom P, Marshall E, Jain T, Orgambide G, Philip-Hollingsworth S, Triplett E, Malik KA, Maya-Flores J, Hartmann A, Umali-Garcia M, Izaguirre-Mayor ML (2000) Progress in multinational collaborative studies on the beneficial association between *Rhizobium leguminosarum* bv. *trifolii* and rice. In: Ladha JK, Reddy PM (eds) The quest for nitrogen fixation in rice. IRRI, Los Banos, pp 167–189

Dazzo FB, Yanni YG, Rizk R, Zidan M, Abu-Baker, Gomaa M, Squartini A, Jing Y, Chi F, Shen SH (2005) Recent studies on the *Rhizobium* cereal association. In: Wang YP, Lin M, Tian ZX, Elmericj C, Newton WE (eds) Biological nitrogen fixation: sustainable agriculture and the environment. Proceedings of the 14th International Nitrogen Fixation Congress, Springer, Dordrecht, pp 379–380

De Bruijn FJ, Jing Y, Dazzo FB (1995) Potential and pitfalls of trying to extend symbiotic interactions of nitrogen-fixing organisms to presently non-nodulated plants such as rice. Plant Soil 174:225–240

De Jong AJ, Heidstra R, Spaink HP, Hartog MV, Meijer EA, Hendrisk T, Schiavo FL, Terzi M, Bisseling T, van Kammen A, de Vries SC (1993) A plant somatic embryo mutant is rescued by rhizobial lipo-oligosaccharides. Plant Cell 5:615–620

Dell'Amico E, Cavalca L, Andreoni V (2008) Improvement of *Brassica napus* growth under cadmium stress by cadmium-resistant rhizobacteria. Soil Biol Biochem 40:74–84

Depret G, Houot S, Allard MR, Breuil MC, Nouaim R, Laguerre G (2004) Long-term effects of crop management on *Rhizobium leguminosarum* biovar *viciae* populations. FEMS Microbiol Ecol 51:87–97

Derylo M, Choma A, Puchalshi B, Suchanek W (1994) Siderophore activity in *Rhizobium* species isolated from different legumes. Acta Biochim Pol 41:7–11

Deshwal VK, Dubey RC, Maheshwari DK (2003) Isolation of plant growth promoting strains of *Bradyrhizobium* (*Arachis*) sp. with biocontrol potential against *Macrophomina phaseolina* causing charcoal rot of peanut. Curr Sci 84:443–444

Dobbelaere S, Vanderleyden J, Okon Y (2003) Plant growth-promoting effects of diazotrophs in the rhizosphere. Crit Rev Plant Sci 22:107–149

Duan J, Muller KM, Charles TC, Vesely S, Glick BR (2009) 1-Aminocyclopropane-1carboxylate (ACC) deaminase gene in *Rhizobium* from Southern Saskatchewan. Microbial Ecol 57:423–436

Dudeja SS, Suneja S, Khurana AL (1997) Iron acquisition system and its role in legume *Rhizobium* symbiosis. Indian J Microbiol 37:1–12

Dullaart U (1970) The bioproduction of indole-3-acetic acid and related compounds in root nodules and roots of *Lupinus luteus* L. and by its rhizobial symbiont. Acta Bot Neerl 19:573–618

Dyachok JV, Tobin AE, Price NPJ, von Arnold S (2000) Rhizobial Nod factors stimulated somatic embryo development in *Picea abies*. Plant Cell Rep 19:290–297

Egamberdiyeva D, Juraeva D, Poberejskaya S, Myachina O, Teryuhova P, Seydalieva L, Aliev A (2004) Improvement of wheat and cotton growth and nutrient uptake by phosphate solubilizing bacteria. In: The "26th Southern Conservation Tillage Conference for Sustainable Agriculture", June 8–9, 2004. J.S. Mckimmon Centre, North Carolina State University, Raleigh, NC, pp 58–66

Ehteshamul-Haque S, Ghaffar A (1992) Use of *Bradyrhizobium japonicaum* and fungicides in the control of root rot disease of sunflower. In: Ghaffar A, Shahzad S (eds) Proceedings of Status of Plant Pathology in Pakistan, Department of Botany, University of Karachi, Karachi – 75270, Pakistan, pp 261–266

Ehteshamul-Haque S, Ghaffar A (1993) Use of rhizobia in the control of root rot diseases of sunflower, okra, soybean and mungbean. J Phytopathol 138:157–163

Ehteshamul-Haque S, Abid M, Sultana V, Ara J, Ghaffar A (1996) Use of organic amendments on the efficacy of biocontrol agents in the control of root rot and root knot disease complex of okra. Nematol Mediterran 24:13–16

El-Tarabily KA, Soaud AA, Saleh ME, Matsumoto S (2006) Isolation and characterization of sulfur-oxidising bacteria, including strains of *Rhizobium*, from calcareous sandy soils and their effects on nutrient uptake and growth of maize (*Zea mays* L.). Aust J Agric Res 57:101–111

Engelhard M, Hurek T, Reinhold-Hurek B (2000) Preferential occurrence of diazotrophic endophytes, *Azoarcus* spp., in wild rice species and land races of *Oryza sativa* in comparison with modern races. Environ Microbiol 2:131–141

Ernesten A, Sabdverg G, Crozier A, Wheeler CT (1987) Endogenous indoles and the biosynthesis and metabolism of indole-3-acetic acid in cultures of *Rhizobium phaseoli*. Planta 171:422–428

Etesami H, Alikhani HA, Jadidi M, Aliakbari A (2009a) Effect of superior IAA producing rhizobia on N, P, K uptake by wheat grown under greenhouse condition. World Appl Sci J 6:1629–1633

Etesami H, Alikhani HA, Aliakbari A (2009b) Evaluation of plant growth hormones production (IAA) ability by Iranian soils rhizobial strains and effects of superior strains application on wheat growth indexes. World Appl Sci J 6:1576–1584

Feng Y, Shen D, Song W (2006) Rice endophyte *Pantoea agglomerans* YS19 promotes host plant growth and affects allocations of host photosynthates. J Appl Microbiol 100:938–945

Fernandez LS, Zalba P, Gomez MA, Sagardoy MA (2007) Phosphate-solubilization activity of bacterial strains in soil and their effect on soybean growth under greenhouse conditions. Biol Fertil Soils 43:805–809

Figueiredo MVB, Vilar JJ, Burity HA, de Fanca FP (1999) Alleviation of water stress effects in cowpea by *Bradyrhizobium* spp. inoculation. Plant Soil 207:67–75

Fyson A, Oaks A (1990) Growth promotion of maize by legume soils. Plant Soil 122:259–266

Galal YGM (2003) Assessment of nitrogen availability to wheat (*Triticum aestivum* L.) from inorganic and organic N sources as affected by *Azospirillum brasilense* and *Rhizobium leguminosarum* inoculation. Egypt J Microbiol 38:57–73

Galal YG, El-Gandaour JA, El-akel FA (2001) Stimulation of wheat growth and N fixation through *Azospirillum* and *Rhizobium* inoculation. A field trial with ^{15}N techniques. In: Walter H, Schenk MK, Bürkert A, Claassen N, Flessa H, Frommer WB, Goldbach HE, Olfs H-W, Römheld V, Sattelmacher B, Schmidhalter U, Schubert S, von Wirén N, Wittenmayer L (eds) Plant nutrition-food security and sustainability of agroecosystems through basic and applied research. Kluwer Academic Publishers, Dordrecht, pp 666–667

Galleguillos C, Aguire C, Barea JM, Azcon R (2000) Growth promoting effect of two *Sinorhizobium meliloti* strains (A wild type and its genetically modified derivative) on a non-legume plant species in specific interaction with two arbuscular mycorrhizal fungi. Plant Sci 159:57–63

Gantar M, Elhai J (1999) Colonization of wheat para-nodules by the N_2-fixing cyanobacterium *Nostoc* sp. strain 2S9B. New Phytol 141:373–379

Garbisu C, Hernández-Allica J, Barrutia O, Alkorta I, Becerril JM (2002) Phytoremediation: a technology using green plants to remove contaminants from polluted areas. Rev Environ Health 17:173–188

Garcia-Rodriguez T, Alvarez C, Perez-Silva J (1984) Indol-3-acetic acid production by cell-free extracts of *Rhizobium trifolii*. Soil Biol Biochem 16:677–678

Garrette SD (1965) Towards the biological control of soil-borne plant pathogens. In: Baker KF, Snyder W (eds) Ecology of soil-borne plant pathogens. University of California Press, Los Angeles, pp 4–17

Gaur Y, Sen A, Subba Rao N (1980) Improved legume-*Rhizobium* symbiosis by inoculating preceding cereal crop wirth *Rhizobium*. Plant Soil 54:313–316

Gerhardt KE, Huang X-D, Glick BR, Greenberg BM (2009) Pytoremediation and rhizoremediation of organic soil contaminants: potential and challenges. Plant Sci 176:20–30

Glick BR (2010) Using soil bacteria to facilitate phytoremediation. Biotechnol Adv 28:367–374

Glick BR, Pasternak JJ (2003) Molecular biotechnology: principles and applications of recombinant DNA, 3rd edn. ASM, Washington, DC

Glick BR, Karaturovic DM, Newell PC (1995) A novel procedure for rapid isolation of plant growth promoting pseudomonads. Can J Microbiol 41:533–536

Goel AK, Sindhu SS, Dadarwal KR (2002) Stimulation of nodulation and plant growth of chickpea (*Cicer arietinum* L.) by *Pseudomonas* spp. antagonistic to fungal pathogens. Biol Fertil Soils 36:391–396

Gopalaswamy G, Kannaiyan S, O'Callaghan KJ, Davey MR, Cocking EC (2000) The xylem of rice (*Oryza sativa*) is colonized by *Azorhizobium caulinodans*. Proc R Soc Land B 267:103–107

Gough C, Webster G, Vasse J, Galera C, Batchelor C, O'Callaghan K, Davey M, Kothari S, Denarie J, Cocking J (1996) Specific flavonoids stimulate intercellular colonization of non-legumes by *Azorhizobium caulinodans*. In: Staceyh G, Mullin B, Gresshoff PM (eds) Biology of plant-microbe interactions. International Society for Plant-Microbe Interactions, Saint Paul, pp 409–415

Grichko VP, Glick BR (2001) Amelioration of flooding stress by ACC deaminase containing plant growth promoting bacteria. Plant Physiol Biochem 39:11–17

Guerinot ML (1991) Iron uptake and metabolism in the rhizobia/legume symbioses. Plant Soil 130:199–209

Guerinot ML (1994) Microbial iron transport. Annu Rev Microbiol 48:743–772

Gunasekaran S, Balachandar D, Mohanasundaram K (2004) Studies on synergism between *Rhizobium*, plant growth promoting rhizobacteria (PGPR) and phosphate solubilizing bacteria in blackgram. In: Kannaiyan S, Kumar K, Govimdarajan K (eds) Biofertilizer technology for rice based cropping system. Scientific Publishers, Jodhpur, pp 269–273

Gupta A, Saxena AM, Gpal M, Tilak KVBR (1998) Effect of plant growth promoting rhizobacteria on competitive ability of introduced *Bradyrhizobium* sp. (Vigna) for nodulation. Microbiol Res 153:113–117

Gutierrez-Zamora ML, Martinez-Romero E (2001) Natural endophytic association between *Rhizobium etli* and maize (*Zea mays* L.). J Biotechnol 91:117–126

Hafeez FY, Safdar ME, Chaudhry AU, Malik KA (2004) Rhizobial inoculation improves seedling emergence, nutrient uptake and growth of cotton. Aust J Exp Agric 44:617–622

Hafeez FY, Hassan Z, Naeem F, Basher A, Kiran A, Khan SA, Malik KA (2008) *Rhizobium leguminosarum* bv.*viciae* strain LC-31: Analysis of novel bacteriocin and ACC-deaminase gene(s). In: Dakora FD, Chimphango SBM, Valentine AJ, Elmerich C, Newton WE (eds) Biological nitrogen fixation: towards poverty alleviation through sustainable agriculture. Springer, Dordrecht, pp 247–248

Haggag WM (2002) Sustainable agriculture management of plant diseases. J Biol Sci 2:280–284

Halder AK, Chakraborty PK (1993) Solubilization of inorganic phosphate by *Rhizobium*. Folia Microbiol 38:325–330

Halder AK, Mishra AK, Bhattacharyya P, Chakraborty PK (1990) Solubilization of rock phosphate by *Rhizobium* and *Bradyrhizobium*. J Gen Appl Microbiol 36:81–92

Han HS, Lee KD (2005) Plant growth promoting rhizobacteria effect on antioxidant status, photosynthesis, mineral uptake and growth of lettuce under soil salinity. Res J Agric Biol Sci 1:210–215

Handelsman J, Stabb EV (1996) Biocontrol of soilborne plant pathogens. Plant Cell 8:1855–1869

Hara F, de Oliveira LA (2004) Physiological and ecological characteristics of rhizobio isolated deriving of acid and alic soils of Presidente Figueiredo. Acta Amazonica 34:343–357

Hilali A, Prevost D, Broughton WJ, Antoun A (2001) Effects de I'inoculation avec des souches de *Rhizobium leguminosarum* bv. *trifolii* sur la croissance du bl'e dans deux sols du Marco. Can J Microbiol 47:590–593

Hiltner L (1904) Uber neure Erfahrungen und Probleme auf dem Gebiete der Bodenbakterriologie under besonderer berucksichtigung der Grundungung und brache. Arbeiten der Deutschen Landwritschafrtlichen Gesellschaft 98:59–78

Hirsch AM, Fang Y, Asad S, Kapulnik Y (1997) The role of phytohormones in plant-microbe symbioses. Plant Soil 194:171–184

Hoflich G (2000) Colonization and growth promotion of non-legumes by *Rhizobium* bacteria. Microbial Biosystems: New Frontiers. In: Bell CR, Brylinsky M, Johnson-Green P (eds) Proceedings of the 8th international symposium on microbial ecology. Atlantic Canada Soci. Microbial Ecol, Halifax, pp 827–830

Hoflich G, Wiehe W, Kohn G (1994) Plant growth stimulation by inoculation with symbiotic and associative rhizosphere microorganisms. Experienca 50:897–905

Hoflich G, Wiehe W, Hecht-Buchholz C (1995) Rhizosphere colonization of different crops with growth promoting *Pseudomonas* and *Rhizobium* bacteria. Microbiol Res 150:139–147

Hossain MS (2007) Potential use of *rhizobium* spp. to improve growth of non-nitrogen fixing plants. Masters thesis. Department of Soil Sciences, Swedish University of Agricultural Sciences, Uppsala, Sweden

Hossain MS, Martensson A (2008) Potential use of *Rhizobium* spp. to improve fitness of non-nitrogen-fixing plants. Acta Agric Scand B Soil Plant Sci 58:352–358

Humphry DR, Andrews M, Santos SR, James EK, Vinogradova LV, Perin L, Reis VM, Cummings SP (2007) Phylogenetic assignment and mechanism of action of a crop growth promoting *Rhizobium radiobacter* strain used as a biofertilizer on *graminaceous* crops in Russia. Antonie Van Leeuwenhoek 91:105–113

Hunter W, Kuykendall LD (1990) Enhanced nodulation and nitrogen fixation by a revertant of a nodulation-defective *Bradyrhizobium japonicum* tryptophan auxotroph. Appl Environ Microbiol 56:2399–2403

Hussain MB, Mehboob I, Zahir ZA, Naveed M, Asghar HN (2009) Potential of *Rhizobium* species for improving growth and yield of rice. Soil Environ 28:49–55

Iruthayathas EE, Gunasekaran S, Vlassak K (1983) Effect of combined inoculation of *Azospirillum* and *Rhizobium* and N_2-fixation of winged bean and soybean. Sci Hortic 20:231–240

Jadhav RS, Thaker NV, Desai A (1994) Involvement of the siderophores of cowpea *Rhizobium* in the iron nutrition of the peanut. World J Microbiol Biotechnol 10:360–361

James EK, Gyaneshwar P, Barraquio WL, Mathan N, Ladha JK (2000) Endophytic diazotrophs associated with rice. In: Ladha JK, Reddy PM (eds) The quest for nitrogen fixation in rice. International Rice Research Institute, Los Banos, Laguna, pp 119–140

Janisiewicz WJ (1996) Ecological diversity, niche overlap, and coexistence of antagonists used in developing mixtures for biocontrol of postharvest diseases of apples. Phytopathology 86:473–479

Jayasinghearachchi HS, Seneviratne G (2006) Fungal solubilization of rock phosphate is enhanced by forming fungal-rhizobial biofilms. Soil Biol Biochem 38:405–408

Ji K-X, Chi F, Yang M-F, Shen S-H, Jing Y-X, Dazzo FB, Cheng H-P (2010) Movement of rhizobia inside tobacco and lifestyle alternation from endophytes to free-living rhizobia on leaves. J Microbiol Biotechnol 20:238–244

Jing Y, Li G, Jin G, Shan X, Zhang B, Guan C, Li J (1990) Rice root odules with acetylene reduction activity. In: Gresshoff PM, Roth LE, Stacey G, Newton WE (eds) Nitrogen fixation: achievements and objectives. Chapman and Hall, London, p 829

Jing Y, Li G, Shan X (1992) Development of nodule-like structure on rice roots. In: Khush GS, Bennett J (eds) Nodulation and nitrogen fixation in rice. International Rice Research Institute, Manila, pp 123–126

Jing Y-D, He Z-L, Yang X-E (2007) Role of soil rhizobacteria in phytoremediation of heavy metal contaminated soils. J Z Univ Sci B 8:192–207

Kaci Y, Heyraud A, Barakat M, Heulin T (2005) Isolation and identification of an EPS-producing *Rhizobium* strain from arid soil (Algeria): Characterization of its EPS and the effect of inoculation on wheat rhizosphere soil structure. Res Microbiol 156:522–531

Kaneko T, Nakamura Y, Sato S, Asamizu E, Kato T, Sasamoto S, Watanabe A, Idesawa K, Ishikawa A, Kawashima K, Kimura T, Kishida Y, Kiyokawa C, Kohara M, Matsumoto M, Matsuno A, Mochizuki Y, Nakayama S, Nakazaki N, Shimpo S, Sugimoto M, Takeuchi C, Yamada M, Tabata S (2000) Complete genomic sequence of nitrogen-fixing symbiotic bacterium *Mesorhizobium loti*. DNA Res 7:331–338

Kaneko T, Nakamura Y, Sato S, Minamisawa K, Uchiumi T, Sasamoto S, Watanabe A, Idesawa K, Iriguchi M, Kawashima K, Kohara M, Matsumoto M, Shimpo S, Tsuruoka H, Wada T, Yamada M, Tabata S (2002) Complete genome sequence of nitrogen fixing symbiotic bacterium *Bradyrhizobium japonicum* USDA110. DNA Res 9:189–197

Kaneshiro K, Slodki ME, Plattner RD (1983) Tryptophan catabolism of indolepyruvic and indole acetic acids by *Rhizobium japonicum* L-259 mutants. Curr Microbiol 8:301–306

Kavimandan SK (1985) Root nodule bacteria to improve yield of wheat. Plant Soil 86:141–144

Kennedy IR, Tchan YT (1992) Biolocial nitrogen fixation in non-leguminous field crops: Recent advances. Plant Soil 141:98–118

Khan MS, Zaidi A, Wani PA, Oves M (2009) Role of plant growth promoting rhizobacteria in the remediation of metal contaminated soils. Environ Chem Lett 7:1–19

Khokhar SN, Qureshi A (1998) Interaction of *Azorhizobium caulinodans* with different rice cultivars for increased N_2 fixation. In: Malik KA, Mirza MS, Ladha JK (eds) Nitrogen fixation with non-legumes. Kluwer Academic Publishers, London, pp 91–93

Khush GS, Bennett J (1992) Nodulation and nitrogen fixation in rice: potential and prospects International Rice Research Institute Press, Manila

Kibblewhite MG, Ritz K, Swift MJ (2008) Soil health in agricultural systems. Philos Trans R Soc B Biol Sci 363:685–701

Kittell BL, HGelinski DR, Ditta GS (1989) Aromatic aminotransferase activity and indole acetic acid production in *Rhizobium meliloti*. J Bacteriol 171:5458–5466

Kloepper JW (1993) Plant growth promoting rhizobacteria as biological control agents. In: Metting FB Jr (ed) Soil microbial ecology. Marcel Dekker, New York, pp 255–273

Kloepper JW, Schroth MN (1978) Plant growth promoting rhizobacteria on radishes. In: Proceedings of the 4th international conference on plant pathogenic bacteria, vol 2, Station de Pathologie Vegetale et Phytobacteriologie, INRA, Angers, France

Kumar BSD, Berggren I, Martensson AM (2001) Potential for improving pea production by co-inoculation with fluorescent *Pseudomonas* and *Rhizobium*. Plant Soil 229:25–34

Ladha JK, Barraquio WL, Watanabe I (1982) Immunological techniques to identify *Azospirillum* associated with wetland rice. Can J Microbiol 28:478–485

Ladha JK, Padre AT, Punzulan GC, Watanabe I (1987) Nitrogen-fixing (C2H2-reducing) activity and plant growth characters of 16 wetland rice varieties. Soil Sci Plant Nutr 33:91–106

Ladha JK, Garcia M, Miyan S, Padre AT, Watanabe I (1989) Survival of *Azorhizobium caulinodans* in the soils and rhizosphere of wetland rice under *Sesbania rostrata* rice rotation. Appl Environ Microbiol 55:454–460

Ladha JK, de Brujin FJ, Malik KA (1997) Introduction: Assessing opportunities for nitrogen fixation in rice-a frontier project. Plant Soil 194:1–10

Law IJ, Strijdom BW (1989) Inoculation of cowpea and wheat with strains of *Bradyrhizobium* sp. that differ in their production of indole acetic acid. South Afr J Plant Soil 6:161–166

Lemanceau P (1992) Effects benefiques de rhizobacteries sur les plantes: exemple des *Pseudomonas* spp. fluorescents. Agromomie 12:413–437

Li WX, Kodama O, Akatsuka T (1991) Role of oxygenated fatty acids in rice phytoalexin production. Agric Biol Chem 55:1041–1147

Loper JE, Schroth MN (1986) Influence of bacterial sources of indole-3-acetic acid on root elongation of sugar beet. Phytopathology 76:386–389

Lopez-Lara IM, van Der Drift KKMGM, van Brussel AAN, Haverkamp J, Lugtenberg BJJ, Thomas-Oates JE, Spaink HP (1995) Induction of nodule primordial on *Phaseolus* and *Acacia* by lipo-chitin oligosaccharide nodulation signals from broad-host-rang *Rhizobium* strain GRH2. Plant Mol Biol 29:465–477

Lugtenberg B, Kamilova F (2009) Plant-growth-promoting rhizobacteria. Annu Rev Microbiol 63:541–556

Lupwayi NZ, Rice WA, Clyton GW (2000) Endophytic rhizobia in barley and canola in rotation with field peas. In: Proceedings of the 17th North American conference on symbiotic nitrogen fixation. Laval University, Quebec, Canada, p 80

Lupwayi NZ, Clyton GW, Hanson KG, Rice WA, Biederbeck VO (2004) Endophytic rhizobia in barley, wheat and canola roots. Can J Plant Sci 84:37–45

Lynch JM (1990) The rhizosphere. Wiley-Interscience, Chichester

Lynch JM, Clark SJ (1984) Effects of microbial colonization of barley (*Hordeum vulgare* L.) roots on seedling growth. J Appl Bacteriol 56:47–52

Lynch JM, Whipps JM (1991) Substrate flow in the rhoizsphere. In: Keister DL, Cregan PB (eds) The rhizosphere and plant growth. Kluwer, Dordrecht, pp 15–24

Ma W, Guinel FC, Glick BR (2003a) *Rhizobium leguminosarum biovar viciae* 1-aminocyclopropane-1-carboxylate deaminase promotes nodulation of pea plants. Appl Environ Microbiol 69:4396–4402

Ma W, Sebestianova SB, Sebestian J, Burd GI, Guinel FC, Glick BR (2003b) Prevalence of 1-aminocyclopropane-1-carboxylate deaminase in *Rhizobium* spp. Antonie Van Leeuwenhoek 83:285–291

Ma W, Carles TC, Glick BR (2004) Expression of an exogenous 1-aminocyclopropane – 1-carboxylate deaminase gene in *Sinorhizobium meliloti* increases its ability to nodulate alfalfa. Appl Environ Microbiol 70:5891–5897

Madhaiyan M, Poonguzhali S, Ryu JH, Sa TM (2006) Regulation of ethylene levels in canola (*Brassica campestris*) by 1-aminocyclopropane-1-carboxylate deaminase-containing *Methylobacterium fujisawaense*. Planta 224:268–278

Mano H, Morisaki H (2008) Endophytic bacteria in rice plant. Microbes Environ 23:109–117

Martens R (1982) Apparatus to study the quantitative relationships between root exudates and microbial populations in the rhizosphere. Soil Biol Biochem 14:315–317

Martinez-Romero E, Wang ET, Lopez-Merino A, Caballero-Mellado J, Rogel MA, Gandara B, Toledo I, Martinez-Romero J (2000) Ribosomal gene based phylogenies on trial: the case of *Rhizobium* and related genera. In: de Wild PJGM, Bisseling T, Stiekema WJ (eds) Biology of plant-microbe interactions, vol 2. International Society for Molecular Plant-Microbe Interactions, St. Paul, pp 59–64

Matiru VN, Dakora FD (2004) Potential use of rhizobial bacteria as promoters of plant growth for increased yield in landraces of African cereal crops. Afr J Biotechnol 3:1–7

Matiru VN, Dakora FD (2005a) Xylem transport and shoot accumulation of lumichrome, a newly recognized rhizobial signal, alters root respiration, stomatal conductance, leaf transpiration and photosynthetic rates in legumes and cereals. New Phytol 165:847–855

Matiru VN, Dakora FD (2005b) The rhizosphere signal molecule lumichrome alters seedling development in both legumes and cereals. New Phytol 166:439–444

Matiru VN, Jaffer MA, Dakora FD (2000) Rhizobial colonization of roots of African Landraces of sorghum and millet and the effect of sorghum growth and P nutrition. In: Proceedings of the 4th congress of the African Association for biological nitrogen fixation: imperatives for BNF research and application in Africa for the 21st century. African Association for Biological Nitrogen Fixation, Nairobi, Kenya, pp 99–100

Matiru VN, Jaffer MA, Dakora FD (2005) Rhizobial infection of African landraces of sorghum (*Sorghum bicolor* L.) and finger millet (*Eleucine coracana* L.) promotes plant growth and alters tissue nutrient concentration under axenic conditions. Symbiosis 40:7–15

Matzanke BF, Muller-Matzanke G, Raymond KN (1989) Siderophore mediated iron transport. Phys Bioinorg Chem Ser 5:1–121

Mayak S, Tirosh T, Glick BR (2004a) Plant growth-promoting bacteria confer resistance in tomato plants to salt stress. Plant Physiol Biochem 42:565–572

Mayak S, Tirosh T, Glick BR (2004b) Plant growth-promoting bacteria that confer resistance to water stress in tomato and pepper. Plant Sci 166:525–530

McInroy JA, Kloepper JW (1995) Survey of indigenous bacterial endophytes from cotton and sweet corn. Plant Soil 173:337–342

Mehboob I, Zahir ZA, Mahboob A, Shahzad SM, Jawad A, Arshad M (2008) Preliminary screening of *rhizobium* isolates for improving growth of maize seedlings under axenic conditions. Soil Environ 27:64–71

Mehboob I, Naveed M, Zahir ZA (2009) Rhizobial association with non-legumes: mechanisms and applications. Crit Rev Plant Sci 28:432–456

Mei C, Flinn BS (2010) The use of beneficial microbial endophytes for plant biomass and stress tolerance improvement. Recent Pat Biotechnol 4:81–95

Mendelsohn R, Rosenberg NJ (1994) Framework for integrated assessments of global warming impacts. Clim Chang 28:15–44

Mia MAB, Shamsuddin ZH (2009) Enhanced emergence and vigor seedling production of rice through growth promoting bacterial inoculation. Res J Seed Sci 2:96–104

Mia MAB, Shamsuddin ZH (2010) *Rhizobium* as a crop enhancer and biofertilizer for increased cereal production. Afr J Biotechnol 9:6001–6009

Mikanova O, Novakova J (2002) Evaluation of the P-solubilizing activity of soil microorganisms and its sensitivity to soluble phosphate. Rostl Výroba 48:97–400

Minamisawa K, Ogawa KI, Fukuhara H, Koga J (1996) Indolepyruvate pathway for indol-3 acetic acid biosynthesis in *Bradyrhizobium elkanii*. Plant Cell Physiol 37:449–453

Miransari M, Smith D (2009) Rhizobial lipo-chitooligosaccharides and gibberellins enhance barley (*Hordeum vulgare* L.) seed germination. Biotechnology 8:270–275

Mishra RPN, Singh RK, Jaiswal HK, Kumar V, Maurya S (2006) *Rhizobium*-mediated induction of phenolics and plant growth promotion in rice (*Oryza sativa* L.). Curr Microbiol 52:383–389

Monteleone M, Thuar A, Olmedo C (2003) Efecto de la promoción del crecimiento con PGPRs en un cultivo de trigo en presencia de *B. japonicum*. Actas de la IV Reunión Nacional Científico Técnica de Biología de Suelos, Santiago del Estero, Argentina. CD-ROM. ISBN 097-99083-6-8

Nadeem SM, Zahir ZA, Naveed M, Arshad M (2007) Preliminary investigations on inducing salt tolerance in maize through inoculation with rhizobacteria containing ACC deaminase activity. Can J Microbiol 53:1141–1149

Nadeem SM, Zahir ZA, Naveed M, Arshad M (2009) Rhizobacteria containing ACC-deaminase confer salt tolerance in maize grown on salt-affected fields. Can J Microbiol 55:1302–1309

Nadeem SM, Zahir ZA, Naveed M, Asghar HN, Arshad M (2010) Rhizobacteria capable of producing ACC-deaminase may mitigate salt stress in wheat. Soil Sci Soc Am J 74:533–542

Naher UA, Othman R, Shamsuddin ZH, Saud HM, Ismail MR (2009) Growth enhancement and root colonization of rice seedlings by *Rhizobium* and *Corynebacterium* spp. Int J Agric Biol 11:586–590

Naidu VSGR, Panwar JDS, Annapurna K (2004) Effect of synthetic auxins and *Azorhizobium caulinodans* on growth and yield of rice. Indian J Microbiol 44:211–213

Natalia K, Gennady K, Vitaly K (1994) Endophytic association of nitrogen-fixing bacteria with plants. In: Proceedings 1st European N_2-fixation conference, Szeged, Hungary, p 104

Nautiyal CS (1997) A method for selection and characterization of rhizosphere-competent bacteria of chickpea. Curr Microbiol 34:12–17

Nautiyal CS (2000) Biocontrol of plant diseases for agricultural management. In: Upadhyay RK, Mukerji KG, Chamola BP (eds) Biocontrol potential and its exploitation in sustainable agriculture. Kluwer Academic/Plenum Publishers, New York, pp 9–23

Newcomb W, Syono K, Torrey JG (1977) Development of an ineffective pea root nodule: morphogenesis, fine structure and cytokinin biosynthesis. Can J Bot 55:1891–1907

Nie L, Shah S, Rashid A, Burd GI, Dixon DG, Glick BR (2002) Phytoremediation of arsenate contaminated soil by transgenic canola and the plant growth-promoting bacterium *Enterobacter cloacae* CAL2. Plant Physiol Biochem 40:355–361

Nieuwenhove CV, Holm LV, Kulasooriya SA, Vlassak K (2000) Establishment of *Azorhizobium caulinodans* in rhizosphere of wet land rice. Biol Fertil Soils 31:143–149

Noel TC, Sheng C, Yost CK, Pharis RP, Hunes MF (1996) *Rhizobium leguminosarum* as a plant growth promoting rhizobacterium: Direct growth promotion of canola and lettuce. Can J Microbiol 42:279–283

O'Callaghan KJ, Stone PJ, Hu X, Griffiths DW, Davey MR, Cocking EC (2000) Effects of glucosinolates and flavonoids on colonization of the roots of *Brassica napus* by *Azorhizobium caulinodans* ORS571. Appl Environ Microbiol 66:2185–2191

O'Sullivan DJ, O'Gara F (1992) Traits of *fluorescent Pseudomonas* spp. involved in suppression of plant root pathogens. Microbiol Rev 56:662–676

Okazaki S, Yuhashi K, Minamisawa K (2003) Quantitative and time-course evaluation of nodulation competitiveness of rhizobitoxine-producing *Bradyrhizobium elkanii*. FEMS Microbiol Ecol 45:155–160

Okazaki S, Sugawara M, Minamisawa K (2004) *Bradyrhizobium elkanii* rtxC gene is required for expression of symbiotic phenotypes in the final step of rhizobitoxine biosynthesis. Appl Environ Microbiol 70:535–541

Ozkoc I, Deliveli MH (2001) *In vitro* inhibition of the mycelial growth of some root rot fungi by *Rhizobium leguminosarum* biovar *phaseoli* isolates. Turk J Biol 25:435–445

Pandey P, Maheshwari DK (2007) Two-species microbial consortium for growth promotion of *Cajanus cajan*. Curr Sci 92:1137–1141

Parveen S, Ghaffar A (1991) Effect of microbial antagonists in the control of root rot of tomato. Pak J Bot 23:187–190

Parveen S, Ehteshamul-Haque S, Ghaffar A (1993) Biological control of *Melodidogyne javanica* on tomato and okra in soil infested with *Fusarium oxysporum*. Pak J Nematol 11:151–156

Peix A, Rivas-Boyero AA, Mateos PF, Rodriguez-barrueco C, Martinez-Molina E, Velaqez E (2001) Growth promotion of chickpea and barley by a phosphate solubilizing strain of *Mesorhizobium mediterraneum* under growth chamber conditions. Soil Biol Biochem 33:103–110

Pena HB, Reyes I (2007) Nitrogen fixing bacteria and phosphate solubilizers isolated in lettuce (*Lactuca sativa* L.) and evaluated as plant growth promoters. Interciencia 32:560–565

Pena-Cabriales JJ, Alexander M (1983) Growth of *Rhizobium* in unamended soil. Soil Sci Soc Am J 47:81–84

Peng S, Biswas JC, Ladha JK, Gyaneshwar P, Chen Y (2002) Influence of rhizobial inoculation on photosynthesis and grain yield of rice. Argon J 94:925–929

Pereira JAR, Cavalcante VA, Baldani JI, Dobereiner J (1988) Field inoculation of sorghum and rice with *Azorhizobium* spp. and *Herbaspirillum seropedicae*. Plant Soil 110:269–274

Perrine FM, Prayito J, Frank JJ, Dazzo B, Rolfe BG (2001) *Rhizobium* plasmids are involved in the inhibition or stimulation of rice growth and development. Aust J Plant Physiol 28:923–937

Perrine FM, Hocart CH, Hynes MF, Rolfe BG (2005) Plasmid-associated genes in the model micro-symbiont *Sinorhizobium meliloti* 1021 affect the growth and development of young rice seedlings. Environ Microbiol 7:1826–1838

Perrine-Walker FM, Gartner E, Hocart CH, Becker A, Rolfe BG (2007a) *Rhizobium*-initiated rice growth inhibition caused by nitric oxide accumulation. Mol Plant Microbe Interact 20:283–292

Perrine-Walker FM, Prayitno J, Rolfe BG, Weinman JJ, Hocart CH (2007b) Infection process and the interaction of rice roots with rhizobia. J Exp Bot 58:3343–3350

Perrine-Walker FM, Hynes MF, Rolfe BG, Hocart CH (2009) Strain competition and agar affect the interaction of rhizobia with rice. Can J Microbiol 55:1217–1223

Phillips DA, Torrey JG (1970) Cytokinin production by *Rhizobium japonicum*. Plant Physiol 23:1057–1063

Phillips DA, Torrey JG (1972) Studies on cytokinin production by *Rhizobium*. Plant Physiol 49:11–15

Phillips DA, Joseph CM, Yang G-P, Martinez-Romero E, Sanborn JR, Volpin H (1999) Identification of lumichrome as a *Sinorhizobium* enhancer of alfalfa root respiration and shoot growth. Proc Nat Acad Sci USA 96:12275–12280

Pineda JA, Kipe-Nolt JA, Rojas E (1994) *Rhizobium* inoculation increases of bean and maize yields in intercrops on farms in the Peruvian Sierra. Exp Agric 30:311–319

Plazinski J, Rolfe BG (1985a) Influence of *Azospirillum* strains on the nodulation of clovers by *Rhizobium* strains. Appl Environ Microbiol 49:984–989

Plazinski J, Rolfe BG (1985b) Interaction of *Azospirillum* and *Rhizobium* strains leading to inhibition of nodulation. Appl Environ Microbiol 49:990–993

Plessner O, Klapatch T, Guerinot ML (1993) Siderophore utilization by *Bradyrhizobium japonicum*. Appl Environ Microbiol 59:1688–1690

Polonenko DR, Scher FM, Kloepper JW, Singleton CA, Laliberte M, Zaleska I (1987) Effects of root colonizing bacteria on nodulation of soybean roots by *Bradyrhizobium japonicum*. Can J Microbiol 33:498–503

Prasad H, Chandra R (2003) Growth pattern of urdbean *Rhizobium* sp. with PSB and PGPR in consortia. J Indian Soc Soil Sci 51:76–78

Prayitno J, Stefaniak J, McIver J, Weinman JJ, Dazzo FB, Ladha JK, Barraquio W, Yanni YG, Rolfe BG (1999) Interactions of rice seedlings with bacteria isolated from rice roots. Aust J Plant Physiol 26:521–535

Prevost D, Saddiki S, Antoun H (2000) Growth and mineral nutrition of corn inoculated with effective strains of *Bradyrhizobium japonicum*. http://www.ag.auburn ed/~mlowens/argentina/pdf%20 manuscripts/prevost.pdf. Accessed on 25 Dec 2010

Prithiviraj B, Zhou X, Souleimanov A, Smith DL (2000) Nod Bj V (C18: 1 MeFuc) a host specific bacterial-to-plant signal molecule, enhances germination and early growth of divers crop plants. In: 17th North American conference on symbiotic nitrogen fixation, 23–28 July, 2000, Quebec, University of Laval, Canada p 6

Puente ME, Bashan Y, Li CY, Lebsky GK (2004) Microbial populations and activities in the rhizoplane of rock-weathering desert plants. 1. Root colonization and weathering of igneous rock. Plant Soil 6:629–642

Puppo A, Riguard J (1978) Cytokinins and morphological aspects of French-bean roots in the presence of *Rhizobium*. Physiol Plantarum 42:202–206

Quispel A (1991) A critical evaluation of the prospects for nitrogen fixation with non-legumes. Plant Soil 137:1–11

Raaijmakers JM, Paulitz TC, Steinberg C, Alabouvette C, Moenne-Loccoz Y (2009) The rhizosphere: a playground and battle field for soilborne pathogens and beneficial microorganisms. Plant Soil 32:341–361

Raupach GS, Kloepper JW (1998) Mixtures of plant growth promoting rhizobacteria enhance biological control of multiple cucumber pathogens. Phytopathology 88:1158–1163

Ray PK, Jana AK, Maitra DN, Saha MN, Chaudhury J, Saha S, Saha AR (2000) Fertilizer prescriptions on soil test basis for jute, rice and wheat in Typic Ustochrept. J Indian Soc Soil Sci 48:79–84

Reddy PM, Ladha JK, So RB, Hernandez RJ, Ramos MC, Angeles OR, Dazzo FB, deBruijn FJ (1997) Rhizobial communication with rice roots: induction of phenotypic changes, mode of invasion and extent of colonization. Plant Soil 194:81–89

Reed MLE, Glick BR (2005) Growth of canola (*Brassica napus*) in the presence of plant growth-promoting bacteria and either copper or polycyclic aromatic hydrocarbons. Can J Microbiol 51:1061–1069

Reimann S, Hauschild R, Hildebrandt U, Sikora RA (2008) Interrelationships between *Rhizobium etli* G12 and *Glomus intraradices* and multitrophic effects in the biological control of the root-knot nematode *Meloidogyne incognita* on tomato. J Plant Disease Protect 115:108–113

Reinhold-Hurek B, Maes T, Gemmer S, Van Montagu M, Hurek T (2006) An endoglucanase is involved in infection of rice roots by the not-cellulose-metabolizing endophyte *Azoarcus* sp. strain BH72. Mol Plant Microbe Interact 19:181–188

Reiter B, Pfeifer U, Schwab H, Sessitsch A (2002) Response of endophytic bacterial communities in potato plants to infection with *Erwinia carotovora* sub sp. Atroseptica Appl Environ Microbiol 68:2261–2268

Reitz M, Rudolph K, Schroder I, Hoffmann-Hergarten S, Hallmann J, Sikora RA (2000) Lipopolysaccharides of *Rhizobium etli* strain G12 act in potato roots as an inducing agent of systemic resistance to infection by the cyst nematode *Globodera pallida*. Appl Environ Microbiol 66:3515–3518

Requena N, Jimenez I, Toto M, Barea JM (1997) Interactions between plant-growth-promoting rhizobacteria (PGPR), arbuscular mycorrhizal fungi and *Rhizobium* spp. in the rhizosphere of *Anthyllis cytisoides*, a model legume for revegetation in Mediterranean semi-arid ecosystem. New Phytol 136:667–677

Reyes VG, Schmidt EL (1979) Population densities of *Rhizobium japonicum* strain 123 estimated directly in soil and rhizosphere. Appl Environ Microbiol 37:854–858

Ridge RW, Bender GL, Bolefe BG (1992) Nodule-like structures induced on roots of wheat seedlings by addition of synthetic auxin 2,4-dichlorophenoxy acetic acid and the effects of microorganisms. Aust J Plant Physiol 19:481–492

Riggs PJ, Chelius MK, Iniguez AL, Kaeppler SM, Triplett EW (2001) Enhanced maize productivity with diazotrophic bacteria. Aust J Plant Physiol 28:829–836

Rodelas B, Salmeron V, Martinez-Toledo MV, Gonzalez-Lopez J (1993) Production of vitamins by *Azospirillum brasilense* in chemically defined media. Plant Soil 153:97–101

Rodelas B, Gonzalez-Lopez J, Salmeron V, Pozo C, Martinez-Toledo MV (1996) Enhancement of nodulation, N_2-fixation and growth of faba bean (*Vicia faba* L.) by combined inoculation with *Rhizobium leguminosarum* bv. *viceae* and *Azospirillum brasilense*. Symbiosis 21:175–186

Rodriguez H, Fraga R (1999) Phosphate solubilizing bacteria and their role in plant growth promotion. Biotechnol Adv 17:319–339

Rodriguez H, Fraga R, Gonzalez T, Bashan Y (2006) Genetics of phosphate solubilization and its potential applications for improving plant growth promoting bacteria. Plant Soil 287:15–21

Roger PA, Watanabe IW (1986) Technologies for utilizing biological nitrogen fixation in wetland rice: potentialities, current usage, and limiting factors. Fert Res 9:39–77

Rolfe BG, Bender GL (1990) Evolving a *Rhizobium* for non-legume nodulation. In: Gresshoff PM, Roth LE, Stacey G, Newton WE (eds) Nitrogen fixation: achievements and objectives. Chapman and Hall, London, pp 779–786

Rosenblueth M, Martinez-Romero E (2004) *Rhizobium etli* maize populations and their competitiveness for root colonization. Arch Microbiol 181:337–344

Sabry RS, Saleh SA, Batchelor CA, Jones J, Jotham J, Webster G, Kothari SL, Davey MR, Cocking EC (1997) Endophytic establishment of *Azorhizobium caulinodans* in wheat. Proc R Soc Lond 264:341–346

Sahin F, Cakmakci R, Kantar F (2004) Sugar beet and barley yields in relation to inoculation with N_2-fixing and phosphate solubilizing bacteria. Plant Soil 265:123–129

Saikia SP, Jain V (2007) Biological nitrogen fixation with non-legumes: An achievable target or a dogma? Curr Sci 92:317–322

Saikia SP, Srivastava GC, Jain V (2004) Nodule-like structures induced on the roots of maize seedlings by the addition of synthetic auxin 2,4-D and its effects on growth and yield. Cereal Res Commun 32:83–89

Santaella C, Schue M, Berge O, Heulin T, Achouak W (2008) The exopolysaccharide of *Rhizobium* sp. YAS34 is not necessary for biofilm formation on *Arabidopsis thaliana* and *Brassica napus* roots but contributes to root colonization. Environ Microbiol 10:2150–2163

Sarig S, Kapulnik Y, Okon Y (1986) Effect of *Azospirillum inoculation* on nitrogen fixation and growth of several winter legumes. Plant Soil 90:335–342

Schloter M, Wiehe W, Assmus B, Steindl H, Becke H, Höflich G, Hartmann A (1997) Root colonization of different plants by plant growth promoting *Rhizobium leguminosarum* bv. *trifolii* R39 studied with mono-specific polyclonal antisera. Appl Environ Microbiol 63:2038–2046

Schwinghamer EA, Belkengren RP (1968) Inhibition of Rhizobia by a strain of *Rhizobium trifolii*: some properties of the antibiotic and of the strain. Arch Microbiol 64:130–145

Schwyn B, Neilands JB (1987) Universal chemical assay for the detection and determination of siderophores. Anal Biochem 160:47–56

Senthilkumar M, Madhaiyan M, Sundaram SP, Sangeetha H, Kannaiyan S (2008) Induction of endophytic colonization in rice (*Oryza sativa* L.) tissue culture plant by *Azorhizobium caulinodans*. Biocontrol Lett 30:1477–1487

Senthilkumar M, Madhaiyan M, Sundaram SP, Kannaiyan S (2009) Intercellular colonization and growth promoting effects of *Methylobacterium* sp. with plant-growth regulators on rice (*Oryza sativa* L. CvCO-43). Microbiol Res 164:92–104

Sessitsch A, Reiter B, Pfeifer U, Wilhelm E (2002) Cultivation-independent population analysis of bacterial endophytes in three potato varieties based on eubacterial and Actinomycetes-specific PCR of 16S rRNA genes. FEMS Microbiol Ecol 39:23–32

Sevilla M, Kennedy C (2000) Colonization of rice and other cereals by *Acetobacter* diazotrophicus, an endophyte of sugarcane. In: Ladha JK, Reddy PM (eds) The quest for nitrogen fixation in rice. International Rice Research Institute, Manila, pp 151–165

Shao H-B, Chu L-Y, Ruan C-J, Li H, Guo D-G, Li W-X (2010) Understanding molecular mechanisms for improving phytoremediation of heavy metal-contaminated soils. Crit Rev Biotechnol 30:23–30

Shaukat SS, Siddqui IA (2003) The influence of mineral and carbon sources on biological control of charcoal rot fungus, *Macrophomina phaseolina* by *fluorescent pseudomonads* in tomato. Lett Appl Microbiol 36:392–398

Sheikh LI, Dawar S, Zaki MJ, Ghaffar A (2006) Efficacy of *Bacillus thuringiensis* and *Rhizobium meliloti* with nursery fertilizers in the control of root infecting fungi on mung bean and okra plants. Pak J Bot 38:465–473

Shimshick EJ, Hebert RR (1979) Binding characteristics of N_2-fixing bacteria to cereal roots. Appl Environ Microbiol 38:447–453

Siddiqui ZA (2007) Biocontrol of *Alternaria triticina* by plant growth promoting rhizobacteria on wheat. Arch Phytopathol Plant Protec 40:301–308

Siddiqui ZA, Mahmood I (2001) Effects of rhizobacteria and root symbionts on the reproduction of *Meloidogyne javanica* and growth of chickpea. Bioresource Technol 79:41–45

Siddiqui IA, Ehteshamul-Haque S, Ghaffar A (1998) Effect of Rhizobia and fungal antagonists in the control of root infecting fungi on sunflower and chickpea. Pak J Bot 30:279–286

Siddiqui IA, Ehteshamul-Haque S, Zaki MJ, Ghaffar A (2000) Effect of urea on the efficacy of *Bradyrhizobium* sp. and *Trichoderma harzianum* in the control of root infecting fungi in mungbean and sunflower. Sarhad J Agric 16:403–406

Sierra S, Rodelas B, Martinez-Toledo MV, Pozo C, Gonzalez-Lopez J (1999) Production of B-group vitamins by two *Rhizobium* strains in chemically defined media. J Appl Microbiol 86:851–858

Sindhu SS, Suneja S, Goel AK, Parmar N, Dadarwal KR (2002) Plant growth promoting effects of *Pseudomonas* sp. on co-inoculation with *Mesorhizobium* sp. Cicer strain under sterile and "wilt sick" soil conditions. Appl Soil Ecol 19:57–64

Singh R, Kumar V, Sharma S, Behl RK, Singh BP, Narula N (2005) Performance and persistence of green fluorescent protein (gfp) marked *Azotobacter chroococcum* in sterilized and unsterilized wheat rhizospheric soil. Chinese J Appl Environ Biol 11:751–755

Singh RK, Mishra RPN, Jaiswal HK, Kumar V, Pandev SP, Rao SB, Annapurna K (2006) Isolation and identification of natural endophytic rhizobia from rice (*Oryza sativa* L.) through rDNA PCR-RFLP and sequence analysis. Curr Microbiol 52:345–349

Smith SE, Dickson S, Smith FA (2001) Nutrient transfer in arbuscular mycorrhizae: how are fungal and plant processes integrated? Aust J Plant Physiol 28:683–694

Smith DL, Prithiviraj B, Zhang F (2002) Rhizobial signals and control of plant growth. In: Finan TM, O'Brian MR, Layzell DB, Vessey JK, Newton W (eds) Nitrogen fixation: global perspectives. CABI Publishing, Wallingford, pp 327–330

Somers E, Vanderleyden J, Srinivasan M (2004) Rhizosphere bacterial signalling: a love prade beneath our feet. Crit Rev Microbiol 30:205–240

Souleimanov A, Prithiviraj B, Smith DL (2002) The major Nod factor of *Bradyrhizobium japonicum* promotes early growth of soybean and corn. J Exp Bot 53:1929–1934

Spaink HP (1992) Rhizobial lipo-oligosaccharide: answers and questions. Plant Mol Biol 20:977–986

Spaink HP, Lugtenberg BJJ (1994) Role of rhizobial lipo-chitin oligosaccharide signal molecules in root nodule organogenesis. Plant Mol Biol 26:1413–1422

Spencer D, James EK, Ellis GJ, Shaw JE, Sprent JI (1994) Interactions between rhizobia and potato tissues. J Exp Bot 45:1475–1482

Sturz AV, Christie BR, Matheson BG, Arsenault WJ, Buchanan NA (1999) Endophytic communities in the periderm of potato tubers and their potential to improve resistance to soil-borne plant pathogens. Plant Pathol 48:360–369

Sullivan JT, Trzebiatowski JR, Cruickshank RW, Gouzy J, Brown SD, Elliot RM, Fleetwood DJ, McCallum NG, Rossbach U, Stuart GS, Weaver JE, Webby RJ, De Bruijin FJ, Ronson CW (2002) Comparative sequence analysis of the symbiosis island of *Mesorhizobium loti* strain R7A. J Bacteriol 184:3086–3095

Tan ZY, Hurek T, Vinuesa P, Muller P, Ladha JK, Reinhold-Hurek B (2001a) Specific detection of *Bradyrhizobium* and *Rhizobium* strains colonizing rice (*Oryza sativa*) roots by 16S-23S ribosomal DNA intergenic spacer-targeted PCP. Appl Environ Microbiol 67:3655–3664

Tan ZY, Kan FL, Peng GX, Wang ET, Reinhold-Hurek B, Chen WX (2001b) *Rhizobium yanglingense* sp. nov., isolated from arid and semi-arid regions in China. Int J Syst Evol Microbiol 51:909–914

Terouchi N, Syono K (1990) *Rhizobium* attachment and curling in asparagus, rice, and oat plants. Plant Cell Physiol 31:119–127

Trinick MJ (1979) Structure of nitrogen fixing nodules of *Rhizobium* on roots of *Parasponia andersonii*. Can J Microbiol 25:565–578

Trinick MJ, Habdodas PA (1995) Formation of nodular structures on the non-legumes *Brassica napus, B. campestris, B. juncea* and *Arabidopsis thaliana* with *Bradyrhizobium* and *Rhizobium* isolated from *Parasponia* spp. on legumes grown in tropical soils. Plant Soil 172:207–219

Trott S, Bauer R, Knackmuss HJ, Stolz A (2001) Genetic and biochemical characterization of an enantioselective amidase from *Agrobacterium tumefaciens* strain d3. Microbiology 147:1815–1824

Tu JC (1978) Protection of soybean from severe *Phytophthora* root rot by *Rhizobium*. Physiol Plant Pathol 12:233–240

Tu JC (1979) Evidence of differential tolerance among some root rot fungi to rhizobial parasitism *in vitro*. Physiol Plant Pathol 14:171–177

Uchiumi T, Oowada T, Itakura M, Mitsui H, Nukui N, Dawadi P, Kaneko T, Tabata S, Yokoyama T, Tejima T, Saeki K, Oomori H, Hayashi M, Maekawa T, Sriprang R, Murooka Y, Tajima S, Simomura K, Nomura M, Suzuki A, Shimoda S, Sioya K, Abe M, Minamisawa K (2004) Expression islands clustered on symbiosis island of *Mesorhizobium loti* genome. J Bacteriol 186:2439–2448

Ueda T, Suga Y, Yahir N, Matsuguchi T (1995) Remarkable N_2-fixing bacterial diversity detected in rice roots by molecular evolutionary analysis of nifH gene sequences. J Bacteriol 177:1414–1417

Upadhyaya NM, Letham DS, Parker CW, Hocart CH, Dart PJ (1991) Do rhizobia produce cytokinins? Biochem Int 24:123–130

Valdes M, Reza-Aleman F, Furlan V (1993) Response of *Leucaena esculenta* to endomycorrhizae and *Rhizobium* inoculation. World J Microbiol Biotechnol 9:97–99

van Loon LC (2007) Plant responses to plant growth-promoting rhizobacteria. Eur J Plant Pathol 119:243–254

Vargas LK, Lisboa BB, Schlindwein G, Granada CE, Giongo A, Luciane AB, Passaglia MP (2009) Occurrence of plant growth-promoting traits in clover-nodulating rhizobia strains isolated from different soils in Rio Grande do sul state. R Bras Ci Solo 33:1227–1235

Velazquez E, Peix A, Zurdo-Pin˜eiro JL, Palomo JL, Mateos PF, Rivas R, Munoz-Adelantado E, Toro N, García-Benavides P, Martínez-Molina E (2005) The coexistence of symbiosis and pathogenicity-determining genes in *Rhizobium rhizogenes* strains enables them to induce nodules and tumours or hairy roots in plants. Mol Plant Microbe Interact 18:1325–1332

Venieraki A, Dimou M, Pergalis P, Kefalogianni I, Chatzipavlidis I, Katinakis P (2011) The genetic diversity of culturable nitrogen-fixing bacteria in the rhizosphere of wheat. Microbial Ecol 61:277–285

Verma SC, Singh A, Chowdhury SP, Tripathi AK (2004) Endophytic colonization ability of two deep-water rice endophytes, *Pantoea* sp. and *Ochrobactrum* sp. using green fluorescent protein reporter. Biotechnol Lett 26:425–429

Vessey JK (2003) Plant growth promoting rhizobacteria as biofertilizers. Plant Soil 255:571–586

Vidhyasekaran P, Muthamilan M (1995) Development of formulations of *Pseudomonas fluorescence* for biocontrol of chickpea wilt. Plant Dis 79:782–786

Vidhyasekaran P, Muthamilan M (1999) Evaluation of powder formulation of *Pseudomonas fluorescens* Pf1 for control of rice sheath blight. Biocontrol Sci Technol 9:67–74

Wang TI, Wood EA, Brewin NJ (1982) Growth regulators, *Rhizobium* and nodulation in peas: Indole-3-acetic acid from the culture medium of nodulating and non-nodulating strains of *R. leguminosarum*. Planta 155:345–349

Wang C, Knill E, Glick BR, Defago G (2000) Effect of transferring 1- aminocyclopropane 1-carboxylic acid (ACC) deaminase genes into *Pseudomonas fluorescens* strain CHA0 and its gacA derivative CHA96 on their growth promoting and disease suppressive capacities. Can J Microbiol 46:898–907

Webster G, Gough C, Vasse J, Batchelor CA, O'Callaghan KJ, Kothari SL, Davey MR, Denarie J, Cocking EC (1997) Interactions of rhizobia with rice and wheat. Plant Soil 194:115–122

Wei G, Kloepper JW, Tuzun S (1996) Induced systemic resistance to cucumber diseases and increased plant growth by plant growth promoting rhizobacteria under field conditions. Phytopathology 86:221–224

Wenzel WW, Lombi E, Adriano DC (1999) Biogeochemical processes in the rhizosphere: Role in phytoremediation of metal-polluted soils. In: Prasad NMV, Hagemeyer J (eds) Heavy metal stress in plants – from molecules to ecosystems. Springer, Heidelberg, pp 273–303

Werner D (1992) Symbiosis of Plants and Microbes. Chapman and Hall, London

West PM, Wilson PW (1938) Synthesis of growth factors by *Rhizobium trifolii*. Nature 142:397–398

West SA, Kiers ET, Simms EL, Denision RF (2002) Sanctions and mutualism stability: why do rhizobia fix nitrogen? Proc R SocLond 269:685–694

Weyens N, van der Lelie D, Taghavi S, Vangronsveld J (2009) Phytoremediation: plant-endophyte partnerships take the challenge. Curr Opin Biotechnol 20:248–254

Wheeler CL, Crozier A, Sandberg G (1984) The biosynthesis of indole-3-acetic acid by *Frankia*. Plant Soil 78:99–104

Wiehe W, Hoflich G (1995) Survival of plant growth promoting rhizosphere bacteria in the rhizosphere of different crops and migration to noninoculated plants under field conditions in northeast Germany. Microbiol Res 150:201–206

Wiehe W, Hecht-Buchholz C, Hoflich G (1994) Electron microscopic investigations on root colonization of *Lupinus albus* and *Pisum sativum* with two associative plant growth promoting rhizobacteria. *Pseudomonas fluorescens* and *Rhizobium leguminosarum* bv. *trifolii*. Symbiosis 17:15–31

Wood M, Cooper JE (1984) Aluminum toxicity and multiplication of *Rhizobium trifolii* in a defined growth medium. Soil Biol Biochem 16:571–576

Xavier LJC, Germida JJ (2002) Response of lentil under controlled conditions to co-inoculation with arbuscular mycorrhizal fungi and rhizobia varying in efficacy. Soil Biol Biochem 34:181–188

Xie Z-P, Staehlin C, Vierheilig H, Wiemken A, Jabbouri S, Broughton WJ, Vogeli-Lange R, Boller T (1995) Rhizobial nodulation factors stimulate mycorrhizal colonization of nodulating and nonnodulating soybeans. Plant Physiol 108:1519–1525

Yagi K (1962) Chemical determination of flavins. In: Glick D (ed) Methods of biochemical analysis, vol X. Interscience Publishers, London, pp 319–356

Yanagita T, Foster JW (1956) A bacterial riboflavin hydrolase. J Biol Chem 221:593–607

Yang G, Bhuvaneswari TV, Joseph CM, King MD, Phillips DA (2002) Roles for riboflavin in the *Sinorhizobium*-alfalfa association. Mol Plant Microbe Interact 5:456–462

Yanni YG, Rizk RY, Corich V, Squartini A, Dazzo FB (1995) Endorhizosphere colonization and growth promotion of *indica* and *japonica* rice varieties by *Rhizobium leguminosarum* bv. *trifolii*. In: Proceedings of the 15th symbiotic nitrogen fixation conference, North Carolina State University, Raleigh, NC, p 17

Yanni YG, Rizk RY, Corich V, Squartini A, Ninke K, Philip-Hollingsworth S, Orgambide G, De Bruijn F, Stoltzfus J, Buckley D, Schmidt TM, Mateos PF, Ladha JK, Dazzo FB (1997) Natural endophytic association between *Rhizobium leguminosarum* bv. *trifolii* and rice roots and assessment of its potential to promote rice growth. Plant Soil 194:99–114

Yanni Y, Rizk R, Maya-Flores J, Dazzo F (2000) Potential of the rice root occupant *Rhizobium leguminosarum* bv. *trifolii* as plant growth promoting biofertilizer for rice. In: Proceedings of the 17th North American conference on symbiotic nitrogen fixation, Universitie Laval., Quebec, Canada, p 79

Yanni YG, Rizk RY, Abd El-Fattah FK, Squartini A, Corich V, Giacomini A, De Bruijn F, Rademaker J, Maya-Flores J, Ostrom P, Vega-Hernandez M, Hollingsworth RI, Martinez-Molina E, Mateos P, Velazquez E, Wopereis J, Triplett E, Umali-Gracia M, Anarna JA, Rolfe BG, Ladha JK, Hill J, Mujoo R, Ng PK, Dazzo FB (2001) The beneficial plant growth-promoting association of *Rhizobium leguminosarum* bv. *trifolii* with rice roots. Aust J Plant Physiol 28:845–870

Yasutu T, Okazaki S, Mitrsui H, Yuhashi K, Ezura E, Minamisawa K (2001) DNA sequence and mutational analysis of rhizobitoxine biosynthesis gene in *Bradyrhizobium elkanii*. Appl Environ Microbiol 67:4999–5009

You M, Nishiguchi T, Saito A, Isawa T, Mitsui H, Minamisawa K (2005) Expression of the nifH gene of a *Herbaspirillum* endophyte in wild rice species: daily rhythm during the light–dark cycle. Appl Environ Microbiol 71:8183–8190

Yu D, Kennedy IR (1995) Nitrogenase activity (C2H2 reduction) of *Azorhizobium* in 2,4-D-induced root structures of wheat. Soil Biol Biochem 27:459–462

Zahir ZA, Arshad M, Frankenberger WT Jr (2004) Plant Growth Promoting Rhizobacteria: Applications and Perspectives in Agriculture. Adv Agron 81:97–168

Zahir ZA, Ghani U, Naveed M, Nadeem SM, Arshad M (2009) Comparative effectiveness of *Pseudomonas* and *Serratia* sp. containing ACC-deaminase for improving growth and yield of wheat under salt-stressed conditions. Arch Microbiol 191:415–424

Zhang F, Smith DL (2001) Interorganismal signaling in suboptimum environments: the legume-rhizobia symbiosis. Adv Agron 76:125–161

Zhang X, Nick G, Kaijalainen S, Lonnqvist C, Tas E, Tighe SW, Graham PH, Lindstrom K (1997) Diversity of peanut *Rhizobium* isolated from Sichuan, China. In: Proceedings of the 11th international congress on nitrogen fixation, Institut Pasteur, Paris, France, p 49

Zhuang X, Chen J, Shim H, Bai Z (2007) New advances in plant growth promoting rhizobacteria for bioremediation. Environ Int 33:406–413

Chapter 27
Effect of Drip and Subsurface Drip Irrigation with Saline Water on Tomato Crop

Besma Kahlaoui, Mohamed Hachicha, Saloua Rejeb, and Mohamed Néjib Rejeb

Contents

Abstract With the purpose of improving salinity management and water use efficiency in agriculture, an experiment was carried out to study the effect of surface drip irrigation (DI) and subsurface drip irrigation (SDI) on a tomato crop (*Lycopersicon esculentum* Mill, cv. Heinz-2274) in a silty clayey soil with three irrigation water qualities: 3.0, 6.0 and 8.3 dS m^{-1}. The results did not show any difference in the crop response of the two irrigation systems, whereas, the effect of the water quality was manifested. Saline water irrigation affected the tomato growth, in particular leaf area, dry matter, as well as the shoot/root ratios and the mineral composition of leaves, stems and roots. The accumulation of Na^+ and Cl^- was associated with a

B. Kahlaoui (✉) • M. Hachicha • S. Rejeb • M.N. Rejeb
National Institute of Research in Rural Engineering, Waters and Forests,
17 Hédi Karray Street, 10 Ariana 2080, Tunisia
e-mail: besma.kahlaoui@gmail.com; hachicha.mohamed@iresa.agrinet.tn; rejeb.saloua@iresa.agrinet.tn; rejeb.nejib@iresa.agrinet.tn

M. Ashraf et al. (eds.), *Crop Production for Agricultural Improvement*,
DOI 10.1007/978-94-007-4116-4_27,

decrease in the contents of Ca^{2+}, K^+ and Mg^{2+}, whereas, the P content in different organs remained constant. The more the salinity of the irrigation water the more marked was the decrease in yield parameters of tomato (setting, size, and yield). This experiment shows that at normal water management (100% of crop water requirement), the SDI does not present additional advantages as compared with the DI for water whose electrical conductivity is not higher than 8.3 dS m^{-1}.

Keywords Drip irrigation • Subsurface drip irrigation • Saline water • Tomato • Silty clay soil • Tunisia

1 Introduction

Many studies addressed to more suitable irrigation systems to face water salinity problem. Indeed, aridity and water scarcity compel several countries to use saline water for irrigation, to try to optimize the use of their water resources in particular in agriculture, and to reduce water used at the farm scale. In Tunisia, despite the primary soils salinization, which concerns about 1.5 millions of hectares (about 10% of the country area), about 30% of irrigated area (400,000 ha) is affected by the secondary salinization. This salinization is generally induced by irrigation with saline water; as a matter of fact, more than 30% of the Tunisian water contains more than 3 g of salt L^{-1} (Hachicha 2007).

Moreover, as in the other Mediterranean regions, the tomato crop is economically the most important vegetable. In 2006, the cultivated area of tomato in Tunisia was about 16,800 ha. This crop is generally irrigated by submersion, but drip irrigation has become more and more widespread since the beginning of the 1990s. Drip irrigation (DI) allows a low and frequent application of water directly at the root zone (Tariq et al. 1990) and a better efficiency in comparison with surface irrigation (Bogle et al. 1989). However, irrigation with saline water by DI produces an accumulation of salt at the soil surface (Ayers and Westcot 1985; Rajak et al. 2006). According to Hanson and May (2004), salts in DI accumulate at the surface soil before migrating and reaching the root zone. To reduce this effect of salts, the subsurface drip irrigation (SDI) was developed. According to Phene et al. (1990) and Oron et al. (1998), the SDI decreases the accumulation of salts within the root zone, generating an improvement in yield and fruit quality. This was observed for tomato (Ayars et al. 2000; Hanson and May 2004), onion (Enciso et al. 2007), cotton (DeTar 2007) and bean (Gençoğlan et al. 2006). SDI improves water use efficiency (Taylor et al. 2005) by allowing the application of a small quantity of water directly at the root zone and maintaining this layer at a suitable soil moisture (Enciso et al. 2007), thus reducing the effect of the saline stress (Incrocci et al. 2005).

Salinity is a constraint which affects the growth, flowering and fructification (size reduction and apical necrosis) of several crops, which has repercussions on their yield and quality of their fruits. This was observed for tomato (Karlberg et al. 2007; Tuna et al. 2005; Flowers et al. 2005). According to Flowers et al. (2005), improvement of salinity tolerance of tomato irrigated with saline water by DI results from the increase in irrigation frequency, which reduces the accumulation of salt in the soil, as well as from the vacuolar compartmentalization of the toxic ions, in particular of Na^+ and Cl^-. The increase in salinity involves an increase in Na^+ concentration. This toxic element competes with the major elements like, Ca^{2+} and K^+ (Juan et al. 2005; Maggio et al. 2006) and causes a nutritional disturbance inhibiting the synthesis of proteins, as well as an inactivation of enzymes (Gaines and Shennan 1999; Fernández-García et al. 2002). The objective of this work was to compare the effect of the DI and SDI systems of irrigation in relation to three different salt concentrations of irrigation water on open field cultivated tomato.

2 Material and Methods

2.1 *Material and Treatments*

The experiment was carried out at the Agricultural Experimental Station of Cherfech, situated in the Mejerda Low Valley, 25 km North of Tunis. The climate is Mediterranean with an average annual rainfall average and an annual evapotranspiration average 470 and 1,370 mm, respectively (Penman method). The soil has a silty-clay texture, xerofluvet with 45% of clay, 45% of silt, 10% of sand, about 1% of organic matter and 20% of calcareous. The irrigation network comprised a water basin and a main pipeline. The DI equipment was constituted of 30 cm depth buried pipelines with filter protected emitters every 40 cm. The same pipelines were used for SDI. The crop was a seasonal tomato (*Lycopersicon esculentum* Mill) variety 'Heinz-2274' adapted to the mechanical harvest. The tomato seedlings were planted at the 4-leaf stage, in simple lines, on May 27, 2005. The lines were 1 m spaced out. The parcel was divided into three blocks, one for each quality of water (Fig. 27.1). There were six planting rows in each block (three lines per treatment). Every line included 27 plants: lines 1, 3 and 5 were for the DI treatment, whereas lines 2, 4 and 6 for the SDI treatment. The irrigation waters came from Mejerda River (3 dS m^{-1}), a shallow well (8.3 dS m^{-1}) and a mixture of these two waters at the rate of 50% for each (6 dS m^{-1}). These waters are rich in NaCl. Their SAR was 7, 10.7 and 6 respectively. The chemical characteristics are presented in Table 27.1.

Soil salinity analysis: Soil salinity was determined before installation and at the end of the crop cycle at the roots level and every 20 cm up to 60 cm depth. The soil/water ratio was 1/5. A significant relationship permitted to convert the electrical conductivity EC (1/5) to the electrical conductivity of the saturated paste extract (ECe) (Jammazi and Hachicha 2002) (Table 27.2).

$$\mathrm{ECe} = 5.853 * \mathrm{EC}(1/5) - 0.262 \quad \mathrm{n} = 134 \text{ and } \mathrm{R}^2 = 0.91$$
$$\text{with } 1.2 < \mathrm{ECe} < 8.3 \text{ dS m}^{-1} \text{and } 0.2 < \mathrm{EC}(1/5) < 1.4 \text{ dS m}^{-1}$$

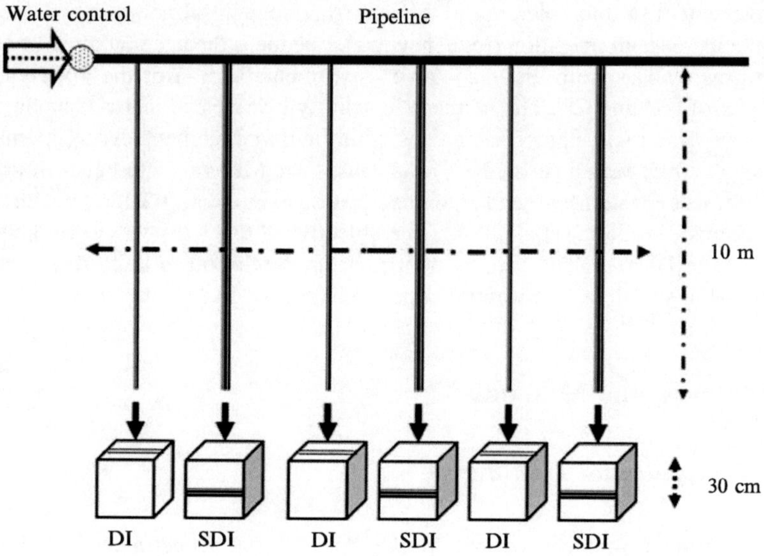

Fig. 27.1 Representation of the experimental device for every quality of irrigation water

Table 27.1 Chemical analysis of three qualities of waters used for the irrigation of tomato

				Anions (me L^{-1})			Cations (me L^{-1})				
Water quality	pH	EC (dS m^{-1})	TDS (g L^{-1})	HCO^{3-}	SO^{4-}	Cl^-	Ca^{2+}	Mg^{2+}	K^+	Na^+	SAR
Mejerda	7.5	3	2	0.8	4.1	19	8.1	5.6	0.2	18.3	7
Well	7.2	8.6	6	1.2	27.6	56.8	23	19.1	0.8	49	10.7
Mixture	7.4	6	4	2	8.4	44	18.6	13.5	1.7	34	6

EC electrical conductivity, *TDS* total dissolved solids

Table 27.2 Analysis of variance (ANOVA) of the soil salinity at the end of the cycle of irrigation

	Variables	ECe (dS m^{-1})	STD
System of irrigation	DI	4.9 a	1.72
	SDI	5.1 a	1.85
Water quality	Mejerda	2.9 c	0.42
	Well	4.9 b	0.44
	Mixture	7.1 a	0.35
Depth (cm)	0–20	4.9 a	1.78
	20–40	5.1 a	1.77
	40–60	4.9 a	1.85

2.2 *Crop Determinations and Analyses*

The following determinations were made on the tomato crop:

- Length growth: values are the average measured on 15 plants.
- Setting rate: ratio of the number of fruits set to number of flowers bloomed. (A fruit was considered set when its diameter reached 3 cm at least).
- Fresh and dry biomass: plants were harvested, divided into roots, stems and leaves. Roots were cleared of the soil particles and washed with water. Every part was weighed to determine the fresh matter. The dry matter was determined after a drying at 80°C for 48 h.
- Shoot/root ratio: allowed estimation of the allocation of assimilates between absorbing organs and photosynthetic organs.
- Leaf area: it was measured by a planimeter (Delta-T Devices Ltd). It was expressed in cm^2.
- Fruit yield: 90 plants were used to measure yield parameters. The yield was expressed per plant.
- Fruit size: Seven fruit- grade classes were adopted: C0<30 mm, 30 mm<C1< 35 mm, 35 mm<C2<40 mm, 40 mm<C3<45 mm, 45 mm<C4<50 mm, 50 mm<C5<55 mm, 55 mm<C6<60 mm.

Besides these determinations, chemical analyses were done on roots, leaves and stems for six types of elements. The extraction of ions was performed by HNO_3 at room temperature for at least 48 h:

- Dosage of Cl: by colorimetry with Buchler chloridimeter in the presence of acetic acid-nitric acid tampon and gelatin destined to agglomerate the precipitate of AgCl;
- Dosage of P: by colorimetry with vanado-molybdate;
- Dosage of K and Na: by flame- emission photometry;
- Dosage of Ca and Mg: by atomic absorption spectrophotometry;

Data analysis: The software STATITCF Version 5 (Beaux et al. 1991) was used for the statistical analyses of the data. The two-factor analysis of variance was computed by Neuman and Keuls' comparison for all pair of means.

3 Results

The soil study was extensively presented in a separate paper by Ben Necib (2008). Before irrigation, the salinity of soil was lower than 1 dS m^{-1}, and it increased at the end of the irrigation cycle. We observed significant mean differences between the qualities of irrigation water and none between DI and SDI. For the tomato crop, three aspects were studied: growth, yield and mineral composition of the plant.

3.1 Effect on Crop Growth

During the experiment, symptoms attributed to salinity were observed on the plants of tomato. From 4 gL^{-1} of salt, there were tanning as well as, necrosis and coil of leaves, revealing a disruption of the mineral nutrition and this ocurred at the same time for DI and SDI. Leaf area was affected strongly by the salinity: compared to the control (Mejerda water, 2 gL^{-1}), the reduction being about 50% for 4 gL^{-1} and 60% for 6 gL^{-1}. This effect was thus similar for DI and SDI (Fig. 27.3). Whatever the irrigation system, the salinity had a depressive effect on the fresh matter yield (FM) of different organs of the tomato (Fig. 27.2). This effect was higher as the salinity of the irrigation water increased. The yield reduction became significant from 4 gL^{-1}. Compared to the control, the reduction concerning 6 gL^{-1} was about 26% for leaves, 34% for stems and 22% for roots, under all irrigation systems. The same effects were observed for the dry matter (DM). More differences were found for leaves and stems than for roots (Fig. 27.2). The reduction become significant at 6 gL^{-1} for DI and SDI, it was 53% for leaves and 23% for stems. The shoots/roots ratio showed a significant reduction at the significance level of 5% when the salt concentration of the irrigation water was high (Fig. 27.3), passing from 14 for 2 gL^{-1} to 12 for 4 gL^{-1} and 10 for 6 gL^{-1}. So, the shoot, in particular leaves, was more sensible to the salinity than the roots. However, no significant difference was observed between DI and SDI.

3.2 Effect of the Salinity on Tomato Yield

Unlike the marked effect of the salinity of the irrigation water, the crop yield did not show any significant difference between DI and SDI. The rate of fruit setting (Fig. 27.4) was reduced by salt. The reduction was significant at the significance level of 5%; it was about 20% at 4 gL^{-1} and 40% at 6 gL^{-1}. Concerning the size of fruits (Fig. 27.5), the irrigation with Mejerda water gave the best proportions of marketable fruits of good size. Sizes C4, C5 and C6 were dominant, their proportions being 17%, 23% and 22% respectively for DI, and 20%, 22% and 21% for SDI. Sizes C1, C2 and C3 were more represented in plants irrigated by the mixture of waters (4 gL^{-1}). At 6 gL^{-1}, the size C0 was the most dominant, with about 40%. In comparison with those irrigated by the well water and the mixture, plants irrigated with Mejerda water had a better fruit setting and a good vegetative development, indicating that an important proportion of assimilates was allocated to fruits. As regards the yield in fresh weight/plant (Fig. 27.4), it was the result of the number of flowers by plant and the percentage of fruit setting. Therefore, factors that affect these two parameters had an impact on yield. The effect of the salinity was significant at the significance level of 5%; an increase in the salinity of the irrigation water induced a reduction in the crop yield.

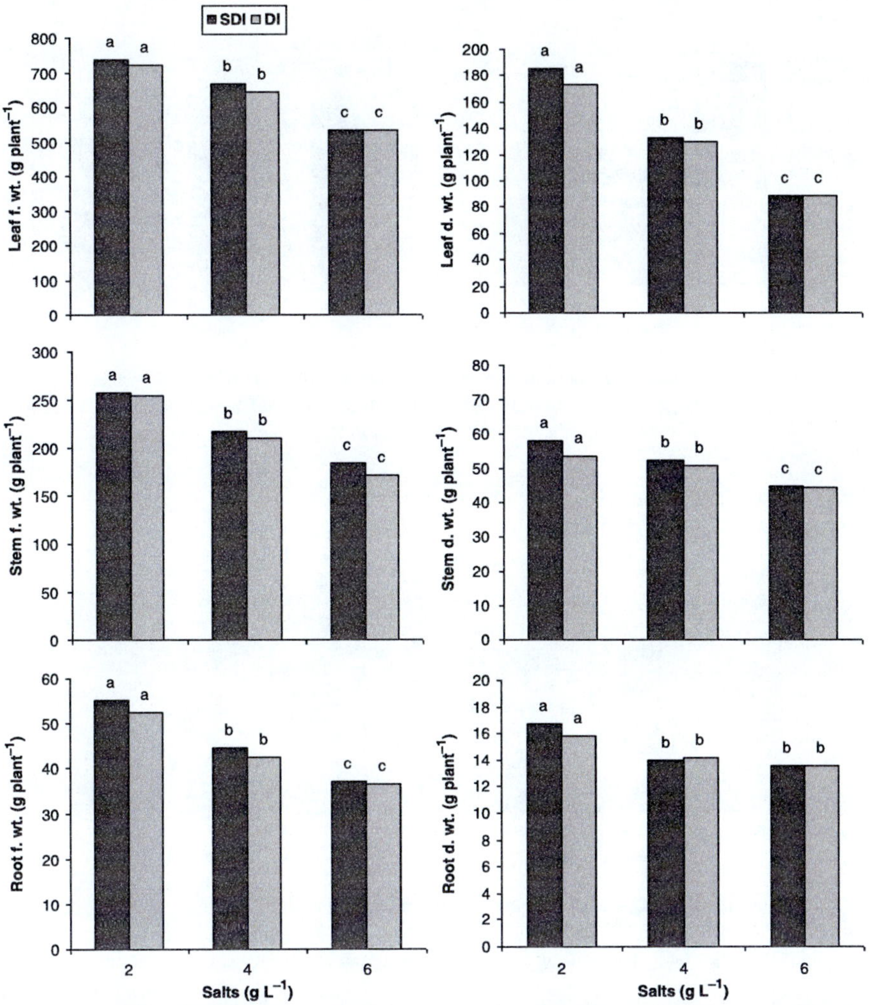

Fig. 27.2 Fresh and dry weight of different organs of irrigated tomato as a function of three salt concentrations of water and two irrigation systems (DI and SDI). Values followed by the same letter are not significant at the 5% level of significance according to the test of Neuman and Keuls

3.3 Effect on the Mineral Composition

Chemical analyses did not show any significant difference between DI and SDI. Conversely, the salinity effect of the irrigation water was manifest. For either irrigation system, DI or SDI, the presence of salt in the irrigation water resulted in an increase of the contents of Na^{+} and Cl^{-} in different organs of tomato (Fig. 27.6).

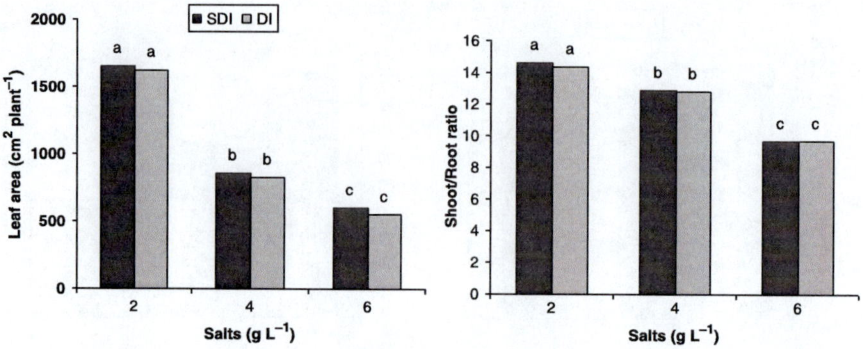

Fig. 27.3 Leaf area and shoot/root ratio of irrigated tomato as a function of three salt concentrations of water and two irrigation systems (DI and SDI). Values followed by the same letter are not significant at the 5% level of significance according to the test of Neuman and Keuls

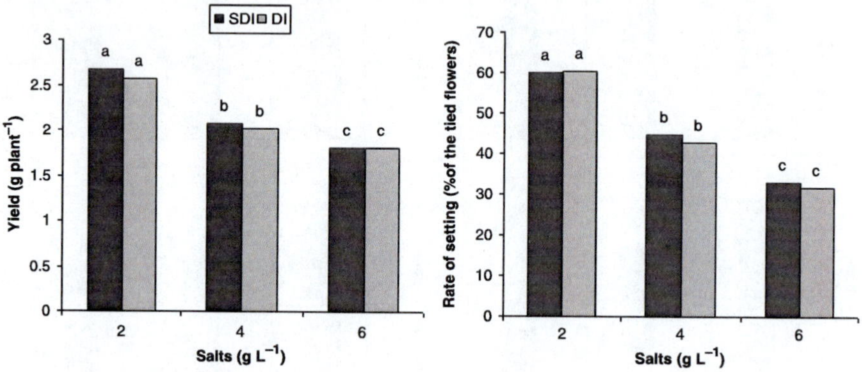

Fig. 27.4 Setting rate and yield of irrigated tomato in relation to three salt concentrations of water and two irrigation systems (DI and SDI). Values followed by the same letter are not significant at the 5% level of significance according to the test of Neuman and Keuls

This increase was higher when the concentration in salts was increased. The presence of salt in the irrigation water induced a reduction of potassium contents for all crop organs (Fig. 27.7). This reduction was proportional to the water salinity. At 6 $g L^{-1}$, the reduction was 40% for leaves, 11% for stems and 51% for roots. The presence of salt in the irrigation water induced also a reduction in calcium contents for the different organs (Fig. 27.7). This reduction was 34% for leaves and reached 69% for roots. For magnesium, stems and leaves were relatively not affected by the water salinity (Fig. 27.8). Only roots were affected by 4 $g L^{-1}$. The reduction was manifest at 6 $g L^{-1}$ about 82%. Concerning phosphorus, a small reduction of contents in leaves, stems and roots was observed, but this effect was not significant at the significance level of 5% (Fig. 27.8).

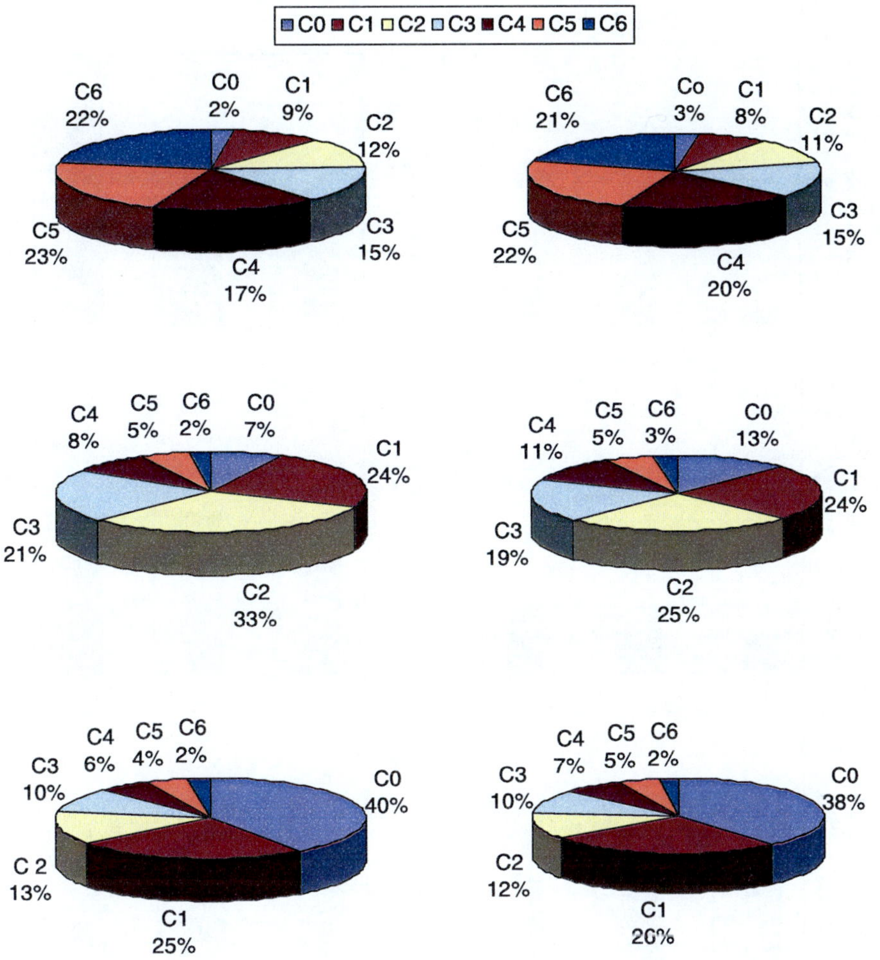

Fig. 27.5 Effect of waters of three salt concentrations (2, 4 and 6 $g\,L^{-1}$) and two irrigation systems (DI and SDI) on the fruit diameter of irrigated tomato

4 Discussion

The growth of tomato is affected by the salinity at different degrees, depending on the salt concentration in the irrigation water and the crop organ (Zhang et al. 2004). In the present study, leaf area decreased greatly under saline stress for SDI and DI. This reduction was more marked at 6 $g\,L^{-1}$, and it resulted from a general reduction of the crop growth and from the fall of leaves from the bottom part of plants, induced

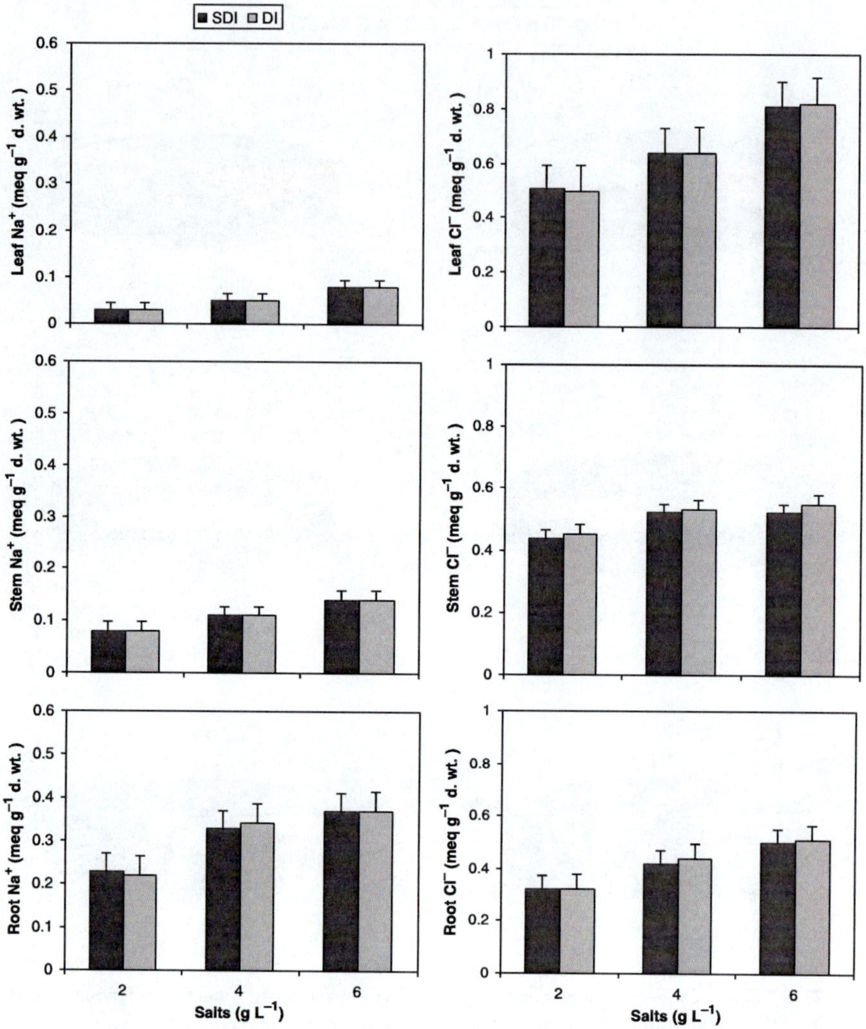

Fig. 27.6 Na^+ and Cl^- content in the dry matter of different organs of irrigated tomato in relation to three salt concentrations of water and two irrigation systems (DI and SDI)

by a premature senescence. These results are analogous to those observed by Taleisnik (1987). However, Karlberg et al. (2007) showed that the increase of salinity in irrigation water (6 gL^{-1}) produced an increase of leaf area of the tomato cultivar Daniella when DI system was used. With regard to the biomass, the results revealed that the salinity limited the growth, whatever the applied treatment may have been. This effect depends on the salt concentration and the type of organ. So, the high salt concentrations (6 gL^{-1}) reduced the biomass significantly. The reduction in growth

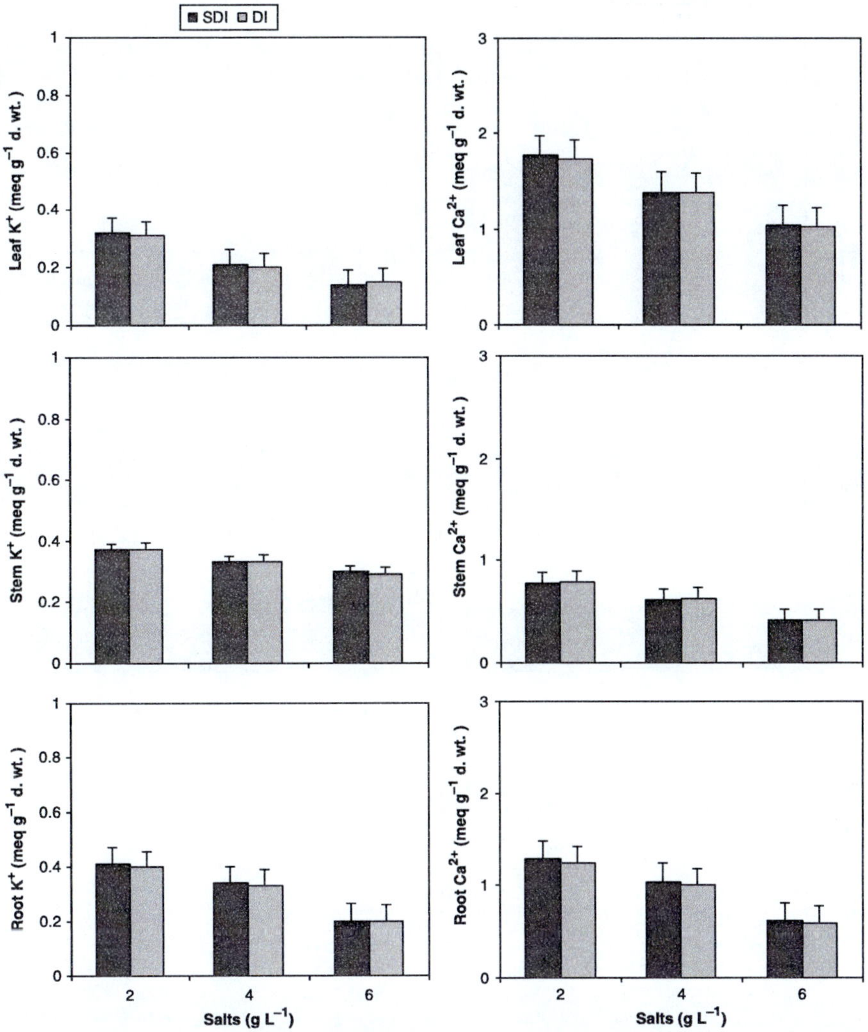

Fig. 27.7 K and Ca content in the dry matter of different organs of irrigated tomato in relation to three salt concentrations of water and two irrigation systems (DI and SDI)

affected all crop organs. Similar results were earlier observed for lentil (Lachaâl et al. 1996). The shoot/root ratio decreased progressively with the increase in salinity. The reduction resulted from the depressive effect of salt on the aerial and root parts. These results are in agreement with those observed by Cornillion et al. (1995) and Maggio et al. (2006). Engel and Marschner (1992) consider that the reduction of this ratio is an adaptive component of plants to environmental constraints; it serves to maintain a functional balance between the shoots and roots by increasing the root

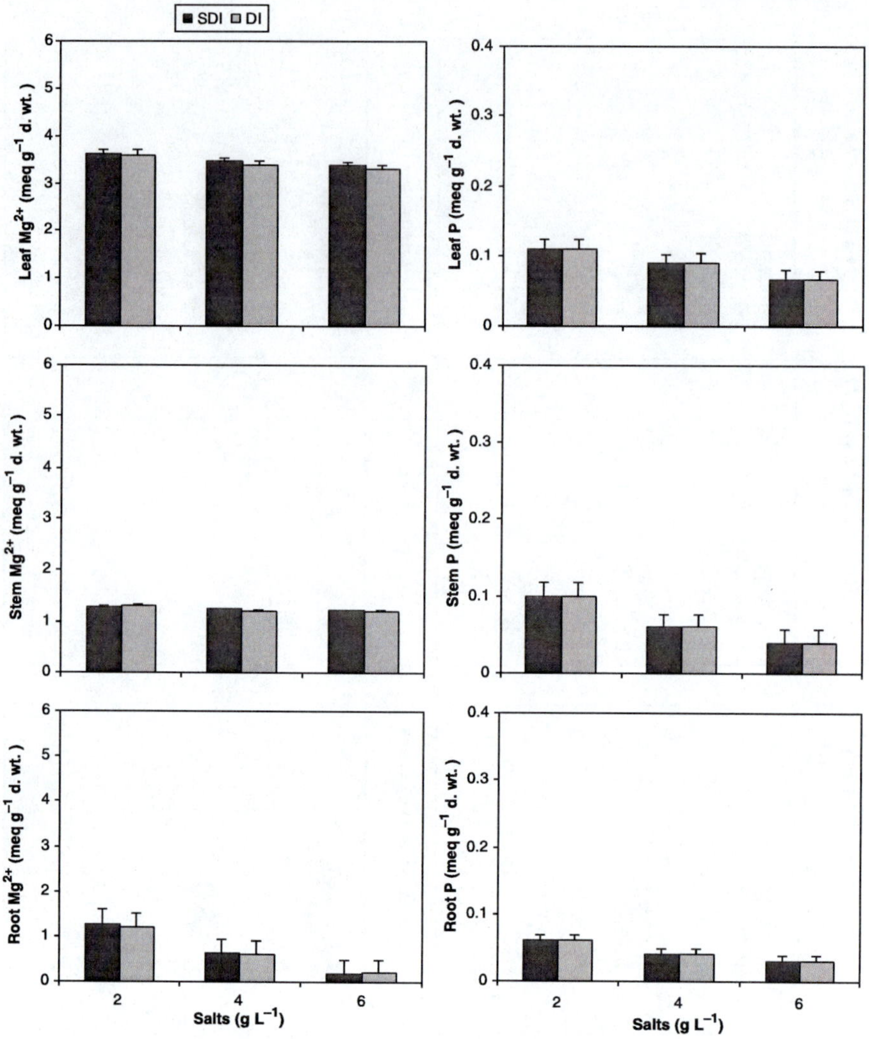

Fig. 27.8 Mg and P content in the dry matter of different organs of irrigated tomato in relation to three salt concentrations of water and two irrigation systems (DI and SDI)

length when their physiological activity is inhibited by external factors. The study of the saline stress effects for the two irrigation systems on flowering and yield of tomato shows that an increase of the irrigation water salinity decreased the agronomic characters at all crop stages (flower number, setting rate, fruits number as well as fruits size). The final consequence is a reduction of the yield. These results are in agreement with those observed for tomato by Perez-Alfocea et al. (1996), Karlberg et al. (2007) and Flowers et al. (2005). No significant difference between

DI and SDI was observed for the pepper cultivar Beldi irrigated by the water of the Mejerda (Hachicha et al. 2006) and for tomato (Incrocci et al. 2005).

The mineral nutrition study of Heinz-2274 tomato variety shows nutritional disruptions in major cations (K^+, Ca^{2+}, Mg^{2+}) and in the P anion according to the salinity for DI as well as for SDI. The significant decrease of all ions content per tissue especially under high salinity level (6 $g\,L^{-1}$) could probably be attributed to the high accumulation of toxic ions (Na^+, Cl^-) in tissue subjected to severe osmotic stress caused by high salinity level (Botia et al. 1998; Roussos et al. 2006). Results concerning the ionic contents of different organs show that Heinz-2274 excluded Na^+ from its leaves because the contents of this ion were low as compared to those of stems or roots. Such results have also been observed by Taleisnik and Grunberg (1994) and Reina-Sánchez et al. (2005) in tomato. They showed that most cultivars of tomato excluded Na^+ from leaves when they were exposed to salt stress. Like Na^+, Cl^- concentration increased in the same way with the increasing salt concentration in the irrigation water. Salt stimulates the absorption of Na^+ and Cl^-. This stimulation induces a reduction of contents in Mg^{2+}, K^+, Ca^{2+} and P (Maggio et al. 2006). Roots being the first organs in direct contact with salts, showed a reduction in Ca^{2+} (about 55-69%) for the studied variety. This significant decrease causes a reduction in Ca^{2+} content at the shoots level (Bernstein et al. 1995). Such reduction is associated with a respiration rate increase and a competitive effect with Na^+ (Cuartero and Fernández-Muñoz 1999). In the same way, a deficiency in K^+ and Mg^{2+} is also in correlation with an excessive accumulation of Na^+, suggesting a competitive effect between these ions and the toxic ion, Na^+ (Juan et al. 2005; Maggio et al. 2006). Therefore, the growth decrease observed was due to a low accumulation of K^+, Ca^{2+}, $Mg^{2+}$and P rather than to a strong accumulation of Na^+ and Cl^- in the tissues.

5 Conclusion

The experiment on tomato crop irrigated with three water qualities by DI and SDI permitted to show a significant decrease of growth parameters particularly with the more saline water (6 $g\,L^{-1}$). This significant decrease refers to leaf area (60%) and biomass of all aerial and root organs, as well as, the shoot/root ratio. The reduction of this ratio was similar for SDI and DI. The depressive effect of salt was more important on the growth of aerial parts than of roots. Concerning mineral nutrition, tomato accumulated more Na^+ and Cl^- in different crop organs in the presence of salt. This high accumulation of these ions induced a reduction in the contents of K^+, Ca^{2+} and Mg^{2+} in different organs, whereas the P content remained constant. The effect of the two irrigation systems on tomato crop was almost uniform.

Acknowledgement This study was carried out under the INRGREF/ACSAD project for saline water use in agriculture and PISEAU II project for saline water use in the Centre of Tunisia. We would like also to thank E. Misle from the Fac. Agr. and Forestry Sci., Univ. Católica del Maule, Chile for his comments on the manuscript.

References

Ayars JE, Schoneman RA, Dale F, Meso B, Shouse P (2000) Managing subsurface drip irrigation in the presence of shallow ground water. Agric Water Manage 47:243–264

Ayers RS, Westcot DW (1985) Water quality for agriculture irrigation and drainage paper 29. FAO, Rome

Beaux MF, Gouet H, Gouet JP, Morghem P, Philippeau G, Tranchefort J, Verneau M (1991) Manuel d'utilisation du logiciel STATITCF. Impri. ITCF, France, p 190

Ben Necib (2008) Effet de l'irrigation goutte à goutte souterraine avec différentes salinités d'eaux sur une culture de tomate de saison. Mastère Génie hydraulique et Aménagement Rural, INAT, pp 42

Bernstein N, Silk WK, Läuchli A (1995) Growth and development of sorghum leaves under conditions of salt stress: possible role of some mineral elements in growth inhibition. Planta 196:699–705

Bogle CR, Hartz TK, Nunez C (1989) Comparison of subsurface trickle and furrow irrigation on plastic-mulched and bare soil for tomato production. J Am Soc Hortic Sci 144:40–43

Botia P, Carvajal M, Cerda A, Martinez V (1998) Response of eight *Cucumis melo* cultivars to salinity during germination and early vegetative growth. Agronomy 18:503–513

Cornillion P, Auge M (1995) Salinité adaptation du piment. Dossier: PHM Revue Hort 4:241–243

Cuartero J, Fernández-Muñoz R (1999) Tomato and salinity. J Sci Hortic 78:83–125

DeTar WR (2007) Yield and growth characteristics for cotton under various irrigation regimes on sandy soil. Agric Water Manage 92:497–505

Enciso J, Jifon J, Wiedenfeld B (2007) Subsurface drip irrigation of onions: effects of drip tape emitter spacing on yield and quality. Agric Water Manage 92:126–130

Engel C, Marschner H (1992) Adaptation of potassium translocation into the shoot of maize (*Zea mays*) to shoot demand: evidence for xylem loading as a regulating step. Physiol Plant 86:263–268

Fernández-García N, Martínez V, Cerdá A, Carvajal M (2002) Water and nutrient uptake of grafted tomato plants grown under saline conditions. J Plant Physiol 159:899–905

Flowers TJ, Ragab R, Abdel Gawad NMG, Cuartero J, Arslan A (2005) Sustainable strategies for irrigation in salt-prone Mediterranean: SALTMED. Agric Water Manage 78:3–14

Gaines AM, Shennan C (1999) Interactive effects of Ca^{2+} and NaCl salinity on the growth of two tomato genotypes differing in Ca^{2+} use efficiency. Plant Physiol Biochem 37:569–576

Gençoğlan C, Altunbey H, Gençoğlan S (2006) Response of green bean (*P. vulgaris* L.) to subsurface drip irrigation and partial root zone-drying irrigation. Agric Water Manage 84:274–280

Hachicha M (2007) Les sols salés et leur mise en valeur en Tunisie. Sécheresse 18:45–50

Hachicha M, Nahdi H, Rejeb S (2006) Effet de l'irrigation au goutte à goutte souterraine avec l'eau salée sur une culture de piment. Ann De l'INRAT 79:85–103

Hanson B, May D (2004) Effect of subsurface drip irrigation on processing tomato yield, water table depth, soil salinity, and profitability. Agric Water Manage 68:1–17

Incrocci L, Malorgio F, Bartola AD, Pardossi A (2005) The influence of drip irrigation or subirrigation on tomato grown in closed-loop substrate culture with saline water. Sci Hortic 107:365–372

Jammazi M, Hachicha M (2000) Application et validation de nouvelles méthodes pour la détermination électromagnétique de la teneur en eau et de la salinité : la réflectométrie dans le domaine temporel (TDR) et la conductivimétrie électromagnétique, INRGREF, rapport de technique – Tunisia, pp 30

Juan M, Rivero RM, Romero L, Ruiz JM (2005) Evaluation of some nutritional and biochemical indicators in selecting salt-resistant tomato cultivars. Environ Exp Bot 54:193–201

Karlberg L, Rockstrom J, Annandale JG, Steyn JM (2007) Low-cost drip irrigation – a suitable technology for southern Africa? An example with tomatoes using saline irrigation water. Agric Water Manage 89:59–70

Lachaâl M, Abdelly C, Grignon C, Soltani A, Hajji M (1996) Variation de la sensibilité au sel en fonction du stade de développement chez la lentille (*Lens culinaris* L). Agronomy 16:81–390
Maggio A, Raimondi G, Martino A, De Pascale S (2006) Salt stress response in tomato beyond the salinity tolerance threshold. Environ Exp Bot 59:276–282
Oron G, DeMalach Y, Gillerman L, David I, Rao VP (1998) Improved saline-water use under subsurface drip irrigation. Agric Water Manage 39:19–33
Perez-Alfocea F, Balibrea ME, Santa Cruz A, Estan MT (1996) Agronomical and physiological characterization of salinity tolerance in commercial tomato hybrid. Plant Soil 180:251–257
Phene CJ, Davis KR, Hutmacher RRB, Meek DW (1990) Effect of high frequency surface and subsurface drip irrigation on root distribution of sweet corn. Irrig Sci 12:135–140
Rajak D, Manjunatha MV, Rajkumar GR, Hebbara M, Minhas PS (2006) Comparative effects of drip and furrow irrigation on the yield and water productivity of cotton (*Gossypium hirsutum* L.) in a saline and waterlogged vertisol. Agric Water Manage 38:30–36
Reina-Sánchez A, Romero-Aranda R, Cuartero J (2005) Plant uptake and water use efficiency of greenhouse tomato cultivars irrigated with saline water. Agric Water Manage 78:54–66
Roussos PA, Tsantili E, Pontikis CA (2006) Responses of jojoba explants to different salinity levels during the proliferation stage *in vitro*. Ind Crops Prod 23:65–72
Taleisnik E, Grunberg K (1994) Ion balance in tomato cultivars differing in salt tolerance. I. Sodium and potassium accumulation and fluxes under moderate salinity. Physiol Plant 92:528–534
Taleisnik EL (1987) Salinity effects on growth and carbon balance in *Lycopersicon esculentum* and *L. pennellii*. Physiol Plant 71:213–218
Tariq M, Tariq JA, Ali M (1990) The effect of trickle and furrow irrigation on the yield of tomatoes. Sarhad J Agric 6:231–233
Taylor KS, Anda M, Sturman J, Mathew K, Ho G (2005) Subsurface dripline tubing-an experimental design for assessing the effectiveness of using dripline to apply treated wastewater for turf irrigation in Western Australia. Desal 187:375–385
Tuna LA, Kaya C, Ashraf M, Altunlu H, Yokas I, Yagmur B (2005) The effects of calcium sulphate on growth, membrane stability and nutrient uptake of tomato plants grown under salt stress. Environ Exp Bot 59:173–178
Zhang X, Zang R, Li C (2004) Population differences in physiological and morphological adaptations of *Populus davidiana* seedlings in response to progressive drought stress. Plant Sci 66:791–797

Chapter 28
Lipid and Carbohydrate Metabolism of Cowpea (*Vigna unguiculata* L. Walp) Cultivars in Relation to Temperature Stress

Shahidul Islam, Carmen Rafaela Carvajal Rebanales, and James Otis Garner Jr.

Contents

S. Islam (✉) • J.O. Garner, Jr.
Department of Agriculture, University of Arkansas at Pine Bluff,
1200 North University Drive, Mail Slot 4913, Pine Bluff, AR 71601, USA
e-mail: islams@uapb.edu; garnerj@uapb.edu

C.R.C. Rebanales
Plant and Soil Sciences, Missippi State University, Starkville, MS, USA
e-mail: ccarvajal@inia.gob.ve

M. Ashraf et al. (eds.), *Crop Production for Agricultural Improvement*,
DOI 10.1007/978-94-007-4116-4_28, © Springer Science+Business Media B.V. 2012

Abstract Cowpea (*Vigna unguiculata* L. Walp) is a legume adapted to subtropical to tropical conditions. It is susceptible to chilling temperatures, primarily during the early stages of germination. Twenty five cultivars were screened for germination at low (10°C), moderate (30°C), and at high (40°C) temperatures. Temperatures held at 10°C, <40°C, <30°C had a negative effect on germination percent and coefficient of velocity of germination (CVG). Three cultivars (Texas Cream 40, Black Crowder and Mississippi Purple) were chosen such as 'Texas Cream 40' was able to germinate at very high and low temperatures. 'Black Crowder' demonstrated acceptable germination at high temperatures but negatively affected at low temperature. Cultivar 'Mississippi Purple' obtained low germination percentage and CVG at all temperatures studied. Lipid, sugar compositions, and peroxidase activity were determined in whole ungerminated seed, cotyledon and embryo tissues of the cowpea cultivars. The main sugars present in cowpea seed were sucrose, raffinose, and stachyose. Sucrose contents were higher in the embryo tissue of cultivars with low percent germination, and reduced in the cultivar with higher percent germination suggesting the use of sucrose for germination. Raffinose and stachyose contents were higher in ungerminated seed. In germinated seed, raffinose and stachyose contents were found only in cotyledon tissues at 10°C. The most abundant fatty acids in cowpea seed were palmitic acid, palmitoleic acid, stearic acid, oleic acid, linoleic acid, linolenic acid, and arachidic acid. The results showed that the long-chain fatty acids appear to be important in the cowpea seed germination process. Peroxidase activity was affected by cultivars, type of tissue and temperature. The highest peroxidase activity was found at low temperature (10°C) in embryo tissue of the cultivar with the highest germination. High peroxidase activity was related to ability of seed to germinate at low temperature. Thus, the information provided by this research will facilitate future plant physiological and genetic studies of cowpea as well as other plant species.

Keywords Carbohydrate • Lipid • Peroxidase • Temperature • Seed • Germination • Cowpea

1 Introduction

Cowpea (*Vigna unguiculata* L. Walp), an indigenous African annual legume, is commonly known as southern pea, blackeye pea, crowder pea, lubia, niebe, coupe or frijole, (Craufurd et al. 1996; Hall et al. 1997). It is a high protein (25%), fiber (6.3%) and low fat (1.9%) vegetable crop having good nutritional qualities (Ricardo 1985). It is a chilling sensitive crop (Ehlers and Hall 1997; Hall et al. 1997) and is adapted to warm weather and humid conditions (Craufurd et al. 1996; Hall et al. 1997; Terao et al. 1997; Singh et al. 2003). It is more tolerant than common beans to drought, waterlogging, infertile soils, and acid stress. Cowpea is cultivated for food (Hall et al. 1997), silage, green manure for soil improvement, livestock feed and pasture. Due to its nutritional content, versatility, adaptability, and high yield, cowpea was chosen by the U.S. National Aeronautical and Space Administration as one of the few crops to be tested for cultivation in the space station (Ehlers and Hall 1997). Furthermore, it is one of the mandated crops by the International Institute of Tropical Agriculture (Smartt and Hymowitz 1985; Quin 1997; Fery 2002). Cowpea was a major agronomic crop in the United States during the early part of the twentieth century, with production peaking at 2.4 million ha in 1937 (Fery 1990). By the early 1980s, however, annual cowpea production in the United States was estimated at 80,000 ha (Fery 1981). Cowpea has long been valued in the southern United States as a vegetable crop, and an extensive industry currently distributes fresh, canned, frozen, and dry-pea products nationwide. Additionally, cowpea has long been a popular item with home gardeners throughout the south. There is a broad range of characteristics among cowpea cultivars grown for horticultural use in the United States (Fery 1990, 2002; Islam et al. 2005, 2006).

Cowpea has been reported to have greater adaptation to high temperatures than any other crop species. Germination greater than or equal to 80% has been observed in cowpea at temperatures ranging from 10°C to 40°C (Ismail et al. 1997, 1999; Terao et al. 1997). Temperatures outside this range adversely affect germination of cowpea. Variation in germination percentage and rate of germination under stress temperatures have been observed among cowpea cultivars (Marsh 1993; Craufurd et al. 1996). Temperature plays an important role during seed germination and initial root growth. The rate of seed germination and seedling emergence is determined by temperature (Craufurd et al. 1996). As a chilling-sensitive crop, cowpea is most susceptible during germination and early seedling development. If cowpea is planted late in the spring when temperatures are still low in subtropical areas, seed germination is hindered. Good stand establishment is necessary to avoid loss of yield, thus it would be beneficial to plant a cultivar able to tolerate low temperatures.

A wide range of cellular responses occur when seeds are exposed to different temperatures, including adjustments in the level of fatty acids. Lipid contents have been related to seed vigor in some seed (Perl et al. 1987). Seed of broad bean (*Vicia faba* L.), evening primrose (*Oenothera Lamarckiana* De Varies), and carrot (*Daucus carota* L.) showed a drop of approximately 20–30% in the lipid fraction when subjected to accelerated aging; however, in cucumber (*Cucumis sativus* L.), squash

[*Praecitrullus fistulosus* (Stocks) Pang], and pea (*Pisum sativum* L.) seed, an increase in lipid content was observed. In pepper (*Capsicum annum* L.) seed, the lipid content did not change with seed vigor level, but noticeable changes in fatty acid composition were observed. In the labeling of the lipids present in pea seed, phospholipids (70%), triglycerides (20%), waxes (8%), monoglycerides (2%), and diglyceride were detected (Harwood and Stumpf 1970). Free fatty acids were seldom detected. The most common fatty acids in the phospholipids were palmitic (C16:0), palmitoleic (C16:1), oleic (C18:1), linoleic (C18:2), and linolenic (C18:3) acids. The longer chain fatty acids were more common in the wax fraction. The endogenous fatty acids present in the seed were not always related to those synthesized during germination (Harwood and Stumpf 1970). In germinating seed, *de novo* synthesis of long chain fatty acids occurs at early stages of germination. Fatty acids including palmitic (C16:0), stearic (C18:0), oleic acid (C18:1), linoleic (C18:2), and linolenic (C18:3), have been detected in cotyledons and axes of water oak (*Quercus nigra* L.), white oak (*Quercus alba* L.), americam muskwood [*Guarea guidonia* (L.) Sleumer and *Carapa guianensis* Aubl.], silver maple (*Acer saccharinum* L.), and red buckeye (*Aesculus pavia* L.) (Connor et al. 1996, 1998, 2000; Connor and Bonner 1998, 2001; Connor and Sowa 2000). Significant changes in the amounts of individual fatty acids in the seed, and the subsequent changes in the total fatty acid content have been reported in these species.

Seed vigor of seed has also been associated with carbohydrate content. In corn (*Zea mays* L.), the decline of sucrose and raffinose contents during aging was closely correlated with the loss of ability to germinate and rate of germination (Bernal-Lugo and Leopold 1992). Furthermore, the quantity of oligosaccharides larger than sucrose ($C_{12}H_{22}O_{11}$) decreased as the period of imbibition during the germination process increased (Koster and Leopold 1988). As a chilling-sensitive crop, cowpea is affected during germination and early seedling development if planted late in the spring when temperatures are still low in subtropical areas. Good stand establishment is necessary to avoid yield loss, thus it would be beneficial to plant a cultivar able to tolerate low temperatures. The objectives for this research were to determine the carbohydrate and fatty-acid metabolism of the cowpea seed prior to and during germination and evaluate the peroxidase enzyme activity in cowpea seed and its effect on temperature tolerance.

2 Materials and Methods

2.1 Cultivars/Genotypes Used in This Study

Twenty five cowpea cultivars were evaluated for germination percentage and coefficient of velocity of germination (CVG): (1) Pinkeye Purple Hull (PPH), (2) California Black Eye (CBE), (3) Tahoro O' Adhan (TA), (4) Elite (E), (5) Big Boy Purple Hull (BBPH), (6) Texas Cream (TC), (7) Tennessee White Crowder (TWC),

(8) Cream 8 (C-8), (9) Guajirio Yoro Mami (GYM), (10) Mississippi Cream (MC), (11) C.T. Pinkeye Purple Hull (CTPPH), (12) Tetapeche Grey (TG), (13) Mayo Colina (MAC), (14) Speckled Purple Hull (SPH), (15) Mayo Colina Pinto (MCP), (16) Black Crowder (BC), (17) Whippoorwill (W), (18) Knuckle Purple Hull (KPH), (19) Corriente (C), (20) Frijol Reata (FR), (21) Mississippi Purple (MP), (22) Mississippi Pinkeye (MPE), (23) Zipper Cream (ZC), (24) Mississippi Silver (MS), and (25) Bisbee Black (BB). The seeds were obtained from Native Seed, Search, Tucson, AZ., USA. To get the seeds from the same lot the cultivars/genotypes were grown in the same year under a common set of environmental conditions, handled using the same harvesting/seed processing equipment, and stored under the same conditions. The seeds from the above lot were used in two different years experiment. Trials were carried out over 2 years using the above cultivars/genotypes grown to initially screen for germination and coefficient of variability of germination tests (Islam et al. 2005, 2006). Three cultivars (Texas Cream 40, Black Crowder and Mississippi Purple) were chosen for further study for biochemical characteristics such as Texas Cream 40 which showed ability to germinate at very high (40°C) and low (10°C) temperature; Black Crowder which had acceptable high germination at 40°C, but reduced germination at 10°C; and Mississippi Purple which exhibited lower germination at all temperatures tested (Islam et al. 2005, 2006).

2.2 *Germination Test*

Germination percentages of all cultivars were evaluated adopting the methods described by the Rules for Testing Seed (AOSA 1991), except for the number of seed used per replication and the inclusion of different temperatures. Seeds were germinated using the "between-paper" method, in which two sheets of germination paper (Anchor brown germination paper, Anchor Paper Co., St. Paul, Minnesota) were placed in the bottom of 48 × 50 cm trays and an additional sheet of the same paper was used to cover the seed. The paper sheets were previously saturated with 2.5 times their weight of water. Four replications of 20 seed of each cultivar were randomly assigned to trays. Each experiment repeated for four times.

2.3 *Environmental Conditions*

The germination tests were conducted in germination chambers each of which was set at a different temperature: 10°C (low temperature), 30°C (moderate temperature), and 40°C (high temperature). All chambers were maintained at 100% ± 1 relative humidity. Light/dark periods were maintained at 8/16 h d^{-1} at all three temperatures.

2.4 Evaluation of Germination Percentage

Seeds were counted as germinated when radicle protrusion was approximately 1 cm. The final count for germination was done 8 days after planting the seed in the trays. Our preliminary observation shows that within 8 days we had maximum germination. The germination percentage for each cultivar was calculated as the mean of the four 20-seed replications. A germination percentage over 75% was considered as reasonable for cowpea seed. Each experiment was repeated four times.

2.5 Coefficient of Velocity of Germination (CVG)

The same procedures and conditions as used in the germination study were used in CVG evaluation. The numbers of germinated seeds were evaluated every day for a period of 8 days. We have conducted a preliminary observation using the seeds of the above cowpea cultivars. This showed that no significant germination occurred after 8 days. The CVG was calculated using the Kotowski's (1926) modified method proposed by Islam et al. (2005):

$$CVG = \frac{\sum G1 + G2 + G3 + \ldots\ldots\ldots + Gn}{(T1 \times G1) = (T2 \times G2) + (T3 \times G3) + \ldots\ldots + (Tn \times Gn)}$$

where, G = number of germinated seeds and T = number of days after incubation corresponding to G.

2.6 Extraction of Carbohydrates

Seed carbohydrates were extracted and identified following the methodology suggested by Connor and Sowa (2000). To determine the initial content of carbohydrates in cowpea seed, whole dry seeds were ground using a hand mill machine and 0.5 g of the resulting flour was used for carbohydrate extraction. For the extraction from the germinating seed, after radical protrusion and before first leaf expansion, the seeds were dissected into cotyledons and axes. Tissues from each cultivar germinated at each temperature were separately dried in a freeze-dry system (Laconic Freezone 4.5 LABCONCO Corporation, Kansas City, MO). After removing all moisture, the tissue samples were finely-ground using a mortar and pestle. A 0.2–0.5 g dry tissue sample was used for each carbohydrate extraction. The tissue samples were placed in test tubes with 10 ml of 80% ethanol solution and then boiled in a water bath at 75°C for 1 h. The samples were then passed through a piece of filter paper in filter funnel. Additional ethanol solution was used to rinse the samples. The extracts were poured into evaporation flasks and then roto-evaporated to dryness in a flask-evaporator (Buchler Instruments. Fort Lee, NJ). The evaporation flasks were rinsed with 10 ml of distilled water, and the samples were freeze-dried overnights in a stoppered vacuum flask attached to the freeze-dry apparatus. When all the moisture was removed from the flasks, the samples were dissolved in 1 ml of trimethylsilylimidiazole (TMSI), then heated in a water bath at 75°C for 30 min, and aspirated to dryness. The samples were then resuspended in 2 ml of chloroform and stored refrigerated until analysis.

2.7 Analysis of Carbohydrates

The analysis of carbohydrates was performed in a Hewlett Packard (HP) gas chromatograph (GC) equipped with a flame ionization detector and using a 125-5037 DB-5 column (30 m length × 0.53 mm I.D. and 0.5 μm film thickness) (J & W Scientific Folsom, CA). The program in the GC preset a detector and injector temperature of 230°C. An initial oven temperature of 210°C was held for 7 min and then increased 15°C min^{-1} until a temperature of 270°C was reached and held for 25 min. The flow was kept at 1.76 nn ml$^{-1.}$ A graph and numerical map with the respective percentage peak area of each sugar was obtained. The carbohydrates were identified by comparing with standards of pure sugars prepared in a similar manner to the tissue samples and injected in the GC.

2.8 Extraction of Lipids

Fatty acids in the seeds were extracted and identified through a modification of the methodology suggested by Whitaker (1986). To determine the content of fatty acids in cowpea seed, whole dry seeds were ground using a glass homogenizer and pestle. For extractions from germinating seeds, after radicle protrusion and before first leaf expansion, the seed were dissected into cotyledons and embryos. Two to five grams of each type of tissue from each cultivar germinated at each temperature were finely chopped and transferred to Pyrex culture tubes and were then covered with isopropanol using a Pasteur pipette. Tubes were then boiled in a water bath for 3–5 min, removed from the water bath, and allowed to cool at ambient temperature. The supernatant (isopropanol fraction), was transferred to another tube and then dried under nitrogen to prevent airoxidation. The crude remains of the extract were resuspended with approximately 2 ml of 2:1 chloroform: methanol. To prevent autooxidation, the supernatants were then aspirated under nitrogen atmosphere until dry. Three ml of chloroform and 1.5 ml of methanol were added to these fractions, followed by the addition of 1.5 ml of 0.85% sodium chloride. This wash procedure described by Folch et al. (1957) was used to remove non-lipid contaminants from the samples. The tubes were then sealed under nitrogen and centrifuged at 1,500 *g* for 5 min to obtain better separation. The pellet parts were removed with a Pasteur pipette and dried under nitrogen. The residues were taken up in 1.5 ml of 2:1 chloroform: methanol. The extracts from the sample were combined. A second addition of 1.5 ml of 0.85% sodium chloride and a second centrifugation at the same speed was performed to improve separation. The pellets were then removed and dried under nitrogen. The remaining portions were taken up in 2 ml of chloroform.

2.9 *Transesterification of Lipids*

The individual polar fractions were dried under nitrogen at 40°C. The samples were then resuspended in 0.5 ml of chloroform followed by the addition of 0.5 ml of 0.6 N KOH in dry methanol. The tubes were sealed under nitrogen and placed on a rototorque rotator in the dark at room temperature for 2 h. After the addition of 0.5 ml of distilled water and 50 μl 6 N HCl, the fatty acid methyl esters (FAME) were recovered by extraction with 2 ml of hexane.

2.10 *Analysis of Lipids*

Fatty acids were identified as described by Whitaker (1986). Lipids were analyzed in a Varian 3300 gas chromatograph (Gas Chromatography, Varian Associates, Sugarland, TX) equipped with a flame ionization detector and using a Supelcowax 10 fused-silica wide-bore capillary column (30 m length × 0.53 mm i.d. and 1.0 μl film thickness) (Supelco, Inc., Bellenfonte, PA). The program used established an injector temperature of 250°C and a detector temperature of 300°C. The initial column temperature was set at 190°C and a holding time of 3 min. The column temperature was increased 3°C/min until the final temperature of 220°C was reached, and a holding time of 20 min. Data were transferred to a Varian 4290 integrator (Varian Instrument Division, Walnutcreek, CA), where they were expressed as percentage (%) area under the peak. Identification of the FAMEs was achieved by comparing them with the standards of FAME mixture reference # 625006, 625007, and 625009 (Alltech Association Inc., Deerfield, IL).

2.11 *Peroxidase Extraction*

Peroxidase was extracted using a slight modification of the methodology of Silva et al. (1990). Seeds were separated into cotyledons and axes and 3.0 g of each tissue were used for extraction. Each sample was ground using a pre-chilled mortar and pestle, and 3.0 ml of citrate phosphate buffer (pH 6.0) were added as a grinding solution. This material was then placed in a centrifuge tube. An additional 3.0 ml of the buffer were used to wash the mortar and pestle and this plus a small amount of polyvinyl-polypyrrolidone (PVP) was added to the tube. The samples were centrifuged at 6,000 × *g* at 5°C for 10 min and the supernatant was saved and maintained at freezing temperature for the next step of enzyme analysis.

2.12 Peroxidase Activity

Peroxidase activity was measured using the methodology of Silva et al. (1990). The enzymatic activity of peroxidase was measured spectrophotometrically in a Spectronic Genesys 5 spectrophotometer (Spectronic Instruments Inc., Rochester, NY). One ml of crude peroxidase extract was added to a reaction mixture consisting of 2 ml of 0.05 M citrate phosphate buffer (pH 6.0) and 2 ml of 4.0×10^{-4} guaiacol. The reaction was initiated by adding 2 ml of 3% H_2O_2. The change in the absorbance at 470 nm was used to measure the enzyme activity.

2.13 Statistics

Each experiment was arranged in a completely randomized design (CRD) with five replicates of 20 seed each for each cultivar at each temperature (for germination and CVG). The data were analyzed as a combined series of CRDs. Data were subjected to analysis of variance (ANOVA) using the general linear models (GLM) procedure of SAS version 8.1 (SAS Instit., Inc. Cary, North Carolina, USA). Mean separations were done using Fisher's protected least significant difference (LSD) test.

3 Results and Discussion

3.1 Seed Germination and CVG

The germination percent and coefficient of velocity of germination (CVG) is shown in Table 28.1. The above three cowpea cultivars were chosen for biochemical evaluation based on physiological performance of the seed namely performance in the germination and CVG tests under temperature stress conditions (Islam et al. 2005, 2006). Texas Cream 40 (TC-40) performed well at 10°C, 30°C and 40°C temperatures; Black Crowder (BC) was more sensitive to chilling temperature than high temperature (Islam et al. 2005, 2006), and Mississippi Purple (MP) with low germination and CVG at all temperatures such as 10°C, 30°C and 40°C (Islam et al. 2005, 2006).

Table 28.1 Germination percentage and coefficient of velocity of germination (CVG) of three selected cowpea cultivars evaluated at different temperatures

	Germination (%)				CVG			
Temperature	10°C	30°C	40°C	LSD (0.05)	10°C	30°C	40°C	LSD (0.05)
MP	C	B	B	8.52	B**	A	A	0.014
	17.5b[a]	63.5a**	65.8a		0.110b	0.152b	0.151a	
TC-40	A*	A	A	8.98	C	A	B	0.011
	89.8b	93.4a	88.3b		0.125a	0.169a	0.155a	
BC	B	A	A	7.75	B	A	A	0.011
	27.0c	93.8a	87.5b		0.111b	0.154b	0.154a	
LSD(0.05)	5.04	9.27	10.11		0.013	0.012	0.010	

MP Mississippi Purple, *TC* Texas Cream, *BC* Black Crowder

*A–C Means within a cultivar not followed by a common capital letter are significantly ($p<0.05$) different according to Fisher's protected LSD test

**Means within a temperature not followed by a common letter are significantly ($p<0.05$) different according to Fishers's procted LSD test

[a]Analysis performed on arcsine transformed data

3.2 Sugar Compositions

Carbohydrates were determined in whole ungerminated seed and in cotyledon and embryo tissues of germinating seeds at low (10°C), moderate (30°C), and high (40°C) temperature. The main sugars present in cowpea seed were sucrose, raffinose, and stachyose. The sugar content was affected by temperature, cultivar, and types of tissue.

3.3 Sucrose Content

There was a significant temperature × cultivar × tissue interaction effect for sucrose content of cowpea seed (Table 28.2). The initial sucrose content in ungerminated seed was highest in BC, followed by TC-40, and lowest in MP. At low temperature (10°C), the higher mean sucrose content was in the embryo tissue of MP followed by BC, and the lowest sucrose content was in TC-40. In cotyledon tissue, the highest mean sucrose content was in TC-40, and the lowest in MP, and BC (Table 28.2). At moderate temperature (30°C), the highest sucrose content was in cotyledon of TC-40 (53.51 mg g^{-1}) and the lowest in MP. In embryo tissues, highest mean sucrose content was in TC-40, and the lowest in MP and BC. At high temperature (40°C), the highest mean sucrose content in cotyledon tissues was in TC-40 and MP, followed by BC. In embryo tissue there were no significant differences among cultivars for mean sucrose content. Sucrose is a primary nutrient in the cells of higher plants (Xu et al. 1989). During seed development, changes in carbohydrates occur in such a way that monosaccharide content decreases whereas oligosaccharide content increases, but during the germination process this pattern is reversed

Table 28.2 Sucrose content of whole seed, cotyledon (Cot.) and embryo (Emb.) tissues of seed of three cultivars of cowpea germinated at three different temperatures

	Temperature							
	Ambient	10°C		30°C		40°C		
Cultivars	Whole seed	Cot.	Emb.	Cot.	Emb.	Cot.	Emb.	LSD (0.05)
MP	E*	C	A	C	D	B	B	3.58
	9.9c	40.8c	67.9a	41.3b	29.6d	47.7ab	47.2ab	
TC-40	E	AB	D	A	C	AB	B	4.69
	14.8b	49.8b	26.7d	53.5a	37.3c	49.1a	46.4ab	
BC	E	C	A	BC	D	BC	B	4.16
	18.3a	39.9c	65.8a	42.9b	28.2d	43.6b	45.9ab	
LSD (0.05)	2.06	4.72		3.22		4.02		

MP Mississippi Purple, *TC*-40 Texas Cream-40, *BC* Black Crowder
a–d Means within a temperature not followed by a common lowercase letter are significantly ($p<0.05$) different according to Fisher's protected LSD test
*A–E Means within a cultivar not followed by a common capital letter are significantly ($p<0.05$) different according to Fisher's protected LSD test

(Vertucci and Farrant 1995). At low temperature, TC-40 seed had faster and higher percent germination than MP and BC (Islam et al. 2006); thus, the higher sucrose content in the cotyledon and lower sucrose contents in embryos of the above cultivar suggest faster and more efficient use of sucrose in germination process and seedling development in this cultivar. Xu et al. (1989) indicated that during germination seeds have a dual role: first as a source of sucrose produced in cotyledons and secondly as sink when the young seedling uses sucrose from development.

3.4 *Raffinose Content*

There was a significant temperature × cultivar interaction for raffinose content (Table 28.3). In whole ungerminated seed, the highest raffinose content mean was in BC (2.5 mg g^{-1}), and lowest in TC-40 (1.5 mg g^{-1}) and MP (1.3 mg g^{-1}). In germinating seed at low temperature (10°C), raffinose contents in cotyledon tissues were 0.67 mg.g^{-1} (MP), 0.62 mg g^{-1} (BC), and 0.35 mg g^{-1} (TC-40). No raffinose was detected in embryo tissues at low temperature or in any tissue at 30°C and 40°C.

3.5 *Stachyose Content*

There was a significant temperature × cultivar interaction for stachyose content (Table 28.4). The highest stachyose mean content in the whole ungerminated seed was in BC (10.4 mg g^{-1}), followed by TC-40 (7.8 mg g^{-1}), and MP (5.8 mg g^{-1}). Stachyose was also detected in cotyledon tissues of germinating seed at low

Table 28.3 Raffinose content (mg g^{-1} d. wt.) of whole seed, cotyledon (Cot.) and embryo (Emb.) tissues of seed of three cultivars of cowpea germinated at three different temperatures

	Temperature							
	Ambient	10°C		30°C		40°C		
Cultivars	Whole seed	Cot.	Emb.	Cot.	Emb.	Cot.	Emb.	LSD (0.05)
MP	A*	B	C	C	C	C	C	0.03
	1.3b**	0.67a	0.00	0.00	0.00	0.00	0.00	
TC-40	A	B	C	C	C	C	C	0.05
	1.3b	0.53b	0.00	0.00	0.00	0.00	0.00	
BC	A	B	C	C	C	C	C	0.10
	2.5a	0.62a	0.00	0.00	0.00	0.00	0.00	
LSD (0.05)	0.30	0.07		ns		ns		

MP Mississippi Purple, *TC*-40 Texas Cream-40, *BC* Black Crowder
*A–C Means within a cultivar not followed by a common capital letter are significantly ($p<0.05$) different according to Fisher's protected LSD test
** a–b Means within a temperature not followed by a common lowercase letter are significantly ($p<0.05$) different according to Fisher's protected LSD test

Table 28.4 Stachyose content (mg g^{-1} d. wt.) of whole seed, cotyledon (Cot.) and embryo (Emb.) tissues of seed of three cultivars of cowpea germinated at three temperatures

	Temperature							
	Ambient	10°C		30°C		40°C		
Cultivars	Whole seed	Cot.	Emb.	Cot.	Emb.	Cot.	Emb.	LSD (0.05)
MP	A*	B	C	C	C	C	C	0.03
	5.8c**	2.2a	0.00	0.00	0.00	0.00	0.00	
TC-40	A	B	C	C	C	C	C	0.05
	7.8b	0.79b	0.00	0.00	0.00	0.00	0.00	
BC	A	B	C	C	C	C	C	0.10
	10.4a	2.35a	0.00	0.00	0.00	0.00	0.00	
LSD (0.05)	1.51	0.27		ns		ns		

MP Mississippi Purple, *TC*-40 Texas Cream-40, *BC* Black Crowder
*A-C Means within a cultivar not followed by a common capital letter are significantly ($p<0.05$) different according to Fisher's protected LSD test
** a-b Means within a temperature not followed by a common lowercase letter are significantly ($p<0.05$) different according to Fisher's protected LSD test

temperature (10°C) in BC (2.4 mg g^{-1}), MP (2.2 mg g^{-1}), and TC-40 (0.8 mg g^{-1}). No stachyose was detected in tissues of germinating seeds at moderate or high temperatures, which suggest that larger chain sugars had been reduced to sucrose or other monosaccharides. Koster and Leopold (1988) demonstrated that raffinose as well as stachyose have an important role in dessication tolerance of pea, corn, and soybean seed, probably helping to prevent sucrose crystallization. The increase in sugar content when comparing the dry ungerminated seed to the sugar content in cotyledon tissue of germinating seed at low temperature indicated either "de novo" synthesis of or convection of other products to sucrose. Miguel and Browse (1995) pointed out that seed storage components such as lipids, proteins and carbohydrates are converted to sucrose in order to feed the developing embryo.

Sucrose, raffinose and stachyose contents were significantly different among the cultivars studied. Black Crowder had the highest sucrose, raffinose, and stachyose contents, and this cultivar also exhibited high germination percentage. The lowest sucrose, raffinose, and stachyose contents were in Mississippi Purple, which had low germination percentage. The sucrose content in ungerminated seed was highest in BC, followed by TC-40, and lowest in MP. Sucrose is a primary nutrient in cells of higher plants (Xu et al. 1989). During seed development, changes in carbohydrates occur in such a way that monosaccharide content decreases whereas oligosaccharide content increases, but during the germination processes this pattern is reversed (Vertucci and Farrant 1995). In this study, BC seed had faster and higher percent germination than TC-40 and MP; thus, the higher sucrose content in the above cultivar suggests, faster and more efficient use of sucrose in germination process and seedling development in this cultivar. Xu et al. (1989) indicated that during germination, seeds have a dual role: first as a source of sucrose produced in cotyledons and secondly as a sink when the young seedling uses sucrose for development. The highest raffinose content was in BC (2.46 mg g^{-1}), and lowest in TC-40 (1.53 mg g^{-1}) and MP (1.25 mg g^{-1}). The highest stachyose content in the whole ungerminated seed was in BC (10.54 mg g^{-1}), followed by TC-40 (7.78 mg g^{-1}), and MP (5.77 mg g^{-1}). The increase in sugar content when comparing the dry ungerminated seed to the sugar content in cotyledon tissue of germinating seed at low temperature indicated either "*de novo*" synthesis of or convection of other products to sucrose. Miguel and Browse (1995) pointed out that seed storage components such as lipids, proteins and carbohydrates are converted to sucrose in order to feed the developing embryo.

3.6 *Fatty Acid (Lipid) Contents*

The main fatty acids present in cowpea seed and seedlings were palmitic acid (C16:0), palmitoleic acid (C16:1), stearic acid (C18:0), oleic acid (C18:1), linoleic acid (C18:2), linolenic acid (C18:3), and arachidic acid (C20:0). The feature of the cowpea seed fatty acids composition was presented in Table 28.5. The fatty acid content was affected by temperature, cultivar, and types of tissue.

3.7 *Palmitic Acid (PA)*

There was a significant temperature × cultivar × tissue interaction for PA (C16:0) (Table 28.6). At low temperature (10°C), the highest mean PA content was in cotyledon tissue of TC-40, followed by MP and BC. In embryo tissue, TC-40 had higher PA content than MP. At moderate temperature (30°C) in cotyledons, the highest

Table 28.5 Fatty acid composition in cowpea seeds

Common name	Scientific name	Carbon atoms	Double bonds	Structures
Saturated				
Palmitic acid	Hexadecanoic acid	16	0	$CH_3(CH_2)_{14}COOH$
Stearic acid	Octadecanoic acid	18	0	$CH_3(CH_2)_{16}COOH$
Arachidic acid	Eicosanoic acid	20	0	$CH_3(CH_2)_{18}COOH$
Unsaturated				
Palmitoleic acid	9-Hexadecenoic acid	16	1	$CH_3(CH_2)_5CH{=}CH(CH_2)_7COOH$
Oleic acid	9-Octadecenoic acid	18	1	$CH_3(CH_2)_7CH{=}CH(CH_2)_7COOH$
Linoleic acid	9, 12-Octadecadienoic acid	18	2	$CH_3(CH_2)_4CH{=}CHCH_2CH{=}CH CH_2(CH_2)_6COOH$
Linolenic acid	9, 12, 15-Octade-catrienoic acid	18	3	$CH_3CH_2CH{=}CHCH_2CH{=}CHCH _2CH{=}CH(CH_2)_7COOH$

Table 28.6 Palmitic acid (PA) (C16:0) concentrations (%) of cotyledon (COT) and embryo (EMB) tissues of seeds of three cowpea cultivars germinated at different temperatures

	10°C		30°C		40°C		
Cultivar	Cot.	Emb.	Cot.	Emb.	Cot.	Emb.	LSD (0.05)
MP	B*	D	A	B	A	C	2.55
	28.04b**	21.20d	34.99a	27.00d	35.28a	25.91c	
TC-40	BC	D	A	BC	B	C	2.64
	30.99a	23.45c	36.08a	30.56bc	33.28a	28.48bc	
BC	C	D	A	BC	B	BC	2.14
	27.52b	22.43cd	32.09b	28.85cd	29.83b	29.44b	
LSD (0.05)	1.96		2.17		3.30		

MP Mississippi Purple, *TC*-40 Texas Cream-40, *BC* Black Crowder
*A–D Means within a cultivar not followed by a common capital letter are significantly ($P<0.05$) different according to Fisher's protected LSD test
**a–d Means within a temperature not followed by a common letter are significantly ($P<0.05$) different according to Fisher's protected LSD test

mean PA content was in TC-40 and MP, and the lowest in BC. In embryo tissues at 30°C, the highest PA content was in TC-40, followed by BC and MP. At high temperature (40°C), the highest mean PA content was in cotyledon tissue of MP and TC-40 followed by BC. In embryo tissue, MP was significantly lower than BC (Table 28.3). PA content was lower at low temperature for all cultivars, suggesting that it does not play an important role in seed tolerance to chilling temperature. Harwood and Stumpf (1970) pointed out that the initial production of fatty acids in germinating pea seed showed a "lag" which depended on temperature. At higher temperature, lag time was shorter. They also observed that production of fatty acids during the first 30 h of germination initiated with palmitate and stearate, followed later by oleic acid. No polyunsaturated acids were detected during this period of pea seed germination. Higher PA contents were found in cotyledons where it is deposited or synthesized.

Table 28.7 Palmitoleic acid (PAA) (C16:1) concentrations (%) of cotyledon (COT) and embryo (EMB) tissues of seeds of three cowpea cultivars germinated at different temperatures

	10°C		30°C		40°C		
Cultivar	Cot.	Emb.	Cot.	Emb.	Cot.	Emb.	LSD (0.05)
MP	A*	A	B	B	B	B	1.22
	4.33a**	3.49a	0.00	0.00	0.00	0.00	
TC-40	B	A	C	C	C	C	1.64
	3.65a	5.60a	0.00	0.00	0.00	0.00	
BC	A	A	B	B	B	B	2.20
	4.28a	5.56a	0.00	0.00	0.00	0.00	
LSD (0.05)	ns		ns		ns		

MP Mississippi Purple, *TC*-40 Texas Cream-40, *BC* Black Crowder

* A–C Means within a cultivar not followed by a common capital letter are significantly ($P<0.05$) different according to Fisher's protected LSD test

** Means within a temperature not followed by a common letter are significantly ($P<0.05$) different according to Fisher's protected LSD test

3.8 Palmitoleic Acid (PAA)

There was a significant effect of temperature on PAA (C16:1) content (Table 28.7). In germinating seed, PAA was evident only at low temperature (10°C). The highest PAA contents in cotyledons were MP (4.79%), and BC (4.28%), and the lowest was TC-40 (3.65%). There were no significant differences among cultivars in PAA content of embryo tissue. De Santis et al. (1999) reported that PAA, a component of the mitochondrial inner membrane isolated from shoots of maize seedlings, had no significant role in chilling tolerance.

3.9 Stearic Acid (SA)

There was a significant temperature × cultivar × tissue interaction for SA (C18:0) content (Table 28.8). At low temperature (10°C), the SA content means differed only between embryo tissue of MP and cotyledon tissue of BC. At moderate temperature (30°C), the highest SA content was in embryo tissue of TC-40, followed by BC and MP, whereas in cotyledon tissue, the highest mean was in BC followed by TC-40 and MP. At high temperature (40°C), TC-40 had higher SA content in embryo tissue than BC. In cotyledon tissues, BC (4.36%) and TC-40 (3.87%) showed higher SA content than MP (2.77%). The content of this fatty acid in embryos of all cultivars was higher at 30°C than at 10°C. Stearic acid is found in fatty acid composition of important oilseeds such as soybean (*Glycine max* L. Merr.), rapeseed (*Brassica napus* L.), sunflower (*Helianthus annuus* L.), palm (*Elais guinenis* L.), cotton (*Gossypium hirsutum* L.), peanut (*Arechis hypogaea* L.), and coconut (*Cocus nucifera* L.) in proportions of approximately 4% (Miguel and Browse 1995). Thus, this long-chain fatty acid appears to be important in germination process.

Table 28.8 Stearic acid (SA) (18:0) concentrations (%) of cotyledon (COT) and embryo (EMB) tissues of seeds of three cowpea cultivars germinated at different temperatures

	10°C		30°C		40°C		
Cultivar	Cot.	Emb.	Cot.	Emb.	Cot.	Emb.	LSD (0.05)
MP	CD*	B	CD	AB	D	A	0.56
	3.23ab**	4.13a	2.87c	4.49b	2.77c	4.72ab	
TC-40	BCD	D	D	A	CD	B	1.28
	4.05ab	3.77ab	3.60c	6.63a	3.87b	5.17a	
BC	D	CD	B	A	B	BC	0.61
	3.18b	3.36ab	4.60b	5.22b	4.36ab	4.19b	
LSD (0.05)	0.94		0.83		1.02		

MP Mississippi Purple, *TC*-40 Texas Cream 40, *BC* Black Crowder
*A–D Means within a cultivar not followed by a common capital letter are significantly ($P<0.05$) different according to Fisher's protected LSD test
**a–c Means within a temperature not followed by a common letter are significantly ($P<0.05$) different according to Fisher's protected LSD test

Table 28.9 Oleic acid (OA) (C18:1) concentrations (%) of cotyledon (COT) and embryo (EMB) tissues of seeds of three cowpea cultivars germinated at different temperatures

	10°C		30°C		40°C		
Cultivar	Cot.	Emb.	Cot.	Emb.	Cot.	Emb.	LSD (0.05)
MP	3.64	3.18	4.97	4.73	4.97	4.98	0.66
TC-40	3.47	2.86	3.82	6.54	6.46	7.40	2.06
BC	3.82	2.99	5.26	4.97	5.79	4.24	0.72
LSD(0.05)	0.76		0.73		2.04		

MP Mississippi Purple, *TC* Texas Cream, *BC* Black Crowder

3.10 Oleic Acid (OA)

There was a significant cultivar × tissue interaction effect for OA (18:1) content (Table 28.9). Higher OA content was observed in the cotyledon tissues of BC and MP, followed by TC-40 at low temperature (10°C). In embryo tissue, OA content was in descending order: MP, BC, and TC-40. At moderate temperature (30°C), the highest OA content was in embryo tissue of TC-40, followed by BC and MP. In cotyledon tissue, BC and MP had higher OA content, and TC-40 (3.82%) had the lowest. At 40°C, the highest mean oleic acid content in embryo tissues was observed in TC-40, followed by MP and BC. In cotyledon tissue, no significant differences were found among cultivars. For all cultivars, OA content was lower at low temperature. OA is an unsaturated fatty acid which is thought to play an important role in chilling tolerance during germination and in other physiological processes. In TC-40, the OA content, which was higher in ungerminated seed, decreased drastically when exposed to chilling temperature. This decrease was more evident in cotyledons, which could be explained by the effect of temperature on enzyme activity.

Table 28.10 Linoleic acid (LA) (C18:2) concentrations (%) of cotyledon (COT) and embryo (EMB) tissues of seeds of three cowpea cultivars germinated at different temperatures

	10°C		30°C		40°C		
Cultivar	Cot.	Emb.	Cot.	Emb.	Cot.	Emb.	LSD (0.05)
MP	B*	C	A	C	A	C	2.02
	39.46a**	37.00b	41.85a	36.24cd	41.93a	36.71b	
TC-40	A	B	A	B	A	B	2.00
	39.60a	34.91b	38.01bc	34.45d	38.55b	34.38c	
BC	A	C	AB	C	A	B	1.75
	40.04a	35.10b	39.40b	35.55d	41.38a	37.9b	
LSD(0.05)	2.15		1.99		2.25		

MP Mississippi Purple, *TC* Texas Cream, *BC* Black Crowder
*A-C Means within a cultivar not followed by a common capital letter are significantly ($P<0.05$) different according to Fisher's protected LSD test
**a-d Means within a temperature not followed by a common letter are significantly ($P<0.05$) different according to Fisher's protected LSD test

Miguel and Browse (1995) suggested that temperature controls the rate of OA production by the fatty acid synthase, which decreases as the temperature decreases, whereas "de novo" synthesis and activity of oleate desaturase is stimulated at low temperature. Results of this study agree with Miguel and Browse (1995), as OA content increased the temperature at which cowpea seed germination increased. Additionally, the higher initial content of OA in ungerminated seed of TC-40, along with the activity of oleate desaturase could help increase the content of linolenic acid. Higher OA content was observed in cotyledon tissue, emphasizing its importance at the beginning of germination.

3.11 *Linoleic Acid (LA)*

There was a significant temperature × cultivar interaction and a significant effect of tissue on the mean LA (18:2) content (Table 28.10). At low temperature (10°C), no significant differences in LA content were detected among cultivars, but LA was more abundant in cotyledon than in embryo tissue. At moderate temperature (30°C), the highest mean linoleic acid content in cotyledon tissue was in MP, followed by BC and TC-40, whereas in embryo tissue, no significant differences were found. At high temperature (40°C), the highest LA content in cotyledon tissue was in MP and BC, and the lowest in TC-40. There was a similar trend in the embryo tissue. This unsaturated fatty acid is thought to play an important role in seed germination at low temperature. In this experiment, the increase in LA content from dry seed (35.98%) to cotyledons of germinating seed (39.60%) at low temperature was highest in TC-40, which was the cultivar with the higher germination percentage.

Table 28.11 Linolenic acid (LIA) (C18:3) concentrations (%) of cotyledon (COT) and embryo (EMB) tissues of seed of three cowpea cultivars germinated at different temperatures

	10°C		30°C		40°C		
Cultivar	Cot.	Emb.	Cot.	Emb.	Cot.	Emb.	LSD (0.05)
MP	C*	A	D	B	D	B	1.28
	21.27b**	30.95a	15.06d	27.51a	15.06d	27.52a	
TC-40	D	A	D	C	D	B	2.15
	18.18c	29.62a	18.06c	21.75b	17.78c	24.54b	
BC	C	A	D	B	D	B	1.65
	21.18b	30.71a	18.19c	22.96b	19.26c	24.24b	
LSD(0.05)	2.43		1.44		1.70		

MP Mississippi Purple, *TC* Texas Cream, *BC* Black Crowder

* A-D Means within a cultivar not followed by a common capital letter are significantly ($P<0.05$) different according to Fisher's protected LSD test

** a-d Means within a temperature not followed by a common letter are significantly ($P<0.05$) different according to Fisher's protected LSD test

3.12 Linolenic Acid (LIA)

There was a significant temperature × cultivar × tissue interaction effect for LIA (18:3) content of cowpea seed (Table 28.11). At low temperature (10°C), the highest LIA content was in the embryo tissue of MP, BC, and TC-40, while in the cotyledon tissue, the means were MP, BC, and TC-40. The same trend was observed at moderate and high temperatures. This unsaturated fatty acid is important in maintaining membrane permeability and in chilling tolerance. In this study, the seed of TC-40 had a slightly higher initial LIA content, which could be related to its ability to germinate under cool conditions. The highest content of LIA was in the embryo tissue and it was higher at low temperature than at higher temperatures. This suggests a more important role for this fatty acid in protecting the seedlings from chilling injury than in low temperature germination tolerance acquisition.

3.13 Fatty Acids Content in Ungerminated (Whole Seed)

Figure 28.1 shows the fatty acid composition in ungerminated (whole seed) seed of three cowpea cultivars. The C18:2/C18:1 ratio was higher in BC followed by MP and TC-40. The 18-carbon unsaturated to 18-carbon saturated fatty acid ratio was lower in TC-40 than in BC and MP. The ratios of unsaturated to saturated fatty acids were higher in BC, TC-40 than in MP (Fig. 28.1). In ungerminated seed, PA (C16:0) content was highest in MP. This cultivar had a lower germination percentage at all temperatures, suggesting that PA was not essential to the germination process. PAA (C16:1) was extracted in small quantity in seed of TC-40. SA (C18:0) content was highest in the TC-40 which was the cultivar with highest germination percentage, and was lowest in MP which showed low germination. This long-chain fatty acid would appear to be important in germination process. OA (C18:1) content was highest

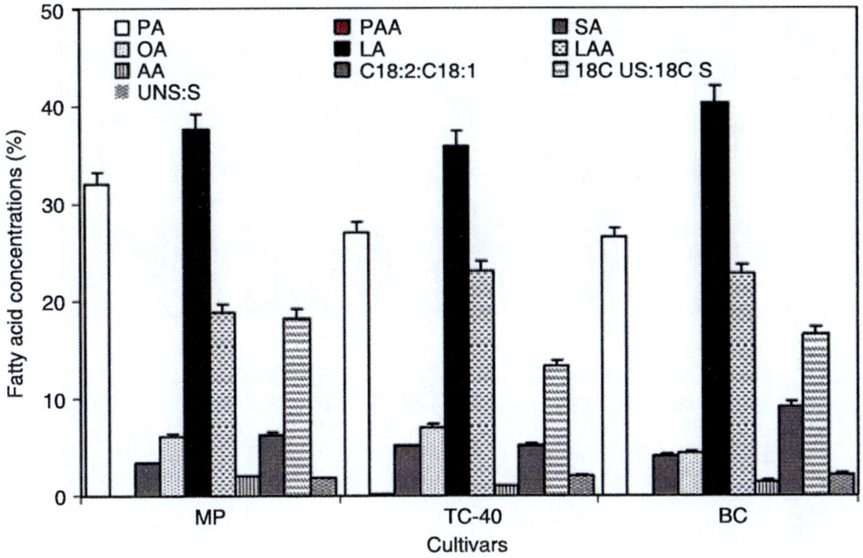

Fig. 28.1 Fatty acid concentrations (%) of ungerminated (whole seed) of three cowpea cultivars. Data are means of four individual experiments. Each experiment consists of four replications of 20 seeds each. Bars represent the standard error of the mean when absent concealed by the symbol. *PA* palmitic acid, *PAA* palmitoleic acid, *SA* stearic acid, *OA* oleic acid, *LA* linoleic acid, *LAA* linolenic acid, *AA* arachidic acid, C18:2/C18:1 polyunsaturated fatty acids: mono-unsaturated fatty acids, 18 US: 18 S 18-carbon unsaturated fatty acid: 18-carbon saturated fatty acid; *UNS: S* unsaturated: saturated, *MP* Mississippi Purple, *TC*-40 Texas Cream 40, *BC* Black Crowder

in TC-40, the cultivar with highest germination and lowest in BC, a cultivar with low germination. LA (C18:2) content was highest in the cultivars with low germination (Fig. 28.1). LAA (18:3) content was higher in ungerminated seed of TC-40 and lower in MP. There was a significant effect of temperature on the AA (C20:0) content in cowpea seeds. This large chain fatty acid was detected only in the whole seed. The highest mean arachidic acid contents were in MP (1.97%), a cultivar with low germination, and TC-40 (1.43%), and the lowest in BC (1.05%). Miguel and Browse (1995) reported that long-chain fatty acids are common in seed from the *Brassicaceae* and other species such as meadowfoam (*Limnanthes alba* Benthan) and jojoba (*Simmondsia chinensis* L.).

3.14 Ratios of the Fatty Acids in Cotyledon and Embryo Tissue

There were significant differences in the 18:2/18:1 ratios (polyunsaturated: monounsaturated) at moderate temperature [TC-40 (10.04 cot. and 5.28 emb.), MP (8.43 cot. and 7.74 emb.0), and BC (7.51 cot. and 7.30 emb.)], and also at high temperature (Fig. 28.2). Temperature affects not only fatty acid and triacylglycerol synthesis

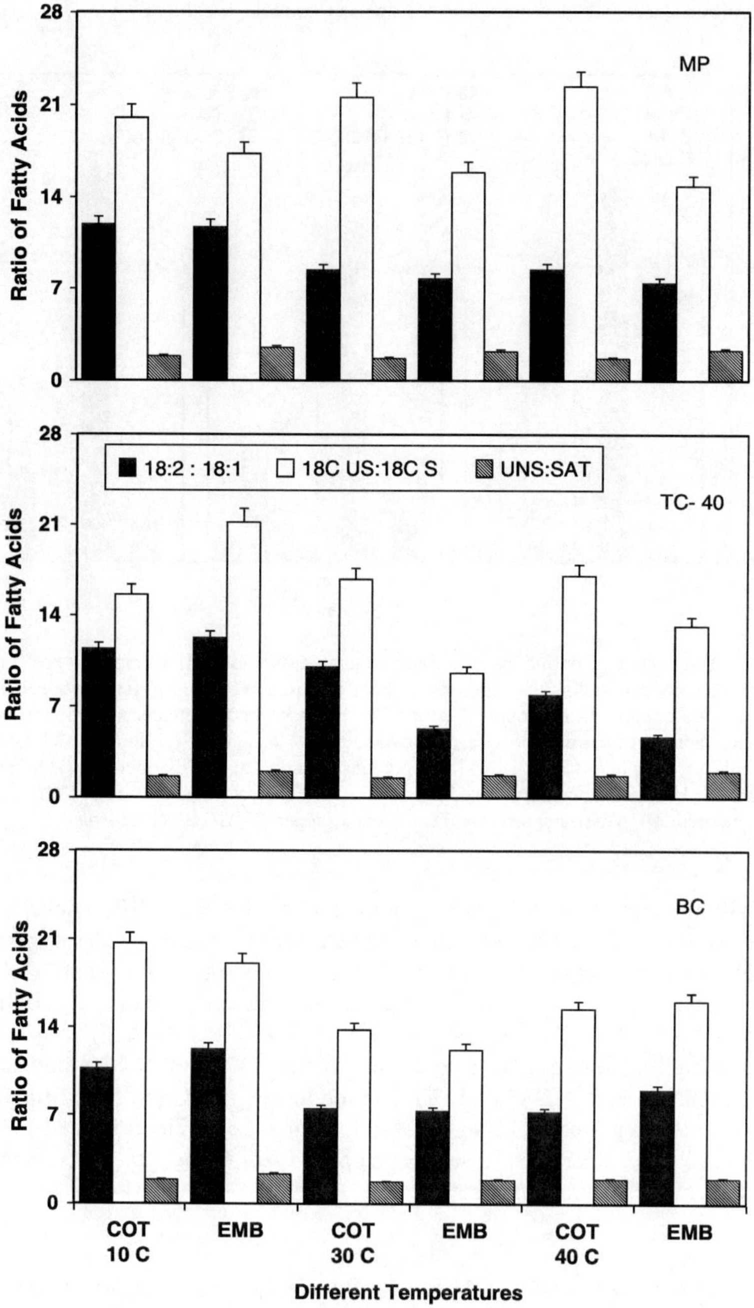

Fig. 28.2 Ratio of fatty acids of cotyledon (COT) and embryo (EMB) tissues of seeds of three cowpea cultivars germinated at different temperature. Data are means of four individual experiments. Each experiment consists of four replications of 20 seeds each. Bars represent the standard error of the mean when absent concealed by the symbol. 18:2: 18:1 polyunsaturated: mono-unsaturated, 18 US: 18 S 18-carbon unsaturated: 18-carbon saturated, *UNS: SAT* unsaturated: saturated, *MP* Mississippi Purple, *TC*-40 Texas Cream 40, *BC* Black Crowder

but also the response of oleate desaturase. Low temperature controls the 18:2/18:1 ratio of lipids by stimulating *de novo* synthesis and activity of oleate desaturase. It also controls the rate of oleate production by fatty acid synthase, which decreases as the temperature decreases. Thus 18:2 content increases as temperature decreases (Miguel and Browse 1995; Murata and Los 1997). At low temperature (10°C), the 18-carbon unsaturated to 18-carbon saturated fatty acid ratio was slightly higher in embryos of TC-40 than in embryos of BC and MP. In cotyledon tissue this ratio was higher in BC than in MP and TC-40. The 18-carbon unsaturated to 18-carbon saturated fatty acid ratio decreased when seeds were exposed to moderate temperature for embryos of TC-40 and BC. A small increase in this ratio, compared to moderate temperature, was observed at high temperature in the embryos of TC-40 and BC. BC had better germination at higher temperature than MP (Table 28.10). At low temperature (10°C), means were higher in embryo tissue than in cotyledon tissue as follows, MP, BC and TC-40. This ratio was higher in embryo tissue at low (10°C) than at moderate (30°C) temperature. Miguel and Browse (1995) reported that in membrane and storage lipids of plants grown at lower temperatures contained more polyunsaturated 18-carbon fatty acids than mono-unsaturated fatty acids. Membrane lipids contain mainly unsaturated fatty acids such as 18:3 and 16:3. When plants or plant parts are exposed to low temperature, the desaturation of fatty acids occurs mainly from 18:2 to 18:3 (Murata and Los 1997). The importance of the desaturation of fatty acid of membrane lipids in tolerance to low temperature has been demonstrated in transgenic systems. When relative levels of saturated molecular species of phosphatidylglycerol are reduced by transformation with glycerol-3-phosphate acyltransferase from a chilling resistant plant, plants become more tolerant to low temperature (Murata and Los 1997). Usually when cold temperature sensitive organisms are subjected to temperatures lower than optimal for their normal development, the degree of unsaturation of fatty acids in their membranes decreases significantly (Murata and Los 1997). The presence and composition of fatty acids in the seed affected its ability to germinate under different temperatures. The fatty acid composition of tomato seed, for example, accounts strongly for its ability to germinate at low temperature. Cold germinating genotypes (CGG) showed a higher proportion of LA and lower proportion of OA than non cold germinating genotypes (NCG). This relationship was not affected by the period of incubation at 10°C. This suggests that the genes controlling germinating ability affect the overall fatty acid composition towards increased unsaturation, even in the storage lipids (Maluf and Tigchelaar 1982). Harwood and Stumpf (1970) pointed out that the initial production of fatty acids in germinating pea seed showed a "lag" which depended on temperature. They also observed that production of fatty acids during the first 30 h of germination initiated with palmitate and stearate, followed later by oleic acid. No polyunsaturated acids were detected during this period of pea seed germination. Stearic acid is found in fatty acid composition of important oilseed such as soybean (*Glycine max* L. Merr.), rapeseed (*Brassica napus* L.), sunflower (*H. annuus* L.), palm (*Elais guinenis* L.), cotton (*G. hirsutum* L.), peanut (*A. hypogaea* L.), and coconut (*Cocus nucifera* L.) in proportions of approximately 4% (Miguel and Browse 1995). Thus, this long-chain fatty acid appears to be important in germination process.

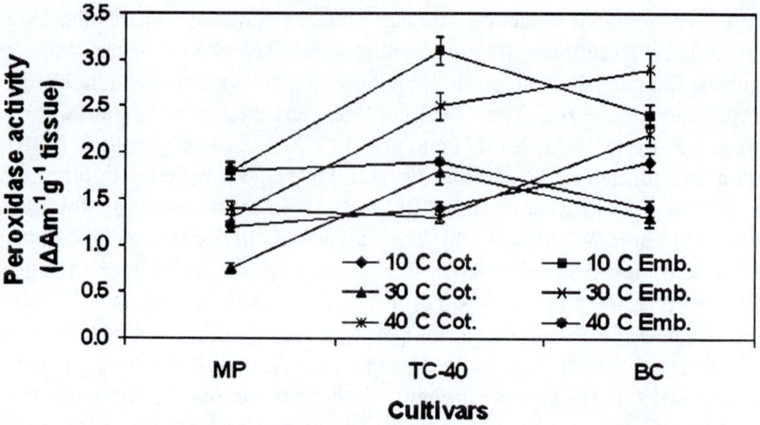

Fig. 28.3 Peroxidase activity (ΔA m^{-1} g^{-1} tissue) cotyledon (COT.) and embryo (EMB.) tissues of seed of three cultivars of cowpea germinated at three temperatures. *MP* Mississippi Purple, *TC*-40 Texas Cream 40, *BC* Black Crowder

3.15 Peroxidase Activity

There was a significant temperature × cultivar × tissue interaction for peroxidase activity in cowpea seed tissues (Table 28.4). At low temperature (10°C) which is considered a chilling temperature for cowpea cultivars, the highest peroxidase activity was in embryo tissue of TC-40 (3 ΔAm^{-1} g^{-1}), followed by embryo tissue of BC (2.7 ΔAm^{-1} g^{-1}). The lowest peroxidase activity was in cotyledon tissues of TC-40 (1.4 ΔAm^{-1} g^{-1}) and MP (1.9 ΔAm^{-1} g^{-1}). At moderate temperature (30°C) the higher peroxidase activity was in embryo tissues of BC (2.9 ΔAm^{-1} g^{-1}) and TC-40 (2.5 ΔAm^{-1} g^{-1}) and the lowest in cotyledon tissue of MP (0.8 ΔAm^{-1} g^{-1}). As at low temperature, higher peroxidase activity was observed in embryo tissue than cotyledon tissue. At high temperature (40°C) the highest peroxidase activity was in cotyledon tissue of BC (2.2 ΔAm^{-1} g^{-1}), followed by embryo tissues of TC-40 (1.9 ΔAm^{-1} g^{-1}) and MP (1.8 ΔAm^{-1} g^{-1}). The lower peroxidase activity was in cotyledon tissues of MP (1.38 ΔAm^{-1} g^{-1}) and TC-40 (1.4 ΔAm^{-1} g^{-1}), and in embryo tissues of BC (1.4 ΔAm^{-1} g^{-1}) (Fig. 28.3). Peroxidase activity in plant tissue increases as a response to stress treatment (Wagih and Coutts 1982a, b). The enzyme activity of cowpea leaves was increased by mechanical abrasion or as a response to infection by Tobacco Necrosis Virus. Chilling temperatures is thought to increase the levels of active oxygen species, which could increase the risk of chilling injury, thus increase in the activity of the active oxygen species scavenging system is desirable to increase tolerance to chilling temperature. The cowpea cultivar with the higher germination percentage, at low temperature, had higher peroxidase activity, suggesting the peroxidase acted in scavenging free radicals, which if accumulated could cause damage to cellular components and alters metabolic function. This is in agreement with Anderson et al. (1995) who mentioned that peroxidase activity increased in chilling acclimated maize seedlings.

4 Conclusions

In conclusion, it is apparent that the main sugars present in cowpea seeds were sucrose, raffinose, and stachyose. Raffinose and stachyose are stored carbohydrates present only in dry ungerminated seed and in cotyledon tissues of germinating seeds at low temperature. No "*de novo*" synthesis of these sugars was detected. Sucrose "de novo" synthesis was found at higher temperatures. An accumulation of this sugar was evident in embryo tissues of cultivars with reduced ability to germinate at low temperature. The results also indicated that high peroxidase activity was related to ability of seeds to germinate at low temperature. The cowpea cultivars with high germination percentage showed higher sucrose, raffinose, and stachyose, as well as peroxidase activity in ungerminated seed. The differences in germination capacity may be related to over expression or inhibition of genes encoding synthesis of relevant molecules that reflect the composition differences demonstrated in this study (Khan and Ungar 1997; Castellon et al. 2003). Furthermore, the cowpea cultivars with high germination percentage showed higher palmitoleic acid (PAA), stearic acid (SA), and oleic acid (OA) contents in ungerminated seed. Cultivars with high germination percentage at low temperature showed higher 18:2/18:1 and 18-carbon unsaturated to 18-carbon saturated ratios. Furthermore, unsaturated fatty acid contents were found higher at lower temperatures. The results suggest that different cultivars obtained by genetic combination may present relevant differences in their fatty acid composition. The results also show significant changes in the amount of individual fatty acids in the cowpea seed during the germination processes. These changes could be explained by the possible translocation of fat-like substances from reserves (cotyledons) to the sink where the respiratory process is more active, or due to lipid peroxidation. The results may help in future breeding for specific constituents to enhance or reduce the fatty acid and sugar contents, and for the improvements of desired quality criteria of cowpea seeds. Thus, the information provided by this research will facilitate future genetic as well as plant biological studies of cowpea cultivars.

References

Anderson MD, Prasad TK, Stewart CR (1995) Changes in Isozyme profiles of catalase, peroxidase, and glutathione reductase during acclimation to chilling in mesocotyls of maize seedlings. Plant Physiol 109:1247–1257

Association of Official Seed Analysts (1991) Rules for testing seed. J Seed Technol 12:18–19

Bernal-Lugo I, Leopold AC (1992) Changes in soluble carbohydrate during seed storage. Plant Physiol 98:1207–1210

Castellon RER, Araujo FMM, Ramos D, Marcio V (2003) Biochemical composition and characterization of lipid fraction of six cowpea cultivars. Rev Bras Agric Ambient 7:149–153

Connor KF, Bonner FT (1998) Physiology and biochemistry of recalcitrant *Guarea guidonia* (L.) sleumer seeds. Seed Technol 20:32–42

Connor KF, Bonner FT (2001) The effects of dessication on seeds of *Acer saccharinum* and *Aesculus pavia*: recalcitrance in temperate tree seeds. Tree 15:131–136

Connor KF, Sowa S (2000) Sucrose accumulation as a response to desiccation in *Quercus alba* L. Acorn: a natural Glycoprotectant. In: The ecological society of America 85th annual meeting, Snowbird, Utah, P 235, 06–10 August 2000

Connor KF, Bonner FT, Vozzo JA (1996) Effect of desiccation of temperate recalcitrant seeds: differential scanning, calorimetry, gas chromatography, electron microscopy and moisture studies on *Quercus nigra* and *Quercus alba*. Can J For Res 26:1813–1821

Connor KF, Kossmann-Ferraz ID, Bonner FT, Vozzo JA (1998) Effect of desiccation on the recalcitrant seeds of *Carapa guinensis* Aubl. and *Carapa procera* DC. Seed Technol 20:71–82

Connor KF, Bonner FT, Vozzo JA, Kossmann-Ferraz ID (2000) Changes in desiccating seeds of temperate and tropical forest tree species. In: Michael AK (ed) Frontiers of forest biology. Proceedings of joint meeting of North American forest Biology Workshop and the Western Forest Genetics Association 1998. Food products press Inc, New York, pp 319–326

Craufurd Q, Ellis H, Summerfield J, Menin L (1996) Development in cowpea (*Vigna unguiculata*). I. The influence of temperature on seed germination and seedling emergence. Exp Agric 32:1–12

De Santis A, Landi P, Genchi G (1999) Changes of mitochondrial properties in maize seedlings associated with selection for germination at low temperature, fatty acid composition, cytochrome c oxidase, and adenine nucleotide translocase activities. Plant Physiol 119:743–754

Ehlers D, Hall A (1997) Cowpea (*Vigna unguiculata* L. Walp.). Field Crops Res 53:187–204

Fery L (1981) Cowpea production in the United States. Hortic Sci 16:473–474

Fery L (1990) The cowpea: production, utilization, and research in the United States. Hortic Rev 12:197–222

Fery L (2002) New Opportunities in *Vigna*. In: Janick J, Whipkey A (eds) Trends in new crops and new uses. ASHS press, Alexandria, pp 424–428

Folch J, Lees M, Stanley S (1957) A simple method for the isolation and purification of total lipids from animal tissues. J Biol Chem 226:497–509

Hall A, Singh B, Ehlers D (1997) Cowpea Breeding. In: Janick J (ed) Plants breeding. Reviews, vol 15. Wiley and Sons, Inc, Hoboken, pp 215–274

Harwood JL, Stumpf PK (1970) Fat metabolism in higher plants. XL. Syntheisis of fatty acids in the initial stage of seed germination. Plant Physiol 46:500–508

Islam S, Carmen CR, Garner JO (2005) Genotypic variation of germination percentage of southern peas (*Vigna unguiculata* L. Walp) according to temperature stress. J Ark Agric Rural Dev 6:7–11

Islam S, Carmen CR, Garner JO (2006) Screening for tolerance of stress temperature during germination of twenty five cowpea (*Vigna unguiculata* L. Walp) cultivars. J Food Agric Environ 4:191–195

Ismail A, Hall A, Close T (1997) Chilling tolerance during emergence of cowpea associated with a dehydrin and slow electrolyte leakage. Crop Sci 37:270–277

Ismail A, Hall A, Close T (1999) Allelic variation of a dehydrin gene cosegregates with chilling tolerance during seedling emergence. Proc Natl Acad Sci 96:13566–13570

Khan MA, Ungar IA (1997) Effect of thermo period on recovery of seed germination of halophytes from saline conditions. Am J Bot 84:279–283

Koster KL, Leopold AC (1988) Sugar and dessication tolerance in seeds. Plant Physiol 88:829–832

Kotowski F (1926) Temperature relations to germination of vegetable seed. Proc Am Soc Hortic Sci 23:176–184

Maluf WR, Tigchelaar EC (1982) Relationship between fatty acid composition and low temperature seed germination in tomato. J Am Soc Hortic Sci 104:620–623

Marsh L (1993) Moisture affects cowpea and okra seed emergence and growth at low temperatures. Hortic Sci 28:774–777

Miguel M, Browse J (1995) Lipid biosynthesis in developing seeds. In: Kigel J, Galili G (eds) Seed development and germination. Marcel Dekker, Inc, NY

Murata N, Los AD (1997) Membrane fluidity and temperature perception. Plant Physiol 115:875–879

Perl M, Yaniv A, Feder Z (1987) The effect of natural and accelerated againg on the lipid content and the fatty acid composition of seeds. Acta Hortic 215:61–67

Quin F (1997) Advances in cowpea research. In: Singh B, Mohan R, Dashiell E, Jackai N (eds) Copublication of International Institute of Tropical Agriculture and Japan International Research Center for Agricultural Sciences. Sayce Publishing, Devon, UK

Ricardo B (1985) Nutritive value of cowpea. In: Singh R, Rachie O (eds) Cowpea research, production and utilization. John Willey and Sons Ltd., Chichester, p 448

Silva JL, Huang JX, Garner JO, Bardwell T, Newwll L (1990) Enzymatic activity and quality changes in refrigerated southern peas. J Miss Acad Sci 35:45–48

Singh BB, Ajeigbe HA, Tarawali SA, Fernandez-Rivera S, Musa A (2003) Improving the production and utilization of cowpea as food and fodder. Field Crops Res 84:167–177

Smartt J, Hymowitz T (1985) Domestication and evolution of grain legumes. In: Summerfield J, Roberts H (eds) Grain legume crops. William Collins Sons and Co. Ltd, London

Terao T, Watanabe I, Matsunaga R, Hakoyama S, Singh BB (1997) Agro-physiological constraints in intercropped cowpea: an analysis. In: Dashiell KE, Jackai LEN, Singh BB, Mohan Raj DR (eds) Advances in Cowpea Research. Co publication of IITA and JIRCAS, Ibadan

Vertucci CW, Farrant JM (1995) Acquisition and loss of desiccation tolerance. In: Kigel J, Galili G (eds) Seed development and germination. Marcel Dekker, Inc, New York

Wagih EE, Coutts RHA (1982a) Peroxidase, polyphenoloxidase and ribonuclease in tobacco necrosis virus infected or mannitol osmotically-stressed cowpea and cucumber tissue. I. Quantitative alterations. Phytopath Z 104:1–12

Wagih EE., Coutts RHA (1982b) Peroxidase, polyphenoloxidase and ribonuclease in tobacco necrosis virus infected or mannitol osmotically-stressed cowpea and cucumber tissue. II. Qualitative alterations. Phytopath Z 104:124–137

Whitaker BD (1986) Fatty-acid composition of polar lipids in fruits and leaf chloroplast of "16:3"- and "18:3"- plant species. Planta 71:118–121

Xu DP, Shi-Jean S, Black CC (1989) Sucrose metabolism in Lima been seeds. Plant Physiol 89:1106–1116

Chapter 29
A Review on Barley Yellow Dwarf Virus

Syed Jawad Ahmad Shah, Muhammad Bashir, and Naheed Manzoor

Contents

S.J.A. Shah (✉)
Nuclear Institute for Food and Agriculture,
Tarnab, Peshawar, Pakistan
e-mail: jawadshah@hotmail.com

M. Bashir • N. Manzoor
Department of Biology, Allama Iqbal Open University,
H-8, Islamabad, Pakistan
e-mail: fodder47@hotmail.com; naheed_manzoor@aiou.edu.pk

M. Ashraf et al. (eds.), *Crop Production for Agricultural Improvement*,
DOI 10.1007/978-94-007-4116-4_29, © Springer Science+Business Media B.V. 2012

Abstract Barley yellow dwarf (BYD) is an economically important, most widely distributed, and destructive viral disease of cereals. The disease is caused by barley yellow dwarf virus (BYDV). The virus is phloem-limited pathogen and causes variable symptoms depending upon the virus isolate, crop species, time of infection and environmental conditions. In general, yellowing and reddening of leaves starting from upper tips to downward, dwarfing and reduction in size and number of ears and grains are observed in infected plants. BYDV is an isometric Luteovirus containing ssRNA genome and is transmitted in a persistent manner by more than 20 species of aphids. The disease has a serious economic impact and yield losses are variable and may range from 5% to 50%. Presently, five isolates of the virus such as PAV, MAV, RMV, RPV and SGV have been identified based on epitope profile and aphid vector specificity. The five biologically well defined isolates are sub-divided in two subgroups; subgroup 1 including PAV, MAV and SGV and subgroup 11 including RMV and RPV. Although the disease can be managed by integrated approaches such as cultural practices, use of pesticides and host plant resistance, but planting of crop varieties with tolerance or resistant is the most economical and practical approach. Development of host plant resistance through genetic engineering is another possible option to combat the problem. Recombinant DNA technology has opened up the prospect of increasing genetic diversity in crop plants by their transformation. This will further help to increase the genetic resistance of bread wheats while using BYDV-resistant genes already identified in wild relatives of wheat and other species. Given the economic significance of the disease and its world wide occurrence, this review has been presented to update the knowledge of various aspects of BYDVs.

Keywords BYDV of cereals • Symptomatology • Properties • Serology • Transmission • Diagnosis • Control

1 Introduction

Barley yellow dwarf (BYD) is a cereal disease caused by an aphid transmitted virus belonging to the family Luteoviridae and genus Luteovirus, the members of which cause yellows disease. The virus which causes this disease is known as barley yellow dwarf virus (BYDV). Barley yellow dwarf is also called cereals yellow plague, cereal yellow dwarf, yellow dwarf and red leaf (Wiese 1987; Burnett 1990). It is distributed worldwide, infects practically all members of the cereals and grasses and possesses all the attributes of destructive plant viruses (Plumb 1983; Rochow et al. 1986). It has been reported in most areas of the world and is believed to damage cereals long before 1951, when it was first described on barley (Singh et al. 1993). The disease occurs on most cereals and numerous grasses, but is not known to affect dicotyledonous plants. The reduction in yield over large production areas is up to 25% (Wiese 1987).

BYDVs are ubiquitous across the globe where Poaceae (wild or cultivated) are grown. These viruses occur wherever cereals are grown. Incidence studies, particularly in the developing world, have been conducted only in some countries. It is evident that worldwide distribution of BYDVs is uneven. As a whole, it appears that PAV-like variants are the most common, but other strains can be predominant in particular agro-ecological zones. The distribution of BYDVs has been shown in Fig. 29.1. The disease was first identified in Netherlands by Oswald and Houston (1951), and later was confirmed in the United Kingdom (Watson and Mulligan 1957). Before the European identification of the disease, however, cereal diseases characterized by yellowing, dwarfing and decreased yields had been observed in North America, sometimes in epidemic proportions. Widespread outbreaks with significant yield losses were probably caused by BYDVs in 1907 and 1949 (Hewings and Eastman 1995). The first reports of barley yellow dwarf viruses in Australia and New Zealand were made more than 47 years ago by Smith (1955, 1957). In Hungary, BYDV was first identified and described in winter barley by Szirmai (1967). Five years later, it was observed in winter wheat during the spring of 1972 by Szunics and Szunics (1980). The virus was later recorded in maize by Milinkó et al. (1984), in rice by Pocsai et al. (1985), in Hungary by Pocsai et al. (1995).

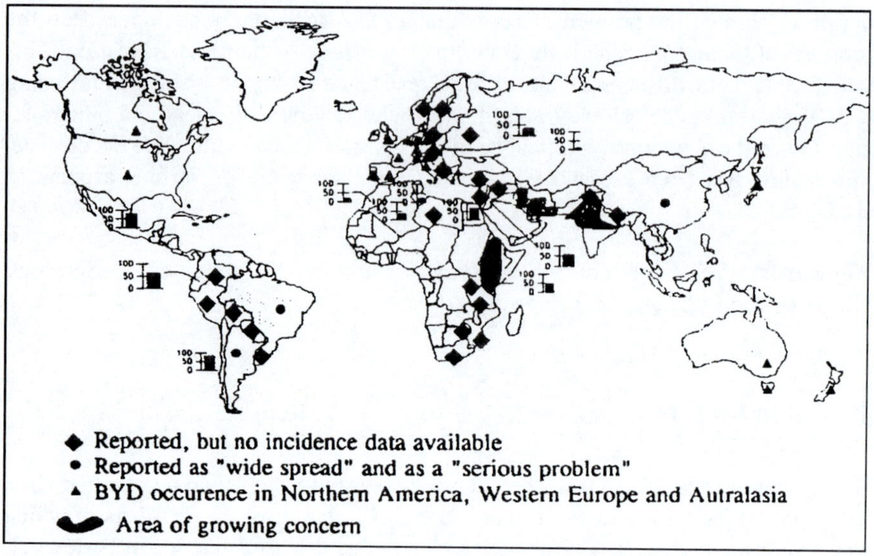

Fig. 29.1 Worldwide presence and incidence of the barley yellow dwarf disease (BYD) in bread wheats as compiled from Burnett (1990), Comeau and Makkouk (1992) and other references. For industrialized countries in Northern America, Western Europe and Australasia, only presence is displayed even if incidence data is available. For Hungry, incidence data was available only for barley

Occurrence of BYDV in some developing countries is well documented (Lister and Ranieri 1995) and was first reported on wheat and barley from India by Nagaich and Vashisth (1963). In Pakistan BYDV was first observed in 1964 near Pakistan-Afghan border on a single plant of local wheat and after that BYDV was noticed in NWFP on all commercial wheat varieties (Aslam and Iftikhar 1990). Later on, the occurrence of BYDV in various provinces of Pakistan was confirmed by several authors (Khalid et al. 1992a, b; Bashir et al. 1994, 1997). Two workshops organized by CIMMYT, Mexico are good sources of information on the distribution and importance of the disease (CIMMYT 1984; Burnett 1990). According to Plum (1983) BYD is truly a global problem.

BYDVs can have a serious impact on, and be an important limiting factor for grain production wherever cereals are grown. However, global yield losses due to the BYDVs are difficult to estimate due to the availability of insufficient information. Average yield losses attributable to natural BYDV infection can range from 11% to 33% (Lister and Ranieri 1995). In some areas the losses have been reported to reach up to 87% (Gildow and Frank 1988). Yield losses caused by BYDV ranged from 27% to 100% in different barley varieties (Pocsai and Kobza 1983). The relationship between the disease incidence and yield loss was found to be linear in wheat and oats. A 1% increase in BYD disease incidence causes yield reduction to increase from 20 to 50 kg/ha in wheat and from 30 to 60 kg/ha in oats. Hewings and

Eastman (1995) calculated that hypothetical 5% losses caused by BYDVs in the United States in 1989 would result in crop losses valued at $847.0 million for corn, $387.1 million for wheat, $48.5 million for barley, and $28.0 million for oats. The greatest yield loss is seen if infection occurs prior to stem elongation (Mann et al. 1997), and severe infections at this growth stage may kill the plants. A PAV-like virus may also cause sugarcane yellow leaf disease in Brazil, Hawaii, and Australia (Vega et al. 1997). In Denmark severe infection of BYDV was seen in 2000. Yield losses were recorded as high as 80% (Gron Viden 2000). Thus the range of economic losses caused by BYDVs may be greater and increasing than previously thought. In Pakistan although there is no specific report on the losses caused by BYDV, but the incidence of the disease has been reported from 0.5% to 5% (Aslam and Ahmad 1990; Bashir et al. 1997).

Keeping in view the world wide occurrence and economic significance of the disease, the present review is being presented to update the information on BYDV globally available, so that future strategies be developed to prevent its spread and better management based on available scientific knowledge.

2 Host Range and Symptoms

BYDV was first reported in oats *(Avena sativa),* barley (*Hordeum vulgare*) and wheat (*Triticum aestivum*) from California, USA by Oswald and Houston (1951, 1953a, b). (*Given above*) Oswald and Houston (1951) reported that wheat, oats and barley were susceptible to BYDV.

BYDVs can infect more than 150 species in the family Poeaceae (D'Arcy 1995). Wheat (*Triticum aestivum, T. durum*), barley (*Hordeum vulgare*), oats (*Avean sativa*), rice (*Oryza sativa*), rye (Secale cereale) and sorghum (*Sorghum vulgare*) are all susceptible to BYDV infection and suffer different degrees of damage, depending on interaction between virus isolate and host cultivar. The agriculturally important grasses in genera Lolium, Poa, Bromus, Festuca, Phleum, and *Cynodon* are also susceptible. Bruehl (1961) summarized many studies and produced a list of 97 susceptible species from 34 genera within the family Poaceae. Slykhuis (1967) reported that about 100 species of grasses were susceptible to BYDV. He also stated that no dicotyledonous plants were known to be susceptible. Bruehl (1961) pointed out that because of the extremely wide host range of this virus among long-lived grass species, there was very little chance of its being eradicated as a result of the elimination of the host. Nevertheless, in most areas where BYDV occurs in cereals, non-cereal grasses ply a role in the epidemiology of the disease. The extensive host range and numerous aphid species that can transmit the BYDV enable the virus to survive in many different environments (van Riessen et al. 1998) but no BYDV isolate is known that can infect dicot plants.

Barley yellow dwarf (BYD) sometimes is difficult to diagnose under field conditions and may go unrecognized in some cases. Field symptoms are easily confused with nutrient or water deficiencies. The disease resembles aster yellows MLO in small grains (Banttari 1965). The symptoms of BYD vary with the crop cultivar, the age of the plant at the time of infection, the strain of virus, the number of aphids present and environmental conditions. It appears that virus interferes with transloca-tion by partially plugging the phloem. It can cause sever stunting of plants, inhibi-tion of root formation, delaying or prevention of heading and reduction in yield. The symptoms persist, or vary seasonally, or disappear soon after infection.

Symptoms of barley yellow dwarf are highly variable and can be confused with those of wheat streak mosaic, nutrient deficiency, root and crown diseases, and envi-ronmental stress. Diagnosis is even more difficult when both wheat streak mosaic and barley yellow dwarf are mixed in the same field or even in the same plant. A big stumbling block to laboratory confirmation of barley yellow dwarf is the difficulty in consistently detecting all five strains of barley yellow dwarf virus. This some-times leads to a false negative diagnosis when, in fact, barley yellow dwarf virus is present. On the positive side, developing technologies using new serological, bio-chemical and biotechnological methods are expected to accurately and rapidly detect all five barley yellow dwarf virus strains. Barley yellow dwarf is diagnosed in the field by the presence of yellowish to reddish stunted plants grouped singly or in small patches among normal plants. Early infection of any of the cereal grains may result in severe stunting, excessive or reduced tillering, bright yellowing or redden-ing of older leaves, delayed heading or ripening, increased sterility, and fewer and lighter-weight kernels. Post-seedling infections are progressively less severe to the point where only the upper leaves, or the flag leaves, show discoloration. Mildly infected wheat may not show any discoloration. The leaves of plants infected with barley yellow dwarf virus are shorter than normal, and the flag leaf may be severely shortened. Leaves often are stiffer and more erect. Root systems are reduced and diseased plants are more easily pulled than healthy plants. Stunted plants result from the failure of the stem inter node to elongate. This leads to a "telescoped" plant where the leaves may unfurl before they have fully emerged from the sheath of the previous leaf. Infected plants are "dwarfs" and have lost their normal confirmation. Even the head, or panicle, doesn't emerge fully or properly. Patterns of barley yel-low dwarf in a field either may be seen as random within the crop or as circular or angular patches which reflect the pattern of movement of the aphid vectors or carri-ers. Many of the infected plants ripen prematurely, after which they may be invaded by sooty molds which give a dirty appearance to the plants and may lower germina-tion of harvested seed. Symptoms described in major hosts by (Watkins and Leslie 1997) are as under.

2.1 Wheat

Different kinds of symptoms of BYD on wheat are reported. However, the most common symptoms reported in wheat are bright yellowing of leaf tips and mar-gins, later spreading to entire lamina, distortion of leaves, dwarfing and stunting, excess tillering and production of sterile spikes in some cases (Khetarpal et al. 1994). If winter wheat is infected in the fall, yellowing of leaves usually does not occur

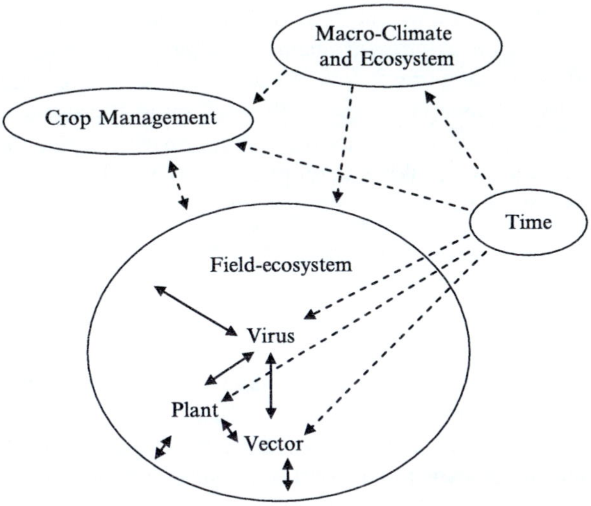

Fig. 29.2 Conceptualization of a pathosystem of a vector-transmitted plant virus

until mid-spring. In severe fall infections, some stunting and reduced tillering may develop. Barley yellow dwarf symptoms start to become obvious at about the jointing stage of growth. Barley yellow dwarf virus does not produce a distinct mosaic pattern, as do wheat streak mosaic virus or soil-borne wheat mosaic virus. The pattern of symptom expression is similar to that in barley or oats. Leaf symptoms begin as blotches near the tip, and with time, these turn various shades of yellow, red or purple. Progression of symptoms is from leaf tip to base and margin to mid-rib. Symptoms are more pronounced under cool temperatures, causing the tips of flag leaves to sometimes become reddish-purple. The yellowing in wheat is not as brilliant as it is in barley, and the reddening of leaves is not as pronounced as it is in oats. In wheat, the pale yellowing of older leaves is the more typical symptom (Fig. 29.2). The extent of yellowing, stunting and yield reduction is contingent on whether the plant is infected as a seedling or during post-seeding development. Leaves of some cultivars under some conditions remain green, but plants are stunted. Under field conditions, barley yellow dwarf first appears in small-localized patches that increase in size as more and more infected plants show symptoms. Generally the plants in the center of these patches show more severe symptoms, with the symptom intensity decreasing toward the perimeter.

2.2 Barley

In barley, a bright yellow discoloration begins at the leaf tip and rapidly progresses down the entire blade. This symptom may sometimes confuse with nitrogen deficiency. Necrotic brown flecking or spotting may accompany the yellowing. The blotchy or uneven discoloration within leaf blades is one of the most characteristic symptoms of barley yellow dwarf and helps to distinguish it from other virus diseases and nutritional or environmental maladies. The brilliant yellowing is seen on older living leaves, and infected plants stand out among surrounding healthy plants in a field. Reddening or purpling of leaves may occur with some barley varieties. The blotchy yellowing begins near the leaf tip; as it progresses towards the leaf base, the green tissues discolor unevenly along the leaf margins, leaving a green stripe along the midrib area. When adult plants become infected, usually only the upper most leaves on the main stem turn yellow.

2.3 *Oats*

Symptoms vary according to the oat variety, the virus strain, the growth stage of the plant at the time of infection, and the general health of the plant. The main color change is to shades of yellow, reddish-orange, reddish-brown, or purple. First symptoms of infection are yellowish-green spots or blotches near the tips of older leaves. Eventually these blotches enlarge and coalesce, turning various shades of yellow, red and brown, entire leaf ultimately becomes a reddish-orange to brown or purple. Severely infected plants are shorter, produce lower test weight grain, and have more blasted florets.

2.4 *Maize and Rice*

The lower leaves show purpling and yellowing. Leaves turn yellow to orange, the discolouration begins at the tips and edges and progresses down the leaves.

3 Properties/Features of BYDV

3.1 *Particle Morphology*

All BYDVs have many common features. The virus particle is ~25 nm icosahedral (T=3) virions. One major (22 kDa) and one minor (50–55 kDa) coat protein. 5.6–5.8 kb positive sense RNA genome with no 5′-cap and no poly(A) tail (Miller et al. 2002).

3.2 *Physical Properties*

Usually one sedimenting component in partially purified preparations; sedimentation coefficient is 115–118S (for BYDV-MAV). 1.4–1.405 g cm in CsCl (for BYDV-PAV). Density is 1.335 g cm^{-3} in Cs_2SO_4 for BYDV-PAV.

3.3 Biochemical Properties

Virion contain 28% nuclei acid (in BYDV-MAV-like isolate), 72% protein and no lipid (Paliwal 1982). Genome consists of single-stranded, linear RNA with molecular weight 2.0 × 10^6d. Total genome size is 5.673 kb (Gerlach et al. 1987). Genome is unipartite, largest (or only) genome part 5.6 kb (BYDV-RPV; 5.677 for BYDV-PAV). Genome nucleic acid was isolated by Brakke and Rochow (1974). Base composition is 24.6% G, 29.6% A, 23.8% C and 22% U. Poly A region is absent. Information on nucleotide sequence on various isolates of BYDV has been reviewed by Miller et al. (1987).

3.4 Features of the Genome and Proteins

Non-genomic nucleic acid found in the virions is satellite RNA (Miller et al. 1987). RNA Sub-genomic mRNA is found in infected cells (Gerlach et al. 1987). Virion protein (s) one; Mr 23500 (MAV), or 24400 (PAV), or 24450 (RPV) (Scalla and Rochow 1977; Hammond et al. 1983). Virus-coded non-virion proteins identified by genomic sequence analysis; six proteins found. Mr. 95000; probably a polymerase, read through protein. Mr of second largest 69000; possibly with a transport function. Mr. of 3rd 34000; probably a polymerase. Mr. of 4th 22000; coat protein. Mr of 5th and smaller 6000; possibly VPg; 17000 internal gene coding for coat protein (Brunt et al. 1996).

3.5 Cytopathology

Virions are found in phloem, cytoplasm, nuclei and around plasmodesmata. Viruses of different BYDV subgroups alter the nucleus in different ways. Other cellular changes are degradation of phloem tissue to which virions are confined (Gill and Chong 1975).

4 Classification and Serological Relationships

4.1 BYDV Isolates

BYDVs present challenges in classification. They are members of the luteovirus group (Randles and Rathjen 1995), which is defined as having icosahedral, T D 3, 25–30 nm virions that are non mechanically transmissible, but rather are transmitted only by aphids in a persistent, circulative manner, and are confined to the phloem tissue in the plant. The studies by Rochow 1969 and Rochow and Muller 1971 have clarified the differences and described five variants or isolates (RPV, RMV, MAV, PAV, and SGV), that were initially classified by their biological properties and have subsequently been shown to be serologically distinct (Lister and Rochow 1979; Torrance et al. 1986). Four aphid vectors were used to distinguish the five isolates of BYDV reported from New York State: *Rhopalosiphum padi, Macrosiphum avena, Rhopalosiphum maidis* and *Schizophis graminum*. The basis for the distinction was the relative efficiency of transmission of the five isolates by these vectors and their vector specificity. The isolates and their major vectors are: RPV transmitted by *Rhopalosiphum padi*, RMV transmitted by *Rhopalosiphum maidis*, MAV transmitted by *Macrosiphum avenae* (now called *Sitobion avenae*), SGV transmitted by *Schizophis graminum* and PAV transmitted by *Rhopalosiphum padi, S. avenae* and others aphids. These are distinguishable serologically (Waterhouse et al. 1988). Laboratories worldwide use antibodies to classify local isolates into one of the above serotypes. However, aphid-transmission properties do not always correlate with serotype (Lister and Sward 1988; Creamer and Falk 1989). Symptoms can vary widely among different PAV isolates (Chay et al. 1996a, b). Thus, the simple five-serotype scheme may have been over applied.

4.2 Subgroups of Isolates

Numerous observations support division of barley yellow dwarf viruses into two viruses and even into separate genera. The former notion was first proposed based on cytopathological differences (Gill and Chong 1975) and subsequently supported by serological evidence (Waterhouse et al. 1988) and most strikingly, by differences in genome organization (Miller et al. 1995; Mayo and Ziegler-Graf 1996). The BYD viruses (BYDv) are presently sub-divided in two major sub-groups based on serological relationships (Rochow 1970a; Rochow and Duffus 1981), cytopathological ultrastructure of infected cells (Gill and Chong 1975; Randles and Rathjen 1995) and dsRNA profile obtained from infected tissue (Gildow et al. 1983). Sub-group 1 includes the strains PAV, MAV, and SGV and are related to soybean dwarf virus, whereas the strains RPV and RMV are included in Subgroup 11 and are closely related to beet western yellow virus (BWYV), potato leaf roll and carrot red leaf viruses, and more distantly to pea enation mosaic and southern bean mosaic sobemovirus (Duffus and Rochow 1978; Rochow and Duffus 1981; Martin and D'Arcy 1990). The acronyms have originally been chosen according to their principal aphid vector species (e.g. RPV for Rhopalosiphum padi virus by Rochow 1970b). These viruses are persistently transmitted by aphid vectors. The virus is not known to multiply in the insect. Newborn nymphs (young aphids) are virus-free and acquire BYDVs by feeding on infected plants. The pathosystem of BYDVs is extremely complex. Figure 29.2 may be one way of conceptualizing this complexity arising from the interaction between virus, host plant and virus vector with the environment and time.

The International Committee on the Taxonomy of Viruses (ICTV) Working Group on Luteoviruses is considering a reclassification in which Subgroup I serotypes would be called BYDV, and members of Subgroup II would be renamed as cereal yellow dwarf virus (CYDV). This dichotomy extends to other luteoviruses at the level of gene homologies and organization (Miller et al. 1995; Mayo and Ziegler-Graf 1996). Beet western yellows luteovirus (BWYV) and potato leaf roll luteovirus (PLRV) resemble Subgroup II BYDVs. Soybean dwarf luteovirus has a Subgroup I-like organization and replicase, but the structural genes are most similar to those of Subgroup II (Rathjen et al. 1994). The chasm between Subgroups is so deep that the Subgroup II BYDVs are more similar to PLRV and BWYV in genome organization, replication genes, and cis-acting signals than they are to Subgroup I BYDVs. Conversely, other than in the structural genes, Subgroup I BYDVs are more closely related to SDV than to BYDVs in Subgroup II. Hence, the ICTV is also considering raising each subgroup category to the level of virus group or genus. The groups would all be members of the Luteoviridae family.

4.2.1 Comparison of Subgroup 1 with Subgroup II Isolates

PAV is the best studied BYDV, especially at the molecular level. It is also the most widespread (D'Arcy 1995) and usually causes the most severe symptoms. RPV is the best-studied Subgroup II BYDV, but it is much less well characterized than other Subgroup II luteoviruses (BWYV and PLRV), or PAV. Thus, PAV and RPV as representative members of each Subgroup, but often use BWYV or PLRV as additional representatives of Subgroup II. BYDV was also compared with related viruses outside the luteovirus group. The polymerase and translational frameshift signals of Subgroup I luteoviruses are more similar to those in red clover necrotic mosaic (RCNMV) and other diantho viruses than they are to those in Subgroup II luteoviruses (Xiong et al. 1993; Miller et al. 1995). Conversely, the polymerase genes of subgroup II luteoviruses are most closely related to those of sobemoviruses. Pea enation mosaic virus (PEMV), the sole member of the enamovirus group, has two RNAs, each encoding its own polymerase (Demler et al. 1996). The polymerase of RNA1 is Subgroup II-like (Demler and de Zoeten 1991) and that of RNA2 is Subgroup I-like (Demler et al. 1993). PEMV is aphid transmissible, probably by the same mechanism as luteoviruses, but it is also mechanically transmissible (Demler et al. 1996).

4.3 *Current Classification*

BYDV is the sole member of genus Luteovirus and the type member of the Luteoviridae family (formerly luteovirus group) (D'Arcy et al. 2000). BYDV serotypes were divided into two subgroups, which were subsequently reclassified as separate species. Currently, only BYDV-MAV (transmitted primarily by *Sitobion avenae*) and BYDV-PAV (transmitted efficiently by *S. avenae* and *Rhopalosiphum padi*) are barley yellow dwarf viruses. Former BYDV serotype RPV (transmitted primarily by *R. padi*) was given a new name, Cereal yellow dwarf virus -RPV (CYDV-RPV) and placed in genus Polerovirus along with four non-BYDV viruses in the Luteoviridae. A third genus, *Enamovirus*, consists only of RNA-1 of the bipartite Pea enation mosaic virus (PEMV). Its organization resembles poleroviruses, but lacks open reading frame (ORF) 4. Comprehensive reports on all aspects of the Luteoviridae are available (Smith and Barker 1999).

After publication of the BYDV sequence (Miller et al. 1988), the sequences of several other luteoviruses were determined in rapid succession. These revealed a taxonomic dilemma that has continued to this day. Essentially the replication machinery of the Luteoviridae has two different evolutionary histories, whereas the proteins that form the virus particles and interact with the aphid vectors clearly have a common origin. Functional and comparative genomic analyses of BYDV and related viruses indicate that BYDV belongs to Tombusviridae family (Miller et al. 2002). The replication proteins and the RNA sequences that control replication and translation most closely resemble those of viruses in the Tombusviridae family. Yet the coat protein, movement and aphid transmission proteins clearly resemble those of the other Luteoviridae, including genus Polerovirus. Like the Tombusviridae, BYDV RNA lacks a 5′ cap or any other modification (Allen et al. 1999), and terminates at the 3′ end with the sequence CCC, preceded by a conserved stem-loop (Koev et al. 2002). In contrast, the poleroviruses, including CYDV- RPV have a protein (VPg) linked to the 5′ end and terminate in GU. In poleroviruses ORF codes for a suppressor of the post-transcriptional gene silencing defense response (Pfeffer et al. 2002). This ORF is absent in BYDV. ORF 1 of poleroviruses encodes a proteinase and the VPg which also are lacking in BYDV. All of the luteovirus-like genes of BYDV can be deleted and the remaining RNA can still replicate in protoplasts. Thus, the core of the virus, i.e. the gene expression and replication framework is more closely related to the Tombusviridae family than to other members of the Luteoviridae (Miller et al. 2002).

4.4 Gene Function and Expression

BYDV has a positive sense, 5.7 kb genomic RNA that encodes six open reading frames (ORFs) and produces three sub genomic RNAs (Kelly et al. 1994) that serve as mRNAs for downstream genes. Only ORFs 1 and 2 are translated from genomic RNA. ORF 2 is translated only as a fusion with ORF 1 via -1 ribosomal frame shifting (Di et al. 1993). ORF 2 encodes the active site of the viral RNA-dependent RNA polymerase (RdRp). The role of the ORF 1 product alone is unknown. These are the only two ORFs that are essential for RNA replication in plant cells (protoplasts) (Mohan et al. 1995). ORF 3 codes for the major coat protein (CP). ORF 4, which is imbedded in the sequence that codes for ORF 3, but in a different reading frame, codes for a protein required for systemic infection of plants (Chay et al. 1996a). ORFs 3, 4, and 5 all are translated only from subgenomic RNA1 (sgRNA1). ORF 5 is translated as a fusion with CP via in-frame read-through of the CP stop codon (Brown et al. 1996). ORF 4 is translated via leaky ribosomal scanning (Dinesh-Kumar and Miller 1993). ORF 6 is translatable only from sgRNA2 in vitro (Wang et al. 1999), via the TE in its 5′UTR. Its role, if any, is unknown.sgRNA3, consists of the 3′-terminal 300 nucleotides of the BYDV genome. It is scarce in protoplast infections after 48 h, but accumulates to very high level in plants, a week after inoculation (Kelly et al. 1994; Koev et al. 1998) sgRNA3 has no ORFs and no known function

4.5 *Translational Control*

BYDV RNA undergoes a number of unusual events during translation (protein synthesis). It lacks the 5′ cap and poly (A) tail that are necessary for translation of normal host mRNAs. Normally eukaryotic mRNAs must circularize via proteins that bind the 5′cap, the poly (A) tail, and each other, prior to binding of the ribosome (Sachs et al. 1997). Instead, BYDV RNA harbors a sequence in the 3′ untranslated region (UTR) called the 3′ TE that brings about efficient translation initiation at the first AUG at the 5′ end of the genome and sub genomic RNA1 (Wang et al. 1997). The 3′ TE base pairs with a sequence in the 5′ UTR to circularize the mRNA and facilitate translation initiation (Guo et al. 2001). This is the first known case of a functional mRNA formed by base pairing between the UTRs rather than by protein-mediated circularization. It provides an example of how BYDV research has revealed fundamental new insight on the workings of the eukaryotic translation machinery. A similar structure and interaction exists in genus Necrovirus of the Tombusviridae, but not in genus Polerovirus of the Luteoviridae. The translation signals of poleroviruses, which also lack a cap and a poly (A) tail, have not been well characterized, but they bear no resemblance to those of BYDV. The BYDV polymerase is expressed via -1 ribosomal frame-shifting, i.e. in the region of overlap between ORF 1 and ORF 2, a small number of ribosomes translating ORF 1 shift back one nucleotide relative to the mRNA and resume translation in the new ORF (ORF 2). This event, which is common among retroviruses such as HIV (Dinman et al. 2002) and various other RNA viruses, is controlled by the mRNA sequence in and around the frame-shift site. The exact mechanism of frame*shifting is unknown. Because host genes are not known to use frame-shifting, disruption of frame-shifting is a promising target for antiviral research (Dinman et al. 2002). BYDV provides new insight into the mechanisms of ribosomal reading frame maintenance because its RNA has a novel structure for bringing about the frame-shift. Unlike any other known RNA, base pairing between a region four-kilo bases downstream in the 3′ UTR, and the frame-shift site is necessary for frame-shifting (Barry and Miller 2002). In contrast, for the Poleroviruses, a small pseudo knotted RNA structure is sufficient to bring about frame-shifting (Kim et al. 1999). Viruses in genus Dianthovirus (Tombusviridae) have a frame-shift structure very similar to BYDV (Kim and Lommel 1998), although it is not known if long-distance base pairing to a downstream sequence modulates diantho viral frame-shifting. However, it emphasizes yet again how BYDV in many ways is more like the Tombusviridae than other Luteoviridae.

4.6 Relationship Among Serotypes

Sequences of coat proteins of many isolates of BYDV have been determined, revealing much variation within serotypes. Based on these sequences, the most common serotype, PAV, has been unofficially subdivided into two Subgroups, A and B, which have about 90% amino acid sequence homology in the CP gene (Mastari et al. 1998). Coat proteins of PAV isolates within subgroup A have at least 96% amino acid sequence identity to each other. The complete genomes of few isolates have been determined. Completed genome of a severe isolate called PAV-129 was sequenced by its discoverer Gray (1996). This isolate is quite divergent from other completely sequenced PAV isolates. In the replicase ORFs 1 and 2, PAV-129 has only 80 and 88% sequence homology, respectively, to ORFs 1 and 2 of all other PAV isolates and MAV, which are all around 97% identical to each other. The 3′ terminus also differs significantly in sequence but retains a similar secondary structure to the other PAV isolates (Koev et al. 2002). The coat protein sequence of PAV-129, with 86% identity to Subgroup A PAV isolates, falls into Subgroup B (Miller et al. 2002), and PAV-129 is the first Subgroup B isolate for which the complete genome has been determined. If other members of Subgroup B prove to be as different from Subgroup A members throughout their genomes as PAV-129, then there is more taxonomic confusion. For example, BYDV-MAV has the most divergent coat and read-through proteins (76% and 60% homology, respectively, to all PAVs), but the remainder of its genome is more closely related to PAV Subgroup A than Subgroup A is to PAV-129. To map the severe symptom determinants of PAV-129, chimeric isolates of an infectious clone PAV6 (originally derived from the PAV-Ill and PAV-Aus isolates) (Di et al. 1993; Mohan et al. 1995) and PAV-129. A genome with the 5′ half (replication genes) derived from PAV6 and the 3′ half (structural genes) derived from PAV-129, replicated only about 10% as efficiently as full-length PAV6 (Koev et al. 2002). Given the divergence of the replication genes and cis-acting signals recognized by them, it is not surprising that replication was reduced. Despite the reduced virus accumulation, the chimera still caused more severe symptoms than PAV6. The chimera also was more efficiently aphid transmitted (32%) than full-length PAV6 (13%). Thus the 3′ half of the PAV-129 genome, which includes the CP, aphid transmission, and system movement genes, contains symptom determinants and (not surprisingly) aphid transmission efficiency determinants.

5 Transmission of BYDVs

BYDV is only transmitted by aphids. There are some reports of transmission through seed, which have not been established and confirmed (Szirmai 1979; Mills et al. 1980). Fruit fly (Oscinella frit) has been claimed as vector of BYDV but no confirmation and supporting evidence are reported.

5.1 Aphid Vectors

BYDVs are persistently transmitted by aphids to all common small grain cereals (Singh et al. 1993) and are not transmissible through seed, soil or sap. More than 20 aphid species transmit BYDVs and the most important are *Rhopalosiphum padi* L., the oat bird-cherry aphid; R. maidis (Fitch), the corn leaf aphid; *Macrosiphum avenae* Fabr., the English grain aphid; and *Shizophis graminum* (Rondani), the greenbug. In some areas, *Metopolophium (Acyrthosiphon) dirhodum* (Walker) is an important vector. In South Africa, the Russian aphid (*Diuraphis noxia*) is a reported as vector (Wiese 1987).

5.2 Aphid Transmission

BYDV isolates are transmitted in the persistent or circulative manner. The acquisition and infection feeding periods of 48 h or more are required to obtain maximum transmission rate. Once acquired, the virus is retained for many days, often the rest of the vector's life. As the virus is phloem limited, the aphids that transmit BYDV must feed to acquire virus and usually multiply on susceptible hosts. After feeding, the virions pass through at least three barriers in the aphid by specific uptake. They do not replicate in the aphid. Each BYDV serotype is transmitted efficiently by only a limited number of aphid species (Power and Gray 1995). The vector specificity of BYDVs does not always correlate with serotypes. (Lister and Sward 1988; Halbert et al. 1992; Lei et al. 1995). For example, Creamer and Falk (1989) described an RPV isolate from California (RPV-CA) that is transmitted efficiently by Sitobion avenae, which gives it the transmission phenotype of PAV. The genomes of RPV-CA and the type RPV-NY isolate exhibit sequence homology in the 30 halves but are unrelated in their 50 halves, based on the northern blot hybridization. The transmission phenotypes of BYDVs may be altered transiently by heterologous encapsidation during mixed infections (Rochow 1970a, b; Creamer and Falk 1990; Wen and Lister 1991). Aphid transmission has been discussed in detail (Wen and Lister 1991; Power and Gray 1995).

5.2.1 Role of the Readthrough Domain

Aphid transmission of luteoviruses requires that the genome be encapsidated (Chay et al. 1996a, b). BYDV virions contain 180 subunits of CP (Randles and Rathjen 1995). A few copies of CP in the virion also contain the RTD (Filichkin et al. 1994; Wang et al. 1995). BYDV mutants deficient in the RTD can form virus particles (Filichkin et al. 1994; Mohan et al. 1995) but are not aphid transmissible (Chay et al. 1996a, b). These experiments can be difficult to interpret because a portion of the RTD may also be required for efficient virus movement within the plant

(Brault et al. 1995). Chay et al. (1996a, b) avoided this problem by using a complementation experiment in which an RTD mutant that replicated better than wild-type RNA in protoplasts was heterologously encapsidated in wild-type CP and CP-RTD that were provided by a co-infecting PAV transcript containing a mutation in another gene. The RTD mutant RNA could then be transmitted by aphids to plants, in which it replicated and spread. The virus in these plants could not be transmitted by aphids. Thus the absence of aphid transmission of the RTD mutant that was observed originally was not due to an inability to replicate or spread in the plant. Similarly, mutations in ORF 5 of BWYV knocked out its ability to be transmitted by aphids to plants (Brault et al. 1995; Bruyere et al. 1997). In these experiments, infectious DNA clones of the virus were transmitted to *Nicotiana clevelandii* plants by *Agrobacterium*-mediated inoculation. Those with mutations in ORF 5 were not transmittable by aphids to other plants. BYDV virions purified from plants contain an approximately 50-kDa, C-terminally truncated form of the RTD (Cheng et al. 1994; Filichkin et al. 1994; Wang et al. 1995), suggesting that the amino terminal half of the RTD provides the aphid transmission function.

5.2.2 Interactions Within the Aphid

How do the CP and RTD interact within the aphid? After acquisition from the phloem, virions are transported to the aphid hindgut. The virus must then cross three distinct transmission barriers to complete the transmission process (Gildow 1993; Gildow and Gray 1993). The first barrier is the hindgut epithelium. The virus is transported into the hemocoel in coated vesicles (Gildow 1993). The recognition at the hindgut membrane appears to be luteovirus-specific but not serotype-specific, as most BYDVs can be acquired into hemolymph of both vectors and non-vectors, whereas unrelated viruses, such as BMV, that are not transmitted in a circulative manner are excluded. It is likely that a hindgut receptor interacts with CP domain(s) shared among most BYDVs (Gildow 1993; Chay et al. 1996a, b). The readthrough protein appears not to be required for the translocation of BYDV virions across the hindgut membrane because PAV mutants lacking the RTD still accumulate in the hemolymph (Chay et al. 1996a, b).

5.2.3 In the Hemolymph

Infectious virus can remain in the hemolymph for the life of the aphid if the aphid is not continuously feeding (Power and Gray 1995). Virions may persist by interacting with the most abundant protein in the aphid, called symbionin, that is produced by a bacterial endosymbiont. Symbionin is closely related to heat-shock proteins in the GroEL family. GroEL is a chaperonin, i.e. a protein that facilitates proper folding of other proteins. Van den Heuvel et al. (1994) reported in vitro binding of pea leaf roll virus (PLRV) virions to symbionin. In vitro, PAV virions, and RTD in particular, specifically bind symbionin (SymL) isolated from R. padi and S. avenae, but not GroEL (Filichkin et al. 1997). PLRV was reported to bind symbionin-like molecules from both vectors and non vectors (van den Heuvel et al. 1994). Therefore, it is unlikely that symbionin determines vector specificity of luteoviruses. Van den Heuvel et al. (1994) speculated that the interaction between virions and symbionin is involved in maintaining virus integrity and thus infectivity. Symbionin binding may also permit the virus to evade the aphid immune system (Gray 1996). Filichkin et al. (1997) proposed that the interaction occurs with the N terminus of the RTD that is conserved among luteoviruses. If the purpose of the interaction is indeed to stabilize virions, localization of binding to the RTD appears to be inconsistent with the observations of Chay et al. (Chay et al. 1996a, b) who detected virions lacking RTD in hemolymph. However, the level of virus lacking RTD was possibly reduced to a level below a threshold required for aphid transmission, but still detectable by the PCR method they used (Chay et al. 1996a, b). Treatment of aphids with antibiotics to kill the endo-symbiotic bacteria and purge symbionin from the hemolymph inhibited the ability of aphids to transmit luteoviruses (van den Heuvel et al. 1994). However, this treatment made the aphids sick and thus may have inhibited their ability to transmit virus only indirectly. The biological role of symbionin remains to be demonstrated.

5.2.4 Entering the Accessory Salivary Gland

Once the virus reaches the accessory salivary gland (ASG) it must penetrate the ASG basal lamina and plasmalemma to be released into salivary canals and ducts. The virus is then excreted in the aphid saliva during the feeding. Both basal lamina and basal plasmalemma of ASG have been implicated in vector-specific transmission of BYDVs (Gildow and Gray 1993; Power and Gray 1995). The association of BYDVs with basal lamina probably requires specific interaction between the BYDV capsid and binding sites on basal lamina (Gildow and Gray 1993). PAV and MAV virions did not attach to the basal lamina of non-vectors. Similarly, BMV virions failed to accumulate in the basal lamina when injected into the aphid's hemocoel. However, RPV did concentrate in basal lamina of a non-vector but was unable to cross the ASG basal plasmalemma (Gildow and Rochow 1980). The transport of BYDVs across the ASG plasmalemma occurs in coated vesicles, probably via a receptor-mediated endocytosis (Gildow 1993). The receptor on the ASG plasmalemma as well as domains on BYDV virions that interact with the receptor are unknown, and identification of these entities is a major goal of current research. Because the RTD is not required for virions to enter the hemocoel, but is required for aphid transmission, it may be required for transport of luteoviruses across the ASG membranes (Chay et al. 1996a, b). The coat protein may be the major determinant of vector specificity. By examining exchanges of regions of the CP gene between infectious genomic clones from serotypes with different vector specificities, it was found that a portion of the CP itself seemed to harbor the vector-specificity determinant. Thus, even though the RTD is required for transmission, it may not be the component that determines the precise vector specificity that distinguishes BYDV isolates.

5.2.5 Dependent Transmission

Some times aphids are able to transmit viruses from a plant only if the plant is also infected by a second virus called helper virus. Rochow (1973) reported that R. padi does not regularly transmit MAV isolate from infected oats; it often transmits MAV together with RPV, a serologically unrelated isolate, from plants doubly infected by both isolates. Rochow and Gill (1978) reported that dependent transmission occurs with a range of BYDV isolates. Dependent transmission by R. padi was 81% from doubly infected plants with either MAV or 6407 together with RPV or 6524 isolates of BYDV. Furthermore, it is also reported that dependent transmission of five isolates of BYDV occurred in tests with aphid species (Rochow 1982). R. padi transmitted RMV, MAV and SGV isolate in the presence of RPV from the mixed infections. The PAV isolate was equally effective in enabling R. padi to transmit RMV. The RMV helped in the transmission of PRV and MAV isolates by R. maidis.

5.2.6 Effect of Temperature on Virus Transmission

Temperature affects the specificity of BYDV transmission. RPV and PAV transmission by S. graminum was higher at high temperature, while the transmission of PAV by Sitobion avenae was greater at low temperature (Rochow 1969). Epidemic of BYD is expected to occur at low temperature with optimum humidity. Both the factors are favorable for cereal's growth and aphid multiplication and migration (Mathre 1985).

6 Interaction of BYDV with Other Factors

6.1 Biotic Factors

6.1.1 Interaction with Other Diseases

Comeau and Makkouk (1988) reported that in case of certain fungal diseases, most of the damage is BYDV-dependent and the damage could be almost entirely attributed to the virus. For example, Fusarium root rot of oats disappeared almost entirely from the province of Quebec, Canada, when very susceptible oat cultivars were replaced by line with moderate tolerance to the virus. Fusarium root rot is generally not very aggressive in farmers' fields, however, in experiments between 1972 and 1974, 50–65% of the plants became infected in plots artificially inoculated with BYDV. This fungal disease is therefore an example of one that has apparently been controlled by breeding for tolerance to BYDV which was the hidden causal factor. Alternaria species are other organisms that are essentially non-aggressive in absence of BYDV.

Most cereal diseases possess some aggressiveness in the absence of BYDV. However, their is field evidence on the ecological interaction of BYDV with many fungi, for example, epidemics of Septoria avanae in Canada were considered a serious disease of oats when BYDV-sensitive cultivars were grown, from the early days until about 1976. These cultivars were replaced by new ones that were less BYDV-sensitive during the period 1977–1980. Since that time only limited damage by BYDV and Septoria has been observed. However, the new cultivars are not different from the old ones when sprayed with Septoria inoculum under controlled conditions. The only logical explanation is that BYDV most likely predisposed oats to Septoria by increasing the reproductive rate of the fungus (Pelletier et al. 1974). Similar phenomena exist in bread wheat, durum wheat, and barley, but the research data has not yet been published in most cases. In Australia, BYDV was shown to have a synergistic effect with take-all disease in reducing yield (Sward and Kollmorgen 1989). In USA and Canada, results showed that BYDV induced more root rot in winter wheat. In Quebec and Ontario, it was shown that the winter-kill of winter barley was partly explained by the level of BYDV sensitivity in more than 3 years out of 10. In Quebec, the recent cultivar OAC Elmira from Ontario show better winter-survival than all checks. No significant cold tolerance was detected in this line, but it was tolerant to BYDV, which could explain its better survival.

In winter cereals, Comeau and Makkouk (1988) observed interactions between snow mold and BYDV in eastern Canada. One of the best survivors in OAC Winter triticale in northern areas, it seems that its high BYDV tolerance enhances its survival even in the presence of high snow mold levels that kill other winter triticales and winter wheats. It was observed, "hot spots' of BYDV coincide with 'hot spots' of Septoria tritici in spring durum and in winter bread wheat while similar phenomenon was also observed in recent cereal nurseries distributed by International Centre for Agricultural Research in the Dry Areas (ICARDA) (Comeau and Makkouk 1988). However, there is generally no correlation between Septoria tritici symptoms and BYDV symptoms on current cultivars which are not BYDV tolerant. Under natural Septoria epidemics it is very difficult to take symptoms score on BYDV except in cases where BYDV infection was early, severe, and uniform. Such conditions are rare except in special trials where seedling dates and other practices have been altered to increase BYDV damage. If symptoms of BYDV infection occur after the boot stage, they may be mistaken for normal plant senescence; where as invisible damage and predisposition to fungal disease may occur 10–15 days earlier. That is why samples of Septoria infected plants should always be taken for BYDV detection by ELISA. The same procedure could be applied to epidemics of other fungi suspected to develop faster on BYDV-infected plants.

Trials on the interaction of BYDV with head scab (*Fusarium roseum*) showed that when BYDV infection occurs early, there is significantly more scab by kernels in BYDV-susceptible bread wheat than in BYDV-tolerant ones. It seems more than mere coincidence that the best BYDV-tolerant wheat IAS-20, was recently used with success as a parent for scab resistance (Bekele et al. 1988) although this line is not in itself highly scab resistant (Luzzardi 1985).

In one trial in Quebec, the incidence of oat leaf rust was slightly decreased on oats artificially inoculated with BYDV (Pelletier et al. 1974). In contradiction to this result, the important leaf rust strain PC-59 was often abundant on the most BYDV-sensitive oat lines in northeastern USA (Comeau 1988b). Considering this new evidence, there is now a need for more research on rust BYDV interactions in all cereal species.

6.1.2 Effect on Root Length

Semi-dwarf cultivars usually have short roots. In normal growing conditions, it has been shown that oat plant height was correlated with root depth ($r = 0505, p < 0.001$), and weight of shoots was correlated with weight of roots ($r = 0.87$, $p < 0.001$) (Mackey 1988). Similar relationship was also reported in bread wheat (Mackey 1973). In preliminary trials conducted in Canada, BYDV was found to have devastating effect on root length of susceptible barley cultivars, confirming similar observations on various cereals in USA (Kainz and Hendrix 1981). Little success was achieved in Canada to create BYDV-tolerant semi-dwarf cereal cultivars. This may be related to the cumulative root-dwarfing effects due in part to the dwarfing genes and in part to the root starving effect of the virus (Comeau and Jedlinski 1990).

6.1.3 BYDV and Aphid Damage

BYDV-infected sensitive cereal plants often have higher levels of amino acids in the phloem sap because of impaired translocation (Comeau 1988a). As a result the insect reproduces much more rapidly on BYDV-infected plants than on healthy plants. In response to crowding, alates are produced and migration follows. In humid, cool, and windy days, daily migration may go beyond 100 km. Under controlled conditions, scientists from the UK showed that large numbers of virus-free aphids could cause damage by themselves. For example, 10 or more aphids per head impede grain filing, however, aphids often come from BYDV infected fields some distance away. These can cause direct aphid damage plus BYDV damage to the root system. BYDV-tolerant crops could theoretically reduce the importance of aphid migration, but this is a long-term hope. Other aphid flights may originate from virus-free grasses. BYDV-tolerant crops could at least reduce in situ aphid multiplication and prolong root activity (Comeau and Makkouk 1988).

6.2 *Abiotic Factors*

6.2.1 Drought

Comeau and Makkouk (1988) reported that BYDV invades roots as well as aerial part, but the damage on roots can be enormous in some lines such as barley line 85 0L303.

Not only is roots growth reduced but their ability to function is also impaired. Therefore, BYDV causes reduction of uptake of water and minerals, which in turn reduces resistance to other abiotic stresses. Part of the drought damage should be attributed to BYDV which reduces the root system. The relative importance of BYDV versus drought may vary but the "drought plus BYDV" combination is always devastating. Moderate BYDV tolerance is not enough to protect plants when drought reaches a critical level; while, plants combining enough BYDV tolerance with some true drought tolerance will continue to grow.

It is worthwhile to note that some of the best BYDV-tolerant lines appear to possess drought resistance. In triticale, Muskox 658 and Merino 274 possessed this useful dual resistance while other examples of multiple resistance in bread wheat and barley have been reported by Comeau and Makkouk (1988). The barley cultivar Biirka is very sensitive to BYDV in Quebec and yields poorly when BYDV is accompanied by lack of rain. The ICARDA barley cultivar Tadmor, known for its drought tolerance, showed better BYDV tolerance than most other lines that were devoid of the Yd2 gene in Syria (Tel Hadya) in 1987–1988. In oats, the widely grown cultivar Ogle displayed BYDV tolerance and some drought tolerance; both traits are heritable but not entirely linked (Comeau and Makkouk 1988). From all available evidence, simultaneous selection for drought and BYDV tolerance would seem to be a promising filed of research. Selection for tolerance to both stresses may favor better root systems and increased water-use efficiency.

6.2.2 Mineral Stress

Mineral deficiencies observed in BYDV-infected plants usually are the result of BYDV infection rather than the cause of damage. The damaged root system of BYDV-infected oat plants did not allow proper assimilation of N, P and K (Comeau and Barnett 1979). Some wheat plants possesses a better than average ability to extract minerals from the soil. Especially phosphorus in the case of some Brazilian wheats. Many Brazilian wheat cultivars also have roots resistant to acid soil and excess of aluminum. It is striking that a large number of the best BYDV-tolerant lines in the Quebec project were found to have resistance to acid soil (Comeau and Makkouk 1988), although initial selection was made at a soil pH of about 6.0, which is good for bread wheat. The aggressive root system of these lines may have contributed to improved BYDV tolerance by better root growth throughout the soil profile. The correlation between BYDV tolerance and vigorous root system may have the strongest practical implications.

6.3 Complex Interactions

BYDV may increase the reproductive rate of aphids by modifying plant sap and also interferes with the deposition of cell wall constituents (Harper et al. 1976), making the plants more fragile and subject to seed shattering. Triple interactions involving virus, aphids and fungi are common; climate should be considered as the fourth component. Aphids can carry fungal spores around (Fuentes and Exconde 1969; Huang et al. 1981). Moreover, aphids move to floral parts at flowering and the sticky honey dew may act as a "spore trap" to further increase fungal attack. Plant height reduction by BYDV may be in itself a further cause of increased fungal attack, for example by Fusarium and Septoria (Tavella 1987; Couture 1982).

7 Serology and Diagnosis

Enzyme linked immunosorbent assay (ELISA) using polyclonal and monoclonal antibodies are routinely used in numerous laboratories for BYDV detection and virus titre measurement. (Bashir et al. 1997; Khalid et al. 1992a, b; Khetarpal et al. 1994). The presence of plant viruses has also been detected in plant tissue by other diagnostic techniques, including immunosorbent electron microscopy (Derrick 1973), tissue blot immunoassay (TBIA) (Makkouk and Comeau 1994) and cDNA probes using virus group-specific primers and polymerase chain reaction (PCR) (Waterhouse et al. 1986). ELISA is being replaced by TBIA due to its several advantages over ELISA. TBIA is simple, quick, sensitive and cheaper than ELISA. Researchers are increasingly preferring these new techniques. They may be advantageous compared to ELISA because of specificity, sensitivity, targeting of determined sequence etc. Mostly detection methods are based on polyclonal antisera produced in rabbits. To overcome the problems of cross reaction with polyclonal antisera, production of monoclonal antibodies is the other option. Monoclonal antibodies of BYDV isolates have been produced and tested effectively for diagnosis (Pead and Torrance 1988; Torrance et al. 1986). cDNA probes have been made to clone sequences from BYDV isolates and dot blot assay using 32P-labelled probes readily detected BYDV (Waterhouse et al. 1986; Barbara et al. 1987; Eweida and Oxelfelt 1989). However, the detection on radioactive labeling and the use of probe is restricted to some laboratories in which the safety measures are available.

7.1 Antiserum Production

Production of monoclonal antibodies of hybridoma clones were reported by Hsu et al. (1984) when mice were injected with either RPV and MAV isolates which were tested for evaluation with five previously known isolates of BYDV. The sometic cells hybridization technique was used. In experiment, five cell lines were from mice injected with PAV and eight lines were from MAV injected mice. Out of 13, seven virus specific antibodies were produced three of which related with PAV, one with MAV. Two additional MAV derived antibodies reacted with both RPV and MAV. Hu et al. (1985) reported that an antiserum against SGV was obtained from the hen's egg yolk, which provided useful immunoglobulins for indirect ELISA of SGV.

In indirect ELISA, SGV was coated with a monoclonal antibody mAB-MAV4 (1: 1,000 dilutions) that reacted with SGV and MAV but not with others.

7.2 *Enzyme Linked Immunosorbent Assay (ELISA)*

The application of enzyme linked immunosorbent assay (ELISA) created new interests in serological detection of viruses especially viruses having low titre in their respective hosts. It is simple, rapid and more sensitive than any other serological tests. Duffus and Rochow (1978) reported that antisera produced against MAV, RPV isolates of BYDV, when tested against four isolates of beet western yellow virus (BWYV) for virus neutralization, this antisera reduced/eliminated viruses in the normal BYDV bearing areas. Usefulness of ELISA for BYDV diagnoses has been reported by (Rochow 1979; Lister and Rochow 1979). While Rochow and Carmichall (1979) and Rochow (1982) used biological and serological tests for comparing viruses collected from small grain field samples with five characterized isolates of BYDV i.e. RPV, RMV, MAV, PAV and SGV. Of 265 isolates identified, 164 were similar to PAV, 69 resembled RPV, 20 were SGV like. The PAV and MAV like strains of BYDV in ryegrasses were detected by using ELISA techniques. The plants showing foliar symptoms were found to contain large proportions of MAV like strains of BYDV. However, there was a poor association between PAV like virus and foliar symptoms (Holmes 1985). It has been reported by Rochow et al. (1987) that four isolates of luteoviruses including BYDV were identified with a modified indirect ELISA for SGV isolate to permit simultaneous use with direct ELISA procedure.

7.3 *Serologically Specific Electron Microscopy (SSEM)*

Serologically specific electron microscopy (SSEM) has been used successfully for diagnoses of BYDV (Paliwal 1977). It was reported by Diaco et al. (1986) that SSEM was highly sensitive procedure and can detect very small quantity of BYDV. When SSEM procedure were performed on a mixture of BYDV and morphologically different soybean mosaic virus (SW) detected only BYDV. SSEM proved very useful in detecting a wide range of BYDV isolates, even in mixed infection.

8 Control of BYD

The most appropriate method for the control of BYD will depend on the conditions under which a crop is grown. The losses caused by BYD can be reduced by adopting the following three approaches:

1. **Crop husbandry** largely by changing sowing dates and adopting cultural methods to avoid risk of infection.
2. **Killing of aphid vectors** by the use of pesticides or preventing them feeding on susceptible hosts
3. Breeding for resistant/tolerant cultivars

The selection of method depends on many local circumstances, not the least of which is the economics of growing the susceptible crop. Although some insecticides have been reported very effective but are not economical. Breeding resistant or tolerant cultivars is the most economical and practical method for controlling BYDVs.

8.1 Cultural Methods

The usual way to control BYDV is to avoid early sowing in September to October, or very late sowing in April to May. As aphids transmit BYDV, it is important to minimize the time where young plants are exposed to active aphids populations (Plumb and Johnstone 1995) or by carefully timed planting when aphid populations are monitored (Plumb 1995). Altering the sowing time does not prevent infection by BYDV, but it will put off the time of infection so that the plants are more tolerant when they are attacked. In spring planting it is usually the latest sown crops that suffer most damage, as this increases the risks of younger, very susceptible seedlings being exposed to aphids carrying virus. In spring there is usually no conflict between sowing date, potential yield, and BYDV infection, as early sowing increases yield potential and decreases virus infection. Ensuring clean seed beds, especially for autumn-sown crops, and avoiding carryover of virus on volunteers and previous grass crop, are essential precautions to avoid virus infection. Although late planting has proven to be successful for avoiding exposure of the crop high aphid population, but potential yield might be reduced as well due to the shorter growing cycle (Plumb 1984). In general, it appears to be complex to make such recommendations to crop production in developing countries. Promoting resistant or tolerant germplasm still appears to be the most feasible method for BYD management.

8.2 Use of Pesticides

Pesticides either applied as granules or sprays have been widely used to control the aphid vectors of BYDV. Insecticidal sprays should be applied only if they become essential and are economical (Eastop 1983). Pesticides have been widely used in Europe to control BYDV. When insecticides are applied, the time of application is very critical. Spread of the disease can be controlled by aphicides (McKirdy and Jones 1996) and supply of the chemical pirimicarb prior to stem elongation has momentary shown effective in controlling the vector (Mann et al. 1997). The fact that many grass species are alternative hosts for the vector can complicate the control of BYDV as a small population of aphids can over winter here and attack new sown crops (Hewings and Eastman 1995). Due to the acquisition time and the latent period needed before new transmission occur, BYDV is actually one of the viruses where control of the vector actually makes sense.

8.3 Cross Protection

Classical cross protection between BYDV isolates has been demonstrated (Ranieri et al. 1992). Cross-protection occurs between BYDVs of Subgroup-I but not between BYDVs of Subgroup-II or between Subgroup-I and II BYDVs (Wen et al. 1991). While similar BYDVs can cross-protect against one another, mixed infections with unrelated BYDVs, i.e. one from each subgroup, can have the opposite effect. Mixed infections of PAV and RPV give more severe symptoms, including more stunting and higher virus titer, than do infections with either virus alone (Baltenberger et al. 1987; Miller et al. 1997). This is consistent with the phenomenon observed among many luteo and luteo-like viruses (Miller et al. 1997). In all cases, the synergistic interaction involves a replicating RNA with a Subgroup I-like polymerase and one with a Subgroup II-like polymerase, suggesting a positive interaction between the two polymerases or the RNAs that they recognize. This interaction has important practical applications. First, it is unlikely that transgenic plants expressing a Subgroup-I BYDV polymerase will be resistant to a Subgroup-II BYDV. More importantly, such transgenic plants could be more susceptible than untransformed plants to a virus of the opposite subgroup if they express the synergy-conferring gene at high enough levels (Vance et al. 1995; Miller et al. 1997). Genetically engineered cross protection was evaluated as a means of controlling BYD (Vincent et al. 1992) with some success. However, it will take time until transformation techniques in wheat become routine. Until such time, use of transgenic expressing viral coat proteins or other viral genes will not be feasible.

8.4 Breeding for Resistance

Genetic resistance is generally considered as being the most practical means of reducing losses inflicted by virus. In order to breed resistant cultivars, sources of resistance must be identified and available. So far no major gene has been identified in bread wheats that confer resistance to infection by BYDVs. Qualset et al. (1990) summarized the genetics of host plant resistance to BYDV and considered some of the problems facing planting breeding in identifying and utilizing the genes that are beneficial. Most success in identifying specific genes conferring resistance to BYDV has been achieved in barley which a major gene conferring tolerance has been identified. Schaller (1984) summarized the genetic background and the recognition of the Yd2 (Rasmusson and Scheller 1959) gene that conditions resistance, which is located on chromosome 3 (Schaller et al. 1964). In California, where most of early selection work was done, five tolerant cultivars have been released and comprise the majority of the acreage. Other tolerant cultivars of barley have been released but do not appear to contain Yd2 gene (Grafton et al. 1982). BYDV infects triticale (a wheat x rye hybrid) and can be damaging. BYDV-resistant lines have been identified (Comeau and Pierre 1990) but none are agronomically acceptable. Recent work on both spring and winter triticale and methods of evaluating its response to BYDV are described by Collin et al. 1990. No resistance to BYDV in maize has been reported. However, some tolerance to BYDV has been reported in maize hybrids in Italy (Lorenzoni et al. 1990).

8.4.1 Natural Tolerance to BYDVs

Well-characterized resistance genes to BYDVs are few. Tolerance is conditioned by one to four genes in oats (Burnett et al. 1995a, b). In general, tolerance or resistance is specific only against certain isolates of a serotype or subgroup (Gray et al. 1993). A single partially dominant gene, Bdv1, confers tolerance to BYDV in some wheat varieties (Singh et al. 1993). High levels of resistance to BYDVs (PAV and RMV) in Agropyron species were reported in early 1980. Derivatives of the hybrid between Thinopyron intermedium (Host) and hexaploid wheat were tested for resistance to BYDVs in the 1980 and providing promising, i.e. resistance. The resistance derived from the partial amphiploid TAF 46 has been transferred to hexaploid wheat through recombination during cell culture (Xin et al. 1991; Griggs 1992). Several researchers have reported desirable traits in the wild relatives of cereals (Sharma et al. 1984; Brettel et al. 1988; Ceoloni et al. 1988). These traits constitute an important genetic resources for cereal improvement (Cauderon 1979). Aegilops species are often used in interspecific hybridization with durum (*Triticum durum*) and bread wheat (*Triticum aestivum*). The gene bank of the International Centre for Agricultural Research in Dry Areas (ICARDA) and CIMMYT have a large collection of Aegilops species. The genes of resistant wild wheat grasses (Thinopyrum or Agronpyron species) have been introgressed into wheat in interspecific crosses to produce wheat lines containing a portion of the wheat grass chromosome that confers substantial resistance to BYDVs (Banks et al. 1995; Sharma et al. 1995). The best characterized resistance to BYDV is conditioned by the Yd2 gene in barley, which confers resistance only to Subgroup-I BYDVs (Baltenberger et al. 1987; Burnett et al. 1995a, b). Barley lines containing the Yd2 gene express a unique polypeptide that is tightly linked to Yd2 (Holloway and Heath 1992). It encodes a putative subunit of a vacuolar proton-translocating ATPase (Ford et al. 1996). A high-resolution map of chromosome 3H around the Yd2 locus has been generated, and cloning of the Yd2 gene is imminent (Ford et al. 1996). This will provide a major breakthrough in our understanding of BYDV-plant interactions. At present, there is only one marker known for resistance/tolerance to BYDVs which could assist in selection. Leaf tip necrosis is a marker for tolerance to gene (Bdv1) in bread wheats (Singh et al. 1993).

8.4.2 Transgenic Resistance

The most cost-effective and environmentally desirable form of disease control is genetic resistance. However, natural resistance genes to BYDV are few (Burnett et al. 1995a, b). Yet, because of the threat of BYDV, breeders are limited to lines that have significant natural BYDV resistance or tolerance. To allow breeders to expand beyond these limitations, transgenic oats tolerant to BYDV were developed (Koev et al. 1998). Oats were transformed with a gene designed to express the 5′ half of the BYDV genome driven by a CaMV 35S promoter. The most resistant line of transgenic oats initially showed mild symptoms but then recovered and grew to maturity. In laboratory growth conditions, yield was slightly reduced compared to un inoculated controls and virus was sometimes detectable, but the yield was infinitely greater than in the inoculated non-transgenic controls which were actually killed by virus infection long before flowering. The transgene was stable inherited in a Mendelian fashion. Field trials were less promising, mainly because the only line of oats that could be transformed (genetically engineered) at the time was not agronomically useful (Somers et al. 1992). It was derived by artificial hybridization of Avena fatua with a cultivar of oat (*A. sativa*), followed by back crosses to oat. These plants (with or without a transgene) were smaller and less robust than agronomic cultivars, in the presence or absence of virus infection. Fortunately, recent improvements in technology now permit transformation of such agronomic lines as Bell. Scientists greatly through improved design of transgenes to engineer resistant plants with high efficiency. Barley plants transformed with inverted sequences of BYDV genes, causing the transcripts to form long, double-stranded hairpin RNAs were immune to BYDV infection (Wang et al. 2000). Presumably the double-stranded RNA induces the host's post-transcriptional gene silencing system (Waterhouse et al. 2001). To the best of our knowledge, no plants engineered for BYDV resistance have been released for use by growers. Unfortunately, due to the low value of oats as a profit-making enterprise, and perhaps due to concern about consumer acceptance of food derived from GMO crops, corporate interest in funding transgenic oat research has waned. Perceived risks imposed by transgenic BYDV-resistant oats drew attention in a Science magazine article about an unpublished poster presentation at a scientific conference (Kaiser 2001). The experiments alleging that pollen escape from transgenic BYDV-resistant oats could lead to "super weeds " were confined to the greenhouse and used no transgenic plants. Yet the benefits of new resistance genes, such as reduced pesticide inputs and increased yields, are clear (Miller et al. 1997).

Application of insecticides on wheat in UK and Australia to control the aphid vectors of BYDV often results in substantial yield increases (Plumb and Johnstone 1995) that are attributable to the absence of BYDV infection. For more details on the economic costs and worldwide occurrence of BYDV diseases, and the economic and environmental costs of controlling them, we refer the reader to other reports in these proceedings, and to the book BYDV: Forty Years of Progress (D'Arcy and Burnett 1995). One legitimate concern with regard to the applicability (but not safety) of virus-derived transgenes for resistance to BYDV is the wide range of sequence variation among isolates. BYDV isolates that lack high homology to the transgene would not be hindered by transgene-induced post-transcriptional gene silencing (Miller et al. 1997; Wang et al. 2000; Waterhouse et al. 2001). Thus, virus-derived transgenes may confer resistance to only a subset of BYDV isolates in the field.

Recombinant DNA technology has opened the prospects of increasing genetic diversity in crop plants by their transformation. However, no suitable systems yet exist for regenerating small grains cereal plants. Such techniques will soon be developed and will make available genetic information to overcome the present difficulties. No information has yet made of decreasing BYDV infection through resistance to the aphid vector. Although some differences in host reaction to aphid vectors have been reported but no evidence to differentiate in infection of different cultivars (Lowe 1978).

9 Conclusion

A large number of studies have been conducted since the disease BYD has been identified. Although a lot of information are available on the biology, pathology, epidemiology and disease management, but still the problem of crop losses due to BYDV infection exists in many parts of the world. Due to continuous expansion of wheat in the absence of BYDV-resistant cultivars, it is likely to increase yield losses related to BYDV. In regions of low inputs where primary infection is extensive and use of pesticides is un-economical, the only option for disease management is through manipulation of sowing dates and host resistance or tolerance. With the emergence of more knowledge on host genotype x virus isolate interaction, the development of cultivars with resistance or tolerance to virus or vectors will be possible. The use of modern molecular techniques will help in proper diagnosis, to understand the virus-vector interaction and epidemiology of the disease. In high input agriculture, pesticides are likely to remain important but their use can be decreased with the introduction of disease resistant or tolerant cultivars.

There is a continuing need to characterize BYDV isolates. Sound knowledge of a pathogen's epidemiology and ecology and of the pathogen's interaction with the host and its environment will help for better disease management and reduce crop losses. Recombinant DNA technology be used to create genetic diversity in host plant and to utilize Yd1 resistant gene already identified in wild relatives of bread wheat and other hosts. Host resistance to aphid vectors should also be explored in wheats or wild relatives and included in breeding programme.

References

Allen E, Wang S, Miller WA (1999) Barley yellow dwarf virus RNA requires a cap-independent translation sequence because it lacks a 5′ cap. Virology 253:139–144

Aslam M, Iftikhar A (1990) Barley yellow dwarf in Pakistan. In: Burnett PA (ed) World perspectives on barley yellow dwarf. In: Proceeding of the international workshop, Udine, Italy, CIMMYT, Mexico, DF Mexico, p 511

Baltenberger DE, Ohm HW, Foster JE (1987) Reactions of oat, barley and wheat to infection with barley yellow dwarf isolates. Crop Sci 27:195–198

Banks PM, Larkin PJ, Bariana HS, Lagudah RA (1995) The use of cell culture for sub chromosomal introgressions of barley yellow dwarf virus resistance from *Thinopyrum* inter-medium to wheat. Genome 38:395–405

Banttari EE (1965) Occurrence of aster yellows in barley in the field and its comparison with BYD. Phytopathology 55:838–843

Barbara DJ, Kawata EE, Veng PP, Lister RM, Larkins BA (1987) Production of cDNA clones from the MAV isolate of barley yellow dwarf. J Gen Virol 68:2419

Barry JK, Miller WA (2002) A programmed ribosomal frameshift that requires base pairing across four kilobases suggests a novel mechanism for controlling ribosome and replicase traffic on a viral RNA. Proc Natl Acad Sci 99:11133–11138

Bashir M, Aftab M, Khan S, Hussain A, Muhammad D, Bhatti MB (1994) Screening of barley and oats genotypes against barley yellow dwarf Luteovirus under natural infection conditions in Pakistan. Barley yellow dwarf virus newsletter no: 5, CIMMYT, Mexico, pp 13–14

Bashir M, Berchinger L, Kisana NS, Mujahid MY, Hashmi NI (1997) Detection of five serotypes of barley yellow dwarf virus in Pakistan. Rachis 16:47–49

Bekele G, Singh RP, Alcala M (1988) Results of the first international scab resistance screening nursery (SRSN), 1985–86, CIMMYT, Mexico, D.F. Mexico

Brakke MK, Rochow WF (1974) Ribonucleic acid of barley yellow dwarf virus. Virology 61:240–248

Brault VJ, Van den Heuvel FJM, Verbeek M, Ziegler-Graff V, Reutenauer A (1995) Aphid transmission of beet western yellows luteovirus requires the minor capsid read-through protein P74. EMBO J 14:650–659

Brettel RIS, Banks PM, Cauderon Y, Cheng XZM, Larkin PJ, Waterhouse PM (1988) A single wheatgrass chromosome reduces the concentration of barley yellow dwarf virus in wheat. Ann Appl Biol 13:599–603

Brown CM, Dinesh-Kumar SP, Miller WA (1996) Local and distant sequences are required for efficient read-through of the barley yellow dwarf virus-PAV coat protein gene stop codon. J Virol 70:5884–5892

Bruehl GW (1961) Barley yellow dwarf, a virus disease of cereals and grasses. Monograph no-1. Am Phythopathol Soc, USA

Brunt AA, Crabtree K, Dallwitz MJ, Gibbs AJ, Watson L, Zurcher EJ (1996) Plant viruses online-barley yellow dwarf Luteovirus, Description and lists from the VIDE database. Version 16th January, URL http://biology.annu.edu. Au/Groups/MES/Vide

Bruyere A, Brault V, Ziegler-Graff V, Simonis MT, van den Heuvel JFJM (1997) Effects of mutations in the beet western yellows virus readthrough protein on its expression and packaging, and on virus accumulation, symptoms and aphid transmission. Virology 230:323–334

Burnett PA (ed) (1990) World perspectives on barley yellow dwarf virus. In: Proceedings of the international workshop CIMMYT, Mexico, DF Mexico

Burnett PA, Comeau A, Qualset CO (1995a) Host plant tolerance or resistance for control of barley yellow dwarf. In: D'Arcy CJ, Burnett PA (eds) Barley yellow dwarf: 40 years of progress. APS Press, St. Paul, pp 321–343

Burnett PA, Comeau A, Qualset CO (1995b) Host plant tolerance or resistance for control of barley yellow dwarf. In: D'Arcy CJ, Burnett PA (eds) Barley yellow dwarf: 40 years of progress. APS Press, St. Paul, pp 9–28

Cauderon Y (1979) Use of Agropyron species for wheat improvement. In: Proceeding of the conference broadening genetic base of crops, Wageningen, The Netherlands, pp 129–139

Ceoloni C, Del Signore G, Pasquini M, Testa A (1988) Transfer of mildew resistance from Triticum longissimum into wheat by ph1 induced homoeologous recombination. In: Miller TE, Koebner RMD (eds) Proceedings of the 7th international wheat genetic symposium, Institute of Plant Science Research, Cambridge, UK, pp 221–226

Chay C, Smith DM, Vaughan R, Gray SM (1996a) Diversity among isolates within the PAV serotype of barley yellow dwarf virus. Phytopathology 86:370–377

Chay CA, Gunasinge UB, Dinesh-Kumar SP, Miller WA, Gray SM (1996b) Aphid transmission and systemic plant infection determinants of barley yellow dwarf luteovirus-PAV are contained in the coat protein readthrough domain and 17-kDa protein, respectively. Virology 219:57–65

Cheng SL, Domier LL, D'Arcy CJ (1994) Detection of the readthrough protein of barley yellow dwarf virus. Virology 202:1003–1006

CIMMYT (1984) Barley yellow dwarf: a proceeding of the workshop CIMMYT, Mexico, DF Mexico

Collin J, St-Pierre CA, Comeau A (1990) Analysis of genetic resistance to barley yellow dwarf virus in Triticale and evaluation of various estimators of resistance. In: Burnett PA (ed) World perspectives on barley yellow dwarf virus: Proceedings of the international workshop CIMMYT, Mexico, DF Mexico, p 404

Comeau A (1988a) La resistance aux purcerons; aspects theoriques et pratiques [Aphid resistance; theoretical and practical aspects]. In: Vincent C, Coderre D (eds) Principles et applications de lutte biologique. Lutte biologique, Canada, pp 433–449

Comeau A (1988b) Tolerance of oats to barley yellow dwarf. In: Mattsson B, Lyhagen R (eds) Proceedings of the third international oat conference, Lund, Sweden, 4–8 July 1988, Svaloev (Sweden): Svaloef AB, 1989.- ISBN 91-971323-0-6, pp 279–286

Comeau A, Barnett G (1979) Effect of barley yellow dwarf virus on N, P, K fertilizer efficiency and on the harvest index of oats. Can J Plant Sci 59:43–54

Comeau A, Jedlinski H (1990) Successful breeding for barley yellow dwarf virus resistance: a systematic approach related to other agronomic characteristic. In: Burnett PA (ed) World perspectives on barley yellow dwarf. Proceeding of the international workshop, Udine, Italy, CIMMYT, Mexico, DF Mexico, p 511

Comeau A, Makkouk KM (1988) Recent progress in barley yellow dwarf virus research: interactions with diseases and other stresses. Rachis 7:5–11

Comeau A, Makkouk KM (1992) Recent progress in barley yellow dwarf virus research: interactions with diseases and other stresses. Rachis 7:5–11

Comeau A, Pierre CA (1990) The genetics of resistance to barley yellow dwarf virus. In: Burnett PA (ed) World perspectives on barley yellow dwarf. Proceeding of the international workshop, Udine, Italy, CIMMYT, Mexico, DF Mexico, pp 107–112

Couture L (1982) Receptivite de cultivars de cereales de printemps a la contamination des graines sur inflorescence par les *Fusarium* spp. Can J Plant Sci 62:29–34

Creamer R, Falk BW (1989) Characterization of a nonspecifically aphid-transmitted CA-RPV isolate of barley yellow dwarf virus. Phytopathology 79:942–946

Creamer R, Falk BW (1990) Direct detection of transcapsidated barley yellow dwarf luteoviruses in doubly infected plants. J Gen Virol 71:211–217

D'Arcy CJ (1995) Symptomatology and host range of barley yellow dwarf. In: D'Arcy CJ, Burnett PA (eds) Barley yellow dwarf: 40 years of progress. American Psychopathological Society Press, St. Paul, pp 9–28

D'Arcy CJ, Burnett PA (eds) (1995) Barley yellow dwarf, 40 years of progress. APS Press, St. Paul, p 374

D'Arcy CJ, Domier LL, Mayo MA (2000) Family Luteoviridae. In: van Regenmortel MHV, Fauquest CM, Bishop DHL, Carstens EB, Estes MK, Lemon SM, Maniloff J, Mayo MA, McGeoch DJ, Pringle CR, Wickner RB (eds) Virus taxonomy: seventh report of the international committee on the taxonomy of viruses. Academic, San Diego, pp 775–784

Demler SA, de Zoeten GA (1991) The nucleotide sequence and luteovirus-like nature of RNA1 of an aphid non-transmissible strain of pea enation mosaic virus. J Gen Virol 72:1819–1834

Demler SA, Rucker DG, de Zoeten GA (1993) The chimeric nature of the genome of pea enation mosaic virus: the independent replication of RNA 2. J Gen Virol 74:1–14

Demler SA, de Zoeten GA, Adam G, Harris KF (1996) Pea enation mosaic enamovirus: properties and aphid transmission. In: Harrison BD, Murant AF (eds) The plant viruses, polyhedral virions and bipartite RNA. Genome 5:303–344

Derrick KS (1973) Detection and identification of plant viruses by serologically specific electron microscopy. Phytopathology 63:441–end page

Di R, Dinesh-Kumar SP, Miller WA (1993) Translational frameshifting by barley yellow dwarf virus RNA (PAV serotype) in *Escherichia coli* and in eukaryotic cell-free extracts. Mol Plant Microb Interact 6:444–452

Diaco R, Lister RM, Hill JH, Durand DP (1986) Detection of homologous and heterologous barley yellow dwarf virus isolates with monoclonal antibodies in serologically specific electron microscopy. Phytopathology 76:225–230

Dinesh-Kumar SP, Miller WA (1993) Control of start codon choice on a plant viral RNA encoding overlapping genes. Plant Cell 5:679–692

Dinman JD, Richter S, Plant EP, Taylor RC, Hammell AB, Rana TM (2002) The frameshift signal of HIV-1 involves a potential intra-molecular triplex RNA structure. Proc Natl Acad Sci 99:5331–5336

Duffus JE, Rochow WF (1978) Neutralization of beet western yellow virus by antisera against barley yellow dwarf virus. Phytopathology 68:45–49

Eastop VF (1983) The biology of the principal aphid virus vector. In: Plum RT, Thresh JM (eds) Plant virus epidemiology. Blackwell Scientific Publications, Oxford, UK, pp 115–132

Eweida M, Oxelfelt P (1989) Production of cloned cDNA from a Swedish barley yellow dwarf isolate. Ann Appl Boil 114:61–69

Filichkin SA, Lister RM, McGrath PF, Young MJ (1994) In vivo expression and mutational analysis of the barley yellow dwarf virus readthrough gene. Virology 205:290–299

Filichkin SA, Brumfield S, Filichkin TP, Young MJ (1997) In vitro interactions of the aphid endosymbiotic SymL chaperonin with barley yellow dwarf virus. J Virol 71:569–577

Ford C, Collins N, Rathjen J, Shams-Bakhsh M, Paltridge N, Symons RH (1996) Barley Yd2, a naturally occurring gene that provides resistance against barley yellow dwarf virus. Presented at International congress of Virology, Jerusalem, p 185

Fuentes FD, Exconde OR (1969) Etiology of spot blotch of barley in the Philippines. Philippines Agric 52:483–504

Gerlach W, Miller A, Waterhouse PM (1987) Barley yellow dwarf virus newsletter 1:17

Gildow FE (1993) Evidence for receptor – mediated endocytosis regulating luteovirus acquisition by aphids. Phytopathology 83:270–277

Gildow FE, Frank JA (1988) Barley yellow dwarf virus in Pennsylvania: effect of the PAV isolate on yield components of Nobel spring oats. Plant Dis 77:254–256

Gildow FE, Gray SM (1993) The aphid salivary gland basal lamina as a selective barrier associated with vector-specific transmission of barley yellow dwarf luteoviruses. Phytopathology 83:1293–1302

Gildow FE, Rochow WF (1980) Role of accessory salivary glands in aphid transmission of barley yellow dwarf virus. Virology 104:97–108

Gildow FE, Ballinger ME, Rochow WF (1983) Identification of double stranded RNAs associated with barley yellow dwarf virus infection in oats. Phytopathology 73:1570–1572

Gill CC, Chong J (1975) Cytopathological evidence for the division of barley yellow dwarf virus isolates into two subgroups. Virology 95:59–69

Grafton KF, Poehlman JM, Sechler DT, Sehgal OP (1982) Effect of barley yellow dwarf virus infection on winter survival and other agronomic traits in barley. Crop Sci 22:596–600

Gray SM (1996) Plant virus proteins involved in natural vector transmission. Trends Microbiol 4:259–264

Gray SM, Smith D, Altman N (1993) Barley yellow dwarf virus isolate-specific resistance in spring oats reduced virus accumulation and aphid transmission. Phytopathology 83:716–720

Griggs T (1992) Smart genetic beat a devastating wheat virus. In: Griggs T (ed) Partners in research for development. ACIAR, Canberra, pp 17–19

Gron V (2000) Havrerodsot. Nielsen SL, Nicolaisen M, Hansen LM, Nielsen GC. Markbrug nr. 228. Ministeriet for Fodevarer, Landbrug og Fiskeri. Danmarks JordbrugsForskning

Guo L, Allen E, Miller WA (2001) Base-pairing between untranslated regions facilitates translation of uncapped, non-polyadenylated viral RNA. Mol Cell 7:1103–1109

Halbert SE, Cornelly J, Lister RM, Klein RE, Bishop GW (1992) Vector specificity of barley yellow dwarf virus stereotypes and variants in southwestern Idaho. Ann Appl Biol 121:123–132

Hammond J, Lister RM, Foster JE (1983) Purification, identity and some properties of an isolate of barley yellow dwarf virus from Indiana. J Gen Virol 64:667–676

Harper AM, Atkinson TG, Smith AD (1976) Effect of *Rhopalosiphum padi* and barley yellow dwarf virus on forage yield and quality of barley and oats. J Econ Entomol 69:383–385

Hewings AD, Eastman CE (1995) Epidemiology of barley yellow dwarf in North America. In: D'Arcy CJ, Burnett PA (eds) Barley yellow dwarf: 40 years of progress. APS Press, St. Paul, pp 75–106

Holloway PJ, Heath R (1992) Identification of polypeptide markers of barley yellow dwarf virus resistance and susceptibility genes in non-infected barley (*Hordeum vulgare*) plants. Theor Appl Genet 85:2–3

Holmes SI (1985) Barley yellow dwarf virus in ryegrass and detection by ELISA. Plant Pathol 34:214–220

Hsu HT, Aebig J, Rochow WF (1984) Differences among monoclonal antibodies to barley yellow dwarf virus. Phytopathology 74:600–605

Hu JS, Rochow WF, Dietert RR (1985) Production and use of antibodies from hen eggs for the SGV isolate of barley yellow dwarf virus. Phytopathology 75:914–919

Huang HC, Harper AM, Kokko EG, Howard RI (1981) Aphid transmission of *Verticillium alboatrum* to alfalfa. Can J Plant Pathol 5:141–147

Kainz M, Hendrix W (1981) Response of cereal roots to barley yellow dwarf virus infection in a mist culture. Phytopathology 71:229

Kaiser J (2001) Breeding a hardier. Weed Sci 293:1425–1427

Kelly L, Gerlach WL, Waterhouse PM (1994) Characterisation of the sub genomic RNAs of an Australian isolate of barley yellow dwarf luteovirus. Virology 202:565–573

Khalid S, Aftab M, Ahmad I, Aslam M (1992a) Detection of barley yellow dwarf virus in Pakistan. Pak J Bot 24:225–225

Khalid S, Ahmad I, Aftab M, Mohsin A (1992b) Barley yellow dwarf disease in Pakistan. In: 5th international Plant virus symposium on viruses, vectors and the environment, 27–31 July 1992, Valenzano (BARI), Italy, p 201

Khetarpal PK, Kumar J, Beuve M, Parakh DB, Nath R (1994) Outbreak of MAV-type barley yellow dwarf virus on wheat in the Garwal Hills in India. Plant Pathol 43:415–416

Kim KH, Lommel SA (1998) Sequence element required for efficient -1 ribosomal frameshifting in red clover necrotic mosaic diantho virus. Virology 250:50–59

Kim YG, Su L, Maas S, O'Neill A, Rich A (1999) Specific mutations in a viral RNA pseudo-knot drastically change ribosomal frameshifting efficiency. Proc Natl Acad Sci 96:14234–14239

Koev G, Mohan BR, Dinesh-Kumar SP, Torbert KA, Somers DA, Miller WA (1998) Extreme reduction of disease in oats transformed with the 5′ half of the barley yellow dwarf virus-PAV genome. Phytopathology 88:1013–1019

Koev G, Liu S, Beckett R, Miller WA (2002) The 3′-terminal structure required for replication of barley yellow dwarf virus RNA contains an embedded 3′ end. Virology 292:114–126

Lei CH, Lister RM, Vincent JR, Karan-jkar MN (1995) SGV serotype isolates of barley yellow dwarf virus differing in vectors and molecular relationships. Phytopathology 85:820–826

Lister RM, Ranieri R (1995) Distribution ands economic importance of barley yellow dwarf. In: D'Arcy CJ, Burnett PA (eds) Barley yellow dwarf: 40 years of progress. APS Press, St. Paul, pp 29–53

Lister RM, Rochow WF (1979) Detection of barley yellow dwarf virus by enzyme-linked immunosorbent assay. Phytopathology 69:649–654

Lister RM, Sward RJ (1988) Anomalies in serological and vector relationships of MAV-like isolates of barley yellow dwarf virus from Australia and the USA. Phytopathology 78:766–770

Lorenzoni C, Bertolini M, Loi N, Osler R, Snidaro M (1990) Tolerance to barley yellow dwarf virus in maize. In: Burnett PA (ed) World perspectives on barley yellow dwarf virus: Proceedings of the international workshop CIMMYT, Mexico, DF Mexico, pp 402–403

Lowe HJB (1978) Detection of resistance to aphids in cereals. Ann Appl Biol 88:401–403

Luzzardi GC (1985) Wheat breeding for scab resistance. In: Wheat for more tropical environments: a proceedings of the international symposium UNDP/CIMMYT, 24–28 Sept 1984, CIMMYT, Mexico, DF Mexico, pp 158–168

Mackey J (1973) The wheat roots. In: Proceedings of the fourth international wheat genetics symposium, Missouri Agricultural Experiments Station, Columbia, USA, pp 827–842

Mackey J (1988) Shoot:root interrelations in oats. In: Proceedings of the third international oat conference, Lund, Sweden

Makkouk KM, Comeau A (1994) Evaluation of various methods for the detection of barley yellow dwarf virus by tissue blot immunoassay and its use for virus detection in cereals inoculated at different growth stages. Eur J Plant Pathol 100:71–80

Mann JA, Harrington R, Carter N, Plumb RT (1997) Control of aphids and barley yellow dwarf virus in spring-sown cereals. Crop Prot 16:81–87

Martin RR, D'Arcy CJ (1990) Relationships among luteoviruses based on nucleic acid hybridization and serological studies. Intervirology 31:23–30

Mastari J, Lapierre H, Dessens JT (1998) A symmetrical distribution of barley yellow dwarf virus PAV variants between host plant species. Phytopathology 88:818–821

Mathre DE (ed) (1985) Compendium of barley diseases. American Phytopathology Society (APS), St. Paul, p 78

Mayo MA, Ziegler-Graf V (1996) Molecular biology of luteoviruses. Adv Virus Res 46:413–460

McKirdy SJ, Jones RAC (1996) Use of imidacloprid and newer generation synthetic pyrethroids to control the spread of barley yellow dwarf luteovirus in cereals. Plant Dis 80:895–901

Milinkó I, Nagy P, Rakk Z, Dezséry M (1984) Elõzetes közlemény egy hazánkban új kukorica patogén vírusról. Növényvédelem 8:350–352

Miller WA, Waterhouse PM, Gerlach WL, Helms K (1987) Genome organization of barley yellow dwarf virus. Phytopathology 77:1704–1704

Miller WA, Waterhouse PM, Kortt AA, Gerlach WL (1988) Sequence and identification of the barley yellow dwarf virus coat protein gene. Virology 165:306–309

Miller WA, Dinesh-Kumar SP, Paul CP (1995) Luteovirus gene expression. Crit Rev Plant Sci 14:179–211

Miller WA, Koev G, Mohan BR (1997) Are there risks associated with transgenic resistance to luteoviruses? Plant Dis 81:700–710

Miller WA, Liu S, Beckett R (2002) Barley yellow dwarf virus: Luteoviridae or Tombusviridae? Mol Plant Pathol 3:1173–1183

Mills PR, Mercer PC, McGimpsey HC (1980) Barley yellow dwraf virus. Annual Report Department of Agriculture, Ireland

Mohan BR, Dinesh-Kumar SP, Miller WA (1995) Genes and cis-acting sequences involved in replication of barley yellow dwarf virus-PAV RNA. Virology 212:186–195

Nagaich BB, Vashisth KS (1963) Barley yellow dwarf: a new viral disease for India. Indian Phytopathol 16:318–319

Oswald JW, Houston BR (1951) A new virus disease of cereals transmissible by aphids. Plant Dis Rep 35:471–475

Oswald JW, Houston BR (1953a) Host range and epiphytology of the cereal yellow dwarf virus. Phytopathology 43:128–136

Oswald JW, Houston BR (1953b) Host range and epiphytology of cereal yellow dwarf. Phytopathology 43:309–313

Paliwal YC (1977) Rapid diagnosis of barley yellow dwarf virus in plants using serologically specific electron microscopy. Phytopathology 89:25–30

Paliwal YC (1982) Role of perennial grasses, winter wheat and aphid vectors in the disease cycle and epidemiology of barley yellow dwarf virus. Can J Plant Pathol 4:367–374

Pead MT, Torrance L (1988) Some characteristics of monoclonal antibodies to a British MAV-like isolate barley yellow dwarf virus. Ann Appl Biol 113:215–218

Pelletier GJ, Comeau A, Couture L (1974) Interaction entre le virus de la feuille rouge de l'avoine (BYDV), Septoria avenae et *Puccinia coronate* sur *Avena sativa*. Phytoprotection 55:9–12

Pfeffer S, Dunoyer PF, Heim KER, Jonard G, Ziegler-Graff V (2002) P0 of Beet western yellows virus is a suppressor of posttranscriptional gene silencing. J Virol 76:6815–6824

Plumb RT (1983) Barley yellow dwarf virus a global problem. In: Plum RT, Thresh JM (eds) Plant virus epidemiology. Blackwell Scientific Publications, Oxford, pp 185–198

Plumb RT (1984) Chemical and cultural control of barley yellow dwarf virus. In: Burnett PA (ed) Barley yellow dwarf: Proceedings of the workshop CIMMYT, Mexico, DF Mexico, pp 52–57

Plumb RT (1995) Epidemiology of barley yellow dwarf in Europe. In: Burnette PA (ed) World perspectives on barley yellow dwraf virus: Proceeding of the international workshop CIMMYT, Mexico, DF Mexico, pp 107–127

Plumb RT, Johnstone GR (1995) Cultural, chemical, and biological methods for the control of barley yellow dwarf. In: D'Arcy CJ, Burnett PA (eds) Barley yellow dwarf: 40 years of progress. APS Press, St. Paul, pp 307–319

Pocsai E, Kobza S (1983) Epidemiological occurrence of barley yellow dwarf virus in Hungary. In: Proceedings of the international conference of integrated plant protection, 4–9 July Budapest, Hungary, vol 1, pp 50–57

Pocsai E, Kiss-Simonné I, Basky Z, Dezséry M (1985) Árpa sárga törpeség virus fellépése rizsen. Abstr Növényvédelem 7:308

Pocsai E, Kovács G, Murányi I, Orosz Á, Papp M, Szunics L (1995) Differentiation of barley yellow dwarf luteovirus serotypes infecting cereals and maize in Hungary. Agronomy 15:401–408

Power AG, Gray SM (1995) Aphid transmission of barley yellow dwarf viruses: interactions between viruses, vectors, and host plants. In: D'Arcy JC, Burnett PA (eds) Barley yellow dwarf: 40 years of progress. APS Press, St. Paul, pp 259–289

Qualset CO, Lorens GF, Ullman DE, McGuire PE (1990) Genetics of host plant resistance to barley yellow dwarf virus. In: Burnett PA (ed) World perspectives on barley yellow dwarf virus: Proceedings of the international workshop. Mexici, CIMMYT, Mexico, DF Mexico, p 338

Randles JW, Rathjen JP (1995) Luteovirus. In: Murphy FA, Fauquet CM, Bishop DHL, Ghabrial SA, Jarvis AW, Martelli GP (eds) Virus taxonomy: sixth report of the international committee on taxonomy of viruses. Springer, Wien/New York, pp 379–383

Ranieri R, Lister RM, Shaner G, Burnett PA, Vallejo J (1992) Cross protection among Mexican barley yellow dwarf virus isolates APS/MSA Joint Meeting, 8–12 Aug, Portland, USA

Rasmusson DC, Scheller CW (1959) The inheritance of resistance in barley to the barley yellow dwarf virus. Agron J 51:661–664

Rathjen JP, Karageorgos LE, Habili N, Waterhouse PM, Symons RH (1994) Soybean dwarf luteovirus contains the third variant genome type in the luteovirus group. Virology 198:571–579

Rochow WF (1969) Biological properties of four isolates of barley yellow dwarf virus. Phytopathology 59:1580–1589

Rochow WF (1970a) Barley yellow dwarf virus. Descriptions of plant viruses, no 32, Kew, Surrey, UK, CMI/AAB

Rochow WF (1970b) Barley yellow dwarf virus: phenotype mixing and vector specificity. Science 167:875–878

Rochow WF (1973) Selective virus transmission by *Rhopalosiphum padi* exposed sequentially to two barley yellow dwarf viruses. Phytopathology 63:1317–1322

Rochow WF (1979) Comparative diagnosis of barley yellow dwarf by serological and aphid transmission tests. Plant Dis Rep 63:426–430

Rochow WF (1982) Identification of barley yellow dwarf virus: comparison of biological and serological methods. Plant Dis 66:381–384
Rochow WF, Carmichael LE (1979) Specificity among barley yellow dwarf viruses in enzyme linked immunosorbent assays. Virology 55:415–420
Rochow WF, Duffus JE (1981) Luteoviruses and yellow diseases. In: Kurstak E (ed) Handbook of plant virus infections and comparative diagnosis. Elsevier, Dordrecht
Rochow WF, Gill CC (1978) Dependent virus transmission by *Rhopalosiphum padi* from mixed infection of the various isolates of barley yellow dwarf virus. Phytopathology 68:451–456
Rochow WF, Muller I (1971) A fifth variant of barley yellow dwarf virus in New York. Plant Dis 55:874–877
Rochow WF, Muller I, Tufford LA, Smith DM (1986) Identification of luteoviruses of small grains from 1981 through 1984 by two methods. Plant Dis 70:461–464
Rochow WF, Hu JS, Foster RL, Hsu HT (1987) Parallel identification of five luteoviruses that cause barley yellow dwarf. Plant Dis 71:272–275
Sachs AB, Sarnow P, Hentze MW (1997) Starting at the beginning, middle, and end: translation initiation in eukaryotes. Cell 89:831–838
Scalla R, Rochow WF (1977) Protein component of two isolates of barley yellow dwarf virus. Virology 78(2):576–580
Schaller CW (1984) The genetics of resistance to barley yellow dwraf virus in barley. In: Burnett PA (ed) Barley yellow dwarf: a proceedings of the international workshop, CIMMYT, Mexico, DF Mexico, pp 93–99
Schaller CW, Qualset CO, Rutger JN (1964) Inheritance and linkage of the Yd2 gene conditioning resistance to barley yellow dwarf virus disease in barley. Crop Sci 4:544–548
Sharma HC, Gill BS, Uyemoto JK (1984) High levels of resistance in *Agropyron* species to barley yellow dwarf and wheat streak mosaic viruses. Phytopathology 110:143–147
Sharma H, Ohm H, Goulart L, Lister R, Appels R, Benlhabib O (1995) Introgression and characterization of barley yellow dwarf virus resistance from *Thinopyrum intermedium* into wheat. Genome 38:406–413
Singh RP, Burnett PA, Albarran M, Rajaram S (1993) Bdv1: a gene for tolerance to barley yellow dwarf virus in bread wheats. Crop Sci 33:231–234
Slykhuis JT (1967) Virus disease of cereals. Rev Appl Mycol 46:401–429
Smith HC (1955) Notes on new diseases in New Zealand. Commonwealth Phytopathol Newsl No 1:10–11
Smith HC (1957) Cereal yellow dwarf virus seen in Australia. Commonwealth Phytopathol Newsl 3:10–11
Smith HG, Barker H (1999) The Luteoviridae. CABI Publishing, Wallingford
Somers DA, Rines HW, Gu W, Kaeppler HF, Bushnell WR (1992) Fertile, transgenic oat plants. Biotechnology 10:1589–1594
Sward RJ, Kollmorgen JF (1989) The effects of the interaction of barley yellow dwarf virus and take-all fungus on the growth and yield of wheat. In: Burnette PA (ed) World perspectives on barley yellow dwarf virus: Proceeding of the international workshop CIMMYT, Medico, DF Mexico, pp 305–312
Szirmai J (1967) Új virusbetegség gabonaföldjeinken: A sárga törpeség. Magyar Mezõgazdaság 22:19
Szirmai J (1979) Seed transmission experiments with barley yellow dwarf virus. Novenytermeles 28:147–150
Szunics L, Szunics L (1980) Vírus a búzán (Virus on the wheat). Magyar Mezõgazdaság 35:9
Tavella CM (1987) Date of heading and plant height of wheat varieties, as related to *Septoria* leaf blotch damage. Euphytica 27:577–580
Torrance L, Pead MT, Larkins AP, Butcher GW (1986) Characterization of monoclonal antibodies to a UK isolate of barley yellow dwarf virus. J Gen Virol 67:569–572
van den Heuvel JFJ, Verbeek MM, van der Wilk F (1994) Endosymbiotic bacteria associated with circulative transmission of potato leaf roll virus *by Myzus persicae*. J Gen Virol 75:2559–2565

van Riessen HW, Paul LM, Kathy LF, John FM, William SG (1998) Barley yellow dwarf in small grains Alabama Cooperative Extension System, CIRCULAR ANR-1082, USA
Vance VB, Berger PH, Carrington JC, Hunt AG, Shi XM (1995) 50 proximal potyviral sequences mediate potato virus X/potyviral synergistic disease in transgenic tobacco. Virology 206:583–590
Vega J, Scagliusi SMM, Ulian EC (1997) Sugarcane yellow leaf disease in Brazil: evidence of association with a luteovirus. Plant Dis 81:21–26
Vincent JR, Lister RM, Ueng PP, Wen F, Lei CH, Klein RE, Larkin BA (1992) Biotechnology: a new weapon against barley yellow dwarf virus. In: Comeau A, Makkouk KM (eds) Barley yellow dwarf in West Asia and North Africa Proceedings of a Workshop ICARDA, Aleppo, Syria, pp 189–196
Wang JY, Chay C, Gildow FE, Gray SM (1995) Readthrough protein associated with virions of barley yellow dwarf luteovirus and its potential role in regulating the efficiency of aphid transmission. Virology 206:954–962
Wang S, Browning KS, Miller WA (1997) A viral sequence in the 3′- untranslated region mimics a 5′ cap in facilitating translation of uncapped mRNA. EMBO J 16:4107–4116
Wang S, Guo L, Allen E, Miller WA (1999) A potential mechanism for selective control of cap-independent translation by a viral RNA sequence in cis and in trans. RNA 5:728–738
Wang MB, Abbot DC, Waterhouse PM (2000) A single copy of a virus-derived transgene-encoding hairpin RNA gives immunity to barley yellow dwarf virus. Mol Plant Pathol 1:347–356
Waterhouse PM, Gerlach WL, Miller WA (1986) Sero-specific and general luteovirus probes from cloned cDNA sequences of barley yellow dwarf virus. J Gen Virol 68:2419–2420
Waterhouse PM, Gildow FE, Johnstone GR (1988) Luteoviruses. In: Descriptions of plant viruses no 339. Kew, Surrey, England: Commonwealth Mycological Institute Assoc Appl Biol
Waterhouse PM, Wang MB, Lough T (2001) Gene silencing as an adaptive defense against viruses. Nature 411:834–842
Watkins JE, Leslie CL (1997) Barley yellow dwarf disease of barley, oats, and wheat C-28. In: Field crops Published by Cooperative Extension, Institute of Agriculture and Natural Resources, University of Nebraska-Lincoln, USA
Watson MA, Mulligan TE (1957) Cereal yellow dwarf virus in Great Britain. Plant Pathology 6:12–14
Wen F, Lister RM (1991) Heterologous encapsidation in mixed infections among four isolates of barley yellow dwarf virus. J Gen Virol 72:2217–2223
Wen F, Lister RM, Fattouh FA (1991) Cross-protection among strains of barley yellow dwarf virus. J Gen Virol 72:791–799
Wiese MV (1987) Compendium of wheat diseases, 2nd edn. APS Press, St. Paul, p 112
Xin ZY, Hu HJ, Chen X, Lin ZS, Zhou GH, Qian YT, Cheng ZM, Larkin PJ, Banks P, Apples P, Glarke B, Brettel RIS (1991) Development of common wheat germplasm resistant to barley yellow dwarf virus vY biotechnology. Sci China 34:1055–1062
Xiong Z, Kim KH, Kendall TL, Lommel SA (1993) Synthesis of the putative red clover necrotic mosaic virus RNA polymerase by ribosomal frameshifting *in vitro*. Virology 193:213–221

Index

M. Ashraf et al. (eds.), *Crop Production for Agricultural Improvement*,
DOI 10.1007/978-94-007-4116-4, © Springer Science+Business Media B.V. 2012

Q

R

S

T